MEDICAL GEOLOGY

IMPACTS OF THE NATURAL ENVIRONMENT ON PUBLIC HEALTH

ESSENTIALS OF
MEDICAL GEOLOGY
IMPACTS OF THE NATURAL ENVIRONMENT ON PUBLIC HEALTH

Editor-in-Chief

OLLE SELINUS
Geological Survey of Sweden

Associate Editors

BRIAN J. ALLOWAY
The University of Reading, UK

JOSÉ A. CENTENO
The Armed Forces Institute of Pathology, USA

ROBERT B. FINKELMAN
United States Geological Survey, USA

RON FUGE
University of Wales, UK

ULF LINDH
Research in Metal Biology, Uppsala University, Sweden

PAULINE SMEDLEY
British Geological Survey, UK

ELSEVIER
ACADEMIC
PRESS

Amsterdam • Boston • Heidelberg • London • New York • Oxford
Paris • San Diego • San Francisco • Singapore • Sydney • Tokyo

Elsevier Academic Press
30 Corporate Drive, Suite 400, Burlington, MA 01803, USA
525 B Street, Suite 1900, San Diego, California 92101-4495, USA
84 Theobald's Road, London WC1X 8RR, UK

This book is printed on acid-free paper.

Library of Congress Cataloging-in-Publication Data
Application Submitted

British Library Cataloguing in Publication Data
A catalogue record for this book is available from the British Library

ISBN: 0-12-636341-2

For all information on all Elsevier Academic Press publications visit our Web site at
www.books.elsevier.com

Printed in China
04 05 06 07 08 09   9 8 7 6 5 4 3 2 1

CONTENTS

Peter W. Abrahams, Institute of Geography and Earth Sciences, University of Wales, Aberystwyth, UK

Brian J. Alloway, Department of Soil Sciences, The University of Reading, Berkshire, UK

J. Donald Appleton, British Geological Survey, Nottingham, UK

Peter Bobrowsky, Geological Survey of Canada, Natural Resources Canada, Ottawa, Canada

Charlotte Bowman, School of Earth and Ocean Sciences, University of Victoria, British Columbia, Canada

Mark W. Bultman, GD Southwest Field Office, United States Geological Survey, Tucson, AZ, USA

Joseph E. Bunnell, United States Department of the Interior, United States Geological Survey, Reston, VA, USA

Allen P. Burke, Department of Cardiovascular Pathology, The Armed Forces Institute of Pathology, Washington, D.C., USA

Sergio Caroli, Community Reference Laboratory, Instituto Superiore di Sanita, Roma, Italy

José A. Centeno, Department of Environmental and Toxicologic Pathology, The Armed Forces Institute of Pathology, Washington, D.C., USA

M. George Cherian, Department of Pathology, University of Western Ontario, Ontario, Canada

Gerald F. Combs, Jr., Grand Forks Human Nutrition Center, USDA Agricultural Research Service, Grand Forks, ND, USA

Angus Cook, School of Population Health, University of Western Australia, Crawley, Australia

Brian E. Davies, Clemson University, Clemson, SC, USA

Theo C. Davies, Moi University, Kenya

Edward Derbyshire, Royal Holloway, University of London, Egham, Surrey

Mike Edmunds, British Geological Survey, Wallingford, UK

Robert B. Finkelman, United States Department of the Interior, United States Geological Survey, Reston, VA, USA

Frederick S. Fisher, Department of Geosciences, University of Arizona, Tucson, AZ, USA

Fiona Fordyce, British Geological Survey, Nottingham, UK

Jefferson Fowles, Institute of Environmental Science and Research, Ltd., Porirua, New Zealand

Teri J. Franks, Department of Environmental and Toxicologic Pathology, The Armed Forces Institute of Pathology, Washington, D.C., USA

Ron Fuge, Institute of Earth Sciences, University of Wales, Aberystwyth, UK

Robert G. Garrett, Applied Geochemistry and Mineralogy Subdivision, Geological Survey of Canada, Natural Resources Canada, Ottawa, Canada

Pierre D. Glynn, United States Department of the Interior, United States Geological Survey, Reston, VA, USA

Tee L. Guidotti, Department of Environmental Health, The George Washington University, Washington, D.C., USA

Stephen C. Guptill, United States Department of the Interior, United States Geological Survey, Reston, VA, USA

Kevin M. Hiscock, School of Environmental Sciences, University of East Anglia, England

Kamal G. Ishak, Department of Hepatic Pathology, The Armed Forces Institute of Pathology, Washington, D.C., USA

Tina Kold Jensen, Institute of Public Health, Research Unit of Environmental Medicine, University of Southern Denmark, Odense, Denmark

Wayne B. Jonas, East Coast U.S. Office, Samueli Institute for Information Biology, Alexandria, VA, USA

Bernt Jones, Department of Clinical Chemistry, Faculty of Veterinary Medicine, Swedish University of Agricultural Sciences, Uppsala, Sweden

Alexander W. Karlsen, United States Department of the Interior, United States Geological Survey, Reston, VA, USA

David G. Kinniburgh, British Geological Survey, Oxfordshire, UK

Leonard F. Konikow, United States Department of the Interior, United States Geological Survey, Reston, VA, USA

Michael Koss, Medical Laboratories, University of Southern California Medical Center, Los Angeles, CA, USA

Gerald Lalor, International Centre for Environmental and Nuclear Sciences, University of the West Indies, Kingston, Jamaica

Ulf Lindh, Biomedical Radiation Sciences, Rudbeck Laboratory, Research in Metal Biology, Uppsala University, Uppsala, Sweden

Stephen Macko, Department of Environmental Sciences, University of Virginia, Charlottesville, VA, USA

Bernhard Michalke, GSF National Research Center for Environment and Health, Munich, Neuherberg, Germany

Chester G. Moore, Colorado State University, Fort Collins, CO; and Division of Vector-Borne Infectious Diseases, Centers for Disease Control and Prevention, Fort Collins, CO, USA

Florabel G. Mullick, Department of Environmental and Toxicologic Pathology, The Armed Forces Institute of Pathology, Washington, D.C., USA

Jesper B. Nielsen, Institute of Public Health, University of Southern Denmark, Denmark

Monica Nordberg, Institute of Environmental Medicine, Karolinska Institute, Stockholm, Sweden

Demosthenes Pappagianis, School of Medicine, Medical Microbiology & Immunology, University of California-Davis, Davis, CA, USA

Roger Paulen, Alberta Geological Survey, Alberta Energy and Utilities Board, Edmonton, Canada

Daniel P. Perl, Mount Sinai School of Medicine, New York, NY, USA

Joseph P. Pestaner, Brody School of Medicine, East Carolina University, Greenville, NC, USA

Eva Rubenowitz-Lundin, Department of Environmental Medicine, Göteborg University, Göteborg, Sweden

Olle Selinus, Geological Survey of Sweden, Uppsala, Sweden

Timothy M. Shields, Bloomberg School of Public Health, The Johns Hopkins University, Baltimore, MD, USA

H. Catherine W. Skinner, Department of Geology and Geophysics; and Department of Orthopaedics and Rehabilitation, Yale University, New Haven, CT, USA

Pauline Smedley, British Geological Survey, Oxfordshire, UK

Paul B. Tchounwou, College of Science, Engineering and Technology, Jackson State University, NIH-Center for Environmental Health, Jackson, MS, USA

Todor Todorov, Department of Environmental and Toxicologic Pathology, The Armed Forces Institute of Pathology, Washington, D.C., USA

Chian-Hsiao Tseng, Department of Internal Medicine, National Taiwan University Hospital, Taipei, Taiwan

Mitko Vutchkov, International Centre for Environmental and Nuclear Sciences, University of the West Indies, Kingston, Jamaica

Philip Weinstein, School of Population Health, University of Western Australia, Crawley, Australia

Robert J. P. Williams, Inorganic Chemistry Laboratory, Oxford University, Oxford, UK

1H																	2He
3Li	4Be											5B	6C	7N	8O	9F	10Ne
11Na	12Mg											13Al	14Si	15P	16S	17Cl	18Ar
19K	20Ca	21Sc	22Ti	23V	24Cr	25Mn	26Fe	27Co	28Ni	29Cu	30Zn	32Ga	32Ge	33As	34Se	36Br	36Kr
37Rb	38Sr	39Y	40Zr	41Nb	42Mo	43Tc	44Ru	45Rh	46Pd	47Ag	48Cd	49In	50Sn	51Sb	52Te	53I	54Xe
55Cs	56Ba	57-71	72Hf	73Ta	74W	75Re	76Os	77Ir	78Pt	79Au	80Hg	81Tl	82Pb	83Bi	84Po	85At	86Rn
87Fr	88Ra	89-103	104Db	105Jo	106Rf	107Bh	108Hn	109Mt	110	111							

57La	58Ce	59Pr	60Nd	61Pm	62Sm	63Eu	64Gd	65Tb	66Dy	67Ho	68Er	69Tm	70Yb	71Lu
89Ac	90Th	91Pa	92U	93Np	94Pu	95Am	96Cm	97Bk	98Cf	99Es	100Fm	101Md	102No	108Lr

FIGURE 1 Periodic table illustrating major elements (pink), minor elements (blue), trace elements (yellow), and noble gases (gray) in the biosphere. Those in green are essential trace elements. Known established toxic elements are shown in red.

All substances are poisons; there is none which is not a poison. The right dose differentiates a poison and a remedy.

Paracelsus (1493–1541)

Medical geology is the science dealing with the relationship between natural geological factors and health in humans and animals and with understanding the influence of ordinary environmental factors on the geographical distribution of such health problems. It is a broad and complicated subject that requires interdisciplinary contributions from various scientific fields if its problems are to be understood, mitigated, or resolved. Medical geology, which focuses on the impacts of geologic materials and processes (i.e., the natural environment) on animal and human health, can be considered as complementary to environmental medicine. The field of medical geology brings together geoscientists and medical and public health researchers to address health problems caused or exacerbated by geologic materials such as rocks, minerals, and water and geologic processes such as volcanic eruptions, earthquakes, and dust.

Paracelsus defined a basic law of toxicology: Any increase in the amount or concentration of elements causes increasing negative biological effects, which may lead to inhibition of biological functions and, eventually, to death. However, despite the harmful effects of some elements, others are essential for life. Therefore, deleterious biological effects can result from either increasing or decreasing concentrations of various trace elements. Thus, as with many aspects of life, either too much or too little can be equally harmful. All of the elements that affect health are found in nature and form the basis for our existence as living creatures. The periodic table of elements, as an indicator of the roles played by the elements in the biosphere, is the basis for our understanding (Figure 1).

The writings of Hippocrates, a Greek physician of the Classical Period, demonstrate how far back our basic knowledge extends:

Whoever wishes to investigate medicine properly, should proceed thus. . . . We must also consider the qualities of the waters, for as they differ from one another in taste and weight, so also do they differ much in their quality.

Hippocrates (460–377 BC)

Hippocrates held the belief that health and "place" are related to ancient origin. Knowledge of specific animal diseases also originated long ago. Even in Chinese medical texts of the third century BC, cause-and-effect relationships are found. Unfortunately, most such observations were lost because they were never written down. As the science grew, many previously unknown relationships began to be understood and a new scientific field evolved: medical geology. This book covers the essentials of our knowledge in this area.

GEOLOGY AND HEALTH

Geology may appear far removed from human health. However, rocks and minerals comprise the fundamental building blocks of the planet and contain the majority of naturally occurring chemical elements. Many elements are essential to plant, animal, and human health in small doses. Most of these elements are taken into the human body via food, water, and air. Rocks, through weathering processes, break down to form the soils on which crops and animals are raised. Drinking water travels through rocks and soils as part of the hydrological cycle and much of the dust and some of the gases contained in the atmosphere are of geological origin. Hence, through the food chain and through the inhalation of atmospheric dusts and gases, human health is directly linked to geology.

The volcanic eruption of Mount Pinatubo is a splendid example of the dramatic effects of geology. Volcanism and related activities are the principal processes that bring elements to the surface from deep within the Earth. During just two days in June 1991, Pinatubo ejected 10 billion metric tonnes of magma and 20 million tonnes of SO_2; the resulting aerosols influenced global climate for three years. This single event introduced an estimated 800,000 tonnes of zinc, 600,000 tonnes of copper, 550,000 tonnes of chromium, 100,000 tonnes of lead, 1000 tonnes of cadmium, 10,000 tonnes of arsenic, 800 tonnes of mercury, and 30,000 tonnes of nickel to the surface environment. (Garrett, R.G., 2000). Volcanic eruptions redistribute many harmful elements such as arsenic, beryllium, cadmium, mercury, lead, radon, and

Garrett, R.G., 2000. Natural sources of Metals in the Environment. *Human and Ecological Risk Assessment*, Vol. 6, No. 6, pp 954–963.

uranium. Many other redistributed elements have undetermined biological effects. At any given time, on average, 60 volcanoes are erupting on the land surface of the Earth, releasing metals into the environment. Submarine volcanism is even more significant than that at continental margins, and it has been conservatively estimated that at least 3000 vent fields are currently active along the mid-ocean ridges.

GOAL AND APPROACH

Because of the importance of geological factors on health and the widespread ignorance of the importance of geology in such relationships, in 1996 the International Union of Geological Sciences (IUGS) commission COGEOENVIRONMENT (Commission on Geological Sciences for Environmental Planning) established an International Working Group on Medical Geology with the primary goal of increasing awareness of this issue among scientists, medical specialists, and the general public. In 2000 the United Nations Educational, Scientific, and Cultural Organization (UNESCO) became involved through a new International Geological Correlation Programme (IGCP) project 454 Medical Geology. Project 454 brings together, on a global scale, scientists working in this field in developing countries with their colleagues in other parts of the world and stresses the importance of geoscientific factors that affect the health of humans and animals. In 2002 the International Council for Science (ICSU) made it possible to put together international short courses on this subject, a cooperation involving the Geological Survey of Sweden, US Geological Survey, and the US Armed Forces Institute of Pathology in Washington DC. The aim of these short courses, which are offered all over the world, is to share the most recent information on how metal ions and trace elements impact environmental and public health issues. The scientific topics of the courses include environmental toxicology; environmental pathology; geochemistry; geoenvironmental epidemiology; the extent, patterns, and consequences of exposures to metal ions; and analysis. Areas of interest include metal ions in the general environment, biological risk-assessment studies, modern trends in trace-element analysis, and updates on the geology, toxicology, and pathology of metal ion exposures.

Because of this increasing activity and interest in this field we decided to write a book that could be used both as a reference and as a general textbook. Our goal is to emphasize the importance of geology in health and disease in humans and animals. The audience of the book consists of upper division undergraduates, graduate students, environmental geoscientists, epidemiologists, medics, and decision makers, but, we have also strived to make the book interesting and understandable to environmentally conscious members of the general public.

There are important relationships between our natural environment and human health. Our approach is to integrate these two fields to enable better understanding of these often complex relationships. All chapters have numerous cross-references not only among the other chapters but also to related reading.

SECTIONAL PLAN

Chapter 1 gives a brief history of medical geology. It is not intended to be an exhaustive overview; instead our overview highlights some important cases in the development of the science of medical geology.

The subsequent material is presented in four sections, each describing different aspects of the subject.

The first section (Chapters 2–8) covers environmental biology. Environmental biology may be characterized by interactions between geological and anthropogenic sources and the kingdoms of life. The geological sources provide life with essential major, minor, and trace elements. In addition, geology provides access to nonessential elements. To influence life, both beneficially and adversely, elements have to be in the environment as well as, in most cases, bioavailable. Therefore this section gives an introduction to the different aspects of environmental biology and provides a foundation for the following sections.

The second section (Chapters 9–20), on pathways and exposures, covers many of the myriad different aspects of medical geology. It has long been said that "we are what we eat"; however, in terms of medical geology we are in fact what we eat, drink, and breathe. The major pathways into the human body of all elements and compounds, whether beneficial or harmful, derive from the food and drink we consume and the air we breathe. The twelve chapters of this section concentrate on the interrelationships among our natural environment, geology, and health. Numerous examples from all over the world are presented on topics ranging from element toxicities and deficiencies, to geophagia,

to global airborne dust and give a clear view of the vast importance of the natural environment on our health. After reading these chapters, you should have no doubt that geology is one of the most important, although often neglected, factors in our well-being.

The third section (Chapters 21–25), on environmental toxicology, pathology, and medical geology, covers the medical aspects of medical geology. In recent decades there has been an increasing awareness of the importance of the interaction of mammalian systems with their natural environment. The primary focus has been on understanding exposure to hazardous agents in the natural environment through air, water, and soil. Such appreciation has led to myriad investigations focused on identifying those natural (and sometimes anthropogenic) environmental risk factors that may be involved in the development of human and other animal diseases. These five chapters describe the different effects of elements in our bodies, how geology affects us, and how we can recognize these effects.

The fourth section (Chapters 26–31), on techniques and tools, brings together in a very practical way our knowledge of the different relevant disciplines. Geoscientists and medical researchers bring to medical geology an arsenal of valuable techniques and tools that can be applied to health problems caused by geologic materials and processes. Although some of these tools may be common to both disciplines, practitioners of these disciplines commonly apply them in novel ways or with unique perspectives. In this section we look at some of these tools and techniques.

Finally, we have included three appendices. Appendix A covers international and some national reference values for water and soils. Appendix B lists numerous Web links from Chapters 19 and 26. Appendix C is a large glossary to be used whenever you need a term explained. We have tried to make this glossary as comprehensive as possible but there will of course be some shortcomings. However, the glossary can also be found and downloaded from the Internet (books.elsevier.com/companions); therefore it can be completed with more explanations when needed.

ACKNOWLEDGMENTS

A volume like this does not come into being without the efforts of a great number of dedicated people. We express our appreciation to the sixty authors who wrote chapters. In addition to writing chapters, the authors have carried out revisions. To ensure the quality and accuracy of each contribution, at least two independent reviewers scrutinized each chapter from a scientific point of view. However, we have gone even one step further. An interdisciplinary book like this must be written at a level that makes it accessible to workers in many different professions and also to members of the general public interested in environmental sciences. Therefore each chapter has also been read by additional reviewers. The geoscientific chapters have been read by those from the medical profession and the medical chapters have been read by geoscientists. We wish to thank all these people for making this book possible.

We are indebted to the following reviewers: Gustav Åkerblom, James Albright, Neil Ampel, Arthur C. Aufderheide, Stefan Claesson, Rick Cothern, Gerald Feder, Peter Frisk, Arne Fröslie, Mark Gettings, Larry Gough, Steve Guptill, Michalan Harthill, Anders Hemmingsson, Brenda Houser, John Hughes, John M. Hunter, Nils Gunnar Ihlbäck, G. V. Iyengar, Erland Johansson, Karen Johnson, Chris Johnson, Andre Kajdacsy-Balla, David G. Kinniburgh, Allan Kolker, Orville Levander, Jan Luthman, Bo L. Lönnerdal, Jörg Matschullat, Chester G. Moore, Maria Nikkarinen, Björn Öhlander, James Oldfield, Rolf Tore Ottesen, Joseph P. Pestaner, Geoffrey Plumlee, Clemens Reimann, Reijo Salminen, Ulrich Siewers, Håkan Sjöström, David Smith, Barry Smith, Alex Stewart, David Templeton, and Paul Younger.

In addition to these, all editors and several of the authors have also acted as reviewers for different chapters.

We also thank the dedicated team of Academic Press—our executive editor Frank Cynar, project manager Jennifer Helé, and production editor Daniel Stone—for their outstanding work and for encouraging us and helping us when needed.

Lastly, we also want to remember professor Valentin K. Lukashev, of Minsk, Belarus, our good colleague and friend who attended the first planning meeting of this book. He died among us of a heart attack in Uppsala, Sweden, on June 8, 1998, shortly after having given a presentation on medical geology in Belarus and the former Soviet Union and after having contributed valuable suggestions for the contents of this book at the first planning discussions. The logotype of medical geology will always serve as a fond remembrance of him, since he had suggested it just before he passed away.

Olle Selinus
Brian Alloway
José A. Centeno
Robert B. Finkelman
Ron Fuge
Ulf Lindh
Pauline Smedley

MEDICAL GEOLOGY: PERSPECTIVES AND PROSPECTS

BRIAN E. DAVIES
Clemson University

CHARLOTTE BOWMAN
University of Victoria

THEO C. DAVIES
Moi University

OLLE SELINUS
Geological Survey of Sweden

CONTENTS

This chapter provides a brief history of medical geology—the study of health problems related to "place." This overview is not exhaustive; instead, it highlights some important cases that have arisen during the development of the science of medical geology. An excess or deficiency of inorganic elements originating from geological sources can affect human and animal well-being either directly (*e.g.*, effect of iodine on the goiter) or indirectly (*e.g.*, effect on metabolic processes such as the supposed protective effect of selenium in cardiovascular disease). Such links have long been known but were unexplained until attempts at alchemy became successful in the seventeenth century, when medicine ceased to be the art of monks versed in homeopathic remedies and modern geology was forged by Lyell and Hutton.

I. THE FOUNDATIONS OF MEDICAL GEOLOGY

A. Ancient Findings

Various ancient cultures made reference to the relationship between environment and health. In many cases, the health problems were linked to occupational environments, but close links to the natural environment were also noted. Chinese medical texts dating back to the third century BC contain several references to relationships between environment and health. During both the Song Dynasty (1000 BC) and the

Ming Dynasty (14th–17th century AD), lung problems related to rock crushing and symptoms of occupational lead poisoning were recognized. Similarly, the Tang Dynasty alchemist Chen Shao-Wei stated that lead, silver, copper, antimony, gold and iron were poisonous (cited in Liang *et al.*, 1998).

Contemporary archaeologists, osteologists, and historians have provided us with evidence that the poor health often revealed by the tissues of prehistoric cadavers and mummies can commonly be linked to detrimental environmental conditions of the time. Goiter, for example, which is the result of severe iodine deficiency, was widely prevalent in ancient China, Greece, and Egypt, as well as in the Inca state of Peru. The fact that this condition was often treated with seaweed, a good source of iodine, indicates that these ancient civilizations had some degree of knowledge with regard to the treatment of dietary deficiencies with natural supplements.

As early as 1500 years ago, certain relationships between water quality and health were also known:

Whoever wishes to investigate medicine properly, should proceed thus.... We must also consider the qualities of the waters, for as they differ from one another in taste and weight, so also do they differ much in their quality. (Hippocrates, 460–377 BC)

Hippocrates, a Greek physician of the Classical period, recognized that health and place are causally related and that environmental factors affected the distribution of disease (Låg, 1990; Foster, 2002). Hippocrates noted in his treatise *On Airs, Waters, and Places* (Part 7) that, under certain circumstances, water "comes from soil which produces thermal waters, such as those having iron, copper, silver, gold, sulphur, alum, bitumen, or nitre," and such water is "bad for every purpose." Vitruvius, a Roman architect in the last century BC, noted the potential health dangers related to mining when he observed that water and pollution near mines posed health threats (cited in Nriagu, 1983). Later, in the first century AD, the Greek physician Galen reaffirmed the potential danger of mining activities when he noticed that acid mists were often associated with the extraction of copper (cited in Lindberg, 1992).

An early description linking geology and health is recounted in the travels of Marco Polo and his Uncle Niccoló. Journeying from Italy to the court of the Great Khan in China in the 1270s they passed to the south and east of the Great Desert of Lop:

At the end of the ten days he reaches a province called Su-chau.... Travelers passing this way do not venture to go among these mountains with any beast except those of the country, because a poisonous herb grows here, which makes beasts that feed on it lose their hoofs; but beasts born in the country recognize this herb and avoid it. (Latham, 1958)

The animal pathology observed by Marco Polo that resulted from horses eating certain plants was similar to a condition that today we know is caused by the consumption of plants in which selenium has accumulated, and this explorer's account may be the earliest report of selenium toxicity. Marco Polo also described goiter in the area around the oasis city of Yarkand (Shache) and ascribed it to a peculiarity of the local water. Earlier, near Kerman on the Iranian eastern frontier, he commented on a lack of bellicosity in the tribesmen that he attributed to the nature of the soil. In what could be considered the first public health experiment, Marco Polo imported soil to place around the tribe's tents in an effort to restore their bellicosity. His approach proved to be effective (see also Chapter 15).

Health problems resulting from the production of metal have been identified in many parts of the world. The common use of heavy metals in ancient societies revealed their toxicity. Although the relationship between lead and a variety of health risks is now well documented in modern society, the relationship was less well known in the past. Lead has been exploited for over six millennia, with significant production beginning about 5000 years ago, increasing proportionately through the Copper, Bronze, and Iron Ages, and finally peaking about 2000 years ago (Hong *et al.*, 1994; Nriagu, 1998). Several descriptions of lead poisoning found in texts from past civilizations further corroborate the heavy uses of lead. Clay tablets from the middle and late Assyrian periods (1550–600 BC) provide accounts of lead-poisoning symptoms, as do ancient Egyptian medical papyri and Sanskrit texts dating from over 3000 years ago (Nriagu, 1983). About 24% of discovered lead reserves were mined in ancient times (Nriagu, 1998).

It has been estimated that during the time of the Roman Empire the annual production of lead approached 80,000 tonnes (Hong *et al.*, 1994; Nriagu, 1998), and copper, zinc and mercury were also mined extensively (Nriagu, 1998). Lead usage exceeded 550 g per person per year, with the primary applications being plumbing, architecture, and shipbuilding. Lead salts were used to preserve fruits and vegetables, and lead was also added to wine to stop further fermentation and to add color or bouquet (Nriagu, 1983). The use of large amounts of lead in the daily life of Roman aristocracy had a significant impact on their health, including epidemics of plumbism, high incidence of sterility and stillbirths and mental incompetence. Physiological profiles of Roman emperors who lived between 50 and 250 BC suggest that the majority of these individuals suffered from lead poisoning (Nriagu, 1983). In turn, it is generally believed that a contributing factor to the fall of the Roman Empire, in AD 476, may have been the excessive use of lead in pottery and other sources (Hong et al., 1994).

Mercury was used during the Roman Empire to ease the pain of teething infants, as well as to aid in the recovery of gold and silver. Such applications were also widely found in Egypt in the twelfth century and in Central and South America in the sixteenth century (Eaton & Robertson, 1994; Silver & Rothman, 1995). Mercury was used to treat syphilis during the sixteenth century and in the felting process in the 1800s (Fergusson, 1990).

Copper was first used in its native form approximately 7000 years ago, with significant production beginning some 2000 years later and eventually peaking at a production rate of about 15,000 tonnes annually during the Roman Empire, when it was used for both military and civilian purposes, especially coinage. A significant drop in the production of copper followed the fall of the Roman Empire, and production remained low until about 900 years ago when a dramatic increase in production occurred in China, reaching a maximum of 13,000 tonnes annually and causing a number health problems (Hong et al., 1994).

Arsenic was used for therapeutic purposes by the ancient Greeks, Romans, Arabs, and Peruvians, because small doses were thought to improve the complexion; however, it has also long been used as a poison (Fergusson, 1990). In the sixteenth century, George Agricola described the symptoms of "Schneeberger" disease among miners working in the Erzgebirge of Germany to mine silver in association with uranium. That disease has since been identified as lung cancer deriving from metal dust and radon inhalation.

B. More Recent Findings

The industrial revolution in Europe and North America encouraged people to quit the poverty of subsistence agriculture in the countryside to live in increasingly crowded cities where they found work in factories, chemical plants, and foundries; however, such occupations exposed the workers to higher levels of chemical elements and compounds that, as rural dwellers, they would rarely have encountered. Friedrich Engels wrote graphic descriptions of the ill health of the new English proletariat in his politically seminal book, *The Conditions of the Working Class in England*, published in 1845. He described the plight of children forced to work in the potteries of Staffordshire: "By far the most injurious is the work of those who dip . . . into a fluid containing great quantities of lead, and often of arsenic. . . . The consequence is violent pain, and serious diseases of the stomach and intestines . . . partial paralysis of the hand muscles . . . convulsions" (Engels, 1845). Engels further characterized the conditions of workers in mid-nineteenth century industrial England as "want and disease, permanent or temporary."

The sciences of toxicology and industrial medicine arose in response to the health problems caused by unregulated industrialization. These sciences have provided the clinical data that allow us to understand the consequences of excess exposure to elements in the natural environment, whether it be due to simple exposure to particular rocks or the exploitation of mineral resources. The emergence of modern geological sciences coupled with increasingly powerful analytical techniques laid the foundation for determining the nature and occurrence of trace elements in rocks and sediments. Scientific agriculture has focused attention on inorganic element deficiencies in plants and animals, and modern medicine has provided reliable descriptions of diseases and more accurate diagnoses through internationally recognized nomenclatures.

Rural people have always recognized that the health of their animals is influenced by their diet and, therefore, soil properties. These observations could not be explained until the advent of scientific agriculture in the nineteenth century, when it required only a small step to suggest that humans may also be caught up in similar relationships. Diseases now known to be caused by a lack or excess of elements in soil and plants were given names that reflected where they occurred, such as Derbyshire neck in the iodine-deficient areas of the English Midlands or Bodmin Moor sickness over the granites of southwest England where cobalt deficiency is endemic in sheep unless treated. It is interesting to note that in Japan, before the 1868 Meiji Restoration, meat was rarely eaten so there was no tradition of animal husbandry. Japanese authors have suggested that this lack of animal indicators largely contributed to the failure to recognize the significance of metal pollution until it became catastrophic.

Archaeologists have also noted links between health and environmental factors. Analysis of bone material has provided an excellent tool for studying the diet and nutritional status of prehistoric humans and animals (Krzysztof & Glab, 2001). For example, the transition from a hunter–gatherer society to an agriculturally based economy resulted in a major dietary change and an accompanying iron deficiency. Iron in plants is more difficult to absorb than iron from a meat source; hence, it has been proposed that this new reliance on a crop diet may have resulted in iron deficiency and anemia among the general population (Roberts & Manchester, 1995).

Skeletal remains found in Kentucky have provided prime examples of the relationship between geology and ancient human health. The area is blanketed by mineral-deficient soils. Native Americans, however, established permanent settlements in the area and began normal crop cultivation practices. As a result of the soil mineral deficiency, the maize produced had extremely low levels of zinc and manganese. These deficiencies led to a range of diet-related health effects that have been clearly documented through the study of dental and skeletal pathology in human remains (Moynahan, 1979).

Several landmark discoveries in medical geology have been made in Norway. For a long time, Norwegian farmers have been aware of the unusually frequent occurrence of osteomalacia among domestic animals in certain districts, and to combat the disease they initiated the practice of adding crushed bones to the feed of the animals. Some farmers suspected that a particular pasture plant caused osteomalacia, and a Norwegian official named Jens Bjelke (1580–1659), who had an interest in botany and a knowledge of foreign languages, gave the suspected plant the Latin name *Gramen ossifragum* ("the grass that breaks bones"). The name has also been *written Gramen Norwagicum ossifragum*. One hundred years ago, the geologist J. H. L. Vogt learned of the practice of adding crushed bones to the diets of farm animals and investigated a region where osteomalacia was common. When he found very small amounts of the mineral apatite in the rocks, he drew the logical and correct conclusion that a deficiency of phosphorus was the cause of the osteomalacia. Another Norwegian geologist had previously pointed out that vegetation was extraordinarily sparse over the bedrock which was found by Vogt to be very poor in apatite. Once the cause of the osteomalacia was determined, it became a relatively simple matter to prevent the damage by adding phosphorus fertilizer to the soil (Låg, 1990) (see also Chapter 14, this volume).

A significant publication that must be mentioned here is André Voisin's book, *Soil, Grass and Cancer* (1959), especially in light of today's interest in the dangers of free radicals in cells and the protective effects of antioxidant substances and enzymes. Over 40 years ago, Voisin stressed the protective role of catalase and observed that copper deficiency was accompanied by low cytochrome oxidase activity.

Oddur Eiriksson and Benedikt Pjetursson provided detailed descriptions of the damage to teeth of domestic animals that resulted from the eruption of the Icelandic volcano Hekla in 1693. Of course, at that time it was not known that the cause was fluorosis. The relationship between the incidence of fluorine deficiency and dental caries has been carefully studied in Scandinavia since World War II, with attention being particularly centered around the need for fluoridation of water. Analyses of the fluoride content of natural waters from various sources and their relationships to the frequency of caries have been reported from several districts (see also Chapter 9, this volume).

II. GEOCHEMICAL CLASSIFICATION OF THE ELEMENTS

It is humbling to realize that the principles of geochemistry and, hence, medical geology were established at a time when few had access to advanced analytical facilities and most scientists relied on the very laborious classical chemical approaches. Despite the limitations imposed by a lack of rapid analysis of rocks and soils, the basic principles of geochemistry were known by the start of the twentieth century. In 1908, Frank W. Clarke, of the U.S. Geological Survey, published the original edition of *The Data of Geochemistry*, in which he adopted a systems approach to present his information. Clarke's book was the forerunner of several texts published during the first half of the twentieth century that have helped us understand how geochemistry is linked to health. Arguably the most important text of the period was V. Goldschmidt's *Geochemistry* (1954), which was based on work by Linus Pauling; it was completed by Alex Muir in Scotland and published after Goldschmidt's death in 1947. Two of Goldschmidt's ideas are of special relevance to medical geology: his geochemical classification of the elements and his recognition of the importance of ionic radii in explaining "impurities" in natural crystals.

Goldschmidt's geochemical classification groups elements into four empirical categories (Table I). The *siderophilic* elements are those primarily associated with

TABLE I. Geochemical Classification of Elements

Group	Elements
Siderophile	Fe, Co, Ni, Pt, Au, Mo, Ge, Sn, C, P
Atmophile	H, N, O
Chalcophile	Cu, Ag, Zn, Cd, Hg, Pb, As, S, Te
Lithophile	Li, Na, K, Rb, Cs, Mg, Ca, Sr, Ba, Al, rare earths (REE)

the iron–nickel (Fe–Ni) core of the Earth; these elements may be found elsewhere to some extent, but this classification explains why, for example, platinum and associated metals are normally rare and dispersed in crustal rocks. This fundamental geochemical observation allowed Alvarez *et al.* (1980) to recognize the significance of the high iridium contents of clays found at the Cretaceous/Tertiary (K/T) boundary. They proposed the persuasive idea that the impact of an asteroid (Fe–Ni type) on the surface of the Earth could explain the massive species extinctions that define the K/T boundary, including the demise of the dinosaurs. Was this an example of medical geology on a global scale?

The *atmophilic* elements are those dominating the air around us, and *lithophilic* elements are common in crustal silicates (Alvarez *et al.*, 1980). Of special interest are the *chalcophilic* elements, which derive their name from a geochemical grouping of these elements with copper (Greek χαλκός). These elements are encountered locally in high concentrations where recent or ancient reducing conditions (and hydrothermal conditions) have led to the reduction of sulfate to sulfide, resulting in the formation of sulfide minerals such as pyrite (FeS_2) and the ores of lead (galena, PbS) or zinc (sphalerite, ZnS). This same thiophilic tendency underlies the toxicity of lead, mercury, and cadmium because they readily link to the —SH groups of enzymes and thereby deactivate them. Goldschmidt's empirical classification of chalcophilic elements is now reinterpreted in terms of hard and soft acids and bases; soft bases (*e.g.*, R—SH or R—S$^-$) preferentially bind to soft acids (*e.g.*, Cd^{2+} or Hg^{2+}).

Goldschmidt's (and Pauling's) second important concept was the importance of ionic size in explaining both the three-dimensional structures of silicate crystals and how other elements can become incorporated in them. The rules are now generally known as *Goldschmidt's rules of substitution*:

1. The ions of one element can replace another in ionic crystals if their radii differ by less than about 15%.
2. Ions whose charges differ by one unit can substitute provided electrical neutrality of the crystal is maintained.
3. For two competing ions in substitution, the one with the higher ionic potential (charge/radius ratio) is preferred.
4. Substitution is limited when competing ions differ in electronegativity and form bonds of different ionic character.

These rules of substitution and the geochemical classification of elements are fundamental to our growing understanding of medical geology, for they explain many environmental occurrences of toxic elements and allow scientists to predict where such occurrences might be found.

III. CONTRIBUTIONS TO MEDICAL GEOLOGY FROM PUBLIC HEALTH AND ENVIRONMENTAL MEDICINE

Although most public health problems involve diseases caused by pathogens, inorganic poisons can also affect public health; among these poisons are arsenic, cadmium, and mercury. Currently much concern exists about environmental levels of mercury, especially in Amazonia, where the amalgamation of gold by mercury in small-scale mining operations has caused widespread mercury pollution. The effects of this metal on human health can be traced back several centuries. For example, in the sixteenth century and later, mercury and its compounds were widely used to treat syphilis despite its known toxicity (D'itri & D'itri, 1977), and mercuric nitrate solution was used to soften fur for hat making. Long-term exposure caused neurological damage in workers handling mercury and gave rise to expressions such as "mad as a hatter" or the "Danbury shakes." In Birmingham, England, buttons were gilded by exposing them to a gold–mercury (Au–Hg) amalgam followed by vaporization of the mercury. By 1891, many tons of mercury had been dissipated around Birmingham, to the great detriment of that city's inhabitants, many of whom suffered from Gilder's palsy. Neurological damage due to exposure to inorganic mercury compounds was well understood by the end of the nineteenth century.

Modern concerns, however, are focused on lipid-soluble organic compounds that tend to concentrate as one proceeds up the food chain. Recognition of such a problem resulted from the well-known outbreak of methylmercury poisoning in 1956 in Minamata city in Japan, thus the name used today—Minamata disease. Subsequently, methylmercury poisoning has been observed in, for example, Niigata (Japan), Sweden, Iraq, and the United States.

Concern about environmental cadmium can be traced back to the outbreak of itai itai disease in Japan earlier in the twentieth century (Chaney et al., 1998). The disease resulted in severe bone malformations in elderly women, and a zinc mine in the upper reaches of the Jintsu river was found to be the source of the cadmium that caused the disease. Later, cadmium was found to be linked to kidney damage, and the element was found to build up in soil following the application of some sewage sludges. Many countries now control the land application of sludge and have set limits in terms of permissible cadmium additions (Friberg et al., 1974).

The colored compounds of arsenic were used as pigments as early as the Bronze Age, and knowledge of its toxicity is just as old. Of concern today are the skin lesions and cancers observed among the millions of people drinking arsenic-rich well water, especially in West Bengal and Bangladesh. As with mercury, links between arsenic and certain cancers were identified early on. Fowler's solution, which contained potassium arsenite, was widely prescribed as a tonic. Patients who believed that if a little of something (a few drops) would do them good then a lot of it must do them a lot of good tended to overdose on the solution. By the late eighteenth century, it was recognized that injudicious use of Fowler's solution led first to peripheral neuritis, which was followed by skin lesions and cancer (see also Chapters 11 and 23, this volume).

Coal is a sedimentary rock formed by the diagenesis of buried peats, which, in turn, form from organic debris under wet, reducing conditions. This process favors the precipitation of the sulfides of chalcophilic metals (especially pyrite, FeS_2). Pyrite can contain significant concentrations of arsenic as well as mercury, thallium, selenium, nickel, lead, and cobalt. Incineration of coal releases mercury to the atmosphere; sulfur gases, which cause acid precipitation, and arsenic compounds may also be released or remain in the ash.

In the autumn of 1900, an epidemic of arsenic poisoning occurred among beer drinkers in Manchester, Salford, and Liverpool in England. The poisoning was first traced to the use of sulfuric acid to make the glucose required for the brewing process; apparently, the breweries had unknowingly switched from de-arsenicated acid (sulfuric acid is a valuable by-product of smelting industries, including those dealing with arsenic ores). Additionally, however, malted barley was dried over coal fires, which contributed to the problem. Even moderate beer drinkers suffered from peripheral neuritis and shingles (herpes zoster), which can be induced by arsenic exposure. Arsenic poisoning has recently emerged again in China, where severe arsenic poisoning has been reported in recent years as a result of consumption of vegetables dried over coal fires (Finkelman et al., 1999).

IV. DEVELOPMENT OF MEDICAL GEOLOGY

A. The Knowledge Gained from Single-Element Studies

Over the course of the twentieth century, geoscientists and epidemiologists gained a greater understanding of the many ways in which the environment of Earth can affect the health of its inhabitants. Incidents of metal poisoning and the identification of specific relationships between dietary constituents and health became representative examples of more general human reactions to exposures to the geochemical environment. The clearest example of the relationship between geology and health is when the presence of too much or too little of a single element in the environment is found to cause or influence disease as a result of being transferred into the body through dust in the soil or air or via water or food.

Iodine remains the classic success story in medical geology as far as human health is concerned. The most common health effect associated with an iodine deficiency is goiter, a swelling of the thyroid gland. Late in the nineteenth century, it was determined that iodine concentrates in the thyroid gland, but the iodine concentrations were reduced in the thyroids of patients from endemic goitrous areas. Iodine deficiency disorders (IDDs) remain a major threat to the health and development of populations the world over. Clinical treatment of IDDs is, of course, the prerogative of medical doctors; nonetheless, a greater understanding of the conditions leading to IDDs has resulted from the work of geoscientists. (Iodine is described in detail in Chapter 16.)

The study of arsenic remained the province of toxicology and forensic medicine until the middle twentieth century. A paper on arsenic in well water in the Canadian Province of Ontario stated: "The occurrence of arsenic in well water is sufficiently rare to merit description" (Wyllie, 1937). Pictures accompanying the text illustrate keratoses on the feet and the palm of a hand. It was concluded in the article that the occurrence of arsenic poisoning from well water was infrequent. Less than 40 years later, however, the scientific world learned of "blackfoot disease" in the Republic of China (Taiwan), and skin disorders and cancer due to arsenic-polluted well water have been described in Chile, Mexico, and Argentina. Serious problems are currently being reported in West Bengal and Bangladesh. In all cases, the geological link is clear (described in detail in Chapters 11 and 23).

Cobalt deficiency provides a good example of the relationship between animal health and the geological environment. In New Zealand, cobalt deficiency was known as "bush sickness" or Morton Mains disease; in Kenya, as *nakuruitis*; in England, as pining; in Denmark, as *vosk* or *voskhed*; and in Germany, as *hinsch*. The underlying cause was discovered by Dr. Eric Underwood, an early expert in the medical geology field (Underwood & Filmer, 1935). His discovery in 1935 of the essentiality of cobalt is an example of triumph over analytical difficulty. Underwood and Filmer showed that "enzootic marasmus" could be cured by treatment with an acid extract of the iron oxide limonite, from which all but negligible quantities of iron had been removed. In all cases, the problem can be traced back to a low cobalt content of the soil parent material. Inadequate cobalt is passed up the food chain for microflora in the gut of herbivores to use in the synthesis of the essential cobalt-containing cobalamin or vitamin B_{12}. Only one case of human cobalt deficiency appears to have been published (Shuttleworth *et al.*, 1961). A 16-month-old girl on an isolated Welsh hill farm was a persistent dirt eater and suffered from anemia and behavioral problems. The cattle on the farm were being treated for cobalt deficiency, and the child recovered her health after oral administration of cobaltous chloride.

Lead poisoning has dominated the environmental agenda for several decades. It is interesting to note that geologists were aware of the potential health problems associated with lead when medical opinion on the subject was still mixed. In mid-nineteenth century Britain, residents expressed growing concern about the unregulated disposal of mine and industrial wastes in rivers. In west Wales, farmers complained that lead mining was ruining their fields as a result of the deposition of polluted sediment when rivers flooded. A Royal Commission in 1874 evaluated their complaints, and legislation soon followed (River Pollution Prevention Act, 1878); however, it was too late. Well into the twentieth century, cattle poisoning in the Ystwyth valley of west Wales continued to occur due to the earlier contamination by mines in the previous century. As late as 1938, the recovery of these rivers was monitored, and even in the 1960s evidence of past pollution was still evident. It was the late Professor Harry Warren in Vancouver, Canada, who first recognized the important implications of high levels of environmental lead. He devoted the last 30 years of his professional life to arguing for the significance of lead uptake by garden vegetables and its possible role in the etiology of multiple sclerosis. Warren had pioneered the use of tree twigs in prospecting for mineral ores in British Columbia, Canada, and he was surprised to observe that lead contents were often higher in forests bordering roads and concluded that "industrial salting" was a widespread and serious problem. Nonetheless, until the 1960s, environmental lead remained a mere curiosity. Health problems were thought to occur only from industrial exposure or due to domestic poisoning from lead dissolved by soft water from lead pipes.

Over the past 20 years, the removal of lead from gas, food canning, and other sources has reduced population blood lead levels by over 80%. Milestones along the way included evidence that dust on hands represented a major pathway of lead exposure, and the phasing out of lead in gasoline in the United States was accompanied by a general reduction in blood lead levels. Adding to the debate was the contention that even relatively low levels of lead exposure could harm the development of a child's brain (Davies & Thornton, 1989; Nriagu, 1983; Ratcliffe, 1981; Warren & Delavault, 1971).

The medical geology of selenium provides a good example of the interaction between geology and medicine. In the late 1960s, selenium was shown to be essential for animals and to be an integral part of glutathione oxidase, an enzyme that catalyzes the breakdown of hydrogen peroxide in cells (Prasad, 1978). In sheep and cattle, a deficiency in selenium accounted for "white muscle disease" (especially degeneration of the heart muscle), and glutathione peroxidase activity was found to be a good measure of selenium status. The problem was particularly widespread among farm animals in Great Britain (Anderson *et al.*, 1979). Humans have also been shown to suffer from selenium deficiency, and in China this condition is referred to as Keshan disease (Rosenfeld & Beath, 1964; Frankenberger & Benson, 1994; Frankenberger & Enberg, 1998). The disease has occurred in those areas of China where dietary intakes of selenium are less than 0.03 mg/day because the selenium content of the soils is low. The condition is characterized by heart enlargement and congestive heart failure. The disease has been primarily seen in rural areas and predominantly among peasants and their families. Those most susceptible have been children from 2 to 15 years of age and women of child-bearing age (Chen *et al.*, 1980; Jianan, 1985). Also, it has been suggested that adequate selenium intake may be protective for cancers (Diplock, 1987), and self-medication with selenium supplements has become widespread with the belief that a lack of selenium is a risk factor in heart diseases. (Selenium is described in greater detail in Chapter 15.)

B. The Importance of Element Interactions Is Recognized

The number of productive single-element studies has obscured two fundamental geochemical principles: First, from a geochemistry perspective, elements tend to group together, and, second, the study of physiology recognizes that elements can be synergistic or antagonistic. Cadmium is a good example of both principles. In some environments, soil cadmium levels are high because of rock type (such as black shales) or from mining contamination. A highly publicized polluted environment is that of the village of Shipham, which in the eighteenth century was a thriving zinc mining village in the west of England. It has been speculated that an adequate selenium intake may be protective for cancers (Diplock, 1987). The belief that a lack of selenium is a risk factor in heart diseases has caused self-medication with selenium supplements to become widespread. A study in 1979 suggested that 22 out of 31 residents showed signs of ill health that could be traced to cadmium. As a result, the health of over 500 residents was subsequently assessed and compared with that of a matching control population from a nearby non-mining village, but "there was no evidence of adverse health effects in the members of the population studied in Shipham" (Thornton *et al.*, 1980). Chaney *et al.* (1998) have commented on the disparity between the reports of ill health in Japan and no-effect observations from other parts of the world: "research has shown that Cd transfer in the subsistence-rice food-chain is unique, and that other food-chains do not comprise such high risk per unit soil Cd" and "Evidence indicates that combined Fe and Zn deficiencies can increase Cd retention by 15 fold compared to Fe and Zn adequate diets . . . it

is now understood that rice grain is seriously deficient in Fe, Zn, and Ca for human needs".

Copper and molybdenum taken individually and together demonstrate the importance of not relying upon simple single-cause relationships. In Somerset (England) there is an area in which pasture causes scouring in cattle. The land is known locally as "teart" and was first reported in the scientific literature in 1862 (Gimmingham, 1914), but the cause of the disorder (molybdenum) was not ascertained until 1943 (Ferguson *et al.*, 1943), when it was shown that the grass contained 20–200 mg molybdenum per kg (d.m.) and that the disorder could be cured by adding cupric sulfate to the feed. The origin of the excess molybdenum was the local black shales (Lower Lias) (Lewis, 1943). Over 20 years later, geochemical reconnaissance of the Lower Lias throughout the British Isles showed that elevated molybdenum contents in soils and herbage were a widespread problem over black shale, regardless of geological age, and that this excess molybdenum was the cause of bovine hypocuprosis (Thornton *et al.*, 1966, 1969; Thomson *et al.*, 1972). A moose disease in Sweden provides another example of the effects of molybdenum, in this case resulting from the interaction of molybdenum with copper. This disease is covered in detail in Chapter 20 (see also Kabata-Pendias and Pendias, 1992; Kabata-Pendias, 2001; Adriano, 2001).

C. Mapping Diseases as a Tool in Medical Geology

For some important groups of diseases (*e.g.*, cancers, diseases of the central nervous system, and cardiovascular disease), the causes are by and large unknown and cure and control is uncertain. When the incidence or prevalence of these diseases has been mapped, especially in countries of western Europe, significant differences from place to place have been reported that are not easily explained by genetic traits or social or dietary differences. Environmental influences appear to be involved in the etiologies, and a role for geology has been suggested by many authors (see, for example, Chapter 13). Association is not necessarily evidence for cause and effect. For mapping approaches to be reliable, two conditions must be satisfied. First, it is essential to be able to show a clear pathway from source (*e.g.*, soil) to exposure (*e.g.*, dirt on hands) to assimilation (*e.g.*, gastric absorption) to a target organ or physiological mechanism (*e.g.*, enzyme system). The second condition, rarely satisfied, is that the hypothetical association must be predictive: If the association is positive in one area, then it should also be positive in a geologically similar area; if not, why not? This condition is well illustrated by fluoride and dental caries—environments where fluoride is naturally higher in drinking water have consistently proved to have lower caries rates.

Over the years much attention has been paid to the geographical variability of cancer occurrences and it has been speculated that this variability may be influenced by soil or water quality. An early study of gastrointestinal cancer in north Montgomeryshire, Wales (Millar, 1961) seemed to show an association with environmental radioactivity because local black shales were rich in uranium. There was no direct evidence to support the hypothesis, and the study was marked by a problem of earlier work—namely, an indiscriminate use of statistics. Work in 1960 in the Tamar valley of the west of England appeared to show that mortality from cancer was unusually low in certain villages and unusually high in others (Davies, 1971). Within the village of Horrabridge, mortality was linked to the origin of different water supplies: The lowest mortality was associated with reservoir water from Dartmoor, whereas the highest mortality was associated with well or spring water derived from mineralized rock strata. Although this study was again statistically suspect, it stimulated a resurgence of interest in the link between cancer and the environment.

Stocks and Davies (1964) sought direct associations between garden soil composition and the frequency of stomach cancer in north Wales, Cheshire, and two localities in Devon. Soil organic matter, zinc, and cobalt were related positively with stomach cancer incidence but not with other intestinal cancer. Chromium was connected with the incidence of both. The average logarithm of the ratio of zinc/copper in garden soils was always higher where a person had just died of stomach cancer after 10 or more years of residence than it was at houses where a person had died similarly of a nonmalignant cause. The effect was more pronounced and consistent in soils taken from vegetable gardens, and it was not found where the duration of residence was less than 10 years.

A possible link between the quality of water supply, especially its hardness, was the focus of much research in the 1970s and 1980s. This was noticed, for example, in Japan in 1957. A statistical relationship was found between deaths from cerebral hemorrhage and the sulfate/carbonate ratio in river water which, in turn, reflected the geochemical nature of the catchment area. In Britain, calcium in water was found to correlate inversely with cardiovascular disease, but the presence of magnesium did not; thus, hardwater may exercise some protective effect. Attention has also been paid to

a possible role for magnesium, because diseased heart muscle tissue is seen to contain less magnesium than healthy tissue. Still, it has to be pointed out that hardwaters do not necessarily contain raised concentrations of magnesium; this occurs only when the limestones through which aquifer water passes are dolomitized, and most English limestones are not. More details can be found in Chapter 13.

Mapping diseases has also been a valuable tool for a long time in China, where pioneering work has been done by Tan Jianan (1989). Modern mapping techniques are now widely used in medical geology; mapping and analytical approaches to epidemiological data are covered in Cliff and Haggett (1988), while discussions on using GIS and remote sensing, as well as several examples, are offered in Chapters 26 and 27.

D. Dental Health Provides an Example of the Significance of Element Substitutions in Crystals

Dental epidemiology has provided some of the most convincing evidence that trace elements can affect the health of communities (Davies & Anderson, 1987). Dental caries is endemic and epidemic in many countries, so a large population is always available for study. Because diagnosis relies upon a noninvasive visual inspection that minimizes ethical restrictions, a high proportion of a target population can be surveyed. Where the survey population is comprised of children (typically 12 year olds), the time interval between supposed cause and effect is short, and it is possible to make direct associations between environmental quality and disease prevalence. In the case of fluoride, a direct link was established over 50 years ago that led to the successful fluoridation of public water supplies. This is an example of medical geology influencing public health policy. The relationship between dental caries and environmental fluoride, especially in drinking water, is probably one of the best known examples of medical geology. So strong is the relationship that the addition of 1 milligram of fluoride per liter to public water supplies has been undertaken regularly by many water utilities as a public health measure.

The history of the fluoride connection is worth recounting. In 1901, Dr. Frederick McKay opened a dental practice in Colorado Springs, Colorado, and encountered a mottling and staining of teeth that was known locally as "Colorado stain." The condition was so prevalent that it was regarded as commonplace but no reference to it could be found in the available literature. A survey of schoolchildren in 1909 revealed that

87.5% of those born and reared locally had mottled teeth. Inquiries established that an identical pattern of mottling in teeth had been observed in some other American areas and among immigrants coming from the volcanic areas of Naples, Italy. Field work in South Dakota and reports from Italy and the Bahamas convinced McKay that the quality of the water supply was somehow involved in the etiology of the condition. He found direct evidence for this in Oakley, Idaho, where, in 1908, a new piped water supply was installed from a nearby thermal spring and, within a few years, it was noticed that the teeth of local children were becoming mottled. In 1925, McKay persuaded his local community to change their water supply to a different spring, after which stained teeth became rare.

A second similar case was identified in Bauxite, Arkansas, where the water supply was analyzed for trace constituents, as were samples from other areas. The results revealed that all the waters associated with mottled teeth had in common a high fluoride content ($2–13\,mg\,F/L^{-1}$). In the 1930s, it was suggested that the possibility of controlling dental caries through the domestic water supply warranted thorough epidemiological-chemical investigation. The U.S. Public Health Service concluded that a concentration of 1 mg fluoride per liter drinking water would be beneficial for dental health but would not be in any way injurious to general health. Fluoride was first added to public water supplies in 1945 in Grand Rapids, Michigan. Fluoridation schemes were subsequently introduced in Brantford, Ontario (1945); Tiel, The Netherlands (1953); Hastings, New Zealand (1954); and Watford, Anglesey, and Kilmarnock in Great Britain (1955). There is no doubt that whenever fluorides have been used a reduction in the prevalence of dental caries follows (Davies & Anderson, 1987; Leverett, 1982) (see also Chapter 12, this volume).

V. AN EMERGING PROFESSION

The field of medical geology (or geomedicine) has developed around the world over the last few decades. The development of activities and the organizational structure of medical geology in a number of regions will be discussed in this section, including the United States, Great Britain, Scandinavia, some African countries, and China.

As research interest in medical geology grew during the 1960s, the desire emerged for conference sessions or even entire conferences dedicated to the subject. The late Dr. Delbert D. Hemphill of the University of Missouri organized the first Annual Conference on Trace Substances in Environmental Health in 1967, and these meetings continued for a quarter of a century. Early in the 1970s, several countries took the initiative to organize activities within the field of medical geology, and a symposium was held in Heidelberg, West Germany, in October 1972. In the United States, Canada, and Great Britain, research on relationships between geochemistry and health were carried out, and the Society for Environmental Geochemistry and Health was established. Geochemistry has for a long time maintained a strong position in the former Soviet Union, and basic knowledge of this science is routinely applied to medical investigations. Medical geology has a long tradition in northern Europe, and the development of this emerging discipline in Scandinavia has been strong. In Norway, too, geochemical research has been regarded as important for quite some time.

In North America in the 1960s and 1970s, a number of researchers made important contributions to our understanding of the role of trace elements in the environment and their health effects; among these are Helen Cannon (1971), H. T. Shacklette et al. (1972), and Harry V. Warren (1964). A meeting on environmental geochemistry and health was held and sponsored by the British Royal Society in 1978 (Bowie & Webb, 1980). Another landmark date was 1979, when the Council of the Royal Society (London) appointed a working party to investigate the role in national policy for studies linking environmental geochemistry to health. This was chaired by Professor S. H. U. Bowie of the British Geological Survey (Bowie & Thornton, 1985). In 1985, the International Association of Geochemistry and Cosmochemistry (IAGC) co-sponsored with the Society for Environmental Geochemistry and Health (SEGH) and Imperial College, London, the first International Symposium on Geochemistry and Health. Also in 1985, the journal *Environmental Geochemistry and Health* first appeared (Thornton, 1985). In 1987, a meeting on geochemistry and health was held at the Royal Society in London, and in 1993 a meeting on environmental geochemistry and health in developing countries was conducted at the Geological Society in London (Appleton et al., 1996).

Traditionally, the terms *geomedicine* and *environmental geochemistry and health* have been used. Formal recognition of the field of geomedicine is attributed to Ziess, who first introduced the term in 1931 and at the time considered it synonymous with *geographic medicine*, which was defined as "a branch of medicine where geographical and cartographical methods are used to present medical research results." Little changed until the 1970s, when Dr. J. Låg, of Norway, redefined the term as the "science dealing with the influence of ordinary environmental factors on the geographic distribution of health problems in man and animals" (Låg, 1990).

The Norwegian Academy of Science and Letters has been very active in the field of medical geology and has arranged many medical geology symposia, some of them in cooperation with other organizations. The proceedings of 13 of these symposia have been published. Since 1986, these symposia have been arranged in collaboration with the working group Soil Science and Geomedicine of the International Union of Soil Science. The initiator of this series of meetings was the late Dr. Låg, who was Professor of Soil Science at the Agricultural University of Norway from 1949 to 1985 and who was among the most prominent soil scientists of his generation, having made significant contributions to several scientific disciplines. During his later years, much of Dr. Låg's work was devoted to medical geology, which he promoted internationally through his book (Låg, 1990).

The countries of Africa have also experienced growth in the field of medical geology. The relationships between the geological environment and regional and local variations in diseases such as IDDs, fluorosis, and various human cancers have been observed for many years in Africa. Such research grew rapidly from the late 1960s, at about the same time that the principles of geochemical exploration began to be incorporated in mineral exploration programs on the continent. In Africa, evidence suggesting associations between the geological environment and the occurrence of disease continues to accumulate (see, for example, Davies, 2003), but in many cases the real significance of these findings remains to be fully appreciated. The reasons are threefold: (1) the paucity of reliable epidemiological data regarding incidence, prevalence, and trends in disease occurrence; (2) the lack of geochemists on teams investigating disease epidemiology and etiology; and (3) a shortage of analytical facilities for measuring the contents of nutritional and toxic elements at very low concentration levels in environmental samples (Davies, 1996). Confronting these challenges, however, could prove to be exceedingly rewarding, for it is thought that the strongest potential significance of such correlations exists in Africa and other developing regions of the world. Unlike the developed world, where most people no longer eat food grown only in their own area, most of the people in Africa live close to the land and are exposed in their daily lives, through food and water

intake, to whatever trace elements have become concentrated (or depleted) in crops from their farms (Appleton et al., 1996; Davies, 2000).

The first real attempt to coordinate research aimed at clarifying these relationships took place in Nairobi in 1999, when the first East and Southern Africa Regional Workshop was convened, bringing together over 60 interdisciplinary scientists from the region (Davies & Schlüter, 2002). One outcome of this workshop was the constitution of the East and Southern Africa Association of Medical Geology (ESAAMEG), establishing it as a chapter of the International Medical Geology Association (IMGA). The Geomed 2001 workshop held in Zambia testified to the burst of interest and research activities generated by that first workshop (Ceruti et al., 2001). As a result of this increasing awareness of medical geology problems around the continent, membership and activities of the ESAAMEG have continued to grow. This is a welcome sign on both sides of what has hitherto been an unbridged chasm between geology and health in Africa.

China has a long history of medical geology. Chinese medical texts dating back to the third century BC contain several references to relationships between geology and health. During both the Song Dynasty (1000 BC) and the Ming Dynasty (14th–17th century), lung ailments related to rock crushing and symptoms of occupational lead poisoning were recognized. Similarly, as noted earlier, the Tang Dynasty alchemist Chen Shao-Wei stated that lead, silver, copper, antimony, gold, and iron were poisonous.

In the twentieth century, much research has been carried out in China (for example, on the selenium-responsive Keshan and Kashin Beck diseases) that has resulted in clarification of the causes of a number of diseases, including endemic goiter and endemic fluorosis. One of the centers for this research has been the Department of Chemical Geography at the Chinese Academy of Sciences. At this institute, several publications have been produced, such as *The Atlas of Endemic Diseases and Their Environments in the People's Republic of China* (Jianan, 1985). Also the Institute of Geochemistry in Guiyang in Southern China is known for its studies in the field that is now referred to as medical geology.

VI. PROSPECTS

As we progress into the early years of the twenty-first century, it can be safely claimed that medical geology has emerged as a serious professional discipline. If respect for medical geology as a discipline is to continue to grow, then future studies must go well beyond simplistic comparisons of geochemical and epidemiological data. Dietary or other pathways must be traced and quantified and causative roles must be identified with regard to target organs or body processes. Moreover, studies must become predictive. Occasionally, simple direct links between geochemistry and health may be identified, but even in these instances confounding factors may be present (for example, the possible role of humic acids in arsenic exposure or the established role of goitrogenic substance in goiter). Ordinarily, geochemistry will provide at best only a risk factor: Unusual exposures, trace element deficiencies, or elemental imbalances will contribute toward the disturbance of cellular processes or activation of genes that will result in clinical disease. The problem of geographical variability in disease incidence will remain.

Rapid growth in the field of medical geology is predicted, as it is a discipline that will continue to make valuable contributions to the study of epidemiology and public health, providing hyperbole is avoided and a dialogue is maintained among geochemists, epidemiologists, clinicians, and veterinarians.

The structure of all living organisms, including humans and animals, is based on major, minor, and trace elements—given by nature and supplied by geology. The occurrence of these gifts in nature, however, is distributed unevenly. The type and quantity of elements vary from location to location—sometimes too much, sometimes too little. It is our privilege and duty to study and gain knowledge about natural conditions (*e.g.*, the bioavailability of elements essential to a healthy life), and the field of medical geology offers us the potential to reveal the secrets of nature.

SEE ALSO THE FOLLOWING CHAPTERS

Chapter 9 (Volcanic Emissions and Health) • Chapter 11 (Arsenic in Groundwater and the Environment) • Chapter 12 (Fluoride in Natural Waters) • Chapter 13 (Water Hardness and Health Effects) • Chapter 14 (Bioavailability of Elements in Soil) • Chapter 15 (Selenium Deficiency and Toxicity in the Environment) • Chapter 16 (Soils and Iodine Deficiency) • Chapter 20 (Animals and Medical Geology) • Chapter 23 (Environmental Pathology) • Chapter 26 (GIS in Human Health Studies) • Chapter 27 (Investigating Vector-Borne and Zoonotic Diseases with Remote Sensing and GIS)

FURTHER READING

Adriano, D. C. (2001). *Trace Elements in Terrestrial Environments: Biogeochemistry, Bioavailability and Risk of Metals*, Springer-Verlag, Berlin.

Agricola, G. (1950). *De Re Metallica* (1556), translated by Hoover, H., and Hoover, L. H., Dover Publications, New York.

Alvarez, L. W., Alvarez, W., Asaro, F., and Michel, H. (1980). Extraterrestrial Cause for Cretaceous-Tertiary Extinction, *Science*, 208, 1095–1108.

Anderson, P. H., Berrett, S., and Patterson, D. S. P. (1979). The Biological Selenium Status of Livestock in Britain as Indicated by Sheep Erythrocyte Glutathione Peroxidase Activity, *Vet. Rec.*, March 17, pp. 235–238.

Appleton, J. D., Fuge, R., and McCall, G. J. H. (Eds.) (1996). *Environmental Geochemistry and Health with Special Reference to Developing Countries*, Geological Society Special Pub-lication No. 113, Geological Society, London.

Bowie, S. H. U., and Thornton, I. (Eds.) (1985). *Environmental Geochemistry and Health*, D. Reidel Publishing, Dordrecht .

Bowie, S. H. U., and Webb, J. S. (1980). *Environmental Geochemistry and Health*, Royal Society, London.

Cannon, H. L. (1971). Environmental Geochemistry in Health and Disease, *Geol. Soc. Am. Mem.*, p. 123.

Carruthers, M., and Smith, B. (1979). Evidence of cadmium toxicity in a population living in a zinc mining area-Pilot survey of Shipham residents: The Lancet, v. April 21, 1979, pp. 845–847.

Ceruti, P. O., Davies, T. C., and Selinus, O. (2001). Geomed 2001—Medical Geology, The African Perspective, *Episodes*, 24(4), 268–270.

Chaney, R. L., Ryan, J. A., Li, Y.-M., and Brown, S. I. (1998). Soil Cadmium as a Threat to Human Health. In *Cadmium in Soils, Plants and the Food Chain* (M. J. McLaughlin and B. R. Singh, Eds.), Kluwer Academic, Dordrecht, pp. 219–256.

Chen, X., Yang, G., Chen, J., Chen, X., Wen, Z., and Ge, K. (1980). Studies on the Relations of Selenium and Keshan Disease, *Biol. Trace Element Res.*, 2, 91–107.

Cliff, D. C., and Haggett, P. (1988). *Atlas of Disease Distributions*, Blackwell Reference, Oxford, England.

Davies, B. E. (1971). Trace Element Content of Soils Affected by Base Metal Mining in the West of England, *Oikos*, 22, 366–372.

Davies, B. E., and Anderson, R. J. (1987). The Epidemiology of Dental Caries in Relation to Environmental Trace Elements, *Experientia*, 43, 87–92.

Davies, B. E., and Thornton, I. (1989). *Environmental Pathways of Lead into Food: A Review*, The International Lead Zinc Research Organization, Research Triangle, NC.

Davies, T. C. (1996). Geomedicine in Kenya. In *Environmental Geology of Kenya* (T. C. Davies, Ed.), Special Issue, *Journal of African Earth Sciences*, 23(4), 577–591.

Davies, T. C. (2000). Editorial, *Environ. Geochem. Health*, Special Issue, 24(2), 97.

Davies, T. C., and Schlüter, T. (2002). Current Status of Research in Geomedicine in East and Southern Africa, *Environ. Geochem. Health*, 24(2), 99–102.

Davies, T. C. (2003). Historical Development of Medical Geography in Africa. Unpublished report, Moi University, Kenya.

Diplock, A. T. (1984). Biological Effects of Selenium and Relationships with Carcinogesis Toxicological and Environmental Chemistry. v8. pp. 305–311.

D'itri, P. A., and D'itri, F. M. (1977). *Mercury Contamination: A Human Tragedy*, John Wiley & Sons, New York.

Eaton, D. L., and Robertson, W. O. (1994). Toxicology. In *Textbook of Clinical, Occupational and Environmental Medicine* (L. Rosenstick and W. R. Cullen, Eds.), W.B. Saunders, Philadelphia, pp. 116–117.

Engels, F. (1845) *The Condition of the Working Class in England* (Penguin Classics, edited by Victor Kernan), Penguin Books, London.

Ferguson, W. S., Lewis, A. H., and Watson, S. J. (1943). The Teart Pastures of Somerset. I. The Cause and Cure of Teartness, *J. Agric. Sci. (Cambridge)*, 33, 44–51.

Fergusson, J. E. (1990). *The Heavy Elements: Chemistry, Environmental Impact and Health Effects*, Pergamon Press, New York.

Finkelman, R. B., Belkin, H. E., and Zheng, B. (1999). Health Impacts of Domestic Coal Use in China, *Proc. Natl. Acad. Sci.*, 96, 3427–3431.

Foster, H. D. (2002). The Geography of Disease Family Trees: The Case of Selenium. In *Geoenvironmental Mapping: Methods, Theory and Practice* (P. Bobrowsky and A. A. Balkema, Eds.), pp. 497–529.

Frankenberger, W. T., and Benson, S. (Eds.) (1994). *Selenium in the Environment*, Marcel Dekker, New York.

Frankenberger, W. T., and Engberg, R. A. (Eds.) (1998). *Environmental Chemistry of Selenium*, Marcel Dekker, New York.

Friberg, L., Piscator, M., Nordberg, G., and Kjellstrom, T. (1974). *Cadmium in the Environment*, 2nd ed., CRC Press, Boca Raton, FL.

Gimingham, C. T. (1914). The Scouring Lands of Somerset and Warwickshire, *J. Agric. Sci. (Cambridge)*, 6, 328–336.

Goldschmidt, V. M. (1954). *Geochemistry*, Clarendon Press, London.

Hong, S., Candelone, J. P., Patterson, C. C., and Boutron, C. F. (1994). Greenland Ice Evidence of Hemisphere Lead Pollution Two Millennia Ago by Greek and Roman Civilizations, *Science*, 265, 1841–1843.

Jianan, T. (1985). *The Atlas of Endemic Diseases and Their Environments in the People's Republic of China*, PR China, Science Press, Beijing.

Kabata-Pendias, A. (2001). *Trace Elements in Soils and Plants*, 3rd ed., CRC Press, Boca Raton, FL.

Kabata-Pendias, A., and Pendias, H. (1992). *Trace Elements in Soils and Plants*, 2nd ed., CRC Press, Boca Raton, FL.

Krzysztof, K., and Glab, H. (2001). Trace Element Concentrations in Human Teeth From Neolithic Common Grave at Nakonowo (Central Poland), *Variabil. Evol.*, 9, 51–59.

Låg, J. (1990). General Survey of Geomedicine. In *Geomedicine* (J. Låg, Ed.), CRC Press, Boca Raton, FL, pp. 1–24.

Latham, R. (1958). *The Travels of Marco Polo*, Penguin Books, London.

Leverett, D. H. (1982). Fluorides and the Changing Prevalence of Dental Caries, *Science*, 217, 26–30.

Lewis, A. H. (1943). The Teart Pastures of Somerset. II. Relation Between Soil and Teartness, *J. Agric. Sci. (Cambridge)*, 33, 52–57.

Lindberg, D. C. (1992). *The Beginnings of Western Science*, University of Chicago Press, Chicago.

Mahaffey, K. R., Annest, J. L., Roberts, J., and Murphy, R. S. (1982). National Estimates of Blood Lead Levels: United States, 1976–1980, *N. Engl. J. Med.*, 307, 573–579.

Millar, I. B. (1961). Gastrointestinal Cancer and Geochemistry in North Montgomeryshire, *Br. J. Cancer*, 15(2), 176–199.

Morgan, H., and Simms, D. L. (1988). The Shipham Report: Discussion and Conclusions: The Science of the Total Environment, v. 75, pp. 135–143.

Moynahan, E. J. (1979). Trace Elements in Man, *Philos. Trans. R. Soc. London*, 288, 65–79.

Nriagu, J. O. (1983). Lead Exposure and Lead Poisoning. In *Lead and Lead Poisoning in Antiquity*, John Wiley & Sons, New York, pp. 309–424.

Nriagu, J. O. (1998). Tales Told in Lead, *Science*, 281, 1622–1623.

Prasad, A. S. (1978). *Trace Elements and Iron in Human Metabolism*, John Wiley & Sons, Chichester, England.

Ratcliffe, J. M. (1981). *Lead in Man and the Environment*, Ellis Horwood, New York.

Roberts, C., and Manchester, K. (1995). Metabolic and Endocrine Disease. In *The Archaeology of Disease*, Alan Sutton Publishing, London, pp. 163–185.

Rosenfeld, I., and Beath, O. A. (1964). *Selenium: Geobotany, Biochemistry, Toxicity, and Nutrition*, Academic Press, New York.

Shacklette, H. T., Sauer, H. I., and Miesch, A. T. (1972). Distribution of Trace Elements in the Environment and the Occurrence of Heart Disease in Georgia, *Bull. Geol. Soc. Am.*, 83, 1077–1082.

Shuttleworth, V. S., Cameron, R. S., Alderman, G., and Davies, H. T. (1961). A Case of Cobalt Deficiency in a Child Presenting as "Earth Eating," *Practitioner*, 186, 760–766.

Silver, C. S., and Rothman, D. S. (1995). Toxics and Health: The Potential Long-Term Effects of Industrial Activity, World Resources Institute, Washington, D.C.

Skinner, H. C. W., and Berger, A. (Eds.), (2003). *Geology and Health: Closing the Gap*, Oxford University Press, New York.

Stocks, P., and Davies, R. I. (1964). Zinc and Copper Content of Soils Associated with the Incidence of Cancer of the Stomach and Other Organs, *Br. J. Cancer*, 18, 14–24.

Thomson, I., Thornton, I., and Webb, J. S. (1972). Molybdenum in Black Shales and the Incidence of Bovine Hypocuprosis, *J. Sci. Food Agric.*, 23, 879–891.

Thornton, I., (Ed.) (1983). *Applied Environmental Geochemistry*, Academic Press, San Diego, CA.

Thornton, I. (Ed.) (1985). *Proceedings of the First International Symposium on Geochemistry and Health*, Science Reviews, Ltd., St. Albans, U.K.

Thornton, I. (Ed.) (1988). *Geochemistry and Health*, Science Reviews, Ltd., St. Albans, U.K.

Thornton, I., Atkinson, W. J., Webb, J. S., and Poole, D. B. R. (1966). Geochemical Reconnaissance and Bovine Hypocuprosis in Co. Limerick, Ireland, *Irish J. Agric. Res.*, 5(2), 280–283.

Thornton, I., John, S., Moorcroft, S., and Watt, J. (1980). Cadmium at Shipham—A Unique Example of Environmental Geochemistry and Health, *Trace Substances Environ. Health Symp.*, 14, 27–37.

Thornton, I., Moon, R. N. B., and Webb, J. S. (1969). Geochemical Reconnaissance of the Lower Lias, *Nature (London)*, 221, 457–459.

Underwood, E. J., and Filmer, J. F. (1935). The Determination of the Biologically Potent Element (Cobalt) in Limonite, *Aust. Vet. J.*, 11, 84–92.

Voisin, A. (1959). *Soil, Grass and Cancer* (translated by C. T. M. Herriot and H. Kennedy), Crosby, Lockwood, and Son, London.

Warren, H. V. (1964). Geology, Trace Elements and Epidemiology, *Geogr. J.*, 130(Pt. 4), 525–528.

Warren, H. V., and Delavault, R. E. (1971). Variations in the Copper, Zinc, Lead, and Molybdenum Contents of Some Vegetables and Their Supporting Soils, *Geol. Soc. Am. Mem.*, 123, 97–108.

Wyllie, J. (1937). An Investigation of the Source of Arsenic in a Well Water, *Can. Publ. Health J.*, 28, 128–135.

The Norwegian Academy of Science and Letters has published several proceedings from symposia on medical geology with Professor Jul Låg as editor. These include:

Geomedical Aspects in Present and Future Research, 1980
Geomedical Research in Relation to Geochemical Registrations, 1984
Geomedical Consequences of Chemical Composition of Freshwater, 1987
Commercial Fertilizers and Geomedical Problems, 1987
Health Problems in Connection with Radiation From Radioactive Matter in Fertilizers, Soils, and Rocks, 1988
Excess and Deficiency of Trace Elements in Relation to Human and Animal Health in Arctic and Subarctic Regions, 1990
Human and Animal Health in Relation to Circulation Processes of Selenium and Cadmium, 1991
Chemical Climatology and Geomedical Problems, 1992
Geomedical Problems Related to Aluminum, Iron, and Manganese, 1994
Chemical Data as a Basis of Geomedical Investigations, 1996
Some Geomedical Consequences of Nitrogen Circulation Processes, 1997
Geomedical Problems in Developing Countries, 2000
Natural Ionizing Radiation and Health (Björn Bölviken, Ed.), 2001

"*Environmental Geochemistry and Health*", published by Kluwer Academic Publishers, is an international quarterly journal focussed on the interaction of the geochemical environment and health and is the official journal of the Society for Environmental Geochemistry and Health.

SECTION I

ENVIRONMENTAL BIOLOGY

INTRODUCTION: ULF LINDH

Environmental biology may be characterized by interactions between geological and anthropogenic sources and life. Geological sources provide biological systems with major, minor, and trace elements. Elements present in soils are influenced by a variety of geological processes. If environmental conditions permit the elements to be available to plants, some will be taken up while others will be rejected. What is taken up becomes available to grazing animals and humans.

Anthropogenic sources provide both essential and nonessential elements. In some cases, elements do not have to be biologically available to present health problems. Some elements or compounds may impact the epithelial cells in the respiratory system merely by mechanical irritation and cause damage. Often, human activities may lead to the movement of elements from places where they reside outside of biological systems to places where their inherent chemical nature is realized.

Chapter 2 provides a comprehensive discussion of what is termed natural background. The chapter emphasizes and illustrates the importance of the biogeochemical cycle, which is intimately related to the concept of bioavailability. Numerous interpretations of bioavailability are presented.

During the past few decades, a number of environmental problems have been attributed, rightly or wrongly, to anthropogenic activities. A fundamental goal of medical geology is to provide a foundation for discussion, in which the anthropogenic sources can be distinguished from natural sources. Chapter 3 describes a variety of anthropogenic sources and reviews the known and potential hazards associated with them. In addition, current and future issues surrounding waste disposal are described, as are agricultural practices and transport of contaminants and the importance of maintaining potable water resources.

Chapter 4 reviews the chemistry of life, beginning with the unique properties of water. The chemical behavior of various elements within living cells is outlined. This chapter highlights the role of elements as chemical messengers and the requirements of multicellular organisms, while introducing two new designations, metallome and metabollome, concerning metals and nonmetals.

The biological mechanisms of element uptake into living organisms are described in Chapter 5. Following a review of some fundamental biochemical principles, this chapter highlights the uptake of iron, zinc, and copper as examples.

The essentiality of elements and particularly trace elements is often inadequately defined. Chapter 6 offers a working definition of major, minor, and trace elements. The biological functions of the major elements are reviewed, followed by an in-depth discussion of the minor elements (including calcium and magnesium) and an emphasis on the biological functions of trace elements.

The discussion in Chapter 7 takes a more physiological approach to geological sources of elements. This chapter reviews sources of essential elements and discusses their bioavailability.

Toxicity is inherent in all elements. For many essential elements, nutritional deficiency may be the common issue. Chapter 8 discusses the concepts of nutritional deficiency and toxicity, beginning with an introduction to biological responses. Various aspects of elements as toxins, and carcinogens are featured in this chapter.

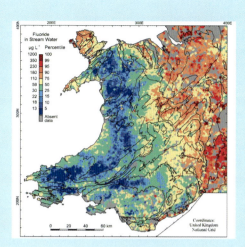

CHAPTER 2

NATURAL DISTRIBUTION AND ABUNDANCE OF ELEMENTS

ROBERT G. GARRETT
Geological Survey of Canada

CONTENTS

I. NATURAL BACKGROUND

A widely recognized biological characteristic of a healthy and sustainable environment is diversity—as with biology, so with geology. Regions characterized by the presence of different bedrock units, and different surficial materials in areas affected by recent (geologically speaking) glaciation, develop varied landscapes that support differing ecosystems. Examples of varied landscapes range from Alpine and Cordilleran mountains, through gentler landscapes of rolling hills, to the glacial plains of Northern Europe and North America,

or similarly from the high Himalayas, through verdant foothills, across fertile plains to the desert of Sind. In the parts of the world characterized by stable geological platforms, where mountain building has not taken place for many hundreds of millions of years and there have been long periods of landscape development, peneplains are the eventual outcome. Their topography is gentle without mountains. High relief areas are largely limited to inselbergs or ravines and river valleys where, due to crustal uplift, modern rivers and streams are cutting down into and eroding the old land surfaces. These are the physical expressions of the underlying geology, but there is another changing characteristic that cannot be seen directly—the chemistry of the underlying rocks and sediments and the soils that lie upon them.

It is the soils that either directly, or indirectly, sustain the vast majority of life on terrestrial parts of Planet Earth. The plants people eat (cereals and vegetables) or use (e.g., wood for construction, fibers for fabric and line, maize or sugar cane for ethanol production) grow in the soil. Furthermore, soils interact with precipitation as it moves from surface to groundwater storage; they are vital to sustaining life.

Soils have developed over very different time spans, from those on the peneplains of Africa, Australia, and South America that are hundreds of millions of years old, to soils developed over the last few decades on recent volcanic material, and on freshly deposited silts from rivers that have overflowed their banks. Soils that have developed on glacial sediments are somewhat older. As the last ice retreated, about 8–12 thousand years ago, at the close of the Wisconsin (North America) and Weischelian or Würm (Northern Europe) Ice Ages, a bare landscape was exposed. What lay underfoot was poorly sorted glacial till, a mixture of eroded rock and sometimes previous soil, containing material from cobbles and large "rocks" down to finely ground mineral fragments. In places where glacial rivers had flowed under the ice, sinuous sand ridges called eskers were deposited. Where the rivers emerged from under the ice outwash fans were formed, and these became deltas when they flowed into glacial lakes. Sand dunes often formed near these glacial river outlets and back from lake shores as there was no vegetative cover to anchor the newly deposited sediments and save them from wind erosion as they dried. The soil cover had yet to form.

The soils that sustain life develop as an interaction between the solid rock or unconsolidated surface material, the climate, and biological and other physical processes. Over time vertical zonations called profiles develop as a function of the interaction of these processes (see also Plant et al., 2001, Figure 6). Many soils are characterized by an organic, carbon-rich black upper layer (the L, F, and A horizons); sometimes a sandy textured light-colored layer (the A_e horizon); commonly a brownish or reddish layer richer in iron and some other elements, organic matter, and minerals (the B horizon); and finally, the weathered soil parent material (the C horizon). Other characteristics develop where the soils are wet; in arid (desert) regions; or frozen in high northern and southern latitudes. In extreme northern and southern latitudes, polar deserts may form, or where there is sufficient moisture, permafrost may form. In the tropics the upper organic-rich A horizons are often thin due to the rapid degradation of the leaf litter and other organic materials present; below these iron-rich B horizons develop. In very old soils the B horizons may become cemented with iron oxides to form hard carapaces—variously named duricrust, ferricrete, or canga. One of the key outcomes of soil formation is that chemical elements commonly become vertically redistributed by the pedological (soilforming) processes acting in the biogeochemical cycle. Within this major cycle many smaller cycles exist, such as that from soil to plant, back to soil, and soil to plant (see Section V for further discussion).

Natural backgrounds characterize the chemistry of rocks and surface materials, including soils, river and lake sediments, and biological tissues. Differences in natural backgrounds arise due to landscape-forming processes, which in turn are influenced by diversity in the underlying geology. There is no one natural background level for any solid material in or on the Earth as the Earth is far too inhomogeneous (diverse). For there to be a single natural background for any substance it would have to be homogeneously distributed throughout the planet, and that situation is only approached in the atmosphere, where the major weather systems of the globe keep the atmosphere relatively well mixed in each hemisphere. Therefore, natural backgrounds are variable, and this chapter discusses and illustrates that reality.

Natural background concentrations of elements provide the pool of essential chemical elements required by biological processes; therefore, they are vitally important. Life on Planet Earth has developed in the presence of all the 97 naturally occurring elements of the periodic table. To varying extents biological processes employ these elements to fulfill specific biochemical tasks which ensure the continuation of life. However, in addition to essentiality there is toxicity (see also Chapter 8, this volume). A few elements, e.g., mercury, lead, and thallium, have no known essential role in sustaining life. On the contrary, at high levels in biota they may be toxic and cause dysfunction and eventually death. In this context, the case of mercury in fish is interesting. Although fish appear to be able to bioaccumulate mercury dominantly as highly toxic methyl-mercury species, without harm to themselves, the consumption of these fish by mammals leads to elevated mercury levels that can be cause for concern. Others, such as cadmium, are toxic at high levels in most animal life, but may be essential for metabolic processes that support life in some species (this is an area of current research). Other elements appear benign, for example, bismuth and gold; the latter is even used for dental reconstruction. Finally, a great number of elements are bioessential at some level. Calcium is necessary for building bones and shells; and iron is important in blood in higher mammals and vanadium and copper for similar roles in marine biota. Other major and trace elements, e.g., sodium, potassium, magnesium, copper, nickel, cobalt, manganese, zinc, molybdenum, sulfur, selenium, iodine, chlorine, fluorine, and phosphorus are also essential for a variety of biotic processes. For most elements it is a question of balance, enough to ensure the needs of essentiality and good health, but not too much which might cause toxicity. As Paracelsus stated 450 years ago and paraphrased to modern English: "The dose makes the poison." It is the imbalance between amounts available naturally and

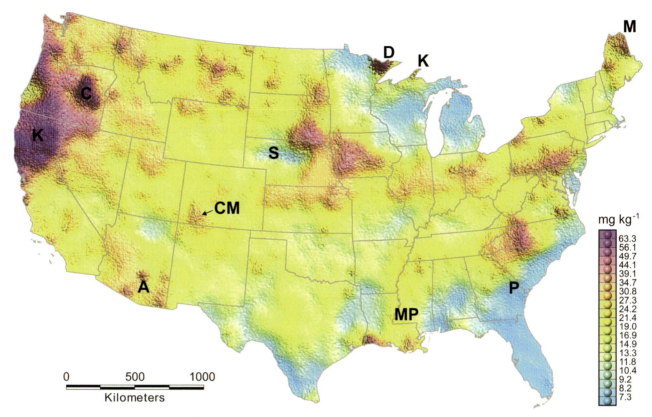

FIGURE 1 Copper content (mg kg⁻¹) of soils in the conterminous United States. (After Gustavsson et al., 2001 and Shacklette & Boerngen, 1984. Reproduced with the permission of the United States Geological Survey.)

those needed to sustain a healthy existence that poses the issues of medical geology.

II. A CHEMICALLY VARIABLE EARTH

An impressive way to demonstrate the chemical variability of the Earth's surface is with maps. Figure 1 displays the distribution of copper in the soils across the conterminous United States (7.84 million km²), which is about 5.3% of the Earth's land surface. What is important to know when using such a map and data is how the soil samples were collected, processed, and analyzed. These are critical facts that influence the conclusions drawn from geochemical data. In this instance the soils, characterized as natural supporting native vegetation or agricultural, were collected from 20 cm below the surface at sites generally over 100 m from roads. The soils were dried, disaggregated, and the fraction that passed a 2-mm stainless steel sieve was pulverized and directly analyzed by optical emission spectroscopy.

This method of analysis, which does not involve a chemical dissolution step, measures all of the copper present in the sample and is referred to as a "total" analysis. The samples used to prepare the map were collected by the U. S. Geological Survey between 1961 and 1975, and although over 25 years old, they still represent one of the few continental-scale depictions available (Shacklette & Boerngen, 1984; Gustavsson et al., 2001).

What are the noticeable features of these data? First, the map indicates they range from 7 to 63 mg kg⁻¹. This is almost an order of magnitude; however, in reality the individual 1323 sample analyses ranged from <1 to 700 mg kg⁻¹, almost three orders of magnitude, yet they were all collected from uncontaminated, background sites. The reduction in range in the scale from three to one order of magnitude is due to the smoothing process used to prepare the map (Gustavsson et al., 2001). Spatially, striking features are the high levels in the northwest versus the low levels in the southeast. An applied geochemist would state, the high background levels, etc. and Figure 1 is a graphic example of how natural background levels vary spatially. The high copper background in the west is associated with the Columbia River basalts (C on Figure 1) and basaltic volcanic rocks, e.g., the Klamath Mountains (K), in northern California and adjacent Oregon. The low copper backgrounds in the southeast in Florida and extending northward are associated with limestones in Florida and old beach sand Piedmont deposits (P) at the foot of the Appalachian mountains through Georgia and the Carolinas. Similarly low levels occur in the area of the Nebraska Sand Hills (S). Other notable features are high background levels in Minnesota associated with the Duluth gabbro (D), in Arizona (A) where major porphyry copper deposits have been mined, in southwestern Colorado associated with the Central Mineral Belt (CM), and in northern Maine (M) adjacent to a mineral-rich region in adjoining New Brunswick, Canada. More subtle, but recognizable features are the locally elevated levels along the lower Mississippi River valley due to an abundance of overbank levee sediments deposited when the river overflowed its banks (MP). Similarly, the native (metallic) copper deposits of the Keweenaw Peninsula (K) are reflected by locally elevated values. Other areas of elevated or depressed background can be related to a variety of geological and pedological features. Although sites were avoided that may have been directly contaminated, some may have been influenced by airborne transport from local or major remote sources.

Clearly to speak of a single background level for copper in United States soils does the reality a great injustice. There is no one average background level. Backgrounds need to be regional and reflect contiguous areas where the processes influencing background levels are similar. Secondly and most importantly, background levels are not single average values but are ranges reflecting the natural heterogeneity of the entity being characterized.

However, having said this, average values are frequently published, e.g., Wedepohl (1995), Reimann and de Caritat (1998), and Kabata-Pendias (2000), for different sample media such as rocks, soil, and waters. These provide a useful service in establishing order of magnitude levels for the abundance of elements in various materials, and compilations such as Reimann and de Caritat provide a great amount of useful information. As a historical note, global averages are sometimes referred to as "Clarkes," after F. W. Clarke who was the chief chemist at the U. S. Geological Survey from 1884 to 1925. Clarke and Washington (1924) were the first persons to attempt to characterize geochemistry on a global scale with publication of an average composition for igneous rocks based on a collection of 5159 "superior" analyses.

Applied geochemists apply two descriptors to data in the context of background distributions: level and relief. Level is the central tendency of concentrations or measures of the amount of some property for a sampled unit. The unit could be rock, soils, waters, and vegetable matter or any discrete sample type for a specified geographic area. The central tendencies are most frequently expressed as a mean, geometric mean, or median. The median is widely used as it is unaffected by the occurrence of a high proportion of abnormally high or low values; it is robust in a statistical sense. Relief has no formal accepted numerical measure, but is an expression of the homogeneity of the data. Data with a small range, tight about the central tendency, are said to have low relief, whereas data with a large range, or exhibiting skewness, are said to have high relief. This term is also used in a spatial context, when a discrete area of a geochemical map is characterized by locally variable, "noisy" data, that area is said to have "high relief." High relief data are characteristics of areas of complex geology and/or multiple processes, and these could be associated with the formation of mineral occurrences, weathering, soil formation, etc. Statistical measures such as coefficient of variation (relative standard deviation), standard deviation, or median absolute deviation have been used to objectively quantify relief. The key issue is that background is not sufficiently characterized by a single number. Background is characterized by a range of values and some quantification is desirable, whether it be quoting percentiles of the data distribution or the computation of some statistic. The advantage of quoting percentiles is that they involve no assumptions as to the statistical distribution of the data. These data are very often mixtures of several distributions related to different bedrock units or materials derived from them and different processes, e.g., the presence of mineral occurrences or weathering and pedological factors. The greater the diversity of an area described, the more likely is it that there are multiple data populations present.

It is common for data from areas not characterized by the presence of anthropogenic contamination, mineral deposits, or a particularly diverse geology to span in excess of an order of magnitude. Figure 2 presents box-and-whisker plots for trace element concentrations in the <2 mm fraction of 973 surface (0–20 cm) soils collected from the Canadian Prairies in 1992. These soils developed on glacial sediments derived dominantly from the sedimentary rocks—i.e., limestones and dolomites (carbonates), shales and sandstones—of the Western Canadian Sedimentary Basin, and to a lesser extent from Canadian Shield rocks to the

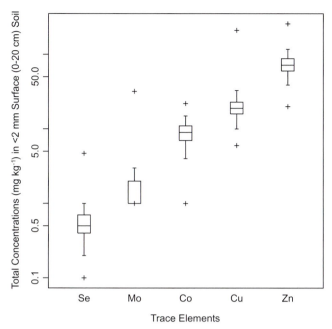

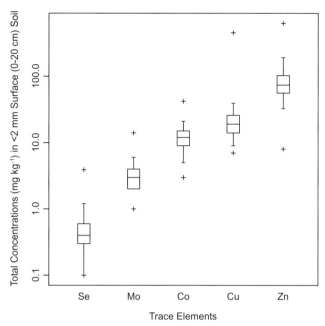

FIGURE 2 Selenium, molybdenum, cobalt, copper, and zinc contents (mg kg⁻¹) of the <2 mm fraction of soils (N = 973) from the Canadian Prairies. These were chosen because of biological importance. Note: Crosses indicate maximum and minimum values, ends of whiskers are the 5th and 95th percentiles, the box is bounded by the 1st and 3rd quartiles, and the bar indicates the median. If a notch is present rather than a bar, the notch indicates the 95% confidence bounds around the median, which is at the narrowest point.

FIGURE 3 Selenium, molybdenum, cobalt, copper, and zinc contents (mg kg⁻¹) of the <2 mm fraction of soils (N = 294) from southern Ontario.

northeast and north, and material from the Rocky Mountains to the west. As can be seen, the data span between one and one-and-a-half orders of magnitude. Most of the elements exhibit a positive skew, i.e., a greater abundance of higher values than lower. The cobalt data are an exception and exhibit a negative skew. The element with the greatest skew is molybdenum. Shales with abnormally high molybdenum levels occur in the region, and the distribution reflects this fact. The molybdenum data are characterized as having a greater relief than the other trace element data.

A soil survey undertaken using similar field sampling, sample preparation, and analytical protocols was undertaken in Ontario, Canada, in 1994. This survey included the Sudbury region which contains some of the largest nickel deposits in the world. For comparison, the same trace elements are plotted in Figure 3. The differences between the mid-50% of the data are small—molybdenum and cobalt levels in surface soil are higher in Ontario, while selenium levels are lower. What are different are the ranges and skewness of the data or their

relief, which reflect the differences in geological diversity between the two survey areas. The range of the molybdenum data is greater in the Prairies, and the skew is greater reflecting the shales mentioned above. In the Ontario data, the ranges of the selenium, copper, and zinc data are larger which reflects the greater geological diversity (both older Shield rocks and younger Phanerozoic sediments) relative to the Canadian Prairies. Of particular significance is the increased positive skew or higher relief of the copper data, which reflects the presence of copper in the nickel deposits of the Sudbury basin.

The Sudbury ore deposits contain a wide range of metals present as sulfides and arsenides, and many are recovered commercially. Figure 4 presents the Ontario soil data for arsenic, cobalt, copper, and nickel. The deposits influence a small number of the survey sample sites, so the central parts of the distributions are not affected by their presence. Only the extremes for the major metals produced, i.e., nickel and copper are affected. The impact of the Sudbury basin, both as a geological and anthropogenic source (stack emissions), on what would be described as the background data distribution can be seen in Figure 5. The main part of the data spans one order of magnitude between 6 and 50 mg kg⁻¹ nickel; two individuals (3 mg kg⁻¹) that fell

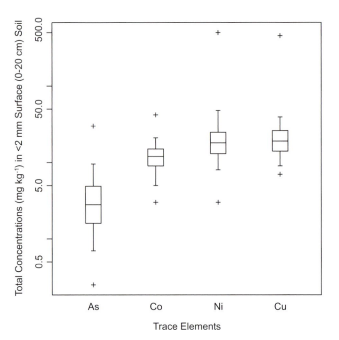

FIGURE 4 Arsenic, cobalt, nickel, and copper contents (mg kg⁻¹) of the <2 mm fraction of soils (N = 294) from southern Ontario. These were chosen as they represent the Sudbury copper–nickel ore deposits.

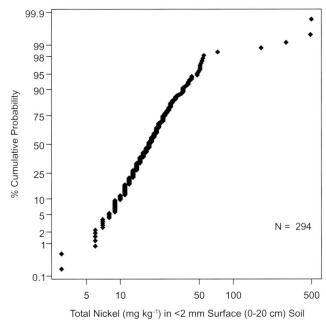

FIGURE 5 Cumulative probability plot of nickel contents (mg kg⁻¹) of the <2 mm fraction of soils (N = 294) from southern Ontario, which demonstrate how mineral deposits are reflected by high outlying concentrations of an element in the ore.

below the detection limit of the analytical procedure (6 mg kg⁻¹) were arbitrarily set to half that limit; and most interestingly, a group of five individuals extend the range of the data a further order of magnitude to 500 mg kg⁻¹ nickel. It is this latter group of samples that causes the data to have high relief and reflect the presence of the Sudbury basin, its mineral deposits, and its smelting facilities.

Background distributions may be influenced by naturally occurring high concentrations of trace elements and metals sometimes referred to as "natural contamination", with the resultant data exhibiting a high relief. Such natural processes that lead to the accumulation of elements at specific sites in the Earth's crust are what make them available to society for use. They raise concentrations to a level described as "ore", i.e., that which can be extracted from the ground at a profit (noting that there are many different economic and social models to define profit). Often in such regions there are also areas of unusually low trace element and metal concentrations called alteration zones, which reflect where natural processes have removed metals to transport and concentrate them elsewhere. When this occurs the relief of data associated with ore elements can be very high. In general, "ore grades" exceed average crustal

abundance levels by two to four orders of magnitude (McKelvey, 1960), which results in natural ranges of trace element and metal concentrations in areas characterized by mineral deposits of four or more orders of magnitude. This depends on how much "ore-grade" material was incorporated in the samples collected. Two examples are provided below.

First, and to demonstrate that spatial scale has no effect on natural backgrounds per se, data for the nickel, copper, and zinc content of 292 glacial till samples collected from the walls of two adjacent trenches cutting across the Nama Creek copper–zinc deposit at Manitouwadge, Ontario, are presented in Figure 6. These trenches, about 300 m long and up to 4 m deep, were dug and sampled prior to the development of the deposit. The nickel distribution, as a measure of background—there are no nickel sulfides in the deposit—spans one-and-a-half orders of magnitude exhibiting low relief similar to the background distributions of most metals in the earlier examples. However, this is not the case for copper and zinc, which span over two-and-a-half orders of magnitude, exhibiting high relief, each due to the incorporation of ore-minerals containing these elements into the glacial till by the erosion of the mineral deposit. The additional order of magnitude

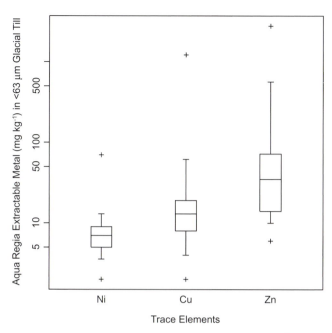

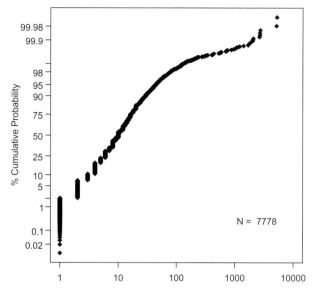

FIGURE 6 Nickel, copper, and zinc contents (mg kg⁻¹) of the <63 μm fraction of glacial tills (N = 292) from the Nama Creek deposit, Manitouwadge, Ontario. This demonstrates the impact of a copper–zinc deposit on the data with anomalously high copper and zinc levels in till samples derived from the erosion of the ore deposit.

FIGURE 7 Cumulative probability plot of nickel (mg kg⁻¹) in the <177 μm fraction of stream sediments (N = 7778) from Goias State, Brazil. The flexure above 100 mg kg⁻¹ reflects the presence of nickel-rich ultramafic rocks, and the outlying values above 1500 mg kg⁻¹ reflect the nickeliferous pyrrhotite mineral occurrences within them.

reflects the presence of the mineral deposit, a factor which increases the geological and geochemical diversity in the area.

The second example is a regional stream sediment survey of approximately 80,000 km² from Goias State in central Brazil. The area is extremely diverse geologically with a wide range of sedimentary, igneous, and metamorphic rocks present, and it is blanketed by residual soils that have developed *in situ*. Of importance to the example is that the rocks range from "metal-poor" limestones and sandstones and their metamorphic derivatives to ultramafic igneous intrusives containing nickel and copper mineral occurrences. The distribution of nickel in the <177 μm fraction of the stream sediments is presented in Figure 7, where data span almost four orders of magnitude. These data are largely uninfluenced by anthropogenic activities, and therefore reflect natural processes. Four features are noteworthy: (1) 2% of these data were below the analytical detection limit of 2 mg kg⁻¹; (2) these and data up to 132 mg kg⁻¹ (99% of all the data) reflect the variation of a wide group of different rock types and soil-forming processes active in the region; (3) the upper tail of the distribution reflects samples collected from areas underlain by nickel-rich ultramafic rocks; and (4) the highest

levels, in excess of 1000 mg kg⁻¹, reflect the presence of nickel sulfide occurrences.

Thus the controlling factor in determining the range of natural background is not the size of an area, but the diversity of the geology present. High diversity, due to some combination of contrasting rock types and/or the presence of mineral deposits, leads to geochemical data that are similarly diverse, i.e., they are characterized by high relief. Another example of the presence of interesting patterns at widely different scales can be seen in Plate 3-1 of Darnley et al. (1995). This plate displays the copper stream sediment geochemistry of the island of St. Lucia, approximately 40 km², juxtaposed to the internal nickel chemistry of a grain of a platinum-bearing mineral 10 μm² in area. The difference in scale (area) is of the order of 10¹², yet well-designed sampling and analytical procedures at both scales reveal patterns of interest and geochemical significance.

Due to the diversity of geology and secondary environmental conditions, a vast number of regional and local backgrounds exist. This can be problematic when natural background distributions are used to establish national reference levels for regulatory purposes. For these to be effective they need to be very clearly defined as to what "environment" they represent, and they need to be based on adequately large sample sets. It is most important that data are not used out of the context of their collection. This issue can be exacerbated if measures of central tendency are used as national reference levels, because this immediately implies that approximately 50% of all measurements relevant to the reference level will fall above the quoted value.

To avoid this problem, reference levels associated with environmental regulations are sometimes quoted at some other level, e.g., mean plus two standard deviations—notionally the 97.5th percentile of data derived from the estimated mean and standard deviation. In using this procedure an assumption has to be made, often implicitly with no discussion of the ramifications, as to the distribution of the data—normal, lognormal, or some other model. This can be fraught with problems, especially as the "geographic units" from which reference levels are derived get larger and more geologically and environmentally diverse. In such cases these data are likely drawn from a number of different distributions and agglomerated into a "mixture". Often these mixtures appear to have lognormal distributions, despite the fact that many of the underlying components are more likely to have normal distributions (Vistelius, 1960). An alternative is to use a percentile of the natural background distribution as a reference value. An example of such a procedure is the use of an Ontario Typical Range 98 (OTR^{98}) value by the Department of Environment of the government of the Province of Ontario, Canada, which corresponds to the 98th percentile of the background data for a specific entity, e.g., residential lands. Every distribution has a 98th percentile; natural processes may be the cause of higher observed levels, and anthropogenic contamination may result in levels lower than the 98th percentile. Acceptable numbers of false positives or negatives, i.e., type I and II statistical errors, are chosen during the selection of any particular percentile. However, the OTR^{98} level is used to trigger an investigation into whether the excedance is due to natural phenomena or is the result of anthropogenic contamination. If it is the latter, appropriate actions are taken on a site-specific basis.

III. MINERAL CHEMISTRY— THE KEY TO THE DISTRIBUTION OF ELEMENTS IN ROCKS

A natural question is: Why is there such diversity in the chemistry of surface materials? The answer lies in the composition of the individual minerals that compose rocks. Their properties are carried forward to other materials through erosional, weathering, and soil-forming processes, and are transferred to varying extents to waters that pass through these solid phase materials. In some respects the chemical diversity is self-fulfilling, as the main criteria that geologists use to "name" a rock, particularly in the field, are its mineralogy and texture (the shapes and interactions of and between the individual minerals). To have different names rocks must be visibly different from each other.

The major components of the common, abundant rock types are silicates. The exceptions are rocks such as limestones and dolomites and their metamorphic derivatives (marbles), which are composed of calcium and magnesium carbonates. Other exceptions include sedimentary rocks containing phosphates and iron carbonates. Oxide, hydroxide, sulfide, and other minerals can also host the trace elements found in biological systems.

Silicates and aluminosilicates are important minerals in geochemistry, particularly ferromagnesian minerals, feldspars, and phyllosilicates (e.g., micas and clays). These minerals all contain silicon and sometimes aluminum as major components. When they occur alone in a mineral they are present, respectively, as ubiquitous quartz and the rare corundum. Corundum occurs as the gemstones ruby and sapphire where the colors are induced by trace amounts of chromium (ruby), iron, or titanium (sapphire). The ferromagnesian minerals and feldspars are important as they, respectively, contain iron and magnesium, and calcium, sodium, and potassium as major components. The presence of these elements establishes a situation where other physically similar elements may enter the lattices of the mineral crystals. It is this phenomenon that results in the wide range of trace element concentrations observed in rocks.

Minerals are rarely "pure" and are commonly contaminated with a wide range of other elements present at "trace" concentrations. Pure minerals are so rare, and often beautiful, that they are only seen in museums and mineral collections. The key to understanding which trace elements enter different mineral crystal forms, by a process known as substitution, is through knowledge of the physical properties of ionic radius (Figure 8) and electronegativity (Figure 9).

One example is the ferromagnesian mineral olivine, which is a major component of many ultramafic and mafic rocks and forms the essentially monomineralic rock dunite. Its composition is $(Mg, Fe)_2SiO_4$, which like many minerals is an intermediate form on a con-

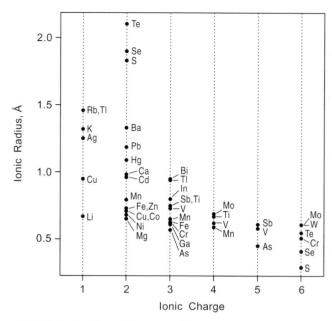

FIGURE 8 The relationship of ionic radius to ionic charge (valence) for major and trace elements of mineralogical and geochemical interest.

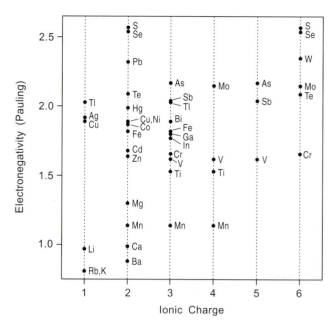

FIGURE 9 The relationship of electronegativity to ionic charge (valence) for major and trace elements of mineralogical and geochemical interest.

tinuous solid solution of two other minerals—pure Mg_2SiO_4, forsterite, and pure Fe_2SiO_4, fayalite—which themselves are rare. The ionic radii of iron and magnesium in their two-valent states are 0.74 and 0.66 Å, respectively. Thus in the solid solution form, olivine, there are crystal lattice sites that can just as easily be occupied by other two-valent ions, such as nickel (0.69 Å), copper (0.72 Å), cobalt (0.72 Å), zinc (0.74 Å), and with a bit more crystal stretch, manganese (0.8 Å). The inclusion of manganese is facilitated by the fact that there is a further solid solution between olivine and monticellite, $(Ca, Mg)_2SiO_4$, and the ionic radius of two-valent manganese lies between that for calcium (0.99 Å) and magnesium (0.66 Å).

When iron occurs in the three-valent form it can be incorporated into garnet group minerals, $Fe_3Al_2(SiO_4)_3$, which are common accessory minerals in many metamorphic rocks, and to a lesser extent, igneous rocks. Garnet chemistry can also be expressed as $3R^2O.R^3_2O_3.3SiO_2$, where the superscript number indicates the valence and the letter R indicates a metal. In this latter form it can be seen that three-valent aluminum (0.51 Å), iron (0.64 Å), chromium (0.63 Å), and titanium (0.76 Å) can enter the garnet crystal lattice. Garnets are truly remarkable in this fashion, and can host a wide range of divalent and trivalent metal ions. It is known that they occur widely at the base of the continental crust, as they come to the surface in rocks

named eclogites entrained in certain volcanic extrusives, and they are believed to be the host and "reservoir" for many of the trace elements stored deep within the crust.

Similar examples can be provided for the other important ferromagnesian minerals, and their capacity to host trace elements by substitution is most easily understood when their formulae are expressed in the same way as garnets. Thus, pyroxenes expressed as $R^2O.R^3_2O_3.SiO_2$ and amphiboles expressed as $R^2O.SiO_2$ can also contain aluminum and ferric iron and have very complex chemistries. They may also include hydroxyl groups and fluorine, epidotes expressed as $2R^2O.R^3OH.R^3_2O_3.3SiO_2$, and micas, e.g., biotite, expressed as $K_2O.3(Mg,Fe)O.3(Al,Si)O_2.(OH)_2$. These examples again can have very complex chemistries. Because of the abundant sites for divalent and trivalent metal ions in sixfold coordination, and the ability for small cations such as aluminum to replace silicon in fourfold coordination, these ferromagnesian minerals are hosts for a wide range of trace elements.

The aluminosilicate feldspars also play an important role as hosts for larger ionic radius metal ions. Feldspar chemistry lies between three end members: anorthite $(CaO.Al_2O_3.2SiO_3)$, albite $(Na_2O.Al_2O_3.6SiO_2)$, and orthoclase $(K_2O.Al_2O_3.6SiO_2)$. The physical structure of feldspars consists of SiO_4 and AlO_4 tetrahedra (silicon and aluminum in fourfold coordination) in a three-dimensional network. This network is elastic and

TABLE I.   Classification of Elements as Lithophile, Chalcophile, Siderophile, or Atmophile

Lithophile	Chalcophile	Siderophile	Atmophile
C, O, P, H, F, Cl, Br, I, Si, Al, Fe, Mg, Ca, Na, K, Ti, Sc, Cr, V, Mn, Th, U, Nb, Ta, Sn, W, Be, Li, Rb, Cs, Ba, Sr, B, Y, Zr, Hf, rare earths (REEs), Ga, (Cd), (Zn), (Pb), (Cu), (Ni), (Co), (Mo), (Tl)	S, Se, Te, As, Sb, Bi, Ag, In, Ge, Tl, Hg, Cd, Zn, Pb, Cu, Ni, Co, Mo, Re, (Fe), (Sn), (Au)	Pt, Ir, Os, Ru, Rh, Pd, Au, (Fe)	N, O, C (as CO_2), H, He, Rn, and other noble gases, (S as oxides), (Hg)

accommodates not only the large positively charged cations, calcium (0.99 Å), sodium (0.97 Å), and potassium (1.33 Å), but also strontium (1.12 Å), lead (1.2 Å), barium (1.34 Å), rubidium (1.47 Å), and thallium (1.47 Å), within its interstices.

Phyllosilicates are an important group of minerals in both rocks and their weathering products. They include silicate (e.g., talc) and aluminosilicate minerals commonly known as micas (e.g., phlogopite) and clays (e.g., montmorillonite), and may contain sodium, potassium, calcium, iron, and magnesium. These minerals have a sheeted mineral structure with pairs of sheets of SiO_4 tetrahedra held together tightly by cations; these pairs are held together loosely by other cations. In the case of the mica biotite, iron and magnesium provide the tight bonds and potassium the looser bonds. Muscovite mica is similar, but the tight bonds are provided by aluminum. A wide variety of cations may replace the iron, magnesium, and aluminum in the tight binding sites, and other elements may substitute for the potassium that loosely binds the sheets together. The breakdown of silicates and aluminosilicates due to alteration or weathering leads to the formation of a wide range of clay minerals, which host a wide range of cations in addition to silicon and/or aluminum. The mica and clay minerals with the greatest ability to support substitution with metal cations are generally those that employ iron and magnesium at interlayer sites, e.g., montmorillonite. In contrast, kaolinite or gibbsite only contains aluminum and silicon or aluminum, respectively.

The elements discussed above are commonly called lithophile (rock loving) and are distinct from other elements referred to as chalcophile, siderophile, or atmophile (Table I). The lithophile elements may occur as silicates, aluminosilicates, oxides, carbonates, sulfates, halides, phosphates, and vanadates, among other mineral forms in the natural environment.

An important second group are the chalcophile elements (Table I), which are characterized by forming sulfides, arsenides, antimonides, selenides, and tellurides. It is these compounds that form the ore minerals that are the source of the nonferrous metals used by society. Some of these minerals, particularly iron sulfides such as pyrite and marcasite (FeS_2), pyrrhotite ($Fe_{(1-x)}S$), and the sulfarsenide arsenopyrite (FeAsS), and to a lesser extent copper, zinc, lead, and molybdenum sulfides like chalcopyrite ($CuFeS_2$), sphalerite ($ZnFeS_2$), galena (PbS), and molybdenite (MoS_2), occur in many igneous and metamorphic and some sedimentary rocks. Due to the large amounts of these trace elements that can be held in sulfide and related minerals, it is not necessary to have abundant sulfides, etc., present in order to raise the levels of the chalcophile trace elements in rocks to quite high levels. As with silicate and aluminosilicate minerals, the chalcophile trace elements are often present as substitutions in commonly found minerals rather than in their own unique minerals. Again this is the result of fundamental physical properties, in this case electronegativity (Figure 9). As examples, silver and mercury replace copper in many copper minerals, cadmium and indium replace zinc in sphalerite, selenium and tellurium replace sulfur, and arsenic and antimony occur interchangeably in others and with sulfur. In iron sulfides copper, cobalt, and nickel commonly substitute for iron. In igneous and metamorphic rocks sulfides exist as blebs and crystals along the boundaries between the majority silicate and aluminosilicate minerals. In addition to this, they may occur within the rock-forming minerals along fracture planes. This is the result of a process known as exsolution, which occurs as rocks cool down and the individual rock-forming crystals are less able to accommodate incompatible components. The offending substances are then rejected to form discrete minerals along

TABLE II. Common Geochemical Associations

Group	Associations	
Generally associated elements	K-Rb	Ca-Sr
	Al-Ga	Si-Ge
	Zr-Hf	Nb-Ta
	Rare earths (REEs), La, Y	
	Pt-Ru-Rh-Pd-Os-Ir	Au-Ag
Plutonic rocks		
Generally associated elements	Si-Al-Fe-Mg-Ca-Na-K-Ti-Mn-Cr-V	
	Zr-Hf-REEs-Th-U-Sr-Ba-P	
	B-Be-Li-Sn-Ga-Nb-Ta-W-Halides	
Specific associations		
Felsic igneous rocks	Si-K-Na	
Alkaline igneous rocks	Al-Na-Zr-Ti-Nb-Ta-F-P-Ba-Sr-REEs	
Mafic igneous rocks	Fe-Mg-Ti-V	
Ultramafic igneous rocks	Mg-Fe-Cr-Ni-Co	
Some pegmatites	Li-Be-B-Rb-Cs-REEs-Nb-Ta-U-Th	
Some contact metasomatic deposits	Mo-W-Sn	
Potassium feldspars	K-Rb-Ba-Pb	
Many other potassium-rich minerals	K-Na-Rb-Cs-Tl	
Ferromagnesian minerals	Fe-Mg-Mn-Ni-Co-Cu-Zn	
Sedimentary rocks		
Fe-oxide rich	Fe-As-Co-Ni-Se	
Mn-oxide rich	Mn-As-Ba-Co-Mo-Ni-V-Zn	
Phosphatic limestones	P-F-U-Cd-Ag-Pb-Mo	
Black shales	Al-As-Sb-Se-Mo-Zn-Cd-Ag-U-Au-Ni-V	

After Rose, Hawkes, and Webb, 1979.

internal lines of crystal weakness or completely to a discrete mineral grain boundary.

Siderophile and atmophile elements are less important in the following discussion. The siderophile elements form alloys with iron and these are important sources of platinum group metals, together with gold, to society. The atmophile elements are ubiquitous in relatively hemispherically homogeneous atmospheres. Mercury is the only metal that occurs as a gas at "normal temperatures and pressures," and this permits its transport over long distances independent of fluvial systems.

Although not strictly a crystalline mineral phenomenon, many elements are associated with organic matter in sedimentary rocks and metamorphic rocks derived from them. This is due to two general processes: (1) the ability of organic compounds to sequester and adsorb trace elements, e.g., copper and mercury; and (2) the actual formation of metallo-organic compounds to fulfill particular biochemical functions such as copper and vanadium in the heme of marine invertebrates. In the geological context, organic matter is only preserved in rocks under anoxic conditions, which due to the prevalent redox conditions are also sulfur-reducing environments that lead to the presence of sulfides. This is particularly important in the formation of rocks described as "black shales" that can become enriched in many trace elements.

As a result of these relationships geochemists have observed consistent patterns in the distribution of many elements. Some of the more interesting of these elements are presented in Table II. Several of the associations are related to mineral deposits, which are major natural sources of elements to the Earth's surface environment, and the processing of the ores can be major anthropogenic sources of contamination if appropriate emissions controls are not installed at processing plants and smelters.

In the secondary, weathering, environment most of the ferromagnesian and aluminosilicate minerals are unstable and break down to more hydrated forms, e.g., phyllosilicates, oxides, and hydroxides and residual silica (quartz). During this process the trace elements held in the rocks are liberated: some are removed in solution as surface runoff or enter groundwater, and others are incorporated into new minerals or sequestered by organic matter. Two key mineral forms capable of retaining trace elements are the phyllosilicates, minerals such as smectites and chlorites, and the oxides and hydroxides of iron and manganese. The phyllosilicates sequester trace elements by two processes: by cation exchange to constant electrical-charge sites on the tabular surfaces of the minerals, and by adhering to the broken edges of the clay particles where variable charge sites occur. Smectites are particularly effective in this role as they have large cation exchange capacities. The oxyhydroxides of iron and manganese formed during weathering are also effective in sequestering cations. This ability is enhanced when the oxyhydroxides are linked to humic or fulvic acids, which raises the charge on the oxyhydroxide surfaces. This effect is even more pronounced with the formation of humic colloids.

In contrast to the trace elements that are associated with minerals that break down in the weathering zone, those associated with resistate minerals that do not weather to any significant extent are retained in that mineral form in the soil and subsequent erosion products. Examples of resistate minerals are chromium in the mineral chromite; tin in cassiterite; niobium and tantalum in columbite-tantalite; zirconium and hafnium in zircons; and cerium, lanthanum, yttrium, and thorium in monazite.

IV. Diversity in the Chemistry of Rocks

The combination of the chemistry of minerals and their abundances in different rock types, which are defined upon a mineralogical and textural basis, leads to a varied rock geochemistry. Table III provides examples of estimated average values for the trace elements in the Earth's crust and different rock types. These estimated averages give no indication of the actual range observed in these individual rock types; such an estimate has never been made on a global basis, but it is likely at least one or two orders of magnitude. The variability behind the Continental Crust estimates can be implied from

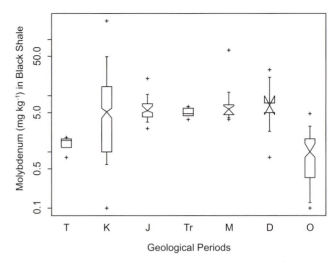

FIGURE 10 Distribution of molybdenum (mg kg⁻¹) by age in black shales, Manitoba, Canada. From youngest to oldest: T = Tertiary, K = Cretaceous, J = Jurassic, Tr = Triassic, M = Mississippian, D = Devonian, and O = Ordovician.

the variability for the estimates of individual rock types. These range from less than an order of magnitude for mercury to almost three orders of magnitude for nickel.

As an example of the variability associated with a single rock type, the example of black shales from Manitoba, Canada, is presented. These data come from 54 surface outcrop sites and drill holes in an area approximately 300 km wide (ENE–WSW) and 500 km long (NNW–SSE) along the eastern margin of the Western Sedimentary Basin. This area represents shales varying in age from Ordovician to Tertiary, spanning some 360 million years (Ma). Figure 10 displays the molybdenum data for the 476 samples subdivided by geologic period, oldest to the right and youngest to the left. The oldest (Ordovician, O) and youngest (Tertiary, T) black shales have lower molybdenum contents than the generally similar median valued Devonian (D) to Cretaceous (K) shales. However, what is outstanding is the variability of the Cretaceous (K) black shales, which extend over three orders of magnitude.

The Cretaceous shales span approximately 85 Ma of deposition in a sea that went through various transgressive (deepening) and regressive (shallowing) stages. The 333 black shales have been subdivided by stratigraphic formation (except for undivided Cretaceous rocks, K) in Figures 11–13. Figure 11 displays the by-age distribution of manganese (mg kg⁻¹), which forms a bowl shape with lowest manganese levels in the Favel (uKf) and Morden (uKm) Formations when the Western Interior Seaway was at it deepest. Manganese

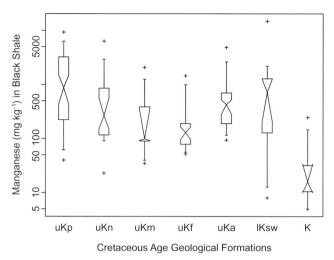

FIGURE 11 Distribution of manganese (mg kg⁻¹) in Cretaceous Age black shales, Manitoba, Canada, subdivided by Formation. Formations from youngest to oldest, where u as a prefix indicates upper Cretaceous and l indicates lower Cretaceous: uKp = Pierre, uKn = Niobrara, uKm = Morden, uKf = Favel, uKa = Ashville, lKsw = Swan River, K = undifferentiated, mostly lowermost Cretaceous, black shale.

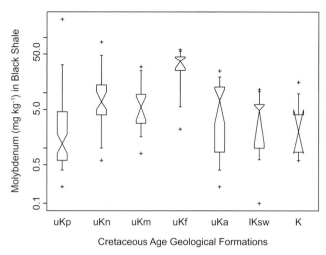

FIGURE 12 Distribution of molybdenum (mg kg⁻¹) in Cretaceous Age black shales, Manitoba, Canada, subdivided by formation. See Figure 11 for legend.

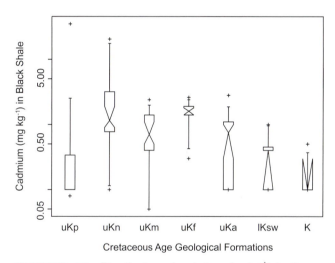

FIGURE 13 Distribution of cadmium (mg kg⁻¹) in Cretaceous age black shales, Manitoba, Canada, subdivided by formation. See Figure 11 for legend.

levels increase in older Ashville (uKa) and Swan River (lKsw) and younger Niobrara (uKn) and Pierre (uKp) shales, which were deposited in shallower water. These variations reflect the fact that in the deep anoxic carbonate-poor waters of the Western Interior Seaway manganese was preferentially retained in the seawater. Figure 12 illustrates the distribution of molybdenum (mg kg⁻¹) in shales. Levels are highest in the deep water shales of the Favel Formation (uKf) where anoxic sulfide- and organic-rich sediments favored the sequestration of molybdenum, and the median molybdenum level is almost two orders of magnitude higher than for the youngest Pierre shales (uKp). Some of the Cretaceous shales, particularly the Favel (uKf), Morden (uKm), and Niobrara (uKn), are enriched in cadmium (Figure 13). At the surface, these shales were eroded from the Manitoba Escarpment and transported westward across the Canadian Prairies about 400 km during the Wisconsin glaciation at the close of the last Ice Age, and they have contributed to higher levels of cadmium in the agricultural soils of the region.

In some instances, large segments of the Earth's crust may exhibit consistent patterns of trace element enrichment. These features are known as geochemical provinces. Rose, Hawkes, and Webb (1979) provide the definition of these geochemical provinces as "a relatively large segment of the Earth's crust in which the chemical composition is significantly different from the average." They go on to state: "One of the criteria of a bona fide geochemical province is that the characteristic chemical peculiarities should be recognizable in rocks covering a substantial period of time." Examples of geochemical provinces are the Bear uranium province in the northwestern part of the Canadian Shield and the central European uranium province that includes parts of Germany, Poland, and the Czech Republic; the great tin province that spans from eastern

Australia, through Indonesia, western Malaysia, Thailand, and into China where rocks of many ages are enriched in tin; and the lesser known manganese province in northeastern North America including Maine and New Brunswick.

A second type of province, metallogenic, is of particular economic importance, and may have environmental and health consequences. Metallogenic provinces are regions of the Earth's crust that are characterized by the presence of mineral deposits and occurrences for particular metals, and they often are of a particular mineral deposit type. In any mineral district they are many more small mineral occurrences than economically exploitable mineral deposits. A distinction is made here between mineral occurrences and mineral deposits. Mineral deposits contain ore, i.e., that which can be extracted from the ground at a profit, using a sufficiently broad definition of profit to meet various societal needs. Metallogenic provinces may also be considered "mineral" provinces, as it is the presence of the metals in specific mineral forms at high concentrations that makes deposits exploitable. Examples are the copper deposits that occur in the western Americas extending from the southwestern United States through Central America to Peru, Bolivia, Chile, and Argentina; the prolific tin deposits coincident with the tin geochemical province of southeast Asia in Indonesia, western Malaysia, Thailand, and China; the Copper Belt deposits of Zambia and Zaire in central Africa; and the gold and base-metal deposits of the Abitibi Greenstone Belt in Quebec and Ontario, Canada.

The mineral forms for base- and precious-metals, apart from native (metallic) gold and platinum alloys, are sulfides, arsenides, antimonides, tellurides, and selenides that are far less stable in the surface weathering environment than the silicates, aluminosilicates, oxides, etc., which host the metals in common rocks. As a result, metals in mineral occurrences and deposits weather more easily and are transferred to soils, sediments, and waters. From a mineral exploration viewpoint their dispersion away from point sources facilitates exploration geochemical surveys. However, from a biological viewpoint this may be a good or bad outcome. If the metals are essential for life this may be beneficial; high metal levels may permit species that require such high levels, or are resistant to them, to flourish. If the metal species present is toxic, this will inhibit some life forms which may lead to an absence of biota or permit only certain hardy species to be present. It was well known that in the central African Copper Belt the high copper in the stream waters derived from the weathering of copper deposits was sufficient to be toxic to the snails that were critical in the *Bilharzia* cycle. As a result, schistosomiasis was largely absent in the region. For those interested in the interaction of geology, geochemistry, and botany, readers are referred to Brooks (1972 and 1998).

The relationship between geochemical and metallogenic provinces is of some interest. By their nature geochemical provinces are low-entropy phenomena. In contrast, metallogenic provinces are high-entropy phenomena with extreme segregation of metals into spatially small discrete high concentration zones. Despite these contrasting characteristics, many metallogenic provinces lie within geochemical provinces, and it is assumed that the increases in regional background levels of the metals have provided part of the pool of metals that have been concentrated into the mineral deposits and occurrences. Reimann and Melezhik (2001) provided a discussion of the relationships between metallogenic and geochemical provinces in the context of a large (188,000 km^2), low sampling density (1 site per 300 km^2) regional geochemical study of surficial materials in Arctic Europe called the Kola Ecogeochemistry Project (Reimann et al., 1998). The authors noted that some significant metallogenic provinces were not recognized in the project they undertook, and also cast doubt on the value of the term "geochemical province." However, considering the very different nature of the two province types in terms of entropy, and the interaction of sampling density through search theory to the probability of recognition of the mineral occurrences of a metallogenic province with low-density field sampling programs (see Section V and Garrett, 1983), the death of these provinces as useful concepts is premature.

It is this underlying variability in rock chemistry, that is in turn due to the mineralogical and compositional variability of the rocks composing the Earth's crust, let alone the mineral deposits and occurrences that occur in them, that causes the geochemical variability in natural background levels in the surficial materials and weathering products discussed in the first section of this chapter.

Although tables of averages (Table III) are useful as general indicators of the element content of rocks, the variability behind them must never be forgotten. As noted and demonstrated above, that variability is considerable and can easily span up to three orders of magnitude for trace elements. Readers requiring average estimates for composition of both the Earth's crust, individual rock types, and other materials are referred to Wedepohl (1995), Reimann and de Caritat (1998), and Kabata-Pendias (2000).

TABLE III. Compilation of Average Geochemical Background Data for the Earth's Crust and Selected Rock Types

	Hg ($\mu g\,kg^{-1}$)	Pb ($mg\,kg^{-1}$)	Cd ($mg\,kg^{-1}$)	Cr ($mg\,kg^{-1}$)	Ni ($mg\,kg^{-1}$)	As ($mg\,kg^{-1}$)	Cu ($mg\,kg^{-1}$)	Zn ($mg\,kg^{-1}$)	Ref.
Earth's crust									
	80	13	0.2	100	75	2	55	70	Taylor, 1964
	90	12	0.2	110	89	2	63	94	Lee & Yao, 1970
Upper continental crust									
		20	0.1	35	20	1.5	25	71	McLennan, 1992
	80	13	0.2	77	61	1.7	50	81	Lee & Yao, 1970
Igneous rocks									
Ultramafic	4	1	0.1	1600	2000	1	10	50	Turekian & Wedepohl, 1961
Mafic	13	6	0.2	170	130	2	87	105	Turekian & Wedepohl, 1961
Intermediate	21	15	0.1	22	15	2	30	60	Turekian & Wedepohl, 1961
		10		55	30		60		McLennan, 1992
Felsic (4)	39	19	0.1	4	5	1	10	39	Turekian & Wedepohl, 1961
Sedimentary rocks									
Sandstone	57	14	0.02	120	3	1	15	16	Faust & Aly, 1981
Limestone	46	16	0.05	7	13	2	4	16	Faust & Aly, 1981
Shale	270	80	0.2	423	29	9	45	130	Faust & Aly, 1981
Black shale		15	4.0	18	68	22	50	189	Dunn, 1990
		100		700	300		200	1500	Vine & Tourtelot, 1970

V. THE BIOGEOCHEMICAL CYCLE

The discussion in this chapter thus far has concerned the mineral kingdom. However, mineral-related processes only form one part of what is known as the biogeochemical cycle—the sum of the biotic and abiotic processes that move elements from rocks, to soils, sediments, and waters where they are incorporated into plants and animals and become parts of food chains. As these processes proceed elements are returned to soils, sediments, and waters and, given sufficient geological time, are incorporated into deposits that will be transformed to newly formed rocks. However, of more immediate interest is the small scale cycling that occurs at local levels.

Plants play a key role in the biogeochemical cycle; they are critical to soil formation in all but desertic regions. The acids that their roots produce to liberate nutrients from the minerals in the soil contribute to the breakdown of those minerals. Another source of organic acids that contributes to mineral decomposition is the decay of litter-fall as plant material decomposes in the surface layers of soil. As these acids percolate downward they solubilize and carry many elements with them to lower levels in the soil. This process is vividly demonstrated in the podzolic soils (Spodosols) of humid temperate zones. Organic matter accumulates in the topmost layers of the soils (the L, F, H, and A_h horizons), giving them their characteristic dark brown-black color. Immediately below this there is a "sandy," colorless, leached (eluviated, A_e horizon) zone composed of mineral grains that have resisted corrosion and decomposition. Below the eluviated zone the soils are enriched in iron-oxyhydroxides, clay-sized materials, and to a lesser extent organic substances. Known as the B horizon, it is well developed in podzols and exhibits a rich red-brown coloration. Together, the humic-rich (A_h) top layer of the soil and the B horizon are sites of trace element, especially metal, accumulation due to the abundance of organic matter, smectites, clay minerals, and iron- and manganese-oxyhydroxides. The retention of trace elements in these horizons introduces a barrier in the biogeochemical cycle that halts circulation through the cycle for varying periods of time. Goldschmidt (1937) introduced the term "geochemical barrier" specifically for the retention of trace elements in humic-rich surface soil as it was a barrier to "flow" in the biogeochemical cycle.

As a result of the biogeochemical cycle, trace elements are dispersed into different materials, three examples are provided below, i.e., one for stream waters (Wales), one for organic stream plant material (Sweden), and one for terrestrial plant material and a foodstuff called hard red spring wheat (Canada). A fourth example of mineral rich stream sediment is provided in Section V.

In recent years the British Geological Survey has applied new ultrasensitive water analysis procedures in the preparation of hydrogeochemical maps. Figure 14, drawn from Simpson et al. (1996), is an example of the distribution of fluoride in stream waters from Wales and adjacent parts of England as revealed by a suite of 17,416 analyses. There is a clear boundary between high F^- waters draining dominantly Permo-Triassic rocks in the east and low F^- waters draining older Paleozoic rocks to the west. These two rock types were laid down in very different environments: the older rocks in a variety of marine environments, and the Permo-Triassic rocks in a terrestrial environment that was desertic to the east along the western margins of the Zechstein Sea and later along the northwestern margin of the Tethyan Ocean. What is important here is that the different environments that either favored the retention of F^- in terrestrial environments or its retention in seawater have survived in excess of 200 Ma to influence ground and surface water chemistry today and have epidemiological consequences (Simpson et al., 1996).

The Swedish Geological Survey is unique among geological surveys in employing organic stream material for its national-scale geochemical survey program (Fredén, 1994). The reason for this choice of sample material, as distinct from the mineral-rich stream sediments commonly collected in other countries (see for example Figure 17), is that it provides data that better estimate the bioavailable amounts of trace elements present in the environment. Aquatic plants such as aquatic mosses and sedge (*Carex* L.) roots are in equilibrium with the stream sediments and waters in which they grow, and their composition reflects the available amount of trace elements. Although stream water compositions may vary seasonally, the composition of the mosses and sedge roots varies more slowly, smoothing out temporal variations in the water chemistry. This makes the survey data particularly effective for monitoring anthropogenic impacts that result in the dispersion of trace elements into the surficial environment, and detecting natural geological sources that may have an impact on the ecosystem. Figure 15 presents the stream plant chemistry for chromium in part of southern Sweden. The elevated levels in the western part of the map around Vänersborg and Trollhättan are associated with pollution from a smelter. Yet in the central

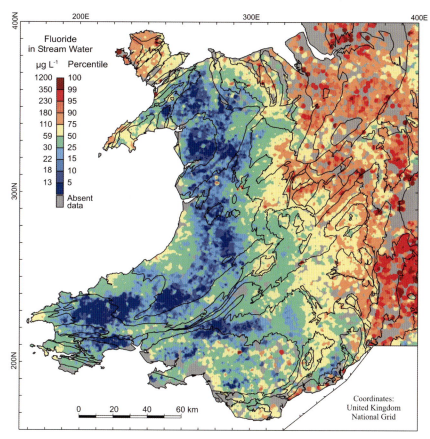

FIGURE 14 Fluoride ($\mu g/L^{-1}$) in stream waters (N = 17,416) from Wales and adjacent parts of England. (Reproduced with the permission of the British Geological Survey and Pergamon Press.)

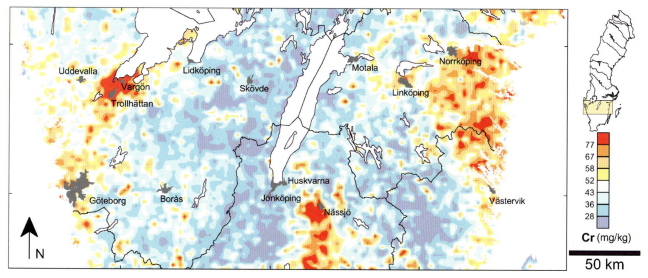

FIGURE 15 Distribution of chromium ($mg\,kg^{-1}$ in ash material) in organic stream sediments in southern Sweden. (Reproduced with the permission of the Geological Survey of Sweden.)

part of the area near Nässjö equally high levels are observed which relate to the presence of dolerites, a mafic rock that is naturally enriched in chromium relative to most other rocks (see Table III). To the east-southeast elevated chromium levels occur due to the presence of small gabbro and ultramafic rock bodies. Another notable pattern occurs in a triangular area east of a line joining Norrköping, Linköping, and Västervik where elevated chromium levels characterize an area underlain by old basement granitoid, likely reflecting even older rocks that were incorporated into the granitoids by the processes of granitization. These chromium elevated areas lie within an extensive area of low chromium rocks dominantly composed of orthogneisses and granites. Thus it can be seen that the organic stream sediment geochemistry reflects both recent anthropogenic processes and a variety of regional geological features. In this context the high chromium patterns east of Linköping and Norrköping are of particular interest. These rocks are granitic, like so much of the area characterized by low chromium levels; however, a relict geochemistry reflecting earlier rocks is retained and indicates that these rocks are "different." This demonstrates one of the strengths of regional geochemistry. Although rocks may look the same, i.e., granitic, they may be geochemically different in their trace element composition, which demonstrates their different geological histories.

As a last example, the regional distribution of selenium in Canadian Prairie hard red spring wheat (*Triticum aestivem* L.) is displayed in Figure 16. These data plotted are averages of the selenium content of harvested grain from 1996 to 1998 (Gawalko et al., 2001). This is an unusual geochemical map, but it represents an important end pause, rather than an end point, in the biogeochemical cycle. In terms of the full biogeochemical cycle involving human populations the next step is milling and incorporation into foodstuffs. Unlike the data for copper, zinc, iron, and manganese, the data for selenium and cadmium show much greater variability (Gawalko et al., 2001). The reason for this variability is that the wheat plants interact with their soil environment to ensure uptake of the essential micronutrients copper, zinc, iron, and manganese. They do not appear to regulate cadmium and selenium that are taken up with the regulated micronutrients to be sequestered in various plant tissues. The spatial distribution of these data is strongly influenced by soil properties, and to a lesser extent the varying selenium content of the soil parent material. The area of highest regional selenium levels is in southwestern Saskatchewan and adjoining Alberta where the soils are dominantly Brown Chernozems (Mollisols, Aridic Borolls), with some Solonetzic (Natric) soils. These are relatively organic carbon poor in comparison with the Dark Brown, Black, and Dark Gray Chernozems (Typic, Udic, and Boralfic

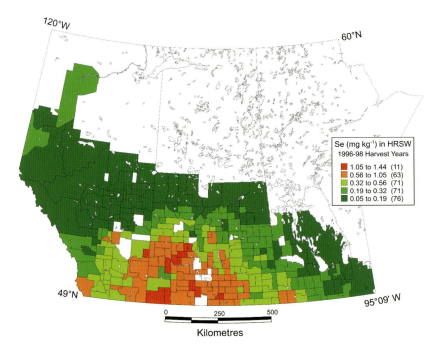

FIGURE 16 Distribution of selenium (mg kg⁻¹) in hard red spring wheat for the 1996–1998 harvests in the Canadian Prairies.

Borolls) and Gray Luvisols (Boralfs) that underlie most of the remainder of the Canadian Prairie agricultural region. Organic compounds in soils form seleno-complexes with labile selenium, thus restricting its availability to the plants. The likely cause of the spatial pattern of high selenium is low organic carbon content. The absence of seleno-complexes favors selenium partitioning into soil pore waters where it is available to the plants. A similar situation has been reported from China (Fordyce et al., 2000; Johnson et al., 2000; Wang and Gao, 2001), where higher soil organic-carbon contents are associated with lower selenium levels in rice and the increased incidence of Keshan disease. Interestingly, a link was made by MacPherson et al. (1997) between the selenium content of Canadian wheat and the selenium status of the Scottish population. The authors related this lowering of selenium status over time to the shift to importing lower selenium European wheat in preference to higher selenium Canadian bread wheat.

The examples above represent the many regional geochemical and biogeochemical studies that have been undertaken in the 20th century. In some cases these data are presented spatially, and in others temporally. They clearly demonstrate the spatial and temporal chemical variability of the natural environment. Life is supported on the Earth's chemically inhomogeneous surface, and there are ecosystem consequences to that reality. To fully understand those consequences and manage any associated risks, a sound knowledge of baseline geochemistry is required.

VI. ESTABLISHING GEOCHEMICAL BASELINES

Applied geochemical surveys are undertaken for one of two basic reasons: (1) to detect geochemical anomalies both natural, e.g., related to mineral occurrences, and anthropogenic, e.g., related to industrial releases; and (2) to map and establish natural background levels or baselines. In the first instance, surveys are designed as search exercises so that a feature of known size and geochemical contrast from the local natural background concentrations can be detected. The sampling is undertaken over an area of fixed extent, perhaps a map sheet, a particular geological terrain, or a particular jurisdiction or economic zone. Where surveys are undertaken to establish natural background and baselines they are designed to sample the area of interest, a particular jurisdiction or terrain (e.g., an eco-district), in an unbiased manner so that an average may be estimated together with a measure of the data variability. Depending on the size of the area, determining if there is systematic spatial variability across the survey area may be required. The sampling considerations for these two types of survey have been discussed by Garrett (1983).

To be able to reliably estimate the geochemical level and relief (e.g., mean and variance) for a study area a sufficient number of samples must be collected and analyzed. If the study area can be treated as a homogeneous entity, a useful rule of thumb based on the formula for the standard error of the mean (SE = $s.n^{-0.5}$) is that a minimum of 30 sites should be sampled, 60 would be better, and it is probably a waste of resources to sample more than 120. In order to obtain unbiased estimates the samples sites should be distributed randomly across the study area. A common strategy used by geoscientists is to use a square grid with a cell size such that the required number of cells, e.g., a minimum of 30, are present in the study area, and then use a random number generator to locate an x–y position (site) in each cell to be sampled. An alternate strategy is to use prior knowledge of rock or soil type distribution and collect material from sufficient sites from these postulated homogeneous units in an unbiased manner to reliably estimate their geochemical level and relief (e.g., mean and standard deviation). For national-scale surveys many more sites are sampled as the desire is to reveal broad-scale regional geochemical variability.

Clearly, factors such as resource availability influence decisions concerning sample design. Concern is sometimes expressed that widely spaced sampling will not yield reliable estimates. Examples from many parts of the world have demonstrated the ability of low-density sampling to map natural backgrounds on a regional scale, e.g., Darnley et al. (1995), Xie and Cheng (2001), and Reimann et al. (2003). It has to be remembered that if the sampling is unbiased (random) all parts of the study area have an equal opportunity to be sampled, and there is a finite probability that small features that do not form a "significant" proportion of the study area will be missed. If these are important, resources are required to increase the sample density. The cruel reality for surveys is triangular, the apices are survey area, detail of information generated, and resources—time, staff, and funds. One can fix any two of these apices, but one cannot fix all three. To minimize survey costs and maximize the return in information, Xie and Cheng (2001) recommended the widely spaced sampling of major river floodplain sediments. They demonstrated that sampling densities between 1 site per 1000–6000 km², i.e., about 520 samples representing all of mainland China, can provide reliable estimates of regional background levels. The field work for this survey was completed in one year (1992–1993) by a sampling team of three people. On this basis, they recommend that surveys of this type be completed before more detailed, time-consuming, and expensive surveys are undertaken.

In any baseline survey consistent protocols are essential for both field and laboratory work. The "target population" needs to be specifically defined, e.g., river/stream water collected midstream 10 cm below surface or surface soils collected from 0–25 cm (0–10 inches) at sites at least 100 m from a road that are visibly uncontaminated. Adequate field notes and location (this is now easy with global positioning systems, GPS) information need to be recorded. The procedures for on-site treatment such as filtering and acidification of waters, and storage, shipping, and preparation of samples like drying and screening of soil (2 mm or in the range 0.1–0.18 mm), drainage sediment (in the range 0.1–0.18 mm), and the retention of the fine fraction for analysis need to be clearly laid out. The difference in retained fraction for soils is historical, and care needs to be taken in selecting an appropriate size fraction. Traditionally soil (as prescribed by the United Nations Food and Agriculture Organization, UN-FAO) and environmental scientists have employed the <2 mm fraction for analysis, while applied geochemists have tended to employ a finer fraction of <0.150 or <0.177 mm, as used for drainage sediment surveys in mineral exploration, or <0.063 mm (<63 μm) as used for glacial sediments and the soils developed on them. Compatibility with prior data sets is a major consideration, as are considerations of plant–soil relationships. Appropriate analytical protocols and QA/QC procedures need to be in place. Procedures and considerations for baseline surveys are discussed in Darnley et al. (1995), Salminen and Gregorauskienė (2000), and by the Forum of European Geological Survey (FOREGS) in Salminen et al. (1998).

An important issue to consider in planning baseline surveys is temporal variability. For soils this is not a major consideration; however, for surface and groundwater variations in flow rates resulting from climatic variations such as seasonal rains (e.g., monsoons) or snow-melt events affect the elemental levels observed in the waters. Therefore, care has to be taken to sample a region under similar conditions, and only to subsequently use the data for comparison with data sets collected under similar conditions. In the case of stream and river sediments, strong seasonal differences in water flow can modify the bed load composition, and therefore its chemistry (Chork, 1977; Rose et al., 1979; Steenfelt & Kunzendorf, 1979). In general, if an effect is present, seasonal rains or spring freshets mass waste bank material into streams and rivers, and fine sediments are then winnowed from the streambed in higher energy (faster flow) environments and deposited in lower energy (slower flow) environments over the period of the subsequent dry season or lower flow period. The result is that levels of trace elements that occur in the finer fraction decrease in higher energy stream environments and increase in lower energy environments. The converse is true for trace elements that occur in coarser or heavier fractions of the sediment. Thus, if seasonal variation is expected, sampling programs should be restricted to longer periods of steady stream flow. If severe weather events occur, e.g., cyclones or hurricanes, the complete bed load may be changed which results in major changes in sediment geochemistry (Ridgway & Dunkley, 1988; Ridgway & Midobatu, 1991; Garrett & Amor, 1994). Thus, if a specific long-term study is undertaken where a knowledge of baseline levels is important, catastrophic weather events will likely require a post-event survey to determine if baseline levels have changed significantly.

Once the analytical data are in hand the accuracy and precision of the data need to be estimated, consistently with international standards (Darnley et al., 1995), to determine if they are adequate. If the field sampling has been structured so that analysis of variance (Garrett, 1983) or geostatistical procedures (e.g., Issaks & Srivastava, 1989) can be applied, the presence of significant spatial trends across the study area can be investigated.

As discussed earlier, geochemical data are often drawn from multiple populations. Therefore it is prudent to summarize the data as percentiles (e.g., minimum, 2nd, 5th, 10th, 25th (first quartile), 50th (median), 75th (3rd quartile), 90th, 95th, 98th percentiles, and maximum), as well as arithmetic means and standard deviations, robust estimates such as the median and mean absolute deviation, and possibly some estimates in logarithmic units. Other properties of the data set to be reported are number of samples analyzed, the lower quantification limit, and how many samples were below the limit. The reporting of the data as percentiles is a nonparametric procedure that avoids any assumption concerning the distribution of the data, and their inspection quickly reveals whether the distribution is skewed. In addition, their availability can assist in setting realistic, in the sense of the natural distribution of an element, regulatory levels as described in Section II.

Often for jurisdictional or regulatory reasons the geographical entities over which baseline surveys are undertaken have no direct, or only an indirect, relation to geology and pedology. The Commission for Environmental Cooperation of the North American Free Trade Agreement (NAFTA-CEC) has prepared a geographic eco-classification for North America (see http://www.cec.org/pubs_info_resources/publications/enviro_conserv/ecomap.cfm and Marshall et al., 1996), which is likely to see increasing use as a way to subdivide natural background data into entities of ecological and environmental relevance. The eco-district boundaries are strongly influenced by soil (soil series) properties that reflect the underlying geological and biological processes, which in turn reflect climate. An example of presentation of natural background data using this framework is presented in Figure 17 for reverse aqua-regia soluble zinc determined in the <0.177 mm fraction of stream sediments from the Yukon Territory, Canada. The eco-district medians vary by a factor of 6, again demonstrating the spatial variability in natural background levels. There are sound geological reasons for the spatial patterns: the highest levels relate to zinc-rich black shales in the Selwyn basin in the northeast; low levels relate to the Yukon crystalline terrain in Central Yukon with generally higher

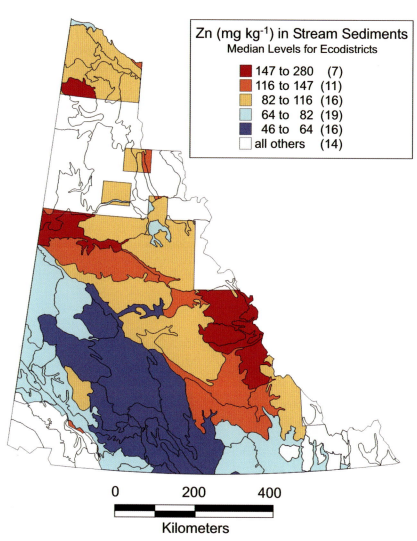

FIGURE 17 Distribution of zinc (mg kg^{-1}) in the <0.177 μm fraction of 26,862 stream sediments displayed as eco-district median values, Yukon Territory, Canada.

levels to the southwest and northeast; and numerous smaller details may be explained by the presence of particular rock units.

VII. TOTAL ANALYSES AND BIOAVAILABILITY

Geoscientists traditionally determine the total amount of elements in the samples they collect, except for specific geochemical exploration procedures where a wide range of protocols are employed. The total amount of metal in a rock, soil, or sediment is a poor estimator of what may become phyto- or bioavailable and be able to cross cellular barriers.

A wide variety of protocols have been developed by agronomists and ecotoxicologists to better estimate the phyto- and bioavailable amounts of an element in soils, waters, plants, etc. An additional issue is: Bioavailable to what? (Allen, 2002). This has led to a great variety of protocols, some of which are locally adequate for specific studies (see, for example, Sauvé, 2002). However, none is universal, though the estimation of free-ion concentrations in free or pore waters (the Free Ion Activity Model, FIAM) approaches that requirement (see Campbell, 1995). Another important factor influencing the availability of metals in solutions to biota is the presence of dissolved organic carbon compounds, many of which are capable of sequestering trace elements so that they remain unavailable. This situation has been addressed by Tipping (1994) in the WHAM model which includes recognition of humic acid complexes. Finally, it must be remembered that the interaction between the biotic and abiotic realms is not passive. Biota are capable of mediating the passage of ions across cellular barriers. In the case of plants, they are capable of acidifying or releasing low molecular weight organic acids, e.g., phytosiderophores (deoxymugineic acid), to the immediate root environment (rhizosphere) to liberate or chelate micronutrients and other trace elements into the proximal pore waters (moisture) so they are available to the plant.

As a result, much geoscience-driven baseline work is still undertaken using total or near-total aqua-regia or hot nitric acid digestions, and for many sample materials these procedures overestimate the amount of metal that may be available to biota. A notable exception is the recent 1:2,500,000 scale soils found in the geochemical atlas of Poland (Lis & Pasieczna, 1995),

which employed a 1:4 HCl mixture. Although there are sound grounds for this choice, these data are no longer comparable to other international data sets. In an ideal world the best procedure would be to have a total/near-total determination and an appropriate weaker extraction that better estimated the phyto- or bioavailable amount in the sample material. A further point that has to be remembered is that all biota are not created equal, and an extraction that may be good for agricultural crops might be quite inappropriate for soil invertebrates.

The prime controls on phyto- and bioavailability are the mineralogical or material form of an element in question and its mobility in the aqueous environment, e.g., soil or sediment pore waters, fresh stream or lake water, or marine sediments and water. Table IV provides an indication of the mobility of many trace elements of interest as a function of pH and redox conditions (see also Plant et al., 2001, Figure 6).

With reference to solid phases, elements tend to be bioavailable when they are loosely held on mineral surfaces or present as metallo-organic complexes. Thus weak extractants—acetic acid, sodium acetate, calcium chloride, potassium or ammonium nitrate—and chelating agents EDTA, DTPA, and sodium pyrophosphate are commonly used in analytical protocols to estimate phytoavailable amounts of elements in soils. Sauvé (2002) has provided a thorough and extensive review of these methods in the context of determining metal speciation (chemical bonding) in soils. In contrast to the soil protocols, dilute hydrochloric acid with various additions has been used to simulate gastric fluids in estimating the amount of trace elements that could be solubilized from soils in the digestive tract of an animal.

VIII. FUTURE CHALLENGES

The great challenge of the future is to provide consistent, relevant, high-quality geochemical data to support epidemiological research, environmental regulation, and other studies such as agricultural and forestry resource management that concern the transfer of trace elements through potable water and the food chain into human and animal populations.

Attention has been drawn to this by publication of the report on International Geological Correlation Projects 259 and 360 (Darnley et al., 1995) and by Plant et al.

TABLE IV. Mobility of Elements in the Surface Environment

	Oxidizing (pH 5–8)	Oxidizing (pH < 4)	Reducing
Relative mobility			
Highly mobile	Cl, Br, I, S, Rn, He, C, N, Mo, B, Se, Te	Cl, Br, I, S, Rn, He, C, N, B	Cl, Br, I, Rn, He
Moderately mobile	Ca, Na, Mg, Li, F, Zn, Ag, U, V, As, Sb, Sr, Hg	Ca, Na, Mg, Sr, Li, F, Zn, Cd, Hg, Cu, Ag, Co, Ni, U, V, As, Mn, P	Ca, Na, Mg, Li, Sr, Ba, Ra, F, Mn
Slightly mobile	K, Rb, Ba, Mn, Si, Ge, P, Pb, Cu, Ni, Co, Cd, In, Ra, Be, W	K, Rb, Ba, Si, Ge, Ra	K, Rb, Si, P, Fe
Immobile	Fe, Al, Ga, Sc, Ti, Zr, Hf, Th, Sn, REEs, Pt metals, Au, Cr, Nb, Ta, Bi, Cs	Fe, Al, Ga, Sc, Ti, Zr, Hf, Th, Sn, REEs, Pt metals, Au, As, Mo, Se	Fe, Al, Ga, Ti, Zr, Hf, Th, Sn, REEs, Au, Cu, Pt metals, Ag, Pb, Zn, Cd, Hg, Ni, Co, As, Sb, Bi, U, V, Se, Te, Mo, In, Cr, Nb, Ta

After Rose, Hawkes, and Webb, 1979.

(2001). Funding of such regional- and continental-scale mapping activities poses a major challenge. To date significant progress has only been made in China, where there is a national commitment to monitor the surficial environment in sufficient detail to yield useful maps (Xie & Cheng, 2001). Progress has also been made in Europe through the collaborative efforts of nations working through the Forum of European Geological Surveys. The challenge for the future is to create the interdisciplinary teams that can generate the critical mass to organize and execute systematic baseline surveys at continental scales with the support of agencies with the resources and vision to appreciate the value of a global geochemical background database.

The advent of rapid global change will stress the world's resource base, and make sustainable development an even more important issue than it is now. The role of human activity as a causative factor may be argued by some, but it remains that global economic development has radically increased the rate of change in the environment (Fyfe, 1998). As Plant et al. (2001) noted: "The problem is most acute in tropical, equatorial, and desert regions where the surface environment is particularly fragile because of its long history of intense chemical weathering over geological time scales." Change needs to be monitored, but how can it be monitored if the baseline is not known? Concerted international action is required to acquire the data essential for managing the risks that the natural environment poses to the world's population.

SEE ALSO THE FOLLOWING CHAPTER

Chapter 3 (Anthropogenic Sources)

FURTHER READING

Allen, H. E. (Ed.) (2002). *Bioavailability of Metals in Terrestrial Ecosystems: Importance of Partitioning for Bioavailability to Invertebrates, Microbes and Plants*, Society of Environmental Toxicological and Chemistry (SETAC), Pensacola, FL.

Brooks, R. R. (1972). *Geobotany and Biogeochemistry in Mineral Exploration*, Harper & Row, New York.

Brooks, R. R. (Ed.) (1998). *Plants that Hyperaccumulate Heavy Metals: Their Role in Phytoremediation, Microbiology, Archaeology, Mineral Exploration, and Phytomining*, CAB International, Oxford, UK.

Campbell, P. G. C. (1995). Interactions between Trace Metals and Aquatic Organisms: A Critique of the Free-Ion Activity Model. In *Metal Speciation and Bioavailability in Aquatic Systems* (A. Tessier and D. R. Turner, Eds.), John Wiley & Sons, New York, pp. 45–102.

Chork, C. Y. (1977). Seasonal Sampling and Analytical Variations in Stream Sediment Surveys, *J. Geochem. Explor.*, 7, 31–47.

Clarke, F. W., and Washington, H. S. (1924). The Data of Geochemistry, *U. S. Geol. Surv. Bull.*, p. 770.

Darnley, A. G., Björklund, A., Bølviken, B., Gustavsson, N., Koval, P. V., Plant, J. A., Steenfelt, A., Tauchid, M., and Xie, X. with contributions by Garrett, R. G., and Hall, G. E. M. (1995). *A Global Geochemical Database for Environmental and Resource Management*, Earth Science Series No. 19, UNESCO, Paris, France, p. 122.

Dunn, C. E. (1990). Lithogeochemical study of the Cretaceous in central Saskatchewan—perliminary report. *Sask. Geol. Surv. Misc. Rep.*, 90–4, 193–197.

Faust, S. D., and Aly, O. M. (1981). *Chemistry of Natural Waters*, Ann Arbor Science, Ann Arbor, MI, p. 400.

Fordyce, F. M., Zhang, G., Green, K., and Liu, X. (2000). Soil, grain and water chemistry in relation to selenium-responsive diseases in Enshi District, China, *Appl. Geochem.*, 15, 117–132.

Fredén, C. (Ed.) (1994). *Geology. National Atlas of Sweden Series* (L. Wastenson, Ed.), SNA Publishing, Stockholm, Sweden, p. 208.

Fyfe, W. S. (1998). Towards 2050; the Past is not the Key to the Future; Challenges for Environmental Geology, *Environ. Geol.*, 33, 92–95.

Garrett, R. G. (1983). Sampling Methodology. In *Handbook of Exploration Geochemistry, Vol. 2, Statistics and Data Analysis in Geochemical Prospecting* (R. J. Howarth, Ed.), Elsevier, Amsterdam, The Netherlands, pp. 83–110.

Garrett, R. G., and Amor, S. D. (1994). Temporal Variations in Stream Sediment Data from Jamaica. In *Prospecting in Tropical and Arid Terrains* (L. Bloom, Ed.), Prospectors and Developers Association of Canada, Toronto, Canada, Section 2, 117–137.

Gawalko, E. J., Garrett, R. G., and Nowicki, T. W. (2001). Trace Elements in Western Canadian Hard Red Spring Wheat (*Triticum aestivem* L.): Levels and Quality Assurance, *J. Assoc. Anal. Intl.*, 84, 1953–1963.

Goldschmidt, V. M. (1937). The Principles of Distribution of Chemical Elements in Minerals and Rocks, *J. Chem. Soc. (Lond.)*, 1937 (Pt. 1), 655–673.

Gustavsson, N., Bolviken, B., Smith, D. B., and Severson, R. C. (2001). Geochemical Landscapes of the Conterminous United States—New Map Presentations for 22 Elements, *U. S. Geol. Surv. Bull.*, 1645, 38.

Issaks, E. H., and Srivastava, R. M. (1989). *An Introduction to Applied Geostatistics*, Oxford University Press, New York.

Johnson, C. C., Ge, X., Green, K. A., and Liu, X. (2000). Selenium Distribution in the Local Environment of Selected Villages of the Keshan Disease Belt, Zhangjiakou District, Hebei Province, People's Republic Of China, *Appl. Geochem.*, 15, 385–401.

Kabata-Pendias, A. (2000). *Trace Elements in Soils and Plants*, 3rd edition, CRC Press, Boca Raton, FL.

Lis, J., and Pasieczna, A. (1995). *Geochemical Atlas of Poland*, Polish Geological Institute, Warsaw, Poland, p. 36, 7 Tables, and 74 Plates.

Lee, T., and Yao, C.-L. (1970). Abundance of chemical elements in the earth's crust and its majortectonic units. *Intl. Geol. Rev.*, 12(7), 778–786.

MacPherson, A., Barclay, M. N. I., Scott, R., and Yates R. W. S. (1997). Loss of Canadian Wheat Imports Lowers Selenium Intake and Status of the Scottish Population. In *Trace Elements in Man and Animals-9* (P. W. F. Fischer, M. R. L. Abbé, K. A. Cockell, and R. S. Gibson, Eds.), National Research Council Press, Ottawa, Canada, pp. 203–205.

Marshall, I. B., Smith, C. A. S., and Selby, C. J. (1996). A National Framework for Monitoring and Reporting Environmental Sustainability in Canada, *Environ. Monitor. Assess.*, 39, 25–38.

McKelvey, V. E. (1960). Relation of Reserve of the Metals to their Crustal Abundance, *Am. J. Sci.*, 258–A, 234–241.

McLennan, S. M. (1992). Continental Crust. In *Encyclopedia of Earth Sciences, Vol. 1* (W. A. Nierenberg, Ed.), Kluwer, Dortrecht, The Netherlands, pp. 581–592.

Plant, J., Smith, D., Smith, B., and Williams, L. (2001). Environmental Geochemistry on a Global Scale, *Appl. Geochem.*, 16, 1291–1308.

Reimann, C., and de Caritat, P. (1998). *Chemical Elements in the Environment: Fact Sheets for the Geochemist and Environmental Scientist*, Springer-Verlag, Berlin p. 398.

Reimann, C., and Melezhik, V. (2001). Metallogenic Provinces, Geochemical Provinces and Regional Geology—What Causes Large-Scale Patterns in Low Density Geochemical Maps of the C-horizon of Podzols in Arctic Europe, *Appl. Geochem.*, 16, 963–983.

Reimann, C., Äyräs, M., Chekushin, V. A., Bogatyrev, I. V., Rognvald, B., Caritat, de P., Dutter, R., Finne, T. E., Halleraker, J. H., Jæger, Ø., Kashulina, G., Niskavaara, H., Lehto, O., Pavlov, V. A., Räisänen, M. L., Strand, T., and Volden, T. (1998). *Environmental Geochemical Atlas of the Central Barents Region*, Geological Survey of Norway, Trondheim, Norway p. 745.

Reimann, C., Siewers, U., Tarvainen, T., Bityukova, L., Erikson, J., Gilucis, A., Gregorauskiene, V., Lukashev, V. K., Matinian, N. N., and Pasieczna, A. (2003). Agricultural soils in northern Europe: A geochemical atlas. Geologisches Jahrbuch: Sonderhefte Reihe D, Hefte SD5, Hannover, Germany.

Ridgway, J., and Dunkley, P. N. (1988). Temporal Variations in the Trace Element Content of Stream Sediments: Examples from Zimbabwe, *Appl. Geochem.*, 3, 609–621.

Ridgway, J., and Midobatu, C. (1991). Temporal Variations in the Trace Element Content of Stream Sediments: An

Example from a Tropical Rain Forest Regime, Solomon Islands, *Appl. Geochem.*, 6, 185–193.

Rose, A. W., Hawkes, H. E., and Webb, J. S. (1979). *Geochemistry in Mineral Exploration*, 2nd edition, Academic Press, London.

Salminen, R., and Gregorauskienė, V. (2000). Considerations Regarding the Definition of a Geochemical Baseline of Elements in the Surficial Materials in Areas Differing in Basic Geology, *Appl. Geochem.*, 15, 647–653.

Salminen, R., Tarvainen, T., Demetriades, A., Duris, M., Fordyce, F. M., Gregorauskiene, V., Kahelin, H., Kivisilla, J., Klaver, G., Klein, H., Larson, J. O., Lis, J., Locutura, J., Marsina, K., Mjartanova, H., Mouvet, C., O'Connor, P., Odor, L., Ottonello, G., Paukola, T., Plant, J. A., Reimann, C., Schermann, O., Siewers, U., Steenfelt, A., Van der Sluys, J., and Williams, L. (1998). FOREGS Geochemical Mapping: Field Manual, *Geol. Surv. Finland Guide*, 47, 42.

Sauvé, S. (2002). Speciation of Metals in Soil. In *Bioavailability of Metals in Terrestrial Ecosystems: Importance of Partitioning for Bioavailability to Invertebrates, Microbes and Plants* (H. E. Allen, Ed.), Society for Environmental Toxicology and Chemistry (SETAC), Pensacola, FL, pp. 7–58, chap. 2.

Shacklette, H. T., and Boerngen, J. G. (1984). Element Concentrations in Soils and Other Surficial Materials of the Conterminous United States, *U. S. Geol. Surv. Bull.*, 1270, 105.

Simpson, P. R., Breward, N., Flight, D. M. A., Lister, T. R., Cook, J. M., Smith, B., and Hall, G. E. M. (1996). High Resolution Regional Hydrogeochemical Baseline Mapping of Stream Waters in Wales, the Welsh Borders and West Midlands Region, *Appl. Geochem.*, 11, 621–632.

Steenfelt, A., and Kunzendorf, H. (1979). Geochemical Methods in Uranium Exploration in Northern East Greenland. In *Geochemical Exploration 1978* (J. R. Watterson and P. K. Theobald, Eds.), Association of Exploration Geochemists Spec. Vol. 7, Toronto, Canada, 429–442.

Taylor, S. R. (1964). *Geochim. Cosmochim. Acta*, Abundance of Chemical Elements in the Continental Crust, A New Table., 28(8), 1273–1285.

Tipping, E. (1994). WHAM—A Chemical Equilibrium Model and Computer Code for Water, Sediments, and Soils Incorporating a Discrete Site/Electrostatic Model of Ion-Binding by Humic Substances, *Comput. Geosci.*, 21, 973–1023.

Turekian, K. K., and Wedepohl, K. H. (1961). Distribution of the elements in some major units of the Earth's crust. *Bull. Geol. Soc. Am.*, 72(1), 175–192.

Vine, J. D., and Tourtelot, E. B. (1970). Geochemistry of black shale deposits; a summary report. *Econ. Geol.*, 65(3), 253–272.

Vistelius, A. B. (1960). The Skew Frequency Distributions and the Fundamental Law of the Geochemical processes, *J. Geol.*, 68, 1–22.

Wang, Z., and Gao, Y. (2001). Biogeochemical Cycling of Selenium in Chinese Environments, *Appl. Geochem.*, 16, 1345–1351.

Wedepohl, K. H. (1995). The Composition of the Continental Crust. *Geochim. Cosmochim. Acta*, 59, 1217–1232.

Xie, X., and Cheng, H. (2001). Global Geochemical Mapping and its Implication in the Asia-Pacific Region, *Appl. Geochem.*, 16, 1309–1321.

CHAPTER 3

ANTHROPOGENIC SOURCES

RON FUGE
University of Wales

CONTENTS

I. INTRODUCTION

As outlined in Chapter 2, the geochemistry of environmental media is largely dependent on the chemistry of the natural sources from which they have been derived, or with which they have interacted. Thus soil and surficial sediment chemistry are strongly influenced by the composition of their parent materials. Similarly stream and river waters, derived initially from precipitation, depend on the rocks, sediments, and soils from which they come into contact and interact with for their chemical composition. However, with the evolution of humans in the relatively recent geological past there have been anthropogenic impacts on the environment, which have increased dramatically with increasing population, urbanization, and industrialization (Fyfe, 1998). Thus humans have contaminated or polluted the once pristine environment, and this impact is manifested in the chemistry of environmental materials that reflect anthropogenic signals superimposed on the natural composition.

Many human activities have resulted in environmental contamination and these include:

1. Mineral extraction and processing
2. Smelting and refining of mineral ores and concentrates
3. Power generation—fossil fuel, nuclear, geothermal, and hydroelectric
4. Other industrial and manufacturing activities— metallurgical and chemical industries, brick and pipe manufacture, cement manufacture, the ceramics and glass industry, plastics and paint manufacture, and fertilizer manufacture
5. Waste disposal—household refuse, fly ash, sewage, nuclear, and the open burning of refuse
6. Agricultural practices—application of mineral-based fertilizers and manure together with sewage sludge, application of pesticides and herbicides,

farmyard runoff including such materials as sheep dip, etc., and deforestation, which has contributed to problems of mercury contamination of Brazilian rivers

7. Transportation,—motor vehicle derived contamination; this is particularly important in urban environments

8. Treatment and transport (through metal pipes and fittings) of potable water

Nowhere is the impact of environmental pollution more apparent than in the urban environment. Road dust and soils in the urban environment can be heavily contaminated, particularly by metals. Road dust and fine soil particles are the major sources of house dust and as such represent a potential pathway into the human body from inhalation and inadvertent ingestion from hands, etc., particularly for children.

There are many examples of anthropogenically derived substances having marked effects on human health; for example, in 1956 the mercury poisoning experienced in the Minimata region of Japan, known as "Minimata disease." This resulted from a factory situated on the coast releasing mercury, which was used as a catalyst in plastics manufacture, into Minimata Bay. From there it was passed through the marine food chain in a methylated form $[CH_3Hg^+$ or $(CH_3)_2Hg]$, and into humans via consumption of fish. Mercury attacks the central nervous system and causes irreversible brain damage. The methyl form of mercury represents a particularly serious toxic threat, as it is able to cross membranes and accumulate in the central nervous system. It has been estimated that over 20,000 people were affected.

Although the main thrust of this book concerns the impact of the natural environment on human health, this chapter deals briefly with anthropogenic impacts on the composition of environmental media together with some brief considerations of health impacts. This is included for the sake of completeness and as such it is not a detailed account of the topic; indeed, such a detailed account would require a second book.

II. IMPACT OF MINERAL EXTRACTION AND REFINING

A. Mining and Mineral Extraction

Mineral deposits represent concentrations of an element or elements to a level at which they can be profitably extractable. As such these deposits represent concentra-

tions of several elements well above crustal abundance. Such naturally occurring high concentrations are reflected in the chemistry of the soils, waters, sediments, plants, etc., in the immediate vicinity of the deposit. Indeed, measurement of the concentrations of various metals and non-metals in media such as soils, sediments, and waters have been used to locate mineral deposits. This practice is described as geochemical exploration.

Although there are likely to be significant natural enrichments of several elements in the vicinity of mineral deposits, mining and extraction of the deposit will add greatly to these enrichments. The mining and subsequent beneficiation of minerals and the separation and refining of their various components is one of the most serious sources of contamination of soils, waters, and the biosphere.

Humans have extracted minerals, particularly the metalliferous ores, since ancient times, and the extraction and refining of metals have played a major role in human development. The mining and processing of minerals have increased through time, due to population growth and the greater utilization of raw materials for manufacture. Many areas of past mining activity, in both the Old and New Worlds, bear witness to these extractions in the form of abandoned workings and extensive waste tips. Modern mineral extraction technology is generally far more efficient than past practices, and in many countries such processes are heavily regulated to limit the degree of contamination from extractive industries. However, historical mineral extraction involved less efficient technologies, and in those times virtually no environmental regulations were in place. Long-abandoned mineral workings are currently the cause of serious environmental pollution in many countries.

A large number of different materials are extracted from the Earth ranging from fuels such as oil and coal, industrial minerals such as clays and silica, aggregates for building and roadstone, and minerals for fertilizers as well as sources of non-metals. However, the major cause of concern are the metalliferous ores that are used as sources of metals and metalloids. Some of the more important metalliferous ore minerals are listed in Table I.

The extraction and subsequent processing of ores can be summarized as follows:

$$mining \rightarrow crushing/grinding$$
$$\rightarrow concentration\ of\ ore\ mineral$$
$$\rightarrow smelting/refining$$

Ores are extracted from the Earth by either subsurface mining, open pit surface techniques, or in a few

TABLE I. Some of the Important, Mostly Metalliferous Minerals

Mineral	Composition	Comments
Arsenopyrite	FeAsS	Frequently occurs as a gangue mineral
Barite	$BaSO_4$	Major use in drilling muds
Bauxite	Mainly $Al(OH)_3$	Only ore of aluminum
Bornite	Cu_5FeS_4	Important ore of copper
Carrollite	$Cu(Co,Ni)_2S_4$	Important ore of cobalt
Cassiterite	SnO_2	Main ore of tin
Chalcocite	Cu_2S	Important ore of copper
Chalcopyrite	$CuFeS_2$	Major ore of copper
Chromite	$FeCr_2O_4$	Main ore of chromium
Cinnabar	HgS	Main ore of mercury
Galena	PbS	Main ore of lead
Gold (native)	Au	Main source of metallic gold
Haematite	Fe_2O_3	Major ore of iron
Ilmenite	$FeTiO_3$	Important ore of titanium
Magnetite	Fe_3O_4	Important ore of iron
Molybdenite	MoS_2	Main ore of molybdenum
Pentlandite	$(Fe,Ni)_9S_8$	Major ore of nickel
Platinum	Pt (with other metals)	Main source of platinum
Pyrite	FeS_2	Common gangue mineral
Pyrrhotite	$Fe_{(1-x)}S_2$	Common gangue mineral
Rutile	TiO_2	Important ore of titanium
Scheelite	$CaWO_4$	Important ore of tungsten
Sphalerite	ZnS	Major ore of zinc
Stibnite	Sb_2S	Main ore of antimony
Tetrahedrite	$(Cu,Fe)_{12}Sb_4S_{13}$	Copper ore—often contains silver
Uraninite (Pitchblende)	UO_2	Main ore of uranium
Wolframite	$(Fe,Mn)WO_4$	Important ore of tungsten
Zircon	$ZrSiO_4$	Main ore of zirconium

cases by solution mining, which carries with it risks of groundwater pollution. Both subsurface and surface extraction result in waste material, which is generally piled on the surface in the vicinity of the mine. However, while such waste piles, which frequently contain ore minerals, are sources of environmental contamination, it is the subsequent processing of the ores that results in the greatest environmental problems. The crushing and grinding (comminution) of mineral processing has the objective of separating the ore minerals from the waste, generally referred to as gangue. To effect separation, the mined ore is finely crushed to liberate individual ore mineral grains to enable concentration of the sought after ore minerals. The very fine waste material left after this concentration process is referred to as tailings, and this material can contain, along with the gangue minerals, residual amounts of the ore minerals and can be a serious source of pollution. The tailings are very fine and are subject to wind ablation and can easily be transported by surface runoff. At many long-abandoned mine sites tailings have been left open to the environment resulting in serious contamination of surrounding soils and waters. In more recent mining operations, tailings are stored wet in tailings ponds often behind artificial dams. However, leakage of metal ions into both surface and subsurface waters has, in some cases, resulted in serious contamination of these waters, some of which have been sources of potable water. In addition, after mine closure tailings ponds will dry out unless arrangements are made to keep them permanently wet, which renders the fine material susceptible to spreading across the neighboring area. In modern mining operations, upon closure, remedial action such as isolation of the tailings material by covering serves to limit subsequent environmental pollution.

A further problem of tailings dams is the possibility of dam failure releasing large quantities of highly contaminated sediments and waters into the local environment. There have been several such dam failures in the last decade such as at the Omai gold mine in Guyana in 1995 and at the Mar copper mine on Marinduque Island, Philippines. A major failure of the Los Frailes tailings dam at Aznalcóllar, Spain, in 1998, spilled 6.8 million m^3 of water and pyrite-rich tailings that covered approximately 2000–3600 ha of agricultural land. In 2000 a gold mine tailings dam at Borsa in northwestern Romania released large quantities of cyanide and metals into local rivers, the Vaser and Tizla, which ultimately drain into the Danube River. This caused the death of many fish and birds that ate the fish.

In addition to the major metal components of the various ore minerals, many trace constituents are

TABLE II. Some Trace Constituents of Selected Sulfide Minerals (values in $mg\,kg^{-1}$)

Element	Normal range	Maximum found
Galena (PbS)		
Ag	500–5000	30000
As	200–5000	10000
Bi	200–5000	50000
Cu	10–200	3000
Sb	200–5000	30000
Tl	<10–50	1000
Sphalerite (ZnS)		
As	200–500	10000
Cd	1000–5000	44000
Cu	1000–5000	50000
Hg	10–50	10000
Sn	100–200	10000
Tl	10–50	5000
Chalcopyrite (CuFeS$_2$)		
Ag	10–1000	2300
Co	10–50	2000
Ni	10–50	2000
Sn	10–200	770
Pyrite (FeS$_2$)		
As	500–1000	50000
Co	200–5000	>25000
Cu	10–10000	60000
Ni	10–500	25000
Pb	200–500	5000
Sb	100–200	700
Tl	50–100	100
Zn	1000–5000	45000

From Levinson, 1980.

included and these frequently cause as much environmental concern as the major elements in the ores (see Table II). Of the trace elements perhaps the most notorious is cadmium. It is ubiquitous in zinc ores with concentrations of up to 4.4% having been recorded in some ores. Weathering of the ore minerals within the abandoned mines and in waste and tailings tips results in the release of the trace constituents along with the major components of the ore. Soils and waters in the vicinity of disused metal mines are frequently heavily contaminated. Figure 1 is a geochemical map for cadmium in stream plants in Sweden. The high cadmium concentrations in the central area of the country reflect metal mining in the area.

The most important ores of several base metals such as lead, zinc, and copper are sulfide minerals. The sulfide ore minerals represent the most serious threat of environmental contamination, because they are fairly easily oxidized in the presence of air to the considerably more soluble sulfates such as

$$ZnS + 2O_2 \rightarrow ZnSO_4$$

As a result surface and groundwaters in the vicinity of the weathering sulfide minerals can be seriously impacted.

A particular problem concerning weathering of sulfide minerals is that of pyrite and marcasite (both FeS$_2$). These iron sulfides oxidize to give various iron oxides and hydroxides together with some sulfates; these oxidation products are collectively called ocher. In addition to ocher, a by-product of the oxidation process is sulfuric acid; the resultant runoff from mines and waste piles is called acid mine drainage or acid rock drainage. Such acidic solutions can chemically attack other ore minerals and rocks to produce a cocktail of elements that can have a serious environmental impact on receiving rivers and streams. The reactions resulting in acid drainage can be summarized in the following equations:

$$FeS_2 + 7/_2 O_2 + H_2O \rightarrow FeSO_4 + H_2SO_4$$

$$2FeSO_4 + H_2SO_4 + 1/_2 O_2 \rightarrow Fe_2(SO_4)_3 + H_2O$$

$$FeS_2 + Fe_2(SO_4)_3 + 2H_2O + 3O_2 \rightarrow 3FeSO_4 + 2H_2SO_4$$

In acid mine drainage impacted rivers, lakes, and estuaries the contaminants can cause extreme damage to the biosphere. This is exacerbated by the precipitation of ocher, which is enriched in metals such as cadmium, copper, zinc, and aluminum, and metalloids such as arsenic. The ocher is frequently fine enough to be ingested by fish. With decreasing acidity which results from dilution and neutralization of the acid by interaction with rocks, the concentration of contaminant metals in the ocher increases. This results in sediments in the impacted streams becoming heavily enriched in potentially harmful elements.

In addition to its occurrence in metalliferous ore deposits, pyrite is frequently associated with coal, with drainage from coal mines and associated waste piles being heavily ocherous, but generally not as strongly enriched in environmentally harmful elements.

As stated above, historical mineral extraction processes were much less efficient than current methods and were not subject to strict environmental regulation.

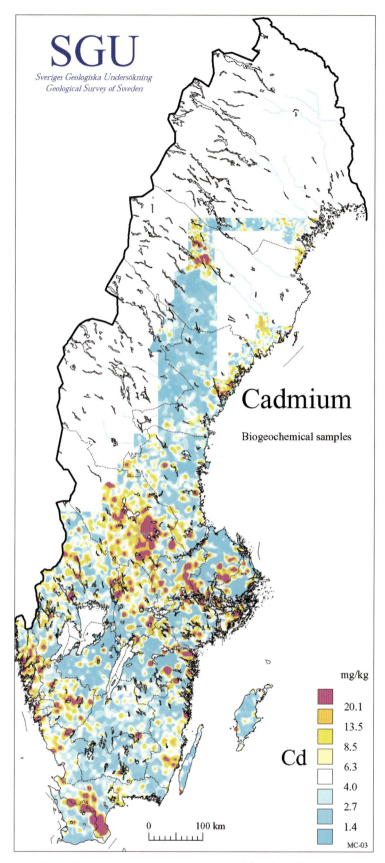

FIGURE 1 Geochemical map for cadmium in stream plants in part of Sweden. (Reproduced with permission of the Swedish Geological Survey.)

The high cadmium concentrations in south central Sweden are due to anthropogenic sources, those on the east coast are due to long range aerial deposition, those in the central area are derived from metal mining, and those in the central eastern area are derived from fertilizer. The high cadmium concentrations in the extreme south of the country are natural and are derived from cadmium-rich sediments.

In many orefields where mining occurred in river valleys, the plentiful supply of water was used to separate ore minerals by gravity. As a result of this inefficient process, considerable amounts of highly contaminated sediment and fine ore grains were released into rivers. Due to subsequent flooding, they were also released onto the floodplains of the river valleys concerned. Thus many floodplain soils in old mining areas are highly enriched with contaminant metals.

One problem concerning gold mining, which deserves a special mention, is that of associated mercury pollution. Mercury has long been used to extract gold from ores and gold-rich sediments due to its ability to amalgamate with gold. As a result of this process many areas where gold has been extracted, particularly from river sediments, have suffered serious mercury contamination.

B. Smelting and Refining

Following the mining and processing of the ores, the resultant concentrate is transported to a smelter. Pyrometallurgical smelting involves roasting of the ore concentrate at high temperatures with the consequent emission of large quantities of potentially harmful elements. The smelter emissions can be in the form of gases, such as sulfur dioxide, aerosols, and larger particulates. Modern smelter stacks are fitted with electrostatic precipitators and other dust recovery mechanisms, which results in the retention of most of the particulates, but some gaseous and aerosolic emissions are still released into the atmosphere. Although any larger particulates released are likely to be deposited close to the source, aerosols and gases can be transported long distances and as a result the smelting of ores has far wider aerial impact than the mining and processing of these ores.

As many of the ores are sulfides and the smelting process is designed to be oxidizing, the gases sulfur dioxide (SO_2) and sulfur trioxide (SO_3) are produced. These gases react with water vapor in the atmosphere to produce sulfuric acid (H_2SO_4), which results in acid rain with potentially serious consequences for the environment.

During the smelting of metalliferous ores many metals and metalloids are released to the environment. The elements released depend on the ore or smelted ores and include antimony, arsenic, bismuth, cadmium, chromium, cobalt, copper, lead, mercury, nickel, thallium, selenium, and zinc. Extremely high concentrations of some of these elements have been recorded in close proximity to the smelters. Rieuwerts and Farago

(1996) showed that soils within 500 m of a lead smelter in the Czech Republic contained up to 3.73% lead and 2.76% zinc. It has also been shown that high concentrations of elements can be transported considerable distances from the smelter. Reimann *et al.* (1998) demonstrated that nickel from smelters in the Kola Peninsula of Russia was transported over 100 km in the direction of the prevailing wind. These same authors state that one of the smelters in the Kola Peninsula (Monchegorsk) emitted 1619 tonnes of nickel during 1994.

Although present day smelters emit potentially harmful elements, areas of past smelting operations continue to be sites of environmental contamination. In the area of the lower Swansea Valley, South Wales, UK, a major smelting center for copper from 1717 until 1925, soils contain up to 200 mg kg^{-1} copper with the affected area extending 25 km north and 20 km east, the prevailing wind direction being from the southwest. The contamination is thought to have resulted from smelter fumes and wind ablation of smelter waste piles.

Many specialist refining facilities have also been shown to cause environmental contamination. Perhaps the best documented is the release of fluorine during aluminum production. Aluminum is derived from alumina (aluminum oxide), which is recovered from bauxite ore by high temperature electrolysis with the alumina in a fluorine-rich electrolyte. The fluorine released during the process has been found to cause toxicity in plants in the vicinity of the refineries, and high fluorine occurs in the skeletons of animals in the same areas.

Smelter contaminants have been shown to have a detrimental effect on human and animal health. Copper toxicity was found to occur in cattle in the vicinity of a copper smelter in Mpumalanga (eastern Transvaal), South Africa (Gummow *et al.*, 1991).

III. POWER GENERATION

A. Fossil Fuel

Globally, fossil fuel (coal, oil, peat) combustion provides most of the power generated for industrial and domestic use. Burning of these fuels has achieved notoriety in recent years due to the large volumes of carbon dioxide (CO_2) produced, the consequent buildup of this gas in the atmosphere, and its possible contribution to the greenhouse effect on the Earth. In addition, combustion of high sulfur-containing fuels in some areas has resulted in production of sulfur dioxide and sulfur trioxide that, as in the case of smelter emissions (see Section II.B above), results in acid rain.

An additional environmental consequence of the use of fossil fuels for power generation derives from the many trace elements contained in the fuels. Although fossil fuels are predominantly made up of organic matter—the combustion of the carbon in these fuels being the source of energy—they also contain variable amounts of inorganic constituents retained in the ash left after the combustion process with some emitted in fine combustion products into the atmosphere. The ash residue, called fly ash, can contain many potentially harmful elements and therefore needs to be carefully disposed of (see Section V). However, some of the inorganic components are emitted during combustion and can impact the soils, water, and biosphere in the vicinity of the power plant.

Many trace elements have been detected in fossil fuels. The actual concentrations of individual elements are variable and depend on the source of the fuel. However, it has been suggested that in Europe oil and coal combustion contribute significantly to atmospheric deposition of arsenic, cadmium, chromium, copper, nickel, and vanadium (Rühling, 1994). Coal combustion is thought to have made a significant contribution to atmospheric lead deposition in the UK, with the coals containing up to 137 mg kg^{-1} (Farmer et al., 1999). It has also been suggested that coal combustion is the major anthropogenic source of selenium in the environment, and United States coals contain up to 75 mg kg^{-1} (Coleman et al., 1993). Enhanced concentrations of uranium in many coals have resulted in enrichments of this element around coal-fired power stations. Elevated concentrations of mercury occur in some oils with values up to 72 mg kg^{-1} recorded (Al-Atia, 1972). Elements enriched in oils such as vanadium have also been found to be elevated in the environment in the vicinity of oil refineries (Rühling, 1994).

The occurrence of fluorine in coal has been shown to impact the biosphere. Fluorine is highly phytotoxic and combustion of fluorine-rich coals has caused toxicity in vines downwind of a power station in New South Wales, Australia (Leece et al., 1986). Dental fluorosis occurring in cattle in the vicinity of large power stations in Yorkshire, England, was suggested by Burns and Allcroft (1964) to be due to the use of high-fluorine coals. Human fluorosis from burning fluorine-rich coals as a domestic fuel occurs in Guizhou Province, China (Zheng & Hong, 1988).

Arsenic is enriched in coals from various parts of the world; however, in Guizhou Province, China, extreme enrichments have been found with up to 3.5% arsenic found in some samples (Ding et al., 2001). More than 3000 cases of arsenic poisoning have occurred from the combustion of these coals for domestic heating (Ding et al., 2001). Abandoned coal gasification sites are another source of contamination related to fossil fuels. Such sites have residual waste piles that are sources of elements such as arsenic, cadmium and copper, and cyanide and organic compounds such as phenols and tars which may have leaked into the subsoil.

B. Nuclear

Nuclear power generation has been utilized since the mid-1950s and accidental leakages and permitted effluent releases have impacted the environment. The nuclear industry is now strictly regulated, but in the early years this was not so and authorized discharges of radioactivity were considerably larger. Radionuclides released during these early years still pose a problem. For example, radioactive elements such as americium and plutonium released from Sellafield nuclear power station in Cumbria, northwestern England, are still retained in nearby marshy areas.

Although much of the contamination released from nuclear power plants affects only the immediate environment of the nuclear installation in question, the catastrophic explosion at Chernobyl in April 1986 caused widespread contamination, which seriously affected the Ukraine with radioactivity spreading over much of Europe and many other parts of the world. Some of the more important radioactive species released in the explosion are listed in Table III. Radioactive cesium from the Chernobyl accident rained out over upland areas of the UK and high concentrations were found in sheep in the area.

Anthropogenic radioactivity in the environment poses a serious threat to human health. Of particular concern is radioactive iodine, mainly ^{131}I, which has been found to move through the food chain rapidly. As a result of exposure to radioactive iodine, humans are prone to increased incidences of thyroid cancer, as evidenced in the aftermath of Chernobyl (see Soils and Iodine Deficiency, Chapter 16, this volume).

C. Geothermal

Geothermal energy has often been assumed to be a "clean" form of power generation. However, many geothermal areas are associated with volcanic activity and many of the hot springs actively precipitate arsenic,

TABLE III. Some of the More Important Radioactive Isotopes Released by the Chernobyl Accident

Isotope	$T_{1/2}$
Relatively short half-life isotopes	
^{132}Te	78 hours
^{133}Xe	5.27 days
^{131}I	8.07 days
^{95}Nb	35.2 days
^{89}Sr	52 days
Isotopes with longer half-lives	
^{134}Cs	2.05 years
^{85}Kr	10.76 years
^{90}Sr	28.1 years
^{137}Cs	30.23 years
Actinides (some of the more problematic)	
^{241}Pu	13.2 years
^{244}Pu	8×10^7 years
^{241}Am	458 years

antimony, mercury, and thallium, whereas some geothermal waters contain very high concentrations of boron. Thus spent waters from geothermal production in areas such as New Zealand contain very high concentrations of arsenic (see Arsenic in Groundwater and the Environment, this volume), which pose serious environmental problems for the receiving rivers.

D. Hydroelectric

Hydroelectric power generation has led to problems resulting from flooding of areas for dams. In flooded areas where soils have been inundated it has been found that fish frequently contain elevated concentrations of mercury. The source of the mercury has been found to be the waterlogged soil where this element becomes converted to a methylated form which is bioavailable (see also Section I of this chapter).

IV. OTHER INDUSTRIAL ACTIVITIES

There are a large number of other industrial activities that have the potential to cause environmental con-

tamination. In this section only a small selection typifying such processes is discussed.

A. Metallurgical

Many metals and metalloids are used in the manufacture of alloys, with steel being the major product. The manufacture and recycling of steel results in the release of many elements into the atmosphere with subsequent deposition and contamination of the local environment. Figures obtained for atmospheric pollutants in the north of the former Soviet Union show that 79% of the chromium, 76% of the manganese, and 75% of the zinc together with significant quantities of antimony, arsenic, cadmium, lead, nickel, and vanadium are derived from the iron and steel industry (Pacyna, 1995). More modern electric arc furnaces processing steel scrap give rise to a large amount of metal-rich flue dusts.

In the proximity of a steel plant in Mpumalanga (eastern Transvaal), South Africa, chronic vanadium toxicity occurs in cattle as a result of air pollution from the steelworks (Gummow et al., 1994). Another health problem related to steel production could possibly stem from the use of the mineral fluorite (calcium fluoride) as a flux in the process. An occurrence of fluorosis in cattle in northern England was suggested by Burns and Allcroft (1964) to be due to fluorine releases from local steel plants. In addition the waste, or slag, from steel plants can contain many metals that can have a detrimental effect on the environment, and this needs to be disposed of appropriately. Some slags have been used in fertilizers and Frank et al. (1996) reported vanadium poisoning in cattle after grazing on pasture land that had been amended with basic slag.

B. Brick and Pipe Manufacture

Bricks and pipes are very important in the construction industry. Both are produced from naturally occurring clays and clay-containing rocks. Manufacture of these materials involves shaping the clays and then roasting them at temperatures up to 1200°C. As a result of this roasting, elements, which are contained in the clay minerals such as copper, lead, and zinc, are released to the environment (Fuge & Hennah, 1989). However, a more significant health problem is the release of fluorine, which is frequently enriched in the minerals of the clays and rocks used during the manufacture. Very serious fluorosis in farm animals occurring in the East Anglia region of the UK in the 1950s was due to the extensive brick manufacture in this area (Burns & Allcroft, 1964).

C. Cement Manufacture

Cement is also manufactured by a process involving the roasting of clay-rich rocks. A mixture of limestone and shale rocks is heated in a furnace at temperatures around 1450°C. Shale is a clay-rich rock and can contain elevated concentrations of potentially harmful elements, and as in brick and pipe manufacture, these can be released at high temperatures.

D. Others

Many other industrial activities have been shown to result in environmental contamination. The production of enamel, ceramics, and glass releases appreciable concentrations of fluorine into the environment (Koritnig, 1972). Another industry shown to release large quantities of fluorine into the environment is the manufacture of phosphate fertilizer from the fluorine-containing mineral apatite [$Ca(PO_4)_3(OH,F,Cl)$] (Dabkowska et al., 1995).

Of the many other industrial processes that impact the environment, it is perhaps relevant to mention the serious impact of chromium in the environment from tanning plants (Davis et al., 1994). In addition, the chlor-alkali industry, which manufactures sodium hydroxide and chlorine from seawater by electrolysis with mercury electrodes has resulted in serious environmental contamination of the atmosphere and waterways with mercury. Spillages and leakage from the electrodes is also an environmental problem. Although the chlor-alkali process is being phased out, disused chlor-alkali works are a continuing source of contamination.

V. Waste Disposal

Large quantities of waste materials are produced on a daily basis from urban and industrial sources. Although there is a move to recycle and reuse waste products, the vast majority of waste has to be treated and subsequently disposed of. Many of the waste products created in modern society have a detrimental effect on the environment.

A. Refuse

Most household refuse together with waste materials from industrial sources is consigned to landfill sites. Such landfill sites produce leachate that contaminates soils, surface waters, and groundwaters in the vicinity. Landfill leachate chemistry is variable and depends on the components of the landfill plus varying conditions within the buried material, such as the differing redox conditions of various parts of the landfill and the percolation and flow of water into and through the site. Modern landfill sites are lined with low permeability materials such as high-density plastics to eliminate leachate leakage. However, old landfill sites made use of suitable depressions, such as disused quarries, and were not lined. These sites have become serious sources of contamination. Because landfills can contain a wide range of waste materials, many elements and organic compounds can be released, which include toxic metals such as cadmium and mercury and substances such as ammonia, nitrate, and sulfate. Due to the generally anaerobic nature of landfills, methane gas is generated and in modern facilities this is collected and used as fuel.

Incineration is the other commonly used method of refuse disposal. This process, which is utilized to provide a source of energy in some instances, results in the production of ash that contains many elements that are potentially harmful to the biosphere so necessitating disposal in a landfill. Additionally, incinerators can vent various elements and organic compounds such as dioxins into the atmosphere which impact local soils and waters. In this context it is perhaps relevant to note that uncontrolled burning of refuse, particularly in the developing world, and the consequent emissions into the atmosphere are of particular concern because they are likely to increase in the future.

B. Fly Ash

The residue from the combustion of coal is called fly ash and large quantities are created which result in disposal problems. This ash is composed of finely powdered glass-like particles and is highly reactive. It contains high concentrations of many potentially harmful elements, some of which could be readily mobilized in the environment. The composition of fly ash depends on the composition of the original coal, and it can contain relatively high concentrations of arsenic, boron, barium, cadmium, chromium, copper, fluorine, germanium, lead, molybdenum, nickel, selenium, and zinc together with high quantities of sulfur in the form of the sulfate anion.

Generally, fly ash is disposed of as a water-based slurry into lagoons or as a dry powder in landfills. In the case of slurry disposal, water from the lagoon is drained to local rivers after treatment and dilution. The soluble components of the ash are then transferred to local drainage systems.

A modern approach to fly ash disposal is to reuse it economically. Thus fly ash has been used in the manufacture of blocks for use in building; it has also been added to agricultural soils to neutralize acidity and to add essential elements to the soils.

C. Sewage

Sewage effluent is transported to treatment plants where the solid material, sewage sludge, is separated and the remaining liquid portion is discharged into rivers and seas. The liquid effluent can greatly impact the aqueous environment. The effluent can be enriched in nitrate, which results in algal blooms and eutrophication in coastal waters (see Section VI for discussion of problems of eutrophication). This problem has been exacerbated due to effluent enriched in phosphate due to the addition of detergents. These additions have since been limited or banned in many countries.

Although liquid effluent can cause serious environmental problems, it is the solid waste or sludge that is the larger problem. Sewage sludge is produced in very large quantities and its disposal can have serious environmental implications. The methods used for its disposal include application to agricultural land (see Section VI), burial in landfills, and incineration. Disposal at sea, which was once a common practice, has been banned in most countries since the late 1990s. While several methods of sludge disposal are used, it is its use on agricultural land that causes the greatest concern.

The composition of sewage sludge is variable because it is derived from many sources such as domestic effluent, surface runoff that includes soil and street dust, and from different industrial effluents. Thus sludge derived from a predominantly urban area will be very different compositionally from that derived from highly indus-

trialized areas. The composition of sewage sludge is variable and has been found to contain metals such as cadmium, chromium, copper, lead, mercury, molybdenum, nickel, and zinc along with elements such as arsenic and fluorine and several organic contaminants such as polyaromatic hydrocarbons (PAHs) and chlorophenols. Nicholson *et al.* (2003) noted that in the UK controls on discharges from industrial sources mean that domestic sources of metals have become relatively more important. Thus copper and zinc from water pipes (see Section X) and elements such as zinc and selenium, which are contained in many shampoos, are mainly derived from urban sources.

The impact of sewage sludge application to agricultural land is discussed in Section VI.

D. Nuclear

The disposal of nuclear waste is a relatively modern issue that is and will continue to be a serious problem for mankind. This waste is classified as high-, medium-, and low-level waste depending on its radioactivity. The favored method for disposal of high- and medium-level waste is long-term storage in underground vaults, and the potential for leakage and transport from these is currently the subject of much research.

Medium- and high-level waste needs to be isolated in vaults, but low-level waste is frequently disposed of in surface locations. It is of note that drainage from one such site in northwestern England, which is the low-level waste disposal site for the nuclear installation at Sellafield, contains $61 \mu g L^{-1}$ uranium with associated stream sediments containing $48 mg kg^{-1}$ uranium, far higher concentrations than are found in surrounding streams (British Geological Survey, 1992).

VI. AGRICULTURAL PRACTICES

The well-documented environmental contaminants from agricultural practices are the nitrate and phosphate derived from fertilizer and farmyard manure that are applied to soils. These soluble ions are washed into surface and groundwaters. Nitrate and phosphate are essential nutrients and in lakes and coastal marine waters they can cause algal blooms and ultimately eutrophication when massive algal growths cause all of the oxygen in the water to be depleted. This results in the death of the algae and other biota, which leads to masses of rotting organic matter. In addition, the leaching of nitrates into groundwater in some agricultural areas has caused problems when the groundwaters are sources of potable waters. Such nitrate-rich waters have been shown to have caused methemoglobinemia in young babies due to nitrate interfering with the iron in blood which results in defective transport of oxygen. This is the cause of so-called "blue blood," and methemoglobinemia is sometimes referred to as "blue baby syndrome." It has also been suggested that ingestion of nitrates can contribute to gastric cancer when some of the nitrates are converted to nitrosamines, which are thought to be carcinogens (Mirvish, 1991).

In addition to the problems of nitrate and phosphate derived from fertilizers, many other contaminants are added to soil, surface, and groundwaters due to agricultural activities. Thus many substances applied to soil as fertilizers contain trace elements known to be detrimental to the biosphere and potentially toxic to humans. These substances can accumulate in soils and be leached into local waters (Table IV). Of the inorganic fertilizers used, phosphate fertilizers manufactured from apatite-rich rocks are of particular interest because they can contain elevated concentrations of several potentially harmful elements. High cadmium in some soils and surface waters in agricultural areas has been linked to the application of phosphate fertilizer, as shown in Figure 1 by stream plant chemistry for an area

of Sweden, where the high concentrations in the east central part of the country reflect fertilizer sources. Phosphate fertilizers have also been linked to high uranium concentrations in waters draining agricultural areas and have been found to be the source of elevated concentrations of the element in surface peats and waters in the Florida Everglades (Zielinski et al., 2000). It should be mentioned that modern phosphate fertilizers contain less contaminant metals than the earlier varieties such as superphosphate.

The application of sewage sludge has been shown to be a source of elemental and organic contamination of agricultural soils. As mentioned in Section V.D, one of the methods of disposal of sewage sludge is its application to agricultural land as a fertilizer. The composition of sewage sludges is extremely variable depending on their geographical source, and their application to soils is generally controlled with limits that are a function of their elemental compositions. However, some sludge applied to agricultural land has, in the past, resulted in health effects in livestock; for example, Davis (1980) found that an outbreak of fluorosis in cattle on a farm in the UK was due to the application of sewage sludge containing 33,000 mg kg^{-1} of fluorine.

The composition of livestock manure is also variable with its elemental composition essentially derived from their food and water intake. However, in some cases feeds are supplemented with metals, as in pig and poultry diets where copper and zinc are added; thus, pig and poultry manure are particularly rich in these two metals. In the past arsenic was added to pig diets, but this has recently ceased. It is worth noting also that both sewage sludge and animal manure can add pathogens to soil and subsequently to surface and groundwaters.

From a study of metal inputs to agricultural soils in the UK, Nicholson et al. (2003) suggested that of the fertilizer sources, livestock manure represents the greatest source of zinc and copper. Sewage sludge was the most important source for lead and mercury, and inorganic fertilizers (mainly phosphate fertilizer) were the major source of cadmium. However, Nicholson et al. (2003) concluded that the major input of metals to agricultural soils in the UK was atmospheric deposition (see Section VIII).

The application of pesticides—insecticides, herbicides, and fungicides—contributes to contamination of soils and surface and groundwaters. Many organic and inorganic compounds are utilized in pesticides. Some of the insecticides are chlorinated hydrocarbons (organochlorines) that include DDT and lindane, both of which were found to be toxic to humans. The use of

TABLE IV. Concentrations of Some Trace Elements in Fertilizers Added to Agricultural Land (mg kg^{-1})

	Phosphate fertilizer	Nitrate fertilizer	Manure	Sewage sludge
As	1–1200	2–120	3–25	2–30
Cd	0.1–190	0.05–8.5	0.1–0.8	<1–3400
Cr	66–245	3.2–19	1.1–55	8–41000
Cu	1–300	—	2–172	50–8000
Hg	0.01–2	0.3–3	0.01–0.4	<1–55
Mo	0.1–60	1–7	0.05–3	1–40
Ni	7–38	7–34	2–30	6–5300
Pb	4–1000	2–27	11–27	30–3600
Se	0.5–25	—	0.2–2.4	1–10
U	20–300	—	—	<2–5
V	2–1600	—	—	—
Zn	50–1450	1–42	15–570	90–50000

Data from several sources but predominantly from Alloway (1995).

these insecticides has been banned for some years, but they still persist in the environment as they decay slowly. Other insecticides such as organophosphates have replaced the organochlorines and, together with other organic compounds, are extensively used. There are a large number of organic herbicides and fungicides that show varying degrees of persistence in the environment ranging from a few weeks to many years. Many of these compounds have been found in groundwater, some of which is used as a source of drinking water. Several pesticides are based on inorganic compounds containing elements such as arsenic, lead, manganese, and zinc. Copper salts are used extensively as pesticides, for example, the fungicide Bordeaux mixture.

Another agricultural practice that can impact soils is the addition of lime, which is used to reduce acidity. Lime derived from limestone can contain many trace elements that can influence the composition of the soils. However, the major influence of lime is in increasing soil pH, which in turn can reduce the uptake of some elements by plants. Lime is sometimes added to soils to limit uptake of potentially harmful elements; thus, in sewage sludge amended soils, where high concentrations of some metals occur in the sludge, liming of soils is used to limit the uptake of metals. In some cases liming of soils has a detrimental effect because it limits plant uptake of elements that are essential to grazing animals. Some animals have suffered copper deficiency when grazing on recently limed pasture. In addition, copper deficiency is exacerbated if there are high amounts of molybdenum in the diet, which can arise from liming soils. This increases alkalinity (pH) and enhances molybdenum uptake by plants. Molybdenum-induced copper deficiency (molybdenosis) has been suggested to be the cause of unexplained moose deaths in Sweden, which have occurred since the mid-1980s in an area that had previously been heavily limed (Frank, 1998) (see also Chapter 20, this volume).

Many other agricultural processes impact soils locally, for example, the application of steelworks slag (see Section IV.A) and fly ash (see Section V.B). The use of wood preservative, both in agricultural and urban areas, can add elements such as copper, chromium, and arsenic to soils and runoff. In addition, metal fences and gates are frequently coated with zinc (galvanized) and can act as a source of this metal in the environment. In arid areas the practice of irrigation can sometimes lead to problems. High selenium in waters in the San Joaquin Valley, California, which has lead to toxicity in birds in the area, was found to be due to leaching of natural selenium from soils by irrigation waters (Kharaka *et al.*, 1996). Kharaka *et al.* (1996) also suggested that several

other western states in the United States have similar leaching problems with selenium-rich irrigation waters.

A particularly interesting problem resulting from an agricultural practice has been manifested in Brazil. Elevated mercury concentrations occurring in tributaries of the Amazon were originally ascribed to artisanal gold mining in river sediments, which result from the use of mercury to amalgamate gold (see Section II.A). However, it has subsequently been shown that the mercury in the rivers is derived from soil as a result of deforestation.

VII. TRANSPORTATION-DERIVED CONTAMINATION

It has been suggested that motor vehicles represent the greatest single source of atmospheric contamination. Motor vehicles are powered by gasoline or diesel, and their combustion in vehicle engines results in the production of exhaust gases in which there are large quantities of carbon dioxide. However, due to incomplete combustion, carbon monoxide, hydrocarbons, and nitrogen oxides are also evolved. These gases pollute the urban atmosphere and are thought to contribute, along with emitted particulates, to respiratory diseases. In addition the nitrogen oxide gases destroy ozone in the stratosphere, which aids in thinning of the ozone layer.

Vehicle exhaust gases can also contain metallic elements such as lead, manganese, nickel, and vanadium. Nickel and vanadium are derived from diesel fuel and these metals can be very rich in diesel exhaust fumes, which represent a major component of atmospheric nickel and vanadium loads. Roadside soils have been found to be markedly enriched in nickel, and diesel locomotives can emit significant quantities of nickel and vanadium.

Lead has been added to gasoline since the early 1920s as tetraethyl and tetramethyl lead to make its combustion more efficient. However, research showed that the lead emitted through the exhaust system had a detrimental effect on the environment with estimates that over 75% of environmental lead was derived from this source. As a result the use of lead-containing fuel has been phased out in many countries. This is not true of all countries and gasoline with lead additives is still used in many developing countries. Because of the phasing

TABLE V. Vehicular Sources of Metals

Metal	Source
Lead	Gasoline
Manganese	Gasoline
Nickel	Diesel + alloys
Vanadium	Diesel + alloys
Zinc	Tires + galvanized items
Cadmium	Tires + lubricating oils (minor amounts)
Chromium	Chrome plating, brake linings, etc.
Copper	Electrical wiring, thrust bearings, etc.
Platinum group metals	Catalytic converters

out and subsequent ban on the use of leaded gasoline in the UK, lead concentrations in the atmosphere have decreased markedly. Although lead concentrations in roadside soils have also decreased, these soils together with road dusts still contain high concentrations that continue to be pathways into the biosphere. In some countries where lead additives have been banned methylcyclopentadienyl manganese tricarbonyl (MMT) is used as an alternative which causes emissions of manganese, a suspected neurotoxin, in exhaust emissions.

Many other pollutant metals in roadside soils are derived from motor vehicles (see Table V). Zinc is added to the rubber used for tires and some steel components of cars are coated with zinc (galvanized) to minimize rusting. Cadmium, a component of zinc ores (see Section II.A), finds its way into tires and is also used as an antioxidant in lubricating oils. Chromium is used in steel alloys and in chrome plating, and nickel is also used in some steel alloys, which provides a second potential source of nickel. Motor vehicles are also thought to be a source of copper pollution, it being released from copper wiring, thrust bearings, and brakes. Similarly, metals such as tungsten used in high-temperature alloys in turbojet components and airplanes may be found in the vicinity of airfields.

A fairly modern group of pollutant metals derived from motor vehicles are the platinum group elements mainly platinum, palladium, and rhodium, which have been used in catalytic converters. These converters were first introduced in the 1970s in Japan and the United States and subsequently in other countries to clean up motor vehicle exhaust. The catalytic converter essentially converts carbon monoxide to carbon dioxide, nitrogen oxides to nitrogen, and hydrocarbons to

carbon dioxide and water vapor. These converters were originally assumed to be fairly indestructible, but a study by Hodge and Stallard (1986) showed that roadside dust in San Diego, California, was enriched in platinum and palladium. Subsequently there have been strong enrichments of the platinum group elements in roadside dust, urban river sediments, roadside drain sediments, sewage sludges, etc. The concentration of these elements in some of these sediments is similar to that found in platinum group metal ores.

VIII. ATMOSPHERIC DEPOSITION OF CONTAMINANTS

Many contaminants are released into the atmosphere and carried as gases, aerosols, and particles, and they are subsequently deposited on the surface where they are incorporated into soils and waters, absorbed by plants, or inhaled by animals and humans. Contaminant sources augment the many natural sources of elements and compounds contributing to the atmospheric load, such as volcanic and hot spring activity, forest and grassland fires, and wind-ablated soil particles. The sources of the atmospheric pollutants are listed in the proceeding sections of this chapter (II, III, IV, V, VII) but include mining, smelting and refining, power generation, various industrial processes, waste incineration, and transportation related activities. In addition, fine soil particles carried into the atmosphere can have contaminants adsorbed on their surfaces.

The degree of atmospheric contamination is geographically variable and depends markedly on meteorological conditions and potential sources. The degree of transport of contaminants varies depending on the form in which they are held. Many contaminants from vehicles are carried as fairly large particles and are not carried very far from their source, which accounts for much of the contamination of road dusts (see Section VII and IX). However, gaseous and aerosol transport can carry contaminants very great distances, and such forms of atmospheric transport are the major causes of contamination of the Arctic regions where metals such as lead have been found to be enriched in ice cores.

Deposition of atmospheric metals and arsenic in Europe has been estimated using moss analysis (Rühling, 1994). This has demonstrated the importance of various industries and power generation on the distribution of the elements, with coal and oil combustion and mining and smelting accounting for the majority of the deposition of arsenic, cadmium, and copper and much of the nickel and zinc. Almost all of the vanadium is derived from coal and oil combustion and oil refining, whereas the steel industry accounted for most of the chromium and iron and some of the nickel. Lead was also derived from industrial and power generation sources, but vehicular contamination was found to be a major source. Pacyna (1995) estimated that in the heavily industrialized northern part of the former Soviet Union, fossil fuel combustion accounted for 75% of the lead in the atmosphere, 87% of the vanadium, and 34% of the selenium with 79% of the chromium and 75% of the zinc derived from the steel industry. Eighty-five percent of the arsenic, 53% of the cadmium, and 51% of the antimony came from the mining and smelting industries.

Deposition from the atmosphere is a major source of elements for the surface environments. Nicholson *et al.* (2003) showed that for UK agricultural soils atmospheric deposition was the major source of several metals and accounted for 85% of the total mercury input, 53% of the cadmium, 55–77% of the arsenic, nickel, and lead, and 38–48% of the zinc.

In Figure 1 the high cadmium on the west coast of Sweden reflects long-range atmospheric transport.

IX. CONTAMINATION OF URBAN ENVIRONMENTS

Contamination of the urban environment is of major importance because a large majority of the world's population live in this environment. Urban contamination is derived from many sources and motor vehicles are a major present-day source (see Section VII); however, many urban areas also house industrial activities with consequent contamination. Even in modernized cities in many countries, even though industries have moved out of the urban area, historical industrial activities have left large areas of contaminated land. Additionally, in many urban areas fossil fuels are still burned as sources of household heating. Contamination can also arise from fertilizers applied in parks and gardens, pesticide use, and garden fires and bonfires adding to atmospheric and soil pollution. A further source of urban contamination is the disposal of waste and refuse in gardens and open areas, and uncontrolled burning in the latter.

TABLE VI. Metals in the Urban Environment—Birmingham, England

Urban area—street dust	Range (mg kg^{-1})
Cadmium	0.4–25
Chromium	9–228
Copper	36–3160
Lead	32–4820
Nickel	11–683
Tin	3–332
Zinc	79–5210
Urban area—soil	
Cadmium	0.7–1.6
Copper	38–715
Lead	75–350
Zinc	53–450
Rural area—soil	
Cadmium	0.2–0.5
Copper	5–57
Lead	14–74
Zinc	10–180

Note: Data for dust was obtained with a strong acid attack and represents a virtually total extraction of metals. Soil data is for a weak acid attack and represents metals which are easily extractable.

Data for street dust are from Brothwood (2001) and data for soils are from Davies and Houghton (1984).

Many contaminants have derived from urban sources with many elements noted as enriched in urban soils relative to their rural counterparts (Table VI). Thus arsenic, antimony, boron, cadmium, chromium, copper, lead, mercury, molybdenum, nickel, tin, vanadium, and zinc, among others, have been listed as serious contaminants of the urban environment. The soils and street dusts of the urban environment are enriched in many potentially harmful metals (Table VI) and are the major sources of household dust. They represent a major pathway into the human body via inhalation and inadvertent ingestion from hands, which is a particular problem for young children. Although many of these elements are potentially harmful to humans, it is lead that has attracted the most attention because it has been shown to cause serious health problems in children from urban environments.

Lead in inner city areas is derived from a variety of sources. The use of leaded fuel in vehicles (see Section II) either has been or is being phased out in most developed countries; however, it is still used in many less developed countries. Even in countries where leaded fuel is banned, much of the urban dust and soil lead is a legacy of its former usage (Table VI). Even though the major source of urban lead is likely to be the past or present use of leaded fuel, Mielke (1994) in a study of New Orleans, Louisiana, found that a major source of lead in soils of inner city areas was lead-containing paint. Although such paints are no longer used, the exteriors of older houses have layers of lead-based paint which have been painted over. Weathering of the paint or renovation of the houses where the old paintwork is scraped off results in the addition of lead to the soils. Old lead paintwork has also been shown to be a source of lead in children who ingest paint flakes that have been either picked off the paintwork or scraped off during renovation.

Lead is an extremely toxic element and attacks the central nervous system. Young children are particularly vulnerable. In extreme cases of lead toxicity, although now very rare, the brain swells (encephalopathy) and death can result. The obvious symptoms of lead toxicity are rare, but the subclinical effects are a cause for concern as they have been shown to impair learning and cognitive response. Surveys of children residing in inner cities have shown that relatively low intelligence quotients (IQ) correlate with high body burdens of lead. These levels are sometimes estimated from the lead content of the milk teeth of young children and monitored by blood lead levels.

In addition to the elemental contaminants mentioned above, it is worth pointing out that many organic contaminants, such as dioxins and PAHs, are added to the urban environment from industry, fossil fuel combustion, and refuse burning.

X. TREATMENT AND TRANSPORT OF POTABLE WATERS

Drinking water for humans comes from surface and groundwater sources and, hence, its initial chemistry is governed mainly by the chemistry of the surface sediments and rocks with which it interacts. The initial water is generally subjected to various treatments to ensure that it is fit for human consumption and this process modifies its composition. Subsequently, the water is distributed through a network of pipes that are frequently metallic; this transport can also modify the water composition.

Surface water frequently carries solid matter that needs to be removed. Although coarse material can be removed easily by filtering, finer material poses more of a problem and a coagulant is added to cause the flocculation and precipitation of this material. As a result of the coagulation process traces of the coagulant remain in the water. Several chemicals are used as coagulants and one of the most important is aluminum salts. However, concern has been expressed concerning elevated aluminum in drinking water, because some studies suggest that it is a contributory factor in Alzheimer's disease (premature senility). Iron compounds have also been used as coagulants, but their use can lead to subsequent strong leaching of metals from pipe surfaces.

Following removal of solid material the water is disinfected to remove any potentially harmful organisms. The most common disinfectant is chlorine. However, it has been shown that chlorine can react with organic matter in the water to produce chlorine-containing organic compounds (chlorophenols) that have been shown to be carcinogenic.

Following treatment the water is distributed through pipes, which are frequently metallic, and as a result traces of the pipe metal are transferred to the water. For example, where copper pipes are used copper and zinc are found in the water. The concentrations of these metals are generally low and Fuge and Perkins (1991) recorded $<65\,\mu g\,L^{-1}$ copper and $<40\,\mu g\,L^{-1}$ zinc for most samples analyzed in the main drinking water in north Ceredigion, Wales. However, in low-calcium containing (soft) waters from springs and wells where the drinking waters were untreated, much higher copper and zinc levels were present, with up to $500\,\mu g\,L^{-1}$ of copper and $125\,\mu g\,L^{-1}$ of zinc recorded.

Lead pipes have been extensively used in the past for transporting water, but in view of the known serious health effects of lead (see Section IX), this metal is no longer used and in many cases older lead piping has been replaced. However, due to its historical use some lead pipes are still in place and Fuge and Perkins (1991) record $29\,\mu g\,L^{-1}$ lead in a supply with lead pipes. Lead in drinking water is highly bioavailable, and it has been suggested that this is a major pathway into the human body in some areas where lead piping is still in use.

TABLE VII. Important Anthropogenic Sources of Some Elements Known to Have Detrimental Effects on the Biosphere

Element	Sources
Antimony	Mining, smelting, fossil fuel combustion
Arsenic	Mining, smelting, steel making, fossil fuel combustion, geothermal energy production, phosphate fertilizer, pesticides
Cadmium	Mining, smelting, fossil fuel combustion, incineration, phosphate fertilizer, sewage sludge, motor vehicles
Chromium	Smelting, steel making, fossil fuel combustion, phosphate fertilizer, sewage sludge
Cobalt	Mining, smelting, fossil fuel combustion
Copper	Mining, smelting, fossil fuel combustion, manure, sewage sludge, pesticides
Fluorine	Mining, aluminum refining, steel making, fossil fuel combustion, brick making, glass and ceramic manufacture, phosphate fertilizer
Lead	Mining, smelting, fossil fuel combustion, sewage sludge, pesticides, motor vehicles
Mercury	Smelting, fossil fuel combustion, incineration, sewage sludge
Nickel	Mining, smelting, steel making, fossil fuel combustion, oil refining, sewage sludge, motor vehicles
Selenium	Smelting, fossil fuel combustion
Thallium	Smelting, fossil fuel combustion
Uranium	Fossil fuel combustion, phosphate fertilizer
Vanadium	Steel making, fossil fuel combustion, oil refining
Zinc	Mining, smelting, steel making, fossil fuel combustion, phosphate fertilizer, manure, sewage sludge, pesticides, motor vehicles, galvanized metal

To this end the World Health Organization (WHO) suggests that drinking water contain no more than $10\,\mu g\,L^{-1}$ of lead.

XI. SUMMARY

The geochemistry of environmental media is strongly dependent on the chemistry of the natural sources from which they have been derived or with which they have interacted. However, the chemistry of environmental materials demonstrates a superimposed anthropogenic signal. It has been suggested that no environment on Earth is free of contamination and even the polar ice caps have elevated lead concentrations. Anthropogenic impacts on the environment result from many human activities that range from the extraction of raw materials from the Earth and manufacturing the many products needed by society that provide the food and generate the power for the world's population and subsequent disposing of the resultant waste materials. With much of the world's population living in major conurbations, nowhere is the impact of anthropogenic activities more evident and much of this is derived from the use of motor vehicles.

The anthropogenic influence on the environment has resulted in many elements and inorganic and organic compounds impacting soils, waters, and the atmosphere. As humans and animals are dependent on these media for food, drinking water, and air, any contaminants can enter the biosphere and can potentially have a serious impact with resultant health problems. A summary of the major sources of elements that can have serious effects on the biosphere is given in Table VII.

SEE ALSO THE FOLLOWING CHAPTERS

Chapter 2 (Natural Distribution and Abundance of Elements) · Chapter 11 (Arsenic in Groundwater and the Environment) · Chapter 20 (Animals and Medical Geology)

FURTHER READING

Al-Atia, M. J. (1972). Trace Elements in Iraqi Oils and their Geological Significance. In *Mineral Exploitation and Economic Geology*, University of Wales, Aberystwyth.

Alloway, B. J. (Ed.) (1995). *Heavy Metals in Soils*, 2nd edition, Blackie Academic & Professional, London.

British Geological Survey, 1992. Regional Geochemistry of the Lake District and Adjacent Areas, Keyworth, Nottingham, UK.

Brothwood, S. (2001). Vehicle Related Emissions of Platinum Group Elements and Other Heavy Metals in the Urban Environment, PhD thesis, University of Wales, Aberystwyth.

Burns, K. N., and Allcroft, R. (1964). Fluorosis in Cattle: Occurrence and Effects in Industrial Areas of England and Wales 1954–57, Industrial Disease Surveys, Reports 2, Part 1, Ministry of Agriculture Fisheries and Food, London.

Coleman L., Bragg, L. J., and Finkelman, R. B. (1993). Distribution and Mode of Occurrence of Selenium in US Coals, *Environ. Geochem. Health*, 15, 215–227.

Dabkowska, E., Machoy-Mokrzynska, A., Straszko, J. Machoy, Z., and Samujlo, D. (1995). Temporal Changes in the Fluoride Levels of Jaws of European Deer in Industrial Regions of Western Pomerania, Poland, *Environ. Geochem. Health*, 17, 155–158.

Davies, B. E., and Houghton, N. J. (1984). Distance-Decline Patterns in Heavy Metal Contamination of Soils and Plants in Birmingham, England, *Urban Ecol.*, 8, 285–294.

Davis, A., Kempton, J. H., Nicholson, A., and Yare, B. (1994). Groundwater Transport of Arsenic and Chromium at a Historical Tannery, Woburn, Massachusetts, U. S. A., *Appl. Geochem.*, 9, 569–582.

Davis, R. D. (1980). The Uptake of Fluoride by Ryegrass Grown on Soil Treated with Sewage Sludge, *Environ. Pollut. (Series B)*, 1, 277–284.

Ding, Z., Zheng, B., Long, J., Belkin, H. E., Finkelman, R. B., Chen, C., Zhou, D., and Zhou, Y. (2001). Geological and Geochemical Characteristics of High Arsenic Coals from Endemic Arsenosis Areas in Southwestern Guizhou Province, China, *Appl. Geochem.*, 16, 1353–1360.

Farmer, J. G., Eades, L. J., and Graham, M. C. (1999). The Lead Content and Isotopic Composition of British Coals and their Implications for Past and Present Releases of Lead to the UK Environment, *Environ. Geochem. Health*, 21, 257–272.

Frank, A. (1998). "Mysterious" Moose Disease in Sweden. Similarities to Copper Deficiency and/or Molybdenosis in Cattle and Sheep. Biochemical Background of Clinical Signs and Organ Lesions, *Sci. Total Environ.*, 209, 17–26.

Frank, A., Madej, A., Galgan, V., and Petersson, L. R. (1996). Vanadium Poisoning of Cattle with Basic Slag. Concentrations in Tissues from Poisoned Animals and from a Reference, Slaughter House Material, *Sci. Total Environ.*, 181, 73–92.

Fuge, R., and Hennah, T. J. (1989). Fluorine and Heavy Metals in the Vicinity of Brickworks, *Trace Substances Environ. Health*, 23, 183–197.

Fuge, R., and Perkins, W. T. (1991). Aluminum and Heavy Metals in Potable Waters of the North Ceredigion Area, Mid-Wales, *Environ. Geochem. Health*, 13, 56–65.

Fyfe, W. S. (1998). Towards 2050; the Past is not the Key to the Future; Challenges for Science and Chemistry, *Environ. Geol. Health*, 33, 92–95.

Gummow, B., Bastianello, S. S., Botha, C. J., Smith, H. J. C., Basson, A. T., and Wells, B. (1994). Vanadium Air Pollution: A Cause of Malabsorption and Immunosuppression in Cattle, *Onderstepoort J. Vet. Res.*, 61, 303–316.

Gummow, B., Botha, C. J., Basson, A. T., and Bastianello, S. S. (1991). Copper Toxicity in Ruminants: Air Pollution as a Possible Cause, *Onderstepoort J. Vet. Res.*, 58, 33–39.

Hodge, V. F., and Stallard, M. O. (1986). Platinum and Palladium in Roadside Dust, *Environ. Sci. Technol.*, 20, 1058–1060.

Kharaka, Y. K., Ambats, G., Presser, T. S., and Davis, R. A. (1996). Removal of Selenium from Contaminated Agricultural Drainage Water by Nanofiltration Membranes, *Appl. Geochem.*, 11, 797–802.

Koritnig, S. (1972). Fluorine. In *Handbook of Geochemistry* (K. H. Wedepohl, Ed.), Springer, Berlin, chap. 9.

Leece, D. R., Scheltema, J. H., Anttonen, T., and Weir, R. G. (1986). Fluoride Accumulation and Toxicity in Grapevines *Vitus vinifera* L. in New South Wales, *Environ. Pollut. (Series A)*, 40, 145–172.

Levinson, A. A. (1980). *Introduction to Exploration Geochemistry*, 2nd edition, Applied Publishing, Wilmette, IL.

Mielke, H. W. (1994). Lead in New Orleans Soils: New Images of an Urban Environment, *Environ. Geochem. Health*, 16, 123–128.

Mirvish, S. S. (1991). The Significance for Human Health of Nitrate, Nitrite and N-Nitroso Compounds. In *Nitrate Contamination Exposure, Consequences and Control* (I. Bogárdi, R. D. Kuzelka, and W. G. Ennenga, Eds.), Springer, New York, pp. 253–266.

Nicholson, F. A., Smith, S. R., Alloway, B. J., Carlton-Smith, C., and Chambers, B. (2003). An Inventory of Heavy Metals Inputs to Agricultural Soils in England and Wales, *Sci. Total Environ.*, 311, 205–219.

Pacyna, J. M. (1995). The Origin of Arctic Air Pollutants: Lessons Learned and Future Research, *Sci. Total Environ.*, 160/161, 39–53.

Reimann C., Äyräs M., Chekushin V. A., Bogatyrev, I. V., Boyd, R., de Caritat, P., Dutter, R., Finne, T. E., Halleraker, J. H., Jæger, Ø, Kashulina, G., Lehto, O., Niskavaara, H., Pavlov, V. A., Räisänen, M. L., Strand, T, and Volden, T. (1998). Environmental Geochemical Atlas of the Central Barents Region. Geological Survey of Norway.

Rieuwerts, J. S., and Farago, M. E. (1996). Heavy Metal Pollution in the Vicinity of a Secondary Lead Smelter in the Czech Republic. *Appl. Geochem.*, 11, 17–23.

Rühling, Å. (Ed.) (1994). Atmospheric Heavy Metal Deposition in Europe—Estimations Based on Moss Analysis, Nordic Council of Ministers, Copenhagen.

Zheng, B., and Hong, Y. (1988). Geochemical Environment Related to Human Endemic Fluorosis in China. In *Geochemistry and Health* (I. Thornton, Ed.), Science Reviews Limited, Northwood, U.K., pp. 93–96.

Zielinski, R. A., Simmons, K. R., and Orem, W. H. (2000). Use of ^{234}U and ^{238}U Isotopes to Identify Fertilizer-Derived Uranium in the Florida Everglades, *Appl. Geochem.*, 15, 369–383.

CHAPTER 4

UPTAKE OF ELEMENTS FROM A CHEMICAL POINT OF VIEW

ROBERT J. P. WILLIAMS
Oxford University

CONTENTS

I. INTRODUCTION: THE ESSENTIAL CHEMISTRY OF ALL LIVING CELLS

In order to appreciate the initial link between the geosphere and the biosphere about 4×10^9 years ago, certain characteristic, essential features of the *organic chemistry* of all known living organisms need to be noted and put into the context of the nature of primitive Earth. The chemicals concerned are in the *cytoplasm*, which is the central compartment of all cells (Figure 1). The common features of cells of this cytoplasmic chemistry are the syntheses of polymers such as the production of lipids (fats), polysaccharides (sugars), proteins, and nucleotides (DNA and RNA). They are all products of a few non-metal elements extracted from the environment; namely carbon, hydrogen, nitrogen, oxygen, sulfur, and phosphorus. Now the very nature of the polymers and their small molecule building blocks is that all of these six elements except phosphorus have to be in reduced states relative to $CO(CO_2)$, H_2O, N_2, (O_2), and elemental sulfur. This means that they are bound not just to one another but are bound, except for phosphorus, mainly by hydrogen. There is then a requirement in all cells for the cytoplasm to have a reductive synthesis capacity (below $-0.5\,V$). Because the required reductions cannot be brought about by organic chemicals, a transition metal ion redox catalyst, such as iron and copper, is also essential. This cytoplasmic chemistry is demanding in other ways because the initial small building blocks, such as simple aliphatic acids, sugars, amino acids, and bases, need to be polymerized. There is then a second requirement for a suitable Lewis acid/base catalyst and again metal ions, especially divalent ions magnesium and zinc, are extremely valuable in reactions such as condensation polymerization.

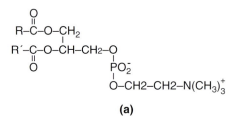

(a)

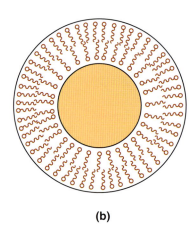

(b)

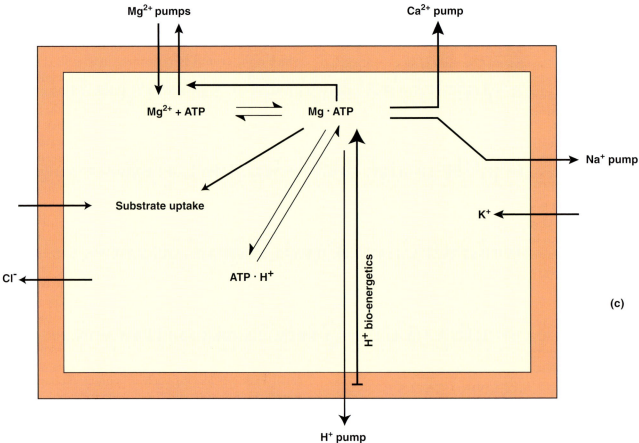

FIGURE 1   The simplest possible cell had a single membrane that enclosed an aqueous solution called the cytoplasm. The membrane was made from lipids; see (a) and (b). The cytoplasm had to have a supply of energy shown as adenosine triphosphate (ATP) in (c) which is bound pyrophosphate (see text). As far as simple ions were concerned, the cell in the sea had to maintain relatively low Na^+, Ca^{2+}, and Cl^- while accepting K^+ and Mg^{2+} using energy (ATP) to pump them. The pH had to be kept at about 7.0. Note that ATP is also used to provide energy for organic synthesis, and it is made from a proton gradient due to redox reactions.

$$XOH + YH \rightarrow XY + H_2O$$

A pH close to neutral is also necessary to avoid hydrolysis and the solution must therefore be buffered by small molecules. Of course all the activities of a cell require energy, so the cytoplasm must have a way of capturing energy. Energy is used in synthesis, uptake and rejection of elements and compounds as required. All environmental energy sources such as light and unstable chemicals also require several inorganic elements in the capture machinery of cells. Many steps in this capture require one-electron redox reactions, which are almost invariably aided by transition metal catalysts. Condensation of phosphate, for example, in adenosine triphosphate (ATP) is also necessary. Energy-containing ATP is made by reaction of adenosine diphosphate (ADP) with phosphate giving a pyrophosphate, which on hydrolysis is used to remove water

$$XOH + YH + ATP \rightarrow X\text{-}Y + ADP + P.$$

Finally, all of this organic chemistry occurs in water and to be reproducible it has to be carried out in a fixed ionic medium.

Along with the absolute requirements of the organic chemistry of the cytoplasm of all cells, there is then an absolute need for metal elements. Next the limitations imposed upon the possible selection of these elements will be reviewed as well as the nature of the original inorganic environment of the Earth from which these elements had to be obtained and which allowed life's chemistry to start. Once the system started, certain unique features were unalterable as is seen in the cytoplasm of cells in all life. The most important environments for our consideration are the primitive sea (Table I and Figure 2) and the atmosphere from 4×10^9 years ago, because it was in the sea (or at least in water) and the interface with the atmosphere that life began. There will be no attempt to describe the origin of life in this chapter, instead it will try to show the chemical limitations of the primitive sea and atmosphere for any development of a chemical system such as life before going forward to evolutionary considerations as this environment changed.

II. An Elementary Introduction to Primitive Earth

The formation of the Earth with its excess of metals and hydrogen gas over non-metals such as oxygen, due to

TABLE I. Available Concentrations in the Sea as They Changed with Time (Estimates)

Metal ion	Original conditions (molar)	Aerobic conditions (molar)
Na^+	$>10^{-1}$	$>10^{-1}$
K^+	$\sim 10^{-2}$	$\sim 10^{-2}$
Mg^{2+}	$\sim 10^{-2}$	$>10^{-2}$
Ca^{2+}	$\sim 10^{-3}$	$\sim 10^{-3}$
V	$\sim 10^{-7.5}$	$\sim 10^{-7.5}$ (VO_4^{3-})
Mn^{2+}	$\sim 10^{-7}$	$\sim 10^{-9}$
Fe	$\sim 10^{-7}$ (Fe^{II})	$\sim 10^{-17}$ (Fe^{III})
Co^{2+}	$<10^{-13}$	$\sim (10^{-11})$
Ni^{2+}	$<10^{-12}$	$<10^{-9}$
Cu	$<10^{-20}$ (very low), Cu^I	$<10^{-10}$, Cu^{II}
Zn^{2+}	$<10^{-12}$	$<10^{-8}$
Mo	$<10^{-10}$ (MoS_4^{2-}, $Mo(OH)_6$)	$\sim 10^{-7}$ MoO_4^{2-})
W	$\sim 10^{-9}$ (WS_4^{2-})	$\sim 10^{-9}$ (WO_4^{2-})
H^+	pH ~ 7	pH 8 (sea)
H_2S	$\sim 10^{-2}$	10^{-2} (SO_4^{2-})
HPO_4^{2-}	$<10^{-5}$	$<10^{-5}$ (HPO_4^{2-})

Note: The values for the original primitive conditions are estimates based on a pH of 8.0, an H_2S concentration of $\sim 10^{-2}$M, and a CO_2 pressure of 1 atm.

The concentrations in today's aerobic condition are taken from Cox (1995).

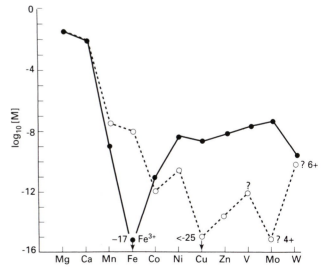

FIGURE 2 Contents of the sea in profile. The dashed line gives the molar concentration in the most primitive sea and the full line gives the concentration in the sea today.

the abundance of elements in the universe and the way Earth was formed, meant that from its beginning the whole planet was a reducing system. Although much of this metal content was isolated in the Earth's core and much hydrogen gas was lost, the different zones as they formed were made from reduced forms of many elements. Thus the atmosphere contained carbon as CO/CO_2 and some CH_4, nitrogen as N_2 and NH_3, oxygen and hydrogen as H_2O, sulfur as H_2S, and chlorine as HCl. The mantle contained lower valent metal oxides such as complex silicates and sulfides. (Note that all the metals could not be made into their preferred thermodynamic states as oxidic minerals so some were formed as sulfides.) It was seen that HCl reacted immediately with the metal oxides to give simple metal chlorides such as NaCl, which is virtually the only form of chlorine on Earth. Its solubility made NaCl the dominant chemical in the sea. The excess of basic metal over acidic non-metal oxides, which is partly due to abundances in the universe and partly to the volatility of non-metal compounds, gave rise to reactions that neutralized the acidity of the non-metal compounds to an ocean pH of about 7–8. At this pH carbonates, phosphates, and hydroxides form a buffered solution. Further solution of many elements was prevented by the insolubility of mixed oxides such as silicates and carbonates and sulfides. Full equilibrium between all elements in the core, the mantle, and the sea was not reached; otherwise there would have been much less of an *inner* metal core and no aqueous layer. However, it is suspected that approximate equilibrium was reached between the sea and its immediate neighbor phases with occasional injections from remote zones.

At that time there would have been and there still is a large temperature gradient from core to atmosphere. It is the high temperature of the isolated core plus energy from the sun that allowed water to remain as a condensed surface *liquid* sea, separate from the core and mantle. Its temperature range from 0 to 100°C, a temperature demanded for life to exist, has been maintained fortuitously for about 4×10^9 years. It was these energy gradients and the lack of chemical equilibrium between the sea and the core that also allowed a small flow of unstable reduced materials from the mantle to the sea. This also allowed a flow in cells that needed energy to reduce the oxides of carbon. Earth thus provided all the chemical ingredients as well as the energy for the cytoplasm of the most primitive cells. (Clearly the situation is not stable and a major eruption from the core could return the present atmosphere to its primitive state.)

III. CHEMICAL ELEMENT RESTRICTIONS ON PRIMITIVE LIFE

These considerations show that there are clearly strong restrictions upon the elements which could form a reactive system and become living cells. These reactions are due to the limitations of abundance and availability of the elements in the primitive sea. Turning to specific groups of elements, abundance severely restricted the amounts of lithium, beryllium, boron, and fluorine and all elements beyond atomic number 36. Abundance also restricted the amount of chlorine so that it was effectively all in the form of NaCl with some KCl and $MgCl_2$ in the sea or trapped in salt deposits. (There was no sulfate, only sulfide.) The sea was and is not saturated in such soluble salts, because the chloride supply is limited. Other availabilities in the early sea were restricted by the water solubility of oxides—sometimes as mixed oxides such as silicates, phosphates, and carbonates—and sulfides. The obvious restrictions due to oxide solubilities, apart from the mixed oxides of Si and P, were of the elements of Group 3 (Al group), 4 (Ti group), 13 (Ga group), and part of Group 14 (Sn). Restrictions due to sulfide solubilities generally applied to transition metals (Groups 3–12) and later elements in Groups 14 (Pb) and 15 (Sb, Bi), though some remained as insoluble metals (Au). The concentration of hydrogen as H^+ was controlled by the variety of buffering systems described above. Of great interest is the restriction on the availability of the reduced states M^{2+} and M^+ of the first transition metal series, but molybdenum and tungsten will also be looked at among metals and also selenium of the non-metals.

The precipitation of their sulfides limited concentrations [M], of the reduced (divalent) ions of the first transition metal series as follows:

Mn^{2+}	Fe^{2+}	Co^{2+}	Ni^{2+}	$CU+/(Cu^{2+})$	Zn^{2+}
$<10^{-6}$	$<10^{-7}$	$<10^{-10}$	$<10^{-12}$	$<10^{-20}$	$<10^{-15}$

The order is close to the order of solubility of many other of the salts of these elements and close to the inverse stability and extractability of their complexes. This order is called the Irving-Williams series. Outside this series the most interesting elements are from Group 5, 6, and 16. Molybdenum was reduced by H_2S to Mo(IV) and precipitated as MoS_2, whereas tungsten remained somewhat soluble as WO_4^{2-} or WS_4^{2-}. Note the geological states of these elements. Vanadium from Group 5 would also have been reduced, but its VO^{2+} ion remained somewhat soluble. Very few metallic elements other than Mn^{2+} and Fe^{2+} could have been present at greater than $10^{-8}M$ due to abundance and solubility restrictions. (The dominance of iron is in part due to its universal abundance.)

Among non-metals—other than carbon, nitrogen, hydrogen, oxygen, chlorine, and sulfur—which were all available as lower oxides or hydrides in the reducing atmosphere and/or in the sea, phosphorus and silicon remained as somewhat soluble elements HPO_4^{2-} and $Si(OH)_4$ in the presence of divalent metal ions in the sea. All halogens remained as halides usually seen in the sea. Selenium as H_2Se was the only other non-metal element of significant abundance and availability, because all others are rare or precipitated as sulfides.

Finally it must observed that these concentrations of elements in the sea, (Table I) are close to equilibrium values of solubility and redox potential and complex ion formation among themselves because on the time scales of interest the elements in different compounds are in fast exchange.

IV. Elements Presumed (Known) to Be Required by Primitive Cells

In the light of these geochemical data about the composition of the primitive sea and atmosphere, it must be asked which elements were not only available but have been found to be of significance for the required primitive chemistry of cells described above. The most primitive cells had a single compartment, the cytoplasm, encircled by a membrane. These cells are called prokaryotes. Rather than make further remarks about the environment, attention is now turned to the composition of the anaerobic prokaryote cells in sulfide media that are known today and assumed to be similar to the earliest forms of life. The elements in such cells are listed together with their free ion concentrations in Table II. Note that quantitative knowledge of *concentrations* of particular compounds, e.g., *free* ions, is necessary as we wish to appreciate the nature of a *reaction system* that cannot be understood by looking at qualitative considerations such as the total amounts of elements or of such molecules as DNA. The profile of metal elements in each compartment of a cell (or the sea; Figure 2) is called a metallome, which has two concentration parts: the free metallome and the bound metallome. It is also important to distinguish the elements that are in fast exchange within the lifetime of a cell (see

TABLE II. Composition of Cytoplasmic Fluid of All Cells

Element	Form	Concentration (M)
H	H^+	10^{-7}
Na	Na^+	$<10^{-3}$
K	K^+	10^{-1}
Cl	Cl^-	$<10^{-3}$
Mg	Mg^{2+}	10^{-3}
Ca	Ca^{2+}	$<10^{-6}$
Mn	Mn^{2+}	$\sim10^{-6}$
Fe	Fe^{2+}	$\sim10^{-7}$
Co	Co^{2+}	$<10^{-9}$
Ni	Ni^{2+}	$<10^{-10}$
Cu	Cu^+	$<10^{-15}$
Zn	Zn^{2+}	$<10^{-11}$
Mo	MoO_4^{2-}	$\sim10^{-7}$

TABLE III. Exchange Rates of Water from Metal Ions

Ions	Exchange rate (sec^{-1})
Na^+, K^+ many anions	10^9
H^+	10^{11}
Ca^{2+}, Cu^+, Mn^{2+}, Zn^{2+}	10^8
Fe^{2+}, Co^{2+}	$>10^7$
Mg^{2+}	10^6
Ni^{2+}	10^4
Fe^{3+}, Al^{3+}	$<10^3$

Table III). It is immediately noted that there is a general agreement between the concentrations of the available elements free in the primitive sea and those free in the cytoplasm of cells (Tables I and II). This is still true despite the changes in the sea. Just as the essential organic chemicals of the cytoplasm have remained fixed, so the free cytoplasmic inorganic element concentrations have not changed. In fact the two may well have initially formed an *inevitable, locally energized, system of reactivity* in some restricted compartment based on the composition of the primitive sea and atmosphere before there were coded molecules such as DNA. The suggestion is that long before there was Darwinian pressure on reproductive evolution there was an energized chemical drive to a particular system. This has not been proven, but it is very difficult to see how life began from any other standpoint.

At the quantitative level there are still some striking divergences between the cellular and the environmental sea free ion concentrations. Most notable are the differences in free sodium, chlorine, and especially calcium. Why is the system energized in these gradients much as organic chemistry is energized by covalent binding?

Before further consideration of the free and combined forms of the elements in the environment and in primitive cells and then the ways in which evolution has generated new cellular systems, and given that the cell concentrations under discussion are not all under thermodynamic control but can have strong kinetic management, a brief outline of the methods of study of the element concentrations and controls will be given in the next section. It is believed that a thorough knowledge of the element content of every compartment of living organisms is equally essential to the knowledge of its protein content, the proteome. In consideration of evolution, it is necessary to follow the analysis of elements in free and combined forms in cellular compartments with that of those available in the environment with time. This may give new insight into what is taken here to be the inevitability of the development of life, although its timing is quite uncertain. It may also help us to set up criteria for possible life forms on other planets.

V. Methods of Study

The knowledge that is required to understand the ways in which different elements function in organisms can be divided into:

1. Determination of their total analytical concentration
2. Appreciating the precise location of each element
3. Deciphering the differently combined forms and concentrations including the free ions in different places
4. Determining the kinetics of exchange, including uptake and rejection, and the associated activity of free and combined forms, including restrictions on diffusion in chemical and compartmental traps
5. The link to organic syntheses

This chapter cannot handle these extensive topics in any depth and the interested reader is directed to a variety of methods described in the literature. Under (1)

see above destructive micro-analysis is sufficient. To appreciate location (2) the best methods employ sectional slice analysis using dark-field EM methods and low temperatures where possible to determine atomic masses. Note that modern tomography and image reconstruction procedures can give high resolution below 10 Å in all three-space dimensions for certain materials. The free ion concentrations locally are determined using optical or fluorescent binding reagents, and total local element concentrations are determined by staining or SIMS in which a focused laser beam volatilizes a small local volume into a mass spectrometer. Proton bean microscopy is useful for many elements too. Often the only methods available under (3) depend upon careful extraction because the molecules are unstable. Fractionation into units such as vesicles or of molecules is then required. Traditional analytical methods as in (1) can then be used to quantify element composition. The determination of uptake or rejection (4) can use radio-isotopes or mass spectrometry (isotope mass fractionation) or of course the traditional methods of enzymology. Using all the methods together generates knowledge of the concentrations of selected elements, free and combined, in many separate compartments. However, the collection of analytical knowledge is far from complete.

It is difficult to understand the link under (5) between the concentration of the combined elements and organic synthesis, but we especially need to know about the organic synthesis of the reagents with which metals are combined. There are two very different final situations: thermodynamic equilibration of metal ion, M, and ligand, L; and irreversible insertion of a metal ion into a ligand. Note that the ligand can be a small chelating agent, a protein, or even DNA. In equilibrated binding the uptake of free metal ion and the synthesis of the ligand must be limited by feedback controls over uptake and synthesis, i.e., to expression via DNA/RNA, so that a complex ML is present in a fixed concentration. In the second situation formation of [ML] is not related to free [M] but to control the kinetics of insertion and the accessibility of L feedback controls rest upon free [ML] as well as upon [M] and [L]. The methods of study now depend not on analysis alone but on knowledge of organic synthetic pathways and their controls all the way back to the gene. Genetic control of L synthesis resides in feedback to transcriptional and translational regulation by [M], [L] and [ML] (see Figure 6). Study is now needed of DNA and its regulated response to inorganic elements. Note again that study of evolution through DNA can lead to evolutionary connections, but it says nothing about concentra-

TABLE IV.   Sources of Energy

Period (years ago)	Energy sources
Initial (4.5×10^9)	(a) The sun
	(b) Basic unstable chemicals in the crust
	(c) Chemicals stored at high temperature in the core
After say 1 billion years (3.5×10^9)	(a) As above
	(b) Some oxidized materials, some SO_4^{2-}; very little O_2, H_2O_2
After say 2 billion years (2.5×10^9)	(a) As above
	(b) Further oxidized materials, e.g. Fe^{3+}; modest O_2, H_2O_2
After say 3 billion years (1.5×10^9)	(a) As above
	(b) Further oxidized materials, almost 1% of final O_2 pressure
Today	(a) As above
	(b) Man's fuels
	(c) Atomic energy

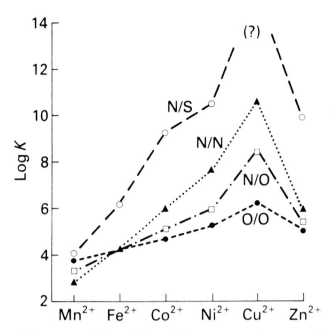

FIGURE 3   A plot of the stability constants of some metal ion complexes. The constant is for the equilibrium

$$M + L \rightleftharpoons ML.$$

Mg^{2+} (not shown) binds less to all ligands except those labeled O/O such as ATP (see Figure 1). The ligands here donate electrons to the metal ions. Typically O/O is a pyrophosphate, N/O is an amino acid, N/N is a di-amine, and N/S contains an amine and a thiolate (see Figure 4). Note that the inverse of these plots are the profiles of the free metal concentrations in the presence of certain ligands.

tions and causation of systems. It is not possible to appreciate the developments of chemical systems through the analysis of single molecules. Systems are to be seen through quantitative intensity factors such as temperature and concentrations of components in local zones. Finally, all the processes of uptake, rejection, and synthesis require energy (see Table IV), which have changed with time.

VI. Equilibrium Binding and Exchange

As indicated above, equilibrium is established rapidly between most organic ligands or inorganic anions with lower valent cations even when binding is quite strong. This feature of inorganic chemistry contrasts strongly with the kinetic control over the reactions of organic elements in compounds. Although there are exceptions, the rule is that thermodynamic constants (see Figure 3) are limiting factors over the binding of many metal elements in cell compartments. These include at least Na^+, K^+, Mg^{2+}, Ca^{2+}, Mn^{2+}, and much of Fe^{2+}, but only some of the bindings in biological cells of other elements found in cells. Molybdenum, cobalt, nickel, copper, and

zinc can be considered in this way too. The pattern of anion reactions is often different because they can be linked covalently in organic molecules, but there are frequent occasions when it is necessary to refer to the binding equilibrium constants of Cl^-, HCO_3^-, NO_3^-, SO_4^{2-}, HPO_4^{2-}, MoO_4^{2-}, WO_4^{2-}, and $Si(OH)_4$ in cells. Here our concern is with the metals, referring only now and then to non-metals.

It has already been seen that the equilibrated binding in oxides and sulfides restricted the primitive environmental availability of the metal elements. The plot of insolubility of the sulfides in the Irving-Williams series is very steep, rising to beyond $10^{-20}\,M$ at copper at pH = 8. We have to consider these insolubility data against the ability of the organic molecules, which cells could produce from H, C, N, O, P, and S, in considerable variety, to bind reversibly to the metal ions at pH = 7 (Figure 4). The resulting organic molecules have frameworks with pendant side-chains that can bind to

Selective ligands based on different donor atoms

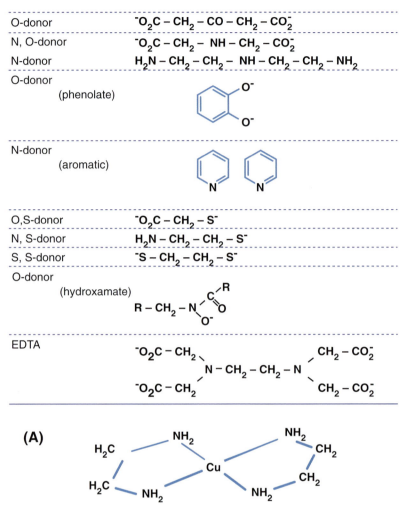

FIGURE 4   The formulae of some organic ligands and a complex ion with organic ligands (A).

metal ions. Single side-chains such as —OH, —NH₂, —SH, —COOH, and —OPO₃H₂ and their anions bind weakly. The vast majority of bindings therefore involve from two to six of such groups in the form of looped structures called chelates. These binding ligands are divided according to the binding (donor) atom into oxygen, nitrogen, and sulfur donors. Model studies have given the optimal conditional binding to various such ligands, and are therefore to be expected of proteins at pH = 7 which form the majority of the frameworks. Proteins have sets of such O-, N-, or S-donor binding centers. The binding constant profiles for the different combinations in small model complexes (and proteins) are shown in Figure 3. It is clear that not only did and

does availability restrict the possibility of a cell obtaining an element from the environment, but binding constants indicate the limits at which metal ions could have been captured by ligands in primitive cells against the competition from the environmental oxides and sulfides even though uptake could be energized. Certain general statements can be made based upon these effective equilibria and knowledge of cellular organic chemicals, largely proteins. (Effective refers to the binding under the conditions of temperature and competitive reactions in the cell.) These statements include:

1. Na⁺ and K⁺ could only be retained at the concentration of the free ions in cells, 10^{-1} to 10^{-2}

M, by physical barriers, because binding constants are very small. The binding by special cyclic or rigid frame O-donor ligands can retain only a small percentage of the ions.

2. Mg^{2+} and Ca^{2+} could be retained considerably in bound forms at the observed concentration, 10^{-3} and $<10^{-6}$ M respectively, of these free ions, by a variety of O-donor ligands but not by S- or N-donors.

3. In marked contrast with (1) and (2), it was hard to retain almost any transition metal ion by O-donor ligands alone, but the increased binding by N- and S-donors did allow binding at the ion concentrations seen in cells for Mn^{2+} and Fe^{2+} ($\sim 10^{-6}$ M) even in the presence of H_2S, which gave rise to external precipitation of sulfides.

4. There were very few ways of retaining Co^{2+}, Ni^{2+}, Cu^+ (Cu^{2+}), and Zn^{2+} because of the removal of these ions by the sulfide of the primitive sea.

5. Molybdenum was more strongly held in MoS_2 than were tungsten and vanadium in their sulfides so that tungstate and VO^{2+} would have been captured in primitive cells by organic chemical ligands to a greater degree than molybdenum.

6. Among anions that were not covalently retained, chloride could only be retained at the concentration found in cells (10^{-2} M) by physical membrane constraints because it is very weakly bound. Sulfate did not exist in primitive times but some phosphate could be retained at equilibrium binding at 10^{-4} M and also kinetically by covalent binding. Bicarbonate (carboxylate of small ligands) is weakly held but organic phosphates are more strongly retained.

It is very important to understand that all equilibria in all systems in an isolated compartment are restrictions on the variables of the system that are not biological but purely chemical, contrast a coded synthesis of a protein, and each unavoidable effective equilibrium constant is then an enforced characteristic of life or of any other organization. The profile of free exchanging metal ions at effective equilibrium in the cell cytoplasm is a characteristic of all life and is fundamental to it (Figure 5). Any code had to work with this profile just as it had to produce only certain organic molecules.

7. Because all ligands bind all the metal ions when both are in sufficient concentration, it is important to devise the ligands such that their binding constants are as discriminating as possible opposite the free ion concentrations allowed in cells. This permits each metal to be used in a selected

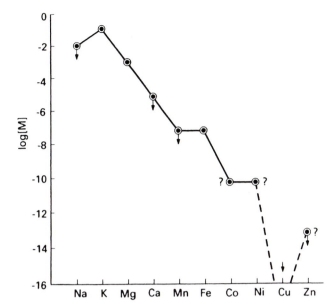

FIGURE 5 The profile of free metal ion concentrations in primitive cells. The values are not very different in the present-day cytoplasm of all cells.

functional way (see below). It is necessary therefore to limit the ligand concentration so that there is little excess free ligand after binding its selected element. The latter avoids binding of the wrong metal to a site. This was achieved in cells by limiting the production of the binding ligand through feedback inhibition by [M], [L], or [ML] via DNA of the synthesis of the ligands. This possibility could hardly have been present in the most primitive system but is a dominant feature in coded cells (see Figure 6).

Before continuing this analysis, it is necessary to be careful not to overgeneralize. Equilibrium considerations are important, but in two ways cellular chemistry escapes their limitations:

1. Elements can be energized as they pass through physical, membrane barriers so that free concentrations can be increased or reduced by up to 10^4 relative to environmental values while to some degree all ions will leak into cells. This prevents equilibrium with the environment but not internally to a compartment.

2. Binding can be made as irreversible inside cells as in the uptake of non-metals.

Only those elements that exchange reasonably readily inside cells can be examined. A factor of 10^4 to 10^{-4} has been added in Figure 5 as the variation away from equilibrium based upon known values of free ions in the sea

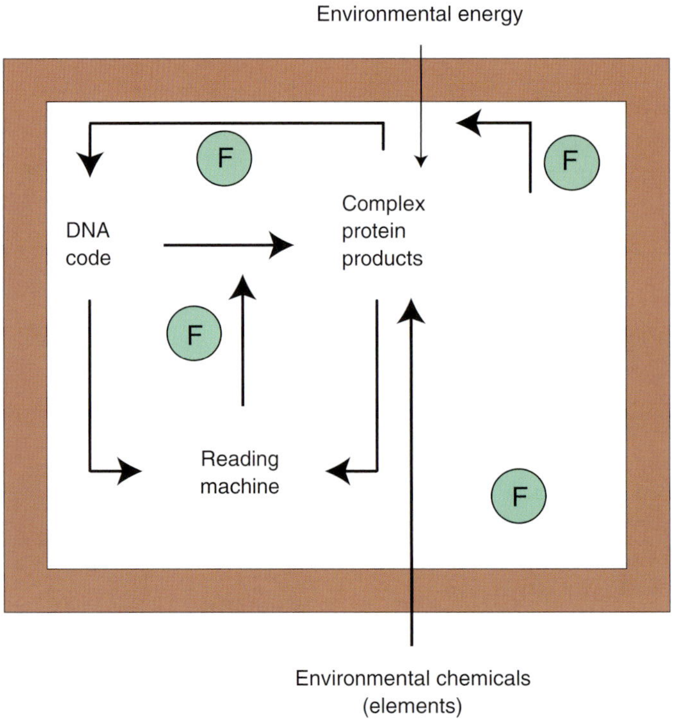

FIGURE 6   A simple picture of feedback, F. In each case a molecule or ion in a cell builds up to a certain level by pumping (Figure 1) or by metabolism. At this level the molecule or ion feeds back to the pump or the metabolic path and blocks it. This ensures that cells work at a fixed concentration of many small components, including both free metal ions, the metallome, and free mobile coenzymes, which are part of the metabollome.

and the known limitations of uptake or rejection into or out of cells. This gives a picture of free ion levels based on model chemical and real biological complex ion studies (shown in Figure 5), which applies to the cytoplasm of all cells. These profiles have been referred to as the free metallome because there are parallels in all cells. Moreover, the elements that were bound by selected organic ligands, given the variety of O-, N-, and S-donors which the cell could make, are close to expectation (Figures 3 and 4). This leads again to the suggestion that life started as a system of energized chemical reactions in one compartment which required the presence of an environmental zone on Earth containing about 20–30 available elements. Of these, about 20 are utilized in cells and a few are rejected. The zone that became a cell was then self-sustaining but unable to reproduce. These are general features of life which are not due to competitive Darwinian pressure but arise from geological and chemical limitations. In this light it is possible to see how cells use elements in an energized way.

VII. THE FUNCTIONAL VALUE OF ELEMENTS IN CELLS

A. Osmotic and Electrical Balance

The most basic requirement of the inside of a cell is to avoid excess concentration and charge, which could destabilize a cell by osmotic or electric pressures (see Figure 1). By necessity a cell is rich in organic molecules and, hence, it must have had a lower content of inorganic ions than the sea. The organic molecules of the cell are mostly anions so the cell must have also rejected the major anions of the sea, but even then it had to maintain approximate charge neutrality. These objectives are achieved by pumping out Na^+ and Cl^- and allowing K^+ to come in. Another essential feature is that Ca^{2+} must be strongly rejected because it forms precipitates and aggregations with organic anions. Most other free and smaller divalent ions, with the possible exception of the larger Mn^{2+}, can be retained up to the limits seen in the primitive sea. Mn^{2+} forms some insoluble salts such as oxalates. The resultant energized gradients of Na^+, Cl^-, and Ca^{2+}, the main ions rejected, allowed the favorable uptake of K^+ (to neutralize charge) and an internally equilibrated intake of all other cations (and anions) in small free amounts but with a high percentage bound. Although this description of the primitive cell system is certainly a necessity, it is still unknown how this energized system evolved. (This puzzle parallels that of the origin of synthesis of the particular organic compounds, which came to form a coordinated set in a cell and no solution to the problems is offered.) However, given fast exchange it is clear that the two surfaces of the energized pumps, in and out, must have binding constants for each selected ion close to the inverse of concentrations of the ion in the cytoplasm and in the sea, respectively.

B. Simple Condensation and Energy-Transfer Reactions

Given our lack of understanding of the basic initial organic chemistry in life, we will assume that all cells can select the organic chemicals, L, they produce and control their amounts. This step needs to be taken because the next metal elements, M, for consideration bind strongly to these organic chemicals. It is essential that [M] and [L] are fixed. [ML] is then fixed selectively by equilibrium. Hence, it should be taken into account that a coded cell can limit production of an organic chemical that binds metal ions by a feedback control to the production center, DNA/RNA (Figure 6). In feedback control the production of [L] occurs when [M] is high, but [M] itself is limited by feedback to the pumps which stop entry at certain values. For elements in fast exchange, *e.g.*, Mg^{2+} and Ca^{2+}, this implies that the feedback control must have a binding constant close to the inverse of the free metal concentration. This is true for all the metal ion complexes that operate under free metal ion control including the pumps. The equilibrated metal ion, therefore, binds equally to all proteins that are active: enzymes, some small molecules, the inner sites of membrane pumps, any carrier molecules, and transcription factors which control the production of all these molecules either directly or ultimately. The case of Mg^{2+} is typical. Its binding constant is close to $10^3 M^{-1}$ to all its sites (Figure 7), which implies that all binding sites are roughly half-saturated because free $[Mg^{2+}]$ is close to

10^{-3} M. This is ideal for response to variations in $[Mg^{2+}]$ because a system acts most sensitively as a dynamic buffer at half-saturation. Thus the activity of the cytoplasm of cells is closely dependent on the value of free $[Mg^{2+}]$ in the sea (10^{-2} M). This includes all reactions of nucleotide tri- and diphosphates (ATP)—the major energy distributing entity of cells for metabolism including pumping and synthesis. The only available sites that can bind Mg^{2+} because of its chemical nature (see Section VI) are O-donor centers. None of these Mg^{2+} O-donor sites can bind any other metal ion at the concentrations available inside a cell, which again are limited by those in the sea. Many of the condensation reactions of the organic syntheses depend on O-donor groups brought together for synthesis with energy-giving ATP which also has only O-donor groups and also requires activation by metal ions. Limitations are imposed on the availability of other elements by solubility or energized rejection (Ca^{2+}). Hence, it is clear that Mg^{2+} was the only possible initial acid catalyst for much cytoplasmic activity. The $[Mg^{2+}]$ concentration is actively maintained by pumps, but this may not have been initially necessary. Pumps have to operate to ensure 10^{-3} M cytoplasmic Mg^{2+} against any external concentration. It is seen that some cellular chemistry arises from chemical (thermodynamic) necessity, not from coded chance.

The corresponding necessary binding constants for K^+, Na^+, and Cl^- in the cell are $10^1 M^{-1}$ (K^+) and $>10^2$ M^{-1} (Na^+ and Cl^-) opposite their free ion concentrations in the cytoplasm of 10^{-1} M (K^+) and $<10^{-2}$ M (Na^+, Cl^-). These binding factors also arise from the nature of the primitive sea and not from any coded imposition. In fact the code molecules, DNA and RNA, are still absolutely dependent for activity upon binding constants of about $10^{-3} M^{-1}$ for Mg^{2+} and 10^{-1} M for K^+.

Next the free Ca^{2+} ion levels in the cytoplasm are considered. Here outward pumping was and is essential and reduces free $[Ca^{2+}]$ in the cytoplasm of all cells generally to $<10^{-6}$ M, although advanced cells have $[Ca^{2+}]$ at 10^{-8} M. This means that the range of useful Ca^{2+} equilibrated binding sites inside cells must be around $10^6 M^{-1}$ to $10^8 M^{-1}$. Later in eukaryotic cells, which have a huge variety of Ca^{2+} binding proteins (Figure 8), Ca^{2+} became the major messenger giving the cell, upon fast Ca^{2+} pulsed entry and exit, information about its external world. On entry Ca^{2+} rises to 10^{-7} or 10^{-6} M and binds to all the response proteins of about ten different kinds. These have known binding constants of about $10^7 M^{-1}$ and this includes the exit pump which rejects the pulsed Ca^{2+} immediately. $[Ca^{2+}]$ at 10^{-7} M does not usually connect directly to DNA to give transcription;

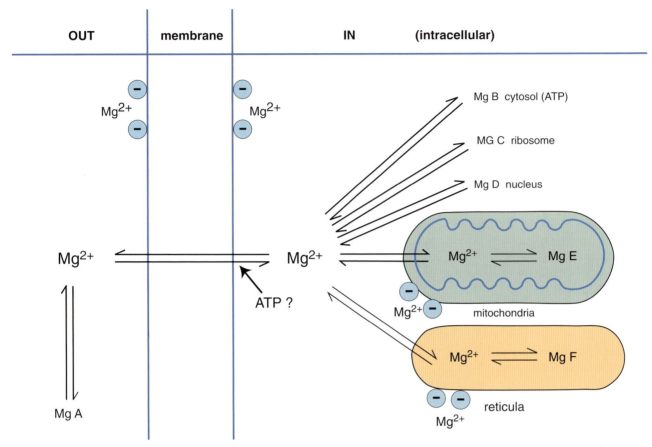

FIGURE 7 The general way in which Mg^{2+} interacts in equilibrium with many sites in a cell. Note that the free Mg^{2+} concentration in the cytoplasm of all cells is close to 10^{-3} M, and this is controlled by feedback and energy input at pumps (Figures 1 and 6).

instead it activates Mg^{2+}-dependent sensors such as kinases, which are proteins dependent on ATP. In a passing note none of these Ca^{2+} selective proteins bind Mg^{2+} with a constant $>10^3 M^{-1}$ so that Ca^{2+} and Mg^{2+} do not compete. These effective binding constants are necessities even though the syntheses of the molecules are coded. Note that Ca^{2+} pumps bind Ca^{2+} reversibly on the outside at $<10^3 M^{-1}$ and on the inside at $>10^6$ M. It is also found that the reversible binding of Ca^{2+} to CO_3^{2-} (carbonate) and PO_4^{3-} (phosphate) gave rise to precipitates at pH = 7 at 10^{-3} M Ca^{2+}, close to the value in the sea.

C. Electron Transfer

When the binding constants and the availability of elements for the catalysis of the other essential reactions of primitive cells such as reduction and energy capture by electron flow are considered, it is seen from solubility data (Table I) that only Fe^{2+} could have been used initially. It could not have been retained selectively by O-donors but only by S- or N-donors of the organic molecules available (Figure 3). The binding constants to all those of its proteins, which are in fast exchange, must be about $10^7 M^{-1}$ if all are to operate together in a controlled fashion. This limitation, built into feedback controls, is due to the nature of Fe^{2+} and organic ligands of the types found in proteins.

D. Oxygen-Atom Transfer

Iron was essential for one-electron reactions $Fe^{2+} \rightleftarrows Fe^{3+}$ especially in a sulfur matrix. The major primitive iron proteins are small $[FeS]_n$ units. Such units cannot do

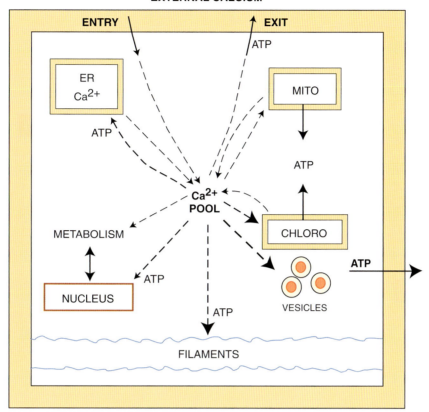

FIGURE 8 The general way in which Ca^{2+} can act as a messenger in eukaryotes. The initial effect is to allow Ca^{2+} entry to increase due to an environmental change. This input raises $[Ca^{2+}]$ above $10^{-7}M$ so that it activates a multitude of cellular responses including energy available (ATP), synthesis, and import/export of chemicals. Subsequently and rapidly the $[Ca^{2+}]$ returns to $10^{-7}M$.

another required oxidation/reduction reaction, the transfer of oxygen (a two-electron reaction), such as $M + RCOOH \rightleftarrows RCHO + MO$. This reaction is catalyzed readily at low redox potential by molybdenum and tungsten. In primitive conditions it could only have been that tungsten was used in two-electron, O-atom transfer because it was the only available element with a low two-electron oxidation-reduction potential for such transfer. Molybdenum was precipitated as a sulfide. It does look as if life's system in the cytoplasm arose as an inevitable consequence of the nature of the primitive environment, and it was repeatable in form but not reproducible. The basis of this assertion is that much of the inorganic chemistry was in fast exchange unlike the organic chemistry, which could only occur in the presence of these elements. The use of tungsten, a very rare element in essential reactions, was a necessity not an accident. Free tungstate, later molybdate, is approximately $10^{-7}M$ in the cytoplasm of cells (see Tables I and II).

VIII. EVOLUTION

A. New Forms of Old Elements

There are three possible modes for the evolution of such a chemical system. These are logistical requirements, independent from coding, and all three depend upon isolating the initial *cytoplasmic, partly equilibrated, organic/inorganic system* from alteration, while evolving in other ways. The central chemistry has to be so maintained almost unchanged in order to keep the basic reducing and acid/base metabolism, which generates all the essential organic polymers common to all known life. The first possible mode of evolution is to develop new organic chemical pathways in the cytoplasm, but now using irreversible chemical traps for the already used or any other newly available elements. This is a change of the bound not the free metallome. The devel-

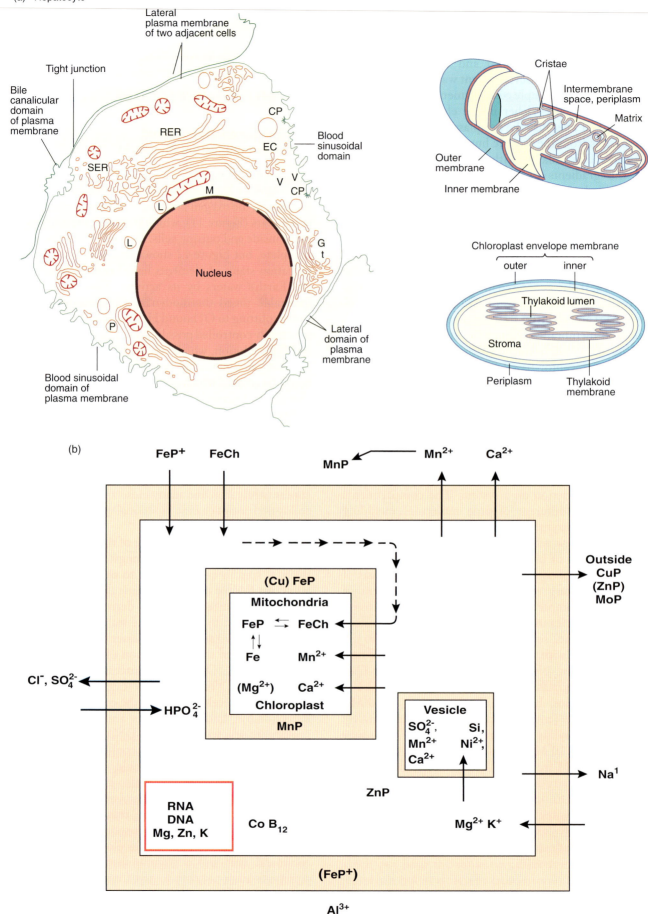

FIGURE 10(A): The nature of a modern cell with compartments. M, mitochondria (see side Figure); RER, rough endoplasmic reticulum; L, lysozome; V, vesicle; SER, smooth endoplasmic reticulum; G, Golgi apparatus; CP, chloroplast (see side Figure). 10(b): The distribution of elements in the new compartments; compare with Figure 1.

are independently energized and controlled (Figure 10b). For bacteria the development was the introduction of the periplasm, a simple compartment surrounding the cytoplasm, which is richer in the elements Na^+, Ca^{2+}, Mn^{2+}, and Cl^-, because the periplasm was open to diffusion of ions from the outside but did not allow proteins to escape. This isolation with new external binding sites and the gradients of ions produced novel catalytic energy-producing and communication possibilities which have evolved over 3×10^9 years. The rejected Mn^{2+} (plus Ca^{2+} and Cl^-) produced a periplasmic facing enzyme in the cytoplasmic bacterial membrane capable of generating external O_2, and in turn, this has enabled all future evolution of chemicals and compartments as seen below. It also generated increased reducing capacity internally. Both features were aided by the new chlorophyll (Mg) and electron transfer (heme iron) complexes mainly in membranes (Figure 9). At the same time the binding of Ca^{2+} on the outside of the cell became the basis for the later receptor-based messages giving inward Ca^{2+} pulses (Figure 8). There was, also later, the creation of internal vesicles, membrane-contained space separated from the cytoplasm. Cells containing such vesicles are called *eukaryotes*.

Kinetically controlled vesicular compartments created the second opportunity to multiply the variables of the system, in contrast with a phase at equilibrium which is a restriction on a system. The compartments could contain different organic molecules including proteins and enzymes. In this way the new compartments create new activities, the most obvious of which is the synthesis of biological minerals such as carbonate, silica, and phosphate shells. These are the basis of many sedimentary rocks. It was the development of such vesicle compartments that caused the explosion of separate species in eukaryotes. The origin of vesicular compartments, separate but kinetically linked to the cytoplasm, lies with the coming of dioxygen as a new chemical species. Note that the equilibrium for Ca^{2+} outside the cytoplasm, now also in vesicles where it binds to proteins, has a binding constant of $10^3 M^{-1}$ opposite the free (environmental) Ca^{2+} of approximately $10^{-3}M$. The binding constant for external Mn^{2+} may well be around $10^7 M^{-1}$ opposite the level of the primitive sea of $10^{-7}M$ free Mn^{2+}. The production of dioxygen linked to energized Mn chemistry using light made it possible first to make the new membrane structures found in eukaryotes and then to do much more novel chemistry in ever extending organization. This is the third evolutionary step.

IX. Dioxygen and Evolution: Single Cell Eukaryotes

A. New Compartments and New Communication Networks

Elsewhere we have described the slow progression of the increase in the oxygen pressure in the atmosphere over 4×10^9 years (Figure 11), which is a new kinetic compartment inside and outside cells. Its final effects are not yet complete, but Figure 12 shows that with time different oxidation states of elements have been introduced into the Earth's waters by its activity. This is seen as an inevitable timed thermodynamic succession following water splitting to O_2^+ bound H by cells and produces new kinetically controlled possibilities. The consequences for cellular evolution, which created the oxygen in the first place, are also inevitable: only certain new reaction paths could be created which give new chemicals in cells, and they had to be largely confined in the new compartments and controlled by novel proteins. The timing of change is unknown, but once oxygen was produced the general direction of use of higher redox potential chemicals is unavoidable. Note that it is mainly acid-base (Zn^{2+}) and redox catalysis (copper, molybdenum, selenium) which appear, while iron and manganese are reduced in availability and do not change functional value significantly in the cytoplasm.

Subsequent to the appearance of ring chelates and the coenzymes of molybdenum, tungsten, and vanadium, the first step in evolution due to oxygen was the creation of a flexible outer membrane through the synthesis of an oxidized squalene derivative, cholesterol. Simultaneously, to maintain the stability of a cell, filamentous internal structure was essential. Together these changes allowed the construction of much larger cells or eukaryotes. Once this structure appeared, new compartments, vesicles, and organelles could be permanently developed within the cell. The organelles, mitochondria, and chloroplasts are, in fact, captured prokaryote cells with their own DNA. Within the structures new reactions pathways could evolve and some pathways in the cytoplasm could be safely dropped because the organelles could also manage them. The evolution of the organic chemistry should not be stressed here, but it must be remembered that any organization that develops as a system of compartments, which is the general way to develop, needs a central command unit and a communication network. The communication network to and from the cytoplasm, to

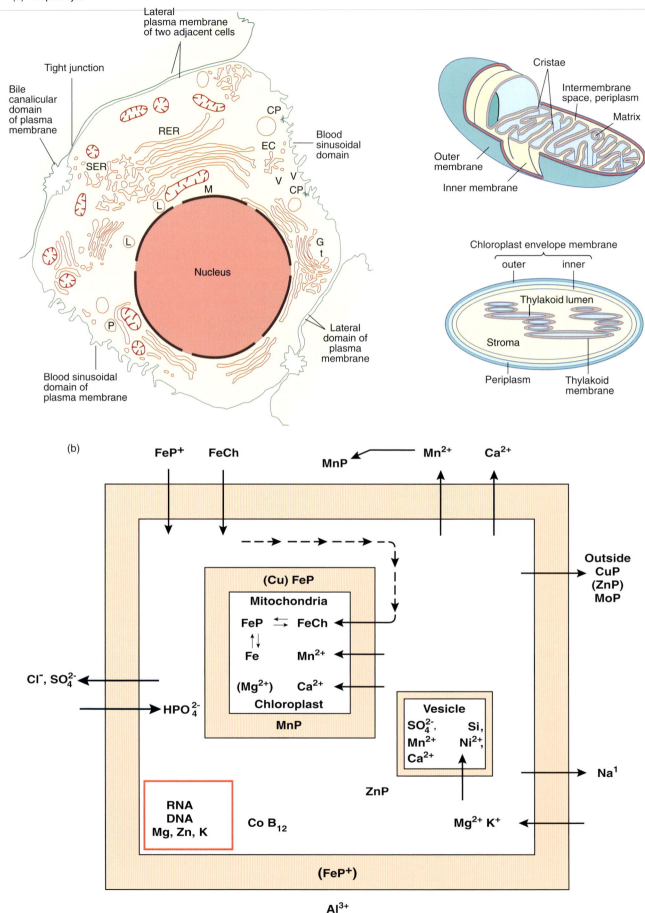

FIGURE 10(A): The nature of a modern cell with compartments. M, mitochondria (see side Figure); RER, rough endoplasmic reticulum; L, lysozome; V, vesicle; SER, smooth endoplasmic reticulum; G, Golgi apparatus; CP, chloroplast (see side Figure). 10(b): The distribution of elements in the new compartments; compare with Figure 1.

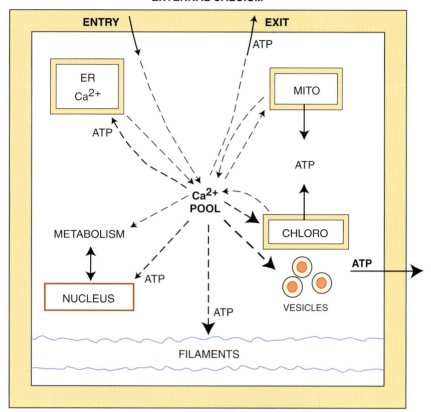

EXTERNAL CALCIUM

FIGURE 8 The general way in which Ca^{2+} can act as a messenger in eukaryotes. The initial effect is to allow Ca^{2+} entry to increase due to an environmental change. This input raises $[Ca^{2+}]$ above $10^{-7}M$ so that it activates a multitude of cellular responses including energy available (ATP), synthesis, and import/export of chemicals. Subsequently and rapidly the $[Ca^{2+}]$ returns to $10^{-7}M$.

another required oxidation/reduction reaction, the transfer of oxygen (a two-electron reaction), such as M + RCOOH $\rightleftarrows$ RCHO + MO. This reaction is catalyzed readily at low redox potential by molybdenum and tungsten. In primitive conditions it could only have been that tungsten was used in two-electron, O-atom transfer because it was the only available element with a low two-electron oxidation-reduction potential for such transfer. Molybdenum was precipitated as a sulfide. It does look as if life's system in the cytoplasm arose as an inevitable consequence of the nature of the primitive environment, and it was repeatable in form but not reproducible. The basis of this assertion is that much of the inorganic chemistry was in fast exchange unlike the organic chemistry, which could only occur in the presence of these elements. The use of tungsten, a very rare element in essential reactions, was a necessity not an accident. Free tungstate, later molybdate, is approximately $10^{-7}M$ in the cytoplasm of cells (see Tables I and II).

VIII. EVOLUTION

A. New Forms of Old Elements

There are three possible modes for the evolution of such a chemical system. These are logistical requirements, independent from coding, and all three depend upon isolating the initial *cytoplasmic, partly equilibrated, organic/inorganic system* from alteration, while evolving in other ways. The central chemistry has to be so maintained almost unchanged in order to keep the basic reducing and acid/base metabolism, which generates all the essential organic polymers common to all known life. The first possible mode of evolution is to develop new organic chemical pathways in the cytoplasm, but now using irreversible chemical traps for the already used or any other newly available elements. This is a change of the bound not the free metallome. The devel-

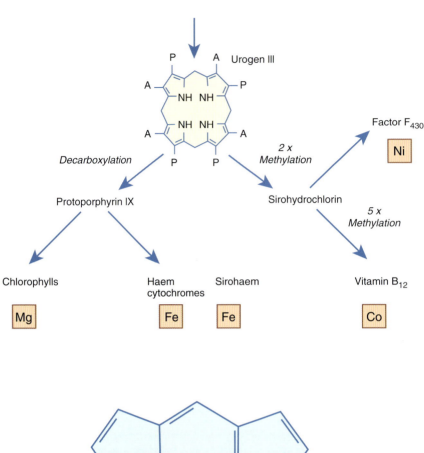

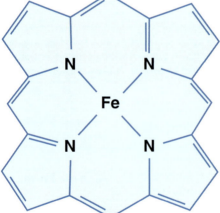

FIGURE 9   The development of porphyrin-like chelating agents to generate new variables in the cytoplasm of primitive cells.

opments of protein-independent ring chelates for iron, cobalt, nickel, and magnesium, all related in synthesis to uroporphyrin (Figure 9), and of the organic part of the coenzymes for molybdenum (or tungsten or vanadium) are clear examples. In effect these chelates, which do not exchange their central metal ions, are new "elements," i.e., new independent variables in the chemical system of cells. They avoid thermodynamic competition with the environment or cell's internal equilibria, as does the cell's organic chemistry. However, their inter-

nal concentrations are controlled by feedback like all other cell components.

B. New Compartments

The second possible evolutionary mode is the creation of new physical compartments, *e.g.*, vesicles (Figure 10a), which can have the same or different elements as the cytoplasm but at different concentrations because they

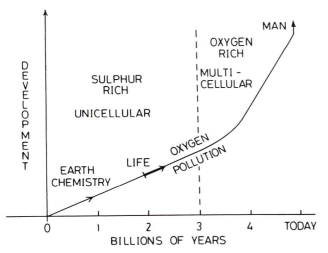

FIGURE 11 The rise of dioxygen with time and some of the enforced cellular changes.

and from vesicles, and the external environment together with its relationship to the central command unit of the cell, DNA/RNA, is described next.

DNA remains the central command unit in the cytoplasm of all cells. Not only is it linked to the activity of cells through proteins and the input of inorganic elements by feedback, but it is also limited by membrane activities. To maintain activity in a systematic fashion the command structure, DNA, receives information from the cytoplasm by sensing its contents, and this very sensing sends back instructions to maintain the system by synthesis. The system also sends signals to and from the cytoplasm and to and from the membrane to adjust the input or output of chemical elements to or from its contents, including those from the environment. The major communication for the organic chemistry connected to the DNA directly in primitive cells were free

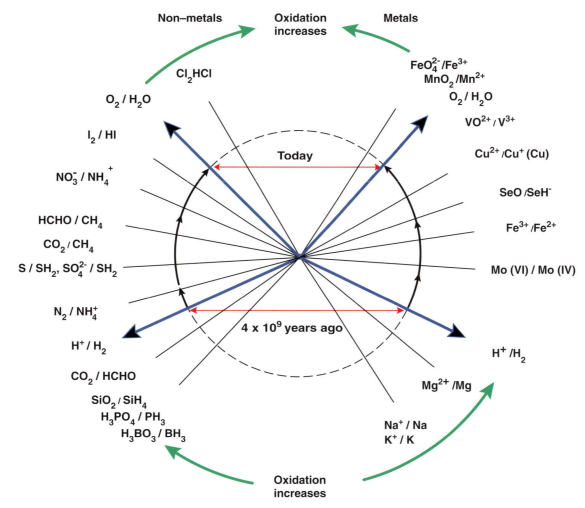

FIGURE 12 The enforced changes in the oxidation states of elements in the environment due to the equilibration with the gradual rise in the oxygen pressure.

ions, the metallome, substrates, and coenzymes acting on transcription factors, while similar chemicals acting on receptors and pumps are the major communication to the membrane. Here particular attention is again drawn to the cytoplasmic network principally of the inorganic elements hydrogen, phosphorus, sulfur, sodium, potassium, chlorine, magnesium, and iron in primitive cells. These must be closely involved with protein synthesis. This network could not be greatly changed throughout evolution because it maintains the central cytoplasmic system including the membrane uptake. (Much of it is derivative from the nature of the sea, as explained above.) Thus new vesicle compartments, after O_2 rose, clearly needed communication modes, which would not be confusing to this network but would connect to it, linking to the central command of the DNA of each cell. One almost dominant new messenger in the evolution of eukaryotes mentioned above was the Ca^{2+} ion, which was previously just rejected from the prokaryote cytoplasm to the external environment and now also into the vesicles of eukaryotes. This gave the necessary energized large gradient using the binding capability of pulsed Ca^{2+} to activate cytoplasmic systems (Figure 8). Once again this is not the place for details, but it is now known that the new Ca^{2+} message from outside the cytoplasm interacts directly with the old phosphate messenger system in the cytoplasm. The phosphate systems required iron for energy transformation and magnesium to catalyze many reactions. It is hard to see how eukaryote organization could have arisen without the utilization of the calcium gradient originally formed by prokaryotes as a way of sensing. These developments needed little use of oxygen. Note that even the earlier eukaryote cells were informed about the environment in this way.

B. New Oxidized Elements

One of the first chemicals to be oxidized by molecular oxygen was sulfide to sulfate (see Figure 12). Sulfur isotope studies indicate the sulfate produced by organisms appeared some 3×10^9 years ago. Sulfate has value like oxygen itself. Although both new components can be introduced into organic compounds, they are also sources of energy. Do not forget that light produced oxygen (and then sulfate) simultaneously with reduced carbon compounds. The new source of energy for cells was then the reaction of the reduced organic molecules with oxygen or sulfate. Sulfate bacteria are well-known and are virtually anaerobes.

The element oxidized at about the same time was molybdenum from its lower oxidation state MoS_2 (see Figure 12). In solution molybdate replaced tungstate forming Moco, and assisting O-transfer reaction, while possibly it replaced reduced vanadium in FeMoco, the coenzyme for nitrogen fixation. Molybdate became established at about 10^{-7} M in the cytoplasm. Subsequently zinc was released from sulfide and became a major signaling element and acid-base catalyst, but it is kept at or below 10^{-11} M as a free ion in the cytoplasm. Other elements with new uses mainly outside cells were selenium in the detoxification of peroxides and in assisting metabolism of the most easily oxidized halide, iodide. Iodide as iodine was incorporated into tyrosine derivatives and found value as the hormone, thyroxine. Vanadium as vanadate, which was able to bind peroxide, was involved in halide oxidation and halogen (bromium, chlorine) incorporation in organic molecules, for example, in toxins. These changes in element use in cells, especially in their vesicles and extracellular spaces, follow in time the succession of availabilities in the sea which is the succession of redox potentials (Figure 12). The sea is closely at equilibrium at a given time with a redox potential increasing toward 0.8 V. One non-metal, which is now oxidized toward nitrate, is nitrogen. Today O_2 is above the NO^-_3/N_2 potential. Nitrate is already an energy source and a nitrogen source for many organisms, but it is dangerous in man.

Some oxidations of metal ions have been saved until the last. The first, which took place some 2×10^9 years ago, was the virtual removal of ferrous to be replaced by ferric Fe^{3+} iron. Fe^{3+} is extremely insoluble at pH = 7.0, and this oxidation is a great problem for the required 10^{-7} M Fe^{2+} in the cell cytoplasm. All cells have devised intricate methods for scavenging of iron from the environment. A second oxidation from an increase in availability was the release of copper as Cu^{2+} from cuprous sulfide. Copper as Cu^{2+} was and is an extreme poison in the cytoplasm of cells and is always maintained at $<10^{-15}$ M. Its use is outside the cytoplasm.

Finally, although nickel (Ni^{2+}) and cobalt (Co^{2+}) became more available in the sea, the substrates for their enzymes largely disappeared, e.g., CH_4, H_2, CO. Note that plants require no cobalt and all higher life requires very little of it as well as very little nickel.

Although all this basic chemistry is largely to be found in the periplasm of prokaryotes (bacteria) and in eukaryotes in their vesicles, the biggest change which oxidation finally generated some 1×10^9 years ago was of multicellular animals with a new large dependence on copper. The further developments of oxidized organic and inorganic chemistry is therefore best described in

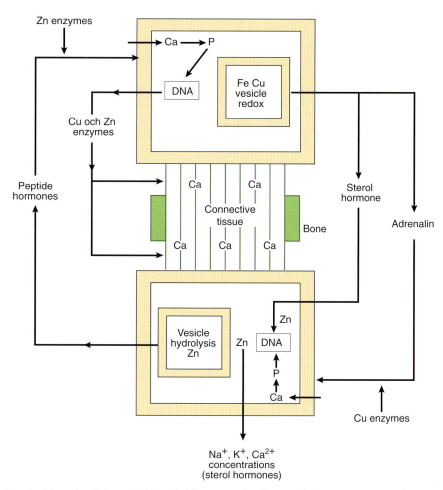

FIGURE 13 The interlocking of cellular activities of differentiated cells as in higher organisms utilizing a greater variety of elements than in primitive cells.

more detail under the heading of multicellular eukaryotes, although much of the new chemistry described is common to the prokaryote periplasm and the vesicles of single cell eukaryotes.

X. MULTICELLULAR ORGANISMS

A. Cell/Cell Organization

Further development of internal organization from simple cells (prokaryotes) and then single cells with internal compartments (eukaryotes) must clearly be the introduction of novel chemistry, mainly oxidative, in new *external compartments*, as cells came together. Once cells became kept in a spatial organization they could then have separate functions in one organism. Such a development needed fixed external filaments and, because communication is central to organization, messengers external to and between the cell (Figure 13). The filament connective tissue had to be open to adjustment with growth making growth signals essential. Hence new chemistry external to a cell had to provide the ability to make strong filaments, ways to make and break the filaments, and new messengers for communication, hormones, and transmission factors (Figure 13). The parallel with the earlier evolution of internal compartments and filaments plus new internal messengers (Ca^{2+}) is obvious. Again the old chemical systems of the cytoplasm could not be changed much because the basic primitive metabolism in all cells had to be left as it was. It is not a coincidence that the appearance of this mul-

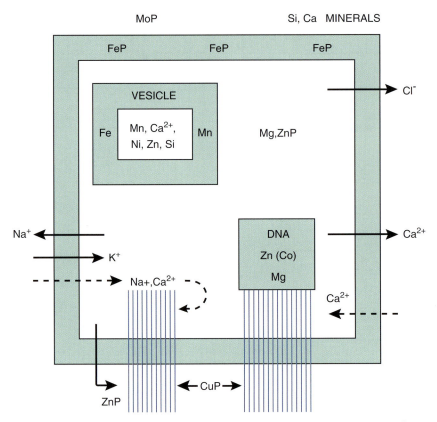

FIGURE 14 An increase in cell-cell organization due to control by zinc and copper enzymes over the extracellular filaments and the use of messengers (see Figure 13 and Table V).

ticellular life coincided with a higher oxygen content of the atmosphere and the extensive use not only of higher oxidation states of iron (especially heme), molybdenum, manganese, and selenium, but also the use of zinc and copper. Zinc was released from its sulfide before copper but, by one billion years ago, both were freely available in the sea.

Let us consider the new filaments external to cells (Figure 14). In order to gain rigidity they had to be cross-linked by oxidation. At first heme iron and manganese lignases, using peroxide, were employed in plants, but use of peroxide would have been dangerous in animals where the risk from incorrect oxidation is greater. Because external ferrous iron was not available, the obvious metal to use was copper, and it was found that copper became the major oxidase for cross-linked external filament production. Does the appearance of copper herald the coming of animals? Use was also made of sulfate in the sulfation of polysaccharides to keep an open network of connective tissue. Individual steps in the creation of this new external network and

its activity cannot be timed, but a major feature became cell differentiation and then organ development in multicellular constructs. Zinc became the major hydrolase center for external breakdown of filaments to allow growth (Figure 14).

B. Cell/Cell Communication: New Messengers

It has been seen that communication was based first inside the cytoplasm upon organic substrates and coenzymes and a few non-metal and metal ions such as phosphate, ferrous iron, and magnesium. This prokaryote organization was augmented by the use of the calcium gradient between outside and inside when larger cells (eukaryotes) with inner vesicles evolved. The possibility of triggering reactions due to simple ion messenger signals was then exhausted until oxidizing conditions arose. Oxidation allowed two new kinds of messenger: oxidized organic molecules such as adrenaline, sterols, thyroxine, and amidated peptides as seen

TABLE V. Organic Messengers Produced by Oxidation

Messenger	Production	Reception	Destruction
NO	Arginine oxidation (heme)	G-protein	? Heme oxidation
Sterols	Cholesterol oxidation (heme)	Zn-fingers[*]	Heme oxidation
Amidated peptides	Cu oxides	(Ca^{2+} release)	Zn peptidases
Adrenaline	Fe/Cu oxidase	(Ca^{2+} release)	Cu enzyme?
δ-OH Tryptamine	Fe/Cu oxidase	(Ca^{2+} release)	Cu enzyme?
Thyroxine	Heme (Fe)	Zn finger?[*]	Se enzyme
Retinoic acid	peroxidase	Zn finger?[*]	?
	Retinol (vitamin A) oxidation		

(*) In the nuclear receptor super family of transcriptional receptors.

in Table V. The role of copper, iron, and selenium must not be missed both in their production and destruction. A messenger has to be removed as soon as its effect has been attained. An old way of making and destroying a messenger quickly was by hydrolysis. Hydrolysis allowed preparation of small peptides from larger units in vesicles, but it also allowed the destruction of the molecules from the vesicles once their message task had been completed outside cells. It was the increased availability of zinc that allowed this activity. The two elements dominating, respectively, hydrolysis (Mg^{2+}) and oxidation/reduction (Fe^{2+}) in the cytoplasm throughout evolution became augmented by the two elements Zn^{2+} for hydrolysis and Cu^{+}/Cu^{2+} for oxidation/reduction in vesicles, periplasm, and outside cells. The metals cobalt and nickel almost dropped out of use in higher organisms because their redox chemistry is much less valuable in oxidative as opposed to reductive chemistry. Of course the redox chemistry of non-metals also changed, and now the value of selenium in peroxide chemistry generally and opposite de-iodination developed with the ability to use the iodinated organic hormone thyroxine as a messenger in animal life. In plants, molybdenum liberated from sulfide became valuable in oxidation as well as in the creation of the hormone absiscic acid from its aldehyde.

Just as the very early use of kinetically isolated new compounds was possible in the cytoplasm of prokaryotes, e.g., heme and chlorophyll, such a development was now possible in eukaryotes though it could not easily relate to oxidative chemistry. The major change seen is of zinc in transcription factors linked to messengers such as sterols at very low free concentrations of <10^{-10} M. It is our belief that in long-lasting cell

systems zinc acts like Mg^{2+}, but without interference due to its high binding constants to very different binding sites and much slower exchange. Zinc could be said to have become an internal connector of cell growth and development.

Thus oxidative chemistry outside the cytoplasm added on to reductive chemistry inside the cytoplasm was forced by the "accidental" conversion of the chemical nature of sea and atmosphere by molecular oxygen. There was no other way organization (evolution) could arise except by utilizing novel oxidative chemistry in new compartments.

XI. THE NON-METAL BALANCE: THE METABOLLOME

The handling of non-metals by organisms is more difficult to describe than that of metals for the very obvious reason that much of the establishment of their steady state levels is through kinetic feedback controls as seen in Figure 6. However, they also have the common problem of the sharing of energy and chemicals between pathways so that different pathways must have almost equal access to the non-metals. Thus there is a network comparable but more extensive than those for magnesium and calcium, for example, Figures 7 and 8. Consider the case of phosphate in nucleotide triphosphates like ATP. These coenzymes transfer phosphate and energy in many chemical steps leading to nucleotides, proteins, saccharides, and lipids. Their networks have to communicate with one another. As a consequence binding constants for them, usually with Mg^{2+} attached, are all close to 10^3 M and their concentrations, in the locality where they are used, are also close to 10^{-3} M. The suggestion is that the non-metal concentrations in certain compounds are relatively fixed. The profile of these compounds is called the metabollome.

Now the other free non-metal compounds which must be of relatively fixed concentration are the other mobile coenzymes that distribute carbon, hydrogen, nitrogen, and sulfur. They are among the common vitamins. Of especial interest here and associated with energy distribution, much as ATP is thus associated, are the hydrogen/electron carriers. There are two distinct cytoplasmic networks to consider just as there were two kinetically distinct networks for free Fe^{2+} and iron in heme. Here the two networks are at a potential close to -0.45 V NAD(H), (NADP(H)) and at around 0.0 V, glutathione. Note that the running of these two H(e) carriers about 0.5 V apart is close to the running of the two primitive prebiotic non-metal reactants, the H_2/H^+ and the H_2S/S_n redox couples. It was these elementary couples and then the corresponding coenzymes that supplied the initial feedstock of cellular life. These coenzyme pods connect not just to an energy supply but also to the distribution of hydrogen. They are all held close to millimolar in concentration in the cytoplasm of all active cells.

It would appear that the feedstock of non-metals from the environment may be variable, unlike that of metals, but for certain components of the non-metal system there is a fixed profile. This profile is the most intriguing part of the metabollome. Again this is suggestive of a selective quantitative requirement for a chemical system of reactions on which a code was superimposed to make reproductive organisms. Maybe the metaphoric original "watchmaker" of Dawkins was blind in some sense, but he must have had some other chemical sense which drove the beginnings and even evolution as if in a tunnel.

Once the code was in place and the cytoplasmic activity established for all time, new chemistry of the non-metals appeared in the vesicles. Examples are the handling and use of sulfate in connective tissue, of the dinitrification processes for nitrate metabolism in the periplasm, peroxide chemistry in defense mechanisms inside cells, the oxidation of halides to create a range of halogenated organic molecules in vesicles, and the development of organic messengers (Table V). How did these changes (see Figure 10) drive evolution? Are there ways by which geological changes interact more directly with the genetic apparatus than through blind mutation of the whole genome? All novel chemicals are

poisons, but they affect localized genes more than others. Do these genes mutate most rapidly to give rise to useful novel chemical features?

XII. THE NEW COMMAND CENTER: THE BRAIN

Once a multicellular structure and then organs appeared, increasing the size of organisms enormously, there were more problems than just producing extracellular matrices and their management and a message system between cells. With gross differentiation there was the logistical need to evolve a central organization of the parts, which were now organs. This need was the greatest in animals, which are scavengers rather than synthesizers such as are plants. Strikingly, the answer found was the brain as an organ. Initially there was development from a new physical type of compartment or cell, a nerve, which grew as a fast communicating compartment linking, in long cylindrical structures (Figure 15), one set of organs to another. The message system pressed into service in this construct was the Na^+/K^+ current. Although it utilizes a new Na^+/K^+ ATPase pump, this message system is derived directly from the required rejection of Na^+ and Cl^- by primitive cells. However, $Na^+/K^+/Cl^+$ flows are not a complete answer because when these ions cross membranes they produce electric potential switches but no chemical trace as they bind so poorly. The solution found was to connect the potential switch to a calcium message and the calcium message to a release of new organic chemical messenger molecules at nerve cell junctions. Note that the use of Na^+, K^+, and Cl^- to bring about fast electric potential switches of Ca^{2+} over long distances is the only possible mechanism open to cell systems. Cells cannot use electronic circuits. The new command center, the brain, is baced upon a coordinated set of such nerve cells.

It did not matter that this new command center, the brain, relies on DNA genes for its basic units, proteins. After its creation an animal was no longer dominated by genes but by the interaction with the environment. In fact because it can sense the environment, the machinery of the brain is capable of responding to control it. The further appearance of organization outside the most complex biological organism is seen in man's society and his industry, which can capitalize on the whole 92 elements of the periodic table in new external compartments using new reducing conditions to produce metals in what is the logical conclusion to evo-

(a)

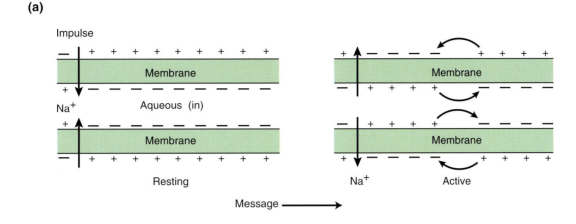

(b)

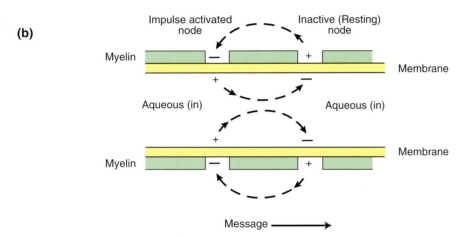

FIGURE 15 A nerve cell in schematic representation.

lution on Earth. The metals have new functional value in electronics. Further advances of materials allow the use of light as a messenger. Clearly the computer becomes the next new electronic command center. All these are but logical developments of organization. However, while rejoicing in this end product of chemical evolution, it must be remembered that the cytoplasm of all cells is a remnant of the primitive environment and cannot tolerate many of the new chemicals man can generate. The chemical system called living is robust but it needs protection. Many new protective devices have been introduced during the course of evolution, but there is no space to discuss them here. Note, however, that rare elements—platinum, gold, mercury, and bismuth—never used by cells become, in man's hands, the basis of medicines, though they are, in fact, poisons.

XIII. SURVIVAL: ECOLOGY AND SYMBIOSIS

Now we have dealt with the progression of compartments from single cell prokaryotes, to single cell eukaryotes, to multicell organisms, to organisms with organs and brains. We must pause now to ask if these systems are in competition, or are they really a natural way for cooperation to evolve in which one kind of cell system is best for one kind of task, with the ultimate objective of maximal use of the environmental chemicals in an overall ecosystem. In such an ecosystem symbiosis reigns supreme. Can we look at these developments of the chemistry and physics of cells (Figure 16) and see them as an inevitable progression, dependent on the changes of environmental chemistry and the logistics of organization? Maybe there was no

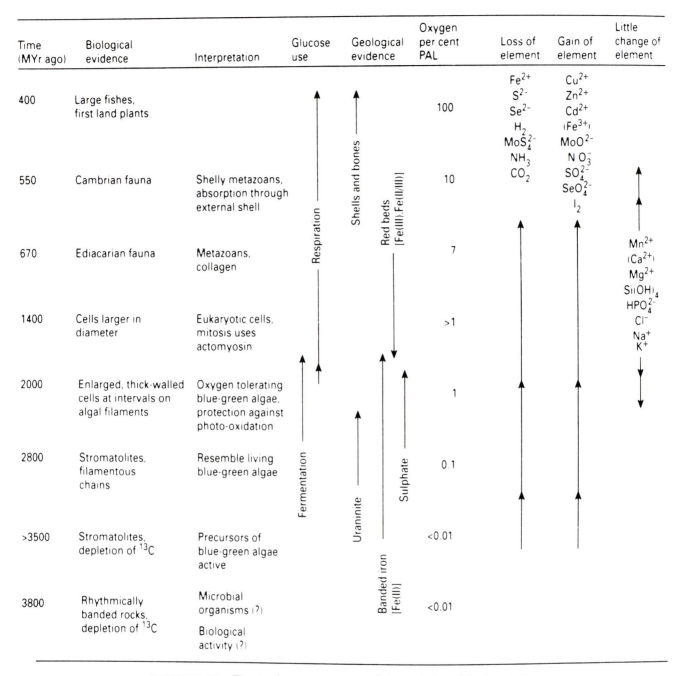

FIGURE 16   The simultaneous sequence of the evolution of Earth and life.

way to evolve novelty from an organic chemical cell except to utilize the possible oxidative chemistry and increase in compartments. However, there is a problem of the economy of such a system. The ability of one organism to evolve in an environment becomes increasingly difficult the more complicated the chemistry and physical constructs it uses, because it must keep all that came before and all that is new in harmony. Intrinsically a prokaryote is a better simple synthesis unit and a better stripped-down energy transducer than a eukaryote, and in turn better than a multicellular organism, because it does little else except synthesize. A eukaryote is a better consumer, even of prokaryotes, and has greater potential. Moreover, the eukaryote has a longer lifetime and needs greater protection. It follows that cooperative life-forms will have a greater survival chance than competitive styles, and this is seen in the successive development of symbiosis and of ecosystems. We see prokaryotes (organelles) in all eukaryotes and a gross dependence of eukaryotes on prokaryotes for N_2 fixation. Man as the last product (thus far) of evolution is not able to synthesize many of the basic requirements of the cytoplasm and requires essential amino acids, fats, coenzymes (vitamins), and saccharides from lower organisms. All minerals are collected by plants for animals. This again is a logical outcome of cooperative organization, not of competition. In the context of this chapter lies the requirement for the supply of twenty essential elements, and while we may concentrate upon the dependence of one phyla upon another for their supply, we must not miss the dependence of the biological ecosystem upon the geosphere. The environment is not separable from the overall chemical systems of life and its evolution.

XIV. Conclusion

The central theme of this chapter is that the primitive sea greatly influenced the way the chemical elements could be used by a chemical system which became primitive life (Figure 5). We consider that the profile of free elements in that sea self-generated a chemical system of reactions that became the one we see in the cytoplasm of all cells even today. Hence, fundamental to life before there was a code was an organized set of reactions that can be characterized by the profile of its free ion exchanging inorganic content—its cytoplasmic free metallome and metabollome. Metallomes and metabollomes, free and bound, have evolved largely due to new compartments and in oxidative possibilities. The way in which this was done is then the major story of cell evolution even though it relied on codes and cooperation to secure survival. The appearance of species is a decoration on this logical progression. The final twist is the turn around to extreme reductive chemistry which has allowed man to evolve in the new industrial world around us.

See Also the Following Chapters

Chapter 5 (Uptake of Elements from a Biological Point of View) · Chapter 6 (Biological Functions of the Elements)

Further Reading

Anderson, G. M., and Crerar, D. A. (1993). *Thermodynamics and Geochemistry: the Equilibrium Model*, Oxford University Press, New York.

Berkner, I. V., and Marshall, L. C. (1965). On the Origin and Rise of Oxygen Concentration in the Earth's Atmosphere, *J. Atm. Sci.*, 22, 225–61.

Caporale, L. H. (Ed.) (1999). Molecular Strategies in Biological Evolution, *Ann. NY Acad. Sci.*, Volume 870.

Cox, P. A. (1995). *The Elements on Earth*, Oxford University Press, New York.

Depew, D. J., and Weber, B. H. (1995). *Darwinism Evolving*, MIT Press, Cambridge, MA.

Fraústo da Silva, J. J. R., and Williams, R. J. P. (2001). *The Biological Chemistry of the Elements*, 2nd edition, Oxford University Press, Oxford.

Margulis, L. (1999). *Symbiotic Planet*, Basic Books, New York.

Martell, A. E., and Hancock, R. D. (1996). *Metal Complexes in Aqueous Solution*, Plenum Press, New York.

Neidhardt, F. C. (Ed.) (1996). *Escherichia and Salmonella* (2nd edition), American Society for Microbiology Press, Washington DC. (This is a detailed compilation of the properties of prokaryotes.)

Stryer, L. (1995). *Biochemistry* (4th edition), W. H. Freeman, New York.

Turner, D., Whitfield, M., and Dickson A. G. (1981). Inorganic Speciation of Trace Metals in the Sea, *Geochim. Cosmochim. Acta*, 45, 855–890.

Williams, R. J. P. (1970). The Biochemistry of Sodium, Potassium, Magnesium and Calcium, *Q. Rev. Chem. Soc. Lond.*, 24, 331–360.

Williams, R. J. P., and Fraústo da Silva, J. J. R. (1996). *The Natural Selection of the Chemical Elements*, Oxford University Press, Oxford.

Williams, R. J. P., and Fraústo da Silva, J. J. R. (1999). *Bringing Chemistry to Life*, Oxford University Press, Oxford.

Iron lost with sloughed cell
Tip of villus
Unaccepted iron
Dietary iron
Accepted by cell
Absorptive surface of villus
Crypt of Lieberkühn
Body iron

UPTAKE OF ELEMENTS FROM A BIOLOGICAL POINT OF VIEW

ULF LINDH
Uppsala University

CONTENTS

The utilization of an element in biology is intimately dependent on its uptake into the living organism. A lot is known of the qualitative aspects of uptake; for example, common sense tells us that what originates in the geological background has to be transported through the soils and presented to plants in a convenient form for uptake. These processes are affected by physicochemical factors due to Nature itself and the increasing pressures from human activities.

From a biologist's perspective, the uptake process is extremely fascinating. Current knowledge tells us that organisms need about 20 of the naturally occurring elements found in the periodic table. Over the course of evolution, mechanisms have developed for the uptake and utilization of elements in organisms that are more or less specific for each. Additionally, those elements that are not essential or even detrimental to the organ-

ism are excluded, usually in an efficient manner. This chapter deals with the general aspects of element uptake as well as principles of exclusion. In addition, examples will be given pertaining to some major, minor, and trace elements collectively described as essential. This chapter begins with a discussion of the periodic table and what is meant by *essentiality*.

I. ESSENTIALITY OF ELEMENTS

A usually ignored fact in biology is that explanations for the behavior of elements can be found in the periodic table itself. Among the elements known to be involved in biology, 11 appear to be approximately constant and predominant in all biological systems. The human body is comprised of about 99.9% of the 11 elements, but surprisingly only 4 of them—hydrogen, carbon, nitrogen, and oxygen—account for 99% of the total. These 4 elements, the major elements, comprise the bulk of living organisms. In addition to these elements, there are the minor elements—sodium, magnesium, phosphorus, sulfur, chlorine, potassium, and calcium (also called *electrolytes*). The minor elements appear in much lower concentration than major elements. One group of elements has still to be defined: the trace elements.

From an analytical chemistry standpoint, trace elements would be described as elements appearing in low concentrations in living systems (*i.e.*, <100 mg/kg). In biology, however, trace elements would be defined by exclusion; for example, a biological approach begins by excluding the major elements as well as the minor elements. Furthermore, group 18 elements (the noble gases) are excluded due to their disinclination for chemical reactions, a property that makes these elements less likely to be a factor in biological functions. Depending on how many elements are considered naturally occurring, the trace elements thus constitute the remainder of the periodic table (*i.e.*, 73 or 75 elements). Most of the elements of the periodic table, then, are trace elements in the eyes of a biologist. Surprisingly, this exclusion definition coincides closely with the one of analytical chemistry. Most of the trace elements appear in biology at concentrations below or well below 100 mg/kg.

Essentiality is usually defined in an operational way, based on early protein chemistry. More stringent criteria have evolved as our knowledge has improved. A trace element can be considered essential if it meets the following criteria: (1) it is present in all healthy tissues of all living things, (2) its concentration from one animal to the next is fairly constant, (3) its withdrawal from the systems induces reproducibly the same physiological and structural abnormalities regardless of the species studied, (4) its addition either reverses or prevents these abnormalities, (5) the abnormalities induced by deficiencies are always accompanied by specific biochemical changes, and (6) these biological changes can be prevented or cured when the deficiency is prevented or cured. It is obvious that the number of elements recognized as essential depends on the sophistication of experimental procedures and that proof of essentiality is technically easier for those elements that occur in reasonably high concentrations than for those with very low requirements and concentrations. Thus, it can be expected that, with further improvement of our experimental techniques, additional elements may be deemed essential.

These six criteria might appear to be too stringent, and in some cases they must be modified. Most of the trace elements essential to both plants and animals are found in the first row of the transition (redox) metals. Zinc is not included in the transition metals, and it does not take part in redox reactions, which is an important property in biology; however, zinc is a good Lewis acid. All of the bulk elements are non-metals. The minor elements include metals as well as non-metals, with only one oxidation state available. Metals are the dominant

components of the essential trace elements, but some very important trace elements are non-metals, such as selenium and iodine. Boron and silicon are non-redox non-metals, and both of them are acknowledged as being essential. Boron, in fact, has been shown to be essential to plants, although it is found in appreciable concentrations in animals as well. The functions of these elements will be discussed in greater detail in Chapter 6.

II. GENERAL ASPECTS OF ELEMENT UPTAKE

The uptake of elements is a process that may vary considerably depending on the complexity of the living system being considered. Unicellular organisms account for the simplest processes, but in complex organisms several aspects of the uptake process must be considered. In humans, for example, the primary uptake process takes place in the gastrointestinal tract, predominantly in the duodenum and first part of the jejunum. Elements taken up have to be transported across the mucosal cells of the intestines to reach the bloodstream, from which they are transported to the liver, where the elements are isolated and delivered into the main bloodstream. After being transported to the organs that will utilize them, these elements must then enter the cells of these organs. If the final target is not found inside the cell, then further transport across additional membranes may be required. Let us review the general principles of transport across cell membranes. A cell or an organelle cannot be entirely open or entirely closed to its surroundings. Although the cell interior must be protected from certain toxic compounds, metabolites must be taken in and waste products removed. Because the cell must contend with thousands of substances, it is not surprising that much of the complex structure of membranes is devoted to the regulation of transport.

III. THE THERMODYNAMICS OF TRANSPORT

Before considering specific mechanisms of transport, it is useful to review some general ideas. The free energy change, ΔG, for transporting one mole of a substance from one place with concentration C_1 to another place with concentration C_2 is

$$\Delta G = RT \ln \frac{C_1}{C_2} \qquad (1)$$

where R is the gas constant and T the temperature. According to Eq. (1), if C_2 is less than C_1, then ΔG is negative and the process is thermodynamically favorable. When more substance is transferred (between two finite compartments), C_1 decreases and C_2 increases, until $C_2 = C_1$. Now, the system is at equilibrium, and $\Delta G = 0$. This equilibrium is the ultimate state approached by transport across any membrane. The concentration of any substance traversing the membrane will end up the same on both sides. In kinetic terms, if the molecules enter the membrane randomly, the number entering from each side will be proportional to the concentration on that side. Once the concentrations are equal, the rates of transport in the two directions will be the same; consequently, no net transport occurs. This equalization can be sidestepped under three conditions, each of which is important in the behavior of membranes.

A substance may be bound by macromolecules restricted to one side of the membrane. It could also be chemically modified once it has crossed the membrane. Compound M could be more concentrated inside a cell than outside, but much of M could be bound to cellular macromolecules or could have been modified. This part of M is not included in Eq. (1), which only deals with the free M, implying, then, that the concentration would be equal on both sides at equilibrium. We can use oxygen in red blood cells to illustrate this principle. Measurements would indicate that the total concentration of molecular oxygen is lower in the blood plasma than in the red cells. Included in the total concentration of the red cells, however, is the portion bound to hemoglobin. The concentration of free dioxygen is still the same in the red cells and plasma at equilibrium.

A membrane is often characterized by an electrical potential governing ion distribution. The well-known principle saying that equal charges repel each other and unequal charges attract each other may now be used to show a simple example. The negatively charged inside of a cell tends to attract cations and to drive out anions. Mathematically, the free energy change for transport over a membrane is

$$\Delta G = RT \ln \frac{C_1}{C_2} + ZF\Delta\Psi \qquad (2)$$

where Z is the charge of the ion, F is the Faraday constant ($96,485\, J\, mol^{-1}\, V^{-1}$) and $\Delta\Psi$ is the transmembrane electric potential (in V). If Z is positive and $\Delta\Psi$ negative (with the inside negative relative to the outside), transport of cations into the cell is favored. The concentration difference of ions across the membrane in most cells is kept at more than ten times, implying that active transport is a major energy-requiring process in biology.

Usually Eq. 1 does not reflect the real situation well enough and must be modified accordingly:

$$\Delta G = RT \ln \frac{C_1}{C_2} + \Delta G' \qquad (3)$$

where $\Delta G'$ may correspond to a thermodynamically favored reaction. Adenosine triphosphate (ATP) hydrolysis coupled to the transport might be such a situation. Equation (3) is clearly a generalization of Eq. (2) that now allows a variety of processes to participate in the transport.

Equations (2) and (3) convey the message that transport across membranes (in and out of cells) can take place against unfavorable concentration gradients. The sodium–potassium pump provides continuous import of potassium and export of sodium, thereby maintaining the concentration difference between inside and outside. Following is a review of the mechanisms by which substances are passed through membranes.

A. Passive Transport: Diffusion

Passive transport or passive diffusion occurs due to the random walk of molecules through membranes. The process is the same as the Brownian motion of molecules in any fluid and is called *molecular diffusion*. During passive transport, the diffusing substance ultimately reaches the same free concentration inside and outside the membrane. The net rate of transport, J ($mol\, cm^{-2}\, s^{-1}$), is proportional to the concentration difference ($C_2 - C_1$) over the membrane:

$$J = -\frac{KD_1(C_2 - C_1)}{l} \qquad (4)$$

where l is the thickness of the membrane, D_1 is the diffusion coefficient of the diffusing substance in the membrane, and K is the partition coefficient for the diffusing material between lipid and water (the ratio of solubilities of the material in lipid and water). For ions and other hydrophilic substances, K is a very small number. Diffusion of such substances through lipid membranes is thus extremely slow. In agreement with Eq. (1), Eq.

TABLE I. Permeability Coefficients from Some Ions and Molecules Through Membranes

P (cm s^{-1})	Membrane Phosphatidylserine	Human Erythrocyte
K$^+$	$<9 \times 10^{-13}$	2.4×10^{-10}
Na$^+$	$<1.6 \times 10^{-13}$	10^{-10}
Cl$^-$	1.5×10^{-11}	1.4×10^{-4a}
Glucose	4×10^{-10}	2×10^{-5a}
Water	5×10^{-3}	5×10^{-3}

[a] Facilitated transport. Note that whenever facilitated transport is encountered, the permeability coefficient rises dramatically.

(4) says that net transport stops when $C_2 = C_1$. If C_1 and C_2 are expressed in mol cm^{-3} and l in centimeters, then D_1 has the dimension cm^2 s^{-1}. Note that D_1 is not the same as the diffusion coefficient (D) that the same molecule would have in aqueous solution. D_1 depends not only on the size and shape of the molecule but also on the viscosity of the membrane lipid.

K, D_1, or the exact thickness of the membranes involved are not usually known, so the rate of passive transport is often described in terms of the permeability coefficient, P, which can be measured by direct experiment:

$$\mathcal{J} = -P(C_2 - C_1) \tag{5}$$

By comparing Eqs. (5) and (4), we see that P is expressed by:

$$P = \frac{KD_1}{l} \tag{6}$$

with the dimensions cm s^{-1}.

Table I shows the permeability coefficients for a number of small molecules and ions in membranes. The low P values of the ions are expected, because ions, as already mentioned, have low values of K; however, the relatively large permeability value for water is conspicuous. Biological membranes are not, in fact, very good barriers against water, the reasons of which are not entirely clear, but they have the obvious advantage of allowing the ready exchange of water with their surroundings.

B. Facilitated Transport: Accelerated Diffusion

The functional and metabolic needs of cells often require transport that is more efficient than passive

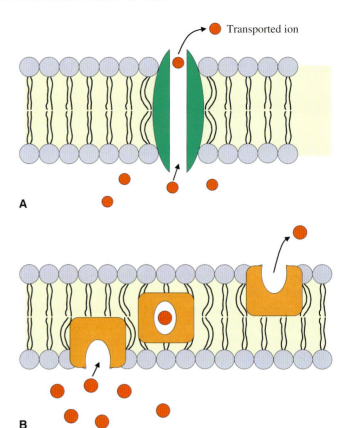

FIGURE 1 (A) Facilitated transport by pores. (B) Facilitated transport by carrier molecules.

diffusion. The adequate handling of catabolically produced CO$_2$ is coupled to the red blood cell exchange of HCO$_3^-$ and Cl$^-$. In respiring tissues, CO$_2$ enters the red cell and is converted to HCO$_3^-$ by carbonic anhydrase, an enzyme that is zinc dependent. HCO$_3^-$ leaves the red cell, and Cl$^-$ enters. The HCO$_3^-$ is transported to the lungs in the plasma where it enters the red cell, and Cl$^-$ is driven out. Inside the red cell, HCO$_3^-$ is again converted to CO$_2$ by carbonic anhydrase, leaves the cell, and is exhaled. The permeability coefficients for Cl$^-$ and HCO$_3^-$ in red cell membranes are about 10^{-4} cm s^{-1}, or about 10 million times greater than the permeability coefficients for such ions in pure lipid bilayers such as the artificial phosphatidylserine membrane described in Table I. Some special mechanism, then, is required to account for this difference. Two general types of facilitated transport, or facilitated diffusion, are known. One type, which is responsible for the rapid transport of Cl$^-$ and HCO$_3^-$ through red cell membranes, involves pores formed by transmembrane proteins (Figure 1A). The other type is mediated by transmembrane carrier molecules (Figure 1B).

Pore-facilitated transport is an important process. An example is the band 3 integral protein that exists as a dimer in the red cell membrane and serves as an anion channel that exchanges Cl^- and HCO_3^-. This protein probably spans the membrane 12 times. Exit of the HCO_3^- is balanced by an influx of Cl^-, which means that, in the absence of Cl^- ions, the transport of HCO_3^- stops. The band 3 protein does not simply form a hole in the membrane for the passage of ions; rather, the pore is very selective and exchanges HCO_3^- and Cl^- on a 1:1 basis. Such facilitated transport, however, is not necessary for O_2 or CO_2. These small, nonpolar molecules are allowed to move rapidly through the membrane by passive diffusion.

A common example of carrier-facilitated transport is the antibiotic valinomycin (from *Streptomyces*), which is a polymer with an approximately spherical shape. Its outer layer has numerous methyl groups; thus it is hydrophobic. Inside the sphere are collections of nitrogen and oxygen that can bind (chelate) a potassium ion. This structure, however, cannot chelate other cations, so it is specific for potassium. Due to its hydrophobic exterior, valinomycin easily passes through a membrane, in contrast to the ion itself. In mathematical terms, the valinomycin increases the factor K in Eq. (6).

The measurable difference between passive and facilitated diffusion is, of course, the transport rate. Another measure is the phenomenon of saturation, which is a characteristic feature of facilitated transport. There is a finite number of carriers or pores in a membrane, and each of them can handle only one ion at a time. Saturation will occur when all carriers are occupied (Figure 2). Equations (4) and (5) imply that the rate of passive diffusion increases linearly with the concentration difference because there are no sites to saturate. The carrier-facilitated transport can be described by:

$$V_0 = \frac{V_{max}[S]_{out}}{K_{tr} + [S]_{out}} \qquad (7)$$

Equation (7) looks familiar and, with some knowledge of enzyme kinetics, it is recognized as being similar to the Michaelis–Menten equation. V_0 represents the initial rate of transport into the membrane at an external concentration of $[S]_{out}$. V_{max} is the maximum transport rate of the substrate, and K_{tr} is analogous to the Michaelis constant K_m. This means that K_{tr} is the substrate concentration when the transporter is half-saturated. The transport rate approaches the maximum value at a high substrate concentration. In Figure 2, the straight line illustrates passive diffusion with, theoretically, no saturation.

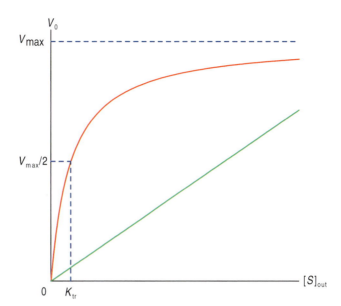

FIGURE 2 Facilitated and passive transport.

It still has to be remembered that passive and facilitated transport are diffusion processes, and as such they do not require energy. Pores are more effective because they offer open gates. Carriers increase the solubility in membranes by offering a hydrophobic outer surface. Irrespective of the kind of diffusion transport, the final free concentration on both sides of the membrane will be equal at equilibrium.

C. Three General Classes of Transport

With regard to transport across membranes, three different types have been identified, all of which depend on the number of substances and the direction in which each is transported. When only one substrate is transported, it is referred to as a *uniport transport process*—for example, the transport of glucose into red blood cells. The band 3 protein (the anion-exchange protein) is an example of an *antiport process*, in which one ion is transported out of and another into the cell. *Symport* is the transfer of two substrates in the same direction. Glucose and certain amino acids are transported via symport with Na^+. In this case, use is made of the gradient caused by the Na^+, K^+-ATPase in the plasma membrane. Figure 3 illustrates these general classes of transport. This characterization, however, takes into account only the direction of transport and does not show whether or not energy is required.

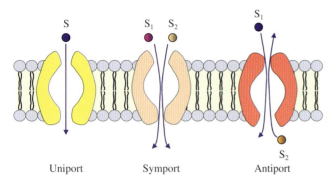

S   S₁ S₂   S₁

Uniport   Symport   Antiport

FIGURE 3 The three general classes of transport systems differ in the number of substrates transported and the direction in which each is transported. In uniport systems, only one substrate is transported. Symport and antiport systems are characterized by the transport of two substrates in the same and in opposite directions, respectively.

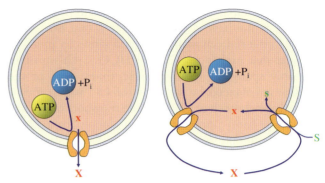

FIGURE 4 Active transport. In primary active transport (left), the energy released by ATP hydrolysis drives solute transport against an electrochemical gradient. In secondary active transport (right), a gradient of ion X (often Na^+) has been secured by primary active transport. Transport of X down its electrochemical gradient accordingly provides the energy to drive cotransport of a second solute (S) against its electrochemical gradient.

D. Active Transport: Transport Against a Concentration Gradient

In many situations, transport must be carried out even against concentration gradients. This requires a type of transport other than facilitated or passive diffusion. A calcium ion ratio of 30,000 must be maintained across membranes of the sarcoplasmic reticulum in muscle fibers. Using Eq. (1), such a ratio corresponds to $\Delta G = 25.6\,kJ\,mol^{-1}$, which indicates an insurmountable impediment. An active transport is necessary in such a scenario, but such a process is thermodynamically unfavorable and can only take place when coupled to a thermodynamically favorable process, such as absorption of light, an oxidation reaction, the breakdown of ATP, or an accompanying flow of some other chemical species down its electrochemical gradient. We can differentiate between *primary* and *secondary active transport* (Figure 4). In the former process, accumulation is coupled directly to a thermodynamically favorable chemical reaction, such as the conversion of ATP to ADP + P_i. When uphill transport of one solute is coupled to the downhill flow of a different solute that has originally been transported uphill by primary active transport, the process is secondary active transport.

The energy required to export one Ca^{2+} ion from the inside of the cell is 9.1 kJ. The energetic cost of moving an ion depends on the electrochemical potential or the sum of the chemical and electrical gradients (see, for example, Eq. (2)). Most cells maintain more than tenfold differences in ion concentrations across their plasma or intracellular membranes; therefore, for many

cells and tissues active transport is a major energy-consuming process.

E. Types of Transport ATPases

The four known types of ATPases are P-type, V-type, F-type, and multidrug transporter. The last type will not be dealt with in this chapter because it is not involved in the transport of elements. P-type ATPases are the most versatile, at least from an elemental point of view. They all transport cations and are reversibly phosphorylated by ATP in the transport cycle. They are all integral proteins with multiple membrane-spanning regions, although they are only single polypeptides. This type of transporter is very widely distributed. The Na^+, K^+-ATPase, an antiporter for Na^+ and K^+, and the Ca^{2+}-ATPase, a uniporter for Ca^{2+}, are ubiquitous, well-understood P-type ATPases in animal tissues. They maintain disequilibrium in the ionic composition between the cytosol and the extracellular media. P-type ATPases are responsible for pumping H^+ and K^+ over the plasma membrane in parietal cells lining the mammalian stomach, thereby acidifying the contents. Bacteria use P-type ATPases to export toxic metal ions such as Cu^{2+}, Cd^{2+}, and Hg^{2+}. Table II provides a summary of the properties of the transport ATPases.

V-type ATPases act as proton pumps and are not structurally similar to the F-type ATPases. The name V-type derives from their role of keeping the pH of vacuoles of fungi and higher plants at 3 to 6. In addi-

TABLE II.   Four Classes of ATPases

P-type ATPases	Organism or Tissue	Type of Membrane
Na^+, K^+	Animal tissues	Plasma
H^+, K^+	Acid-secreting (parietal) cells of mammals	Plasma
H^+	Fungi (Neurospora)	Plasma
H^+	Higher plants	Plasma
Ca^{2+}	Animal tissues	Plasma
Ca^{2+}	Myocytes of animals	Sarcoplasmic reticulum (ER)
Cd^{2+}, Hg^{2+}, Cu^{2+}	Bacteria	Plasma
V-type ATPases		
H^+	Animals	Lysosomal, endosomal, secretory vesicles
H^+	Higher plants	Vacuolar
H^+	Fungi	Vacuolar
F-type ATPases		
H^+	Eukaryotes	Inner mitochondrial
H^+	Higher plants	Thylakoid
H^+	Prokaryotes	Plasma

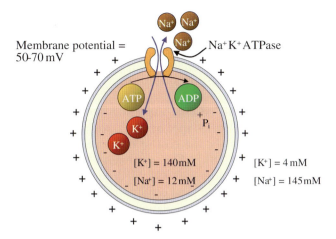

Membrane potential = 50-70 mV

Na^+ K^+ ATPase

$[K^+] = 140\,mM$
$[Na^+] = 12\,mM$

$[K^+] = 4\,mM$
$[Na^+] = 145\,mM$

FIGURE 5   In animal cells, this active transport system is primarily responsible for establishing and maintaining the intracellular concentrations of Na^+ and K^+ and for generating the transmembrane electrical potential.

F. Ion Pumps

Anyone having been in the slightest contact with biochemistry or physiology has not been able to avoid the sodium–potassium pump (Figure 5). This remarkable ion pump maintains concentration gradients for sodium from the inside of cells to the outside of 12 and 145 mM, respectively. At the same time, the concentration of potassium is kept at 140 mM intracellularly, in contrast to 4 mM outside. This situation pertains to almost all animal cells and would not be possible to maintain based solely on passive diffusion. The motor of this pump is the Na^+, K^+-ATPase that couples ATP breakdown to the concomitant movement of both Na^+ and K^+ against their electrochemical gradients.

The sodium–potassium pump transports two K^+ ions into the cell and exports three Na^+ ions at the cost of one molecule of ATP converted into ADP and P_i. Na^+, K^+-ATPase is an integral protein comprised of two subunits of approximate molecular weight 50,000 and 110,000, both of which span the membrane. This transporter is a P-type ATPase. The mechanism seems simple; however, the import of two potassium ions and the simultaneous export of three sodium ions are still not fully understood. It is generally assumed, though, that the ATPase cycles between two forms, one of which is phosphorylated with a high affinity for K^+ and a low affinity for Na^+, as well as one that is dephosphorylated with a high affinity for Na^+ and low affinity for K^+. The breakdown of ATP to ADP and P_i by hydrolysis takes place in two steps catalyzed by the enzyme: formation

tion to acidification of vacuoles, the same occurs for lysosomes, endosomes, the Golgi complex, and secretory vesicles in animal cells. The complex structures of V-type ATPases are similar throughout the family and possess an integral transmembrane domain that comprises the proton channel and a peripheral domain containing the ATP-binding site and the ATPase activity.

The F-type ATPase is so called because it has been identified as an energy-coupling factor. F-type ATPases catalyze uphill as well as downhill transport of protons. The uphill process is propelled by ATP hydrolysis, whereas the downhill reaction drives ATP synthesis. In this case, we may call them ATP synthases rather than ATPases. The F-type ATPases are multi-subunit complexes. They provide a transmembrane pore for protons and a molecular machine using the energy release by downhill proton flow to form the phosphoanhydride bonds of ATP.

of the phosphoenzyme and hydrolysis of the phospho-enzyme with the overall net reaction:

$$ATP + H_2O \rightarrow ADP + P_i$$

In this way, energy is supplied to cover the expenditure of the pump.

Calculating the cost of exporting three moles of sodium from 12 to 145 mM at 37°C, we arrive at 39.5 kJ. Correspondingly, the cost of importing two moles of potassium is 4.8 kJ. The net energy necessary to perform the transport is 44.3 kJ. Hydrolyzing one mole of ATP under physiological conditions to ADP gives 31 kJ. According to this calculation, more than 44 kJ is required for the transport, which does not seem reasonable. The trick is that in most cells the concentration of ATP is much higher than the concentrations of ADP and P_i. The energy available in real life is thus enough to pay for the transport.

Free calcium ions in cytosol are usually kept at a concentration of about 100 nM, which is far below what is found outside cells; thus, it is a significant finding that the total concentration of calcium in cells is much higher. One reason is that inorganic phosphates such as P_i and PP_i occur at millimolar concentrations. The concentration of free calcium ions must be kept low because inorganic phosphates readily combine with calcium, and relatively insoluble calcium phosphate precipitates will form. Maintaining the concentration of free calcium ions requires effective pumping out of the cytosol. This is accomplished by a P-type ATPase, which is the plasma membrane Ca^{2+} pump. Another calcium ion pump of the P-type resides in the endoplasmic reticulum (ER) and moves Ca^{2+} into the ER lumen, which is separated from the cytosol. In myocytes (muscle cells), Ca^{2+} is usually sequestered in a specialized form of ER called the *sarcoplasmic reticulum*. These pumps are closely related in structure and mechanism and are collectively called *sarcoplasmic and endoplasmic reticulum calcium* (SERCA) pumps. In contrast to the plasma membrane Ca^{2+} pump, the SERCA pumps are inhibited by the tumor-promoting agent thapsigargin.

These different pumps—the plasma membrane Ca^{2+} pump and SERCA pumps—share similarities in that both are integral proteins cycling between two conformations in a mechanism not very different from that for Na^+, K^+-ATPase. The calcium-ion pump of the sarcoplasmic reticulum has been thoroughly characterized and has been identified as a prototype for Ca^{2+} pumps of the P-type. It is built from a single polypeptide (M $\approx$ 100,000) spanning the membrane ten times. In a large cytosolic domain, there is a site for ATP binding, as well as an Asp residue undergoing reversible phos-phorylation by ATP. This process favors a conformation with a high-affinity Ca^{2+}-binding site exposed at the cytosolic side, and the opposite process, dephosphory-lation, favors a conformation with a low-affinity Ca^{2+}-binding site at the luminal side. One consequence of these cyclic changes of conformation is that the trans-porter binds Ca^{2+} on the side of the membrane where the calcium ion concentration is low and releases it on the side where the concentration is high. The energy released by hydrolysis of ATP to ADP and P_i during one cycle of phosphorylation and dephosphorylation drives Ca^{2+} across the membrane against the steep electro-chemical gradient.

IV. UPTAKE AND REGULATION OF IRON

Iron is vital for all living organisms because it is essential for many metabolic processes, the most well known being oxygen transport; however, further examples include DNA synthesis and electron transport. Although iron is abundant in nature, the metal is most commonly found as the virtually insoluble Fe^{3+} hydroxide $Fe(OH)_3$. Thus, iron-uptake systems require strategies to solubilize Fe^{3+}. Many organisms use siderophores (low-molecular-weight molecules secreted by bacteria, some fungi, and plants), which can solubilize Fe^{3+} for uptake by siderophore-specific transport systems. Genetic and biochemical evidence has demonstrated the presence of multiple pathways for iron uptake by eukaryotic cells. In mammals, changes in iron absorption are the major control point for altering the iron content of the body and of individual cells.

The intestine is the major site of iron regulation with regard to controlling the uptake of dietary iron across the brush border and the release of absorbed iron across the basolateral membrane to the circulation. Cells responsive to iron uptake are born in the crypt of Lieberkühn, located in the duodenum and jejunum. These cells differentiate and move toward the absorptive surface of the villus, where they are referred to as *enterocytes*. Gradually, mature enterocytes move toward the tip of the villus and are sloughed into the intestinal lumen (Figure 6).

In the intestinal lumen, iron exists in the forms of ferrous (Fe^{2+}) and ferric (Fe^{3+}) iron salts. Because ferric iron becomes insoluble at pH values above 3, ferric ions must be reduced or chelated by amino acids or sugars to be efficiently absorbed. Most ferrous iron remains soluble even at pH 7, so absorption of ferrous iron salts

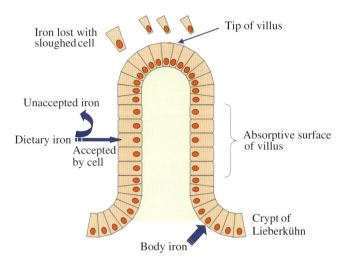

FIGURE 6 Iron is taken up into the enterocytes of the proximal small intestine from both the diet and blood plasma. The enterocytes are born in the crypts of Lieberkühn and move toward the villous tip to be discarded into the intestinal lumen at the end of a 2- to 3-day life span. (Adapted from Conrad and Umbreit, 2000.)

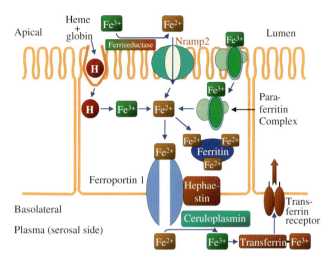

FIGURE 7 Pathways of heme and non-heme uptake of iron and transport in the intestine. The majority of the dietary ferric iron is reduced to ferrous iron or solubilized by mucin, ascorbic acid, or other reducing agents. The ferric iron in the lumen of the intestine is reduced by ferrireductase or in the cytoplasm by monooxygenase. An iron transporter, Nramp2, situated at the apical cell surface, transports most of the ferrous iron into the enterocyte. Yet another pathway for ferric and ferrous iron into the cell is the paraferritin complex, comprised of β integrin, mobilferrin, and flavin monooxygenase. Heme is taken up in the enterocytes as an intact metalloporphyrin (H). The uptake process is probably mediated by endocytosis. In the cytoplasm, heme is degraded by heme oxidase to release its inorganic iron. Two possibilities exist for the intracellular iron; either it is stored in ferritin or it is transported over the basolateral membrane by ferroportin I into plasma. Iron export seems to be facilitated by hephaestin concomitant with ferroportin I. Ferroportin I could also load iron onto transferrin assisted by the plasma ferroxidase, ceruloplasmin. (Adapted from Lieu *et al.*, 2001.)

is more efficient than absorption of ferric iron salts; however, most dietary inorganic iron is in the form of ferric iron. In most industrialized countries, two-thirds of the iron in the diet is present as ferric iron and one-third as heme iron (Carpenter & Mahorey, 1992). Reduction of ferric irons becomes necessary for efficient dietary iron absorption and is mediated by a mucosal ferrireductase that is present in the intestines. Inhibition of ferrireductase activity in intestinal cells reduces iron absorption, which demonstrates the importance of ferric iron reduction in dietary iron import. Alternatively, uptake of ferric irons might be mediated by the paraferritin pathway, though less efficiently. In addition to ferrireductase activity, the presence of dietary ascorbate provides a reduction of ferric iron to ferrous, whereby absorption is enhanced. Figure 7 illustrates the intestinal absorption and balance of iron.

A. Non-Heme Iron Uptake

Absorption of both heme and non-heme iron occurs predominantly in the crypt cells of the duodenum and jejunum (Wood & Han, 1998). Enterocytes, the specialized cells located on the intestinal villus, control the passage of dietary iron in the lumen of the intestine and the transfer of iron into the circulation of the body. To enter the circulation, dietary iron has to cross three

cellular barriers: iron absorption over the apical membrane, intracellular iron transport through the cell, and iron export over the basolateral membrane and into the circulation. The enterocytes, however, have no transferrin receptors on the surface exposed to the lumen (Pietrangelo *et al.*, 1992); thus enterocytes differ from other nuclear-bearing cells. Consequently, there has to be a mechanism of absorption other than the usual transferrin–transferrin receptor pathway. Absorption of iron across the apical membrane of the enterocytes is mediated by a divalent cation transporter, which is called *Nramp2* or *divalent cation transporter 1* (DCT1) (Fleming *et al.*, 1997). Nramp is an acronym for *natural resistance-associated macrophage protein*; Nramp2 is highly homologous to Nramp1, a molecule that is important in host defense against pathogen infection. Evidence speaks in favor of Nramp2 being responsible for iron transport from the duodenum lumen into the cytoplasm of enterocytes.

B. The Iron Importer: Nramp 2

The gene coding for Nramp2 in humans contains more than 36,000 bases. At least two different forms of mRNA are coded (Lee *et al.*, 1998), as this is the first step in the expression of the gene. One region of the Nramp2 isoform I is the 3′ untranslated region, which contains an iron-responsive element similar to the iron-responsive element present in the 3′ untranslated region of the mRNA transferrin receptor 1 (Tandy *et al.*, 2000). The isoform II, however, lacks the iron-responsive element. Expression of Nramp2 isoform I is, therefore, upregulated in iron-deficient animals and human intestinal cells, whereas expression of Nramp2 isoform II is not due to the lack of the iron-responsive element (Fleming *et al.*, 1999). The importance of the presence of a functional iron-responsive element in the Nramp2 isoform I is that its expression probably is controlled by intracellular iron concentration. Such a control should not be functional in the case of Nramp2 isoform II.

As might be expected, Nramp2 is highly expressed at the duodenum brush border, which corroborates its important role in intestinal iron absorption (Cannone-Hergeaux *et al.*, 1999), because enterocytes lack the transferrin receptor system at the absorptive surface. Nramp2 is located on the plasma membrane as well as on subcellular vesicular compartments. The Nramp2 protein is thought to consist of 12 transmembrane domains, and studies show that Nramp2 acts as a proton-coupled divalent cation transporter (Gunshin *et al.*, 1997). Nramp2 is, therefore, capable of transporting not only ferrous iron but also a number of divalent cations such as Zn^{2+}, Mn^{2+}, Co^{2+}, Cd^{2+}, Ni^{2+}, and Pb^{2+} (Gunshin *et al.*, 1997). In addition, Nramp2 function is pH dependent, being optimal at a pH of <6 (Gunshin *et al.*, 1997). Nramp2 has been shown to be colocalized at the subcellular level with transferrin, and Nramp2 might be involved in transporting transferrin-bound iron across the membrane of endosomes into the cytoplasm (Gruenheid *et al.*, 1999).

C. Heme Iron Uptake

Hemoglobin iron from food is absorbed more efficiently than inorganic iron; therefore, absorption of iron from myoglobin and hemoglobin is different from the way in which inorganic iron is absorbed. Hemoglobin is enzymatically digested in the intestinal lumen, and the heme molecule is internalized by the enterocytes as an intact metalloporphyrin (Majuri & Grasbeck, 1987). It might be that the heme molecule enters the cell through a receptor-mediated process. Once inside the enterocyte, heme is metabolized by heme oxygenase, and inorganic iron is released. This is either stored as ferritin or transported across the basolateral membrane to enter the bloodstream (Figure 7). When the enterocyte ends its life cycle, iron in the form of ferritin will be sloughed with the aged cells and leave the body through the gastrointestinal tract. Humans have a limited means of eliminating iron; therefore, this process is an important mechanism of iron loss (Lieu *et al.*, 2001).

D. Paraferritin-Mediated Iron Uptake

Nramp2 is a much better transport agent for ferrous iron than ferric iron. In addition, ferrous and ferric iron can also be internalized by enterocytes in different pathways. Paraferritin is a membrane complex with a molecular weight of 520 kDa that contains β-integrin, mobilferrin (a homolog of calreticulin, which is a lectin-like chaperone promoting efficient folding of proteins in the ER), and flavin monooxygenase. It participates in the mucin-mediated iron uptake in the intestinal lumen (Figure 7) (Umbreit *et al.*, 1998). Experiments with erythroleukemia cells show that an anti-$β_2$-integrin monoclonal antibody blocks 90% of ferric citrate uptake. Little effect, however, was observed on the uptake of ferrous iron. Consequently, it seems that ferric iron is absorbed via the paraferritin-mediated pathway (Conrad *et al.*, 1999). A possible mechanism is that ferric iron is solubilized by mucin in the intestinal lumen, transferred to the mobilferrin- and β-integrin-containing paraferritin complexes, and then internalized (Conrad *et al.*, 1999). Having been internalized, flavin monooxygenase is associated with the complexes and ferric iron is reduced to ferrous iron in parallel with the activity of NADPH. The β-integrin- and mobilferrin-containing paraferritin complex also interacts with $β_2$ microglobulin. In addition, mobilferrin and $β_2$ microglobulin have been shown to play critical roles in the development of iron overload in hemochromatosis in animals (Rothenberg & Voland, 1996).

E. The Iron Exporter: Ferroportin1

A novel iron transporter gene, ferroportin1, has recently been identified (Donovan *et al.*, 2000). Sequence analysis of ferroportin1 shows that it has a stem-loop structure, typical of iron-responsive elements, in the 5′ untranslated region (Donovan *et al.*, 2000). It has been shown that the iron-responsive element binds to iron regulatory proteins 1 and 2. This indicates that expression of ferroportin1 is regulated by intracellular iron levels (McKie *et al.*, 2000). Studies of ferroportin1 demonstrate that ferroportin1 mediates iron efflux across membranes by a mechanism requiring an auxiliary ferroxidase activity (Donovan *et al.*, 2000; McKie *et al.*, 2000). Ferroportin1 is expressed highly in the placenta, liver, spleen, macrophages, and kidneys. In the cell, ferroportin1 is located on the basolateral membrane of duodenal enterocytes (McKie *et al.*, 2000). This suggests that ferroportin1 probably functions as an iron exporter in the enterocytes (Figure 7). Ferroportin1 is located on the basal surface of placental syncytiotrophoblasts, which probably suggests a role for ferroportin1 in iron transport into the embryonic circulation (Donovan *et al.*, 2000).

The function of ferroportin1 is believed to be in parallel with the membrane-resident ferroxidase hephaestin and serum ceruloplasmin (McKie *et al.*, 2000). Hephaestin has a high degree of similarity to ceruloplasmin, which is a multi-copper oxidase possessing ferroxidase activity, which is required for the release of iron into blood and the binding to transferrin. Hephaestin does not transport iron, which is similar to ceruloplasmin; however, it facilitates the transport of iron from enterocytes into the body's circulation (Harris *et al.*, 1998). Sex-linked anemic mice with defective hephaestin show normal dietary iron absorption into the enterocytes, but they suffer from a defect in the transport of iron from duodenum to the blood (Vulpe *et al.*, 1999). Unfortunately, the mechanism by which ferroportin1 mediates the transport of iron across the basolateral membrane and by which it interacts with hephaestin and ceruloplasmin is still unknown (Lieu *et al.*, 2001).

F. Regulation of Dietary Iron Absorption

The regulation of iron absorption by enterocytes is exercised in various ways. In the first place, it may be modulated by the amount of iron in recently consumed food. This mechanism is referred to as the *dietary regulator* (Andrews, 1999). Enterocytes are resistant to acquiring additional iron for several days after consumption, a phenomenon referred to as *mucosal block* (Andrews, 1999). A second regulatory mechanism, which monitors the iron levels stored in the body rather than the dietary iron status, is referred to as the *stores regulator*. When there is iron deficiency, the stores regulator can modify the amount of iron uptake by a factor of approximately two to three (Finch, 1994). Saturation of plasma transferrin with iron is also thought to influence the dietary absorption of iron, at least indirectly. A detailed mechanism of the stores regulator remains to be defined. The *erythropoietic regulator* is a third regulatory mechanism that has a greater capacity to increase iron absorption than the stores regulator. The erythropoietic regulator does not respond to the cellular iron levels; however, it does modulate iron absorption in response to the requirements for erythropoiesis. Further studies are required to increase our knowledge of the molecular mechanisms of intestinal iron absorption (Lieu *et al.*, 2001).

G. Transferrin Receptor-Mediated Iron Uptake

In the blood, iron is transported by the plasma glycoprotein transferrin, which has a molecular weight of about 80 kDa and a high affinity for ferric iron. Most cells in the body, except enterocytes in the intestine, get iron from transferrin. The uptake of iron in cells begins with the binding of transferrin to a receptor on the cell surface known as the *transferrin receptor*. It binds transferrin only when it carries iron. The maximum capacity of transferrin is two iron ions. The transferrin–receptor complex is then internalized through the endocytic pathway. Transferrin receptors do not interact directly with iron, yet they control iron uptake and storage by most cells in the organism. There are at least two types of transferrin receptors. Transferrin receptor 1 is a membrane-resident glycoprotein that is expressed in all cells, with the exception of mature erythrocytes. The other type, transferrin receptor 2, is a homolog of transferrin receptor 1. It is specifically expressed in the liver, particularly in the hepatocytes. Following internalization, the endosome is covered with clathrin. This is a protein complex of three large and three small polypeptide chains that is thought to help bend the membrane in the internalization process. The endosome is then uncoated by an uncoating ATPase. Then protons are pumped into the endosome, causing iron to be released from the transferrin. Iron then passes

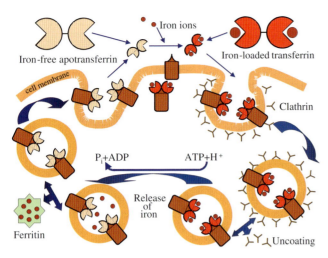

FIGURE 8 A schematic view of the extracellular binding of iron to apotransferrin, receptor-mediated endocytosis, release of iron into the endosome (driven by ATP), and the loading of iron into ferritin.

through the endosomal membrane and enters the intracellular labile pool. Intracellular iron in the labile pool can then be utilized for the synthesis of heme- and non-heme-containing proteins or stored within ferritin with a molecular weight of 474 kDa. The storage capacity of ferritin is as high as 4500 iron atoms. The receptor-bound transferrin is recycled back to the cell surface for reuse after completing a cycle of highly specific and efficient cellular iron uptake (Figure 8).

H. The Iron-Binding and Iron-Transport Protein: Transferrin

Transferrin is a single polypeptide chain of a glycoprotein that consists of two globular domains. Both domains offer a high-affinity binding site for one iron ion. Iron affinity of transferrin is pH dependent, and iron is released from transferrin when the pH is below 6.5. Transferrin might also be involved in the transport of a number of metals, such as aluminum, manganese, copper, and cadmium (Davidsson *et al.*, 1989; Moos *et al.*, 2000), but iron has the highest affinity to transferrin and will drive other metals out.

The liver is the primary site of synthesis of transferrin (Takeda *et al.*, 1998); however, it is synthesized in significant amounts in the brain, testis, lactating mammary gland, and some fetal tissues during development. The three different forms of transferrin, existing as a mixture, are iron-free (apotransferrin), one-iron (monoferric), and two-iron (transferrin diferric). The ratio between these forms depends on the concentration of iron and transferrin in blood plasma. Under normal conditions, most of the iron molecules in blood plasma are bound to transferrin (Lieu *et al.*, 2001). The main function of transferrin is to capture iron from plasma and to transport it to various cells and tissues in the organism.

1. The Transferrin-Binding and Transferrin-Transport Protein: Transferrin Receptor 1

Transferrin receptor 1 is a dimer comprised of two identical subunits and having a molecular weight of approximately 90 kDa. The receptor crosses the plasma membrane. The monomers are joined by two disulfide bonds and consist of three domains: a 61-residue amino-terminal domain, a 28-residue transmembrane region that helps to anchor the receptor into the membrane, and a large extracellular carboxyl terminus of 671 amino acid residues (McClelland *et al.*, 1984). As a type II membrane protein, the carboxyl terminal ectodomain of the transferrin receptor 1 is critical for transferrin binding. Indeed, replacement of the carboxyl-terminal, 192-amino-acid residues of the human transferrin receptor 1 with the corresponding region of the chicken transferrin receptor dramatically reduces or completely abolishes its binding affinity for transferrin. Because each ectodomain contains a binding site for the transferrin molecule, a homodimer of transferrin receptor 1 can bind up to two molecules of transferrin simultaneously.

Transferrin receptor 1 is synthesized intracellularly in the ER. Additionally, it undergoes a number of posttranslational modifications. Its ectodomain is comprised of three nitrogen-linked glycosylation sites and one oxygen-linked glycosylation site (Kohgo *et al.*, 2002). Correct folding is strongly dependent on the nitrogen-linked glycosylation sites of transferrin receptor 1. If the oxygen-linked glycosylation at threonine 104 is eliminated, the cleavage of transferrin is enhanced, which, in turn, promotes the release of its ectodomain (Rutledge & Enns, 1996). The segment of transferrin receptor 1 that crosses the membrane consists of 18 hydrophobic amino acids and also undergoes posttranslational modifications. The hydrophobic membrane-crossing segment is covalently bound to fatty acids and is subjected to acylation with palmitate. This probably helps to fasten the receptor to the plasma membrane (Kohgo *et al.*, 2002).

The part of transferrin receptor 1 that is resident in the cytoplasm is important for the clustering of the receptor into the chlatrin-coated pits of the plasma membrane and, subsequently, for endocytosis (Iacopetta *et al.*, 1988). A conserved internalization signal (YTRF; tyrosine, threonine, arginine, phenylalanine) within the 61-amino-acid residues in the cytoplasmic part of transferrin receptor 1 is critical for efficient endocytosis of the receptor (Collawn *et al.*, 1993).

Human transferrin receptor 1 is a tightly associated homodimer. Each transferrin receptor 1 monomer consists of three distinct globular domains (Rolfs & Hediger, 1999). The general form of the homodimer suggests that transferrin could bind to either side with no contact between the two transferrin molecules. The extracellular part of transferrin receptor 1 is separated from the membrane by a stalk, which presumably includes residues involved in disulfide bond formation and oxygen-linked glycosylation (Rolfs & Hediger, 1999; Lieu *et al.*, 2001).

Transferrin receptor 1 is the general mechanism for cellular uptake of iron from plasma transferrin. The current model of iron uptake from transferrin via receptor-mediated endocytosis in mammals is shown in Figure 8 and in greater detail in Figure 9. The first step, the binding of transferrin to transferrin receptor 1, is accomplished by a physical interaction that does not require an increase in temperature or energy (Conrad & Umbreit, 2000). Transferrin can bind one or two ferric ions, and the iron status of transferrin affects its affinity for its receptor. Diferric transferrin has the highest affinity, followed by monoferric transferrin; apotransferrin (without iron) has the lowest affinity. An estimated dissociation constant for diferric transferrin is about $2-7\,nM$ (Lieu *et al.*, 2001). The plasma concentration of diferric transferrin is about $5\,\mu M$ under physiological conditions; consequently, most surface transferrin receptors become saturated with transferrin. Thus, the homodimeric transferrin receptor can mediate a maximum uptake of four atoms of iron at the same time.

The complexes consisting of transferrin receptor–transferrin–iron interact with adaptor proteins in the clathrin-coated pit and are then internalized by the cells through an endocytic pathway mediated by the receptor. The tyrosine internalization motifs located on the parts of the transferrin receptors that reside in the cytoplasm seem to be necessary for a high-affinity binding to the adaptor protein complexes on the plasma membrane. Importantly, but not surprisingly, this process is temperature and energy dependent (Lieu *et al.*, 2001). Inside the endosome, an ATPase proton pump causes acidification of the endosome and results in the release of iron from transferrin. The apotransferrins remain attached to the transferrin receptors and return to the cell surface, where they are released from the cells. The binding between transferrin and the transferrin receptor is dependent on the pH, which is critical to membrane uptake and the release of transferrin. The release of apotransferrin from its receptor occurs at neutral pH at the cell surface. Both the ligand and receptor, in this way, become available for recycling the absorption of iron. After its release from the transferrin, iron passes through the endosomal membrane into the cytoplasm via the iron transporter Nramp2 (also known as the DCT1). Iron that enters the cell can be utilized in the synthesis of heme or incorporated in iron-containing molecules. Intracellular iron can also be stored in the ferritin complexes or can modulate the activity of iron regulatory proteins (Lieu *et al.*, 2001).

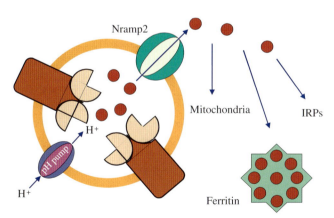

FIGURE 9 Release of iron from transferrin in the endosome and following events. A proton pump (energy-requiring) decreases the pH within the endosome with the consequence that iron is released from transferrin. The iron transporter Nramp2 carries out the subsequent transport of iron over the endosome membrane into the cytoplasm. IRP stands for iron-responsive protein.

2. Second Transferrin-Binding and Transferrin-Transport Protein: Transferrin Receptor 2

Transferrin receptor 2, which is a homolog of transferrin receptor 1, was recently identified. Its gene is located on chromosome 7q22 and gives rise to two transcripts approximately 2900 and 2500 bases in length. Like transferrin receptor 1, transferrin receptor 2 is a type II transmembrane glycoprotein, and it shares 66% similarity in its ectodomain with transferrin receptor 1. Although the cytoplasmic portion of transferrin receptor 2 is very different from transferrin receptor 1, transferrin receptor 2 also contains an internalization motif

(YQRV; tyrosine, glutamine, arginine, valine), which is similar to the YTRF motif in transferrin receptor 1. Transferrin receptor 2 does not possess iron-responsive elements. It seems that expression of transferrin receptor 2 is not regulated by an iron-regulatory, protein-mediated feedback regulatory mechanism in response to cellular iron status (Kawabata *et al.*, 1999).

In contrast to transferrin receptor 1, transferrin receptor 2 is primarily expressed in the liver. Like transferrin receptor 1, binding of transferrin receptor 2 to transferrin is also pH dependent. The binding of apo-transferrin to transferrin receptors 1 and 2 only takes place at acidic pH. Expression levels of both transferrin receptors 1 and 2 correlate with stages of the cell cycle which, in turn, are related to requirements for iron during DNA synthesis. Nevertheless, transferrin receptor 2 differs from transferrin receptor 1 in its binding properties with transferrin and regulation of expression. Holotransferrin has a lower affinity for transferrin receptor 2 than for transferrin receptor 1.

Transferrin receptors 1 and 2 are likely not only to be regulated through distinct pathways but also to mediate iron uptake and storage by a different, yet unidentified, mechanism. Transferrin receptor 1 seems to play a general role in cellular iron uptake; however, transferrin receptor 2 appears to play a specific role in iron uptake and storage in the liver, due to its high expression in hepatocytes (Lieu *et al.*, 2001).

I. Control of Iron Metabolism

Not surprisingly, animal cells differ somewhat from plants and lower eukaryotes with regard to the way in which they control iron metabolism. Transcription is the preferred method by which plants and lower eukaryotes maintain iron homeostasis. In the yeast *Saccharomyces cerevisiae*, for example, the iron-regulated transcription factor AFT1 controls production of multiple gene products that are needed to make up the high-affinity iron transport systems. Similarly, in plants, the iron storage protein ferritin is transcriptionally regulated by iron, which differs from the way in which it is regulated in animal cells, as animal cells utilize post-transcriptional control of iron metabolism in most cell types in the body. Apparently, regulation of gene transcription has a more important role in cell-type-specific modulation of iron homeostasis. Tissue-specific regulation of the expression of H- and L-ferritin, an erythroid-specific isoform of 5-aminolevulinate synthase, is one example, as well as control of the relative expression of iron regulatory protein 1 (IRP1) and IRP2 between tissues. This means that mammalian iron homeostasis is maintained through integrated use of sensory and regulatory systems operating at multiple levels of gene regulation (Eisenstein, 2000).

Although iron is an essential trace element, it might be detrimental if it is available as a free ion; consequently, besides providing storage, it is also necessary to prevent toxicity. Alteration of ferritin gene transcription provides an important means by which the relative abundance of the ferritin subunits can be modified to meet the unique iron storage and/or detoxification needs of specific tissues. The ratio of the abundance of the heavy-chain (H) and light-chain (L) subunits varies among tissues, and this variation is probably due to tissue-specific differences in the rates by which the ferritin genes are transcribed (Tsuji *et al.*, 1999; Eisenstein, 2000). Gene transcription of ferritin can be modulated by both iron-dependent and iron-independent factors. Experiments show that an excess of iron can cause a selective increase in L-ferritin gene transcription in the liver. In other systems, though, H- and L-ferritin transcription is altered in parallel to iron. Transcription of the ferritin genes is also modulated by a number of iron-independent signaling pathways (Eisenstein, 2000). Synthesis of transferrin and transferrin receptor takes place in fewer tissues than is the case for the ferritins. Transcription thus seems to dictate their expression in specific tissues. Iron deficiency causes induction of the transcription of both genes (Testa *et al.*, 1989). In many cases, however, transferrin receptor expression in iron deficiency is controlled by the regulation of mRNA stability. Seemingly contradictory is the fact that transcription appears to be a more critical factor in the iron regulation of transferrin expression in a deficiency, although translational regulation may be important in reducing transferrin synthesis when there is excess iron (Eisenstein, 2000). The gene transcription of transferrin and transferrin receptor is enhanced during hypoxia to increase iron delivery to the erythron, which is necessary to advance erythropoiesis and increase oxygen-carrying capacity (Tacchini *et al.*, 1999; Eisenstein, 2000). During erythropoiesis, transcription plays a greater role in modulating transferrin receptor expression than in many non-erythroid cells. This makes increases in transferrin receptor expression possible without maximal induction of IRP activity and allows for simultaneous expression of the erythroid isoform of delta-aminolevulinate synthase (ALAS, eALAS). The growth state of cells also influences transferrin receptor gene transcription. Clearly, regulation of the transcription of the H- and L-ferritins, transferrin, and transferrin receptor genes contributes greatly to the maintenance of cell and organ iron homeostasis (Eisenstein, 2000).

J. Iron Regulatory Proteins and the Coordination of Iron Homeostasis

Iron regulatory proteins play a significant role in maintaining iron homeostasis by coordinating many of the mRNAs that encode proteins for the control of uptake or the metabolic fate of iron. Furthermore, changes in rates of transcription help establish the final level of these mRNAs. IRPs are considered central regulators of mammalian iron metabolism because they regulate the synthesis of proteins that is required for the uptake, storage, and use of iron by cells (Eisenstein, 2000). Important factors with regard to this regulation include the following:

- It is well accepted that IRPs are critical factors of the posttranscriptional regulation of transferrin receptor expression.
- IRPs play a major role in determining the iron storage capacity of cells by regulating translation of both H- and L-ferritin mRNA.
- Translation of the mRNA for the eALAS also seems to be regulated by IRPs.

In this role, IRPs may coordinate the formation of protoporphyrin IX with the availability of iron. IRPs may thus be important modulators of iron cycling in the body. In addition, the mRNA-encoding divalent metal transporter 1 (DMT1) and ferroportin1/iron-regulated gene 1 (IREG1) contain iron-responsive element (IRE)-like sequences. This suggests that IRPs might possibly affect the use of these mRNAs. DMT1 expression is iron regulated in some but not all situations; however, there are indications that the abundance of ferroportin1/IREG1 mRNA responds to changes in iron status. If the IREs in DMT1 and ferroportin1/IREG1 mRNA are functional, as is the case for transferrin receptor mRNA, then IRPs probably are major modulators of the transmembrane transport of transferrin and non-transferrin iron (Eisenstein, 2000).

V. Uptake and Regulation of Zinc

A number of physiologic systems contribute to zinc homeostasis under different conditions. Central to the maintenance of zinc homeostasis, however, is the gastrointestinal systems, especially the small intestine, liver, and pancreas. Specifically, the processes of absorption of exogenous zinc and gastrointestinal secretion and excretion of endogenous zinc are critical to zinc homeostasis throughout the body. During evolution, cells developed efficient uptake systems to allow for the accumulation of zinc even when it is scarce. These uptake systems use integral membrane transport proteins to move zinc across the lipid bilayers of the plasma membrane. Once inside a eukaryotic cell, a portion of the zinc must be transported into intracellular organelles to serve as a cofactor for various zinc-dependent enzymes and processes present in those compartments; therefore, transporter proteins must be present in organelle membranes to facilitate this flux of zinc. Zinc can also be stored in certain intracellular compartments when supplies are high and used later if zinc deficiency ensues. Again, zinc transporters are required to facilitate this transport in and out of organelles.

A. Families of Zinc Transporters in Eukaryotes

Many types of transporters have been found to be involved in zinc transport. In prokaryotes, transporters of the ATPase binding cassette (ABC) family have been demonstrated to work in zinc uptake. The zinc ABC proteins of *Escherichia coli*, for example, are a major source of zinc incorporation for these cells. A family of P-type ATPases functions as zinc efflux transporters; the ZntA protein in *E. coli* is one such transporter. Interestingly, this protein is important for zinc detoxification by pumping the metal ion out of the cell when intracellular zinc levels get too high (Hantke, 2001).

Eukaryotes have been found not to use ABC transporters or P-type ATPases; instead, zinc transport apparently is accomplished by two other families of transporters. The uptake of zinc and transport from the extracellular space to the cytoplasm have been found to be associated with the ZIP (Zrt-, Irt-like proteins) family. Additionally, the mobilization and transport of stored zinc from an organelle to the cytoplasm have been shown to be carried out by ZIP transporters. The CDF (cation diffusion facilitator) family does the opposite of the ZIP proteins; namely, it pumps zinc from the cytoplasm out of the cell or into the lumen of an organelle. All of the known members of these families play roles in metal ion transport, and zinc is often the substrate; consequently, it might be that several other

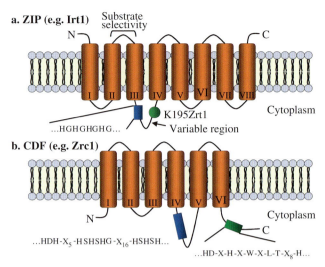

FIGURE 10 A sketch of the predicted membrane topology of ZIP and CDF proteins. (a) ZIP proteins as well as Irt1 are predicted to have eight membrane-crossing domains (I–VIII). Indicated as the variable region are the conserved and functionally important residues in domains VI and V, the ubiquitinated K195 in Zrt1, and the extracellular loop region affecting Irt1 substrate specificity. (b) The majority of CDF transporters as well as Zrc1 are predicted to have six membrane-crossing domains (I–VI). Conserved polar or charged residues within the membrane-crossing domains I, II, and V are indicated. H, histidine; G, glycine; D, aspartate; S, serine; L, lysine; T, threonine; W, tryptophan; and X, any amino acid. (Adapted from Gaither and Eide, 2001.)

members work in zinc transport (Gaither & Eide, 2001).

Two of the first members of the ZIP family to be discovered were Zrt1 of *Saccharomyces cerevisiae* and Irt1 of *Arabidopsis thaliana* (Gaither & Eide, 2001); thus the name ZIP transporters. Zrt1 is a zinc uptake transporter in yeast, and Irt1 is an iron transporter in plants. Currently, 86 ZIP members can be found in the protein sequence database at the National Center for Biotechnology Information (NCBI). This list includes proteins from eubacteria, archaea, fungi, protozoa, insects, plants, and mammals.

The degree of sequence conservation can be used to split the ZIP family into subfamilies. Most proteins in the ZIP family are predicted to have eight membrane-crossing domains; however, some may have as few as five. The majority of ZIP proteins share a similar predicted topology where the amino and carboxyl termini are located in the extracellular space (Figure 10a). Parts of this topology have been corroborated for some members of the family. Examples include the amino

terminus of Zrt1 and the carboxyl terminus of hZip2, which have been shown to be on the outside surface of the plasma membrane. Many of the ZIP proteins have a long loop region located between membrane-crossing domains III and IV. This region is called the "variable region," because both its length and sequence show little conservation among the family members. Many of the ZIP proteins are characterized by the presence of many histidine residues. In Zrt1, this sequence is histidine–aspartate–histidine–threonine–histidine–aspartate–glutamate . . . , and in Irt1 the sequence is histidine–glutamate–histidine–glutamate–histidine–glutamate–histidine. . . . The function of this region is not known; however, it is acknowledged to be a potential metal binding domain. Consequently, its conservation in many of the ZIP proteins implies a role in metal ion transport or its regulation (Gaither & Eide, 2001).

The mechanism of transport used by the ZIP proteins has yet to be unveiled. Conspicuously, the zinc uptake by human hZip2 zinc transporter has been shown to be energy independent (Gaither & Eide, 2000); however, this finding does not correspond with studies of the yeast zinc transporters Zrt1 and Zrt2, which demonstrated strict energy dependence (Zhao & Eide, 1996). Fungal and human ZIP proteins may, consequently, use different mechanisms. Zinc uptake by hZip2 was stimulated by HCO_3^- but was not dependent on K^+ or Na^+ gradients (Gaither & Eide, 2000); it has been suggested that hZip2 functions *in vivo* by a Zn^{2+}–HCO_3^- symport mechanism. Another possibility is that zinc uptake by these proteins may be driven by the concentration gradient of the metal ion substrate. Although the total level of zinc in cells can be as high as several hundred micromoles (Mantzoros *et al.*, 1998), only small amounts of that zinc are present in a "free" or labile form. Estimates of the labile pool of zinc in cells are in the nanomolar range (Suhy & O'Halloran, 1995). A concentration gradient of labile zinc across the plasma membrane may thus be an important driving force for Zn^{2+} uptake. The negative-inside membrane potential existing in cells could also be a driving force for the uptake of zinc (Gaither & Eide, 2001; Zhang & Allen, 1995).

The CDF (cation diffusion facilitator) proteins were early recognized to often play roles in metal ion transport (Nies & Silver, 1995). They are similar to the ZIPs found in organisms at all phylogenetic levels. Many members of this family have been implicated specifically in the transport of zinc from the cytoplasm out of the cell or into organellar compartments (Gaither & Eide, 2001). The CDF family was recently said to comprise 13 members; however, more sequence data and better tools have increased the number of members to 101 (Gaither & Eide, 2001).

The CDF family can be divided into three different subfamilies (I, II, and III) based on clusters or proteins with greater sequence similarities. CDF subfamily I is found mostly in prokaryota, including both eubacteria and archaea. Subfamilies II and III are comprised of about equal numbers of species of eukaryotic and prokaryotic origin. Six membrane-crossing domains seem to be common in a majority of the members of the CDF family. Their predicted membrane topology is similar to that shown for one such protein, Zrc1, from *Saccharomyces cerevisiae*, shown in Figure 10b (Gaither & Eide, 2001).

B. Zinc Transport and Its Regulation in Plants

Both ZIP and CDF family genes have been discovered in many plant species and have contributed to our increasing understanding of zinc transport and regulation in plants. The number of ZIP family members in plants is remarkable. The *Arabidopsis* genome contains 18 ZIP family genes from three of the four subclasses of ZIP proteins. ZIP subfamily II, however, does not have a plant representative. Plants and animals are multicellular organisms, as reflected by the high number of potential metal ion transport proteins, probably due to the greater diversity of tissue-specific roles played by these proteins (Gaither & Eide, 2001). The first ZIP protein to be discovered in any organism was Irt1 (iron-regulated transporter 1) (Eide *et al.*, 1996). Its gene (*IRT1*) was cloned because its expression in a yeast mutant with an impaired iron uptake suppressed the growth defect of this strain when growth media contained low amounts of iron. Irt1 expression indeed increased iron uptake in this yeast strain, as confirmed by biochemistry (Eide *et al.*, 1996). Later studies showed that Irt1 could also transport Zn^{2+}, Mn^{2+}, and Cd^{2+} (Korshunova *et al.*, 1999). Iron accumulation, rather than the transport of other metals such as zinc, seems to be the main function of Irt1 in plants. In addition, Irt1 is expressed only in the roots of plants for which iron access is restricted. If Irt1 takes part in the accumulation of metals other than iron, such as zinc, this probably occurs only under iron-limiting conditions (Gaither & Eide, 2001). It has, indeed, been observed that iron-limited plants accumulate higher levels of other metals such as zinc, manganese, and cadmium (Cohen *et al.*, 1998). These findings corroborate the prediction of Gaither and Eide (2001).

Zip1 through Zip4, ZIP transporters in *Arabidopsis*, may play roles in zinc transport. In *Saccharomyces cerevisiae*, the expression of Zip1, Zip2, or Zip3, each with distinct biochemical properties, results in increased zinc uptake. These proteins, consequently, most probably are zinc transporters. Zip4 expressed in yeast, however, does not result in increased zinc uptake. This may be due to poor expression or mislocalization of the protein in the yeast cell. *ZIP1* is expressed predominantly in roots while *ZIP3* and *ZIP4* mRNA could be found in both roots and shoots. The induction of *ZIP1*, *ZIP3*, and *ZIP4* mRNA takes place under zinc-limiting conditions. A role for these proteins in zinc transport is thus further confirmed. Neither subcellular localization of these proteins nor tissue-specific expression has been determined, so their exact roles cannot be assessed as yet. It is quite clear that some mechanism of regulation exists in plants because there is a zinc-responsive regulation of mRNA levels in response to zinc availability (Gaither & Eide, 2001).

If regulation of the expression of zinc transporters was altered in any way, zinc accumulation in plants would probably be greatly impacted. This presumption may in part explain the physiology of an unusual group of plants called *metal hyperaccumulators*. These are plants that take up large quantities of metal ions from the soil. They are of great interest because of their potential to remove metal pollutants from surface soils in a process called *phytoremediation* (Raskin, 1995; Gaither & Eide, 2001). A well-known hyperaccumulator is *Thlaspi caerulescens*, a member of the Brassicaceae family that also includes *Arabidopsis*. Certain ecotypes of *T. caerulescens* are capable of accumulating zinc in their shoots at levels up to as much as $30{,}000\,\mu g\,g^{-1}$ without evident toxic effects (Gaither & Eide, 2001). Plants that are not hyperaccumulators normally accumulate only 0.1% of that level. A salient ability to accumulate and detoxify metal ions should therefore be a significant property of hyperaccumulators. In studies of *T. caerulescens*, it was found that the maximum velocity, V_{max}, was elevated almost fivefold compared to a non-hyperaccumulating ecotype, *T. arvense*, although there was no difference in the Michaelis-Menten constant K_m (Lasat *et al.*, 2000). Expression of zinc uptake transporters should thus be higher in *T. caerulescens*. *ZNT1*, a ZIP family member, has been cloned from *T. caerulescens* and *T. arvense*. In *T. arvense*, Znt1 is expressed at a low level and regulated by zinc status. In *T. caerulescens* this gene is expressed at a much higher level and is unaffected by zinc availability. The increased zinc accumulation in this and perhaps other metal hyperaccumulating plant species can thus be explained by Znt1 expression.

Many members of the CDF family are also contained in the genomes of plant species. For example, *Arabidopsis* alone encodes ten CDF member genes. The proteins expressed by these genes are likely to function in subcellular zinc compartmentalization as well as in zinc efflux. The Zat protein of *Arabidopsis* is the only plant CDF member to have been studied. This protein seems to be a zinc transporter; however, *ZAT* mRNA expression is not zinc regulated. Transgenic plants overexpressing the *ZAT* gene demonstrate increased zinc resistance. In roots of these transgenic plants, the zinc content was also found to be increased, which suggests that Zat transports zinc into an intracellular compartment (*e.g.*, the vacuole or root cells). In any multicellular organism, zinc transporters are required for both cellular zinc uptake as well as efflux to allow utilization of the metal. Plants, for example, need a zinc efflux transporter to pass zinc from the root tissue into the xylem for distribution to aerial portions of the plant. CDF proteins such as Zat probably perform this function as well (Gaither & Eide, 2001).

C. Zinc Transporters and Their Regulation in Mammals

Both the ZIP and CDF families are represented by several zinc transporters found in mammalian organisms. Fourteen ZIP genes have been identified in humans, and three have been found in the mouse. Functional data are available for three of the human genes (*hZIP1*, *hZIP2*, and *hZIP4*). Recently, a subfamily of mouse zinc transporter genes was characterized (Dufner-Beattie *et al.*, 2003). The proteins hZip1 and hZip2 appear to play roles in zinc uptake across the plasma membrane. Expression of *hZIP2* mRNA has been detected in prostate and uterine tissue as well as monocytes, indicating restricted tissue specificity (Kambe *et al.*, 2004). Overexpressed *hZIP2* in cultured K562 erythroleukemia cells resulted in an increased accumulation of zinc compared to control cells. Furthermore, the hZip2 protein was localized to the plasma membrane of these cells (Kambe *et al.*, 2004). These results indicated that *hZIP2* might serve in zinc uptake in the few tissues where it is expressed (Gaither & Eide, 2001).

Endogenous uptake of zinc was shown to be biochemically different from uptake mediated by hZip2 in the K562 cell line in a number of ways. For example, HCO_3^- treatment stimulated zinc uptake mediated by hZip2, whereas the endogenous system did not react. Moreover, several other metal ions (*e.g.*, Co^{2+}, Fe^{2+}, and

Mn^{2+}) significantly inhibited zinc uptake by hZip2; however, the endogenous uptake was far less sensitive. It has recently been demonstrated that another ZIP transporter, hZip1, represents the endogenous zinc uptake system in K562 cells (Gaither & Eide, 2001). Three important observations support this hypothesis. First, K562 cells express *hZIP1* mRNA, and the functional hZip1 protein is localized to the plasma membrane of these cells. Second, a twofold increase in zinc uptake activity was a consequence of a twofold overexpression of hZIP1 mRNA. It was not possible to distinguish, by biochemical means, the increased uptake of zinc in hZip1-overexpressing cells from the endogenous system. Last, but not least, antisense oligonucleotides targeted to inhibit *hZIP1* expression also inhibited the endogenous zinc uptake activity. The hypothesis that hZip1 is the endogenous transporter in K562 cells is thus strongly supported. The antisense *hZIP1* oligonucleotide treatment reduced zinc uptake to 10–20% of control levels, again corroborating the idea that hZip1 is the major pathway of zinc uptake in these cells (Gaither & Eide, 2001).

A wide variety of different cell types demonstrates expression of hZIP1, in sharp contrast to the *hZIP2* gene. The results of Gaither and Eide (2001), therefore, suggest that hZip1 is an important candidate for being the primary factor for zinc uptake in many human tissues. A recent study by Franklin et al. (2003) in which a correlation was found between *hZIP1* expression levels and zinc uptake in human malignant cell lines derived from the prostate provided significant support for this conclusion. Prostate cell lines LNCap and PC-3 possess high levels of zinc uptake activity that is stimulated by prolactin and testosterone. It was found that *hZIP1* is expressed in LNCap and PC-3 cells, and this expression is increased by prolactin and testosterone treatment (Franklin *et al.*, 2003). Expression of *hZIP1* was also repressed by adding zinc to the medium, which suggested that some regulation of zinc uptake occurs in response to cellular zinc status. Lioumi *et al.* (1999) recently reported a closely related ortholog of hZip1 obtained from the mouse; this protein was named Zirtl for zinc-iron regulated transporter-like protein. The *ZIRTL* gene is expressed in a wide variety of tissues as is *hZIP1*.

A conspicuous finding is that the transporters hZip1 and hZip2 have a surprisingly low affinity for their substrate. The K_m values of both proteins are about $3\,\mu M$ for free Zn^{2+} ions. Additionally, zinc transporters in a wide variety of mammalian cells have K_m values of the same order. We are faced with an apparent paradox that arises when considering the free Zn^{2+} concentration in mammalian serum. The total zinc concentration of serum is about 15 to $20 \mu M$, and very little of that amount is present in an unbound form (Zhang & Allen, 1995). About 75% of Zn^{2+} is bound to albumin, and 20% is bound to α_2-macroglobulin. What does not exist in free form is complexed with amino acids such as histidine and cysteine. The serum has a high binding capacity for metals; thus, the free Zn^{2+} concentration in serum is estimated to be in the low nanomolar range. It is difficult to understand how such a low concentration of substrate would allow these transporters to contribute to zinc accumulation by mammalian cells under physiological conditions. The cellular requirements of zinc have to be considered and compared to transporter capacity. Recent studies showed that the capacity, expressed as V_{max}, for uptake is so high relative to the cellular demand for zinc that adequate levels can be obtained despite the apparent low affinity of the transporters (Gaither & Eide, 2001).

Curiously, a ferrous iron transporter, the DCT1/DMT1/Nramp2 Fe^{2+}, may be involved in zinc uptake. This transporter is a member of the Nramp family of transporters and is not related to ZIP or CDF proteins. Experiments with *Xenopus* oocytes suggested that cation influx currents could indicate Zn^{2+} movement across the membrane; however, more recent results have indicated that the currents recorded in these oocytes result from Zn^{2+}-induced proton fluxes rather than transport of the metal ion (Sacher et al., 2001).

Export of zinc from the cell as well as transport into intracellular organelles is related to mammalian CDF family members. Seven CDF genes in humans and six in the mouse genome have been identified plus a small number of others from the rat and other mammals. Four of the mammalian genes, *ZnT-1*, *ZnT-2*, *ZnT-3*, and *ZnT-4*, have been functionally characterized to such an extent that their roles in zinc metabolism are not in doubt. ZnT-1 is a zinc export transporter in the plasma membrane of mammalian cells; consequently, ZnT-1 may play a role in the cellular detoxification of zinc by exporting unnecessary metal ions out of the cell. An observation that cells overexpressing this transporter show higher zinc resistance than control cells further corroborates this role. ZnT-1 may also be involved in the dietary absorption of zinc in the intestine as well as in the reabsorption of zinc from urine in the renal tubules of the kidney. The intestinal enterocytes of the duodenum and the jejunum (*i.e.*, the primary sites of zinc absorption) express ZnT-1, and the protein is found localized to the basolateral membrane. This indicates a role in transporting zinc out of the enterocyte

and into the bloodstream. The protein is also found on the basolateral surface of renal tubule cells, where it would be expected to appear to be involved in transporting zinc that has been reabsorbed from the urine back into the circulation. It is well established that the loss of zinc in urine is very low because of an efficient renal reabsorption (Gaither & Eide, 2001).

Intracellular zinc sequestration and storage may be dependent on ZnT-2, a role similar to that proposed for Zrc1 and Cot1 in yeast. ZnT-2 is located in the membrane of the late endosome that accumulates zinc when cells are grown under high zinc conditions (Palmiter et al., 1996).

The third CDF transporter, ZnT-3, has a role similar to that of ZnT-2. It also transports zinc into an intracellular compartment where the metal may play a role in neuronal signaling. Messenger RNA of *ZnT-3* has been detected only in the brain and testis and is most abundant in the neurons of the hippocampus and the cerebral cortex. The protein is localized in membranes of synaptic vesicles in these neurons, which suggests that the protein transports zinc into this compartment. This hypothesis is further supported by the fact that a subset of glutamatergic neurons contains histochemically reactive zinc in their synaptic vesicles. The ZnT-3 protein was colocalized with these zinc-containing vesicles. Furthermore, a mouse line lacking *ZnT-3* did not accumulate zinc in these vesicles; therefore, the protein ZnT-3 must be required for the transport of zinc into synaptic vesicles in some types of neurons where it may play a neuromodulatory role (Cousins & McMahon, 2000).

The protein ZnT-4 is expressed in the mammary gland, brain, and small intestine. In the mammary gland, it is responsible for zinc transport into milk. In fact, mutations in the *ZnT-4* gene produced a mutant mouse referred to as the lethal milk (lm) mouse. This mutant gene is the *lm* gene. Pups of any genotype suckled on *lm/lm* dams die before weaning, and the cause of death is zinc deficiency from an insufficient supply of zinc in the milk. In intestinal enterocyte ZnT-4 is localized in endosomal vesicles concentrated at the basolateral membrane. It seems that ZnT-4, in a manner similar to ZnT-1, may facilitate transport of zinc into the portal blood (Kambe et al., 2004).

It is becoming increasingly clear that regulation of zinc export in many cell types is managed by zinc. It has been demonstrated that *ZnT-1* mRNA is upregulated during ischemia, which is known to cause zinc influx into neurons. Cultured neurons transiently increased *ZnT-1* mRNA when exposed to zinc, a finding that is in accordance with *ZnT-1* regulation as a result of zinc influx (Gaither & Eide, 2001). Transcriptional control of *ZnT-1* could therefore contribute to zinc detoxification by stimulating its export. As *ZnT-1* is expressed in many cell types, this could be a general mechanism of cellular zinc homeostasis. Zinc absorption may also be dependent on the transcriptional control of *ZnT-1*. Messenger RNA levels were found to be increased in enterocytes following an oral dose of zinc (Cousins & McMahon, 2000). The location of the ZnT-1 protein on the basolateral membrane of these cells suggests that upregulation of ZnT-1 promotes zinc absorption by facilitating transport into the portal blood.

Regulation of zinc uptake transporters in mammals is less well known. Evidence suggests that the activity of these transporters is controlled by the levels of zinc. Zinc uptake in brush border membrane vesicles, for example, has been found to increase in zinc-deficient rats. Additionally, cultured endothelial cells grown under low-zinc conditions displayed a higher rate of zinc uptake than zinc-replete cells. Zinc deficiency may thus increase the expression or activity of zinc uptake transporters in some cell types. A hypothetic mechanism of this regulation could be similar to that described in yeast. This is supported by the finding that *hZIP1* mRNA levels in cultured malignant prostate cells were reduced when treated with zinc. This suggests a transcriptional control mechanism. Such a mechanism would play a critical role in mammalian zinc homeostasis (Gaither & Eide, 2001; Kambe *et al.*, 2004).

VI. Uptake and Rejection of Copper

The transport and cellular metabolism of copper depends on a series of membrane proteins and smaller soluble peptides that comprise a functionally integrated system for maintaining cellular copper homeostasis. Inward transport across the plasma membrane appears to be a function of integral membrane proteins that form the channels that select copper ions for passage. Two membrane-bound, copper-transporting ATPase enzymes—ATP7A and ATP7B (the products of Menkes' and Wilson's disease genes, respectively)—catalyze an ATP-dependent transfer of copper to intracellular compartments or expel copper from the cell. ATP7A and ATP7B work in concert with a series of smaller peptides, the copper chaperones, which exchange copper at the ATPase sites or incorporate the copper directly into the structure of copper-dependent enzymes such as cytochrome *c* oxidase and Cu–Zn

superoxide dismutase (CuZnSOD). This enzyme is found in the cytoplasm and scavenges the superoxide anion, a reactive oxygen species. These mechanisms come into play in response to a high influx of copper or during the course of normal copper metabolism.

A. Accessing the Intracellular Pool

In a defined culture medium (an artificial environment), cells can get copper ions from a great number of suitable donors. The uptake process is usually rapid and, curiously enough, does not depend on the ATP status of the cell, so it seems that the uptake process is not energy demanding. It is tempting to conclude that a passive copper transport system exists in the membrane (Tong & McArdle, 1995). Plasma factors seem to have a similar capability to influence cellular access *in vivo*. The capacity of amino acids and chloride and bicarbonate ions to stimulate copper uptake gives some support for this idea (Harris, 2000). Inhibitors of protein biosynthesis seem to influence increased uptake of copper. This indicates that the transport system has a two-sided character and accounts for both import and export (Harris, 2000).

To work properly, a transport mechanism requires an easy exchange of copper. There has been some discussion about whether albumin- and ceruloplasmin-bound copper actually could be sources of copper for the tissues; however, a reducing environment, be it in plasma or in the membrane, can compromise binding strengths, making release of copper possible. These can be compromised by reducing systems in the membrane. Even amino acids could interact with the copper–protein complex (Harris, 2000).

B. Albumin as a Copper Transport Factor

Albumin is the most abundant protein in plasma; therefore, it is a good candidate for copper transport because it can bind copper in several sites on the protein (Masuoka & Saltman, 1994). The best candidates for transportation seem to be the sites at the extreme N terminal region or certain cysteine residues within the protein. At the N terminal, a histidine at position 3 may bind copper (Cu^{2+}); however, this site does not reach a rapid equilibrium with unbound copper. Histidine forms stable complexes with Cu^{2+}, and the histidine concentration in plasma is about $135\,\mu M$. These factors favor histidine as a transport ligand for copper. The interaction is reflected by a ternary complex with albumin–Cu(II) and histidine. Furthermore, the interaction induces the protein to release copper as a histidine–Cu(II) complex. It seems, however, that the histidine ligand is not transported across the membrane (Hilton *et al.*, 1995). This would suggest that the role of histidine ends at the cell surface (Harris, 2000).

C. Ceruloplasmin as Copper Transporter

Ceruloplasmin was first isolated from plasma and characterized as a copper-containing protein by Holmberg and Laurell (1948). This protein is a member of a class of proteins known as multicopper oxidases, which are characterized by three distinct copper sites. Ceruloplasmin is also a ferroxidase and plays an important role in oxidizing Fe^{2+} to Fe^{3+} for incorporation in apotransferrin. As a copper transporter, it contains about 95% of the copper in plasma. Multicopper oxidases utilize the electron chemistry of bound copper ions to couple substrate oxidation with the four-electron reduction of dioxygen. Electrons pass from the substrate to the type I copper, then to the trinuclear copper cluster, and subsequently to the oxygen molecule bound at this site (Garrick *et al.*, 2003). Human ceruloplasmin is encoded in 20 exons encompassing about 65 kb of DNA localized to chromosome 3q23–q24. The human ceruloplasmin gene in hepatocytes is expressed as two transcripts of 3.7 and 4.2 kb. These come from use of alternative polyadenylation sites within the 3′ untranslated region. When these transcripts are expressed in the liver, the 1046-amino-acid protein ceruloplasmin is found in plasma (Bielli & Calabrese, 2002).

D. Membrane Transport of Copper

Some years ago, a major area of investigation was agents that transported copper in plasma. Lately, however, interest has been focused on interactions at the membrane surface and transport intracellularly. What brought about this change in focus was the identification of specific copper-transporting proteins in the membrane as well as within the cell. Yeast has both low- and high-affinity systems for copper uptake, and these mediate copper transport (Eide, 1998). In yeast, the copper transport 1 (*CTR1*) gene was the first to be identified. Surprisingly, this gene was not directly for copper transport but was essential for iron transport in *Saccharomyces cerevisiae*. In fact, the copper was required for

Fet3. This is a multicopper ferroxidase that catalyzes the oxidation of Fe^{2+} to Fe^{3+} to make absorption by a ferric transport protein possible (Dancis *et al.*, 1994). The protein Ctrp1 is expressed by this gene and is a membrane-crossing protein. In addition, it is heavily glycosylated, with a serine- and methionine-rich composition in which the structural motif methionine–X–X–methionine is repeated 11 times. Ctrp1 transports Cu^+ and not Cu^{2+} or any other metal ions. The identification of this protein establishes a mechanistic link between copper and iron uptake (Harris, 2000). Strains of *S. cerevisiae* are equipped with a second high-affinity transporter gene, *CTR3* (Knight *et al.*, 1996). The expressed protein, Ctrp3, restores copper-related functions in strains lacking the *CTR1* gene. A structurally similar transporter gene in *Arabidopsis thaliana*, *COPT1*, was later discovered as well as the human transporter *hCTR1*. In HeLa cells, the *hCTR1* gene is located on chromosome 9 (9q31/32, to be more precise) (Harris, 2000). The human hCtr1 protein is much smaller than the yeast Ctrp1 protein, based on cDNA sequence data. All these copper transport proteins recognize only Cu^+; thus, there must be a reductase in the membrane to reduce Cu^{2+} to Cu^+ at the moment of membrane penetration. In yeast, the expression of *FRE1* and *FRE2* reduces both Cu^{2+} and Fe^{3+} for transport. Expression of a third gene, *FRE7*, increases when extracellular copper becomes very limiting (Martins *et al.*, 1998).

The first important factor to take into consideration is that the cytosolic environment is highly reducing. This means that reducing Cu^{2+} to Cu^+ is a straightforward process. The nitrogen-containing molecule glutathione (γ-glutamylcysteinglycine, or GSH) is the most common intracellular thiol. It has a concentration in mammalian cells of 0.5 to 10 mM and is a reducing agent in many reactions. Glutathione was one of the earliest intracellular components identified with copper transport (Denke & Farburg, 1989). Glutathione may also play the role of a general transporter of copper ions by delivering copper to Ctr1 in the plasma membranes. Glutathione reduces and binds Cu^+ and delivers it to metallothioneins and to some copper-dependent apoenzymes such as superoxide dismutase and hemocyanin (Tapiero *et al.*, 2003). Formation of Cu(I)–GSH is a spontaneous reaction apparently independent of enzyme involvement. When cellular GSH is low, cells are slower to take up copper from the medium and have a lower cellular concentration at steady state (Harris, 2000).

A decline in GSH levels in a cell impairs the subsequent binding of copper to apo-CuZnSOD or the delivery of copper to the cytosolic enzymes; thus, Cu–GSH complexes have the capacity to mediate Cu(I) transfer to a variety of binding sites on macromolecules. The function of GSH, then, extends beyond that of preventing copper toxicity to playing an important role in internal copper metabolism (Harris, 2000).

A new family of soluble metal receptor proteins acting in the intracellular trafficking of metal ions is the metallochaperones. These metal receptors do not act as scavengers or detoxifiers. On the contrary, they act in a "chaperone-like" manner by guiding and protecting metal ions while facilitating appropriate partnerships (O'Halloran & Culotta, 2000). Copper chaperones are a family of cytosolic peptides that form transient complexes with Cu⁺. An invariant methionine–X–cysteine–X–X-cysteine metal-binding motif in the N-terminal region is a structural feature of most chaperones. Their function is to guide copper ions in transit to specific proteins that require copper (Harris, 2000). Significant examples of metallochaperones are the gene products of the mercury resistance (*mer*) operon of the Gram-negative transposon Tn*21*, including MerP. MerP is a small, soluble periplasmic protein that transports Hg^{2+} to a membrane transporter and eventually to a reductase that reduces the Hg^{2+} to the volatile Hg^0 as part of a detoxifying mechanism (Hamlett *et al.*, 1992). Chaperones for copper act in a similar manner, moving copper from one location in the cell to another, often crossing membrane boundaries. In contrast to nuclear activating factors, chaperones demonstrate no capacity to enter the nucleus or interact with DNA. ATX1 and copZ are copper chaperones in *Saccharomyces cerevisiae* and *Enterococcus hirae*, respectively, and both have ferredoxin-like folds with the βαββαβ motif in the folded chains. Copper as Cu(I) is bound to two cysteine sulfur groups, forming a linear bidentate ligand. This structure probably allows for easy exchange of the bound copper to a structurally similar copper-binding site on the receiving protein (Harris, 2000). Of utmost significance is that other structural features allow the peptide to specify the target proteins with virtually no possibility for mismatch. The targets for these copper chaperones are cytochrome *c* oxidase in the mitochondria, ATP7B in the trans-Golgi, and the apo form of copper- and zinc-dependent superoxide dismutase (CuZnSOD) in either peroxisomes or the cell cytosol. The obvious advantage of a chaperone is that it selects the receiving molecule with high precision. This is not the case with Cu(I)–GSH, which has no target-specifying property. A conceptual disadvantage is that chaperones force a partitioning of copper into multiple pools in order to replenish copper enzymes (Harris, 2000). To date, three chaperones have been described in yeast, and all three are known to share structural features with mammalian and plant counterparts. The following is a brief description of known chaperones.

In yeast cells lacking superoxide dismutase activity (*sod1Δ*) and auxotrophic for lysine, *ATX1* (antioxidant 1) was identified as an antioxidant gene that suppressed oxygen toxicity (Lin *et al.*, 1997). The human homolog of *ATX1*, named *ATOX1*, was shown to complement yeast lacking *ATX1*. Like ATX1, the human ortholog contains one methionine–threonine–cysteine–X–glycine–cysteine copper-binding domain (Harris, 2000). The copper chaperone (CCH) in *Arabidopsis* has a 36% sequence identity with ATX1. Chaperones with the ATX1 structural domain target P-type ATPases. The ATPase in yeast is Ccc2p, a membrane-bound protein that mediates the transfer of copper to a late or post Golgi compartment. By the same reasoning, ATOX1 in mammals is thought to target ATP7B, which is a P-type ATPase that occurs in Wilson's disease. The ATPase ATP7B transfers copper to apo-ceruloplasmin or forces its extrusion into the bile, whereas Ccc2p transports copper to Fet3p, which is a multicopper oxidase that oxidizes Fe^{2+} to Fe^{3+} for incorporation into the ferric ion transporter and subsequent delivery (Harris, 2000).

Another copper chaperone, COX17, carries out the transport of copper to cytochrome oxidase in the mitochondria of the yeast *Saccharomyces cerevisiae*. A human homolog of COX17 has also been reported. Cox17, acting as a mitochondrial copper shuttle, is the only known chaperone that violates the glycine–methionine–X–cysteine–X–X-cysteine consensus motif. The copper binding sites on Cox17 are cysteine residues occurring in tandem (Cys14 and Cys16) and positioned near the N terminus. A unique property is the binding of Cu(I) to Cox17, which is a binuclear cluster that is similar to the copper cluster in metallothionein, with the exception that the Cox17–Cu(I) complex is more labile. In *S. cerevisiae* the delivery of copper to cytochrome oxidase apparently involves two inner mitochondrial membrane proteins, SCO1 and SCO2, which are penultimate receivers of the copper (Harris, 2000).

LYS7 is a 27-kDa copper chaperone that delivers copper to the apo-SOD1. There are mutants of yeast LYS7 mutants; they are defective in SOD1 (the gene encoding CuZnSOD) activity and are unable to incorporate copper into the protein. A single methionine–histidine–cysteine–X–X–cysteine consensus sequence is present in the N-terminal region of the protein. The copper chaperone for SOD (CCS) is a human counterpart of comparable size and 28% sequence identity (Harris, 2000).

The most significant finding with regard to copper transport in cells is without a doubt the discovery, cloning, and sequencing of the genes responsible for Menkes' and Wilson's diseases. These diseases have constituted important models of abnormal copper metabolism in humans. The etiology of these diseases was unknown for a long time. Classical Menkes' disease is an X-linked (*i.e.*, strikes mainly males) copper deficiency. The incidence is about 1 in 200,000. Boys with the disease usually do not survive past 10 years of age. Manifestations of the disease are a series of enzyme defects. Hypopigmented hair is caused by a deficiency of the tyrosinase required for melanin synthesis. Connective tissue abnormalities, including aortic aneurysms, loose skin, and fragile bones, result from reduced lysyl oxidase activity and consequent weak cross-links in collagen and elastin. Severe neurological defects are a predominant feature of the classical form of the disease and possibly result from the reduced activities of cytochrome oxidase; however, effects due to the reduced activity of superoxide dismutase, peptidyl-glycine-α-amidating monooxygenase, and dopamine-β-monooxygenase may also contribute to brain abnormalities (Suzuki & Gitlin, 1999). All of these enzymes are copper dependent. A diagnostic feature is the unusual steely or kinky hair caused by reduced keratin cross-linking, a process that also is copper dependent. The copper deficiency in Menkes' disease is caused by reduced uptake of copper across the small intestine (Figure 11) catalyzed by the Menkes' protein ATP7A, compounded by defective distribution of copper within the body wherever ATP7A is required for copper transport. ATP7A is involved in the transport of copper across the blood–brain barrier, which explains the marked brain copper deficiency and consequent severe neurological abnormalities in patients (Suzuki & Gitlin, 1999).

Wilson's disease, or hepatolenticular degeneration, is an autosomal recessive copper toxicosis condition with an incidence of 1 in 50,000 to 1 in 100,000, depending on the population. Very high concentrations of copper accumulate in the liver because of impaired biliary excretion of copper or failure to incorporate copper into ceruloplasmin, the major copper-binding protein in the circulation. This will ultimately cause the death of hepatocytes. The disease has a variable age of onset but is rarely observed in children younger than five and can present as a hepatic or neurologic disease. Copper may be released from damaged hepatocytes and accumulates in extrahepatic tissues, including the central nervous system. The diagnostic copper deposits that can sometimes be seen in the cornea of the eyes are known as the

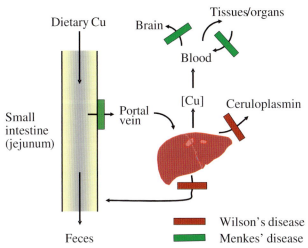

FIGURE 11 A schematic view of copper pathways and the blocks in Menkes' disease and Wilson's disease. Copper is taken up by enterocytes in the small intestine and exported over the basolateral membrane into the portal circulation. The export mechanism is defective in Menkes' disease; consequently, an overload of copper occurs in the enterocytes and subsequent copper deficiency in the organism. The liver normally takes up most of the copper in the portal circulation. If there is a copper overload, excess copper is excreted in the bile. This process is blocked in Wilson's disease. Furthermore, the delivery of copper to ceruloplasmin is also blocked. (Adapted from Mercer, 2001.)

Kayser-Fleischer rings disorder, which results from pathological accumulations of copper, predominantly in the liver and brain tissues. The dominant symptoms relate to a failure to release liver copper into bile. Both diseases have provided unprecedented molecular insights into genetic factors that regulate copper transport and bioavailability to organs and tissues (Suzuki & Gitlin, 1999).

Menkes' and Wilson's diseases are caused by mutations in genes on the X chromosome (Xq13) and chromosome 13 (13q14.3). The isolation and sequencing of these disease genes have revealed that both code for P-type Cu-ATPases. The Wilson's (ATP7A) and Menkes' (ATP7B) proteins are specific copper transporters; furthermore, ATP7B and ATP7A have a 57% sequence homology to one another. In addition, they have remarkable parallels to copper-binding proteins in bacteria. The gene for Menkes' disease spans about 150 kb. Its mRNA is 8.3 to 8.5 kb, encompassing 23 exons that range in size from 77 to 4120 bp, with a single open reading frame and an ATG start codon in the second exon. Exon 23 contains the TAA stop codon, 274 bp that are translated, and a 3.8-kb untranslated region that has the polyadenylation site. When analyzed as a cDNA,

ATP7A mRNA encodes a protein of exactly 1500 amino acids; however, it may have additional nucleotide sequences at the 5′ end. A 22-amino-acid presequence generated by an in-frame ATG site upstream occurs in some ATP7A transcripts (Harris, 2000). Strong expression of Menkes' disease mRNA is observed in muscle, kidney, lung, and brain. In placenta and pancreas, the expressions are weaker, and liver shows only traces. The Wilson's disease transcript is 7.5 kb and encodes a protein of 1411 amino acids. In contrast to the Menkes' disease gene, the Wilson's disease gene is strongly expressed in the liver and kidney (Harris, 2000).

The biological functions of ATP7A and ATP7B are different, although their structures are similar. The Menkes' disease protein (ATP7A) seems to be responsible for the regulation of copper-ion release at the outer membrane. Experiments with Chinese hamster ovary cells have shown that overexpression of ATP7A makes the cells tolerate highly toxic amounts of copper in their immediate environment (Camakaris et al., 1995). Superior tolerance is manifested by forced expulsion that prevents copper accumulation. The similarity in overall appearance between ATP7A and ATP7B might lead to the conclusion that they are similarly distributed. ATP7B, in contrast to the membrane association of ATP7A, resides within an internal organelle of the cell, where it functions to incorporate copper into apo-ceruloplasmin. This process takes place in either the ER or a Golgi compartment. Additionally, the protein works to force the release of copper into the bile (Harris, 2000).

The structure of ATP7A contains a comparatively large, heavy-metal binding domain (Hmb). This domain is comprised of six metal-binding cysteine clusters within the structural motif glycine–methionine–threonine/histidine–cysteine–X–serine–cysteine which contain eight transmembrane (Tm) regions. The purpose of these Tm regions is to guarantee anchorage and orientation of the protein. The correct assembly defines the channel through which copper ions pass. ATP7A is a type II membrane protein and is thus defined because both the —NH$_2$ and the —COOH termini are on the cytosolic side of the membrane. Two flexible loops, one on a 135-residue chain and the second on a 235-residue chain, extend into the cytosol. The smaller one is between Tm 4 and 5 and the larger one is between Tm 6 and 7 (Harris, 2000). The bacterial CopA protein contains a smaller Hmb with a single glycine–methionine–threonine/histidine–cysteine–X–X–X motif (Solioz et al., 1994). A yeast Cu-ATPase has the motif at most twice, which leads one to speculate whether the 650-residue Hmb region performs some

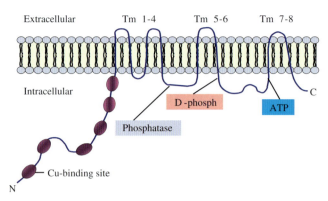

FIGURE 12 An illustration of the Menkes' and Wilson's Cu-ATPases. Eight membrane-crossing (Tm) domains are predicted for both proteins. Most of the protein is localized in the cytoplasm. In addition to the ATP-binding site, phosphatase domain and phosphorylated aspartic acid (D-phosph) is common to all P-type ATPases. The N terminal contains six copper-binding motifs that interact with copper chaperones. (Adapted from Mercer 2001.)

transport-related function other than binding copper for export. A cysteine–proline–cysteine motif in Tm 6 is thought to build the channel that allows copper in the cytosol to be transported across the membrane (Figure 12). Copper(I) is the preferred ion to bind to ATP7A and ATP7B. There is little or no affinity of the Hmb to bind to Fe(II), Fe(III), Ca(II), Mg(II), Mn(II), or Ni(II); however, Zn(II) shows some binding (Harris, 2000).

The perinuclear area within the region of the cell thought to represent the Golgi seems to be the localization of ATP7A. ATP7A-loaded vesicles have been postulated to continually be moving between the Golgi and plasma membrane. Experiments show that high concentrations of copper in the exterior of the cell induce movement of the marked vesicles to the cell boundary. The Wilson's disease ATP7B is also localized in the Golgi. A truncated homolog of ATP7B that lacks four of the eight membrane-spanning domains seems to reside in the cytosol. A structural analog of the Wilson's disease gene or Menkes' disease gene, the *CCC2* gene in yeast, also encodes a P-type ATPase that exports cytosolic copper to the extracytosolic domain of Fet3p, a copper oxidase required for iron uptake (Harris, 2000).

The mouse homolog of ATP7A, the Atp7a, is expressed in all tissues but is particularly strong in the choroid plexus of the brain. The localization in the brain, more specifically the blood–brain barrier, places the ATPase in a strategic position to control the flow of copper into the ventricles of the brain. An animal model of Menkes' disease is the macular mutant mouse. A gradual erosion of cytochrome *c* oxidase activity in the brain has been observed in this model. The effect can be partially prevented by a single injection of copper in an early perinatal period (Megura *et al.*, 1991). Copper shows a propensity to accumulate in brain blood vessels and in astrocytes which apparently hinders its movement to neurons (Kodama, 1993); furthermore, embryonic mouse liver expresses Atp7a mRNA in contrast to adults. The rat homolog of ATP7B, Atp7b, is expressed early only in the central nervous system, heart, and liver. With development, Atp7b appears in intestine, thymus, and respiratory epithelia (Kuo *et al.*, 1997). Transfection assays in yeast have demonstrated that the two ATPases share biochemical functions. Atp7b in hepatocytes from (LEC) rats, a model for Wilson's disease, mimics the sequestration of ceruloplasmin in cotransfected cells. The data support a metabolic connection between the plasma copper protein and Wilson's disease ATPase (Cox & Moore, 2002). ATP7B may also be localized on the apical surface of hepatocytes, a location that allows the protein to expel copper ions into the bile (Fuentealba & Aburto, 2003).

SEE ALSO THE FOLLOWING CHAPTER

Chapter 6 (Biological Functions of the Elements)

FURTHER READING

Andrews, N. C. (1999). Disorders of Iron Metabolism, *N. Engl. J. Med.*, 341, 1986–1995.

Bielli, P., and Calabrese, L. (2002). Structure to Function Relationships in Ceruloplasmin: A "Moonlighting" Protein, *Cell. Mol. Life Sci.*, 59, 1413–1427.

Cannone-Hergeaux, F., Gruenheid, S., Ponka, P., and Gros, P. (1999). Cellular and Subcellular Localization of the Nramp2 Iron Transporter in the Intestinal Brush Border and Regulation by Dietary Iron, *Blood*, 93, 4406–4417.

Carpenter, C. E., and Mahorey, A. W. (1992). Contributions of Heme and Non-Heme Iron to Human Nutrition, *CRC Crit. Rev. Food Sci. Nutr.*, 31, 333–367.

Cohen, C. K., Fox, T. C., Garvin, D. F., and Kochian, L. V. (1998). The Role of Iron-Deficiency Stress Responses in Stimulating Heavy-Metal Transport in Plants, *Plant Physiol.*, 116, 1063–1072.

Collawn, J. F., Lai, A., Domingo, D., Fitch, M., Hatton, S., and Trowbridge, I. S. (1993). YTRF is the Conserved Inter-

nalisation Signal of the Transferrin Receptor, and a Second YTRF Signal at Position 31–34 Enhances Endocytosis, *J. Biol. Chem.*, 268, 21686–21692.

Camakaris, J., Petris, M. J., Bailey, L., Shen, P. Y., Lockhart, P., Glover, T. W., Barcroft, C., Patton, J., and Mercer, J. F. (1995). Gene Amplification of the Menkes (MNK: ATP7A) P-type ATPase Gene of CHO Cells is Associated with Copper Resistance and Enhanced Copper Efflux, *Hum. Mol. Genet.*, 4, 2117–2123.

Conrad, M. E., and Umbreit, J. N. (2000). Iron Absorption and Transport—An Update, *Am. J. Hematol.*, 64, 287–298.

Conrad, M. E., Umbreit, J. N., and Moore, E. G. (1999). Iron Absorption and Transport, *Am. J. Med. Sci.*, 318, 213–229.

Cousins, R. J., and McMahon, R. J. (2000). Integrative Aspects of Zinc Transporters, *J. Nutr.*, 130, 1384S–1387S.

Cox, D. W., and Moore, S. D. (2002). Copper Transporting P-Type ATPases and Human Disease, *J. Bioenerg. Biomembr.*, 34, 333–338.

Dancis, A., Yuan, D. S., Haile, D., Askwith, C., Eide, D., Moehle, C., Kaplan, J., and Klausner, R. D. (1994). Molecular Characterization of a Copper Transport Protein in *S. cerevisiae*: An Unexpected Role for Copper In Iron Transport, *Cell*, 76, 393–402.

Davidsson, L., Lönnerdal, B., Sandström, B., Kunz, C., and Keen, C. L. (1989). Identification of Transferrin as the Major Plasma Carrier Protein for Manganese Introduced Orally or Intravenously or After *In Vitro* Addition in the Rat, *J. Nutr.*, 119, 1461–1464.

Denke, S. M., and Farburg, B. L. (1989). Regulation of Cellular Glutathione, *Am. J. Physiol.*, 257, L163–L173.

Donovan, A., Brownile, A., Zhou, Y., Shepard, J., Pratt, S. J., Moynihan, J., Paw, B. H., Drejer, A., Barut, B., Zapata, A., Law, T. C., Brugnarall, C., Lux, S. E., Pinkus, G. S., Pinkus, J. L., Kingsley, P. D., Pails, J., Fleming, M. D., Andrews, N. C., and Zon, L. I. (2000). Positional Cloning of Zebrafish Ferroportin Identifies a Conserved Vertebrate Iron Exporter, *Nature*, 403, 776–781.

Dufner-Beattie, J., Langmade, S. J., Wang, F., Eide, D., and Andrews, G. K. (2003). Structure, Function, and Regulation of a Subfamily of Mouse Zinc Transporter Genes, *J. Biol. Chem.*, 278, 50142–50150.

Eide, D. J. (1998). The Molecular Biology of Metal Ion Transport in Saccharomyces Cerevisiae, *Annu. Rev. Nutr.*, 18, 441–469.

Eide, D., Broderius, M., Fett, J., and Guerinot, M. L. (1966). A Novel Iron-Regulated Metal Transporter from Plants Identified by Functional Expression in Yeast, *Proc. Natl. Acad. Sci. U.S.A.*, 93, 5624–5628.

Eisenstein, R. S. (2000). Iron Regulatory Proteins and the Molecular Control of Mammalian Iron Metabolism, *Annu. Rev. Nutr.*, 20, 627–662.

Finch, C. (1994). Regulators of Iron Balance in Humans, *Blood*, 84, 1697–1702.

Fleming, M. D., Trenor, C. C., Su, M. A., Foernzler, D., Beier, D. R., Dietrich, W. F., and Andrews, N. C. (1997). Microcytic Anemia Mice Have a Mutation in Nramp2, a Candidate Iron Transporter Gene, *Nature Genet.*, 16, 383–386.

Fleming, R. E., Miqas, M. C., Zhou, X. Y., Jiang, J., Britton, R. S., Brunt, E. M., Tomatsu, S., Waheed, A., Bacon, B. R., and Sly, W. S. (1999). Mechanism of Increased Iron Absorption in Murine Model of Hereditary Hemochromatosis: Increased Duodenal Expression of the Iron Transporter DMT1, *Proc. Natl. Acad. Sci. U.S.A.*, 96, 3143–3148.

Franklin, R. B., Ma, J., Zou, J., Guan, Z., Kykoyi, B. I., Feng, P., and Costello, L. C. (2003). Human ZIP1 is a Major Zinc Uptake Transporter for the Accumulation of Zinc in Prostate Cells, *J. Inorg. Biochem.*, 96, 435–442.

Fuentealba, I. C., and Aburto, E. M. (2003). Animal Models of Copper-Associated Liver Disease, *Comp. Hepatol.*, 2, 5–16.

Gaither, L. A., and Eide, D. J. (2000). Functional Characterization of the Human hZIP2 Zinc Transporter, *J. Biol. Chem.*, 275, 5560–5564.

Gaither, L. A., and Eide, D. J. (2001). Eukaryotic Zinc Transporters and Their Regulation, *Biometals*, 14, 251–270.

Garrick, M. D., Nunez, M. T., Olivares, M., and Harris, E. D. (2003). Parallels and Contrasts Between Iron and Copper Metabolism, *Biometals*, 16, 1–8.

Gruenheid, S., Cannone-Hergeaux, F., Gauthier, S., Hackam, D. J., Grinstein, S., and Gros, P. (1999). The Iron Transport Protein NRAMP2 is an Integral Membrane Glycoprotein That Colocalizes with Transferrin in Recycling Endosomes, *J. Exp. Med.*, 189, 831–841.

Gunshin, H., Mackenzie, B., Berger, U. V., Gunshin, Y., Romero, M. F., Boron, W. F., Nussberger, S., Gollan, J. L., and Hediger, M. A. (1997). Cloning and Characterization of a Mammalian Proton-Coupled Metal-Ion Transporter, *Nature*, 388, 482–488.

Hamlett, N. V., Landale, E. C., Davis, B. H., and Summers, A. O. (1992). Roles of the *Tn21*, *merT*, *merP*, and *merC* Gene Products in Mercury Resistance and Mercury Binding, *J. Bacteriol.*, 174, 6377–6485.

Hantke, K. (2001). Bacterial Zinc Transporters and Regulators, *Biometals*, 14, 239–249.

Harris, E. D. (2000). Cellular Copper Transport and Metabolism, *Annu. Rev. Nutr.*, 20, 291–310.

Harris, Z. L., Klomb, L. W., and Gitlin, J. D. (1998). Aceruloplasminemia: An Inherited Neurodegenerative Disease with Impairment of Iron Homeostasis, *Am. J. Clin. Nutr.*, 96, 10812–10817.

Hilton, M., Spenser, D. C., Ross, P., Ramsey, A., and McArdle. H. J. (1995). Characterisation of the Copper Uptake Mechanism and Isolation of Ceruloplasmin Receptor/Copper Transporter in Human Placental Vesicles, *Biochim. Biophys. Acta*, 1245, 153–160.

Holmberg, C. G., and Laurell, C. B. (1948). Investigations in Serum Copper. II. Isolation of the Copper-Containing Protein and a Description of Some of Its Properties, *Acta Chem. Scand.*, 2, 550–556.

Iacopetta, B. J., Rothenberger, S., and Kuhn, L. C. (1988). A Role for the Cytoplasmic Domain in Transferrin Receptor Sorting and Coated Pit Formation During Endocytosis, *Cell*, 54, 485–489.

Kambe, T., Yamaguchi-Iwai, Y., Sasaki, R., and Nagao, M. (2004). Overview of Mammalian Zinc Transporters, *Cell. Mol. Life Sci.*, 61, 49–68.

Kawabata, H., Yang, R., Hirama, T., Vuong, P. T., Kawano, S., Gombart, A. F., and Koeffler, H. P. (1999). Molecular Cloning of Transferrin Receptor 2. A New Member of the Transferrin Receptor-Like Family, *J. Biol. Chem.*, 274, 20826–20832.

Knight, S. A. B., Labbé, S., Kwon, L. F., Kosman, D. J., and Thiele, D. J. (1996). A Widespread Transposable Element Marks Expression of a Yeast Copper Transport Gene, *Genes Dev.*, 10, 1917–1929.

Kodama, H. (1993). Recent Developments in Menkes' Disease, *J. Inherited Metab. Dis.*, 16, 791–799.

Kohgo, Y., Torimoto, Y., and Kato, J. (2002). Transferrin Receptor in Tissue and Serum: Updated Clinical Significance of Soluble Receptor, *Int. J. Hematol*, 76, 213–218.

Korshunova, Y. O., Eide, D., Clark, W. G., Guerinot, M. L., and Pakrasi, H. B. (1999). The IRT1 Protein from *Arabidopsis thaliana* Is a Metal Transporter with a Broad Substrate Range, *Plant Mol. Biol.*, 40, 37–44.

Kuo, Y. M., Gitschier, J., and Packman, S. (1997). Developmental Expression of the Mouse Mottled and Toxic Milk Genes Suggests Distinct Functions for the Menkes and Wilson Disease Copper Transporters, *Hum. Mol. Genet.*, 6, 1043–1049.

Lasat, M. M., Pence, N. S., Garvin, D. F., Ebbs, S. D., and Kochian, L. V. (2000). Molecular Physiology of Zinc Transport in the Zn Hyperaccumulator *Thlapsi caerulescens*, *J. Exp. Biol.*, 51, 71–79.

Lee, P. L., Gelbart, T., West, C., Halloran, C., and Beutler, E. (1998). The Human Nramp2 Gene Characterization of the Gene Structure, Alternative Splicing, Promoter Region and Polymorphisms, *Blood Cells Mol. Dis.*, 24, 199–215.

Lieu, P. T., Heiskala, M., Peterson, P. A., and Yang ,Y. (2001). The Roles of Iron in Health and Disease, *Mol. Aspects Med.*, 22, 1–87.

Lin, S. J., Pufahl, R. A., Dancis, A., O'Halloran, T. V., and Culotta, V. C. (1997). A Role for the *Saccharomyces cerevisiae*

ATX1 Gene in Copper Trafficking and Iron Transport, *J. Biol. Chem.*, 272, 9215–9220.

Lioumi, M., Ferguson, C. A., Sharpte, P. T., Freeman, T., Marenholz, I., Mischke, D., Heizmann, C., and Ragoussis, J. (1999). Isolation and Characterization of Human and Mouse ZIRTL, a Member of the IRT1 Family of Transporters, Mapping Within the Epidermal Differentiation Complex, *Genomics*, 62, 272–280.

Majuri, R., and Grasbeck, R. (1987). A Rosette Receptor Assay with Haem-Microbeads. Demonstration of a Haem Receptor on K562 Cells, *Eur. J. Haematol*, 38, 21–25.

Mantzoros, C. S., Prasad, A. S., Beck, F. W. J., Grabowski, S., Kaplan, J., Adair, C., and Brewer, G. J. (1998). Zinc May Regulate Serum Leptin Concentrations in Humans, *J. Am. Coll. Nutr.*, 17, 270–275.

Martins, L. J., Jensen, T. L., Simons, J. R., Keller, G. L., and Winge, D. R. (1998). Metalloregulation of *FRE1* and *FRE2* Homologs in Saccharomyces Cerevisiae, *J. Biol. Chem.*, 273, 23716–23721.

Masuoka, J., and Saltman, P. (1994). Zinc(II) and Copper(II) Binding to Serum Albumin, *J. Biol. Chem.*, 269, 25557–25561.

McClelland, A., Kuehn, L. C., and Ruddle, F. H. (1984). The Human Transferrin Receptor Gene: Genomic Organization, and the Complete Primary Structure of the Receptor Deduced from a cDNA Sequence, *Cell*, 39, 267–274.

McKie, A. T., Marciani, P., Rolfs, A., Brennan, K., Wehr, K., Barrow, D., Miret, S., Bomford, A., Peters, T. J., Farzaneh, F., Hediger, M. A., Hentze, M. W., and Simpson, R. J. (2000). A Novel Duodenal Iron-Regulated Transporter, IREG1, Implicated in the Basolateral Transfer of Iron to the Circulation, *Mol. Cell*, 5, 299–309.

Meguro, Y., Kodama, H., Abe, T., Kobayashi, S., Kodama, Y., and Nishimura, M. (1991). Changes of Copper Level and Cytochrome *c* Oxidase Activity in the Macular Mouse with Age, *Brain Dev.*, 13, 184–186.

Mercer, J. F. B. (2001). The Molecular Basis of Copper-Transport Diseases, *Trends Mol. Med.*, 7, 64–69.

Moos, T., Trinder, D., and Morgan, E. H. (2000). Cellular Distribution of Ferric Iron, Ferritin, Transferrin and Divalent Metal Transporter 1 DMT1 in Substantia Nigra and Basal Ganglia of Normal and Beta 2-Microglobulin Deficient Mouse Brain, *Cell. Mol. Biol.*, 46, 549–561.

Nies, D. H., and Silver, S. (1995). Ion Efflux Systems Involved in Bacterial Metal Resistance, *J. Ind. Microbiol.*, 14, 186–199.

O'Halloran, T. V., and Culotta, C. V. (2000). Metallochaperones: An Intracellular Shuttle Service for Metal Ions, *J. Biol. Chem.*, 275, 25057–25060.

Palmiter, R. D., Cole, T. B., and Findley, S. D. (1996). ZnT-2, a Mammalian Protein that Confers Resistance to Zinc by Facilitating Vesicular Sequestration, *EMBO J.*, 15, 1784–1791.

Pietrangelo, A., Rocchi, E., Casalgrandi, G., Rigo, G., Ferrari, A., Perini, M., Ventura, E., and Cairo, G. (1992). Regulation of Transferring, Transferrin Receptor, and Ferritin Genes in Human Duodenum, *Gastroenterology*, 102, 802–809.

Raskin, I. (1995). Plant Genetic Engineering May Help with Environmental Cleanup, *Proc. Natl. Acad. Sci. U.S.A.*, 93, 3164–3166.

Rolfs, A., and Hediger, M. A. (1999). Metal Ion Transporters in Mammals: Structure, Function and Pathological Implications, *J. Physiol.*, 518, 1–12.

Rothenberg, B. E., and Voland, J. R. (1996). Beta2 Knockout Mice Develop Parenchymal Iron Overload: A Putative Role for Class I Genes of the Major Histocompatibility Complex in Iron Metabolism, *Proc. Natl. Acad. Sci. U.S.A.*, 93, 1529–1534.

Rutledge, E. A., and Enns, C. A. (1996). Cleavage of the Transferrin Receptor Is Influenced by the Composition of the O-Linked Carbohydrate at Position 104, *J. Cell. Physiol.*, 168, 284–293.

Sacher, A., Cohen, A., and Nelson, N. (2001). Properties of the Mammalian and Yeast Metal-Ion Transporters DCT1 and Smf1p Expressed in *Xenopus laevis* Oocytes, *J. Exp. Biol.*, 204, 1053–1061.

Solioz, M., Odermatt, A., and Krapf, R. (1994). Copper Pumping ATPases: Common Concepts in Bacteria and Man, *Fed. Eur. Biochem. Soc. Lett.*, 346, 44–47.

Suhy, D., and O'Halloran, T. V. (1995). Metal Responsive Gene Regulation and the Zinc Metalloregulatory Model. In *Metal Ions in Biological Systems* (H. Siegel, Ed.), Marcel Dekker, New York, Vol. 32, pp. 557–558.

Suzuki, M., and Gitlin, J. D. (1999). Intracellular Localisation of the Menkes' and Wilson's Disease Proteins and Their Role in Intracellular Copper Transport, *Pediatr. Int.*, 41, 436–442.

Tacchini, L., Bianchi, L., Bernelli-Zazzera, A., and Cairo, G. (1999). Transferrin Receptor Induction by Hypoxia: HIF-1-Mediated Transcriptional Activation and Cell-Specific Post-Transcriptional Regulation, *J. Biol. Chem.*, 274, 24142–24146.

Takeda, A., Devenyi, A., and Connor, J. R. (1998). Evidence for Non-Transferrin-Mediated Uptake and Release of Iron and Manganese in Glial Cell Cultures from Hypotransferrinemic Mice, *J. Neurosci. Res.*, 51, 454–462.

Tandy, S., Williams, M., Leggett, A., Lopez-Jimenez, M., Dedes, M., Ramesh, B., Srai, S. K., and Sharp, P. (2000). Nramp2 Expression Is Associated with pH-Dependent Iron Uptake by the Brain: Effects of Altered Iron Status, *J. Biol. Chem.*, 275, 1023–1029.

Tapiero, H., Townsend, D. M., and Tew, K. D. (2003). Trace Elements in Human Physiology and Pathology: Copper. *Biomed. Pharmacother.*, 57, 386–398.

Testa, U., Petrini, M., Quaranta, M. T., Pelosi-Testa, E., Mastroberardino, G., Camagna, A., Boccoli, G., Sargiacomo, M., Isacchi, G., Cozzi, A., Arosio, P., and Peschle, C. (1989). Iron Up-Modulates the Expression of TfRs During Monocyte–Macrophage Maturation, *J. Biol. Chem.*, 264, 13181–13197.

Tong, K. K., and McArdle, H. J. (1995). Copper Uptake by Cultured Trophoblast Cells Isolated from Human Term Placenta, *Biochim. Biophys. Acta*, 1269, 233–236.

Tsuji, Y., Moran, E., Torti, S. V., and Torti, F. M. (1999). Transcriptional Regulation of the Mouse Ferritin H Gene: Involvement of p300/CBP Adaptor Proteins in FER-1 Enhancer Activity, *J. Biol. Chem.*, 274, 7501–7507.

Umbreit, J. N., Conrad, M. E., Moore, E. G., and Latour, L. F. (1998). Iron Absorption and Cellular Transport: The Mobilferrin/Paraferritin Paradigm, *Sem. Hematol.*, 35, 13–26.

Vulpe, C. D., Kuo, Y.-M., Murphy, T. L., Cowley, L., Askwith, C., Libina, N., Gischier, J., and Anderson, G. J. (1999). Hephaestin, a Ceruloplasmin Homolog Implicated in Intestinal Iron Transport, Is Defective in the sla Mouse, *Nature, Genet.*, 21, 195–199.

Wood, R. J., and Han, O. (1998). Recently Identified Molecular Aspects of Intestinal Iron Absorption, *J. Nutr.*, 128, 1764S–1765S.

Zhang, P., and Allen, J. C. (1995). A Novel Dialysis Procedure Measuring Free Zn^{2+} in Bovine Milk and Plasma, *J. Nutr.*, 125, 1904–1910.

Zhao, H., and Eide, D. (1996). The Yeast ZRT1 Gene Encodes the Zinc Transporter of a High Affinity Uptake System Induced by Zinc Limitation, *Proc. Natl. Acad. Sci. U.S.A.*, 93, 2454–2458.

BIOLOGICAL FUNCTIONS OF THE ELEMENTS

ULF LINDH
Uppsala University

CONTENTS

I. ESSENTIALITY OF ELEMENTS

Much discussion has taken place with regard to how to define the essentiality of elements, particularly trace elements. The earliest definition was actually borrowed from protein chemistry. In this definition, an element is essential if:

- It is present in living tissues at a relatively constant concentration.
- It provokes similar structural and physiological anomalies in several species when removed from the organism.

- These anomalies are prevented or cured by supplementation of the element.

The current definition, suggested by an expert consultation of the World Health Organization/Food and Agricultural Organization/International Atomic Energy Agency (Mertz, 1998), states:

An element is considered essential to an organism when reduction of its exposure below a certain limit results consistently in a reduction in a physiologically important function, or when the element is an integral part of an organic structure performing a vital function in the organism.

The concept of essentiality has the practical consequence that it is necessary to supply an organism with adequate amounts of the concerned elements. An immediate question raised by this consequence is how much is adequate. For most elements, ranges of safe and adequate intakes have been defined. In some cases, however, there is considerable uncertainty regarding the limits of such ranges. Adequate intakes do vary substantially among elements, in both amount and width of the range. In very general terms, the range may be visualized as shown in Figure 1. For a detailed discussion of deficiencies and toxicities, the reader is referred to Chapter 8.

TABLE I. Abundance by Mass of Major and Minor Elements in the Human Body

Element	Mass Percent	Element	Mass Percent
Oxygen	65.0	Magnesium	0.50
Carbon	18.0	Potassium	0.34
Hydrogen	10.0	Sulfur	0.26
Nitrogen	3.0	Sodium	0.14
Calcium	1.4	Chlorine	0.14
Phosphorus	1.0	—	—

TABLE II. Abundance of Certain Trace Elements in the Human Body by Mass ($\mu g\ g^{-1}$)

Element	Mass Fraction	Element	Mass Fraction
Arsenic	0.26	Manganese	0.17
Bromine	2.9	Molybdenum	0.08[a]
Cobalt	0.021	Nickel	0.14
Chromium	0.094	Selenium	0.11
Copper	1	Silicon	260
Fluorine	37	Tin	0.24
Iron	60	Tungsten	0.008[a]
Iodine	0.19	Vanadium	0.11[a]
Lithium	0.009*	Zinc	33

[a]Estimated from Iyengar et al., (1978) and Li (2000).

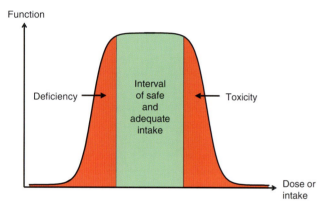

FIGURE 1 Dose–response of essential trace elements.

II. Major, Minor, and Trace Elements in Biology

Eleven elements seem to be consistently abundant in biological systems: hydrogen, oxygen, carbon, nitrogen, sodium, potassium, calcium, magnesium, phosphorus, sulfur, and chlorine. In humans, these elements comprise 99.9% of the atoms. Usually, these elements are divided into two groups of major and minor elements. The major elements—hydrogen, oxygen, carbon, and nitrogen—make up 99% of the atoms, or just over 96% of the body mass. Sodium, potassium, calcium, magnesium, sulfur, and chlorine comprise 3.78% of the body mass (Table I).

The entire group of noble gases is excluded from consideration because their chemical properties make them unlikely to fulfill any biological function. The remaining elements are considered to be trace elements. There are 90 naturally occurring elements in the periodic table; thus, 73 are trace elements. Of these 73, 18 are considered to be essential or possibly essential trace elements: lithium, vanadium, chromium, manganese, iron, cobalt, nickel, copper, zinc, tungsten, molybdenum silicon, selenium, fluorine, iodine, arsenic, bromine, and tin. Their abundance in the human body is reflected in Table II.

Several problems are associated with proving the essentiality of trace elements. Experiments are based upon the general acceptance that, if an essential trace element is completely withdrawn from the diet of experimental animals, signs and symptoms of a deficiency should occur, such as growth retardation and loss of hair. When a state of deficiency has been established, supplementation of the trace element should alleviate these symptoms and reverse the deficiency state. The first basic problem is that it is not possible to completely eliminate every bit of an element in food. Even if this were possible, the analytical techniques are inadequate due to their limits of detection. A second problem is that when essentiality is being evaluated, there is usually no well-grounded hypothesis for a possible biological function. Withdrawal of one essential trace element from the diet may result in altered uptake patterns for other trace elements, which makes results ambiguous. Most results have been obtained on plants and rodents. Veterinary medicine has contributed with information on the essentiality of elements in domestic animals. When it comes to humans, however, our knowledge of essential trace elements is less advanced for obvious reasons.

The essentiality of 12 of the trace elements in Table II is generally agreed upon, although perhaps not for all biological species. Without exhausting the available data on vanadium biology, a conspicuous property of ascidians must be pointed out. The blood cells of some ascidians accumulate vanadium at a degree of 10^7 as compared to the sea concentration. *Ascidia gemmata* has been shown to have the highest vanadium concentration—350 mM—which corresponds to about 1.7% (Michibata *et al.*, 2002).

III. BRIEF DESCRIPTION OF THE FUNCTION OF MAJOR ELEMENTS

Because the human body is roughly 71% water, it is not surprising that the quantities of hydrogen and oxygen are so substantial. Water makes up more than two-thirds of the weight of the human body, and without water a human would die in just a few days. All the cell and organ functions depend on water for functioning. It serves as a lubricant and forms the base for saliva and the fluids that surround the joints. Water regulates body temperature, as cooling and heating occur through perspiration. Water helps to alleviate constipation by moving food through the intestinal tract, thereby eliminating waste. Water also contributes to the high contents of oxygen in the body.

A. A Few Important Points About Hydrogen

Hydrogen is a very special element in biology. It appears in three states, H^+ (cation), H— (a covalently bound state), and H^- (anion). It is a very strong acid as a proton (H^+). In the H— form, it takes part in stable non-metal bonds such as C—H and N—H. Even in the presence of dioxygen, these bonds are kinetically stable. Hydrogen can also be transferred from a non-metal, not just as H^+ or H— but also as H^- (Fraústo da Silva & Williams, 2001). This makes it possible for hydrogen atoms to take part in one- or two-electron processes. Many biological redox reactions are based upon this property.

B. Carbon: The Backbone of Organic Chemistry and Biochemistry

In the chemical sense, use of the term "organic" means that carbon is involved; thus, an organic compound contains carbon atoms, and organic chemistry is the chemistry of carbon compounds. There are a few exceptions, however, as oxides, carbonates, and cyanides are considered inorganic compounds. Biochemistry, the chemistry of life, is a special branch of organic chemistry. Slightly contradictory is the fact that biochemistry involves both organic and inorganic compounds. To be more precise, it is appreciated in this field of study that the large molecules found in cells all contain carbon but many of the small molecules may be inorganic. A carbon atom is capable of combining with up to four other atoms, but in some cases a carbon atom combines with fewer than that. These bonds are covalent. Not only can a carbon atom form covalent bonds with four other atoms, but it can also combine with other carbon atoms; thus, carbon atoms can form chains and rings onto which other atoms can be attached. The atomic number of carbon is six; it has two electrons in the K shell and four in the L shell. Carbon must gain or lose four electrons to be ionized, but this process is difficult so instead it shares electrons to fill its L shell. In summary, carbon turns out to be a very versatile atom. The organic compounds in biology are represented by carbohydrates, lipids, proteins, and nucleic acids.

1. Carbohydrates

Carbon, hydrogen, and oxygen comprise most of the carbohydrates. For each carbon and oxygen there are two hydrogen atoms (*i.e.*, CH_2O). Carbohydrates perform a series of important functions in biology, such as short-term energy storage (*e.g.*, monosaccharides), long-term energy storage (*e.g.*, starches and glycogen), and structural support (*e.g.*, cellulose found in all plant cell walls), as well as important components of DNA and RNA.

2. Lipids

Lipid molecules are insoluble in polar solvents such as water. They dissolve in nonpolar solvents, and they are nonpolar. Lipids work as energy storage molecules, as insulation and protection for internal organs, as lubricants, and as hormones. Additionally, phospholipids are the major structural elements of membranes that are composed of a bilayer of phospholipids.

3. Proteins

Amino acids can be combined to form peptides, in which case the order of the amino acids is significant. The amino acids consist of an amino (–NH) group and a carboxylic acid (–CO_2H) group bonded to a central carbon atom. When the combination of amino acids exceeds more than about ten amino acids, the resulting peptide is referred to as a *polypeptide*. If the number of amino acids in a combination is more than 50, the molecule is referred to as a *protein*. The most important function of proteins is to maintain and drive the reactions of cells; however, additional properties include acting in supportive tissue like cartilage, and they are involved in muscle movement. Many proteins are enzymes, and their association with trace elements is described later in this chapter.

4. Nucleic Acids

Ribose, which is a monosaccharide, interacts with nitrogenous bases to form nucleosides such as adenosine. When nucleosides exist in cells mainly in the form of esters with phosphoric acid they are called *nucleotides*. If the sugar is ribose, the molecule is called *ribonucleotide*. The two nucleic acids, DNA and RNA, are polymers of nucleotides. In DNA, the bases are adenine, cytosine, guanine, and thymine; in RNA they are the same, except that uracil takes the place of thymine. The nucleotides in both DNA and RNA are joined by covalent bonds between the phosphate of one nucleotide and the sugar of the next one. A chain of many nucleotides is a polymer forming a nucleic acid.

C. Oxygen: The Savior and Reactionist

Without oxygen, humans and other mammals would not survive. Structures with a high degree of organization require specific transport mechanisms, and it is well known that in mammals oxygen is transported in the blood by hemoglobin; however, other oxygen-transporting molecules, such as hemerythrin, are used by some marine invertebrates, and hemocyanin is found, for example, in snails. In muscles, the diffusion of oxygen is facilitated by myoglobin. Oxygen is intimately involved in the production of energy-rich molecules (adenosine triphosphate, or ATP) which takes place in the mitochondrial membrane. The reduction of dioxygen to water requires the transfer of four electrons onto dioxygen at the same time:

$$O_2 + 4e^- + 4H^+ \rightarrow 2H_2O$$

Such a process, however, is not chemically possible; instead, on the way to becoming water, dioxygen passes through stages of aggressive power: the superoxide anion ($O_2^{-\bullet}$), hydrogen peroxide (H_2O_2), and the hydroxyl radical ($OH^\bullet$). All three intermediates are

extremely reactive. Even hydrogen peroxide is very reactive, although it is not a radical. Electron- and proton-transfer reactions are extremely rapid, so the reactive intermediates are generally kept in the enzyme. In summary, the reduction of dioxygen to water involves four one-electron transfers:

$$O_2 \rightarrow O_2^{-\bullet} \rightarrow H_2O_2 \rightarrow OH^\bullet \rightarrow H_2O$$

Nature has evolved protective functions to take care of leaking intermediates. The superoxide anion is metabolized by superoxide dismutase to hydrogen peroxide, which is metabolized to water by catalase or glutathione peroxidase without releasing the extremely reactive hydroxyl radical. This balance, however, is delicate, and attacks from metals may result in the release of radicals. The plethora of reactions and functions that could be described for oxygen are too voluminous to present in this chapter, and the reader is referred to standard biochemistry textbooks.

D. Nitrogen Fixation

The supply of nitrogen to biological systems comes from gaseous N_2, which is very abundant in the atmosphere (80%). The bond between the nitrogen atoms is very strong; hence dinitrogen is chemically unreactive. Similar to dioxygen, dinitrogen has to be reduced in order to be biologically available. Reduction to ammonia requires a very specific and sophisticated enzyme in a process known as *nitrogen fixation*.

In the biosphere, most nitrogen fixation is carried out by a few species of bacteria that synthesize the enzyme nitrogenase. This enzyme is found in the species *Rhizobium*, living in symbiosis with root nodules of many leguminous plants, such as beans, peas, alfalfa, and clover. Nitrogen fixation also takes place in free-living soil bacteria such as *Azobacter*, *Klebsiella*, and *Clostridium* and by cyanobacteria found in aquatic environments.

Two proteins comprise nitrogenase. One of them contains a [4Fe–4S] cluster, and the other has two oxidation–reduction centers. Iron is involved in one of these centers, molybdenum in the other. The net reaction is

$$N_2 + 8H^+ + 8e^- + 16ATP \rightarrow$$
$$2NH_3 + H_2 + 16ADP + 16P_i$$

To obtain the reducing power and ATP required for this process, symbiotic nitrogen-fixing microorganisms rely on nutrients obtained through photosynthesis carried out by the plant with which they are associated.

IV. BRIEF DESCRIPTION OF THE FUNCTIONS OF MINOR ELEMENTS

The minor elements in biology are comprised of sodium, magnesium, phosphorus, sulfur, chlorine, potassium, and calcium. The biological functions of the minor elements are discussed in three groups: (1) sodium, potassium, and chlorine; (2) magnesium and phosphorus; and (3) calcium.

A. Sodium, Potassium, and Chlorine: Interactions and Ion Properties

Life evolved from water. Do concentrations of abundant elements in sea- or freshwater reflect this evolution? Table III makes a crude comparison between seawater and blood serum. It shows that intracellular concentrations are in most cases much lower than extracellular, the exception being potassium, which is an abundant intracellular ion. To function properly, organisms must actively pump sodium and chloride out of cells and actively take in potassium. It is not quite clear why life must reject the most abundant anion (Cl^-) and the most abundant cation (Na^+), although this is probably due to a need for the cellular stability that comes from maintaining an osmotic balance. Maintaining the intracellular and extracellular balance requires active transport processes that require considerable energy. For details regarding uptake and transport processes, the reader is referred to Chapter 5.

TABLE III. Concentrations of Free Cations and Anions of Calcium, Magnesium, Potassium, and Sodium in Seawater, Human Blood Serum, and Human Red Cells (mmol/L)

Ion	Seawater	Serum	Red Cells
Calcium	10.25	2.20–2.55	10^{-4}
Magnesium	53.60	0.76–1.10	2.5
Potassium	9.96	3.5–5.1	92
Sodium	471	136–146	11
Chloride	549	98–106	50
Bicarbonate	—	22–29	10?
Phosphate	2×10^{-3}	0.74–3.07	10?

1. Biological Functions of Sodium, Potassium, and Chloride

Circuits maintained by K^+, Na^+, and Cl^- generally control the following properties in all cells of all organisms:

- Osmotic pressure
- Membrane potentials
- Condensation of polyelectrolytes
- Required ionic strength for activity

The first two properties are quite easy to understand; however, condensation of polyelectrolytes may have to be clarified. Biopolymers are polyelectrolytes; DNA, for example, possesses a linear series of charges. Other polyelectrolytes such as fats show two-dimensional arrays, whereas proteins have curved surfaces not seldom almost spherical. The surfaces of polyelectrolytes are stabilized by the surrounding ionic environment mainly due to sodium, potassium, and chloride ions.

Absorption of glucose, or any molecule for that matter, entails transport from the intestinal lumen, across the epithelium and into blood. The transporter that carries glucose and galactose into the enterocyte is the sodium-dependent hexose transporter, known more formally as SGLUT-1. As the name indicates, this molecule transports both glucose and sodium into the cell and, in fact, will not transport either alone.

The essence of transport by the sodium-dependent hexose transporter involves a series of conformational changes caused by the binding and release of sodium and glucose. It can be summarized as follows:

- The transporter is initially oriented facing into the lumen, and at this point it is capable of binding sodium, but not glucose.
- Sodium binds, which induces a conformational change that opens the glucose-binding pocket.
- Glucose binds, and the transporter reorients in the membrane such that the pockets holding sodium and glucose are moved inside the cell.
- Sodium dissociates into the cytoplasm, which causes glucose binding to destabilize.
- Glucose dissociates into the cytoplasm, and the unloaded transporter reorients back to its original, outward-facing position.

Sodium is intimately involved in vitamin transport. For example, the uptake of biotin, a member of the vitamin B complex, is dependent on the sodium-dependent multivitamin transporter, or SMVT (Stanley *et al.*, 2002). There are also indications that the transport in brain parenchyma of *N*-acetylaspartate (the second most abundant amino acid in the adult brain) is accomplished by a novel type of sodium-dependent carrier that is present only in glial cells (Sager *et al.*, 1999).

Sodium is also closely connected to the transport of vitamin C. It has been shown that the sodium-dependent vitamin C transporter (SVCT) is responsible for an age-dependent decline of ascorbic acid contents in tissues (Michels *et al.*, 2003). The transport of vitamin C into the brain relies heavily on the sodium-dependent ascorbic-acid transporter Slc23a1 (Sotiriou *et al.*, 2002). A conspicuous finding of Handy *et al.* (2002) was that copper uptake across epithelia seems to be sodium dependent.

Experiments have shown the presence of an ouabain-insensitive, potassium-dependent *p*-nitrophenylphosphatase in rat atrial myocytes. This enzyme is suggested to be an isoform of an H-transporting, potassium-dependent adenosine triphosphatase (Zinchuk *et al.*, 1997). The enzyme pyruvate kinase requires potassium for maximal activity (Kayne, 1971); however, it was recently shown that in the absence of potassium dimethylsulfoxide induces active conformation of the enzyme (Ramirez-Silva *et al.*, 2001).

There is an interesting coupling of sodium and chloride in the transport of the neurotransmitter gamma-aminobutyric acid (GABA). The removal of GABA from the extracellular space is performed by sodium- and chloride-dependent high-affinity plasma membrane transporters (Fletcher *et al.*, 2002). It has been suggested that chloride is involved in the regulation of proteolysis in the lysosome through cathepsin C. This enzyme is a tetrameric lysosomal dipeptidyl–peptide hydrolase that is activated by chloride ion (Cigic & Pain, 1999). Alpha-Amylases have also been shown to be chloride dependent (D'Amico *et al.*, 2000).

B. Magnesium and Phosphate: Close Connections

In contrast to many metal and non-metal ions, magnesium is rather homogeneously distributed in organisms with about the same intra- and extracellular concentration of 10^{-3} *M*. It is also unique among the biological cations due to its size, charge density, and structure in aqueous solution, as well as its aqueous chemistry. These properties make magnesium generally different from all other cations, monovalent or divalent. Mg^{2+} has the largest hydrated radius of any common cation; in contrast, its ionic radius (*i.e.*, minus waters of hydration)

is among the smallest of any divalent cation (Kehres & Maguire, 2002). Magnesium is an essential element, a fact that was first demonstrated 77 years ago (Leroy, 1926); however, its general role in cellular function is poorly understood.

Although there does not seem to be any significant chemical gradient, the net electrochemical gradient for Mg^{2+} is markedly directed inward because of the negative membrane potential inside; consequently, mechanisms must be in place for maintaining a low intracellular free Mg^{2+} concentration and to regulate Mg^{2+} homeostasis. At least two transport processes have been discovered; one is sodium dependent and the other works even in the absence of sodium. The sodium-dependent exchanger operates with a $3Na^+_{in} : 1Mg^{2+}_{out}$ ratio (Romani & Maguire, 2002). Various monovalent and divalent cations may replace the Na^+ when sodium is absent. In that case, the stoichiometry is $1:1$ (Romani & Maguire, 2002). The physiological significance of this mechanism is, however, unclear as low extracellular concentrations of sodium are unlikely.

1. Hormonal Regulation of Magnesium Homeostasis

In physiological terms, it is reasonable to envision that, because hormones stimulate Mg^{2+} extrusion, other hormones or agents must also operate in eukaryotic organisms to promote Mg^{2+} accumulation and the maintenance of Mg^{2+} homeostasis (Romani & Maguire, 2002). The mechanism behind hormonal regulation seems to be hormones or agents acting on the production of cyclic AMP. Another possibility is that activating a protein kinase C pathway decreases cyclic AMP in many tissues. An exiting finding was identifying the protein family Mrs2p in mitochondria; this is the first molecularly identified metal ion channel protein in the inner mitochondrial membrane.

2. Magnesium Binding and Magnesium Enzymes: Magnesium and Phosphates

The best known example of the very strong binding of magnesium ions is the Mg^{2+} in chlorophyll. A few other cases are ATP-synthetases in thylakoids and mitochondria and in the ATPases of muscles. Nucleoside diphosphates and triphosphate, both in aqueous solution and at the active site of enzymes, usually are present as complexes with magnesium (or sometimes manganese) ions. These cations coordinate with oxygen atoms of the phosphate groups and form six-membered rings with ADP or ATP (Figure 2). A magnesium ion can form several different complexes with ATP. In solution, formation of the β,γ complex is favored.

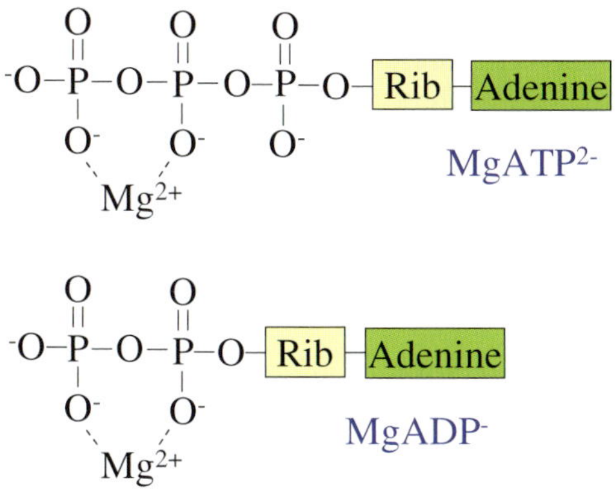

FIGURE 2 Mg^{2+} complexes with ADP and ATP.

TABLE IV. Some Examples of Magnesium-Dependent Enzymes and Their Functions

Enzymes	Function
Kinases	G-transfer reactions
Adenylate cyclase	cAMP formation from ATP
ATPases	Hydrolysis of ATP
Alkaline phosphatase	Splitting off phosphorus
Isocitrate lyase	Formation of succinate and glyoxylate in the citric acid cycle
Methyl aspartase	Glutamate receptor
Ribulose bisphosphate carboxylase	Carboxylation and oxygenation of ribulose bisphosphate
Myosin ATPase	Hydrolysis of ATP in muscles
Nucleases	Hydrolysis of phosphodiesters in nucleic acids
GTP-dependent enzymes	Restriction enzymes cleaving DNA, for example

Glycolysis is one of two important pathways in carbohydrate metabolism. The other is the pentose phosphate pathway. Both pathways provide energy to cells and both are also involved in the formation and degradation of other molecules such as amino acids and lipids. The glycolysis pathway has ten enzyme-catalyzed steps. In one of them, 2-phosphoglycerate is dehydrated to phosphoenolpyruvate in a reaction catalyzed by enolase. This enzyme requires Mg^{2+} for activity. Two magnesium ions participate in this reaction: A "conformational" ion binds to the hydroxyl group of the substrate and a "catalytic" ion participates in the dehydration reaction. Some other magnesium-dependent enzymes of the general metabolism are shown in Table IV together with their biological functions.

3. Magnesium and Nucleic Acid Biochemistry

Monovalent metal ions such as Na^+ and K^+ act more or less as bulk electrolytes to stabilize surface charge. Divalent magnesium, on the other hand, interacts with nucleic acids with higher affinity. The role of Mg^{2+} is to neutralize negative charges from phosphates, either electrostatically or by forming hydrogen bonding networks from waters of solvation. Mg^{2+} may lower the pK_a of coordinated water, thereby facilitating phosphate ester hydrolysis. It is becoming increasingly clear that many drug molecules interact with DNA in a specific and Mg^{2+}-dependent manner. An understanding of the latter may aid in the design of other novel DNA binding drugs (Sreedhara & Cowan, 2002).

In addition to the enzymes noted in Table IV, it is worth mentioning a newly discovered class of enzymes requiring magnesium. Several categories of RNA have been found to catalyze reactions autonomously without protein or with only secondary protein assistance. Ribozymes are RNA molecules adopting three-dimensional structures that allow them to catalyze a variety of chemically important reactions, including but not restricted to phosphate ester hydrolysis, amide bond formation, and ligation (Sreedhara & Cowan, 2002). Divalent magnesium seems to be involved in both the structure of RNA and the catalytic mechanism of ribozymes. Other metals such as manganese and calcium might take part in this structural and catalytic mechanism. Due to the intracellular abundance of magnesium, this would be the preferred metal ion. Cleavage of RNA is yet another exciting Mg^{2+}-dependent process of ribozymes.

4. Magnesium and Photosynthesis

As living organisms became abundant on the primitive Earth, their consumption of organic nutrients produced by geochemical processes outpaced production. Developing alternative sources of organic molecules that provided energy and the raw materials required for biosynthetic processes became critical for survival. The abundant CO_2 in the Earth's early atmosphere was a natural carbon source for organic synthesis; hence, photosynthesis became the pragmatic solution to the problem. Organisms capable of photosynthesis include certain bacteria, cyanobacteria (blue-green algae), algae, nonvascular plants, and vascular (higher) plants. Photosynthesis is also the source of the Earth's molecular oxygen. With the exception of anaerobic bacteria, all organisms capable of photosynthesis give off O_2 as an end product.

The net reaction of photosynthesis is:

$$CO_2 + H_2O \xrightarrow{\text{light}} (CH_2O) + O_2$$

where (CH_2O) represents carbohydrate. The oxidation of water, a thermodynamically unfavorable reaction, is driven by solar energy. Electrons from this oxidation pass through electron-transport systems that resemble the mitochondrial electron-transport chain. Photosynthesis encompasses two major processes that can be described by two partial reactions:

$$H_2O + ADP + P_i + NADP^+ \xrightarrow{\text{light}}$$
$$O_2 + ATP + NADPH + H^+$$

$$CO_2 + ATP + NADPH + H^+ \rightarrow$$
$$(CH_2O) + ADP + P_i + NADP^+$$

$$\text{Sum:} \ CO_2 + H_2O \xrightarrow{\text{light}} (CH_2O) + O_2$$

Photosynthesis takes place in chloroplasts. Chlorophyll is the most abundant pigment involved in the light harvesting, and the tetrapyrrole ring of chlorophylls, called *chlorine*, is similar to heme but contains Mg^{2+} chelated to the nitrogen atoms of the ring (Figure 3). Chlorophylls a and b are the most abundant types. Magnesium ions are inserted into the chlorin ring through the action of a chelatase enzyme. Without the enzyme, magnesium could not combine with chlorins. Other metal ions, such as Zn^{2+}, Cu^{2+}, and Fe^{2+}, are more likely to combine with chlorins.

Mg^{2+} has an additional function in the chloroplast. When light shines on the thylakoids, pH drops to about 4.0. The protons force Mg^{2+} out of the thylakoid, and some chloride enters. The consequence is increased levels of Mg^{2+} in the stroma, where magnesium activates the carboxylase rubisco for incorporation of CO_2 into ribulose bisphosphate. This reaction is a step in the Calvin cycle of photosynthesis.

C. Calcium: Messenger and Support

Calcium has a plethora of functions in biology, and new roles are still being discovered. Calcium triggers new life at fertilization. It controls several developmental processes, and when cells have differentiated it functions to control such diverse cellular processes as metabolism, proliferation, secretion, contraction, learning, and memory (Jaiswal, 2001). The best known function of calcium is its being an integral component of bone and teeth phosphates. Of the approximately 2.14 kg of calcium in the entire body, the skeleton contains 1.1 kg and the soft tissues 13.8 g. The bulk of the calcium, therefore, is contained in the skeleton as more or less crystalline calcium phosphate (called *bone apatite*), and it promotes the stability and rigidity of bones. Although only a minor portion is contained in soft tissues, the role of calcium here should not be overlooked. Control over metabolic processes probably involves a calcium-dependent step (Table V).

TABLE V.   Calcium-Controlled Events in Cells

Activity	Controlled Events or Systems
Photosynthesis	Dioxygen release
Oxidative phosphorylation	Dehydrogenases
Receptor responses	Nerve synapse
	IP3-linked reactions
Contractile devices	Muscle triggering (actomysin)
	Cell filament controls
Phosphorylation	Activation of kinases (e.g., in fertilization)
Metabolism	Numerous enzymes inside cells
Membrane/filament organization	Annexin-like proteins modulate tension
Cell division	S-100 proteins, immune system
Cell death (apoptosis)	Internal proteases
Hormone/transmitter release	Homeostasis
Binding to membranes	C-2 domains of enzymes
Cross-linking	Outside cells
Enzyme-activation	Outside cells; in membranes

Source: Adapted from Fraústo da Silva and Williams (2001).

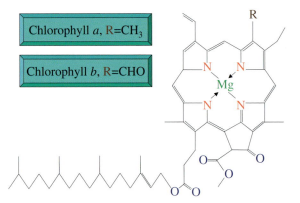

Chlorophyll *a*, R=CH₃

Chlorophyll *b*, R=CHO

FIGURE 3   Magnesium in chlorophyll.

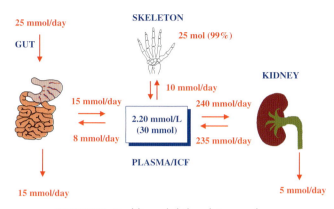

FIGURE 4 Normal daily calcium exchange.

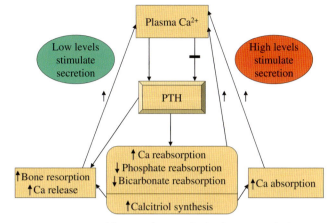

FIGURE 5 Hormonal control of calcium metabolism.

1. Calcium Homeostasis

Only a minor proportion of skeletal calcium (about 1%) is rapidly exchangeable with plasma, although remodeling of bone results in the turnover of nearly 20% of skeletal calcium each year. Approximately 1% of body calcium is present in the extracellular fluid, where functions include the regulation of neuromuscular excitability and acting as a cofactor for clotting enzymes. The gradient of extracellular to intracellular free calcium is around 10,000:1, although cellular calcium is also found as insoluble complexes. Within cells, free calcium ions (Ca^{2+}) regulate the activity of various enzymes directly and also exert second-messenger hormonal functions by interaction with calcium-binding proteins such as troponin C and calmodulin.

The normal daily calcium exchange is represented in Figure 4. Calcium in the gastrointestinal tract originates from the diet and also from secretions. Approximately half is absorbed, mainly in the upper small intestine (jejunum and duodenum), by active transport. Up to 250 mmol calcium is filtered daily by the kidney. With the majority reabsorbed in the proximal tubule and loop of Henle, urinary excretion is normally 2.5 to 7.5 mmol L^{-1}, depending on uptake. The small amounts lost in sweat are usually insignificant, unless profuse sweating occurs for a prolonged period.

Parathyroid hormone (PTH) is a key regulatory hormone of calcium metabolism whose secretion is stimulated by low plasma Ca^{2+} concentrations and by low plasma magnesium concentrations. The physiological importance of regulation by magnesium is unclear, although magnesium depletion can cause hypocalcemia. The secretion of PTH is inhibited by increased Ca^{2+} levels. Within the parathyroid glands, PTH is transcribed as an 115-amino-acid polypeptide and processed to form an 84-amino-acid polypeptide, in which form it

is stored prior to secretion. The biological activity of PTH is contained within a 32- to 34-amino-acid fragment of the molecule located at the N-terminal end. Following secretion, PTH is cleaved, mainly in the liver, to produce two fragments, one of which is an inactive (C-terminal) fragment. PTH is usually measured in blood by assays that depend on immunoreactivity rather than biological activity (immunoassays). In renal failure, the rate of metabolism of the C-terminal fragment is reduced as the kidney normally removes this. If assays for measuring PTH are based on the immunoreactivity of the C-terminal end of the molecule, levels may appear higher than would be apparent if biological activity were determined. This is important, because renal failure causes secondary hyperparathyroidism.

The main effect of PTH is to raise plasma Ca^{2+} concentrations through actions on bones, the kidney, and, indirectly, the gastrointestinal tract (Figure 5). In the bones, PTH stimulates osteoclast activity, while in the kidney it increases the reabsorption of calcium and reduces the rate of transport of phosphate and bicarbonate. PTH also stimulates the hydroxylation of 25-hydroxycholecalciferol to form calcitriol, which then acts on the gut to increase calcium and phosphate absorption.

Vitamin D is converted to its biologically active form, 1,25-dihydroxycholecalciferol (calcitriol) by successive hydroxylations in the liver and kidney. Calcitriol stimulates intestinal absorption of calcium and phosphate by regulating the synthesis of a protein that transports calcium across the enterocyte. In addition, calcitriol is required for normal mineralization of bone, which is defective in deficiency states. Weakness of skeletal muscles also occurs in vitamin D deficiency, which responds to supplements. This suggests that vitamin D is important for normal skeletal function, although the basis of this is not understood.

2. Calcium Signaling

The development and improvement of analytical methods to qualitatively and quantitatively determine the presence of specific elements are of profound importance for understanding their biological roles. Calcium is no exception in this respect. The ability of cells to precisely regulate the cellular concentrations of free and bound calcium both in time and space is a feature that adds to the versatility of the calcium ion. Calcium plays an important role in cellular signaling and therefore has been classified as one of the major second-messenger molecules. Calcium signals control a vast array of cellular functions, ranging from short-term responses such as contraction and secretion to longer term control of transcription, cell division, and cell death (Berridge et al., 2000; Venkatchalam et al., 2002). Eukaryotic cells have developed specialized internal Ca^{2+} stores that are localized in the sarco- or endoplasmic reticulum (Carafoli & Klee, 1999; Sorrentino & Rizzuto, 2001), at least for higher vertebrate cells. These stores represent a significant contribution to intracellular signaling, as the stores can release Ca^{2+} either in conjunction with or independently of the Ca^{2+} entry pathways localized on the plasma membrane after stimulation by agonists (Sorrentino & Rizzuto, 2001). Molecular studies have both identified the genes and characterized the proteins participating in intracellular Ca^{2+} homeostasis (Carafoli & Klee, 1999; Sorrentino & Rizzuto, 2001).

Intracellular Ca^{2+} stores have a specific molecular organization with regard to the way in which Ca^{2+} is released through specific intracellular Ca^{2+} release channels (ICRCs). Accumulation and storage, however, are mediated by dedicated Ca^{2+} pumps and Ca^{2+} binding proteins, respectively.

Ca^{2+} release from intracellular stores is mediated by channels encoded by a superfamily of genes, which includes three genes encoding channels capable of binding the vegetal ryanodine (RY) receptors and three genes encoding channels that bind inositol (1,4,5)-trisphosphate (IP3 receptors). These should be common for vertebrates (Sorrentino & Rizzuto, 2001). In most nonexcitable cells, the generation of receptor-induced cytosolic calcium signals is complex and involves two interdependent and closely coupled components: the rapid, transient release of calcium from stores in the endoplasmic reticulum and then the slow and sustained entry of extracellular calcium (Putney & McKay, 1999; Berridge et al., 2000; Venkatchalam et al., 2002). Through the activation of phospholipase C subtypes, G-protein-coupled receptors and tyrosine-kinase-coupled receptors generate the second messenger inositol (1,4,5)-trisphosphate and diacylglycerol. Doing so functions as a chemical message that diffuses rapidly within the cytosol and interacts with inositol-trisphosphate receptors located on the endoplasmic reticulum lumen and generates the initial calcium signal phase (Sorrentino & Rizzuto, 2001).

3. Structural Role of Calcium

Bone is not a static tissue, although it would be easy to think that when growth is finished bone does not change anymore; however, bone is a very dynamic tissue undergoing continuous destruction and remodeling. This process is important for the maintenance of bone volume and calcium homeostasis. The cells responsible for these dynamics are osteoblasts and osteoclasts, respectively. Osteoblasts produce bone matrix proteins, of which type I collagen is the most abundant extracellular bone protein. In addition, osteoblasts are critical to mineralization of the tissue (Aubin & Triffitt, 2002; Katagiri & Takahashi, 2002). Undifferentiated mesenchymal cells become osteoblasts, chondrocytes, myocytes, and adipocytes. Progenitor cells acquire specific phenotypes that are under the control of regulatory factors during the differentiation. Bone morphogenetic proteins (BMPs) play critical roles in the differentiation of the mesenchymal cells to osteoblasts (Katagiri & Takahashi, 2002).

Osteoclasts are multinucleated cells responsible for bone resorption. They are differentiated from hematopoietic cells of monocyte/macrophage lineage under the control of bone microenvironments. The presence of ruffled borders and a clear zone is characteristic for these cells (Väänänen & Zhao, 2001). Osteoblasts or bone marrow stromal cells have been shown to regulate osteoclast differentiation, thus providing a microenvironment similar to bone. The resorbing area under the ruffled border is acidified by vacuolar H^+–ATPase. A clear zone surrounds the ruffled border. This zone allows for the attachment of osteoclasts to the bone surface and maintains a microenvironment favorable for bone resorption. The recent discovery of the tumor necrosis factor (TNF) receptor–ligand family has clarified the molecular mechanism of osteoclast differentiation (Katagiri & Takahashi, 2002).

D. Sulfur Bioinorganic Chemistry

About 10 nonmetallic elements are essential for life. Of these, sulfur stands out due to its astounding chemical versatility. This versatility is reflected by several papers dealing with the biology of sulfur—for example, the chemistry of sulfane sulfur, disulfide as a transitory covalent bond and its significance in protein folding, the role of protein sulfenic acids in enzyme catalysis and redox regulation, and sulfur protonation in $[3Fe–4S]^{0,2-}$ clusters. In addition, it has been suggested that sulfur plays a role with regard to the μ_3 sulfide, where μ_3 indicates the number of metal ions connected to a multivalent ligand of Fe–S clusters in the catalysis of homolytic reactions, such as in pyruvate–formate lyase, anaerobic ribonucleotide reductase, 2,3-lysine amino mutase, biothin synthase, and the thioredoxin reductase (Beinert, 2000). A close neighbor in the periodic table, phosphorus, works in biological systems by ionic reactions in the form of its oxo-anions. In contrast, sulfur in its most oxidized form, sulfate, is of limited use to higher organisms except for sulfation and detoxification reactions; rather, it is the chemical versatility of the lower oxidation states of sulfur that is vital for anabolic reactions (Beinert, 2000).

1. Inorganic Systems

A conspicuous example of Fe–S clusters is cytoplasmic c-aconitase/iron regulatory protein (IRP). The aconitases lose one Fe readily on exposure to O_2, or even faster to O_2^-, with formation of the enzymatically inactive 3Fe form $[3Fe–4S]^+$, which can be reconstituted to form active $[4Fe–4S]^{2+}$ enzyme (Figure 6a). In addition, c-aconitase alternates as aconitase and, after complete loss of its Fe–S cluster, as an IRP (Figure 6b) that can bind to iron-responsive elements (IREs) in the untranslated regions of mRNAs and thereby regulate translation (Haile *et al.*, 1992; Beinert *et al.*, 1996; Beinert, 2000). There is some spontaneous breakdown of aconitase; however, this process is probably too slow to be used in regulation but for subtle adjustments.

c-aconitase $[4Fe–4S]^{2+}$(active) $\xrightarrow{O_2,\,O_2^-}$ c-aconitase [3Fe–4S] (inactive)

a)

c-aconitase $[4Fe–4S]^{2+}$(active) $\longrightarrow$ apo-c-aconitase $\rightarrow$ IRP (iron regulatory protein)

b)

FIGURE 6   Possibilities for aconitases.

FNR $[4Fe–4S]^{2+}$(active) $\xrightarrow[\text{Fast}]{O_2}$ FNR $[2Fe–2S]^{2+}$ $\xrightarrow[\text{Slow}]{O_2}$ apo-FNR (inactive)

FIGURE 7   Transition from anaerobic to aerobic metabolism in *E. coli.*

Agents such as NO or H_2O_2 can speed up the process significantly. Transition of the organism from anaerobic to aerobic metabolism (*Escherichia coli*) is governed by the global transcription regulator fumarate nitrate reduction (FNR). FNR alternates between the active, O_2-sensing, holoform $[4Fe–4S]^{2+}$ and the inactive form $[2Fe–2S]^{2+}$ and, finally, apoform during O_2 sensing (Bates *et al.*, 2000) by means of its very O_2-sensitive Fe–S cluster (Figure 7). Obviously, in all these cases, there must be a continuous cycle of disassembly and reassembly or repair of clusters *in vivo.*

2. Organic Systems

If the processes in which Fe–S clusters are involved are considered primarily inorganic, it is interesting to look for sulfur processes without the involvement of these clusters. It appears that at least one process is devoid of Fe–S clusters—the biosynthesis of thiamine. This pathway is complicated and not yet fully understood (Begley *et al.*, 1999). Thiamine synthesis involves at least six peptides (ThiF, S, G, H, I, J). Cystein is the original source of the sulfur, and the immediate donor, which completes the heterocycle synthesis, is thiocarboxylate formed at the carboxyl terminus of ThiS in an ATP-demanding reaction.

V. THE FUNCTIONAL VALUE OF TRACE ELEMENTS

Trace elements are found in minute amounts in tissues and body fluids. Nevertheless, they play an extremely important role in biology. The paramount function is to be necessary for the structure and function of significant biomolecules, mainly enzymes. An intriguing property of many of the trace elements is that they are found among the transition elements, and most of them are also in the fourth row of the periodic table. Only boron, selenium, molybdenum, iodine, and tungsten are found at other sites.

A. Vanadium, Cobalt, and Nickel Are Not Used Extensively

These trace elements are treated more or less collectively without any specific chemical argument. They can be found in the transition metal area in the periodic table and are known to have or are suspected to have biological functions.

1. Vanadium

The rich aqueous chemistry of vanadium suggests that its metabolism is complex. Both the anion vanadate (VO_3^-) and the cation vanadyl (VO^{2+}) can complex with molecules of physiological significance. Vanadium species such as the oxyanion and oxycations VO_4^{2-}, VO^{2+}, and VO_2^+ are oxidizing agents capable of reacting by one-electron transfer or by two-electron steps when V acts as an O-atom donor. In addition, vanadium can form sulfur-containing anionic centers (e.g., VS_4^{3-}) and sulfur-containing cationic centers such as VS^{2+} and $VSSH^+$ (Fraústo da Silva & Williams, 2001). Apart from the strikingly high concentrations of vanadium in some ascidians, for which there still is no clear biological function, vanadium concentrations in tissues and body fluids generally are low, as can be seen from Table VI.

Most of the vanadium in the vanadocytes of ascidians, which are known to be the signet ring cells, has been shown to be in the V(III) state, with a minor part occurring in the V(IV) state; consequently, reducing agents have to participate in the accumulation of vanadium in vanadocytes. Among the proposed agents are tunichromes. Earlier it was thought that the vanadium incorporated by ascidians was dissolved as ionic species or associated with low-molecular-weight substances rather than proteins. This is in contrast to other metals that generally bind to macromolecules such as proteins and are incorporated into the tissues of living organisms. Michibata *et al.* (2002), however, demonstrated the presence of a vanadium-associated protein in vanadocytes. This protein was estimated to associate with vanadium in an approximate ratio of 1:16. The protein was comprised of three peptides of estimated molecular weights of 12.5, 15, and 16kDa; however, the physiological roles of vanadium still remain to be elucidated.

Other functions of vanadium are associated with various defense systems, such as peroxidase and catalase activity or haloperoxidases, as well as dinitrogen fixation. The peroxidase and catalase activity is exerted by the vanadium-containing substance amavadine (V(IV) *bis*-complex of *N*-hydroxyimino-di-a-propionate), the mechanism of which is illustrated in Figure 8 (Matoso *et al.*, 1998). Amavanadine is found in some *Amamita* toadstools. It seems that amavadine could be a kind of primitive protective substance able to use H_2O_2 for self-regeneration of damaged tissues or to defend against foreign pathogens and predators, decomposing it when not necessary (Fraústo da Silva & Williams, 2001).

An intriguing and exciting property of vanadium is that it mimics insulin both *in vitro* and *in vivo*. Even more fascinating is that vanadium as sodium vanadate ($NaVO_3$) was used to treat patients with diabetes mellitus as early as 1899 (Lyonnet *et al.*, 1899), before insulin was even discovered. The proposed mechanism by which vanadyl acts is on at least three sites (Sakurai, 2002), as illustrated in Figure 9. Vanadate behaves in a manner similar to that of phosphate; therefore, the effects of vanadium are thought to inhibit the production of protein phosphotyrosine phosphatase that follows stimulation of protein tyrosine phosphorylation. Vanadate also activates autophosphorylation of solubi-

---

TABLE VI. Concentrations of Vanadium in the Tissues of Several Ascidians Compared with Human Serum (mmol/L)

Species	Tunic	Mantle	Branchial Basket	Serum	Blood Cells
Ascidia gemmata	ND	ND	ND	ND	347.2
A. ahodori	2.4	11.2	12.9	1.0	59.9
A. sydneiensis	0.06	0.7	1.4	0.05	12.8
Phallusia mammillata	0.03	0.9	2.9	ND	19.3
Ciona intestinalis	0.003	0.7	0.7	0.008	0.6
Homo sapiens	—	—	—	0.000003–0.000018	—

lized insulin receptors in a way similar to the insulin activation. Vanadate also stimulates the tyrosine kinase activity of the insulin receptor β-subunit. Both vanadate and vanadyl are effective in stimulating glucose metabolism in adipocytes. Vanadate restores the expression of the insulin-sensitive glucose transporter and induces the recruitment of the GLUT4 glucose transporter to the plasma membrane. The 3′,5′-cyclic adenosine monophosphate (cAMP)-mediated protein phosphorylation cascade in adipocytes is activated during diabetes

or in the presence of adrenalin. Both glucose and vanadyl, which are incorporated in the adipocytes in response to vanadyl treatment, lead to the restored regulation of this cascade. Free fatty acid (FFA) release is thought to be inhibited by vanadyl. Vanadyl, thus, acts on at least three cell sites—phosphatidyl inositol 3-kinase, glucose transporter, and phosphodiesterase—to normalize both glucose and FFA levels in diabetes.

It has been demonstrated that vanadium compounds can be used in the treatment of diabetes mellitus, and vanadium also seems to be able to prevent the onset of diabetes. A novel hypothesis has been proposed in which nitric oxide (NO) production can be attributed to macrophages (Mφ). The possible mechanism is shown schematically in Figure 10 (Sakurai, 2002). In the macrophages of normal animals treated with $VOSO_4$, incorporation of vanadium and responses to enhanced NO production are low. In the prediabetic phase of mice treated with streptozotocin (STZ), activated Mφ exudes through pancreatic islets, and the NO produced concomitantly by activated Mφ destroys normal islet β cells. The onset of diabetes by STZ administration is proposed to be based on the enhancement of the generation of superoxide anions ($\cdot O_2^-$) in β cells. NO reacts with $\cdot O_2^-$ to produce peroxynitrite ($ONOO^-$). One of the degradation products of peroxynitrite is a hydroxyl radical ($\cdot OH$). The radicals $\cdot O_2^-$ and $\cdot OH$ destroy the β

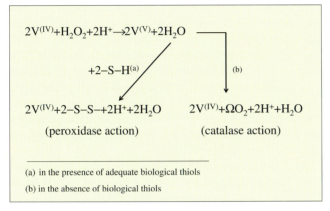

$$2V^{(IV)}+H_2O_2+2H^+\rightarrow2V^{(V)}+2H_2O$$

$$+2-S-H^{(a)}$$

$$(b)$$

$$2V^{(IV)}+2-S-S-+2H^++2H_2O \qquad 2V^{(IV)}+\Omega O_2+2H^++H_2O$$

(peroxidase action) (catalase action)

(a) in the presence of adequate biological thiols
(b) in the absence of biological thiols

FIGURE 8 Peroxidase and catalase activity of amavadine. (Adapted from Matoso *et al.*, 1998.)

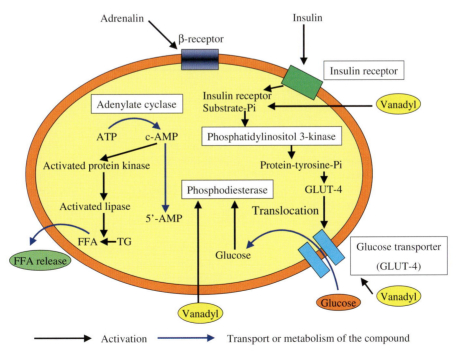

FFA release

FIGURE 9 Proposed mechanisms of vanadyl action. (Adapted from Sakurai, 2002.)

cells. Treatment with $VOSO_4$ suppresses NO production and consequently formation of $\cdot OH$ that damages the β cells.

2. Cobalt

A spontaneous association with cobalt biological functions is most likely with vitamin B_{12}, as cobalt is an integral cofactor in this vitamin. Although the biological

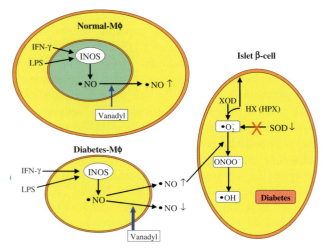

FIGURE 10 A possible mechanism of vanadium prevention of diabetes. (Adapted from Sakurai, 2002.)

functions of vitamins are slightly out of the scope of this book, a few comments are necessary here. Vitamin B_{12} is involved in four types of important reactions:

1. Reduction of ribose to deoxyribose
2. Rearrangement of diols and similar molecules
3. Rearrangement of malonyl to succinyl
4. Transfer of methyl groups

Worth mentioning is the enzyme ribonucleotide reductase that requires vitamin B_{12} for the reduction of ribonucleotides to the corresponding deoxyribonucleotides. Not only is this reaction dependent on the enzyme ribonucleotide reductase and vitamin B_{12}, but the reaction is also considered to proceed via the radical pathway. This constitutes an interesting example of how radicals can have important functions in biological processes that are not destructive in nature. In vitamin B_{12}, cobalt is associated with a corrin ring, a relative of porphyrin. Non-corrin cobalt is receiving increased interest, and ten non-corrin-cobalt-containing enzymes have been isolated (Kobayashi & Shimizu, 1999) and characterized (Table VII).

Metionine aminopeptidase cleaves the N-terminal methionine from many newly translated polypeptide chains in both prokaryotes and eukaryotes. It is an important catalyst for N-terminal modification involved in functional regulation, intracellular targeting, and protein turnover. The *E. coli* methionine aminopeptidase is a monomering protein of 29 kDa and consisting

---

TABLE VII.   Cobalt-Containing Proteins

Enzyme or Protein	Source	Cofactor content	Postulated role of cobalt
Methionine aminopeptidase	Animals, yeast, bacteria	2 Co per subunit	Hydrolysis
Prolidase	Archaea	1–2 Co per subunit	Hydrolysis
Nitrile hydratase	Actinomycetes and bacteria	1 Co in each α-subunit	H_2O activation, CN-triple-bond hydration and protein folding
Glucose isomerase	Actinomycetes	1 Co per 4 subunits	Isomerization
Cobalt transporter	Actinomycetes and yeast	—	Cobalt uptake
Methylmalonyl-CoA carboxytransferase	Bacteria	1 Co, 1 Zn per subunit	Carboxytranserfation
Aldehyde decarbonylase	Algae	1 Co-porphyrin per αβ-subunit	Decarbonylation for aldehyde
Lysine-2,3-aminomutase[a]	Bacteria	0.5–1 Co per subunit	Mutation
Bromoperoxidase	Bacteria	≈0.35 Co per 2 subunits	Bromination
Cobalt-porphyrin-containing protein	Bacteria	1 Co-porphyrin per protein	Electron carrier

[a]Lysine-2,3-aminomutase also contains an iron–sulfur cluster, zinc, and PLP as cofactors.
Source: Adapted from Kobayashi and Shimizu (1999).

of 263 residues that bind two Co^{2+} ions in its active site. The two subfamilies of cobalt-containing methionine aminopeptidase are a prokaryotic class (type I) and a human class (type II).

Prolidase (or praline dipeptidase) specifically cleaves Xaa–Pro dipeptides. In concert with other endopeptidases and exopeptidases, prolidase is thought to be involved in the terminal degradation of intracellular proteins and may also function in the recycling of proline (Ghosh *et al.*, 1998).

Nitrile hydratase catalyzes hydration of nitriles to amides (see below) and is a key enzyme involved in the metabolism of toxic compounds. The presence of cobalt as an essential cofactor may be explained by the effective catalysis of CN-triple bond hydration as well as a requirement for the protein folding.

Glucose isomerase catalyzes the reversible isomerization of D-glucose to D-fructose. Although this enzyme requires divalent cations for the activity, its specific requirement depends on the source of the enzyme.

Most metals that play essential roles as cofactors in biological processes have to be actively incorporated into cells against concentration gradients. COT1 acts as a cobalt toxicity suppressor in *Saccharomyces cerevisiae* by sequestration or compartmentalization within the mitochondria of cobalt ions that cross the plasma membrane (Conklin *et al.*, 1992). The cobalt transporter NhlF mediates cobalt incorporation into the cell in an energy-dependent manner. The sequence of this transporter shows eight putative hydrophobic membrane-spanning domains. NhLF contains nine histidine residues and two cysteine residues. One histidine (His306) and one cysteine (Cys301) residue are located on the fourth outside loop and may comprise a cobalt-binding site for the initial fixation of the metal. The transmembrane segments could form a cobalt channel in which the four histidines might function as transient cobalt-binding ligands.

The biotin-containing enzyme methylmalonyl–CoA carboxytransferase (transcarboxylase) is a complex multisubunit enzyme that catalyzes the transfer of a carboxyl group from methylmalonyl–CoA to pyruvate to form propionyl–CoA and oxaloacetate.

Aldehyde decarboxylase converts a fatty aldehyde to hydrocarbon and carbon monoxide. This enzyme is responsible for a key step in the biosynthesis of hydrocarbon compounds. Lysine-2,3-aminomutase catalyzes the reversible isomerization of l-lysine to l-β-lysine, a reaction in which the hydrogen on the 3-pro-R position of lysine is transferred to the 2-pro-R position of β-lysine and the 2-amino group of lysine migrates to carbon-3 of β-lysine. The enzyme contains three

cofactors: pyridoxal phosphate, Fe–S centers, and cobalt or zinc.

Bromoperoxidase catalyzes the formation of a carbon–bromine bond in the presence of peroxides. More studies, however, are needed to elucidate the function of bromoperoxidase. Other proteins contain cobalt–porphyrins in plants and in sulfate-reducing bacteria. The cobalt–porphyrins are not covalently bound to the proteins. The prosthetic groups of these cobalt-containing proteins are found to be cobalt isobacteriochlorins.

3. Nickel

Nickel-containing enzymes are involved in at least five metabolic processes, including the production and consumption of molecular hydrogen, hydrolysis of urea, reversible oxidation of carbon monoxide under anoxic conditions, methanogenesis, and detoxification of superoxide anion radicals. The active sites of the relevant enzymes harbor unique metallocenters, which are assembled by auxiliary proteins. Different ligand environments are involved in the coordination of nickel in the various metalloenzymes (Eitinger, 2000). The reactions catalyzed by nickel-dependent enzymes are summarized in Figure 11.

Some transporters are potentially able to transport nickel; however, CorA, MgtA, and MgtB have affinities that are too low to be of physiological importance (Ragsdale, 1998). Two different types of high-affinity nickel transporters have been identified. One is a mul-

a. Superoxide dismutase

$$2H^+ + 2O_2^{\bullet -} \rightarrow H_2O_2 + O_2$$

b. Urease

$$H_2N - CO - NH_2 + 2H_2O \rightarrow 2NH_3 + H_2CO_3$$

c. Hydrogenase

$$2H^+ + 2e^- \leftrightarrow H^+ + H^- \leftrightarrow H_2$$

d. Methyl-CoM reductase

$$CH_3 - CoM + CoB - SH \rightarrow CH_4 + CoM - S - S - CoB$$

e. CO dehydrogenase

$$CO + H_2O \rightarrow 2H^+ + CO_2 + 2e^-$$

f. Acetyl-CoA synthase

$$CH_3 - CFeSP + CoA - SH + CO \rightarrow CH_3 - CO - SCoA + CFeSP$$

FIGURE 11 Reactions catalyzed by nickel-dependent enzymes.

ticomponent ATP-binding cassette (ABC) transporter system such as NikABCDE that uses ATP. The other is a one-component transporter such as NixA, UreH, HupN, and HoxN. NixA and HoxN are integral membrane proteins that have eight transmembrane-spanning helices and a sequence motif that is essential for function. HypB can sequester nickel and release it for incorporation into apoproteins when nickel becomes limiting (Ragsdale, 1998).

A nickel–superoxide dismutase (Ni–SOD) has been isolated from *Streptomyces*. The protein is a homotetramer of four 13-kDa subunits with little sequence similarity to earlier known SODs. Nickel induces its expression, represses the FeZn–SOD, and is involved in maturation of a precursor polypeptide (Ragsdale, 1998).

Urease plays a key role in the nitrogen metabolism of plants and microbes and acts as a virulence factor for some human and animal pathogens (Mobley *et al.*, 1995). There are two nickel ions in the enzyme, and CO_2 is required to generate a carbamylated lysine bridge between the two nickel ions.

Hydrogenases are of different types. Some hydrogenases involve Ni, Fe, and Se; some Ni and Fe. In addition, some depend only on Fe. In fact, one hydrogenase has no association with metals at all (Hartmann *et al.*, 1996).

Methyl-coenzyme M reductase is a remarkable enzyme composed of six subunits forming a heterohexamer. Three unusual coenzymes are embedded in a long channel between the subunits. Of particular interest is the binding of coenzyme F430, a Ni–porphinoid that occurs exclusively in this enzyme. As substrates, the enzyme binds methyl-coenzyme M (methyl-thioethane sulfonate) and coenzyme B (7-thioheptanoyl threoninephosphate). Structural studies of this enzyme have been hampered by the low activity of the purified enzyme; however, substantial improvements in this respect have been seen in recent years (Finazzo *et al.*, 2003).

Carbon monoxide (CO) dehydrogenase oxidizes CO to CO_2 in the half reaction shown in Figure 12 using a Ni–Fe cluster. Another enzyme, a hydrogenase, also contains a Ni–Fe cluster and uses the electrons in the first half reaction to reduce protons to hydrogen gas: $2H^+ + 2e^- \leftrightarrow H_2(g)$ (Watt & Ludden, 1999). Carbon monoxide dehydrogenases are bifunctional enzymes that perform reversible CO oxidation and also function to synthesize or degrade acetyl–CoA. The bifunctional characteristics of this enzyme led to changing its name to CO–dehydrogenase/acetyl–CoA synthase (Hausinger, 1993). Both catalytic sites for the individual reactions require nickel for catalysis. Methanogenic

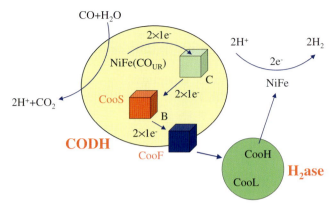

FIGURE 12   CO-dehydrogenase oxidizes CO to CO_2 using a NiFe cluster and a nickel-dependent hydrogenase (H2ase) to produce hydrogen gas. (Adapted from Watt and Ludden, 1999.)

bacteria convert acetate to methane and CO_2 and couple acetate degradation to ATP synthesis. This metabolic process uses the CO–dehydrogenase/acetyl–CoA synthase (Watt & Ludden, 1999).

Nickel compounds are recognized as human carcinogens. It has been suggested that the molecular mechanism of the genotoxicity underlying the induction of carcinogenesis is the delivery of nickel into a cell from particulates outside the cell. Water-soluble nickel salts penetrate cells poorly. Following phagocytosis, the particles are contained in vacuoles that become highly acidified, and this greatly enhances the dissolution of soluble nickel from the particles. This increases the intracellular load of nickel, which can subsequently attack chromatin and particularly histones and produce effects associated with carcinogenic activity (Zoroddu *et al.*, 2002).

B. Chromium, Molybdenum, and Tungsten

This triad of trace elements comprises group 6 in the periodic table; consequently, they should have some chemical properties in common. An intriguing property of these trace elements is their relationship with the non-metal elements of group 16; hence, the trace elements of group 6 often appear covalently bonded and not as simple metal ions.

1. Chromium

Chromium(III) was already established as an essential dietary component in the 1950s. Many studies have been conducted to elucidate the biological role of chromium but have had a low degree of success. A major

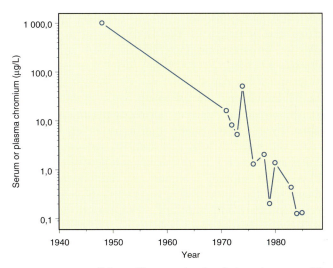

FIGURE 13 "Normal" serum levels of chromium over 30 to 40 years.

problem is the low concentration in tissues and body fluids and the high risk of contamination of samples. It was not until the 1980s that reliable analytical methods were developed. Over the course of 30 to 40 years of analytical efforts (Figure 13), the so-called normal serum level of chromium in humans has been decreasing by orders of magnitude. It has been suggested that chromium is integral to the glucose tolerance factor (GTF). Efforts to purify this factor have led to the detection of nicotinic acid, glycine, glutamic acid, and cysteine, as well as chromium (Mertz *et al.*, 1974).

Recent years have witnessed a plethora of activity related to the elucidation of a potential role for trivalent chromium in mammalian carbohydrate and lipid metabolism at a molecular level. In the 1980s, the isolation and characterization of a unique chromium-binding oligopeptide known as low-molecular-weight chromium-binding substance (LMWCr) or chromodulin were reported (Yamamoto *et al.*, 1987). The oligopeptide has a molecular weight of about 1500 Da and is comprised of only four types of amino acid residues (*i.e.*, glycine, cysteine, glutamate, and aspartate). Despite its small molecular weight, it binds four equivalents of chromic ions, apparently in a tetranuclear assembly, as necessitated by charge balance arguments (Vincent, 2000). Chromodulin has an intriguing ability to potentiate the effects of insulin on the conversion of glucose into carbon dioxide or lipid. No other naturally occurring chromium-containing species potentiates insulin action in this manner (Vincent, 2000).

How chromium is absorbed and transported is still uncertain. It appears that transport is mediated by trans-ferrin, the main iron-transporting protein (molecular weight, 80 kDa). This may be due to the fact that transferrin usually carries only about a 30% load of iron so it has unused transportation capacity. Transferrin is also thought to be a transporter of various trace elements. Recent reports on the effects of insulin on iron transport and the relationship between hemochromatosis and hepatic iron overload and diabetes suggest that transferrin may actually be the major physiologic chromium transport agent (Vincent, 2000).

Hexavalent chromium compounds have been established as being carcinogenic. Chromate easily enters cells through the sulfate channel and is quickly reduced by, for example, glutathione. The ultimate step of the metabolic pathway yields Cr(III) inserted within the cell nucleus, where it cross-links DNA to proteins. Recent results indicate that glutathione is not only a primary target for oxidation by chromate but also acts as an efficient ligand-stabilizing Cr(V) in a dimeric bridged cluster (Gaggelli *et al.*, 2002).

2. Molybdenum

Molybdenum is found in the second row of transition metals in the periodic table. It is the only metal in this row that is required by most living organisms. Although only a minor constituent of the Earth's crust, molybdenum is readily available to biological systems because of the solubility of molybdate salts in water. In fact, molybdenum is the most abundant transition metal in seawater. It is not, therefore, surprising that molybdenum has been incorporated widely in living organisms. Molybdenum is redox active under physiological conditions (ranging between oxidation states VI and IV). The V valence state is also available, and molybdenum can act as a transducer between obligatory two- and one-electron oxidation–reduction systems such as the hydroxylation of carbon centers under more moderate conditions than are required by other systems (Hille, 2002).

An important feature of molybdenum-containing enzymes is the molybdenum cofactor (Moco) that is able to associate with different apoenzymes to form the Mo–holoenzymes where, depending on the type of apoenzymes, molybdenum catalyzes redox reactions on C, N, and S atoms. The only exception is bacterial nitrogenases that contain an FeMo cofactor not related to Moco (Mendel, 1997). It has been shown that in Moco molybdenum is complexed by a pterin with a four-carbon alkyl side chain containing a Mo-coordinating dithiolene group and a terminal phosphate ester (Mendel, 1997). This pterin has been named molybdopterin (Figure 14).

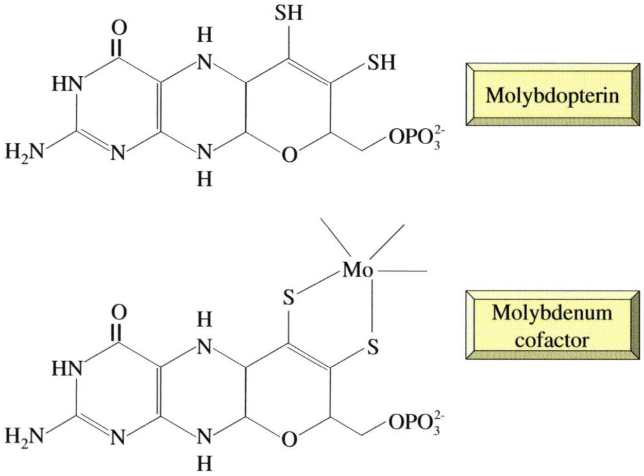

FIGURE 14   Molybdopterin and molybdenum cofactor.

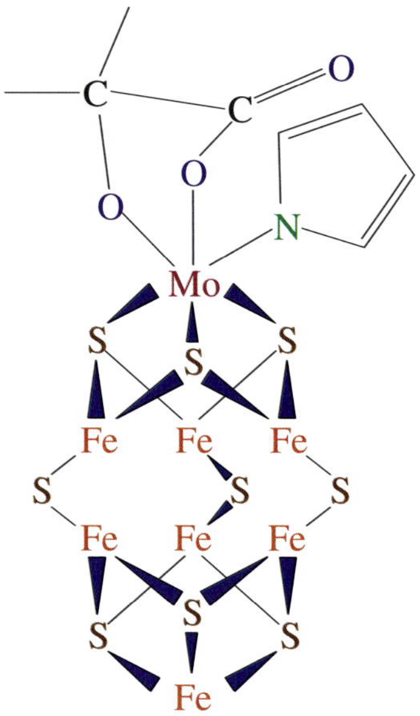

FIGURE 15   Multinuclear center with iron, sulfur, and molybdenum in nitrogenase.

There are two different kinds of molybdenum enzymes. One is exemplified by the nitrogenase enzyme family, which is characterized by a multinuclear center with iron, sulfur, and molybdenum (Figure 15). In fact, molybdenum can be replaced by vanadium or iron. Nitrogenases catalyze the reaction from dinitrogen to ammonia and are the basic components in nitrogen fixation. Indeed, most cycling of nitrogen in the biosphere depends on the trace element molybdenum. The second type of molybdenum enzyme is characterized by a dependence on a mononuclear center, which is associated with molybdopterin. Three enzyme families are

included in this group: xanthine oxidases, sulfite oxidases, and dimethylsulfoxide DMSO reductases. Examples of molybdenum enzymes of these families are discussed below.

Xanthine oxidases comprise the largest family of molybdenum enzymes, with up to 20 members. The xanthine oxidase catalyzes purine or pyrimidine catabolism by inserting oxygen and/or removing hydrogen from the substrates. It is a well-studied enzyme that exists as a dimer with a molecular weight of about 300 kDa. Each subunit contains one molybdenum, one flavin adenine dinucleotide, and two Fe_2S_2 centers. The reaction catalyzed is xanthine + H_2O → uric acid + $2H^+$ + $2e^-$.

Sulfite is a highly reactive and potentially toxic compound. Like other reduced inorganic sulfur compounds such as hydrogen sulfide or thiosulfate, it occurs in nature as a consequence of geological and industrial processes and the anaerobic mineralization of organic matter by dissimilatory sulfate reduction (Kappler & Dahl, 2001). There is no doubt that sulfite oxidases have developed during evolution as a response to sulfite reactivity. Sulfite oxidase contains only a single prosthetic group, a heme group, in addition to the Moco-derived center. The holoenzyme is a dimer with a molecular weight of 120 kDa. The enzyme can oxidize sulfite by using oxidized cytochrome c, ferricyanide, or dioxygen as an electron acceptor. Sulfite oxidation occurs at the molybdenum center, with the heme center serving to couple this two-electron oxidation to the reduction of two molecules of cytochrome c. DMSO reductase catalyzes the reduction of dimethylsulfoxide to dimethylsulfide and liberates the oxygen atom of DMSO as water. The reducing equivalents come from a specific pentaheme cytochrome. In addition to the enzymes described here are a few other molybdenum enzymes, such as pyridoxal oxidase, xanthine dehydrogenases, and pyropallol transhydrolases.

3. Tungsten

Tungsten is by far the heaviest metal with a biological function (*i.e.*, an essential trace element). The three classes of tungsten enzymes are aldehyde ferredoxin–oxidoreductase, formate dehydrogenase, and acetylene hydratase. Formate is a common metabolite in most life forms. In most cases, its production and consumption involve formate dehydrogenase (FDH), which catalyzes the reversible two-electron conversion of CO_2 to formate according to:

$$CO_2 + 2H^+ + 2e^- \rightleftharpoons HCOO^-$$

Not all FDHs are tungsten dependent. On the other hand, most FDHs of aerobic organisms do not contain metals or other cofactors. The tungsten-dependent FDHs have mostly been identified in Chlostridiae (e.g., *C. thermoaceticum*, *C. formioaceticum*, and *C. acidiurici*) (Kletzin & Adams, 1996).

Formyl methanofuran dehydrogenase (FMDH) catalyzes the reversible formation of *N*-formylmethanofuran from CO_2 and methanofuran (MFR). The first step in CO_2 utilization by methanogens is thus:

$$CO_2 + MFR^+ + H^+ + 2e^- \rightleftharpoons CHO\text{-}MFR + H_2O$$

However, FMDH was found to be a molybdoenzyme in the first methanogens examined (Kletzin & Adams, 1996). Tungsten-dependent FMDHs have been isolated from moderate thermofils such as *Methanobacterium wolfei* and *M. thermoautotrophicum*.

Aldehyde-oxidizing enzymes catalyze oxidation of aldehydes of one type or another and again have been found in microorganisms. The reaction catalyzed is

$$CH_3CHO + H_2O \rightleftharpoons CH_3COO^- + 3H^+ + e^-$$

Among the aldehyde-oxidizing enzymes are carboxylic acid reductase (CAR) from acetogens which was the second tungsten-dependent enzyme isolated. Another tungsten-dependent enzyme is aldehyde ferredoxin oxidoreductase (AOR), from *Pyrococcus furiosus*, which catalyzes the oxidation of a range of aliphatic and aromatic aldehydes and reduces ferredoxin. Formaldehyde ferredoxin oxidoreductase (FOR) has also been demonstrated in *P. furiosus*. These enzymes also use ferredoxin as the physiological electron carrier and are maximally active at temperatures above 95°C. In contrast to AOR, however, they oxidize only C1–C3 aldehydes and are of much lower specific activity (Kletzin & Adams, 1996).

The enzyme glyceraldehyde-3-phosphate ferredoxin oxidoreductase (GAPDH) has been found in the peculiar hyperthermophilic archaeon *P. furiosus* and is thought to play a role in gluconeogenesis. A fourth tungsten-containing enzyme from *P. furiosus* has recently been characterized (Roy & Adams 2002). This enzyme, preliminarily named WOR 4, is thought to play a role in SO reduction. It has an interesting structure in that it contains approximately one W atom, three Fe atoms, three or four acid-labile sulfides, and one Ca atom per subunit (Roy & Adams, 2002). Aldehyde dehydrogenases (ADH) have been shown in *Desulfovibrio gigas*.

Acetylene hydratase (AH) converts acetylene to acetaldehyde according to:

$$H\text{-}C \equiv C\text{-}H + H_2O \rightarrow H_3C\text{-}HC = O$$

which is a hydration reaction.

In addition to availability a key factor in tungsten utilization appears to be its redox properties relative to molybdenum. When tungsten substitutes for molybdenum in molybdoenzymes, some will be inactive, probably because the tungsten site has a lower reduction potential as compared to the molybdenum site. Conversely, to catalyze a reaction of extremely low potential, tungsten should be preferred over molybdenum (Kletzin & Adams, 1996).

C. Manganese: Photosynthesis and Defense Against Oxygen

Although fairly abundant in the biosphere, manganese is found only in trace amounts in living organisms. Manganese exhibits the widest range of oxidation states of any of the first row *d*-block metals. The lowest states are stabilized by π-acceptor ligands, usually in organometallic complexes. Most of the biochemistry of manganese can be explained by two properties: It is redox active, and it is a close but not exact analog of Mg^{2+}. Manganese plays many roles in biological systems ranging from acting as a simple Lewis acid catalyst to being an element that can transverse several oxidation states to carry out water oxidation. The presence of Mn^{2+} in the cytoplasm in significant concentrations would pose a serious problem for aerobic eukaryotes. One reason is that manganese is considered to be mutagenic, and it binds relatively weakly to proteins. Prokaryotes, on the other hand, may make use of this property to promote mutations to generate variation. Manganese in aerobic eukaryotic cells, then, has to be pumped into various organelles. Examples of such organelles are presented in Table VIII.

TABLE VIII. Manganese-Containing Vesicles

Vesicle	Known Mn protein
Chloroplast	Superoxide dismutase
Golgi apparatus	Glycosyl-transferases
Lysosome (acid)	Acid phosphatase (Mn(III))
Mitochondria	Superoxide dismutase
Thylakoid (acid)	Generation of molecular oxygen
Vacuole (acid)	Mn(II) is free

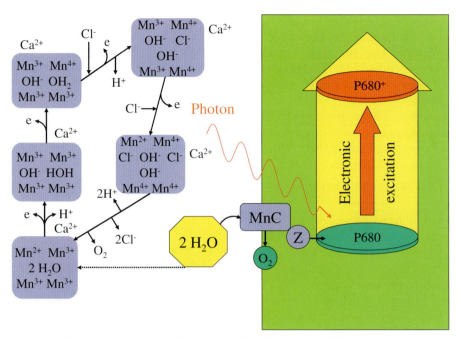

FIGURE 16 A partly hypothetical model of the splicing of water.

1. Generation of Dioxygen from Water

Higher plants and algae complete oxygenic photosynthesis whereby water is oxidized by a cluster of manganese ions. This process takes place in Photosystem II (PS II), which is a multisubunit complex embedded in the thylakoid membranes. This photosynthetic enzyme catalyzes the most thermodynamically demanding reaction in biology, the splitting of water into dioxygen and reducing equivalents (Barber, 2003). Light is captured by antennae chlorophyll and funneled to the primary reaction center, which contains chlorophyll (P680). The oxidized P680 first oxidizes a redox-active tyrosine, hydrogen bonded to a histidine (Yz). Subsequent oxidation of the manganese cluster occurs via this generated tyrosine radical. The enzyme can exist in five oxidation levels named S_n states, where S_0 is the most reduced and S_4 the most oxidized. Oxygen is liberated upon the $S_3 \rightarrow S_4 \rightarrow S_0$ transition (Yocum & Pecoraro, 1999). A partly hypothetical model of this process is depicted in Figure 16. It is interesting to note that the oxidation of water to dioxygen takes place in the thylakoid membranes of plants and algae. These membranes correspond to the mitochondrial membrane of eukaryote cells, in which the reverse process of reducing dioxygen to water takes place. Even more intriguing is that manganese plays important roles in both processes.

2. Defense for a Life with Oxygen

The enzyme most commonly associated with manganese is superoxide dismutase, which is a scavenger of the very reactive superoxide anion produced during reduction of oxygen in cellular respiration. As the name suggests, the enzyme acts on the superoxide anion produced in the first step in the reduction of dioxygen to water. It catalyzes the following reaction:

$$Mn(III) + O_2^- \rightleftharpoons [Mn(III)\text{-}O_2^-] \rightarrow Mn^{2+} + O_2$$

$$Mn^{2+} + O_2^- \rightleftharpoons [Mn^{2+}\text{-}O_2^-] + 2H^+ \rightarrow Mn(III) + H_2O_2$$

The end product, hydrogen peroxide, although not a radical, is a reactive oxygen species. MnSODs are widespread in bacteria, plants, and animals. In most animal tissues and yeast, MnSOD is largely confined to mitochondria (Fridovich, 1995). It is conspicuous that, although manganese is deeply involved in the generation of oxygen from water, it is also necessary in scavenging reactive oxygen species when dioxygen is metabolized in mitochondria. MnSOD is the most phylogenetically widespread manganese-dependent enzyme (Kehres & Maguire, 2003).

There are some aspects of redox biochemistry that have to be taken into account to understand the unexpected tolerance of relatively high concentrations of

Mn^{2+}. Similar to iron, manganese can cycle readily *in vivo* between the 2+ and 3+ oxidation states; however, the reduction potential of any molecule depends on its ligand environment. Another important feature is that manganese is less reducing than iron under most biological conditions (Kehres & Maguire, 2003). "Free" (solvated) Mn^{2+} has a reduction potential too high to reduce H_2O_2 in aqueous solution; however, replacing one or two inner hydration shell waters with hydrogen carbonate results in rapid catalase activity by a Mn^{2+}-dependent disproportionation reaction. That the reducing potential of manganese is smaller than that of iron can be explained by the different $3d$ electron occupancies of the two cations. The electrochemical consequence is that $Mn^{3+} + e^- \rightarrow Mn^{2+}$ has a standard reduction potential of $+1.51\,V$, while $Fe^{3+} + e^- \rightarrow Fe^{2+}$ has a standard reduction potential of $+0.77\,V$.

Two consequences of this redox chemistry are obvious. The similarity between Mn^{2+} and Fe^{2+} is that their intrinsic reduction potentials are close enough to those of many common biological molecules that each metal can take part in biologically important redox catalysis. In addition, the critical difference between the two metals is that the higher reduction potential of Mn^{2+} renders the free (*i.e.*, solvated) Mn^{2+} innocuous under conditions (notably in an aerobic environment) where free Fe^{2+} would actively generate toxic radicals. Cells can thus tolerate very high cytoplasmic concentrations of Mn^{2+} with essentially no negative redox consequences. This is not the case with iron or with any other biologically relevant redox-active metal (Kehres & Maguire, 2003).

3. Similarity to Magnesium

The greatest similarity between these two metals is in the context of structure. Mg^{2+} is an ideal structural cation for biological molecules, especially phosphorylated ones such as nucleic acids and many intermediary metabolites. This is due to a complete lack of d electrons in Mg^{2+}, and its $2s^2 2p^6$ electron configuration confines it to strict octahedral liganding geometry with liganding bond angles very close to $90°$. This geometry is useful in organizing the conformations of complex compounds or macromolecules (Kehres & Maguire 2003). The lack of d electrons is important because there is very little covalent interaction between Mg^{2+} and its ligands; therefore, Mg^{2+} is a labile and rapidly exchangeable cation that does not interpose itself in the way of other close intermolecular interactions. Mn^{2+}, with its relatively similar ionic radius and relatively

minor involvement of its stable symmetric $3d^5$ shell electrons in bonding, readily exchanges with Mg^{2+} in most structural environments and exhibits much of the same octahedral, ionic, labile chemistry. However, less similarity is seen in catalysis. The $3d^5$ electrons of Mn^{2+} do interact to some extent with electrophilic ligands. Thus, Mn^{2+}-ligand bonds are generally much more flexible than Mg^{2+}-ligand bonds, in both length and angle (Kehres & Maguire, 2003).

4. Manganese and Reactive Oxygen Species

The manganese-dependent superoxide dismutase has already been mentioned. In addition to this enzyme is a set of enzymes involved in the detoxification of reactive oxygen species that are dependent on manganese. There is a family of manganese-dependent catalases that often are referred to as "non-heme catalases." They are structurally and mechanistically unrelated to conventional catalases that are cofactored by iron in a prosthetic heme group. Both classes operate on hydrogen peroxide. In addition to catalases, a family of catabolic heme enzymes known as manganese peroxidases couples the redox activity of H_2O_2 to the degradation of nutrients such as lignin via oxidation of a manganese bound to a propionate side-chain of the heme group. An intriguing phenomenon is that salens, synthetic chelators that are derivatives of N,N'-*bis*(salicylidene) ethylendiamine chloride, form complexes with Mn^{2+} that show efficacy as combined superoxide dismutase/catalase mimics. They have been shown, in their oxo forms, to oxidize nitric oxide and nitrite to the more benign nitrate *in vitro* (Kehres & Maguire, 2003). A summary of manganese-related enzymes and proteins can be found in Table IX.

D. Iron: Savior and Threat

Iron is the most important of all metals and is the fourth most abundant element in the Earth's crust. It is a d-block element that can exist in oxidation states ranging from -2 to $+6$. In biological systems, however, these oxidation states are limited primarily to the ferrous ($+2$), ferric ($+3$), and ferryl ($+4$) states. Fe^{3+} is quite water insoluble and significant concentrations of water-soluble Fe^{3+} species can be attained only by strong complex formation. The interconversion of iron oxidation states is not only a mechanism whereby iron participates in electron transfer but also a mechanism

TABLE IX.   Manganese-Dependent Enzymes and Proteins

Enzyme/Protein	Function	Occurrence
MnSOD	Detoxify superoxide radical anion	Bacteria, archaea, and eukaryotes
Non-heme Mn-catalase	Detoxify hydrogen peroxide	Bacteria
Transcription factor (*mntR*)	Repress Mn^{2+} uptake transporter expression	Homologs in diverse bacteria; extent unknown
ppGpp hydrolase (*spoT*)	Hydrolyze RNA synthesis regulator ppGpp	Practically ubiquitous in bacteria
Protein phosphatases	Dephosphorylate many cellular proteins	All cells; highly conserved between prokaryotes and eukaryotes
Agmatinase	Synthesize osmoprotectant putrescine from angmatine (decarboxylated arginine)	Many Gram-negative bacteria
Aminopeptidase P	Hydrolyze typical X-Pro sequence	Enterobacteria, most other bacteria
Phosphoglyceromutase	Catalyzes interconversion of 3-phosphoglycerate and 2-phosphoglycerate	Enterobacteria, other bacteria, and plants
Fructose-1,6-BP phosphatase	Convert fructose-1,6-BP to fructose-6-phosphate	Enterobacteria; extent otherwise unknown
Adenyl cyclase	Synthesize cyclic AMP	*Mycobacterium tuberculosis*
Aromatic hydrocarbon metabolism	Oxidation of catechols and other aromatics	*Arthrobacter globiformis* and similar enzymes in many soil bacteria
Lipid phosphotransferases	Modify or remove polar headgroups on lipids	Enterobacteria, Gram-positive bacteria
Polysaccharide polymerases	Synthesize capsular or secreted polysaccharide	Some Gram-positive and -negative bacteria
Protein kinases	Phosphorylation of unknown proteins	Extent unknown
Pyruvate carboxylase	Catalyze carboxylation of pyruvate to oxaloacetate	Eukaryotes, *Bacillus licheniformis*, and *Mycobacterium smegmatis*
Ribonucleotide reductase	Convert ribonucleotides to deoxyribonucleotides	Most bacteria, eurkaryotes
Arginase	Convert arginine to urea+ornithine	Higher eukaryotes, liver, and macrophages/monocytes; *Bacillus* sp.
Concavalin A	Plant lectin binding	Plants
Mn-lipoxygenase	Synthesize lipoxins from fatty acids	Fungi
Mn-peroxidase	Degrade lignin	White- and brown-rot fungi
Photosynthetic reaction center	Convert H_2O to O_2	Photosynthetic bacteria and plants

Source: Adapted from Kehres and Maguire (2003).

whereby iron can reversibly bind ligands. Iron can bind to many ligands by virtue of its unoccupied *d* orbitals. The preferred biological ligands for iron are oxygen, nitrogen, and sulfur atoms. Iron(III) is a hard acid that prefers hard oxygen ligands, while iron(II) is on the borderline between hard and soft and favors nitrogen and sulfur ligands. The electronic spin state and biological redox potential (from +1000 mV for some heme proteins to −550 mV for some bacterial ferredoxins) of iron can change according to the ligand to which it is bound. By exploiting the oxidation state, redox potential, and electron spin state of iron, nature can precisely adjust iron's chemical reactivity (Beard, 2001). Biologically important iron-containing proteins carry out oxygen transport and storage, electron transfer, and substrate oxidation–reduction. Four major classes of protein carry out these reactions (Beard, 2001) in mammalian systems: (1) iron-containing, nonenzymatic proteins (hemoglobin and myoglobin); (2) iron–sulfur enzymes; (3) heme-containing enzymes; and (4) iron-containing enzymes that are non-iron-sulfur, non-heme enzymes (Figure 17).

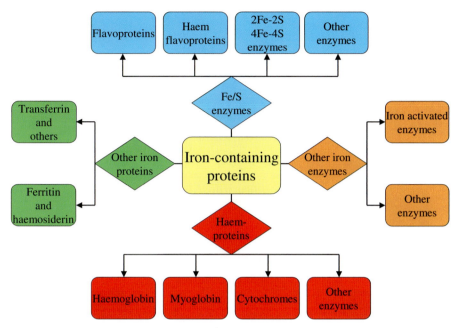

FIGURE 17 Iron-containing proteins.

1. Hemeproteins

Proteins with heme as the prosthetic group carry out important functions in biological systems; thus the biosynthesis of heme is very important. It begins with the synthesis of tetrapyrroles, where the term *tetrapyrrole* indicates compounds containing four linked pyrrole rings. Four classes of such compounds are very common in biology: (1) the widely distributed iron porphyrin, heme; (2) the chlorophylls of plants and photosynthetic bacteria; (3) the phycobilins, photosynthetic pigments of algae; and (4) the cobalamins, especially vitamin B_{12} and its derivatives. All tetrapyrroles are synthesized from a common precursor, δ-aminolevulinic acid (ALA). Figure 18 summarizes the relationships between the synthetic pathways.

2. Hemoglobin and Myoglobin

Evolution has provided animals with hemoglobin and myoglobin for oxygen transport and storage. Aerobic metabolism requires some kind of oxygen transporter because relying on diffusion would be adequate for only very small animals. Insects, however, have solved the problem with oxygen transport through their networks of tubes (tracheae) leading from the body surface to the inside tissues.

Oxygen transport and storage are tricky problems to solve. The molecules used must bind dioxygen without allowing oxidation of other substances, thereby reducing dioxygen. Are proteins suited for direct oxygen binding? The answer is no; however, *d*-block metals such as iron and copper in their lower oxidation states readily bind oxygen. Proteins can bind Fe(II) in various ways. In hemoglobin and myoglobin, iron is bound by the tetrapyrrole ring protoporphyrin IX (see Figure 19). When iron is bound to protoporphyrin IX, the system is referred to as ferroprotoporphyrin or heme. In addition to its use in hemoglobin and myoglobin, the heme group is a prosthetic group in a variety of proteins (see below). This makes the biosynthesis of heme very important. Myoglobin is a smaller molecule relative to hemoglobin and is the principal oxygen storage protein.

An interesting property of hemoglobin, in addition to its oxygen-binding ability, is its enzymatic activity. It has been thought that the oxygen carrier function was so specialized that globins were not recruited to new tasks; however, it has recently been found that the globin of some marine worms (*Amphitrite ornata*) has evolved into a powerful peroxidase, more precisely dehaloperoxidase. This enzyme catalyzes the oxidative dehalogenation of polyhalogenated phenols in the presence of hydrogen peroxide at a rate at least ten times faster than all known halohydrolases of bacterial origin. *A. ornata* can thus successfully survive in an environment where other species secrete brominated aromatics and other halocompounds as repellents (Lebioda, 2000).

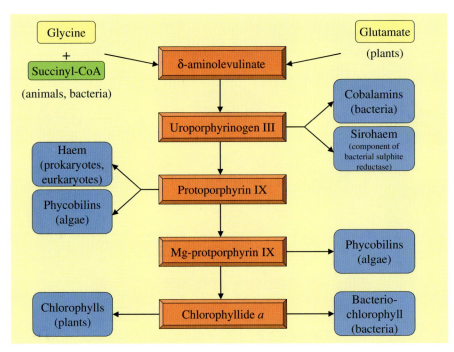

FIGURE 18   The principal steps of heme biosynthesis.

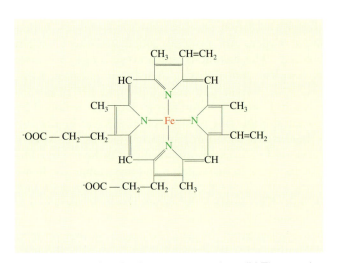

FIGURE 19   Iron binding in protoporphyrin IX. The complex is called *heme*.

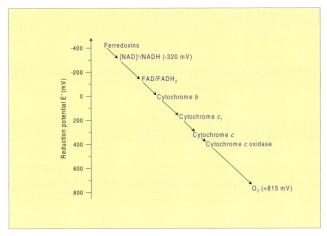

FIGURE 20   The mitochondrial electron-transfer chain.

3. Cytochromes

Cytochromes constitute a group of hemeproteins with distinctive visible-light spectra that function as electron carriers from biological fuels to oxygen. The major respiratory cytochromes are classified as *b*, *c*, or *a*, depending on the wavelengths of the spectral absorption peaks. They are vital members of the mitochondrial electron-transfer chain. Figure 20 shows a schematic representa-

tion of part of this electron-transfer chain. In addition, the cytochromes are also essential components in plant chloroplasts for photosynthesis. It is the ability of the iron center to undergo reversible Fe(III) ↔ Fe(II) changes to allow them to act as electron-transfer centers. In the mitochondrial electron-transfer chain, cytochrome *c* accepts an electron from cytochrome c_1 and then transfers it to cytochrome *c* oxidase. The electron is ultimately used in the four-electron reduction of dioxygen.

4. Cytochrome P-450 Enzymes

Cytochrome P-450 enzymes are hemeproteins that function as monooxygenases to catalyze the insertion of oxygen into a C—H bond of an aromatic of aliphatic hydrocarbon (i.e., the conversion of RH to ROH). Examples of the biological functions of cytochrome P-450 are drug metabolism and steroid synthesis. Carbon monoxide adducts of cytochrome P-450 absorb at 450 nm; thus its name. One of the many interesting aspects of cytochrome P-450 enzymes is that some are inducible, which means that following exposure of the cell to an inducing chemical enzyme activity increases, in some cases several orders of magnitude. It has been proposed that the catalytic cycle begins with the enzyme in a resting state with iron present as Fe(III). The hydrocarbon substrate then binds, followed by one-electron transfer to the iron porphyrin. The resulting Fe(II) complex with bound substrate proceeds to bind O_2. A key reaction is the reduction of the porphyrin ring of the oxygen complex by a second electron, which produces the ring radical anion. Uptake of two H^+ ions then leads to the formation of the Fe(IV) oxo-complex, which attacks the hydrocarbon substrate to insert oxygen. Loss of ROH and uptake of an H_2O molecule at the vacated coordination position bring the cycle back to the resting state (Ortiz de Monetallano & De Voss, 2002).

5. Iron-Activated Enzymes

The most prominent example of this class of enzymes is heme oxygenase, which has evolved to carry out the oxidative cleavage of heme, a reaction essential in several physiological processes as diverse as iron reutilization and cellular signaling in mammals and synthesis of essential light-harvesting pigments in cyanobacteria and higher plants, as well as the acquisition of iron by bacterial pathogens (Wilks, 2002). Heme oxygenase (HO) is concentrated in both blood vessel endothelium and adventitial neurons which suggests that HO subserves functions that are handled by NO-generating enzymes. The gene of this cellular stress protein, mediating the catabolism of heme to biliverdin in brain and other tissues, is strongly induced by dopamine, oxidative stress, and metal ions. In the brain, it is primarily expressed in the astroglia, and, when upregulated, HO promotes mitochondrial iron deposition in these cells (Schipper, 1999). HO protects cells by degrading prooxidant metalloporphyrins and appears to facilitate iron efflux from the cell (Berg et al., 2001). HO is responsible not only for iron export but also for the intracellular sequestration of iron by glial mitochondria (Schipper, 1999).

6. Fe/S Enzymes

The porphyrin ligand environment of iron that occurs in hemoglobin and myoglobin is also important in redox enzymes. Thus, in the large class of biochemically important heme proteins iron is coordinated to a porphyrin ligand. All other iron proteins are defined as non-heme. Those that contain iron in a tetrahedral environment of four sulfur atoms are particularly important. Iron–sulfur clusters, however, were not familiar to inorganic chemists prior to recognition of their biochemical importance. Iron–sulfur clusters are simple inorganic groups that are contained in a variety of proteins having functions related to electron transfer, gene regulation, environmental sensing, and substrate activation. Biological Fe-S clusters, however, are not formed spontaneously, but a consortium of highly conserved proteins is required for both the formation of Fe-S clusters and their insertion into various protein partners. The formation or transfer of Fe-S clusters appears to require an electron-transfer step (Frazzon et al., 2002). Table X provides examples of proteins with Fe-S clusters and their function.

Although not all of the enzymes involved in the mitochondrial respiratory chain are iron–sulfur proteins, it is enlightening to look through the various parts of the chain. Electrons move from reduced nicotinamide adenine dinucleotide (NADH), succinate, or some other primary electron donor, through flavoproteins, ubiquinone, iron–sulfur proteins and cytochromes, and finally to O_2. Table XI summarizes the protein components of the mitochondrial electron-transfer chain.

In the context of iron–sulfur proteins, we can observe an intriguing link to another essential metal (namely, molybdenum) through xanthine dehydrogenase and xanthine oxidase. These are the most studied of the small but important class of molybdenum-containing iron–sulfur flavoproteins. Xanthine oxidoreductase catalyzes the hydroxylation of a wide variety of purine, pyrimidine, pterin, and aldehyde substrates. The active form of the enzyme is a homodimer of molecular mass 290 kDa, and each of the monomers acts independently in catalysis. Each subunit contains one molybdopterin group, two identical 2Fe-S centers, and one flavin adenine dinucleotide cofactor (Nishino & Okamoto, 2000).

7. Transferrin, Lactoferrin, and Hemopexin

Transferrin originally was the name of the serum protein that binds and transports iron for delivery to cells. Today, it is the name applied to a wider family of homologous proteins that includes serum transferrin,

TABLE X.   Some Examples of Proteins and Enzymes with [Fe–S] Clusters and Their Function

Enzyme or Protein	Function	Cofactors
Rubredoxin	Electron transfer	Fe
Ferredoxin, Rieske ferredoxin	Electron transfer	[2Fe–2S]
Phthalate dioxygenase reductase	Electron transfer	[2Fe–2S], FAD
Naphthalene dioxygenase	O_2-activation	[2Fe–2S], Fe(II)
Adenine glycosylase, glutamine PRPP amidotransferase, endonuclease III	Structural stabilization	[4Fe–4S]
Aconitase	Electrophilic catalysis	[4Fe–4S]
IRP-1,[a] FNR,[b] soxRS[c]	Regulation of gene expression	[4Fe–4S], [2Fe–2S]
Dinitrogenase reductase, dinitrogenase	Nitrogen fixation	[4Fe–4S], FeMoCo, P-clusters

[a] Iron regulatory protein-1.
[b] Ferredoxin:NADP+oxidoreductase.
[c] Superoxide regulatory system.

TABLE XI.   Protein Components of the Mitochondrial Electron-Transfer Chain

Enzyme Complex	Mass (kDa)	Number of Subunits	Prosthetic Group(s)
I NADH dehydrogenase	850	42	FMN,[a] [Fe–S]
II Succinate dehydrogenase	140	5	FAD,[b] [Fe–S]
III Ubiquinone: cytochrome *c* oxidoreductase	250	11	Hemes, [Fe–S]
Cytochrome *c*	13	1	Heme
IV Cytochrome oxidase	160	13	Hemes, Cu_A, Cu_B

[a] Flavin mononucleotide.
[b] Flavin adenine dinucleotide.

lactoferrin, ovotransferrin, and melanotransferrin (Baker *et al.*, 2003). Only serum transferrin has a proven transport function; however, transferrins seem to be involved in the homeostatic control of free iron in all the places where it might be found. The serum transferrin has two structurally similar lobes. Both lobes contain one iron-binding center that is very specific for iron(III) and has a binding constant of about 10^{20}. It is only when transferrin is loaded with two ferric ions that it binds strongly to the receptor for internalization. In addition to iron(III), transferrin can bind strongly to a range of other metal ions. Many such complexes are still recognized by the transferrin receptor, and some bacteria have transferrin receptors. Transferrin, therefore, holds promise for use in the development of antimicrobial therapies (Andrews *et al.*, 2003).

Lactoferrin is an iron-binding protein that binds iron even more tightly than transferrin. It is present in milk, many other exocrine secretions, and white blood cells (Baker *et al.*, 2003). One of the first functions attributed to lactoferrin was the ability to inhibit bacterial growth and viral infection. It is thought that lactoferrin is able to sequester iron from certain pathogens, which inhibits their growth. Another important function is its ability to stimulate the release of the neutrophil-activating polypeptide interleukin-8 (IL-8). This suggests that lactoferrin may function as an immunomodulator for activating the host defense system (Kruzel & Zimecki, 2002).

Hemopexin is a recycler and transporter of heme. Turnover of heme proteins, notably hemoglobin, leads to the release of heme into extracellular fluids with potentially severe consequences. Like free iron, heme is a source of essential iron for invading bacterial pathogens and is highly toxic because of its ability to catalyze free-radical formation. Protection is given by hemopexin, a 60-kDa serum glycoprotein that sequesters heme with very high affinity from the bloodstream; transports it to specific receptors on liver cells, where it undergoes receptor-mediated endocytosis; and releases the bound heme into cells. It thus serves both to protect against heme toxicity and to conserve and recycle iron (Baker *et al.*, 2003).

8. Ferritin and Hemosiderin

Ferritins constitute a class of iron storage proteins found in bacterial, plant, and animal cells. They form hollow, spherical particles in which up to 4500 iron atoms can be stored as iron(III). Although ferritins are quite small molecules about 8 to 12 nm in diameter, they are very effective iron stores. Additionally, they can provide iron on demand. The biosynthesis of ferritin is controlled by the level of iron in the cell via the iron regulatory protein (Andrews et al., 2003). Hemosiderin is another iron-storage complex; however, knowledge of its structure is minimal. It is found solely in cells, in contrast to ferritin, which can also be found in the circulation. It has been suggested that hemosiderin could be a complex of ferritin, denatured ferritin, and some other material. Iron present in hemosiderin deposits is poorly available to provide iron on demand. The storage complex is found in macrophages and appears to be especially abundant following hemorrhage; thus, its formation might be related to phagocytosis of red blood cells and hemoglobin (Trinder et al., 2000).

G. Copper: The Master of Oxidases

Historical records show that copper and copper compounds had been used medicinally at least as early as 400 BC (Mason, 1979). Many copper compounds were used to treat a variety of diseases during the nineteenth century, and the presence of copper in plants and animals was recognized more than 150 years ago. For quite some time it has been widely accepted that copper is an essential trace element required for survival by all organisms, from bacterial cells to humans (Linder & Hazegh-Asam, 1996). Copper ions undergo a unique chemistry due to their ability to adopt distinct redox states, either oxidized [Cu(II)] or in the reduced state [Cu(I)]. Consequently, copper ions serve as important catalytic cofactors in redox chemistry for proteins that carry out fundamental biological functions; however, copper provides a challenge to biological systems. The very properties that make copper indispensable to biology become toxic when copper is present in excess. Copper's outstanding redox properties, however, such as the transitions between Cu(II) and Cu(I), can in certain circumstances result in the generation of reactive oxygen species such as superoxide radicals and hydroxyl radicals. Susceptible cellular components can be damaged by these reactive species if an effective scavenging mechanism is not in operation. Copper can also bind with high affinity to histidine, cysteine, and methionine residues of proteins. This may result in the

inactivation of the proteins (Camarakis et al., 1999). Consequently, there is a great need for effective homeostatic mechanisms controlling the cellular concentration of copper.

1. Copper Proteins

Copper is present in three different forms in proteins: (1) blue proteins without oxidase activity (e.g., plastocyanin), which function in one-electron transfer; (2) non-blue proteins that produce peroxidases and oxidize monophenols to diphenols; and (3) multicopper proteins containing at least four copper atoms per molecule and acting as oxidases (e.g., ascorbate oxidase and laccase). Table XII provides an arbitrary selection of copper-dependent proteins to emphasize the versatility of copper proteins (Peña et al., 1999; Fraústo da Silva & Williams, 2001).

2. Intracellular Distribution of Copper

To be distributed in the cell, copper has to be taken up in the cell. This is accomplished by copper transport 1 (Ctr1p), which is a membrane-spanning protein. Ctr1p specifically transports Cu(I), not Cu(II) or any other metals. A conspicuous finding was that the gene *CTR1* (Cu transport 1) was first discovered not as a gene for copper but as a gene essential for iron transport in *Saccharomyces cerevisiae* (Harris, 2000). The copper was required for Fet3, a multicopper ferroxidase that catalyzed the oxidation of Fe(II) to Fe(III).

The intracellular environment is generally reducing. This means that Cu(II) is rapidly reduced to Cu(I). Glutathione was identified early as being involved in copper transport. Glutathione (GSH) is a cysteine-containing tripeptide present more or less exclusively within cells at concentrations in the millimolar range. Copper(I) reacts directly with the internal cysteine sulfhydryl group of glutathione; however, it is less likely that Cu(II) binds to glutathione because of the propensity of Cu(II) to catalyze the oxidation of sulfhydryl groups. The formation of Cu(I)–GSH is a spontaneous reaction probably independent of enzyme involvement. Besides transferring copper to metallothionein, glutathione is required for biliary excretion of copper. In addition, there is evidence that a Cu–GSH complex can mediate stable Cu(I) binding to apocuproteins (Harris, 2000).

Copper chaperones are a family of cytosolic peptides. They are usually referred to as metallochaperones and serve as an intracellular shuttle service for metal ions. In the case of copper, they form transient complexes with Cu(I). These chaperones escort copper ions on

TABLE XII.   Examples of Copper-Dependent Proteins

Protein	Function
Cytochrome oxidase	Reduction of O_2 to H_2O
Laccase	Oxidation of phenols
Ceruloplasmin	Oxidation of Fe(II) to Fe(III), Cu transport
Hemocyanin	Transport of O_2
Lysine oxidase	Cross-linking of collagen
Ascorbate oxidase	Oxidation of ascorbate
Galactose oxidase	Oxidation of primary alcohols to aldehydes in sugars
Amine oxidase	Removal of amines and diamines
Blue proteins	Electron-transfer (many kinds)
Superoxide dismutase	Superoxide dismutation (defense)
Nitrate reductase	Reduction of NO_2^- to NO
Nitrous oxide reductase	Reduction of N_2O to N_2
Metallothionein	Cu(I) storage
Dopamine monooxygenase	Hydroxylation of Dopa
Co-proporphyrin decarboxylase	Production of protoporphyrin IX
Ethylene receptor	Hormone signaling
Methane oxidase	Oxidation to methanol
Terminal glycine oxidases	Production of signal peptides
Tyrosinase	Melanin production
Clotting factors V and VII	Blood clotting
Angiogenin	Induction of blood vessel formation
Hephaestin	Iron egress from intestines
CP-x type ATPase	Copper pump
Atx-I (Lys 7)	Copper transfer

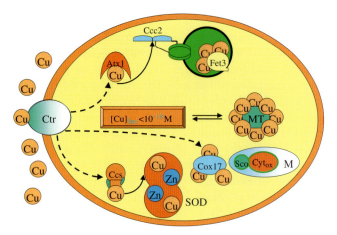

FIGURE 21   A schematic view of copper localization in the cell.

three copper atoms for proper function. Two are situated in a dinuclear site in one subunit and one in another subunit next to heme a_3. It has been suggested that Cox17, a small 8-kDa protein, functions in a copper trafficking pathway of cytochrome oxidase (Huffman & O'Halloran, 2001). The CCS metallochaperone seems to be necessary for the incorporation of copper into the radical scavenging enzyme Cu,Zn–superoxide dismutase. A schematic view of copper uptake and cellular localization is shown in Figure 21. A few intriguing examples of copper-binding proteins are presented in the next section.

3. Ceruloplasmin

This protein was first isolated from plasma and characterized as a copper-containing protein as early as 1948 (Holmberg & Laurell). It was later discovered that the concentration of this protein was low in patients with Wilson's disease. The first proposed physiological role of the protein was in iron homeostasis and as a ferroxidase (Hellman & Gitlin, 2002). Ceruloplasmin belongs to the group of multicopper oxidases. This group of proteins has three distinct copper sites, which are type I copper sites. Charge transfer between the cysteine ligand sulfur and the copper at these sites results in strong absorption at 600 nm. This is why they are called blue proteins. A single copper of type II is coordinated by four imidazole nitrogens and is in close proximity to two antiferromagnetically coupled type III copper ions absorbing at 330 nm. The type II and type III coppers form a trinuclear copper cluster, which is the site of oxygen binding during the catalytic cycle (Hellman & Gitlin, 2002).

Multicopper oxidases use the smooth electron chemistry of bound copper ions to couple substrate oxidation

their way into proteins that require copper. The prototype of metallochaperones is MerP, a small soluble mercury-binding protein that transports Hg^{2+} to a membrane transporter and eventually to a reductase that reduces Hg^{2+} to the volatile Hg^0 as part of a detoxifying process (Lund & Brown, 1987). The best known metallochaperone protein is Atx1 (antioxidant 1), which was originally isolated as an antioxidant protein in *Saccharomyces cerevisiae* and functions in a high-affinity iron uptake pathway in yeast. It works together with Ccc2, a copper-transporting P-type ATPase. The cooperation of these two proteins is necessary for copper loading of the multicopper oxidase Fet3, which is required for the high-affinity iron uptake (Huffman & O'Halloran, 2001).

The human homologs of Ccc2 are the Wilson's (Atp7ab) and Menkes' (Atp7a) disease proteins, and the human homolog of Fet3 is ceruloplasmin. Cytochrome oxidase is essential for cellular respiration and requires with the four-electron reduction of molecular oxygen. Electrons pass from the substrate to the type I copper and then to the trinuclear copper cluster and subsequently to the oxygen molecule bound at this site (Messerschmidt *et al.*, 1989). In addition to ceruloplasmin, several multicopper oxidases have been identified that play a critical role in iron homeostasis. Fet3 is a ferroxidase essential for iron uptake in yeast, and hephaestin is a ceruloplasmin homolog required for efficient iron efflux from the placenta and enterocytes in mammals (Vulpe *et al.*, 1999). Ceruloplasmin is an acute-phase reactant, and its serum concentration increases during inflammation, infection, and trauma largely as a result of increased gene transcription in hepatocytes mediated by the inflammatory cytokines (Hellman & Gitlin, 2002).

The liver largely exercises copper homeostasis, and the hepatocytes are the primary site of copper metabolism. Hepatocytes are highly polarized epithelial cells that regulate copper excretion in the bile dependent on the intracellular copper concentration. The copper chaperone Atx1 is required for the delivery of copper to ceruloplasmin; however, the mechanism of copper incorporation into ceruloplasmin is not well understood. Studies of *Saccharomyces cerevisiae* show that both the H^+-transporting V-type ATPase and the CLC chloride channel Gef1 are necessary for copper incorporation in the homologous multicopper oxidase Fet3 (Gaxiola *et al.*, 1998). Ceruloplasmin is capable, *in vitro*, of catalyzing oxidation of a number of different substrates. This has caused some confusion as to the physiologic role of this protein. Ceruloplasmin from human serum has considerable ferroxidase activity necessary for the oxidation of the ferrous iron and incorporation of ferric iron in apotransferrin (Osaki *et al.*, 1971).

4. Superoxide Dismutase

Superoxide dismutase (SOD) is a member of the oxidoreductase family of enzymes. There are several forms of superoxide dismutase, including Mn–SOD, Fe–SOD, and Cu,Zn–SOD. The Mn–SOD is found exclusively in the mitochondria, whereas the Fe–SOD is generally found in prokaryotes. The Cu,Zn–SOD is active in the cytoplasm of eukaryotes and is the most abundant form of SOD. Cu,Zn–SOD is a dimer of identical subunits with a molecular weight of 16 kDa. One Cu(II) and one Zn(II) are included in each dimer. The role of zinc ions in SOD is structural, and the copper ions take part in the catalytic process. It is based on a redox process, and zinc does not take part in those reactions. SOD catalyzes the disproportion of the superoxide anion ($O_2^{-\bullet}$) to a less dangerous reactive oxygen species. SOD cat-

alyzes the conversion of the superoxide anion according to the following process:

$$Cu^{2+} + O_2^{\bullet-} \rightarrow Cu^+ + O_2$$
$$Cu^+ + O_2^{\bullet-} + 2H^+ \rightarrow Cu^{2+}H_2O_2$$
$$\text{Net reaction: } 2O_2^{\bullet-} + 2H^+ \rightarrow H_2O_2 + O_2$$

Recently, a Ni-containing SOD was found in *Streptomyces griseus* and *S. coelicolor* (Youn *et al.*, 1996). The subunits of Cu,Zn–SOD are stabilized by an intrachain disulfide bond but associated by noncovalent forces. This enzyme requires copper and zinc for its biological activity, and the loss of copper results in its complete inactivation, which can lead to the development of human diseases.

F. Zinc: The Ubiquitous Trace Element

Zinc has been known to be essential to life since 1869, when it was discovered that it was required by *Aspergillus niger*. It differs chemically from its neighbors in the transition metal area of the periodic table. Zinc does not take part in redox reactions, but it is a good Lewis acid. In fact, it could be called Nature's Lewis acid. It has a hard metal center and is ideally suited for coordination of N- and O-donors. It is also highly polarizing, and the activity of Zn(II)-dependent enzymes is due to the Lewis acidity of the metal center. In addition, it is characterized by fast ligand exchange. The first enzyme to be recognized as a zinc metalloenzyme was carbonic anhydrase, an enzyme essential for respiration in mammals. At present, zinc metalloenzymes have been recognized in all classes of enzymes. Today, more than 300 enzymes are known to be dependent on zinc for catalytic, structural, regulatory, and noncatalytic functions. Examples of enzymes in which zinc plays a catalytic role include carbonic anhydrase, carboxypeptidase, thermolysin, and aldolases. Zinc stabilizes the quaternary structure of oligomeric holoenzymes. It dimerizes *Bacillus subtilus* α-amylase without affecting its enzymatic activity and stabilizes the pentametric quaternary structure of asparatate–transcarbamylase. Zinc acts as an activator of bovine lens leucine aminopeptidase and inhibits the activity of porcine kidney leucine aminopeptidase and fructose-1,6-bisphosphatase.

1. Zinc and Enzymes

Zinc has three types of roles with regard to enzymes: catalytic, cocatalytic, and structural (see Table XIII). Carbonic anhydrases are a widely expressed family of enzymes that catalyze the reversible reaction $CO_2 + H_2O \rightleftharpoons HCO_3^- + H^+$. These enzymes therefore both

TABLE XIII.   Examples of Zinc Metalloproteins

Enzyme	Role of Zinc	Enzyme	Role of zinc
Class I: Oxidoreductases		Carboxypeptidase (other)	Catalytic
Alcohol dehydrogenase	Catalytic, non-catalytic	Carboxypeptidase A	Catalytic
D-Lactate cytochrome reductase	?	Carboxypeptidase B	Catalytic
D-Lactate dehydrogenase	Catalytic	Collagenase	Catalytic
Superoxide dismutase	Structural (copper catalytic)	Creatinase	?
		Cytidine deaminase	Catalytic
		D-Carboxypeptidase	Catalytic
Class II: Transferases		DD carboxypeptidase	Catalytic
Aspartate transcarbamylase	Structural	Dihydropyrimidine aminohydrolase	?
Cobalamin-dependent methionine synthase	Catalytic	Dipeptidase	Catalytic
		Elastase	?
Cobalamin-independent methionine synthase	Catalytic	Fructose-1,6-bisphosphatase	Regulatory
		Neutral protease	Catalytic
DNA polymerase	Catalytic	Nuclease P_I	?
Mercaptopyruvate sulphur transferase	?	Nucleotide pyrophosphatase	Catalytic
Nuclear poly(A) polymerase	Catalytic	Phosphodiesterase (exonuclease)	Catalytic
Phosphoglucomutase	?	Phospholipase C	Catalytic
Protein farnesyltransferase	Catalytic	Procarboxypeptidase A	Catalytic
Reverse transcriptase	Catalytic	Procarboxypeptidase B	Catalytic
RNA polymerase	Catalytic	Thermolysine	Catalytic
Terminal dNT transferase	Catalytic		
Transcarboxylase	?	Class IV: Lyases	
		δ-Aminolevulinic acid dehydratase	Catalytic
Class III: Hydrolases		Carbonic anhydrase	Catalytic
α-Amylase	Structural	Fructose-1,6-bisphosphatase adolase	Catalytic
α-D-Mannosidase	?	Glycoxalase	Catalytic
β-Lactamase II	Catalytic	L-Rammulose-1-phosphate adolase	Catalytic
Adenosine deaminase	Catalytic		
Alkaline phosphatase	Catalytic, noncatalytic	Class V: Isomerases	
Aminocyclase	?	Phosphomannose isomerase	?
Aminopeptidase	Catalytic, regulatory		
Aminotripeptidase	Catalytic	Class VI: Ligases	
AMP deaminase	?	Pyruvate carboxylase	?
Angiotensin-converting enzyme	Catalytic	TRNA synthetase	Catalytic
Astacin	Catalytic		

Source: Data from Prasad (1995) and McCall *et al.* (2000).

produce HCO_3^- for transport across membranes and consume HCO_3^-, which has been transported across membranes (Sterling *et al.*, 2001). Erythrocytes of mammals have two isoenzymes of carbonic anhydrase: CAI, which has a low activity, and CAII, which has high activity. The molecular weight of human CAI is 30,000 Da, and it contains one atom of zinc per molecule. Human CAII has 259 amino acids, while human CAI has 260 amino acids, and the two isoenzymes share 60% sequence homology. In carbonic anhydrase, zinc is catalytic.

An example of the structural role of zinc in enzymes is aparatate transcarbamylase (ATCase). This enzyme catalyzes the first step in pyrimidine biosynthesis, condensation of aspartate, and carbamyl phosphate. It is an allosterically regulated enzyme, and its activity is inhibited by cytidine triphosphate (CTP) and activated by ATP. Both responses seem to make sense from a physiological perspective. If CTP levels are already high, additional pyrimidines are not needed, and high ATP signals offers both a purine-rich state and an energy-rich cell condition under which DNA and RNA synthesis can be active. There are similarities with hemoglobin in that the allosteric regulation of ATCase involves changes in the quaternary structure of the molecule. A change in the molecule from the tense state to the relaxed state involves a major rearrangement of subunit positions (Purcarea et al., 1997).

Leucine aminopeptidase (LAP) is a prototypic dizinc peptidase that has been studied intensely. The enzyme is present in animals, plants, and bacteria and has various tissue-specific physiological roles in the processing or degradation of peptides. Human LAP has been shown to catalyze postproteosomal trimming of the N terminus of antigenic peptides for presentation on major histocompatibility complex class I molecules (Beninga et al., 1998; Sträter et al., 1999). The two zinc atoms bound to LAPs work in two different ways: One has a catalytic function, and the other regulates the activity induced by the zinc atom at the first site.

Alcohol dehydrogenase is a zinc-dependent enzyme. At least in some species, zinc may be essential for protection against oxidative damage (Tamarit et al., 1997). In humans, there are at least nine different forms of the enzyme, most of which are found in the liver. It should be noted that an unusual iron- and zinc-containing alcohol dehydrogenase has been identified in the hyperthermophilic archaeon *Pyrococcus furiosus* (Ma & Adams, 1999). Alcohol dehydrogenase is our primary defense against alcohol intoxication. It catalyzes the following reaction:

$$\text{Ethanol} \underset{\text{dehydrogenase}}{\overset{\text{Alcohol}}{\rightleftarrows}} \text{Acetaldehyde} \underset{\text{hydrogenase}}{\overset{\text{Aldehydede}}{\rightleftarrows}} \text{Acetic acid}$$

In fact, alcohol dehydrogenase catalyzes a transformation to a yet more toxic product, acetaldehyde. So, this toxic molecule is transformed in the next step by aldehyde dehydrogenase to acetic acid and other molecules (Duester, 1998) that can be used by the cells. Alcohol dehydrogenase also catalyzes the transformation of retinol in the eye to retinaldehyde and by aldehyde dehydrogenase to retinoic acid. This first line of defense against alcohol, however, is beset with some dangers because alcohol dehydrogenase also modifies other alcohols, often producing dangerous products. For instance, methanol is converted into formaldehyde, which causes damage to proteins and possibly cancer. Small amounts of methanol cause blindness when the sensitive proteins in the retina are attacked, and larger amounts lead to widespread damage and death (Barceloux et al., 2002).

The alcohol dehydrogenases are NAD(H) dependent and have two subunits. Each subunit contains two zinc atoms and binds one molecule of NAD(H). One zinc atom is essential for the catalytic effect, and the role of the second is largely still unknown; however, it does not seem to be necessary for structure in many cases. There is evidence that zinc may be of importance for the conformational stability of yeast alcohol dehydrogenase (Yang & Zhou, 2001).

2. Zinc and Gene Expression

It is by now well established that zinc plays a very important role in gene expression. The importance can be appreciated from the fact that about 25% of the zinc content of rat liver is found in the nucleus, and a significant amount of zinc is incorporated into nuclei *in vitro* (Cousins, 1998). One subject of major importance is genetic stability. The correct sequence of nucleotides in DNA is essential for proper replication, gene expression, and protein synthesis. Zinc is involved in the processes of genetic stability and gene expression in a variety of ways, including the structure of chromatin, replication of DNA, and transcription of RNA through the activity of transcription factors and RNA and DNA polymerases, as well as playing a role in DNA repair and programmed cell death (Falchuk, 1998).

Until the late 1980s, DNA-binding proteins were not well represented among the nearly 300 zinc-containing proteins known at that time. The multisubunit bacterial RNA polymerases were found to be zinc dependent in the early 1970s. Following these findings, the eukaryotic RNA polymerases, containing many more subunits than the bacterial enzymes, were also found to be zinc enzymes. Replicative DNA polymerases are essential for the replication of the genome of all living organisms. They catalyze the chemical reaction of DNA synthesis:

$$\text{template} - \text{primer} - (\text{dNMP})_n + \text{dNTP} \rightarrow$$
$$\text{template} - \text{primer} - (\text{dNMP})_{n+1} + \text{PP}_i$$

where dNMP and dNTP are deoxynucleoside 5'-monophosphate and 5'-triphosphate, respectively. During the reaction, inorganic pyrophosphate is released. The reaction requires a 3'-hydroxyl group of the primer for the nucleophilic attack on the α-phosphate of the incoming dNTP. The released pyrophosphate is hydrolyzed, thus providing energy for the reaction.

The sequence similarities may be used to classify them into three types. Type A polymerases are homologous to bacterial polymerases, type B includes archaeal DNA polymerases and eukaryotic DNA polymerase α, and type C is made up of the bacterial polymerase III class. The catalytic mechanism of all three types involves two metal-binding acidic residues in the active site.

One very important aspect of zinc is the so-called zinc fingers, which were first discovered as the transcription factor IIIA (TFIIIA). TFIIIA is a site-specific DNA-binding protein that plays a central role in controlling the transcription of 5S ribosomal RNA genes in the African toad *Xenopus laevis*. This protein is slightly unusual because not only does it recognize the internal control region of about 45 bp in the center of the 5S RNA gene, but also TFIIIA itself is bound to the product. The name "zinc fingers" was introduced because of the specific interaction between the amino acids cysteine and histidine and a zinc ion responsible for the formation of the characteristic loop structure (Figure 22). Zinc fingers are generic protein motifs that can mediate DNA-binding and are both widespread and multifunctional. Since first being discovered in the early 1980s, several more zinc-finger proteins have been identified. More than 50 zinc-finger proteins are known today.

Zinc-finger domains are common, relatively small protein motifs that fold around one or more zinc ions. In addition to their role as DNA-binding modules, zinc-finger domains have recently been shown to mediate protein:protein and protein:lipid interactions (Ladomery & Dellaire, 2002). This small zinc-ligating domain, often found in clusters containing fingers with different binding specificities, can facilitate multiple, often independent intermolecular interactions between nucleic acids and proteins. Classical zinc fingers, typified by TFIIIA, ligate zinc via pairs of cysteine and his-

tidine residues, but there are at least 14 different classes of zinc fingers, which differ in the nature and arrangement of their zinc-binding residues. Some types of zinc fingers can bind to both DNA and a variety of proteins. Thus, proteins with multiple fingers can play a complex role in regulating transcription through the interplay of these different binding selectivities and affinities (Ladomery & Dellaire, 2002). Other zinc fingers have more specific functions, such as DNA-binding zinc-finger motifs in the nuclear hormone receptor proteins and small-molecule-binding zinc fingers in protein kinase C. Some classes of zinc fingers appear to act exclusively in protein-only interactions. It has also been suggested that zinc fingers, in addition to the functions described above, play a protective role through their prevention of chemical attack by, for example, radicals or reactive oxygen species (Dreosti, 2001).

3. Zinc and Metallothionein

Metallothionein is a generic name for a superfamily of ubiquitous proteins or polypeptides possessing sulfur-based metal clusters. These proteins consist of a single polypeptide chain of 61 to 62 amino acids that contains 20 cysteine residues that, in turn, contain several bivalent cations such as zinc bound through metal–thiolate linkages. These clusters are usually formed through the preferential coordination of d^{10} metal ions by the cysteine thiolate ligands. Currently, four isoforms of metallothionein have been identified. Not all of them are expressed in all organs of mammals. Quite substantial differences exist among species. Metallothionein isoforms I and II have a ubiquitous tissue distribution, with particular abundance in liver, pancreas, intestine, and kidney, whereas isoforms III and IV are found principally in brain and skin (Davis & Cousins, 2000).

The zinc coordination in metallothionein is exceptional. The structure is a dumbbell-shaped molecule with two domains, in each of which zinc is bound in a cluster. In one domain, three zinc atoms are bound to nine cysteines; in the other domain, four zinc atoms are bound to 11 cysteines. In this way, each zinc atom is bound tetrahedrally to four cysteines, but overall there are fewer than the maximum number of possible ligands for the seven metals; consequently, some of the cysteines form ligand bridges that form an extensive zinc–sulfur network. The protein envelops the zinc atoms in a manner that effectively shields them from the environment and leaves only a few of the sulfur ligands partially exposed to solvent. Because both protein structure and tetrahedral zinc coordination preclude access of ligands to zinc, it would seem that a conformational

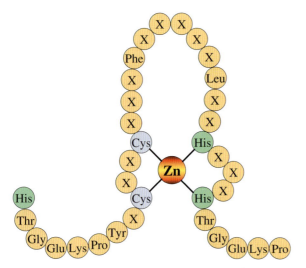

FIGURE 22 General structure of zinc fingers.

change of the protein is necessary to release zinc (Maret, 2000).

It is challenging that very few suggestions as to the biological functions of metallothionein have emerged, although the protein has been extensively studied for decades. The main consensus seems to be that it has a role in the detoxification of metal. The biological function of metallothionein is likely related to the physiologically relevant metals that it binds. Recent studies have produced strong evidence to support the idea that metallothionein functions as a metal chaperone for the regulation of gene expression and for the synthesis and functional activity of metalloproteins and metal-dependent transcription factors (Xun & Kang, 2002). Vital roles for this pleiotropic protein in more primitive life forms involve the sequestration of environmental metals such as cadmium and mercury.

It has been suggested recently that a biological function of metallothionein is to provide redox functions to the cells. The association of zinc with only sulfur ligands and the biological significance of the peculiar cluster of metallothionein and its purpose have not been elucidated in detail; however, Maret (2000) provided a chemical solution to this challenge: The cluster unit operates via a novel mechanism that allows the cysteine sulfur ligands to zinc to be oxidized and reduced with the concomitant release and binding of zinc. This results in an oxidoreductive process exercised by the ligands of the otherwise redox-inert zinc atom. Thus, metallothionein can become a redox protein in which the redox chemistry originates not from the metal atom but rather from its coordination environment.

Although zinc, copper, cadmium, mercury, gold, and bismuth are all metals that induce metallothionein, zinc is the primary physiological inducer. Zinc and copper are essential trace elements, and the other metals are environmental toxicants; however, copper in nontoxic concentrations does not induce metallothionein (Coyle & Phicox, 2002). Metal regulation of metallothionein genes is a complex process involving several steps. In short, the binding of zinc to metal transcription factor 1 (MTF-1) allows the protein to bind to metal-responsive elements (MREs) in the promotor region, which subsequently initiates metallothionein gene transcription. It has been proposed that MTF-1 regulates the free zinc concentration by controlling the expression of metallothionein as well as that of a zinc-transporter protein, ZnT-1. Basal expression of MTF-1 may be controlled by a zinc-sensitive inhibitor, which prevents MTF-1 binding to MREs. Zinc dissociates the inhibitor from MTF-1, thereby promoting transcription of metallothionein. MTF-1 is important in the regulation of a number of genes that play a role in cellular response to various stresses (Coyle & Phicox, 2002). Figure 23 illustrates a model of the complex interactions controlling the expression of metallothionein.

4. Zinc and Inflammation

There is no single factor regulating metallothionein synthesis in inflammation; instead, a complex interrelationship exists between factors that, in combination and in different tissues, act synergistically on metallothionein gene transcription. Nucleotide sequences other than MREs respond to glucocorticoids, interleukin-6 (IL-6), phorbol esters, and hydrogen peroxide. Many of the acute-phase proteins appear to be regulated by combinations of the same factors that include catecholamines and glucocorticoids as well as the cytokines IL-6, IL-1, TNF-α, and γ-interferon. Unlike other acute-phase proteins, metallothionein induction by inflammatory mediators has been found to be conditional upon the presence of zinc. Reactive oxygen species generated during the inflammatory response may induce metallothionein through multiple pathways, including directly stimulating an antioxidant response element and specific MREs in the promoter region as well as by events associated with various second-messenger protein kinase pathways.

5. Zinc, Insulin, and Diabetes

Several forms of disordered glucose metabolism are collectively referred to as diabetes. Although all diabetes mellitus (mellitus = "sweet as honey") syndromes have some degree of hyperglycemia in common, this is a symptom rather than the metabolic error itself. In insulin-dependent diabetes mellitus (IDDM), there is a destruction of the beta cells of the islets of Langerhans in the pancreas, most often on an autoimmune basis, which results in no insulin being produced. Without insulin, muscle, fat, and liver cells cannot transport glucose from the blood to the intracellular space. Intracellular starvation ensues, with fats becoming the primary intracellular energy source. This form of energy generation results in the production of ketone bodies and organic acids, primarily acetoacetic and beta hydroxybutyric acids, resulting in the development of severe metabolic acidosis (Chausmer, 1998).

With non-insulin-dependent diabetes mellitus (NIDDM), the pancreatic islet cells are capable of producing large quantities of insulin, at least at the beginning of the disease. In the healthy individual, insulin binds to a cell membrane receptor and, through several pathways, results in the transport of glucose across the

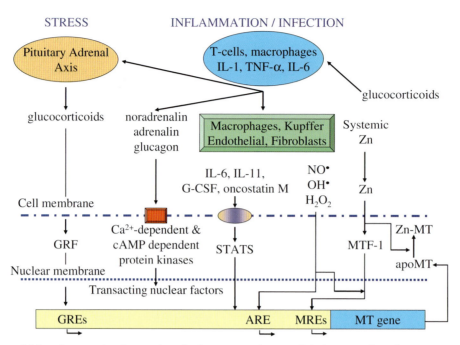

FIGURE 23 A model for the complex interactions both extra- and intracellularly controlling the expressions of MT. (Adapted from Coyle *et al.*, 2002.)

membrane. The intracellular events associated with the activation of glucose transport, after the signal from the insulin–receptor complex is received, are referred to as postreceptor events. To a great extent, it is the failure of the postreceptor events that results in hyperglycemia. In response to the hyperglycemia, the pancreatic islets produce greater and greater quantities of insulin, resulting in downregulation of the number of insulin receptors on the cell membrane and compounding the problem. This results in both hyperglycemia and hyperinsulinemia (Chausmer, 1998).

There is an intriguing relation between zinc and insulin regarding the storage of insulin in granulae of the pancreatic islets. Although insulin circulates in the blood and binds to its receptor as a monomer, it forms dimers at micromolar concentrations, and in the presence of zinc ions it further assembles to hexamers. The insulin monomer itself consists of two chains: an A chain of 21 amino acids and a B chain of 30 amino acids. Insulin is synthesized in the β cells of the islets of Langerhans. The β cells are characterized by two features associated with cells that export proteins: rough endoplasmatic reticular surfaces and well-defined storage vesicles. In this case, the vesicle typically contains microcrystals of packaged insulin. The presence of zinc is crucial for the stability of this storage system. On release into the blood, the insulin microcrystals experi-

ence a change of pH from about 5.5 to 7.4. This causes the hexamers and therefore the crystal to disintegrate rapidly (Dodson & Steiner, 1998).

The predominant effect on zinc homeostasis of diabetes is hypozincemia, which may be the result of hyperzincuria or decreased gastrointestinal absorption of zinc or both. Whereas the evidence for increased zinc excretion is uniform, the data supporting decreased absorption of zinc are less clear cut. It appears that hyperzincuria is more a result of hyperglycemia than of any specific effect of endogenous or exogenous insulin on the renal tubule (Chausmer, 1998).

The zinc–metallothionein complex in the islet cells may provide protection against radicals produced in the cell from any cause, and certainly the immune-mediated, cytokine-provoked oxidative stress would be a significant oxidative stress. The more depleted the intracellular zinc stores, the less able the cell is to defend itself against this oxidative load. This provides a potential mechanism for zinc deficiency to affect the progress of IDDM. With NIDDM, there is no good evidence for oxidative stress as a major factor in the development of either insulin deficiency or islet cell damage; however, there is clear evidence for increased secretion of insulin, at least early in the progress of the disease. Because zinc leaves the cell with insulin, the greater secretion of insulin causes depletion of zinc. The cell can make more insulin, but it cannot make more zinc. With hypozincuria and decreased retention, the zinc is more likely to be excreted and not available for reuptake into the cellular pool. Zinc deficiency may therefore negatively affect the progress of NIDDM (Chausmer, 1998).

6. Zinc and the Immune System

It is by now well recognized that nutritional factors are important for the function of the immune system. In the case of zinc, the situation currently of interest is its deficiency and subsequent adverse effects on immune functions. Studies in young adult mice have shown greatly depressed responses to both T-lymphocyte-dependent and -independent antigens. Both primary and secondary antibody responses have been reported to be lowered in zinc-deficient mice. Declines in *in vivo*-generated cytotoxic T-killer activity as well as decreased natural killer (NK) cells have been reported in zinc-deficient mice. All these effects of zinc deficiency on immune functions in mice can be reversed with zinc supplementation (Dardenne, 2002).

Malabsorption of zinc occurs in the hereditary disease acrodermatitis enteropathica, in which patients experience thymic atrophy, anergy, reduced lymphocyte proliferative response to mitogens, selective decrease in T4+ helper cells, and deficient thymic hormone activity. All of these symptoms may be corrected with zinc supplementation. Less severe cellular immune defects have been reported in patients who become zinc-deficient while receiving total parenteral nutrition. Controversial and partly contradictory results have been obtained when zinc intakes were high. In experimental models, high-zinc diets have been shown to reinforce immune functions above basal levels. Other studies have demonstrated the adverse effects of zinc excess; therefore, caution should be exercised when taking large zinc supplements for prolonged periods of time (Dardenne, 2002).

Several possible hypotheses can be offered regarding the mode of action of zinc on immune function. Zinc may be necessary for the activity of some immune system mediators such as thymulin, a nonapeptidic hormone secreted by thymic epithelial cells that requires the presence of zinc for its activity. This peptide promotes T-lymphocyte maturation, cytotoxicity, and IL-2 production. Thymulin activity in both animals and humans is dependent on plasma zinc concentrations. Zinc could also be critical for the activity of some cytokines; for example, it has been demonstrated that the production or activity of IL-1, IL-2, IL-3, IL-4, IL-6, IFN-γ, and TNF-α are affected by zinc

deficiency. Zinc could contribute to membrane stabilization by acting at the cytoskeletal level. Additionally, zinc is a major intracellular regulator of lymphocyte apoptosis. Thus, it is becoming evident that the thymic atrophy and lymphopenia that accompany zinc deficiency are primarily due to an alteration in the production of lymphocytes and the loss of precursor cells via an apoptotic mechanism (Dardenne, 2002).

G. Selenium and Iodine: Young and Old Trace Elements

These trace elements are found in the non-metal area of the periodic table. Although they are neighbors, significant differences exist in their chemical behavior. Selenium is a nonmetal with semiconductor properties, and iodine is a halogen. The biological history of iodine can be traced back to the beginning of the nineteenth century when a physician named Jean-Francois Coindet (1821) used various iodine solutions to treat goiter. Another halogen, although more reactive than iodine, fluorine was also quite early connected to goiter. It was shown that feeding a dog with sodium fluoride caused goiter to appear (Maumené, 1854). The human essentiality of iodine was established in 1850.

The scientific community's appreciation of the trace element selenium has more or less undergone a metamorphosis. The toxic effects of selenium were first discovered in the 1930s when livestock ate certain plants with unfortunate results. This problem, mistakenly called "alkali disease," occurred in an acute form following the consumption by range animals of some wild vetches of the genus *Astragalus*, which accumulated toxic amounts of selenium from the soil (Moxon, 1937). An historical aside is that it is thought that General Custer might have survived his trip to the Little Bighorn if reinforcements had not been delayed by pack animals that were apparently suffering from selenium-induced lameness. In 1943, selenium was even considered to be a carcinogenic element (Nelson, 1943). It was some years before selenium was recognized as an essential trace element (Schwartz & Foltz, 1957). In the late 1960s, research suggested an anticarcinogenic effect of selenium (Shamberger & Frost, 1969).

1. Iodine Biochemistry

The thyroid, the largest endocrine gland in the body, is located in the neck. The normal gland consists of two lobes connected by a narrow isthmus and is composed of numerous functional units called *follicles*. Only one

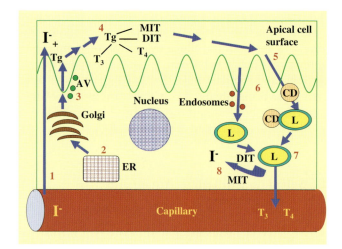

FIGURE 24 An overview of iodine metabolism.

clearly established function has been demonstrated: to synthesize its hormones, thyroxine (T_4) and 3,5,3'-triiodothyronine (T_3). Iodine is an integral part of T_4 and T_3 that contributes 65 and 59% to their respective molecular sizes (Dunn & Dunn, 2001). Figure 24 provides an overview of iodine metabolism, which includes thyroglobulin (Tg) iodination, hormone formation, hormone release, and factors that influence these processes. Details regarding iodine metabolism have been known for some time; however, new information still appears quite often in the literature. Of special interest, of course, are the cell trafficking and regulatory aspects.

Iodine provides the raw material for hormone synthesis. In order to accomplish this function, iodine must be transported across the basal membrane of the thyroid by the sodium iodide symporter (Shen *et al.*, 2001). The major factors involved in this process are thyroperoxidase (TPO), hydrogen peroxide, nicotinamide adenine dinucleotide phosphate (NADPH), pendrin, cell-trafficking proteins (the molecular chaperones), and Tg itself (Dunn & Dunn, 2001).

Iodine is ingested in a variety of chemical forms. Most ingested iodine is reduced in the gastrointestinal tract and absorbed almost completely. Some iodine-containing compounds (*e.g.*, thyroid hormones) are absorbed intact. Iodate, widely used in many countries as an additive to salt, is rapidly reduced to iodide and completely absorbed. Once in the circulation, iodide is removed by the thyroid gland and the kidney. The thyroid selectively concentrates iodide (see above and Figure 24) in amounts required for adequate thyroid hormone synthesis. Most of the remaining iodine is excreted in the urine. Several other tissues can also concentrate iodine, including salivary glands, breast tissue, choroid plexus,

and gastric mucosa. Other than the lactating breast, these are minor pathways of uncertain significance.

The NIS sodium iodide symporter in the thyroidal basal membrane is responsible for iodine concentration. It transfers iodide from the circulation into the thyroid gland at a concentration gradient of about 20 to 50 times that of the plasma to ensure that the thyroid gland obtains adequate amounts of iodine for hormone synthesis.

Iodine in the thyroid gland participates in a complex series of reactions to produce thyroid hormones. Thyroglobulin, a large glucoprotein weighing 666 kDa, is synthesized within the thyroid cell and serves as a vehicle for iodination. Iodide and Tg meet at the apical surface of the thyroid cell. There, TPO and H_2O_2 promote the oxidation of the iodide and its simultaneous attachment to tyrosyl residues within the Tg molecule to produce the hormone precursors diiodotyrosine and monoiodotyrosine. Thyroperoxidase further catalyzes the intramolecular coupling of two molecules of diiodotyrosine to produce T_4. A similar coupling of one monoiodotyrosine and one diiodotyrosine molecule produces T_3. Mature iodinated Tg is stored extracellularly in the lumen of thyroid follicles, with each consisting of a central space rimmed by the apical membranes of thyrocytes (Dunn & Dunn, 2001).

Thyroglobulin, which contains the thyroid hormones, is stored in the follicular lumen until needed. The endosomal and lysosomal proteases digest Tg and release the hormones into the circulation. About two-thirds of Tg iodine is in the form of the inactive precursors monoiodotyrosine and diiodotyrosine. The iodine is not released in the circulation but instead is removed from the tyrosine moiety by a specific deiodinase (see below) and then recycled within the thyroid gland. This process is an important mechanism for iodine conservation, and individuals with impaired or genetically absent deiodinase activity risk iodine deficiency.

Once in the circulation, T_4 and T_3 rapidly attach to several binding proteins synthesized in the liver, including thyroxine-binding globulin, transthyretin, and albumin. The bound hormone then migrates to target tissues where T_4 is deiodinated to T_3, which is the metabolically active form. The responsible deiodinase contains selenium, and selenium deficiency may impair T_4 conversion and hormone action. The iodine of T_4 returns to the serum iodine pool and follows the cycle of iodine again or is excreted in the urine.

Thyroid-stimulating hormone (TSH) is the major regulator of thyroid function. The pituitary secretes this protein hormone, which has a molecular weight of

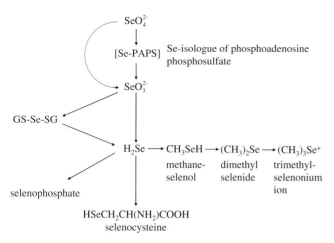

FIGURE 25 The assimilation of selenate.

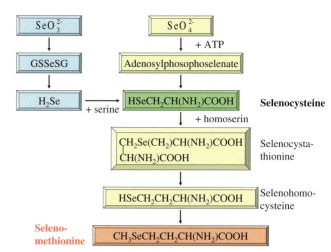

FIGURE 26 The major pathway of selenomethionine synthesis by plants, marine algae, and yeast.

about 28 kDa, in response to circulating concentrations of thyroid hormone: TSH secretion increases when levels of circulating thyroid hormone decrease. TSH affects several sites within the thyrocyte. The principal actions are to increase thyroidal uptake of iodine and to break down Tg in order to release thyroid hormone into the circulation. An elevated serum TSH concentration indicates primary hypothyroidism and a decrease in TSH concentration reflects hyperthyroidism.

2. Selenium Biochemistry

Selenium is primarily taken up from the soil by plants as selenate (SeO_4^{2-}) or selenite (SeO_3^{2-}). The assimilation of selenate appears to follow the sulfate reduction pathway common to higher plants (Figure 25). Analogous to sulfur metabolism, selenate is presumed to be activated by ATP sulfurylase to adenosine phosphoselenate, which then is reduced to selenite. Selenite reacts nonenzymatically with glutathione to form selenodiglutathione, which is readily reduced to selenide by flavine-dependent disulfide reductases, such as glutathione or thioredoxin reductase. In mammals, selenite may also be directly reduced by thioredoxin reductase. Selenide is the central metabolite in the utilization and excretion of selenium. It serves as a substrate for the biosynthesis of selenocysteine by cystein synthases, and it is transformed into selenophosphate, which is required for selenoprotein biosynthesis. Alternatively, it can be methylated by S-adenosylmethionine-dependent methyl transferases, which leads to excretion or volatilization of selenium (Ellis & Salt, 2003).

Cereals and forage crops convert selenium into mainly selenomethionine and incorporate it into protein in place of methionine because tRNA^{Met} does not discriminate between methionine and selenome-

thionine. The major pathway of selenomethionine synthesis by plants, marine algae, and yeast is shown in Figure 26. There are some doubts, however, that marine algae can reduce selenate. It is also unclear whether higher animals can make use of selenate without the support of intestinal flora (Birringer et al., 2002). Selenomethionine is not required for growth by plants but is produced along with methionine in quantities depending on the amount of selenium available (Schrauzer, 2000). Practically all small organic selenium compounds in plants, bacteria, yeast, and animals are the isologs of corresponding sulfur compounds (Table XIV). Most of the known selenoproteins contain one or more selenocysteine residues integrated into the main polypeptide chain. The incorporation of selenocysteine is a cotranslational process that makes use of the microsomal machinery of protein synthesis; however, the process is unique in several respects (Birringer et al., 2002).

An intriguing phenomenon, usually referred to as the hierarchy of selenoproteins, has been observed in mammals. It recognizes that the individual selenoproteins respond differently to selenium availability when it is limiting (Birringer et al., 2002).

3. Selenoproteins

The selenium-containing proteins identified thus far can be divided into three groups: proteins into which selenium is incorporated nonspecifically, specific selenium-binding proteins, and specific proteins that contain selenium in the form of genetically encoded selenocysteine and that have been defined as seleno-

TABLE XIV. Distribution (%) of Inorganic and Organic Selenium Compounds in Garlic and Yeast Extracts, Determined by HPLC-ICP-MS

Compound	Garlic (296 mg kg^{-1} Se)	Yeast (1922 mg kg^{-1} Se)
Selenate	2	ND
Selenite	ND	1
Selenocystine	0.5	0.5
Selenocystathione	0.5	1
Se-methylated selenocysteine	3	0.5
γ-Glutamyl-Se—methylselenocysteine	73	0.5
Selenomethionine	13	85
γ-Glutamylselenomethionine	4	ND
Se-Adenosylselenohomocysteine	ND	3
Selenolanthionine	ND	1.5
Total selenium (%)	96	93

Note: ND = not determined.
Source: From Kotrebai, M. *et al.*, *Analyst*, 125, 71–78, 2000.

proteins. In addition, there are proteins in which selenium has been demonstrated but for which no information regarding function is yet available (Behne & Kyriakopoulos, 2001).

Although there are intriguing differences between the selenoproteins found in prokaryotes and eukaryotes, especially mammals (Tables XV and XVI), this chapter deals exclusively with mammalian proteins in this context. In fact, at present it seems that selenophosphate synthetase is the only protein belonging both to prokarya and eukarya.

Selenium exerts its biological function through certain proteins containing the element. In this situation, it is in the form of covalently bound selenocysteine. The incorporation of selenium into these proteins requires a set of specific factors. Among others, incorporation of sulfur instead of selenium must be prevented. These elements share similar chemical and physical properties, and sulfur is much more abundant in the biosphere than selenium.

All selenoproteins identified thus far are enzymes in which the selenocysteine residues are responsible for their catalytic functions. Their metabolic importance is based on the fact that in contrast to the thiol in cysteine-containing enzymes, the selenol is fully ionized at normal physiological pH and that under comparable conditions it is of much higher reactivity than the thiol group (Behne & Kyriakopoulos, 2001).

4. Glutathione Peroxidases

Glutathione peroxidases (GSH-Px) catalyze the reduction of hydrogen peroxide and organic hydroperoxides. This is an important component of the cellular defense against reactive oxygen species. As can be expected, glutathione usually serves as the electron donor; however, in some cases other thiols are oxidized to fulfill a specific biological function. At present, four selenocysteine-containing GSH-Pxs have been identified (Table XVI):

- **Cytosolic or classical glutathione peroxidase**—As the name indicates, it is found in the cytosol of cells. It is present in almost all tissues but is not homogeneously distributed. Cytosolic GSH-Px consists of four identical selenocysteine-containing subunits of about 22 kDa. It catalyzes the reduction of hydrogen peroxide and various soluble organic peroxides. In this way, it contributes to the antioxidant defense against reactive molecules and complements the effects of vitamin E. This system seems to be quite insensitive to low activities of cGSH-Px, and it seems that the protective effects of cGSH-Px are of particular importance when the system is exposed to additional stress factors (Behne & Kyriakopoulos, 2001).

- **Gastrointestinal glutathione peroxidase**—This enzyme is thought to protect mammals from the toxicity of ingested lipid hydroperoxides. It is similar to cGSH-Px in that it is a cytosolic enzyme consisting of

TABLE XV. Selenoproteins in Prokaryotes

Protein	Function
Glycine reductase	Formation of a selenoether
Glycine/sarcosine/betaine reductase	Redox function
Glycine reductase selenoprotein B	Formation of a selenoether
Sarcosine reductase selenoprotein B	Formation of a selenoether
Betaine reductase selenoprotein B	Formation of a selenoether
Proline reductase	Redox function, formation of a selenoether
Heterosulfide reductase	Redox function
Selenoperoxiredoxin	Redox function (peroxidase)
Putative redox active selenoprotein	Redox function
Formate dehydrogenase	Hydrogen donor
Formylmethanofuran dehydrogenase	Redox function
NiFeSe-hydrogenase	Hydrogen donor
F420 nonreducing hydrogenase	Redox function
F420 reducing hydrogenase	Redox function
Selenophosphate synthetase	Selenoprotein synthesis
CO dehydrogenase	Formation of a carbon oxide selenide
Nicotinic acid hydroxylase	Unknown
Xanthine dehydrogenase	Unknown

Source: Data from Köhrle *et al.* (2000) and Birringer *et al.* (2002).

TABLE XVI. Mammalian Selenoproteins

Glutathione peroxidases (GSH-Px)
    Cytosolic or classical GSH-Px (cGSH-Px, GSH-Px-1)
    Gastrointestinal GSH-Px (GI-GSH-Px, GSH-Px-GI, GSH-Px-2)
    Plasma GSH-Px (pGSH-Px, GSH-Px-3)
    Phospholipid hydroperoxide GSH-Px (PhGSH-Px, GSH-Px-4)
Iodothyronine deiodinases
    5'-deiodinase, type I (5'DI)
    5'-deiodinase, type II (5'DII)
    5-deiodinase, type III (5-DIII)
Thioredoxin reductases
    Thioredoxin reductase (TrxR)
    Testicular thioredoxin reductase (TrxR-2)
    Mitochondrial thioredoxin reductase (TrxR-3)
    Thioredoxin reductase homologues
Selenophosphate synthetase-2
Functionally undefined
    15-kDa selenoprotein of T cells
    Selenoprotein P10
    Selenoprotein P12
    Selenoprotein W
    Selenoprotein R
    Selenoprotein T
    Selenoprotein X
    Selenoprotein N

Source: Data from Köhrle *et al.* (2000) and Birringer *et al.* (2002).

four identical selenocysteine-containing subunits weighing slightly below 22 kDa. In animal studies, selenium deficiency decreases the enzyme activity; however, no effect of human gastrointestinal (GI) GSH-Px has been reported. Gastrointestinal glutathione peroxidase is the most important selenoprotein antioxidant in the colon. Oxidative stress is a critical event in tumorigenesis. It is therefore likely that the antioxidant function of GI-GSH-Px is important in the early defense against colon cancer (Brown & Arthur, 2001).

• **Phospholipid hydroperoxide glutathione peroxidase**—This protein has been shown to have a membrane-associated as well as a cytosolic location. It is responsible for the reductive destruction of lipid hydroperoxides, and it was the second mammalian selenoenzyme to be discovered. The enzyme is a monomer of 19.7 kDa. The activity of the enzyme is preserved in preference to cGSH-Px when dietary selenium is low. It reacts with phospholipid hydroperoxides as well as

small soluble hydroperoxides. It is also capable of metabolizing cholesterol and cholesterol ester hydroperoxides in oxidized low-density lipoprotein. Consequently, it is well recognized as being essential to the destruction of fatty acid hydroperoxides, which, if not reduced to hydroxyl fatty acids, will lead to uncontrolled radical chain reactions that are deleterious to the integrity of membranes (Brown & Arthur, 2001). It has also been suggested that the enzyme may have important functions in the redox regulation of a variety of processes, such as inflammation and apoptosis, although it is not known to what extent the other GSH-Pxs are involved in these reactions (Behne & Kyriakopoulos, 2001). Of note is that a significant role in spermatogenesis is exclusively fulfilled by this enzyme.

• **Plasma glutathione peroxidase**—Extracellular glutathione peroxidase (pGSH-Px) is another protein with antioxidant potential. Again, we are confronted with a tetrameric GSH-Px with subunits of about 23 kDa. The significant difference between pGSH-Px and cGSH-Px, as well as GI-GSH-Px, is that pGSH-Px is a glucoprotein and is present in extracellular fluids. It is secreted into the extracellular fluids from tissues where it is expressed. The kidney is the main site of production for this enzyme. Similar to other tetrameric GSH-Pxs, pGSH-Px catalyzes the reduction of hydrogen peroxide and various organic peroxides when glutathione is used as a substrate. The biological function of this enzyme still has not been unveiled.

5. Thioredoxin Reductases

Thioredoxin reductases are a recently identified family of selenoproteins that catalyze the NADPH-dependent reduction of thioredoxin and therefore play a regulatory role in its metabolic activity. This is a family of homodimeric flavoenzymes present in various tissues. In addition to the flavin and the active site of the prokaryotic homologs with their redox-active disulfide, they also contain selenocysteine as the penultimate C-terminal amino acid residue, which is indispensable for their enzymatic activity (Gromer et al., 1998). A description of the thioredoxin reductases identified so far follows:

- **Thioredoxin reductase 1**—TrxR1 was the first thioredoxin reductase to be identified. It is a dimer with two identical 50-kDa subunits. It is a ubiquitous cytosolic enzyme in contrast to the other types.

- **Thioredoxin reductase 2**—The second thioredoxin reductase to be discovered, it was named mitochondrial thioredoxin reductase 2 (TrxR2). The biological role of TrxR2 in the mitochondria is not known, but it seems to be involved primarily in the protection against mitochondria-mediated oxidative stress (Behne & Kyriakopoulos, 2001).

- **Thioredoxin reductase 3**—A third selenocysteine-containing thioredoxin reductase, here named thioredoxin reductase 3 (TrxR3), is preferentially expressed in the testes. The deduced sequence of the human enzyme shows 70% identity to that of TrxR1. It contains a long N-terminal extension and has a higher molecular mass (about 65 kDa) than the two other isozymes (Behne & Kyriakopoulos, 2001).

6. Selenophosphate Synthetase 2

Selenophosphate synthetase catalyzes the reaction of selenide with AMP. The product, selenophosphate, acts as the selenium donor for the biosynthesis of selenocysteine. In addition to selenophosphate synthetase 1, which contains threonine in its active center, a selenocysteine-containing homolog of about 50 kDa has also been identified in various human and mouse tissues. The detection of a selenoenzyme that is involved in the production of the selenoproteins is of special interest with regard to regulation of mammalian selenium metabolism (Behne & Kyriakopoulos, 2001).

7. Iodothyronine Deiodinases

Comparable to the prokaryotic and the methanococcus world, most of the eukaryotic selenocysteine-containing proteins with identified function are also involved in redox reactions (Köhrle, 1999). The deiodinases were discovered in the 1990s. The iodothyronine deiodinases are a large group of selenoproteins. Three iodothyronine deiodinases regulate the conversion of thyroxine (T_4) to 3,3′,5-triiodothyronine (T_3): the active thyroid hormone or reverse triiodothyronine (rT_3), the inactive thyroid hormone. Type I deiodinase (IDI) is expressed in liver, kidney, brain, pituitary, and brown adipose tissue of ruminants. Type II deiodinase (IDII) has been present in the brain and pituitary of all species examined thus far and in brown adipose tissue of humans. IDII catalyzes conversion of T_4 to T_3 within tissues that cannot utilize circulating T3. Type III deiodinase (IDIII) converts T_4 to rT_3 and T_3 to diiodothyronine; is found in brain, skin, and placenta; and functions to deactivate thyroid hormones. The role of selenium in iodothyronine deiodinases implies that some of the consequences of selenium deficiency may be directly attributed to disturbances in thyroid hormone metabolism. The type I enzyme deiodinates 4′-O-sulfates of T_4 and T_3 at the tyrosyl ring by 5-deiodination. The type II enzyme acts as a heterotrimeric complex of about 200 kDa containing a 29-kDa subunit that interacts with filamentous actin (Köhrle, 2000). In contrast to IDI, IDII is strictly a phenolic-ring 5′-deiodinase that, as a substrate, prefers T_4 to rT_3. The type III enzyme inactivates T_4 and its metabolites by removal of iodine atoms at the tyrosyl ring. The holoenzyme structure of this enzyme, containing a 32-kDa substrate binding subunit, is not yet known (Köhrle, 2000).

8. Selenoprotein P

The existence of selenoprotein P was reported in 1973 (Burk, 1973); however, it was not until ten years later that it was recognized as a selenoprotein. The P stands for its presence in blood plasma. Selenoprotein P has been purified from humans and rats by a procedure that includes immunoaffinity chromatography. Six typical glycosylation sites in the deduced amino acid sequence of the human proteins have been detected; however, no characterizations of the bound carbohydrates has been reported. Selenoprotein P is the first and so far the only protein described to contain more than one selenium atom per polypeptide chain. Ten selenocysteine residues were predicted (Persson Moschos, 2000). Selenoproteins with known enzymatic activity are redox enzymes that contain selenocysteine in their active sites. A key issue concerning selenoprotein P that remains to be revealed is its catalytic function. It has been suggested that it acts as an extracellular oxidant defense or as a transport protein. It seems less likely to be a transporting protein because the selenium is covalently bound in the protein (Persson Moschos, 2000). It is currently unknown whether selenoprotein P is able to react with certain phospholipid hydroperoxides under physiological conditions (Saito & Takahashi, 2000). Recent work suggests that selenoprotein P in plasma diminishes the oxidizing and nitrating reactivity of peroxynitrite, a reactive intermediate formed by the reaction of nitrogen monoxide and superoxide anion. Due to the association with endothelial membranes, it has been speculated that endothelial cells are protected against peroxynitrite toxicity by selenoprotein P (Saito & Takahashi, 2000).

9. Selenoprotein W

Selenoprotein W contains both a selenocysteine residue that is encoded by a UGA codon in the open reading frame of the mRNA as well as a bound glutathione molecule. The protein is localized predominantly in the cytoplasm (Whanger, 2000; Dae-won et al., 2002); however, a small portion of the total selenoprotein W is associated with the cell membrane. Selenoprotein W was expressed in all tissues examined in selenium-supplemented animals including muscle, heart, testis, spleen, kidney, intestine, tongue, brain, lung, and liver (Gu et al., 1999). The loss of the protective effect of selenoprotein W against hydrogen-peroxide-induced cytotoxicity in cells treated with an inhibitor of glutathione synthesis indicates that the protein is a glutathione-dependent antioxidant in vivo (Dae-won et al., 2002). Glutathione and its redox cycle play a critical role in catabolizing hydrogen peroxide and other peroxides through enzymatic coupling reactions. Additionally, glutathione is important for the detoxification of electrophiles and for protection of the thiol groups from oxidation. Glutathione is also required for regeneration of the glutathione peroxidase and glutaredoxin system (Dae-won et al., 2002).

10. Selenoprotein R

Selenoprotein R is a small protein with a molecular mass of about 12 kDa. It contains selenocysteine. Homologs of this protein have been identified in bacteria, archaea, and eukaryotes, and, with the exception of vertebrate selenoprotein R, all homologs contain cysteine in place of selenocysteine (Kryukov et al., 2002).

Bioinformatic analyses have suggested a functional linkage of selenoprotein R to a pathway of methionine sulfoxide reduction as well as a role of selenoprotein R in protection against oxidative stress and/or redox regulation of cellular processes (Kryukov et al., 2002). Methionine sulfoxide reduction is an important process by which cells regulate biological processes and cope with oxidative stress (Hoshi & Heinemann, 2001). Methionine sulfoxide reductase is a protein that has been known for decades. It is involved in the reduction of methionine sulfoxides in proteins. It has been shown that methionine sulfoxide reductase is only specific for methionine-S-sulfoxides. The fact that oxidized methionines occur in a mixture of R and S isomers in vivo raised the question of how methionine sulfoxide reductase could be responsible for the reduction of all protein methionine sulfoxides. The study of Kryukov and coworkers (2002) explained this puzzle. It appeared that a second methionine sulfoxide reductase exists specific for methionine-R-sulfoxides. This reductase was selenoprotein R, and, in addition, these researchers showed that it contains zinc.

11. Selenium and Cancer

Chemoprevention is a recently introduced and strongly growing area within oncology. The number of studies of chemoprevention has increased drastically during recent years. The reason for this is, of course, that the possibility to prevent or hinder cancer is generally attractive. A great number of agents have been tested for prevention of different forms of cancer (Decensi & Costa, 2000). Tamoxifen has been used as a chemoprevention agent against breast cancer as well as raloxifen and synthetic retinoides and combinations. Several nonsteroidal antiinflammatory preparations have been used against colorectal cancer, as have cyclooxygenase-2 and prostanoides. Even calcium has been used in this way, as well as α-difluoromethylornithine, which is an irreversible inhibitor of the enzyme ornithine decarboxylase; also, beta-carotene and retinol have been used against lung cancer.

The first intervention study of prevention of human cancer with selenium was performed in Qidong, an area north of Shanghai with a high incidence of primary liver cancer. In an urban area, 20,847 inhabitants received table salt supplemented with 15 ppm of selenium as sodium selenite. Those individuals were thus supplemented with about 30 to 50 μg of selenium per day over the course of 8 years. Supplementation resulted in a decrease of the incidence of primary liver cancer to 27.2 per 100,000 inhabitants, while it remained at 50.4 per 100,000 inhabitants in four surrounding areas. When selenium was no longer added to the table salt, the incidence began to increase (Yu *et al.*, 1991, 1997). In another trial, 2474 members of families with a high risk of primary liver cancer were administered 200 μg of selenium as yeast with a high concentration of selenium or placebo. During the 2-year trial, 1.26% of the controls developed primary liver cancer in contrast to 0.69% in the treated group ($p < 0.05$). Out of 226 bearers of HBVsAg, 7 of 113 individuals in the placebo group developed primary liver cancer, while none of the 113 individuals in the treated group developed cancer during the same period ($p < 0.05$).

The effects of supplementation of vitamins and trace elements on cancer incidence and mortality were investigated in an intervention study in Linxian, China (Li *et al.*, 1993). This area is known to have the highest mortalities of esophagus and ventricular cancer. Among those who received supplementation with beta-carotene, vitamin E, and selenium, total mortality was reduced by 9% and cancer mortality by 13%. Mortality due to ventricular cancer decreased significantly (20%), while mortality in other forms of cancer decreased by 19%.

Within this context, prevention of skin cancer assumes a prominent position, although the outcome of chemoprevention has not always been positive. Two excellently designed studies evaluated the preventive effects of selenium or retinol on skin cancer (Clark *et al.*, 1996; Moon *et al.*, 1997). The selenium study (Clark *et al.*, 1996) showed negative results for the prevention of squamous cell cancer and basilioma. This study involved 1312 patients that had a history of skin cancer of the nonmelanoma type and lived in areas of the United States with a naturally low intake of selenium. The investigation, presented as a post hoc observation, showed a significant preventive effect on prostate, lung, and colon cancer. The cancer incidences were as follows: prostate cancer (selenium, $n = 13$; placebo, $n = 35$; RR (relative risk) = 0.37; 95% CI (confidence interval); 0.18–0.71; $p = 0.002$), lung cancer (selenium, $n = 17$; placebo, $n = 31$; RR = 0.54; 95% CI; 0.30–0.98; $p = 0.04$), and colorectal cancer (selenium, $n = 8$; placebo, $n = 19$; RR = 0.42; 95% CI; 0.18–0.95; $p = 0.03$). The study also showed a decreased total mortality without statistical significance, although the total mortality in cancer decreased significantly.

SEE ALSO THESE CHAPTERS

Chapter 5 (Uptake of Elements from a Chemical Point of View) · Chapter 8 (Biological Responses of Elements)

FURTHER READING

Andrews, S. C., Robinson, A. K., and Rodriquez-Quinones, F. (2003). Bacterial Iron Homeostasis, *FEMS Microbiol. Rev.*, 27, 215–237.

Aubin, J. E., and Triffitt, J. T. (2002). Mesenchyman Stem Cells and Osteoblast Differentiation. In *Principles of Bone Biology*, 2nd edition (J. P. Bilezikian, L. G. Raisz, and G. A. Rodan, Eds.), Academic Press, San Diego, CA, 59–81.

Baker, H. M., Anderson, B. F., and Baker, E. (2003). Dealing with Iron: Common Structural Principles in Proteins That Transport Iron and Heme, *Proc. Natl. Acad. Sci. U.S.A.*, 100, 3579–3583.

Barañano, D. E., and Snyder, S. H. (2001). Neural Roles for Heme Oxygenase: Contrasts to Nitric Oxide Synthase, *Proc. Natl. Acad. Sci. U.S.A.*, 98, 10,996–11,002.

Barber, J. (2003). Photosystem II: The Engine of Life, *Q. Rev. Biophys.*, 36, 71–89.

Barceloux, D. G., Bond, G. R., Krenzelok, E. P., Cooper, H., and Vale, J. A. (2002). American Academy of Clinical Toxicology Practice Guidelines on the Treatment of Methanol Poisoning, *J. Toxicol. Clin. Toxicol.*, 40, 415–446.

Bates, D. M., Popescu, C. V., Khoroshilova, N., Vogt, K., Beinert, H., Münck, E., and Kiley, P. J. (2000). Substitution of Leucine 28 with Histidine in the *Escherichia coli* Transcription Factor FNR Results in Increased Stability of the [4Fe–4S]²⁺ Cluster to Oxygen, *J. Biol. Chem.*, 275, 6234–6240.

Beard, J. L. (2001). Iron Biology in Immune Function, Muscle Metabolism and Neuronal Functioning, *J. Nutr.*, 131, 568S–580S.

Begley, T. P., Xi, J., Kinsland, C., Taylor, S., and MacLafferty, F. (1999). The Enzymology of Sulphur Activation During Thiamine and Biotin Biosynthesis, *Curr. Opin. Chem. Biol.*, 3, 623–629.

Behne, D., and Kyriakopoulos, A. (2001). Mammalian Selenium-Containing Proteins, *Annu. Rev. Nutr.*, 21, 453–473.

Beinert, H. (2000). A Tribute to Sulphur, *Eur. J. Biochem.*, 267, 5657–5664.

Beinert, H., Kennedy, M. C., and Stout, C. D. (1996). Acontiase as Iron–Sulfur Protein, Enzyme and Iron-Regulatory Protein, *Chem. Rev.*, 96, 2335–2373.

Beninga, J., Rock, K. L., and Goldberg, A. L. (1998). Interferon-Gamma Can Stimulate Post-Proteasomal Trimming of the N Terminus of an Antigenic Peptide by Inducing Leucine Aminopeptidase, *J. Biol. Chem.*, 273, 18734–18742.

Berg, D., Gerlach, M., Youdim, M. B. H., Double, K. L., Zecca, L., Riederer, P., and Becker, G. (2001). Brain Iron Pathways and Their Relevance to Parkinson's Disease, *J. Neurochem.*, 79, 225–236.

Berridge, M. J., Lipp, P., and Bootman, M. D. (2000). The Versatility and Universatility of Calcium Signalling, *Nat. Rev. Mol. Cell Biol.*, 1, 11–21.

Birringer, M., Pilawa, S., and Flohé, L. (2002). Trends in Selenium Biochemistry, *Nat. Prod. Rep.*, 19, 693–718.

Brown, K. M., and Arthur, J. R. (2001). Selenium, Selenoproteins and Human Health: A Review, *Pub. Health. Nutr.*, 4, 593–599.

Burk, R. F. (1973). Effect of Dietary Selenium Level on Se Binding to Rat Plasma Proteins, *Proc. Soc. Exp. Biol. Med.*, 143, 719–722.

Camarakis, J., Voskoboinik, I., and Mercer, J. F. (1999). Molecular Mechanisms of Copper Homeostasis, *Biochem. Biophys. Res. Commun.*, 261, 225–232.

Carafoli, E., and Klee, C. (1999). *Calcium as a Cellular Regulator*, Oxford University Press, Oxford, England.

Chausmer, A. B. (1998). Zinc, Insulin and Diabetes, *J. Am. Coll. Nutr.*, 17, 109–115.

Cigic, B., and Pain, R. H. (1999). Location of the Binding Site for Chloride Ion Activation of Cathepsin C, *Eur. J. Biochem.*, 264, 944–951.

Clark, L. C., Combs, G. F., Jr., Trumbull, B. W., Slate, E. H., Chalker, D. K., Chow, J., Davis, L. S., Glover, R. A., Graham, G. F., Gross, E. G., Krongrad, A., Lesher, J. L., Jr., Park, H. K., Sanders, B. B., Jr., Smith, C. L., and Taylor, J. R. (1996). Effects of Selenium Supplementation for Cancer Prevention in Patients with Carcinoma of the Skin. A Randomized Controlled Trial. *National Prevention of Cancer Study Group*, 276, 1957–1963.

Coindet, J. F. (1821). Nouvelles recherches sur les effets d'iode, et sur les precautions à suivre dans le traitement du goitre par ce nouveau remède, *Ann. Chim. Phys. (Paris)*, 16, 252–266.

Conklin, D. S., McMaster, J. A., Culbertson, M. R., and Kung, C. (1992). COT1, a Gene Involved in Cobalt Accumulation in *Saccharomyces cerevisiae*, *Mol. Cell. Biol.*, 12, 3678–3688.

Cousins, R. J. (1998). A Role of Zinc in the Regulation of Gene Expression, *Proc. Nutr. Soc.*, 57, 307–311.

Coyle, P., and Phicox, J. C., Carey, L. C., and Rofe, A. M. (2002). Metallothionein: The Multipurpose Protein, *Cell. Mol. Life Sci.*, 59, 627–647.

Dae-won, J., Tae, S. K., Youn, W. C., Byeong, J. L., and Ick, Y. K. (2002). Selenoprotein W is a Glutathione-Dependent Antioxidant *In Vivo*, *FEBS Lett.*, 517, 225–228.

D'Amico, S., Gerday, C., and Feller, G. (2000). Structural Similarities and Evolutionary Relationships in Chloride-Dependent Alpha-Amylases, *Gene*, 253, 95–105.

Dardenne, M. (2002). Zinc and Immune Function, *Eur. J. Clin. Nutr.*, 56, S20–S23.

Davis, S. R., and Cousins, R. J. (2000). Metallothionein Expression in Animals: A Physiological Perspective on Function, *J. Nutr.*, 130, 1085–1088.

Decensi, A., and Costa, A. (2000). Recent Advances in Cancer Chemoprevention, with Emphasis on Breast and Colorectal Cancer, *Eur. J. Cancer*, 36, 694–709.

Dodson, G., and Steiner, D. (1998). The Role of Assembly in Insulin's Biosynthesis, *Curr. Opin. Struct. Biol.*, 8, 189–194.

Dreosti, I. E. (2001). Zinc and the Gene, *Mutat. Res.*, 475, 161–167.

Duester, G. (1998). Alcohol Dehydrogenase as a Critical Mediator of Retinoic Acid Synthesis from Vitamin A in the Mouse Embryo, *J. Nutr.*, 128, 459S–462S.

Dunn, J. T., and Dunn, A. D. (2001). Update on Intrathyroidal Iodine Metabolism, *Thyroid*, 11, 407–414.

Eitinger, T., and Mandrand-Berthelot, M. A. (2000). Nickel Transport Systems in Microorganisms, *Arc. Microbiol.*, 173, 1–9.

Ellis, D. R., and Salt, D. E. (2003). Plants, Selenium and Human Health, *Curr. Opin. Plant Biol.*, 6, 273–279.

Falchuk, K. H. (1998). The Molecular Basis for the Role of Zinc in Developmental Biology, *Mol. Cell Biochem.*, 188, 41–48.

Finazzo, C., Harmer, J., Jaun, B., Duin, E. C., Mahlert, F., Thauer, R. K., Van Doorslaer, S., and Schweiger, A. (2003). Characterization of the MCRred2 Form of Methyl-Coenzyme M Reductase: A Pulse EPR and ENDOR Study, *J. Biol. Inorg. Chem.*, 8, 586–593.

Fletcher, E. L., Clark, M. J., and Furness, J. B. (2002). Neuronal and Glial Localization of GABA Transporter Immunoreactivity in the Myenteric Plexus, *Cell Tissue Res.*, 308, 339–346.

Fraústo da Silva, J. J. R., and Williams, R. J. P. (2001). *The Biological Chemistry of the Elements: The Inorganic Chemistry Of Life*, 2nd ed., Oxford University Press, London.

Frazzon, J., Fick, J. R., and Dean, D. R. (2002). Biosynthesis of Iron-Sulphur Clusters is a Complex and Highly Conserved Process, *Biochem. Soc. Trans.*, 30, 60–685.

Fridovich, I. (1995). Superoxide Radical and Superoxide Dismutases, *Annu. Rev. Biochem.*, 64, 97–112.

Gaggelli, E., Berti, F., D'Amelio, N., Gaggelli, N., Valensin, G., Bovalini, L., Paffetti, A., and Trabalsini, L. (2002). Metabolic Pathways of Carcinogenic Chromium, *Environ. Health Perspect.*, 110 (Suppl. 5), 733–738.

Gaxiola, R. A., Yuan, D. S., Klausner, R. D., and Fink, G. R. (1986). The Yeast CLC Chloride Channel Functions in Cation Homeostasis, *Proc. Natl. Sci. U.S.A.*, 95, 4046–4050.

Ghosh, M., Grunden, A. M., Dunn, D. M., Weiss, R., and Adams, M. W. (1998). Characterization of Native and Recombinant Forms of an Unusual Cobalt-Dependent Proline Dipeptidase (Prolidase) from the Hyperthermophilic Archaeon *Pyrococcus furiosus*, *J. Bacteriol.*, 180, 402–408.

Gromer, S., Wissing, J., Behne, D., Ashman, K., Schirmer, R. H., Flohe, L., and Becker, K. (1998). A Hypothesis on the

Catalytic Mechanism of the Selenoenzyme Thioredoxin Reductase, *Biochem. J.*, 332, 591–592.

Gu, Q. P., Beilstein, M. A., Barofsky, E., Ream, W., and Whanger, P. D. (1999). Purification, Characterization, and Glutathione Binding to Selenoprotein W from Monkey Muscle, *Arch. Biochem. Biophys.*, 361, 25–33.

Haile, D. J., Rouault, T. A., Harford, J. B., Kennedy, M. C., Blondin, G. A., Beinert, H., and Klausner, R. D. (1992). Cellular Regulation of the Iron-Responsive Element Binding Protein: Disassembly of the Cubane Iron–Sulfur Cluster Results in High-Affinity RNA Binding, *Proc. Natl. Acad. Sci. U.S.A.*, 89, 11735–11739.

Handy, R. D., Eddy, F. B., and Baines, H. (2002). Sodium-Dependent Copper Uptake Across Epithelia: A Review of Rationale with Experimental Evidence from Gill and Intestine, *Biochim. Biophys. Acta.*, 1566, 104–115.

Harris, E. D. (2000). Cellular Copper Transport and Metabolism, *Annu. Rev. Nutr.*, 20, 291–310.

Hartmann, G. C., Klein, A. R., Linder, M., and Thauer, R. K. (1996). Purification, Properties and Primary Structure of H2-Forming N5,N10-Methylenetetrahydromethanopterin Dehydrogenase from *Methanococcus thermolithotrophicus*, *Arch. Microbiol.*, 165, 187–193.

Hausinger, R. P. (1993). Carbon Monoxide Dehydrogenase, In *The Biochemistry of Nickel*, Plenum Press, New York, 107–145.

Hellman, N. E., and Gitlin, J. D. (2002). Ceruloplasmin Metabolism and Function, *Annu. Rev. Nutr.*, 22, 439–458.

Hille, R. (2002). Molybdenum and Tungsten in Biology, *Trends Biochem. Sci.*, 27, 360–367.

Holmberg, C. G., and Laurell, C. B. (1948). Investigations in Serum Copper. II. Isolation of the Copper-Containing Protein and a Description of Some of Its Properties, *Acta Chem. Scand.*, 2, 550–556.

Hoshi, T., and Heinemann, S. (2001). Regulation of Cell Function by Methionine Oxidation and Reduction, *J. Physiol.*, 531, 1–11.

Huffman, D. L., and O'Halloran, T. V. (2001). Function, Structure, and Mechanism of Intracellular Copper Trafficking Proteins, *Annu. Rev. Biochem.*, 70, 677–701.

Iyengar, G. V., Kollmer, W. E., and Bowen, H. J. M. (1978). *The Elemental Composition of Human Tissues and Body Fluids: A Compilation of Values for Adults*, Verlag Chemie, Weinheim.

Jaiswal, J. K. (2001). Calcium—How and Why?, *J. Biosci.*, 26, 357–363.

Kappler, U., and Dahl, C. (2001). Enzymology and Molecular Biology of Prokaryotic Sulfite Oxidation, *FEMS Microbiol. Lett.* 203, 1–9.

Katagiri, T., and Takahasi, N. (2002). Regulatory Mechanisms of Osteoblast and Osteoclast Differentiation, *Oral Dis.*, 8, 147–159.

Kayne, F. J. (1971). Thallium (I) Activation of Pyruvate Kinase, *Arch. Biochem., Biophys.*, 143, 232–239.

Kehres, D. G., and Maguire, M. E. (2002). Structure, Properties and Regulation of Magnesium Transport Proteins, *Biometals*, 15, 261–270.

Kehres, D. G., and Maguire, M. E. (2003). Emerging Themes in Manganese Transport, Biochemistry and Pathogenesis in Bacteria, *FEMS Microbiol. Rev.*, 27, 263–290.

Kletzin, A., and Adams, M. W. W. (1996). Tungsten in Biological Systems, *FEMS Microbiol. Rev.*, 18, 5–63.

Kobayashi, M., and Shimizu, S. (1999). Cobalt Proteins, *Eur. J. Biochem.*, 261, 1–9.

Köhrle, J. (1999). The Trace Element Selenium and the Thyroid Gland, *Biochimie*, 81, 527–533.

Köhrle, J. (2000). The Deiodinase Family: Selenoenzymes Regulating Thyroid Hormone Availability and Action, *Cell. Mol. Life Sci.*, 57, 1853–1863.

Kotrebai, M., Birringer, M., Tyson, J. F., Block, E., and Uden, P. C. (2000). Selenium Speciation in Enriched and Natural Samples by HPLC-ICP-MS and HPLC-ESI-MS with Perfluorinated Carboxylic Acid Ion-Pairing Agents, *Analyst*, 125, 71–78.

Kruzel, M. L., and Zimecki, M. (2002). Lactoferrin and Immunologic Dissonance: Clinical Implications, *Arch. Immunol. Ther. Exp.*, 50, 399–410.

Kryukov, G. V., Kuman, R. A., Koc, A., Sun, Z., and Gladyshev, V. N. (2002). Selenoprotein R Is a Zinc-Containing Stereo-Specific Methionine Sulfoxide Reductase, *Proc. Natl. Acad. Sci. U.S.A.*, 99, 4245–4250.

Ladomery, M., and Dellaire, G. (2002). Multifunctional Zinc Finger Proteins in Development and Disease, *Ann. Hum. Genet.*, 66, 331–342.

Lebioda, L. (2000). The Honorary Enzyme Haemoglobin Turns Out To Be a Real Enzyme, *Cell. Mol. Life Sci.*, 57, 1817–1819.

Leroy, J. (1926). Nécessité du magnesium pour la croissance de la souris. *C. R. Seances Soc. Biol.*, 94, 431–433.

Li, J. Y., Li, B., Blot, W. J., and Taylor, P. R. (1993). Preliminary Report on the Results of Nutrition Prevention Trials of Cancer and Other Common Disease Among Residents in Linxian, China, *Chin. J. Pathol.*, 15, 165–181.

Li, Y. H. (2000). *A Compendium of Geochemistry: From Solar Nebula to the Human Brain*, Princeton University Press, Princeton, NJ.

Linder, M. C., and Hazegh-Azam, M. (1996). Copper Biochemistry and Molecular Biology, *Am. J. Clin. Nutr.*, 63, 797S–811S.

Lund, P. A., and Brown, N. L. (1987). Role of the merT and merP Gene Products of Transposon Tn501 in the Induction and Expression of Resistance to Mercuric Ions, *Gene*, 52, 207–214.

Lyonnet, B., Martz, F., and Martin, E. (1899). L'emploi thérapeutique des derives du vanadium, *Presse Med.*, 1, 191–192.

Ma, K., and Adams, W. W. (1999). An Unusual Oxygen-Sensitive, Iron- and Zinc-Containing Alcohol Dehydrogenase from the Hyperthermophilic Archaeon *Pyrococcus furiosus*, *J. Bacteriol.*, 181, 1163–1170.

Maret, W. (2000). The Function of Zinc Metallothionein: A Link Between Cellular Zinc and Redox State, *J. Nutr.*, 130, 1455S–1458S.

Mason, K. E. (1979). A Conspectus of Research on Copper Metabolism and Requirements of Man, *J. Nutr.*, 109, 1079–2066.

Matoso, C. M. M., Pombeiro, A. J. L., Fraústo da Silva, J. J. R., Guedes da Silva, M. F. C., Silva, J. A. L., Baptista-Ferreria, J. L., and Pinho-Almeida, F. (1998). A Possible Role for Amavadine in Some Amanita Fungi. In *Vanadium Compounds—Chemistry, Biochemistry and Therapeutic Applications* (A. S. Tracey and D. C. Craus, Eds.), ACS Symposium Series, 711, American Chemical Society, Washington, D.C., 186–201.

Maumené, E. (1854). Experiencé pour determiner l'action des fluores sur l'economie animale. *C. R. Acad. Sci. (Paris)*, 39, 538–539.

McCall, K. A., Huang, C., and Fierke, C. A. (2000). Function and Mechanism of Zinc Metalloenzymes, *J. Nutr.*, 130, 1437S–1446S.

Mendel, R. R. (1997). Molybdenum Cofactor of Higher Plants: Biosynthesis and Molecular Biology, *Planta*, 203, 399–405.

Mertz, W. (1998). Review of the Scientific Basis for Establishing the Essentiality of Trace Elements, *Biol. Trace Element Res.*, 66, 185–191.

Mertz, W., Toepfer, E. W., Roginiski, E. E., and Polansky, M. M. (1974). Present Knowledge of the Role of Chromium, *Fed. Proc. Fed. Am. Soc. Exp. Biol.*, 33, 2275–2280.

Messerschmidt, A., Rossi, A., Ladenstein, R., Huber, R., Bolognesi, M., Gatti, G., Marchesini, A., Petruzzelli, R., and Finazzi-Agro, A. (1989). X-Ray Crystal Structure of the Blue Oxidase Ascorbate Oxidase from Zuccnini. Analysis of the Polypeptide Fold and a Model of the Copper Sites and Ligands, *J. Mol. Biol.*, 206, 513–529.

Michels, A. J., Joisher, N., and Hagen, T. M. (2003). Age-Related Decline of Sodium-Dependent Ascorbic Acid Transport in Isolated Rat Hepatocytes, *Arch. Biochem. Biophys.*, 410, 112–120.

Michibata, H., Uyama, T., Ueki, T., and Kanamore, K. (2002). Vanadocytes: Cells Hold the Key to Resolving the Highly Selective Accumulation and Reduction of Vanadium in Ascidians, *Microsc. Res. Tech.*, 56, 421–434.

Mobley, H. L. T., Island, M. D., and Hausinger, R. P. (1995). Molecular Biology of Microbial Ureases, *Microbiol. Rev.*, 59, 451–480.

Moon, T. E., Levine, N., Cartmel, B., Bangert, J. L., Rodney, S., Dong, Q., Peng, Y. M., and Alberts, D. S. (1997). Effect of Retinol in Preventing Squamous Cell Skin Cancer in Moderate-Risk Subjects: A Randomized, Double-Blind, Controlled Trial. Southwest Skin Cancer Prevention Study Group, Cancer Epidemiol. *Biomarkers Prev.*, 6, 949–956.

Moxon, A. L. (1937). Akali Disease, or Selenium Poisoning, Bulletin 331. South Dakota Agricultural Experimental Station, Brookings.

Nelson, A. A., Fitzhugh, O. G., and Calvery, H. O. (1943). Liver Tumors Following Cirrhosis Caused by Selenium in Rats, *Cancer Res.*, 3, 230–236.

Nishino, T., and Okamoto, K. (2000). The Role of the [2Fe–S] Cluster Centers in Xanthine Oxidoreductase, *J. Inorg. Biochem.*, 82, 43–49.

Ortiz de Montellano, P. R., and De Voss, J. J. (2002). Oxidizing Species in the Mechanism of Cytochrome P450, *Nat. Prod. Rep.*, 19, 477–493.

Osaki, S., Johnson, D. A., and Frieden, E. (1971). The Mobilization of Iron from the Perfused Mammalian Liver by a Serum Copper Enzyme, Ferroxidase 1, *J. Biol. Chem.*, 246, 3018–3023.

Peña, M. M. O., Lee, J., and Thiele, D. J. (1999). A Delicate Balance: Homeostatic Control of Copper Uptake and Distribution, *J. Nutr.*, 129,1251–1260.

Persson Moschos, M. (2000). Selenoprotein P, *Cell. Mol. Life Sci.*, 57, 1836–1845.

Prasad, A. S. (1995). Zinc: An Overview, *Nutrition*, 11, 93–99.

Purcarea, C., Hervé, G., Ladjimi, M. M., and Cunin, R. (1997). Aspartate Transcarbamylase from the Deep-Sea Hyperthermophilic Archaeon *Pyrococcus abyssi*: Genetic, Organisation, Structure, and Expression in *Escherichia coli*, *J. Bacteriol.*, 179, 4143–4157.

Putney, J. W., and McKay, R. R. (1999). Capacitive Calcium Entry Channels, *Bioassays*, 21, 38–46.

Ragsdale, S. W. (1998). Nickel Biochemistry, *Curr. Opin. Chem. Biol.*, 2, 208–215.

Ramirez-Silva, L., Ferreira, S. T., Nowak, T., Tuena de Gomez-Puyou, M., and Gomez-Puyou, A. (2001). Dimethylsulfoxide Promotes K+-Independent Activity of Pyruvate Kinase and the Acquisition of the Active Catalytic Conformation, *Eur. J. Biochem.*, 268, 3267–3274.

Romani, A. M. P., and Maguire, M. E. (2002). Hormonal Regulation of Mg^{2+} Transport and Homeostasis in Eurkaryotic Cells, *Biometals*, 15, 271–283.

Roy, R., and Adams, M. W. W. (2002). Characterization of a Fourth Tungsten-Containing Enzyme from the Hyperthermofilic Archaeon *Pyrococcus furiosus*, *J. Bacteriol.*, 184, 6952–6956.

Sager, T. N., Thomsen, C., Valsborg, J. S., Laursen, H., and Hansen, A. J. (1999). Astroglia Contain a Specific Transport Mechanism for *N*-acetyl-L-aspartate, *J. Neurochem.*, 73 807–811.

Saito, Y., and Takahashi, K. (2000). Selenoprotein P: Its Structure and Functions, *J. Health Sci.*, 44, 409–413.

Sakurai, H. (2002). A New Concept: The Use of Vanadium Complexes in the Treatment of Diabetes Mellitus, *Chem. Rec.*, 2, 237–248.

Schipper, H. M. (1999). Glial HO-1 Expression, Iron Deposition and Oxidative Stress in Neurodegenerative Disease, *Neurotox. Res.*, 1, 57–70.

Schrauzer, G. N. (2000). Selenmethionine: A Review of Its Nutritional Significance, Metabolism, and Toxicity, *J. Nutr.*, 130, 1653–1656.

Schwartz, K., and Foltz, C. M. (1957). Selenium as an Integral Part of Factor 3 Against Dietary Necrotic Liver Degeneration, *J Am. Chem. Soc.*, 79, 3292–3293.

Shamberger, R. J., and Frost, D. V. (1969). Possible Protective Effect of Selenium Against Human Cancer, *Can. Med. Assoc. J.*, 100, 682.

Shen, D. H. Y., Kloos, R. T., Mazzaferri, E. L., and Jhiang, S. M. (2001). Sodium Iodide Symporter in Health and Disease, *Thyroid*, 11, 415–425.

Sorrentino, V., and Rizzuto, R. (2001). Molecular Genetics of Ca^{2+} Stores and Intracellular Ca^{2+} Signalling, *Trends Pharmacol. Sci.*, 22, 459–464.

Sotiriou, S., Gispert, S., Cheng, J., Wang, Y., Chen, A., Hoogstraten-Miller, S., Miller, G. F., Kwon, O., Levine, M., Guttentag, S. H., and Nussbaum, R. L. (2002). Ascorbic-Acid Transporter Slc23a1 Is Essential for Vitamin C Transport into the Brain and for Perinatal Survival, *Nat. Med.*, 8, 514–517.

Sreedhara, A., and Cowan, J. A. (2002). Structural and Catalytic Roles for Divalent Magnesium in Nucleic Acid Biochemistry, *Biometals*, 15, 211–223.

Stanely, J. S., Mock, D. M., Griffin, J. B., and Zempleni, J. (2002). Biotin Uptake into Human Peripheral Blood Mononuclear Cells Increases Early in the Cell Cycle, Increasing Carboxylate Activities, *J. Nutr.*, 132, 1854–1859.

Sterling, D., Reithmeier, R. A. F., and Casey, J. R. (2001). Carbonic Anhydrase: In the Driver's Seat for Bicarbonate Transport, *J. Pancreas (Online)*, 2, 165–170.

Sträter, N., Sun, L., Kantrowitz, E. R., and Lipscomb, W. N. (1999). A Bicarbonate Ion as a General Base in the Mechanisms of Peptide Hydrolysis by Dizinc Leucine Aminopeptidase, *Proc. Natl. Acad. Sci. U.S.A.*, 96, 11151–11,155.

Sun, X., and Kang, Y. J. (2002). Prior Increase in Metallothionein Levels Is Required To Prevent Doxorubicin Cardiotoxicity, *Exp. Biol. Med.*, 227, 652–657.

Tamarit, J., Cabiscol, E., Aguilar, J., and Ros, J. (1997). Differential Inactivation of Alcohol Dehydrogenase Isoenzymes in *Zymomonas mobilis* by Oxygen, *J. Bacteriol.*, 179, 1102–1104.

Trinder, D., Macey, D. J., and Olynyk, J. K. (2000). The New Iron Age, *Int. J. Mol. Med.*, 6, 607–612.

Väänänen, K., and Zhao, H. (2001). Osteoclast Function. In *Principles of Bone Biology* (J. P. Bilezikan, L. G. Raisz, and G. A. Rodan, Eds.), 2nd ed., Academic Press, San Diego, CA, 127–139.

Venkatchalam, K., van Rossum, D. B., Patterson, R. L., Ma, H. T., and Gill, D. L. (2002). The Cellular and Molecular Basis of Store-Operated Calcium Entry, *Nat. Cell Biol.*, 4, E263–E272.

Vincent, J. B. (2000). The Biochemistry of Chromium, *J. Nutr.*, 130, 715–718.

Vulpe, C. D., Kuo, Y. M., Murphy, T. L., Cowley, L., Askwith, C., Libina, N., Gitschier, J., and Anderson, G. J. (1999). Hepaestin, a Ceruloplasmin Homologue Implicated in Intestinal Iron Transport, is Defective in the sla Mouse, *Nat. Genet.*, 21, 195–199.

Watt, R. K., and Ludden, P. W. (1999). Nickel-Binding Proteins, *Cell. Mol. Life Sci.*, 56, 604–625.

Whanger, P. D. (2000). Selenoprotein W: A Review, *Cell. Mol. Life Sci.*, 57, 1846–1852.

Wilks, A. (2002). Heme Oxygenase: Evolution, Structure, and Mechanism, *Antiox. Redox Signal.*, 4, 603–614.

Yamamoto, A., Wada, O., and Ono, T. (1987). Isolation of a Biologically Active Low-Molecular-Mass Chromium Compound from Rabbit Liver, *Eur. J. Biochem.*, 165, 627–631.

Yang, Y., and Zhou, H. M. (2001). Effect of Zinc Ions on Conformational Stability of Yeast Alcohol Dehydrogenase, *Biochemistry (Moscow)*, 66, 47–54.

Yocum, C. F., and Pecoraro, V. L. (1999). Recent Advances in the Understanding of the Biological Chemistry of Manganese, *Curr. Opin. Chem. Biol.*, 3, 182–187.

Youn, H. D., Kim, E. J., Roe, J. H., Hah, Y. C., and Kang, S. O. (1996). A Novel Nicket-Containing Superoxide Dismutase from Streptomyces spp, *Biochem. J.*, 318, 889–896.

Yu, S. Y., Zhu, Y. J., and Li, W. G. (1997). Protective Role of Selenium Against Hepatitis B Virus and Primary Liver Cancer in Qidong, *Biol. Trace Element Res.*, 56, 117–124.

Yu, S. Y., Zhu, Y. J., Li, W. G., Huang, Q. S., Zhi-Huang, C., Zhang, Q. N., and Hou, C. (1991). A Preliminary Report of the Intervention Trials of Liver Cancer in High Risk Populations with Nutritional Supplementation in China, *Biol. Trace Element Res.*, 29, 289–294.

Zinchuk, V. X., Kobayashi, T., Garcia del Saz, E., and Seguchi, H. (1997). Biochemical Properties and Cytochemical Localization of Ouabain-Insensitive, Potassium-Dependent *p*-Nitrophenylphosphatase Activity in Rat Atrial Myocytes, *J. Histochem. Cytochem.*, 45, 177–187.

Zoroddu, M. A., Schinocca, L., Jankowska-Kowalik, T., Koslowski, H., Salnikow, K., and Costa, M. (2002). Molecular Mechanisms in Nickel Carcinogenesis: Modelling Ni(II) Binding Site in Histone 4, *Environ. Health Perspect.*, 101(Suppl. 5), 719–723.

CHAPTER 7

GEOLOGICAL IMPACTS ON NUTRITION

GERALD F. COMBS, JR.
USDA Agricultural Research Service

CONTENTS

I. GEOLOGICAL SOURCES OF NUTRIENTS

Humans, like all living organisms, biosynthesize the proteins, nucleic acids, phospholipids, and many of the smaller molecules on which they depend for life functions. The health and well-being of organisms also depend on their ability to obtain from external chemical environments a number of compounds that they cannot synthesize, at least at rates sufficient to support those functions. Thus, of the large set of bioactive compounds and metabolites called "nutrients," some are referred to as "essential" because they must be obtained from the air (oxygen), water, and diet. These include

vitamins, some fatty acids, some amino acids, and several mineral elements. Foods contain essential nutrients as a result of the capacity of plants, and in some cases food animals, to synthesize and/or store them. The human body, therefore, consists of substantial amounts of "mineral elements" (see Table I) obtained mostly from such foods. (The term mineral elements is used by nutritionists and is equal to "elements" in the other chapters of this book.) These elements, of course, cannot be biosynthesized; ultimately, they are obtained from soils and, in turn, from the parent materials from which soils are derived. Therefore, good mineral nutrition is, in part, a geologic issue.

Nutritionally important mineral elements include some (e.g., manganese [Mn]) that occur predominately in silicates, some (e.g., zinc [Zn], selenium [Se]) that occur in silicates and sulfides, some (e.g., copper [Cu], molybdenum [Mo]) that occur in sulfides or as native elements with iron (Fe), and some (e.g., iron) that occur in silicates, sulfides, and as the native metal. The most abundant of these is iron, which is the fourth most abundant element in the Earth's crust (see also Chapter 2, this volume).

About 22 mineral elements are known or suspected to be essential for humans and other animals (see Table II). Some are required in fairly large amounts, grams per kilogram of diet, and are therefore referred to as "macronutrients"; others are required in much smaller amounts, e.g., microgram-to-milligrams per kilogram

TABLE I. Nutritionally Essential Mineral Elements in the Human Body

Element	Typical amount[a]
Ca	1000 g
P	700 g
Mg	20–28 g
Na	1.3 g
K	110–150 g
Mg	20–28 g
Zn	2–2.5 g
Cu	120 mg
Se	20 mg

[a]70-kg reference.

of diet and are referred to as "micronutrients." At least eight mineral elements function physiologically in their simple cationic forms (Ca^{+2}, Mg^+, Na^+, K^+, Fe^{+2}, Cu^{+2}, Zn^{+2}, and Mn^+) and can, therefore, be subject to chelation by either intact proteins or a variety of small, organic molecules. Some chelates (e.g., the heme moiety in hemoglobin and myoglobin) are essential in metabolism; some (e.g., amino acids, EDTA) facilitate the absorption, transport, and tissue storage of mineral ions; and others (e.g., phytic acid, oxalic acid) can interfere with the enteric absorption of certain essential cations. For example, the transition metal ions (Fe^{+2}, Cu^{+2}, and Zn^{+2}) form coordinate covalent bonds with ligands containing the electron-donor atoms N, S, and O, the histidinyl imidazole-N (Cu^{+2}), the cysteinyl sulfhydryl-S (Zn^{+2}), and the aspartyl and glutamyl carboxyl-O (Fe^{+2}, Cu^{+2}, and Zn^{+2}). Three mineral ele-

TABLE II. Mineral Elements Known and Suspected to be Essential for Optimal Health

Accepted essentials[a]	Suspected essentials[b]	Known or implicated functions
Macronutrient elements		
Ca		Bone structure, nervous transduction
P		Bone structure, membrane structure, metabolic regulation
Mg		Bone structure, electrochemical regulation, enzyme catalysis
Na		Electrochemical regulation, acid-base balance, osmotic control of water distribution
K		Electrochemical regulation, acid-base balance, osmotic control of water distribution
Cl		Electrochemical regulation; acid-base balance, osmotic control of water distribution
Micronutrient elements		
Fe		Oxygen transport, electron transport
Cu		Enzyme catalysis
Zn		Enzyme catalysis, protein structure
I		Metabolic regulation
Se	Ni	Enzyme catalysis, antioxidant protection, redox regulation, anti-tumorigenic metabolites
Mn	Pb	Enzyme catalysis
Mo	As	Potentiation of insulin action in the maintenance of glucose tolerance
Cr	B	Enzyme catalysis
F	V	Protects against dental caries
Co[c]	Si	Single carbon metabolism as active center of the vitamin B_{12} molecule
		Fetal survival and anemia in experimental animals
		Anemia in experimental animals
		Reproductive function and growth in experimental animals
		Bone mineralization in experimental animals
		Growth in experimental animals
		Reproductive function and fetal development in animals, calcification in cell culture

[a]Essentiality demonstrated on the basis of specific biochemical functions.
[b]Essentiality indicated by physiological impairment correctable by supplementation.
[c]The element itself can be used only by ruminants with foregut microflora capable of synthesizing that vitamin. For this reason, Co is considered essential only for ruminants, while the essential form for all non-ruminants including humans is vitamin B_{12}.

ments function as anions or in anionic groupings (Cl^-, PO_4^{-3}, MoO_4^{-2}). Two, iodine [I] and selenium, are nonmetals and function in covalent compounds (e.g., iodothyronine, selenocysteine) formed metabolically. The biologic significance of these elements, therefore, tends to be a property of their particular organic species rather than of the element per se.

II. MINERAL ELEMENTS NEEDED FOR GOOD HEALTH

Sixteen mineral elements are established as being essential for good health (Table II). These, collectively, have five general physiological roles:

1. Bone and membrane structure: calcium, phosphorus, magnesium, fluoride
2. Water and electrolyte balance: sodium, potassium, chloride
3. Metabolic catalysis: zinc, copper, selenium, magnesium, molybdenum
4. Oxygen binding: iron
5. Hormone effects: iodine, chromium

Although some of these functions are effected by the mineral ions themselves, many are effected by macromolecules in which one or more minerals are bound, either covalently or otherwise. Because these are all critical life functions, the tissue levels of many nutritionally essential mineral elements tend to be regulated within certain ranges despite varying levels of intake by homeostatic control of enteric absorption and tissue storage and/or excretion. For mineral cations such as Cu^{+2} and Zn^{+2}, regulation occurs primarily at the level of enteric absorption. For mineral elements that tend to be highly absorbed (e.g., selenium, boron), homeostasis is achieved by control at the level of excretion, i.e., through the urine, bile, sweat, and breath. In the case of iron, access to active forms is regulated by altering the storage of the element in an inactive form, e.g., ferritin. The ability to orchestrate these physiological processes to achieve homeostatic control of cellular access to such mineral elements is an important factor in ameliorating the effects of short-term dietary deficiencies or excesses.

Calcium—The human body contains more than 1 kg of calcium [Ca] 99% of which is in the skeleton where it serves as the dominant cationic component (26% of dry weight). Bone mineral consists of a complex matrix of plate-like crystals laid down by osteoblasts in or along collagen fibrils in several solid phases: hydroxyapatite ($Ca_{10}[PO_4]_6[OH]_2$), whitlockite ($[Ca,Mg]_3[PO_4]_2$), amorphous $Ca_9(PO_4)_6X$, octacalcium phosphate ($Ca_8H_2[PO_4]_6 \cdot H_2O$), and brushite ($CaHPO_4 \cdot 2[H_2O]$). Bone calcium is in constant turnover, with mineralization and mobilization of bone minerals occurring continually in the healthy bones of both children and adults. This "remodeling" allows the bone to serve as a source of calcium for noncalcified tissues, which mitigates against irregularities in day-to-day calcium intakes. Such homeostatic control maintains calcium in the ranges of 8.5–10.5 mg dl^{-1} in plasma and of 45–225 mg kg^{-1} intracellularly. These levels are regulated by vitamin D metabolites (e.g., 1,25-dihyroxycholecalciferol), which affect the active transport of calcium across the gut, the recovery of calcium by the renal tubule, and the remodeling of bone in processes also involving parathyroid hormone, calcitonin, and estrogen (Hollick, 1994; Jones et al., 1998). Only a small fraction of intracellular calcium exists in the ionic form, Ca^{+2}, which functions as a second messenger to signal many key cellular events, e.g., cell volume regulation, fertilization, growth-factor-induced cell proliferation, secretion, platelet activation, and muscle contraction. Therefore, a key aspect of cell regulation is the control of the release of the Ca^{+2} signal. In various cells, this process is thought to involve 1,4-inositoltriphosphate or nervous stimulation as triggers and protein kinase C and cyclic AMP as inhibitors (Bronner, 1997). Impaired bone mineralization in young children results in deformities of the growing bones and is called rickets; in adults with formed bone it is called osteomalacia and is characterized by increase fracture risk and loss of stature. Conditions of either type can be caused by insufficient intakes of calcium, vitamin D (or exposure to sunlight which is necessary for the biosynthesis of the vitamin), phosphorus and/or magnesium. Only in very severe deficiency, when bone mineral has largely been exhausted, does calcium deficiency result in impaired nervous conduction or muscular contraction. Excessive calcium intake, which typically occurs due to the inappropriate use of calcium supplements, can lead to renal stone formation, hypercalcemia, and renal insufficiency as well as impaired utilization of iron, zinc, magnesium, and phosphorus.

Phosphorus—The human body contains approximately 700 g of phosphorus, about 85% of which is in bones where it serves a structural function in bone minerals (see also Chapter 28, this volume). Much of the nearly 14% of body phosphorus in noncalcified tissues also serves a structural function in the phospholipids that comprise plasma and subcellular membranes

(Berner, 1997). As phosphate, the element phosphorus is also important in metabolism because it is incorporated into nucleic acids, RNA, DNA, proteins (including transcription factors), ATP, and numerous other high-energy substrates. Intracellular phosphate serves as a regulator of glycolysis, a key pathway for rendering oxygen available to the tissues; and phosphoproteins play essential roles in the electron-transport system of mitochondria, which generates metabolically useful energy from carbohydrates and lipids. Other phosphoproteins serve as cellular growth factors and cytokines. Like calcium, phosphorus homeostasis is affected by the vitamin D hormone system. Deficiency of phosphorus (hypophosphatemia) can result in tissue hypoxia due to the loss of erythrocyte 2,3-diphosphoglucose and ATP, which leads to nervous signs (convulsions, confusion), renal dysfunction, and smooth muscle problems (e.g., dysphagia, gastric atony). Hypophosphatemia can also result in rickets and osteomalacia. Chronic, excessive intakes of phosphorus can cause hyperphosphatemia which leads to interference with calcium homeostasis, bone demineralization, and ectopic calcification of the kidney.

Magnesium—The human body typically contains 20–28 g of magnesium, which is widely distributed: 60–65% of that amount is in bone, 25–30% in muscle, and the balance is in other tissue and extracellular fluid. In fact, Mg^{+2} is second only to K^+ as the most abundant intracellular inorganic cation. Normal plasma Mg concentrations are in the range of 0.65–0.88 mmol L^{-1}. Magnesium tends to be well absorbed (i.e., at rates of 67–70%) by a saturable process as well as simple diffusion. Amounts of the element not retained for tissue growth/turnover are excreted in the urine. The cation, Mg^{+2}, functions as a cofactor in at least 300 enzymatic reactions (Shils, 1997). These include virtually all kinase reactions (in which Mg^{+2} complexes with the negatively charged ATP^{-4} to form the substrate), pyrophosphotransferases, acyl-CoA synthetases, and adenylate cyclase. The cation is also involved in the regulation of ion movements within cells. Magnesium deficiency (hypomagnesemia) is characterized by neuromuscular signs (hyperactivity, muscle spasms, tremor, weakness) and gastrointestinal symptoms (anorexia, nausea, vomiting). Magnesium has a cathartic effect, but adverse effects have been identified only for magnesium ingested from non-food sources.

Sodium—The human body contains approximately 1.3 g of sodium [Na], 90% of which is in the extracellular space where it serves as one of the three (with K^+ and Cl^-) osmotically active solutes in extracellular fluid. Sodium is freely and quantitatively absorbed, but the element is homeostatically regulated within a normal range of 135–145 mmol L^{-1} at the level of renal reabsorption effected by renin, angiotensin, aldosterone, antidiuretic hormone, atrial natriuretic peptide, and other factors affecting renal blood flow (Harper et al., 1997). Intracellular Na^+ is normally maintained at relatively low levels. The Na^+ gradient is used as an energy source for the uphill transport of a variety of solutes (e.g., amino acids, Ca^{+2}, Mg^{+2}, H^+) into the cell. The maintenance of the Na^+ gradient is maintained by several transport systems including the Na^+–K^+ pump which effects the ATP-dependent anti-transport of Na^+ (in) and K^+ (out), the Na^+–H^+ exchanger, Na^+–K^+–Cl^- co-transporters, the Na^+–Ca^{+2} exchanger, Na^+–Mg^{+2} exchangers, and the voltage-regulated Na^+ channel. Sodium deficiency results in muscle cramps, headache, poor appetite, and dehydration, but the main sign is fatigue.

Potassium—Potassium (K) is the most abundant cation in the human body, with total body stores typically in the range of 110–150 g. In contrast to Na^+, K^+ is found primarily (98%) in the intracellular compartment; most cells contain about 150 mM K^+, while the level in extracellular fluid is only about 4 mM. Potassium is freely absorbed, with homeostasis affected by rapid renal excretion. Potassium passes the plasma membrane into cells by the Na^+,K^+-ATPase, the H^+,K^+-ATPase, the Na^+–2Cl^-–K^+ co-transporter, and K^+ conductance channels. Increases in extracellular K^+ concentrations can be caused by vigorous exercise leading to K^+ efflux from myocytes and mediating vasodilation and increased blood flow. Such increases stimulate the release of catecholamines and insulin, which stimulates K^+ uptake via the Na^+,K^+-ATPase (Peterson, 1997). Potassium deficiency (hypokalemia) can be caused by insufficient intake and/or excessive excretion (e.g., due to diarrhea, bulimia) of the element. This is characterized by skeletal muscular weakness; smooth muscle paralysis resulting in anorexia, nausea, vomiting, and constipation; cardiac arrhythmias; carbohydrate intolerance due to diminished insulin secretion; impaired renal function due to reduced blood flow; and altered water balance involving increased water consumption secondary to elevated angiotensin II levels.

Chloride—Chloride (Cl^-) is the major extracellular anion maintained at a concentration of 100–110 mmol L^{-1} in that fluid. Like Na^+, there is no control over Cl^- absorption, and homeostasis is affected by renal reabsorption/elimination. The transport and cellular uptake of Cl^- is effected by a number of transporters including a K^+–Cl^- co-transporter, a Na^+–K^+–2Cl^- co-transporter,

Cl⁻–HCO₃ exchangers, cystic fibrosis transmembrane conductance regulator (mutation of this causes cystic fibrosis), Ca^{+2}-activated Cl⁻ channels, voltage-regulated Cl⁻ channels, and mechanically activated Cl⁻ channels (Harper et al., 1997).

Iron—The human body typically contains approximately 5 g of iron, and its metabolic function is to transport oxygen and electrons. Iron serves as the redox agent in a large number of enzymatic reactions involving substrate oxidation and reduction. These include oxidoreductases (e.g., xanthine oxidase/dehydrogenase), monooxygenases (e.g., amino acid oxidases, cytochrome P-450), and dioxygenases (e.g., amino acid dioxygenases, lipoxygenases, peroxidases, fatty acid desaturases, nitric oxide synthases) (Beard & Dawson, 1997). Iron homeostasis is effected at the level of enteric absorption. Dietary iron generally exists in either heme (from hemoglobin and myoglobin in animal products) or non-heme (i.e., organic and inorganic salts in plant-based and iron-fortified foods) forms, each of which is absorbed by a different mechanism. Heme iron is much better absorbed and less affected by enhancers and inhibitors of absorption than non-heme iron, which is strongly regulated by the intestinal mucosal cells in response to iron stores and blood hemoglobin status. Thus, iron-adequate men and women typically absorb about 6 and 13% of dietary iron, respectively, with non-heme iron absorption as great as 50% under conditions of severe iron-deficiency anemia. Excess absorbed iron is stored as ferritin and hemosiderin in the liver, reticuloendothelial cells, and bone marrow. The loss of iron from the body is very low, about 0.6 mg per day, and is primarily due to losses in the bile and exfoliated mucosal cells eliminated in the feces. Menstrual losses can be significant, as can nonphysiological losses resulting from parasitism, diarrhea, and enteritis, which are thought to account for half of the cases of global iron-deficiency anemia. Iron deficiency is manifested as hypochromic, normocytic anemia; lethargy; apathy; listlessness; fatigue; impaired non-shivering thermogenesis; impaired immune function; impaired cognitive development; and reduced physical performance. In pregnancy, iron-deficiency increases the risk of premature delivery, low birth weight, and infant and maternal mortality.

Epidemiologic observations have linked high dietary iron intakes or high iron stores with increased risk of coronary heart disease (Salonen et al., 1992). The toxic potential of iron arises from its pro-oxidative effects, which yield reactive oxygen species that attack polyunsaturated membrane lipids, proteins, and nucleic acids. An iron-overload disease, hereditary hemochromatosis, is caused by a defect in the regulation of iron absorption, which leads to very high circulating transferrin iron. Clinical signs appear when body iron accumulates to about tenfold excess of normal: these include hepatic cirrhosis, diabetes, heart failure, arthritis, and sexual dysfunction.

Zinc—The human body contains 2–2.5 g of zinc: just over half (55%) of which is in muscle, 30% in bone, with the balance distributed in other tissues. Zinc is absorbed by both carrier-mediated and simple diffusion processes which render the element only moderately absorbed (about 30%) at nutritionally adequate intakes and more efficiently absorbed under deficient conditions (Chesters, 1997). Both processes can be effected by the presence of chelating substances that may promote (e.g., meats) or impair (e.g., phytic acid) zinc absorption. Zinc homeostasis is also affected by the regulation of zinc excretion/reabsorption in and from pancreatic and intestinal secretions; urinary losses of zinc are low and not generally responsive to changes in zinc intake. Plasma zinc levels, comprising 0.1% of total body zinc, are not regulated and are therefore not indicative of overall zinc status except under conditions of marked deficiency. Zinc has been shown to function in at least 50 widely varied enzymes. In each, the element serves either in a catalytic (i.e., at the active site), a co-catalytic (i.e., near the active site), or a structural role bound most commonly to histidinyl, glutamyl, or aspartyl residues. Zinc deficiency is manifested as losses in activities of at least some zinc enzymes (e.g., some dehydrogenases, alkaline phosphatase, superoxide dismutase), although direct links between such losses and the physiologic manifestations of zinc deficiency have not been established. Zinc deficiency is characterized by poor growth and dwarfism, anorexia, parakeratotic skin lesions, diarrhea, impaired testicular development, impaired immune function (including wound healing), and impaired cognitive function. Low zinc status is also thought to increase risk to osteoporosis and susceptibility to oxidative stress. Very high intakes of zinc, which have occurred due to inappropriate use of zinc supplements, can interfere with copper metabolism and deplete the body of copper. Chronic exposure to excess zinc (more than $100 \, mg \, d^{-1}$) can reduce immune function and HDL cholesterol.

Copper—The human body contains approximately 120 mg of copper, which is widely distributed in many tissues and fluids at $mg \, kg^{-1}$ or $\mu g \, kg^{-1}$ concentrations. Copper serves as a cofactor for a number of oxidase enzymes including lysyl oxidase, ferroxidase (ceruloplasmin), dopamine beta-monooxygenase, tyrosinase, alpha-amidating monooxygenase, cytochrome *c* oxidase, and superoxide dismutase (Harris, 1997). These enzymes are involved in generating oxidative energy, stabilizing connective tissue matrices, maintaining iron in the ferrous (Fe^{+2}) state, synthesizing neurotransmitters, pigmenting hair and skin, supporting immune competence, and protecting the body from reactive oxygen species. Copper also functions non-enzymatically in angiogenesis, neurohormone release, oxygen transport, and the regulation of genetic expression. Copper homeostasis is effected at the level of enteric absorption. It is absorbed by facilitated diffusion (involving either specific transporters or nonspecific divalent metal ion transporters on the brush-border surface) and is transported to the liver where it is re-secreted into the plasma bound to ceruloplasmin. The element is excreted in the bile; only very small amounts are lost in the urine. Copper absorption/retention varies inversely with the level of copper intake, and tends to be moderate (e.g., 50–60%) even at low copper intakes. Copper deficiency is manifested as hypochromic, normocytic or macrocytic anemia; bone abnormalities resembling osteoporosis or scurvy; increased susceptibility to infection; and poor growth. The ingestion of high amounts of copper can cause nausea. Chronic high copper intake can lead to the hepatic accumulation of copper, which has been suspected in juvenile cases of hepatic cirrhosis in India.

Iodine—The human body contains approximately 5 mg of iodine, which functions only in the iodine-containing thyroid hormones. These include the tetraiodinated protein thyroxine (T_4) that is converted by a single deiodination to yield the active thyroid hormone triiodothyronine (T_3). The latter functions as a regulator of growth and development by increasing energy (ATP) production and activating or inhibiting the synthesis of various proteins (Hetzel & Wellby, 1997). Organic forms of iodine are converted in the upper gastrointestinal tract to the iodide anion (I^-), which is rapidly and almost completely absorbed. In contrast, when T_3 is ingested, about 80% is absorbed intact. Absorbed I^- circulates in the plasma in the free ionic form and is rapidly removed by the thyroid and kidney. Iodine homeostasis is effected at the level of the kidney, and urinary excretion is the major route of loss (comprising 90% of iodine absorbed by iodine-adequate individuals). The deiodination of T_4 occurs in the thyroid, skeletal muscles, and brain, but the thyroid gland is the only storage site for iodine where it appears mostly as mono- and diiodotyrosine and T_4, with a small amount of T_3. Iodine deficiency in adults is characterized as thyroid hypertrophy or goiter and in children as myxedematous cretinism. Collectively, these iodine deficiency diseases comprise a global health problem, and cretinism is the greatest source of preventable mental retardation (see also Chapter 16, this volume).

Selenium—The human body typically contains approximately 20 mg of selenium. This is widely distributed in all tissue in which the element is bound to proteins. Some selenium-containing proteins contain the element in the form of the amino acid selenocysteine (SeCys). Selenoamino acid cannot be incorporated into peptides during protein synthesis (SeCys cannot charge the tRNA for cysteine); instead it is synthesized by a unique co-translational process encoded by a codon (UGA) that otherwise signals termination (Sunde, 1997). When consumed as selenomethionine (SeMet), a large number of other proteins can incorporate selenium nonspecifically, due to the mimicry of SeMet with its sulfur-containing analogue in general protein synthesis. Because foods contain both of the selenoamino acids, human tissues typically contain both the specific and nonspecific selenoproteins. Only the former have physiological functions and these include: multiple isoforms of glutathione peroxidase, thioredoxin reductase, and iodothyronine 5′-deiodinase, as well as an enzyme involved in SeCys synthesis, selenophosphate synthase, and at least four selenoproteins of unknown function. As essential constituents of these SeCys proteins, selenium functions in antioxidant protections, redox regulation, and thyroid hormone regulation. It is not clear whether uncomplicated selenium deficiency (i.e., not accompanied by other deficiencies or oxidative stress) results in significant physiological impairment. Severely low selenium intakes (greater than $20 \mu g\,d^{-1}$) have been associated with juvenile cardiomyopathy in China (Keshan disease), which also appears to have a viral component to its etiology.

Recent interest in selenium centers around the apparent efficacy of supranutritional intakes of the element to reduce cancer risk. A decade-long, randomized, double-blind, placebo-controlled clinical intervention trial found that supplementation of free-living American adults with $200 \mu g\,d^{-1}$ selenium (in addition to their normal diets) reduced major cancer incidence by half or more (Clark et al., 1996). Similar effects have been shown in animal studies. Current thinking is that a normal metabolite of selenium can stimulate apoptosis in transformed cells (Combs & Lu, 2001). Selenium intakes greater than $1\,mg\,d^{-1}$ can induce dermatological changes, including brittle hair and nails. Chronic intake approaching 5 mg has been reported to lead to skin rash, paresthesia, weakness, and diarrhea (see also Chapter 15, this volume).

Manganese—Manganese (Mn) functions as a cofactor for enzymes in antioxidant defense (mitochondrial superoxide dismutase), gluconeogenesis (pyruvate carboxylase, phosphoenol–pyruvate carboxykinase), glycoprotein biosynthesis (glycosyl transferases), nitrogen metabolism (arginase, glutamine synthase), and cholesterol biosynthesis (farnesyl pyrophosphate synthetase). Little is known about the mechanisms of absorption, transport, or cellular uptake of manganese, although the element is widely distributed in noncalcified tissues with the greatest concentrations in the liver (Leach & Harris, 1997). The greatest route of manganese excretion appears to be the bile in which it is released in bound form. There is little available evidence of manganese deficiency in humans, although studies in experimental animals have shown effects on fetal survival, normal skeletal development (i.e., shortened limbs, twisted legs, lameness), ataxia, glucose tolerance, and hepatic steatosis.

Molybdenum—Molybdenum (Mo) functions as the active center of three enzymes that catalyze oxidative hydroxylations: sulfite oxidase (the last step in the degradation of sulfur amino acids), xanthine dehydrogenase, and aldehyde oxidase (which transfers electrons to other redox cofactors and, ultimately, to cytochrome c, molecular oxygen, or NAD^+). In these enzymes, the element is found in a pterin-containing, molybdenum cofactor, the synthesis of which in eukaryotes remains poorly understood (Johnson, 1997). Molybdenum appears to be efficiently absorbed at all levels of intake, apparently by a passive process. It is transported in the blood attached to proteins in erythrocytes. Whole blood molybdenum levels vary directly with dietary molybdenum intake, although plasma Mo^+ levels are maintained at about $5\,nmol\,L^{-1}$. Molybdenum is widely distributed in the body, with greatest concentrations in liver, kidney, adrenal gland, and bone. One human case of molybdenum deficiency has been described; signs (all of which responded to molybdenum therapy) included tachycardia, tachypnea, severe headache, night blindness, nausea, and vomiting.

Chromium—Chromium (Cr) potentiates the action of insulin and has been shown to restore glucose tolerance in malnourished infants. Several studies have shown that chromium supplementation lowers circulating glucose levels, increases plasma insulin, and produces a favorable profile of plasma lipids (Offenbacher et al., 1997). It has been suggested that these effects may be due to a low molecular weight chromium-binding substance that may amplify insulin receptor tyrosine kinase activity in response to insulin. Chromium can also bind to one of the binding sites of transferrin, and

it has been proposed that excessive iron storage in hemochromatosis may interfere with the transport of chromium to contribute to the diabetes associated with that disorder. The enteric absorption of the element, which is absorbed as Cr^{+3}, is very low (usually no more than 2%) and appears to be regulated. Absorbed chromium accumulates in the liver, kidney, spleen, and bone.

Fluoride—Fluoride (F^-) is the ionic form of fluorine. It is very highly electronegative and reacts reversibly to hydrogen to form hydrogen fluoride which freely diffuses across the intestine, dissolves in the blood, and is taken up by the tissues where its high affinity for calcium causes it to accumulate in calcified tissues. Fluoride can stimulate new bone formation; when present in oral fluids, F^- exerts cariostatic effects due to enhanced remineralization of dental enamel and reduced acid production of plaque bacteria (Chow, 1990; Cerklewski, 1997). Prior to the widespread use of F^- in dental products and water supplies, most studies showed that the incidence of dental caries (in both children and adults) was 40–60% lower in areas with drinking water F^- concentrations of at least $0.7\,mg\,L^{-1}$ when compared to communities with lower F^- levels. Excessive F^- intake can cause fluorosis of the enamel and bone. Although the former is a largely cosmetic effect involving the mottling of the teeth, skeletal fluorosis is associated with joint stiffness, calcification of ligaments, and some osteosclerosis of the pelvis and vertebrae (see also Chapter 12, this volume).

III. DIETARY SOURCES OF ESSENTIAL MINERAL ELEMENTS

Mineral elements are metabolized and, to varying degrees, stored by plants and animals, some of which constitute important sources of those elements in human diets (see Table III). That the mineral elements are not homogeneously distributed among various types of foods is clear: few foods other than dairy products are rich in calcium; sea foods constitute the best sources of iodine and chloride; meats are the most important sources of iron; and protein-rich foods comprise the best sources of zinc, copper, and selenium. Therefore, optimal mineral nutrition, like optimal nutrition in general, is most likely to be obtained from mixed diets based on a diverse selection of foods. Conversely, the monotonous, non-diverse, grain-based diets accessible to the poor of the developing world are likely to provide

TABLE III.   Important Dietary Sources of Essential Mineral Elements

Element	Sources
Ca	Dairy products, fortified juices, kale, collards, mustard greens, broccoli, sardines, oysters, clams, canned salmon
P	Meats, fish, eggs, dairy products, nuts, beans, peas, lentils, grains
Mg	Seeds, nuts, beans, peas, lentils, whole grains, dark green vegetables
Na	Common table salt, seafood, dairy products, meats, eggs
K	Fruits, dairy products, meats, cereals, vegetables, beans, peas, lentils
Cl	Common table salt, seafood, dairy products, meats, eggs
Fe	Meats, seafood
Cu	Beans, peas, lentils, whole grains, nuts, organ meats, seafood (e.g., oysters, crab), peanut products, chocolate, mushrooms
Zn	Meats, organ meats, shellfish, nuts, whole grains, beans, peas, lentils, fortified breakfast cereals
Se	Meats from Se-fed livestock, sea fish, grain products, nuts, garlic, broccoli grown on high-Se soils
I	Iodized salt, sea fish, kelp
Mn	Whole grains, beans, peas, lentils, nuts, tea
Mo	Beans, peas, lentils, dark green leafy vegetables, organ meats
F	Fluoridated water

insufficient energy, protein, and minerals, especially calcium, copper, selenium, and biologically available iron and zinc. At the same time, the increasing use in industrialized countries of non-diverse eating habits is associated with prevalent insufficient intake of such minerals as calcium.

Soil can contribute to the total dietary intake of mineral elements. This can occur through adherent soil particles on foods and suspended soil particles in drinking and cooking water, as well as through the direct consumption of soil. The latter practice of geophagia can be deliberate in some communities in which the eating of clays occurs (see Geophagy and the Involuntary Ingestion of Soil, this volume). Consumption of clays with high cation-exchange capacities can provide substantial supplements of calcium, iron, copper, zinc, and manganese; other clays can interfere with the enteric absorption of iron and zinc. Consumption of iron-rich lateritic soils or waters draining them can provide enough iron to impair the utilization of copper and zinc.

In some areas, fresh water supplies can provide nutritionally important amounts of such minerals as calcium, magnesium, iron, manganese, and arsenic, and industrialized countries have used municipal water as a vehicle for providing fluoride. In a few locales surface runoff from selenium-rich soils has been found to contain biologically significant amounts of selenium, but such cases are rare and most water supplies are very low in that nutrient.

For many people in industrialized countries, fortified foods and nutritional supplements constitute important sources of several of the mineral elements. Various forms of copper, zinc, iron, and selenium are offered in over-the-counter formations, both as individual supplements as well as compounded in multivitamin mineral supplements. Calcium, typically as the carbonate or gluconate salts, is now commonly used to fortify orange and other fruit juices. Consumer response to such nutrient-fortified foods has been very strong, and this aspect of consumer retailing is expected to continue to grow.

IV. MINERAL ELEMENT BIOAVAILABILITY

For several nutrients only a portion of the ingested amount is absorbed and utilized metabolically. Therefore, it is necessary to consider this when evaluating the nutritional adequacy of foods and diets. This concept, bioavailability, is particularly important in mineral nutrition, because some foods are less useful sources of essential minerals than might be expected from their absolute mineral content.

Mineral bioavailability depends on both physiological and exogenous factors. Physiological determinants of mineral bioavailability include:

1. Age-related declines in the efficiency of enteric absorption of copper and zinc
2. Early postnatal lack of regulation of absorption of iron, zinc, and chromium
3. Adaptive increases in the absorptive efficiencies of iron and zinc, copper, manganese, and chromium by receptor upregulation during periods of deficiency
4. Dependence on other nutrients for the physiological functions of selenium and iodine in thyroid hormone metabolism, and copper and iron in catecholamine metabolism
5. Anabolic effects on tissue sequestration of zinc and selenium
6. Catabolic effects on zinc, selenium, and chromium losses

The chemical form of an element as well as the presence/absence of other factors in foods and diets can impair or enhance its absorption and post-absorptive utilization. For example, 25–30% of the heme iron in animal tissues can be absorbed, but 2–5% of the non-heme iron in plant foods is absorbed. The utilization of plant sources of iron can be markedly improved by including in the diet sources of ascorbic acid (e.g., oranges) or meats, both of which promote the utilization of non-heme iron. Similarly, citrate and/or histidine can enhance the absorption of zinc. Dietary ascorbate (vitamin C) can, thus, also enhance the antagonistic effect of iron on copper utilization.

Mineral bioavailability can be reduced by dietary factors that reduce enteric absorption. For example, phytate, phosphorus, and triglycerides can reduce the luminal solubility and, hence, the absorption of calcium. Phytate and other non-fermentable fiber components can bind zinc and magnesium, reducing the absorption of each. Sulfides can reduce the absorption of copper by similar means. Minerals that share transporters can be mutually inhibitory for absorption, e.g., sulfite and selenite, cadmium and zinc, and zinc and copper.

In general, problems related to poor bioavailability are greatest for iron in plant-based containing phytates and/or polyphenols but there are few problems with promotor substances. For calcium there are problems with bioavailability when poorly soluble forms are consumed with vegetables (spinach, rhubarb, beet greens, chard) containing inhibitory oxalates without others (artichokes) containing fructose oligosaccharide promoters; for zinc in diets high in unrefined (>90% extraction), unfermented cereal grains or high-phytate soy products, especially those fortified with inorganic calcium salts; and for selenium consumed as plant foods (containing SeMet much of which is diverted to protein synthesis). For these reasons, the utilization of these minerals as consumed in most diets tends to be moderate at best, though in each case it can be markedly enhanced through appropriate dietary choices.

V. Quantitative Estimates of Mineral Needs and Safe Exposures

Dietary standards have been set for several, but not all, of the nutritionally essential mineral elements. International standards have been developed for only some minerals (FAO-WHO, 2002) (Table IV). The most current and extensive standards are the Dietary Reference Intakes (DRSs) published by the U.S. National Academy of Science (NAS) (Food and Nutrition Board, 1997, 2000, 2001) (Table V). It is important to note that the expert panels of the respective organizations used the same primary data, i.e., the published scientific literature. Also, each based its recommendations on estimates of individual physiological need (i.e., the World Health Organization's "basal requirement" and the "recommended dietary allowance," RDA, from NAS) which was then inflated to accommodate estimated interindividual variation. This approach produced the WHO "normative requirement" and NAS "estimated average requirement" (EAR). The NAS process went further to include estimates of "average intakes" (AIs) in cases where data were not sufficient to support EARs or RDAs. Both groups also estimated safe limits of exposure: WHO created "upper limits of safe ranges of population mean intakes" (Table VI), and NAS created "upper tolerable limits" (ULs) (Table VII).

VI. Clinical Assessment of Mineral Status

The status assessment of the essential minerals, which vary so much in metabolic function, homeostatic regulation, and tissue distribution, calls for a mixed

TABLE IV. International Dietary Recommendations (Units per Day) for Essential Mineral Elements[a,b]

Life stage	Ca (mg)	P (mg)	Na (mg)	K (mg)	Cl (mg)	Mg (mg)	Fe[c] (mg)	Cu (mg)	Zn[d] (mg)	Se (µg)	I (µg)	Mn (mg)	Mo (mg)	Cr (µg)	F (mg)
Children															
0–3 months								0.33–0.55[f]							
3–6 months	300[e]					26[e]		0.37–0.62[f]	2.8[e]	6	[15][g]				
	400[f]					36[f]									
7–12 months	400					53	[9][g]	0.60	4.1	10	130				
1–3 years	500					60	6	0.58	4.8	17	75				
3–6 years	600					73	6	0.57	5.1	21	115				
6–9 years	700					100	9	0.75	5.6	21	110				
10–11 years											140f				
											135m				
10–12 years								0.77f							
								0.73m							
10–14 years							14f				140f				
							15m				135m				
12–15 years								1.00							
12–18 years											110f				
											100m				
15–18 years							31f	1.33f							
							19m	1.15m							
10–18 years	1300					230f			7.8f	26f					
						250m			9.7m	32m					
Adults															
19–50 years	1000					220f	29f	1.35f	4.9f	26f	110f				
						260m	14m	1.15m	7.0m	34m	130m				
51–65 years	1300f					220f	11f	1.35f	4.9f	26f	110f				
	1000m					260m	14m	1.15m	7.0m	34m	130m				
65+ years	1300					190f	11f	1.35f	4.9f	26f	110f				
						230m	14m	1.15m	7.0m	34m	130m				
Pregnancy															
1st trimester						220		1.15	5.5		200				
2nd trimester						220		1.15	7.0	28	200				
3rd trimester	1200					220		1.15	10.0	30	200				
Lactation															
0–3 months	1000					270	15	1.25	9.5	35	200				
3–6 months	1000					270	15	1.25	8.8	35	200				
6–12 months	1000					270	15	1.25	7.2	42	200				

[a]Recommendations for copper are normative dietary requirements, WHO (1996).

[b]Recommendations for calcium, magnesium, iron, zinc, selenium, and iodine, FAO-WHO (1998).

[c]10% bioavailability conditions.

[d]Moderate (30–35%) bioavailability conditions.

[e]Breast-fed infants.

[f]Formula-fed infants.

[g]Value expressed in units $kg^{-1} d^{-1}$.

TABLE V. Dietary Recommendations for Essential Mineral Elements (Units per Day)

Life stage	Ca (mg)	P (mg)	Na (mg)	K (mg)	Cl (mg)	Mg (mg)	Fe (mg)	Cu (μg)	Zn (mg)	Se (μg)	I (μg)	Mn (mg)	Mo (mg)	Cr (μg)	F (mg)
Children															
0–6 months	210[a]	100[a]	120[c]	500[c]	180[c]	30[a]	0.27[d]	200[d]	2[d]	15[e]	110[d]	0.003[d]	0.2[d]	0.2[d]	0.01[a]
7–12 mos.	270[a]	275[a]	200[c]	700[c]	300[c]	75[a]	11[f]	220[f]	3[f]	20[e]	130[d]	0.6[d]	3[d]	5.5[d]	0.5[a]
1–3 years	500[a]	460[b]	225–300[c]	1000–1400[c]	350–500[c]	80[b]	7[f]	340[f]	3[f]	20[g]	90[f]	1.2[f]	17[f]	11[d]	0.7[a]
4–8 years	800[a]	500[b]	300–400[c]	1400–1600[c]	500–600[c]	130[b]	10[f]	440[f]	5[f]	30[g]	90[f]	1.5[f]	22[f]	15[d]	1[a]
9–13 years	1300[b]	1250[b]	400–500[c]	1600–2000[c]	600–750[c]	240[b]	8[f]	700[f]	8[f]	40[g]	120[f]	1.9[d](m) 1.6[d](f)	34[f]	25[d](m) 21[d](f)	2[a]
14–18 years	1300[b]	1250[b]	500[c]	2000[c]	750[c]	410(m)[b] 360(f)[b]	11[f](m) 15[f](f)	890	11[f](m) 9[f](f)	55[g]	150[f]	2.2[d](m) 1.6[d](f)	43[f]	35[d](m) 24[d](f)	3[a]
Adults															
19–30 years	1000[a]	700[b]	500[c]	2000[c]	750[c]	400(m)[b] 310(f)[b]	8[f](m) 11[f](f)	900[f]	11[f](m) 8[f](f)	55[g]	150[f]	2.3[d](m) 1.8[d](f)	45[f]	35[d](m) 25[d](f)	4(m)[a] 3(f)[a]
31–50 years	1000[a]	700[b]	500[c]	2000[c]	750[c]	420(m)[b] 320(f)[b]	8[f](m) 11[f](f)	900[f]	11[f](m) 8[f](f)	55[g]	150[f]	2.3[d](m) 1.8[d](f)	45[f]	35[d](m) 25[d](f)	4(m)[a] 3(f)[a]
51+ years	1200[a]	700[b]	500[c]	2000[c]	750[c]	420(m)[b] 320(f)[b]	8[f]	900[f]	11[f](m) 8[f](f)	55[g]	150[f]	2.3[d](m) 1.8[d](f)	45[f]	30[d](m) 20[d](f)	4(m)[a] 3(f)[a]
>70 years	1200[a]	700[b]	500[c]	2000[c]	750[c]	420(m)[b] 320(f)[b]	8[f]	900[f]	11[f](m) 8[f](f)	55[g]	150[f]	2.3[d](m) 1.8[d](f)	45[f]	30[d](m) 20[d](f)	4(m)[a] 3(f)[a]
Pregnancy															
≥18 years	1300[a]	1250[b]				400[b]	27[f]	1000[f]	13[f]	60[g]	220[f]	2[d]	50[f]	19[d]	3[a]
19–30 years	1000[a]	700[b]				350[b]	27[f]	1000[f]	11[f]	60[g]	220[f]	2[d]	50[f]	30[d]	3[a]
31–50 years	1000[a]	700[b]				360[b]	27[f]	1000[f]	11[f]	60[g]	220[f]	2[d]	50[f]	30[d]	3[a]
Lactation															
≥18 years	1300[a]	1250[b]				360[b]	10[f]	985[f]	14[f]	70[g]	290[f]	2.6[d]	50[f]	44[d]	3[a]
19–30 years	1000[a]	700[b]				310[b]	9[f]	1000[f]	12[f]	70[g]	290[f]	2.6[d]	50[f]	45[d]	3[a]
31–50 years	1000[a]	700[b]				320[b]	9[f]	1000[f]	12[f]	70[g]	290[f]	2.6[d]	50[f]	45[d]	3[a]

[a]Average Intake (AI) value, (Food and Nutrition Board, 1997).
[b]Recommended Dietary Allowance (RDA), (Food and Nutrition Board, 1997).
[c]Estimated Minimum Requirement, (Food and Nutrition Board, 1989).
[d]Average Intake (AI) value, (Food and Nutrition Board, 2001).
[e]Average Intake (AI) value, (Food and Nutrition Board, 2000).
[f]Recommended Dietary Allowance (RDA), (Food and Nutrition Board, 2001).
[g]Recommended Dietary Allowance (RDA), (Food and Nutrition Board, 2000).

TABLE VI. International Estimates of Upper Limits (Units per Day) of Safe Intakes of Essential Mineral Elements

Life stage	Ca (mg)	P (mg)	Na (mg)	K (mg)	Cl (mg)	Mg (mg)	Fe (mg)	Cu (mg)	Zn[a] (mg)	Se (μg)	I (μg)	Mn (mg)	Mo (mg)	Cr (μg)	F (mg)
Children															
0–6 months									13						
0–12 months								(150)[b]							
1–6 years								1.5	23						
6–10 years								3.0	28						
10–12 years								6.0	32f 34m						
12–15 years								8.0	36f 40m						
15–18 years									38f 48m						
Adults															
15–60 years								12.0		400						
18–60 years									55f 45m						
Pregnancy															
General								10.0							
Lactation															
General								10.0							

[a]Moderate (30–35%) bioavailability conditions.
[b]In mcg kg^{-1}.
From WHO (1996).

approach. This approach includes elemental analyses of tissues and/or body fluids, assays of mineral-dependent enzyme activities, and measurement of functional and/or morphological indices. A battery of such tests may be feasible in research settings, but in clinical settings practicality and timeliness dictate approaches based on analyses of a single specimen of blood.

Beyond the obvious issues pertaining to sampling (i.e., number, bias, amount, homogeneity, interindividual variability, etc.), the analysis of minerals, particularly those present in only trace amounts in foods and tissues, calls for special attention to sample integrity and freedom from contamination (Milne, 2000). For example, the iron and zinc contents of plasma or serum can be affected by hemolysis; rubber stoppers and borosilicate glass can contaminate blood with zinc and boron, respectively; and some anticoagulants can produce osmotic shifts that release several elements from erythrocytes. The laboratory, too, can be a significant source of contamination: poorly treated water can contaminate with iron, calcium, magnesium, manganese, zinc, or copper; stainless steel surfaces can contaminate with chromium and nickel; and dust, paper products, wood, skin, hair, and dandruff can also be sources of contamination. For these reasons, a well-monitored laboratory designed for mineral/trace element analyses is a prerequisite for the generation of useful data.

The available methods for the clinical assessment of mineral status are presented in Table VIII, with normative values for the most practically useful of these presented in Table IX.

Status with respect to mineral elements that are active or highly regulated in circulating tissues can be assessed by analyzing their amounts in plasma/serum or blood cells (Sauberlich, 1999). For example, knowledge of plasma/serum potassium or erythrocyte iron levels can be highly informative, because those elements exert their physiological functions in those respective compartments. This is not the case for mineral elements that function in other compartments and/or chemical forms. For example, analyses of chromium, copper, or

TABLE VII. Estimated Upper Tolerable Intakes of Essential Mineral Elements (Units per Day)

Life stage	Ca (mg)	P (mg)	Na (mg)	K (mg)	Cl (mg)	Mg (mg)	Fe (mg)	Cu (mg)	Zn (mg)	Se (μg)	I (mg)	Mn (mg)	Mo (mg)	Cr (μg)	F (mg)
Children															
0–6 months									4[c]	45[b]					0.7[a]
7–12 months									5[c]	60[b]					0.9[a]
1–3 years	2.5[a]	3[a]				65[a]		1[c]	7[c]	90[b]	0.2[c]	2[c]	0.3[c]		1.3[a]
4–8 years	2.5[a]	3[a]				110[a]		3[c]	12[c]	150[b]	0.3[c]	3[c]	0.6[c]		2.2[a]
9–13 years	2.5[a]	4[a]				350[a]		5[c]	23[c]	280[b]	0.6[c]	6[c]	1.1[c]		10[a]
14–18 years	2.5[a]	4[a]				350[a]		8[c]	34[c]	400[b]	0.9[c]	9[c]	1.7[c]		10[a]
Adults															
19–30 years	2.5[a]	4[a]				350[a]		10[c]	40[c]	400[b]	1.1[c]	11[c]	2[c]		10[a]
31–50 years	2.5[a]	4[a]				350[a]		10[c]	40[c]	400[b]	1.1[c]	11[c]	2[c]		10[a]
51+ years	2.5[a]	4[a]				350[a]		10[c]	40[c]	400[b]	1.1[c]	11[c]	2[c]		10[a]
>70 years	2.5[a]	4[a]				350[a]		10[c]	40[c]	400[b]	1.1[c]	11[c]	2[c]		10[a]
Pregnancy															
≥18 years	2.5[a]	3.5[a]				350[a]		8[c]	40[c]	400[b]	.9[c]	9[c]	1.7[c]		10[a]
19–30 years	2.5[a]	3.5[a]				350[a]		8[c]	40[c]	400[b]	1.1[c]	11[c]	2[c]		10[a]
31–50 years	2.5[a]	3.5[a]				350[a]		8[c]	40[c]	400[b]	1.1[c]	11[c]	2[c]		10[a]
Lactation															
≥18 years	2.5[a]	4[a]				350[a]		8[c]	34[c]	400[b]	.9[c]	9[c]	1.7[c]		10[a]
19–30 years	2.5[a]	4[a]				350[a]		10[c]	40[c]	400[b]	1.1[c]	11[c]	2[c]		10[a]
31–50 years	2.5[a]	4[a]				350[a]		10[c]	40[c]	400[b]	1.1[c]	11[c]	2[c]		10[a]

[a] Food and Nutrition Board (1997).
[b] Food and Nutrition Board (2000).
[c] Food and Nutrition Board (2001).

selenium in serum/plasma have inferential value for assessing status only to the extent that those values correlate with the sizes/activities of other physiologically relevant pools. For elements that are not highly regulated in the blood, such as zinc, that parameter has limited, if any, value in assessing status in all except severely deficient individuals.

For mineral elements such as selenium, iodine, zinc, and copper, which exert their physiological functions as essential constituents of macromolecules, assessment of status calls for measurement of the levels/activities of their respective functional forms or metabolite profiles. Thus, zinc adequacy can be determined on the basis of the cytosolic superoxide dismutase, and iodine adequacy can be determined on the basis of circulating levels of triiodothyronine (T_3), thyroid hormone (T_4), and thyroid-stimulating hormone (TSH). Similarly, plasma selenium, because it consists of several components including nonfunctional selenium bound nonspecifically in albumin and other proteins, is best assessed in cases of subadequacy by determining the selenoproteins—extracellular glutathione peroxidase and selenoprotein P.

VII. ECOLOGICAL ASPECTS OF MINERAL NUTRITION

Because the mineral elements are ultimately derived from soils, the mineral status of humans and other animals depends on the minerals available in the soils

TABLE VIII. Clinical Assessment of Mineral Element Status

Element	Elemental analysis[a]	Indicator enzymes/proteins	Indicator metabolites	Physiological indices
		Most useful parameters of status, by general type		
Ca	Serum total Ca–AAS[b], ES[c], MS[d], NAA[e], ICP-MS[f] Serum Ca^{++} – EC[g]			
P	Serum P–AAS, C[h], EC, ES, MS, GFAAS, NAA			
Mg	Serum/plasma Mg—AAS, MS, ICP-MS, NAA Serum Mg^{++}– EC Muscle Mg—AAS, MS, ICP-MS Erythrocyte Mg—AAS, MS, ICP-MS			
Na	Serum/plasma Na—Aas, C, EC, NAA Urinary Na—AAS, C, EC, NAA			
K	Serum/plasma K—AAS, C, EC, NAA Urinary K—AAS, C, EC, NAA			
Cl	Serum/plasma Cl—C, EC, NAA Urinary Cl—C, EC, NAA			
Fe	Serum Fe—AAS, GFAAS, C, EC, ES, MS, NAA, PIXE[i]	Erythrocyte hemoglobin Serum ferritin Serum transferrin Serum transferrin receptor	Free erythrocyte protoporphyrin Zn-protoporphyrin	Hematocrit Serum total Fe-binding capacity
Zn	Serum/plasma Zn—AAS, GFAAS, ES, MS Hair/nail Zn—AAS, GFAAS, ES, MS Urinary Zn–AAS, GFAAS, ES, MS Leukocyte Zn—AAS, GFAAS, ES, MS	Metallothionine I Alkaline phosphatase Carbonic anhydrase Nucleoside phosphorylase Ribonuclease		
Cu	Serum/plasma Cu—AAS, GFAAS, ES, MS, PIXE	Ceruloplasmin activity Superoxide dismutase activity Cytochrome c oxidase activity		
Se	Serum/plasma Zn—GFAAS, HGAAS[j] Hair/nail Se—EAAS	Glutathione peroxidase isoforms Serum selenoprotein P		
I	Urinary I—C, POT[k], NAA		Tetraiodothyronine (T$_4$) Thyroid hormone (T$_3$) Thyroid-stimulating hormone (TSH)	
Mn	Serum/plasma Mn—AAS, GFAAS, ES, MS, NAA			
Mo	Serum/plasma Mo—GFAAS, MS, NAA			
Cr	Serum/plasma Cr—GFAAS, NAA			
F	Serum/plasma F—NAA			

[a]Acceptable analytical method (typically yields CV < 10%).
[b]AAS = atomic absorption spectrophotometry.
[c]ES = emission spectroscopy.
[d]MS = mass spectrometry.
[e]NAA = neutron activation analysis.
[f]GFAAS = AAS with electrothermal atomization using a graphite furnace.
[g]EC = electrochemistry.
[h]Chemical methods.
[i]PIXE = proton-induced x-ray emission.
[j]HGAAS = hydride generation AAS.
[k]POT = potentiometry.

TABLE IX. Reference Values for Key Clinical Parameters of Mineral Element Status

Element	Parameter	Reference value[a]
Ca	Serum total Ca	86–102 µg L^{-1}
	Serum Ca^{++}	46.4–52.8 µg L^{-1}
P	Serum P	Adult: 25–45 µg L^{-1}
		Children: 40–70 µg L^{-1}
Mg	Serum/plasma Mg	16–26 µg L^{-1}
Na	Serum/plasma Na	300–350 µg L^{-1}
	Urinary Na	1950–3400 mg/d
K	Serum/plasma K	130–200 µg L^{-1}
	Urinary K	90–450 mg/d
Cl	Serum/plasma Cl	350–400 µg L^{-1}
	Urinary Cl	>40 mg L
Fe	Serum Fe	Women: 500–1700 µg L^{-1}
	Serum total Fe-binding capacity	Men: 650–1650 µg L^{-1} 2500–4250 µg L^{-1}
	Serum ferritin	Women: 100–1200 µg L^{-1} Men: 200–2500 µg L^{-1}
Zn	NA[b]	plasma Zn normal range: 700–1500 µg L^{-1}
Cu	Serum/plasma Cu	Women: 800–1900 µg L^{-1} Pregnant women: 1180–3020 µg L^{-1} Men: 700–1400 µg L^{-1} Infants: 200–700 µg L^{-1} Children (6–12 y) 80–190 µg L^{-1}
Se	Plasma/serum Se	800–2000 µg L^{-1}
I	Urinary I	>1000 µg L^{-1}.
	Serum T$_4$	60–100 µg L^{-1}
	Serum TSH	1–50 µg L^{-1}
Mn	NA[b]	Serum/plasma Mn normal range: 4–11 µg L^{-1} Whole blood Mn normal range: 77–121 µg L^{-1}
Mo	NA[b]	Serum/plasma Mo normal range: 1–30 µg L^{-1} Whole blood Mo normal range: 8–33 µg L^{-1} Urine Mo normal range: 80–340 µg L^{-1}
Cr	Serum/plasma Cr	1–1.6 µg L^{-1}
F	NA[b]	NA[b]

[a]Normal range for healthy adults, unless otherwise indicated.
[b]Validated method not available.

upon which their foods were grown and through which their drinking and cooking waters drained (see Table III). Therefore, it is not surprising that mineral nutritional status can vary geographically, particularly in cases where the soil-water-plant-animal linkages are fairly direct as in the cases of grazing animals and people in highly localized food systems. Such cases have been described for iodine, copper, zinc, selenium, molybdenum, manganese, iron, boron, and cobalt (see Table X). Soil mineral deficiencies can involve intrinsically low mineral contents of soils (e.g., selenium), inefficient uptake by crops (e.g., zinc deficiency in calcareous soils), and excessive leaching (e.g., iodine, zinc). In at least two general cases, Keshan disease and the iodine deficiency diseases goiter and myxedematous cretinism, endemic distributions of a disease are directly related to the geographic patterns of soil deficiencies in selenium and iodine, respectively. Interregional and international transshipment of foods can be expected to mitigate against such local soil effects, particularly in industrialized countries (see also Chapter 15, this volume).

VIII. SUMMARY

Minerals play essential roles in the normal metabolism and physiological functions of animals and humans. Some (calcium, phosphorus, magnesium, fluoride) are required for structural functions in bone and membranes. Some (sodium, potassium, chloride) are required for the maintenance of water and electrolyte balance in cells. Some (zinc, copper, selenium, manganese, molybdenum) are essential constituents of enzymes or serve as carriers (iron) for ligands essential in metabolism. Some serve as essential components of a hormone (iodine) or hormone-like factor (chromium).

Unlike other essential nutrients, the mineral elements cannot be derived from the biosynthesis of food plants or animals—they must be obtained from soils and pass through food systems to humans in food forms. For this reason, local deficiencies of minerals in soils can produce deficiencies in local food systems which clinically impact the people dependent on those systems. The development of international trade and interregional transportation of foods has ameliorated the impact of such local mineral deficiencies. However, cases still occur in areas where the transshipment of food and, thus, the diversity of the diet are limited.

TABLE X. Examples of Geographic Patterns of Deficiencies of Nutritionally Important Mineral Elements

Element	Known deficient areas
I	A wide range of soils in areas remote from sea coasts; in the United States, the northwestern mountains and upper Midwest lake areas
Cu	Acid histosols in the eastern United States; acid sands in Florida; podzolic soils in Wisconsin; some sandy alkaline soils
Zn	An estimated half of the world's soils with small deficient spots in many areas; most likely in calcareous or leached, acid, sandy soils
Se	Mountainous belt of northeast China to the Tibetan plateau; parts of Africa; Pacific Northwest, northeast, and lower eastern seaboard of the United States
Mo	Acid soils; eastern seaboard, Great Lakes, and Pacific Coast areas of the United States
Mn	Humid, organic soils of the eastern United States
Fe	Seldom a problem for food plants except in arid regions
B	much of the United States, particularly in neutral-to-alkaline soils
Co	In the United States, lower Atlantic coastal plain and lower Maine coast; parts of Australia

SEE ALSO THE FOLLOWING CHAPTERS

Chapter 2 (Natural Distribution and Abundance of Elements) · Chapter 5 (Uptake of Elements from a Biological Point of View) · Chapter 8 (Biological Responses of Elements) · Chapter 15 (Selenium Deficiency and Toxicity in the Environment) · Chapter 16 (Soils and Iodine Deficiency) · Chapter 25 (Speciation of Trace Elements) · Chapter 28 (Mineralogy of Bones)

FURTHER READING

Beard, J. L., and Dawson, H. D. (1997). Phosphorus. In *Handbook of Nutritionally Essential Mineral Elements* (B. L. O'Dell and R. A. Sunde, Eds.), Marcel Dekker, New York, pp. 275–334, chap. 3.

Berner, Y. N. (1997). Phosphorus. In *Handbook of Nutritionally Essential Mineral Elements* (B. L. O'Dell and R. A. Sunde, Eds.), Marcel Dekker, New York, pp. 63–92, chap. 3.

Bogden, J. D., and Klevay, L. M. (Eds.) (2000). *Clinical Nutrition of the Essential Trace Elements and Minerals. The Guide for Health Professionals*, Humana Press, Totowa, NJ.

Bronner, F. (1997). Calcium. In *Handbook of Nutritionally Essential Mineral Elements* (B. L. O'Dell and R. A. Sunde, Eds.), Marcel Dekker, New York, pp. 13–61, chap. 2.

Cerklewski, F. L. (1997). Fluorine. In *Handbook of Nutritionally Essential Mineral Elements* (B. L. O'Dell and R. A. Sunde, Eds.), Marcel Dekker, New York, pp. 583–602, chap. 20.

Chesters, J. K. (1997). Zinc. In *Handbook of Nutritionally Essential Mineral Elements* (B. L. O'Dell and R. A. Sunde, Eds.), Marcel Dekker, New York, pp. 185–230, chap. 7.

Chow, L. C. (1990). Tooth-Bound Fluoride and Dental Caries. *J. Dent. Res.*, 69, 595–600.

Clark, L. C., Combs, Jr., G. F., Turnbull, B. W., Slate, E., Alberts, D., Abele, D., Allison, R., Bradshaw, R., Chalker, D., Chow, J., Curtis, D., Dalen, J., Davis, L., Deal, R., Dellasega, M., Glover, R., Graham, G., Gross, E., Hendrix, J., Herlong, J., Knight, F., Krongrad, A., Lesher, J., Moore, J., Park, K., Rice, J., Rogers, A., Sanders, B., Schurman, B., Smith, C., Smith, E., Taylor, J., and Woodward, J. (1996). The Nutritional Prevention of Cancer with Selenium 1983–1993: A Randomized Clinical Trial. *J. Am. Med. Assoc.*, 276, 1957–1963.

Combs, Jr., G. F., and Lü, J. (2001). Selenium as a Cancer Preventive Agent. In *Selenium: Molecular Biology and Role in Health* (D. Hatfield, Ed.), Kluwer Academic, New York, pp. 205–217.

FAO-WHO (2002). *Human Vitamin and Mineral Requirements: Report of a Joint FAO/WHO Expert Consultation*, Food and Agricultural Organization of the United Nations, World Health Organization, Rome.

Food and Nutrition Board (1989). *Recommended Dietary Allowances*, 10th edition. National Academy Press, Washington DC.

Food and Nutrition Board (1997). *Dietary Reference Intakes for Calcium, Phosphorus, Magnesium, Vitamin D and Fluoride*, National Academy Press, Washington DC.

Food and Nutrition Board (2000). *Dietary Reference Intakes for Vitamin C, Vitamin E, Selenium and Carotenoids*, National Academy Press, Washington DC.

Food and Nutrition Board (2001). *Dietary Reference Intakes for Vitamin A, Vitamin K, Arsenic, Boron, Chromium, Copper, Iodine, Iron, Manganese, Molybdenum, Nickel, Silicon, Vanadium and Zinc*, National Academy Press, Washington DC.

Harper, M. E., Willis, J. S., and Patrick, J. (1997). Sodium and Chloride in Nutrition. In *Handbook of Nutritionally Essential Mineral Elements* (B. L. O'Dell and R. A. Sunde, Eds.), Marcel Dekker, New York, pp. 93–116, chap. 4.

Harris, E. D. (1997). Copper. In *Handbook of Nutritionally Essential Mineral Elements* (B. L. O'Dell and R. A. Sunde, Eds.), Marcel Dekker, New York, pp. 231–273, chap. 3.

Hetzel, B. S., and Wellby, M. L. (1997). Iodine In *Handbook of Nutritionally Essential Mineral Elements* (B. L. O'Dell and R. A. Sunde, Eds.), Marcel Dekker, New York, pp. 557–581, chap. 19.

Hollick, M. F. (1994). Vitamin D—New Horizons for the 21st Century, *Am. J. Clin. Nutr.*, 60, 619–630.

Johnson, J. L. (1997). Molybdenum. In *Handbook of Nutritionally Essential Mineral Elements* (B. L. O'Dell and R. A. Sunde, Eds.), Marcel Dekker, New York, pp. 413–438, chap. 13.

Jones, G., Strugnell, R. A., and DeLuca, H. F. (1998). Current Understanding of the Molecular Actions of Vitamin D, *Physiol. Rev.*, 78, 1193–1231.

Kubota, J., and Allaway, W. H. (1972). Geographic Distribution of Trace Element Problems. In *Micronutrients in Agriculture* (J. J. Mortvedt, P. M. Giodano, and W. L. Lindsay, Eds.), Soil Science Society of America, Madison, WI, pp. 525–554.

Leach, Jr., R. M., and Harris, E. D. (1997). Manganese. In *Handbook of Nutritionally Essential Mineral Elements* (B. L. O'Dell and R. A. Sunde, Eds.), Marcel Dekker, New York, pp. 335–355, chap. 10.

Milne, D. B. (2000). Laboratory Assessment of Trace Element and Mineral Status. In *Clinical Nutrition of the Essential Trace Elements and Minerals: The Guide for Health Professionals*, Humana Press, Totowa, NJ, pp. 69–90.

Offenbacher, E. G., Pi-Sunyer, F. X., and Stoeker, B. J. (1997). Chromium. In *Handbook of Nutritionally Essential Mineral Elements* (B. L. O'Dell and R. A. Sunde, Eds.), Marcel Dekker, New York, pp. 389–411, chap. 12.

Peterson, L. N. (1997). Potassium in Nutrition. In *Handbook of Nutritionally Essential Mineral Elements* (B. L. O'Dell and R. A. Sunde, Eds.), Marcel Dekker, New York, pp. 153–183, chap. 6.

Salonen, J. T., Nyyssonen, K., Korpela, H., Tuomilehto, J., Seppanen, R., and Salonen, R. (1992). High Stored Iron Levels are Associated with Excess Risk of Myocardial Infarction in Eastern Finnish Men, *Circulation*, 86, 803–811.

Sauberlich, H. E. (1999). *Laboratory Tests for the Assessment of Nutritional Status*, 2nd edition, CRC Press, Boca Raton, FL.

Shils, M. E. (1997). Magnesium. In *Handbook of Nutritionally Essential Mineral Elements* (B. L. O'Dell and R. A. Sunde, Eds.), Marcel Dekker, New York, pp. 117–152, chap. 5.

Sunde, R. A. (1997). Selenium. In *Handbook of Nutritionally Essential Mineral Elements* (B. L. O'Dell and R. A. Sunde, Eds.), Marcel Dekker, New York, pp. 493–556, chap. 18.

WHO (1996). *Trace Elements in Human Nutrition and Health*, World Health Organization, Geneva.

CHAPTER 8

BIOLOGICAL RESPONSES OF ELEMENTS

MONICA NORDBERG
Karolinska Institute

M. GEORGE CHERIAN
University of Western Ontario

CONTENTS

I. AN INTRODUCTION TO BIOLOGICAL RESPONSES OF CERTAIN ELEMENTS

Medical geology is defined as the science dealing with the relationship between natural geological factors and health problems in man and animals. The geographical distribution of trace elements and metals in nature can explain "natural deficiency or toxicity," which includes the occurrence of health problems, diseases, and adverse health effects endemically seen. It is a broad subject and it attracts interdisciplinary collaboration and contributions. A relationship between humans and the ecosystem, environment and life, and environment and illness exists, and geological factors are of major interest in the environment in this respect (see also Chapters 7 and 22, this volume).

Biological effects in response to environmental exposure of elements constitute two parts, deficiency and toxicity, which are the themes of this chapter. Under normal conditions human nutritional deficiencies may occur due to environmental, sociological, and genetic influences. Thus deficiency syndromes are an expression of multiple simultaneous effects. Functional abnormalities related to specific nutrients may be identified by the biological role of the nutrient, in controlled physiological experiments and by multivariate analysis of outcomes. Model systems and dose response experiments allow measurements of deficiency effects at different levels of physiological states and at different stages in the life cycle. Such experiments provide a basis for understanding the roles of nutrients in human physiology and the biological consequence of deficiencies. Nutritional deficiencies of certain minerals in soil have resulted in diseases in populations. For example, zinc deficiency in Iran and selenium deficiency (Keshan disease) in China were due to mineral deficiencies in soil. In most western countries, farmers are aware of this problem, and they test the soil every year and add the deficient nutrients before planting their crops. Thus they get better crops and also take care of the nutritional status of the population.

Primary nutritional deficiencies are caused by adverse economic conditions, customs, and food choices that

TABLE I.   Different Forms of Monitoring of Metals and Their Relationship to Exposure, Dose, and Effects

Emission	Exposure	Internal dose indicators	Biological effects
Sources/rates/patterns	Air	Absorbed dose	Bioindicators of effects
	Water	Body burden	Early health effects
	Soil	Target tissue concentration	Overt health impairment
	Food		
Source characterization and emissions monitoring	Environmental monitoring	Biological monitoring	Health monitoring

From Clarkson et al., 1988.

limit dietary variety and thus nutrient availability. Conditioned (secondary) deficiencies are caused by non-dietary factors, especially illnesses and iatrogenic conditions that interfere with absorption, utilization, or retention of nutrients. In many instances, especially among the poor, combined effects of low intakes and conditioning factors result in deficiency disease. The conditioning factors can also give rise to toxicity.

Toxicity can occur either due to high bioavailability of elements or due to interactions of trace elements in the environment. To follow changes in the eco- and biological systems and their relationship to exposure conditions, different forms of monitoring, e.g., environmental and biological, are performed as illustrated in Table I. These may involve mode of exposure, dose, and effects (see also Chapter 24, this volume).

Essentiality and toxicity of trace elements need to be in balance. Otherwise, adverse health effects to the organism will develop with cellular functional changes as the first sign of toxicity, followed by organ damage, and subsequent manifestation of illness as the end point. Most studies on toxicity of metals have been performed with the assumption that intake of essential trace elements are on a "normal basal" level. When the dietary level of essential trace elements is low, toxicity can occur at much lower concentration levels of the non-essential elements than previously regarded as toxic dose levels for these elements. The recommendation for supplementation of daily intake of essential metals in certain cases can exceed the dose that may cause toxicity. With that background a more intense collaboration between researchers in toxicology and nutrition was initiated on an international level (WHO, 2002).

Toxicity of metals depends on bioavailability, species, and exposure conditions. Metals that are essential for life can also cause adverse health effects if exposure is excessive or if the chemical species or exposure route is different from the physiological route. Low iron status, and other nutritional factors such as low intake of protein, calcium, and zinc, can increase the toxicity of toxic metals such as cadmium and lead. Although an element is classified as essential to the organism it can cause toxicity with excessive exposures and interactions. The environmental level of metals in the ecosystem is influenced by industrial emissions and other human activities. Recycling of batteries and modern electronic equipment materials increases concentrations of metals such as cadmium, mercury, and semi- and superconductor materials like gallium and indium in the ecosystem. The interactions between ecosystem and these human activities may result in excessive exposure to metals (see also Chapter 3, this volume).

The acidification of lakes and soils due to emissions of sulfur and nitrogen oxides results in changes in the mobility and promotes chemical conversion of metals and may result in chemical species of certain metals with greater toxicity than the parent compound. Aluminum, arsenic, cadmium, and mercury constitute examples of elements with increased mobility whereas selenium has decreased mobility in an acidified ecosystem. With increased acidification of lakes, increased mercury concentration in the fish has been reported due to increased bioavailability of methylmercury. It is reported that the concentration of metals in fish and crops can vary with change of pH in lakes, sediments, and soils.

Environmental biochemistry of metals provides special emphasis on chemical reactions that transform metals in the environment into species or compounds of the element that are either more toxic or less toxic. For example, chemical reactions like methylation and demethylation can alter the toxicity of mercury and arsenic. It is known that both bacteria and vitamin B_{12} can methylate mercury in sediments.

Medical geology is of global concern not only for humans but also for plants and animals. Tissue samples from penguins and seals from the Antarctic that were

The cycle of mercury

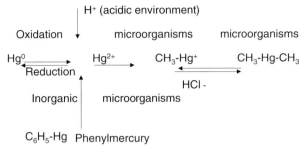

FIGURE 1 The cycle of mercury.

analyzed for cadmium and copper showed increased concentrations in samples from areas that were regarded as non-polluted. This might be explained by the geology of the Antarctic. Oysters from New Zealand can be high in cadmium but binding to various bioligands can vary their intestinal absorption pattern, and thus affect toxicity.

The adverse health effects of mercury depend on the species of mercury. Figure 1 shows the cycle of mercury in the ecosystem. Inorganic mercury, mercury vapor, methylmercury, and various other mercury compounds exist in the ecosystem due to bioconversion, and can be transformed to a stable form of methylmercury by bacteria and vitamin B_{12}. This is the most toxic form of mercury for humans because it is easily absorbed (about 99%) from the gastrointestinal tract and retained in the body with high distribution to the brain.

The aim of this chapter is to illustrate the relationship between deficiency and toxicity of trace elements and to give examples of diseases related to geology-medical geology or medical geology. Special attention is paid to iodine, iron, selenium, cadmium, and mercury and this chapter also includes a few other trace elements. These elements are selected for discussion because of their known role in certain diseases or their interactions that can lead to diseases (see also Chapters 4, 5, 21, and 23, this volume).

II. Metals and Geo-Environment

Human exposure to metals generally occurs through food, drinking water, and air. The drainage of contaminants from farmland is a major source of toxic metals in well water in rural areas. In several cities in develop-ing countries, air pollution with metals has become a major problem. Occupational exposure mostly takes place via inhalation of metal fumes and dust where the trace elements exist as oxides, sulfides, or in the elemental form. Exposure to metals can be monitored in various ways: the exposure sources, the type of indicators, and the biological effects in general can be all documented in such monitoring methods (Table I).

A. Toxicity

Paracelsus (1493–1541) stated that everything is toxic, it is just a matter of dose. Trace elements are important for life. Both deficiency and toxicity can cause adverse health effects. Many metals constitute part of the structure or cofactor of the enzymes and play important roles for the biological activity of enzymes and vitamins. Zinc plays an essential role in zinc-dependent enzymes such as alcoholdehydrogenase and cobalt in vitamin B_{12}.

Upon exposure to high concentrations of non-essential elements clinical disease or death can occur in exposed individuals, and in most acute toxic cases, the effect is directly related to the dose. At a lower dose range, less severe effects might occur after a long latent period. In most cases, the higher the dose, the shorter the latency period for appearance of toxic effects.

Risk assessment in environmental medicine serves as a basis of preventive action in order to avoid adverse health effects in general populations exposed to chemicals. To discuss toxicity it is important to define concepts such as factors influencing uptake (Table II), critical organ, critical dose, critical effects, and biological half-time. The organ or tissue where the exposed chemical can give rise to a critical effect is known as the critical organ. Normally, the critical organ is the organ where the earliest adverse effect occurs in an individual. Critical concentration is the concentration of the chemical, which can cause the earliest adverse effect in the critical organ in an individual or population (Nordberg & Nordberg, 2002) (see also Chapters 21 and 24, this volume).

In order to prevent toxicity and maintain quality of life, it is important to find various sensitive chemical or biological indicators for early detection of effects of metals in the critical organ. Each metal and sometimes each species of metal is unique with regard to its critical organ and toxicity. A typical example for this is the different forms of mercury and their different target organs for toxicity. Mercury vapor and methylmercury exposures can affect the brain while inorganic forms of mercury can affect the kidney. The cornerstone to

TABLE II. Factors Influencing Uptake of Trace Elements and Metals

Concentration of agent
Chemical and physiological properties of the agent and the soil
Particle size
Physiological factors: age and nutritional status
Any existing disease
Genetic makeup

TABLE III. Health Risk Evaluation of Agents in Soil

Agents
Concentration and quantities
Exposure situation
Type of soil/rock
Mining and volcanic activity
Bioavailability
Health effects

TABLE IV. Examples of the Relationship Between Elements, Geology, and Health Effects as Expression of Medical Geology

Metal	Activity	Disease
Arsenic and drinking water	Well-drilling in Bangladesh and feeding cattle in United States	Cancer and keracytoses
Cadmium	Japan	Itai itai
Cesium as 137	Radioactive release, Chernobyl	Cancer
Copper	Genetic defects Copper in kitchen utensils	Menkes' disease, placenta; Wilson's disease, liver; Indian liver cirrhosis, liver
Selenium	Farming in China Farming in Finland	Kashin-Beck; Keshan Cancer and heart disease

develop a metabolic model constitutes knowledge of absorption, distribution, excretion, biotransformation, and biological half-time for specific species of the metal.

In order to understand the toxicity of metals and the role of geology (Tables III and IV), it is important to have information on concentration levels of metals in water, air, and soils in a particular geographical area (see also Part II, this volume).

Geological factors such as availability of trace elements and metals can be reflected in the occurrence of diseases in man and animals. If plants actively take up a metal, it can result in high exposures to people living in that area. The increased amount of cadmium in rice was the cause of "itai itai" disease in Japan, and also in Shipham in the UK. Nutritional deficiencies (iron, calcium, and vitamin D) also might have played a role in itai itai disease.

B. Neurotoxicity

Certain forms of metals such as methylmercury can cause neurotoxicity. Certain forms of mercury such as mercury vapor and methylmercury can easily cross the blood-brain barrier. Mercury can bind with critical ligands and cause a direct toxic effect in the cell. However, for cadmium neurotoxicity, the uptake in brain and the mechanism behind neurotoxicity are more obscure. Lactating pups of dams exposed to cadmium showed changes in the serotonin levels. Because cadmium does not pass the blood-brain barrier, this observation remains controversial. Other factors like interference of cadmium with zinc or calcium metabolism and the presence of a specific form of metallothionein (MT-3) in the brain should also be considered in the neurotoxicity of metals. Lead is known to cause central nervous system (CNS) toxic effects in children, but it may cause damage to peripheral nerves in adults because of the differences in permeability to the blood-brain barrier.

C. Carcinogenicity

A few metals or metalloids (arsenic) have been classified by the International Agency for Research on Cancer/World Health Organization, (IARC/WHO) (IARC 1980, 1990, 1994) to be carcinogenic in humans, i.e., group 1 as presented in Table V. These classifications are based on epidemiological studies and detection of tumors in certain organs depending on the mode of exposures. A number of metals have been shown to be carcinogenic in animals, especially after injection of metal salts, but only a few have been classified as human carcinogens, i.e., group 1.

TABLE V. Human Metal/Metalloid Carcinogens

As and inorganic compounds	1980
Chromium (VI)	1990
Nickel and compounds	1990
Cadmium and compounds	1993
Beryllium and compounds	1993

TABLE VI. Examples of Protective Mechanisms in Metal Toxicology

Metal binding proteins
Localization in the cell
Metallothionein in cytoplasm/nucleus
Lead-binding inclusion bodies in kidney
Detoxification by binding to glutathione or amino acids
Change of pH in lysosomes
Interaction between metals, e.g., selenium-mercury, cadmium-zinc
Protein intake
Iron status
Genetic polymorphism
Methylation
Demethylation

III. PROTECTIVE MECHANISMS

There are several binding sites for metals in the cells, and therefore, a number of mechanisms may be involved in the cell to protect it from developing toxic effects, as shown in Table VI.

A. Importance of Various Proteins

New families of proteins, important for protecting the cell from toxic insults by reactive oxygen species (ROS) and also heat shock proteins, have been identified after exposure to metals. Although these proteins are found in low basal levels in cells, they are inducible at a transcriptional level when exposed to metals (Piscator, 1964). The genes coding for metallothioneins (MTs) and heat shock proteins are present in most of the organisms, and their induction after exposure to metals plays an important role in protection against metal toxicity. There are several review articles published on metals and stress proteins (Goyer & Cherian, 1978; Goering & Fisher, 1995; DelRazo et al., 2001), and they describe the specificity of certain metals to induce various heat shock proteins, and some of the similarities on gene expression for these proteins and MTs. The metal regulatory elements (MRE) present in the metallothionein gene and metal transcription factor (MTF-1) are important factors that may control the mechanisms of induction of these proteins after exposure to certain metals (Seguin, 1991; Radtke, 1993). Both essential (zinc and copper) and non-essential (cadmium and mercury) metals can induce the synthesis of MTs and also bind to them. Thus, these proteins may have a role in the metabolism of essential metals and protection against the toxicity of metals (Cherian & Nordberg, 1983; Cherian, 1995, 1997).

B. Metallothionein

The structure, chemicophysical properties, and the physiological/biological functions of MTs in several organisms have been investigated for the last 55 years. Metallothionein is a family of proteins with low molecular weight (6500 Da) with a unique capacity to bind 7 metal ions through the 20 cysteinyl groups. Of the 61 amino acids of this protein, 20 of them are cysteines. In MT, divalent zinc and cadmium are bound tetrahedrally, with both primary and bridging sulfur bonds. It may bind with 12 monovalent copper trigonally. Two distinct metal binding domains, the α- and β-clusters have been characterized in MT by both NMR solution structure and crystallization. The purified MTs generally contain 5–10% metals as zinc, cadmium, mercury, or copper. The characteristic properties of MT (Kägi & Nordberg, 1979) are listed in Table VII.

The definition of the MT super family follows the criteria set for polypeptides that have common features with equine renal MT. The four major forms of groups of MT consist of MT-1, -2, -3, and -4 isoforms. Mammalian MT-1 and -2 forms are present and expressed in almost all tissues. However, the number of isoforms differs, as MT-1 exists in many subisoforms, whereas only one isoform for MT-2 has been identified. MT-3 has 7 additional amino acids and contains a total of 68 amino acids with differences in charge characteristics when compared to isoforms MT-1 and -2. The MT-3 isoform was identified years after characterization of

TABLE VII.   Characteristics of Metallothionein

1. Molecular weight 6000–7000, 61 amino acids (aa)
2. 20 cysteine (30%), N-acetylmethionine, C-alanine, no aromatics, no histidine
3. Metal content (cadmium, zinc, copper, mercury) 5–10% wet weight
4. Absorption 250 nm (cadmium), 225 nm (zinc), 275 nm (copper), 300 nm (mercury)
5. Induced synthesis by cadmium, zinc, copper, and mercury
6. No disulfide bonds
7. Heat stability
8. Cytoplasmic and nuclear localization
9. Unique amino acid sequence

TABLE VIII.   Functions of Metallothionein

1. Metabolism of essential metals
2. Detoxification of metals
3. Protection from metal toxicity
4. Storage of metals
5. Protection from oxidative stress
6. Cellular proliferation and differentiation

MT-1 and -2 as a growth inhibitory factor (GIF) in the brain. At the N terminal region of MT-3 an additional threonine is inserted and 6 additional amino acids consisting of glutamic acid and alanine are present as a loop at the C terminal region. Thus MT-3 differs from MT-1 and -2 in its amino acid sequence. Another difference in the sequences of the MT-3 form is the presence of a proline close to the N-terminal region, which contributes to the growth inhibitory effects of MT-3. The threonine in MT-3 increases the acidity and the charge surface facilitates the interaction of MT-3 with other biological constituents. MT-4 consists of 62 amino acids with one glutamate inserted and is specific for squamous epithelium and expressed in keratinocytes. There are 14 human MT genes that are localized on chromosome 16q13–22. Of these six are functional, two are not, and six have not been characterized.

MT is often related to toxicokinetics and biochemistry of essential and non-essential metals such as zinc, cadmium, mercury, and copper. *In vivo* binding to other metals/metalloids such as selenium and bismuth is not yet understood. Although it is mainly an intracellular protein, MT has been detected in small amounts in blood and urine.

Several functions have been proposed for MT, and they are listed in Table VIII. They may play an important role in the homeostasis of essential metals such as zinc and copper. It has been demonstrated that mammalian fetal liver contains very high levels of MT bound to these metals, and it may act as a metal storage protein during gestation similarly to ferritin as an iron storage protein. In fetal and newborn liver, MT is localized in the cell nucleus and cytoplasm, but during growth, MT is degraded and the levels of zinc and copper are maintained at a low basal level. In adults, the low levels of

MT in the hepatocytes are detected in areas around the wound after partial hepatectomy and surgery. Thus, MT may serve as a storage protein for zinc during development and other conditions when the requirement for zinc is high. In addition, because of its high affinity for metals, MT can detoxify toxic effects of certain non-essential metals such as cadmium and mercury.

Cellular membranes are targets for metals, and MT may play a role in protection of the cell from metal toxicity. However, the release of Cd-MT from liver and other tissues can cause toxic effects. Membrane damage caused by Cd-MT in the renal tubule is most likely explained by a direct interference of cadmium on calcium transport in renal membranes and might explain chronic toxicity of cadmium on the kidney. Metals like mercury, copper, and cadmium are continuously accumulated in liver and kidneys with a major part bound to MT. Cd-MT is efficiently transported through the glomerular membrane and actively taken up by the renal tubular cells, which causes damage to the cell.

Expression of MT in different tissues is influenced by the exposure to metals and their accumulation. Therefore these proteins could be used as a biomarker of toxic metal contamination at an environmental level and as a biomarker of toxic effects in an individual.

A number of different techniques and methodologies have been used for the quantification of MTs. The detection is based on the redox properties of the thiol groups in the molecule, for electrochemical methods, on saturation with metal ions, for indirect quantification methods, and the immunoreactivity with antibodies for immunochemical methods. These methods are widely used to determine the levels of MT in tissues and biological fluids, depending on their detection limit. Still, certain technical problems exist with the determination of MT in certain biological samples.

When MT is isolated from various tissue samples, the content of metal ions may differ depending on tissue factors and the exposure levels of metals. In most cases, isolated MT is completely saturated with metals. It is well known that atmospheric oxygen can easily oxidize MTs, which results in the formation of disulfide bonds with "free" thiol groups. In isolation methods, a reduction agent, such as 2-mercaptoethanol or dithiothreitol, is used to prevent the disulfide bond (S—S) formation. The immunoassays of MT in biological samples may provide limited values because of the lack of antibodies, which can cross-react with all the forms.

Increased tissue concentration of MT-1 and -2 may indicate increased exposure to certain elements. MT-3 has not been shown to be inducible with metal exposure, but it may play a role in zinc metabolism and other elements in the brain. A mechanistic model for cadmium and MT illustrates the chronic toxic action of cadmium in renal tubular cells and the development of adverse health effects after low-dose chronic exposure.

Because of its low molecular weight and related efficient glomerular filtration in the kidney, the binding of Cd to MT in blood plasma serves as one of the modes (Figure 2) of transport of cadmium from liver and other tissues to the kidney. The binding of cadmium to MT and other low molecular ligands in plasma and tissues may play a role in the tissue distribution of cadmium after uptake from the intestine. The absorption of cadmium in the gastrointestinal tract may involve divalent metal transporters (DMT-1), and it could also be influenced by MT synthesis in the intestine. Because both cadmium and iron are taken up by DMT-1, a direct competition by these metals in the intestinal level cannot be ruled out. Previous studies have shown that the gastrointestinal (GI) absorption of cadmium in humans may depend on the iron status (Flanagan et al.,

1978). Uptake and distribution of cadmium occurs mainly in the initial phase in a form where cadmium is bound to albumin in plasma. A report suggests that iron deficiency can increase levels of MT-1 in bone marrow of rats with hemolytic anemia with unchanged hepatic and reduced renal MT, and it may indicate that MT-1 levels in blood reflect erythropoietic activity (Robertson et al., 1989).

MT levels in the liver and kidneys may serve as potential indicators of environmental exposure to cadmium, zinc, and copper. The extremely long biological half-time of cadmium in the kidney (15–20 years in humans) may be due to its binding to intracellular MT. Thus, MT can immobilize cadmium intracellularly, and can play a role in the kinetics of cadmium and protection from metal toxicity (Nordberg, 1998; Nordberg & Nordberg, 2000).

C. Metabolism and Kinetics of Trace Elements and Toxic Metals

The metabolism and kinetics of trace elements involve binding to various proteins. The specific role of MT in the binding of cadmium and zinc has been described in the previous sections.

A similar role for MT in the kinetics for copper also has been shown (Bremner, 1987). In certain diseases like Wilson's disease, the excretion of copper is impaired and results in accumulation of copper in the liver. Initially, the toxicity of copper may be prevented by induced synthesis of MT, and its binding to copper. At a later stage, the cells are unable to synthesize MT, and binding capacity of MT may get saturated. The excess copper ions or copper-saturated MT may cause toxicity to liver cells.

IV. BIOLOGICAL MONITORING OF METALS AND TRACE ELEMENTS

To study the toxicity of metals in a population and protect public health, biological monitoring of metals has been performed in various populations such as lead in children and pregnant women and mercury in people with dental amalgam.

The use of MT as a potential biochemical indicator for environmental cadmium exposure has been tried by

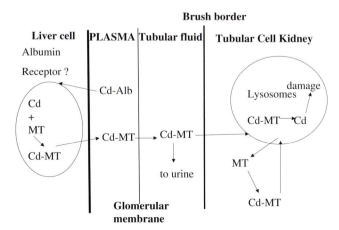

FIGURE 2 Model of transport of cadmium in blood.

analyses of tissue samples from moose and reindeer from the northern part of Sweden and penguins and seals from the Antarctic. Cadmium was found to be a major metal component in MT even from this area that is regarded as non-polluted. Human tissue samples from autopsy have been used as potential indicators of environmental exposure to cadmium in Canadians. For bioindicators for metals and their potential use in risk assessment see Clarkson et al. (1988).

The quality of the analyses in biological monitoring depends on a number of factors. The time of collection and storage of specimens for analysis of metals or species of metal (inorganic or organic) should be controlled. The difficulty in interpretation of such results and accuracy of the method also should be considered in such studies. Among the sources of error in biological monitoring of environmental exposure are physiological factors such as variations in the biological material, age, diet, smoking, alcohol intake, and concurrent exposure to other compounds and drugs that can affect metabolic changes. Other factors may involve species-dependent kinetics of elements and simultaneous exposure to mixed species of an element.

The external contamination is a major problem in trace element analysis in biological samples during collection, and some of the special techniques can be successfully performed only by trained personnel. The contamination might arise from both the collection procedure and the storage container itself. Good hygiene should be practiced when collecting specimens, i.e., smokers should not collect samples collected for cadmium analyses. Contamination might also occur in sampling for zinc and aluminum where powder used as lubricant in disposable gloves can contribute to contamination. The material of the containers also can contribute to errors. Specific care has to be taken upon monitoring of aluminum. Quartz knifes are, for example, recommended for the handling of tissue samples. Metal analysis is usually performed by atomic absorption spectrometry (flame, graphite furnace, and electrothermal), and also by X-ray fluorescence spectrometry.

V. SPECIFIC METALS

Several trace elements are found to be essential to human health, and they are iron, zinc, copper, chromium, iodine, cobalt, molybdenum, and selenium (WHO, 2002). Other elements that might have a beneficial effect and probably are essential for humans are silicon, manganese, nickel, boron, and vanadium, and they may constitute a second group of trace elements of human concern. If they are accepted as essential trace elements, a biological requirement should be demonstrated at appropriate dose levels (WHO, 2002). Some of these elements can be toxic, and those that can interact with others are discussed here.

A. Aluminum

1. Aluminum—Plasma

Aluminum is found in several human organs, including lungs. Pulmonary fibrosis has been reported after inhalation of aluminum in certain industrial operations and in mining. Aluminum is mainly excreted through urine, and kidney damage can increase its retention.

Aluminum administration to experimental animals can produce osteomalacia and cause neurotoxicity in certain animals with neurofibrillary tangle formation. The role of aluminum in Alzheimer's disease is still controversial. Reference values (Commission of the European Community, CEC, recommendations for patients receiving dialysis):

- Normal = $<10\,\mu g\,L^{-1}$ (no history of chronic renal failure; CRF)
- $<60\,\mu g\,L^{-1}$ = desirable in CRF patients
- $>60\,\mu g\,L^{-1}$ = excessive accumulation
- $>100\,\mu g\,L^{-1}$ = cause for concern; high risk of toxicity in children
- $>200\,\mu g\,L^{-1}$ = urgent action required; high risk of toxicity in all

2. Aluminum—Water, Dialysis Fluid

Treatment of CRF by dialysis can give rise to increased aluminum concentration in the patient. Reference values (CEC Recommendations):

- Maximum allowable concentration (MAC) for potable water: $200\,\mu g\,L^{-1}$
- Guideline concentration (GLC) for potable water: $50\,\mu g\,L^{-1}$
- MAC for water for preparation of dialysis fluid: $30\,\mu g\,L^{-1}$

3. Aluminum—Urine

CEC reference range shows that excretion is usually <25 µg/24 hours.

B. Antimony

Antimony exists as V and III valence status with similar chemical characteristics as arsenic. In the geo-environment they occur simultaneously. The daily intake via food is estimated to be around 10 µg/day. Exposure to antimony is rare. For biological monitoring concentration in blood and urine is an indicator of exposure and internal dose.

C. Arsenic

Arsenic is distributed in the Earth's crust with an average concentration of 2 mg kg^{-1}. Rock, soil, water, and air contain trace amounts of arsenic. Arsenic compounds are found in close association with gold as arsenopyrites in mining areas. Wines made from grapes sprayed with arsenic-containing insecticides may also contain high levels of arsenic. The chemistry of arsenic focuses on inorganic forms (−3, 0, +3, and +5) and on organic arsenic compounds of which arsenobetaine occurring in fish is the most prominent one. Inorganic arsenic of geological origin is found in groundwater and used as drinking water it gives rise to adverse health effects in several parts of the world, e.g., in Bangladesh, Hungary, and China.

The development of a variety of instrumental techniques, i.e., atomic absorption spectroscopy (AAS) and inductively coupled plasma mass spectrometry (ICP-MS) with hyphenated methods and element-specific detectors coupled to chromatographic separation techniques makes it possible to study chemical species of arsenic. A test kit that is based on the color reaction of arsine with mercuric bromide allows arsenic to be determined under field conditions, e.g., in the groundwater in Bangladesh with a detection limit of 50–100 µg L^{-1}.

Biochemical condition in the geo-environment enhances biotransformation between arsenite and arsenate reduction and methylation of arsenic including organoarsenic compounds. Similar to other cases, biotransformation follows pH. This is an important matter because toxicity of arsenic depends on the chemical species. Acute poisoning occurs when the arsenic blood level is above the normal range of <10 µg L^{-1} as inorganic arsenic and metabolites. Acceptable occupational exposure (ACGIH value) is <50 µg g^{-1} creatinine as inorganic arsenic and metabolites.

D. Cadmium

Cadmium, discovered in 1817, is a soft, silver-white metal and is similar in appearance to zinc. Cadmium does not have a defined taste or odor. Many radioactive isotopes of cadmium, e.g., 109 and 115 m, are well recognized in experimental toxicology. Cadmium is an element with an average distribution of 0.1 mg kg^{-1} in the Earth's crust. Cadmium is usually found associated with zinc. Particularly high concentrations of cadmium occur in some sulfide ores, but many soils, rocks, coal, and mineral fertilizers contain some cadmium. Cadmium is widely dispersed in the environment. Human exposure to low levels occurs as a result of natural processes and of human activities, e.g., mining, smelting, fossil fuel combustion, and industrial use. Due to the natural occurrence in the geo-environment and its active uptake by plants, some farming products such as tobacco could be high in cadmium content. The metal also remains strongly bound to other compounds in the soil and water (WHO, 1992).

Cadmium causes kidney damage with proteinuria and calciuria. Bone effects have been reported as itai itai disease in Japan and also as low mineral density of the skeleton in studies from Belgium and China. Cadmium is classified as a human carcinogen (IARC, 1994).

A specific environmental exposure to cadmium occurred in Japan. Widespread exposure with both subclinical and clinical effects was found in itai itai patients in areas where water from the Jinzu river was used for irrigation. The cause of this disease was confirmed to be due to excessive exposure to cadmium through rice. A large number of people living in the cadmium-polluted area have renal tubular dysfunction, and people in other polluted areas have the same health effects. Itai itai disease is characterized by osteomalacia and renal tubular dysfunction and is an unusual disease. The opinion is that cadmium can give rise to both osteomalacia and osteoporosis. The age, sex, nutrition, exercise, and number of pregnancies influences the type of bone effect. The kidney damage can also lead to bone damage. The subclinical bone and renal effects in cadmium-exposed populations may be common and undetected, and could contribute to more severe effects such as fractures and kidney stones.

For biological monitoring of environmental exposure, cadmium is measured in blood and urine. It should be noted that smokers have increased cadmium concentrations by a factor of two due to the high content of cadmium in tobacco. The reference range for blood is $<0.2\,\mu g\,L^{-1}$ for nonsmokers and $<1.4\,\mu g\,L^{-1}$ for smokers. Cadmium concentration in urine for nonsmokers is $<1.0\,\mu g\,g^{-1}$ creatinine, and for smokers it is $<3.0\,\mu g\,g^{-1}$ creatinine. The high urinary cadmium may be detected only after renal damage, and by then it may be too late to treat these patients.

E. Chromium

Trivalent chromium is essential for man and animals, and it plays a role in carbohydrate metabolism as a glucose tolerance factor. The daily chromium requirement for adults is estimated to be $0.5–2\,\mu g$ of absorbable chromium(III). However, this is based on the calculation that if a fractional absorption value of 25% for "biologically incorporated" chromium(III) in food is assumed, then this is provided by a daily dietary intake of $2–8\,\mu g$ of chromium(III). Both acute and chronic toxic effects of chromium are caused by hexavalent compounds that are very toxic.

Ingestion of $1–5\,g$ of "chromate" results in severe acute toxic effects such as gastrointestinal disorders, hemorrhagic diathesis, and convulsions. Chromium concentration in serum/plasma and urine has a toxic reference range value in serum/plasma of $<0.5\,\mu g\,L^{-1}$ and for urine $<1.0\,\mu g\,g^{-1}$ creatinine.

F. Cobalt

Cobalt is an essential metal for vitamin B_{12}, which is involved in various methyl transfer reactions. About 25% of cobalt is absorbed from the GI tract, and it is mainly excreted in urine. Addition of cobalt to beer has caused endemic outbreaks of cardiomyopathy among beer drinkers resulting in fatalities.

G. Copper

In the general environment humans are exposed to copper via food and drinking water. Copper in drinking water has for some time been regarded as the cause of diarrhea and stomach problems in certain countries. Copper is found naturally in a wide variety of mineral salts and organic compounds and in the metallic form. It is essential for all biota. It is widely used in cooking utensils, water distribution systems, and in fertilizers.

Bactericides, fungicides, algicides, antifouling paints, and animal feed additives and growth promoters also contain copper. It is also used in a number of industrial applications. Major sources for copper distribution in the environment are mining operations, agriculture, solid waste, and sludge from treatment factories. Copper that is biologically available can accumulate in tissues and give rise to high body burdens in certain animals and terrestrial plants (see also Chapters 14 and 20, this volume).

The highest concentration of copper for the human diet is found in animal flesh, liver, oysters, fish, whole grains, nuts, seeds, and chocolate. Concentrations of approximately $10\,mg\,kg^{-1}$ have been reported. The adult daily intake of copper via food is estimated to be between $1–2\,mg$ and for children 2 years of age it is $0.6–0.8\,mg$. The contribution from drinking water is usually not included in these estimates. Thus the contribution of copper from drinking water can in some cases be high and populations should be alerted.

Also breastfeeding can contribute to an infant's exposure to copper. Concentrations of copper in drinking water exceeding $1–2\,mg\,L^{-1}$ give rise to staining of sanitation porcelain. Also hair turns blue-green as described for people exposed to water in, for example, swimming pools. The taste from copper influences the quality of water as reported for water of $0.3–12.7\,mg\,L^{-1}$. The taste from copper might be hidden due to the content of other additives of the drink and thus a misleading impression of copper content in water can occur. WHO recommends that the daily essential need for copper is $0.6\,mg$ for children from six months to six years.

The risks of copper exposure to human health for the general population are through food and drinking water when contaminated with copper. Thus, the major route of excessive copper exposure is oral. In Sweden, most of the water pipeline system consists of copper. Copper pipes were also used in several other countries. Copper is generally thought to be a good material to use for pipeline systems but under certain circumstances copper is released from the pipes, for example, acid rain can increase the bioavailability of copper.

About 30–40% of intake of copper is absorbed in the intestine and transported by albumin to the liver. After hepatic uptake, copper can be incorporated into copper-containing enzymes or into ceruloplasmin, which is then exported into the blood. Copper in the cytoplasm is predominantly bound to MT, and any excess of copper is excreted into the bile mainly through a lysosome-to-bile pathway, which results in fecal excretion. Normally copper concentration in tissues is regulated by homeostatic control. Under abnormal conditions such as genetic diseases (e.g., Wilson's disease), biliary excretion of copper is impaired and it is accumulated in the liver, which is the critical organ of copper toxicity.

Copper is an essential metal that can be toxic when homeostatic control fails. Adverse health effects of copper can develop both from deficient and excessive intake. Copper deficiency can cause heart diseases. In the general population, copper toxicity occurs due to consumption of contaminated beverages, including drinking water. Copper can catalyze the production of hydroxyradicals and oxyradicals when it is available in Cu+, a redox-active form. Both these radicals are extremely active and can attack many cell constituents, including lipids, nucleic acids, and proteins. Thus the occurrence of apoptotic bodies in the livers of copper-loaded animals is indicative of copper-induced DNA damage.

Special attention should be paid to a sensitive population with genetic disorders such as Menkes' or Wilson's disease. These are two specific inherent genetic disorders that give rise to disturbed copper metabolism. The copper transport protein, P-type ATPase is mutated in Wilson's disease, and copper accumulates in the liver, causing hepatic damage. Menkes' syndrome is characterized by disruption of copper transport from the intestine to the blood, which gives rise to copper deficiency and low activity of copper-dependent enzymes. Increased copper concentration in livers is seen in subjects suffering from Indian liver cirrhosis.

A relationship between MT and copper has been detected in patients with Wilson's disease. Induction of MT in the GI tract by oral zinc administration has been used to treat Wilson's disease by blocking the intestinal uptake of copper and decreasing the toxic tissue accumulation of copper.

MT induction in the intestine by feeding high levels of zinc can decrease copper uptake and tissue concentration. In Wilson's disease, the high hepatic copper is mainly bound to MT. This disease is due to a genetic defect in the transport of Cu into bile. Metallothionein has also been shown to be present in the placenta in patients suffering from Menkes; disease. The cause of Indian liver cirrhosis is still unknown, but high levels of copper have been detected in the liver of these patients.

Another disease is Indian childhood cirrhosis (ICC) reported mainly from India due to cooking utensils that are rich in copper. Nutritional health effects due to low and insufficient copper intake also exist and can affect the heart.

Limit values for health effects are weakly estimated. WHO has a recommended value of copper in drinking water of $2 \, mg \, L^{-1}$ based on an assumption that 10% of copper intake originates from drinking water. A provisional tolerable daily intake of copper exists since 1982; however, it is very close to the dose that causes vomiting. Tap water of $0.05 \, mg \, L^{-1}$ may be caused by corrosion of pipelines. Concentrations execeeding $0.2 \, mg \, L^{-1}$ can cause staining of sanitation porcelain (toilets and bath tubs) and hair. At $10 \, mg$ copper L^{-1} there is an increased risk for taste and odor. Health effects start to occur at $2 \, mg \, L^{-1}$. This concentration increases the risk for diarrhea in children. The practical general recommendation is to allow water to run free from the tap for one minute before using the water for food preparation.

H. Iodine

In 1990, the United Nations and WHO estimated that about one billion people are at risk for iodine deficiency disorders (IDD), 211 million with goiter (enlargement of the thyroid gland), 5.1 million with severe cognitive and neuromotor deficiencies (cretinism), and many more with less severe neuropsychological defects. The loss of human capital contributes to the perpetuation of poverty and its associated social ills. Therefore, the elimination of IDD is a priority for the WHO (see also Chapter 16, this volume).

Goiter is a recognized disease of great antiquity. Some more recent advances in the understanding of iodine deficiency are listed in Table IX. Iodine deficiency may be understood through the context of thyroid function. The normal adult thyroid weighs 20–25 g and contains 8–10 mg of iodine. Goiter is the earliest and most common manifestation of IDD (Table X).

Features of neurological cretinism include a wide range of mental retardation (nearly normal to severe) with associated hearing and speech disorders—the most severe is deaf-mutism—and abnormal neuromotor functions, e.g., proximal spastic rigidity of muscles of the leg with shuffling gait. Disposition is usually equable—most are able to carry out simple tasks of daily living and primitive farming and may have families and children. Growth is similar to that of the indigenous population, plasma T_4 is normal or low-normal, hypothyroidism is infrequent, and goiter may be present. Myxedematous cretinism is characterized by mental retardation, fetal hypothyroidism, persistent myxedema, growth stunting, and musculoskeletal disorders such as scoliosis, and atrophic thyroid gland. Mixed

TABLE IX. Some Historical Advances in Knowledge of Iodine Deficiency

Date	Name	Observation
	Chinese	The ancient Chinese were treating goiter with powdered seaweed and sea urchins several thousand years ago
BCE	Greeks	Burnt sponge used to treat goiter
1811		The discovery of iodine by adding concentrated sulfuric acid to a seaweed of the type that was used to treat goiter
1819	Fyfe	Iodine identified in sponge
1820	Coindet	Treated goiter with iodine
1854	Chatin	Suggested low iodine in soil, water, and food caused goiter
1896	Baumann	Thyroid rich in iodine
1896	Halsted	Maternal thyroid removal caused fetal thyroid hyperplasia (dog)
1908	McCarrison	Endemic cretinism characterized
1909	Marine	Maternal iodine deficiency caused goiter in fetus (dog)
1915	Kendall	Discovery of thyroxin
1917	Smith	Maternal iodine deficiency caused "cretinism" (swine)
1921	Marine	Prevention of goiter by iodide
1927	Harrington	Synthesis of thyroxin
1941	Mackenzie	Sulfanilguanidine inhibits iodide concentration by thyroid (rat)
1943	Mackenzie	Aminobenzene and thiourea inhibit iodine concentration by thyroid (rat)
1943	Mackenzie	Hyperplasia of pituitary gland in hypothyroid state (rat)
1947	Vanderlaan	Thiocyanate inhibits iodide concentration by thyroid (rat)

TABLE X. Iodine Deficiency Disorders

	Malformations	
Fetus	Abortion	
	Perinatal death	
	Infant death	
	Neurological cretinism	Severe mental deficiency
		Deaf-mutism
		Spastic diplegia
		Squint
	Myxedematous cretinism	Growth-stunting
		Severe mental deficiency
	Psychomotor deficiency	
Neonate	Goiter	
	Hypothyroidism	
Child and adolescent	Goiter	
	Hypothyroidism	
	Mental deficiency	
	Low physical development	
Adult	Goiter	Mechanical compression of adjacent organs in the neck
		Endocrine disorders: hyperthyroidism; hypothyroidism;
		Neoplasia: benign tumors; cancer
	Mental deficiency	

endemic cretinism is characterized by a combination of the above manifestations. The two types of cretinism may occur in the same endemia. The prevalence of cretinism in IDD endemias is 0–15% with 5–8% common. Factors that determine the prevalence and type of cretinism are incompletely understood. High-dietary goitrogens and selenium deficiency, noted below, have been postulated.

The physical finding of goiter suggests the presence of iodine deficiency. Proof is provided by the concentration of iodine in urine. Concentrations $<20 \mu g L^{-1}$ are indicative of severe IDD, $20–49 \mu g L^{-1}$ indicates moderate IDD, and $50–99 \mu g L^{-1}$ indicates mild IDD. Concentrations between $100–200 \mu g L^{-1}$ are satisfactory.

Thyroid-stimulating hormone (TSH) concentration in blood plasma reflects iodine nutriture. This test is used to screen newborns for hypothyroidism. In IDD endemic regions such as northern India and Zaire up to 10% of neonates were found to have increased plasma TSH, which implied severe iodine deficiency, hypothyroidism, and a high likelihood of brain damage.

Selenium is required for activity of iodothyronine deiodinase enzymes I–III that deiodinate T_4, T_3, and reverse T_3 and thus regulate the concentration of T_3. Additionally, the selenoproteins glutathione peroxidase and thioredoxin reductase, in concert with glutathione reductase, are believed to protect the thyroid gland from peroxides produced during the synthesis of T_4 and T_3. When soils are low in selenium as well as iodine, simultaneous deficiencies can occur. This phenomenon apparently occurs in central Africa and western China. It has been speculated that selenium deficiency increases the severity of IDD and contributes to the occurrence of myxedematous cretinism.

Limited data suggest zinc nutriture affects human thyroid function. Zinc deficiency inhibits liver type I 5'-deiodinase, and lowers plasma concentrations of T_3. These effects are expressed physiologically by impaired temperature regulation. Theoretically zinc nutriture also affects the resistance of thyroid tissue to peroxidation through glutathione reductase and thioredoxin reductase. Both are flavoenzymes. Zinc is essential for flavokinase and synthesis of flavin adenine dinucleotide (FAD). Some zinc-deficient humans were found to have low plasma concentrations of T_3 and increased plasma concentrations of reverse T_3. It is possible that some zinc-deficient humans are hypothyroid. Because zinc deficiency is common among the world's poor, many of whom are at risk of iodine deficiency, interactions between low zinc status and iodine deficiency are of more than theoretical interest.

The minimal iodine requirement of adults under usual circumstances is 50–75 µg/day. To meet this need and provide a margin of safety the Food and Nutrition Board (FNB), Institute of Medicine, National Academy of Medicine, United States, recommends 150 µg/day for both sexes, and more during pregnancy (220 µg/day) and lactation (290 µg/day).

Diets containing foods of marine origin are rich in iodine. For example, marine fin and shellfish contain 300–3000 µg iodine g^{-1} as contrasted to the 20–40 µg iodine g^{-1} in freshwater fish. Fertilizers that contain marine products can increase the iodine content of plants 10- to 100-fold, and iodine-enriched rations can increase the iodine in eggs and milk 100- to 1000-fold.

Iodine in soil and water determines the iodine content of foods (see also Chapter 16, this volume). For example, in England the average daily intake of iodine is about 220 µg. Similarly in the northeast United States average intakes are about 240 µg. In contrast, intakes in the southwest United States are about 740 µg. In Japan daily intakes are about 300 µg, when little seaweed is eaten. Seaweed consumption can increase iodine intakes to 10 mg daily. Approximate iodine contents of American foods are listed in Table XI. Of note is the wide variability. Foods raised on iodine-depleted soil are poor sources of iodine.

Prevention of primary iodine dietary deficiency requires the administration of iodine. Iodine enrichment of foods commonly consumed by the population at risk is the preferred approach. Salt is the most common vehicle. Salt iodization with KI or KIO4 is

TABLE XI. Approximate Iodine Content of Foods (U.S.)

Food class	Iodine (µg/g w.w.) Mean ± SEM	Number of samples
Fruit	40 ± 20	18
Bread/cereals	100 ± 20	18
Dairy	130 ± 10	18
Eggs	260 ± 80	11
Meat	260 ± 70	12
Vegetables	320 ± 100	13
Marine	660 ± 180	7

used in some areas and where used it is economical and efficacious. However, salt iodization may not be effective when customs or economic reasons cause the population to obtain salt from traditional small producers and distributors, and not from large producers equipped for iodization. In addition, iodine enrichment of salt has no effect on populations that use little or no salt. A second approach is the addition of iodine to bread. Wide variation in bread consumption and the preparation of bread in the home and small bakeries can detract from this approach. The addition of iodine to water supplies has been used in some settings. The small amount of iodized water consumed is a major limitation. When these approaches are not practical, oral administration of iodine-containing tablets, oil, or confections have been used, or iodine-containing oil has been injected. An obvious limitation of these approaches is the need for continuous cooperation by the recipients and the people responsible for delivery of the iodine. Even so, every three years intramuscular injection of iodized oil given to women of child-bearing age has eliminated cretinism in some regions. In addition, the Chinese have used irrigation of crops with iodine-enriched water.

I. Iron

Iron is an essential element in human nutrition but it can be toxic. Estimates of the minimum daily requirement of iron for humans depend on age, sex, physiological status, and iron bioavailability. The range is 10–18 mg per day, 30 mg per day if pregnant (U. S. recommended daily allowance; RDA), and 14 mg per day (EU RDA).

Iron toxicity can occur at high levels of intake. The average lethal dose of iron is 200–250 mg kg^{-1} of body weight. However, even an oral intake as low as 40 mg kg^{-1} of body weight has been lethal. Chronic iron overload results primarily from a genetic disorder (hemochromatosis) characterized by increased iron absorption and from diseases that require frequent transfusions. Intake of 0.4–1 mg kg^{-1} of body weight per day is unlikely to cause adverse effects in healthy people.

Iron deficiency, probably the most common nutritional deficiency, is believed by some authorities to affect 80% of the world's population, or about 5 billion people, 2 billion of whom are anemic. Especially at risk are young children and premenopausal women, both pregnant and non-pregnant. Low socioeconomic status (SES), high dietary phytate, and other inhibitors of iron bioavailability and low consumption of flesh foods and chronic blood loss increase the risk of deficiency. Knowledge of iron deficiency and its effects has a long and rich history. Examples of advances in understanding iron in human nutrition are listed in Table XII.

Iron deficiency exists when body iron stores are completely depleted and iron-dependent processes malfunction. Iron deficiency may be associated with many abnormalities (Table XIII). However, because iron deficiency related to diet seldom occurs alone and is usually part of a syndrome of micronutrient deficiencies, a cause-and-effect relationship between iron status and some clinical phenomena is obscure or nonexistent. Controlled experiments in other species have clarified these issues (Table XIV).

Iron deficiency related to diet and/or chronic blood loss evolves slowly. Initially iron stores are depleted and iron is less available for iron-dependent biochemical functions. This is reflected by decreased synthesis of heme, and the activity of cytochrome enzymes, aconitase, and other iron-sulfur enzymes is decreased. Iron-depleted aconitase (three irons) in intestinal mucosa, whose concentration is effected by iron in plasma, mediates mechanisms that increase intestinal iron absorption. These cellular effects occur when iron stores are depleted and before hemoglobin concentration is significantly decreased. More than half of iron-deficient people have this type of iron deficiency. Most affected individuals are unaware of its morbidity. Research data indicate that abnormalities in cold intolerance, muscle endurance, immunity, and neuropsychological function can be detected. Other associated micronutrient deficiencies depend on the type of diet consumed. For example, diets that exclude red meat increase the likelihood of both iron and zinc deficiencies in young women who menstruate regularly.

TABLE XII. Advances in Understanding of Iron in Human Nutrition

Date	Name	Observation
1554	Lange	Chlorosis: greenish pallor and languor in young women
1661	Doctor Sydenham's practice of physick	Iron treatment for chlorosis
1713	Lemery	Iron present in blood
1832	Foedisch	Low iron content in blood of chlorotics
1832	Blaud	Ferrous sulfate efficacious for chlorosis
1846	Magendi	Dietary iron increased blood iron of experimental animals
1895	Stockman	Dietary iron of men and women and its relation to chlorosis
1897	Cloetta	Dietary iron prevented "milk anemia" in dogs
1910	Dock	Hookworm associated with growth-stunting
1919	International Health Board	Hookworm associated with low intellectual performance in children and young adults
1919	Waite	Hookworm associated with retarded mental development
1925	Whipple	Iron and unidentified nutrients essential for erythropoiesis
1925	Keilin	Cytochrome enzymes discovered
1927	Hart	Copper essential for iron intestinal absorption
1928	Mackay	Iron treatment of infants decreased morbidity from respiratory and gastrointestinal infections
1933	Strauss	Infants born to iron-deficient women at risk of deficiency
1936	Bernhardt	Low maze learning in post-weaning rats
1943	McCance	Dietary phytate impairs iron absorption
1943	Hahn	Mucosal block theory of iron absorption
1963	Prasad	Iron and zinc deficiencies occur together
1966	Oski	Ceruloplasmin oxidizes Fe^{+2} to Fe^{+3} for binding to transferrin
1967	Sandstead	Zinc deficiency, not iron deficiency, causes growth-stunting
1968	Lee	Copper essential for utilization of tissue iron
1973	Webb	Low academic achievement in adolescents
1974	Cantwell	Residual neurological abnormalities in children after iron deficiency as an infant
1978	Oski	Mental development of iron-deficient infants improved by iron

TABLE XIII. Examples of Chemical Effects of Iron Deficiency on Cells

Type	Effect
Heme proteins	Hemoglobin and myoglobin: low
	Cytochromes—cytochrome c: low in skeletal muscle, liver, intestine, and kidney; but little or no change in brain and heart
	Catalase: low in erythrocytes
Metalloflavoproteins	Monoamine oxidase: low in liver and platelets
	Aldehyde oxidase: low in brain mitochondria
	Alpha-glycerophosphate dehydrogenase: low in skeletal muscle
	Succinate dehydrogenase: low in heart and kidney
	Xanthene oxidase: low in heart and kidney
Cofactors	Aconitase: low in kidney and liver; activity in intestinal mucosa directly correlated with iron stores and inversely with iron absorption
	Protocollagen-proline hydroxylase: low
Nucleic acid metabolism	Ribonucleotide reductase: low

TABLE XIV. Clinical Abnormalities Described as Associated with Iron Deficiency in Humans

Skin	Poor growth of hair
	Dry and fissured
	Infections
	Spoon-shaped deformity of nails (koilonychia)
Nose	Atrophy of mucosa (ozena)
Mouth	Bilateral angular stomatitis
	Atrophy of mucosa
Tongue	Atrophy of mucosa with loss of papillae
Esophagus	Dysphagia from hyperplasia of basal cells and parakeratosis of surface cells causing a web-shaped constriction of the mucosa just below the cricoid cartilage, known as Plummer-Vinson syndrome
Gastric mucosa	Atrophy, achlorohydria, and low intrinsic factor secretion
Intestinal mucosa	Atrophy of duodenal mucosa
	Iron absorption regulated by iron-depleted aconitase: increased
Liver	Reticuloendothelial cell iron: absent
Spleen	Reticuloendothelial cell iron: absent
Immune system	Morbidity from infections: increased or decreased (depends on agent)
	Circulating immunoglobulins: no change or increased
	Circulating T cells: decreased
	Lymphocyte subsets: decreased
	Lymphocyte proliferation in vitro: decreased, increased, or unchanged
	Skin hypersensitivity to antigens: decreased
	Phagocyte function: decreased
Bone marrow	Iron: decreased to absent
	Hemoglobin synthesis: decreased
	Erythropoiesis: increased
	Intermedullary hemolysis: increased
Erythrocytes	Normochromic normocytic anemia (early in deficiency)
	Hypochromic microcytic anemia
	Erythrocyte survival: decreased
	Zinc protoporphyrin concentration: increased
	Transferrin receptors: increased
Blood plasma	Iron in ferritin: decreased
	Ferritin protein: decreased when acute-phase stimuli are absent
	Serum iron: decreased
	Transferrin concentration: increased
	Iron bound to transferrin: decreased
	T_4 concentrations: decreased
	Catecholamine concentrations: increased
Muscle	Exercise endurance: decreased
	Physical work productivity: decreased
	Myoglobin synthesis: decreased
Brain	Neuropsychological function: decreased
	Pica (perverse appetite craving for and compulsive ingestion of food and non-food items)
	Affective behaviors: inconsistent symptoms, e.g., fatigue, weakness, headache, irritability, depression
Growth	Stunted
Maturation	Retarded
Fetus	Perinatal death increased; poorly reversible retarded development
Maternal status	Maternal death increased

Iron deficiency anemia occurs after all chemical indicators of deficiency are abnormal. Anemia is relatively mild in most individuals and is poorly related to symptoms or awareness of morbidity. When anemia is more severe other manifestations of the deficiency syndrome are more evident. For example, the anemia may become a "mixed anemia" as deficiencies of other micronutrients such as copper, pyridoxine, retinol, zinc, and folate become manifest. When this occurs treatment with iron alone will not fully restore erythropoiesis to normal. Other examples of conditions that may in part be caused by iron deficiency, but are also caused by associated deficiencies, include stunted growth and delayed sexual development, low immunity, impaired neuropsychological function, and abnormal pregnancy outcomes. In addition, research in animal models suggests that some clinical signs involving epithelia (skin, hair, nails, mouth, tongue, esophagus, stomach, and duodenum) that may be associated with iron deficiency are probably caused by associated micronutrient deficiencies, e.g., zinc, pyridoxine, and riboflavin.

Iron deficiency occurs when the diet provides insufficient bioavailable iron. Bleeding is a conditioning factor. Body iron content is regulated by feedback through aconitase-mediated control of iron absorption. With the exception of surface and menstrual losses, iron is reutilized.

Basal iron losses from the exterior and interior surfaces of men are about $14 \mu g \, kg^{-1} \, day^{-1}$, an amount of iron readily available from usual "western" diets. Under usual circumstances premenopausal women are at risk of iron deficiency because their non-menstrual iron losses are similar to those of men, and they lose about 6–179 mL of blood monthly (intrauterine devices can cause substantially greater losses). About 10% of normal women lose >80 mL of blood monthly. Thus the iron needs of some women for maintenance are 2.84 mg iron per day (95th percentile). Therefore women, who absorb 15% of iron when iron stores are low, must consume about 18.9 mg of food iron daily to be assured of adequate iron status. This is difficult to do. Consequently, a national survey in the United States found that the 25th percentile for serum ferritin of premenopausal women was $14 \mu g \, L^{-1}$, which suggested one out of four American premenopausal women is iron deficient by the criteria of absent iron stores.

Newborns generally have adequate iron stores for the first 4–6 months of life. They then become an increased risk of iron deficiency up to 2 years of age because of the low bioavailability of iron from common cereal-based weaning foods and high growth rate. Their iron requirements are about $100 \mu g \, kg^{-1} \, day^{-1}$, or four times the average needs of menstruating women. Later, after

TABLE XV.   Iron Requirements of Pregnancy and Lactation

	Iron (mg)
Fetus	200–300
Placenta and cord	30–170
Blood lost at delivery	90–310
Maternal milk (6 months)	100–180
Total	420–1030
Daily need: gestation and 6 months lactation	1.0–2.5

age 2, growth of children slows and requirements decrease. Adolescents who are growing rapidly have iron requirements of about 20% (boys) and 30% (girls) more, respectively, than the needs of menstruating women. The requirements of girls are related to the combined effects of growth and menses.

Pregnancy and lactation substantially increase maternal iron requirements (Table XV). Multiple pregnancies without repletion of iron nutriture can result in severe iron deficiency. This phenomenon is especially common in societies where diets are based on unrefined cereals and legumes and flesh foods are infrequently eaten.

Bioavailability is a critical factor affecting iron adequacy. Most food iron (non-heme iron) is bound to proteins and other food constituents and is usually about 1–5% bioavailable. In iron-deficient individuals feedback stimulation may increase absorption of non-heme iron as much as 20% if substances that bind iron and inhibit absorption are absent and facilitators are present. Flesh foods provide iron-protoporphyrin (heme iron) that is 15–35% bioavailable. Red meat is the best source in that nearly 50% of its iron is heme iron. Non-red-flesh foods contain less heme iron. The iron content of foods can be retrieved at Web sites of national agencies.

Dietary non-heme iron is in the ferric state (Fe^{+3}). Organic acids reduce ferric iron to the ferrous state (Fe^{+2}), the form in which most iron is absorbed. Absorption of ferrous iron occurs primarily in the duodenum. Nondigestible substances including phytate (hexaphosphate-inositol), certain dietary fibers, lignins, phenolic polymers, oxalate, products of Maillard browning, and alkaline clays bind ferrous iron and inhibit its absorption. Calcium inhibits absorption of both non-heme and heme iron. Facilitators of non-heme-iron absorption include cysteine from meat, ascorbic acid, and other organic acids.

TABLE XVI. Iron Available to Infants from Human Milk and Cows' Milk Preparations

Source	(mg L⁻¹)	% Bioavailable	Absorbed (mg L⁻¹)
Breast milk	0.5	~50	0.25
Whole cows' milk	0.5	~10	0.05
Non-fortified formula	1.5–4.8[a]	~10	0.15–0.48
Iron-fortified formula[b]	10.0–12.8[a]	~4	0.40–0.51

[a]Common infant formulas in the United States.

[b]Iron-fortified formula contains 1.0 mg iron/100 kcal formula; most iron-fortified formulas contain 680 kcal L⁻¹, equivalent to 6.8 mg iron L⁻¹.

Diets low in heme iron and facilitators of non-heme-iron absorption greatly increase the risk of iron deficiency. Such diets are common in nonindustrialized countries where poverty and/or customs limit consumption of flesh foods and cereals and legumes are the principal sources of protein. In developed countries food choice is an important determinant of the occurrence of dietary iron deficiency. Clay eating, a culturally determined "dietary" practice is the cause of iron deficiency in some populations.

Breast milk is adequate in iron until about 6 months postpartum. Non-human milk, e.g., cows' and goats' milk, contains very little bioavailable iron. Thus infants weaned early to non-human milk are at risk for iron deficiency. Weaning foods based on cereals that are rich in phytate increases the risk of deficiency, as do formulas prepared from phytate-rich soy products. Modern iron-fortified, processed cows' milk formulas that provide 1.0 mg iron/100 kcal substantially decrease the risk of iron deficiency. A comparison between breast milk and cows' milk preparations is seen in Table XVI.

Empirical practice suggests that iron-fortified cereals are the first foods fed infants at weaning. Recent research suggests infants also accept and tolerate weaning foods prepared from animal products rich in heme and sulfur amino acids. This innovation needs further evaluation.

Chronic blood loss is the second major cause of iron deficiency, and hookworm is the primary agent responsible. Hookworm afflicts at least 2 billion people. *Ancylostoma duodenale* affects the Middle East, North Africa, and southern Europe, whereas *Necator americanus* predominates in the Americas and Australia, and both also occur in central and southern Africa. The disease is endemic where the climate is warm and moist and sanitation is rudimentary. Adult worms suck blood from the mucosa of the small intestine and disperse their eggs in human feces. If dropped on appropriate soil larva mature to the filariform stage which can penetrate the skin (feet and legs are the usual points of entry) and migrate to the small intestine where they transform into adults.

J. Lead

Lead occurs in ore and soil as both inorganic and organic compounds with different types of toxicity to humans. Lead has been used in paint, in ceramics, and in home utensils. The main exposure route is via food and drinking water. It can cause neurotoxic effects in children. An early effect in adults is the interference with the hemoglobin syntheses with an increase of erythrocyte zinc protoporphyrin (ZPP) in whole blood. ZPP is sometimes monitored in whole blood as an indicator of lead exposure. This can also be seen in iron-deficiency anemia. Humans suffering from intermittent protoporphyria are considered a vulnerable group for lead toxicity.

In acute and chronic poisoning from environmental exposure to inorganic Pb the reference range in blood for adults is usually <100 µg Pb L⁻¹. For children, the levels should not exceed 40 µg Pb L⁻¹ because of its neurotoxic effects.

The metabolism and distribution of organic Pb differs markedly from the inorganic form, and urine is used as the medium for biological monitoring. Monitoring of inorganic Pb in urine is only recommended during chelation therapy. After the removal of tetraethyl lead from gas, the concentration of lead in blood in children has markedly decreased.

K. Manganese

Manganese is an essential trace element with an estimated daily nutritional requirement of 30–50 µg kg⁻¹ of body weight. Manganese is required for several enzymes involved in carbohydrate metabolism. Absorption rate is influenced by actual intake, chemical form, and the presence of other metals, such as iron and copper, in the diet. In infants and young animals very high absorption rates of manganese have been observed.

The biological half-time of manganese in humans has been determined to be about 12–35 days, depending on the nutritional status. The major route of excretion of manganese is through the biliary system, and a low

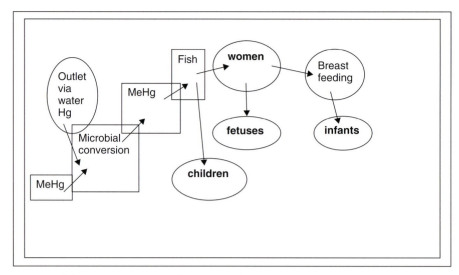

FIGURE 3 Mercury exposure via the nutritional chain with vulnerable groups indicated in bold.

tissue level is maintained by this mechanism. Because of its chemical similarity to iron, manganese also binds to iron-binding proteins such as transferrin. Interactions between manganese, iron, and lead have been shown in rats (Malhotra et al., 1984). Although manganese is not very toxic, its increased tissue accumulation can cause toxic effects in the brain and lung.

Excessive absorption of manganese from lungs can increase its accumulation in the brain. Brain damage has been reported in miners exposed to manganese dioxide dust, and it can progress to irreversible brain injury similar to Parkinson's disease. The most important biochemical effects of manganese are on the metabolism of various neurotransmitter substances such as dopamine. The selective injury of catecholamine neurons by manganese is similar to 6-hydroxydopamine-induced neuronal injury, and it may be related to a generation of toxic free radicals. Certain manganese compounds such as manganese carbonyl compounds have been used as anti-knocking agents to replace lead in several countries. The long-term effects of these compounds on people, especially the elderly, should be monitored because of its known effects on catecholamines in brain and its relation to Parkinson's disease. The interactions of other metals on the neurotoxicity of manganese should also be studied to understand the mechanisms involved in metal-induced neurotoxicity.

L. Mercury

Mercury is found mainly in food, especially fish as methylmercury, and it is the major source of mercury exposure in various parts of the world. Exposure to mercury gives rise to adverse health effects such as neurobehavioral disorders and is an issue of public health concern. Cognitive impairment in children, following exposure of methylmercury during pregnancy, has been reported from several epidemiological studies from New Zealand (Kjellström et al., 1986), from the Seychelles (Clarkson, 1997), and from the Faroe islands (Grandjean et al., 1997). The message of public health concern is that women of child-bearing age should not consume species containing methylmercury (Figure 3).

The major source of exposure to mercury vapor in the general population is from dental amalgam fillings. Dental amalgam consists of about 50% mercury along with other metals such as silver and copper. They have been used widely for the last 150 years because they are long lasting, inexpensive, and easier to use than other types of dental fillings. There are reports of mercury poisoning in dentists and dental technicians who handle amalgam as well as reports of side effects in patients. However, the use of amalgam in dental practice has been controversial because of the assumed toxic effects of mercury. It is known that amalgams can release mercury vapor into the oral cavity and if inhaled up to 80% is absorbed in the lungs. What is absorbed is transported to the bloodstream and part of it may enter the CNS through the blood-brain barrier, but a great part of it may be excreted in the urine. Normally the absorbed mercury vapor is excreted in the urine. About 80% of inhaled mercury vapor is taken up. The normal mercury level in urine is less than 0.05 mg L^{-1}, and the maximum allowable level is 0.15 mg L^{-1}. The threshold limit value (TLV) for mercury vapor is 0.05 mg m$^{3-1}$, and is calculated for a working schedule of 8 hours/day and 40 hours/week.

1. Mercury—Blood

The blood level of mercury is a useful indicator for both acute and chronic exposure to organic mercury compounds; but it may be of limited use for acute exposure to vapor or inorganic salts. Reference range is <4 μg L^{-1}.

2. Mercury—Urine

Urinary levels of mercury are useful to measure exposure to mercury vapor and to inorganic mercury salts. Analysis of mercury in urine has been recommended for people concerned about its release from dental fillings. Reference range: excretion usually <10 μg/g creatinine/<10 μg/24 hours.

According to the IPCS/WHO Health Criteria Document on inorganic mercury (1991), when mercury vapor exposure is above 80 μg m$^{3-1}$, 25–80 μg m^{-3} corresponding to a urinary mercury level of 100 μg/g creatinine, the probability of developing the classical neurological signs of mercury intoxication such as tremor, erethism, and proteinuria is high. Exposure to mercury vapor in the range of 25–80 μg m^{-3}, corresponding to urinary level of 30–100 μg mercury/g of creatinine can lead to increases in the incidence of certain less severe toxic effects.

M. Molybdenum

Molybdenum is considered to be an essential trace element in both animals and humans. Safe and adequate intake levels suggested for the population are 0.015–0.04 mg/day for infants, for children of age 1–10 the level is 0.025–0.15 mg day^{-1}, and 0.075–0.25 mg day^{-1} for humans over 10 years of age.

A disease of Swedish moose is explained by interference of molybdenum with copper; see Chapter 20.

N. Selenium

Selenium is an essential trace element. Recent studies have indicated that selenium exerts a beneficial effect on coronary disease mortality, and that selenium plus garlic produces significant anticancer activity. In the Scandinavian countries the intake of selenium is low due to the fact that the soils are poor in selenium. Also, due to acid rain selenium has less bioavailability. Depending on the soil content selenium concentration varies in grain. Countries like New Zealand, Finland, and Sweden, with a low concentration of selenium in soils, report intakes below 70 μg per day and from China low intakes are reported as below 20 μg per day. High concentrations are reported from Scotland, Venezuela, and from certain parts of China of up to 200–700 μg per day (see also Chapter 15, this volume).

An estimated daily intake in Sweden is about 24–35 μg. A recommended daily intake in the United States is 55 μg for ages 19 and older. For pregnant women the recommended daily intake is 60 μg and during lactation the recommended rate is 70 μg. In some areas this dose is exceeded. Deficiency occurs when the daily intake is 10 μg. Toxicity occurs at intakes of 500 μg/day. Health effects related to deficient intake, which occurs at daily intake below 20–30 μg occur in muscles and the heart. At high intakes above 1000 μg gastrointestinal irritation, hair loss, and nerve damage occur. Acute exposure may give rise to liver damage. Reported increased incidence of asthma in populations in New Zealand is associated with low selenium intakes.

Selenium deficiency gives rise to heart diseases like Keshan disease, which is most frequently seen in children. Another disease associated with selenium deficiency is Kashin-Beck. Cancer is connected to selenium and influences diseases of the muscles and joints and rheumatics and senility. High intakes of selenium influence the occurrence of caries, the garlic smell of breath, and the blue staining of nails. Selenium intoxication of cattle has been treated with arsenic supplement.

A selenium level in serum between 60–120 μg L^{-1} reflects sufficient intake of selenium via food. Selenium concentration in hair exceeding 5 μg g^{-1} reflects selenium poisoning. Selenium in plasma/serum is measured upon indications of deficiency and/or toxicity and can measure recent (months) changes in intake of or exposure to selenium. In humans the activity of the Se-dependent enzyme glutathione peroxidase is of interest. Reported reference values for neonates to 16 years of age are 30–115 μg L^{-1} and are age dependent and for adults ranging from 70 to 130 μg L^{-1}.

O. Thallium

Thallium (Tl), is a soft metal and the oxidation states are I and III. The salts are highly toxic. Environmental samples contain μg kg^{-1} or less. However, the determination level for minerals is 20 μg kg^{-1} and for aqueous solutions it is about 0.1 μg L^{-1}. Thallium can be found in phyllosilicates and in sulfide deposits.

TABLE XVII. Advances in Understanding of the Nutritional Role of Zinc

Date	Name	Observation
1869	Raulin	Essential for *Aspergillus niger*
1887	Lechartier	Presence in living tissues
1905	Mendel	Constituent of respiratory pigment of the snail *Sycotypus*
1910	Mazé	Growth factor for maize
1919	Birckner	Proposed "nutritive function" from its "constant occurrence" in human and cows' milk
1926	Somner	Essential for higher green plants (sunflower, barley)
1927	Hubbell	Improves growth of mouse
1934	Todd	Deficiency stops growth and causes alopecia (rat)
1941	Follis	Deficiency causes dermatitis with acanthosis, parakeratosis, and inflammation; parakeratosis and basal cell hyperplasia of esophagus (rat)
1952	Mawson	Deficiency causes atrophy of testicular germinal epithelium, epididymides, and prostate (rat)
1955	Tucker	Deficiency causes stunting, dermatitis with parakeratosis, diarrhea, and death (swine)
1956	Vallee	Possible zinc deficiency in humans with alcoholic cirrhosis
1958	O'Dell	Deficiency in *Aves* causes dermatitis and retards feathering, osteogenesis and lymphoid tissue
1959	Winder	Facilitates DNA, RNA, and protein synthesis of *Mycobacterium smegmatis*
1960	Blamberg	Deficiency is teratogenic (chicks)
1961	Prasad	Deficiency thought to cause stunting and hypogonadism in teenage Iranian farmers
1962	Lieberman	Facilitates DNA synthesis in cultured mammalian cells
1962	Miller	Deficiency in calf has effects similar to those in swine
1963	Prasad	Stunted teenage farmers in Egypt have zinc metabolism consistent with zinc deficiency
1963	Sandstead	Zinc treatment improved growth and development of stunted teenage farmers in Egypt
1964	Ott	Deficiency in lambs has effects similar to those in swine
1966	Hurley	Deficiency teratogenic for rat

Bioavailability of thallium increases when pH decreases in soil. Thallium can also be leached to ground and surface water. It has a strong tendency to accumulate in aquatic life and plants can easily take up the element by the roots.

Oral intake of (Tl1) 20–60 mg thallium kg^{-1} body weight is lethal within one week. Thallium(III) oxide, which is water soluble, shows a somewhat lower acute toxicity compared to thallium(I) salts. The U. S. Environmental Protection Agency (EPA) suggests that drinking water exceeding action levels can lead to gastrointestinal irritation and nerve damage in the short term, and to changes in blood chemistry, damage to liver, kidney, intestinal, and testicular tissues and hair loss in the long term.

The major symptoms of acute intoxication are anorexia, vomiting, depression, and hair loss. Respiratory failure is lethal. The same symptoms are reported for chronic intoxication. Loss of hair is a typical sign of intoxication caused by thallium. The reference range for acute and chronic poisoning for blood is $<1\,\mu g\,L^{-1}$ and for urine $<1\,\mu g\,L^{-1}$.

P. Zinc

Zinc (Zn) is an essential trace element found in all food and potable water as salt or an organic complex. The principal source of zinc is normally the diet. Zinc in surface and groundwater usually does not exceed 0.01 and 0.05 mg L^{-1}, respectively, and concentrations in tap water can be much higher as a result of dissolution of zinc from pipes. A daily dietary requirement of zinc of 0.3 mg kg^{-1} of body weight was proposed by the Joint FAO/WHO Expert Committee on Food Additives (JECFA) in 1982. At the same time JECFA also suggested that the provisional maximum tolerable daily intake for zinc is 1 mg kg^{-1} of body weight. The dietary reference values for adults range from 6 to 15 mg/day. Drinking-water containing zinc at levels above 3 mg L^{-1} may not be acceptable to consumers.

Zinc deficiency occurs in human beings when intake is low. This depends on the intake of diets that are low in readily bioavailable zinc. Unrefined cereals are rich in phytate and dietary fibers which all bind zinc and prohibit the bioavailability of the metal. This means that even if food is high in zinc content the intake can give rise to deficiency. This has been described as a public health problem in certain countries such as Egypt.

Zinc in serum is measured upon indications of deficiency. Another way of finding out zinc status is simply by asking about number of meals with red meat per week. This can provide a crude estimate as to whether zinc deficiency should be suspected.

1. *Zinc Deficiency*

a. History

Examples of advances in understanding of the nutritional role of zinc are given in Table XVII. Reference values for zinc status are that $<0.5\,\mathrm{mg\,L^{-1}}$ may indicate zinc deficiency and $0.5–0.7\,\mathrm{mg\,L^{-1}}$ might be of no clinical significance. The "normal" range for all ages is regarded to be $0.7–1.6\,\mathrm{mg\,L^{-1}}$ whereas $>1.6\,\mathrm{mg\,L^{-1}}$ might reflect the use of dietary supplements.

Zinc in urine is measured upon indications and metabolic studies or for chelation therapy. It is, however, of little value in assessing deficiency. This analysis may be of value in monitoring the effect on zinc body burden of long-term chelation therapy of other trace elements. The reference value for excretion is usually 0.3–0.6 mg/24 hours.

VI. SUMMARY

To understand the full context of the relationship of trace elements to the geo-environment, it is important to encourage interdisciplinary research and to coordinate the knowledge obtained from various scientific fields. This approach may provide a better understanding of the mechanisms involved in both nutritional deficiency and toxicity of trace elements and metals. The nutritional deficiencies can arise from lack of the essential elements in the drinking water or food. This can be due to lack of these elements in the soil where the food is grown or can be due to the eating habits of the people. Certain genetic defects can also affect the absorption or transport of the elements to required tissues. Toxicity might occur when high concentrations of metals in soil and drinking water lead to high exposure to metals. The potential risk of developing toxicity depends on the bioavailability of the specific trace element. These influencing factors are shown in Table XVIII.

TABLE XVIII. Factors Influencing the Relationship for Trace Elements and Health Effects

Eco– and biological system related to exposure, dose, and effect
Environmental exposure in the ecosystem
Environmental and biological monitoring
Acidification and species of metals and trace elements
Bioavailability, species, and exposure conditions and toxicity of metals
Metals, geo-environment, and environmental biochemistry
Metabolism and kinetics of trace elements
Importance of various proteins such as metallothionein for binding of elements

SEE ALSO THE FOLLOWING CHAPTERS

Chapter 4 (Uptake of Elements from a Chemical Point of View) · Chapter 5 (Uptake of Elements from a Biological Point of View) · Chapter 6 (Biological Functions of the Elements) · Chapter 7 (Geological Impacts on Nutrition)

FURTHER READING

Bremner, I. (1987). Nutritional and Physiological Significance of Metallothionein. In *Metallothionein II* (J. H. R. Kägi and Y. Kojima, Eds.), Birkhauser Verlag, Basel, pp. 81–107.

Cherian, M. G., and Nordberg, M. (1983). *Cellular adaptation in metal Toxicology and metallothionen Toxicology*, 28, 1–15.

Cherian, M. G. (1995). Metallothionein and Its Interaction with Metals. In *Toxicology of Metals* (R. A. Goyer and M. G. Cherian, Eds.), Springer, 121–132.

Cherian, M. G. (1997). Metallothionein and Intracellular Sequestration of Metals. In *Comprehensive Toxicology* (I. G. Sipes, C. A. McQueen, and A. J. Gandolfi, Eds.), Volume 3, Pergamon, pp. 489–500.

Clarkson, T. W. (1997). The Toxicology of Mercury, *Crit. Rev. Crit. Lab. Sci.*, 34(4), 369–403.

Clarkson, T. W., Friberg, L., Nordberg, G. F., and Sager, P. R. (Eds.). (1988). *Biological Monitoring of Toxic Metals*, Plenum Press, New York.

DelRazo, L. M., Quintanilla-Vega, B., Brambila-Colombres, E., Calderon-Aranda, E. S., Manno, M., and Albores, A.

(2001). Stress Proteins Induced by Arsenic, *Toxicol. Appl. Pharmacol.*, 177, 132–148.

Flanagan, P. R., McLellan, J. S., Haist, J., Cherian, M. G., Chamberlain, M. J., and Valberg, L. S. (1978). Increased Dietary Cadmium Absorption in Mice and Human Subjects with Iron Deficiency, *Gastroenterology*, 74, 841–846.

Goering, P. L., and Fisher, B. R. (1995). Metals and Stress Proteins. In *Toxicology of Metals* (R. A. Goyer and M. G. Cherian, Eds.), Springer, pp. 229–266.

Goyer, R. A., and Cherian, M. G. (1978). Metallothioneins and Their Role in the Metabolism and Toxicity of Metals, *Life Sci.*, 23, 1–10.

Grandjean, P., Weihe, P., and White, R. F. (1997). Cognitive Deficit in 7-Year-Old Children with Prenatal Exposure to Methylmercury, *Neurotoxicol. Teratol.*, 19(6), 417–428.

IARC. (1980). *Some Metals and Metallic Compounds*, Volume 23, IARC, Lyon, France.

IARC. (1990). *Chromium, Nickel and Welding*, Volume 49, IARC, Lyon, France.

IARC. (1994). *Beryllium, Cadmium, Mercury and Exposures in the Glass Manufacturing Industry Monograph*, Volume 58, IARC, Lyon, France.

Kägi, J. H. R., and Nordberg, M. (Eds.). (1979). *Metallothionein*, Birkhauser Verlag, Basel, pp. 1–378.

Kjellstrom, T., Kennedy, P., Wallis, S., and Mantell, C. (1986). Physical and Mental Development of Children with Prenatal Exposure to Mercury From Fish, National Swedish Environmental Protection Board Report 3080, Stockholm.

Malhotra, K. M., Murthy, R. C., Srivastrava, R. S., and Chandra, S. V. (1984). Concurrent Exposure of Lead and Manganese to Iron-Deficient Rats: Effect of Lipid Peroxidation and Contents of Some Metals in the Brain, *J. Appl. Toxicol.*, 4, 22–30.

Nordberg, M. (1998). Metallothioneins, 1998. Historical Review and State of Knowledge, *Talanta*, 46, 243–254.

Nordberg, M., and Nordberg, G. F. (2000). Toxicological Aspects of Metallothionein, *Cell. Mol. Biol.*, 46(2), 451–463.

Nordberg, M., and Nordberg, G. F. (2002). Cadmium. In *Handbook of Heavy Metals in the Environment* (B. Sarkar, Ed.), Marcel Dekker, Inc., New York, 231–269.

Piscator, M. (1964). On Cadmium in Normal Human Kidneys Together with a Report on the Isolation of Metallothionein From Livers of Cadmium-Exposed Rabbits, *Nord. Hyg. Tidskv.*, 48, 76.

Radtke, F., Heuchel, R., Georgiev, O., Hergersberg, M., Garglio, M., Dembic, Z., and Schaffner, W. (1993). Cloned Transcription Factor MTF-1 Activates the Mouse Metallothionein 1 Promoter, *EMBO J.*, 12, 1355–1362.

Robertson, A., Morrison, J. N., Woods, A. M., and Bremner, I. (1989). Effects of Iron Deficiency on Metallothionein-1 Concentrations in Blood and Tissues of Rats, *J. Nutr.*, 119, 439–445.

Seguin, C. (1991). A Nuclear Factor Requires Zn^{2+} to Bind a Regulatory MRE Element of the Mouse Gene Encoding Metallothionein-1, *Gene*, 97, 295–300.

WHO/IPCS ENVIRONMENTAL HEALTH CRITERIA DOCUMENTS FOR VARIOUS METALS

WHO/IPCS Environmental Health Criteria Document 118 Inorganic Mercury, WHO, Geneva, 1991.

WHO/IPCS Environmental Health Criteria Document 134 Cadmium, WHO, Geneva, 1992.

WHO/IPCS Environmental Health Criteria Document 182 Thallium, WHO, Geneva, 1992.

WHO/IPCS Environmental Health Criteria Document 224 Arsenic, 2nd ed, WHO, Geneva, 2001.

WHO/IPCS Environmental Health Criteria Document 221 Zinc, WHO, Geneva, 2001.

WHO/IPCS Environmental Health Criteria Document 228 Principles and Methods for the Assessment of Risk from Essential Trace Elements, WHO, Geneva, 2002.

PATHWAYS AND EXPOSURE

INTRODUCTION: RON FUGE

It has long been said that "we are what we eat"; however, in terms of medical geology we are in fact what we eat, drink, and breathe. The major pathways into the human body of all elements and compounds that are needed for well-being, together with those that are harmful, derive from the food and drink we consume and the air we breathe. Although direct absorption through the skin can also serve as a pathway, this is of relatively minor importance. The composition of the air, water, and food is directly influenced by interactions with geological media. It should, however, be stated that air, soil, and water are subject to chemical changes due to anthropogenic activities, as discussed in Chapter 3. Although pollutants from this source can have a dramatic effect on the composition of air, water, and food, which can in turn seriously impact human and animal health, this book is essentially concerned with the influence of natural sources of the elements on human health. Therefore, anthropogenic sources are largely ignored in this section. The chapters in the section discuss the importance of the various pathways, together with examples of the influence of various elements from different sources, on human and animal health.

The atmosphere is of particular importance for life on Earth because it is the source of essential oxygen. However, in addition to the oxygen, the air inhaled can contain substances that are detrimental to health. As outlined in Chapter 9, volcanic eruptions can directly result in large loss of life due to phenomena such as explosive vulcanicity, pyroclastic flows, and lahars. In addition, volcanic eruptions can result in the emission of large quantities of fine ash and dust that can be inhaled and cause serious health problems. Similarly, volcanic gases, such as the sulfur gases and hydrogen fluoride, can be extremely harmful to exposed populations, both from inhalation and from skin contact. The radioactive gas radon, derived from the natural decay of uranium and thorium, can be emitted during volcanic eruptions and can also leak out of crustal rocks that contain uranium and thorium minerals (see Chapter 10). Inhalation of radon can increase the risk of lung cancer.

The trace element chemistry of terrestrial water is essentially controlled by its interaction with rocks and soils. The composition of the water deriving from precipitation depends essentially on geography. Because rain and snow are derived from evaporation of seawater, they contain some remnant salts of marine origin, the content being controlled essentially by the proximity of the coast. Thus, near the coast precipitation is enriched in sea salts, whereas well inland the influence is much weaker. However, once precipitated, the water interacts with soils and surface rocks, whereas some penetrates into the lithosphere to become groundwater, where it comes into more intimate contact with the rocks through which it flows and in which it becomes stored. Thus, the chemistry of the water that animals and humans consume can be greatly modified by interaction with the lithosphere, particularly in the case of groundwater, which can be many millions of years old.

Two elements that can have a marked effect on health are arsenic (Chapter 11) and fluorine (Chapter 12). Chronic exposure to high concentrations of these elements is most often through drinking water. Consumption of arsenic-rich drinking water is having serious and detrimental health effects on large populations in several countries. Fluorine, occurring in groundwater as the fluoride ion, is essential in small quantities and harmful in larger quantities. Again, the harmful effects of drinking fluoride-rich water are a serious problem in many countries.

Water hardness is essentially dependent on the concentrations of calcium and magnesium in the water. It has been suggested that the hardness of drinking water can influence human health (Chapter 13), with harder waters appearing to offer some protection against cardiovascular disease.

Plants take up nutrients from soil through their roots and from the atmosphere through their aerial parts. However, the major pathway for most trace elements is from soil. To that end, it is important to understand soil formation and the chemistry of soils, together with bioavailability of elements from this source (Chapter 14). Soil chemistry is very strongly influenced by the chemistry of the parent materials. Thus, the origin of most trace elements in plants is the lithosphere, from which soils are derived through weathering. Consumption of plants then becomes a major pathway into animals and humans.

Chapters 15 and 16 deal with the geochemistries of selenium and iodine, respectively, with particular reference to their behavior in soils and their bioavailability. These two elements have long been known to be essential in trace quantities. Although excesses of both, particularly selenium, are detrimental to health, it is their deficiencies in many areas of the world that has resulted in large geographically defined areas where endemic diseases related to their low concentrations and bioavailability have been and are prevalent.

In addition, soil can have a more direct pathway into humans through its ingestion intentionally, or geophagy, or by involuntary consumption (Chapter 17). This can also result in the ingestion of soil microbes, some of which can be pathogenic (Chapter 19). Wind-ablated soil and dust particles can also represent a pathway for pathogen ingestion and inhalation. Such dusts, particularly in drylands, can also cause serious respiratory problems and can represent a pathway into the body of many elements, some of which may be harmful (Chapter 18).

The final chapter in this section (Chapter 20) deals with medical geology in relation to animals. This is a particularly important topic because animals, whose movements and sources of food are more limited than humans, are more frequently affected by deficiency and excess diseases related directly to geology. They are also part of the food chain.

CHAPTER 9

VOLCANIC EMISSIONS
AND HEALTH

PHILIP WEINSTEIN AND ANGUS COOK
University of Western Australia

CONTENTS

I. AN INTRODUCTION TO VOLCANIC TOXICOLOGY

Volcanic vents and fissures provide a conduit by which magma—the molten rock, gases, and water within the earth—may interact with human biological systems (Figure 1). Because of the range of materials that are ejected during eruptions, the consequent effects on human health are diverse. Contact may occur dramatically and immediately for people living close to the vent, such as during pyroclastic flows or the emission of large tephra projectiles. Alternatively, effects on health may occur slowly or at great distances from the eruptive site as a result of dispersal of volcanic material such as ash, dust, and aerosols.

The vast majority of volcanogenic fatalities in the past few centuries have resulted from "proximal" events—such as pyroclastic flows, lahars, and suffocation or building collapse from ash or debris—and "distal" events, such as tsunamis, which may spread for hundreds of miles from the active site, and indirect consequences of eruptions, such as famine or infectious disease outbreaks. Simkin et al. (2001) placed the death toll over the last 500 years at above 250,000, with seven eruptions dominating the historical record (including Tambora in 1815, Krakatau in 1883, Pelee in 1902, and Ruiz in 1985). Apart from the obvious thermal and physical injuries resulting from an eruption, ejecta may also contain toxic elements and compounds that disrupt biological systems. These compounds may be released in the form of volcanic gases or carried with volcanic matter falling from eruptive columns or plumes. Some of the material ejected may induce disease by undergoing radioactive decay: among the best documented of such products is radon. Although these toxins are often not the major causes of mortality in volcanic eruptions, they may persist and have the potential to cause long-term morbidity.

Considering the pathological consequences of exposure to volcanic toxic compounds is a complex task. There is no simple, predictable path between the "emergence" of a toxin from the magma to the eventual health consequences it imparts to a particular individual. This chapter reviews the varied mechanisms by

FIGURE 1   Vent of an active volcano: White Island, New Zealand. (Photo: Michael Durand).

FIGURE 2   Krafla fissure eruption, Iceland. (Photo: Olle Selinus).

which the emissions associated with volcanism may compromise health and generate disease. In general, these mechanisms may be influenced by one or more of the following factors:

1. Eruptive variables
2. Toxin-specific properties
3. Patterns of toxin dispersal and persistence
4. Biological variables

Eruptive variables—The nature of the eruption (or other volcanic event) influences the duration of emissions, the chemical composition of the toxic compounds expelled, and the range of dispersal. For example, eruptions may be broadly grouped as *explosive* (releasing large quantities of gas, hot ash, and dust, as with Mount St. Helens), *effusive* (associated with large lava flows but less dramatic outpourings of gas and dust, as with the basaltic volcanoes of Hawaii), or *mixed* (a combination of the two patterns). Activity may be measured using the

Volcano Explosivity Index, which incorporates many variables including the volume of tephra, the eruption type, and duration (Newhall & Self, 1982).

Toxin-specific properties—These primarily pertain to the chemical and physical properties of toxic compounds. Volcanic products vary in terms of particle size, concentration, pH, and water solubility. All of these factors can influence the bioavailability of toxins, and thus their patho-physiological effects.

Patterns of toxin dispersal and persistence—In terms of evaluating possible effects on human populations, physical proximity to the vent or eruptive site is an important component of risk assessment. Populations who fall within the "near-vent" range may be exposed to the full array of ejected materials, often at high concentrations (Figure 2). By contrast, areas that are distal from the volcanic site (or are less vulnerable to volcanic products for some other reason, such as the presence of a natural barrier) tend to be exposed to a smaller range of toxic compounds, and at concentrations less likely to result in injury. There are exceptions, however, with toxins such as fluoride adhering to fine, wind-dispersed ash particles and thereby occurring in highest concentrations some distance from the vent. In this discussion, near-vent is used loosely to refer to an area extending tens of kilometers from the eruptive site.

The mode of toxin dispersal should also be considered. Eruptive products may travel along many routes, and in a variety of chemical forms, before finally appearing in human biological systems. Carriage in the atmosphere and hydrosphere are the most common modes of dispersal, but poisoning may also occur as a result of volcanic products entering the soil and food chain.

The duration of exposure plays one of the most crucial roles in determining health outcomes. For example, some insults may be short-lived and reversible, as with conjunctival irritation from ash particles, or may be chronic, as with inhalation of silica particles resulting in the life-long respiratory problems of silicosis. Some toxic compounds, such as radon, may persist in volcanic products (and continue to cause injury) long after the eruptive event ceases. These patterns of injury, and the body systems that are predominantly affected, are discussed in detail in Sections III, IV, and V.

The properties of the environment also affect the pattern of dispersal and settlement of volcanic toxins, including features of the physical environment (geography, air pressure, and climate) or man-made surroundings (for example, the opportunities for asphyxiant gases to enter low, enclosed spaces such as cellars).

Biological variables—The mechanism of damage must be considered at a histopathological level. Volcanic products may produce injury in human tissues and cells, either individually or in combination, in the following ways: (1) direct physical interaction (e.g., skin contact with acidic gases); (2) initiation of a chronic process of damage and repair (e.g., fibrosis from the deposition of respirable silica particles in lung tissue); (3) metabolic disruption (e.g., with carbon monoxide toxicity); or (4) genotoxicity and genetic alteration (e.g., from exposure to carcinogenic agents, such as radon).

Finally, the characteristics of the affected individuals play an important role in determining health outcomes. Important parameters include age, the presence of coexisiting cardiac or respiratory diseases, and access to appropriate diagnostic services. There is also an inherent response spectrum, both phenotypic and genotypic, in any human population and some individuals are more susceptible than others. Such variation in susceptibility is discussed in Section III.C.

Figure 3 illustrates the major determinants of health outcomes in a simplified schema that shows: (1) the spatial range of health effects (in relation to proximity to the vent), (2) the primary mode of human exposure, and (3) potential duration of health effects following contact with the toxic compound. A comparison is provided for three common volcanic products: free silica, a respirable mineral; sulfur dioxide, a noxious gas (and its associated sulfuric acid aerosol); and fluoride compounds (including the highly acidic gas, hydrogen fluoride, HF).

II. THE HEALTH EFFECTS OF TEPHRA DISPERSAL

A. Atmospheric Dispersal of Toxic Compounds

Tephra dispersal is a major cause of morbidity following eruptions. Tephra thrown into the atmosphere may cause disease through the fallout of particles from eruption columns or plumes on human populations, or through the movement of individuals into eruptive clouds (such as aircraft passengers and crew). The emission of large fragments of rock, such as "blocks" and "bombs," may cause severe physical injury including lacerations and fractures. Heavy fallouts (especially of pumice) can lead to burial and asphyxiation, either directly or through a roof collapse. Smaller particles of pumice, scoria, and ash may be distributed over a wide area around the eruptive site, and in some cases plumes may affect settlements situated hundreds of kilometers away. This chapter considers the health effects of tephra by primarily examining the effects of ash and dust (the constituents of tephra less than 2 mm in diameter). The pathophysiological effects of the compounds discussed are summarized in Table I.

The eyes are particularly vulnerable to the emission of fine tephra particles. Common ocular injuries include abrasions of the cornea and conjunctivitis from accumulation of ash in the conjunctival sac (Blong, 1984). Ocular irritation has been reported in people using contact lenses, because of the interposition of matter between the contact lens and the eye. Swelling of the eyelids and other facial tissues around the eyes have been reported less frequently. Symptomatically, ash produces higher reported rates of ocular redness, itchiness, throbbing pain, and discharge. Superficial tissues such as the skin, lips, mouth, and other mucous membranes may also be exposed. Nasal and throat irritation occur more frequently, and higher rates of nasopharyngeal irritation and nasal stuffiness were observed in a group of loggers following the Mount St. Helens eruption (Baxter et al., 1986). A less well-documented effect of tephra is skin irritation, including in the axillary area, following deposition of volcanic ash particles on the skin—the so-called "ash-rash." Irritation of superficial mucous membranes, eyes, and exposed skin commonly resolves shortly after the exposure to the ash ceases, and longer term toxic injury to such body structures following ash-fall is unusual.

The lungs may be exposed to any particulate matter able to penetrate into the respiratory passages. This

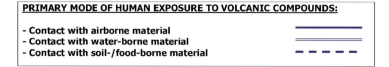

PRIMARY MODE OF HUMAN EXPOSURE TO VOLCANIC COMPOUNDS:

- Contact with airborne material
- Contact with water-borne material
- Contact with soil-/food-borne material

a. Free silica

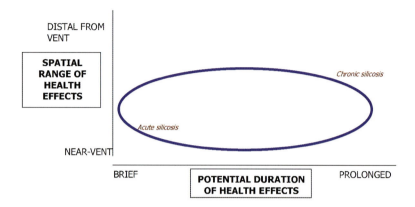

b. Sulphur dioxide / sulphuric acid

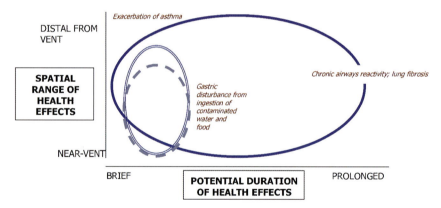

c. Fluoride compounds

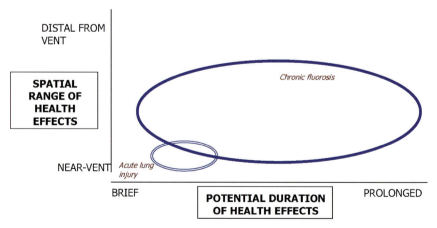

FIGURE 3   Major determinants of health outcomes for silica, sulfur dioxide, and fluoride.

TABLE I. A Review of Major Toxic Compounds of Volcanic Origin and Their Potential Pathophysiological Effects

Toxin (active form)	Mode of dispersal to human populations	Mechanism of injury	Acute systemic effects	Chronic or recurrent systemic effects
Sulfur compounds				
Sulfur dioxide, SO_2; sulfur trioxide, SO_3; sulfuric acid, H_sSO_4	Gas emissions during eruptions, lava flows, degassing episodes	Acidic irritant	**RESP**: upper airway irritation, pneumonitis, pulmonary edema, acute ARDS $\rightarrow \leq$ **HEENT**: nose and throat irritation, conjunctivitis **DERM**: skin irritation	**RESP**: recurrent or prolonged exacerbation of respiratory disease, bronchiolitis obliterans $\rightarrow \leq$
Hydrogen sulfide, H_2S	Gas emissions during eruptions	Irritant, asphyxiant, inhibition of metabolic enzymes	**GENERAL**: headache nausea, vomiting, confusion, collapse, paralysis of respiratory centers $\rightarrow \leq$ **GIT**: diarrhea **GU/REPRO**: pain on urination **RESP**: cough, shortness of breath, pulmonary edema **HEENT**: eye and throat irritation	
Fluoride compounds				
Fluoride compounds (including related acidic gases, aerosols and liquids)	Gas emissions during eruptions, ash leaching	Acidic irritant on inhalation or contact with skin, conjunctiva, or mucous membranes	**GENERAL**: hypocalcemia (low serum calcium), low serum magnesium, collapse and shock $\rightarrow \leq$ **RESP**: (e.g., HF inhalation) coughing, laryngeal spasm, bronchitis, pneumonitis, pulmonary edema, acute ARDS $\rightarrow \leq$ **GIT**: nausea, vomiting, gastrointestinal hemorrhage $\rightarrow \leq$ **GU/REPRO**: nephritis **HEENT**: eye and throat irritation **DERM**: severe, slow healing burns; may be absorbed through skin causing internal effects	**RESP**: permanent lung injury from toxic inhalation **HEENT**: mottling and/or pitting of teeth **MSS**: osteoporosis, osteosclerosis, calcification of ligaments and tendons, kyphosis of spine with bony exostoses

continues

Continued

Chlorine compounds				
Hydrochloric acid, HCl	Gas emissions during eruptions, lava (e.g., with seawater contact)	Acidic irritant	**GENERAL**: collapse → ≤ **RESP**: coughing, laryngeal spasm, pneumonitis, pulmonary edema, acute ARDS → ≤ **HEENT**: eye and throat irritation	**RESP**: permanent lung injury from toxic inhalation
Carbon compounds				
CO	Gas emissions during eruptions	Noxious asphyxiant; binds to hemoglobin	**GENERAL**: collapse, coma, "cherry red" skin → ≤ **NEURO**: headache, impaired dexterity	**NEURO**: permanent neurological impairment from brain injury
CO_2	Gas emissions during eruptions	Inert asphyxiant	**GENERAL**: asphyxia, collapse → ≤	
Free silica and silicates				
Free silica	Ash plumes	Mineral dusts initiate inflammatory response and fibrosis (scarring)	**RESP**: acute exacerbation of respiratory disease (e.g., asthma)	**RESP**: chronic silicosis
Metals				
Mercury vapor, Hg	Gas emissions during eruptions	Oxidant irritant	**RESP**: bronchitis, pneumonitis, pulmonary edema **NEURO**: neurotoxicity, (may lead to acute or chronic mercury intoxication → ≤)	**NEURO**: neurotoxicity

Key to major body systems:
CVS: Cardiovascular system
DERM: Dermatological system (skin and adnexa)
GENERAL: Includes multiorgan, metabolic, and endocrine effects
GIT: Gastrointestinal system
GU/REPRO: Genitourinary and reproductive systems
HEENT: "Head, eyes, ears, nose and throat," including the scalp, face, eyes, and adnexa, ears, nasal cavities, sinuses, pharynx, oral cavity, or dentition
MSS: Musculoskeletal system
NEURO: Neurological system
RESP: Respiratory system
→ ≤ = Potential cause of mortality

respirable portion of tephra refers to particles less than 10 μm in diameter, and those under 2.5 μm may penetrate further into the lungs into the terminal bronchi and alveoli. The proportion of respirable ash varies greatly across eruptions. Higher levels of total suspended particles (TSP) caused by ash-fall may escacerbate some pre-existing respiratory complaints including asthma and bronchitis (Baxter et al., 1981; Baxter, 1983; Yamo et al., 1990). An increase in the frequency of bronchitis was also detected after the 1996 Mount

Ruapehu eruption in New Zealand, despite the small amount of ash and the relatively low respirable portion (Hickling et al., 1999). The probable mechanism by which ash produces such respiratory symptoms is by provoking hypersecretion of mucus and bronchoconstriction (narrowing of the air passages): both are reversible, however, and diminish once exposure ceases (Buist, 1988).

Dispersal of tephra may also produce health effects over a longer duration, and respiratory diseases are among the most common subacute, recurring, and chronic problems in this context. Baxter et al. (1983) reported higher rates of asthma and bronchitis exacerbations for over 3 months after the 1981 Mount St. Helens eruption. There appeared to be a clear correlation with respirable ash, which persisted in the ambient air of surrounding settlements long after the eruption. This effect was less prominent following the Mount Pinatubo eruption, despite the high volume of tephra. This lower degree of respiratory morbidity may have occurred because subsequent rainfall acted to settle volcanic dust and limit opportunities for respiration of particles. Alternatively, the chemical nature of the respirable portion of ash may have differed significantly between the two eruptions: such differences are difficult to predict and quantify, but may be important with respect to health protection (see Section VII, this volume).

In terms of chronic lung pathogenesis, one of the most troublesome compounds produced by volcanic activity is silica. Certain forms of silica, such as cristobalite and tridymite, occur in lava and may be formed when amorphous silica or quartz is heated to high temperatures. Inhalation of fine particles of crystalline silica, including quartz, is a well-established cause of both acute and chronic inflammatory reactions in lung tissue (see also Chapter 18, this volume). If large quantities of silica are inhaled over a short period, exposure may result in the appearance of inflammatory cells in the interstitium and filling of alveoli with proteinaceous material. If exposure to silica is prolonged, the particles may evoke a chronic inflammatory reaction which recurs until the lung tissue displays extensive signs of fibrosis (tissue repair with collagen fibers). In the chronic phase of silicosis, the classical pathological features of the disease are lung nodules, which are masses of concentric collagen fibers with a rim of silica dust and macrophages (white cells) containing trapped silica particles. These whorled masses slowly enlarge and may impinge on airways and the vasculature of the lung.

The mechanisms by which silica evokes this reaction are probably related to the chemical reactivity of the particle surface. SiOH groups form bonds with, and ultimately damage, proteins and phospholipids embedded in cellular membranes. Cell death may result. Furthermore, silica interacts with macrophages, causing the activation of these cells and the release of inflammatory mediators, such as interleukin-1. This process triggers the migration of other inflammatory cells to the alveoli, which amplifies the body's response and encourages intense deposition of collagen fibers.

During the Mount St. Helens eruption, the free silica content was 3–7% of the "sub-10 μm" (that is, potentially respirable) fraction of ash. Due to the relatively low free silica content, it was estimated that the chances of silicosis developing even in high-exposure groups (such as cleanup crews or loggers) was negligible (Baxter et al., 1986). The risk of silicosis is thought to be much greater for the residents remaining on the island of Montserrat, which has experienced many years of eruptions starting from 1995. At this location, the free silica content has been much higher (10–24% by weight of the sub-10 μm fraction), and the exposure is much more prolonged (Baxter et al., 1999; Baxter, 2001). Populations living in close proximity to Mount Pinatubo are also thought to be vulnerable following exposure to high levels of respirable silica.

Other particles of volcanic origin, including silicates such as talc, may also cause prolonged fibrotic changes, and these are included in a broader classification of lung diseases known as pneumoconioses. (This term encompasses inflammatory lung reactions that occur from contact with respirable mineral dusts, but that are distinct from asthma or bronchitis that may follow mineral dust exposure.) Following the Mount St. Helens eruption, workers exposed to tephra over a long duration, such as loggers and road gangs around the eruptive site, were examined for the advent of pneumoconiosis. One deceased victim appeared to have been suffering from distinctive granulomatous lesions of the lungs, a chronic inflammatory reaction similar to that observed in rats exposed to respirable tephra (Green et al., 1982). However, a longitudinal study of survivors found that respiratory symptoms largely disappeared three years after exposure, and no permanent changes in chest x-rays were observed relating to ash exposure (Buist et al., 1986). Animal studies indicated that volcanic ash from various eruption sites (including Mount St. Helens ash) displayed less cytotoxicity than free silica, and was more readily phagocytosed by macrophages in the alveoli (Vallyathan et al., 1984).

It is has also been suggested that respirable dusts from volcanic eruptions may have carcinogenic effects. Apart from the example of radon (described below), ash may contain fibrous zeolites (including erionite). These asbestiform minerals have been linked to endemic pleural disease and a high rate of mesotheliomas (a highly lethal malignancy which typically infiltrates the pleura or peritoneum) (Rohl et al., 1982).

B. Hydrospheric Dispersal of Toxic Compounds

During and following volcanic activity, tephra particles may affect water supplies in two principal ways. First, tephra particles (such as ash or dust) may be deposited onto bodies of water, which include irrigation or filtration plants, thus rendering the water highly turbid and unusable. Secondly, tephra may carry a variety of adsorbed chemicals. Plumes of ash and dust, which may cover hundreds of kilometers, may effectively disperse such "stowaway" toxic agents. After the ash settles to the ground, these toxins may be dissolved—often by rain water—and thus leach into the environment. The dominant chemicals which may adsorb to tephra, and thus act as leachates, include chlorine (Cl), sulfur (S) compounds, sodium (Na), calcium (Ca), potassium (K), magnesium (Mg), and fluorine (F). Other elements that are present in smaller quantities are manganese (Mn), zinc (Zn), copper (Cu), barium (Ba), selenium (Se), bromine (Br), boron (B), aluminum (Al), silicon (Si), and iron (Fe). Furthermore, even light tephra-falls may have significant effects on water pH. During the 1953 eruption of Mount Spurr (Alaska), a 3–6 mm tephra fall on Anchorage caused the pH of the town water supply to fall to 4.5, returning to 7.9 after a few hours (Blong, 1984). Such excursions in pH and turbidity can alter the chlorine demand significantly at treatment plants, adding a significant, albeit indirect, microbial water hazard.

Some of these elements and compounds have safety levels established in drinking water and could potentially cause harm if ingested in quantities exceeding these concentrations. There is a paucity of conclusive studies relating to the health effects of ash leachates, although some effects may be inferred from accidental or occupational exposures. An example is provided by fluoride, a relatively common volcanic product, for which numerous cases of toxicity have been reported in various settings.

Fluoride is rapidly and effectively absorbed following ingestion (up to 90% absorption from the gastrointestinal tract), and most of the ionic fluoride remaining in the body has a high affinity for calcified tissues, particularly bones and developing teeth. Despite its use-fulness in reducing dental caries, it may produce toxicity if ingested in high concentrations. Acute fluoride poisoning can result in death, such as following the accidental contamination of public water systems (Gessner, 1994). Acute symptoms of excessive fluoride ingestion include nausea, vomiting, excessive sweating, and abdominal discomfort (probably from the formation of the corrosive hydrofluoric acid in the stomach) (Grandjean, 1982). Clinical and laboratory evidence indicates that kidney damage (nephritis) may also result, thus disturbing water and sodium balances in affected patients. Higher concentrations of fluoride may produce cardiovascular convulsions, collapse, and shock partly as a result of fluid loss from vomiting and gastrointestinal bleeding. Fluoride may also lower levels of calcium in the blood, possibly by a process of precipitation in which insoluble CaF_2 is formed. Painful and involuntary muscle contractions may result from the depletion of available calcium. Although not recorded in humans following volcanic eruptions, these signs and symptoms are well-established in grazing livestock in Iceland (see below) (see also Chapter 12, this volume).

Chronic ingestion of fluoride may cause a number of low-grade symptoms, including loss of appetite, headache, vertigo, and joint pain. The most characteristic effects, however, involve the dentition and skeletal system. Although not directly a result of recent ash leaching, elevated fluoride levels in potable water adjacent to the Furnas volcano in the Azores have resulted in dental fluorosis, a condition characterized by mottling of the teeth (Baxter & Coutinho, 1999b). Mottling—the appearance of chalky white patches distributed irregularly over the teeth—develops if chronic fluorosis occurs during the development phase of permanent dentition. The enamel is structurally weak, and may eventually become pitted, in which case the affected teeth acquire a corroded appearance. Such dental changes are often the only signs of chronic fluorosis.

In areas where waters are highly fluoridated, however, more serious problems may emerge. For example, teeth and gums may degenerate. Skeletal symptoms start to appear, which include back pain and limitation of spinal movement. Studies of fluoride exposure in various contexts have indicated that, as a result of the fixation of bone calcium by fluorine, bones undergo a number of changes: increased hardening and rigidity (sclerosis) of bones; the development of bony "projections" (osteophytes and exostoses) on the ribs, pelvis, and vertebrae; increased thickness of long bones; and calcification of the ligaments (Grandjean, 1982). The results of long-term fluoride exposure may thus be crippling.

Other evidence relating to emission of ash rich in soluble fluorine have been obtained from studies of effects on other mammals. It is known that the ingestion of high concentrations of fluoride is highly toxic to livestock (Baxter, 1983). Following the 1947 eruption of the Icelandic volcano Hekla, a mere 1-mm deposition of fluoride-rich ash was sufficient to kill thousands of sheep. Icelanders in the eighteenth century also suffered from the long-term consequences of fluorosis following the eruption of Laki (1783). During this eruption, in which over 140 cones formed along a 27-km fissure, it is estimated that eight megatons of fluoride compounds were discharged. (Fluoride may also be expelled from volcanic vents as unadsorbed gas, including fluorine, F_2, and HF; volcanic gases are discussed subsequently.) The deposition of high concentrations of fluorine on pastures and waterways proved fatal for numerous animals. Death occurred from at least two causes. First, the volcanic discharge resulted in extensive crop damage, and thus stock starved from loss of feed. In addition, ingestion of fluoride from pastures and waterways caused direct toxicity (see also Chapter 20, this volume).

Records of the period indicate that fluoride may also have directly poisoned the human population. A contemporary account describes a curious disorder affecting farmers living around the eruption site. These symptoms included severe swelling in many joints that was associated with painful cramps. The most striking effects, however, appeared to affect the oral cavity: sores appeared on the palates and tongues of the victims, and eventually teeth and blackened portions of gum fell from their mouths. Although it is difficult to exclude the effects of scurvy, Blong (1984) suggested that the illness may be the result of toxic fluorosis. Recent studies of meltwater around ash particles from Mount Hekla, Iceland, show fluoride levels up to 2000 ppm. Although rapidly diluted to about 200 ppm in standing water and 20 ppm in running surface water, these fluoride concentrations could readily produce symptomatic fluorosis if ingested by humans (N. Oskarsson, personal communication).

Determining areas at risk from fluoride leachates remains difficult, and the relationship between concentrations and distance from the volcanic vents is highly variable. For example, during the Mount Ruapehu eruption in New Zealand in 1995, it was concluded that the levels of leachable fluoride in adjacent water supplies did not pose a threat to human health (Weinstein & Patel, 1997). Following the Popocatepetl eruption on May 12, 1997, however, levels of fluoride were maximal at a distance of 13.4 km from the eruptive site. It was predicted that had such ash entered the water supplies, fluoride levels would have exceeded the national standard of 1.5 mg/L (Armienta et al., 2001). Such results suggest that the possibility of significant water contamination by fluoride should be considered during and following the eruptive phase.

C. Toxic Compounds in the Soil and Food Chain

Volcanic material may also enter the human body by direct ingestion of contaminated foods or soils. As with the hydrospheric effects, toxic chemicals may be carried into the soil or food supplies by adsorption to tephra particles and subsequent leaching. (This is discussed in Section III. Emissions of volcanic gases also contribute to chemical deposition.) Usually ingestion of these compounds is unintentional, but humans may consume volcanic products such as mineral-rich muds for therapeutic or nutritional reasons.

Selenium is an example of a volcanic compound that may potentially produce health effects through food-borne exposure. This element may be emitted in substantial concentrations (e.g., 630 kg of aerosolized selenium per day during a 1976 eruption of Mount Etna [Faivre-Pierret & Le Guern, 1983]), thus providing unusually large Se loads for soils and edible plants. Chronic toxicity from the consumption of foods with high levels of selenium have been reported which include cases with probable associated fluorosis in some districts in China (Yang et al., 1988). Signs of chronic exposure include loss of hair and nails, skin lesions, and abnormalities of the nervous system. Elevated selenium levels may increase the risk of dental caries. The exact mechanism of selenium toxicity remains uncertain, although it has been suggested that the element may cause disruption of sulfur metabolism, inhibition of protein synthesis, or oxidation of sulfhydryl groups. The role of selenium and cancer is debated. Some studies indicate that there is a protective effect against cancer, and others have argued that selenium displays pro-oxidant—and thus possibly carcinogenic—effects (Spallholz et al., 1994) (see also Chapter 15, this volume).

The effects of tephra on the food chain and the hydrosphere intersect through the damage caused to aquatic life. During the 1979 eruption of Karkar, Papua New Guinea, heavy tephra-falls (and possibly small lahars) tainted the headwaters of surrounding rivers, which resulted in the death or contamination of aquatic life. However, edible fish, eels, and prawns continued to be consumed by local farmers, despite complaints of the sulfurous taste (Blong, 1984). The health effects of this form of food contamination are unknown.

III. The Health Effects of Volcanic Gas Emissions

A. Description of Volcanic Gases

The gaseous substances produced by volcanic activity are varied, and may be classified as follows:

1. Gases and vapors: The gaseous state of an element which normally exists in a liquid or solid form and can be readily reverted to this form by decreases in temperature or increases in pressure (such as mercury)
2. Aerosols: Droplets or particles suspended in a gaseous medium
3. Fumes: Aerosols of solid particles, usually less than 0.1 μm in size, usually formed by escape of volatiles from molten materials
4. Smoke: Volatile gases and particles, usually less than 0.5 μm in size, produced by combustion (Kizer, 1984).

The pathophysiological effects of the compounds discussed are summarized in Table I.

Steam, from both magmatic and superficial sources (such as overlying lakes or groundwater), is the most common volcanic gas. Other, often very toxic, gases are also emitted during eruptive events and may impact on human health (Figure 4). There are numerous accounts

FIGURE 4   Volcanic gas emissions. (Photo: Michael Durand).

of volcanic gases causing death. In Japan, for example, 49 people have been killed over the last 50 years by becoming overwhelmed by volcanic gas emissions (Hirabayashi, 1999).

Among the most notorious gases are those which are heavier than air, such as CO_2 and H_2S. These may pool at ground level and result in asphyxia. HF, hydrochloric acid (HCl), hydrogen (H), helium (He), carbon monoxide (CO), and radon may also be produced in considerable quantities. Vaporized metals, such as mercury (Hg), are also found in plumes. Gases may be released even in the absence of obvious volcanic activity (degassing).

In assessing the impact of volcanic gases, it is not always possible to separate the toxicological effects of each gas released during a particular eruption. Numerous gases may be emitted simultaneously, or the gaseous components or concentrations may vary over time as the eruptive process evolves. For example, the sequence of activity around Rabaul, Papua New Guinea, over the last seventy years included: relatively quiet solfataric emissions of CO_2 and H_2S (prior to 1937); a violent steam explosion resulting in over 500 fatalities (between May and June 2, 1937); the appearance of vigorous vents producing large clouds of sulfur dioxide, with periodic emissions of hydrogen chloride, gradually increasing until another major eruption commenced (1937–1941); a release of CO_2 in a pit crater at the Tavurvur site, causing the death of six persons who had been collecting eggs (1990); the eruption of the two major vents, producing large quantities of ejecta, widespread building collapse, the evacuation of over 50,000 residents from Rabaul, and an accompanying SO_2 plume with an estimated size of 45,000 km², with a maximum SO_2 mass of 80 kilotons (1994); and the emission of a large bubble of flammable gas, presumed to be methane, which ignited and initiated bush fires (1997).

Although not as dramatic as full-scale eruptions, other types of geothermal activity (including hot springs and fumaroles) also generate a variety of toxic gases (Sparks et al., 1997). Because such activity may often continue unabated over prolonged periods, and because the benefits of hot springs may encourage the presence of humans, there is a risk of toxic exposure. For example, fumaroles are very numerous in some areas (such as Yellowstone Park) and may emit high levels of numerous gases including CO_2, SO_2, HCl, and H_2S. A full discussion of geothermal toxins is beyond the scope of this chapter.

In terms of adverse impacts on human health, volcanic gases may be classified into the following groups: gases with *irritant effects* on the respiratory system; those which act as *inert asphyxiants*; and those which combine both properties and act as *noxious asphyxiants*.

B. Inert Asphyxiants

Carbon dioxide, CO_2, illustrates the effect of an inert asphyxiant gas: it replaces oxygen, but does not have a directly toxic effect on biological tissue. Concentrations of CO_2 are particularly high near emission vents (Faivre-Pierret & Le Guern, 1983). The degassing of volcanic soil may result in the collection of carbon dioxide in cellars, huts, and in low-lying areas (Baubron & Toutain, 1990; Baxter & Coutinho, 1999). Low concentrations (e.g., under 5%) produce accelerated breathing, and often feelings of discomfort, by direct activation of the respiratory centers in the brain. Headache and vertigo are early symptoms. If sufficient concentrations are reached (for example, concentrations of 7–10% for a few minutes), fainting occurs. Elevated levels of CO_2 in the bloodstream (hypercapnia) eventually result in circulatory failure and death from acidosis.

In Java's Dieng Volcanic Complex, it is believed that emissions of CO_2 from a fissure resulted in the deaths of 149 people (Le Guern & Faivre-Pierret, 1982). The dramatic effect of CO_2 emissions is also illustrated by two disasters in the Cameroon at Lake Monoun in 1984 and at Lake Nyos in 1986. It is thought that large volumes of CO_2 emerged from these crater lakes and, carried downward by gravity, engulfed whole villages in their path. As a result, 37 people were asphyxiated near Lake Monoun, and the death toll reached 1700 at Lake Nyos. The survivors from the Lake Nyos disaster reported falling in a deep state of unconsciousness for up to 36 hours. It appears that no long-term respiratory effects occurred in the survivors, although some sustained burns by falling into cooking fires during a period of CO_2-induced coma (Wagner et al., 1988; Afane et al., 1996). In areas with the highest CO_2 concentrations, a variety of animal life ranging from insects to livestock also perished. Although the exact mechanism of release is unclear, it is probable that the gas release was not in fact secondary to underlying volcanic activity. A widely accepted hypothesis suggests that soda springs deep in these crater lakes create accumulated pockets of CO_2, which may be abruptly emitted at periodic intervals.

C. Irritant Gases

Volcanic gases which have primarily *irritative* (directly injurious) *effects* include the hydrogen halides, HF and HCl. At low doses, eye and throat irritation may occur. At higher levels, both may cause ulceration of the respiratory tract upon inhalation and corrosive burns upon contact with the skin or mucous membranes. Cutaneous burns from contact with HF are particularly severe and notoriously slow to heal. Fatalities may occur from pulmonary edema and laryngeal spasm (discussed below).

Although there are few clinical accounts of direct toxicity from these gases in a volcanic context, an indication of likely health effects is provided by fumes from Hawaii's fumaroles and basaltic lava, which contain high concentrations of both HCl, and, to a lesser degree, HF (Murata, 1966; Kullmann et al., 1994). As discussed in Section V, HCl and HF concentrations are often highest in dense plumes arising close to the ocean, from which sea winds carry mist clouds to adjacent villages of Kalapana and the Hawaii Volcanoes National Park. The HCl-/HF-acidic aerosol may exacerbate pre-existing lung disease (Ostro et al., 1991), even at great distances from volcanic vents (Mannino et al., 1996). On the island of Ambrym, Vanuatu, cutaneous burns resulted from acid rain following a 1979 emission of eruptive gases high in HCl and sulfur compounds. Gastric upsets were also reported from the ingestion of acidic cistern water (Scientific Event Alert Network, 1989).

At a pathophysiological level, irritant gases affect the respiratory tract in a number of ways. The relatively more soluble gases, which include HF and HCl, tend to be removed by mucus linings before they reach the alveoli. Therefore, such gases predominantly cause inflammation in the upper airways, which results in symptoms such as cough and reactive bronchoconstriction. These effects, although often relatively short-lived, may be severe, particularly in people with hyperreactive airways (such as asthma sufferers). Less soluble volcanic gases, including hydrogen sulfide and mercury vapor (both discussed below), are less likely to be "cleared" by the mucus secretions of the upper airways. Thus, there is greater risk of penetration to the level of the smaller airways and alveoli.

As the concentration of respired irritant gases rises, there is a greater chance that alveolar damage will result. The most vulnerable targets are the epithelial cells lining the airways, the mucosal tissues underlying these cells, and endothelial cells lining the pulmonary blood vessels (including the capillaries of the alveoli). Damage to these cells results in an increase in vessel permeability. Ultimately, the alveoli fill with protein-rich fluid from plasma, which interrupts the process of effective gas exchange. In addition, there may be denudation and sloughing of alveolar epithelium and the mucous membranes of bronchiolar walls.

During the acute phase, the spectrum of disease caused by irritant gases is highly variable and includes pneumonitis, which results in a dry cough, shortness of breath, and evidence of patches of inflammation on chest radiography. The condition may often resolve without long-term sequelae such as pulmonary edema (the accumulation of unwanted fluid in the respiratory organs, for which there are numerous other etiologies). In other cases, pulmonary edema emerges within a day of exposure, and may be associated with the formation of mucus plugs and collapse of areas of the lung. Because of compromised oxygen intake, patients may experience breathlessness and acute (or adult) respiratory distress syndrome (ARDS). This syndrome is associated with severe alveolar injury, edema, and hemorrhage. It has a mortality rate of around 50%, and represents the end point for a number of disease processes apart from toxic inhalation.

If the victim survives the acute stage of gas exposure, a number of further clinical outcomes may occur. Often the symptoms may rapidly reverse once the gas is removed and full recovery occurs. If the gas has produced severe epithelial damage, however, cellular regeneration may take weeks. A secondary effect of this alveolar disruption is that the lung tissues become more vulnerable to bacterial invasion, and thus chest infections may intervene in the post-exposure period. Occasionally, toxic gas exposure results in permanent fibrosis (scarring) of lung tissue. Another potential outcome from toxic inhalation (described in greater detail below) is prolonged airway hyperreactivity, although the specific cause of this disorder remains uncertain.

Sulfur dioxide, SO_2, is well-established as a cause of acute and chronic disease. Both the gas, and the sulfuric acid aerosols into which it forms, are highly irritant, particularly to the eyes, nasal passages, throat, and respiratory tract. High SO_2 concentrations may also act to cause asphyxia, although in volcanic contexts the effects appear primarily irritative.

Sulfur dioxide exposure may provoke exacerbations of asthma, even at low concentrations. Increased airway resistance has been noted in asthmatics exposed to SO_2 at concentrations of 0.5 ppm when exercising, and at 1 ppm during rest. In non-asthmatic controls, an increase in airway resistance has been provoked at 5 ppm (Bethel et al., 1983). Cough and eye irritation occurs at concentrations of 20 ppm or above. At higher doses, pneumonitis, pulmonary edema, ARDS, and ultimately death may result.

In the context of volcanically active sites, SO_2 concentrations of 1 ppm have been recorded far downwind (30 km) from the Masaya Volcano in Nicaragua (Baxter et al., 1982a). The effects of long-term emissions have been examined at Kilauea, Hawaii, which has been erupting for 15 years and (in 2001) continues to eject 1500 tons of sulfur dioxide per day. Episodes of increased SO_2 in the ambient air have exceeded health standards 80 times in the last 15 years and appear correlated to ongoing eye irritations, throat pain, and respiratory problems, including asthma exacerbations (Elias & Sutton, 2001).

At the histopathological level, sulfur dioxide has a diverse range of respiratory effects. Although larger aerosols are filtered out by the nose and nasopharynx and are rapidly removed by mucociliary clearance, smaller SO_2 particles (particularly those less than 10 μm in diameter) may be deposited in deeper airways. Sulfur dioxide also acts to increase levels of mucus secretion and viscosity. Animal studies also indicate that the gas impairs the ability of macrophages to destroy bacteria in the alveoli, thus increasing the risk of respiratory infection.

Apart from affecting mucus production, SO_2 may produce airway narrowing through other mechanisms. As mentioned, people with high levels of airway hyperresponsivness, such as those with asthma and/or atopy (tendencies to allergic responses), are particularly vulnerable. Sulfur dioxide may provoke the recruitment of inflammatory cells, including histamine-secreting mast cells and eosinophils, which persist long after exposure ceases. Such cells contribute to the risk of recurrent airway constriction, particularly if provocation with gases (or other agents and allergens) recurs. This bronchoconstrictive response may enter a chronic phase (sometimes after only one exposure of a particular irritant gas). Other irritant gases and aerosols, such as nitrogen dioxide (discussed below), may precipitate a similar response. If such sequelae occur without a previous history of asthma or allergy, some authorities apply the term reactive airways dysfunction syndrome (RADS). Other changes from SO_2 exposure include chronic neutrophil infiltration and edema (fluid accumulation) in the mucosa, which is more distinctive of chronic bronchitis.

Two related, but less common, syndromes resulting from exposure to SO_2 (and other volcanic products, such as mercury vapor) are bronchiolitis obliterans and bronchiolitis obliterans-organizing pneumonia (BOOP). Both conditions are related to a delayed inflammatory and reparative process occurring some time after the initial injury. Bronchiolitis obliterans is associated with the appearance of plugs of granulation tissue (capillary-rich tissue associated with healing) in small airways, together with the presence of fibrous scarring. A similar process occurs in BOOP, except that the alveoli and their adjacent tissues are also involved. The different patterns may be associated with variation in host responses to the toxic insult (Epler et al., 1985).

In addition to acting directly, irritant gases may create health problems by their interaction with other atmospheric processes. For example, droplets of rainfall or mud may pass through the irritant gases present in toxic clouds or ash plumes, thus creating acid rain. The health effects of this process were experienced by people on Kodiak Island, who suffered from falls of both ash and sulfuric acid rain following the Katmai eruption on the Alaskan mainland in 1912. A number of Kodiak islanders, 160 km downwind from the active vent, suffered from "stinging burns" when this rain contacted their lips or skin (Blong, 1984). Acid rains in other sites have also resulted in eye irritation (e.g., with Masaya in Nicaragua [Baxter, 1982a]) and apparently hair loss (during the 1917 eruption of Boqueron). A secondary health risk is posed by the interaction between acid rains and heavy metals: acid rain has reacted with zinc in galvanized roofs and tainted water supplies as a result (Baxter et al., 1982b).

As with tephra fallout, chemicals dispersed with volcanic gas emissions may affect water and food supplies. During the 1783 eruption of Laki in Iceland, apart from the high levels of expelled fluorine, an estimated 150 megatons of sulfur dioxide was discharged. This vast quantity of gas, and the aerosolized sulfuric acid it formed, had destructive consequences for vast tracts of surrounding pasture lands. The event tipped the balance in the already marginal farming environment of Iceland. The ensuing period of crop damage resulted in massive livestock losses: half of all the island's cattle and horses and four-fifths of the sheep perished. One-fifth of Iceland's population, around 10,000 people, died as a consequence of famine (called the haze famine from the persistent presence of sulfur compounds in the atmosphere) with toxic fluorosis probably also contributing to some fatalities.

D. Noxious Asphyxiants

The pungent gas hydrogen sulfide, H_2S, is a noxious asphyxiant: that is, it acts as both an asphyxiant and a powerful irritant. Its metabolic effect is to inhibit cytochrome oxidase, one of the enzymatic drivers of cellular metabolism (Jappinen et al., 1990).

At low concentrations, H_2S may cause irritation of the conjunctivae and mucous membranes. Short exposures at concentrations of 2 ppm have not shown any statistically significant effects on respiratory function (Jappinen et al., 1990), and other effects of chronic exposure at low levels form the basis for ongoing studies (P. Shoemack, personal communication). However, once the H_2S concentration in ambient air increases, the effects become more definitive, and exposure even for a few seconds may be fatal. For example, at the Kusatsu-Shirane crater in Japan, a high school teacher and two students were overcome and killed by H_2S emitted from nearby fumaroles that had concentrated in a bowl-shaped depression (Scientific Event Alert Network, 1989). Early signs of poisoning include headaches, ocular and respiratory irritation, and loss of smell (anosmia). Apart from these effects, inhalation of the gas also directly damages the respiratory tract and precipitates pulmonary edema in the lungs. At 1000 ppm, fainting occurs. Ultimately, H_2S causes cessation of breathing by direct action on the respiratory centers of the brain. Those who recover from poisoning may suffer from long-term neuropsychological effects. These sequelae are probably related to acute hypoxic effects on the brain either during poisoning or subsequently from pulmonary damage.

E. Health Risks from Other Volcanic Gases

Information relating to the effects of exposures to other volcanic gases is limited. After symptomatic (or fatal) episodes of inhalation, it may be difficult to determine the nature of the gas or combinations of gases that were present, and whether the concentrations were high enough to cause illness. Other gases produced during eruptive or degassing events that may *potentially* reach concentrations sufficient to produce symptoms in humans include: carbon monoxide, CO; nitrogen dioxide, NO_2; carbon disulfide, CS_2; methane, CH_4; and ammonia, NH_4.

Carbon monoxide—Like hydrogen sulfide, CO is an example of a noxious asphyxiant. Although concentrations in plumes are usually low, the gas may reach high concentrations in certain eruptions and near fumaroles (Tazieff & Sabroux, 1983; Williams & Moore, 1983). CO is toxic to humans in small doses. Once inhaled and absorbed, it rapidly permeates across the membrane of red blood cells and binds to the iron component of hemoglobin. Its affinity for hemoglobin is profound, over 200 times greater than oxygen, and it rapidly reduces the oxygen-carrying capacity of blood. Symptomatically, low levels of blood saturation with CO produce headache, nausea, and impaired judgment and dexterity. As the percentage saturation exceeds 30–40%, severe nausea, vomiting, confusion, hyperventilation, and collapse ensue. Coma and death occur if treatment is not provided rapidly. The characteristic "cherry red" skin associated with CO poisoning occurs because of the persistent saturation of both arterial and venous blood with this "hemoglobin-loving" gas (which forms carboxyhemoglobin). Survivors of CO poisoning may suffer from permanent neurological and behavioral symptoms, which include disorientation, mood changes, and movement disorders.

Nitrogen dioxide—Significant concentrations of NO_2 have been detected in plumes from the Mount St. Helens eruption (Olsen & Fruchter, 1986). If high concentrations are inhaled, hydrolysis of NO_2 results in the production of nitric acid, particularly in the lower respiratory tract (including the alveoli). Nitric acid in turn produces nitrates and nitrites, both of which are capable of producing cytotoxicity and disruption of cellular membranes. The clinical outcomes of NO_2 inhalation are in some respects comparable to those of contact with SO_2. Acutely, victims may suffer from chemical pneumonitis, pulmonary edema, ARDS, or fatal lung injury. Susceptible individuals may suffer from exacerbations of asthma, increased need for bronchodilators, and decreased pulmonary function (Moseholm et al., 1993). As with SO_2, emergency room visits and hospital admissions for asthma are also increased by exposure to NO_2. Delayed or chronic outcomes of toxic inhalation, which may appear even after apparent "recovery," include chronic airway obstruction and bronchiolitis obliterans.

Carbon disulfide—CS_2 has been detected in volcanic plumes, although concentrations are usually low. This gas acts as a powerful toxin and produces headaches, muscular weakness, and delirium.

Methane—Methane, together with CO and CO_2, was among the gases which hampered rescue efforts during the 1973 eruption on the island of Heimaey, Iceland (Williams & Moore, 1983). It may also be concentrated around fumaroles. Primarily, it is an inert asphyxiant and acts by replacing oxygen in breathed air.

Ammonia—Exposure to this highly water-soluble gas results in severe irritation to the eyes, skin, and upper airways. Injuries are produced by thermal and chemical (alkali) burns as ammonia reacts to form hydroxyl ions on exposed mucosal surfaces.

IV. VOLCANIC DISPERSAL OF METALS AND TRACE ELEMENTS

The dispersal of toxins such as fluoride and selenium have been discussed above. In addition, there may be widespread dispersal of other elements and metals with identified risks for humans. For example, using a plume dispersal model for a 1976 eruption of Mount Etna, a variety of toxic aerosols was estimated to have exceeded permissible air concentrations at a distance of 10 km from the vent. These included lead (Pb), mercury (Hg), copper (Cu), zinc (Zn), selenium (Se), and cadmium (Cd). Arsenic (As) exceeded recommended concentrations at a distance of 5 km (Faivre-Pierret & Le Guern, 1983). Daily aerosol outputs were also estimated, which included 360 kg of lead, 110 kg of arsenic, 75 kg of mercury, and 28 kg of cadmium per day. These last four elements, which have well-established toxic effects even in small doses, are discussed briefly below as well as in greater detail in other chapters in this book.

Metal vapors, such as those produced by mercury, may act directly as an irritant gas. It has been suggested that mercury is sometimes present in concentrations sufficient to be hazardous if emissions were inhaled over a prolonged period of time (Baxter, 1983). Mercury vapors act to cause pulmonary edema and bronchitis and are neurotoxic once absorbed.

Arsenic can be deposited in soil or water (including seepage into subterranean wells) following volcanic or low-level geothermal activity (Welch et al., 1988). Chronic arsenic poisoning may affect many organ systems. For example, in affected populations in Taiwan and India, ingestion of drinking water high in arsenic has been associated with a variety of skin lesions, which include increased or decreased pigmentation and keratosis. Gangrene has also been reported (blackfoot disease). The carcinogenic role of arsenic has been suggested in a range of studies: exposure increases the risk of cancer of the skin, lung, liver, bladder, and kidneys (see also Chapter 11, this volume).

Lead produced by volcanic processes may potentially be inhaled or ingested in contaminated water and foodstuffs. The health effects, particularly on the neurological development of children, have been well-documented.

Like lead, cadmium enters the human body through inhalation or ingestion. Cadmium deposited in the soil crosses readily into plants and thus the food chain. Having entered the body, the toxin tends to be retained in the liver and kidneys and is excreted extremely slowly. The effective half-life of cadmium in humans may exceed 30 years. Cadmium thus has the potential to result in prolonged illness, and toxic effects may occur in virtually any organ. The more common manifestations of acute ingestion or inhalation include vomiting, cramps, respiratory difficulties, and ultimately loss of consciousness. Chronic effects include anemia and renal disorders. More controversial is the connection between elevated cadmium and hypertension.

It is important to note that the valency state (species) of these metals and metalloids can be a key determinant of their potential biological activity. Although we have no further details about the role of valency in relation to volcanic emissions, the issue is discussed in more detail for some elements in other chapters.

FIGURE 5 Lava from the Krafla fissure, Iceland. (Photo: Olle Selinus).

V. VOLCANISM AND RADIATION HAZARDS

Radioactive decay of volcanic material may also have consequences for human health. Ash may have high uranium content and carry adherent particles of radon, an alpha-radioactive gas that has been linked to the development of lung cancer (Baubron & Toutain, 1990). Exposure may also occur from use of volcanic material, such as for building. In the Azores, radon has been found in high concentrations in dwellings within the Furnas caldera. Radon is discussed in detail in Radon in Air and Water, this volume.

VI. TOXIC EXPOSURE WITH OTHER ERUPTIVE EVENTS

A. Lava Flows

One of the more visually dramatic outcomes of volcanism is the ejection of fluid or semi-fluid material, such as basaltic lava (Figure 5). In some locations (for example, Hawaii), eruptions may be associated with fountaining of molten material, in which globules of plastic lava are sprayed over a kilometer high. These may feed into lava lakes and lava flows that course away from the volcano. The direct threats to health posed by lava flows are primarily thermal injuries. Often fatalities occur because of unexpectedly rapid flows because

escape routes have been cut off, or from steam explosions created when the lava strikes a water source.

Lava flows may result in illness less directly by exposing humans to toxic chemicals. For example, the basaltic lava flows in Hawaii are often associated with the release of sulfur dioxide and aerosolized droplets of sulfuric acid. As discussed, HCl and (a lesser degree) HF may also be formed, particularly when molten lava strikes the ocean, thus creating falls of acid rain from the steam plume (Mannino et al., 1996). Lava may also act to taint subterranean wells by the process of leaching. The toxicity and health effects of compounds released from lava flows do not differ from those already discussed, although the exposure may at times be intense, such as in the area around an erupting crater. Emergency crews often work in close proximity to lava flows. On Heimaey in 1973, for example, crews spent many days on or near lava flows, applying cooling waters in a successful attempt to solidify and direct flows away from the main town.

B. Pyroclastic Flows

Pyroclastic flows are intensely hot ionized gas flows that contain dispersed fragments of debris, which may travel at speeds up to 200 km/h (Houghton et al., 1999). The exact composition and temperature varies greatly, but usually some of the fragmented rock within such flows is within the respirable range. The gas content will usually include H_2O (which may be superheated), CO_2, SO_2, and H_2S. Such flows present an immediate risk to humans close to the vent. With their considerable kinetic energy, these deadly "volcanic hurricanes" simultaneously sear and blast objects in their path. The fatality rate of those caught in such flows is usually extremely high, and the common causes of death include asphyxiation (often from burial), trauma, and severe burns (especially of the respiratory system). For example, during the 1902 eruption of Mount Pelée on the island of Martinique, such a superheated gas cloud rapidly enveloped the city of St. Pierre, resulting in over 30,000 fatalities.

Pyroclastic flows result in varying degrees of thermal injury to the skin ranging from superficial erythema, to deep penetration into the subcutaneous tissues, to the extreme of complete incineration. Victims are commonly described as appearing dried and "mummified," rather than charred (the outcome usually observed with fire injuries). Respiratory effects appear to occur as a result of intense heat, oxygen deficiency, ash inhalation, and toxicity of the gas. Asphyxia from plugs of ash in the upper airways was described as the cause of death in those caught in the flow of the Mount St. Helens eruption. Survivors of the devastating flows from Lamington, Papua New Guinea, in 1953, have suffered from symptoms suggestive of pharyngeal burns, including throat pain, shortness of breath, and inability to swallow (Taylor, 1958). Health effects subsequent to the acute injury include pneumonia, tracheobronchitis, and ARDS, presumably from irritation and secondary infection of injured respiratory tissues (Eisele et al., 1981).

Flows, surges, and debris avalanches also impact human health because of their capacity to disperse toxic compounds. For example, during the Mount Pinatubo eruption in 1991, pyroclastic flows contributed to the volcanic material that covered a wide area and filled surrounding valleys. Subsequent erosion, often triggered from monsoon rains, then acted to mobilize the volcanic chemicals for many years subsequently.

C. Volcanic Activity and Aquatic Environments

Apart from the interaction between tephra and the hydrosphere described above, volcanic and aquatic processes may intersect in other ways (Figure 6). Crater lakes, for example, can act as a reservoir of toxic compounds which, in some circumstances, may affect human populations. Some lakes, such as the Poas volcano, Costa Rica, sit atop a degassing system. At this site, steady emissions of sulfur dioxide pass through a shallow lake, which is often hot and intensely acidic (pH <1). These emissions, together with particles of

FIGURE 6 Crater lake formation adjacent to a volcanic vent. (Photo: Michael Durand).

rock dissolved in the acid lake water, are periodically dispersed out of the crater and have been linked to respiratory problems in downwind communities (Baxter, 1997). At Poas and Kiwa Ijen in Java, crater lakes have contaminated water supplies with fluoride and other elements (Baxter, 2001).

A fast-moving and potentially lethal consequence of volcanic eruptions is the lahar. These torrential flows of mud, water, and debris wash down the sides of the volcano, and may occur in association with crater lakes or the melting of snow and ice during or after eruptive events. For communities situated in the path of lahars, the opportunities for timely warnings may be limited—sometimes with lethal consequences. Lahars from some volcanic lakes, such as Kelut in Indonesia, may be hot and often acidic. Five thousand people died from a lahar generated from the 1919 eruption of Kalut. Those caught in the flow often suffer from drowning, suffocation while entrapped, or severe trauma from penetrating wounds and fractures (Baxter, 1990). Following the Nevado Del Ruiz lahar in Colombia, burns were also noted and may have been acidic in origin (Lowe, 1986).

For submarine eruptions, rafts of pumice and areas of discolored water on the surface may be the only evidence volcanic activity has occurred. At the other end of the hazard continuum, such events may produce tsunamis, which pose a major threat to coastal populations even at considerable distances from the vent. Submarine activity may manifest itself in other ways, as illustrated by the destruction of a Japanese research vessel *No. 5 Kaiyo-Maru* in 1952. The vessel had been traveling to monitor a submarine eruption 420 km south of Tokyo when radio contact with the vessel was lost.

Examination of the debris (which included boat fragments containing embedded rock particles) suggested that an explosion beneath the ship had caused the disaster, in which all 31 crew members were killed.

Apart from such infrequent events, little is known about the threat to health posed by submarine emissions of gases and particles. It is probable that most ejected material is thoroughly diluted in seawater and thus poses little risk for humans. However, in some regions, activity may occur near coastal populations. In the Antilles, for example, recently active submarine sites include Kick 'Em Jenny, situated 8 km north of Grenada, and other vents near Martinique, where periodic episodes of "boiling water" have been reported (Roobol & Smith, 1989). Along the shallow Reyjkanes oceanic ridge near Iceland, bubbles rich in methane and CO_2 may rise to the sea surface (German et al., 1994). In Papua New Guinea, weak tremors and increased fumarole emissions of SO_2 and HCl around the Kadovar volcano were associated with reddish discoloration of the surrounding seawater, which was possibly associated with iron hydroxide. Coastal villagers reported that this discoloration persisted from June 1976 to 1978 (Mori et al., 1989). These eruptions and emissions, although under water, are in effect near vent phenomena. The potential for toxicity remains, albeit tempered by the aquatic environment.

VII. VOLCANIC MONITORING AND HEALTH PROTECTION

HEKLA, perpetuis damnata estib. er nivib. horrendo boatu lapides evomit (Hekla, cursed with eternal fires and snow, vomits rocks and a hideous sound)

—an early surveillance report of the volcano Hekla (from 1585 map of Iceland)

A. Introduction

From the perspective of health protection, the purposes of volcanic monitoring are threefold: (1) to provide an early warning system of potential health hazards, thus providing a preparation period for resident populations, health services, transport services, electricity utilities, etc., that are located in the hazard zone; (2) to minimize illness from contaminated water supplies, tainted or fouled foodstuffs, and air pollutants, particularly for susceptible groups; and (3) to provide accurate records of adverse health events for use in epidemiological and clinical research.

The phases of volcanic monitoring may be usefully classified as follows:

- Pre-eruptive phase: This period extends from the elevation of alert levels above the baseline state, hence indicating increased likelihood of volcanic activity, to the actual start of the eruption.
- Eruptive phase: The period surrounding the eruptive event (or degassing episode), including event imminence, the primary volcanic event, and periods of ongoing volcanic activity.
- Post-eruptive phase: The period over which volcanic activity wanes, allowing recovery and rehabilitation measures to take precedence.

The current discussion will be based on the assumption that these phases provide the loci of intervention designed to protect the public health.

B. The Pre-Eruptive Phase

This phase is primarily directed at preparing at-risk populations for an impending volcanic event. Early volcanic activity is usually monitored by geologists and involves assessments of seismic activity, ground deformation, gas emissions, geophysical variables, and hydrology. Using hazard maps, the primary aim during the pre-eruptive stage is to predict the temporal and spatial pattern of an eruption, together with the scale and path of any pyroclastic flows, lahars, or lava flows. Disaster scenarios will usually include the need for, and possible extent of, evacuation from the danger zone. Secure locations and safe travel routes for displaced populations should also be considered.

An important strategy during the pre-eruptive phase is the dissemination of information regarding potential health effects and strategies for minimizing exposure. Ideally, this process of training and educating should be continuous (even in periods of quiescence), and only escalated in the event of increased alert levels. (In Japan, for example, evacuation drills are conducted routinely in areas at risk.) Baseline tests of water and air quality should be conducted.

From an epidemiological perspective, this period may be an opportunity for selection of groups for prospective studies (for example, pre-eruptive baseline screening of people with asthma who will be followed through and subsequent to the eruption and ash-fall).

C. The Eruptive Phase

Early warning systems should be established for affected communities in order to ensure that the public, and organizations designed to preserve the societal infrastructure, are prepared for the arrival of the eruptive phase. For example, hospital staff and general practitioners should be informed of possible consequences of tephra fallout, including an increased frequency of motor vehicle accidents, ocular problems, and exacerbations of asthma.

Where risks are minimal (such as from light ash-fall), it may not always be necessary to recommend major changes in daily routines. Some groups of susceptible individuals, such as those with pre-existing respiratory disease, may benefit from consultation with health professionals. Instruction of disease management in the event of ash-fall or increased levels of gaseous emissions should be initiated if appropriate. For some patients, such as those with asthma who may react to raised levels of sulfuric acid aerosol, medical advice may be as simple as recommending that bronchodilator medications (such as inhalers) are always readily at hand. In those with chronic illness, strategies for ongoing medical management are particularly important should evacuation become necessary. More general advice may be required each day, including avoidance of areas of likely ash-fall. Ongoing status reports of hazards and strategies to avoid risk should be provided using newspapers, radio and television broadcasts, and public notices. In at-risk areas, it is imperative that such communication plans *pre-exist* the volcanic event to avoid the need to generate the information on an ad hoc basis.

Table II summarizes health impacts of eruptive events and relevant health preservation strategies. The hazards are separated according to proximity to the eruptive site using the general categories near vent and distal from vent.

Given the range of health consequences from exposure to volcanic ash and dust, monitoring tephra dispersal is an important component of eruption-phase management. The size and composition of tephra constituents vary between volcanoes and from eruption to eruption; this makes it difficult to predict the chemical composition of any given volcanic product. Although the health effects of tephra fallout are often not immediately life-threatening for populations distant from the eruption site, resulting illness may be both widespread and chronic in nature. As discussed, ash may cause ocular injury and exacerbations of respiratory disease. Prior warnings of ash-fall, together with regular announcements on air quality, will enable susceptible individuals to minimize their exposure or to seek medical advice where appropriate. Rapid analysis of particulate levels is also necessary to address or alleviate public concerns regarding other chronic outcomes such as silicosis. As with communication strategies, it is imperative that protocols for these analyses pre-exist the eruption.

As described above, tephra may also have disruptive effects on water supplies. Following some eruptions there is a direct risk of toxicity, which includes acute fluorosis. Water quality may also be rapidly degraded by increased turbidity and fluctuations in pH levels. Apart from direct testing and pH monitoring, it may often be possible to anticipate potential effects on water quality by estimating volumes of ash falling in water catchment areas and by using results from rapidly performed leachate studies. Such analyses are of importance not only to populations drawing water from large catchment areas, but also to those using non-public water supplies (such as for crop irrigation) that are not routinely monitored for quality.

Other health hazards of ash include the mechanical effects (particularly of wet ash), which cause roofs to collapse resulting in trauma. Heavy ash-falls may also result in traffic accidents, a product of both poor visibility and ash coating on roads (Dent et al., 1995). Cleanup crews are particularly at risk, and falls from roof cleaners often cause greater morbidity than any direct effect of the eruption.

Table III summarizes the strategies for tephra monitoring, which indicates how such information may assist in health protection for vulnerable populations.

As discussed in the previous section, volcanic gas emissions—and the aerosols that are derived from them—also pose a major threat to health. Therefore, as with tephra, gas production at the vent should be monitored in order to alert downwind communities of the potential hazards on a daily basis. Information of gas (and tephra) dispersal production will usually be combined with weather predictions. Combining "climatic" and "volcanic" forecasts—including predictions for acid rain or combined ash and fog contributing to poor visibility—will allow communities to implement appropriate precautions on a day-by-day (or hour-by-hour) basis.

The monitoring of gases and pH often requires specialized equipment that may not be available for permanent use in all settlements at risk of exposure (Environment m.f.t., 1994). A possible solution may be the provision of a mobile air quality team that responds to predictions based on geologic and climatic data.

TABLE II. Comparison of Proximal (Near Vent) and Distal Health Impacts and Corresponding Health Preservation Strategies

Eruptive event	Consequence	Health impact	Health preservation strategies
Near vent			
Explosion	Blast, rock fragments, shock waves	Trauma, skin burns, lacerations	Evacuation, movement to secure shelters
	Lightning	Electrocution	
	Forest and bush fires, combustion of buildings and vehicles	Burns, smoke inhalation	
Pyroclasts and other thermal emissions	Pyroclastic flows	Skin and lung burns	Evacuation
	Ash flows and falls	Asphyxiation	
Drainage of crater lakes, melting ice, snow, or rain accompanying eruption	Mudflows, floods	Engulfing, drowning	Evacuation, diversion barriers
Lava	Lava flow	Engulfing and burns (rare)	Evacuation, diversion barriers
	Forest/bush fires		
		Burns	Evacuation
Gas Emissions H_2O, SO_2, CO, CO_2, H_2S, HF	Pooling in low lying areas and inhalation	Asphyxiation	Evacuation, avoidance
			Respiratory protective
		Airways constriction (exacerbation of asthma, COPD)	equipment
Radon	Radiation exposure	Lung cancer	Evacuation
Distal from vent			
Ash-fall	Ash/dust less than 10 μm in diameter	Asthma, exacerbation of pre-existing lung disease	Wear high-efficiency masks, minimize exposure, protect homes and offices from ash infiltration
	Siliceous dust	Silicosis if free silica content high and exposure prolonged	Respiratory protective equipment
	Water contamination with fluoride, possibly also heavy metals (e.g., cobalt, arsenic)	Gastrointestinal upset and electrolyte disturbance (may be fatal)	Avoid water that has not yet been approved, avoid surface water, use water from wells
	Food contamination	As above	
		Foreign bodies in eyes	Goggles for when heavily exposed
		Conjunctivitis, corneal abrasions	
Gas Emissions	Acid rain	Eye and skin irritation; possible toxic contamination	Protection during rainfall, avoid collection of rain water for drinking, especially from metal roofs, etc.
	Dispersal of irritant and/or asphyxiant gases	Exacerbation of pre-existing lung disease (especially asthma) exacerbation from acidic aerosols)	Prophylactic use of medication
		Mucosal and conjunctival irritation; suffocation	Avoidance/evacuation

Adapted from Baxter et al., 1986.

TABLE III.   Strategies for Monitoring Tephra Dispersal: A Health Protection Perspective

Monitored component of tephra fallout	Location of monitoring	Optimal time for monitoring	Health-related objectives of monitoring
Ash production and ash-fall predictions	At the vent	Immediately	a. To alert areas downwind of the probability and degree of ash-fall b. To anticipate and minimize the effects of traffic disruption and mechanical injury (e.g., roof collapse)
Mapping of ash-fall (isopach)	Throughout ash-fall area	As soon as possible after ash-fall in order to determine load of ash in water catchment areas, pastures, etc.	a. To maintain water quality, including monitoring of turbidity and pH b. To minimize the risk of ingesting toxic substances in drinking water (e.g., fluoride)
Proportion of particle size which is respirable ($<10\,\mu m$)	Priority to plume-vulnerable areas with greatest population density	Immediately after an eruption	To quantify the level of respirable ash to assess the risk of acute and chronic illness (e.g., silicosis)
PM-10 (overall measure of particulate matter $<10\,\mu m$, i.e., from volcanogenic and other sources, such as man-made pollutants)	a. Priority to plume-vulnerable areas with greatest population density b. Personal monitors on individuals exposed to much ash	a. Continuously and preferably at locations where humans are likely to receive maximum exposure b. Whenever exposed to ash	a. To monitor daily levels and provide information to individuals with respiratory disease (e.g., to stay indoors on certain days); also for research purposes in linking the respirable portion of ash to health effects b. To monitor total exposure and any associated health effects; this is especially important if the ash has a high free silica content
Free silica content of respirable portion	a. Priority to plume-vulnerable areas with greatest population density b. Individuals that will be exposed to high levels (e.g., cleanup workers)	a. Preliminary studies immediately after an eruption b. More detailed studies at later date	a. To assess, and hopefully alleviate, public anxiety concerning silicosis b. To determine the relative risk of silicosis for individuals exposed to varying concentrations and types of free silica
Particle morphology	At a range of distances from the vent	At a later date	To assess the possible requirements for long-term health monitoring and/or to initiate prospective studies (e.g., relationship between exposure to fibrous zeolites and pleural disease)
Leachate studies	a. In water catchment areas supplying public drinking water b. In water catchment areas of private water supplies (e.g., irrigation, wells)	a. Immediately after ash-fall b. Ideally as soon as possible; however, the assessment based on ash depths and leaching studies may be all that is practicable	a. To assess risk to water quality in the catchment region, and to initiate clearance of toxins and particles; this is often required urgently because water demand is likely to be high during the process of cleanup b. To assess potential effects on water quality

D. The Post-Eruptive Phase

As the post-eruptive phase commences, and recovery becomes possible, it is critical to continue monitoring for as long as toxic compounds remain present in the environment. The conclusion of the "disaster phase" does not indicate the cessation of monitoring requirements. Collection of geological and health-related data should be ongoing for epidemiological reviews in order to anticipate which measurements are important for any subsequent eruptions. It may be necessary to specifically review exposure patterns in high-risk groups (such as cleanup workers) and in areas where dust will be disturbed (e.g., roads, schoolyards, city centers). In areas with persistent or repeated ash-falls, advice and hazard communication similar to that provided during the eruptive phase is important.

E. Utilizing and Integrating Data from Eruptions

In a coordinated response to volcanic hazards, ideally a number of information sources should be utilized. Geological monitoring may often help predict when and where an eruption will take place, as well as the type and scale of activity expected (Figure 7). After appropriate assessment of hazards, the information should be disseminated to all relevant organizations, including territorial authorities or regional councils responsible for identifying risk areas and likely effects upon their constituent populations. Regional authorities are usually also responsible for ensuring that the public are,

FIGURE 7 Monitoring volcanic emissions. (Photo: Michael Durand).

and remain, informed and educated throughout each of the eruption phases. Finally, local public health representatives and medical providers should participate in the assessment of the possible eruptive effects on human welfare.

Problems may arise, however, with the interpretation and integration of geological and medical information to ensure beneficial health outcomes. A number of failures to obtain and integrate health-related data have emerged during recent volcanic events, including the 1995–1996 Ruapehu eruptions in New Zealand (Davies et al., 1998). For example, during the height of Ruapehu's activity, numerous measurements of ash-fall and gas emissions were undertaken, but these data were used almost exclusively for geological purposes. There were few avenues available to usefully relay such information or to discuss its implications with public health officials or medical practitioners. Furthermore, air monitoring of poisonous gases following eruptive activity was difficult, because many areas had no existing facilities in place to test pollutant levels during non-eruptive periods. Data regarding water contamination from falling ash was often unavailable because testing was completed at supply level, thus excluding many catchment areas that were vulnerable to ash-fall (such as small streams serving remote communities).

A review of recent disaster responses has therefore indicated a degree of mismatching between the acquisition of eruption data and its health-related utilization. Given the diverse range of effects of volcanism on humans and ecosystems, full and accurate risk assessment requires integration of data collected by numerous ministries, departments, and regional authorities. Protocols for sharing information and achieving a consensus should be established *prior to* a volcanic event. For an effective response, joint consultation and action should involve organizations responsible for agriculture, fisheries, forestry, water provision, power generation, environmental management, and health. Furthermore, consensus needs to be achieved regarding the process of information transfer to the public, which includes the nature and frequency of health-related information, the organization that should impart the message, and the most appropriate media to use.

Geological data which may be generated before, during, and subsequent to eruptions therefore represent only the first step in the process of decreasing "volcanogenic morbidity." The ideal end result would involve information collected from monitoring (e.g., gas production at the vent) being disseminated to all relevant regional organizations and health providers, and then to vulnerable communities. Recommendations

TABLE IV. Strategies for Monitoring Gas (and Aerosol) Emissions: A Health Protection Perspective

Monitored component of gas emission	Location of monitoring	Optimal time for monitoring	Health-related objectives of monitoring
CO_2	a. At vent	a. At regular intervals preferably daily	a. To minimize risk of asphyxiation from downhill flow of CO_2 into populated areas
	b. In areas where CO_2 (heavier than air) can pool	b. If the areas need to be accessed	b. To minimize risk of asphyxiation from CO_2 collecting in low-lying areas
	c. Soil degassing	c. Continuously	c. Same as for a
SO_2	a. At vent	a. At regular intervals preferably daily	a. To provide early warnings to vulnerable populations
	b. Priority to emission-vulnerable areas with greatest population density	b. Continuously during eruption or in response to early warning	b. To minimize risks of respiratory disease, including exacerbations of asthma
H_2S	a. At vent	a. At regular intervals preferably daily	a. To provide early warnings to vulnerable populations
	b. Downwind in built up areas	b. Continuously during eruption or in response to early warning	b. To minimize risks of respiratory disease, including exacerbations of asthma
	c. In areas where H_2S (heavier than air) can pool	c. If the areas need to be accessed	c. To minimize risk of noxious asphyxiation
HCl and HF	a. At vent	a. At regular intervals preferably daily	a. To provide early warnings to vulnerable populations
	b. Priority to emission-vulnerable areas with greatest population density	b. Continuously during eruption or in response to early warning	b. To minimize risks of respiratory disease, including toxic lung injury
	c. Areas where much water is collected from roofs	c. Continuously during eruption or in response to early warning	c. To minimize effects of acid rain dissolution of metal roof components (which may release heavy metals into drinking water)
Air pH?	a. At vent	a. At regular intervals preferably daily	a. To give indication of expected pH of ambient air downwind
	b. Priority to emission-vulnerable areas with greatest population density	b. Continuously during eruption	b. To minimize the effects of acidic aerosols on respiratory disease, including asthma
Radon	a. At volcano	a. Continuously	a. To minimize possible long-term sequelae of radon exposure, such as lung cancer

communicated to the public should take into account practical realities (including economic or geographical constraints) and should be simple, coherent, noncontradictory, and delivered in a prompt fashion. From a medical perspective, geologic monitoring has a pivotal, but often underutilized, role in helping those living in the shadows of volcanoes. In keeping with the integra-

tive aims of this book, the authors hope that this chapter will help to encourage the cross-disciplinary use of geological and medical data from locations where populations are exposed to volcanic eruptions. Such an approach offers an opportunity not only to advance scientific understanding, but ultimately to better protect public health.

SEE ALSO THE FOLLOWING CHAPTERS

Chapter 2 (Natural Distribution and Abundance of Elements) · Chapter 10 (Radon in Air and Water) · Chapter 12 (Fluoride in Natural Waters) · Chapter 15 (Selenium Deficiency and Toxicity in the Environment) · Chapter 20 (Animals and Medical Geology) · Chapter 23 (Environmental Pathology)

FURTHER READING

Afane, Z., Atchou, G., Carteret, P., and Huchon, G. J. (1996). Respiratory Symptoms and Peak Expiratory Flow in Survivors of the Nyos Disaster, *Chest*, 110(5), 1278–1281.

Armienta, M., de la Cruz-Reyna, S., and Morton O. (2001). Compositional Variations of Ash Deposits as a Function of Distance at Popocatepetl Volcano. Cities on Volcanoes 2 Conference Proceedings, Auckland, New Zealand.

Baubron, J.-C., and Toutain, J. P. (1990). Diffuse Volcanic Emissions of Carbon Dioxide from Vulcano Island, Italy, *Nature*, 344, 51–53.

Baxter, P. (1983). Health Hazards of Volcanic Eruptions, *J. Roy. Coll. Phys. Lond.*, 17(3), 180–182.

Baxter, P. (1990). Medical Effects of Volcanic Eruptions, *Bull. Volcanol.*, 52, 532–544.

Baxter, P. J. (1997). Volcanoes. In *The Public Health Consequences of Disasters* (E. K. Noji, Ed.), Oxford University Press, Oxford, UK.

Baxter, P. J. (2001). Human Health and Volcanoes: Recent Developments. Cities on Volcanoes 2 Conference Proceedings, Auckland, New Zealand.

Baxter, P. J., and Coutinho R. (1999). Health Hazards and Disaster Potential of Ground Gas Emissions at Furnas Volcano, Sao Miguel, Azores, *J. Volcanol. Geotherm. Res.*, 92, 95–106.

Baxter, P. J., Bolyard, M. L., Buist, A. S. et al. (1986). Health Effects of Volcanoes. An Approach to Evaluating the Health Effects of an Environmental Hazard, *Am. J. Public Health*, 76, 1–90.

Baxter, P. J., Dupree, R., Hards, V. L. et al. (1999). Cristobalite Volcanic Ash of the Soufriere Hills Volcano, Montserrat, British West Indies, *Science*, 283, 1142–1145.

Baxter, P. J., Falk R. S., Falk H. et al. (1982b). Medical Aspects of Volcanic Disasters: An Outline of Hazards and Emergency Response Measures, *Disasters*, 6, 268–276.

Baxter, P. J., Ing, R., Falk, H. et al. (1981). Mount St. Helens Eruptions, May 18–Jun 12, 1980: An Overview of the Acute Health Impact, *JAMA*, 246, 2585–2589.

Baxter, P. J., Ing R., Falk H. et al. (1983). Mount St. Helens Eruptions: The Acute Respiratory Effects of Volcanic Ash in a North American Community, *Arch. Environ. Health*, 38, 138–143.

Baxter, P. J., Stroiber R. E., and Williams S. N. (1982a). Volcanic Gases and Health: Masaya Volcano, Nicaragua, *Lancet*, 2, 150–151.

Bethel, R. A., Sheppard, D., Nadel, J. A., and Boushey, H. A. (1983). Sulfur Dioxide-Induced Bronchoconstriction in Freely Breathing, Exercising, Asthmatic Subjects, *Am. Rev. Respir. Dis.*, 128, 987–990.

Blong, R. J. (1984). *Volcanic Hazards: A Sourcebook on the Effects of Eruptions*, Academic Press, Sydney.

Buist, A. S. (1988). Evaluation of the Short and Long Term Effects of Exposure to Inhaled Volcanic Ash from Mt. St. Helens. Kagoshima International Conference on Volcanoes Proceedings, 709–712.

Buist, A. S., Vollmer W. M., Johnson L. R. et al. (1986). A Four-Year Prospective Study of the Respiratory Effects of Volcanic Ash from Mount St. Helens, *Am. Rev. Respir. Dis.*, 133, 526–534.

Davies, H. (1998). Review of the Priority 2 Identification Programme 1997/98, Institute of Environmental Science and Research Limited, Wellington, New Zealand.

Dent, A. W., Barret, P., and de saint Ours, P. J. A. (1995). The 1994 Eruption of the Rabaul Volcano, Papua New Guinea: Injuries Sustained and Medical Response, *Med. J. Aust.*, 163, 635–639.

Eisele, J. W., O'Halloran, R. L., Reay, D. T. et al. (1981). Deaths During the May 18, 1980, Eruption of Mount St. Helens, *N. Engl. J. Med.*, 305, 931–936.

Elias, T., and Sutton, A. J. (2001). Volcanic Air Pollution Creates Health Concerns on the Island of Hawai'i, Cities on Volcanoes 2 Conference Proceedings, Auckland, New Zealand.

Environment, m. f. t. (1994). Ambient Air Quality Guidelines, Ministry for the Environment, Wellington, New Zealand.

Epler, G. R., Colby, T. V., McLoud, T. C. et al. (1985). Bronchiolitis Obliterans Organizing Pneumonia, *N. Engl. J. Med.*, 312, 152–158.

Faivre-Pierret, R., and Le Guern, F. (1983). Health Risks Linked with Inhalation of Volcanic Gases and Aerosols. In *Forecasting Volcanic Events* (H. Tazieff, and J.-C. Sabroux, Eds.), Elsevier Press, Amsterdam.

German, C. R., Briem, J., Chin, C. et al. (1994). Hydrothermal Activity on the Reykjanes Ridge: The Steinholl Vent-Field at 63°06′N, *Earth Planet. Sci. Lett.*, 121, 647–654.

Gessner, B. (1994). Acute Fluoride Poisoning from a Public Water System, *N. Engl. J. Med.*, 330(2), 95–99.

Grandjean, P. (1982). Occupational Fluorosis Through 50 Years: Clinical and Epidemiological Experiences, *Am. J. Ind. Med.*, 3, 227–236.

Green, F. H. Y. et al. (1982). Health Implications of the Mount St. Helens Eruption, Laboratory Investigation, *Ann. Occup. Hyg.*, 26(1–4), 921–933.

Hickling, J., Weinstein, P., and Woodward, A. (1999). Acute Health Effects of the Mount Ruapehu (New Zealand) Volcanic Eruption of June 1996, *Int. J. Environ. Health Res.*, 9, 97–107.

Hirabayashi, J.-I. (1999). Personal communication.

Houghton, B., Johnston, D., Hill, D. et al. (1999). Volcanoes and Society: Planning for a Volcanic Crisis in New Zealand, Wairakei Research Centre, New Zealand.

Jappinen, P., Marttila, O., and Haahtela, T. (1990). Exposure to Hydrogen Sulphide and Respiratory Function, *Br. J. Ind. Med.*, 47, 824–828.

Kizer, K. W. (1984). Toxic Inhalations. *Emerg. Clin. N. Am.*, 2, 649–666.

Kullmann, C. J., Jones, W. G., Cornwell, R. J. et al. (1994). Characterization of Air Contaminants Formed by the Interaction of Lava and Sea Water, *Environ. Health Perspect.*, 102(5), 478–482.

Le Guern, P. L., and Faivre-Pierret, R. (1982). An Example of Health Hazard: People Killed by Gas During a Phreatic Eruption: Dieng Plateau (Java Indonesia), February 20th 1979, *Bull. Volcanol.*, 45, 153–156.

Lowe, D. (1986). Lahars Initiated by the 13 November 1985 Eruption of Nevado del Ruiz, Colombia, *Nature*, 324, 51–53.

Mannino, D. M., Holschuh, F. C., Holscuh, T. C. et al. (1996). Emergency Department Visits and Hospitalizations for Respiratory Disease on the Island of Hawaii, 1981 to 1991, *Hawaii Med. J.*, 55, 48–54.

Mori, J., McKee, C., and Talai, B. (1989). A Summary of Precursors to Volcanic Eruptions in Papua New Guinea. In *Volcanic Hazards: Assessment and Monitoring* (J. H. Latter, Ed.), Springer-Verlag, Berlin.

Moseholm, L., Taudorf, E., and Frosig, A. (1993). Pulmonary Function Changes in Asthmatics Associated with Low-Level SO_2 and NO_2 Air Pollution, Weather, and Medicine Intake. An 8-month Prospective Study Analyzed by Neural Networks, *Allergy*, 48, 334–344.

Murata, K. J. (1966). The 1959–60 Eruption of Kilauea Volcano, Hawaii; An Acidic Fumarolic Gas from Kilauea Iki, U. S. Geological Survey professional paper, 537-C, 1–6.

Newhall, C. G., and Self, S. (1982). The Volcanic Explosivity Index (VEI), *J. Geophys. Res.*, 87, 1231–1238.

Olsen, K. B., and Fruchter, J. S. (1986). Identification of Hazards Associated with Volcanic Emissions, *Am. J. Publ. Health*, 76(Suppl.), 45–52.

Oskarsson, N. (2002). Personal communication.

Ostro, B. D., Wiener, M. B., and Selner, J. C. (1991). Asthmatic Responses to Airborne Acid Aerosols. *Am. J. Publ. Health.*, 81(6), 694–702.

Rohl, A. M., Langer A. M., Moncure, G. et al. (1982). Endemic Pleural Disease Associated with Exposure to Mixed Fibrous Dust in Turkey, *Science*, 216, 518–520.

Roobol, M. J., and Smith, A. L. (1989). Volcanic and Associated Hazards in the Lesser Antilles. In *Volcanic Hazards: Assessment and Monitoring* (J. H. Latter, Ed.), Springer-Verlag, Berlin.

Scientific Event Alert Network (SEAN) (1989). *Global Volcanism 1975–1985: The First Decade of Reports from the Smithsonian Institution's Scientific Event Alert Network*, Prentice Hall, Upper Saddle River, New Jersey.

Shoemack, P. (1999). Personal communication.

Simkin, T., Siebert, L., and Blong, R. (2001). Volcano Fatalities—Lessons from the Historical Record, *Science*, 291, 255.

Spallholz, J. E. (1994). On the Nature of Selenium Toxicity and Carcinostatic Activity, *Free Radical Biol. Med.*, 17, 45–64.

Sparks, R. S. J., Bursik, M. I., Carey, S. N. et al. (1997). *Volcanic Plumes*, John Wiley & Sons, Chichester, England.

Taylor, G. A. M. (1958). The 1951 Eruption of Mt. Lamington, Papua. Bureau of Mineral Resources of Australia, Bulletin 38.

Tazieff, H., and Sabroux, J.-C. (1983). *Forecasting Volcanic Events*, Elsevier Press, Amsterdam.

Vallyathan, V., Robinson, V., Reasor, M. et al. (1984). Comparative *in vitro* Cytotoxicity of Volcanic Ashes from Mount St. Helens, El Chichon, and Galunggung, *J. Toxicol. Environ. Health*, 14, 641–654.

Wagner, G. N., Clark, M. A., Koenigsberg, E. J. et al. (1988). Medical Evaluation of the Victims of the 1986 Lake Nyos Disaster, *J. Forensic Sci.*, 33, 899–909.

Weinstein, P., and Patel, A. (1997). The Mount Ruapehu Eruption, 1996: A Review of Potential Health Effects, *Austr. N. Z. J. Publ. Health*, 21(7), 773–778.

Welch, A. H., Lico, M. S., and Hughes, J. L. (1988). Arsenic in Ground Water of the Western United States, *Ground Water*, 26, 333–347.

Williams, S. N., and Moore J. G. (1983). Man Against the Volcano: The Eruption on Heimaey, Vestmannaeyjar, Iceland, U. S. Geological Survey, Washington DC.

Yamo, E., Yokohama, Y., Higashi, H. et al. (1990). Health Effects of Volcanic Ash: A Repeat Study, *Arch. Environ. Health*, 45(6), 367–373.

Yang, G., Ge, K., Chen, J., and Chen, X. (1988). Selenium-Related Endemic Diseases and the Daily Nutritional Requirements of Humans, *World Rev. Nutr. Diet*, 55, 98–152.

CHAPTER 10

RADON IN AIR
AND WATER

J. DONALD APPLETON
British Geological Survey

CONTENTS

I. INTRODUCTION

Radon is a natural radioactive gas that you cannot see, smell, or taste and that can only be detected with special equipment. It is produced by the radioactive decay of radium, which in turn is derived from the radioactive decay of uranium. Uranium is found in small quantities in all soils and rocks, although the amount varies from place to place. Radon decays to form radioactive particles that can enter the body by inhalation. Inhalation of the short-lived decay products of radon has been linked to an increase in the risk of developing cancers of the respiratory tract, especially of the lungs. Breathing radon in the indoor air of homes contributes to about 20,000 lung cancer deaths each year in the United States and 2000–3000 in the UK. Only smoking causes more lung cancer deaths.

Geology is the most important factor controlling the source and distribution of radon. Relatively high levels of radon emissions are associated with particular types of bedrock and unconsolidated deposits, for example some, but not all, granites, phosphatic rocks, and shales rich in organic materials. The release of radon from rocks and soils is controlled largely by the types of minerals in which uranium and radium occur. Once radon gas is released from minerals, its migration to the surface is controlled by the transmission characteristics of the bedrock and soil; the nature of the carrier fluids, including carbon dioxide gas and groundwater; meteorological factors such as barometric pressure, wind, relative humidity, and rainfall; and soil permeability, drainage, and moisture content (see also Chapter 9, this volume).

Radon levels in outdoor air, indoor air, soil air, and groundwater can be very different. Radon released from rocks and soils is quickly diluted in the atmosphere. Concentrations in the open air are normally very low

British Geological Survey. © NERC.

and probably do not present a hazard. Radon that enters poorly ventilated buildings, caves, mines, and tunnels can reach high concentrations in some circumstances. The construction method and the degree of ventilation can influence radon levels in buildings. A person's exposure to radon will also vary according to how particular buildings and spaces are used.

The concentration of radon in a building primarily reflects (1) the detailed geological characteristics of the ground beneath the building, which determines the potential for radon emissions and (2) the structural detail of the building and its mode of use, which determines whether the potential for radon accumulation is fulfilled. The radon potential of the ground may be assessed from a geologically based interpretation of indoor radon measurements in conjunction with permeability, uranium, soil gas radon, and ground and airborne gamma spectrometric data. The categorization of a group of rocks or unconsolidated deposits as having known or suspected high levels of radon emissions does not imply that there is any problem. That would depend on whether pathways, locations for accumulation, and protracted exposure occur. Whereas geological radon potential maps do not give a direct guide to the level of radon in specific buildings or cavities, there is, in general, a higher likelihood that problems may occur at specific sites within areas of potentially high radon emissions.

Radon potential maps have important applications, particularly in the control of radon through environmental health and building control legislation. They can be used (1) to assess whether radon protective measures may be required in new buildings, (2) for the cost-effective targeting of radon monitoring in existing dwellings and workplaces, (3) to provide a radon potential assessment for homebuyers and sellers, and (4) for exposure data for epidemiological studies of the links between radon and cancer. Whereas a geological radon potential map can indicate the relative radon hazard, it cannot predict the radon risk for an individual building. This can only be established by having the building tested.

Radon dissolved in groundwater migrates over long distances along fractures and caverns depending on the velocity of fluid flow. Radon is soluble in water and may thus be transported for distances of up to 5 km in streams flowing underground in limestone. Radon remains in solution in the water until a gas phase is introduced (e.g., by turbulence or by pressure release). If emitted directly into the gas phase, as may happen above the water table, the presence of a carrier gas, such as carbon dioxide, would tend to induce migration of the radon. This appears to be the case in certain limestone formations, where underground caves and fissures enable the rapid transfer of the gas phase. Radon in water supplies can result in radiation exposure of people in two ways: by ingestion of the water or by release of the radon into the air during showering or bathing, allowing radon and its decay products to be inhaled. Radon in soil under homes is the biggest source of radon in indoor air, and it presents a greater risk of cancer than radon in drinking water.

This chapter explains how radon forms, the associated health risks, the kinds of rocks and soils it comes from, and how it moves through the ground and into buildings. It also explains how the radon potential of an area can be estimated.

II. NATURE AND MEASUREMENT

A. Radioactivity and Radiation

All matter, including the materials that constitute the Earth's crust, consists of atoms, which are usually combined in various chemical compounds. Each atom comprises a nucleus, made up of protons, neutrons, and electrons, which orbit around the nucleus. Nuclei identified by the name of the element and the number of protons and neutrons are referred to as nuclides. All nuclei of the same chemical element have the same number of protons, but they can have different numbers of neutrons, and these are then called isotopes of that element. Many atoms are unstable and will change quite naturally into atoms of another element accompanied by the emission of ionizing radiation. This process is called radioactivity and the change is called radioactive decay. Unstable atoms that change through radioactive decay to form other nuclides are said to be radioactive and are referred to as radionuclides or radioisotopes. The rate of change or decay of an unstable radionuclide is indicated by its half-life, which is the period of time during which half the original number of atoms will have decayed.

The radiations most commonly emitted by radionuclides are alpha particles, beta particles, and gamma rays. The principal geological sources of radiation are *gamma radiation* from the ground and buildings and *radon* gas, which is derived mainly from uranium minerals in the ground. *Terrestrial gamma rays* originate chiefly from the radioactive decay of the natural potassium, uranium, and thorium, which are widely

TABLE I. The Uranium-238 Decay Series

Nuclide	Principal mode of decay	Half-life
^{238}U	α	4.5×10^9 years
^{234}TH	β	24.1 days
^{234}Pa	β	1.2 minutes
^{234}U	α	2.5×10^5 years
^{234}TH	α	7.5×10^4 years
^{226}Ra	α	1,602 years
^{222}Rn	α	3.8 days
^{218}Po	α	3.1 minutes
^{214}Pb	β	26.8 minutes
^{218}At	α	1.5 seconds
^{214}Bi	α	19.9 minutes
^{214}Po	α	$1.6 - 10^{-4}$ seconds
^{210}Tl	β	1.3 minutes
^{210}Pb	β	22.6 years
^{210}Bi	β	5.0 days
^{210}Po	α	138.4 days
^{206}Tl	β	4.2 minutes
^{206}Pb	Stable	Stable

TABLE II. The Thorium-232 Decay Series

Nuclide	Principal mode of decay	Half-life
^{232}Th	α	1.4×10^{10} years
^{228}Ra	β	5.8 years
^{228}Ac	β	6.1 hours
^{228}TH	α	1.9 years
^{224}Ra	α	3.7 days
^{220}Rn	α	55.6 seconds
^{226}Po	α	0.15 seconds
^{212}Pb	β	10.6 hours
^{212}Bi	α 36%	60.5 minutes
	β 64%	
^{212}Po	α	3.0×10^{-7} seconds
^{208}Tl	β	3.1 minutes
^{207}Pb	Stable	Stable

lung with alpha particles and this may increase the risk of developing lung cancer.

distributed in terrestrial materials including rocks, soils, and building materials extracted from the Earth.

There are three naturally occurring radon (Rn) isotopes: ^{219}Rn (actinon), ^{220}Rn (thoron), and ^{222}Rn, which is commonly called radon. ^{219}Rn (actinon) has a very short half-life of about 3 seconds and this, together with its occurrence in the decay chain of ^{235}U (which is only present as 0.7% of natural uranium), restricts its abundance in gases from most geological sources. Actinon does not escape to air in significant quantities. ^{222}Rn (radon) is the main radon isotope of concern to man. It occurs in the uranium-238 decay series (Table I), has a half-life of 3.82 days and provides about 50% of the total radiation dose to the average person. ^{222}Rn is produced by the radioactive decay of solid radium (^{226}Ra). ^{220}Rn (thoron) is produced in the thorium-232 decay series (Table II). ^{220}Rn has been recorded in houses, and about 4% of the average total radiation dose for a member of the UK population is from this source.

Most of the radon that is inhaled is exhaled again before it has time to decay and irradiate tissues in the respiratory tract. Radon (^{222}Rn), however, decays to form very small solid radioactive particles, including polonium-218, that become attached to natural aerosol and dust particles. These may remain suspended in the air or settle onto surfaces. When these particles are inhaled, they irradiate the lining of the bronchi in the

B. Measurement of Radioactivity

There are a number of different ways to measure radioactivity. These include (1) the *radioactivity* of a radioactive material, such as radon gas; (2) the *dose* to living tissue, e.g., to the lungs from solid decay products of radon gas; and (3) the *exposure* caused by the presence of radioactivity. There are also environmental or safety thresholds of radioactivity such as *dose limit*, *action level*, and *reference level*, which are used in legislation and advice. The units of radioactivity and dose are summarized in Table III. In the United States, radioactivity is commonly measured in pico curies (pCi), named after the French physicist Marie Curie, who was a pioneer in the research on radioactive elements and their decay. In most other countries, and throughout this chapter, radioactivity is measured using the SI unit becquerel (Bq). One becquerel represents one atomic disintegration per second. The level of radioactivity in the air due to radon is measured in becquerels per cubic meter (Bq m^{-3}) of air. The average radon concentration in houses in Great Britain is 20 Bq m^{-3}, that is, 20 radon atoms disintegrate every second in every cubic meter of air. The average in the United States is 46 Bq m^{-3}. A 1000-square foot house with 46 Bq m^{-3} of radon has nearly 2 million radon atoms decaying in it every minute.

TABLE III. Units of Measurement of Radioactivity and Dose

Quantity	Unit	Purpose	Comments
Activity	Becquerel (Bq)	Measure activity of a radioactive material (solid or gas); the International System of Units (SI) definition of activity	1 Bq = 1 atomic disintegration per second
	Curie	In the United States, the activity (rate of decay) of ^{222}Rn is expressed in units called curies	The curie is based on the rate of decay of one gram of ^{226}Ra or 3.7×10^{10} disintegrations per second
	Pico curies (pCi)		1 pCi = one trillionth of a curie; 0.037 disintegrations per second, or 2.22 disintegrations per minute
Radioactivity in air or water	Becquerels/m^{-3} (Bq m^{-3})	Measure average concentration of radon gas in building or in soil air Bq/L used to measure radon in water	Average level of radon in houses in Great Britain is 20 Bq m^{-3}; in Sweden 108 Bq m^{-3}
	pico curies/L^{-1} (pCi L^{-1})	Unit used in the United States	Average level of radon in houses in the United States is 1.24 pCi L^{-1} equivalent to 46 Bq m^{-3}
Absorbed dose	Gray (Gy)	Measure energy per unit mass absorbed by tissue	1 joule of energy absorbed by 1 kg of tissue
	rad	Old unit of absorbed dose	1 rad = 0.01 Gy
Dose equivalent	Sievert (Sv)	Measure of absorbed doses caused by different types of radiation	Absorbed dose weighted for harmfulness of different radiations
	Roentgen equivalent man (rem)	Old measure of absorbed dose	The rem is being replaced by the Sievert, which is equal to 100 rem

Radon levels in outdoor air, indoor air, soil air, and groundwater can be very different. Radon in outdoor air is generally low (4–8 Bq m^{-3}) but may be as high as 100 Bq m^{-3} in some valleys when measured in the early morning. Radon in indoor air ranges from less than 20 Bq m^{-3} to about 110,000 Bq m^{-3} with a population-weighted world average of 39 Bq m^{-3} and country averages ranging from 9 in Egypt, 20 in the UK, 46 in the United States, 108 in Sweden to 140 in the Czech Republic (UNSCEAR, 2000). Radon in soil air (the air that occupies the pores in soil) ranges from less than 1 to more than 2500 Bq L^{-1}; most soils in the United States contain between 5 and 55 Bq L^{-1} radon in soil air. The amount of radon dissolved in groundwater ranges from about 3 to nearly 80,000 Bq L^{-1}.

The *absorbed dose* is the energy absorbed by a unit mass of tissue whereas the *dose equivalent* takes account of the relative potential for damage to living tissue of the different types of radiation. The dose equivalent is the absorbed dose multiplied by a "quality factor," which is 1 for beta and gamma rays and 20 for alpha particles. This is because alpha particles deposit their energy more densely. In addition, alpha particles transfer all their energy in short distances so that a relatively small volume of tissue receives a high dose of radiation. The commonly used unit for dose equivalent is the *sievert* (1 Sv = 1000 millsieverts; mSv).

The dose equivalent indicates the potential risk of harm to particular tissues by different radiations, irrespective of their type or energy. *Risk weighting factors* are an approximate measure of the risk to particular parts of the body for a given dose equivalent. Some parts are more susceptible to radiation damage (e.g., lungs, bone marrow, or gonads). These have higher risk-weighting factors than other parts of the body. An overall *effective dose equivalent* for the whole body can be calculated from the organ dose equivalents and risk-weighting factors. The annual effective dose equivalent for the average member of the UK population arising from all sources is 2.5 mSv, to which exposure of the lungs by radon and its daughters contributes about half. Exposure in the home to a radon gas concentration of 48 Bq m^{-3} would lead, in the course of a year, to an effective dose equivalent of 1 mSv (ICRP, 1993). In the United States, the average person is exposed to an effective dose equivalent of approximately 3.6 mSv (whole-body exposure) per year from all sources.

Governments set occupational *dose limits* in order to ensure that individuals are not exposed to an unacceptable degree of risk from artificial radiation. Occupational levels are conventionally expressed in working level (WL) units. A WL is any combination of short-lived radon daughters (decay products or progeny ^{218}Po, ^{214}Pb, ^{214}Bi, and ^{214}Po) in one liter of air that will result in the emission of 1.3×10^5 MeV of potential alpha energy. Exposures are measured in working level months (WLM). A WLM is the cumulative exposure equivalent to 1 WL for a working month (170 hours). In SI units, a WLM is defined as 3.54 mJ h m^{-3} (ICRP, 1993). One WL is approximately equal to a radon exposure of 7500 Bq m^{-3} and 1 WLM to an average radon exposure of about 144 Bq m^{-3} y (on the assumption that people spend most of their time indoors) (NRPB, 2000).

The International Commission for Radiological Protection (ICRP) has recommended that all radiation exposures should be kept as low as reasonably achievable taking into account economic and social factors. In the UK, statutory regulations apply to any work carried out in an atmosphere containing ^{222}Rn gas at a concentration in air, averaged over any 24-hour period, exceeding 400 Bq m^{-3} except where the concentration of the short-lived daughters of ^{222}Rn in air averaged over any 8-hour working period does not exceed 6.24×10^{-7} Jm^{-3}. The limit on effective dose for any employee of 18 years of age or above is 20 mSv in any calendar year. This dose limit may be compared with the dose to the average person in the UK of 2.5 mSv, the dose of 7.5 mSv to the average person living in the high radon area of Cornwall, UK, and 4.5 mSv to the average nuclear worker in the UK.

In the United States, exposure limits vary by regulating agency and type of worker. The Miners Safety and Health Act (MSHA) covers underground miners, whereas the Occupational Safety and Health Act (OSHA) regulates exposure to ^{222}Rn gas and ^{222}Rn progeny for workers other than miners. The MSHA sets limits so that no employee can be exposed to air containing ^{222}Rn progeny in excess of 1 WL (100 pCi L^{-1}) in active work areas. The MSHA also limits annual exposure to ^{222}Rn progeny to less than 4 WLM per year. OSHA limits exposure to either 30 pCi L^{-1} or 0.33 WL based on continuous workplace exposure for 40 hours per week, 52 weeks per year.

A number of occupations have the potential for high exposure to ^{222}Rn progeny: mine workers, including uranium, hard rock, and vanadium; workers remediating radioactive contaminated sites, including uranium mill sites and mill tailings; workers at underground nuclear waste repositories; radon mitigation contractors and testers; employees of natural caves; phosphate fertilizer plant workers; oil refinery workers; utility tunnel workers; subway tunnel workers; construction excavators; power plant workers, including geothermal power and coal; employees of radon health mines; employees of radon balneotherapy spas (waterborne ^{222}Rn source); water plant operators (waterborne ^{222}Rn source); fish hatchery attendants (waterborne ^{222}Rn source); employees who come in contact with technologically enhanced sources of naturally occurring radioactive materials; and incidental exposure in almost any occupation from local geologic ^{222}Rn sources.

III. HEALTH EFFECTS OF RADIATION AND RADON

Radiation can interact with the electrons in surrounding molecules in the cells and induce changes such as ionization. Ionization of water molecules in organic tissues can alter important molecules in that tissue. Radiation can also ionize and produce chemical changes in DNA molecules, the basic material that controls the structure and function of the cells that make up the human body. This can lead to biological effects, including abnormal cell development, some of which may not be seen for some time after radiation exposure. Alpha particles are considered to be the most dangerous type of radiation. Although they do not penetrate very far, the mass and charge of the particles is so high that it can cause intense ionization. Whereas alpha radiation cannot penetrate the surface layer of the skin (stratum corneum), the interior of the lungs lacks a protective epidermis so that alpha decay particles emitted by radon progeny can damage important molecules in the cells. Gamma rays are very penetrative and can cause ionization and tissue damage comparable in effect to x-radiation, but are usually much more energetic. Provided the radioactive sources remain outside the body, gamma radiation is the greatest problem because it is so penetrating, whereas alpha particles are stopped by clothing and the outer layers of the skin. Beta particles are intermediate in penetrating power.

When radioactive sources are taken into the body however, the situation changes markedly. The major pathways by which alpha activity enters the human body are the ingestion of radioactive elements and inhalation

TABLE IV.　Principal Decay Properties of Radon (^{222}Rn) and Short-Lived Decay Products

Radionuclide		Half-life	Main radiation energies and intensities					
			α		β		γ	
			MeV	%	MeV	%	MeV	%
^{222}Rn	Gas	3.824 day	5.49	100	—	—	—	—
^{218}Po	Solid	3.11 min	6.00	100	—	—	—	—
^{214}Pb	Solid	26.8 min	—	—	1.02	6	0.35	37
					0.70	42	0.30	19
					0.65	48	0.24	8
^{214}Bi	Solid	19.7 min	—	—	3.27	18	0.61	46
					1.54	18	1.77	16
					1.51	18	1.12	15
^{214}Po	Solid	1.64×10^{-4} s	7.69	100	—	—	—	—

After Green et al., 1992.

of radon, and more importantly its daughter products, some of which are alpha particle emitters (^{218}Po and ^{214}Po). Alpha particles give up their energy to a very small volume of tissue and can thus cause intensive damage, which has been shown to result in cancers. Much of the inhaled radon is exhaled and relatively few alpha particles are emitted by it within the body. However, the four immediate decay products of ^{222}Rn have short half-lives and are all radioactive isotopes of solid elements (Table IV). The decay products, which remain in suspension attached to the surface of aerosols, dust, smoke, or moisture particles, or are unattached, may remain in the respiratory system where they may become trapped in the lungs and irradiate the cells of mucous membranes, bronchi, and other pulmonary tissues. Overall doses are due largely to irradiation of the bronchial epithelium and secretory cells by alpha particles from the short-lived decay products of ^{222}Rn. It is believed that the ionizing radiation energy affecting the bronchial epithelial cells initiates carcinogenesis. As a consequence the main danger is an increased risk of developing cancers of the respiratory tract, especially the lungs. Whereas radon-related lung cancers occur primarily in the upper airways, radon increases the incidence of all histological types of lung cancer, including small cell carcinoma, adenocarcinoma, and squamous cell carcinoma. The contribution to both lung dose equivalent and effective dose equivalent by the beta and gamma radiations may be ignored, as they are small compared to those from alpha radiation.

It is interesting to note that radon remained a chemical curiosity for decades, even promoted at times as a "health-giving" gas at various spas. Initially radon was regarded as a fairly innocuous or even benign component of geological gases, and its importance as the major contributor to the radioactive dose received by the general population has been recognized relatively recently. In contrast to the early dramatic effects of high radiation doses on humans, which can cause death in a few days or weeks, or obvious skin damage when a limited area of the body is exposed to a high radiation dose, the effects of the relatively low doses of natural radiation (e.g., cancer) usually occur a long time after exposure.

The overall hazard to human health from gamma radiation, either indoors or outdoors, is negligible compared with the hazard associated with radon.

IV. RADON EPIDEMIOLOGY

Evidence linking the exposure to high levels of radon and an increase in the risk of lung cancer is becoming overwhelming. Indeed more is known about the health risks of radon exposure than about most other human carcinogens. A large body of epidemiological data has accumulated over several decades relating to studies of the incidence of lung cancer in miners and risk estimates

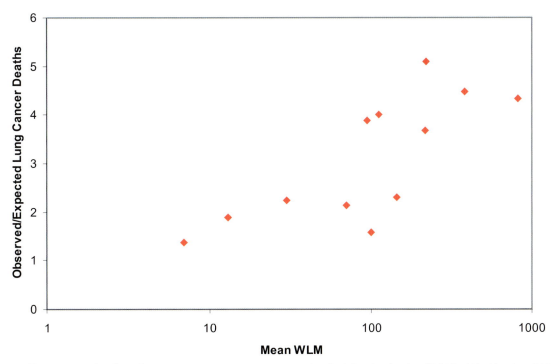

FIGURE 1 Excess mortality from lung cancer among miners exposed to high levels of radon. WLM = Working Level Month, a unit of exposure to radon. (Data from NRPB, 2000.)

have been derived from these data (NAS, 1998; NRPB, 2000). Supporting evidence comes from experimental studies of animals and from radiobiology.

A. Cohort Studies of Miners

High death rates from lung cancer recorded in the Middle Ages among miners in Germany and the Czech Republic are now recognized as radon-induced. Studies of thousands of miners, some with follow-up periods of more than thirty years, have been conducted in uranium, iron, tin, and fluorspar mines in Australia, Canada, China, Europe, and the United States. These studies consistently demonstrated an increase in lung cancer incidence with exposure to radon decay products, despite several differences in study populations and methodologies. The miner studies demonstrated that (1) at equal cumulative exposures, low exposures in the range of the U. S. Environmental Protection Agency's (EPA) $4 \, pCi \, L^{-1}$ ($148 \, Bq \, m^{-3}$) action level over longer periods produced greater lung cancer risk than high exposures over short periods; (2) increased lung cancer risk with radon exposure was observed even after controlling for, or in the absence of, other potentially

confounding mine exposures such as asbestos, silica, diesel fumes, arsenic, chromium, nickel, and ore dust; (3) increased lung cancer risk has been observed in miners at relatively low cumulative exposures in the range of the U. S. EPA's $4 \, pCi \, L^{-1}$ ($148 \, Bq \, m^{-3}$) action level; and (4) nonsmoking miners exposed to radon have been observed to have an increased risk of lung cancer.

A major reassessment of health information mainly on uranium miners from Czechoslovakia, Colorado, Sweden, and Canada by the IRCP (1993) demonstrated a pronounced excess of lung cancers. More recently the Sixth Committee on the Biological Effects of Ionizing Radiation (BEIR VI) of the American National Academy of Sciences (NAS, 1998) re-analyzed the data for miner cohorts and used the most recent information available to estimate the risks posed by exposure to radon in homes.

Results from twelve major epidemiological studies involving a total of more than 60,000 miners clearly indicate a correlation between excess mortality from lung cancer and radon exposure (Figure 1). A combined analysis of studies on underground miners revealed an increase in relative risk from about 2% at a mean exposure of 250 WLM to 10% at 2500 WLM (Lubin et al., 1994).

Differences between mine and home environments could limit the validity of extrapolating risk estimates based on miner data to the home environment. These include generally higher ^{222}Rn gas concentrations, more airborne dust, and larger dust particle diameter in mines compared with homes as well as different activity size distributions of radon progeny and rates of attachment. Other toxic pollutants present in mine air may act as confounders. In addition, there are age and sex differences between miners and the general population; higher levels of physical activity among miners, which affects respiration rates; and more oral (as opposed to nasal) breathing in the miners, which leads to increased deposition of larger particles into the lung. Miners have shorter term high exposure compared with the lifelong lower concentration exposure for the general population. Finally, most miners were smokers compared with a minority of the general population. Even allowing for these factors, the evidence for a causal association between lung cancer and occupational radon exposure in underground miners is overwhelming (see also Chapter 12, this volume).

B. Case Control Studies of the Effects of Domestic Exposure

Comparison of radon exposures among people who have lung cancer with exposures among people who have not developed lung cancer is the most direct way to assess the risks posed by radon in homes. Many factors must be considered when designing a domestic case control radon epidemiology study. These factors include

1. Mobility: People move a lot over their lifetime and it is virtually impossible to test every home where an individual has lived; estimates of radon exposure have to be used to fill in gaps in exposure history.
2. Housing stock changes: Over time, older homes are often destroyed or reconstructed so that radon measurements will be not available or vary dramatically from the time of occupancy by the case; a home's radon level may increase or decrease over time if new ventilation systems are installed, the occupancy patterns may change substantially, or the home's foundation may shift and cracks appear.
3. Inaccurate histories: Often a majority of the lung cancer cases (individuals) studied are deceased or too sick to be interviewed by researchers. This

requires reliance on second-hand information, which may not be accurate.

These inaccuracies primarily affect

1. Residence history: A child or other relative may not be aware of all residences occupied by the patient, particularly if the occupancy is distant in time or of relatively short duration. Even if the surrogate respondent is aware of a residence he or she may not have enough additional information to allow researchers to locate the home.
2. Smoking history: Smoking history historically has reliability problems. Individuals may underestimate the amount they smoke. Conversely, relatives or friends may overestimate smoking history.
3. Other factors: Complicating factors other than variations in smoking habits include an individual's genetics, lifestyle, exposure to other carcinogens, and home heating, venting, and air conditioning preferences.

Several such case control (or cohort) studies have been completed but they have not produced a definitive answer, principally because the risk is small at the low exposure of most domestic environments. In addition, many people involved in the studies moved a number of times so it was difficult to estimate the radon exposures that people had received over their lifetimes. The greatest problem, however, was caused by the fact that far more lung cancers are caused by smoking than are caused by radon (NAS, 1998).

Residential epidemiological case-control studies examining the relationship between contemporary ^{222}Rn gas concentrations in homes and lung cancer have been performed in Canada, China, Finland, Germany, Sweden, the UK, and the United States. These studies indicate that higher lung cancer rates occur in those exposed to higher levels of radon, although in most cases this did not reach a statistical level of significance (Lubin & Boice, 1997; Darby et al, 1998).

Meta-analysis is a statistical attempt to analyze the results of several different studies to assess the presence or absence of a trend or to summarize results. Meta-analysis of the largest case control studies produced a positive risk estimate that was statistically significant and close to that derived from the miner data (Lubin & Boice, 1997). The meta-analysis of eight studies using weighted linear regression found a summary excessive risk of 14% at an average indoor ^{222}Rn gas concentration of $4\,pCi\,L^{-1}$ ($148\,Bq\,m^{-3}$). The excess risk at $4\,pCi\,L^{-1}$ ($148\,Bq\,m^{-3}$) in recent studies in Germany and the UK was in close agreement with risk estimates obtained from the meta-analysis. Lubin and Boice (1997) concluded that the results of their meta-analysis are consistent with the current miner-based estimates of lung cancer risk from radon, which place the number of radon-related deaths at approximately 15,000 per year in the United States. Because meta-analysis has several inherent limitations (such as the inability to adequately explore the consistency of results within and between studies and to control for confounding factors), meta-analysis is not able to prove that residential radon causes lung cancer. But it does provide additional good suggestive evidence. It is one more link in the "chain of evidence" connecting residential radon exposure to increased lung cancer risk. Because the investigators performing a meta-analysis do not have access to the raw data on the individual study subjects, the analysis is based on the published relative risks and confidence intervals of the individual studies. Frequently, the impact of each study is weighted based on some factor considered relevant to the reliability of each study's data. In the Lubin and Boice (1997) meta-analysis, the results of each individual study were weighted so that each study contributed in relation to the precision (i.e., relative lack of random or sampling errors) of its estimate.

An exposure-rate effect is the alteration of an effect by intensity of an exposure. An inverse exposure-rate effect would be the enhancement of an effect as the intensity of the exposure decreases (i.e., low-level chronic exposures would be riskier than high-level more acute exposures). An "inverse exposure-rate effect" was observed in the miner data. This means that for miners who received the same exposure, those that received it over a longer period of time had a greater risk of lung cancer. The inverse exposure-rate effect diminished in miners exposed below 50–100 WLM. The finding that the inverse exposure-rate effect does not seem to apply in residential situations will not change the EPA's risk assessment since the EPA had not included the inverse exposure-rate effect in their latest, 1992, risk estimate.

C. Ecological (Geographical) Epidemiological Studies

Ecological epidemiological studies of the associations between average lung cancer rates and radon concentrations in geographical areas are considered to be much less reliable than case-control studies that consider individual radon exposure and smoking histories. A negative correlation between mean radon and lung cancer

rates in counties in the United States (Cohen, 1997) is not well understood and the study methodology has been criticized by epidemiologists (e.g., Lubin, 1998; Smith et al., 1998) (see also Chapter 21, this volume).

D. Extrapolation from Mines to Homes

Whereas it has been suggested that the dose response relationship seen in miners may not extend to the much lower levels present in most homes, there appears to be sufficient evidence to suggest that a dose threshold for radiation carcinogenesis does not apply to lung cancer. Ionizing radiation is thought to induce specific gene mutations in DNA in single target cells in tissue and, as such, act principally at the initial stage of cancer. The number of cells hit by alpha particles will be broadly proportional to the dose (i.e., the radon concentration in a dwelling). The general consensus is that low dose (i.e., domestic radiation) cancer risk rises in proportion to the dose and there is not a threshold below which risk may be discounted (NRPB, 2000).

More information about residential exposure to radon is needed to answer important questions about radon's effect on women and children—groups not included in the occupational studies of miners. Although children have been reported to be at greater risk than adults for developing certain types of cancer from radiation, currently there is no conclusive evidence that radon exposure places children at any greater risk. Some miner studies and animal studies indicate that for the same total exposure, a lower exposure over a longer time is more hazardous than short, high exposures. These findings increase concerns about residential radon exposures.

E. Experimental Studies with Animals

Results from animal experiments conducted in the United States and France are generally consistent with the human epidemiological data. Health effects observed in animals exposed to radon and radon decay products include lung carcinomas, pulmonary fibrosis, emphysema, and a shortening of life span (USDOE, 1988). The incidence of respiratory tract tumors increased with an increase in cumulative exposure and with a decrease in rate of exposure (NAS, 1998). Increased incidence of respiratory tract tumors was observed in rats at cumulative exposures as low as

20 WLM (NAS, 1998). Exposure to ore dust or diesel fumes simultaneously with radon did not increase the incidence of lung tumors above that produced by radon progeny exposures alone (USDOE, 1988). Lifetime lung tumor risk coefficients observed in animals are similar to the lifetime lung cancer risk coefficients observed in human studies (USDOE, 1988). In a study of rats simultaneously exposed to radon progeny and uranium ore dust, it was observed that the risk of lung cancer was elevated at exposure levels similar to those found in homes. The risk decreased in proportion to the decrease in radon-progeny exposure (Cross, 1992). Confounding factors such as smoking are more readily controlled in animal experiments and qualitatively confirmed that radon can indeed induce lung cancer in the absence of smoking.

F. Other Cancers and Radon Exposure

No consistent association has been observed between radon exposure and other types of cancer. A combined analysis of the data from 11 miner cohort studies involving more than 60,000 miners did not find convincing links (Darby et al., 1995). No clear association between radon and childhood cancer (especially leukemia) emerged from a number of ecological studies, and a review of ecological, miner, and cohort studies did not find an association between radon and leukemia (Laurier et al., 2001).

Radon ingested in drinking water may lead in some circumstances to organs of the gastrointestinal tract receiving the largest dose. Ingested radon is absorbed by the blood. Most of the radon is lost quickly from the bloodstream through the lungs but some will deliver a dose to other body organs, especially those with a high fat content due to the higher solubility of radon in fat compared with water. Other body organs may be irradiated to some extent although the doses involved will be much smaller. Comparative estimated doses to various organs from exposure to radon are indicated in Table V. Apart from lung cancer, there is no epidemiological proof of radon causing any other type of cancer.

V. RADON HEALTH RISKS

Because a valid risk estimate could not be derived only from the results of studies in homes, the BEIR VI com-

TABLE V. Estimated Annual Absorbed Doses to Adult Tissues From ^{222}Rn and Its Short-Lived Progeny for Domestic ^{222}Rn Concentration of 20 Bq m^{-3}

Tissue	Annual dose (μGy y^{-1})
Lung	500
Skin[a]	50–1000
Red bone marrow	0.5–6
Bone surface	0.4–4.4
Breast	1.2–1.5
Blood	1.1
Liver	2.5
Kidney	14.4

[a] Basal cells at 50 μm in exposed skin.
From NAS, 1998.

mittee chose to use data from studies of miners to estimate the risks posed by radon exposures in homes (NAS, 1998). The committee statistically analyzed the data from 11 major studies of underground miners, which together involved about 68,000 men, of whom 2700 have died from lung cancer. A range of models was used to try to explain the relationship between radon and smoking. In the multiplicative model it is assumed that a specific radon exposure will multiply the base risk rate for smokers and non-smokers by the same factor. BEIR VI models take into account total exposure, age and duration of exposure or total exposure, and age and average radon exposure with predicted risks at about 50% higher under the first of these two models. In general, the risk of developing lung cancer increases linearly as the exposure increases.

The number of lung cancer cases due to residential radon exposure in the United States was estimated to be 15,400 (exposure-age-duration model) or 21,800 (exposure-age-concentration model), which is 10–15% of lung cancer deaths. Radon causes 11% of lung cancer deaths among smokers (most of whom die of smoking) but 23% of never-smokers. The BEIR VI committee's uncertainty analyses using the constant relative risk model suggested that the number of lung cancer cases could range from about 3000 to 32,000. The 95% upper confidence limit for the exposure-age-concentration model was approximately 38,000, but it was considered that such a high upper limit was highly unlikely given the uncertainty distributions. The major shortcomings in the existing data relate to estimating lung cancer risks near 148 Bq m^{-3} (4 pCiL^{-1}) and down to the average U. S. indoor level of 46 Bq m^{-3} (1.24 pCiL^{-1}), especially the risks to never-smokers.

Most of the radon-related deaths among smokers in the United States would not have occurred if the victims had not smoked. Whereas there is evidence for a synergistic interaction between smoking and radon, the number of cancers induced in ever-smokers by radon is greater than one would expect from the additive effects of smoking alone and radon alone. The estimated number of deaths attributable to radon in combination with cigarette smoking and radon alone in never-smokers constitutes a significant public-health problem and makes indoor radon the second leading cause of lung cancer after cigarette smoking.

The BEIR VI committee suggested that about one-third of the radon-attributed cases (about 4% of the total lung cancer deaths) would be avoided if all homes had concentrations below the EPA's action guideline of 148 Bq m^{-3} (4 pCiL^{-1}). Of these deaths, about 87% would be in ever-smokers. Deaths from radon-attributable lung cancer in smokers could be reduced most effectively through reduction in smoking, because most of the radon-related deaths among smokers would not have occurred if the victims had not smoked. Whereas the relative risks for smokers and nonsmokers is still disputed, evidence from miners who never smoked demonstrates a clear relationship between cumulative exposure and relative risk. Existing biologic evidence indicates that even very low exposure to radon might pose some risk but that a threshold level of exposure, below which there is no effect of radon, cannot be excluded.

BEIR VI risk models have been used to estimate fatal lifetime lung cancer risk for lifetime exposure at 200 Bq m^{-3} (Table VI). This implies that the 2000–3500 fatal

lung cancers in the UK are attributable to the mean domestic radon concentration of 20 Bq m^3 of which 500–1300 would be nonsmokers. The risk could be an order of magnitude higher in houses with radon concentrations at the current action level of 200 Bq m^{-3}, and up to 50 times higher from lifetime residence in the worst affected houses. To put this in perspective, risk of death from accidents in the home is 0.7%; risk of premature death from accidents on the road is 2.5%, while there is a 5–7.5% overall risk of lung cancer in the UK. A recent review (Darby et al., 2001) demonstrated that over 80% of the radon-related deaths in the UK occur at ages of less than 75 and over 80% in smokers or ex-smokers. Controversially, Darby et al. (2001) estimated that around 90% of radon-induced deaths in the UK probably occur in response to exposure to radon concentrations below the currently recommended action level of 200 Bq m^{-3}, of which 57.3% (1304 deaths) can be attributable to residential radon below 50 Bq m^{-3}. This has major implications for the cost-effectiveness of government intervention strategies designed to manage exposure to radon in the domestic environment. The total number of deaths from lung cancer in the UK is about 34,000, most of them due directly to smoking.

Duport (2002) questioned whether radon risk is overestimated because only the exposure to inhaled radon decay products is generally taken into account in the determination of risk of radiogenic lung cancer in uranium miners, whereas the risk actually reflects the total dose of radiation received by the lung. Radiation dose from sources other than ^{222}Rn decay products may account for 25–75% of the total effective dose, absorbed dose, or equivalent lung dose and this varies between mines. Neglecting these doses would lead to overestimation of risk both through dose underestimation and misclassification. Correction for neglected doses and dose misclassification would reduce the risk per unit of radon exposure by a factor of at least two or three and bring the overall dose-effect relationship toward the no-effect null hypothesis. This would increase the likelihood of a radon exposure threshold for lung cancer risk at current indoor exposure levels (Duport, 2002).

The U. S. EPA estimates that radon in drinking water causes about 168 cancer deaths per year, 89% from lung cancer caused by breathing radon released from water, and 11% from stomach cancer caused by drinking radon-containing water. In general, radon released from tap water and inhaled will present a greater risk than radon ingested through drinking water (NRPB, 2000).

TABLE VI. Fatal Lifetime Lung Cancer Risks for Lifetime Radon Exposure at 200 Bq m^{-3} Based on BEIR VI Models

	Risk (%)
General population	3–5
Smokers	10–15
Non-smokers	1–3

From NRPB, 2000.

VI. SOURCES OF NATURAL RADIATION

A. Introduction

The average person in the UK receives an annual effective radiation dose of 2.8 mSv, of which about 85% is from natural sources: cosmic rays, terrestrial gamma rays, the decay products of ^{220}Rn and ^{222}Rn, and the natural radionuclides in the body ingested through food and drink. Of these the major proportion is from geological sources. About 60% of the total natural radiation dose is from radon isotopes (mostly due to alpha particle activity) while about 15% is thought to be due to gamma radiation from the U, Th, and K in rocks and soils and from building products produced from geological raw materials. X-rays and radioactive materials used to diagnose disease are the largest source of artificial exposure to people. The average dose due to anthropogenic isotopes (radioactive fallout, fuel cycle, etc.) is less than 1% of the total annual dose (Figure 2). Similar average annual effective doses apply worldwide (UNSCEAR, 2000). In European countries, the average annual dose from natural sources is 2 mSv in Denmark and the UK rising to 3 mSv in Finland and Sweden where indoor radon concentrations and gamma radiation are much higher (NRPA, 2000).

On an individual basis, the dose would be dependent upon where one lived, one's lifestyle, and the nature and extent of any medical treatment. Most of the exposures to terrestrial gamma rays and to ^{220}Rn and ^{222}Rn decay products result from living indoors. Building materials are the main source of thoron (^{220}Rn) in room air although a minor contribution comes from soil gas. Radon contributes by far the largest variation in the average dose from natural radiation sources.

B. Radon

The average annual dose to the UK population from radon is 1.2 mSv with a range of 0.3 to more than 100 mSv. In the most radon-prone area in Great Britain, the average person receives a total annual radiation dose of 7.8 mSv of which 81% is from radon. The production of radon by the radioactive decay of uranium in rock, overburden, and soil is controlled primarily by the amount of uranium within the rock-forming minerals and their weathering products. The ^{238}U decay chain may be divided into two sections separated by ^{226}Ra (radium), which has a half-life of 1622 years (Table I). Earlier isotopes mostly have long half-lives, while the later isotopes, including radon (^{222}Rn), have relatively short half-lives. Outdoors, radon normally disperses in the air whereas in confined spaces such as buildings, mines, and caves it may accumulate. Radon in indoor air comes from soil gas derived from soils and rocks

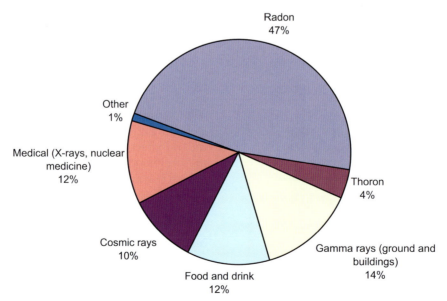

FIGURE 2   Sources of radiation exposure contributing to average effective dose in the UK. Other = occupation 0.3%, fallout 0.2%, nuclear discharges <0.1%, and products <0.1%. (UK NRPB data.)

beneath a building with smaller amounts from the degassing of domestic water into the indoor air and from building materials. Soil gas represents the predominant source of indoor ^{222}Rn gas. Outdoor radon concentrations may occasionally reach potentially hazardous levels. For example, air escaping from an open uranium mine gallery in the town of Schneeberg, Germany, contained up to 10,000 Bq m^{-3} radon and ventilation facilities had to be installed to prevent ingress of this air into an adjacent factory.

Water in rivers and reservoirs usually contains very little radon so homes that use surface water do not have a radon problem from their water. Water processing in large municipal systems aerates the water, which allows radon to escape and also delays the use of water until most of the remaining radon has decayed. However, in many areas of the United States, for example, groundwater is the main water supply for homes and communities. These small public water works and private domestic wells often have closed systems and short transit times that do not remove radon from the water or permit it to decay. In such situations, radon from the domestic water could add radon to the indoor air.

Radon from degassing of domestic water accounts for about 5% of the total indoor radon for homes that use groundwater sources in the United States. In some cases, radon from this source may account for a higher proportion of indoor radon. In Maine (U. S.), radon concentrations in domestic water wells sometimes exceed 37,000 Bq L^{-1} and more than 10% of private well water supplies exceed the action level (740 Bq L^{-1}). Radon from the domestic water supply is inhaled when it is released from the water during showering, washing clothes, and washing dishes. It is estimated that 370 Bq L^{-1} of radon in the domestic water supply contributes about 37 Bq m^{-3} to the indoor air of a home. Areas most likely to have problems with radon from domestic water supplies include those with high levels of uranium in the underlying rocks, such as uraniferous granites. This association has been observed both in the United States and the UK. In Maine, for example, the average for well water in granite areas is over 500 Bq L^{-1}.

In a study of private water supplies in southwestern England, a high proportion derived from granite areas exceeds the draft European Union action level of 1000 Bq L^{-1}. It was also found that radon concentrations varied significantly over the course of a week and between samples taken several months apart. For water from groundwater sources, mean values (by source type) at the tap were generally lower than those at the source. This is consistent with loss of radon due to degassing as a result of water turbulence within the supply system

and natural radioactive decay while the water is resident in the household supply system. All the water sources sampled showed large variability in radon concentration over the summer sampling period, whereas less pronounced variability was observed during the winter sampling. Maximum values were observed during the summer.

Building materials generally contribute only a very small percentage of the indoor air ^{222}Rn concentrations. However, in some areas, concrete, blocks, or wallboard made using radioactive shale or waste products from uranium mining will make a larger contribution to the indoor radon. High radium content and radon exhalation rates in concrete and bricks used in some Hong Kong buildings with high indoor radon concentrations are probably caused by the granitic composition of aggregates (Man & Yeung, 1998). In Sweden, 300,000 houses constructed with radioactive alum shale form the world's largest stock of buildings that have used building materials with enhanced radiation. The houses have radon concentrations of 100–400 Bq m^{-3} and gamma radiation levels of 0.3–1.2 μSv h^{-1} (NRPA, 2000). High effective dose rates (7.1–16.7 mSv y^{-1}) for ^{222}Rn and ^{220}Rn and their progenies have been estimated for cave dwellings excavated in loess in the Yan'an area of China, reflecting both exhalation from the loess and relatively poor ventilation (Weigand et al., 2000).

C. Gamma Rays From the Ground and Buildings (Terrestrial Gamma Rays)

Everyone is irradiated by gamma rays emitted by the radioactive materials in the Earth. Building materials extracted from the Earth may also emit gamma radiation, so people can be irradiated indoors as well as outdoors. Terrestrial gamma rays originate principally from the radioactive decay of the natural potassium, uranium, and thorium. These elements are widely distributed in terrestrial materials which include rocks and soils. The average annual gamma radiation dose from all these sources to the population in Great Britain is about 350 μSv with a range of 120–1200 μSv.

In general, the gamma radiation dose is proportional to the amount of U, Th, and K in the ground and in building materials. Within a masonry building, most of the gamma radiation is received from the building materials, whereas in wooden buildings a larger part of the dose is contributed from gamma radiation from the ground. The average person in the UK spends only 8% of his or her time outdoors so the contribution to total radiation dose from the ground is relatively small. The bulk of the radiation above the ground surface is derived from only the top 30 cm or so of soil or rock. Soils developed upon radioactive rocks generally have a much lower gamma radioactivity than the rock substrate. Whereas one can predict or identify areas of high geological gamma radioactivity, the resultant dose to the population depends on additional factors such as soil type, house construction, and lifestyle.

A single K isotope, ^{40}K, comprising only 0.0119% of the total K, is radioactive. ^{40}K undergoes branched decay producing ^{40}Ca and ^{40}Ar, the latter reaction producing high-energy gamma rays. Uranium consists of two main isotopes (^{238}U and ^{235}U). Because ^{235}U comprises only 0.72% of the total U it may, for practical purposes, be ignored. Isotopes in the later section of the ^{238}U decay series (Table I) include the bismuth isotope (^{214}Bi), which contributes most of the gamma activity of the decay series. As ^{226}Ra is chemically very different from U, it is possible in natural processes for the two to become separated so that the ^{226}Ra, and its daughter products are unsupported by the parent U. In addition, radioactive elements in the rock fragments and derived minerals in the weathered overburden are diluted with organic matter and water. Thus there may not be a simple relationship between the measured gamma ray flux and the U content.

Many of the daughter isotopes of ^{238}U and ^{232}Th are gamma active. ^{208}Tl is the main gamma active daughter product derived from ^{232}Th, and this takes 70 years to reach secular equilibrium in the ^{232}Th decay series. Potassium gives rise to a prompt gamma ray in which the intensity is directly related to the potassium concentration. On average potassium is much more abundant than thorium which, in turn, is more abundant than uranium. However, the specific gamma activities are such that, on average, approximately equivalent gamma emissions are observed from potassium and the decay series of uranium and thorium.

Areas of high natural radiation include areas of monazite sands in Brazil, China, Egypt, and India; volcanic rocks in Brazil and Italy; uranium mineralization in France, the UK, and the United States; and radium-enriched karst soils developed over limestones in Switzerland, the UK, and the United States (UNSCEAR, 2000).

D. Food and Drink

Radioactive materials even occur in food. Potassium-40 in particular is a major source of internal irradiation. Natural radioactivity in the human diet gives an average annual dose of 300 μSv each year of which 180 μSv is from ^{40}K. The range for all internal radiation sources in Great Britain is 100–1000 μSv per annum. Shellfish concentrate radioactive materials so that, even when there is no man-made radioactivity, people who consume large quantities of mussels, cockles, or winkles can receive a dose from natural radioactivity in food that is about 50% higher than average. Apart from restricting intake of shellfish, there is very little possibility of reducing the small exposure to natural radioactivity from food.

E. Cosmic Rays

Approximately 10% of the average annual radiation dose is from cosmic rays (Table VII), although this increases with latitude and altitude. The average dose from cosmic radiation received each hour rises from 0.03 μSv at sea level, to 0.1 μSv in Mexico City (altitude of 2250 m), and 5 μSv at the cruising altitude for commercial jet aircraft (10,000 m). Polar and mountain dwellers, aircrews, and frequent air travelers therefore receive higher doses of cosmic radiation. Little can be done about cosmic radiation because it readily penetrates ordinary buildings and aircraft. The average

TABLE VII. Sources of Radiation for Average Person in the UK

Source	Annual dose (%)
Natural sources	87.0
Radon (^{222}Rn) gas from the ground	47.0
Thoron (^{220}Rn) from the ground	4.0
Gamma rays from the ground and building materials	14.0
Food and drink	12.0
Cosmic rays	10.0
Artificial sources	13.0
Medical	12.0
Nuclear discharges	0.1
Work	0.2
Fallout	0.4
Miscellaneous	0.4

From NRPB, 1989.

annual dose from cosmic rays in Great Britain is 250 μSv, with a range of 200–300 μSv.

VII. Measurement of Radon

A. Radon Testing in the Home

Although radon cannot be seen or smelled, it can be measured relatively easily with the proper equipment. The most common procedures for measuring radon make use of the fact that it is the only natural gas that emits alpha particles, so if a gas is separated from associated solid and liquid phases any measurements of its radioactive properties relate to radon or its daughter products. In the United States, radon in homes is usually measured using inexpensive do-it-yourself radon test kits, which are available by mail order and in many retail outlets or by hiring a U. S. EPA qualified or state-certified radon tester. The EPA recommends that all homes be tested for radon below the third floor.

Common short-term test devices are charcoal canisters, alpha track detectors, liquid scintillation detectors, electret ion chambers, and continuous monitors. A short-term testing device remains in the home for 2–90 days, depending on the type of device. Because radon levels tend to vary from day-to-day and season-to-season, a long-term test is more likely than a short-term test to measure the home's year-round average radon level. If results are needed quickly, however, a short-term test followed by a second short-term test may be used to determine the severity of the radon problem. Long-term test devices, comparable in cost to devices for short-term testing, remain in the home for more than three months. A long-term test is more likely to indicate the home's year-round average radon level than a short-term test. Alpha track detectors and electret ion detectors are the most common long-term test devices.

Charcoal canister and liquid scintillation detectors contain small quantities of activated charcoal. Radon and its decay products are adsorbed onto the charcoal and are measured by counting with a sodium iodide detector or a liquid scintillation counter. Radon adsorbed at the beginning of the exposure decays away after a few days so the duration of the measurement is restricted and the device does not measure the true average exposure. Charcoal detectors are suitable only for short-term tests when results are required urgently and a less accurate measurement is acceptable. The result should be well below the action level before it can be concluded that the annual average concentration will also be below the action level. Ambiguous short-term measurements should be followed up by a long-term measurement (NRPB, 2000).

Alpha (etched) track detectors contain a small sheet of plastic that is exposed for a period of one to three months. Alpha particles etch the plastic as they strike it. These marks are then chemically treated and are usually counted automatically under a microscope to determine the radon concentration. Etched track detectors are relatively cheap and suitable for long-term measurement and are usually deployed for a period of three months.

Electret ion detectors contain an electrostatically charged Teflon disk. Ions are generated by the decay of radon strike and reduce the surface voltage of the disk. By measuring the voltage reduction, the radon concentration can be calculated. Allowance must be made for ionization caused by natural background radiation. Different types of electret are available for measurements over periods of a few days to a few months. The detectors must be handled carefully for accurate results.

Continuous monitors are active devices that need power to function. They require operation by trained testers and work by continuously measuring and recording the amount of radon in the home. These devices sample the air continuously and measure either radon or its decay products (NRPB, 2000).

A rigorous procedure must be followed for short-term tests if relatively reliable results are to be obtained. For example, doors and windows must be closed 12 hours prior to testing and throughout the testing period. The test should not be conducted during unusually severe storms or periods of unusually high winds. The test kit is normally placed in the lowest lived-in level of the home, at least 50 cm above the floor, in a room that is used regularly, but not in the kitchen or bathroom where high humidity or the operation of an exhaust fan could affect the validity of the test. At the end of the test period, the kit is mailed to a laboratory for analysis; results are mailed back in a few weeks. If the result of the short-term test exceeds $100\,Bq\,m^{-3}$ then a long-term test is normally recommended. Remediation of the home is recommended if the radon concentration exceeds certain levels ($150\,Bq\,m^{-3}$ in Luxembourg and the United States; $200\,Bq\,m^{-3}$ in Australia, Israel, Syria, and the UK; $400\,Bq\,m^{-3}$ in Austria, Belgium, Denmark, Finland, Greece, and Sweden).

Radon levels are highest in winter so seasonal corrections have to be applied to estimate the average annual radon level. In workplaces, consideration needs to be taken of work practices and the building design and use. For small premises at least one measurement should made in the two most frequently occupied ground floor rooms. In larger buildings at least one measurement is required for every $100\,m^2$ floor area.

B. Indoor Radon Measurement Validation Scheme

Great care is required in the interpretation of radon monitoring results because the rate at which radon is released into buildings is controlled by a complex series of factors, which requires monitoring equipment to be located in the right place over a prolonged period to take account of temporal variations. Validation schemes are required to (1) ensure organizations measure radon within an acceptable degree of uncertainty; (2) determine that detectors are handled in an appropriate fashion both before and after the detectors have been with householders; and (3) ensure minimum standards in how results are interpreted and presented, which includes requiring the use of the seasonal variation factors. In the United States, the EPA operates a voluntary National Radon Proficiency Program that evaluates radon measurement companies and the test services they offer. Both the UK NRPB and the U. S. EPA recommend that testing services be purchased from certified organizations.

C. Measurement of Radon in Soil Gas

Measurement of radon in soil gas using pumped monitors is recommended as the most effective method for assessing the radon potential of underlying rocks, overburden, and soil. Instruments for the determination of soil gas radon are generally based upon either an extraction method, using a "pump monitor" device for transferring a sample of the soil gas to a detector, or simply emplacing the detector in the ground (passive methods). In the former method, a thin rigid tapered hollow tube is usually hammered into the ground to a convenient depth, which causes minimum disturbance to the soil profile. Detection of radon is usually based upon the zinc sulfide scintillation method or the ionization chamber. Alpha particles produce pulses of light when they interact with zinc sulfide coated on the inside of a plastic or metal cup or a glass flask (Lucas cell). These may be counted using a photomultiplier and suitable counting circuitry. Because the radon isotopes are the only alpha-emitting gases, their concentration may be determined accurately using relatively simple equip-

ment. Because of the different half-lives of these isotopes and their immediate daughter products, it is possible to calculate the activities of radon and thoron. The equipment is relatively robust for field use and is designed for rapid changing of the cell when it becomes contaminated. The large number of instruments produced attests to its suitability for field use. The concentration of radon in soil gases is usually sufficient that the level may be determined relatively fast; a matter of a few minutes generally suffices.

Radon can also be measured by emplacing alpha track detectors in the ground. Holes may be dug with an auger or drill to a depth of at least $0.5\,m$ and preferably $1.0\,m$. Holes are normally lined with plastic piping in which the detector is emplaced and the top of the pipe sealed. The detectors are normally taped to the bottom of a plastic cup, which is inverted before burial. The detector is then recovered 3–4 weeks later. This procedure is used when long-term monitoring is required to overcome problems of short-term variation in radon concentration.

Although alpha track detectors overcome many of the problems associated with temporal variation in radon fluxes, they are time-consuming to emplace, requiring two visits to each site, with all the problems in reoccupying the site. More important, they require a laboratory processing stage. In practice they are generally not favored for primary investigations, although they do have an important role to play at later stages. They are also sensitive to thoron, but the presence of a polyethylene film seal, at a distance of about $5\,cm$ from the detector, reduces the amount of short half-lived thoron while having a negligible effect on radon. The polyethylene film allows radon but not water vapor to diffuse. Water droplets on the surface of the film may also affect the recorded alpha counts and water vapor absorbers may need to be introduced into the sampling device.

C. Measurement of Radon in Water

Radon has a high partition coefficient (gas to water) so that the passage of fine gas bubbles through water provides an efficient means of extraction. The gas may be drawn into an evacuated Lucas cell. Alternatively a sealed re-circulating system may be set up. Very careful attention must be paid to the timing of both degassing and counting and careful calibration of the procedure with standardized radon solutions is required. Other methods require expensive equipment and laboratory processing. For example, the radon daughter [214]Bi pho-

topeak emission can be measured using either a sodium iodide scintillation crystal or a high-resolution, lithium-drifted germanium semiconductor detector.

D. Measurement of Radon in Solid Materials

One of the solid daughter products of radon is ^{214}Bi. This emits high-energy gamma radiation at 1.76 MeV. Gamma spectrometric determinations of uranium in the field and laboratory often make use of this photopeak on the assumption that the decay chain is in equilibrium and therefore this measurement provides an effective total radon determination. If the parent uranium mineral is resistant to weathering (e.g., thorium- and REE-rich uranium oxides and silicates, monazites, zircons, etc.) then the radium will tend to be in secular equilibrium with the uranium. In such minerals the radon loss is normally low and gamma spectrometric measurements give a good indication of the uranium contents. The measurement of radon release from solid samples requires an alternative method. Radon release from disaggregated samples (soils, stream sediments, and unconsolidated aquifer sands) may be determined by agitating a slurry of the material with distilled water in a sealed glass container, allowing a period of about 20–30 days for the generation of radon from radium, and then measuring the radon in the aqueous phase using a liquid scintillation counter. Emanation of radon from solid rock samples can be determined using a similar method.

VIII. Factors Controlling Release and Transfer of Radon Gas

Most radon remains in rocks and soil and only some of that near a free surface is released. Soil generally releases more radon than rock, as its constituents are more comminuted. The rate of release of radon from rocks and soils is largely controlled by the uranium concentration and by the types of minerals in which the uranium occurs. Once radon gas is released from minerals, the most important factors controlling its migration and accumulation in buildings include (1) transmission characteristics of the bedrock including porosity and permeability; (2) the nature of the carrier fluids, including carbon dioxide gas, surface water, and groundwater; (3) weather; (4) soil characteristics including permeability; (5) house construction characteristics; and (6) lifestyle of house occupants.

A. Mineralogical Effects

The main mineralogical factors affecting the release of radon are the solubility, internal structure, and specific surface area of uranium-bearing minerals. Uranium is very seldom homogeneously distributed throughout rocks and soils. Most of the uranium in rocks can be attributed to discrete uranium-bearing minerals, even when there is only a few mg kg^{-1} of uranium present. Because radon is a gas with a limited half-life, its chances of escaping from the parent mineral are much greater if it is generated from grain margins. Other important controls are the openness of and imperfections in the internal structure of the mineral and the specific surface area of the mineral grains.

The release of radon is generally controlled by alpha particle recoil mechanisms, which tend to expel radon from radium derived from uranium-bearing minerals. Most of the radon remains within the mineral to decay again to solid products. Only a very small proportion of the radon generated can be released by recoil. The location of the radium atoms in the mineral grains and the direction of the recoil of the radon atoms will determine whether the newly formed radon atoms enter pore spaces between mineral grains. Factors such as the specific surface area, the shape, degree of fracturing, imperfections, and even radiation-induced damage of the host uranium-bearing mineral affect the efficiency of radon expulsion. Because uranium minerals have high densities, the recoil range is usually low. However, if radium is present in intergranular films then recoil ranges varying between 20 and 70 µm would occur. The fraction of radon, produced by radium decay, that escapes from rock or soil (called the emanation coefficient) is dependent on the surface area of the source material. Emanation coefficients are greater for rocks than minerals, whereas soils usually have the highest values. If water is present in the pore space, however, the moving radon atom slows very quickly and is more likely to stay in the pore space.

Differences in the uranium-bearing minerals, and especially in the solubility of the major uranium-bearing minerals, control the amount of radon released. In some granites, for example, much of the uranium is found in the mineral uraninite (uranium oxide), which is easily weathered, especially near the surface. Uranium is more soluble in water so it is removed from the original mineral site, but the relatively insoluble radium, which is the immediate parent of radon gas, remains in a mixture of iron oxides and clay minerals. This material is a highly efficient radon generator because of the high specific surface area of the radium-bearing phase.

Radon generated from the radium can escape into the fluid phase with high efficiency thus facilitating its rapid migration to the surface. An inert gas, radon is relatively unaffected by chemical buffering reactions that often control the generation of other gases in rocks and their weathering products. In contrast, uranium in other granites may occur in chemically resistant high-thorium uraninite, zircon, monazite, and apatite, all of which liberate less radon.

The mineral associations typically found in sedimentary rocks differ significantly from those in granites. In carboniferous limestone of northern England, for example, uranium is relatively uniformly distributed and associated with finely divided organic matter in the matrix of bioclastic limestones (usually <10 mg kg^{-1} U), although it may also be concentrated in stylolites, which typically contain 20–60 mg kg^{-1} U. Even though the overall concentration of U in the limestones is below 2 mg kg^{-1}, high radon emissions are probably derived from radium deposited on the surfaces of fractures and cavities. The high specific surface area of the radium permits efficient release of radon and high migration rates are promoted by the high permeability of the limestone. In addition, uranium and radium are concentrated in residual soil overlying limestone. Radium is sometimes preferentially concentrated in soil organic material, which has a high emanation coefficient (Greeman & Rose, 1996). In black shales in the UK, uranium is located mainly in the fine-grained mud matrix, where it may be present at levels up to 20 mg kg^{-1} U, and also in organic-rich bands at concentrations up to 40 mg kg^{-1} U. Much higher uranium concentrations have been reported from the Chattanooga shale in the United States (20–80 mg kg^{-1}), the Dictyonema shale in Estonia (30–300 mg kg^{-1}), and the alum shale in Sweden and Norway (50–400 mg kg^{-1}). Uranium is rare in detrital phases and may also be remobilized and adsorbed on iron oxides. In sandstones, uranium is concentrated in primary detrital minerals, such as apatite and zircon, which can contain high concentrations of U (>100 mg kg^{-1}). Uranium may also be adsorbed onto Fe oxides in the matrix of sandstone or its weathering products. Emission of radon from sandstones is restricted by the relatively low specific surface area of the uranium minerals and appears to be more dependent upon fracturing of the rock.

B. Transmission Characteristics of Bedrock

Although the generation of high levels of radon is ultimately dependent upon the concentration of uranium and upon the nature of the parent mineral, the transmission of radon gas to the surface is largely independent of these characteristics (Åkerblom & Mellander, 1997). Once the radon is released from the parent mineral into the space between mineral grains (the intergranular region) other factors take over. The most notable of these are (1) the fluid transmission characteristics of the rock including permeability, porosity, pore size distribution, and the nature of any fractures and disaggregation features and (2) the degree of water retention (saturation) of the rocks. Faults and other fractures permit the efficient transmission of radon gas to the surface. The presence of faults with their enhanced fluid flow frequently results in high radon in soil gases (Ball et al., 1991).

C. Carrier Fluids

Radon readily diffuses into pores and cavities from mineral surfaces. However, its relatively short half-life (3.8 days) limits the distance over which diffusion may occur. In highly permeable dry gravel, radon has decayed to 10% of its original concentration over a diffusion length of 5 m (UNSCEAR, 2000). In more normal soils, which are generally moist, this distance would be substantially less. Diffusive ^{222}Rn in soil gas can be determined from the specific ^{226}Ra activity, specific density, effective porosity, and radon emanation coefficients of soils and rocks (Washington & Rose, 1992). In caves, radon concentrations of approximately 100 Bq m^3 would be expected if radon were generated by diffusion from solid limestone with 2.2 mg kg^{-1} U. However, the enhanced concentration of radon in caves suggests that structurally controlled convective transport of radon in fluids along faults, shear zones, caverns, or fractures is more significant than diffusive transport. Transportation of radon in this way may exceed 100 m.

Following radon release, migration in carrier fluids, such as carbon dioxide gas or water, is considered to be the dominant means of gas transmission to the surface. Radon release is higher in rocks that have a high surface area in contact with groundwater. Once released from the uranium minerals, radon migration is dependent upon the fluid flow characteristics of the rock and soil. In water, convective or pressure gravity flow mechanisms can influence migration of the radon, whereas in a gas the transport may be controlled by the diffusion characteristics of the carrier gas. Water flow below the water table is generally relatively slow as is groundwater transport in the soil aquifer (<1–10 cm per day).

Thus all hydraulically transported radon will have decayed over a distance of less than 1–2 m. Radon is likely to be carried away more quickly by fluids in areas of permeable rocks such as limestones. The carrier effect may also be important for other rock types. Carbon dioxide may collect radon gas in the unsaturated zone and transport it along fractures, fissures, and faults. In situations where the carbon dioxide flux is high, such as in active volcanic areas, radon may be either diluted or enhanced because of rapid transport from the generation zone to the surface.

Radon in surface water is not generally accompanied by dissolved radium. In surface stream waters, the radon concentration appears to be more closely related to the radium concentration of the stream sediment. However, the radon concentration in surface streams is usually far too low for more than a very small degree of transfer across the air/water interface to occur, unless a gas phase is introduced. Radon dissolved in subsurface fluids migrates over long distances along fractures and caverns depending on the velocity of fluid flow. Radon is soluble in water and may thus be transported for distances of up to 5 km in streams flowing underground in limestone. Radon remains in solution in the water until a gas phase is introduced, for example, by turbulence or by pressure release. If emitted directly into the gas phase, as may happen above the water table, the presence of a carrier gas, such as carbon dioxide, would tend to induce migration of the radon. This appears to be the case in certain limestone formations, where underground caves and fissures enable the rapid transfer of the gas phase.

D. Weather

The principal climatic factors affecting radon concentrations are barometric pressure, rainfall, and wind velocity. In the absence of a less permeable humic or clay rich topsoil, radon concentration in soil gas varies directly with barometric pressure, and to a lesser extent, inversely with wind speed. Where the topsoil is finer grained and more humic, the effects of barometric pressure and wind velocity are reduced to a marginal role. The extent to which rainfall affects radon concentrations depends on the permeability of the soil. For permeable soils, radon concentrations are only affected during precipitation when saturation of small pore spaces with moisture effectively prohibits the rapid outgassing of radon from the soil. This causes the buildup of radon below the moisture-saturated surface layer and

increases of an order of magnitude are sometimes observed. Prolonged rainfall may penetrate deeply and seal the pore spaces in the soils to a considerable depth.

A similar buildup of radon is often observed during the night when dew forms on the surface and this can result in a twofold increase in soil gas alpha activity. Sealing of the pore spaces by near-surface moisture can result in temporary entrapment of radon in soil gases, with a significant increase in total gamma activity from radon daughters (^{214}Bi). Dry conditions cause clay-rich soils to dry out and to fracture, allowing easier egress for the soil gases and hence an increase in radon activity at the soil surface. Seasonal variation in soil pore radon concentration was observed by Rose et al. (1990) who found that the radon concentration tended to be lower in the winter and higher in the summer, often varying by a factor of 3–10. The variation was attributed largely to changes in the soil moisture content with more radon held in solution in the soil pore water during the winter. The variation is greater in the soil above 70 cm depth than below this depth, presumably due to greater short-term fluctuations in soil moisture content. This suggests that radon in soil gas measurements should be taken at depths greater than 70 cm in order to reduce the effects of temporal variations caused by rainfall.

Although barometric pressure and rainfall obviously cause temporal variation in radon concentration (indicated by alpha activity), it is encouraging to note that the soil gas radon fluxes in areas that are not mineralized appear to be relatively uniform. Various rock types have been tested and the site variation is often less than that between adjacent rock types. This is particularly important for radon potential mapping based on the measurement of radon in soil gas.

E. Soil Characteristics

The principal soil properties that influence the concentration of radon in soil gas, including the rate of release of radon and its transfer through soils, are soil permeability and soil moisture. In general soil permeability depends on such factors as soil texture, structure, median pore diameter, pore size distribution, pore volume, packing density, soil bulk density, and grain size. Soil mineralogy is an important factor controlling soil gas radon concentrations; in some cases, organically bound ^{226}Ra can be a principal source of ^{222}Rn in soil gas (Greeman et al., 1990; Greeman & Rose, 1996). Radon volume activity increases with the percentage of coarse material in the soil thus confirming the general correlation between radon fluxes and soil permeability. In general, coarse gravelly soils will tend to have higher radon fluxes than impermeable clay soils. However, humic and clay soils may be impermeable in the winter when saturated with water or filled with ice if the ground is frozen, and during very dry periods they may crack and behave in a permeable manner. Soil permeability and rainfall (soil saturation) exert a considerable control on radon concentrations in houses. Soil permeability generally closely reflects the permeability of the underlying rocks and superficial deposits such as glacial till, alluvium, or gravel. Radon diffuses more slowly through water than air, so water-saturated soils impede the diffusion of radon enough for it to decay to harmless levels before it has diffused more than 5–10 cm. Consequently, radon from water-saturated soils is unlikely to enter buildings unless it is transported in other gases such as carbon dioxide or methane.

It is important to remember that whereas the top meter of the soil profile is generally removed during the construction of foundations for a dwelling, only a few centimeters of topsoil are removed from the remainder of the subfloor space. Indeed, in many cases the soil profile beneath a dwelling will not be unduly influenced by temporal variations in rainfall. The influence of the geochemistry and permeability of the bedrock or overburden beneath a dwelling on the potential for radon emissions from the ground may be greater than near-surface soil properties.

Only 10–50% of the radon produced in most soils escapes from the mineral grains and enters the pores. Soils in the United States generally contain between 5 and 80 Bq L^{-1} of radon. Drier, highly permeable soils and bedrock—such as limestones, coarse glacial deposits, and fractured or cavernous bedrock, and hill slopes—are usually associated with relatively high levels of indoor radon. The permeability of the ground permits radon-bearing air to move greater distances before it decays, and thus contributes to high indoor radon even if the radon content of soil gas is in the normal range (5–50 Bq L^{-1}).

IX. RADON MIGRATION PATHWAYS

When considering *natural migration pathways*, it should be noted that although the general direction and position of planar discontinuities and openings including bedding planes, joints, shear zones, and faults can be determined by detailed structural mapping, the precise location of such migration pathways is often difficult to establish, especially if the area is covered with soil or drift. In the United States, high radon is associated with U-enriched shear zones in granites, which are characterized by high radon in soil gas and groundwater. Indeed, some of the highest indoor radon levels in the United States are associated with sheared fault zones. Similar observations have been made in southwest England. Radon and other gases are known to concentrate and migrate upward along faults and through caves and other solution cavities. However, natural cavities such as potholes and swallow holes in limestone would also be difficult to locate precisely due to their irregular and relatively unpredictable disposition.

Artificial pathways underground include mine workings and disused tunnels and shafts. Radon concentrations in old uranium mine workings are commonly 10,000–60,000 Bq m^{-3} and can be as high as 7,100,000 Bq m^{-3} (Gilmore et al., 2001). High radon is known to be associated with gassy ground overlying coal-bearing rock strata. In addition, relatively randomly orientated and distributed blasting and subsidence fractures will affect areas underlain by mined strata. The sites and disposition of recent coal mine workings in some countries may be obtained from mine records, although these may not be reliable. Other artificial pathways related to near-surface installations include electricity, gas, water, sewage, and telecommunications services, the location of which may be obtained from the local service agencies. Land drains provide another potential migration pathway. The detection and prediction of migration pathways is difficult and may be imprecise, although a detailed geological and historical assessment together with appropriate radon gas monitoring and a detailed site investigation should provide a reasonable assessment of the source and radon gas migration pathways. Information on the local geology may be obtained from maps, memoirs, boreholes, and site investigation records.

X. FACTORS AFFECTING RADON IN BUILDINGS

The design, construction, and ventilation of the home affect indoor radon levels. Radon can enter a home through cracks in solid floors and walls below construction level; through gaps in suspended concrete and timber floors and around service pipes; and through

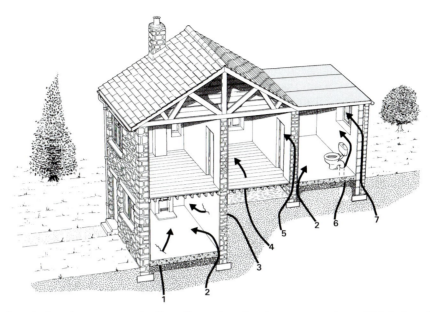

FIGURE 3 Routes by which radon enters a dwelling. (Reproduced with permission from CRC Ltd., publishers of BR211, BRE, 1999.)

crawl spaces, cavities in walls, construction joints, and small cracks or pores in hollow-block walls (Figure 3). Radon concentrations are generally highest in basements and ground floor rooms that are in contact with the soil or bedrock. Air released by well water during showering and other household activities may also contribute to indoor radon levels, although this generally makes a relatively small contribution to the total radon level.

When constructing a house with a basement in the United States, a hole is dug, footings are set, and coarse gravel is usually laid down as a base for the basement slab. The gap between the basement walls and the ground outside is backfilled with material that often is more permeable than the original ground. Radon moves into this permeable material and the gravel bed underneath the slab from the surrounding soil. The backfill material is typically rocks and soil from the foundation site but may be imported material with different radon emanation characteristics to the local rocks and soils. Therefore, the amount of radon in the permeable material depends on the amount of uranium in the local or imported rock as well as the type, permeability, and moisture content of the soil. The backfill layer will need to have a thickness of at least 1 m or have a very high Ra^{226} concentration for it to be a significant source of radon in the building.

In a typical masonry building in which radon occurs at the UK national average level of $20\,Bq\,m^3$,

approximately 60% of radon comes from the ground on which the building stands, 25% from building materials, 12% from fresh air, 2% from the water supply, and 1% from the gas supply. These figures apply to the average house in the UK, but can vary substantially, and the proportion of radon entering a home from the ground will normally be much higher in homes with high radon levels. The dominant mechanism of radon ingress is pressure-induced flow through cracks and holes in the floor. Slightly negative pressure differences between indoor and outdoor atmospheres caused by wind outside and heating inside the building draw radon contaminated air into the building, especially through the floor. Energy-conserving measures such as double-glazing restrict the fresh supply of air and lessen the dilution of radon indoors. Conversely, they may also reduce the pressure difference between indoors and outdoors and thus reduce the influx of radon from the ground. Poor ventilation may increase radon concentrations, but it is not the fundamental cause of high indoor radon levels.

Indoor radon concentrations are generally about 1000 times lower than radon in the soil underlying the house. Most houses draw less than one percent of their indoor air from the soil with the remainder from outdoors where the air is generally quite low in radon. In contrast, houses with low indoor air pressures, poorly sealed foundations, and several entry points for soil air may draw as much as 20% of their indoor air from the soil. Consequently, radon levels inside the house may be very high even in situations where the soil air has only moderate amounts of radon.

Clavensjö and Åkerblom (1994) suggested that the ^{222}Rn concentration in a building or room that results from the transport of soil air may be calculated by the following formula:

$$C_{building} = \frac{C_t \cdot L}{V_{building} \cdot (n + \lambda)}$$

$C_{building}$ = ^{222}Rn concentration in the building/room (Bq m^{-3})

C_t = ^{222}Rn concentration in the soil air entering from the ground (Bq m^{-3})

L = volume of soil air entering from the ground (m^3 h^{-1})

n = rate of air change in the building (air changes h^{-1})

λ = decay constant, for ^{222}Rn, $7.55 \cdot 10^{-3}$ (h^{-1})

$V_{building}$ = building/room volume (m^3)

XI. RADON POTENTIAL MAPPING METHODS

Accurate mapping of radon-prone areas helps to ensure that the health of occupants of new and existing dwellings and workplaces is adequately protected. Radon potential maps have important applications, particularly in the control of radon through planning, building control, and environmental health legislation. Radon potential maps can be used (1) to assess whether radon protective measures may be required in new buildings, (2) for the cost-effective targeting of radon monitoring in existing dwellings and workplaces, and (3) to provide a radon assessment for homebuyers and sellers. It is important, however, to realize that radon levels often vary widely between adjacent buildings due to differences in the radon potential of the underlying ground as well as differences in construction style and use. Whereas a radon potential map can indicate the relative radon risk for a building in a particular locality, it cannot predict the radon risk for an individual building. In the UK, radon potential maps generally indicate the probability that new or existing houses will exceed a radon reference level, which in the UK is called the action level (200 Bq m^{-3}). In other countries, geological radon potential maps predict the average indoor radon concentration (United States) or give a more qualitative indication of radon risk (Germany and the Czech Republic).

Two main procedures have been used for mapping radon-prone areas. The first uses radon measurements in existing dwellings to map the variation of radon potential between administrative or postal districts or grid squares. The second is geological radon potential mapping in which each geological feature is assigned to a radon potential class based on the interpretation of one or more of the following types of data: (1) radon concentrations in dwellings (indoor radon), (2) concentration, mineralogical occurrence, and chemical state of uranium and radium in the ground (radiometric and geochemical data), (3) rock and soil permeability and moisture content, (4) concentration of radon in soil gas, and (5) building architecture (construction characteristics). Because the purpose of maps of radon-prone areas is to indicate radon levels in buildings, maps based on actual measurements of radon in buildings are generally preferable to those based on other data.

Procedures for monitoring and surveys of radon in dwellings are described in Nazaroff (1988) and Miles (2001). In the UK, measurements are made with passive integrating detectors over a period of three months whereas short-term "screening" measurements taken over a 2- to 7-day period are commonly used for mapping in the United States. Measurements carried out over less than a year should be corrected for seasonal variations. In the United States, houses with basements typically have higher indoor radon than those with slab-on-grade construction because basements tend to have more entry points for radon and a lower internal pressure relative to the soil than non-basement homes. Architecture type is one factor within the Radon Index Matrix used to estimate geological radon potential in the United States (Gundersen & Schumann, 1996).

Bungalows and detached houses tend to have higher indoor radon than terraced houses or flats in the same area of the UK. Building material, double-glazing, draught-proofing, date of building, and ownership also have a significant impact on indoor radon concentrations. Radon potential mapping is sometimes based on indoor radon data that have been normalized to a mix of houses typical of the housing stock as this removes possible distortion caused by construction characteristics. Maps based on results corrected for temperature but not normalized to a standard house mix reflect such factors as the greater prevalence of detached dwellings in rural areas, and hence the higher risk of high radon levels in rural areas compared with cities where flats are usually more prevalent. Radon potential estimates based on radon levels in the actual housing stock are more appropriate for the identification of existing dwellings with high radon.

Requirements for mapping radon-prone areas using indoor radon data are similar whether the maps are made on the basis of grid squares or geological units. These requirements include (1) accurate radon measurements made using a reliable and consistent protocol, (2) centralized data holdings, (3) sufficient data evenly spread, and (4) automatic conversion of addresses to geographical coordinates. It appears that Great Britain is the only country that currently meets all of these requirements for large areas. In countries where lesser quality or quantity of indoor radon data are available, there is greater reliance on proxy data for radon potential mapping (e.g., Czech Republic, Germany, Sweden, and United States). Where there are no existing houses or indoor radon measurements, proxy data (such as soil gas radon concentrations) are required to map radon potential.

Mapping levels of radon in administrative areas has the advantage of simplifying any subsequent administrative action. Use of grid squares allows an appropriate size of area to be chosen and simplifies the analysis. Use of geological boundaries may help to delineate differences in radon potential with greater spatial accuracy than other types of boundary. Whereas a wide variety of factors affect the concentration of radon in buildings, regional variations are related principally to the geological characteristics of the ground. Indoor radon surveys in the UK have confirmed the association of high levels of radon in dwellings with uraniferous granites, uraniferous sedimentary rocks, permeable limestones, and phosphatic ironstones, as well as fault and shear zones. Similar observations have been made in the Czech Republic, Germany, Luxembourg, Sweden, and the United States.

It is important to remember that however indoor radon data are grouped (whether by grid square or geological unit), a wide range of indoor radon levels is likely to be found. This is because there is a long chain of factors that influence the radon level found in a building, such as radium content and permeability of the ground below it, and construction details of the building (Miles & Appleton, 2000). Radon potential does not indicate whether a building constructed on a particular site will have a radon concentration that exceeds a reference level. This can only be established through measuring radon in the building.

A. Non-Geological Radon Potential Mapping

Radon measurements in existing dwellings are used to map the radon potential of countries (Figure 4), administrative districts, or grid squares without taking into consideration the geological controls on radon in dwellings. Because the factors that influence radon concentrations in buildings are largely independent and multiplicative, the distribution of radon concentrations is usually lognormal, so lognormal modeling can be used to produce accurate estimates of the proportion of homes above a reference level (Miles, 1998). In the UK, maps show the fraction of the housing stock above the action level in each 5-km grid square (Figure 5) (Lomas et al., 1996). Where house radon data are plentiful, maps using grid squares smaller than 5 km can be made. In some cases, this method can show up variations that are obscured by general geological grouping, such as variations in radon potential within a geological unit. Investigations in southwestern England revealed that the finer the grid, the closer the correlation with the geological controls of radon in dwellings.

Radon potential mapping using indoor radon measurements has been carried out in other European countries which include Ireland, Luxembourg, and France, but the maps are not as detailed as the NRPB maps of the UK. This is mainly due to the relatively low measurement density and restricted coverage.

In the United States most measurements of indoor radon have been made using short-term charcoal monitors, so these cannot be used directly to estimate long-term average radon levels. Although individual short-term measurement results are poor indicators of radon potential, aggregations of them can be corrected for bias and can provide useful information where no long-term results are available. A statistical technique known as Bayesian analysis improves estimates of mean radon level in areas where the data are sparse. A U. S. EPA survey of radon in homes covered about 6000 homes across the United States, all measured using long-term etched track detectors. The data distribution is very sparse, given the size of the country, but it can provide estimates of the mean radon levels and distributions for the whole country and states, although not for smaller areas.

B. Geological Radon Potential Mapping

The most accurate and detailed radon potential maps are generally those based on house radon data and geological boundaries provided that the indoor radon data can be grouped by sufficiently accurate geological boundaries. In the absence of an adequate number of high quality indoor radon measurements, proxy indicators such as soil gas radon data or information on U content may be used to assess geological radon

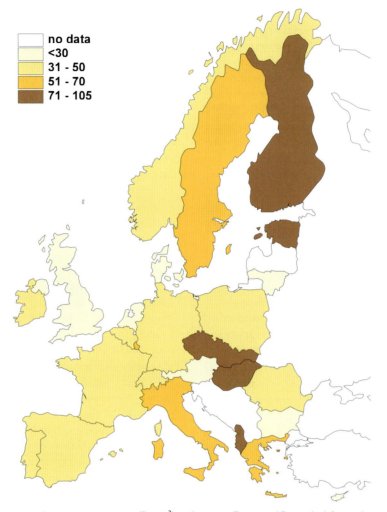

FIGURE 4 Geometric mean radon concentrations ($Bq\,m^{-3}$) indoors in Europe. (Compiled from data in UNSCEAR, 2000.)

potential. The reliability of maps based on proxy data increases with the number of classes as well as the quantity and quality of available data. Radon potential maps based on indoor radon data grouped by geological unit have the capacity to accurately estimate the percentage of dwellings affected together with the spatial detail and precision conferred by the geological map data (Miles & Ball 1996). The reliability and spatial precision of mapping methods is, in general, proportional to the indoor radon measurement density. It is, however, reassuring that even when the measurement density is as low as the minimum for 5-km grid square mapping (i.e., 0.2–0.4 per km^2), geological radon potential mapping discriminates between geological units in a logical way. These relationships can be explained on the basis of the petrology, chemistry, and permeability of the rock units and are confirmed in adjoining map sheets with higher measurement densities (Miles & Appleton, 2000).

Geological radon potential maps of the UK have been produced at 1:625,000, 1:250,000, and 1:50,000. Each geological unit within a map sheet or smaller area, such as a 5-km grid square, has a characteristic geological radon potential that is frequently very different from the average radon potential for the grid square shown (Figure 6). Lithological variations within geological units can cause geological radon potential mapping to miss significant areas of higher radon potential identified by 1-km grid square mapping. Geological and grid square mapping are likely to be most powerful when used in a complementary fashion by integrating maps produced by the two methods and by grouping results both by geological unit and by grid square (Appleton & Miles, 2002).

Uranium and radium concentrations in surface rocks and soils are useful indicators of the potential for radon emissions from the ground. Uranium can be estimated

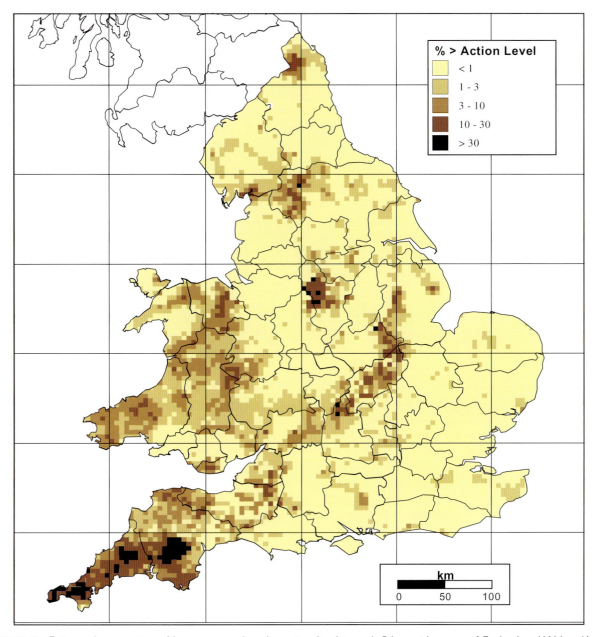

FIGURE 5 Estimated proportion of homes exceeding the action level in each 5-km grid square of England and Wales. (Adapted from Figure 2-2 in Appleton et al., 2000a.)

by gamma spectrometry either in the laboratory or by ground, vehicle, or airborne surveys. The close correlation between airborne radiometric measurements and indoor radon concentrations has been demonstrated in the United States in Virginia and New Jersey, Nova Scotia in Canada, and also in parts of England (Appleton & Ball, 2001). Duval and Otton (1990) identified a linear relationship between average indoor radon levels and surface radium content for soils of low

to moderate permeability. However, areas with high permeability ($>50\,cm\,h^{-1}$) had significantly higher indoor radon levels than would otherwise be expected from the ^{226}Ra concentrations, which reflects an enhanced radon flux from permeable ground. Grasty (1997) demonstrated that any estimate of natural gamma ray flux from the uranium decay series (i.e., radium) in the ground must take into consideration the radon coefficient of the soil as well as its radon

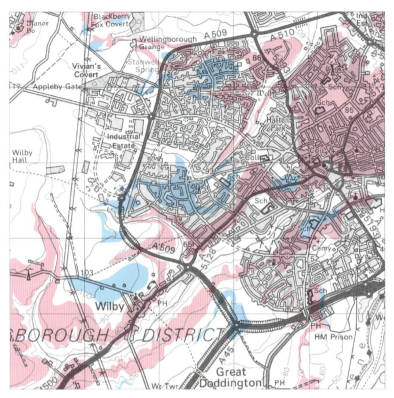

FIGURE 6 Geological radon potential map of the 5-km grid square (485265) that encompasses the western sector of Welling-borough, England. The 1:50,000 scale map illustrates the distribution of geological units with <3% (white), 3–5% (blue), and 10–20% (pink) of dwellings above the UK radon action level. The 5-km grid square has an average radon potential of 3.9% (NRPB 1998 data). (Topography based on Ordnance Survey 1:50,000 Scale Colour Raster data with permission of The Controller of Her Majesty's Stationery Office Crown Copyright. Ordnance Survey Licence number GD272191/2004.)

diffusion coefficient, which depends largely on soil moisture. Clay soils tend to have higher eU when wet whereas sandy soils have lower eU (Grasty, 1997).

Sweden was the first country to make use of airborne gamma ray spectrometry data to produce maps of radon potential. Radon potential is estimated and mapped on the basis of available data including (1) geology, (2) airborne radiometric surveys (covering 65% of Sweden), (3) results from radiometric surveys of the ground, (4) results from radon surveys in buildings, (5) results from earlier geotechnical investigations (e.g., permeability and groundwater level), (6) field surveys including gamma spectrometry, and (7) orientation soil gas radon measurements. Åkerblom (1987) established a simple threefold radon risk classification based on geology, permeability, and soil gas radon (Table VIII). These criteria are used at a mapping scale of 1:50,000 or larger in conjunction with airborne gamma spectrometry surveys to produce provisional radon risk maps.

Radon risk mapping of the Czech Republic at a scale of 1:500,000 (Figure 7) is based upon a number of data

TABLE VIII. Criteria Used in Sweden for Classifying High- and Low-Radon Ground

Bedrock or overburden	^{226}Ra $(Bq\,kg^{-1})$	^{222}Rn in soil gas $(Bq\,L^{-1})$
High radon ground		
Bare rock	>200	Not relevant
Gravel, sand, coarse till	>50	>50
Sand, coarse silt	>50	>50
Silt	>70	>60
Clay, fine till	>110	>120
Low radon ground		
Bare rock	<60	
Gravel, sand, till	<25	<20
Silt	<50	<20
Clay, fine till	<80	<60

After Clavensjö and Åkerblom, 1994.

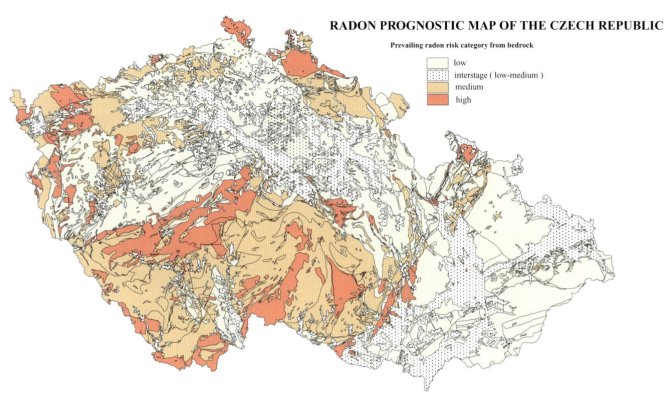

FIGURE 7 Radon prognostic (risk) map of the Czech Republic. (Reproduced with permission from the Czech Geological Survey.)

sets for airborne radiometry, geology, pedology, hydrogeology, ground radiometry, and soil gas radon. Rock and soil permeability were obtained from hydrogeological and pedological maps and reports. Radon risk categories (low, medium, and high) were established for geological and lithological units and were based upon a rigid set of rules accepted by the Ministry of Environment. Radon risk maps are currently produced at the 1:50,000 scale, and these can be used for the identification of dwellings exceeding the guidance level to an accuracy of 70–80% (Mikšová & Barnet, 2002). However, the maps are not recommended for the prediction of the requirement for radon protective measures in new buildings for which soil gas radon site assessments using the Czech Radon Risk Classification For Foundation Soils (Table IX) are required.

Airborne radiometric survey data were used to produce the first radon potential maps in the United States. The U. S. EPA radon map was developed using five factors to determine radon potential (indoor radon measurements, geology, aerial radioactivity, soil permeability, and foundation type). Radon potential assessment is based on geologic provinces adapted to county

TABLE IX. Czech Republic Radon Risk Classes Based on Radon in Soil Gas and Rock-Overburden Permeability

	Rock-overburden permeability		
	High	Medium	Low
Radon risk	Radon concentration in soil gas $(Bq L^{-1})$		
High	>30	>70	>100
Medium	10–30	20–70	30–100
Low	<10	<20	<30

After Barnett, 1994.

boundaries for the Map of Radon Zones (Figure 8). The purpose of the map is to assist national, state, and local organizations to implement radon-resistant building codes. In the United States, high geological radon potential is associated with granites, limestones, black

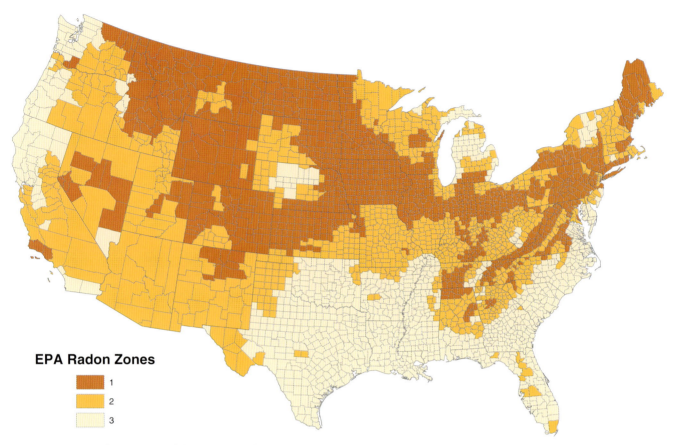

FIGURE 8 U. S. EPA Map of Radon Zones (excluding Alaska and Hawaii). Zone 1, 2, and 3 counties have a predicted average indoor radon screening concentration of >148, 74–148, and <74 Bq m^{-3}, respectively. (Map based on state radon potential maps available at http://www.epa.gov/radon/zonemap.html; state and county boundaries SRI ArcUSA 1:2M.)

shales, and glacial tills and gravels derived from these bedrocks in the Appalachians; sandy and clay tills derived from sandstones, limestones, and black shales in the northern Great Plains and Great Lakes areas; and uraniferous granites, permeable limestones, sedimentary, and metamorphic rocks together with derived colluvial and alluvial deposits in the Rocky Mountains and parts of the western Great Plains.

After uranium and radium concentration, the permeability and moisture content of rocks and soils is probably the next most significant factor influencing the concentration of radon in soil gas and buildings. Radon diffuses farther in air than in water, so in unsaturated rocks and overburden with high fluid permeability, higher radon values are likely to result from a given concentration of uranium and radium than in less permeable or water-saturated materials. Weathering processes can also affect permeability. Enhanced radon in soil gas is also associated with high-permeability features such as fractures, faults, and joints. The fracturing of clays,

resulting in enhanced permeability, combined with their relatively high radium content and their emanation efficiency may also result in higher radon concentrations in dwellings. The permeability of glacial deposits exerts a particularly strong influence on the radon potential of underlying bedrock.

It has been demonstrated in a number of countries, including Canada, Germany, the UK, the United States, and Sweden, that soil gas radon measurements combined with an assessment of ground permeability can be used to map geological radon potential in the absence of sufficient indoor radon measurements. Significant correlations between average indoor and soil gas radon concentrations, grouped according to geological unit, have been recorded in the Czech Republic (Figure 9), Germany, the UK, and the United States. Where low correlations have been measured between radon in soil gas and radon in adjacent houses, the probable causes include: (1) the small number of houses with variable design in the study, (2) single rather than multiple soil

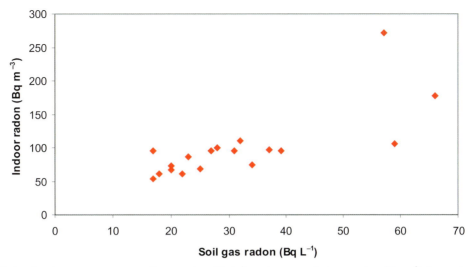

FIGURE 9 Relationship between average soil gas radon (Bq L^{-1}) and average indoor radon (Bq m^{-3}) for major rock types of the Czech Republic. (Based on data in Table 1, Barnet et al., 2002.)

gas measurements, (3) short-term indoor radon measurements, and (4) a mixture of summer and winter measurements. Spot measurements of soil gas radon and short-term indoor data are known to be relatively unreliable and it is now generally accepted that 10–15 soil gas radon measurements are required to characterize a site or geological unit.

XII. RADON SITE INVESTIGATION METHODS

Radon migrates into buildings as a trace component of soil gas. Therefore the concentration of radon in soil gas should provide a good indication of the potential risk of radon entering a building if its construction characteristics permit the entry of soil gas. There is a growing body of evidence that supports the hypothesis that soil gas radon is a relatively reliable indirect indicator of indoor radon levels at the local as well as the national scale (Figures 9 and 10).

Soil gas radon data may be difficult to interpret due to the effects of large diurnal and seasonal variations in soil gas radon close to the ground surface and variations in soil gas radon on a scale of a few meters. The former problem may be overcome by sampling at a depth greater than 70 cm or by the use of passive detectors with relatively long integrating times, although this may not be a practical option if site investigation results are

required rapidly. Small-scale variability in soil gas radon may be overcome by taking 10–15 soil gas radon measurements on a 5- to 10-m grid to characterize a site. Radon in soil gas varies with climatic changes including soil moisture, temperature, and atmospheric pressure. Weather conditions should be as stable as possible during the course of a soil gas radon survey. A range of methods such as controlled gas extraction, air injection procedures, or water percolation tests can be used to estimate gas permeability at a specific site. In the absence of permeability measurements, more qualitative estimates of permeability can be based on visual examination of soil characteristics, published soil survey information, or on the relative ease with which a soil gas sample is extracted.

In some areas and under some climatic conditions, site investigations using soil gas radon cannot be carried out reliably, for example, when soil gas cannot be obtained from waterlogged soils or when soil gas radon concentrations are abnormally enhanced due to the sealing effect of soil moisture. These conditions are particularly common in winter. Problems with the determination of permeability and its incorporation into a radon site investigation procedure have been encountered in the Czech Republic where the quality of the permeability classification obtained at a site is very reliant on the personal experience of the technical staff carrying out the site investigation. If soil gas radon concentrations cannot be determined because of climatic factors, measurement of radon emanation in the laboratory or gamma spectrometric measurement of eU can

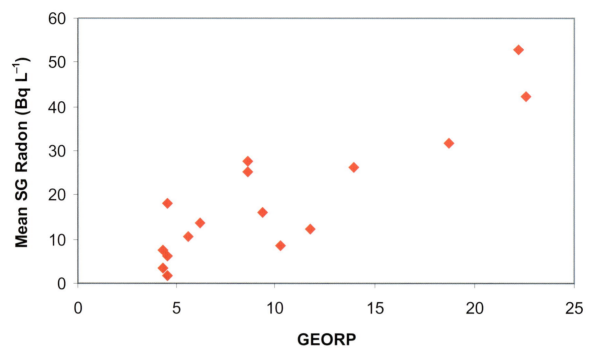

FIGURE 10   Relationship between average soil gas radon concentration (BqL⁻¹) and the geological radon potential (GEORP = estimated proportion of dwellings exceeding the UK radon action level, 200 Bqm⁻³. Data for dwellings sited on the Jurassic Northampton Sand Formation grouped by 5-km grid square). (Reproduced from Appleton et al, 2000a.)

be used as radon potential indicators in some geological environments. However, few data are available and the methods have not been fully tested.

The Swedish National Board of Housing, Building, and Planning has adopted a ground classification based on geology, permeability, and soil gas radon measurement. This procedure is used to predict radon emissions expected on a particular construction site. Finland has adopted a similar radon risk classification of building ground based on radioactivity and permeability. Measurement of radon emanation coefficients and radium concentrations by gamma spectrometry is also used to investigate radon characteristics of the ground in new building areas where buildings are to be constructed on unconsolidated sediments or directly onto bedrock (Table VIII) (Clavensjö & Åkerblom, 1994). In Germany an empirical ranking classification has been developed for radon potential based on median soil gas radon and permeability measured by air injection through the soil gas probe. All new development sites in the Czech Republic require a site investigation comprising a geological survey and measurement of radon in soil gas. The radon risk classification (Table IX) is based upon soil gas radon concentration limits and is

broadly similar to classifications used in Finland, the UK, the United States, and Sweden.

XIII. GEOLOGICAL ASSOCIATIONS

Relatively high levels of radon emissions are associated with particular types of bedrock and unconsolidated deposits, for example, some, but not all, granites, uranium-enriched phosphatic rocks, and shales rich in organic materials; soils over some limestones; and some permeable sandstones. Rock types that are high radon sources in the United States include:

1. Uraniferous metamorphic rocks and granites—some of the highest indoor levels in the United States, particularly in the Rocky and Appalachian ranges and the Sierra Nevada are associated with fault shear zones in these rocks
2. Marine black shales are sources of high radon throughout the United States and especially in the central region from Ohio to Colorado

3. Glacial deposits derived from uranium-bearing rock and sediment, especially in the northwestern Midwest, where high radon emanation reflects large surface area and high permeability caused by cracking when dry

4. Soils derived from carbonate, especially karstic terrain, which are high in uranium and radium

5. Uranium mining residues and mine tailings in the states of the western United States (e.g., Colorado)

6. Phosphate ore close to the surface and in mining waste on the surface, can result in high radon concentrations, especially in Polk County, Florida.

The maximum ^{226}Ra concentration in phosphate ores is typically about 50 times greater than the average concentration in soil. Releases from coal residues and the burning of natural gas and coal complete the list of major contributors to atmospheric radon (Gundersen et al., 1992).

Enhanced levels of radon in houses and soil gas in the UK are associated with the following geological associations. The *uranium* association comprises rocks and their weathering products containing enhanced levels or uranium or radium. The *permeable rock* association comprises permeable rocks, unconsolidated overburden, and their weathering products. Areas underlain by less permeable rocks, unconsolidated overburden, and soils, especially where these have low uranium concentrations, are generally characterized by low radon in houses and soil. Low radon is also associated with permeable sandstones containing low quantities of uranium.

The uranium association comprises granites in southwestern England characterized by high uranium concentrations, a deep weathering profile, and uranium in a mineral phase that is easily weathered. Although the uranium may be removed through weathering, radium generally remains *in situ* (Ball & Miles, 1993). Radon is easily emanated from the host rock and high values of radon have been measured in groundwaters and surface waters (110–740 Bq L^{-1}) and also in soil gas (frequently >400 Bq L^{-1}). There is a clear correspondence between areas where more than 30% of the house radon levels are above the action level and the major granite areas (Ball & Miles, 1993).

The depositional and diagenetic environment of many black shales leads to enrichment of uranium. For example, some Carboniferous shales in northern England contain 5–60 mg kg^{-1} uranium. Weathering and secondary enrichment can substantially enhance U levels in soils derived from these shales. It is found that 15–20% of houses sited on uraniferous shales with >60 mg kg^{-1} U and high soil gas radon (32 Bq L^{-1}; Ball et al., 1992) are above the UK radon action level.

Uranium-enriched phosphatic horizons occur in the Carboniferous Limestone, the Jurassic oolitic limestones, and in the basal Cretaceous Chalk in the UK and these sometimes give rise to high radon in soil gases and houses. Many iron deposits are phosphatic and slightly uraniferous and a large proportion (>20%) of houses underlain by the Northampton Sand Formation (NSF) ironstone in England are affected by high levels of radon (Figures 5 and 10). Phosphatic pebbles from the Upper Jurassic, and Lower and Upper Cretaceous phosphorite horizons in England contain 30–119 mg kg^{-1} U. Radon in dwellings is a significant problem in areas where these phosphatic rocks occur close to the surface, especially if the host rocks are relatively permeable. The NSF consists of ferruginous sandstones and oolitic ironstone with a basal layer up to 30 cm thick containing phosphatic pebbles. Whereas the ferruginous sandstones and ironstones mainly contain low concentrations of U (<3 mg kg^{-1}), the phosphatic pebbles contain up to 55 mg kg^{-1}. It is, however, probable that the mass of the NSF, which in many cases contains disseminated radium, may contribute more to the overall level of radon emissions than the thin U-enriched phosphate horizons.

High levels of radon occur in both soil gas and houses underlain by Carboniferous Limestone in the UK as well as in caves and mines. There are 10% to more than 30% of houses built on the limestones that have radon concentrations greater than the UK action level (Appleton et al., 2000a). Much of the radon is thought to emanate from uranium- and radium-enriched residual soils that overlie the highly permeable limestones.

Chalk is a particularly abundant limestone in the south of England, but its radon emanation characteristics are different from the Carboniferous and Jurassic limestones. Chalk still retains its primary porosity, although most of the water and gas flow is through fissures. The proportion of dwellings with radon above the action level is much lower than over the Carboniferous Limestone, but higher levels of radon occur where the chalk is covered with congeliturbate and residual clay-with-flint deposits.

Thick, permeable Cretaceous sand formations in southwestern England, including the glauconitic Lower and Upper Greensand and the Upper Lias Midford Sands, all emanate high levels of soil gas radon (mean values 20–48 Bq L^{-1}), and are characterized by a high proportion of houses above the action level (13 and 22% for the Upper Greensand and Midford Sands, respectively). In contrast, impermeable mudstones and clays in England and Wales are generally characterized by low to moderate soil gas radon (about $20\,Bq\,L^{-1}$) and less than 1% of homes exceed the action level.

Similar associations between high radon and Lower Carboniferous limestones, Namurian uraniferous and phosphatic black shales, and some granites and highly permeable fluvioglacial deposits have also been recorded in Ireland (Cliff & Miles, 1997).

In the Czech Republic, the highest indoor and soil gas radon levels are associated with the Variscan granites, granodiorites, syenites, and phonolites of the Bohemian massif. Syenites contain $12–20\,mg\,kg^{-1}\,U$ and the phonolites have $10–35\,mg\,kg^{-1}\,U$ and soil gas radon levels up to more than $450\,Bq\,L^{-1}$. High radon is also associated with Paleozoic metamorphic and volcanic rocks and also with uranium mineralization in the Příbram area (Barnet et al., 2002; Mikšová & Barnet, 2002).

In Germany the highest radon occurs over the granites and Paleozoic basement rocks. Median soil gas radon for some granites ranges from 100 to $200\,Bq\,L^{-1}$ (Kemski et al., 2001). In contrast, the highest radon potential in Belgium is associated with strongly folded and fractured Cambrian to Lower Devonian bedrocks in which uranium preferentially concentrated in ferric oxyhydroxides in fractures and joints is considered to be the main source of radon (Zhu et al., 2001). In France some of the highest radon levels occur over peraluminous leucogranites or metagranitoids in a stable Hercynian basement area located in South Brittany (western France). These rocks are derived from uraniferous granitoids with average uranium contents of over $8\,mg\,kg^{-1}$ (Ielsch et al., 2001). Soil gas and indoor radon concentrations were found to be controlled by lithology, structure, and uranium mineralization in India (Singh et al., 2002). High radon is associated with alum shale in both Sweden (Tell et al., 1993) and Belgium (Poffijn et al., 2002). In Korea, the mean values of soil gas radon concentrations were highest in granite gneiss and banded gneiss and lowest in soils over shale, limestone, and phyllite schist (Je et al., 1999).

The impact of unconsolidated deposits mainly reflects their permeability. For example, peat and lacustrine clays strongly reduce radon potential associated with the underlying bedrock, whereas permeable sand and gravel and river terrace deposits tend to enhance radon potential. In Sweden fragments and mineral grains of uranium-rich granites, pegmatites, and black alum shales are dispersed in till and glaciofluvial deposits leading to high radon in soils and dwellings, especially when the glaciofluvial deposits are highly permeable sands and gravels (Clavensjö & Åkerblom, 1994).

XIV. Administrative and Technical Responses

A. Environmental Health

Responses include provision of guidance for radon limitation including recommendations for dose limits and action levels, establishment of environmental health standards for houses and workplaces, and enforcement of Ionizing Radiations Regulations to control exposure to radon in workplaces (Åkerblom, 1999; Appleton et al., 2000b; NRPA, 2000). There are substantial variations in action levels (or their equivalents) in countries that perceive a radon problem. International and national recommendations for radon limitation in existing and future homes, given as the annual average of the gas concentration in $Bq\,m^{-3}$, range from 150 to 1000 for existing dwelling and from 150 to 250 for new dwellings (Åkerblom, 1999). The majority of countries have adopted 400 and $200\,Bq\,m^{-3}$, respectively, for the two reference levels.

The reasons for these different reference levels appear largely historical but are also due to a combination of environmental differences, different construction techniques, and varying levels of political and environmental concern. There would be advantages in harmonizing standards because the existence of different levels may lead to confusion among the public. The ICRP considers that one common international standard is unlikely to be achieved, and that this is less important than achieving reasonable reductions in radon levels in radon prone areas.

In addition to variations in house radon standards between countries, there also appears to be some variation in standards applied within the field of radiological protection. For example, some observers have suggested that if the highest natural radon levels recorded in some houses in Cornwall and Devon were reached, for example, at a nuclear installation, it would be closed down immediately. Equally, it is reported that all houses in the Chernobyl area had to be evacuated under Soviet law if they were contaminated to a level equivalent to the current UK action level of $200\,Bq\,m^{-3}$. However, international and national radiological protection authorities are united in acknowledging the need for a distinction in the ways radiation is approached in these different circumstances.

Recommendations differ from country to country. In the UK, testing for radon is recommended by government in radon affected areas where more than 1% of dwellings exceed the action level of $200\,Bq\,m^{-3}$. In contrast, the U. S. EPA recommends that all homes be tested for radon because (1) high levels of indoor radon have been found in every state, (2) radon levels vary so much from place to place, and (3) because dwellings differ so radically in their vulnerability to radon. The U. S. EPA estimates that 1 in every 15 homes has radon levels higher than $4\,pCi\,L^{-1}$ ($148\,Bq\,m^{-3}$), the level above which the EPA recommends that corrective action is taken. The U.S. EPA recommends that no action is required below $150\,Bq\,m^{-3}$, action within a few years between 170 and $750\,Bq\,m^{-3}$, urgent action between 750 and $7500\,Bq\,m^{-3}$, and immediate action above $7500\,Bq\,m^{-3}$.

The European Commission Recommendation (2001/928/Euratom) on the protection of the public against exposure to radon in drinking water supplies recommends $1000\,Bq/L$ as an action level for public and commercial water supplies above which remedial action is always justified on radiological protection grounds. Water supplies that support more than 50 people or distribute more than $10\,m^{-3}$ per day, as well as all water that is used for food processing or commercial purposes, except mineral water, are covered by the Europe Commission Recommendation. The $1000\,Bq/L$ action level also applies to drinking water distributed in hospitals, residential homes, and schools and should be used for consideration of remedial action in private water supplies. The U. S. EPA recommends that states develop multimedia mitigation (MMM) programs to address the health risks from radon in indoor air while individual water systems should reduce radon levels in drinking water to $148\,Bq\,L^{-1}$ ($4000\,pCi\,L^{-1}$) or lower. The EPA is encouraging states to adopt this option because it is the most cost-effective way to achieve the greatest radon risk reduction. If a state chooses not to develop an MMM program, individual water systems would be required to either reduce radon in their system's drinking water to $11\,Bq\,L^{-1}$ ($300\,pCi\,L^{-1}$) or develop individual local MMM programs and reduce levels in drinking water to $148\,Bq\,L^{-1}$ ($4000\,pCi\,L^{-1}$). The regulations will not apply to private wells, because the EPA does not regulate them. A guideline value of $2\,\mu g\,L^{-1}$ uranium (equivalent to approximately $0.02\,Bq\,L^{-1}$) is recommended by the World Health Organization (WHO), although this is based on its toxicity, which is more detrimental to health than its radioactivity (WHO, 1996). The U. S. EPA established maximum contaminant levels (MCL) of $30\,\mu g\,L^{-1}$ uranium, and $0.185\,Bq\,L^{-1}$ for radium-226 and radium-228 in community water supplies. No specific value for uranium is given in the EU Directive for drinking water (CEC,

1998), which establishes a $0.1\,mSv\,y^{-1}$ total indicative dose guidance level for radionuclides, excluding tritium, ^{40}K radon, and radon decay products. Action levels ranging from 7.4 to $160\,Bq\,L^{-1}$ uranium in drinking water have been reported for Austria, Finland, and France.

B. Radon Monitoring

The overall aim of most countries that have identified a radon problem is to map radon-prone areas and then identify houses and workplaces with radon concentrations that exceed the radon reference level. In the UK, for example, radon affected areas are delineated by measuring radon in a representative sample of existing dwellings. Householders are then encouraged to have radon measured in existing and new dwellings in affected areas and local authority environmental health departments are generally responsible for ensuring that radon in workplaces is monitored in appropriate areas.

C. Protective Measures

Provisions have been made in the building regulations to ensure that new dwellings are protected against radon where a significant risk of high radon concentrations in homes has been identified on the basis of house radon surveys. Nine European countries (Czech Republic, Denmark, Finland, Ireland, Latvia, Norway, Slovak Republic, Sweden, and the UK) have regulations and guidelines for construction requirements to prevent elevated radon concentrations in new buildings. Austria, Germany, Greece, and Switzerland plan to introduce such regulations. In most of the countries with regulations, enforced radon protection in new buildings is specified in the national building codes. Implementation of regulations is normally shared by national and local authorities. Eight European countries (Czech Republic, Denmark, Finland, Ireland, Norway, Slovak Republic, Sweden, and the UK) have regulations and guidelines for radon prevention in the planning stages of new development (e.g., where construction permits are applied for dwellings, offices, and factories). Austria and Germany are considering the introduction of guidance and/or regulations for dealing with radon at the planning stage. In the Czech Republic, Ireland, Slovak Republic, and Sweden regulations require an investigation of radon risk at construction sites before building is permitted (Åkerblom, 1999).

Remedial Measures

In the UK, owners of workplaces may be forced to carry out remedial measures whereas householders in dwellings with radon above the action level are generally only advised to take action to reduce the radon level. Guidance on reducing radon in dwellings is provided, but the cost of installing remedial measures in a dwelling is normally the householder's responsibility. Grant aid may be available.

The principal ways of reducing the amount of radon entering a dwelling are similar to those used for protective measures in new dwellings. These are

1. Install an airtight barrier across the whole of the ground floor to prevent radon getting through it and also seal voids around service inlets
2. Subfloor ventilation of underfloor cavities, i.e., drawing the air away from underneath the floor so that any air containing radon gas is dispersed outside the house
3. Subfloor depressurization (radon sump)
4. Positive pressurization (i.e., pressurize the building in order to prevent the ingress of radon)
5. Ventilation (i.e., avoid drawing air through the floor by changing the way the dwelling is ventilated)

In the United States, the cost of radon mitigation in a typical home ranges from about $500 to about $2500. Fitting radon resistant measures at the time of construction would cost $350–$500. Similar costs apply in the UK.

Radon gas may be easily removed from high-radon groundwaters by aeration and filter beds will remove daughter products. Various aeration technologies are available including static tank, cascade, or forced aeration in a packed tower. Radon removal technologies used in the United States include removal of ^{222}Rn by spray jet aeration, packed tower aeration, and multistage bubble aeration. Packed tower aeration is simple and cheap and is recommended for large drinking water supplies. Removal of ^{222}Rn by granular activated carbon is efficient but ^{238}U decay products, including U, Po, Bi, and Pb (^{210}Pb), are adsorbed onto the activated carbon, which produces a disposal problem. The U. S. EPA recommends that the most practical treatment methods for radionuclide removal are ion exchange and lime-soda softening for radium, aeration and granular activated carbon for radon, and anion exchange and reverse osmosis for uranium.

SEE ALSO THE FOLLOWING CHAPTERS

Chapter 2 (Natural Distribution and Abundance of Elements) · Chapter 9 (Volcanic Emissions and Health) · Chapter 21 (Environmental Epidemiology)

FURTHER READING

Åkerblom, G. (1987). Investigations and Mapping of Radon Risk Areas. In *Proceedings of International Symposium on Geological Mapping, Trondheim, 1986: In the Service of Environmental Planning*, Oslo, Norges Geologiske Undersoekelse, pp. 96–106.

Åkerblom, G. (1999). Radon Legislation and National Guidelines 99:18, Swedish Radiation Protection Institute.

Åkerblom, G., and Mellander, H. (1997). Geology and Radon. In *Radon Measurements by Etched Track Detectors* (S. A. Durrani, and R. Ilić, Eds.), World Scientific, New Jersey, pp. 21–49.

Appleton, J. D., and Ball, T. K. (2001). Geological Radon Potential Mapping. In *Geoenvironmental Mapping: Methods, Theory and Practice* (P. T. Bobrowsky, Ed.), Balkema, Rotterdam, pp. 577–613.

Appleton, J. D., and Miles, J. C. H. (2002). Mapping Radon-Prone Areas Using Integrated Geological and Grid Square Approaches. In *Radon Investigations in the Czech Republic IX and the Sixth International Workshop on the Geological Aspects of Radon Risk Mapping* (I. Barnet, M. Neznal, and J. Mikšová, Eds.), Czech Geological Survey, Prague, pp. 34–43.

Appleton, J. D., Miles, J. C. H., and Talbot, D. K. (2000a). Dealing with Radon Emissions in Respect of New Development: Evaluation of Mapping and Site Investigation Methods for Targeting Areas Where New Development May Require Radon Protective Measures, British Geological Survey Research Report, RR/00/12.

Appleton, J. D., Miles, J. C. H., Scivyer, C. R., and Smith, P. H. (2000b). Dealing with Radon Emissions in Respect of New Development: Summary Report and Recommended Framework for Planning Guidance, British Geological Survey Research Report, RR/00/07.

Ball, T. K., Cameron, D. G., and Colman, T. B. (1992). Aspects of Radon Potential Mapping in Britain, *Radiat. Prot. Dosimetry*, 45, 211–214.

Ball, T. K., Cameron, D. G., Colman, T. B., and Roberts, P. D. (1991). Behavior of Radon in the Geological Environment—a Review, *Q. J. Eng. Geol.*, 24(2), 169–182.

Ball, T. K., and Miles, J. C. H. (1993). Geological and Geochemical Factors Affecting the Radon Concentration in Homes in Cornwall and Devon, UK, *Environ. Geochem. Health*, 15(1), 27–36.

Barnet, I. (1994). Radon Risk Classification for Building Purposes in the Czech Republic. In *Radon Investigations in Czech Republic V* (I. Barnet and M. Neznal, Eds.), Czech Geological Survey, Prague, pp. 18–24.

Barnet, I., Mikšová, J., and Fojtíková, I. (2002). The GIS Analysis of Indoor Radon and Soil Gas in Major Rock Types of the Czech Republic. In *Radon Investigations in the Czech Republic IX and the Sixth International Workshop on the Geological Aspects of Radon Risk Mapping* (I. Barnet, M. Neznal, and J. Mikšová, Eds.), Czech Geological Survey, Prague, pp. 5–11.

BRE (1999). Radon: Guidance on Protective Measures for New Dwellings, Building Research Establishment Report, BR 211.

CEC (Council of the European Community) (1998). Council Directive 98/83/EC on the Quality of Water Intended for Human Consumption.

Clavensjö, B., and Åkerblom, G. (1994). *The Radon Book*, The Swedish Council for Building Research, Stockholm.

Cliff, K. D., and Miles, J C H. (Eds.) (1997). Radon Research in the European Union, EUR 17628, National Radiological Protection Board, Chilton, UK.

Cohen, B. L. (1997). Problems in the Radon vs. Lung Cancer Test of the Linear No-Threshold Theory and a Procedure for Resolving Them, *Health Phys.*, 72, 623–628.

Cross, F. T. (1992). A Review of Experimental Animal Radon Health Insights and Implications. In *Radiation Research, a Twentieth Century Perspective* (D. C. Dewey et al., Eds.), Academic Press, New York, pp. 333–339.

Darby, S., Hill, D., and Doll, R. (2001). Radon: A Likely Carcinogen at All Exposures, *Ann. Oncol.*, 12(10), 1341–1351.

Darby, S. C. et al. (1995). Radon and Cancers Other Than Lung Cancer in Underground Miners—a Collaborative Analysis of 11 Studies, *J. Natl. Cancer Inst.*, 87(5), 378–384.

Darby, S. C., Whitely, E., Silcocks, P. et al. (1998). Risk of Cancer Associated with Residential Radon Exposure in South-West England: A Case-Control Study, *Br. J. Cancer*, 78, 394–408.

Duport, P. (2002). Is the Radon Risk Overestimated? Neglected Doses in the Estimation of the Risk of Lung Cancer in Uranium Underground Miners, *Radiat. Prot. Dosimetry*, 98(3), 329–338.

Duval, J. S., and Otton, J. K. (1990). Radium Distribution and Indoor Radon in the Pacific Northwest, *Geophys. Res. Lett.*, 17(6), 801–804.

Gates, A. E., and Gundersen, L. C. S. (Eds.) (1992). Geologic Controls on Radon, Special Paper 271, Geological Society America.

Gilmore, G. K., Phillips, P., Denman, A., Sperrin, M., and Pearce, G. (2001). Radon Levels in Abandoned Metalliferous Mines, Devon, Southwest England, *Ecotoxicol. Environ. Safety*, 49(3), 281–292.

Grasty, R. L. (1997). Radon Emanation and Soil Moisture Effects on Airborne Gamma-Ray Measurements, *Geophysics*, 62(5), 1379–1385.

Greeman, D. J., and Rose, A. W. (1996). Factors Controlling the Emanation of Radon and Thoron in Soils of the Eastern U. S. A, *Chem. Geol.*, 129, 1–14.

Greeman, D. J., Rose, A. W., and Jester, W. A. (1990). Form and Behavior of Radium, Uranium, and Thorium in Central Pennsylvania Soils Derived From Dolomite, *Geophys. Res. Lett.*, 17(6), 833–836.

Green, B. M. R, Lomas, P. R., and O'Riordan, M. C. (1992). Radon in Dwellings in England, NRPB-R254, National Radiological Protection Board, UK.

Gundersen, L. C. S., and Schumann, E. R. (1996). Mapping the Radon Potential of the United States: Examples from the Appalachians, *Environ. Int.*, 22(Suppl. 1), S829–S844.

Gundersen, L. C. S., Schumann, E. R., Otton, J. K, Dubief, R. F., Owen, D. E., and Dickenson, K. E. (1992). Geology of Radon in the United States. In *Geologic Controls on Radon* (A. E. Gates and L. C. S. Gundersen, Eds.), Special Paper 271, Geological Society America, pp. 1–16.

ICRP (International Committee on Radiological Protection) (1993). Protection Against Radon-222 at Home and at Work, ICRP Publication 65, *Ann. ICRP*, 23(2).

Ielsch, G. et al. (2001). Radon (Rn-222) Level Variations on a Regional Scale: Influence of the Basement Trace Element (U, Th) Geochemistry on Radon Exhalation Rates, *J. Environ. Radioact.*, 53(1), 75–90.

Je, H. K., Kang, C. G., and Chon, H. T. (1999). A Preliminary Study on Soil-Gas Radon Geochemistry According to Different Bedrock Geology in Korea, *Environ. Geochem. Health*, 21(2), 117–131.

Kemski, J., Siehl, A., Stegemann, R., and Valdivia-Manchego, M. (2001). Mapping the Geogenic Radon Potential in Germany, *Sci. Total Environ.*, 272(1–3), 217–230.

Laurier, D., Valenty, M., and Tirmarche, M. (2001). Radon Exposure and the Risk of Leukemia: A Review of Epidemiological Studies, *Health Phys.*, 81(3), 272–288.

Lomas, P. R., Green, B. M. R., Miles, J. C. H., and Kendall, G. M. (1996). Radon Atlas of England, NRPB-290, National Radiological Protection Board, UK.

Lubin, J. H. (1998). On the Discrepancy between Epidemiologic Studies in Individuals of Lung Cancer and Residential Radon and Cohen's Ecologic Regression, *Health Phys.*, 75(1), 4–10.

Lubin, J. H., and Boice, J. D. (1997). Lung Cancer Risk from Residential Radon: Meta-Analysis of Eight Epidemiologic Studies, *J. Natl. Cancer. Inst.*, 89(1), 49–57.

Lubin, J. H. et al. (1994). Radon Exposure in Residences and Lung-Cancer Among Women—Combined Analysis of 3 Studies, *Cancer Causes Control*, 5(2), 114–128.

Man, C. K., and Yeung, H. S. (1998). Radioactivity Contents in Building Materials used in Hong Kong, *J. Radioanal. Nucl. Chem.*, 232(1–2), 219–222.

Mikšová, J., and Barnet, I. (2002). Geological Support to the National Radon Programme (Czech Republic), *Bull. Czech Geol. Surv.*, 77(1), 13–22.

Miles, J. C. H. (1998). Mapping Radon-Prone Areas by Log-Normal Modeling of House Radon Data, *Health Phys.*, 74(3), 370–378.

Miles, J. C. H. (2001). Temporal Variation of Radon Levels in Houses and Implications for Radon Measurement Strategies, *Radiat. Prot. Dosimetry*, 93(4), 369–376.

Miles J. C. H., and Appleton, J. D. (2000). Identification of Localised Areas of England Where Radon Concentrations are Most Likely to Have >5% Probability of Being Above the Action Level, Department of the Environment, Transport and the Regions Report, DETR/RAS/00.001.

Miles, J. C. H., and Ball, T. K. (1996). Mapping Radon-Prone Areas Using House Radon Data and Geological Boundaries, *Environ. Int.*, 22(Suppl. 1), 779–782.

NAS (1998). *Health Effects of Exposure to Radon (BEIR VI)*, National Academy of Sciences, Washington DC.

Nazaroff, W. W. (1988). Measurement Techniques. In *Radon and Its Decay Products in Indoor Air*, John Wiley & Sons, New York, pp. 491–504.

NRPA (Nordic Radiation Protection Authorities) (2000). Naturally Occurring Radioactivity in the Nordic Countries—Recommendations, The Radiation Protection Authorities in Denmark, Finland, Iceland, Norway, and Sweden.

NRPB (1989). Living with Radiation, National Radiological Protection Board, UK.

NRPB (2000). Health Risks from Radon, National Radiological Protection Board, UK.

Poffijn, A., Goes, E., and Michaela, I. (2002). Investigation of the Radon Potential of an Alum Deposit. In *Radon Iinves-tigations in the Czech Republic IX and the Sixth International Workshop on the Geological Aspects of Radon Risk Mapping* (I. Barnet, M. Neznal, and J. Miksova, Eds.), Czech Geological Survey, Prague.

Rose, A. W., Hutter, A. R., and Washington, J. W. (1990). Sampling Variability of Radon in Soil Gases, *J. Geochem. Explor.*, 38, 173–191.

Singh, S., Kumar, A., and Singh, B. (2002). Radon Level in Dwellings and Its Correlation with Uranium and Radium Content in Some Areas of Himachal Pradesh, India, *Environ. Int.*, 28(1–2), 97–101.

Smith, B. J., Field, R. W., and Lynch, C. F. (1998). Residential Rn-222 Exposure and Lung Cancer: Testing the Linear No-Threshold Theory with Ecologic Data, *Health Phys.*, 75(1), 11–17.

Tell, I. et al. (1993). Indoor Radon-Daughter Concentration and Gamma-Radiation in Urban and Rural Homes on Geologically Varying Ground, *Sci. Total Environ.*, 128(2–3), 191–203.

UNSCEAR (United Nations Scientific Committee on the Effects of Atomic Radiation) (2000). Sources, Effects, and Risks of Ionizing Radiation, United Nations, New York.

USDOE (United States Department of Energy) (1988). Radiation Inhalation Studies of Animals, DOE/ER-0396, Washington DC.

Washington, J. W., and Rose, A. W. (1992). Temporal Variability of Radon Concentrations in the Interstitial Gas of Soils in Pennsylvania, *J. Geophys. Res.*, 97(B6), 9145–9159.

WHO (1996). *Guidelines for Drinking-Water Quality*, 2nd edition, Vol. 2, World Health Organization, Geneva.

Wiegand, J. et al. (2000). Radon and Thoron in Cave Dwellings (Yan'an, China), *Health Phys.*, 78(4), 438–444.

Zhu, H. C., Charlet, J. M., and Poffijn, A. (2001). Radon Risk Mapping in Southern Belgium: An Application of Geostatistical and GIS Techniques, *Sci. Total Environ.*, 272(1–3), 203–210.

CHAPTER 11

ARSENIC IN GROUNDWATER AND THE ENVIRONMENT

PAULINE SMEDLEY AND DAVID G. KINNIBURGH
British Geological Survey

CONTENTS

I. INTRODUCTION

Arsenic is a ubiquitous element in the environment. It may be mobilized through a combination of natural processes such as weathering and erosion, biological activity, and volcanic emissions (see Volcanic Emissions and Health, this volume), as well as through the activities of man. Although most environmental arsenic problems are the result of mobilization under natural conditions, anthropogenic impacts have been significant in places due to activities such as mining; fossil fuel combustion, use of arsenical pesticides, herbicides, and crop desiccants, and arsenic-based additives in livestock feed. Although such pesticides and herbicides have been used much less over the last few decades, arsenic is still used widely in wood preservation and such sources may still pose a localized threat to the environment.

Human exposure to arsenic may be through a number of pathways, including air, food, water, and soil (Cullen & Reimer, 1989; NRC, 1999). The relative impacts of these vary depending on local circumstances, but of the potential sources of arsenic available, drinking water poses one of the greatest threats to human health and has been shown to have direct detrimental effects in many parts of the world. Drinking water may be obtained from a number of sources (surface water, rainwater, groundwater) depending on local availability. The concentrations of arsenic in these sources are highly variable and the observed ranges vary over several orders of magnitude. Excepting localized sources of anthropogenic contamination, the highest aqueous arsenic concentrations tend to be found in groundwaters because of natural water–rock interaction processes and the high solid/solution ratios found in aquifers. Groundwaters therefore pose the greatest overall threat to health. Groundwaters with arsenic concentrations sufficiently high to be detrimental to humans, or with already detectable health impacts, have been reported in Bangladesh, India, Taiwan, Thailand, China, Hungary, Vietnam, Nepal, Myanmar, Mexico, Argentina, and Chile, and it is likely that occasional problems will be found in many other countries.

Some of the groundwater arsenic problems have been recognized for a considerable time. Probably the earliest cases of health effects from arsenic contamination of drinking water were recognized in a mining area of Poland in the 1890s. Here, contamination of water supplies by oxidation of arsenic-bearing sulfide minerals produced localized health problems (see Tseng et al., 1968). In central Argentina, arsenic-related health problems were first documented in 1917 (Círculo Médico del Rosario, 1917) and problems in Taiwan and Chile were first identified in the 1960s (Tseng et al., 1968; Smith et al., 1998). However, in each of these cases the problems were solved primarily by engineering solutions, which for the most part provided alternative supplies of surface water or of treated water. Hence, geochemical investigations into the processes controlling the arsenic mobilization in these areas have generally not been carried out until relatively recently. Even today, many of the arsenic-affected areas have received little attention and much remains unknown about the precise mechanisms involved. Hence, the problems observed in seriously impacted regions such as the Bengal Basin and Vietnam were not anticipated by water providers or the scientific community and have emerged only recently.

Both advisory guideline values and national standards for arsenic in drinking water have been reduced in recent years following mounting evidence of its chronic toxic effects. The World Health Organization (WHO) guideline value for arsenic in drinking water was provisionally reduced in 1993 from $50 \mu g L^{-1}$ to $10 \mu g L^{-1}$. Many other regulatory authorities in the western world have subsequently reduced their limits for arsenic in line with this recommendation. The EC maximum permissible value for arsenic in drinking water was reduced to $10 \mu g L^{-1}$ in 2000. The U. S. Environmental Protection Agency (EPA) maximum contaminant level (MCL) was also reduced to $10 \mu g L^{-1}$ in 2001, although the revision has been the subject of long debate over the last few years, largely because of the major cost implications to the U. S. water supply industry. While many national authorities are seeking to reduce their limits in line with the WHO guideline value, many countries and indeed all affected developing countries, still adopt the former WHO value of $50 \mu g L^{-1}$, in part because of a lack of adequate testing facilities for measuring lower concentrations. Concentrations of $10 \mu g L^{-1}$ and $50 \mu g L^{-1}$ are therefore both often used as yardsticks for the testing and reporting of arsenic.

This chapter reports the current state of knowledge of the sources and distributions of arsenic in natural waters and their host rocks, and attempts to describe what is currently known of the main geochemical processes that control its mobilization in the environment (see also Chapters 2, 22, and 23, this volume).

II. Sources of Arsenic in the Natural Environment

A. Minerals

Arsenic occurs as a major constituent in more than 200 minerals including elemental arsenic, arsenides, sulfides, oxides, arsenates, and arsenites. Most are ore minerals or their alteration products. These minerals are relatively rare in the natural environment. Among the most common occurrences in ore zones are arsenian pyrite $(Fe(S,As)_2)$, arsenopyrite (FeAsS), realgar (AsS), orpiment (As_2S_3), cobaltite (CoAsS), niccolite (NiAs), and scorodite $(FeAsO_4.2H_2O)$. Arsenian pyrite $(Fe(S,As)_2)$ is probably the most important source of arsenic in ore zones (Nordstrom, 2000). The arsenic ore minerals also often contain high concentrations of transition metals, as well as cadmium, lead, silver, gold, antimony, phosphorus, tungsten, and molybdenum.

Arsenic is often present in varying concentrations in other common rock-forming minerals. As the chemistry of arsenic follows sulfur closely, the greatest concentrations tend to occur in sulfide minerals, of which pyrite (FeS_2) is the most abundant. Arsenic concentrations in pyrite, chalcopyrite, galena, and marcasite can be highly variable, even within a given grain, but in some cases exceed $10 wt\%$ (Table I). Pyrite is an important component of ore bodies and is also formed in low-temperature sedimentary environments under reducing conditions. Such authigenic pyrite plays a very important role in present-day geochemical cycles. It is present in the sediments of many rivers, lakes, and the oceans as well as in many aquifers. During pyrite formation, it is likely that some of the soluble arsenic will also be incorporated into the pyrite. Pyrite is not stable in aerobic systems and oxidizes to iron oxides with the release of significant amounts of sulfate, and acidity and associated trace constituents, including arsenic. The presence of pyrite in mineralized veins is responsible for the production of acid mine drainage and for the pres-

TABLE I. Typical Ranges of Arsenic Concentrations in Common Rock-Forming Minerals

Mineral	Arsenic concentration range (mg kg^{-1})
Sulfide minerals	
Pyrite	100–77,000
Pyrrhotite	5–100
Marcasite	20–126,000
Galena	5–10,000
Sphalerite	5–17,000
Chalcopyrite	10–5000
Oxide minerals	
Hematite	up to 160
Fe(III) oxyhydroxide	up to 76,000
Magnetite	2.7–41
Ilmenite	<1
Silicate minerals	
Quartz	0.4–1.3
Feldspar	<0.1–2.1
Biotite	1.4
Amphibole	1.1–2.3
Olivine	0.08–0.17
Pyroxene	0.05–0.8
Carbonate minerals	
Calcite	1–8
Dolomite	<3
Siderite	<3
Sulfate minerals	
Gypsum/anhydrite	<1–6
Barite	<1–12
Jarosite	34–1000
Other minerals	
Apatite	<1–1000
Halite	<3–30
Fluorite	<2

From sources summarized in Smedley and Kinniburgh, 2002.

ence of arsenic problems around coal mines and areas of intensive coal burning.

High arsenic concentrations are also found in many oxide minerals and hydrous metal oxides, either as part of the mineral structure or as adsorbed species. Concentrations in Fe oxides can also reach several weight percent (Table I), particularly where they form as the oxidation products of primary iron sulfide minerals, which have an abundant supply of arsenic. Adsorption of arsenate to hydrous iron oxides is particularly strong and adsorbed loadings can be great even where con-

centrations in solution are low (e.g., Goldberg, 1986). Adsorption of arsenic is complicated because the two most common oxidation states of dissolved arsenic, As(III) and As(V), behave quite differently and because the amount of adsorption of these varies greatly with the concentration of other dissolved species. Adsorption to hydrous aluminum and manganese oxides may also be important if these oxides are present in quantity (Goldberg, 1986; Stollenwerk, 2003). Sorption to the edges of clays may also occur, although the loadings are much smaller on a weight basis than for the iron oxides (Smedley & Kinniburgh, 2002). Adsorption reactions are responsible for the relatively low concentrations of arsenic found in most natural waters.

Arsenic concentrations in phosphate minerals can also reach high values, for example, up to 1000 mg kg^{-1} in apatite (Table I). However, phosphate minerals are much less abundant than oxide minerals and thus make a correspondingly small contribution to the arsenic concentration in most sediments. Arsenic tends to be present at much lower concentrations in the rock-forming minerals, although low concentrations are invariably present. Most common silicate minerals (including quartz, feldspar, micas, amphiboles) contain around 1 mg kg^{-1} or less (Table I). Carbonate minerals usually contain concentrations less than 10 mg kg^{-1}.

B. Rocks, Sediments, and Soils

Arsenic occurs ubiquitously but at variable concentrations in rocks, unconsolidated sediments, and soils. Average crustal abundance is around 1.5 mg kg^{-1}. In igneous rocks concentrations are generally low. Ure and Berrow (1982) quoted an average value of 1.5 mg kg^{-1} for undifferentiated igneous rocks. Other averages quoted are generally slightly higher than this value but usually less than 5 mg kg^{-1}. Volcanic glasses are slightly higher with an average of around 6 mg kg^{-1} (Smedley & Kinniburgh, 2002). Overall, there is relatively little difference between the different igneous rock types (see also Chapter 2, this volume).

Arsenic concentrations in metamorphic rocks tend to reflect the concentrations in their igneous and sedimentary precursors. Most contain around 5 mg kg^{-1} or less. Pelitic rocks (slates, phyllites) typically have the highest concentrations (Table II), with an average of around 18 mg kg^{-1}.

Concentrations in sedimentary rocks are typically in the range of 5–10 mg kg^{-1} (Webster, 1999). Average sediments are enriched in arsenic relative to igneous and

TABLE II. Typical Ranges of Arsenic Concentrations in Rocks, Sediments, Soils, and Other Superficial Deposits

Rock/sediment type	As concentration range (mg kg^{-1})
Igneous rocks	
Ultrabasic rocks	0.03–15.8
Basic rocks	0.06–113
Intermediate rocks	0.09–13.4
Acidic rocks	0.2–15
Metamorphic rocks	
Quartzite	2.2–7.6
Hornfels	0.7–11
Phyllite/slate	0.5–143
Schist/gneiss	<0.1–18.5
Amphibolite and greenstone	0.4–45
Sedimentary rocks	
Marine shale/mudstone	3–15 (up to 490)
Shale (mid-Atlantic ridge)	48–361
Non-marine shale/mudstone	3.0–12
Sandstone	0.6–120
Limestone/dolomite	0.1–20.1
Phosphorite	0.4–188
Iron formations and Fe-rich sediment	1–2900
Evaporites (gypsum/anhydrite)	0.1–10
Coals	0.3–35,000
Unconsolidated sediments	
Alluvial sand (Bangladesh)	1.0–6.2
Alluvial mud/clay (Bangladesh)	2.7–14.7
River bed sediments (Bangladesh)	1.2–5.9
Lake sediments	0.5–44
Glacial till	1.9–170
World average river sediments	5
Stream and lake silt	<1–72
Loess silts, Argentina	5.4–18
Continental margin sediments	2.3–8.2
Soils	
Mixed soils	0.1–55
Peaty and bog soils	2–36
Peat	up to 9T
Acid sulfate soils	1.5–45
Soils near sulfide deposits	2–8000
Contaminated superficial deposits	
Mining-contaminated lake sediment	80–1104
Mining-contaminated reservoir sediment	100–800
Mine tailings	396–2000
Soils and tailings-contaminated soil	120–52,600
Industrially polluted intertidal sediments	0.38–1260
Soils below chemicals factory	1.3–4770

From sources summarized in Smedley and Kinniburgh, 2002.

metamorphic rocks because they contain greater quantities of minerals with high adsorbed arsenic loads. Of the sediments, sands and sandstones tend to have the lowest concentrations which reflect the low arsenic concentrations in their dominant minerals: quartz and feldspars. Average sandstone arsenic concentrations are around 4 mg kg^{-1}. Ure and Berrow (1982) gave a lower average value of 1 mg kg^{-1}.

Argillaceous deposits have a broader range of concentrations than sandstones (Table II), with a typical average of around 13 mg kg^{-1} (Ure & Berrow, 1982). The higher values reflect the larger proportion of sulfide minerals, oxides, organic matter, and clays present. Black shales have arsenic concentrations at the upper end of the range, principally because of their enhanced pyrite content. Marine clays appear to have higher concentrations than non-marine equivalents. This may also be a reflection of the grain-size distributions with potential for a higher proportion of fine material in offshore pelagic sediments. Marine shales also tend to contain more sulfur and hence are likely to contain more pyrite. High arsenic concentrations have been found in mid-ocean ridge shales (Table II).

Arsenic concentrations in coals and bituminous deposits are often high, in part because they are closely associated with sulfide minerals. Organic-rich shales from Germany have arsenic concentrations of 100–900 mg kg^{-1} (Riedel & Eikmann, 1986). Some coal samples have concentrations up to 35,000 mg kg^{-1} (Belkin et al., 2000) (Table II), although lower concentrations of 2.5–17 mg kg^{-1} are more typical (e.g., Palmer & Klizas, 1997). Some of the highest observed arsenic concentrations are found in ironstones and Fe-rich rocks. James (1966) collated data for ironstones from various parts of the world and reported arsenic concentrations up to 800 mg kg^{-1} in a chamosite-limonite oolite from the former USSR. Boyle and Jonasson (1973) reported concentrations for iron-rich rocks up to 2900 mg kg^{-1} (Table II). Phosphorites are also relatively enriched in arsenic with concentrations up to around 400 mg kg^{-1} reported. Carbonate rock types typically have low arsenic concentrations (around 3 mg kg^{-1}) as a result of the low concentrations of their constituent minerals (Tables I and II).

Unconsolidated sediments have arsenic concentrations that do not differ significantly from their indurated equivalents. As noted above, muds and clays usually have higher concentrations than sands and carbonates. Values are typically 3–10 mg kg^{-1}, depending on texture and mineralogy (Table II). High concentrations tend to reflect the amounts of pyrite or iron oxides present. There is often a significant positive correlation between the iron and arsenic concentrations in sediments. High arsenic concentrations are also common in mineralized areas. Placer deposits in streams can have very high concentrations as a result of the abundance of sulfide minerals. Average arsenic concentrations for stream sediments in England and Wales are in the range of 5–8 mg kg^{-1} (AGRG, 1978). Similar concentrations have also been found in river sediments where groundwater arsenic concentrations are high: Datta and Subramanian (1997) found concentrations in sediments from the River Ganges averaging 2.0 mg kg^{-1}, from the Brahmaputra River averaging 2.8 mg kg^{-1}, and from the Meghna River averaging 3.5 mg kg^{-1}.

Cook et al. (1995) found concentrations in Canadian lake sediments ranging between 0.9 and 44 mg kg^{-1} (median 5.5 mg kg^{-1}) but noted that the highest concentrations were present up to a few kilometers downslope of mineralized areas. They also found concentrations in glacial till of 1.9–170 mg kg^{-1} (median 9.2 mg kg^{-1}) (Table II) and noted the highest concentrations down-ice of mineralized areas. Relative arsenic enrichments have been observed in reducing sediments in both near-shore and continental shelf deposits (Peterson & Carpenter, 1986; Legeleux et al., 1994). Legeleux et al. (1994) noted concentrations increasing with depth (up to 30 cm) in continental shelf sediments as a result of the generation of increasingly reducing conditions. Concentrations varied between sites, but generally increased with depth in the range 2.3–8.2 mg kg^{-1} (Table II).

Concentrations of arsenic in uncontaminated soils are generally of the order of 5–15 mg kg^{-1}. Boyle and Jonasson (1973) quoted an average in world soils of 7.2 mg kg^{-1} (Table II) and Shacklette et al. (1974) quoted an average of 7.4 mg kg^{-1} for American soils. Ure and Berrow (1982) gave an average of 11.3 mg kg^{-1}. Peats and bog soils can have higher concentrations (average 13 mg kg^{-1}, Table II), but this is principally because of an increased prevalence of sulfide mineral phases under the reduced conditions. Shotyk (1996) found a maximum of 9 mg As kg^{-1} in two Swiss peat profiles and in the profile with the lower mineral content, i.e., the purer peat, the As content was 1 mg kg^{-1} or lower.

Acid sulfate soils that are generated by the oxidation of pyrite in sulfide-rich terrains such as pyritic shales, mineral veins, and dewatered mangrove swamps can also be relatively enriched in arsenic. Acid sulfate soils from the weathering of pyrite-rich shales in Canada have arsenic concentrations up to 45 mg kg^{-1} (Dudas, 1984). Gustafsson and Tin (1994) found similarly high concentrations (up to 41 mg kg^{-1}) in acid sulfate soils from the Mekong delta of Vietnam.

Additional arsenic inputs to soils may be derived locally from industrial sources such as smelting and fossil fuel combustion products and agricultural sources such as pesticides and phosphate fertilizers. Ure and Berrow (1982) gave concentrations in the range 366–732 mg kg^{-1} in orchard soils following historical application of arsenical pesticides to fruit crops. Long-term use of phosphate fertilizers may also add significant arsenic to soil.

Concentrations of arsenic in sediments and soils contaminated by the products of mining activity, including tailings and effluent, can be orders of magnitude higher than under natural conditions. Concentrations in tailings piles and tailings-contaminated soils up to several thousands of mg kg^{-1} have been reported (Table II). The values reflect not only increased abundance of primary arsenic-rich sulfide minerals, but also secondary iron arsenates and iron oxides formed as reaction products of the original ore minerals. The primary sulfide minerals are susceptible to oxidation in the tailings pile and the secondary minerals have varying solubility under oxidizing conditions in groundwaters and surface waters. Scorodite ($FeAsO_4.2H_2O$) is a common sulfide oxidation product and its solubility is likely to control arsenic concentrations in such environments. Secondary arsenolite (As_2O_3) is also often represented in such environments. Arsenic is also often strongly bound to iron oxides (see below) and is relatively immobile, particularly under oxidizing conditions.

III. ARSENIC IN GROUNDWATER

A. Aqueous Speciation

Compared to many other toxic trace elements, arsenic is relatively mobile at the pH values typically found in natural waters (pH 6.5–8.5) and under both oxidizing and reducing conditions. Arsenic can occur in the environment in a number of oxidation states (−3, −1, 0, +3, and +5), but under natural conditions it is mostly found in inorganic form as trivalent arsenite (As(III)) or as oxyanions of pentavalent arsenate (As(V)). Organic arsenic species may be produced by biological activity, mostly in surface waters, but are rarely quantitatively important in groundwater. Exceptions potentially occur in cases of industrial pollution.

Redox potential (Eh) and pH are the most important factors controlling arsenic speciation in aqueous systems.

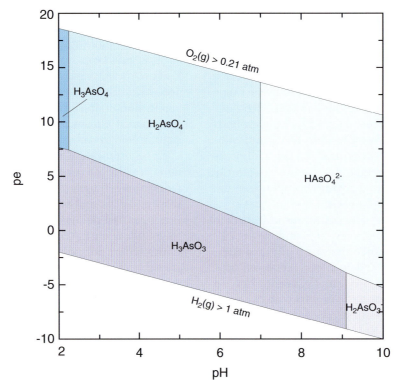

FIGURE 1   pe-pH diagram for soluble arsenic species in the system As-O_2-H_2O (As$_T$ = 10^{-5} M, NaCl = 10^{-2} M; 25°C).

Under oxidizing conditions, $H_2AsO_4^-$ is dominant at low pH (less than about pH 6.9), while at higher pH, $HAsO_4^{2-}$ it dominates (Figure 1). In extremely acidic conditions, $H_3AsO_4^0$ is important whereas AsO_4^{3-} may be present in alkaline conditions. Under reducing conditions where pH is less than about 9.2, the uncharged arsenite species $H_3AsO_3^0$ predominates. Native arsenic may be present under extremely reducing conditions.

As(III) and As(V) may form aqueous complexes with reduced sulfur (e.g., thioarsenite) and carbonate ligands, and these may be significant in some groundwaters. In the presence of extremely high concentrations of reduced dissolved sulfur at low pH, dissolved arsenic sulfide species can be formed. Reducing, acidic conditions also favor precipitation of orpiment (As_2S_3), realgar (AsS), or other sulfide minerals containing co-precipitated arsenic (Cullen & Reimer, 1989). As a result, high-arsenic waters are not expected where there is a high concentration of free sulfide (Moore et al., 1988).

The ratio of As(III) to As(V) in groundwaters varies with the redox status of the aquifer, which in turn depends upon the abundance of the redox-active solids,

especially organic carbon, and the flux of potential oxidants (oxygen, nitrate, and sulfate). Microorganisms play a key role in these redox reactions, and in environments of high microbial activity can be responsible for lack of redox equilibrium between the arsenic species. In strongly reducing aquifers, typified by Fe(III) and sulfate reduction, As(III) typically dominates, as expected from the redox sequence. In oxidizing systems, As(V) is typically dominant. The extent of redox equilibrium with respect to arsenic in natural waters has been a matter of considerable debate. Although observations of the rate of oxidation of As(III) in groundwater are difficult under field conditions, the rates are generally believed to be slow. Microbial activity is also generally low in groundwaters, but this is compensated for to some extent by the long time scales usually involved in groundwater flow that are typically decades and longer.

Reducing arsenic-rich groundwaters from Bangladesh have As(III)/As$_T$ ratios varying between 0.1 and 0.9 but are typically around 0.5 and 0.6 (Smedley et al., 2001). Ratios in reducing groundwaters from Inner Mongolia are typically 0.6–0.9 (Smedley et al., 2003).

Concentrations of organic forms are generally low or negligible in groundwaters (Del Razo et al., 1990; Chen et al., 1995).

B. Arsenic Abundance and Distribution in Groundwater

Aquifers have high solid/solution ratios, typically 3–20 kg L^{-1} and as a result, groundwaters within them are especially vulnerable to the buildup of arsenic in solution. In addition, aquifers more often have the physical and chemical conditions favorable for arsenic mobilization and transport than is the case in surface waters. Despite this, the occurrence of high arsenic concentrations in groundwaters is the exception rather than the rule. Concentrations in groundwater are in most countries less than 10 μg L^{-1} and often substantially lower. However, values quoted in the literature show a very large range from <0.5 to 5000 μg L^{-1}, i.e., more than four orders of magnitude (Smedley & Kinniburgh, 2002).

Mobilization of arsenic in solution is favored, especially under oxidizing conditions at high pH and under strongly reducing conditions. Evaporative concentration can also increase arsenic (and other element) concentrations substantially and may be important in some arid areas. Additional arsenic problems are encountered in some geothermal areas and in many areas of sulfide mineralization and mining. Cases of industrial arsenic pollution (including those from agriculture) have also been reported. Although these may be severe locally, occurrences are relatively rare and can usually be anticipated.

Investigations worldwide have revealed a number of major aquifers with significant groundwater arsenic problems (concentrations exceeding 50 μg L^{-1}). The hydrogeological and geochemical conditions in these affected aquifers vary, although some common unifying features are apparent. Aquifers at greatest risk appear to be those found in large alluvial and delta plains as well as large inland basins, the latter especially in arid and semi-arid areas. In each case, geologically young (Quaternary) aquifers are particularly prone to developing and preserving high-arsenic groundwater. Aquifers in alluvial and delta plains with recognized arsenic problems include Bangladesh, West Bengal (India), Nepal, northern China, Taiwan, Hungary, Romania, and Vietnam. Those from inland and closed basins include parts of Argentina, Chile, Mexico, and areas of the United States, particularly the southwest (Figure 2). In many of these areas, significant numbers of wells have arsenic concentrations of several hundreds of micrograms per liter, with occasional sources in the milligram per liter range. Arsenic-related health problems have been recognized in the resident populations in some areas as a result. Recent reconnaissance surveys of groundwater quality in other areas such as parts of Myanmar, Pakistan, and Cambodia have also revealed concentrations of arsenic in some wells exceeding 50 μg L^{-1}. However, documentation of the affected aquifers in these areas is thus far limited. Little is known about other large deltas, for example, that of the Nile (Egypt), Chao Phraya (Thailand), Niger (Nigeria), Huang Ho (China), and Yangste (China), although they have many geological characteristics similar to other arsenic-affected areas. The regions of the world having major aquifers with recognized arsenic problems are outlined below and categorized in terms of their environmental conditions. The distributions of arsenic in the environment related to geothermal activity as well as mining and mineralization are also described.

1. Alluvial Plains and Deltas

a. Bangladesh and India (West Bengal)

Promoted by the presence of fertile land and plentiful supplies of water, the Bengal delta of Bangladesh and West Bengal (India) is one of the most densely populated regions on Earth. Traditional sources of water for the region included rivers and ponds as well as shallow hand-dug wells. These have long been the source of significant problems from life-threatening diarrheal diseases. As a response, the last few decades have seen a proliferation in the development of groundwater by the installation of boreholes, both for domestic supply and irrigation. This has been aided by the presence of abundant groundwater at shallow depths in the unconsolidated sediments and the ease of drilling. Today, there are estimated to be up to 11 million boreholes in Bangladesh alone (BGS & DPHE, 2001), and the vast majority are private hand-pumped boreholes for domestic use. Although the discovery of arsenic problems in the aquifers of West Bengal was made by local physicians as early as the 1980s, the problem was not widely recognized until the mid-1990s and the scale of the problem has emerged only recently.

Arsenic problems in groundwater of the Bengal Basin occur within young, mainly Holocene, shallow aquifers (<150 m depth), composed of alluvial and deltaic sediments deposited by the vast river systems of the Ganges, Brahmaputra, and Meghna and their tributaries. These alluvial and deltaic sediments typically include in the

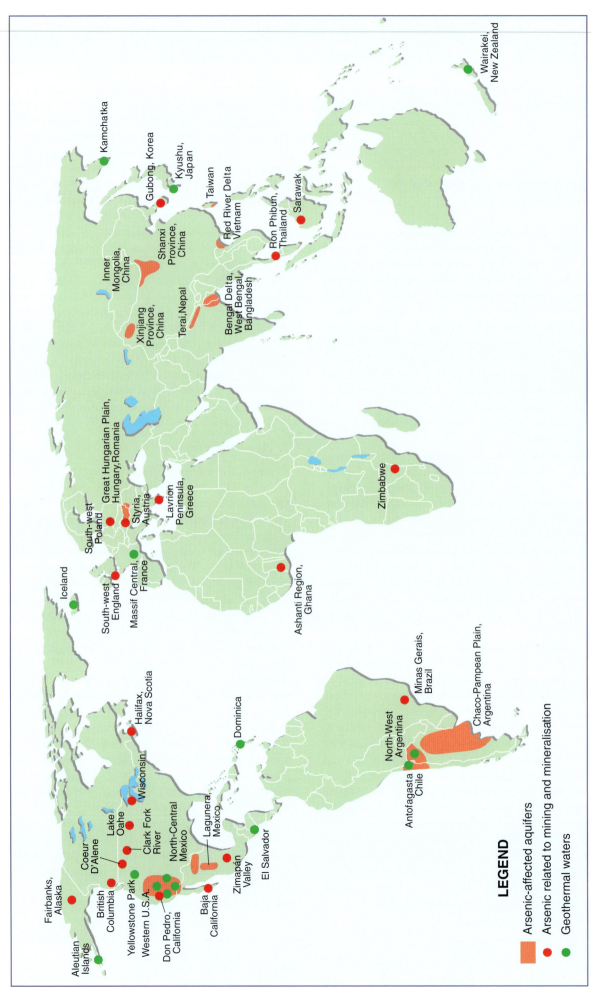

FIGURE 2 Occurrence of documented arsenic problems in groundwater (arsenic >50 µg L^{-1}) in major aquifers and (Smedley and Kinniburgh, 2002). Related to mining and geothermal sources.

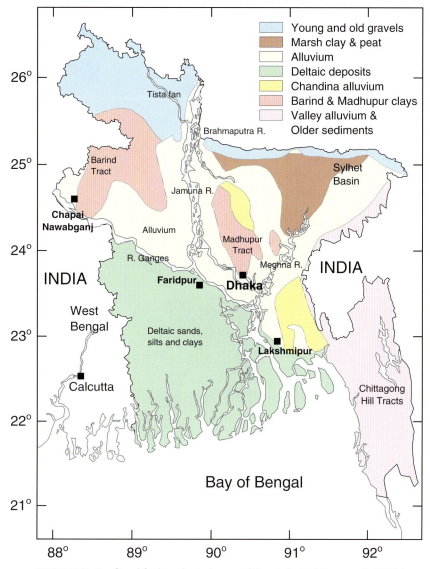

FIGURE 3 Simplified geological map of Bangladesh. (Alam et al., 1990.)

upper reaches coarse alluvial fan deposits (e.g., the Tista Fan; Figure 3) with dominant sand and gravel. In the middle parts of the basin, meander belts consist of levee, backswamp, ox-bow lake, and abandoned channel deposits. In the lower reaches of the delta, marsh and tidal flat deposits mainly consist of fine silts and clays, but with some sand horizons. Buried channel deposits in particular tend to contain sand and gravel and frequently make good aquifers. The sediments are highly heterogenous both laterally and with depth, although fining-upwards sequences are commonly observed. The sediments include some contemporaneous disseminated organic matter and occasional peat horizons. The lower parts of the basin have a surface cover of fine alluvial

overbank silts and clays of variable thickness (around 10–80 m; BGS & DPHE, 2001; Chakraborti et al., 2001). These are in large part responsible for restricting the diffusion of oxygen to the underlying aquifers and, together with the presence of organic matter, have led to the widespread development of reducing conditions, particularly in the distal parts of the basin. These alluvial and deltaic sediments cover the majority of Bangladesh (Figure 3) and a large proportion of West Bengal.

In terms of the populations at risk, groundwater arsenic problems in Bangladesh and West Bengal represent the most serious occurrences identified globally. Observed concentrations have a very large range from

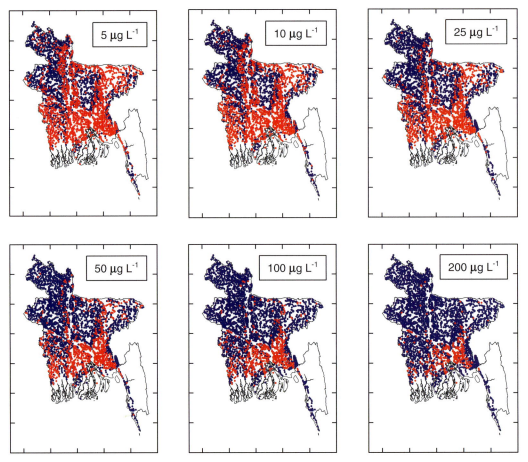

FIGURE 4 Map of Bangladesh showing the distributions of arsenic in groundwater from shallow (<150 m) tubewells at various threshold concentrations. Blue symbols denote values less than the indicated threshold; red symbols denote values above. (From BGS & DPHE, 2001.)

<0.5 μg L⁻¹ to around 3200 μg L⁻¹. Chakraborti et al. (2001) reported that, of 90,000 water analyses measured by them in West Bengal, some 34% exceeded the national standard for arsenic in drinking water of 50 μg L⁻¹ and 55% exceeded the WHO guideline value of 10 μg L⁻¹. These problems occur mainly to the east of the Bhagirathi River in a relatively narrow strip bordering Bangladesh. They also reported that of 27,000 analyses carried out in Bangladesh, 59% exceeded 50 μg L⁻¹ while 73% exceeded 10 μg L⁻¹. These percentages differ somewhat from the survey by BGS and DPHE (2001) of around 3500 wells sampled randomly across Bangladesh, where 27% of samples from shallow wells (<150 m depth) contained arsenic concentrations exceeding 50 μg L⁻¹ and 46% contained more than 10 μg L⁻¹. The differences most probably relate to differing sampling strategies and sample populations. The worst-affected area of Bangladesh is the southeast (Figure 4) where in the worst-affected district, BGS and DPHE (2001) identified more than 90% of wells having arsenic concentrations >50 μg L⁻¹. The indications are that the degree of contamination is not as severe in West Bengal as in the worst districts of Bangladesh. Certainly, the overall areal extent of contamination in West Bengal is less than in Bangladesh. Around 35 million people in Bangladesh and 6 million people in West Bengal were estimated to be at risk from arsenic in drinking water at concentrations above 50 μg L⁻¹ by Smedley & Kinniburgh (2002).

Recognized health problems resulting from chronic exposure to arsenic in drinking water consist mainly of skin disorders, notably pigmentation changes (melanosis) and keratosis, although skin cancer has also been identified. Around 5000 patients have been identified with arsenic-related health problems in West Bengal (including skin pigmentation changes), although some estimates put the number of patients with arsenicosis at more than 200,000 (Smith et al., 2000). The number with recognized problems in Bangladesh is of the order of 14,000 though many more may be unrecognised or in a preclinical stage. Thus far, data do not show a strong relationship between groundwater arsenic concentrations and health problems in Bangladesh, but epidemiological studies are ongoing. The prevalence of internal arsenic-related health problems, including cancers, is also not known, but could be appreciable given the growing evidence of the toxic effects of chronic arsenic exposure (see also Chapters 22 and 23, this volume).

The high-arsenic groundwaters of the Bengal Basin have variable chemical compositions, but a number of features are commonly observed and highlight the strongly reducing nature of the affected groundwaters. Measured redox potentials are typically less than 100 mV, dissolved oxygen is usually very low or absent (<1 mg L^{-1}), and high concentrations of dissolved iron (often several milligrams per liter, up to 60 mg L^{-1}), manganese (often >0.5 mg L^{-1}, up to 8 mg L^{-1}), bicarbonate (often >500 mg L^{-1}; up to about 1100 mg L^{-1}), and ammonium-N (>1 mg L^{-1}) are typical (Table III) (BGS & DPHE, 2001). Correlations between these parameters have been found in some studies of localized areas such as villages, but they are generally poor on a regional scale. High concentrations of phosphorus (often >0.5 mg L^{-1}; up to 20 mg L^{-1}) are also a common feature. The dissolved iron and manganese most likely derive from reductive dissolution of iron and manganese oxides in the sediments. Bicarbonate is in part derived from reaction of carbonate minerals and partly from the oxidation of organic matter. Phosphorus is likely to be derived from both iron oxides and organic matter, although dissolution of detrital apatite may also be involved. Derivation of phosphate by leaching of fertilizers from overlying soils has been suggested by some workers (Acharyya et al., 1999), although this is considered unlikely given the widespread occurrence of phosphate in the groundwater and its presence in deep groundwaters (>150 m) as well as in shallower groundwaters. The groundwaters are generally fresh, but salinity has increased in the southern coastal region as a result of saline intrusion.

Additional indicators of reducing conditions in the arsenic-affected aquifers include low concentrations of nitrate, except for a number of shallow polluted wells, and low to very low concentrations of sulfate (typically <1 mg L^{-1}). A general negative correlation observed between arsenic and sulfate concentrations (BGS & DPHE, 2001) suggests that arsenic mobilization occurs under the most strongly reducing conditions—

alongside sulfate reduction and iron oxide dissolution. Arsenic speciation studies have revealed a large range in the relative proportions of dissolved arsenate and arsenite in the groundwaters. However, arsenite is generally dominant; the modal proportion in pumped groundwaters from Bangladesh is around 60% of the total arsenic (BGS & DPHE, 2001). The variable ratios may reflect lack of redox equilibrium in the aquifers or the presence of mixed groundwater from a strongly stratified aquifer. Some of the groundwaters of Bangladesh are sufficiently reducing for methane generation to have taken place.

The origin of the arsenic in the Bengal groundwaters has been in dispute in recent years, but it is generally accepted to be of natural origin. The widespread development of strongly reducing conditions in the affected aquifers is likely the main factor controlling the mobilization. This has probably occurred in a complex combination of redox changes brought on by rapid burial of the alluvial and deltaic sediments, which include reduction of the solid-phase arsenic to As(III), desorption of arsenic from iron oxides, reductive dissolution of the oxides themselves, and likely diagenetic changes in iron-oxide structure and surface properties following the onset of reducing conditions (BGS & DPHE, 2001; Smedley & Kinniburgh, 2002). Competition for adsorption sites between dissolved arsenic species and other constituents such as phosphate and bicarbonate may also be involved. It has recently been suggested from a study of groundwaters in the Munshiganj district, southern Bangladesh, that arsenic mobilization in the aquifers has resulted from increased drawdown of organic carbon of anthropogenic origin as a result of recent irrigation (Harvey et al., 2002). However, the evidence for this is scant and not altogether convincing, and it remains to be seen whether this process is of widespread significance.

The distribution of arsenic in the groundwaters of Bangladesh and West Bengal is known to be spatially highly variable, with large differences in arsenic concentration in wells over short lateral distances. This is likely a reflection of the considerable heterogeneity in composition, texture, and permeability of the sediments. At a practical level, the variability makes prediction of arsenic concentrations from well to well extremely difficult, although where the density of sampling is very high, distinct patterns within villages have been observed. Despite this, on a regional scale the distribution shows a clear geological control (Figure 4). Wells in older uplifted Plio-Pleistocene sediments of the Barind and Madhupur Tracts of north and central Bangladesh (including those of the city of Dhaka) have

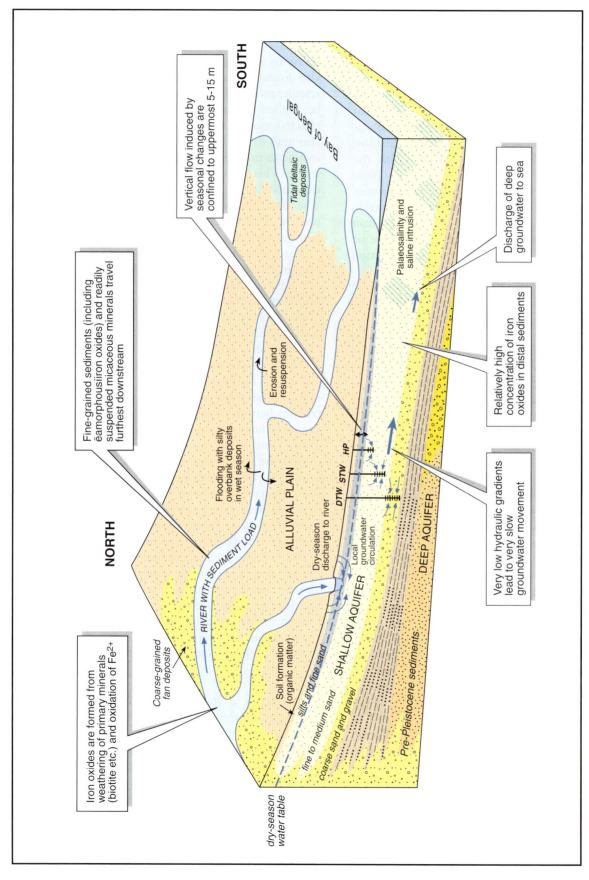

FIGURE 5(A) Block diagrams showing schematically (A) simplified geology and hydrogeology and (B) principal geochemical processes involved in the generation of high-arsenic groundwaters in Bangladesh. (From BGS & DPHE, 2001, from which further details of geochemical processes and the groundwater flow model can be found.)

low arsenic concentrations (less than $10\,\mu g\,L^{-1}$ and usually $<0.5\,\mu g\,L^{-1}$; Figure 4).

Arsenic concentrations also vary with depth as shown by both individual wells at a given site and by the distribution of arsenic with depth for wells from different sites. The concentration vs. depth profile typically is a bell-shaped curve with low concentrations in shallow ($<10\,m$) and deep wells ($>150–200\,m$) and with the largest concentrations in wells from the 10- to 70-m depth range. Typically wells are drilled to the minimum depth required for an adequate yield and acceptable quality (in terms of salinity and sometimes iron content). This often corresponds with a high-arsenic depth interval. Little is known about the exact variation of arsenic concentration at greater depths.

In the BGS and DPHE (2001) survey, practically all deep wells, tapping depths greater than 100–200 m, had low arsenic concentrations. These were mostly from the extreme south and north-east of the country. Although the stratigraphy of Bangladesh aquifers at depth is poorly understood, these are also likely to be Plio-Pleistocene aquifers of comparable facies and composition to those of the Barind and Madhupur Tracts. The source of arsenic in the few identified deep wells with increased arsenic concentrations is uncertain. However, the deep aquifer itself is not necessarily the dominant source. Downward leakage of groundwater from the shallow aquifer(s) and multilevel screening of wells are likely alternative explanations. Dug wells in Bangladesh and West Bengal are also observed to have generally low arsenic concentrations, typically $<10\,\mu g\,L^{-1}$ (BGS & DPHE, 2001; Chakraborti et al., 2001).

Several workers have suggested that the source of the arsenic in the Bengal sediments derives from discrete high-arsenic mineralized zones upstream of the affected areas (e.g., Acharyya, et al., 1999). However, the high-arsenic groundwaters occur in sediments with total arsenic concentrations typically in the range of $<2–20\,mg\,kg^{-1}$, i.e., not exceptionally high concentrations (Table II). Many studies have shown that fine-grained sediments (silts and clays) tend to have greater arsenic concentrations than coarse-grained sediments. BGS and DPHE (2001) found concentrations of 1.3–$10\,mg\,kg^{-1}$ in sediments from three arsenic-affected areas of Bangladesh. Chakraborti et al. (2001) also reported low concentrations in most of their samples analyzed from West Bengal. In one borehole from 24 South Parganas, 108 samples had an arsenic range of 8–$12\,mg\,kg^{-1}$. The near-average concentrations observed are not surprising considering the scale of the Bengal arsenic problem, with the affected aquifer sediments derived from a very large catchment area of the

Himalayas. They are therefore by definition likely to be "close to average." However, BGS and DPHE (2001) also found the concentrations of readily extractable arsenic, probably largely associated with iron oxides, to be higher in Bangladesh sediments from the badly affected aquifers than from elsewhere. The iron oxides also tend to be more concentrated in the finer-grained sediments in the lower, distal part of the delta than elsewhere. This is probably a further contributory factor in the distribution of arsenic problems across the basin. The distribution may reflect the high biotite mica content of these sediments, because the weathering of biotite provides a potential source of iron oxides.

The reasons for the distinction between groundwater arsenic concentrations in the shallow and deep aquifers of the Bengal Basin are not yet well-understood. Differences between the sediments at depth may occur in terms of absolute arsenic concentrations and in the oxidation states, structure, and binding properties of their iron oxides (BGS & DPHE, 2001; van Geen et al., 2003). It is also possible that the history of groundwater movement and aquifer flushing in the Bengal Basin has been important in generating the differences in dissolved concentrations between the shallow and deep aquifers. Older, deeper sediments have been subject to longer periods of groundwater flow, aided by greater hydraulic heads during the Pleistocene period when glacial sea levels around the Bangladesh landmass were as much as 130 m lower than today (Umitsu, 1993). Flushing of the deeper older aquifers with groundwater is therefore likely to have been much more extensive than in the shallow Holocene sediments deposited during the last 5000–10,000 years. Hence, much of the arsenic released during diagenesis of the deep sediments may have already been flushed away. Models of the hydrogeology and geochemical processes believed to be controlling the arsenic mobilization in the Bengal Basin aquifers are shown schematically in Figure 5. This model is discussed more fully by BGS and DPHE (2001).

b. Nepal

Groundwater is abundant in the Quaternary alluvial sediments of the lowland Terai region of southern Nepal and is an important resource for domestic and agricultural use. The region is estimated to have around 200,000 boreholes which supply groundwater for some 11 million people (Chitrakar & Neku, 2001). Both shallow and deep aquifers occur throughout most of the Terai region. The shallow aquifer appears to be mostly unconfined and well-developed, although it is thin or absent in some areas. The deep aquifer (precise depth uncertain) is artesian. Quaternary alluvium also infills

TABLE III. Summary of Documented High-Arsenic Groundwater Provinces and Their Typical Chemical and Hydrogeological Characteristics

Groundwater environment/aquifer type	Examples	Typical aquifer conditions	Typical chemical features of high-arsenic groundwaters	Likely mechanisms of arsenic mobilization
Strongly reducing groundwater	Alluvial/deltaic aquifers of Bangladesh, West Bengal, northern China, Taiwan, Myanmar, Nepal, Hungary, Romania	Young (Quaternary) sediments, slow groundwater flow, low-lying parts of aquifers; rapidly accumulated sediments	High Fe ($>1 \, mg \, l^{-1}$), Mn ($>0.5 \, mg \, l^{-1}$), NH_4-N ($>1 \, mg \, l^{-1}$), high HCO_3 ($>500 \, mg \, L^{-1}$); low NO_3-N ($<1 \, mg \, l^{-1}$), low SO_4 ($<5 \, mg \, L^{-1}$); often high P ($>0.5 \, mg \, l^{-1}$); sometimes high concentrations of dissolved organic matter, including humic acid; arsenic dominated by As(III)	Reductive desorption of As from metal oxides, reductive dissolution of Fe, Mn oxides; possible competition between arsenic and other anionic species (including P, HCO_3)
Aerobic groundwater, high pH	Inland basins or closed basins (arid and semi-arid areas): Argentina, Mexico, parts of the western United States, parts of Chile	Young (Quaternary) sediments, slow groundwater flow, low-lying parts of aquifers	pH typically >8, high HCO_3 ($>500 \, mg \, L^{-1}$), low Fe, Mn, often correspondingly high F, U, B, V, Mo, Se; some groundwaters have high salinity due to evaporation; arsenic dominated by As(V)	Desorption of As and other oxyanion-forming elements from metal oxides, especially of Fe and Mn; evaporative concentration
Geothermally influenced groundwater	Parts of Kamchatka, Chile, Argentina, the western United States, Japan, New Zealand	Any aquifers affected by geothermal inputs, especially in rift zones	High Si, B, Li, often high salinity (Na, Cl); high pH (>7); increased groundwater temperature	Mixing of fresh groundwater with geothermal solutions
Groundwater from sulfide mineralized/ mining areas	Parts of the United States, Canada, southwest England, Thailand, South Korea, Poland, Greece, Ghana, Zimbabwe, Brazil	Groundwater in fractures in crystalline rocks or alluvial placer deposits	Oxidizing or mildly reducing conditions possible; high SO_4 concentrations (typically hundreds of $mg \, L^{-1}$ or higher); acidic (unless buffered by carbonate minerals); often increased concentrations of other trace metals (Ni, Pb, Zn, Cu, Cd)	Oxidation of sulfide minerals

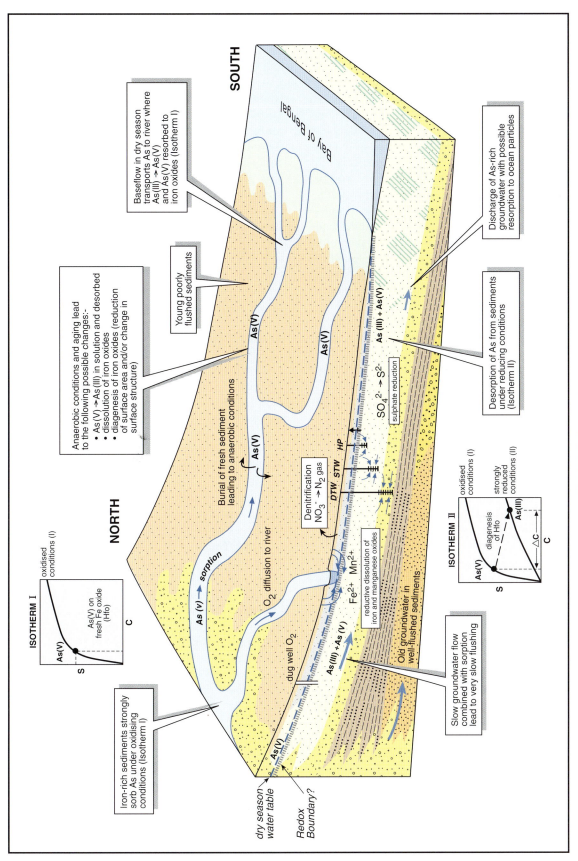

FIGURE 5(B) *Continued*

several intermontaine basins in Nepal, most notably that of the Kathmandu Valley of central Nepal (about 500 km²) where sediment thickness reaches in excess of 300 m.

A number of surveys of groundwater quality in the Terai region have revealed the presence of arsenic at high concentrations in some samples from shallow boreholes (<50 m depth), though most analyzed samples appear to have <10 μg L⁻¹ (Smedley, 2003). The Nepal Department of Water Supply and Sewerage (DWSS) carried out a survey of some 4000 tubewells from the 20 Terai districts, which were mostly analyzed by field-test kits but with laboratory replication of some. Results from the survey indicated that 3% of the samples had arsenic concentrations of >50 μg L⁻¹ (Chitraker and Neku, 2001). The worst affected districts were found to be Rautahat, Nawalparasi, Parsa, and Bara in the central area of the Terai. The highest observed concentration was 343 μg L⁻¹ (Parsa District). From testing in 17 of the 20 Terai districts, the Nepal Red Cross Society (NRCS) also found 3% of groundwater sources sampled to have concentrations above 50 μg L⁻¹; the highest observed concentration was 205 μg L⁻¹. The spatial distribution of the worst affected areas was found to be similar to that reported by Chitrakar and Neku (2001). More recent results indicate that of 25,000 analyses carried out so far in the Terai, 8% exceed 50 μg L⁻¹ and 31% exceed 10 μg L⁻¹ (Shrestha et al., 2004).

The high arsenic concentrations occur in anaerobic groundwaters and are often associated with high concentrations of dissolved Fe. The number of samples with exceedances above 50 μg L⁻¹ are generally small, but are nonetheless cause for further testing and remedial action.

Surveys appear to indicate that deeper boreholes (>50 m depth) in the Terai usually have lower arsenic concentrations (<10 μg L⁻¹). Preliminary investigations also suggest that the Kathmandu Valley does not have an arsenic problem, although more analysis is required for verification.

c. Taiwan

Arsenic problems in groundwaters in Taiwan were first recognized during the 1960s (e.g., Tseng et al., 1968), and related chronic health problems have been well documented since then (e.g., Chen et al., 1985). Black-foot disease (a peripheral vascular disorder similar to gangrene) is a well-publicized health problem of the region and is most likely linked to the high arsenic concentrations, although the high humic acid concentration in groundwaters of the region has also been cited as a possible causal factor. A number of other diseases,

including internal cancers, have also been described. High groundwater arsenic concentrations have been recognized in both the southwest (Tseng et al., 1968) and northeast (Hsu et al., 1997) parts of the island.

Arsenic concentrations in groundwater samples from southwest Taiwan have been found in the range 10–1800 μg L⁻¹ and a significant number of samples are more than 400 μg L⁻¹ (references cited in Smedley & Kinniburgh, 2002). In northeastern Taiwan, arsenic concentrations exceed 600 μg L⁻¹ in some groundwaters and a recent survey of 377 water samples gave an average of 135 μg L⁻¹ (Hsu et al., 1997).

In the southwest, the high arsenic concentrations are found in groundwaters from deep artesian wells (mostly 100–280 m) abstracted from sediments which include fine sands, muds, and black shale (Tseng et al., 1968). Groundwaters abstracted from the northeast of Taiwan are also artesian, but of shallower depth (typically in the range 16–40 m; Hsu et al., 1997). In each area the groundwaters are likely to be strongly reducing, which is supported by the observation that the arsenic is present largely as As(III) (Chen et al., 1994). However, both the geochemistry of the groundwaters and the mineral sources in Taiwan are poorly defined. Groundwater samples taken from shallow, open dug wells are observed to have low arsenic concentrations (Guo et al., 1994).

d. Northern China

Arsenic has been found at high concentrations (in excess of the Chinese national standard of 50 μg L⁻¹) in groundwaters from a number of areas in northern China, including Xinjiang, Shanxi, Jilin, and Liaoning Provinces as well as Inner Mongolia (Figure 2) (Sun et al., 2001; Smedley et al., 2003). The earliest problems were recognized in Xinjiang Province. Wang and Huang (1994) reported concentrations of 40–750 μg L⁻¹ in deep artesian groundwater from the Dzungaria (Junggar) Basin on the north side of the Tianshan Mountains. Wells were up to 660 m deep. The basin has been a zone of subsidence since at least Mesozoic times and is composed of a 10-km thick sequence of sediments, which include a substantial upper sequence of Quaternary age. The high-arsenic zone extends from Aibi Lake in the west to Mamas River in the east (about 250 km). Wang and Huang (1994) found that arsenic concentrations increased with depth in the artesian groundwater. Shallow (non-artesian) groundwaters had arsenic concentrations between <10 μg L⁻¹ (the detection limit) and 68 μg L⁻¹.

In the Datong and Jinzhong Basins of Shanxi Province, arsenic concentrations have been found to exceed 50 μg L⁻¹ in 35% of 2373 randomly selected groundwater samples (Sun et al., 2001). Concentrations in Shanyin county, the worst affected of the regions in Shanxi Province, reached up to $4.4 \, mg \, L^{-1}$. Groundwaters from the Hetao Plain of Inner Mongolia also have high concentrations. Guo et al. (2001) found concentrations up to $1350 \, \mu g \, L^{-1}$ and observed that concentrations were generally much higher in groundwater from boreholes (depth range 15–30 m) than open dug wells (3–5 m depth). Tian et al. (2001) also reported arsenic concentrations between 50 and $1800 \, \mu g \, L^{-1}$ in groundwater from the Ba Men region of west central Inner Mongolia.

In the Huhhot Basin, part of the Tumet Plain of Inner Mongolia, arsenic problems are found in groundwaters from mainly Holocene alluvial and lacustrine aquifers. As with the Bengal Basin, these occur under highly reducing conditions and the problems are worst in the lowest lying parts of the basin (Smedley et al., 2003). The reducing high-arsenic groundwaters have characteristically moderate to high concentrations of dissolved iron, manganese, bicarbonate, and ammonium and low concentrations of nitrate and sulfate (Luo et al., 1997; Smedley et al., 2003). Arsenic concentrations reach up to $1500 \, \mu g \, L^{-1}$ in the groundwaters, with a significant proportion (>60%) of the arsenic present as As(III). The mechanisms of arsenic mobilization are believed to be very similar to those taking place in the Bengal Basin. However, unlike Bangladesh, deep aquifers in the area have relatively high arsenic concentrations (up to $300 \, \mu g \, L^{-1}$). This may relate to differences in sediment chemistry and diagenetic history between the two regions. Alternatively, paleohydrogeological differences may have played a role because hydraulic gradients in Inner Mongolia, an inland basin remote from the sea, would have been less affected by sea-level changes during the Quaternary (Section V.C). Another distinction from Bangladesh is the observation that groundwater in many of the old hand-dug wells in the low-lying part of the Huhhot Basin also has relatively high arsenic concentrations. Concentrations up to $550 \, \mu g \, L^{-1}$ have been reported by Smedley et al. (2003). The high concentrations are believed to be due to the maintenance of relatively reducing conditions in the well waters as a result of high organic-matter content (solid and dissolved) in the shallow aquifer of the region.

The high groundwater arsenic concentrations have resulted in the development of chronic health problems in the affected areas of China, with the best documented cases in Xinjiang Province and Inner Mongolia. The problems are most notably manifested by skin lesions which include melanosis and hyperkeratosis. In the affected region of Inner Mongolia, additional health problems including lung, skin, and bladder cancer have been recognized and are documented by Luo et al. (1997).

e. Vietnam

The aquifers of the large deltas of the Mekong and Red Rivers are now widely exploited for drinking water. The total number of tubewells in Vietnam is unknown but could be of the order of one million, with perhaps 150,000 in the Red River delta region. The majority of these are private tubewells. The capital city, Hanoi, is largely dependent on groundwater for its public water supply. The aquifers exploited are of both Holocene and Pleistocene age. In parts of the Red River delta region (Figure 2), Holocene sediments form the shallow aquifer which may be only 10–15 m deep. Where Holocene sediments are absent, older Pleistocene sediments are exposed at the surface. Unlike Bangladesh, even when the Holocene sediments are present, there is not always a layer of fine silt-clay at the surface. Normally the Holocene sediments are separated from the underlying Pleistocene sediments by a clay layer several meters thick, although "windows" in this clay layer exist where there is hydraulic continuity between the Holocene and Pleistocene aquifers. The total thickness of sediments is typically 100–200 m.

The groundwaters in the delta regions are usually strongly reducing with high concentrations of iron, manganese, and ammonium. Much of the shallow aquifer in the Vietnamese part of the Mekong delta region is affected by salinity and cannot be used for drinking water. Little was known about the arsenic concentrations in groundwater in Vietnam until recently. Results from Hanoi indicate that there is a significant problem in shallow tubewells in the city, particularly in the south (Berg et al., 2001). Concentrations were found in the range of $1–3050 \, \mu g \, L^{-1}$ (average $159 \, \mu g \, L^{-1}$). Early indications suggest that arsenic concentrations are not high in the groundwaters of the Mekong delta of Vietnam, though some groundwaters have been found with concentrations $>50 \, \mu g \, L^{-1}$ further upstream in the Mekong Valley of Cambodia. Investigations are also ongoing in the same aquifer in neighboring Laos.

f. Hungary and Romania

Concentrations of arsenic above $50 \, \mu g \, L^{-1}$ have been identified in groundwaters from alluvial sediments in the southern part of the Great Hungarian Plain of Hungary and neighboring parts of Romania (Figure 2). Concentrations up to $150 \, \mu g \, L^{-1}$ (average $32 \, \mu g \, L^{-1}$, 85 samples) have been found by Varsányi et al. (1991). The

Plain, about 110,000 km^2 in area, consists of a thick sequence of subsiding Quaternary sediments. Groundwaters vary from Ca-Mg-HCO$_3$ type in the recharge areas of the basin margins to Na-HCO$_3$ type in the low-lying discharge regions. Groundwaters in deep parts of the basin (80–560 m depth) with high arsenic concentrations are reducing with high concentrations of iron and ammonium and many have reported high concentrations of dissolved organic matter (humic acid quoted as up to 20 mg L^{-1}; Varsányi et al., 1991). The groundwaters have the largest arsenic concentrations in the lowest parts of the basin where the sediment is fine-grained. Gurzau and Gurzau (2001) reported concentrations up to 176 µg L^{-1} in the associated aquifers of neighboring Romania.

2. Inland Basins in Arid and Semi-Arid Areas

a. Mexico

The Comarca Lagunera of north central Mexico has a well-documented arsenic problem in groundwater with significant resulting chronic health problems. The region is arid and groundwater is an important resource for potable supply. Groundwaters are predominantly oxidizing with neutral to high pH. Del Razo et al. (1990) quoted pH values for groundwaters in the range of 6.3–8.9. They found arsenic concentrations in the range of 8–624 µg L^{-1} (average 100 µg L^{-1}, n = 128), with half the samples having concentrations greater than 50 µg L^{-1}. They also noted that most (>90%) of the groundwater samples investigated had arsenic present predominantly as As(V). The Comarca Lagunera is a closed basin and arsenic concentrations are typically highest in the low-lying parts of the basin. Del Razo et al. (1994) determined the average concentration of arsenic in drinking water from Santa Ana town as 404 µg L^{-1}. The estimated population exposed to arsenic in drinking water with >50 µg L^{-1} is around 400,000 in the Comarca Lagunera (Del Razo et al., 1990). The groundwater also has high concentrations of fluoride (up to 3.7 mg L^{-1}; Cebrián et al., 1994). High arsenic concentrations have also been identified in groundwaters from the state of Sonora in northwest Mexico, where Wyatt et al. (1998) reported concentrations in the range of 2–305 µg L^{-1} (76 samples). The arsenic concentrations were also positively correlated with fluoride, with maximum fluoride concentrations of 7.4 mg L^{-1}, which were significantly greater than the WHO guideline value for fluoride in drinking water of 1.5 mg L^{-1}.

b. Chile

Health problems related to arsenic in drinking water were first recognized in northern Chile in 1962. Typical symptoms included skin pigmentation changes, keratosis, squamous-cell carcinoma (skin cancer), cardiovascular problems, and respiratory disease. More recently, chronic arsenic ingestion has been linked to lung and bladder cancer. It has been estimated that around 7% of all deaths occurring in Antofagasta between 1989 and 1993 were due to past exposure to arsenic in drinking water at concentrations of the order of 500 µg L^{-1} (Smith et al., 1998). Since exposure was chiefly during the period from 1955 to 1970, this pointed to a long latency of cancer mortality. Other reported symptoms include impaired resistance to viral infection and lip herpes (Karcher et al., 1999).

High arsenic concentrations have been reported in surface waters and groundwaters from Administrative Region II (incorporating the cities of Antofagasta, Calama, and Tocopilla) of northern Chile (Cáceres et al., 1992). The region is arid and is part of the Atacama Desert, and water resources are limited. High As concentrations are accompanied by high salinity and high concentrations of boron and lithium. This in part relates to evaporation but is also significantly affected by geothermal inputs from the El Tatio geothermal field. Arsenic concentrations below 100 µg L^{-1} in surface waters and groundwaters are apparently quite rare, and concentrations up to 21,000 µg L^{-1} have been found. Karcher et al. (1999) quoted ranges of 100 µg L^{-1} to 1000 µg L^{-1} in untreated surface waters and groundwaters (average 440 µg L^{-1}).

The affected groundwaters of Chile are taken to be predominantly oxidizing (with dissolved oxygen present), largely because the arsenic is reported to be present in the waters as arsenate. However, the geochemistry of the aquifers of Chile is still poorly understood. The aquifers are composed of volcanic rocks and sediments, but the arsenic sources are not well-characterized. In Antofagasta, concentrations of arsenic in the sediments are about 3.2 mg kg^{-1} (Cáceres et al., 1992). Sediments from the Rio Loa and its tributaries have much higher concentrations (26–2000 mg kg^{-1}) as a result of geothermal inputs from the river system (Romero et al., 2003). Additional exposure to arsenic from the smelting of copper ore has also been noted in northern Chile (Cáceres et al., 1992).

c. Argentina

The Chaco-Pampean Plain of central Argentina constitutes perhaps one of the largest regions of high-arsenic groundwaters known, covering around 1 million km^2. High concentrations of arsenic have been found in the provinces of Córdoba, La Pampa, Santa Fe, Buenos Aires, and Tucumán in particular. Symptoms typical of chronic arsenic poisoning, including skin lesions and some internal cancers, have been recorded in these areas. The climate is temperate with increasing aridity toward the west. The high-arsenic groundwaters are from Quaternary deposits of loess (mainly silt) with intermixed rhyolitic or dacitic volcanic ash (Nicolli et al., 1989, 2001; Smedley et al., 2002). The sediments display abundant evidence of post-depositional diagenetic changes under semi-arid conditions with common occurrences of calcrete in the form of cements, nodules, and discrete layers that are sometimes many centimeters thick.

Nicolli et al. (1989) found arsenic concentrations in groundwaters from Córdoba in the range of 6–11,500 $\mu g L^{-1}$ (median 255 $\mu g L^{-1}$). Smedley et al. (2002) found concentrations for groundwaters in La Pampa Province in the range of <4–5280 $\mu g L^{-1}$ (median 145 $\mu g L^{-1}$). Nicolli et al. (2001) found concentrations in groundwaters from Tucumán Province of 12–1660 $\mu g L^{-1}$ (median 46 $\mu g L^{-1}$). The groundwaters often have high salinity and are also predominantly oxidizing with low dissolved Fe and Mn concentrations. Arsenic is predominantly present as As(V) (Smedley et al., 2002). Under the arid conditions, silicate and carbonate weathering reactions are pronounced and the groundwaters often have high pH values. Smedley et al. (2002) found pH values typically of 7.0–8.7 and Nicolli et al. (2001) found values of 6.3–9.2. Metal oxides in the sediments (especially iron and manganese oxides) are thought to be the main source of the dissolved arsenic with the mobilization promoted by desorption under high-pH conditions (Smedley et al., 2002). Positive correlations between arsenic and pH are apparent in the groundwaters (Figure 6) and the arsenic is also generally well correlated with other anion and oxyanion elements (F, V, HCO_3, Mo, and B; Figure 6), some of these elements having very high concentrations. Dissolution of volcanic glass has been proposed as an alternative potential source of the dissolved arsenic (Nicolli et al., 1989).

d. Southwest United States

Many areas have been identified in the United States with arsenic problems in groundwater. Most of the worst affected and best documented cases occur in the southwestern states of Nevada, California, and Arizona. Within the last decade, aquifers in Maine, Michigan, Minnesota, South Dakota, Oklahoma, and Wisconsin have been found with concentrations of arsenic exceeding 10 $\mu g L^{-1}$ and smaller areas of high-arsenic groundwaters have been found in many other states. Much water analysis has been carried out in the United States, particularly in view of the recent reduction in the U.S.

EPA drinking-water maximum contaminant level. Occurrences in groundwater are therefore found to be widespread, although of those reported, relatively few have significant numbers of sources with concentrations greater than 50 $\mu g L^{-1}$. A review of the analyses of some 17,000 water analyses from the United States concluded that around 40% exceeded 1 $\mu g L^{-1}$ and about 5% exceeded 20 $\mu g L^{-1}$ (Welch et al., 2000). The arsenic appears to be derived from various sources, which include natural dissolution/desorption reactions, geothermal water, and mining activity. The natural occurrences of arsenic in groundwater occur under both reducing and oxidizing conditions in different areas. Concentration by evaporation is thought to be an important process in the more arid areas.

In Nevada, Fontaine (1994) reported at least 1000 private wells with arsenic concentrations in excess of 50 $\mu g L^{-1}$. Welch and Lico (1998) also reported high concentrations that often exceeded 100 $\mu g L^{-1}$ but with extremes up to 2600 $\mu g L^{-1}$ in shallow groundwaters from the southern Carson Desert. These are largely present under reducing conditions. The groundwaters also have associated high pH (>8) and high concentrations of phosphorus (locally >4 $mg L^{-1}$) and uranium (>100 $\mu g L^{-1}$; Welch and Lico, 1998). The high arsenic and uranium concentrations were thought to be due to evaporative concentration of groundwater combined with the influence of redox and desorption processes involving metal oxides.

In California, high arsenic concentrations have been reported in the Tulare Basin of the San Joaquin Valley. A range from <1 to 2600 $\mu g L^{-1}$ was found by Fujii and Swain (1995). Redox conditions in the aquifers are variable and high arsenic concentrations are found in both oxidizing and reducing conditions. The proportion of groundwater arsenic present as As(III) increases with increasing well depth. The groundwaters from the Basin are often strongly affected by evaporation with resultant high concentrations of total dissolved solids. Many also have high concentrations of Se (up to 1000 $\mu g L^{-1}$), U (up to 5400 $\mu g L^{-1}$), B (up to 73,000 $\mu g L^{-1}$), and Mo (up to 15,000 $\mu g L^{-1}$) (Fujii & Swain, 1995).

In Arizona, Robertson (1989) also noted the occurrence of high arsenic concentrations in some groundwaters under oxidizing conditions in alluvial aquifers of the Basin and Range Province. Dissolved oxygen concentrations in the groundwaters were in the range of 3–7 mg L^{-1}. Only limited analysis of arsenic species was carried out, but results obtained suggested that the arsenic was present predominantly as As(V). The dissolved arsenic generally correlated positively with pH, Mo, and V (Figure 6) as well as Se and F. Groundwater pH values

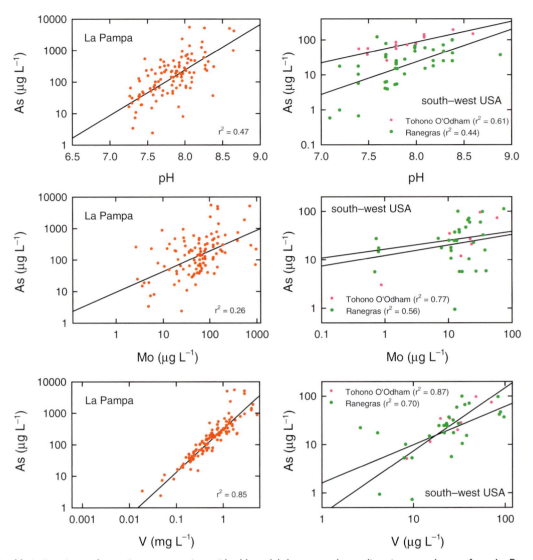

FIGURE 6 Variations in total arsenic concentration with pH, molybdenum, and vanadium in groundwater from La Pampa, Argentina (Smedley et al., 2002) and two areas of southwest USA (from Robertson, 1989). Curves indicate linear regressions for each data set and correlation coefficients (r²) are also given (United States coefficients from Roberston, 1989).

were found to be in the range of 6.9–9.3. These correlations between anionic and oxyanion compounds are comparable with those observed for other oxidizing high-pH groundwater environments such as Argentina (Figure 6). Note, however, that concentrations of the associated trace elements are generally much lower than those found in parts of Argentina. Arsenic concentrations in the aquifer sediments of the Basin and Range Province were in the range of 2–88 mg kg⁻¹. Oxidizing conditions (with dissolved oxygen present) were found to persist in the aquifers down to significant depths (600 m) despite the fact that the groundwaters had very long residence times (up to 10,000 years). The high arsenic (and

other oxyanion) concentrations are a feature of the closed basins of the province.

3. Sulfide Mineralization and Mining-Related Arsenic Problems

Arsenic problems have long been recognized in association with zones of ore mineralization as a result of the high concentrations of arsenic in sulfide minerals (Table I). Oxidation of the sulfide minerals, particularly through mining activity, may lead to the release of substantial quantities of arsenic and a number of transition and heavy metals.

The reaction for pyrite oxidation by oxygen to produce dissolved ferrous iron and sulfate can be described as:

$$FeS_2 + 7/2O_2 + H_2O \rightarrow Fe^{2+} + 2SO_4^{2-} + 2H^+$$

and similarly for oxidation of arsenopyrite as:

$$4FeAsS + 13O_2 + 6H_2O \rightarrow$$
$$4Fe^{2+} + 4AsO_4^{3-} + 4SO_4^{2-} + 12H^+.$$

The reactions can be catalyzed significantly by bacteria, notably *Thiobacillus ferrooxidans*, *T. thiooxidans*, and *Leptospirillum ferrooxidans* (Schrenk et al., 1998). With each of these minerals, besides the release of iron, sulphate, and arsenic into solution, the reactions also generate acidity which keeps the iron and many trace metals in solution. Hence, waters affected by sulfide oxidation commonly have very high concentrations of these constituents (iron and sulfate typically of the order of tens to thousands of milligrams per liter) and many have acidic compositions (pH <6). The highest arsenic concentration reported, 850,000 µg L^{-1}, is from a mine seep at Iron Mountain, California (Nordstrom & Alpers, 1999). Plumlee et al. (1999) also reported arsenic concentrations in mine drainage from various parts of the United States ranging from <1 µg L^{-1} to 340,000 µg L^{-1}.

Although high iron concentrations are common in water affected by sulfide oxidation, under oxidizing conditions iron is usually removed by oxidation and hydrolysis and subsequent precipitation as an iron(III) oxide. Trace elements likewise co-precipitate with, or adsorb onto, these iron oxides. Water is also neutralized in the presence of carbonate minerals. As a result, many mine waters have near-neutral pH values (e.g., Welch et al., 2000) and many have relatively low iron and arsenic concentrations. The considerable capacity of iron oxides to adsorb arsenic means that even in areas where high dissolved arsenic concentrations occur, they are generally restricted to the local area around the zone of oxidation, typically a few kilometers or less. Nonetheless, contamination of soils, sediments, and vegetation in mining and mineralized areas can be substantial. Many documented cases of mining contamination cite very high concentrations of arsenic and other trace elements in soils, stream sediments, and mine tailings (references cited in Smedley & Kinniburgh, 2002).

Although sulfide oxidation causes substantial environmental damage, its consequences for human health are usually indirect as most mining-impacted water sources are not used for potable supply. However, one or two notable exceptions exist and have given rise to localized health problems. Probably the worst case of arsenicosis related to mining activity that has been doc-

umented is that of Ron Phibun District in Nakhon Si Thammarat Province of southern Thailand. This area lies within the Southeast Asian Tin Belt and has been mined for several generations. Arsenic-related health problems were first recognized there in 1987. By the late 1990s, around 1000 people had been diagnosed with skin disorders, particularly in and close to Ron Phibun town (Williams, 1997). Arsenic concentrations up to 5000 µg L^{-1} have been found there in shallow groundwaters from Quaternary alluvial sediments that have been extensively dredged during mining operations. Deeper groundwaters from an older limestone aquifer are apparently less contaminated (Williams et al., 1996), although a few high arsenic concentrations occur, presumably also as a result of contamination from the mine workings. The mobilization of arsenic is related to the oxidation of arsenopyrite, which was exacerbated by the former tin-mining activities. The recent appearance in groundwater occurred during post-mining groundwater rebound (Williams, 1997).

Environmental arsenic damage has also been documented in the gold-mining areas of Ghana, although there is currently little evidence of detrimental effects on human health. Gold reserves have long been mined in the Ashanti Region of central Ghana. The gold there is also associated with sulfide mineralization, particularly arsenopyrite. Arsenic mobilizes in the local environment as a result of arsenopyrite oxidation, and is induced or aided by the mining activity. Around the town of Obuasi, high arsenic concentrations have been found in soils close to the mines and treatment works. Some high concentrations have also been reported in river waters close to the mining activity (Smedley et al., 1996).

Despite the presence of high arsenic concentrations in the contaminated soils and in bedrocks close to the mines, Smedley et al. (1996) found that many of the groundwaters of the Obuasi area had low concentrations of arsenic with a median value in borehole waters of just 2 µg L^{-1}. Some higher concentrations were observed (up to 64 µg L^{-1}), but these were not generally in the vicinity of the mines or related directly to mining activity. Rather, the higher concentrations were found to be present in relatively reducing groundwaters. Oxidizing groundwaters, especially from shallow hand-dug wells, had low arsenic concentrations. This was taken to be due to strong adsorption onto hydrous ferric oxides under the prevailing acidic groundwater conditions (median pH 5.4 in dug wells; Smedley et al., 1996).

Arsenic contamination from mining activities has been identified in numerous areas of the United States, many of which have been summarized by Welch et al. (1988, 1999, 2000). Groundwater from some areas has been reported to have very high arsenic concentrations locally (up to $48,000 \mu g L^{-1}$). Well-documented cases include the Fairbanks gold mining district of Alaska, the Coeur d'Alène Pb-Zn-Ag mining area of Idaho, Bunker Hill mine in Idaho, Leviathan Mine and Mother Lode in California, Kelly Creek Valley in Nevada, Clark Fork river area in Montana, and Lake Oahe in South Dakota.

Many other areas have above-average concentrations of arsenic in water, soils, sediments, and vegetation as a result of local mineralization exacerbated by mining activity. Documented cases include the Zimapán Valley of Mexico, Baja California (Mexico), the Lavrion region of Greece, the Iron Quadrangle of Minas Gerais (Brazil), the Styria region of Austria, the Zloty Stok area of southwest Poland, parts of southwest England, eastern Zimbabwe, South Korea, and Sarawak (Malaysia) (Figure 2) (Smedley & Kinniburgh, 2002 and references cited therein). Doubtless, arsenic problems also exist in many other undocumented mining areas.

Increased concentrations of dissolved arsenic have also been found in parts of the world with local mineralization which has not been mined. In Wisconsin in the United States, arsenic and other trace-element problems in groundwater have arisen as a result of the oxidation of sulfide minerals (pyrite and marcasite) present as a discrete secondary cement horizon in the regional Ordovician sandstone aquifer. Concentrations of arsenic up to $12,000 \mu g L^{-1}$ have been found in the well waters (Schreiber et al., 2000). The oxidation appears to have been promoted by groundwater abstraction which has led to the lowering of the piezometric surface at a rate of around $0.6 m$ $year^{-1}$ since the 1950s, which resulted in a partial dewatering of the aquifer. The high arsenic concentrations were observed where the piezometric surface intersects, or lies close to, the sulfide cement horizon.

Boyle et al. (1998) recorded arsenic concentrations up to $580 \mu g L^{-1}$ in groundwaters from an area of sulfide mineralization in Bowen Island, British Colombia. Heinrichs and Udluft (1999) found many high concentrations in groundwater from the Upper Triassic Keuper Sandstone in northern Bavaria. Out of 500 wells, 160 had arsenic concentrations in the range $10–150 \mu g L^{-1}$. The nature of the mineralization in this aquifer was not clearly identified. As yet unidentified areas of mineralization could be quite widespread, although these are likely to be on a local scale.

4. Geothermal Sources

The common occurrence of high concentrations of arsenic in geothermal fluids has been recognized for a

long time. Geothermal areas with documented high concentrations include the United States, Japan, New Zealand, Chile, Argentina, Kamchatka, France, and Dominica (references cited in Smedley & Kinniburgh, 2002). One of the largest and best documented geothermal systems is that of Yellowstone National Park in the states of Wyoming, Idaho, and Montana in the United States. Arsenic concentrations up to $7800 \mu g L^{-1}$ have been found in hot springs and geysers in the region (Thompson & Demonge, 1996). Geothermal inputs have also given rise to high concentrations of arsenic (up to $370 \mu g L^{-1}$) in waters of the Madison River (Nimick et al., 1998). Other reported geothermal occurrences in the United States include Honey Lake Basin, California (arsenic up to $2600 \mu g L^{-1}$); Coso Hot Springs, California (up to $7500 \mu g L^{-1}$); Imperial Valley, California (up to $15,000 \mu g L^{-1}$); Long Valley, California (up to $2500 \mu g L^{-1}$); Lassen Volcanic National Park, California (up to $27,000 \mu g L^{-1}$); Steamboat Springs, Nevada (up to $2700 \mu g L^{-1}$); and Geyser Bight, Umnak Island, Alaska (up to $3800 \mu g L^{-1}$) (White et al., 1963; Thompson et al., 1985; Welch et al., 1988, 2000). Geothermal inputs from Long Valley, California, are believed to be responsible for the relatively high concentrations ($20 \mu g L^{-1}$) of arsenic in the Los Angeles Aqueduct which provides the water supply for the city of Los Angeles (Wilkie & Hering, 1998). Geothermal inputs also contribute significantly to the high dissolved arsenic concentrations (up to $20 mg L^{-1}$) in Mono Lake, California (Maest et al., 1992).

In New Zealand, arsenic concentrations up to $9000 \mu g L^{-1}$ have been found in geothermal waters (Webster & Nordstrom, 2003). River and lake waters receiving inputs of geothermal water from the Wairakei, Ohaaki, Orakei Korako, and Atiamuri geothermal fields have arsenic concentrations up to $121 \mu g L^{-1}$ (Robinson et al., 1995). Concentrations diminish significantly downstream from the geothermal input areas.

Arsenic concentrations in the range of $45,000–50,000 \mu g L^{-1}$ have been found in geothermal waters from the El Tatio system in the Antofagasta region of Chile (Ellis & Mahon, 1977). The geothermal area lies between the volcanoes of the Andes and the Serrania de Tucle. Romero et al. (2003) found concentrations in waters from the Rio Loa and its tributaries in the range of $120–10,000 \mu g L^{-1}$ as a result of inputs from the El Tatio geothermal system.

Arsenic concentrations in the range of $100–5900 \mu g L^{-1}$ have been found in geothermal waters from Kamchatka (White et al., 1963). Concentrations in the range of $500–4600 \mu g L^{-1}$ were also reported in 26 geothermal water samples from five geothermal fields in Kyushu, Japan (Yokoyama et al., 1993).

White et al. (1963) reported arsenic concentrations in geothermal waters from Iceland. While concentrations were high compared to most groundwaters (range 50–120 $\mu g\,L^{-1}$), these are much lower than the values found in many other geothermal systems described above. This may be related to the fact that the geothermal fluids in the Icelandic volcanic field are associated with magmas of predominantly basaltic composition derived from oceanic mantle, hence having low arsenic concentrations (Webster & Nordstrom, 2003). Typical high-arsenic geothermal fluids are associated with acidic volcanic systems in continental settings. In this case, higher arsenic concentrations may be derived from the interaction of geothermal fluids with the continental crust, particularly argillaceous sediments (Nordstrom, 2002) in which the element is known to be preferentially partitioned. High arsenic concentrations have not been documented in geothermal systems associated with other volcanic provinces of dominantly basaltic composition, whether oceanic or continental (e.g., Hawaii, East African Rift).

Hot springs commonly show a positive correlation between arsenic and chloride. Welch et al. (1988) noted a general relationship between arsenic and salinity in geothermal waters from the United States. Wilkie & Hering (1998) also noted high Cl (and B) concentrations of arsenic-rich geothermal waters in Long Valley, California. High-arsenic geothermal waters are also typically of Na-Cl type in the Kyushu area of Japan (Yokoyama et al., 1993) as are those from the El Tatio geothermal system in Chile. The associations between arsenic and chloride were noted by Webster and Nordstrom (2003) to relate to the similar behavior of these elements during subsurface boiling and phase separation as both partition preferentially into the liquid phase.

IV. Mineral-Water Interactions

A. Adsorbed Arsenic in Sediments

Of the arsenic problem aquifers identified, it appears to be those hosted in young sediments that are most vulnerable to the development of high-arsenic groundwater on a regional scale. In these aquifers, surface reactions are considered to be important controls on the trace-element chemistry of the groundwaters. Arsenic behavior in aquifers, sediments, and soils has commonly been linked to adsorption/desorption reactions (e.g., Korte, 1991; Manning & Goldberg, 1997a; BGS & DPHE, 2001).

The major minerals that adsorb arsenic (as both arsenate and arsenite) in sediments are the metal oxides, particularly those of iron, aluminum, and manganese. Iron oxides are relatively abundant in most sediments and are commonly produced by the weathering of primary minerals. In freshwater sediments and silicate clays, they often constitute about 50% of the iron present (Manning & Goldberg, 1997b). Carbonates may also adsorb arsenic, but their capacity for doing so has not been measured quantitatively and is likely to be relatively small. Of these minerals, the iron oxides are probably the most important adsorbents of arsenic in sandy aquifers because of their relative abundance, strong binding affinity, and high specific surface area, especially the freshly formed amorphous oxides. These oxides are also particularly sensitive to a changing geochemical environment, acid dissolution, and reductive dissolution as well as changes in mineral structure and crystallinity. Aluminum oxides can also be expected to play a significant role in arsenic adsorption when present in quantity (Hingston et al., 1971; Manning & Goldberg, 1997b). Evidence from water treatment suggests that aluminum hydroxides have similar effectiveness to iron hydroxides (on a molar basis) for adsorbing As(V) below pH 7.5, but that iron salts are more efficient at higher pH and for adsorbing As(III) (Edwards, 1994). Iron sulfide minerals such as iron monosulfides and pyrite may also adsorb significant quantities of arsenic.

B. Reduced Sediments and the Role of Iron Oxides

A well-known sequence of reduction reactions occurs when sediments are buried and the environment becomes anaerobic (Berner, 1981; Stumm & Morgan, 1995). Such reactions are common in sediments from a wide variety of environments. The processes causing changes in iron redox chemistry are particularly important because they can directly affect the mobility of arsenic. One of the principal causes of high arsenic concentrations in groundwaters appears to be the reductive dissolution of hydrous iron oxides and/or the release of adsorbed and co-precipitated arsenic. This sequence begins with the consumption of oxygen and an increase in dissolved CO_2 from the decomposition of organic matter. Next, NO_3^- decreases by reduction to NO_2^- and ultimately to the gases N_2O and N_2. Insoluble manganic oxides dissolve by reduction to soluble Mn^{2+} and hydrous ferric oxides are reduced to Fe^{2+}. These processes are followed by SO_4^{2-} reduction to S^{2-}, then CH_4 production from fermentation and methanogenesis, and finally reduction of N_2 to NH_4^+. During sulfate reduction the sulfide produced reacts with any available iron to produce FeS and ultimately pyrite, FeS_2. Iron is often more abundant than sulfur so that there is excess iron beyond that which can be converted to pyrite. Arsenic(V) reduction would normally be expected to occur after Fe(III) reduction but before SO_4^{2-} reduction.

In sulfate-poor environments, iron from free iron oxides is solubilized as Fe^{2+} under reducing conditions. This gives rise to waters characteristically high in Fe. Groundwaters in such conditions tend to have Fe concentrations of $0.1–30\,mg\,L^{-1}$. The reaction is microbially mediated (Lovley & Chappelle, 1995). There is also evidence for solid-state transformations of the iron oxides under reducing conditions. This is reflected by a color change from red/orange/brown colors to gray/green/blue colors. Changes to the magnetic properties have also been documented (Sohlenius, 1996). Direct analysis of the Fe(II) and Fe(III) contents of iron oxides from reduced lake waters and sediments often indicates the presence of a mixed Fe(II)–Fe(III) oxide with an approximate average charge on the iron of 2.5 (Davison, 1993). The fate of iron oxides during reduction is not well understood, in part because they are probably very fine-grained and difficult to observe directly.

"Green rusts" are one possible product of the transformations. These have occasionally been identified or suspected in anoxic soils and sediments but are very unstable under oxidizing conditions (Taylor, 1980; Génin et al., 1998; Cummings et al., 1999). They consist of a range of green-colored Fe(II)–Fe(III) hydroxide minerals with a layered structure and a charge-balancing interlayer counterion, usually carbonate or sulfate. Authigenic magnetite (Fe_3O_4) is another possible product that has been identified in anaerobic sediments (Fredrickson et al., 1998), often with extremely small particle sizes. Magnetite is frequently found in sediments as a residual detrital phase from rock weathering but very fine-grained magnetite is also formed by so-called "magnetotactic" bacteria. Magnetite formation has been established under reducing conditions in the laboratory (Guerin & Blakemore, 1992). However, under strongly reducing conditions magnetite is unstable and in the presence of high concentrations of H_2S, it converts slowly to pyrite on a scale of centuries or more (Canfield & Berner, 1987). At the sediment/water interface in oceans, partial oxidation of primary magnetite can lead to a coating of maghemite, γ-Fe_2O_3. Further burial and reduction leads to the dissolution of the primary magnetite (Torii, 1997).

These studies of iron oxides in reducing environments indicate a lack of understanding of the detailed sequence of events taking place when Fe(III) oxides are subjected to strongly reducing conditions. The changes are evidently substantial and can result in the partial dissolution of the oxides and their transformation to completely new mineral phases. It is not yet clear what impact these transformations have on the adsorbed arsenic load of the original Fe(III) oxides. However, even quite small changes in arsenic binding could have a large impact on porewater arsenic concentrations because of the large solid/solution ratios found in sediments. Therefore, it is likely that understanding the changes to the nature of iron oxide minerals in sedimentary environments is an important part of understanding the processes leading to arsenic mobilization in sedimentary environments.

C. Role of Microbes

It has become increasingly clear in recent years that microbes play an important role in arsenic speciation and release. They can be significant catalysts in the oxidation of arsenite, reduction of arsenate, and methylation and volatilization of arsenic species. The microbiological transformations either provide sources of energy or act as detoxifying mechanisms. A number of chemoautotrophs oxidize As(III) by using oxygen, nitrate, or ferric iron as a terminal electron acceptor and CO_2 as their sole carbon source. Some autotrophs are also capable of As(III) oxidation. Conversely, it has recently been shown that some prokaryotes are capable of the respiratory reduction of As(V) (Oremland et al., 2002). These include several species of eubacteria and a few hyperthermophiles (Archaea).

Arsenic can also be released indirectly as a result of other microbially induced redox reactions. For example, the dissimilatory Fe-reducing bacterium *Shewanella alga* reduces Fe(III) to Fe(II) in scorodite ($FeAsO_4 \cdot 2H_2O$), releasing As(V) in the process (Cummings et al., 1999).

Thermophilic bacteria and cyanobacteria have been recognized in geothermal settings and can affect arsenic speciation and mineral precipitation in such systems (Webster & Nordstrom, 2003). Streams affected by geothermal inputs have also been shown to be influenced by microbial activity. In the geothermally fed waters of Hot Creek, California, Wilkie and Hering (1998) concluded that the oxidation of As(III) was controlled by bacteria attached to macrophytes. In abiotic systems, this oxidation reaction would normally be expected to be slow. In the Hot Creek system, the microbially mediated oxidation proceeded with a pseudo-first-order half-life of just 0.3 hours. Extremophiles such as *Bacillus arsenicoselenatis* have also adapted to the alkaline and saline conditions in the geothermally-influenced Mono Lake of California (Oremland et al., 2002).

V. COMMON FEATURES OF HIGH-ARSENIC GROUNDWATER PROVINCES

A. The Source of Arsenic

In the cases where affected groundwaters are found close to obvious geological or industrial sources rich in arsenic (geothermal springs, drainage from mineralized and mining areas, specific contaminant sources), it is clear that the anomalously high arsenic concentrations in the source region are responsible. The extent of this contamination is usually highly localized because the geochemical conditions within most aquifers do not favor arsenic mobilization on a regional scale. Areas affected by geothermal activity are potentially more widespread because in this case mobilization of arsenic is not required: arsenic is already present in solution and the size of geothermal reservoirs can be large. This probably accounts for why high-arsenic surface waters are normally located in geothermal areas. Perhaps more puzzling is the way in which very high concentrations of arsenic, up to several $mg\,L^{-1}$, are found in groundwaters from areas with apparently near-average source rocks. In aquifers with extensive areas of high-arsenic groundwater, this appears to be the rule rather than the exception. Most of these cases arise in aquifers derived from relatively young sediments, often consisting of alluvium or loess where the total sediment arsenic concentrations are usually in the range of 1–$20\,mg\,kg^{-1}$. Recognition of this fact is a recent development and its late appreciation has delayed the discovery of many high-arsenic groundwater provinces. Hitherto, geochemists had concentrated their investigations on the well-recognized, high-arsenic areas associated with mining and geothermal activity (see also Chapter 2, this volume).

Of critical importance is the fact that drinking-water limits for arsenic are very low in relation to the overall abundance of arsenic in the environment. Fortunately, most arsenic is normally retained in the solid phase and does not constitute a problem for potable water supplies. However, it only takes a very small percentage of this "solid" arsenic to dissolve or desorb to give rise to a serious groundwater problem.

B. Mobilization of Arsenic

There appear to be two key factors involved in the formation of high-arsenic groundwaters on a regional scale. First, there must be some form of geochemical trigger which releases arsenic from the aquifer solid phase into the groundwater. Secondly, the released arsenic must remain in the groundwater and not be flushed away. There are a number of possible geochemical triggers. In mining and mineralized areas, oxidation of sulfide ores may be triggered by influxes of oxygen or other oxidizing agents. This may follow a lowering of the water table or change in hydrogeological regime. However, in most arsenic-affected aquifers, the most important trigger appears to be the desorption or dissolution of arsenic from oxide minerals, particularly iron oxides. An important feature of this process is that the initial adjustment to environmental changes is probably quite rapid because adsorption reactions are surface reactions. The rate-limiting factors are probably those that control the major changes in pH, redox condition, and associated water-quality parameters of the aquifer. These are in part related to physical factors such as the rate of diffusion of gases through the sediment, the rate of sedimentation, the extent of microbiological activity, and the rates of chemical reactions. However, many of these are likely to be rapid on a geological time scale. Dissolution reactions are relatively slow but even oxide dissolution is rapid on a geological time scale and can be observed over the course of weeks or even days in flooded soils (Masscheleyn et al., 1991).

A qualification is that if diagenetic changes to the oxide mineral structure take place or if burial of sediment occurs, then there could be a slow release of arsenic over a much longer time scale. Details of the rate of release of arsenic from these sources are not yet clear. It is likely that the rate will diminish with time, with the greatest changes occurring in the early stages. Natural groundwater flushing means that very slow releases of arsenic are likely to be of little consequence because the arsenic released will be removed and not tend to significantly accumulate. Once the diagenetic readjustment has taken place and the sediments have equilibrated with their new environment, there should be little further release of arsenic. This contrasts with some mineral-weathering reactions which occur in "open" systems and can continue for millions of years—essentially until all of the mineral has dissolved. Seen in this context, the desorption/dissolution of arsenic from metal oxides in young aquifers is essentially a step change responding to a new set of conditions. The geochemical triggers involved could arise for a number of reasons. These are discussed further below.

1. Desorption of Arsenic at High pH Under Oxidizing Conditions

Under the aerobic and acidic to near-neutral conditions typical of many natural environments, arsenic is strongly adsorbed by oxide minerals as the arsenate ion and the concentrations in solution are therefore kept very low. The nonlinear nature of the adsorption isotherm for arsenate (Figure 5) ensures that the amount of arsenic adsorbed is often relatively large, even when dissolved concentrations of arsenic are low. Adsorption protects many natural environments from widespread arsenic toxicity problems. As the pH increases, especially above pH 8.5, arsenic tends to desorb from oxide surfaces thereby increasing the concentration in solution (Dzombak & Morel, 1990). The impact of this is magnified by the high solid/solution ratios typical of aquifers.

Smedley and Kinniburgh (2002) demonstrated the effect of a pH increase on As(V) desorption from hydrous ferric oxide (HFO) using the Dzombak and Morel (1990) diffuse double-layer model. The calculations indicated that for a sandy sediment with 25% porosity containing $1\,g\,Fe\,kg^{-1}$ as HFO, in equilibrium at pH 7 with water having an arsenic concentration of $1\,\mu g\,L^{-1}$ (typical of a river water, for example), increasing the pH under closed-system conditions results in a strong increase in the amount of arsenic desorbed. Above pH 9, the arsenate concentration in solution from this source could exceed $1000\,\mu g\,L^{-1}$.

These calculations assumed that no competing ions existed in the system. In reality, other ions are likely to compete for sorption sites on the HFO and will reduce the arsenic loading on the solid. Smedley and Kinniburgh (2002) demonstrated the reduced arsenic loading on HFO in the presence of phosphorus, but they noted that increasing the pH still had a strong desorbing effect on arsenic. At high pH, phosphate is also

released. Other potential competing ions include bicarbonate, silicate, and dissolved organic carbon.

There are several reasons why the pH might increase, but the most important in the present context is the uptake of protons by mineral-weathering and ion-exchange reactions combined with the effect of evaporation in arid and semi-arid regions. This pH increase is commonly associated with the development of salinity and the salinization of soils. Inputs of geothermal waters with high pH may be important in maintaining high arsenic concentrations in some alkaline lakes. Desorption at high pH is the most likely mechanism for the development of groundwater arsenic problems under oxidizing conditions such as those observed in Argentina, Mexico, and parts of the United States, for example, and would account for the observed positive correlation of arsenic concentrations with increasing pH (Figure 6). Increases in pH also induce the desorption of a wide variety of other oxyanions such as phosphate, vanadate, uranyl, molybdate, and borate. This is indeed observed in many groundwaters with high pH (Figure 6) (Smedley et al., 2002). These specifically adsorbed anions all interact with adsorption sites on the oxides in a competitive way and thus influence the extent of binding of each other. The process is not well understood in a quantitative sense. Phosphate, in particular, may play an important role in arsenic binding because it is usually more abundant than arsenic, often by a factor of 50 or more (in molar terms), and is also strongly bound to oxide surfaces. The infurence of Vanadium may also be significant.

The presence of bicarbonate may also promote desorption of arsenate. However, its role is unclear at present. Experimental evidence for the desorption of anionic compounds in the presence of bicarbonate is contradictory (Smedley & Kinniburgh, 2002). Bicarbonate is often the dominant anion in high-arsenic groundwaters and concentrations can be high, frequently exceeding $500\,mg\,L^{-1}$, and occasionally in excess of $1000\,mg\,L^{-1}$ (Table III). Silica may also exert a control on the adsorption of arsenic (Swedlund & Webster, 1998).

The role of dissolved organic carbon (fulvic and humic acids) is uncertain, at least from a quantitative point of view. Humic substances have been shown to reduce As(III) and As(V) sorption by iron oxides under some conditions (Xu et al., 1991; Bowell, 1994) and high-As groundwaters are associated with high humic-acid concentrations in some aquifers (Varsányi et al., 1991; Smedley et al., 2003). However, direct evidence for a causal link between dissolved organic carbon and arsenic desorption does not yet exist.

By contrast, some cations, because of their positive charge, may promote the adsorption of negatively charged arsenate (Wilkie & Hering, 1996). Calcium and magnesium are likely to be the most important cations in this respect because of their abundance in most natural waters and their +2 charge. Ferrous iron (Fe^{2+}) may be important in reduced waters and Al^{3+} in acidic waters.

The aridity described above enables the high pH values to be maintained and minimizes the flushing of any released arsenic. It also allows the buildup of high chloride and fluoride concentrations. Other environments with a high pH (up to pH 8.3), particularly open-system calcareous environments, are likely to be too well-flushed to allow released arsenic to accumulate.

The pH dependence of arsenic adsorption is important but has not yet been measured in detail for any aquifer materials, especially in the presence of typical groundwater compositions. The pH dependence is likely to depend to some extent on the heterogeneity of the aquifer material. Other specifically adsorbed anions, particularly phosphate and perhaps bicarbonate, may also affect the pH dependence of As(V) and As(III) binding.

2. Arsenic Desorption and Dissolution Due to a Change to Reducing Conditions

The onset of strongly reducing conditions, sufficient to enable iron(III) and probably sulfate reduction to take place, appears to be another trigger for the release of arsenic. The most common cause of this is the rapid accumulation and burial of sediments. This occurs in large alluvial systems, especially broad lowland meander belts and braided channels, and in prograding deltas. The organic carbon content of the buried sediment will largely determine the rate at which reducing conditions are created. Freshly produced soil organic matter readily decomposes and the presence of even small quantities can consume all of the dissolved oxygen, nitrate, and sulfate in the system. Solid-phase Fe(III) in minerals may moderate the rate of reduction of the aquifer. Reducing conditions can only be maintained if the diffusion and convection of dissolved oxygen and other oxidants from the surface is less rapid than the consumption. This is facilitated if there is a confining layer of fine-grained material close to the surface. This often occurs in large deltas where fine-grained overbank deposits overlie coarser-grained alluvial deposits.

A change in the redox state of adsorbed arsenic from As(V) to As(III) upon the onset of reducing conditions is likely to be one of the earliest reactions to take place.

This changing redox state could have wider repercussions because it will also affect a large number of competing reactions. Phosphate-arsenite competition, for example, is likely to be less important than phosphate-arsenate competition. There is also the potential for arsenite-arsenate competition. Although As(V) is normally more strongly bound than As(III), model calculations suggest that adsorbed phosphate can reverse the relative affinity of As(III) and As(V) at near-neutral pH values (BGS & DPHE, 2001). This has yet to be confirmed experimentally.

3. Changes in the Structure of Oxide Minerals

Disordered and fine-grained iron oxides, which may include HFO, lepidocrocite, schwertmannite, and magnetite, are common products of the early stages of weathering. Freshly-precipitated HFO is extremely fine-grained with cluster sizes of about 5 nm in diameter and a specific surface area of $300 \, m^2 \, g^{-1}$ or greater. HFO gradually transforms to more ordered structures such as goethite or hematite with larger crystal sizes and reduced surface areas. Goethite typically has specific surface areas of $150 \, m^2 \, g^{-1}$ or less and those for hematite are even less (Cornell & Schwertmann, 1996). This aging reaction can take place rapidly in the laboratory, but the rate in nature is likely to be inhibited somewhat by the presence of other ions, particularly strongly adsorbed ions such as aluminum, phosphate, sulfate, arsenate, bicarbonate, and silicate (Cornell & Schwertmann, 1996). One consequence of the reduction in surface area is that the amount of As(V) adsorption may decrease on a weight-for-weight basis. If the site density (site nm^{-2}) and binding affinities of the adsorbed ions remain constant, then as the specific surface area of the oxide mineral is reduced, some of the adsorbed ions may be desorbed.

Some of the desorbed ions may also be incorporated into the evolving oxide structure as a solid solution, which results in the reduction of the amount of arsenic released. In addition, if the surface structure changes, it is likely that the binding affinity for both arsenate ions and protons will change because the two are closely related.

Under strongly reducing conditions, it appears that additional processes could operate which may lead to a reduction in the overall adsorption of arsenic. Specifically for iron oxides, some of the surface iron could be reduced from Fe(III) to Fe(II) to produce a mixed-valence oxide perhaps akin to that of a magnetite or a green rust. This would tend to reduce the net positive charge of the surface (or increase its net negative charge) and would thereby reduce the electrostatic interaction between the surface and anions. This could result in the desorption of arsenic and a corresponding large increase in the concentration of arsenic in solution (BGS & DPHE, 2001).

On balance, laboratory and field evidence suggests that at micromolar concentrations of arsenic, freshly formed HFO binds more arsenic than goethite on a mole of Fe basis (De Vitre et al., 1991) and thus a reduction in affinity appears to be more probable. In Bangladesh, areas with high-arsenic groundwaters tend to correspond with those areas in which the sediments contain a relatively high concentration of oxalate-extractable iron (BGS & DPHE, 2001). This provides indirect support for the importance of iron oxides. It is likely that the soils and sediments most sensitive to arsenic release on reduction and aging are those in which iron oxides are abundant, HFO is initially a major fraction of the iron oxides present, and other arsenic-sorbing minerals are relatively scarce. Although the existing evidence is somewhat contradictory, it tends to suggest that a change from aerobic to anaerobic conditions often results in a net release of arsenic.

4. Mineral Dissolution

Mineral dissolution reactions tend to be most rapid under extremes of pH and Eh. For example, iron oxides dissolve under strongly acidic conditions and under strongly reducing conditions. Minor elements, including arsenic, present either as adsorbed (labile) arsenic or as irreversibly bound (non labile) arsenic will also tend to be released during the dissolution. This can explain, in part at least, the presence of high arsenic concentrations in acid mine drainage and in strongly reducing groundwaters. Reductive dissolution of iron(III) oxides accounts for the high Fe(II) content of anaerobic waters. This process also undoubtedly accounts for some of the arsenic found in reducing groundwaters, but it is probably insufficient to account for all, or even most, of the arsenic released (Smedley & Kinniburgh, 2002). It is likely that arsenic release from iron oxides under reducing conditions involves some combination of reductive dissolution and reductive desorption.

Manganese oxides also undergo reductive desorption and dissolution and thus could contribute to the arsenic load of groundwaters in the same way as iron. High concentrations of dissolved Mn(II) are observed in many reducing high-arsenic groundwaters (e.g., Bangladesh). Manganese(IV) oxide surfaces are also thought to catalyze the oxidation of As(III) (Oscarson et al., 1981).

As described above, one of the most pertinent mineral dissolution processes in respect of arsenic mobilization is the oxidation of sulfide minerals. Pyrite, the most abundant of these minerals, can be an important source of arsenic, especially where it is freshly exposed as a result of excavation by mining or by lowering of the water table. In extreme cases, this can lead to highly acidic groundwaters rich in sulfate, iron, and trace metals. As the dissolved iron is neutralized and oxidized, it tends to precipitate as HFO with resultant adsorption and co-precipitation of dissolved As(V). In this sense, pyrite oxidation is not a very efficient mechanism for releasing arsenic into water.

C. Arsenic Transport Through Aquifers

The geochemical triggers described above may release arsenic into groundwater but are not in themselves sufficient to account for the distribution of high-arsenic groundwaters observed in various parts of the world. The released arsenic must also not have been flushed away or diluted by normal groundwater flow. The rate of arsenic release must be set against the accumulated flushing of the aquifer that has taken place during the period of release. The rocks of most aquifers used for drinking water are up to several hundred million years old and yet contain groundwater that may be at most a few thousand years old. Hence, a large number of pore volumes of fresh water will have passed through the aquifer over its history. This is also the case in most young aquifers with actively flowing groundwater. By contrast, many alluvial and deltaic aquifers are composed of relatively young sediments. Where groundwater flow is slow, these contain relatively old groundwater. The relative ages of aquifer rocks and groundwater are important. High groundwater arsenic concentrations only occur on a regional scale when geochemical conditions capable of mobilizing arsenic are combined with hydrogeological conditions which prevent its loss (see also Chapter 31, this volume).

The Quaternary period has seen considerable changes in climate and global sea level. Variation in groundwater piezometric levels over this period would have induced large variations in base levels of erosion and in groundwater flow regimes and rates. During the last glaciation, about 21,000–13,500 years ago, sea levels would have been up to 130 m below the present mean sea level. This was a worldwide phenomenon and would have affected all then existing coastal aquifers. Continental and closed basin aquifers on the other hand would have not been affected. The increased hydraulic gradient in coastal aquifers during the glacial period would have resulted in correspondingly large groundwater flows and extensive flushing. The arsenic in these older aquifers would therefore tend to have been flushed away. The deep unsaturated zone would also have led to more extensive oxidation of the shallower horizons with possible increased sorption of arsenic to Fe(III) oxides. Aquifers younger than around 7000 years old, i.e., of Holocene age, will not have been subjected to this increased flushing that occurred during the most recent glaciation.

In Bangladesh, the age of sediment versus depth relationship is important because this has a direct bearing on the extent of aquifer flushing. Many of the shallow sediments in southern Bangladesh are of Holocene age (less than 13,000 years old and many less than 5000 years old). Hence, they will not have experienced the extensive flushing of the last glacial period. The majority of boreholes abstract groundwater from these sediments. Certainly at present, flushing is slow because of the extremely low hydraulic gradients, especially in the south. By contrast, deeper and older sediments, which probably exceed 13,000 years in age, will have been subjected to more extensive flushing. This may account for the low-arsenic groundwaters found in the deep aquifers of Bangladesh. Geochemical factors may also play a role because the evidence shows that while the deep groundwaters are currently reducing, they are less strongly reducing than the shallow aquifers (BGS & DPHE, 2001; van Geen et al., 2003). Certainly, the aquifers in the Pleistocene uplifted alluvial sediments of the Barind and Madhupur Tracts (Figure 3) will have been well flushed since they are at least 25,000–125,000 years old. These sediments invariably yield low-arsenic groundwaters, typically containing less than $0.5\,\mu g\,L^{-1}$ arsenic (Section 111B1a). A complication is that the Bengal Basin is locally rapidly subsiding and accumulating sediments. This adds to the high degree of local and regional variation.

The process of delta development also favors the separation of minerals based on particle size and produces the characteristic upwardly-fining sequences of sand-silt-clay. These lead to confining or semi-confining layers which aid the development of strongly reducing conditions. The youngest, distal part of the deltas will tend to contain the greatest concentration of fine-grained material and this provides an abundant source of arsenic in the form of colloidal-sized oxide materials. Flocculation of colloidal material, including iron oxides, at the freshwater-sea water interface will tend to lead to

relatively large concentrations of these colloids in the lower parts of a delta. The larger the delta, and the more rapid the infilling, the lower the hydraulic gradient and the less flushing is likely to have occurred. However, some deltas, even large deltas, may be so old and well-flushed that even the existing low hydraulic gradients will have been sufficient to flush away any desorbed or dissolved arsenic.

Regional flow patterns are not the only important factors. At a local scale, small variations in relief or in drainage patterns may dictate local flow patterns and hence the distribution of arsenic-rich groundwater. Evidence from Argentina, for example, suggests that the highest groundwater arsenic concentrations are found in small-scale topographic depressions where seasonal discharge occurs (Smedley et al., 2002). The same is true in Inner Mongolia (Smedley et al., 2003) and may also be true in Bangladesh. In any case, it is a characteristic of groundwater arsenic problem areas to have a high degree of local-scale variation. This reflects the poor mixing and the low rate of flushing characteristic of the affected aquifers. It is clear that flat low-lying areas, particularly large deltas and inland basins, are particularly prone to potentially high-arsenic groundwaters because they combine many of the risk factors identified above.

VI. Mitigation of High-Arsenic Groundwater Problems

Recognizing that an aquifer may have locally high concentrations of arsenic in the groundwater is one problem, mitigating the problem is a more difficult proposition. This is particularly the case in developing countries such as Bangladesh where the degree of short-range variability in arsenic concentrations is large, where the number of wells in use is immense, and where the technical infrastructure and resources are limited. As international drinking-water limits are also very low, arsenic poses a problem even in developed countries because of the expense of removing arsenic to the concentrations required to comply with regulations. While arsenic-removal plants can be built readily for large municipal water supplies, the problem is much greater for smaller rural supplies, many of which may be privately owned. Clearly, the best means of mitigating arsenic-related chronic health problems is to provide alternative low-arsenic water sources on a long-term basis. Options for mitigation are discussed below. The best options available will vary from country to country and from aquifer to aquifer, depending on local geological, climatic, social, and economic factors.

A. Identification and Use of Existing Low-Arsenic Wells

Identification of low-arsenic groundwater in an aquifer that is vulnerable to arsenic contamination demands a large-scale water survey and analysis (Kinniburgh & Kosmus, 2002). Identifying regions where the probability of finding high-arsenic groundwater is increased can be done on a reconnaissance basis, but it requires random sampling. On the other hand, as the concentrations of arsenic in a given aquifer are characteristically variable, identifying individual sources of safe water for potable use demands testing of each well. In a country such as Bangladesh with a large reliance on small private tubewell supplies, this task is enormous. This strategy also requires some consideration of the long-term variability in arsenic concentrations in an individual well and therefore ideally requires some form of monitoring. It is unlikely that the concentrations in most wells will change significantly in the short- to medium-term given natural groundwater flow conditions, but changes may take place more rapidly as a result of extensive pumping. The option to identify and use existing low-arsenic wells is likely to involve well sharing in badly affected areas and may require cultural changes. The suitability of this option will vary depending on the percentage of wells affected in a given region and the scale of variation, e.g., the distance to the nearest safe well.

B. Drilling New Wells in Alternative Low-Arsenic Aquifers

Even in parts of the world with recognized groundwater arsenic problems, there are often other aquifers in the region with low-arsenic sources of water. In the Bengal Basin, for example, groundwater from deeper aquifers (>150 m to 200 m) appears to often have low to very low arsenic concentrations. In many problem areas, development of the alternative aquifer may be a solution, albeit at considerable extra cost. However, in the Bengal Basin issues such as sustainability of supply given increased development and the use of groundwater for irrigation are important. Increased abstraction from deeper levels may induce leakage of high-arsenic groundwater from the overlying aquifer or of saline water in the southern coastal areas. In addition, in Bangladesh at least, the deep aquifer does not appear to exist everywhere in the country (BGS & DPHE, 2001). Hence, use of deeper groundwater is not a universal option. It may also require different equipment and skills for the deepest wells, which incur costs perhaps ten to twenty times greater than for small hand-pump tubewells completed in the shallow aquifer.

In Inner Mongolia (Huhhot Basin), deeper aquifers (>100 m) have been developed for potable supplies of groundwater as an alternative to arsenic-enriched shallow sources. However, these deep sources also often have unacceptably high concentrations of arsenic (Smedley et al., 2003). Clearly, it is difficult to generalize and knowledge of the local geology and hydrogeology of a given area is required.

C. Use of Dug Wells

In a number of strongly reducing high-arsenic aquifers, shallow hand-dug wells have been observed to contain low arsenic concentrations ($<10\,\mu g\,L^{-1}$). The Bengal Basin is a notable example (BGS & DPHE, 2001). The low concentrations observed are likely to be due to the maintenance of locally aerated conditions in the aquifer around the well and due to the presence of very young groundwater that has had little opportunity to react with minerals in the shallow aquifer. As such, dug wells provide a potentially suitable alternative source of drinking water. Indeed, as these were traditional sources of water before the advent of borehole technology in the Bengal Basin, this has often been stated as the reason why arsenic-related health problems in the region are a relatively recent phenomenon. Care must be taken to ensure that the water provided from dug wells is bacterially safe. This is often a major drawback of these shallow sources in many parts of the developing world. Another problem in more arid areas is the potential for dug wells to dry out during dry periods. In the Huhhot Basin of Inner Mongolia, dug wells are also not suitable for potable use as many are relatively reducing and can contain high arsenic concentrations, often greater than $50\,\mu g\,L^{-1}$ (Luo et al., 1997; Smedley et al., 2003).

D. Rainwater Harvesting

Collection and storage of rainwater is another option for providing sources of potable water with very low concentrations of arsenic and is in use in various parts of the world. In Bangladesh, collection of rainwater is currently being piloted. The method requires that the receptacle chosen to store the rainwater be kept clean and safe from the possibility of bacterial contamination for the duration required for storage. The viability of harvesting rainwater will depend to a major extent on the regional climate, as well as the cost of the receptacle, the presence of a suitable place to locate it, and a suitable roof to capture the rainwater. Even in semi-arid areas such as parts of Argentina, it could provide an option for at least part of the year. In the Bengal Basin, the high annual rainfall means that water availability is less of a problem, although there are still 6–7 months with little or no rainfall.

E. Use of Treated Surface Water

Use of surface water can be on a range of scales from small ponds with attached filters for use by a group of families or a village (as installed commonly in Bangladesh), to major piped and treated water supplies from reservoirs for urban areas. Large schemes require major investment and this is a significant drawback in developing countries. Reservoirs also tend to fill up with silt. Treatment systems on any scale also require maintenance. This option may be the best solution in some circumstances, but it is unlikely to be suitable for all high-arsenic affected regions.

F. Treatment of High-Arsenic Groundwater

When other options are unsuitable, a further possibility is to treat water to remove the high arsenic concentrations. This can also be achieved on a range of scales and with varying degrees of technical sophistication and cost. Treatment is achieved by various methods involving adsorption, ion exchange, or coagulation. The suitability and efficacy of each of these depend on factors including pH, arsenic speciation, and concentrations of other ions (Clifford & Ghurye, 2002). Adsorption techniques most commonly make use of metal oxides (mainly of iron or aluminum). Activated alumina is also an effective medium. Reverse osmosis is practiced in some areas. In Argentina, for example, reverse osmosis is commonly used in urban treatment works and is an effective method for reducing salinity as well as the concentrations of fluoride and arsenic. However, this method is expensive and not suitable for small private

supplies in rural areas. Coagulation commonly makes use of alum or ferric chloride, often with an oxidizing agent to oxidize the arsenic to As(V). In developing countries, where treatment has been tried, the methods used are generally the simplest and least expensive, including coagulation with iron salts or alum or adsorption using locally available materials (e.g., brick chips, laterite, iron-oxide coated sand). Various domestic filters have been developed in affected countries such as Bangladesh, India, and China. In reducing high-iron aquifers, passive oxidation is also occasionally used as a partial mitigation measure. This simply involves storing abstracted water for a period to allow oxidation of dissolved Fe(II) and subsequent sedimentation. In the process, arsenic also oxidizes (at least partially) and co-precipitates with, or adsorbs to, freshly forming iron oxides. The effectiveness of this option depends on a number of factors, critically the amount of Fe(II), the ratios of Fe/As and As(III)/As(V), and the time allowed for settling. The treatment methods used in developing countries meet with varying success and require a knowledge of maintenance techniques by the local village community or family. They are, nonetheless, better than no treatment at all and are particularly suitable as short-term measures to remove arsenic and other related elements such as iron and manganese. Clearly, passive oxidation is not an option for groundwaters from oxidizing aquifers such as Argentina or parts of Mexico.

There is also the possibility of *in situ* groundwater treatment whereby the arsenic is removed by promoting its adsorption or precipitation within the aquifer. This could be achieved by both an oxidative route (iron-oxide precipitation) or reductive route (sulfide precipitation) (Welch et al., 2003). These options are currently being explored in Bangladesh and elsewhere.

In summary, the groundwater arsenic problem is complex and in developing countries, where the problem may be large in relation to the resources available to tackle it, there exists no single, simple solution.

VII. Conclusions

The growing interest in arsenic in groundwater developed over the last few years stems from an increased awareness of its severe and detrimental effects on human health and resultant revisions of recommended and regulatory limits for arsenic in drinking water.

Concern has also increased following the recent recognition of the large scale of arsenic problems currently faced in Bangladesh and elsewhere. This account has attempted to characterize the distribution of arsenic in the environment and to describe the main geochemical controls on its speciation and mobilization.

Natural waters have arsenic concentrations varying over more than four orders of magnitude. Although most have low concentrations ($<10 \,\mu g \, L^{-1}$, and often significantly less), where they are higher and the water is used for drinking purposes, they can constitute an important pathway of human exposure to arsenic. Groundwater is probably the source of water most vulnerable to the development of high concentrations because of natural geochemical reactions in aquifers. Well-known high-arsenic groundwater areas occur in Argentina, Chile, China, Hungary, Mexico, Taiwan, and more recently in Bangladesh, West Bengal (India), Nepal, Vietnam, and Myanmar. Many of these areas have developed serious health problems for resident populations and the possibility of more widespread problems in the future is a serious concern. The scale of the problem in terms of populations exposed to arsenic is greatest in the Bengal Basin, with more than 40 million people drinking water containing arsenic above $50 \,\mu g \, L^{-1}$, the national standard for arsenic in many of the affected countries.

Such large-scale "natural" groundwater arsenic problem areas tend to be found in two types of environment: strongly reducing aquifers, often derived from alluvial sediment (e.g., Bangladesh, West Bengal, northern China, Hungary) and inland or closed basins in arid or semi-arid areas (e.g., central Argentina, Mexico). Both environments typically contain geologically young sediments and occur in flat, low-lying areas where groundwater flow is sluggish. Aquifers tend to be poorly flushed and any arsenic released from the sediments following burial has been able to accumulate in the groundwater. The arsenic concentrations of the aquifer materials in such problem aquifers appear not to be anomalously high, being typically in the range of 1–20 $mg \, kg^{-1}$. Arsenic-rich groundwaters may also be found in some geothermal areas and, on a more localized scale, in arsenic-rich mineralized areas, particularly sulfide ore zones and hence areas of mining activity.

The detailed mechanisms leading to the release of arsenic to groundwater are still poorly understood, but there appear to be a number of distinct triggers that can lead to the release of arsenic from mineral sources. The development of strongly reducing conditions at near-neutral pH may lead to the desorption of arsenic from mineral oxides together with the reductive dissolution of iron and manganese oxides, which may also lead to the release of arsenic. These groundwaters typically have abundant Fe(II) and As(III) and in many, sulfate concentrations are low (typically $1 \, mg \, L^{-1}$ or less), signifying probable sulfate reduction. Additional large concentrations of phosphate, bicarbonate, silicate, and possibly organic matter can enhance the desorption of arsenic through competition for adsorption sites on metal (especially iron) oxides. Release of arsenic may also be related to the development of high-pH (>8.5) conditions in semi-arid or arid environments, usually as a result of the combined effects of enhanced mineral weathering and high evaporation rates. The pH increase leads to the desorption of adsorbed arsenic (especially As(V) species) and a range of other anion-forming elements such as vanadium, boron, fluoride, molybdenum, selenium, and uranium from oxide minerals. The oxidation of sulfide minerals may also release arsenic to solution, although its fate following release will be critically dependent on the availability of metal oxides which may effectively scavenge the aqueous arsenic and hence limit its dispersion from the site of release.

Although arsenic problems have now been recognized in a number of regions of the world, there are doubtless other areas, principally aquifers, where problems are yet to be recognized. As more widespread water testing, health awareness, and diagnosis programs are undertaken internationally, these problem areas are likely to emerge gradually. Naturally, the best option in identifying any new areas at risk is to test the groundwater directly for arsenic. Where this is not possible, other water-quality data may provide some indication of the likelihood of arsenic problems. The range of water-quality characteristics identified in various provinces in this chapter should help to categorize areas in terms of likely arsenic risk. No single factor is sufficient to identify a problem area but if collectively many of the environmental characteristics and water-quality indicators point toward a potential problem, then this is an indicator for the need for an urgent testing program. Given the high degree of spatial variability in arsenic concentrations observed in many high-arsenic groundwater provinces, the task involved in screening "at-risk" areas may be large. Randomized reconnaissance surveying of the area is a logical first step for assessing the scale and location of problems.

Groundwater provides an important source of drinking water to many millions of people globally. While this chapter has focused on areas where groundwater arsenic concentrations are often high, it must be borne in mind that these are the exceptions. Groundwater more commonly provides a safe and reliable form of drinking water. Indeed, the proliferation of groundwater use in developing countries over the last few decades has resulted from the need to provide safe sources of drinking water that are protected from the potentially fatal effects of bacterial contamination. In this respect, the use of groundwater for improving community health has been highly successful. With regards to arsenic, most wells in most aquifers are likely to be uncontaminated, even when the groundwaters contain high concentrations of dissolved iron. Therefore it is also important to understand why these groundwaters are not affected. It appears that it is only when a number of critical geochemical and hydrogeological factors are combined that high-arsenic groundwaters occur.

SEE ALSO THE FOLLOWING CHAPTERS

Chapter 22 (Environmental Medicine) · Chapter 23 (Environmental Pathology) · Chapter 31 (Modeling Groundwater Flow and Quality)

FURTHER READING

Acharyya, S. K., Chakraborty, P., Lahiri, S., Raymahashay, B. C., Guha, S., Bhowmik, A., Chowdhury, T. R., Basu, G. K., Mandal, B. K., Biswas, B. K., Samanta, G., Chowdhury, K., Chanda, C. R., Lodh, D., Roy, S. L., Saha, K. C., and Roy, S. (1999). Brief Communications: Arsenic Poisoning in the Ganges Delta. The Natural Contamination of Drinking Water by Arsenic Needs to be Urgently Addressed, *Nature*, 401, 545.

AGRG (1978). *The Wolfson Geochemical Atlas of England and Wales*, Clarendon Press, Oxford, England.

Alam, M. K., Hasan, A. K. M. S., Khan, M. R., and Whitney, J. W. (1990). Geological Map of Bangladesh, Scale 1: 1,000,000. Geological Survey of Bangladesh.

Belkin, H. E., Zheng, B., and Finkelman, R. B. (2000). Human Health Effects of Domestic Combustion of Coal in Rural China: A Causal Factor for Arsenic and Fluorine Poisoning. In Second World Chinese Conference on Geological Sciences, Extended Abstracts, August 2000, Stanford University, 522–524.

Berg, M., Tran, H. C., Nguyen, T. C., Pham, H. V., Schertenleib, R., and Giger, W. (2001). Arsenic Contamination of Groundwater and Drinking Water in Vietnam: A Human Health Threat, *Environ. Sci. Technol.*, 35, 2621–2626.

Berner, R. A. (1981). A New Geochemical Classification of Sedimentary Environments, *J. Sediment. Petrol.*, 51, 359–365.

BGS and DPHE (2001). Arsenic Contamination of Groundwater in Bangladesh. Technical Report, WC/00/19, (D. G. Kinniburgh and P. L. Smedley, Eds.), Four Volumes. British Geological Survey, Keyworth.

Bowell, R. J. (1994). Sorption of Arsenic by Iron-Oxides and Oxyhydroxides in Soils, *Appl. Geochem.*, 9, 279–286.

Boyle, D. R., Turner, R. J. W., and Hall, G. E. M. (1998). Anomalous Arsenic Concentrations in Groundwaters of an Island Community, Bowen Island, British Columbia, *Environ. Geochem. Health.*, 20, 199–212.

Boyle, R. W., and Jonasson, I. R. (1973). The Geochemistry of As and Its Use as an Indicator Element in Geochemical Prospecting, *J. Geochem. Explor.*, 2, 251–296.

Cáceres, L., Gruttner, E., and Contreras, R. (1992). Water Recycling in Arid Regions–Chilean Case. *Ambio*, 21, 138–144.

Canfield, D. E., and Berner, R. A. (1987). Dissolution and Pyritization of Magnetite in Anoxic Marine Sediments, *Geochim. Cosmochim. Acta*, 51, 645–659.

Cebrián, M. E., Albores, M. A., Garciá-Vargas, G., Del Razo, L. M., and Ostrosky-Wegman, P. (1994). Chronic Arsenic Poisoning in Humans. In *Arsenic in the Environment, Part II: Human Health and Ecosystem Effects* (J. O. Nriagu, Ed.), John Wiley, New York, 93–107.

Chakraborti, D., Basu, G. K., Biswas, B. K., Chowdhury, U. K., Rahman, M. M., Paul, K., Chowdhury, T. R., Chanda, C. R., Lodh, D., and Ray, S. L. (2001). Characterization of Arsenic-Bearing Sediments in the Gangetic Delta of West Bengal, India. In *Arsenic Exposure and Health Effects IV*, (W. R. Chappell, C. O. Abernathy, and R. L. Calderon, Eds.), Elsevier, Amsterdam, 27–52.

Chen, C. J., Chuang, Y. C., Lin, T. M., and Wu, H. Y. (1985). Malignant Neoplasms Among Residents of a Blackfoot Disease-Endemic Area in Taiwan: High-Arsenic Artesian Well Water and Cancers, *Cancer Res.*, 45, 5895–5899.

Chen, S. L., Dzeng, S. R., Yang, M. H., Chlu, K. H., Shieh, G. M., and Wal, C. M., (1994). Arsenic Species in Groundwaters of the Blackfoot Disease Areas, Taiwan, *Environ. Sci. Technol.*, 28, 877–881.

Chen, S. L., Yeh, S. J., Yang, M. H., and Lin, T. H. (1995). Trace Element Concentration and Arsenic Speciation in the Well Water of a Taiwan Area with Endemic Blackfoot Disease, *Biol. Trace Element Res.*, 48, 263–274.

Chitrakar, R. L., and Neku, A. (2001). The Scenario of Arsenic in Drinking Water in Nepal. Department of Water Supply & Sewerage, Nepal.

Círculo Médico del Rosario (1917). Sobre la nueva enfermedad descubierta en Bell-Ville, *Rev. Médica del Rosario*, Rosario, Argentina, VII, 485.

Clifford, D. A., and Ghurye, G. L. (2002). Metal-Oxide Adsorption, Ion Exchange, and Coagulation-Microfiltration for Arsenic Removal from Water. In *Environmental Chemistry of Arsenic* (W. T. Frankenberger, Ed.), 217–246.

Cook, S. J., Levson, V. M., Giles, T. R., and Jackaman, W. (1995). A Comparison of Regional Lake Sediment and Till Geochemistry Surveys—A Case-Study from the Fawnie Creek Area, Central British Columbia, *Explor. Mining Geol.*, 4, 93–110.

Cornell, R. M., and Schwertmann, U. (1996). *The Iron Oxides: Structure, Properties, Reactions, Occurrence and Uses*, VCH, Weinheim, Germany.

Cullen, W. R., and Reimer, K. J. (1989). Arsenic Speciation in the Environment, *Chem. Rev.*, 89, 713–764.

Cummings, D. E., Caccavo, F., Fendorf, S., and Rosenzweig, R. F. (1999). Arsenic Mobilization by the Dissimilatory Fe(III)-Reducing Bacterium *Shewanella alga* BrY, *Environ. Sci. Technol.*, 33, 723–729.

Datta, D. K., and Subramanian, V. (1997). Texture and Mineralogy of Sediments from the Ganges-Brahmaputra-Meghna River System in the Bengal Basin, Bangladesh and their Environmental Implications, *Environ. Geol.*, 30, 181–188.

Davison, W. (1993). Iron and Manganese in Lakes, *Earth Sci. Rev.*, 34, 119–163.

Del Razo, L. M., Arellano, M. A., and Cebrián, M. E. (1990). The Oxidation States of Arsenic in Well-Water from a Chronic Arsenicism Area of Northern Mexico, *Environ. Pollut.*, 64, 143–153.

Del Razo, L. M., Hernández, J. L., García-Vargas, G. G., Ostrosky-Wegman, P., Cortinas de Nava, C., and Cebrián, M. E. (1994). Urinary Excretion of Arsenic Species in a Human Population Chronically Exposed to Arsenic Via Drinking Water. A Pilot Study. In *Arsenic Exposure and Health* (W. R. Chappell, C. O. Abernathy, and C. R. Cothern, Eds.), Science and Technology Letters, Northwood, England, pp. 91–100.

De Vitre, R., Belzile, N., and Tessier, A. (1991). Speciation and Adsorption of Arsenic on Diagenetic Iron Oxyhydroxides, *Limnol. Oceanogr.*, 36, 1480–1485.

Dudas, M. J. (1984). Enriched Levels of Arsenic in Post-Active Acid Sulfate Soils in Alberta, *Soil Sci. Soc. Am. J.*, 48, 1451–1452.

Dzombak, D. A., and Morel, F. M. M. (1990). *Surface Complexation Modelling—Hydrous Ferric Oxide*, John Wiley & Sons, New York.

Edwards, M. (1994). Chemistry of Arsenic Removal During Coagulation and Fe-Mn Oxidation, *J. Am. Water Works Assoc.*, 86, 64–78.

Ellis, A. J., and Mahon, W. A. J. (1977). *Chemistry and Geothermal Systems*, Academic Press, New York, p. 392.

Fleet, M. E., and Mumin, A. H. (1997). Gold-Bearing Arsenian Pyrite and Marcasite and Arsenopyrite from Carlin Trend Gold Deposits and Laboratory Synthesis, *Am. Mineralog.*, 82, 182–193.

Fontaine, J. A. (1994). Regulating Arsenic in Nevada Drinking Water Supplies: Past Problems, Future Challenges. In *Arsenic Exposure and Health* (W. R. Chappell, C. O. Abernathy, and C. R. Cothern, Eds.), Science and Technology Letters, Northwood, England, pp. 285–288, chap. 28.

Fredrickson, J. K., Zachara, J. M., Kennedy, D. W., Dong, H., Onstott, T. C., Hinman, N. W., and Li, S.-M. (1998). Biogenic Iron Mineralization Accompanying the Dissimilatory Reduction of Hydrous Ferric Oxide by a Groundwater Bacterium, *Geochim. Cosmochim Acta.*, 62, 3239–3257.

Fujii, R., and Swain, W. C. (1995). Areal Distribution of Selected Trace Elements, Salinity, and Major Ions in Shallow Ground Water, Tulare Basin, Southern San Joaquin Valley, California, U. S. Geological Survey Water-Resources Investigations Report, 95–4048.

Génin, J.-M. R., Bourrié, G., Trolard, F., Abdelmoula, M., Jaffrezic, A., Refait, P., Maitre, V., Humbert, B., and Herbillon, A. (1998). Thermodynamic Equilibria in Aqueous Suspensions of Synthetic and Natural Fe(II)-Fe(III) Green Rusts: Occurrences of the Mineral in Hydromorphic Soils, *Environ. Sci. Technol.*, 32, 1058–1068.

Goldberg, S. (1986). Chemical Modeling of Arsenate Adsorption on Aluminum and Iron Oxide Minerals, *Soil Sci. Soc. Am. J.*, 50, 1154–1157.

Guerin, W. F., and Blakemore, R. P. (1992). Redox Cycling of Iron Supports Growth and Magnetite Synthesis by *Aquaspirillum magnetotacticum*, *Appl. Environ. Microbiol.*, 58, 1102–1109.

Guo, H. R., Chen, C. J., and Greene, H. L. (1994). Arsenic in Drinking Water and Cancers: A Brief Descriptive View of Taiwan Studies. In *Arsenic Exposure and Health* (W. R. Chappell, C. O. Abernathy, and C. R. Cothern, Eds.), Science and Technology Letters, Northwood, England, pp. 129–138.

Gurzau, E. S., and Gurzau, A. E. (2001). Arsenic in Drinking Groundwater in Transylvania, Romania: An overview. In *Arsenic Exposure and Health Effects IV* (W. R. Chappell, C. O. Abernathy, and R. L. Calderon, Eds), Elsevier, Amsterdam, pp. 181–184.

Gustafsson, J. P., and Tin, N. T. (1994). Arsenic and Selenium in Some Vietnamese Acid Sulfate Soils, *Sci. Total Environ.*, 151, 153–158.

Harvey, C. F., Swartz, C. H., Badruzzaman, A. B. M., Keon-Blute, N., Yu, W., Ali, M. A., Jay, J., Beckie, R., Niedan, V., Brabander, D., Oates, P. M., Ashfaque, K. N., Islam, S., Hemond, H. F., and Ahmed, M. F. (2002). Arsenic Mobility and Groundwater Extraction in Bangladesh, *Science*, 298, 1602–1606.

Heinrichs, G., and Udluft, P. (1999). Natural Arsenic in Triassic Rocks: A Source of Drinking-Water Contamination in Bavaria, Germany. *Hydrogeol. J.*, 7, 468–476.

Hingston, F. J., Posner, A. M., and Quirk, J. P. (1971). Competitive Adsorption of Negatively Charged Ligands on Oxide Surfaces. In *Discussions of the Faraday Society*, No. 52. The Faraday Society, London, pp. 334–342.

Hsu, K.-H., Froines, J. R., and Chen, C.-J. (1997). Studies of Arsenic Ingestion from Drinking Water in Northeastern Taiwan: Chemical Speciation and Urinary Metabolites. In *Arsenic Exposure and Health Effects* (C. O. Abernathy, R. L. Calderon, and W. R. Chappell, Eds.), Chapman and Hall, London, pp. 190–209.

James, H. L. (1966). Chemistry of the Iron-Rich Sedimentary Rocks. Data of Geochemistry. USGS Professional Paper, 440-W, p. 61.

Karcher, S., Cáceres, L., Jekel, M., and Contreras, R. (1999). Arsenic Removal from Water Supplies in Northern Chile Using Ferric Chloride Coagulation, *J. Chartered Inst. Water Environ. Manag.*, 13, 164–169.

Kinniburgh, D. G., and Kosmus, W. (2002). Arsenic Contamination in Groundwater: Some Analytical Considerations, *Talanta*, 58, 165–180.

Korte, N. (1991). Naturally Occurring Arsenic in Groundwaters of the Midwestern United States, *Environ. Geol. Water Sci.*, 18, 137–141.

Legeleux, F., Reyss, J. L., Bonte, P., and Organo, C. (1994). Concomitant Enrichments of Uranium, Molybdenum and Arsenic in Suboxic Continental-Margin Sediments, *Oceanologr. Acta*, 17, 417–429.

Lovley, D. R., and Chapelle, F. H. (1995). Deep Subsurface Microbial Processes, *Rev. Geophys.*, 33, 365–381.

Luo, Z. D., Zhang, Y. M., Ma, L., Zhang, G. Y., He, X., Wilson, R., Byrd, D. M., Griffiths, J. G., Lai, S., He, L., Grumski, K., and Lamm, S. H. (1997). Chronic Arsenicism and Cancer in Inner Mongolia—Consequences of Well-Water Arsenic Levels Greater than $50 \mu g L^{-1}$. In *Arsenic Exposure and Health Effects* (C. O. Abernathy, R. L. Calderon, and W. R. Chappell, Eds.), Chapman and Hall, London, pp. 55–68.

Maest, A. S., Pasilis, S. P., Miller, L. G., and Nordstrom, D. K. (1992). Redox Geochemistry of Arsenic and Iron in Mono Lake, California, USA. In *Proceedings of the Seventh International Symposium on Water-Rock Interaction* (Y. K. Kharaka and A. S. Maest, Eds.), A. A. Balkema, Rotterdam, pp. 507–511.

Manning, B. A., and Goldberg, S. (1997a). Arsenic(III) and Arsenic(V) Adsorption on Three California Soils, *Soil Sci.*, 162, 886–895.

Manning, B. A., and Goldberg, S. (1997b). Adsorption and Stability of Arsenic(III) at the Clay Mineral-Water Interface, *Environ. Sci. Technol.*, 31, 2005–2011.

Masscheleyn, P. H., DeLaune, R. D., and Patrick, W. H. (1991). Effect of Redox Potential and pH on Arsenic Speciation and Solubility in a Contaminated Soil, *Environ. Sci. Technol.*, 25, 1414–1419.

Moore, J. N., Ficklin, W. H., and Johns, C. (1988). Partitioning of Arsenic and Metals in Reducing Sulfidic Sediments, *Environ. Sci. Technol.*, 22, 432–437.

Nicolli, H. B., Suriano, J. M., Peral, M. A. G., Ferpozzi, L. H., and Baleani, O. A. (1989). Groundwater Contamination with Arsenic and Other Trace-Elements in an Area of the Pampa, Province of Córdoba, Argentina, *Environ., Geol. Water Sci.*, 14, 3–16.

Nicolli, H. B., Tineo, A., García, J. W., Falcón, C. M., and Merino, M. H. (2001). Trace-Element Quality Problems in Groundwater from Tucumán, Argentina, In *Water-Rock Interaction 2001* (R. Cidu, Ed.), Volume 2, Swets & Zeitlinger, Lisse, The Netherlands, pp. 993–996.

Nimick, D. A., Moore, J. N., Dalby, C. E., and Savka, M. W. (1998). The Fate of Geothermal Arsenic in the Madison and Missouri Rivers, Montana and Wyoming, *Water Res. Res.*, 34, 3051–3067.

Nordstrom, D. K. (2000). An Overview of Arsenic Mass Poisoning in Bangladesh and West Bengal, India. In *Minor Elements 2000: Processing and Environmental Aspects of As, Sb, Se, Te, and Bi* (C. Young, Ed.), Society for Mining, Metallurgy and Exploration, Littleton, CO, pp. 21–30.

Nordstrom, D. K. (2002). Worldwide Occurrences of Arsenic in Ground Water, *Science*, 296, 2143–2144.

Nordstrom, D. K., and Alpers, C. N. (1999). Negative pH, Efflorescent Mineralogy, and Consequences for Environmental Restoration at the Iron Mountain Superfund Site, California, *Proc. Natl. Acad. Sci. U. S. A.*, 96, 3455–3462.

NRC (1999). *Arsenic in Drinking Water*, National Academy Press, Washington DC.

Oremland, R., Newman, D., Kail, B., and Stolz, J. (2002). Bacterial Respiration of Arsenate and its Significance in the Environment, In *Environmental Chemistry of Arsenic* (W. Frankenberger, Ed.), Marcel Dekker, New York, pp. 273–295, chap. 11.

Oscarson, D. W., Huang, P. M., Defosse, D., and Herbillion, A. (1981). Oxidative Power of Mn(IV) and Fe(III) Oxides with Respect to As(III) in Terrestrial and Aquatic Environments, *Nature* 291, 50–51.

Palmer, C. A., and Klizas, S. A. (1997). The Chemical Analysis of Argonne Premium Coal Samples, U. S. Geological Survey Bulletin 2144, U. S. Geological Survey, Reston, VA.

Peterson, M. L., and Carpenter, R. (1986). Arsenic Distributions in Porewaters and Sediments of Puget Sound, Lake Washington, the Washington Coast and Saanich Inlet, B. C, *Geochim. Cosmochim Acta*, 50, 353–369.

Plumlee, G. S., Smith, K. S., Montour, M. R., Ficklin, W. H., and Mosier, E. L. (1999). Geologic controls on the composition of natural waters and mine waters dramine diverse mineral-deposit types. In: EW. Geochemistry of Mineral Deopists. Part B: Case studies pp. 373–432.

Riedel, F. N., and Eikmann, T. (1986). Natural Occurrence of Arsenic and its Compounds in Soils and Rocks, *Wiss. Umwelt*, 3–4, 108–117.

Robertson, F. N. (1989). Arsenic in Groundwater Under Oxidizing Conditions, Southwest United States, *Environ. Geochem. Health*, 11, 171–185.

Robinson, B., Outred, H., Brooks, R., and Kirkman, J. (1995). The Distribution and Fate of Arsenic in the Waikato River System, North Island, New Zealand, *Chem. Speciat. Bioavail.*, 7, 89–96.

Romero, L., Alonso, H., Campano, P., Fanfani, L., Cidu, R., Dadea, C., Keegan, T., Thornton, I., and Farago, M. (2003). Arsenic Enrichment in Waters and Sediments of the Rio Loa (Second Region, Chile), *Appl. Geochem.*, 18, 1399–1416.

Schreiber, M. J., Simo, J., and Freiberg, P. (2000). Stratigraphic and Geochemical Controls on Naturally Occurring Arsenic in Groundwater, Eastern Wisconsin, *Hydrogeol. J.*, 8, 161–176.

Schrenk, M. O, Edwards, K. J., Goodman, R. M., and Banfield, J. F. (1998). Distribution of *Thiobacillus ferrooxidans* and *Leptospirillum ferrooxidans*: Implications for Generation of Acid Mine Drainage, *Science*, 279, 1519–1522.

Shacklette, H. T., Boerngen, J. G., and Keith, J. R. (1974). Selenium, Fluorine, and Arsenic in Superficial Materials of the Conterminous United States, U. S. Geological Survey, Circular 692, U. S. Government Printing Office, Washington DC.

Shotyk, W. (1996). Natural and Anthropogenic Enrichments of As, Cu, Pb, Sb, and Zn in Ombrotrophic versus Minerotrophic Peat Bog Profiles, Jura Mountains, Switzerland, *Water Air Soil Pollut.*, 90, 375–405.

Shrestha, R. R., Shrestha, M. P., Upadhyay, N. P., Pradhan, R., Khadka, R., Maskey, A., Tuladhar, S., Daha, B. M., Shiestha, S., and Shrestha, K. B. (2004). Groundwater arsenic contamination in Nepal: a new challenge for water supply sector. *Env. & Publ. Health Org.* (ENPHO) report.

Smedley, P. L. (2003). Arsenic in Groundwater–South and East Asia. In *Arsenic in Groundwater: Geochemistry and Occurrence* (A. H. Welch, and K. G. Stollenwerk, Eds.), Kluwer, The Netherlands, pp. 179–209, chap. 6.

Smedley, P. L., Edmunds, W. M., and Pelig-Ba, K. B. (1996). Mobility of Arsenic in Groundwater in the Obuasi area of Ghana. In *Environmental Geochemistry and Health* (J. D. Appleton, R. Fuge, and G. J. H. McCall, Eds.), Geological Society Special Publication No 113, Geological Society, London, pp. 163–181.

Smedley, P. L., and Kinniburgh, D. G. (2002). A Review of the Source, Behaviour and Distribution of Arsenic in Natural Waters, *Appl. Geochem.*, 17, 517–568.

Smedley, P. L., Kinniburgh, D. G., Huq, I., Luo, Z., and Nicolli, H. B. (2001). International Perspective on Naturally Occurring Arsenic Problems in Groundwater. In *Arsenic Exposure and Health Effects* IV (W. R. Chappell, C. O. Abernathy, and R. L. Calderon, Eds.), Elsevier, Amsterdam, pp. 9–25.

Smedley, P. L., Nicolli, H. B., Macdonald, D. M. J., Barros, A. J., and Tullio, J. O. (2002). Hydrogeochemistry of Arsenic and Other Inorganic Constituents in Groundwaters from La Pampa, Argentina, *Appl. Geochem.*, 17, 259–284.

Smedley, P. L. Zhang, M., Zhang, G., and Luo, Z. (2003). Hydrogeochemistry of Fluviolacustrine Aquifers and Arsenic Mobilization in the Huhhot Basin, Inner Mongolia, *Appl. Geochem.*, 18, 1453–1477.

Smith, A. H., Goycolea, M., Haque, R., and Biggs, M. L. (1998). Marked Increase in Bladder and Lung Cancer Mortality in a Region of Northern Chile Due to Arsenic in Drinking Water, *Am. J. Epidemiol.*, 147, 660–669.

Smith, A. H., Lingas, E. O., and Rahman, M. (2000). Contamination of Drinking-Water by Arsenic in Bangladesh: A Public Health Emergency, *Bull. WHO*, 78, 1093–1103.

Sohlenius, G. (1996). Mineral Magnetic Properties of Late Weichselian-Holocene Sediments from the Northwestern Baltic Proper, *Boreas*, 25, 79–88.

Stollenwerk, K. G. (2003). Geochemical Processes Controlling Transport of Arsenic in Groundwater: A Review of Adsorption. In *Arsenic in Groundwater: Geochemistry and Occurrence* (A. H. Welch and K. G. Stollenwork, Eds.), Kluwer, The Netherlands, pp. 67–100, chap. 3.

Stumm, W., and Morgan, J. J. (1995). *Aquatic Chemistry: Chemical Equilibria and Rates in Natural Waters*, Wiley Interscience, New York.

Sun, G. F., Pi, J. B., Li, B., Guo, X. Y., Yamauchi, H., Yoshida, T. (2001). Progresses on Researches of Endemic Arsenism in China: Population at Risk, Intervention Actions, and Related Scientific Issues. In *Arsenic Exposure and Health Effects IV* (W. R. Chappell, C. O. Abernathy, and R. L. Calderon, Eds.), Elsevier, Amsterdam, pp. 79–85.

Swedlund, P. J., and Webster, J. G. (1998). Arsenic Removal from Geothermal Bore Waters: The Effect of Monosilicic Acid. In *Proceedings of the 9th International Symposium on Water-Rock Interaction* (G. B. Arehart and J. R. Hulston, Eds.), WRI-9, Taupo, New Zealand, 30 March–3 April 1998, Balkema, Rotterdam, 947–950.

Taylor, R. M. (1980). Formation and Properties of Fe(II)-Fe(III) Hydroxycarbonate and its Possible Significance in Soil Formation, *Clay Minerals*, 15, 369–382.

Thompson, J. M., and Demonge, J. M. (1996). Chemical Analyses of Hot Springs, Pools, and Geysers from Yellowstone National Park, Wyoming, and Vicinity, 1980–1993, USGS Open-File Report, 96–68.

Thompson, J. M., Keith, T. E. C., and Consul, J. J. (1985). Water Chemistry and Mineralogy of Morgan and Growler Hot Springs, Lassen KGRA, California, *Trans. Geotherm. Res. Council*, 9, 357–362.

Tian, D., Ma, H., Feng, Z., Xia, Y., Le, X. C., Ni, Z., Allen, J., Collins, B., Schreinemachers, D., and Mumford, J. L. (2001). Analyses of microunclei in exfoliated epithelial cells from individuals chronically exposed to arsenic via drinking water in Inner Mangolia, China.

Torii, M. (1997). Low-Temperature Oxidation and Subsequent Downcore Dissolution of Magnetite in Deep-Sea Sediments ODP Leg 161, Western Mediterranean, *J. Geomagn. Geoelectr.*, 49, 1233–1245.

Tseng, W. P., Chu, H. M., How, S. W., Fong, J. M., Lin, C. S., and Yeh, S. (1968). Prevalence of Skin Cancer in an Endemic Area of Chronic Arsenicism in Taiwan, *J. Natl. Cancer Inst.*, 40, 453–463.

Umitsu, M. (1993). Late Quaternary sedmentary environments and landforms in the Ganges delta. *Sed. Geol.*, 83, 177–186.

Ure, A., and Berrow, M. (1982). The Elemental Constituents of Soils. In *Environmental Chemistry* (H. J. M. Bowen, Ed.), Royal Society of Chemistry, London, pp. 94–203, chap. 3.

van Geen, A., Zheng, Y., Versteeg, R., Stute, M., Horneman, A., Dhar, R., Steckler, M., Gelman, A., Small, C., Ahsan, H., Graziano, J. H., Hussain, I., and Ahmed, K. M. (2003). Spatial Variability of Arsenic in 6000 Tube Wells in a 25 km^2 Area of Bangladesh, *Water Resour. Res.*, 39, art. -1140.

Varsányi, I., Fodré, Z., and Bartha, A. (1991). Arsenic in Drinking Water and Mortality in the Southern Great Plain, Hungary, *Environ. Geochem. Health*, 13, 14–22.

Wang, L., and Huang, J. (1994). Chronic Arsenism from Drinking Water in Some Areas of Xinjiang, *Arsenic in the Environment, Part II: Human Health and Ecosystem Effects*, (J. O. Nriagu, Ed.), John Wiley, New York, pp. 159–172.

Webster, J. G. (1999). Arsenic. In *Encyclopaedia of Geochemistry* (C. P. Marshall and R. W. Fairbridge, Eds.), Chapman and Hall, London, 21–22.

Webster, J. G., and Nordstrom, D. K. (2003). Geothermal Arsenic. In *Arsenic in Groundwater: Geochemistry and Occurrence* (A. H. Welch, and K. G. Stollenwerk, Eds.), Kluwer, The Netherlands, pp. 101–125, chap. 4.

Welch, A. H., Helsel, D. R., Focazio, M. J., and Watkins, S. A. (1999). Arsenic in Ground Water Supplies of the United States. In *Arsenic Exposure and Health Effects* (W. R. Chappell, C. O. Abernathy, and R. L. Calderon, Eds.), Elsevier, Amsterdam, pp. 9–17.

Welch, A. H., and Lico, M. S. (1998). Factors Controlling As and U in Shallow Ground Water, Southern Carson Desert, Nevada, *Appl. Geochem.*, 13, 521–539.

Welch, A. H., Lico, M. S., and Hughes, J. L. (1988). Arsenic in Ground-Water of the Western United States, *Ground Water*, 26, 333–347.

Welch, A. H., Stollenwerk, K. G., Maurer, D. K., and Feinson, L. S. (2003). In situ Arsenic Remediation in a Fractured, Alkaline Aquifer. In *Arsenic in Ground Water: Geochemistry and Occurrence* (K. G. Stollenwerk, Ed.), Kluwer Academic Publishers, The Netherlands, pp. 403–419.

Welch, A. H., Westjohn, D. B., Helsel, D. R., and Wanty, R. B. (2000). Arsenic in Ground Water of the United States: Occurrence and Geochemistry, *Ground Water*, 38, 589–604.

White, D. E., Hem, J. D., and Waring, G. A. (1963). *Data of Geochemistry*, sixth edition (M. Fleischer, Ed.), Chapter F. Chemical Composition of Sub-Surface waters, Geological Survey Professional Paper 440-F, U. S. Government Printing Office, Washington DC.

Wilkie, J. A., and Hering, J. G. (1996). Adsorption of Arsenic onto Hydrous Ferric Oxide: Effects of Adsorbate/Adsorbent Ratios and Co-Occurring Solutes, *Colloids Surfaces A Physicochem. Eng. Aspects*, 107, 97–110.

Wilkie, J. A., and Hering, J. G. (1998). Rapid Oxidation of Geothermal Arsenic(III) in Streamwaters of the Eastern Sierra Nevada, *Environ. Sci. Technol.*, 32, 657–662.

Williams, M. (1997). Mining-Related Arsenic Hazards: Thailand Case-Study, Summary Report, British Geological Survey Technical Report, WC/97/49.

Williams, M., Fordyce, F., Paijitprapapon, A., and Charoenchaisri, P. (1996). Arsenic Contamination in Surface Drainage and Groundwater in Part of the Southeast Asian Tin Belt, Nakhon Si Thammarat Province, Southern Thailand, *Environ. Geol.*, 27, 16–33.

Wyatt, C. J., Fimbres, C., Romo, L., Mendez, R. O., and Grijalva, M. (1998). Incidence of Heavy Metal Contamination in Water Supplies in Northern Mexico, *Environ Res.*, 76, 114–119.

Xu, H., Allard, B., and Grimvall, A. (1991). Effects of Acidification and Natural Organic Materials on the Mobility of Arsenic in the Environment, *Water Air Soil Pollut.*, 57–8, 269–278.

Yokoyama, T., Takahashi, Y., and Tarutani, T. (1993). Simultaneous Determination of Arsenic and Arsenious Acids in Geothermal Water, *Chem. Geol.*, 103, 103–111.

FLUORIDE IN NATURAL WATERS

MIKE EDMUNDS AND PAULINE SMEDLEY
British Geological Survey

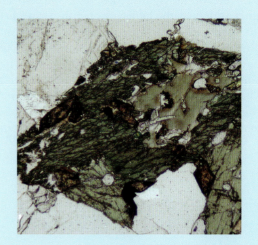

CONTENTS

I. INTRODUCTION

Fluorine is an essential element in the human diet. Deficiency in fluorine has long been linked to the incidence of dental caries, and the use of fluoride toothpastes and mouthwashes has been widely advocated in mitigating dental health problems. Fluoridation of water supplies to augment naturally low fluoride concentrations is also undertaken in some countries. However, despite the essentiality of fluorine in humans, optimal doses appear to fall within a narrow range. The detrimental effects of ingestion of excessive doses of fluorine are also well documented. Chronic ingestion of high doses has been linked to the development of dental fluorosis, and in extreme cases, skeletal fluorosis. High

doses have also been linked to cancer (Marshall, 1990), although the association is not well established (Hamilton, 1992).

Drinking water is particularly sensitive in this respect because large variations in fluoride concentration exist in water supplies in different areas. Concentrations in natural waters span more than four orders of magnitude, although values typically lie in the $0.1–10\,mg\,L^{-1}$ range. Where concentrations are high, drinking water can constitute the dominant source of fluorine in the human diet. Concentrations in drinking water of around $1\,mg\,L^{-1}$ are often thought to be optimal; however, chronic use of drinking water with concentrations above about $1.5\,mg\,L^{-1}$ is considered to be detrimental to health. The World Health Organization (WHO) (2004) guideline value for fluoride in drinking water is $1.5\,mg\,L^{-1}$. Many countries also use this value as a national standard for drinking water, although the standard in both China and India is $1.0\,mg\,L^{-1}$ (Table I). The U. S. Environmental Protection Agency (EPA) has set the primary standard (enforceable limit) at $4\,mg\,L^{-1}$ for fluoride in drinking water, although the secondary standard (non-enforceable) for United States drinking water supplies is $2\,mg\,L^{-1}$. In Tanzania, the national standard is as high as $8\,mg\,L^{-1}$, reflecting the difficulties with compliance in a country with regionally high fluoride concentrations and problems with water scarcity.

High fluoride concentrations are most often associated with groundwaters as these accumulate fluoride

TABLE I. Regulations and Recommendations for Fluoride in Drinking Water From a Number of Organizations or Countries

Institution/ nation	Limit/guideline	Value ($mg\,L^{-1}$)	Comment
WHO	Guideline value (GV)	1.5	2004 guidelines
U.S. EPA	Primary standarad	4	Enforceable
U.S. EPA	Secondary standard	2	Guideline intended to protect against dental fluorosis; not enforceable
EC	Maximum permissible value	1.5	1998 regulations
Canada	National standard	1.5	
India	National standard	1	Lowered from 1.5 $mg\,L^{-1}$ in 1998
China	National standard	1	
Tanzania	National standard	8	Interim standard

from rock dissolution as well as geothermal sources. Many high-fluoride groundwater provinces have been recognized in various parts of the world, particularly northern China, India, Sri Lanka, Mexico, the western United States, Argentina, and many countries in Africa. Fluoride removal by water treatment is carried out in some countries. However, as many of the high-groundwater provinces occur in developing countries, fluoride removal practices vary widely and many high-fluoride water sources are used without treatment. As a result, large populations throughout parts of the developing world suffer the effects of chronic endemic fluorosis. Estimates are not well established, but more than 200 million people worldwide are thought to be drinking water with fluoride in excess of the WHO guideline value. This includes around 70 million in India, 45 million people in China (Wuyi et al., 2002), and about 5 million in Mexico (Diaz-Barriga et al., 1997). The population at risk in Africa is unknown but is also likely to be tens of millions.

Despite the clear evidence for health problems related to fluoride in drinking water and the links between fluoride occurrence and geology, there have

been few reviews on the hydrogeochemistry of fluoride. This chapter addresses the hydrogeochemical aspects of fluoride in water, particularly groundwater, and outlines the links with health impacts. It also characterizes the typical ranges of concentrations of fluoride found in water bodies, along with their distribution, speciation, and mechanisms of mobilization. The principles of fluoride behavior are also illustrated with case studies from Canada, East Africa, Ghana, India, Sri Lanka, and the UK.

II. History of Fluoride Research and Links with Health

Chemists first intimated at the potential for an association between fluorine and health in the 19th century from its variable presence in bones and teeth. In the latter part of that century, fluorine was recommended for administration to pregnant women in the interests of dental health. In 1892, it was suggested that the high incidence of dental caries in England was due to deficiency of fluoride in the diet. However, it was not until the first quarter of the 20th century that a clear association between fluorine in water supplies and dental health was established by medical scientists. The earliest studies in Europe and North America were aided by the fact that populations were less mobile than today, and water supplies in rural areas were generally from wells or springs.

The first studies on fluorosis in the United States were carried out in the 1930s (Dean & Elvove, 1937). These studies established high fluoride concentrations in drinking water as a likely factor in fluorosis disease and led to the adoption (U. S. Public Health Service, 1943, 1946) of an initial upper limit for fluoride in American drinking water supplies of 1.0 $mg\,L^{-1}$, which was later revised to 1.5 $mg\,L^{-1}$. In Britain, the earliest studies (1922) were conducted on school children in Maldon, Essex, where an association was established between mottled teeth and fluoride in Chalk groundwater at concentrations of 4.5–5.5 $mg\,L^{-1}$ (Ainsworth, 1933; Hoather, 1953) (see also Chapter 28, this volume).

The recognition early in the 20th century that certain areas of the United States, the UK, and elsewhere had clearly defined patterns of dental caries and dental fluorosis prompted closer examination of the association with the geology and the underlying groundwater supplies. In the early studies, a strong link between caries and low-fluoride water emerged and this led, as early as 1945, to the fluoridation of fluoride-deficient water supplies in Grand Rapids, Michigan (Maier, 1950). Fluoridation of British water supplies was also initiated in the early part of the 20th century and today, around 10% of the British public water supply is fluoridated, mainly in soft water areas.

Since WW II, the significance of water-derived fluoride has become obscured. This has coincided with greater variety in diets and the increased mobility of populations. A number of other foodstuffs, including tea, have been established as key sources of additional fluoride. The introduction of topical fluorides (e.g., fluoride toothpastes, mouthwashes) has also enhanced the intake of fluoride. Nevertheless, it is clear that exposure to fluoride from drinking water and food remains an important factor. Today, these sources are augmented by exposure from anthropogenic sources such as industrial emissions.

There has been much debate over the alleged benefits of fluoridation of drinking water supplies (Hamilton, 1992), and the issue is still strongly contentious. One of the reasons is that many of the scientific studies, especially the epidemiological studies, are considered to be incomplete or flawed (Marshall, 1990). Studies of water fluoridation also yield conflicting results (Hamilton, 1992). This situation was reviewed by Diesendorf (1986) who called for a thorough re-examination of the fluoridation issue. His studies indicated that large temporal reductions in tooth decay, which could not be attributed to fluoridation, had been observed in both fluoridated and unfluoridated areas of at least eight developed countries over the preceding 30 years. Such arguments have weakened the case for fluoridation considerably. Some studies have even suggested that fluoridation has been detrimental to health (Zan-dao & Yan, 2002).

Despite the uncertain health effects of fluoride in drinking water at low concentrations ($0.7\,mg\,L^{-1}$ or less), the chronic effects of exposure to excessive fluoride in drinking water are well established. The most common symptom is dental fluorosis (mottled enamel), a condition involving interaction of fluoride with tooth enamel, which involves staining or blackening, weakening, and possible eventual loss of teeth. With higher exposure to fluoride, skeletal fluorosis can result. This manifests in the early stages as osteosclerosis, involving hardening and calcifying of bones and causing pain, stiffness, and irregular bone growth. At its worst, the condition results in severe bone deformation and debilitation. Long-term exposure to fluoride in drinking water at concentrations above about $1.5\,mg\,L^{-1}$ can

TABLE II. Health Effects of Fluoride Concentrations in Drinking Water

Fluoride concentration range $(mg\,L^{-1})$	Chronic health effects
Nil	Limited growth and fertility
$0.0–0.5\,mg\,L^{-1}$	Dental caries
$0.5–1.5\,mg\,L^{-1}$	Promotes dental health, prevents tooth decay
$1.5–4.0\,mg\,L^{-1}$	Dental fluorosis (mottled teeth)
$4–10\,mg\,L^{-1}$	Dental fluorosis, skeletal fluorosis
$>10\,mg\,L^{-1}$	Crippling fluorosis

From Dissanayake (1991).

result in dental fluorosis (Table II), whereas values above $4\,mg\,L^{-1}$ can result in skeletal fluorosis, and above about $10\,mg\,L^{-1}$ crippling fluorosis can result (Dissanayake, 1991). However, Nutrition is believed to be an important factor in the onset of fluorosis disease. Dietary deficiencies in calcium and vitamin C are recognized as important exacerbating factors. Young children are particularly at risk as the fluoride affects the development of growing teeth and bones. Once developed, the symptoms of fluorosis are irreversible. Links between high fluoride and other health problems, including birth defects (Hamilton, 1992) and cancer (Marshall, 1990) are less clearly defined (see also Chapter 28, this volume).

III. THE HYDROGEOCHEMICAL CYCLE OF FLUORINE

A. Atmospheric and Surface Water Inputs

The hydrogeochemical cycle of fluorine is illustrated in Figure 1. The cycle involves transfer of fluorine to the atmosphere by volcanic emissions, evaporation, marine aerosols, and industrial pollution. Wet and dry deposition transfer fluorine to the biosphere and geosphere. In the geosphere, uptake and release of fluorine are controlled by various water–rock interactions and by inputs from anthropogenic sources.

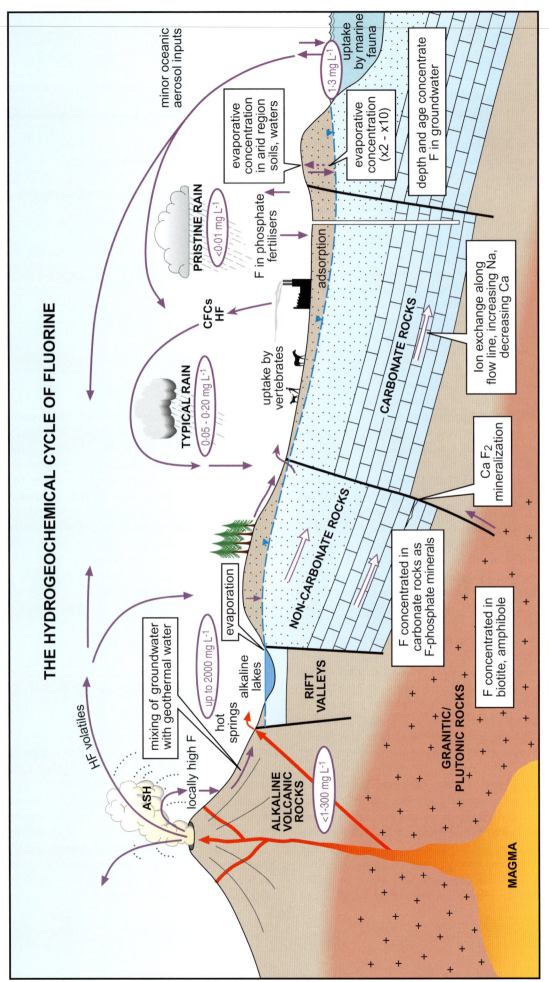

THE HYDROGEOCHEMICAL CYCLE OF FLUORINE

minor oceanic aerosol inputs

uptake by marine fauna

1-3 mg L^{-1}

evaporative concentration in arid region soils, waters

evaporative concentration (x2 - x10)

depth and age concentrate F in groundwater

PRISTINE RAIN

<0.01 mg L^{-1}

F in phosphate fertilisers

adsorption

CARBONATE ROCKS

CFCs HF

Ion exchange along flow line, increasing Na, decreasing Ca

TYPICAL RAIN

0.05 - 0.20 mg L^{-1}

uptake by vertebrates

Ca F$_2$ mineralization

NON-CARBONATE ROCKS

F concentrated in carbonate rocks as F-phosphate minerals

F concentrated in biotite, amphibole

evaporation

HF volatiles

mixing of groundwater with geothermal water

up to 2000 mg L^{-1}

hot springs   alkaline lakes

locally high F

RIFT VALLEYS

ALKALINE VOLCANIC ROCKS

GRANITIC/ PLUTONIC ROCKS

<1-300 mg L^{-1}

ASH

MAGMA

FIGURE 1   Schematic diagram showing the fluorine hydrogeochemical cycle.

Rainfall constitutes an important component of the cycle. Fluorine sources in rainfall include marine aerosols, volcanic emissions, recent introduction of chlorofluorocarbons (CFCs), and industrial emissions. Industrial aerosols are especially produced from coal burning, brick making, and aluminum smelting (Fuge & Andrews, 1988). Fluoride can substitute to a minor extent for OH^- in minerals, including clay minerals, and this is released (as is the OH^- by dehydroxylation) upon heating in industrial processes. Fluoride concentrations in rainfall are low and accurate data are sparse as a result of analytical difficulties. If marine aerosols contribute to rainfall compositions in their seawater proportions, rainfall in coastal areas with $10\,mg\,L^{-1}$ chlorine should have $0.68\,\mu g\,L^{-1}$ fluorine. In many continental areas

where rainfall chlorine concentrations are near or below $1\,mg\,L^{-1}$, fluoride inputs to streams and groundwaters from rainfall should be at or below $0.1\,\mu g\,L^{-1}$. In fact, rainfall fluoride concentrations tend to be higher than these estimates, implying that some fractionation favors greater uptake of the more volatile fluoride at the sea surface or that the ratio is increased by atmospheric fluoride inputs.

The concentrations of fluoride in pristine rainfall are difficult to assess because human impacts are likely to dominate in most areas. Comparative studies at inland and coastal sites in Virginia in the United States found median values of 4 and $9\,\mu g\,L^{-1}$, respectively (Table III) and marine aerosol inputs were considered to be small (Barnard & Nordstrom, 1982). Higher concentrations,

TABLE III. Ranges of Fluoride Concentrations in Various Natural Waters

Country	Region/aquifer	Range of F $(mg\,L^{-1})$	Average $(mg\,L^{-1})$	Number of analyses	Reference
Rainfall					
Norway		0.0–0.253	0.013–0.025		Saether & Andreassen (1989)
USA	Virginia—coastal	0.002–0.02	0.009		Barnard & Nordstrom (1982)
	Virginia—inland		0.004		
Sri Lanka	Anuradhapura	<0.02–0.08	0.03	6	BGS (unpublished data)
UK	Chilton, southeast England		0.096	24	Edmunds et al. (1987)
UK	Mid Wales	0.02–0.22	0.02		Neal (1989)
UK	Loch Fleet, Scotland	<0.05	<0.05	10	Cook et al. (1991)
Surface waters					
India	Ganges River		0.154		Datta et al. (2000)
India	Meghna River		0.066		Datta et al. (2000)
India	Brahmaputra River		0.120		Datta et al. (2000)
UK	Hafren river, Wales	0.03–0.07	0.05		Neal (1989)
Ghana	Rivers, Accra region	0.03–0.14	0.09	5	Botchway et al. (1996)
Ghana	Ponds, Accra region	0.07–0.30	0.18	5	Botchway et al. (1996)
Surface waters in high-fluoride regions					
Tanzania, Kenya, Uganda	Lakes on alkaline volcanic rocks and others	0.2–1627		200	Kilham & Hecky (1973)
Ethiopia	River Awash	0.9–1.3	1.1	2	Ashley & Burley (1994)
Kenya	Streams bordering Lake Magadi	0.1–1.9	0.6	14	Jones et al. (1977)
Kenya	Lake Naivasha	1.7–1.8	1.75	2	Jones et al. (1977)
Tanzania	Lake Magadi	759–1980	1281	13	Jones et al. (1977)
Tanzania	Little Magadi Lake	668–754	711	2	Jones et al. (1977)
Tanzania	Lakes, alkaline volcanic area	60–690		9	Nanyaro et al. (1984)
Tanzania	Rivers, alkaline volcanic area	12–26		9	Nanyaro et al. (1984)
Soil water					
UK	Upland Wales	0.02–0.30	0.05		Neal (1989)

continues

Continued

Country	Region/aquifer	Range of F (mg L^{-1})	Average (mg L^{-1})	Number of analyses	Reference
Geothermal springs					
Tanzania	Mbulu springs	up to 99			Bugaisa (1971)
Tanzania	Lake Natron thermal springs	330			Bugaisa (1971)
Kenya	Hot springs (>50°C) bordering Lake Magadi	141–166	7	155	Jones et al. (1977)
Kenya	Warm springs (<50°C) bordering Lake Magadi	50–146	14	82	Jones et al. (1977)
Tunisia	Hot springs	up to 4			Travi (1993)
USA	Western states	0.8–30.8			Nordstrom & Jenne (1977)
USA	Wister mudpots, Salton Sea	14–15			Ellis and Mahon (1977)
Iceland	Spring, Reykjavik	1.0			Ellis and Mahon (1977)
Iceland	Spring, Hveragerdi	1.1–1.9			Ellis and Mahon (1977)
Kamchatka	Spring	0.8			Ellis and Mahon (1977)
Former USSR	Spring, Makhachkala, Dagestan	0.4			Ellis and Mahon (1977)
New Zealand	Springs	0.3–8.4			Ellis and Mahon (1977)
Taiwan	Spring, Tatun	7.3			Ellis and Mahon (1977)
France	Vichy water	3.8–8.0			Goni et al. (1993)
France	Mont Dore springs	0.07–3.6			Jacob (1975)
France	Plombières springs	1.9–7.0			Fritz (1981)
China	Springs, Shixingsian, Guangdong	up to 45			Fuhong & Shuquin (1988)
Portugal	Rio Vouga hot springs	0.04–20.5	11.0	6	Ten Haven et al. (1985)
Groundwater: crystalline basement rocks					
Norway	Caledonian basic and ultrabasic rocks		1.89		Banks et al. (1998)
	Precambrian granite		1.69		Banks et al. (1998)
	Precambrian anorthosite, charnockite		<0.05		Banks et al. (1998)
Norway	Igneous and metamorphic rocks, Hordaland	0.02–9.48	0.30	1063	Bårdsen et al. (1996)
Ghana	Crystalline basement, including granite and metasediment	0.09–3.8	1.07	118	Smedley et al. (1995)
Senegal	Granitoids, pelites, schists, amphibolites	0.1–3.5			Travi (1993)
India	Crystalline basement, Andhra Pradesh	up to 20			Rao (1974)
India	Archaean granites and gneisses, east and southeast Karnataka	0.80–7.4	3.5	25	Suma Latha et al. (1999)
India	Archaean banded gneiss, southeast Rajasthan	<0.1–16.2	1.28	2649	Gupta et al. (1993)
India	Archaean granite	0.3–6.9		188	Vijaya Kumar et al. (1991)
Sri Lanka	Crystalline basement, including granite and charnockite	<0.02–10			Dissanayake (1991)
South Africa	Western Bushveld Complex	0.1–10	143	3485	McCaffrey (1998)
Groundwater: volcanic rocks					
Ethiopia	Volcanic bedrock (Wonji/Shoa area)	6.1–20.0	12.9	14	Ashley & Burley (1994)
Ethiopia	Pleistocene sediment above volcanic bedrock (Wonji/Shoa)	2.1–4.6	3.4	6	Ashley & Burley (1994)
Ethiopia	Pleistocene sediment above volcanic bedrock (Metahara)	2.7–15.3	5.9	15	Ashley & Burley (1994)

Continued

Country	Region/aquifer	Range of F (mg L^{-1})	Average (mg L^{-1})	Number of analyses	Reference
Tanzania	Ngorongoro Crater and Lemagrut volcanic cone	40–140			Bugaisa (1971)
Tanzania	Kimberlites, Shinyanga	110–250			Bugaisa (1971)
Groundwater: sediments and sedimentary basins					
China	Quaternary sands, Hunchun Basin, northeast China	1.0–7.8		19	Woo et al. (2000)
Argentina	Quaternary loess, La Pampa	0.03–29	5.2	108	Smedley et al. (2002)
India	Quaternary alluvium Agra, Uttar Pradesh	0.11–12.8	2.1	658	Gupta et al. (1999)
UK	Cretaceous Chalk, Berkshire	<0.1–2.4	0.74	22	Edmunds et al. (1989)
	Cretaceous Chalk, London Basin	0.11–5.8	1.44	21	Edmunds et al. (1989)
	Cretaceous Lower Greensand, London	<0.1–0.35	0.17	26	Edmunds et al. (1989)
	Triassic Sandstone, Shropshire	<0.1–0.17	<0.1	40	Edmunds et al. (1989)
	Triassic Sandstone, Lancashire	<0.1–0.14	0.1	16	Edmunds et al. (1989)
	Triassic Sandstone, Cumbria	<0.1–0.26	<0.1	23	Edmunds et al. (1989)
Canada	Carboniferous clastic rocks, Gaspé,	0.02–28	10.9	20	Boyle & Chagnon (1995)
	Alluvial/glacial (<30 m depth)		0.09	39	Boyle & Chagnon (1995)
Canada	Non-marine Upper Cretaceous sediments, Alberta Basin	0.01–22.0	1.83	469	Hitchon (1995)
Germany	Cretaceous Chalk Marls	<0.01–8.9	1.28	179	Queste et al. (2001)
Libya	Miocene, Upper Sirte Basin	0.63–3.6	1.4	11	Edmunds (1994)
Sudan	Cretaceous, Nubian Sandstone (Butana area)	0.29–6.2	1.8	9	Edmunds (1994)
Senegal	Palaeocene sediments	1.5–12.5		26	Travi (1993)
Senegal	Maastrichtian sediments	1.1–5.0		32	Travi (1993)
Tunisia	Cretaceous to Quaternary sediments	0.1–2.3		59	Travi (1993)
USA	Carboniferous sediments, Ohio	0.05–5.9		255	Corbett & Manner (1984)

in the range of <20–80 µg L^{-1} (average 30 µg L^{-1}), have been measured in rainfall from Anuradhapura, central Sri Lanka (BGS, unpublished data). Such values are likely to reflect a large contribution from marine aerosols. Neal (1989) reported typical fluoride concentrations in rainfall from Wales of 20–70 µg L^{-1}, which reflects in part marine influences. Occasional higher values up to 220 µg L^{-1} were interpreted as increased atmospheric inputs, presumably of anthropogenic origins. Saether and Andreassen (1989) found concentrations in Norwegian precipitation up to 253 µg L^{-1}, with values typically of 13–25 µg L^{-1}. They used the correlation between sulfate and fluoride to suggest that anthropogenic contributions amount to as much as 98% of the total fluoride, which is derived principally from industrial aerosols.

Concentrations of fluoride in surface waters are generally much higher than in rainfall, though still typically in the µg L^{-1} range. Stream flow in upland areas of Wales has concentrations in the range of 30–70 µg L^{-1} (Neal et al., 1997). In the Bengal Basin, the Ganges-Brahmaputra river system has mean fluoride concentrations in the range of 66–154 µg L^{-1} (Datta et al., 2000) and rivers from southern Ghana have concentrations of around 30–140 µg L^{-1} (Table III) (Botchway et al., 1996). In most cases, concentrations in surface waters are less than 300 µg L^{-1}.

Despite these generally low concentrations, fluoride in surface waters can be much higher in geothermal areas. Many alkaline lakes in the East African Rift Valley, for example, have concentrations of the order of tens to hundreds of mg L^{-1} (up to 1980 mg L^{-1}; Table III).

B. Mineral Sources of Fluoride

Fluorine is more abundant in the Earth's crust average ($625 \, mg \, kg^{-1}$) compared to its sister halogen element chlorine ($130 \, mg \, kg^{-1}$). Chloride is highly mobile in the aqueous environment and most is found in the oceans. By contrast, fluorine is mostly retained in minerals. Fluorine is the lightest of the halogen elements and is also the most electronegative. It has an ionic radius very similar to that of OH^- and substitutes readily in hydroxyl positions in late-formed minerals in igneous rocks. It is mobile under high-temperature conditions and because it is a light, volatile element it is found along with boron and to a lesser extent chloride in hydrothermal solutions. Thus, concentrations of fluoride are generally localized in their geological occurrence. Highest concentrations are found in acidic igneous rocks, mineralized veins, and sedimentary formations where biogeochemical reactions have taken place.

Fluorine occurs in primary minerals, especially biotites and amphiboles (Figure 2), where it substitutes for hydroxyl positions in the mineral structure. An example is biotite:

$$K_2(Mg,Fe)_4(Fe,Al)_2[Si_6Al_2O_{20}](OH)_2(F,Cl)_2$$

On weathering, the fluorine tends to be released preferentially from these minerals. Where biotites and amphiboles are abundant, such as in granite, a major source of fluoride in water bodies is formed.

Other high-temperature fluorine minerals such as topaz are less soluble. Apatite ($Ca_5(Cl,F,OH)(PO_4)_3$), which may form at both high and low temperatures, is another important source of fluorine (Figure 2). Substituted apatites with high fluorine are more soluble than purer (high-temperature) apatites. Fluorite (CaF_2) is the main fluorine mineral, which occurs in localized secondary hydrothermal vein deposits and as a relatively rare authigenic mineral in sediments.

In marine sediments, fluorine is concentrated both by adsorption onto clays and also by biogeochemical processes involving the removal of phosphorus. Limestones may contain localized concentrations of fluorapatite, especially francolite. Most sandstones contain very low concentrations of fluorine and, hence, the fluoride in resident groundwaters may also be low (see also Chapter 2, this volume).

C. Sources and Reactions in Soils

Typical concentrations of fluoride in soils are around $20–500 \, mg \, kg^{-1}$ (Kabata-Pendias & Pendias, 1984). Similar values of $200–400 \, mg \, kg^{-1}$ were given by Fuge and Andrews (1988), although higher concentrations were found in soils from mineralized areas of the UK. Concentrations up to $3700 \, mg \, kg^{-1}$ were reported in Welsh soils, up to $20,000 \, mg \, kg^{-1}$ in Pennine soils of the English Midlands, and up to $3300 \, mg \, kg^{-1}$ in soils from Cornwall. These are all associated with hydrothermal mineralization that includes the presence of fluorite. Some plant species in the areas of mineralization have also increased concentrations of fluoride which are potentially detrimental to humans and grazing animals (Fuge & Andrews, 1988).

Typically, only a small amount of the element present naturally in soils (e.g., $<10 \, mg \, kg^{-1}$) is easily soluble, the remainder residing within a variety of minerals (Pickering, 1985). Fluoride concentrations have been found in the range of $24–1220 \, mg \, kg^{-1}$ (26 samples) in Argentine soils, and the water-soluble fraction is in the range of $0.53–8.33 \, mg \, kg^{-1}$. The soluble fraction in these may be higher than typical because the loess-derived soils are fine-grained (Lavado & Reinaudi, 1979). Easily soluble fluoride has also been reported in sodic loess-derived soils from Israel and China (Fuhong & Shuquin, 1988; Kafri et al., 1989).

Of the anthropogenic inputs to soils, high fluoride concentrations are found in phosphate fertilizers ($8500–38,000 \, mg \, kg^{-1}$) (Kabata-Pendias & Pendias, 1984) and sewage sludge ($80–1950 \, mg \, kg^{-1}$) (Rea, 1979). Contributions from these inputs can increase the concentrations in agricultural soils considerably.

The amount of fluoride adsorbed by soils varies with soil type, soil pH, salinity, and fluoride concentration (Fuhong & Shuquin, 1988; Lavado & Reinaudi, 1979). Adsorption is favored in slightly acidic conditions and uptake by acid soils can be up to ten times that of alkaline soils. Fluoride adsorption releases OH^-, though the release is non-stoichiometric. Wenzel and Blum (1992) noted that, while minimum mobility occurs at pH 6.0–6.5, it is increased at pH < 6 as a result of the formation of $[AlF]^{2+}$ and $[AlF_2]^+$ complexes in solution. Fine-grained soils also generally retain fluoride better than sandy types. Adsorption is favored strongly by the presence of freshly precipitated $Fe(OH)_3$ or $Al(OH)_3$. Clay minerals are also effective adsorbents (Wuyi et al., 2002) as is soil organic matter (Fuge & Andrews, 1988). In soil profiles containing mainly sand and with little clay, iron, or aluminum, up to half of the infiltrating

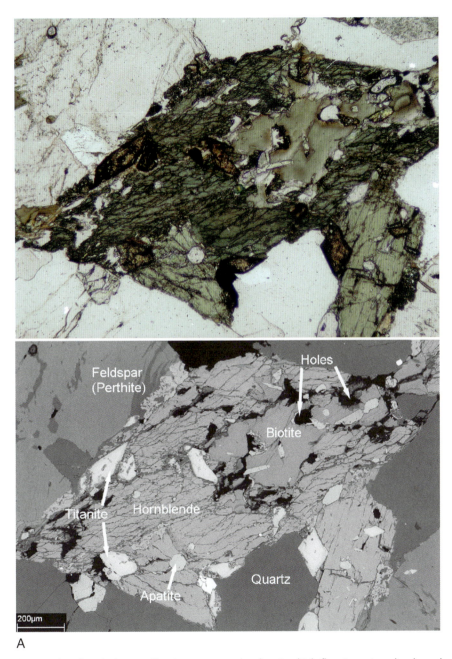

FIGURE 2 Photomicrographs of typical crystalline basement rocks showing high-fluorine minerals, above by plane-polarized light, below by back-scattered SEM. (A) view of hornblende engulfing biotite, sphene (titanite) and euhedral apatite; granite, Bongo, Upper East Region, Ghana.

fluoride in water may pass through the soil profile (Pickering, 1985). Where continued loading of fluoride occurs from anthropogenic sources, it is likely that the soil-retention capacities may be exceeded and fluoride will percolate to the water table. In most cases, however, soils act as a sink rather than a source of fluoride and water reaching the water table is likely to have low fluoride concentrations and be dominated by atmospheric concentrations. Nevertheless, low fluoride concentrations afforded by retention in soils may be offset by evapotranspiration at the soil surface. This may increase fluoride concentrations reaching the water table by up to five times in temperate climates and 10–100 times under semi-arid conditions.

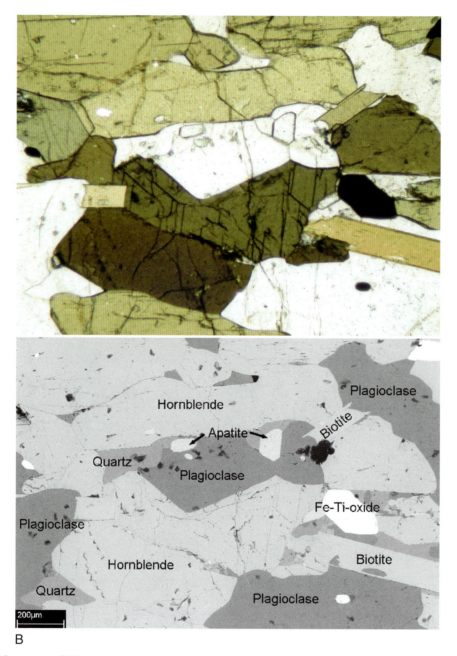

FIGURE 2 *Continued*   (B) View of hornblende, biotite, plagioclase, quartz, and apatite; granite, Anuradhapura, Sri Lanka.

D. Fluoride in Solution

Fluorine occurrence in natural waters is closely related to its abundance in the local minerals and rocks. It is also strongly associated with mineral solubility and in this regard the mineral fluorite, which is least soluble and has favorable dissolution kinetics at low temperature, exerts the main control on aqueous concentrations in the natural environment. An upper limit on fluoride activities in aqueous solution is controlled by the solubility product, $K_{fluorite}$:

$$CaF_2 = Ca^{2+} + 2F^- \tag{1}$$

$$K_{fluorite} = (Ca^{2+}) \cdot (F^-)^2 = 10^{-10.57} \text{ at } 25°C \tag{2}$$

$$or, \log K_{fluorite} = (Ca^{2+}) + 2\log(F^-) = -10.57 \tag{3}$$

This is an important relationship because it shows that in the presence of fluorite, concentrations of fluoride are inversely proportional to Ca^{2+} concentrations. For example, in the presence of $10^{-3}M$ calcium at 25°C, dissolved fluorine$^-$ should be limited to $3.1\,mg\,L^{-1}$. It is therefore likely to be the absence of calcium in solution that allows higher concentrations of fluoride to be stable in solution. Such conditions arise in volcanic regions dominated by alkaline volcanic rocks (e.g., Kilham & Hecky, 1973; Ashley & Burley, 1994;) and also in conditions where cation exchange occurs naturally (e.g., Handa, 1975). Here, removal of Ca^{2+} is achieved by exchange with Na^+ from clay minerals. In both cases, the waters are typically of $Na-HCO_3$ type. Cation exchange occurs in soils and in groundwaters along flow gradients in response to changing chemistry (Edmunds & Walton, 1983) and in zones of saline intrusion.

The solubility of fluorite (Eq. 2) is temperature-dependent. Fluorite solubility decreases with decrease in temperature and this is reflected with a change in the equilibrium constant. For example, the value of $K_{fluorite}$ changes from $10^{-10.57}$ at 25°C to $10^{-10.8}$ at 10°C. Further, the equilibrium constant is affected by the salt content of the waters (the ionic strength) with the solubility increasing with increasing salinity, providing the waters are below saturation with respect to calcite.

Reaction times with aquifer minerals are also important in controlling fluoride buildup. High fluoride concentrations can be built up in groundwaters that have long residence times in the host aquifers, which is probably a result of diagenetic reactions. Surface waters usually have low concentrations, as do most shallow groundwaters from open wells because they represent young, recently infiltrated rainwater. Deeper (older) groundwaters from boreholes are therefore most likely to contain high concentrations of fluoride. Exceptions occur in active volcanic areas where surface water and shallow groundwaters are affected by hydrothermal inputs.

High fluoride concentrations are also a feature of arid climatic conditions (Handa, 1975; Fuhong & Shuquin, 1988; Smedley et al., 2002). Here, groundwater infiltration and flow rates are slow, which allows prolonged reaction times between water and rocks. Fluoride buildup is less pronounced in the humid tropics because of high rainfall inputs and their diluting effect on groundwater chemical composition.

Table IV shows the chemical compositions of a range of typical high-fluoride groundwaters from various parts of the world. These data illustrate the dominance of Na over Ca in most, though not all, waters and the high concentrations of HCO_3 that typify high-fluoride

waters. Table IV also gives the likely dominant species, which is calculated using the program PHREEQCi (Parkhurst, 2002). Free F^- is overwhelmingly the dominant form in most natural waters, with minor additional amounts of complexes with major cations (calcium, magnesium, and sodium). HF^0 may be present in acidic waters and at pH 3.5, and this is likely to be the dominant species (Hem, 1985). Fluorine also readily forms complexes with aluminum, boron, beryllium, vanadium, uranium, silica, and Fe^{3+}, and these may enhance fluorine mobilization if present in solution in significant quantity. The formation of complexes may have ramifications for human health, for example, in aluminum-bearing waters, where the total fluorine may be much higher than the measured ionic fluoride. Thus aluminum fluoride may stabilize the fluoride as a complex ion, but if these complexes are broken down during metabolism they could release both F^- and potentially toxic aluminum.

Table IV gives the saturation indices for fluorite for each of the waters. Fluorite solubility is likely to control the upper limits of fluoride concentrations in most natural waters. This table shows that most are at or near saturation with respect to fluorite, although a couple of samples are significantly oversaturated ($SI_{fluorite} > 0.5$). This presumably reflects the slower kinetics of precipitation of fluorite compared with its dissolution, or discrepancies related to mineral purity or grain size.

E. Fluoride Distribution in Groundwater—Statistical Studies

One of the best ways to understand the controls on the concentrations of fluoride in groundwaters is to examine its distribution according to geology, depth, salinity, and other controlling parameters. Several studies have been made which take large populations of data and examine them statistically, as in Alberta, Canada (Hitchon, 1995), India (Gupta et al., 1999), Germany (Queste et al., 2001), and the UK (Edmunds et al., 1989). A statistical summary of fluorine occurrence in sedimentary basins and other aquifers in various parts of the world is included in Table III.

In the UK, an intercomparison of fluoride in 13 representative major aquifers was made alongside other trace elements by Edmunds et al. (1989). These results are summarized in Figure 3 as a cumulative-frequency plot. A clear distinction is seen between groundwaters in sandstones and those in carbonate aquifers. A very narrow range of concentrations is found in all Triassic sandstones and median values do not exceed $150\,\mu g\,L^{-1}$.

TABLE IV.   Chemical Compositions of a Range of Groundwaters with Naturally High Fluoride Concentrations

Country	Units	Ghana	United Kingdom	Sri Lanka	Argentina	Canada	Tanzania
Location		Yorogu Abagabisi, Bolgatanga	Spalding, Lincolnshire	Paniyankadawala, Anuradhapura	El Cruce, La Pampa	Maria, Gaspé Peninsula	Imalanguzu, Singida
Water type		Unconfined	Confined	Unconfined	Unconfined		Unconfined
Well depth	m	26	99	nd	16	>30	110
pH		6.63	8.32	6.96	8.1	9.2	8.86
Temperature	°C	31.2	11	29.5			27.7
SEC		348		3840	2610	598	1360
Ca	mg L^{-1}	27.6	2.0	173	24.1	4.3	17.6
Mg	mg L^{-1}	13.7	4.1	179	22.9	0.7	1.37
Na	mg L^{-1}	18.8	540	355	616	137	332
K	mg L^{-1}	1.56	4.0	3.02	12.6	0.4	2.07
HCO$_3$	mg L^{-1}	146	506	516	1180	232	845
SO$_4$	mg L^{-1}	1.65	33	15.5	190	13.6	20.8
Cl	mg L^{-1}	5.89	490	1050	113	56.7	13.6
NO$_3$-N	mg L^{-1}	4.92	<0.1	9.00	45.9	nd	7.81
F	mg L^{-1}	3.6	5.6	4.4	15.8	10.9	17.5
Si	mg L^{-1}	34.4	6.0	45.0	28.1	3.5	54.2
Fe	mg L^{-1}	0.37	0.48	0.007	0.055	0.087	0.053
Al	µg L^{-1}	30	14	nd	40	67	81
Be	µg L^{-1}	1	nd		83	nd	2
U	µg L^{-1}	1.3	nd		71	nd	nd
B	mg L^{-1}			0.15	4.58		0.46
Charge imbalance	%	+3.8	+2.01	−0.51	−0.65	−0.06	−2.13
F$^-$	%	95.98	98.56	79.21	96.54	99.58	99.04
NaF	%	0.04	0.98	0.46	1.04	0.29	0.63
MgF$^+$	%	2.91	0.44	18.86	2.22	0.09	0.17
CaF$^+$	%	0.49	0.02	1.46	0.18	0.04	0.16
AlF$_3$	%	0.35					
AlF$_2$$^+$	%	0.18					
HF^0	%	0.04					
AlF$_4$$^-$	%	0.02					
BF(OH)$_3$$^-$	%				0.01		
Total	%	99.99	100.00	99.99	99.99	100.00	100.00
SI$_{fluorite}$	log	−0.32	−1.01	+0.18	+0.75	+0.05	+0.62

Note: This table also gives the modeled dominant species in the waters and fluorite saturation indices, calculated using PHREEQCi.
SEC: specific electrical conductance; nd: not determined.
From British Geological Survey, unpublished data, and Boyle and Chagnon (1995).

For most of the sandstone groundwaters, the fluoride concentrations do not significantly exceed the rainfall input concentrations after allowing for evapotranspiration. The only aquifers to deviate from the general trend are the Moray Sandstone in Scotland in which fluorite is recorded as a cementing phase associated with vein mineralization, and the Carboniferous Sandstone, in Shropshire where 10% of the sample population has higher fluoride, also as a result of local hydrothermal mineralization (Edmunds et al., 1989). Median concentrations in limestones lie between 0.6 and 1.5 mg L^{-1}; the higher values are found in fine-

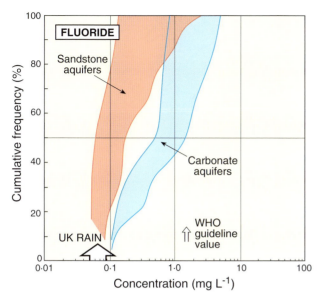

FIGURE 3 Cumulative-frequency diagram showing the distribution of fluoride in groundwater from carbonate and noncarbonate aquifers in the UK. (Modified from Edmunds et al., 1989.)

grained Chalk. Although rainfall concentrations provide a contribution, the main source of the fluoride here is biogenic and closely related to phosphate-enriched horizons. The highest fluoride concentrations in UK groundwaters are found in confined carbonate aquifers where the combination of a geological source of fluorine (CaF_2), sufficient residence time, and some calcium depletion due to Na-Ca exchange leads to fluoride concentrations controlled by fluorite mineral solubility. In the majority of UK groundwaters, fluorite solubility is not achieved.

F. High Groundwater-Fluoride Provinces

Although the concentrations of fluoride are likely to be higher in groundwater than in other water bodies as a result of water–rock interaction, the concentrations in most groundwaters are below the upper concentrations considered detrimental to health, as shown by the UK studies in Section E. However, aquifers with high groundwater fluoride concentrations have been recognized in a number of regions across the world. The distribution of documented cases where fluoride occurs substantially in wells at concentrations >1.5 mg L^{-1} is shown in Figure 4. Endemic fluorosis is a problem in many, though not all, of these regions. High-fluoride

groundwaters are found in many parts of the developing world, and many millions of people rely on groundwater with concentrations above the WHO guideline value for their normal drinking water supply. Worst affected areas are arid parts of China, India, Sri Lanka, West Africa (Ghana, Ivory Coast, Senegal), North Africa (Libya, Sudan, Tunisia), South Africa, the East African Rift Valley (Kenya, Uganda, Tanzania, Ethiopia, Rwanda), northern Mexico, and central Argentina (Figure 4). Problems have also been reported in parts of Pakistan. In the early 1980s, it was estimated that around 260 million people in 30 countries worldwide were drinking water with more than 1 mg L^{-1} of fluoride (Smet, 1990). In India alone, endemic fluorosis is a major problem in 17 out of the 22 states.

High-fluoride groundwaters are typically of Na-HCO$_3$ type with relatively low Ca concentrations (<20 mg L^{-1} or so), and with neutral to alkaline pH values (around 7–9). Fluoride problems are largely found in groundwater from basement aquifers, particularly granites, where fluorine-rich minerals are abundant; in active volcanic zones where fluorine is derived from the volcanic rocks and geothermal sources; and groundwaters in some sediments, particularly in arid areas.

1. Basement Aquifers

Fluoride problems have been found in crystalline basement rocks, particularly those of granitic composition from several areas across the world (Table III). Granitic rocks contain a relatively large proportion of high-fluorine minerals such as micas, apatites, and amphiboles. Fluorite is also an occasional accessory mineral in these rocks.

Basement aquifers in large parts of India and Sri Lanka are known to suffer from severe fluoride and fluorosis problems. In India, the worst affected states are Rajasthan, Andhra Pradesh, Uttar Pradesh, Tamil Nadu, and Karnataka (Handa, 1975; Suma Latha et al., 1999). In Sri Lanka, the Dry Zone in the eastern and north central parts of the country has groundwater with high concentrations of fluoride up to 10 mg L^{-1} with associated dental and possibly skeletal fluorosis. In the Wet Zone in the west, groundwater fluoride concentrations are low, probably as a result of intensive rainfall and long-term leaching of fluorine from rocks (Dissanayake, 1991). Here, the incidence of dental caries is reported to be high. Travi (1993) also found fluoride concentrations in groundwater from basement rocks in eastern Senegal in the range of 0.1–3.5 mg L^{-1} with seven samples (about 7%) exceeding 1 mg L^{-1}.

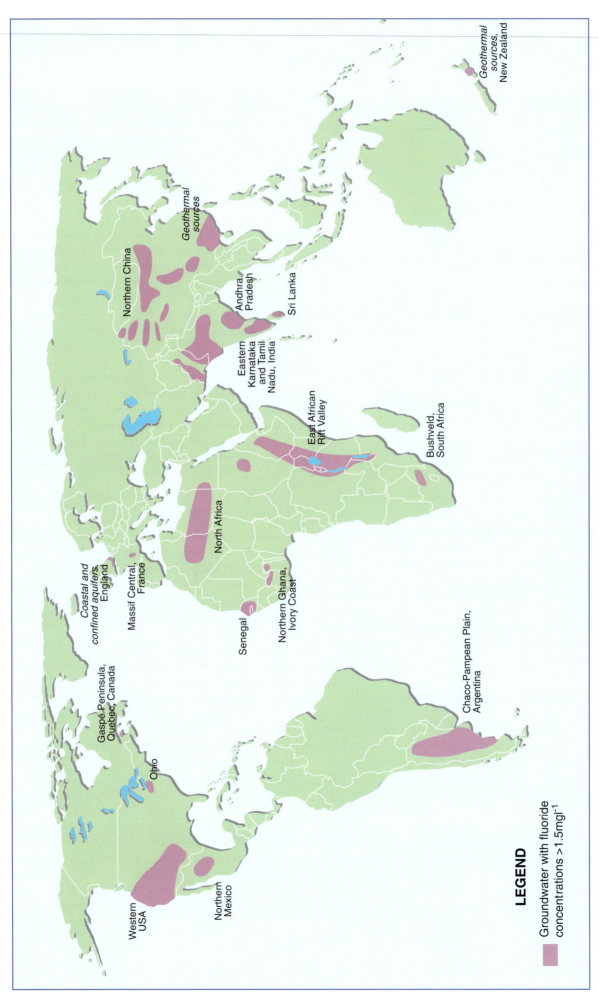

FIGURE 4 World map of documented high-fluoride (>1.5 mg L^{-1}) groundwaters.

Geothermal sources, New Zealand

Northern China

Geothermal sources

Andhra Pradesh

Sri Lanka

Eastern Karnataka and Tamil Nadu, India

East African Rift Valley

Bushveld, South Africa

North Africa

Coastal and confined aquifers, England

Massif Central, France

Senegal

Northern Ghana, Ivory Coast

Chaco-Pampean Plain, Argentina

Gaspé Peninsula, Quebec, Canada

Ohio

Western USA

Northern Mexico

LEGEND

Groundwater with fluoride concentrations >1.5mgl^{-1}

2. Volcanic Areas and Geothermal Sources

High fluoride concentrations have been reported in geothermal sources and active volcanic belts in the western United States (Nordstrom & Jenne, 1977), Iceland, Taiwan, New Zealand, the former Soviet Union (Ellis & Mahon, 1977), France, Algeria, Tunisia (Travi, 1993), and the East African Rift Valley. The most common types of geothermal water are alkali-chloride solutions with near-neutral pH values. In these, fluoride concentrations are typically in the range of $1–10\,mg\,L^{-1}$ (Ellis & Mahon, 1977) and the waters also have increased concentrations of Si and B, as well as high arsenic, NH_3, and H_2S. Under acidic conditions, concentrations in geothermal sources can reach more than $1000\,mg\,L^{-1}$ (Ellis, 1973), with dissolved fluoride in the form of HF, HF_2^-, and SiF_6^{2-} (see also Chapter 9, this volume).

In New Zealand, fluoride concentrations in deep waters have been reported in the range of $1–12\,mg\,L^{-1}$ with maximum concentrations influenced by subsurface temperature and the solubility of fluorite (Mahon, 1964). The main volcanic province with high-fluoride waters, the East African Rift Valley, is outlined as a case study below.

3. Aquifers in Sediments and in Sedimentary Basins

Some sedimentary aquifers also have groundwaters with high fluoride concentrations, particularly in arid and semi-arid regions. In La Pampa Province of central Argentina, high-pH (>8), HCO_3-rich groundwaters have concentrations of dissolved fluoride up to $29\,mg\,L^{-1}$ (Smedley et al., 2002). The concentrations of fluoride are positively correlated with pH and HCO_3 and Smedley et al. (2002) suggested that the fluoride was mobilized at high pH from fluoride-bearing minerals present in the sediments. These contain rhyolitic volcanic ash. Fluoride concentrations were highly variable, but highest concentrations were found in groundwater from shallow depths (top 20 m). High-fluoride groundwaters are also observed in Quaternary sediment aquifers from arid regions of China. Here, soil salinity and pH have been highlighted as key factors in fluoride accumulation in groundwater (Fuhong & Shuquin, 1988).

Groundwater in sedimentary basins can also accumulate potentially detrimental concentrations of fluoride with increasing residence time. As groundwaters flow downgradient in a typical sedimentary aquifer, sodium for calcium exchange takes place and groundwaters evolve from $Ca-HCO_3$ to $Na-HCO_3$ types.

Under low-calcium conditions in old groundwaters, fluoride can be maintained at relatively high concentrations unlimited by the precipitation of CaF_2. Fluoride concentrations therefore often show an increase in groundwaters down the flow line and with increasing depth as a result of increasing residence time (Edmunds & Walton, 1983; Travi, 1993). A case study of the UK Lincolnshire Limestone is given in Section IV (Part C).

Edmunds (1994) reported fluoride concentrations up to $3.6\,mg\,L^{-1}$ in old groundwaters from the Upper Sirte Basin of Libya and up to $6.2\,mg\,L^{-1}$ in groundwaters from the Cretaceous Nubian Sandstone of Sudan (Table III). The concentrations were variable in each aquifer but were found to be higher in marine sediments than in freshwater sediments, which reflected the overall higher concentrations of fluorine in the sediments themselves. Despite the high dissolved concentrations, most were undersaturated with respect to fluorite as a result of low dissolved calcium concentrations.

Groundwater from Cretaceous to Tertiary aquifers of western Senegal has fluoride concentrations ranging between <0.01 and $13\,mg\,L^{-1}$ (Travi, 1993). The sediments form part of an 8000-m thick sequence in a sedimentary basin, which overlies Precambrian basement. In the aquifers, fluoride concentrations generally increase in the direction of groundwater flow. The source of the dissolved fluoride is taken to be phosphatic horizons (containing fluorapatite), particularly in argillaceous deposits. In the Maastrichtian aquifer, fluoride concentrations range between 1.1 and $5.0\,mg\,L^{-1}$ with increases corresponding to a facies change from sandy to more argillaceous deposits (Travi, 1993). The high concentrations are present in groundwater up to 30,000 years old.

In the sequence of Cretaceous to Quaternary sediments of western Tunisia, Travi (1993) recorded fluoride concentrations between 0.1 and $2.3\,mg\,L^{-1}$. Highest concentrations were found in groundwaters from the Cretaceous "Complexe Terminal" aquifer (average $1.7\,mg\,L^{-1}$). The Tunisian sediments are also phosphatic, although the relationships between phosphate occurrence and dissolved fluoride concentrations were found to be less clear than observed in Senegal.

IV. CASE STUDIES TO ILLUSTRATE HIGH-FLUORIDE GROUNDWATER SYSTEMS

A number of aquifers have been selected as case studies to illustrate the different geochemical controls on fluoride mobilization more clearly and where possible associated health problems may also be explained.

A. Rajasthan, India—Precambrian Basement Aquifer

The distribution of fluoride in India has been reviewed by Handa (1975). He recognized Rajasthan as the state most seriously affected by high fluoride, although other high-fluoride areas were also recognized in the country. The distribution of fluoride in Rajasthan (11 districts) was also reviewed by Gupta et al. (1993). Although there are many accounts of the incidence of dental and skeletal fluorosis in the province, it is difficult to find studies linking to regional geology. However, studies in the Sirohi district (one of the administrative districts of Rajasthan) seem typical of the rather complex geology of the region (Maithani et al., 1998) and illustrate the nature of the fluoride occurrence.

Rajasthan has an arid climate with low but highly variable annual rainfall. In the Sirohi district, fluoride concentrations up to $16\,mg\,L^{-1}$ have been found in groundwater from dug wells and boreholes at depths between 25 and 75 m during geochemical exploration for uranium (Maithani et al., 1998). The aquifer comprises Proterozoic metasediments with intrusions of granite and rhyolite. Significant fluorite mineralization associated with the granites and volcanic rocks is also reported from the adjacent Jalore district (Figure 5). This geological setting seems typical of much of Rajasthan where a mixed assemblage of basement rocks outcrops as islands between a highly weathered series of younger Quaternary sediments. The association of fluoride-endemic areas with the bedrock geology is often obscure, and it is unclear whether anomalies are related to primary bedrock or secondary enrichment in the sediments.

In the 150-km^2 area studied by Maithani et al. (1998), 117 samples were collected and over 75% of these contained groundwater with a fluoride concentration in excess of $1.5\,mg\,L^{-1}$. A good correlation with the bedrock geology is observed, with highest concentrations found in association with the granites, acid volcanic rocks, and basic dikes (Figure 5). These dikes act as barriers, which slow down groundwater flow and permit prolonged contact times to raise concentrations of groundwater fluoride. Low-fluoride areas in the east

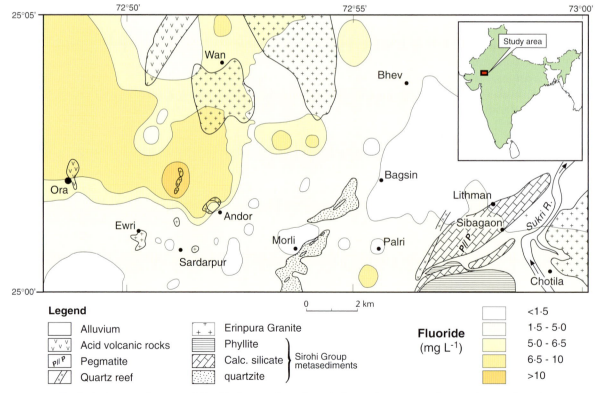

FIGURE 5 Geological map of parts of Sirohi district, Rajasthan, India, with contoured groundwater-fluoride concentrations. (After Maithani et al., 1998.)

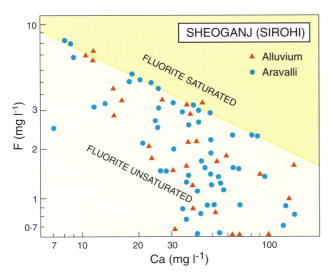

FIGURE 6 Variation of calcium with fluoride in groundwaters from the Sirohi District of Rajasthan (from data provided in Handa, 1975), along with the saturation curve for fluorite. Groundwaters from alluvium and Aravalli Hills (bedrock) are distinguished.

($<1.5 \, mg \, L^{-1}$) are associated with carbonate rocks and higher calcium may inhibit an excess of fluoride. It is not clear from the study to what extent the alluvium acts as a source (or a conduit) for the fluoride-bearing water and no groundwater flow data are shown.

Handa (1975) recognized a negative relationship between calcium and fluoride in the groundwaters of the Sirohi region (Figure 6). Groundwaters from both bedrock and alluvium showed the relationship. The findings demonstrated the solubility control of fluorite on the fluoride concentrations.

The ill effects of high fluoride on human health, including dental fluorosis and a few cases of skeletal fluorosis, were observed in villages in the high-fluoride region of Sirohi, although complete medical statistics have not been reported (Handa, 1975; Maithani et al., 1998). Although the Maithani et al. (1998) study indicates a strong association between geology and groundwater-fluoride occurrence, a full appraisal of the situation is limited by the absence of hydrogeological information, especially water levels and flow data. Additional hydrochemical data would also assist identification of the causes of high and low fluoride.

B. Northern Ghana—Alkaline Granitic Basement

The Bolgatanga area in the Upper East Region of northern Ghana demonstrates the often-strong influ-

ence of geology, specifically certain types of granite, on fluoride concentrations in groundwater. The region experiences maximum temperatures in the range of 20–40°C and average annual rainfall around 1000 mm (Murray, 1960). The climate is semi-arid, largely as a result of northeasterly harmattan (dry, desert) winds, which affect the region from December to March. Geology comprises crystalline basement rocks, including Upper Birimian (Precambrian) rocks of mixed metaigneous and metasedimentary origin. Intruded into these are a suite of coarse granites, the Bongo Granite suite, and associated minor intrusions, the ages of which are unknown. The Bongo Granite typically comprises a pink microcline-hornblende-granite with interstitial quartz, plagioclase, and biotite (Figure 2). The granite is slightly alkaline, with the amphibole appearing to be in the hornblende-arfvedsonite range (Murray, 1960). Accessory mineral phases include abundant sphene as well as magnetite, apatite, zircon, rutile, and more rarely fluorite. The Bongo Granite has been found to contain up to 0.2% fluorine (Smedley et al., 1995). The Tongo Granite just to the south of the study area (Figure 7) is similar in composition (Murray, 1960). To the east of the study area (Sekoti district), a north-south tract of granodiorite has been classed together with other hornblende-biotite granodiorites (Figure 7 and Murray, 1960). However, in outcrop the rock type forms rounded tors resembling those of the Bongo Granite and some workers have grouped the Sekoti Granodiorite with the Bongo suite. The Sekoti Granodiorite comprises mainly plagioclase and quartz with biotite, often replacing amphibole, accessory apatite, and sphene (Murray, 1960).

Groundwater is abstracted for drinking water mostly from shallow hand-pumped boreholes (typically 30–50 m deep), although some traditional shallow hand-dug wells also have depths of less than 10 m. Surveys of groundwater quality (Smedley et al., 1995; Apambire et al., 1997) clearly indicate that highest fluoride concentrations occur in groundwater from the Bongo area in the west, with values reaching up to $3.8 \, mg \, L^{-1}$, and from Sekoti district in the east, with values reaching up to $3.2 \, mg \, L^{-1}$ (Figure 7). The higher concentrations from any given rock type are generally found in borehole waters. Water from shallow hand-dug wells typically (but not always) has very low fluoride concentrations ($<0.4 \, mg \, L^{-1}$) as these have had much shorter reaction times with the host rocks. They also circulate within the superficial weathered overburden layer rather than the fractured granite at greater depth. From this association, it is possible that shallower groundwaters in such fluoride-vulnerable zones are

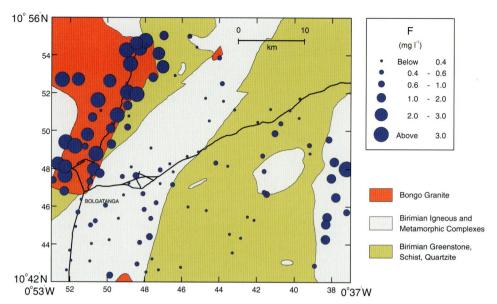

FIGURE 7 Geological map of the Bolgatanga area of Upper East Region, Ghana, which shows the distribution of fluoride concentrations in groundwater. The concentrations are highest where associated with outcrops of biotite- and hornblende-bearing granite and granodiorite (from Smedley et al., 1995).

safer sources of drinking water with respect to fluoride than deeper sources. However, the potential for bacterial contamination is increased in such open, shallow wells.

Villagers in some parts of the Bolgatanga area suffer from dental fluorosis as a result of the high concentrations of dissolved fluoride. The problem is particularly prevalent in Bongo and Sekoti districts, although the number of people affected and the regional extent is not known in detail. The fluorosis is particularly prevalent among children. There is to date no evidence of skeletal fluorosis in the region.

C. The UK Lincolnshire Limestone— Downgradient Evolution

A common situation in many aquifers is to encounter increasing fluoride concentrations in the groundwaters down the flow gradient. This situation arises initially as a result of continuous dissolution of fluoride from minerals in the carbonate aquifers, up to the limit of fluorite solubility, followed by ion-exchange reactions involving removal of calcium. The phenomenon is particularly well illustrated in the Jurassic Lincolnshire Limestone aquifer of eastern England (Edmunds, 1973; Edmunds & Walton, 1983).

The gently dipping oolitic Lincolnshire Limestone is typically 30 m thick and downgradient becomes con-

fined beneath clays and marls. Groundwater flow is predominantly through fractures, although there is also an intergranular porosity (13–18%). The unaltered limestone at depth is gray-green in color due to the presence of Fe^{2+}. However, oxidation of the limestone on a geological time scale has produced an orange-brown oxidized zone through groundwater action. This is also reflected in an oxidation-reduction "barrier" in the groundwaters, with aerobic groundwaters near the outcrop and reducing waters (with some dissolved iron) at depth.

Fluoride concentrations in the limestone at outcrop are quite low (up to $0.2\,mg\,L^{-1}$). The relationship between fluoride and other major ions and iodide is shown in Figure 8. The groundwaters are essentially of calcium-bicarbonate type and are saturated with calcite, buffering the concentrations of Ca and HCO_3 over the first 12 km of the flow line and producing "hard" water. Thereafter, the profile shows distinctive ion exchange, with Ca^{2+} being replaced by $2Na^+$. This produces a "soft," sodium bicarbonate groundwater (with a high pH, up to 8.5) and increased bicarbonate concentrations. In these waters, the Ca^{2+} falls to as low as $1\,mg\,L^{-1}$, yet the groundwater adjusts under the new physicochemical conditions to maintain saturation with calcite.

It can be seen that the fluoride concentration increases progressively across the aquifer and this increase starts in the calcite-buffered part of the aquifer.

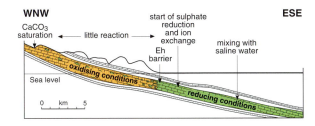

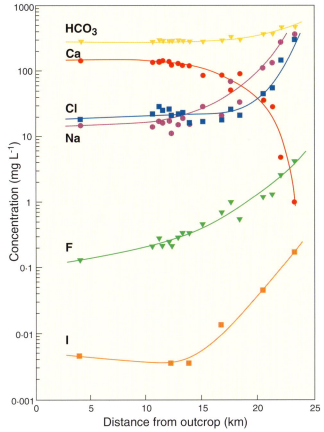

FIGURE 8 Profiles of changing groundwater chemistry with increasing distance from the outcrop zone of the Lincolnshire Limestone, eastern England (after Edmunds & Walton, 1983). A geological cross section across the aquifer is also shown.

The progressive decrease in concentration of Ca^{2+} allows the continuing rise in fluoride concentrations to a maximum of $5.6\,mg\,L^{-1}$. The fluoride increase is not controlled by the increase in salinity downgradient in the aquifer; an increase in chloride concentration only begins about 17 km from outcrop. It is interesting to see that the other halogen measured in this study, iodide, follows the chloride behavior and not the water–rock interaction pathway shown clearly by fluoride. The source of fluoride in this case is thought to be traces of phosphate minerals in the impure limestone.

This situation was utilized by the water engineers in the 1950s who noted "by a fortunate geological accident, it is possible to obtain in the geographical center of the district a pure un-aggressive water which has in effect been softened, fluoridated and transported free of charge by nature" (Lamont, 1958). This strategy has been adopted and large-yielding boreholes are concentrated in the central zone of the limestone aquifer to provide water with naturally-produced fluoride between 0.5 and $1.0\,mg\,L^{-1}$. Today there are additional water supply problems in the Lincolnshire Limestone with the need to manage the high nitrate concentrations from agricultural chemicals. Nature again assists the management of this problem in the central region, because nitrate is reduced naturally through redox reactions.

D. Sri Lanka—Climatic and Geological Controls

Sri Lanka consists of over 90% metamorphic rocks of Precambrian age. These include metasedimentary and metavolcanic rocks and gneisses and granitoids. Within these are abundant fluorine-bearing minerals such as micas, hornblende, and apatite, as well as less common fluorite, tourmaline, sphene, and topaz. As a result, large parts of Sri Lanka can be considered a fluorine province. Concentrations of fluoride up to $10\,mg\,L^{-1}$ have been recorded in the groundwater and the incidence of dental fluorosis among children is high (Dissanayake, 1991). Groundwater compositions across the island indicate that fluoride problems have a strong geographical control linked to both geology and climate.

Dissanayake (1991) compiled a hydrogeochemical atlas for Sri Lanka based on the results of analyses of 1970 water samples from shallow wells. It can be seen from the fluoride map produced (Figure 9) that high concentrations lie in the eastern and north central regions of the country. The central hill country and the southwest coastal regions have relatively low fluoride concentrations. Although rocks containing fluoride-rich minerals underlie almost the whole country (except the Jaffna Peninsula), areas with mean annual rainfall above around 2000 mm experience few fluoride problems. In these wetter areas, there is a tendency for the soluble ions, including F^-, to be leached out under the effects of high rainfall. Indeed, the Wet Zone of Sri Lanka is found to have quite high incidence of dental caries, presumably related to fluoride deficiency. In the Dry Zone, evaporation brings the soluble ions to the

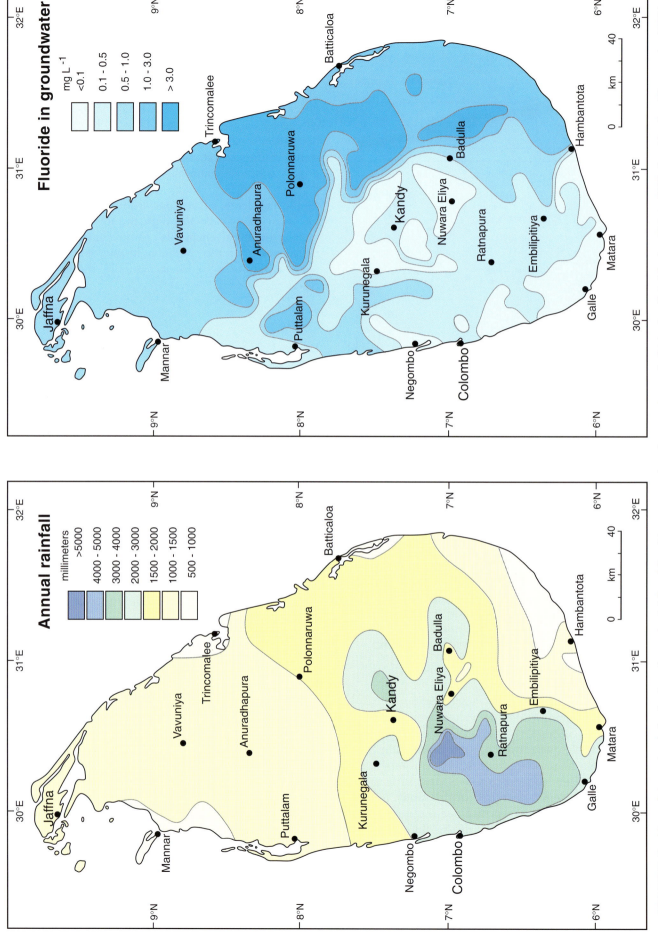

FIGURE 9 Maps of the regional distribution of annual rainfall and fluoride concentrations in groundwater in Sri Lanka. (Rainfall after Zubair, 2003; fluoride after Dissanayake, 1991.)

surface by capillary action. Despite these generalities, the precise composition of the rock types needs to be considered, because even in the Dry Zone, fluoride concentrations vary considerably. In particular, within the Dry Zone the high concentrations coincide with rocks rich in mafic minerals such as hornblende- and biotite-gneisses, charnockites, marble, and calc-gneisses (Dissanayake, 1991).

The high-fluoride groundwaters of Sri Lanka have many characteristics in common with high-fluoride aquifers elsewhere. In an investigation of groundwater chemistry in Anuradhapura, north central Sri Lanka, groundwaters were found to have typically near-neutral pH values (6.0–7.9), with high alkalinity (alkalinity as HCO_3 up to 713 mg L^{-1}; BGS, unpublished data; 123 samples). Groundwater from many wells also has relatively high salinity (SEC up to 3850 µS cm^{-1}, average 1260 µS cm^{-1}), and the high salinity values reflect the importance of water–rock reactions and evapotranspiration in the semi-arid climatic conditions.

The excess or deficiency of fluoride in groundwater is of special concern in Sri Lanka because 75–80% of the population depend on wells without treatment. Unlike iron, for example, fluoride has no taste and a high percentage of the population who suffer from dental fluorosis were unaware of the problem until the later stages of the disease. The use of fluoride toothpaste has alleviated the condition for those suffering from dental caries, but by 1991 at least, no preventative measures were taken against fluorosis. As a result of the studies (Dissanayake, 1991), it was recommended that very detailed maps showing the concentrations of fluoride in groundwater should be made, followed up by rural-water supply schemes to avoid the use of high-fluoride groundwater. The development of relatively inexpensive defluoridation techniques was also recommended coupled with rural education programs.

E. The East African Rift Valley– The Alkaline Volcanic Province

Some of the highest concentrations of fluoride ever recorded have been found in water from the East African Rift Valley. The Rift extends through Eritrea, Djibouti, Ethiopia, Kenya, Tanzania, Uganda, Rwanda, Burundi, and Malawi and excessive fluoride concentrations have been found in groundwaters, hot springs, alkaline lakes, and some river systems in these countries (Table III). The region has well-documented cases of severe dental and skeletal fluorosis (Bugaisa, 1971; Nanyaro et al., 1984; Gaciri & Davies, 1993).

The anomalous fluoride concentrations have been linked to a number of sources and processes. However, they are in large part related to the development of hyperalkaline volcanic rocks in the rift zone, which include nepheline and carbonatite magmas and associated ash deposits. These are capable of accumulating large concentrations of fluoride in melts and volatile fractions. Hence, water bodies in the rift zone can accumulate fluoride directly as a result of weathering of these rocks, as well as from high-fluoride geothermal solutions. The fine-grained and friable ashes are likely to be particularly reactive. Weathering of the silicate minerals in the lavas and ashes by silicate hydrolysis reactions produces Na-HCO_3-rich groundwaters (Jones et al., 1977), which are relatively depleted in Ca and Mg. Hence, high concentrations of fluoride can occur as the solubility of fluorite (CaF_2) is not a limiting factor. Ultimately, villiaumite (NaF) should limit the dissolved concentrations, but because this mineral is very soluble, fluoride can reach very high concentrations before this limit is achieved.

In the area of Mount Meru, a Neogene volcano in northern Tanzania, Nanyaro et al. (1984) reported fluoride concentrations of 12–76 mg L^{-1} in rivers draining the volcano's slopes (Table III) and 15–63 mg L^{-1} in associated springs. They attributed the high concentrations to weathering of fluorine-rich alkaline igneous rocks and to contributions from fumaroles and gases as well as to the re-dissolution of fluorine-rich trona ($Na_2CO_3.NaHCO_3.2H_2O$), which occurs as a seasonal encrustation in low-lying river valleys and lake margins as a result of extreme evaporation. Seasonally the river water chemistry varies significantly as a result of dilution following periods of heavy rainfall.

Both groundwaters and surface waters in the Awash Valley in the Rift zone of Ethiopia also have high concentrations of fluoride. In a study of the Wonji/Shoa and Metahara areas of the Awash Valley, Ashley and Burley (1994) found concentrations up to 26 mg L^{-1} in groundwater from volcanic rocks of the Pliocene to Recent Aden Volcanic Series and up to 4.3 mg L^{-1} in recent alluvium overlying the volcanic rocks. They concluded that the fluoride originated from the volcanic rocks and that high concentrations in solution were maintained by the dominance of sodium ions over calcium in the groundwaters. Concentrations diminished slightly in the overlying alluvium as a result of increased groundwater calcium concentrations following likely reaction with carbonate minerals in the sediments.

Alkaline and crater lakes in the East African Rift have some of the highest recorded dissolved fluoride concentrations. In Lake Magadi in Kenya, Jones et al. (1977) reported concentrations in surface brines from the lake up to 1980 mg L^{-1} (Table III). The lake waters have evolved toward saline compositions, in some cases with dissolved solids in excess of 300 g L^{-1}, principally by surface evaporative concentration in the low-lying lake basin, which has acted as an evaporating pan for the lake waters. Elevated concentrations of fluoride are also achieved by very low calcium concentrations following precipitation of carbonate minerals. The lake waters appear to be largely saturated with respect to fluorite, and the mineral is an abundant accessory authigenic phase in many of the Magadi lake sediments (Jones et al., 1977). Jones et al. (1977) reported extremely high fluoride concentrations (up to 2170 mg L^{-1}) in saline groundwaters from boreholes in Magadi lake sediments. The occurrence of fluoride in the natural waters of Kenya is also further explored by Gaciri and Davies (1993).

Nanyaro et al. (1984) found concentrations of 60–690 mg L^{-1} in alkaline lakes and ponds of the Momella Lakes Group (Mount Meru area) in Tanzania (Table III). The lake waters are also brackish and the most alkaline and saline compositions are found in the largest lakes of the group. Evaporation is also considered a major factor in the development of the extreme fluoride enrichments and the higher concentrations in the larger lakes of the group may be related to a smaller influence of dilution from runoff during wet periods and the long history of evolution of the lakes (Nanyaro et al., 1984). Also in Tanzania, concentrations up to 72 mg L^{-1} have been reported for Lake Natron (Table III). Ashley and Burley (1994) also found 36 mg L^{-1} fluoride in Lake Besaka in the Awash Valley of Ethiopia. Variable concentrations were obtained for lake waters from East Africa by Kilham and Hecky (1973). The highest concentrations (up to 437 mg L^{-1} reported for Lake Tulusia in the Momella Group and up to 437 mg L^{-1} in Lake Magadi) were found in lake waters of Na-HCO$_3$ (low Cl) composition. Kilham and Hecky (1973) postulated that the fluoride most likely originates from weathering of volcanic rocks and/or from geothermal solutions in the rift system. As with other lakes in the Rift, the maintenance of high dissolved concentrations is possible because they have low Ca concentrations resulting from loss of Ca by precipitation of travertine in local streams and springs.

It is an unfortunate fact that this region of anomalously high fluoride concentrations, like many arid areas, also suffers from significant problems with water scarcity. The arid climate, combined with often poor coverage by, or poor maintenance of, abstraction boreholes in many of the affected countries means that the emphasis is usually on water availability and that drinking-water quality is of lower priority.

F. Gaspé Peninsula, Canada–Depth Control Where Glacial Deposits Overlie Bedrock

Increasing fluoride concentrations are found with depth in groundwaters in many parts of the world. Such areas may represent increasing reaction times and decreasing circulation rates or alternatively a change in geology. One such area is the Gaspé Peninsula in Canada where Quaternary glacial till overlies older bedrock. This is likely to be a widespread geological occurrence in North America, Europe, and northern Asia, and this example serves as a type area for other glacial terrain.

A farming community relying on shallow wells for water supply occupies the town of Maria in the Gaspé Peninsula of Quebec, Canada. This community has only recently suffered from cases of chronic skeletal fluorosis, which results from fluoride concentrations up to 28 mg L^{-1} in the water supplies (Boyle & Chagnon, 1995). The area is situated on a flat coastal plain covered by colluvial, alluvial, and glacial sediments that vary in thickness from 10 to 30 m. The coastal plain is underlain by Carboniferous sediments comprising conglomerates, sandstones, mudstones and commonly containing calcareous cements. High-fluoride waters occur throughout the region and there is a clear correlation between fluoride concentrations and depth. Thus, almost all the wells completed in the Carboniferous sediments contain excessive fluoride; the mean concentration in water from the overlying sediments was around 0.1 mg L^{-1}, compared to 10.9 mg L^{-1} in the bedrock. The high fluoride concentrations are explained by the high cation-exchange capacities of the sediments, which give rise to softening of the groundwaters and exchange of Ca^{2+} for $2Na^+$. This is similar to the Lincolnshire Limestone described above. Tritium analyses of groundwater also suggest that the groundwaters in this low-lying area have long residence times.

Farmers relied originally on shallow dug wells, but were advised in the 1970s to deepen these to increase yields (to satisfy the same demands as the townspeople who were supplied by mains water). Two cases of skeletal fluorosis drew attention to the potentially widespread nature of the problem, with the affected individuals only having been exposed to the fluoride-rich water for a period of 6 years. This study draws attention to the need for water quality as well as water quantity assessments during groundwater investigations. There should also be an awareness of the geochemical properties of the sediments and a consideration of groundwater residence times.

V. REMEDIATION OF HIGH-FLUORIDE GROUNDWATER

As a means of combating the fluoride problem, several methods of water treatment using various media have been tested and a number are in use in various parts of the world. Some of the common methods are listed in Table V and a review has been published by Heidweiller (1990). Most low-technology methods rely on precipitation or flocculation or adsorption/ion-exchange processes. Probably the best known and most established method is the Nalgonda technique (Nawlakhe & Bulusu, 1989), named after the Nalgonda District in Andhra Pradesh, India where it was first developed and is in use today. This technique uses a combination of alum (or aluminum chloride) and lime (or sodium aluminate) together with bleaching powder. These are added to high-fluoride water, stirred, and left to settle. Fluoride is subsequently removed by flocculation, sedimentation, and filtration. The method can be used at a domestic scale (in buckets) or community scale (fill-and-draw type defluoridation plants; Nawlakhe & Bulusu, 1989). It has moderate costs and uses materials that are usually easily available. It is therefore preferred for small defluoridation units used exclusively for drinking water. Aluminum polychloride sulfate has been found to have technical advantages over alum as use of alum results in increased concentrations of SO_4 and suspended particles in the treated water (N'dao et al., 1992).

Other precipitation methods include the use of gypsum, dolomite, or calcium chloride. Most methods tested are capable, in principle, of reducing fluoride in treated water to a concentration below the WHO guideline value. Vanderdonck and Van Kesteren (1993) reported that the calcium chloride method is capable of reducing fluoride concentrations up to $20\,mg\,L^{-1}$ to acceptable concentrations.

The most common ion-exchange removal methods tested are activated carbon, activated alumina (Barbier & Mazounie, 1984), ion-exchange resins (e.g., Defluoron 2), plant carbon, clay minerals (Chaturvedi et al., 1988), clay pots, crushed bone, or bone char. Activated alumina and bone materials are among the most effective appropriate-technology removal methods (with highest removal capacity, Table V). These also have drawbacks,

however. Activated alumina may not always be available or affordable and bone products are not readily acceptable in some cultures. Use of fired clay-pot shards has proved a promising approach in some developing countries (Moges et al., 1996). The use of fly ash has also been tested favorably for fluoride removal (Chaturvedi et al., 1990), although the concentrations of other solutes in the water after treatment with fly ash need to be ascertained. Some success has also been demonstrated experimentally using aluminum-rich volcanic soils (Ando soils) from the East African Rift of Kenya (Zevenbergen et al., 1996), although field testing of the technique has not been reported.

Other highly efficient methods of removal include electrodialysis and reverse osmosis (Schneiter & Middlebrooks, 1983). These methods tend to involve higher technology and higher costs (Table V) and are therefore less suitable for many applications in developing countries.

Most methods designed for village-scale fluoride removal have drawbacks in terms of removal efficiency, cost, local availability of materials, chemistry of resultant treated water, and disposal of treatment chemicals. Local circumstances will dictate which methods, if any, are the most appropriate. In addition, many of the defluoridation methods have only been tested at pilot scale or in the laboratory. In practice, remediation techniques such as those outlined above often meet with disappointing success when put into operation. Success rates depend on efficacy, user acceptance, ease of maintenance, degree of community participation, availability, and cost of raw materials. An additional problem lies with monitoring. In most rural communities, it will be almost impossible to monitor the initial fluoride concentrations so that dosing can be accurate. It is also not practicable at village level to chemically monitor the progress of treatment schemes, such as knowing when to recycle or replace the media involved in the treatment.

As an example of problems with the water treatment approach, defluoridation schemes have been in operation in Wonji, Ethiopia, since 1962. In the past, these principally involved bone char, but most recently a resin-activated-alumina adsorption method has been used (Kloos & Haimanot, 1999). Only 2 of 12 plants were still in operation in 1997 and in the years of operation, only 4 of the plants consistently reduced fluoride below $1.5\,mg\,L^{-1}$. The major problems encountered included efficiencies in the operation of the plant, lack of spare parts and materials, and lack of community involvement. The affected areas are currently served by a piped water scheme, but in more remote areas, interest in defluoridation continues.

TABLE V.   Popular Removal Methods for Fluoride from Drinking Water

Removal method	Capacity/dose	Working pH	Interferences	Advantages	Disadvantages	Relative cost
Precipitation						
Alum (aluminum sulfate)	150 mg/mg F	Non-specific	—	Established process	Sludge produced, treated water is acidic, residual Al present	Medium-high
Lime	30 mg/mg F	Non-specific	—	Established process	Sludge produced, treated water is alkaline	Medium-high
Alum+lime (Nalgonda)	150 mg alum+ 7 mg lime/mg F	Non-specific, optimum 6.5	—	Low-tech, established process	Sludge produced, high chemical dose, residual Al present	Medium-high
Gypsum + fluorite	5 mg gypsum + <2 mg fluorite/mg F	Non-specific	—	Simple	Requires trained operators, low efficiency, high residual Ca, SO_4	Low-medium
Calcium chloride	3 mg $CaCl_2$/mg F	6.5–8.0	—	Simple	Requires additional flocculent (e.g., $FeCl_3$)	Medium-high
Adsorption/ion exchange						
Activated carbon	Variable	<3	Many	—	Large pH changes before and after treatment	High
Plant carbon	300 mg F/kg	7	—	Locally available	Requires soaking in potassium hydroxide	Low-medium
Zeolites	100 mg F/kg	Non-specific	—		Poor capacity	High
Defluoron 2	360 g F/m³	Non-specific	Alkalinity		Disposal of chemicals used in resin regeneration	Medium
Clay pots	80 mg F/kg	Non-specific	—	Locally available	Low capacity, slow	Low
Activated alumina	1200 g F/m³	5.5	Alkalinity	Effective, well-established	Needs trained operators, chemicals not always available	Medium
Bone	900 g F/m³	>7	Arsenic	Locally available	May give taste; degenerates, not universally accepted	Low
Bone char	1000 g F/m³	>7	Arsenic	Locally available, high capacity	Not universally accepted; may give adverse color, taste	Low
Other						
Electrodialysis	High	Non-specific	Turbidity	Can remove other ions; used for high salinity	Skilled operators; high cost.; not used much	Very high
Reverse osmosis	High	Non-specific	Turbidity	Can remove other ions; used for high salinity	Skilled operators; high cost	Very high

After Heidweiller (1990); Solsona (1985); and Vanderdonck and Van Kesteren (1993).

Given the potential drawbacks of water treatment, alternative approaches to water quality improvement could prove more effective, such as careful tubewell siting and groundwater management. Factors worth considering in borehole siting are local geology and variations in groundwater fluoride concentration with depth. Groundwater management includes consideration of optimum pumping rates, especially where there exists the possibility of mixing of groundwater with deep fluoride-rich groundwater (e.g., old groundwater or hydrothermal solutions), which would be increasingly drawn upward at high pumping rates (Carrillo-Rivera et al., 2002). Possibilities for enhanced recharge of low-fluoride surface water to aquifers could also give benefits to shallow groundwater quality. Examples include the charco dams of Tanzania. A fully fledged artificial recharge scheme would offer similar benefits.

In view of the site-specific nature of each incidence of high fluoride and also the problems of applying remediation, especially in small communities, the supply of small quantities of imported low-fluoride water should also be considered. This could be achieved by importation of water in bottles or tankers to central distribution points and may prove cost-effective, because the high-fluoride source can still be used for sanitation and possibly also for most forms of agriculture.

In low-rainfall areas with fluoride problems, it is important to assess the hydrogeological situation very carefully. The generation of high-fluoride groundwaters usually requires considerable residence times in the aquifer, for reasons outlined in the preceding sections. Thus, it is likely that younger, shallow groundwaters, for example, those recharged rapidly below wadis or stream channels, may have lower fluoride concentrations than the bulk groundwater. They may be exploitable as a resource overlying older groundwater. Exploitation may require "skimming" the shallow water table rather than abstraction from deeper penetrating boreholes. The harvesting of rainwater, either directly in cisterns or by collection in small recharge dams, offers a potentially attractive alternative solution.

VI. SUMMARY

This review has summarized the occurrence of fluorine in natural water and the environment and outlined its behavior in terms of the hydrogeochemical cycle. The principal sources of fluoride in rainfall are marine aerosols, volcanic emissions, and anthropogenic inputs. Concentrations in rain typically lie in the range of 0.02–0.2 mg L^{-1} with pristine and continental rains lying at the lower end of the range. Surface waters generally also reflect these sources and concentrations in most are low (<300 μg L^{-1}), except where affected by extensive evapotranspiration or geothermal inputs. Surface waters rarely have fluoride concentrations sufficiently high to be detrimental to health and if fluoride-related health problems exist, they are more likely to be linked to deficiency.

Groundwaters are generally more vulnerable to the buildup of fluoride and although concentrations in these are also mostly low (<1 mg L^{-1}), a number of conditions can give rise to detrimentally high concentrations. Geology plays a key role in defining fluoride concentrations. Areas with potentially high-fluoride groundwater include crystalline basement rocks, especially those of granitic composition. These contain relatively high concentrations of fluorine-bearing minerals (e.g., apatite, mica, hornblende, occasionally fluorite) and have low calcium concentrations. Such areas occur over large parts of India, Sri Lanka, and Africa, for example. The active volcanic province of the East African Rift Valley is also an anomalous area with high fluoride concentrations on a regional scale. High concentrations result from both weathering of volcanic rocks and geothermal inputs. In groundwaters from the East African Rift, fluoride concentrations in excess of 1 mg L^{-1} are common, whereas concentrations up to several hundred mg L^{-1} have been recorded in some hot springs and concentrations in excess of 1000 mg L^{-1} have been found in some alkaline lakes. Some sedimentary aquifers also contain high-fluoride groundwaters, especially where calcium concentrations are low and Na-HCO$_3$ waters dominate. These can occur under arid conditions and in aquifers affected by ion exchange.

The influence of low recharge on dissolved fluoride concentrations in groundwater is also clear from documented studies. The tendency for fluoride accumulation in groundwater from arid and semi-arid areas is well illustrated from crystalline basement aquifers of India and Sri Lanka and many sedimentary aquifers. The accumulations under arid conditions are the result of rock weathering, evapotranspiration, and low recharge. Excessive groundwater fluoride concentrations are relatively rare in tropical and even temperate regions.

Groundwaters in a number of aquifers display progressive increases in fluoride along flow lines and demonstrate the importance of residence time and their position in the flow and reaction sequence. This phenomenon has been demonstrated in the English Lincolnshire Limestone aquifer where dissolved fluoride concentrations increase progressively from

$0.11\,\mathrm{mg\,L^{-1}}$ in the unconfined aquifer to $5.6\,\mathrm{mg\,L^{-1}}$ in the deep confined aquifer in response to time-dependent mineral dissolution and ion-exchange reactions. In many aquifers, the evolution of groundwater down the flow gradient has taken place over centuries or millennia, with abstracted water having had significant opportunity for equilibration with host aquifer minerals. The accumulation of fluoride in water is ultimately limited by mineral solubility. In groundwaters where calcium is abundant, fluoride concentrations are limited by saturation with the mineral fluorite. In cases where calcium concentrations are low, or where calcium is removed by ion exchange, fluoride may build up to excessive and dangerous concentrations.

Given these key controls on fluoride occurrence and distribution, it is possible to anticipate broadly where areas of regionally high fluoride concentrations are likely to exist. Such an understanding of the fluoride occurrence is important for the management of the fluoride-related epidemiological problems. Water supply programs in potentially high-fluoride areas should have geological and hydrological guidance, which include chemical analyses and geological maps. High fluoride incidence may not be universal within a given area and remediation strategies may include identification of areas of distinct geology, residence time, or selective recharge where fluoride concentrations are locally lower. Nevertheless, even taking into account all available hydrogeological information, groundwater fluoride concentrations are frequently so variable on a local scale that prediction of concentrations in individual wells is difficult. Hence, for the purposes of compliance, testing of each well used for drinking water should be undertaken in fluoride-prone areas. Testing can be achieved effectively either in the field or laboratory by ion-selective electrodes or colorimetry. Chemical and other forms of intervention and treatment should generally be a last resort, especially because their monitoring in rural communities is virtually impossible.

SEE ALSO THE FOLLOWING CHAPTERS

Chapter 2 (Natural Distribution and Abundance of Elements) · Chapter 9 (Volcanic Emissions and Health) · Chapter 26 (GIS in Human Health Studies) · Chapter 28 (Mineralogy of Bone) · Chapter 31 (Modeling Groundwater Flow and Quality)

FURTHER READING

Ainsworth, N. J. (1933). Mottied Teeth. *Br. Dental J.*, 55, 233–250.

Apambire, W. B., Boyle, D. R., and Michel, F. A. (1997). Geochemistry, Genesis, and Health Implications of Fluoriferous Groundwaters in the Upper Regions of Ghana, *Environ. Geol.*, 33(1), 13–24.

Ashley, P. P., and Burley, M. J. (1994). Controls on the Occurrence of Fluoride in Groundwater in the Rift Valley of Ethiopia. In *Groundwater Quality* (H. Nash and G. J. H. McCall, Eds.), Chapman and Hall, pp. 45–54.

Banks, D., Frengstad, B., Midtgard, A. K., Krog, J. R., and Strand, T. (1998). The Chemistry of Norwegian Groundwaters: I. The Distribution of Radon, Major and Minor Elements in 1604 Crystalline Bedrock Groundwaters, *Sci. Total Environ.*, 222, 71–91.

Barbier, J. P., and Mazounie, P. (1984). Methods of Reducing High Fluoride Content in Drinking Water: Fluoride Removal Methods—Filtration Through Activated Alumina: A Recommended Technique, *Water Supply*, 2, 3–4.

Bårdsen, A., Bjorvatn, K., and Selvig, K. A. (1996). Variability in Fluoride Content of Subsurface Water Reservoirs, *Acta Odontol. Scand.*, 54, 343–347.

Barnard, W. R., and Nordstrom, D. K. (1982). Fluoride in Precipitation—II. Implication for the Geochemical Cycling of Fluorine, *Atmos. Environ.*, 16, 105–111.

Botchway, C. A., Ansa-Asare, O. D., and Antwi, L. A. K. (1996). Natural Fluoride and Trace Metal Levels of Ground and Surface Waters in the Greater Accra region of Ghana, *West. Afr. J. Med.*, 15, 204–209.

Boyle, D. R., and Chagnon, M. (1995). An Incidence of Skeletal Fluorosis Associated with Groundwaters of the Maritime Carboniferous Basin, Gaspé region, Quebec, Canada, *Environ. Geochem. Health*, 17, 5–12.

Bugaisa, S. L. (1971). Significance of Fluorine in Tanzania Drinking Water, Proceedings of a Conference on Rural Water Supply in East Africa, Dar Es Salaam, Tanzania, 107–113.

Carrillo-Rivera, J. J., Cardona, A., and Edmunds, W. M. (2002). Use of Abstraction Regime and Knowledge of Hydrogeological Conditions to Control High-Fluoride Concentration in Abstracted Groundwater: San Luis Potosí Basin, Mexico, *J. Hydrol.*, 261, 24–47.

Chaturvedi, A. K., Pathak, K. C., and Singh, V. N. (1988). Fluoride Removal from Water by Adsorption on China Clay, *Appl. Clay Sci.*, 3(4), 337–346.

Chaturvedi, A. K., Yadava, K. P., Pathak, K. C., and Singh, V. N. (1990). Defluoridation of Water by Adsorption on Fly Ash, *Water, Air Soil Pollut.* 49, 51–61.

Cook, J. M., Edmunds, W. M., and Robins, N. S. (1991). Groundwater Contributions to an Acid Upland Lake (Loch Fleet, Scotland) and the Possibilities for Amelioration, *J. Hydrol.*, 125, 111–128.

Corbett, R. G., and Manner, B. M. (1984). Fluoride in Ground Water of Northeastern Ohio, *Ground Water*, 22, 13–17.

Datta, D. K., Gupta, L. P., and Subramanian, V. (2000). Dissolved Fluoride in the Lower Ganges-Brahmaputra-Meghna River System in the Bengal Basin, Bangladesh, *Environ. Geol.*, 39, 1163–1168.

Dean, H. T., and Elvove, E. (1937). Further Studies on the Minimal Threshold of Chronic Endemic Dental Fluorosis, *Public Health Rep. (Washington)*, 52(37), 1249–1264.

Diaz-Barriga, F., Leyva, R., Quistian, J., Loyola-Rodriguez, J., Pozos, A., and Grimaldo, G. (1997). Endemic Fluorosis in San Luis Potosi, Mexico, *Fluoride*, 30, 219–222.

Diesendorf, M. (1986). The Mystery of Declining Tooth Decay, *Nature*, 322, 125–129.

Dissanayake, C. B. (1991). The Fluoride Problem in the Groundwater of Sri Lanka—Environmental Management and Health, *Int. J. Environ. Stud.*, 38, 137–156.

Edmunds, W. M. (1973). Trace Element Variations Across an Oxidation-Reduction Barrier in a Limestone Aquifer, *Proceedings Symposium on Hydrogeochemistry and Biogeochemistry* (E. Ingerson, Ed.), Clarke Co., Washington DC, pp. 500–526.

Edmunds, W. M. (1994). Characterization of Groundwaters in Semi-Arid and Arid Zones Using Minor Elements. In *Groundwater Quality* (H. Nash and G. J. H. McCall, Eds.), Chapman and Hall, pp. 19–30.

Edmunds, W. M., Cook, J. M., Darling, W. G., Kinniburgh, D. G., Miles, D. G., Bath, A. H., Morgan-Jones, M., and Andrews, J. N. (1987). Baseline Geochemical Conditions in the Chalk Aquifer, Berkshire, UK: A Basis for Groundwater Quality Management, *Appl. Geochem.*, 2(3), 251–274.

Edmunds, W. M., Cook, J. M., Kinniburgh, D. G., Miles, D. G., and Trafford, J. M. (1989). Trace Element Occurrence in British Groundwaters, British Geological Survey Research Report SD/89/3, pp. 424.

Edmunds, W. M., and Walton, N. (1983). The Lincolnshire Limestone—Hydrogeochemical Evolution Over a Ten-Year Period, *J. Hydrol.*, 61, 201–211.

Ellis, A. J. (1973). Chemical Processes in Hydrothermal Systems—A Review. In *Proceedings of a Symposium on Hydrogeochemistry and Biogeochemistry* (E. Ingerson, Ed.), Clarke Co., Washington DC, pp. 1–26.

Ellis, A. J., and Mahon, W. A. J. (1977). *Chemistry and Geothermal Systems*, Academic Press, San Diego, CA.

Fritz, B. (1981). Etude thermodynamique et modélisation des réactions hydrothermales et diagénétiques. Thèse Doc., Université de Strasbourg.

Fuge, R., and Andrews, M. J. (1988). Fluorine in the UK Environment, *Environ. Geochem. Health*, 10, 96–104.

Fuhong, R., and Shuquin, J. (1988). Distribution and Formation of High-Fluorine Groundwater in China, *Environ. Geol. Water Sci.*, 12, 3–10.

Gaciri, S. J., and Davies, T. C. (1993). The Occurrence and Geochemistry of Fluoride in Some Natural Waters of Kenya, *J. Hydrol.*, 143, 395–412.

Goni, J., Greffard, J., Leleu, M., and Monitron, L. (1993). Le fluor dans les eaux de boisson. BRGM Dép. Hydrogéol.

Gupta, M. K., Singh, V., Rajwanshi, P., Agarwal, M., Rai, K., Srivastava, S., Srivastav, R., and Dass, S. (1999). Groundwater Quality Assessment of Tehsil Kheragarh, Agra (India) with Special Reference to Fluoride, *Environ. Monit. Assess.*, 59, 275–285.

Gupta, S. C., Rathore, G. S., and Doshi, C. S. (1993). Fluoride Distribution in Groundwaters of Southeastern Rajasthan, *Ind. J. Environ. Health.*, 35(2), 97–109.

Hamilton, M. (1992). Water Fluoridation: A Risk Assessment Perspective. *J. Environ. Health*, 54(6), 27–32.

Handa, B. K. (1975). Geochemistry and Genesis of Fluoride-Containing Ground Waters in India, *Ground Water*, 13, 275–281.

Heidweiller, V. M. L. (1990). Fluoride Removal Methods. In *Proceedings Symposium on Endemic Fluorosis in Developing Countries: Causes, Effects and Possible Solutions* (J. E. Frencken, Ed.), NIPG-TNO, pp. 51–85.

Hem, J. D. (1985). Study and Interpretation of the Chemical Characteristics of Natural Water, USGS Water Supply Paper 2254, pp. 263.

Hitchon, B. (1995). Fluorine in Formation Waters in Canada, *Appl. Geochem.*, 10, 357–367.

Hoather, R. C. (1953). Fluorides in Water Supplies, *J. R. Sanitary Inst.*, 73, 202–223.

Jacob, B. (1975). Contribution à l'étude géochimique des eaux froides et thermominérales du Massif du Mont-Dore. Thèse troisième cycle, Université de Paris VI.

Jones, B. F., Eugster, H. P., and Reitig, S. L. (1977). Hydrochemistry of the Lake Magadi Basin, Kenya, *Geochim. Cosmochim. Acta*, 41, 53–72.

Kabata-Pendias, A., and Pendias, H. (1984). *Trace Elements Soils Plants*, CRC Press, Boca Raton, FL.

Kafri, U., Arad, A., Halicz, L., and Ganor, E. (1989). Fluorine Enrichment in Groundwater Recharged Through Loess and Dust Deposits, Southern Israel, *J. Hydrol.*, 110, 373–376.

Kilham, P., and Hecky, R. E. (1973). Fluoride: Geochemical and Ecological Significance in East African Waters and Sediments, *Limnol. Oceanogr.*, 18(6), 932–945.

Kloos, H., and Haimanot, R. T. (1999). Distribution of Fluoride and Fluorosis in Ethiopia and Prospects for Control, *Trop. Med. Int. Health*, 4(5), 355–364.

Lamont, P. (1958). Interim Report on an Investigation of the Chemical Characteristics of the Underground Waters from the Lincolnshire Limestone in the Area of Supply, with Special Reference to the Possibility of Locating New Sources Yielding a More Suitable Water for Public Supplies. Unpublished report, Spalding, U.K.

Lavado, R. S., and Reinaudi, N. (1979). Fluoride in Salt Affected Soils of La Pampa (Republica Argentina), *Fluoride*, 12, 28–32.

Mahon, W. A. J. (1964). Fluorine in the Natural Thermal Waters of New Zealand, *N. Z. J. Sci.*, 7, 3–28.

Maier, F. J. (1950). Fluoridation of Public Health Supplies, *J. Am. Water Works Assoc.*, 42, 1120–1132.

Maithani, P. B., Gurjar, R., Banerjee, R., Balaji, B. K., Ramachandran, S., and Singh, R. (1998). Anomalous Fluoride in Groundwater from Western Part of Sirohi District, Rajasthan, and Its Crippling Effects on Human Health, *Curr. Sci.*, 74(9), 773–777.

Marshall, E. (1990). The Fluoride Debate: One More Time, *Science*, 247, 276–277.

McCaffrey, L. P. (1998). Distribution and Causes of High Fluoride Groundwater in the Western Bushveld Area of South Africa, unpublished PhD thesis, University of Cape Town, South Africa.

Moges, G., Zewge, F., and Socher, M. (1996). Preliminary Investigations on the Defluoridation of Water Using Fired Clay Chips, *J. Afr. Earth Sci.*, 22, 479–482.

Murray, R. J. (1960). The Geology of the "Zuarungu" E Field Sheet, Geological Survey Ghana Bulletin No. 25, pp. 118.

Nanyaro, J. T., Aswathanarayana, U., and Mungure, J. S. (1984). A Geochemical Model for the Abnormal Fluoride Concentrations in Waters in Parts of Northern Tanzania, *J. Afr. Earth Sci.*, 2(2), 129–140.

Nawlakhe, W. G., and Bulusu, K. R. (1989). Nalgonda Technique—A Process for Removal of Excess Fluoride from Water, *Water Qual. Bull.*, 14, 218–220.

N'dao, I., Lagaude, A., and Travi, Y. (1992). Défluoruration expérimentale des eaux souterraines du Sénégal par le sulphate d'aluminum et le polychlorosulphate basique d'aluminum, *Sci. Tech. Eau*, 26, 243–249.

Neal, C. (1989). Fluorine Variations in Welsh Streams and Soil Waters, *Sci. Total Environ.*, 80, 213–223.

Neal, C., Wilkinson, J., Neal, M., Harrow, M., Wickham, H., Hill, L., and Morfitt, C. (1997). The Hydrochemistry of the River Severn, Plynlimon, *Hydrol. Earth Sys. Sci.*, 1(3), 583–617.

Nordstrom, D. K., and Jenne, E. A. (1977). Fluorite Solubility Equilibria in Selected Geothermal Waters, *Geochim. Cosmochim. Acta*, 41, 175–198.

Parkhurst, D. L. (2002). PHREECi—A Graphical User Interface for the Geochemical Computer Program PHREEQC, http://wwwbrr.cr.usgs.gov/projects/GWC_coupled/index.html.

Pickering, W. F. (1985). The Mobility of Soluble Fluoride in Soils, *Environ. Pollut. (Ser. B)*, 9, 281–308.

Queste, A., Lacombe, M., Hellmeier, W., Hillerman, F., Bortulussi, B., Kaup, M., Ott, O., and Mathys, W. (2001). High Concentrations of Fluoride and Boron in Drinking Water Wells in the Muenster Region—Results of a Preliminary Investigation, *Int. J. Hyg. Environ. Health*, 203, 221–224.

Rao, A. S. M., Rajayalakshmi, K., Sastry, K. R. K., Rao, P. S., Rao, V. V. K., and Rao, K. N. (1974). Incidence of Fluorides in Drinking Water Resources, in Andhra Pradesh. Proceedings of a Symposition on Fluorosis, Hyderabad, India, 227–235.

Rea, R. E. (1979). A Rapid Method for the Determination of Fluoride in Sewage Sludges, *Water Pollut. Control*, 78, 139–142.

Saether, O. M., and Andreassen, B. T. (1989). Fluoride in Precipitation in Southern Norway: Amounts and Sources, Report 89/106, Geological Survey of Norway.

Schneiter, R. W., and Middlebrooks, E. J. (1983). Arsenic and Fluoride Removal From Groundwater by Reverse Osmosis, *Environ. Int.*, 9, 289–292.

Smedley, P. L., Edmunds, W. M., and Pelig-Ba, K. B. (1995). Groundwater Vulnerability Due to Natural Geochemical Environment: 2. Health Problems Related to Groundwater in the Obuasi and Bolgatanga Areas, Ghana, BGS Technical Report, WC/95/43.

Smedley, P. L., Nicolli, H. B., Macdonald, D. M. J., Barros, A. J., and Tullio, J. O. (2002). Hydrogeochemistry of Arsenic and other Inorganic Constituents in Groundwaters from La Pampa, Argentina, *Appl. Geochem.*, 17(3), 259–284.

Smet, J. (1990). Fluoride in Drinking Water. *Proceedings of the Symposium on Endemic Fluorosis in Developing Countries: Causes, Effects and Possible Solutions*, NIPG-TNO, 51–85.

Solsona, F. (1985). Water Defluoridation in the Rift Valley, Ethiopia, UNICEF, pp. 27.

Suma Latha, S., Anbika, S. R., and Prasad, S. J. (1999). Fluoride Contamination Status of Groundwater in Karnataka, *Curr. Sci.*, 76(6), 730–734.

Ten Haven, H. L., Konings, R., Schoonen, M. A. A., Jansen, J. B. H., Vriend, S. P., Van der Weijden, C. H., and Buitenkamp, J. (1985). Geochemical Studies in the Drainage Basin of the Rio Vouga (Portugal), *Chem. Geol.*, 51, 225–238.

Travi, Y. (1993). Hydrogéologie et hydrochimie des aquifères du Sénégal, Sciences Géologiques, Memoire 95, Université de Paris-Sud.

U. S. Public Health Service (1943). Public Health Drinking Water Standards, *J. Am. Water Works Assoc.*, 35, 93–104.

U. S. Public Health Service (1946). Drinking Water Standards 1946, *J. Am. Water Works Assoc.*, 38, 361–370.

Vanderdonck, P., and Van Kesteren, X. (1993). Utilisation du chlorure de calcium, *Trib. Eau*, 562, 39–40.

Vijaya Kumar, V., Sai, C. S. T., Rao, P. L. K. M., and Rao, C. S. (1991). Studies on the Distribution of Fluoride in Drinking Water Sources in Medchal Block, Ranga Reddy District, Andra Pradesh, India, *J. Fluorine Chem.*, 55, 229–236.

Wenzel, W. W., and Blum, W. E. H. (1992). Fluorine Speciation and Mobility in F-Contaminated Soils, *Soil Sci.*, 153, 357–364.

WHO (2004). *Guidelines for Drinking-Water Quality.* 3rd edition, World Health Organization, Geneva.

Woo, N. C., Moon, J. W., Won, J. S., Hahn, J. S., Lin, X. Y., and Zhao, Y. S. (2000). Water Quality and Pollution in the Hunchun Basin, China, *Environ. Geochem. Health*, 22, 1–18.

Wuyi, W., Ribang, L., Jian'an, T., Kunli, L., Lisheng, Y., Hairong, L., and Yonghua, L. (2002). Adsorption and Leaching of Fluoride in Soils of China, *Fluoride*, 35(2), 122–129.

Zan-dao, W., and Yan, W. (2002). Fluoridation in China: A Clouded Future. *Fluoride*, 35(1), 1–4.

Zevenbergen, C., van Reeuwijk, L. P., Frapporti, G., Louws, R. J., and Schuiling, R. D. (1996). A Simple Method for Defluoridation of Drinking Water at Village Level by Adsorption on Ando Soil in Kenya, *Sci. Total Environ.*, 188(2–3), 225–232.

Zubair, L. (2003). Sensitivity of Kelani Streamflow in Sri-Lanka to ENSO. *Hydrol. Proc.* 17, 2439–2448.

CHAPTER 13

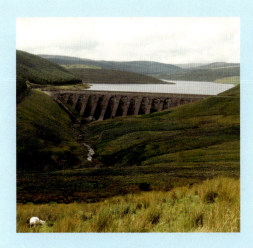

WATER HARDNESS AND HEALTH EFFECTS

Eva Rubenowitz-Lundin
Göteborg University

Kevin M. Hiscock
University of East Anglia

CONTENTS

I. DEFINITION OF WATER HARDNESS

Water hardness is the traditional measure of the capacity of water to react with soap and describes the ability of water to bind soap to form lather, which is a chemical reaction detrimental to the washing process. Hardness has little significance in terms of hydrochemical studies, but it is an important parameter for water users. Today, the technical significance of water hardness is more concerned with the corrosive effects on water pipes that carry soft water.

Despite the wide usage of the term, the property of hardness is difficult to define exactly. Water hardness is not caused by a single substance but by a variety of dissolved polyvalent metallic ions—predominantly calcium and magnesium—although other ions, for example, aluminum, barium, iron, manganese, strontium, and zinc, also contribute. The source of the metallic ions are typically sedimentary rocks, and the most common are limestone ($CaCO_3$) and dolomite ($CaMg(CO_3)_2$).

Hardness is normally expressed as the total concentration of calcium and magnesium ions in water in units of $mg\,L^{-1}$ as equivalent $CaCO_3$. Hardness can be determined by substituting the concentration of calcium and magnesium, expressed in $mg\,L^{-1}$, in the following equation:

$$\text{Total hardness} = 2.5(Ca^{2+}) + 4.1(Mg^{2+}) \qquad (1)$$

Each concentration is multiplied by the ratio of the formula weight of $CaCO_3$ to the atomic weight of the ion; hence, the factors 2.5 and 4.1 are included in the hardness relation (Freeze & Cherry, 1979).

In Europe, water hardness is often expressed in terms of degrees of hardness. One French degree is equivalent to $10\,mg\,L^{-1}$ as $CaCO_3$, one German degree to $17.8\,mg\,L^{-1}$ as $CaCO_3$, and one English of Clark degree to $14.3\,mg\,L^{-1}$ as $CaCO_3$. One German degree of hardness (dH) is equal to 1 mg of calcium oxide (CaO) or 0.72 mg of magnesium oxide (MgO) per 100 mL of water.

Where reported, carbonate hardness includes that part of the total hardness equivalent to the bicarbonate and carbonate (or alkalinity). If the hardness exceeds the alkalinity, the excess is called the non-carbonate hardness, and is a measure of the calcium and magnesium sulfates. In older publications, the terms "temporary" and "permanent" are used in place of carbonate and non-carbonate. Temporary hardness reflects the fact that the ions responsible may be precipitated by boiling, such that:

$$Ca^{2+} + 2HCO_3^- \xrightarrow{\text{Heat}} \underset{\text{"scale"}}{CaCO_3\downarrow} + H_2O + CO_2\uparrow \quad (2)$$

A number of attempts have been made to classify water hardness. Water with hardness values greater than $150\,mg\,L^{-1}$ is designated as very hard. Soft water has values of less than $60\,mg\,L^{-1}$. Groundwaters in contact with limestone or gypsum ($CaSO_4.2H_2O$) rocks can commonly attain levels of $200–300\,mg\,L^{-1}$. In water from gypsiferous formations, $1000\,mg\,L^{-1}$ or more of total hardness may be present (Hem, 1985).

Hardness in water used for domestic purposes does not become particularly troublesome until a level of $100\,mg\,L^{-1}$ is exceeded. Depending on pH and alkalinity, hardness of about $200\,mg\,L^{-1}$ can result in scale deposition, particularly on heating, and increased soap consumption. Soft waters with a hardness of less than about $100\,mg\,L^{-1}$ have a low buffering capacity and may be more corrosive to water pipes resulting in the presence of heavy metals such as cadmium, copper, lead, and zinc in drinking water. This depends on the pH, alkalinity, and dissolved oxygen concentration of the water.

No health-based guideline value is proposed for hardness because it is considered that the available data on the inverse relationship between the hardness of drinking water and cerebrovascular disease (CVD) are inadequate to permit the conclusion that the association is causal (WHO, 2002). However, a concentration of $500\,mg\,L^{-1}$ is at the upper limit of aesthetic acceptability.

II. NATURAL HYDROCHEMICAL EVOLUTION OF GROUNDWATER

The combination of geology and hydrology of a river catchment is important in determining the hardness of water. As illustrated in Figure 1A, catchments underlain by impermeable rocks that are resistant to erosion generate surface runoff with little time for weathering to occur. As a result, the surface water has a chemical com-

A

B

FIGURE 1 Illustration of two types of water resources and the influence of catchment geology on water hardness. In (A) surface water runoff from the impermeable Silurian mudstones and siltstones to the Dinas reservoir in west Wales is a soft water with a hardness of $19\,mg\,L^{-1}$ as $CaCO_3$, pH of 7.0, and electrical conductivity of $73\,\mu S\,cm^{-1}$ (sample date 1998; Hiscock & Paci, 2000). In (B) groundwater discharge from the Fergus River Cave springs developed in Carboniferous limestone in County Clare, Ireland, is a very hard water with a hardness of $256\,mg\,L^{-1}$ as $CaCO_3$, pH of 7.7, and electrical conductivity of $440\,\mu S\,cm^{-1}$ (sample date 2002).

position similar to dilute rainfall and is characterized as soft water. In contrast, and as illustrated in Figure 1B, catchments underlain by permeable rocks allow water to infiltrate below the ground surface such that groundwater in contact with the rock mass promotes solutional weathering, which potentially leads to the development of fissures and conduits. As a result, the groundwater attains a high concentration of dissolved constituents and is characterized as hard water.

In natural aqueous systems, the chemical composition of water is continually changing and is controlled predominantly by geological factors. The geochemistry of rock weathering determines the release of elements into water. In groundwaters, a natural evolution of chemical composition is recognized. Dilute rainwater with a sodium chloride water type and containing CO_2 enters the soil zone whereupon further CO_2, produced by the decay of organic matter, is dissolved in the infiltrating water to form carbonic acid. Within the soil and unsaturated zone of sedimentary rocks, this weak acid dissolves soluble calcium and magnesium carbonates such as calcite and dolomite to give high concentrations of calcium, magnesium, and bicarbonate. In crystalline igneous and metamorphic rocks, slow weathering of silicates by the attack of carbonic acid releases low concentrations of calcium, magnesium, potassium, and sodium in the infiltrating soil water or groundwater and also produces bicarbonate.

Away from the supply of oxygen in the soil and unsaturated zone, infiltrating water becomes increasingly anoxic as a result of progressive bacterial reduction of oxygen. Below the water table and with increasing reducing conditions, iron and manganese become mobilized and then later precipitated as metal sulfides. In the presence of disseminated clay material within an aquifer, ion exchange replaces calcium for sodium as the water evolves to a sodium bicarbonate water type. Hence, the groundwater is naturally softened by ion exchange reactions. In the deeper, confined section of aquifers, mixing with saline water may occur to produce a sodium chloride water type before a region of static water and aquifer diagenesis is reached. Either part or all of this classic sequence of hydrochemical change is identified in a number of aquifers including the Floridan aquifer system.

The Floridan aquifer system occurs in the southeast United States (Figure 2) and is one of the most productive aquifers in the world. The aquifer system is a vertically continuous sequence of Tertiary carbonate rocks of generally high permeability. Limestones and dolomites are the principal rock types, although in southwestern and northeastern Georgia and in South Carolina the limestones grade into lime-rich sands and clays. The Floridan aquifer is composed primarily of calcite and dolomite with minor gypsum, apatite, glauconite, quartz, clay minerals, and trace amounts of metallic oxides and sulfides.

The total hardness of water in the Upper Floridan aquifer varies from <50 to >5000 mg L^{-1} as $CaCO_3$. Generally, where the system is composed only of limestone, the total hardness is equivalent to carbonate hardness and is <120 mg L^{-1}. Groundwater with higher total hardness usually results from (1) dissolution of other aquifer minerals, primarily dolomite and gypsum; (2) mixing of fresh water with residual saline groundwater; (3) encroachment and mixing of modern seawater; or (4) contamination. Natural softening of the groundwater by cation exchange is thought to be responsible for the low hardness in Escambia and Santa Rosa Counties, Florida (Sprinkle, 1989).

A sequence of hydrochemical evolution is identified by Sprinkle (1989), which starts with calcite dissolution in recharge areas that produces a calcium-bicarbonate dominated water type with a total dissolved solids (TDS) concentration of generally less than 250 mg L^{-1}. Downgradient, dissolution of dolomite leads to a calcium-magnesium-bicarbonate water type. Where gypsum is abundant, sulfate becomes the predominant anion. In coastal areas, as shown in Figure 3, seawater increases the TDS concentration and the water type changes to sodium-chloride. In the western panhandle of Florida, cation exchange leads to the development of a sodium-bicarbonate water type and a less hard groundwater.

For additional reading on groundwater modeling, see Modeling Groundwater Flow and Quality, this volume.

III. THE HARD-WATER STORY

The history behind what is considered today to be a commonly accepted fact—that hard water protects against CVD—is often referred to as the "hard-water story." The hard-water story started in 1957 with Jun Kobayashi, a Japanese agricultural chemist. He had for many years been engaged in studies of the nature of irrigation water from an agricultural point of view. In these studies he found a close relation between the chemical composition of river water and the death rate from "apoplexy" (CVD). The death rate of apoplexy in Japan was extraordinarily high compared to other countries and the biggest cause of death in Japan. Kobayashi found that it was the ratio of sulfur to carbonate (SO_4/CaO_3) that was related to the death rate from apoplexy. He suggested that inorganic acid $CaCO_3$ might induce or prevent apoplexy (Kobayashi, 1957).

Three years later a study was presented by Schroeder (1960), comprising the 163 largest cities in the United States. He found inverse correlations between water hardness and CVD in both men and women. He also studied the relation between different water constituents and coronary heart disease (CHD) among 45- to 64-year-old men. He found significant correlations

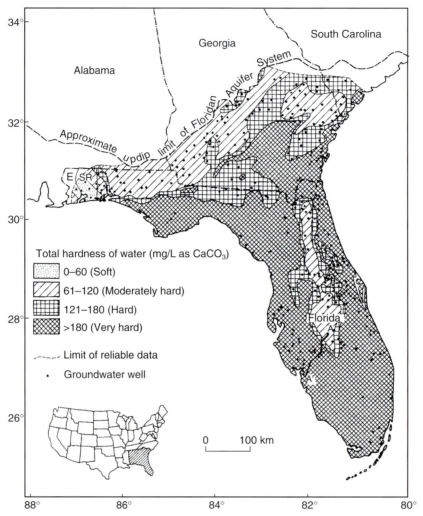

FIGURE 2 Map of the extent of the Upper Floridan Aquifer showing the spatial distribution of total hardness of water. Escambia and Santa Rosa Counties, Florida, are located at positions E and SR, respectively. (After Sprinkle, 1989.)

between death rate from CHD and sulfate and bicarbonate, respectively, but not for the ratio of the agents. He also found significant negative correlations between deaths from CHD and magnesium, calcium, fluoride, and pH. The coefficient of correlation for magnesium in drinking water was $r = -0.30$, $p < 0.01$, and for calcium $r = -0.27$, $p < 0.01$ (Schroeder, 1960).

During the following decades several studies on the relation between water hardness and CVD were presented. Initially the studies dealt with the question of whether there was a toxic effect in soft water or a protective effect in hard water. As described above, soft waters usually have low buffering capacity and are more corrosive, which leads to higher amounts of toxic trace elements. It has been proposed that this could account for the observed increased mortality. However, the results from previous studies have not supported this

hypothesis (Marier, 1986a). For example, in a nationwide survey of more than 500 tap waters in Canada, no significant correlation was found between mortality and trace elements like lead, cadmium, cobalt, lithium, mercury, molybdenum, nickel, or vanadium (Neri et al., 1975). Gradually it became obvious that it was a protective effect from hard water that was responsible for the relations seen. In different parts of the world, studies have been made on the relation between magnesium and calcium in the local drinking water and CVD mortality. These studies were generally based upon death registers and water data at regional or municipality levels. The results of these studies are summarized in Table I.

Even with these studies, the results were not conclusive as to the role of magnesium and calcium in drinking water for CVD. Most of the studies showed a relation

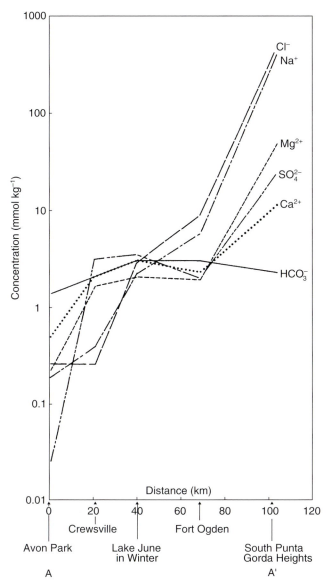

FIGURE 3 Hydrochemical section along line A-A′ (for location see Figure 2) showing the variation in concentrations of major ions downgradient in the direction of groundwater flow. Relative to calcium and magnesium, the groundwater experiences an increasing concentration of sodium as a result of ion ex-change and mixing with saline water in the coastal zone. (After Sprinkle, 1989.)

TABLE I. The Effect of High Magnesium (Mg) or Calcium (Ca) Levels in Drinking Water on the Mortality from Cardiovascular Disease (CVD), Coronary Heart Disease (CHD), or Cerebrovascular Disease (CD) in Different Studies Published since 1975

| | Mg | | | Ca | | |
Studies	CVD	CHD	CD	CVD	CHD	CD
North America						
Dawson et al., 1978	↓			↓		
Neri et al., 1975		↓			—	
Great Britain						
Shaper et al., 1980	—			↓		
Maheswaran et al., 1999	—			—		
Germany						
Teitge, 1990		↓				
Sweden						
Nerbrand et al., 1992	—			↓		
Rylander et al., 1991	↓		—		↓	↓
South Africa						
Leary et al., 1983		↓				

↓ = lower mortality, — = no difference.

between different cardiovascular deaths and either magnesium or calcium, or both, but some studies showed no relation at all. However, most of these studies were ecological meaning that the exposure to water constituents was determined at group levels with a high risk of misclassification. Often, very large groups, for example, all inhabitants in large cities or areas, were assigned the same value of water magnesium and calcium, despite the presence of several waterworks or private wells.

In addition, the disease diagnoses studied were sometimes unspecific, with wide definitions that included both cardiac and cerebrovascular diseases. In some studies, it is also unclear whether the range of magnesium and calcium in drinking water was large enough to allow for appropriate analyses.

One of the most comprehensive studies of the geographic variations in cardiovascular mortality was the British Regional Heart Study. The first phase of this study (Pocock et al., 1980) applied multiple regression analysis to the geographical variations in CVD for men and women aged 35–74 in 253 urban areas in England, Wales, and Scotland from 1969 to 1973. The investigation showed that the effect of water hardness was nonlinear; much greater in the range from very soft to medium-hard water than from medium to very hard water. The geometric mean for the standardized mortality ratio (SMR) for CVD for towns grouped according to water hardness both with and without adjustments (by analysis of covariance) for the effects of four climatic and socioeconomic variables (percentage of days with rain, mean daily maximum temperature, per-

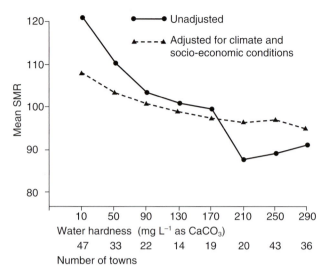

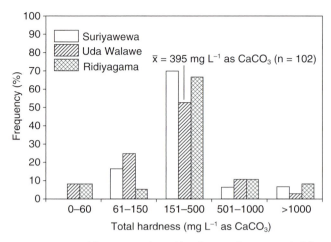

FIGURE 4 Geometric means of the standardized mortality ratio (SMR) (for all men and women aged 35–74 with CVD) for towns in England, Wales, and Scotland grouped according to water hardness (in concentration units of mg L⁻¹ as CaCO₃ equivalent). (After Pocock et al., 1980.)

FIGURE 5 Histogram of total hardness values recorded for groundwaters sampled during the wet season (January and February 2001) from dug wells and tubewells in three subcatchments of the Uda Walawe basin of Sri Lanka. (Data courtesy of L. Rajasooriyar.)

centage of manual workers, car ownership) is shown in Figure 4. The adjusted SMR decreased steadily in moving from a hardness of 10 to 170 mg L⁻¹ but changed little between 170 to 290 mg L⁻¹ or greater. After adjustment, CVD in areas with very soft water, around 25 mg L⁻¹, was estimated to be 10–15% higher than in areas with medium-hard water, around 170 mg L⁻¹, whereas any further increase in hardness beyond 170 mg L⁻¹ did not additionally lower CVD mortality. Hence, it appeared that the maximum effect on CVD was principally between the very soft and medium-hard waters. Adjusting for climatic and socioeconomic differences considerably reduced the apparent magnitude of the effect of water hardness (Pocock et al., 1980).

A problem with correlation studies such as the British Regional Heart Study, as argued by Jones and Moon (1987), is the failure of much of the research to consider the causal mechanism that links independent variables to the disease outcome. Also, many of the calibrated models presented in the literature are socially blind in including only variables pertaining to the physical environment, which often contributes a large number of water quality elements. Even in those better analyses that have included social variables, as in the case of the British Regional Heart Study, the relatively strong correlation found for calcium in England and Wales may be a result of calcium acting as a very good surrogate for social variables. The soft water areas of the north and west of the British Isles equate to the areas of early

industrialization, and today these areas house a disproportionate percentage of the socially disadvantaged (Jones & Moon 1987). Therefore, it is important that further studies undertake the challenge of quantitatively analyzing the separate effects of social variables from those of water hardness.

Fewer studies have been carried out in developing countries but Dissanayake et al. (1982), for example, found a negative correlation between water hardness and various forms of CVD and leukemia in Sri Lanka. More recent studies (Dissanayake, 1991; Rajasooriyar, 2003) have highlighted the problem of high fluoride concentrations and associated dental fluorosis in areas of hard water abstracted from crystalline bedrock aquifers in Sri Lanka.

Rajasooriyar (2003) measured total hardness in dug wells and tubewells in the Uda Walawe basin of southern Sri Lanka in the wide range of 7–3579 mg L⁻¹ as CaCO₃ with an average of 395 mg L⁻¹ as CaCO₃ (Figure 5). Compared with the government water quality limit of 600 mg L⁻¹ as CaCO₃, 12% of the 102 samples collected during the wet season in 2001 were in excess of the limit and are considered too hard to drink (values above 100–150 mg L⁻¹ as CaCO₃ are locally considered too hard as a water supply). Soft waters are found in areas with a dense irrigation network supplied by rain-fed surface reservoirs. Irrigation canal waters in the Suriyawewa and Uda Walawe subcatchments were measured in the dry season to have a total hardness in the range 40–90 mg L⁻¹ as CaCO₃ (Rajasooriyar, 2003), and it is leakage of this water source that leads to the softening of shallow groundwater. The majority of the ground-waters (63%) in the fractured aquifer represent very hard water with carbonate hardness in the range 151–500 mg L^{-1} as CaCO$_3$ contributed by the weathering of ferro-magnesian minerals, anorthite, calcite, and dolomite. The products of this weathering lead to high concentrations of dissolved calcium, magnesium, and bicarbonate in groundwaters. Exceptionally high values of hardness (>1000 mg L^{-1} as CaCO$_3$) typically occur in non-irrigated areas with additional non-carbonate hardness contributed by pyrite oxidation and, in the case of the coastal Ridiyagama coastal catchment, by salt water inputs.

IV. STUDIES AT AN INDIVIDUAL LEVEL

In recent years some studies have been made with a higher precision as regards exposure classification than the previously cited ecological studies. Most of them have been conducted with a case-control design. This means that the exposure to water magnesium and calcium has been estimated for individuals who have suffered from the disease as well as for healthy control persons, and the difference in risk calculated. A similar type of study design is the prospective cohort study, where the subjects are followed over time.

A. Studies in Finland

Punsar and Karvonen made a cohort study in 1979. They compared the death rate from CHD in two rural areas in western and eastern Finland. The cohort comprised men 40–59 years old in 1959, and they were followed for 15 years. The water data were not truly individual but were median values in ten subareas in the western area and 33 subareas in the eastern area. The ranges of subarea medians were 6.9–27.8 mg L^{-1} of water magnesium in the western area and 0.6–7.3 mg L^{-1} in the eastern area. The cohort in eastern Finland had a death rate from CHD which was 1.7 times higher than the western cohort. Calcium was not investigated.

A few years later, Luoma et al. (1983) published a case-control study conducted in the southeastern region of Finland. Cases were men 30–64 years of age with a first acute myocardial infarction (AMI; alive or deceased), who were pair-matched with hospital controls for age and region (rural vs. urban). In addition, population controls were selected and matched for age and municipality. All subjects submitted a sample of

their drinking water. The range of magnesium in the drinking water was 1.0–57.5 mg L^{-1} for the cases and 0.75–30.0 mg L^{-1} and 1.0–16.0 mg L^{-1} for the hospital controls and population controls, respectively. For case-population control comparisons, the relative risk (RR) with 95% confidence limits was 4.7 (95% confidence interval [CI] 1.3–25.3) for magnesium levels lower than 1.2 mg L^{-1}. This means that those with the lowest magnesium levels had a risk of AMI almost five times higher than those with higher magnesium levels. For the case-hospital control comparisons, the RR was 2.0 (95% CI 0.7–6.5), that is double the risk. They also found inverse relations with fluoride levels but no relation to calcium.

B. Studies in Sweden

In Sweden, three case-control studies have been conducted during the last decade. First, using mortality registers, the relation between death from AMI and the level of magnesium and calcium in drinking water was examined among men. A few years later a study with a similar design was made comprising women. The studies were conducted in southern Sweden, in a relatively small geographic area, where there was a great difference between as well as within the municipalities regarding magnesium and calcium content in drinking water. The advantage with this limited study area was that the possible risk of such confounding factors as climate, geographical, cultural, and socioeconomic differences was minimized.

Seventeen municipalities were identified whose water quality with respect to water hardness, acidity, and treatment procedures had been basically unchanged (change of hardness <10% and pH <5%) during the most recent ten years. Figure 6 shows the location of the 17 municipalities that comprised the study area.

Cases were men (n = 854) and women (n = 378) in 17 municipalities in southern Sweden who had died of AMI between ages 50 and 69 years. Controls were men (n = 989) and women (n = 1368) of the same age group who had died of cancer. Individual water data were collected. Table II shows the number of waterworks, the range of magnesium in drinking water, and the amount of magnesium in water supplied to the most densely populated area in each municipality.

The subjects were divided into quartiles according to the levels of magnesium. Odds ratios were calculated in relation to the group with the lowest exposure. Adjustments for age were made in all analyses. The results show odds ratios of 0.65 for men and 0.70 for women in the quartile with highest magnesium levels in the

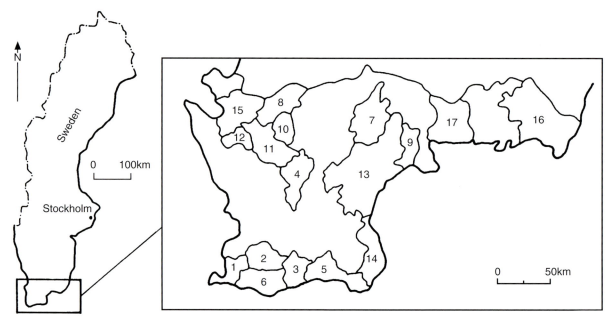

FIGURE 6 The country of Sweden with Skåne and Blekinge counties enlarged. The 17 municipalities in the studies are numbered corresponding to Table II, i.e., 1 Vellinge, 2 Svedala, 3 Skurup, 4 Höör, 5 Ystad, 6 Trelleborg, 7 Ö Göinge, 8 Örkelljunga, 9 Bromölla, 10 Perstorp, 11 Klippan, 12 Åstorp, 13 Kristianstad, 14 Simrishamn, 15 Ängelholm, 16 Karlskrona, and 17 Karlshamn.

TABLE II. Number of Waterworks, Range of Magnesium (Mg) Concentrations in Drinking Water (mg L^{-1}), and Amount of Mg in Water Supplied to the Most Densely Populated Areas (mg L^{-1})

Municipality no.[a]	Number of waterworks in the study	Range of water Mg (mg L^{-1})	Water Mg in the most densely populated area (mg L^{-1})
1	2	5.9–10.0	10.0
2	2	6.8–20.0	20.0
3	4	2.6–10.0	9.0
4	2	8.8–13.0	8.8
5	5	5.1–11.9	7.5
6	4	6.5–18.0	16.0
7	9	3.3–13.5	5.0
8	3	3.0–6.9	6.9
9	5	1.3–7.6	1.3
10	2	4.0–9.0	9.0
11	2	5.5–9.0	8.0
12	1	10.0	10.0
13	15	1.3–14.4	6.7
14	9	3.8–13.4	9.7
15	2	7.0–11.0	11.0
16	9	2.0–13.0	2.0
17	2	1.9–5.0	2.0

[a]Municipality numbers correspond to Figure 6.

drinking water (≥ 9.8–9.9 mg L^{-1}) (Figure 7). This means that the risk of dying from AMI was about 30% lower compared with the risk for those who used drinking water with the lowest levels of magnesium (Rubenowitz et al., 1996, 1999).

A few years later a prospective interview study was conducted in the same area where men and women who suffered from AMI from 1994 to 1996 were compared with population controls (Rubenowitz et al., 2000). The results showed that magnesium in drinking water protected against death from AMI, but the total incidence was not affected. In particular, the number of deaths outside hospitals was lower in the quartile with high magnesium levels. This supports the hypothesis that magnesium prevents sudden death from AMI rather than all CHD deaths. The mechanisms that could explain these findings are discussed below.

V. PHYSIOLOGICAL IMPORTANCE OF MAGNESIUM

A. Physiological Properties of Magnesium in Humans

Magnesium is involved in several important enzymatic reactions. All reactions that involve ATP (adenosine triphosphate), that is energy demanding reactions, have an absolute magnesium requirement (Reinhardt, 1988).

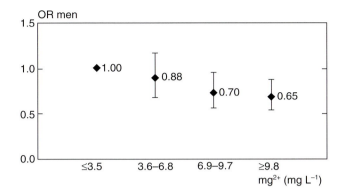

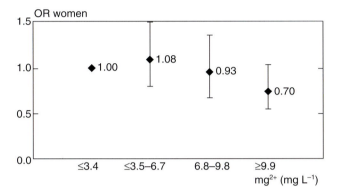

FIGURE 7 Odds ratios (OR) with 95% confidence intervals for death from AMI in relation to magnesium in drinking water (adjusted for age and calcium) in men and women.

Important processes in the body mediated by magnesium are, for example, synthesis of protein, nucleic acid and fat, glucose use, neuromuscular transmission, muscular contraction, and transport over cell membranes (Altura & Altura, 1996) (see also Chapter 30, this volume).

Magnesium is essential to the cardiovascular system, and two properties are especially important: stabilizing the cardiac electric system (preventing cardiac arrhythmia) and regulating vascular tone (Reinhardt, 1991). There are multiple mechanisms behind these properties.

Magnesium is needed to maintain the normal gradient of potassium and calcium over cell membranes (Altura et al., 1981). It is well known that magnesium is necessary to maintain intracellular levels of potassium. This is done by blocking the outward passage of potassium through the cell membrane and by activating the enzyme Na/K-ATPase. It has a similar function in Ca-ATPase. Magnesium also has a direct effect on potassium and calcium channels in the cell membranes (Reinhardt, 1991).

Furthermore, as regards impact on vascular tone, magnesium is a necessary activator for the synthesis of

cyclic adenosine monophosphate (c-AMP), which is a vasodilator. It also acts as a natural calcium antagonist by competing for calcium binding sites in the vascular smooth muscle and thus reducing the constrictive effect of calcium in the blood vessels (Reinhardt, 1991). In addition, the vasoconstrictive actions of hormones such as angiotensin, serotonin, and acetylcholine are enhanced in the case of magnesium deficiency (Altura et al., 1981).

VI. MAGNESIUM INTAKE

A. Magnesium Deficiency

Marginal magnesium deficiency probably affects a large proportion of the population whose dietary intake does not reach the recommended amount. Although severe magnesium deficiency is not common in the population, hypomagnesemia is often present among hospitalized patients (Altura, 1994). A survey showed that hypomagnesemia was the most common electrolyte abnormality in patients entering the intensive care unit and was present in 20% of the patients (Reinhardt, 1988) (see also Chapter 8, this volume).

The causes of hypomagnesemia include reduced intake caused by starvation or intravenous therapy without magnesium supplement; impaired absorption owing to chronic diarrhea or malabsorption syndromes; increased renal loss caused by diseases such as diabetes or renal diseases or by the use of alcohol or such drugs as diuretics or antibiotics (Reinhardt, 1988).

B. Magnesium Intake From Food

The largest part of the total magnesium intake is from foodstuffs. Magnesium is present in many foodstuffs and in large amounts in nuts, beans, green leafy vegetables, and whole grain cereals. Mainly because of increased industrial treatment, which decreases magnesium levels by 80 to 95% (Marier, 1986b), it has been suggested that the majority of people today have a lower magnesium intake than the recommended dietary amount (RDA) of $6\,mg\,kg^{-1}\,day^{-1}$ (Marier, 1986b; Seelig, 1986; Durlach, 1989). Table III shows the magnesium content in some foodstuffs and beverages.

Approximately 40% of magnesium ingested in food is normally absorbed (Hardwick et al., 1990). The proportion absorbed is inversely related to the amount ingested, however, and has been shown to range between approximately 10 and 70% (Fine et al., 1991).

TABLE III. Magnesium (Mg) Content in Food and Beverages

Food items and beverages	Mg (mg $100g^{-1}$)	Food items and beverages	Mg (mg $100g^{-1}$)
Food items		**Food items**	
Bread, white	23	Bananas	33
Bread, fiber-rich	50	Fish	25
Muesli	100	Shrimp	42
Corn flakes	16	Meat	25
Cheese	34	Eggs	13
Common beans, white, dried	184	Almonds	280
Common beans, brown, dried	131	Peanuts	188
Corn	23	Dark chocolate	130
Spinach	79	Milk chocolate	60
Avocado	39		
Broccoli	23	**Beverages**	
Tomato	10	Milk	12
Cucumber	10	Coffee[a]	5
Mushrooms	13	Tea[a]	3
Brown rice	110	Beer, low alcohol content	8
White rice	34	Beer, high alcohol content	11
Potatoes	24	Red wine	12
Apples	5	White wine	3
Oranges	10	Spirits	0

[a]Magnesium content in the water not included.

The absorption rate also depends on the intake of other foodstuffs. Elements such as calcium, phosphorus, fibers, and phytic acids are known to diminish the absorption rate of magnesium (Seelig, 1986). Most magnesium is absorbed in the distal small intestine, mainly by passive diffusion through the paracellular pathway, but also to a smaller extent by solvent drag and active transport (Hardwick et al., 1990).

VII. WATER MAGNESIUM AND BODY MAGNESIUM STATUS

A. Magnesium Intake From Water— Importance for Body Magnesium Status

Magnesium has a natural source in water from the weathering of a range of rock types. In igneous rock, magnesium is typically a constituent of the dark colored ferromagnesian minerals, which include olivine, pyroxenes, amphiboles, and dark colored micas. In altered rocks, magnesian mineral species also occur such as chlorite, montmorillonite, and serpentine. Sedimentary forms of magnesium include magnesite ($MgCO_3$) and dolomite ($CaMg(CO_3)_2$). Magnesium is substantially less abundant than calcium in all rock types and so in most natural waters the magnesium concentration is much lower, usually by 5–10 times, than the calcium concentration. Concentrations of magnesium in fresh waters are controlled by solution and precipitation reactions involving magnesium-bearing silicate and carbonate minerals with concentrations typically less than $50\,mg\,L^{-1}$, although values above $100\,mg\,L^{-1}$ are recorded.

Magnesium intake via water depends on the level in drinking water. An important issue in discussions about the relation between water magnesium and AMI is whether magnesium in drinking water can be critical for the body magnesium status, as the main part of the magnesium intake derives from food (Neutra, 1999). It has been suggested that the quantitative contribution of water magnesium may be crucial for body magnesium status for those who have a low dietary intake and use water with high magnesium levels (Durlach et al., 1989). In addition, cooking food in magnesium-poor water leaches out magnesium, while cooking in magnesium-rich water diminishes this loss (Haring & van Delft, 1981).

Furthermore, it has been suggested that magnesium in water, appearing as hydrated ions, has a higher bioavailability than magnesium in food, which is bound in different compounds that are less easily absorbed (Durlach et al., 1989; Theophanides et al., 1990). However, simultaneous intake of other agents could diminish the absorption rate, as discussed above.

Plants cultivated in areas with magnesium-rich water may have higher magnesium content, especially if the soil is magnesium-rich and the land is irrigated with magnesium-rich water. People living in such areas who eat locally grown vegetables and fruits may also benefit from an addition to the total magnesium intake, especially during summer months. However, genetic factors appear to have a greater effect on plant magnesium composition than do soil and environmental factors (Wilkinson et al., 1987).

Some previous studies have shown relations between water magnesium and body magnesium content. In a study of baboons, tap water was more effective than dietary supplementation in increasing serum levels of magnesium and zinc (Robbins & Sly, 1981). Anderson et al. (1975) found relations between water magnesium and magnesium content in the heart muscle. The myocardial magnesium was approximately 7% lower among residents of cities with soft water. There was, however, no difference in magnesium content in the diaphragm or the pectoralis major muscle. Other studies have shown relations between water magnesium levels and the magnesium content in skeletal muscle and in coronary arteries (Landin et al., 1989).

In a loading test, 10 subjects who normally used water with a magnesium level of $1.6\,mg\,L^{-1}$, instead were given drinking water with $20\,mg\,L^{-1}$ magnesium. After six weeks of supplementation with magnesium-enriched water, the excretion of magnesium was increased, which indicated improved body magnesium status (Rubenowitz et al., 1998).

Another study demonstrated that the ionized serum magnesium level was raised after only six days of an increased dietary load of magnesium, but not the total body magnesium. However, a correlation between ionized serum and ionized intracellular magnesium was shown (Altura & Altura, 1996).

From the available data it is apparent that the quantitative contribution from water may be crucial. The magnesium levels in drinking water in the Swedish case-control studies ranged from 0 to $44\,mg\,L^{-1}$. With a total daily intake of drinking water of two liters, the proportional magnesium contribution from water thus ranges from 0 to $88\,mg\,day^{-1}$. This is a percentage contribution between 0 and 25% of the RDA of $350\,mg\,day^{-1}$. For those who have a daily intake lower than the RDA and use magnesium-rich water, the contribution would be even more important.

In the prospective study, the calculated intake of magnesium from food ranged from 157 to $658\,mg\,day^{-1}$, median $356\,mg\,day^{-1}$. This means that a large number of the subjects had a lower intake than the RDA. One subject with an intake of 157 mg magnesium per day used drinking water with $3.5\,mg\,L^{-1}$, which means a 4.5% addition to the magnesium intake from food. If she instead had used water with $40\,mg\,L^{-1}$, the addition would have been 50%, and the total daily intake $240\,mg\,day^{-1}$.

VIII. CALCIUM IN DRINKING WATER AND CARDIOVASCULAR DISEASE

The majority of the previous ecological studies showed an inverse relation between calcium in drinking water and CVD, as reported above. The physiological mechanisms that could explain the relationship are not clear. There is, however, evidence that calcium deficiency can cause hypertension, which is a well-known risk factor for both stroke and AMI (Lau & Eby, 1985; Moore, 1989; Waeber & Brunner, 1994).

A. Calcium Deficiency

Calcium deficiency is common among the elderly, especially women. The absorption and renal conservation of calcium decreases with age. The absorption of calcium from food varies between 15 and 75% (Schaafsma, 1992), but in menopausal women the absorption is only about 20–30% (Heany & Recker, 1985). Furthermore, calcium intake is often decreased among the elderly (Harlan et al., 1984). The RDA for calcium in Sweden is $800\,mg\,day^{-1}$ for adult women and $600\,mg\,day^{-1}$ for men. In the United States, the RDA is $1000–1500\,mg\,day^{-1}$ for adults.

A study comprising 61,000 women in Sweden aged 40–76 showed that calcium intake decreased with age and that a majority of postmenopausal women had a deficient calcium intake (Michaelsson, 1996). Several studies conducted in the United States have also shown an intake lower than recommended, especially among women (Fleming & Heimbach, 1994). For individuals with a deficiency, the additional calcium from water could be crucial to prevent this. Along with the contribution of drinking water, cooking food in calcium-rich water has been shown not only to prevent leaching, but even to increase calcium levels in the food (Haring & van Delft, 1981).

B. Calcium and Blood Pressure

Several studies have shown an inverse relation between dietary calcium intake and blood pressure. Meta-analysis comprising nearly 40,000 people has shown that a high calcium intake lowered both systolic and diastolic blood pressure (Cappucio et al., 1995). Low serum concentrations of ionized calcium have been measured in patients with hypertension (McCarron, 1982). There are several possible mechanisms that could explain how calcium lowers blood pressure.

One mechanism may be that hypocalcemia inhibits Ca-ATPase activity, which leads to an increase in free intracellular calcium and contraction of vascular smooth muscles (McCarron, 1985). Calcium supplementation has been shown to be efficacious, especially among salt-sensitive hypertensives with a deficient basal calcium intake (Sowers et al., 1991). These individuals often have increased levels of the calcium regulatory hormones parathyroid hormone and active vitamin D $(1,25\text{-}(OH)_2\text{-}D)$, which can cause increased peripheral resistance. Dietary calcium suppresses these hormones which causes the blood pressure to decrease. Calcium and calcium regulatory hormones may also influence blood pressure regulation via the central nervous system. Calcium also induces natriuresis, which has been shown to lower blood pressure in postmenopausal hypertensive women (Johnson et al., 1985).

The case-control studies reported above showed no clear relation between calcium and AMI.

IX. WATER HARDNESS AND OTHER HEALTH EFFECTS

The results of several studies have suggested that a variety of other diseases are also correlated with water hardness, which include various types of cancer. Incidences of a correlation between cancer and water hardness have been reported from studies in Finland and Taiwan. In northern Finland, correlations of age-adjusted incidences of various forms of cancer with the geochemical composition of groundwater were undertaken by comparing geochemical maps showing the hardness, uranium, iron, and nitrate content of water and maps showing the areal distribution of the incidences of ten forms of cancer (Piispanen, 1991). A statistically significant positive correlation was identified between water hardness and several forms of cancer, especially for all forms of cancer combined in the female population ($r = 0.66$). Piispanen (1991) suggested that drinking hard water may be an initiator and promoter of cancer, although it is admitted that a positive correlation between the geochemical and medical variables does not necessarily prove a cause-and-effect relationship between these variables. More recent research in Taiwan also proposes a possible association between water hardness in drinking water and several types of cancer. In a study of esophageal cancer, Yang et al. (1999) set up control groups that consisted of people who had died from causes other than cancer, and the controls were pair-matched to the cancer cases by sex, year of birth, and year of death. For esophageal cancer, the results showed that there was a 42% excess risk of mortality from esophageal cancer in relation to the use of soft water.

In a Japanese study of water hardness, regional geological features, and the incidence of struvite stones, Kohri et al. (1993) found a positive correlation between the magnesium-calcium ratio of tap water and the incidence of struvite stones. The incidence of struvite stones was both high in regions of basalt and sedimentary rock and low in granite and limestone areas.

In an ecological study of the relation between domestic water hardness and the prevalence of eczema among primary-school-age children in Nottingham, England, McNally et al. (1998) found a significant direct relation between a one-year period and lifetime prevalence of eczema and water hardness, both before and after adjustment for confounding factors (sex, age, socioeconomic status, access to health care). Eczema prevalence trends in the secondary-school population were not significant.

X. CONCLUSIONS AND CONSEQUENCES

The significance of earlier studies of the link between water hardness and CVD is unclear, and it is suggested that the reported associations may reflect disease patterns that can be explained by social, climatological, and environmental factors rather than by the hardness of the water. However, from the results of epidemiological and experimental studies, the most significant conclusion is that magnesium in drinking water prevents death from AMI, either by preventing arrhythmia or spasm in coronary blood vessels. It is not difficult to imagine the importance of these findings for public health.

A. A Public Health Perspective

Marier and Neri (1985) attempted to quantify the importance of water magnesium using a number of the abovementioned epidemiological studies. They estimated that an increase in water magnesium level of 6 $mg\,L^{-1}$ would decrease CHD mortality by approximately 10%.

The data collected in the Swedish case-control studies can be used to estimate the impact of water magnesium on the incidence of myocardial infarction in the study population. If everyone in the male study base were to drink water from the highest quartile

($\geq 9.8 \, mg \, L^{-1}$), the decrease in mortality from AMI would be about 19%. This means that the age-specific incidence of death from myocardial infarction in the study area would change from about 350/100,000 $year^{-1}$ to 285/100,000 $year^{-1}$. The decrease of the incidence per $mg \, L^{-1}$ magnesium can be calculated to be approximately 10/100,000 $year^{-1}$, which is an even larger decrease than Marier and Neri (1985) estimated.

For women, the corresponding decrease in mortality from AMI would be about 25% if everyone were to drink water from the highest quartile ($\geq 9.9 \, mg \, L^{-1}$). The age-specific mortality among women in the study area would change from about 92/100,000 $year^{-1}$ to 69/100,000 $year^{-1}$.

B. Consequences

What should be the consequences of today's knowledge of a recommended intake of magnesium? Is there sufficient evidence to recommend an increased intake of magnesium? If so, should the recommendation be applicable to the whole population or only to certain risk groups? Furthermore, what would be the best way to increase magnesium intake?

The consumption of magnesium-rich food and water can be encouraged, and the use of water softeners in areas with hard water discouraged. At least one tap with unfiltered water for drinking should always be left. Although such recommendations are hardly controversial, it is difficult to make them effective, especially if the target is the population as a whole.

Theoretically, a possible way of increasing magnesium intake would be to add magnesium to drinking water, especially in areas with naturally soft water. This is already done in some waterworks in order to reduce corrosiveness. To do so on a larger scale would, however, be expensive and politically difficult.

Recommendations to use oral magnesium supplementation would not be practically feasible for the whole population, but could be considered to be directed toward certain risk groups.

Before a general prevention program can be accepted, large-scale intervention studies must be conducted to accurately evaluate the preventive effect of magnesium. This would require several thousands of subjects, however, and would thus be cumbersome and expensive. Nevertheless, in view of the significant implications for public health, such studies should be conducted. The possibility of a simple and harmless way of reducing AMI mortality rate must not be overlooked.

SEE ALSO THE FOLLOWING CHAPTERS

Chapter 5 (Uptake of Elements from a Biological Point of View) · Chapter 8 (Biological Responses of Elements) · Chapter 22 (Environmental Medicine) · Chapter 23 (Environmental Pathology) · Chapter 31 (Modeling Groundwater Flow and Quality)

FURTHER READING

Altura, B. M. (1994). Introduction: Importance of Magnesium in Physiology and Medicine and the Need for Ion Selective Electrodes, *Scand. J. Clin. Lab. Invest.*, 54, Supplement 217, 5–9.

Altura, B. M., and Altura, B. T. (1996). Role of Magnesium in Pato-physiological Processes and the Clinical Utility of Magnesium Ion Selective Electrodes, *Scand. J. Clin. Lab. Invest.*, 56, Supplement 224, 211–234.

Altura, B. M., Altura, B. T., Carella, A., and Turlapaty, P. D. M. V. (1981). Hypomagnesemia and Vasoconstriction: Possible Relation to Etiology of Sudden Death Ischemic Heart Disease and Hypertensive Vascular Disease, *Artery.*, 9, 212–231.

Anderson, T. W., Neri, L. C., Schreiber, G. B., Talbot, F. D. F., and Zdrojewski, A. (1975). Ischemic Heart Disease, Water Hardness and Myocardial Magnesium, *Can. Med. Assoc. J.*, 113, 199–203.

Cappucio, F., Elliot, P., Allender, P. S., Pryer, J., Follman, P. A., and Cutler, J. A. (1995). Epidemiologic Association Between Dietary Calcium Intake and Blood Pressure: A Meta-Analysis of Published Data, *Am. J. Epidemiol.*, 142, 935–945.

Dawson, E. B., Frey, M. J., Moore, T. D., and McGanity, J. (1978). Relationship of Metal Metabolism to Vascular Disease Mortality Rates in Texas, *Am. J. Clin. Nutr.*, 31, 1188–1197.

Dissanayake, C. B. (1991). The Fluoride Problem in the Groundwater of Sri Lanka – Environmental Management and Health, *Int. J. Environ. Stud.*, 38, 137–156.

Dissanayake, C. B., Senaratne, A., and Weerasooriya, V. R. (1982). Geochemistry of Well Water and Cardiovascular Diseases in Sri Lanka, *Int. J. Environ. Stud.*, 19, 195–203.

Durlach, J. (1989). Recommended Dietary Amounts of Magnesium: Magnesium RDA, *Magnes. Res.*, 2, 195–203.

Durlach, J., Bara, M., and Guiet-Bara, A. (1989). Magnesium Level in Drinking Water: Its Importance in Cardiovascular Risk. In *Magnesium in Health and Disease* (Y. Itokawa and J. Durlach, Eds.), John Libbey & Co., London.

Fine, K. D., Santa Ana, C. A., Porter, J. L., and Fordtran, J. S. (1991). Intestinal Absorption of Magnesium from Food and Supplements, *J. Clin. Invest.*, 88, 396–402.

Fleming, K. H., and Heimbach, J. T. (1994). Consumption of Calcium from the US: Food Sources and Intake levels, *J. Nutr.*, 124, 1426S–1430S.

Freeze, R. A., and Cherry, J. A. (1979). *Groundwater*, Prentice Hall, Englewood Cliffs, New Jersey.

Hardwick, L. L., Jones, M. R., Brautbar, N., and Lee, D. B. N. (1990). Site and Mechanism of Intestinal Magnesium Absorption, *Miner. Electrolyte Metab.*, 16, 174–180.

Haring, B. S. A., and van Delft, W. (1981). Changes in the Mineral Composition of Food as a Result of Cooking in "Hard" and "Soft" waters, *Arch. Environ. Health*, 36, 33–35.

Harlan, W. R., Hull, A. L., Schmouder, R. L., Landis, J. R., Thompson, F. E., and Larkin, F. A. (1984). Blood Pressure and Nutrition in Adults, *Am. J. Epidemiol.*, 120, 17–28.

Heany, R. P., and Recker, R. R. (1985). Estimation of True Calcium Absorption, *Ann. Int. Med.*, 103, 516–521.

Hem, J. D. (1985). Study and Interpretation of the Chemical Characteristics of Natural Water (3rd edition), United States Geological Survey Water Supply Paper 2254.

Hiscock, K., and Paci, A. (2000). Groundwater Resources in the Quaternary Deposits and Lower Paleozoic Bedrock of the Rheidol Catchment, West Wales. In *Groundwater in the Celtic Regions: Studies in Hard Rock and Quaternary Hydrogeology* (N. S. Robins and B. D. R. Misstear, Eds.), Geological Society, London, Special Publications, 182, 141–155.

Johnson, N. E., Smith, I. L., and Freudenhim, J. L. (1985). Effects on Blood Pressure of Calcium Supplementation of Women, *Am. J. Clin. Nutr.*, 42, 12–17.

Jones, K., and Moon, G. (1987). *Health, Disease and Society: A Critical Medical Geography*, Routledge & Kegan Paul, London, 134–140.

Kobayashi, J. (1957). On Geographical Relations Between the Chemical Nature of River Water and Death Rate from Apoplexy, *Berich. Ohara Inst. Landwirtsch. Biol.*, 11, 12–21.

Kohri, K., Ishikawa, Y., Iguchi, M., Kurita, T., Okada, Y., and Yoshida, O. (1993). Relationship Between the Incidence of Infection Stones and the Magnesium-Calcium Ratio of Tap Water, *Urolog. Res.*, 21, 269–272.

Landin, K., Bonevik, H., Rylander, R., and Sandström, B. (1989). Skeletal Muscle Magnesium and Drinking Water Magnesium Level, *Magnes. Bull.*, 11, 177–179.

Lau, K., and Eby, B. (1985). The Role of Calcium in Genetic Hypertension, *Hypertension*, 7, 657–667.

Leary, W. P., Reyes, A. J., Lockett, C. J., Arbuckle, D. D., and van der Byl, K. (1983). Magnesium and Deaths Ascribed to Ischaemic Heart Disease in South Africa, *S. A. Med. J.*, 64, 775–776.

Luoma, H., Aromaa, A., Helminen, S., Murtomaa, H., Kiviluoto, L., Punsar, S., and Knekt, P. (1983). Risk of Myocardial Infarction in Finnish Men in Relation to Fluoride, Magnesium and Calcium Concentration in Drinking Water, *Acta Med. Scand.*, 213, 171–176.

Maheswaran, R., Morris, S., Falconer, S., Grossinho, A., Perry, I., Wakefield, J., and Elliot, P. (1999). Magnesium in Drinking Water Supplies and Mortality from Acute Myocardial Infarction in North West England, *Heart*, 82, 455–460.

Marier, J. R. (1986a). Role of Magnesium in the "Hard-Water Story," *Magnes. Bull.*, 8, 194–198.

Marier, J. R. (1986b). Magnesium Content of the Food Supply in the Modern-Day World, *Magnesium*, 5, 1–8.

Marier, J. R., and Neri, L. L. (1985). Quantifying the Role of Magnesium in the Interrelation Between Human Mortality /Morbidity and Water Hardness, *Magnesium*, 4, 53–59.

McCarron, D. A. (1982). Low Serum Concentrations of Ionized Calcium in Patients with Hypertension, *N. Engl. J. Med.*, 307, 226–228.

McCarron, D. A. (1985). Is Calcium More Important Than Sodium in the Pathogenesis of Essential Hypertension? *Hypertension*, 7, 607–627.

McNally, N. J., Williams, H. C., Phillips, D. R., Smallman-Raynor, M., Lewis, S., Venn, A., and Britton, J. (1998). Atopic Eczema and Domestic Water Hardness, *Lancet*, 352, 527–531.

Michaelsson, K. (1996). *Diet and Osteoporosis*, Uppsala University Medical dissertations.

Moore, T. J. (1989). The Role of Dietary Electrolytes in Hypertension, *J. Am. Coll. Nutr.*, 8, 68S–80S.

Nerbrand, C., Svärdsudd, K., Ek, J., and Tibblin, G. (1992). Cardiovascular Mortality and Morbidity in Seven Counties in Sweden in Relation to Water Hardness and Geological Settings, *Eur. Heart J.*, 13, 721–727.

Neri, L. C., Hewitt, D., Schreiber, G. B., Anderson, T. W., Mandel, J. S., and Zdrojewsky, A. (1975). Health Aspects of Hard and Soft waters, *J. Am. Water Works Assoc.*, 67, 403–409.

Neutra, R. R. (1999). Epidemiology vs. Physiology? Drinking Water Magnesium and Cardiac Mortality (editorial), *Epidemiology*, 10, 4–6.

Piispanen, R. (1991). Correlation of Cancer Incidence with Groundwater Geochemistry in Northern Finland, *Environ. Geochem. Health*, 13, 66–69.

Pocock, S. J., Shaper, A. G., Cook, D. G., Packham, R. F., Lacey, R. F., Powell, P., and Russell, P. F. (1980). British Regional Heart Study: Geographic Variations in Cardiovascular Mortality, and Role of Water Quality, *Br. Med. J.*, 280, 1243–1249.

Punsar, S., and Karvonen, M. J. (1979). Drinking Water Quality and Sudden Death: Observations from West and East Finland, *Cardiology*, 64, 24–34.

Rajasooriyar, L. D. (2003). A Study of the Hydrochemistry of the Uda Walawe Basin, Sri Lanka, and the Factors that Influence Groundwater Quality, PhD thesis, University of East Anglia, Norwich.

Reinhardt, R. A. (1988). Magnesium Metabolism: A Review with Special Reference to the Relationship Between Intracellular Content and Serum Levels, *Arch. Int. Med.*, 148, 2415–2420.

Reinhardt, R. A. (1991). Clinical Correlates of the Molecular and Cellular Actions of Magnesium on the Cardiovascular System, *Am. Heart J.*, 121, 1513–1521.

Robbins, D. J., and Sly, M. R. (1981). Serum Zinc and Demineralized Water, *Am. Journal Clin. Nutr.*, 34, 962–963.

Rubenowitz, E., Axelsson, G., and Rylander, R. (1996). Magnesium in Drinking Water and Death from Acute Myocardial Infarction, *Am. J. Epidemiol.*, 143, 456–462.

Rubenowitz, E., Axelsson, G., and Rylander, R. (1998). Magnesium in Drinking Water and Body Magnesium Status, Measured Using an Oral Loading Test, *Scand. J. Clin. Lab. Invest.*, 58, 423–428.

Rubenowitz, E., Axelsson, G., and Rylander, R. (1999). Magnesium in Drinking Water and Death from Acute Myocardial Infarction Among Women, *Epidemiology*, 10, 31–36.

Rubenowitz, E., Molin, I., Axelsson, G., and Rylander, R. (2000). Magnesium in Drinking Water in Relation to Morbidity and Mortality from Acute Myocardial Infarction, *Epidemiology*, 11, 416–421.

Rylander, R., Bonevik, H., and Rubenowitz, E. (1991). Magnesium and Calcium in Drinking Water and Cardiovascular Mortality, *Scand. J. Work Environ. Health*, 17, 91–94.

Schaafsma, G. (1992). The Scientific Basis of Recommended Dietary Allowances for Calcium, *J. Int. Med.*, 231, 187–194.

Schroeder, H. A. (1960). Relation Between Mortality from Cardiovascular Disease and Treated Water Supplies. Variations in States and 163 Largest Municipalities of the United States, *J. Am. Med. Assoc.*, 172, 1902–1908.

Seelig, M. S. (1986). Nutritional Status and Requirements of Magnesium: With Consideration of Individual Differences and Prevention of Cardiovascular Disease, *Magnes. Bull.*, 8, 170–185.

Shaper, A. G., Packham, R. F., and Pocock, S. J. (1980). The British Regional Heart Study: Cardiovascular Mortality and Water Quality, *J. Environ. Pathol. Toxicol.*, 3, 89–111.

Sowers, J. R., Zemel, M. B., Zemel, P. C., and Standley, P. R. (1991). Calcium Metabolism and Dietary Calcium in Salt Sensitive Hypertension, *Am. J. Hypertension*, 4, 557–563.

Sprinkle, C. L. (1989). Geochemistry of the Florida Aquifer System in Florida and in Parts of Georgia, South Carolina, and Alabama, United States Geological Survey Professional Paper 1403-I.

Teitge, J. E. (1990). Incidence of Myocardial Infarction and the Mineral Content of Drinking Water (in German), *Z. Gesamte Iinn. Med. Ihre Grenzgeb.*, 45, 478–485.

Theophanides, T., Angiboust, J.-F., Polissiou, M., Anastassopoulous, J., and Manfait, M. (1990). Possible Role of Water Structure in Biological Magnesium Systems, *Magnes. Res.*, 3, 5–13.

Waeber, B., and Brunner, H. R. (1994). Calcium Deficiency in the Elderly: A Factor Contributing to the Development of Hypertension, *Eur. J. Endocrinol.*, 130, 433.

WHO (2002). World Health Organization Guidelines for Drinking Water Quality. Hardness. http://www.who.int/water_sanitation_health/GDWQ/Chemicals/hardnfull.htm, July 2002.

Wilkinson, S. R., Stuedemann, J. A., Grunes, D. L., and Devine, O. J. (1987). Relation of Soil and Plant Magnesium to Nutrition of Animals and Man, *Magnesium*, 6, 74–90.

Yang, C. Y., Chiu, H. F., Cheng, M. F., Tsai, S. S., Hung, C. F., and Lin, M. C. (1999). Esophageal Cancer Mortality and Total Hardness Levels in Taiwan's Drinking Water, *Environ. Res.*, 81, 302–308.

CHAPTER 14

BIOAVAILABILITY OF ELEMENTS IN SOIL

BRIAN J. ALLOWAY
The University of Reading

CONTENTS

I. INTRODUCTION

The natural abundance of elements in the soil parent material and the factors controlling their availability to plants and animals provide major links between geology and human medicine and the health of plants and animals in natural and agricultural ecosystems. This is the rock-soil-plant-animal/human pathway, and it is of major importance in the study of medical geology.

The formation of soils (pedogenesis) is closely linked to the weathering of rock-forming minerals and the structural arrangement of organic (mainly humic) and mineral materials to form a soil profile. However, soil formation depends on the involvement of plants, microorganisms, and soil fauna, whereas rock weathering can occur without the biological component and produce a regolith of decomposed rock rather than a soil.

Within the context of soil–plant interactions, the main soil processes controlling the availability of both naturally occurring and contaminant trace and major elements to plants and their leaching down the soil profile to the groundwater are those which influence the sorption and desorption of these elements within the soil. Sorption is the collective term for the retention of metal ions on the surfaces of the solid phase of the soil system. The sorption mechanisms include:

1. Cation or anion exchange in which ions are attracted to oppositely charged sites on the solid surfaces (e.g., negatively charged surfaces in the case of cation exchange and positively charged surfaces in anion exchange)
2. Specific adsorption in which certain metal cations and most anions are held by ligand exchange in the form of covalent bonds
3. Co-precipitation in which ions are precipitated on surfaces simultaneously with other inorganic

compounds, such as iron, aluminum, and manganese oxides, or calcium carbonate

4. Insoluble precipitates of elements on surfaces, including the formation of insoluble carbonates, sulfides, phosphates, and hydroxides

5. Organic complexation with solid-state organic matter ligands (in contrast to the formation of low molecular weight soluble complexes)

Desorption is the term for the release of ions from these sorbed forms due to a change in pH, and redox conditions, or the release of plant root exudates (Alloway, 1995).

The factors controlling adsorption and desorption of ions in soils, include:

1. The properties, speciation, and concentration of specific trace and major elements

2. The composition of the soil, especially the relative abundance of clay minerals of different types, and total contents of iron, aluminum and manganese oxides, free calcium carbonate, and organic matter

3. Soil physicochemical conditions, which include pH, redox status, and concentrations of other cations and anions

Plant factors, especially genotype, also play a major role in determining the extent to which elements are accumulated in plant tissues, including edible parts, and, ultimately, in the food chain.

However, elements differ considerably in the relative extent to which they are taken up from the soil and accumulated in plant tissues even allowing for the differences in soil properties and plant genotype. The soil–plant transfer of different elements varies in orders of magnitude from relatively unavailable metals such as barium, to the more readily accumulated elements such as cadmium. In general, the essential trace elements with relatively low transfer coefficients (Tf = M Plant/M Soil) are likely to be more prone to being deficient, whereas both essential and non-essential elements with high transfer coefficients would be more likely to pose a problem of toxicity.

II. SOIL FORMATION

Pedogenesis is the process by which a thin surface layer of soil develops on weathered rock material, gradually increases in thickness, and undergoes vertical differentiation in morphology to form a soil profile. The soil profile comprises distinct horizontal layers (called horizons), which differ in color, texture, structure, and organic matter content. This soil profile (or solum) is the unit of classification of soils. The basis of soil mapping is to delineate areas of soils with distinct profile characteristics.

Pedogenesis is essentially the processes of chemical weathering of rock fragments in an environment which is generally rich in atmospheric oxygen, moisture, carbon dioxide, humic material, and biochemicals from living and decomposing vegetation (the biosphere). These pedogenic processes are strongly influenced by climatic factors, which include temperature, precipitation, and evaporation. They are also affected by the drainage of the site at which the soil is forming and this depends on the topography (shedding or receiving water) and the permeability of the layers of weathering rock material.

Humus is the name given to the complex organic molecules formed in the soil as a result of microbial action on dead plant material (litter) in the soil. The solid-state humic substances generally have relatively high molecular weights and are distinctly different from biochemical substances found in living and recently dead plant and animal tissues. Low molecular weight, soluble forms of soil organic matter are referred to as dissolved organic compounds and are important in forming soluble complexes containing metals. These soluble organic compounds can desorb metal ions from sorption sites on solid surfaces and thus increase the mobility and plant availability of many elements. Solid-state humus plays several very important roles in the soil. Physically it contributes to the binding of particles to create soil aggregates and a pore system comprising linked voids of varying diameter which are involved in both water transmission and storage and gaseous exchanges with the atmosphere. Chemically, soil humus is a major reserve of carbon, nitrogen, phosphorus, and sulfur which were constituents of the plant tissues that underwent humification. However, humic substances also have relatively strong adsorptive capacities for cations and the presence of humus in soil therefore adds considerably to the sorptive properties of the soil contributed by clay minerals, oxides of iron, and manganese and carbonate minerals.

Soils are formed as a result of the interactions between the geological parent material, climate, vegetation, and topography (especially with regard to drainage status) over time. Jenny (1941) called these the "state factors" of soil formation and expressed them in the form of an equation:

$$\text{Soil} = f(cl, o, r, p, t)$$

where f is a function, cl is climate, o is organisms (vegetation), r is relief (topography), p is parent material, and t is time.

It is important to note that the parent material is the weathered rock at the surface of the Earth and in many cases this may not be the underlying solid geology but the "drift material" that can be fragmented unaltered or chemically weathered rock, which has been transported and deposited on top of the solid geology. This transport may have been by wind (loess), rivers (alluvium), glaciers (glacial sands, gravels, and boulder clay), and downslope movement (colluvium). However, where the solid geology is exposed at the surface the soils will be formed on weathered fragments of this material.

A typical profile of an uncultivated soil comprises a layer of organic litter (desiccated but undecomposed plant material) on the surface (called the L horizon). Below this can occur an A horizon which is a relatively dark colored mixture of humic and mineral material (formed by the actions of earthworms). In cultivated soils where mixing has occurred often over many centuries through plowing and other tillage, the surface layer of dark colored soil is referred to as the plow layer (or Ap horizon). However, under acid conditions in unplowed soils, dark-colored layers of decomposing plant material may exist beneath the litter horizon. These organic layers differ in the extent to which the plant tissues have decomposed and are designated O1 and O2, etc., with O1 only partially decomposed and O2 more decomposed and so on. The O horizons will often overlie an A horizon, but it will not be as thick or deep as in the less acid soils where deep mixing of the soil by large numbers of earthworms or where regular plowing and cultivation has created a thick Ap horizon.

Beneath the A horizon there may be a zone of a lighter color and different texture called an eluvial (E) horizon. This is the zone in the soil profile from which clay minerals and iron have been removed and translocated down the profile. Beneath the E horizon, a B horizon of accumulation can occur. The B horizon can be designated a Bt horizon where clay has accumulated, or a Bs horizon where sesquioxides of iron and aluminum have been deposited and, in podsol profiles, a Bfe and a Bo horizon can also be found. Beneath this B horizon is found the C horizon—the weathering rock material on which the soil is forming. If solid rock is found beneath this C horizon then it is designated the R horizon. Figure 1 shows a diagrammatic soil profile with both the FAO/UNESCO and the USDA Soil Taxonomy horizon nomenclature.

There are two important modifying factors that can affect a soil horizon's appearance and chemical properties, and these are identified in the designation of soil horizons. These are where free calcium carbonate occurs in soil horizons (calcimorphic soils) and where soils are affected by either permanent or intermittent

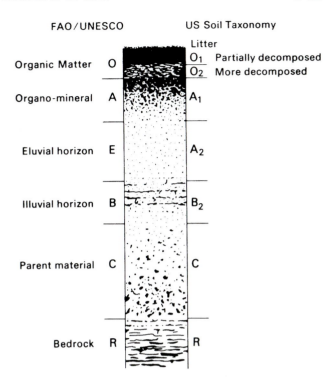

(Not all of these horizons are present in every profile)

FIGURE 1 A diagrammatic soil profile with FAO/UNESCO and USDA Soil Taxonomy Horizon Nomenclature. (From Alloway, 1995.)

waterlogging (hydromorphic soils). In calcimorphic soils (also called calcareous soils, or calcisols), horizons containing visible calcium carbonate are described as calcareous and identified with a lowercase "ca" as in Aca or Bca and so on. This calcium carbonate (calcite) may have originated either from limestone in the parent material or from precipitation of calcite in the pores of soils in semi-arid areas. In hydromorphic soils (usually called gleys) poor drainage results in the onset of reducing conditions which affect both the appearance and the physicochemical properties of the soil. The waterlogging, which gives rise to the creation of reducing conditions in gley soils can be due to impermeability caused by either fine soil texture (where there is a high proportion of clay-size particles), and/or poor structure, or poor site drainage (e.g., in a hollow or at the foot of a slope). Gleying in soils is characterized by pale colors (light brown, gray to bluish green) caused by the reduction of iron oxides. Soils with aerobic, or oxic, conditions normally have darker brown and reddish colors due to the presence of abundant iron oxides.

It should be pointed out that not all soil profiles contain all of the types of distinct horizons mentioned here, but different soil types are characterized by certain combinations of horizons. For example, acidic and strongly leached soils developed on permeable parent material, such as sands or weathered granite, in areas with a high precipitation–evapotranspiration ratio can have a characteristic podsol (or spodosol in the USDA Soil Taxonomy) profile. This podsol comprises a combination of thick organic horizons at the top of the profile—a strongly bleached E horizon, underlain by a Bo horizon of deposited organic matter, over a thin, distinct hard layer of deposited iron (iron pan, Bfe)—which in turn overlies a relatively thick yellowish brown horizon of deposited iron and aluminum sesquioxides (Bs horizon) and this, in turn, overlies the C horizon which is the parent material. This is the distinct profile (L, O1, O2, A, Ea, Bo, Bfe, Bs, C) of a podsol found in many parts of the world with a cool humid climate, such as areas of natural coniferous forest. In contrast, areas underlain by chalk (soft limestone) tend to have a much simpler and shallower profile with a characteristic deep, dark-colored and humus-rich A horizon underlain by a C horizon of chalk fragments. This AC profile belongs to a shallow calcareous soil called a rendzina found on limestones in central and western Europe. In arid and semi-arid areas, soils with a calcium-carbonate-enriched "calcic" horizon (Bca, or Cca) are commonly found. These calcareous soils are called calcisols and a calcic horizon with more than 15% calcium carbonate (and at least 5% more than in the underlying horizon) is their defining characteristic. Calcisols can develop on limestone parent material or on other parent material in low-lying areas where a high water table with calcium-rich groundwater leads to evaporation from the soil surface and precipitation of calcium carbonate within the pore structure of the soil profile.

The chemical weathering of the geological parent material involves the processes of hydrolysis, hydration, dissolution, oxidation, reduction, ion-exchange, and carbonation. These, together with the physical disintegration of rock, bring about the decomposition of rock-forming minerals, which release cations and anions into solution. New secondary minerals (including clay minerals and iron oxides) can be formed from the products of weathering and precipitates of compounds formed when their solubility products are exceeded. Some of the ions released will remain in solution and be leached down the soil profile. These will either reach the groundwater and move to water courses in humid regions or accumulate in low-lying areas and move up through the soil profile as evapotranspiration of water occurs in arid environments. The latter situation results in the formation of calcified and/or saline soils, which are characteristic of arid and semi-arid regions. Soils

with high contents of calcium and a high pH often have low concentrations of plant-available trace elements, such as zinc, and acute deficiencies of this and other essential trace elements can occur in crops growing on them. Low availability of elements to plants can be the result of either low total concentrations in the soil (sometimes referred to as a primary deficiency), to a high degree of sorption leading to low soluble/plant available concentrations, or to antagonistic effects between two or more elements, e.g., copper and zinc, which are both absorbed by the same pathways into plant roots. A relative excess of one element can induce a reduction in the uptake and availability of the other.

The main soil constituents responsible for the adsorption of cations and anions are certain clay minerals, hydrous oxides of iron, manganese and aluminum, calcite, and humus. Therefore, the soil horizons that have relatively high concentrations of these mineral and organic materials will tend to retain ions released from weathering together with ions introduced in fertilizers (especially phosphorus, potassium, and sulfur). Many environmental pollutants (especially trace metals and persistent organic pollutants) will also be retained in these horizons. Nitrogen added to the soil in fertilizers and manures is not retained to any significant extent. It is usually lost from the soil by leaching (mainly as nitrate ions or lost to the atmosphere in gaseous oxide forms [NO_x]). The organic-rich surface horizon, especially in cultivated soils (Ap horizon, topsoil) is usually the horizon with the greatest adsorptive capacity. This horizon is the rooting zone for plants and receives elements cycled through plants and added in fertilizers, manures, and agrichemicals and is also a sink for atmospherically deposited pollutants.

Conversely, horizons of elution where materials, especially clays, have been removed by being washed down the profile, will tend to have lower adsorptive capacities and contents of both nutrient and non-essential elements. In general, sandy textured, well-drained soils tend to have lower adsorptive capacities than soils with a higher proportion of silt and clay-sized mineral particles or iron oxides. However, the humus content can greatly modify this. Intensively cultivated soils generally have lower organic matter contents than soils under permanent grassland or rough grazing in the same climatic zone. Therefore, the texture, organic matter content, and mineral composition of intensively cultivated soils tends to be more important in the dynamics of ions of all types than in grassland soils. In addition to the macronutrients required by plants (nitrogen, phosphorus, potassium, magnesium, calcium, sulfur, and sodium), the trace element content of the parent material will also differ between soils developed on different types of weathered rock. This will include both essential trace elements (boron, cobalt, chlorine chromium, copper, iron, iodine, manganese, molybdenum, nickel, selenium, and zinc) and elements with no known essential function (including arsenic, cadmium, mercury, lead, thallium, and uranium). Therefore, differences in mineralogy and geochemistry of soil parent materials will result in differences in the total concentrations of both major and trace elements found in topsoils and whole soil profiles. Examples of this include soils developed on ultramafic (e.g., serpentinite) rocks where there is a relative excess of magnesium compared to calcium and therefore possible problems with a deficiency of calcium and an excess of magnesium and anomalously high concentrations of cobalt, chromium, and nickel. This combination of high magnesium to calcium ratio (Mg:Ca) and elevated concentrations of certain metals has resulted in the evolution of specialized serpentine flora and if non-adapted plants are grown on these soils they could be affected by nutritional imbalances and possibly toxicity.

Of wider importance with regard to areas of land affected are the soils developed directly on clay or shale strata with distinct sedimentary layers (facies) of marine black shales within them and also soils developed on surface drift deposits containing these materials. The marine black shales generally contain anomalously high concentrations of silver, arsenic, cadmium, copper, molybdenum, lead, selenium, uranium, vanadium, and zinc. There is a distinct possibility of these soils giving rise to elevated concentrations of some or all of these elements in food crops and livestock herbage. Apart from the risk of excess available concentrations of some of these elements, there is also the possibility of antagonistic effects occurring due to interactions between some of the elements present in high concentrations. For example, high molybdenum concentrations can induce copper deficiency in ruminant livestock even though copper itself may be present in elevated total concentrations in the soil and herbage.

In general, soils developed on clay and mudstone formations or drift derived from them tend to have higher concentrations of most essential trace elements than those developed on sandstones and sandy drift deposits. This is due to the clays and mudstones having inherently higher concentrations of these elements and also to the soils having greater adsorptive capacities to retain them against leaching compared with more sandy soils. These geochemical differences, together with variations in soil chemical properties associated with the variations in parent materials and soil forming-factors,

will also help to determine the concentrations of elements (both essential and non-essential) that are available to plants (bioavailable fraction). In addition to these geochemical associations, inputs of fertilizers, agrichemicals, and environmental pollutants will also add to the total concentrations, but the nature of the elements concerned and the physicochemical properties of the soil will determine the bioavailability of these substances.

The global distribution of calcisols (ISSS, 1998) is shown in Figure 2. These soils are important with regard to the mobility and plant availability of trace elements because their alkaline pH and the presence of free calcium carbonate result in the strong adsorption of many cations. This has serious implications in the case of essential trace elements, which include zinc, iron, copper, and manganese because inadequate concentrations of available forms of these elements can cause deficiencies in both crops and livestock. When crops, such as wheat and rice are affected by deficiencies of essential trace elements, such as zinc, there is both the problem of reduced crop yields and also low concentrations of the elements in the human diet. Therefore, the health of regular consumers could also be affected. Zinc deficiency is a major problem in wheat and other staple crops in arid countries such as Turkey, Syria, Iraq, India, and Pakistan which all have large areas of calcisols.

Soils with a high percentage (>65%) of sand grains are called "arenosols," but in contrast to calcisols, they do not necessarily have one particularly distinctive horizon nor do they always have an alkaline soil pH. They are classified by their distinctive sandy texture, which has certain soil physical and chemical properties associated with it. These sandy soils also give rise to deficiencies of essential trace elements, which include zinc, iron, copper, boron, manganese, selenium, and cobalt. In this case the cause of the deficiency is the inherently low total concentrations in the sand grains of the soil parent material and the relatively low sorptive capacity of these soils. Their low clay content results in the loss by leaching of nutrient elements reaching the soil by various pathways. Arenosols are commonly found in deserts, but in areas where they are cultivated (usually where the climate is humid or where irrigation is available) deficiencies of essential trace elements are commonly found. The global distribution of arenosols (ISSS, 1998) is shown in Figure 3.

Soils rich in iron oxides are characteristic of tropical regions of the world. These oxides are the result of the accumulation of iron and aluminum after all the weatherable minerals have decomposed under the aggressive tropical weathering environment with its

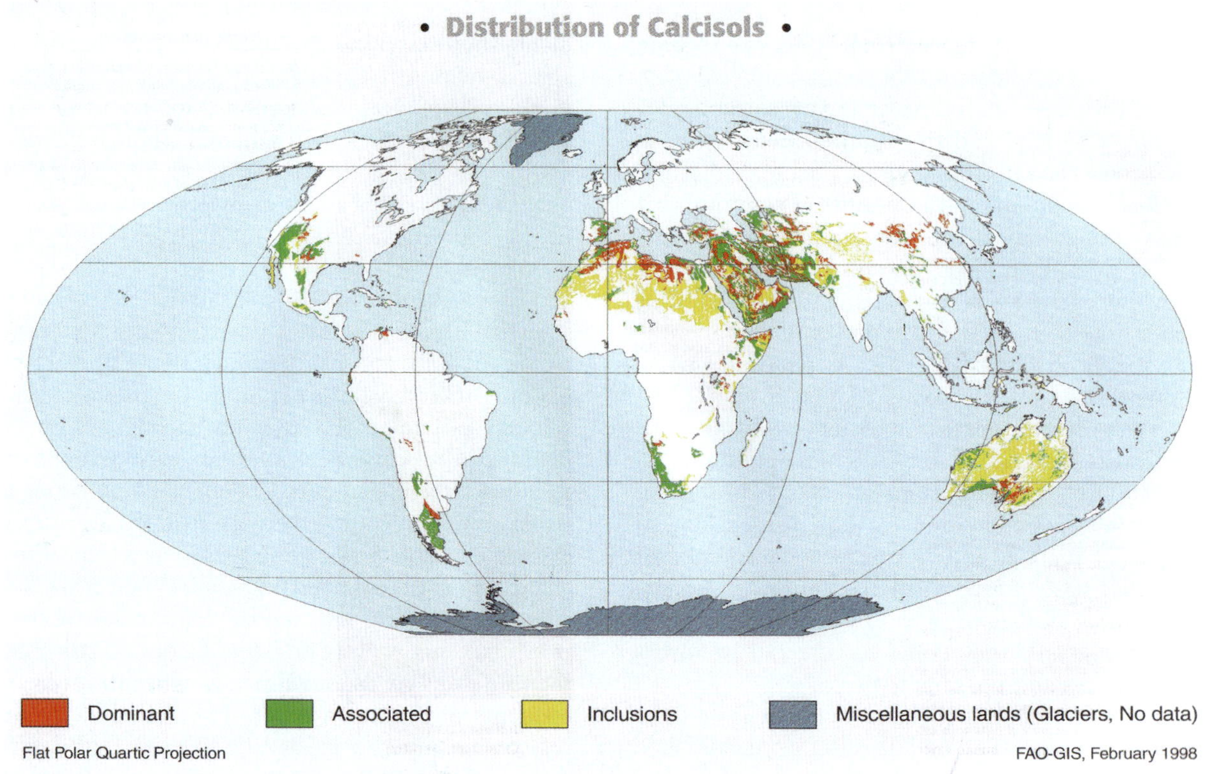

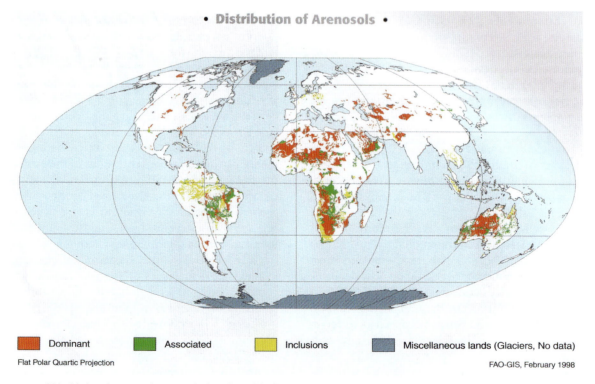

FIGURE 2 World distribution of calcisols (calcareous soils). (Reproduced from the *World Reference Base for Soil Resources Atlas*, ISSS Working Group RB 1998, Bridges et al., Eds., by permission of the editors and publisher.)

FIGURE 3 World distribution of arenosols (sandy soils). (Reproduced from the *World Reference Base for Soil Resources Atlas*, ISSS Working Group RB 1998, Bridges et al., Eds., by permission of the editors and publisher.)

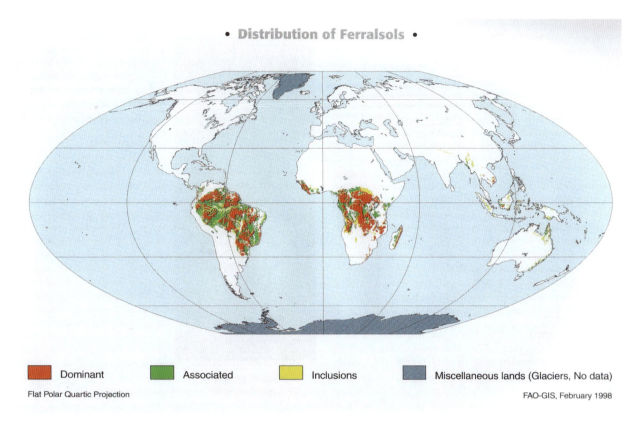

FIGURE 4 World distribution of ferralsols (iron oxide-rich soils). (Reproduced from the *World Reference Base for Soil Resources Atlas*, ISSS Working Group RB 1998, Bridges et al., Eds., by permission of the editors and publisher.)

high temperatures, high precipitation, and low pH, which favor the chemical weathering of geological minerals. These soils are referred to as oxisols in the USDA Soil Taxonomy. They are also sometimes called ferralitic soils, and they are classified as ferralsols in the *World Reference Base for Soil Resources* (ISSS, 1998) and in the FAO/UNESCO World Soil Map. These ferralsols rich in iron oxide are also a potential trace element deficiency problem as a result of both the complete weathering of the original parent material and subsequent leaching and/or the sorption of ions on the iron oxides. The global distribution of ferralsols (ISSS, 1998) is shown in Figure 4.

III. SOIL–PLANT TRANSFER OF TRACE ELEMENTS

The soil–plant transfer coefficient (Tf), also called the bioaccumulation ratio (BR), is a convenient method of expressing the relative ease with which elements in soils (total concentrations) are taken up by plants and accumulated in their above-ground tissues. The coefficient is obtained by dividing the concentration of an element in the plant tissues by the total concentration of the same element in the topsoil (Tf = M in plant/M in soil). Plants differ considerably in their ability to absorb trace elements, but Kabata-Pendias and Pendias (1992) have quoted work that considered the relative ranges of BRs or transfer factors for a large number of plant species from data published in the literature. These range from the least accumulated elements such as barium and titanium to those most readily accumulated (cadmium and boron). However, considerable differences in trace element accumulation occur between plant species and between varieties within a species (inter- and intra-specific variations):

1. Elements lacking accumulation (only slightly available) (Tf < 0.001)

   barium > titanium > scandium > zirconium > bismuth > gallium > iron > selenium

2. Elements showing slight accumulation:
   (Tf 0.001–0.01)

   antimony > beryllium > chromium > iodine
   > vanadium > fluorine > lithium > nickel > manganese

3. Elements showing medium accumulation:
   (Tf 0.01–1.0)

       cobalt > arsenic > germanium > tellurium
       > silver > strontium > lead > copper
       > mercury > molybdenum > zinc

4. Elements which are intensively accumulated
   (Tf 1–10)

   rubidium > cesium > bromine > boron > cadmium

In general, the elements that tend to have relatively low soil–plant transfer coefficients are those that tend to be more strongly sorbed onto the surfaces of soil solids. Those that are more readily accumulated are less strongly adsorbed due to the properties of the ionic forms of these elements. The ions that are strongly sorbed will tend to have lower free ion concentrations in the soil solution and, therefore, be less available for plant uptake and translocation to the aerial tissues of plants.

IV. Soil Chemical Properties and the Bioavailability of Trace and Major Elements

The soil is a dynamic system comprising solid, liquid, and gaseous phases, which is subject to short-term fluctuations such as variations in moisture status, temperature, pH, and redox conditions. In addition to the non-living mineral and organic constituents, the topsoil contains a highly diverse microbial and mesofauna population and a high density of roots of many different plant species. These living organisms are all intricately involved in the physical, chemical, and biological processes taking place in the soil. In addition to short-term changes, soils also undergo gradual alterations in response to changes in management and environmental factors. Examples of these longer term changes in soil properties include a decrease in the content of organic matter with increasing cultivation and/or increasing temperatures and gradual acidification due to acid pre-

cipitation or lack of regular liming in areas of high precipitation relative to evaporation. These short- and long-term changes in soil properties can each have significant effects on the form and bioavailability of trace and major elements (of both indigenous geochemical and external pollution origin). Soils are also inherently heterogeneous at the micro (profile) and macro (field/catchment) scales. For the purposes of soil testing to assess the available concentration of both essential and potentially toxic elements, the spatial variability in soil physical and chemical properties necessitates appropriate and thorough sampling. It is essential to take sufficient samples which include the range of variability in parameters at any site investigated.

A. Key Soil Chemical Properties

1. Soil pH

The soil reaction (pH) is the most important factor controlling the chemical behavior of ions and many other important processes in soils. The pH of a soil applies to the hydrogen ion (H^+) concentration in the solution present in the soil pores. This is in dynamic equilibrium with the predominantly negatively charged surfaces of the soil particles with which it is in contact. Positively charged hydrogen ions are strongly attracted to surface negative charges and they have the power to replace most other cations. There is a diffuse layer close to the negatively charged surfaces, which has a higher concentration of H^+ than the bulk solution. When the soil solution is diluted (as when a suspension of soil in water is made to measure pH or in the field after heavy rain), the diffuse layer expands, causing the pH of the bulk solution to increase. This has important implications for soil testing to measure soil pH. It is the normal practice to mix soil with 2.5 times its weight of distilled water. This generally gives a pH value which is 1–1.5 units higher than that of the soil solution in the diffuse layer near the solid surfaces where the reactions take place. However, this dilution effect can be overcome by measuring the pH in a dilute suspension of a neutral salt such as calcium chloride or potassium chloride, which helps to recreate the ionic strength of the real soil solution. When reporting soil pH values, the method used for measuring it is usually given with the results, but if the solvent used is not named, it is normally assumed that the pH was measured in distilled water.

Soil pH is affected by variations in redox potential that periodically occur when the drainage status of soils changes from waterlogged to more freely drained and vice versa. Reducing conditions (gleying) generally cause an increase in pH and oxidation decreases pH. Fluctuations of up to two pH units can occur over a year in soils prone to periodic waterlogging. Oxidation of iron pyrites (FeS_2) in a soil parent material, such as a weathering black shale or drained marshland, can cause a marked decrease in pH due to the formation of sulfate ions and sulfuric acid.

There are several mechanisms in soils that have the effect of buffering pH changes. These include the formation of hydroxy-aluminum ions ($Al(OH)^{2+}$), partial pressure of CO_2, and formation and dissolution of carbonates and cation exchange reactions. However, even with these buffering mechanisms soil pH can vary by as much as 1 unit diurnally and spatially due to localized conditions in different parts of a field. In humid regions, soil pH usually increases with depth due to the leaching of bases down the soil profile. In arid environments, pH can decrease with increasing depth due to the accumulation of salts of calcium, sodium, and magnesium in the surface horizons through evaporation of the soil solution.

Soils generally have pH values (measured in water) from 4 to 8.5 due to the buffering by aluminum at the lower end and by calcium carbonate at the upper end of the range. Brady (1984) states that the normal pH range of soils in humid regions is 5–7 and 7–9 in the soils of arid regions. The maximum range of pH conditions found in uncontaminated soils is 2–10.5, but some contaminated soils containing large amounts of cement or mortar may have even higher pH values. In a typical temperate environment, such as the UK, the pH range generally found in topsoils is 4–8. The optimum pH for most arable crops is 6.5 and for grassland mineral soils it is 6.0, but for peaty soils, the optimum pH for grassland is 5.5. Soil pH can be relatively easily raised by liming (with calcium carbonate or calcium hydroxide), but it is normally impractical to acidify agricultural soils more alkaline than these values.

In general, most divalent cationic forms of elements are more mobile and more available to plants; therefore, they are less strongly adsorbed on soil solids under acid conditions than they are at neutral and slightly alkaline pH values. However, the pH of the soil also has a major control on the solubility of soil organic matter, which in turn can modify the behavior of trace elements in the soil. As the soil pH increases toward neutrality and beyond, there is an increase in the amount of DOC. This can have a modifying influence on the solubility of several elements which readily form chelate complexes with soluble organic matter. For example, copper and lead are strongly complexed by soil organic matter so

they may become more available with increasing soil pH in soils with a relatively high organic matter content due to the increased concentration of organic ligands in the DOC.

The overall importance of the soil pH in controlling the availability/mobility of elements is due to its effects on (1) the solubility of soil organic matter; (2) the speciation and solubility of elements in the soil solution; (3) the polarity of the charge on solids in soils, such as Fe oxides which have variable charges; (4) the cation exchange capacity of soil solids; (5) the dissolution of certain precipitates and minerals (e.g., calcium carbonate); and (6) effects on soil microorganisms and fauna.

B. Soil Organic Matter

Soil can be distinguished from regolith (decomposed rock) by the presence of living organisms, organic debris (mainly from plants), and humus. All soils (by definition) contain organic matter although the amount and type may vary considerably. Colloidal organic matter has a major influence on soil physical and chemical properties and can be divided into "non-humic" and "humic" substances. The non-humic substances comprise unaltered biochemicals such as amino acids, carbohydrates, organic acids, lignin, lipids, and waxes that have not changed their form since they were synthesized by living organisms. In contrast, humic substances are a series of acidic substances, yellow to black colored, polyelectrolytes of moderately high molecular weight. They are formed by secondary synthesis reactions involving microorganisms and have characteristics that are dissimilar to any compounds in living organisms. They have several functional groups, which include carboxyl, phenolic hydroxyl, carbonyl, ester, and possibly quinone and methoxy groups (Hayes & Swift, 1978; Stevenson, 1979). Although mainly composed of humic substances, soil humus also contains some biochemicals bound to humic polymers. The elemental composition of humus (on an ash-free basis) is 44–53% carbon, 3.6–5.4% hydrogen, 1.8–3.6% nitrogen, and 40–47% oxygen. In general, soil organic matter (humus plus non-humic material) contains about 58–60% organic carbon.

The organic matter contents of soils can vary widely from <1% in intensively cultivated arable soils or soils in semi-arid areas, to more than 10% in permanent grassland soils in cool humid conditions. In poorly drained (gleyed) sites, the soil may be formed on peat and comprise more than 70% organic matter. In general, in hotter drier climates organic matter contents are much lower than those found in cooler humid regions. Within the soil profile, organic matter is always found concentrated in the surface horizon. In cultivated soils this will have been mixed within a deeper layer due to plowing (Ap horizon) and in podsols (usually found in cool humid conditions) and in vertisols (found in hot humid conditions) some organic matter will have been translocated down the profile.

Humus is frequently described as comprising three fractions forming a continuum of compounds varying in molecular weight, carbon, oxygen and nitrogen contents, acidity, and cation exchange capacity in the order: humin > humic acid > fulvic acid. The humic acid fraction has a molecular weight in the range of 20,000–100,000 Da and fulvic acid generally consists of lower molecular weight compounds (some of which may be decomposition products of humic acid) with up to 10% polysaccharides. Humins are humic acid type compounds strongly bound onto minerals.

The contribution of organic matter (humic and non-humic) to the chemical properties of a soil are

1. In the adsorption of cations to negatively charged sites (created by deprotonation of carboxyl and phenol groups)
2. In the mobility and protection of some metal ions from adsorption through the formation of soluble complexes (e.g., chelates) with low molecular weight humic substances (DOC)
3. In the retention of many elements in the higher molecular weight, solid forms of humus by chelation

Soil organic matter is the main reservoir of carbon, nitrogen, phosphorus, and sulfur in soils. These can be gradually released as leachable ionic or gaseous forms (not phosphorus) through the action of soil microorganisms. The carbon locked up in the organic matter of the world's soils is a major consideration in model predictions of global climate change.

Several trace metals show particularly high affinities for soil organic matter. These include cobalt, copper, mercury, nickel, and lead and these are probably held principally in chelated form within humus (Adriano, 2001). Other trace metals, such as cadmium appear to be mainly sorbed in the soil by cation exchange and specific adsorption and are not retained as strongly on solid-state soil organic matter.

Table I gives the summarized results of large-scale surveys of soils in England and Wales, a large area of northern Europe surrounding the Baltic Sea, and the United States for soil pH (measured in water) and per-

TABLE I. Summarized Values for pH and Organic Carbon in Soils in England and Wales, Northern Europe (Baltic Area), and the United States

Country	Number of samples	Minimum	Median	Maximum
England and Wales[a]				
pH (water)	5679	3.1	6.0	9.2
Organic C (%)	5666	0.1	3.6	65.9
Baltic area[b]				
pH (water)	774	3.7	6.1	8.7
Organic C (%)	774	1.8	7.1	97.2
United States[c]				
pH (water)	3045	3.9	6.1	8.9
Organic C (%)	3045	0.09	1.05	63.0

[a]Soils collected on 5 × 5 km grid. (From McGrath & Loveland, 1992.)
[b]Soils collected on a 2500-km grid over an area of 1,800,000 km² from 10 European countries surrounding the Baltic Sea. (From Reimann et al., 2000.)
[c]Soils from sites with healthy crops remote from obvious contamination. (Holmgren et al., 1993).

centage organic carbon. The surveys differed in that the samples from England and Wales were collected on a formal 5 × 5 km grid, the Baltic area on a grid of 1 sample per 2500 km², and the American samples from sites selected as free from obvious contamination. The pH data are remarkably similar but the organic matter data show a much lower median value for the American samples, which is probably a reflection of the hotter climate in many agricultural areas of the United States. In contrast, northern Europe has a cooler, more humid climate which is reflected in more peaty soils and pastures with relatively high organic matter contents.

C. Chemically Active Mineral Constituents

The inorganic constituents of soils usually comprise more than 90% of the mass of soils, and it is the adsorption and desorption of ions on the surfaces of these materials that exerts an important effect on the plant availability and mobility of macro elements and trace elements. The inorganic fraction can comprise a wide range of rock fragments and minerals undergoing weathering; newly synthesized and recycled clay minerals; oxides of iron, aluminum, and manganese; free carbonates of calcium and magnesium; and, in more arid regions, crystals of salts such as calcium sulfate and sodium chloride.

1. Clay Minerals

Clay minerals are either products of rock weathering or are synthesized as new minerals from the products of weathering. They have marked effects on both the physical and chemical properties of soils. Their contribution to soil chemical properties results from their comparatively large surface area and permanent negative charge on their surfaces which adsorb cations. The clay fraction of a soil is defined as the mass of the dispersed inorganic constituents which are less than $2\,\mu m$ in diameter. Although this is based on particle size rather than mineralogy, in most cases it is the mineralogically distinct group of clay minerals that comprise most of the material in this size fraction (together with iron oxides in many cases). Due to space limitations in this book, it is not possible to cover the crystallography of these phyllosilicate minerals; however, they all share two main types of building blocks in their structure. These are a continuous sheet of silicon (Si) oxygen (O) tetrahedra (the silica unit) and another of aluminum (Al) hydroxide (OH) octahedra (the gibbsite unit). In many cases some of the silicon and aluminum ions in the crystal lattice of the minerals may be replaced by other ions.

The most common types of clay mineral include: (1) the kaolinites with one silica sheet and one gibbsite sheet (a 1 : 1 clay); (2) the illites which contain two silica sheets with one gibbsite sheet between them (a 2 : 1 clay); (3) the smectites which also have two silica sheets and one gibbsite sheet (2 : 1 clay); and (4) the vermiculites which have two silica sheets, one gibbsite sheet, and one brucite sheet containing magnesium, which is not found in the other types of clay minerals (2 : 2 clay). In all the clay minerals except kaolinite, isomorphous substitution within the mineral lattice leads to a permanent charge imbalance which gives rise to a net negative charge on the surface of the mineral. For example, this can be caused by Al^{3+} substituting for Si^{4+}, and Mg^{2+} or Fe^{2+} substituting for Al^{3+}.

In kaolinites the 1 : 1 units are tightly bound together by hydrogen bonds between hydrogen and oxygen atoms of adjacent layers. These kaolinites have a smaller surface area than the other clay minerals ($5-40\,m^2\,g^{-1}$) and their cation exchange capacity is relatively low ($3-20\,cmols_c\,kg^{-1}$) because little isomorphous substitution has occurred.

Illites have their 2 : 1 units bonded by potassium ions and their specific surface and cation exchange capacity are larger than those of kaolinites ($100-200\,m^2\,g^{-1}$ and $10-40\,cmols_c\,kg^{-1}$, respectively). Smectites have the largest specific surfaces ($700-800\,m^2\,g^{-1}$) due to relatively weak interlayer bonding, which allows them to expand when they are wetted, and consequently they have high cation exchange capacities ($80-120\,cmols_c\,kg^{-1}$). They shrink on drying and can give rise to cracks during prolonged dry periods in soils in which they predominate. Vermiculites have an intermediate surface area and a high cation exchange capacity ($100-150\,cmols_c\,kg^{-1}$).

2. Oxides of Iron, Manganese, and Aluminum

Oxides of iron, manganese, and aluminum in soils are often referred to as the hydrous oxides, or in the case of iron and aluminum, as sesquioxides. They play important roles in the chemical properties of soils. In temperate regions they generally occur in the clay size fraction ($<2\,\mu m$) mixed with the clay minerals and have a disordered structure. The main iron oxides include gelatinous ferrihydrite ($Fe_2O_3.9H_2O$), goethite ($FeOOH$), hematite ($\alpha\text{-}Fe_2O_3$), and lepidocrocite ($\gamma\text{-}FeOOH$). Precipitation of Fe^{3+} is initially in the form of ferrihydrite, which gradually dehydrates with aging to form more stable minerals. However, ferrihydrite is more likely to be subsequently dissolved again than the other iron oxide minerals when a decrease in redox potential (Eh) or pH occurs. These oxides generally occur as mixtures and are often referred to generically as "iron oxides" or "hydrous iron oxides," but goethite is the iron oxide mineral most frequently found in soils. In tropical regions where there is a more aggressive chemical weathering regime and the soils are usually much older (10^4-10^6 years), the oxides of iron and aluminum are often the predominant soil minerals. This is because all the primary (rock-forming) minerals and most of the clay minerals will have been chemically decomposed (weathered). The characteristic brown color of most soils throughout the world is due to the presence of iron oxides. Nevertheless, the colors of soils vary widely depending on the amounts of iron in the parent material, the presence of other strongly colored minerals, the drainage (redox) status, and the organic matter and calcium carbonate contents.

Gibbsite is the common form of aluminum oxide found in soils, but it is much less abundant than iron oxides. Manganese oxide minerals are generally present in smaller quantities than iron oxides but have stronger adsorptive properties for several cations.

Freshly deposited soil oxides are the most active in adsorbing and co-precipitating trace and major elements. As a result of its large surface area, freshly deposited ferrihydrite acts as a scavenger, sorbing both cations and anions, especially phosphate and arsenate ions (HPO_4^{2-}, H_2PO_4, and AsO_4^{3-}). If oxidizing conditions persist for a long time and/or temperatures remain high, the oxide crystals age and become dehydrated and less strongly charged. Therefore retention of metal ions by oxide surfaces is inversely related to the degree of crystallinity of the oxide minerals (Okazaki et al., 1986).

The adsorptive properties of iron and manganese oxides depend on the soil pH, which determines whether they are positively or negatively charged. Generally speaking, their charge is negative under neutral-alkaline conditions and positive under acid conditions. The pH at which the charge is neutral (called the point of zero charge, PZC) varies for the pure forms of different oxides. The PZC for iron oxides lies in the region pH 7–10, for gibbsite it is in the pH range 8–9.4, and for manganese oxides the pH range is 1.5–4.6. However, when mixed with clay minerals in soils, the PZC values tend to be much lower than these. In acid soils, the positively charged iron oxides are the main adsorptive medium for soil phosphate and arsenic anions. Soil organic matter also has a pH-dependent charge, but its PZC is around pH 2 which is not normally encountered in soils, so this material is nearly always negatively charged. It is, therefore, an important contributor to the cation exchange capacity of a soil.

The presence of chemically active forms of metal oxides in soils is very dependent on the drainage status of the soil at any site. Where waterlogging occurs either as a result of an impermeable soil mineral fraction, such as a high proportion of swelling clays, or topographic position, as in a depression or receiving site, then reducing conditions will predominate. In soils with developing reducing conditions (called gleying), oxides will be reduced and the iron and manganese ions will be mobilized together with the ions of other elements that had been adsorbed on their surfaces. Permanently waterlogged (gleyed) soils have low concentrations of iron and manganese because the oxides of these elements will have been dissolved, transported away, and deposited in an oxygenated environment over a long period of time. Fluctuating redox conditions (such as in paddy rice soils) can result in the periodic mobilization and precipitation of oxides. Certain specialized bacteria, including *Thiobacillus ferrooxidans* and *Metallogenum* spp. are also involved in the precipitation of iron and manganese oxides, respectively.

3. Free Carbonates

The predominant carbonate mineral found in soils is calcite ($CaCO_3$) and more rarely dolomite (Ca, $Mg(CO_3)_2$). These minerals can be present as a result of soils forming on weathering limestone rocks and, in addition, in arid regions from the accumulation of calcite in the pores of soils due to the evaporation of water vapor from calcium-rich ground waters. This latter process is referred to as calcification.

The presence of free carbonate in soils has such a dominant effect on the soil's morphology and chemical properties that they are specially classified as members of the "calcimorphic" group of soils. These include the calcisols (Figure 2) and other soils with relatively high calcium carbonate contents, such as rendzinas, which do not have a horizon of distinct calcium carbonate enrichment within their profile. The carbonate minerals control the soil pH and tend to keep it between 7 and 8.5, and they bind a wide range of cations and certain anions to their surface. Major elements such as phosphorus are strongly bound forming apatite minerals (calcium phosphates) and many trace elements, which include cadmium and zinc are sorbed and rendered less available and less mobile. The agricultural practice of liming soils, where either calcium carbonate or calcium hydroxide is added to soils, has the effect of converting the soil to a near neutral or calcareous soil with a significant proportion of free carbonate surface area. Liming is frequently carried out to remediate soils contaminated with trace metals which include cadmium, copper, lead, and zinc. In naturally calcareous soils, the plant-available concentrations of essential trace elements, including boron, iron, copper, manganese, and zinc, are often inadequate for many agricultural crops. These crops are found to suffer from deficiencies of essential trace elements which can cause major reductions in yield and sometimes crop failure unless appropriate prophylaxis is carried out. This usually involves applying salts of the elements, such as zinc sulfate, copper sulfate, manganese sulfate, and sodium borate to the soil, or other forms, including chelates, as foliar applications to the crops themselves.

D. Redox Conditions

The balance of reducing and oxidizing conditions (redox status) in the soil is important due to its effects on the speciation of several very important elements including: carbon, nitrogen, sulfur, iron, manganese, chromium, copper, arsenic, silver, mercury, and lead. All of these elements can exist in soils in more than one oxidation state. The main factor determining the redox status is the degree of waterlogging which prevents the movement of oxygen through the soil profile in the larger diameter, air-filled pores. Redox equilibria are controlled by the aqueous free electron activity which can be expressed as either a pE value (negative log of the electron activity) or as a redox potential, Eh, (the millivolt difference in potential between a Pt electrode and a standard H electrode). Large positive values of pE or Eh (+300 to +800 mV) indicate the presence of oxidized species and low or negative Eh values (+118 to −414 mV) are associated with reducing conditions. Eh can be converted to pE by the factor: Eh (mV) = 59.2 pE (Lindsay, 1979).

Redox conditions control the precipitation and dissolution of iron and manganese oxides and, therefore, have a direct influence on the soil's adsorptive capacity for anions and cations. In gleyed (periodically or permanently waterlogged) soils, those elements which are normally sorbed to iron oxides tend to be more bioavailable than in freely drained (oxic) soils. These include cobalt, iron, nickel, vanadium, copper, and manganese. However, boron, cobalt, molybdenum, and zinc do not undergo changes in valency themselves when changing redox conditions but are sorbed or desorbed and coprecipitated or dissolved according to the effects of redox on the iron and manganese oxides.

Oxic soils tend to have red and brown colors, whereas gleyed soils with strong reducing conditions tend to have pale colors, such as grayish brown and gray to blue-green colors determined by the presence of ferrous ions (Fe^{2+}). However, strong colored parent materials may mask the color changes to a certain extent. Under severely reducing conditions, several elements may be precipitated as sulfides which renders them insoluble and therefore unavailable to plants. The sulfide ion (S^{2-}) comes from the reduction of sulfate (SO_4^{2-}) or from the degradation of sulfur-containing compounds in soil organic matter. In general, the soil pH in acid soils increases slightly under reducing conditions, but in alkaline soils there is a slight pH decrease. If there are cyclic reducing-oxidizing conditions, such as occur in a rice paddy soil with periodic flooding and drying, sulfides, such as iron pyrites (FeS_2) and cadmium sulfide (CdS) (if the soil has been contaminated by this element) will become oxidized and the metal ions released. However, the formation of sulfate ions (SO_4^{2-}) from the oxidation of sulfide when a waterlogged soil dries out results in the soils becoming strongly acidified

and elements such as cadmium rendered highly mobile and bioavailable to plants. This situation was responsible for the heavily contaminated paddy soils of the Jinzu Valley in Japan, which caused the outbreak of itai-itai disease in women who had subsisted on locally grown rice (see Section VII. C).

E. Adsorption and Desorption of Ions in Soils

The surfaces of the organic and mineral colloidal solids in soils are able to retain ions by several different mechanisms. Frequently, this retention is referred to by the general term "sorption" because in addition to true adsorption by attraction forces, some ions may reside on surfaces as a result of the formation of insoluble precipitates and by chelation.

1. Cation and Anion Exchange

Cation exchange is the term applied to the electrostatic attraction of positively charged cations to negatively charged surfaces. This occurs on several clay minerals, organic matter, and Fe oxides at higher pH values. It is the formation of outer sphere complexes with the surface functional groups to which they are bound electrostatically. The "exchange" part of this mechanism is due to the exchange between counter ions in the soil solution near the charged surface. The ability of an adsorbent (mixtures of colloidal clays, oxides, and organic matter) to attract and retain cations is referred to as the cation exchange capacity (CEC) and its units of measurement are centimoles of charge per kg ($cmols_c kg^{-1}$). The CEC of mineral soils can range from around 3 to 60 $cmols_c kg^{-1}$, but in organic soils this can rise to 200 $cmols_c kg^{-1}$. The importance of pH and PZC in determining the charge of oxides was discussed above. In general, oxides contribute little to the CEC of soils below pH 7 although they can be involved in specific adsorption reactions in acid soils.

Anion exchange occurs where negatively charged anions, such as Cl^-, SO_4^-, and NO_3^- are attracted to positively charged sites on soil solids. These are usually variable charge sites at pH values below the PZC. Soil in which positively charged surfaces predominate, such as ferralsol, at low pH will have the ability to retain various anions against leaching. Maximum anion exchange capacities (AEC) for sesquioxides (iron and aluminum oxides) are 30–50 $cmol_c kg^{-1}$ (White, 1997). The exchange of Cl^- and NO_3^- is straightforward, but the exchange of other anions including sulfates, phosphates, and molybdates is more complicated due to specific reactions between these anions and the adsorbent. The CEC of soils is generally greater than their AEC due to the greater number of negative charges on the colloid surfaces.

Table II gives typical CEC values for colloidal constituents of soils and shows that soil organic matter has a much higher CEC than all other soil constituents except vermiculite clays aluminum.

Cation exchange has the following characteristics: it is reversible, diffusion controlled, stoichiometric, and there is some degree of selectivity for one ion over another by the adsorbing surface. This selectivity gives rise to an order of replacement determined by the concentration of ions, their valency, their degree of hydration, and hydrated radius. The higher the valency of an ion, the greater its replacing power. However, the only exception to this is H^+ which behaves like a polyvalent ion. The greater the degree of hydration, the lower the

replacing power of an ion, other things equal. The commonly quoted order of replaceability on the cation exchange complex (comprising colloidal organic matter, clay, and oxides) is:

$$Lithium(Li^+) = sodium(Na^+) > potassium(K^+)$$
$$= ammonium(NH_4^+) > rubidium(Rb^+)$$
$$> cesium(Cs^+) > magnesium(Mg^{2+})$$
$$> calcium(Ca^{2+}) > strontium(Sr^{2+})$$
$$= barium(Ba^{2+}) > lanthanum(La^{3+}) = hydrogen(H^+)$$
$$= aluminum(Al^{3+}) > thorium(Th^{4+})$$

Examples of replacing power of different trace element ions on specific soil constituents are shown in Table III.

From Table IV it can be seen that metal ions such as cadmium and zinc, which were shown in Section III to have relatively high soil–plant Tfs, tend to have low replacing powers and are therefore not strongly retained on soil surfaces. These ions are therefore more readily available for uptake by plants and are also more easily leached down the soil profile than ions with higher replacing powers such as lead and copper.

2. Specific Adsorption

This mechanism involves the exchange of cations of several elements and most anions with surface ligands on solids to form partly covalent bonds with lattice ions. This mechanism is highly pH-dependent and is related to the hydrolysis of the sorbed ions. The pK (equilibrium constant) values of the reaction $M^{2+} + H_2O = MOH^+ + H^+$ determine the adsorption behavior of different ions. Specific adsorption increases with decreasing pK value. However, where the pK values are

TABLE II. Typical Cation Exchange Capacity Values for Soil Constituents

Soil constituent	CEC ($cmols_c\,kg^{-1}$)
Soil organic matter	150–300
Kaolinite (clay)	2–5
Illite (clay)	15–40
Montmorillonite (clay)	80–100
Vermiculite (clay)	150
Hydrous oxides of iron, manganese and aluminum	4

From Ross, 1989.

TABLE III. Typical Orders of Replacement of Trace Element Cations on Various Soil Constituents

Soil constituent	Selectivity order	Ref.
Montmorillonite	Ca > Pb > Cu > Mg > Cd > Zn	Bittel & Miller (1974)
Illite	Pb > Cu > Zn > Ca > Cd > Mg	Bittel & Miller (1974)
Kaolinite	Pb > Ca > Cu > Mg > Zn > Cd	Bittel & Miller (1974)
Smectite, vermiculite and kaolinite	Zn > Mn > Cd > Hg	Stuanes (1976)
(ferrihydrite)	Pb > Cu > Zn > Ni > Co > Sr > Mg	Kinniburgh et al. (1976)
Fe oxides- (hematite)	Pb > Cu > Zn > Co > Ni	Mackenzie (1980)
(goethite)	Cu > Pb > Zn > Co > Cd	Forbes et al. (1976)
Peat	Pb > Cu > Cd = Zn > Ca	Bunzl et al. (1976)
Fulvic acid	Fe^{3+} > Cu > Zn > Mn > Ca > Mg	Murray & Lindler (1983)
Humic substances	Cu > Pb > Zn = Ni > Co > Cd > Mn > Ca > Mg	Tipping & Hurley (1992)

TABLE IV. Trace Metals Normally Found Co-Precipitated with Secondary Minerals in Soils

Mineral	Co-precipitated trace metals
Iron oxides	V, Mn, Ni, Cu, Zn, Mo
Manganese oxides	Fe, Co, Ni, Zn, Pb
Calcite	V, Mn, Fe, Co, Cd
Clay minerals	V, Ni, Co, Cr, Zn, Cu, Pb, Ti, Mn, Fe

From Sposito, 1983.

the same, the ion with the greater radius will be the more strongly adsorbed. Brummer (1986) gave the order for increasing specific adsorption as:

$$Cd(pK = 10.1) < Ni(pK = 9.9) < Co(pK = 9.7)$$
$$< Zn(pK = 9.0) < Cu(pK = 7.7)$$
$$< Pb(pK = 7.7) < Hg(pK = 3.4)$$

Those ions retained by specific adsorption are held much more strongly than they would be by cation exchange and the CEC of the soil constituent may not reflect the extent of sorption by this different mechanism. For example, it has been shown that the sorptive capacities of iron and aluminum oxides were between 7 and 26 times greater than their CECs at pH 7.6 (Brummer, 1986).

In addition to sorption on colloid surfaces, some ions can diffuse into minerals, such as iron and manganese oxides, illite and smectite clays, and calcite. The rate of diffusion into the minerals increases with pH up to a maximum, which is equal to the pK value for when M^{2+} + H_2O = MOH^+ + H^+ on the mineral surface. Above this pH, the $MOH^+ > M^{2+}$ and diffusion rate decreases. This can be related to the ionic radius of the ions involved. For example, the maximum relative diffusion rates for cadmium, nickel, and zinc decrease in the order: Ni > Zn > Cd where the ionic radii are Ni 0.69 nm, Zn 0.74 nm, and Cd 0.97 nm (Brummer, 1986).

3. Co-Precipitation

Co-precipitation is defined as the simultaneous precipitation of a chemical in conjunction with other elements by any mechanism and at any rate (Sposito, 1983). The types of mixed solids formed include clay minerals, iron and manganese oxides, and calcite in which isomorphous substitution has occurred. In addition to co-precipitation, replacement of Ca^{2+} in $CaCO_3$ by other elements can occur. For example, cadmium can diffuse

into $CaCO_3$ and form cadmium carbonate ($CdCO_3$) (Papadopoulos & Rowell, 1988). Typical co-precipitated elements in different minerals are shown in Table IV.

4. Insoluble Precipitates of Elements in Soils

When the concentrations of cations and anions in the soil solution exceed the solubility products of compounds they can form, then insoluble precipitates of these may be formed that will have an important effect in controlling the concentrations of ions in solution in addition to electrostatic adsorption reactions. Examples of some compounds that can be formed and occur in the solid state in soils include:

1. Phosphates of calcium ($Ca_{10}(OH)_2(PO_4)_6$, cadmium ($Cd(PO_4)_3$), and lead ($Pb_5(PO_4)_3Cl$)
2. Carbonates and bicarbonates of calcium ($CaCO_3$, $Ca(HCO_3)_2$), magnesium ($MgCO_3$), sodium (Na_2CO_3), cadmium ($CdCO_3$), and zinc ($ZnCO_3$)
3. Sulfides of iron (FeS_2), cadmium (CdS), and mercury (HgS, Hg_2S)
4. Chlorides of sodium ($NaCl$) and mercury ($HgCl_2$)
5. Iron (ferrite) forms of copper ($Cu_2Fe_2O_4$), molybdenum ($Fe_2(MoO_4)_3$), and zinc ($ZnFe_2O_4$)

5. Organic Complexation

The solid-phase humic material is involved in the retention of trace elements by forming complexes, such as chelates, in addition to comprising part of the colloidal cation exchange complex. Humic substances with reactive groups, which include hydroxyl, phenoxyl, and carboxyl groups, form coordination complexes with metal and other ions. Carboxyl groups are particularly important in binding by the humic and fulvic acid fractions of humus. The stability constants of chelates with elements tend to be in the following order:

copper > iron = aluminum > manganese = cobalt > zinc

Soluble, low molecular weight organic compounds (DOC) of both humic and non-humic origin can form soluble complexes with many trace elements and thus prevent them from being sorbed onto solid surfaces. This has the effect of making these elements more readily leached down the soil profile and/or more available for plant uptake.

6. Quantitative Description of the Sorption of Ions in Soils

A great deal of research has been conducted on the adsorption of elements by soils, especially trace metals that can pose a potential risk of toxicity in plants, animals, and humans. It has been found that two different adsorption equations are very useful in describing most of the adsorption measured. These are the Freundlich and Langmuir equations.

The Freundlich equation is expressed as:

$$x/m = KC^{1/n}$$

where x/m is the amount of solute adsorbed per unit mass, C is the concentration of solute in solution at equilibrium, and K and n are equation constants. This equation provides an effective means of summarizing adsorption that follows a hyperbolic relationship with the greatest amount occurring at lower concentrations and gradually decreasing with higher concentrations.

The Langmuir equation is expressed as:

$$\frac{C}{x/m} = \frac{1}{Kb} + \frac{C}{b}$$

where C is the concentration of the ion in the equilibrium solution, x/m is the amount of C adsorbed per unit mass, K is a constant related to the bonding energy, and b is the maximum amount of ions that will be adsorbed by a given sorbent.

These and other applicable adsorption isotherm equations can then be used in models of trace element availability to plants.

V. CONCENTRATIONS OF SELECTED TRACE AND MAJOR ELEMENTS IN ROCKS, SOILS, AND CROP PLANTS

The concentrations of macro and trace elements in soils vary widely as a result of differences in the mineralogy of the soil parent material and, in the case of trace elements, the amount of contamination from external sources can also vary widely. The ranges of concentrations of both macro and trace elements given in Table V include soils from around the world developed on highly diverse parent materials that have been subject to varying degrees of contamination. Almost all soils in the technologically advanced regions of the world and also in many parts of developing countries are subject to a certain amount of contamination, often from atmospheric deposition.

The data for elements in soils presented in Table V are for "normal" agricultural soils not considered to be markedly contaminated. Total concentrations of trace elements, such as lead, copper, and zinc can reach very high concentrations in some overtly contaminated soils

such as those near metalliferous mines and smelters, other industrial sources of atmospheric emissions, and on land that has received heavy applications of sewage sludge. The concentrations of metals including cadmium, copper, lead, and zinc in sewage sludges have decreased markedly in most industrialized countries as a result of strict pollution controls and structural changes in industry. Nevertheless, soils that received heavy, repeated applications of sewage sludge in earlier periods when metal contents in sludges were much higher are still likely to retain relatively high total concentrations of several metals for many years (see also Chapter 3, this volume).

Heavily contaminated soils can sometimes contain hundreds of $mg\,kg^{-1}$ of cadmium, when the safe maximum limits are considered to be around $1-3\,mg\,kg^{-1}$; hundreds or thousands of $mg\,kg^{-1}$ of lead and zinc when the safe maximum values for human health is considered to be in the range $125-450\,mg\,kg^{-1}$ for lead, and the safe maximum for plants for zinc is around $200-300\,mg\,kg^{-1}$. However, in most cases the sites with very high concentrations of metals arising from contamination are generally of relatively small extent in comparison with the total area of agricultural land. Perhaps the most insidious contamination problem is with inputs of cadmium in phosphatic fertilizers which are used in most parts of the world, at least where high-yielding crops are grown. Some of the phosphate rock used for making these fertilizers can contain relatively high concentrations of cadmium ($<100\,mg\,kg^{-1}$).

VI. THE BIOAVAILABILITY OF TRACE AND MAJOR ELEMENTS TO PLANTS

Uptake of ions by roots can involve several processes including cation exchange by roots, transport inside cells by chelating agents and other carriers, and rhizosphere effects.

Uptake of ions by roots is controlled by the release of ions and organic compounds which include amino acids (e.g., aspartic, glutamic, and prolinic acids). These exudates vary with plant species, microorganism association, and plant growth conditions (e.g., supply of essential trace elements).

A. Uptake of Trace Elements by Plants

The uptake of trace elements by plants is a key stage in the soil-plant-animal/human pathway and is second

TABLE V. Concentrations of Selected Trace and Major Elements in the Lithosphere, Agricultural Soils, and Food Crops ($mg\,kg^{-1}$ Dry Matter)

Element	Content in lithosphere	Common range for agricultural soils	Selected average for soils	Typical range in food crops
Silver (Ag)	0.10	0.03–0.9	0.05	0.03–2.9
Arsenic (As)	5	<1–95	5.8	0.009–1.5
Barium (Ba)	430	19–2368	500	1–198
Boron (B)	10	1–467	9.5–85	1.3–16
Calcium (Ca)	36000	7000–500,000	13700	1000–50,000
Cadmium (Cd)	0.2	0.01–2.5	0.06–1.1	0.13–0.28
Cobalt (Co)	40	0.1–70	7.9	8–100
Chromium (Cr)	200	1.4–1300	54	0.013–4.2
Copper (Cu)	70	1–205	13–24	1–10
Fluorine (F)	625	10–1360	329	0.2–28.3
Iron (Fe)	51000	5000–50,000	38,000	25–130
Mercury (Hg)	0.1	0.05–0.3	0.03	0.0026–0.086
Iodine (I)	0.3	0.1–10	2.8	0.005–10.4
Potassium (K)	26000	400–30,000	8300	20,000–50,000
Magnesium (Mg)	21000	20–10000	5000	1500–3500
Manganese (Mn)	900	270–525	437	15–133
Molybdenum (Mo)	2.3	0.013–17	1.8	0.07–1.75
Sodium (Na)	28000	750–7500	6300	—
Nickel (Ni)	100	0.2–450	20	0.3–3.8
Phosphorus (P)	1200	200–5000	600	3000–5000
Lead (Pb)	16	3–189	32	0.05–3.0
Sulfur (S)	600	30–10,000	700	1000–5000
Selenium (Se)	0.09	0.005–3.5	0.33	0.001–18.0
Tin (Sn)	40	1–11	—	0.2–7.9
Titanium (Ti)	6000	1000–9000	3500	0.15–80
Vanadium (V)	150	18–115	58	0.5–280
Zinc (Zn)	80	17–125	64	1.2–73.0

Note: The typical plant concentrations for the macronutrients calcium, potassium, phosphorus, and sulfur are those for optimum growth and not the full range that may be actually found in crops around the world.

Compiled from Lindsay (1979), Kabata-Pendias and Pendias (1992), Adriano (2001), and Marschner (1995).

only to intake via drinking water with regard to the link between geochemistry and human health apart from where there is excessive ingestion of soil.

Plants readily take up ionic or soluble complexed forms of trace elements present in the soil solution. The factors affecting the amounts of elements absorbed through the roots are those controlling: (1) the concentration and speciation of the element in the soil solution, (2) movement of the element from the bulk soil to the root surface, (3) transport of the element from the root surface into the root, and (4) its translocation from the root to the shoot. Absorption of mobile ions present in the soil solution is mainly determined by the total quantity of this ion in the soil. However, in the case of

strongly adsorbed ions, absorption into the root is more dependent on the amount of root produced and its ability to explore a large volume of soil. Mycorrhizae are symbiotic fungi which effectively increase the absorptive area of the root and can assist in the uptake of nutrient ions such as orthophosphates and micronutrients when concentrations of these are low. Roots also possess a significant CEC, mainly due to the presence of carboxyl groups, and this probably forms part of the mechanism transporting ions through the outer part of the root to the plasmalemma where active absorption takes place. Evapotranspiration (flow of water into roots, up through the plant and out of leaf canopy) is an important factor controlling root uptake of elements because plants that have more rapid evapotranspiration will absorb greater volumes of soil solution containing ions and complexes in solution.

Absorption of ions by plant roots can be both passive and metabolically active processes. Passive uptake involves diffusion of ions in the soil solution to the root endodermis. In contrast, metabolically active absorption takes place against a diffusion gradient but requires the expenditure of energy so it is vulnerable to inhibition by toxins. The type of mechanism appears to differ with different elements; for example, lead is passive whereas the uptake of copper, molybdenum, and zinc is considered to be metabolically active or a combination of both active and passive mechanisms. Ions which are absorbed into the root by the same type of mechanism are likely to have an antagonistic relationship through competition with each other. For example, zinc absorption is inhibited by copper and hydrogen, but not by manganese and iron. Copper absorption is inhibited by zinc, ammonium ions, calcium, and potassium (Barber, 1984; Graham, 1981).

The rhizosphere is a narrow zone (1–2 mm thick) between a root and the surrounding soil. It is a zone of intense microbiological and biochemical activity because it receives appreciable amounts of organic substances from the roots that provide a substrate for a diverse microbial population. These organic substances include exudates, mucilage, and sloughed-off cells and their lysates (Marschner, 1995). As a result of the processes taking place in the rhizosphere which include acidification, redox changes, and organic complex formation, some ions adsorbed onto the soil in the vicinity of the root may be desorbed and become available for absorption into the root. Phenolic compounds and amino acids are known to be involved in the mobilization of oxidized forms of iron and manganese (Fe^{3+} and Mn^{4+}) (Marschner, 1995).

Cereal plants experiencing a deficiency of iron and/or zinc appear to have exudates containing substances (usually referred to as phytosiderophores) such as phytosiderophore 2'-deoxymugineic acid, which are effective in mobilizing these and other elements coprecipitated with them, including cadmium and copper, from iron and manganese oxides and other sorption sites in the vicinity of the root (Kabata-Pendias & Pendias, 1992). Tobacco plants have root exudates that increase the absorption of cadmium but decrease that of iron (Mench & Martin, 1991).

Kabata-Pendias and Pendias (1992) summarized the main points in the literature relating to the absorption of trace elements from solutions as:

- One of the most important factors determining the biological availability of trace elements is the extent to which they are bound to soil constituents.
- Plants take up the species of trace elements that are dissolved in the soil solution in either ionic or chelated and complexed forms.
- Absorption usually operates at very low concentrations.
- Adsorption depends largely on the concentrations in solution, especially at low ranges.
- The rate of absorption depends on the occurrence of H^+ and other ions.
- The intensity of absorption varies with plant species and stage of development.
- The processes of absorption are sensitive to some properties of the soil environment, such as: temperature, aeration, and redox potential.
- Absorption by a plant may be selective for a particular ion.
- The accumulation of some ions can take place against a concentration gradient.
- Mycorrhizae play an important role in cycling between media (e.g., soil) and roots.
- Root absorption can be by both passive (non-metabolic) and active (metabolic) processes.

In addition, Marschner (1995) stressed that there can be marked differences in ion uptake by different plant species (and cultivars within species; see Section VI.C).

B. Uptake of Major Elements by Plants

The major elements are present in much higher concentrations in plant tissues than trace elements and are referred to as "macronutrients." Trace elements are usually present at concentrations of <100 mg kg^{-1}, whereas macronutrients are present at levels of 1000–40,000 mg kg^{-1} (<4%) in the dry matter.

Wild and Jones(1988) and Marschner (1995) summarized the concentrations of macronutrient elements in crop plants as follows:

1. Calcium has a normal concentration range of 0.1–2.5% in plant dry matter but has a low mobility in plants and thus is not redistributed. Therefore the relatively high concentrations do not necessarily reflect the plants' metabolic requirements. Calcium plays a key role in the maintenance and integrity of membranes.

2. Magnesium generally occurs in the dry matter at lower concentrations than calcium (0.2–0.56%), but this element is more mobile than calcium. Magnesium is a specific constituent of chlorophyll. On serpentine soils, there may be a disproportionately large amount of available magnesium compared with calcium and this has lead to the development of specialized flora.

3. Sulfur is usually present at concentrations of 0.1–1% (dry matter) and is a constituent of the amino acids cysteine, cystine, and methionine, and therefore of proteins containing these. Biochemically, sulfur is a very important element because it is a constituent of enzymes and other key proteins. Crops take up 15–50 kg S ha^{-1} a^{-1}.

4. Phosphorus is present in plants at around 0.2% in the dry matter of shoots and is a key component in metabolic processes involving phosphorylation (e.g., ADP-ATP). It is present in the soil solution mainly as HPO_4^{2-} and $H_2PO_4^-$ but is taken up mainly as $H_2PO_4^-$. It is relatively strongly bound in most soils, in the organic matter, and sorbed to iron oxides (acid soils) or calcium carbonate with which it reacts to form apatite in calcareous and heavily limed soils.

5. Potassium occurs in similar concentrations as nitrogen in plants (1.4–5.6% in the dry matter) and is the most abundant cellular cation, but it can often be in short supply in crops. Crops can take up <500 kg K ha^{-1} a^{-1}, (400–500 kg K ha^{-1} a^{-1} in a 15 t ha^{-1} DM grass crop, 300 kg K ha^{-1} a^{-1} in a normal crop of potatoes, and <300 kg K ha^{-1} a^{-1} in a 10 t ha^{-1} (grain) crop of cereals).

6. Nitrogen can be present at 1.6–4% in the dry matter and is the fourth most abundant element in plants after carbon, hydrogen, and oxygen. Nitrogen has an essential role as a constituent of proteins, nucleic acids, chlorophyll, and growth hormones. It can be absorbed into the roots as either nitrate or ammonium ions but whatever the source, ammonium is the intermediate for the formation of amino acids, amides, and subsequently proteins.

C. Differences in Trace Element Accumulation Between and Within Plant Species

The amount of a trace element taken up from any particular soil will depend on plant factors as well as soil properties. Most important of the plant factors is the genotype or genetic makeup of the plant. Differences between plant families and species and also between varieties (cultivars) within a species are clearly shown in the case of cadmium. Grant et al. (1999) reviewed the literature on cadmium accumulation in crops and reported the general trend between plant families as:

1. Low accumulators: Leguminosae
2. Moderate accumulators: Graminae, Liliacae, Cucurbitacae, and Umbelliferae
3. High accumulators: Chenopodiacae, Cruciferae, Solanacae, and Compositae

However, marked differences can occur within families of plants and within species. For example, many varieties of durum wheat (*Triticum durum*) accumulate significantly more cadmium than common spring or winter (bread) wheats (*T. aestivum*).

In a review, Welch and Norvel (1999) reported that median cadmium concentrations in the seeds of seven different grain crops show a 100-fold variation, ranging from low values in rice and maize, to non-durum wheat, with durum wheat containing higher concentrations than non-durum wheat, and sunflower and flax containing the highest cadmium concentrations. Intraspecific variations of 40-fold have been reported for cadmium in 20 inbred lines of maize.

With regard to the risk to the health of consumers, it is not just the amount of a potentially toxic element, such as cadmium, which occurs in the whole plant that is important, but the concentrations found in the edible portions. Many plants (e.g., oat, soybean, timothy grass, alfalfa, maize, and tomato) tend to concentrate cadmium in their roots which implies that fruits, seeds, or leaves will be less enriched. Other species tend to have higher cadmium concentrations in their leaves, and these include the green leaf crops: lettuce, carrot, and tobacco. Lettuce and tobacco have a higher risk to consumers because of this characteristic distribution within the plant. It is considered that differences between varieties of some crops can be due to variations in the proportions of cadmium translocated around the plant from the roots (e.g., durum wheat).

The finding that varieties of key food crops can vary significantly in their accumulation of potentially hazardous elements such as cadmium does open up the possibility that plant breeders could use this to select new varieties with minimal concentrations in their edible parts. This is a distinct possibility for potatoes, rice, wheat, maize, lettuce, sunflower, and soybean. On the other hand, essential trace elements, such as zinc and copper, also show marked genotypic variations in plant species so plant breeders could exploit this trait. By selecting cultivars of food crops which contain higher concentrations of essential trace elements, such as zinc, it would be possible to match crops to marginally deficient soils and thus reduce the amount of trace element fertilizers needed to rectify the deficiency problems. The diets of people in many developing countries contain marginal to deficient concentrations of zinc and this can have marked effects on health.

VII. TRACE ELEMENTS IN SOILS AND CROPS AND HUMAN HEALTH

Soils used for growing food crops in domestic gardens and commercial horticulture and agriculture can become contaminated with potentially toxic chemicals. In cases where the contamination is not sufficiently great to cause phytotoxicity and/or possible crop failure, there is a possible risk that livestock or people consuming large quantities of crop products grown on contaminated soil could suffer illness and even death from the chronic intake of contaminants or by the direct ingestion of the soil itself.

A. Concentrations of Selected Trace Elements in Soils in Different Parts of the World

Cadmium and lead are the elements in contaminated soils that are generally considered to constitute the widest possible health risk to humans through the plant uptake-dietary route. Surveys of soils in various countries have been carried out to determine the total (and sometimes bioavailable) concentrations of a range of trace elements, some of which are often found to have been elevated from anthropogenic sources of pollution. The concentrations of cadmium, lead, copper, and zinc in England and Wales, the United States, Florida, China, and Poland are shown in Table VI. These data show that all four elements are generally present at higher concentrations in England and Wales than in the other countries and state. This can be explained, at least in part, by the grid sampling (5×5 km cells) for England and Wales which included sites contaminated from various sources including metalliferous mining and heavy applications of sewage sludge. In the United States, the samples were collected on a more selective basis which avoided obvious sources of contamination

either from sewage sludge or atmospheric deposition from nearby industries.

B. Concentrations of Cadmium and Lead in Food Crop Products

Surveys of the concentrations of potentially toxic elements, such as cadmium and lead, in food crops have also been undertaken in various countries and these help to show the extent to which anthropogenic contamination and/or geochemically enriched soil parent materials have influenced the composition of food crops consumed in the countries or areas considered (see also Chapter 7, this volume).

Consumption of grain, potatoes, and leafy vegetables is considered to account for more than 50% of total cadmium intake by people in most countries (Adriano, 2001). In the UK, the largely wheat-based diet gave an intake of $8\,\mu g\,Cd\,day^{-1}$ which is only 11% of the WHO limit of $70\,\mu g\,Cd\,day^{-1}$. The concentration of Cd in wheat grain decreased from $0.052\,mg\,Cd\,kg^{-1}$ in 1982 to 0.042 in 1992 and 0.038 in 1993 due to reductions in atmospheric deposition, change to lower cadmium content phosphorus fertilizers, and increased yields giving rise to a dilution effect (Adriano, 2001).

The European Union (EU) has legally defined maximum permissible concentrations for cadmium and lead in a range of foodstuffs including wheat (*T. aestivum* L.) and barley (*Hordeum vulgare* L.) grain. For cereals, excluding wheat grain, bran, germ, and rice, the maximum permissible cadmium concentration is $0.1\,mg\,kg^{-1}$ (fresh weight), and the limit for the excluded crops mentioned above is $0.2\,mg\,kg^{-1}$ (fresh weight). For lead the maximum permissible concentration in all cereals is 0.2 $mg\,kg^{-1}$. Assuming an 85% dry matter content, the effective limiting concentration on a dry matter basis is $0.118\,mg\,kg^{-1}$ for cadmium in cereals excluding wheat grain, bran, germ, and rice and $0.235\,mg\,kg^{-1}$ for the excluded categories. The dry matter based limiting concentration for lead is $0.235\,mg\,kg^{-1}$ in all crop products.

A survey of the cadmium and lead concentrations in 250 samples of wheat grain and 233 samples of barley grain from throughout the UK after the 1998 harvest is reported by Adams et al. (2001). In the samples of wheat, they found overall mean concentrations (in the dry matter) of $0.063\,mg\,kg^{-1}$ for cadmium and $0.025\,mg\,kg^{-1}$ for lead. Only one sample of wheat had a cadmium concentration above the maximum permissible cadmium concentration. The 230 barley samples showed an overall mean value of $0.022\,mg\,kg^{-1}$ for cadmium and $0.039\,mg\,kg^{-1}$ for lead. Only one sample

TABLE VI. Total Concentrations of Selected Elements in Topsoils in Various Countries or States (mg kg^{-1})

Country	Element	Mean	Minimum	Maximum	Median
England & Wales (n = 5692)[a]	Cadmium	0.8	<0.2	40.9	0.7
	Lead	74.0	3.0	16338	40
	Copper	23.1	1.2	1508	18.1
	Zinc	97.1	5.0	3648	82.0
United States (n = 3045)[b]	Cadmium	0.265	<0.01	2.0	0.2
	Lead	12.3	7.5	135.0	11.0
	Copper	29.6	<0.6	495.0	18.5
	Zinc	56.5	<3.0	264.0	53.0
Florida (n = 448)[c]	Cadmium	0.07	0.004	2.8	0.004
	Lead	11.2	0.18	290	4.89
	Copper	6.1	0.1	318	1.9
	Zinc	8.35	0.9	169	4.6
China (n = 4095)[d]	Cadmium	—	0.2	0.33	—
	Lead	—	9.95	56.0	—
	Copper	—	7.26	55.1	—
	Zinc	—	28.5	161.0	—
Poland (n = 127)[e]	Cadmium	—	0.1	1.7	—
	Lead	—	7.1	50.1	—
	Copper	—	2.0	18.0	—
	Zinc	—	10.5	154.7	—

[a]McGrath and Loveland (1992).
[b]Holmgren et al. (1993).
[c]Chen et al. (1998).
[d]Wei et al. (1990).
[e]Dudka et al. (1993).

of barley exceeded the maximum permissible level for cadmium and two samples exceeded the maximum level for lead (Adams et al., 2001).

Both the wheat and barley samples showed small but significant differences between cultivars in mean concentrations of cadmium. The wheat samples also showed significant cultivar differences in lead concentrations (Adams et al., 2001).

Lead concentrations in samples of rice (*Oryza sativa*) from 17 areas of the world were reported by Zhang et al. (1996). The grand means for different countries ranged from 0.002 mg kg^{-1} in Australia to 0.039 mg kg^{-1} in Indonesia. Samples from within China showed a 76-fold variation in mean lead contents (0.016 mg kg^{-1}–0.152 mg kg^{-1}) in different parts of the country. A later paper (Zhang et al., 1998) reported lead and cadmium concentrations in 59 samples of several types of cereals and 34 samples of pulses from open markets in northeastern

China. Average lead concentrations were 0.031 mg kg^{-1} for the cereals and 0.026 mg kg^{-1} for the pulses. Mean cadmium concentrations were higher in pulses than cereals (0.056 mg kg^{-1} pulses, 0.009 mg kg^{-1} cereals). Foxtail millet (*Stetaria itaica*) was found to contain the highest amounts of lead (0.054 mg kg^{-1}) and cadmium was highest in soya beans (*Glycine max*; 0.074 mg kg^{-1}). There were some possible links between food crop composition and human health effects. It had earlier been found that the consumption of Foxtail millet was a leading determinant of blood lead concentrations in Shandong Province and there was concern about the elevated levels of cadmium in pulses which are an important source of protein and lipids in Asia.

A much larger survey of cadmium and lead in 4113 samples of rice and other cereal products in Japan was reported by Shimbo et al. (2001). They found a grand geometric mean for cadmium in polished, uncooked rice of 0.05 mg kg^{-1} and 0.019 mg kg^{-1} in wheat flour. Mean lead concentrations were much lower for both rice and wheat flour ranging from 0.002 to 0.003 mg kg^{-1}. Rice was therefore shown to be a more important source of cadmium than wheat flour for Japan as a whole.

These four examples show how varying concentrations of elements, such as cadmium and lead in soil, together with genotypic differences between species and cultivars can give rise to variations in the trace element composition of food crops. The soil concentrations are a result of both the geochemical composition of the parent material and inputs from anthropogenic sources.

Examples of significant human health effects occurring as a result of soil contamination include (1) lead poisoning in children ingesting lead-rich garden soils and house dust, (2) skeletal deformity and death due to excessive cadmium intake by multiparous women living in the Jinzu Valley in the Toyama Province of Japan, and (3) tumors due to arsenic poisoning in people drinking contaminated water and eating vegetables and other crops from soils irrigated with arsenic-rich groundwaters in Bangladesh, India, and Taiwan (see Chapter 11).

C. Examples of Contaminated Soils Affecting Human Health

1. Exposure of Children to Lead in Contaminated Domestic Garden Soils

Domestic gardens can contain relatively high concentrations of lead from weathered and scraped painted interior and exterior surfaces (especially important around timber houses), deposition of lead from vehicle exhaust emissions when gasoline used to contain relatively high levels of lead as an anti-knock agent, and other diffuse sources. Prior to the 1950s, some house paints contained more than 50% of lead in the dry matter. It was estimated in the late 1980s that up to 11.7 million children in the United States could have been at risk due to exposure to excessive amounts of lead in contaminated garden soils and house dusts (Millstone, 1997). Culbard et al. (1988) conducted a survey of soil in 3550 urban gardens in the UK and found the following geometric mean concentrations (in mg kg^{-1}): lead 230 (<14,125), cadmium 1.2 (<17), copper 53 (<16,800), and zinc 260 (<14,568). The geometric mean concentrations of metals in 579 gardens in greater London were significantly higher than for the rest of the country: lead 647 mg kg^{-1}, cadmium 1.3 mg kg^{-1}, copper 73 mg kg^{-1}, and zinc 424 mg kg^{-1}. This reveals

that urban gardens in large cities, such as London, can be relatively heavily contaminated with potentially toxic elements including lead and zinc. However, zinc in high concentrations is normally regarded as more of a potential toxicity hazard to plants than to humans or animals.

Lead has a relatively low phytotoxicity compared with most other trace metals: cadmium > copper > cobalt, nickel > arsenic, chromium > zinc > manganese, iron > lead (Chino, 1981). Therefore, crops grown in lead-contaminated gardens are unlikely to show symptoms of toxicity. However, the greatest risk to children and some adults is not through consumption of vegetables which have accumulated lead by uptake through their roots but by direct ingestion of lead-contaminated soil. This soil may have been ingested accidentally on unwashed vegetables and eating with unwashed hands. However, the greatest risk is to children who intentionally eat soil (*pica*) (see also Chapter 17, this volume).

A model has been proposed by Wixson and Davies (1993) to calculate safe guideline values for lead in soils where there may be a risk of soil ingestion by children. This model is:

$$S = \frac{(T/G^n - B)}{(\delta)} 1000$$

where S is the soil guideline value, a geometric mean concentration of lead in g Pb g^{-1} of soil, T is the blood lead guideline or target concentration in μg Pb dl^{-1} whole blood, G is the geometric standard deviation of the blood lead distribution (typically 1.3–1.5 but may be higher, e.g., in mining areas), B is the background or baseline blood lead concentration in the population from sources other than dust, n is the number of standard deviations corresponding to the degree of protection required, and δ is the slope or response of the blood Pb–soil Pb relationship and has units of μg Pb dl^{-1} blood increase per 1,000 μg Pb g^{-1}.

2. Cadmium Contamination in Japan and Great Britain

The Jinzu Valley in the Toyama Province of Japan has metalliferous mining and smelting industries located near the Jinzu River together with intensive paddy rice cultivation around the urban and industrial areas on the alluvial soils of the valley floor. Over many years leading up to WWII the mining and smelting operations had resulted in the contamination of the paddy soils with cadmium, lead, and zinc. Because of the cycle of reducing and oxidizing conditions associated with the flooding and drying out of the soils in the paddy fields,

the contaminants, especially cadmium, underwent marked changes in speciation. In the flooded soils with strong reducing conditions, cadmium is present as an insoluble sulfide precipitate (CdS). As the growing rice approaches maturity, the paddy fields are drained to facilitate harvesting and it is during this period that the cadmium sulfide becomes oxidized and forms Cd^{2+} and SO_4^{2-}, which result in a massive uptake of cadmium ions into the rice plant and translocation into the grain (Asami, 1984). As a consequence of this, rice grain grown in Jinzu Valley had significantly elevated concentrations of cadmium, and the people consuming this rice were exposed to a high risk of cadmium poisoning. It was the women who had had several children who were the worst affected by any kidney damage and an acute skeletal disorder known locally as itai-itai disease, which translated from Japanese means "ouch, ouch" due to the pain experienced by the sufferers when their bodies were touched. More than 200 elderly women living in the Valley during the 1940s who had borne several children were found to have been disabled by the disease and a further 65 women died from its effects. At that time, the rice-based subsistence diets of the peasant farmers were generally deficient in protein, calcium, and vitamin A, which all exacerbated the cadmium poisoning. In addition to the intake of cadmium in the rice, water taken from the river for drinking and cooking was also significantly contaminated with cadmium.

The average concentration of cadmium in the rice grown on the contaminated paddy soils was $0.7\,\mathrm{mg\,Cd\,kg^{-1}}$, which was more than ten times greater than the average cadmium concentration in local control samples of rice ($0.07\,\mathrm{mg\,Cd\,kg^{-1}}$). The maximum content of cadmium found in rice was $3.4\,\mathrm{mg\,Cd\,kg^{-1}}$.

The mean cadmium intake for the residents of the Jinzu Valley was estimated to be around $600\,\mathrm{mg\,day^{-1}}$, which is around ten times greater than the maximum tolerable intake. The Japanese government has set $1\,\mathrm{mg\,Cd\,kg^{-1}}$ in rice as the maximum allowable limit, and it has been found that 9.5% of paddy soils in Japan have been polluted with cadmium to the extent that the concentrations in rice are greater than this value. Even in areas considered to be unpolluted with cadmium, rice is estimated to be the source of more than 60% of cadmium in the diet. Much of this cadmium is probably derived from phosphatic fertilizers.

An interesting comparison can be made with the village of Shipham in Somerset, England, where a relatively large number of houses had been built in the 1950s and 1960s on land contaminated with cadmium, lead, zinc, and other elements from historic metallifer-

ous mining (Morgan & Sims, 1988). Although the total concentrations of cadmium were in some cases more than 100 times those in the Jinzu Valley soils ($<360\,\mathrm{mg\,Cd\,kg^{-1}}$), the cadmium was less bioavailable due to the mining-contaminated soils in Shipham containing appreciable amounts of free calcium carbonate and, consequently, a higher pH and a significant amount of chemisorption (Alloway et al., 1988). The Shipham soils were also contaminated with zinc and lead ($<37200\,\mathrm{mg\,Zn\,kg^{-1}}$ and $6540\,\mathrm{mg\,Pb\,kg^{-1}}$) and antagonistic reactions with these metals may have also played a role in reducing cadmium uptake. Nevertheless, the mean concentration of cadmium in vegetables grown in gardens in Shipham was $0.25\,\mathrm{mg\,Cd\,kg^{-1}}$ DM, which is nearly 17 times higher than the UK national average of $0.015\,\mathrm{mg\,Cd\,kg^{-1}}$ DM. Some samples of leafy vegetables growing on the most highly contaminated soils contained much higher concentrations but these were not common. Health studies on approximately 500 inhabitants of Shipham village revealed some small but significant differences in biochemical parameters, but there was no evidence of adverse health effects in the participants in the investigations.

VIII. WIDESPREAD DEFICIENCIES OF ESSENTIAL TRACE ELEMENTS

Large areas of the world have soils which are unable to supply staple crops, such as rice, maize, and wheat with sufficient zinc. Smaller, but still significant areas of the world are affected by deficiencies of boron, copper, iron, and manganese. In crops, acute deficiencies of essential trace elements result in visible symptoms of stress, such as chlorosis (yellow coloration due to impaired chlorophyll production) and reduced dry matter growth and yield of edible crop products. However, marginally deficient amounts of certain trace elements, such as copper and zinc, can give rise to hidden deficiencies in which yields can be reduced by up to 30% or more without the appearance of obvious visible symptoms of stress. Plants may be smaller but unless normal (sufficient) crops are available nearby for comparison, marginal deficiencies are often difficult to detect. This can result in crop yields which may be very much less than optimal due to a trace element deficiency that could be easily treated by either application of fertilizers to the soil or foliar sprays to the crop.

The impact of deficiencies of essential trace elements on crop yields and quality is much greater overall than that of toxicity due to pollution. In several countries, large proportions of the arable soils are affected by deficiencies, such as in India where around 45% of soils are deficient in zinc, 33% deficient in boron, 8.3% deficient in iron, 4.5% deficient in manganese, and 3.3% deficient in copper (Singh, 2001). Apart from lost food production in terms of tonnes of carbohydrates and protein, the trace element composition of the crop products is also affected. The zinc status of many people in developing countries, such as India and Pakistan, is considered to be suboptimal and therefore low zinc concentrations in food products will exacerbate this problem. Cereal staples, such as rice, wheat, and maize contain relatively large amounts of phytate which tends to make zinc in the diet less available. Phytate concentrations tend to be higher in crops receiving high levels of phosphorus fertilizer, such as the new, high-yielding varieties.

Deficiencies of essential trace elements in crops can be caused by low total concentrations of elements in the soil parent material (such as sandstones and sandy drift deposits), by low availability due to high soil pH, high concentrations of calcium (in calcareous soils on limestones or in semi-arid or arid areas), high organic matter contents (peaty soils), and waterlogged (gleyed) conditions. Zinc deficiency is the most widespread essential trace element deficiency problem in the world and many large areas of land in hot arid climates have severe zinc deficiency problems in calcareous soils (such as in the Middle East) and in rice paddy soils (as in India, Philippines, and Bangladesh).

As with other plant species, crops such as wheat, rice, and maize have been found to vary widely in their ability to tolerate deficiencies of essential trace elements such as zinc. "Zinc-efficient" cultivars of wheat are those that are able to grow and yield reasonably well on soils in which less tolerant cultivars would be acutely affected by zinc deficiency. The zinc efficiency character will enable plant breeders to produce new hybrids that are better suited to zinc-deficient soils. Although any zinc-deficient soil can be fertilized with a zinc compound (such as zinc sulfate) to raise its plant-available concentration of zinc, the use of zinc-efficient cultivars would enable plants to be matched to the soils, rather than the soils fertilized to match them to the crop's requirements. This is particularly important in areas where zinc fertilizers are difficult for farmers to obtain.

As discussed in Selenium Deficiency and Toxicity in the Environment (in this volume), widespread deficiencies of selenium are also related to soil parent materials containing low concentrations of this and other essential trace elements.

IX. SUMMARY

The health of people and animals can be affected by imbalances in their dietary intake of trace and major elements and therefore the rock-soil-plant-animal/human pathway is of major importance in considerations of medical geology. Soils can differ widely in their total concentrations of both macro and trace elements due to variations in the mineralogy of the geological parent material on which the soil has formed, even without inputs from environmental pollution or agricultural husbandry. Concentrations of essential trace elements can be anomalously low on parent materials such as sandstones and sandy or gravelly glacial drift. These can be due to both low concentrations of these elements in the predominant mineral (quartz) and the relatively low adsorptive capacities of sandy soils (arenosols) due to their very low clay contents. On the other hand, anomalously high concentrations of both essential and potentially toxic elements can occur in soils formed on geological materials which themselves have anomalously high concentrations of certain elements. Examples of this are marine black shales enriched in arsenic, cadmium, copper, chromium, mercury, molybdenum, lead, and zinc and serpentinite rocks enriched in magnesium, cobalt, chromium, and nickel.

Given that marked variations can occur in the total concentrations of elements in the soil, the mobility and bioavailability of these elements is going to be controlled by a range of soil factors of which the pH is probably the most important. The adsorptive capacity of a soil is dependent on its mineralogy, especially the amount and type of clays; the amounts of iron, manganese, and aluminum oxides and free carbonates; and the organic matter content. The redox conditions will have a marked control on the oxide content and the free carbonates will have a major effect on the soil pH.

Even after the bioavailable concentrations have been determined by the soil and parent material factors, the plants growing on the soil have both an effect on the soil and also vary markedly in their ability to accumulate trace and major elements due to differences in their genotypes. Therefore, the potential impact of geology on the dietary intake of trace and major elements by humans and livestock is dependent on a range of soil and plant factors. Fortunately many of the mechanisms in soils are reasonably well understood and in due course it will become possible to model their behavior and make more realistic risk assessments of both potential toxicity and deficiency problems. With regard to the plant genotypic factor, with the continual breeding of

new crop varieties, there is an ongoing need to screen these for their ability to accumulate both essential and non-essential elements as they come into use. This would enable the most suitable crop varieties to be grown. This could apply to varieties of crops that have a high efficiency of use of essential trace elements such as zinc in areas where there is a major problem of deficiencies both limiting yields and the dietary available concentrations of trace elements essential for humans consuming the crops. Conversely, where soils have relatively high concentrations of potentially harmful elements, such as arsenic and cadmium, either as a result of anomalous geochemical composition or environmental pollution, then crops which accumulate relatively low concentrations of these elements would be most suitable. Amelioration of soils by manipulation of soil chemical properties, such as pH, can be carried out, but selecting the most appropriate plant species and cultivars would also help to mitigate the problems.

Apart from the uptake of elements by plants followed by the consumption of crop products by people, there are other ways (dealt with elsewhere in this book) by which the composition and properties of soils can affect human health. These range from the inhalation and ingestion of atmospheric dusts containing soil minerals eroded by the wind from land, the intake of elements in drinking water which has been affected by the leaching of elements through soil profiles, and, lastly, to the intentional eating of soil (geophagia). The soil is obviously a key consideration in medical geology.

SEE ALSO THE FOLLOWING CHAPTERS

Chapter 16 (Soils and Iodine Deficiency) · Chapter 17 (Geophagy and the Involuntary Ingestion of Soil) · Chapter 18 (Natural Aerosolic Mineral Dusts and Human Health) · Chapter 19 (The Ecology of Soil-Borne Human Pathogens)

FURTHER READING

Adams, M. L., Zhao, F. J., McGrath, S. P., Nicholson, F. A., Chalmers, A., Chambers, B. J., and Sinclair, A. H. (2001). Cadmium and Lead in British Wheat and Barley: Survey Results and the Factors Affecting their Concentration in Grain. Project Report No. 265, Home Grown Cereals Authority (HGCA), London.

Adriano, D. C. (2001). *Trace Elements in Terrestrial Environments*, (2nd edition), Springer-Verlag, New York.

Alloway, B. J. (Ed.) (1995). *Heavy Metals in Soils* (2nd edition), Blackie Academic & Professional, London.

Alloway, B. J., Thornton, I., Smart, G. A., Sherlock, J., and Quinn, M. J. (1988). Metal Availability (in The Shipham Report), *Sci. Total Environ.*, 75, 41–69.

Asami, T. (1984). Pollution of Soils by Cadmium. In *Changing Metal Cycles and Human Health* (J. O. Nriagu, Ed.), Springer-Verlag, Berlin, 95–111.

Barber, S. (1984). *Soil Nutrient Bioavailability: A Mechanistic Approach*, John Wiley & Sons, New York.

Bittel, J. E., and Miller, R. J. (1974). Lead, Cadmium and Calcium Selectivity Coefficients on Montmorillonite, Illite and Kaolinite, *J. Environ. Qual.*, 3, 243–254.

Brady, N. C. (1984). *The Nature and Properties of Soils*, (9th edition), Collier Macmillan, New York.

Brummer, G. W. (1986). *The Importance of Chemical Speciation in Environmental Processes*, Springer-Verlag, Berlin, pp. 169–192.

Bunzl, K., Schmidt, W., and Sansomi, B. (1976). The Kinetics of Ion Exchange in Soil Organic Matter. IV Adsorption and Desorption of Pb^{2+}, Cu^{2+}, Cd^{2+}, Zn^{2+} and Ca^{2+} by Peat, *J. Soil Sci.*, 27, 154–166.

Chen, M., Ma, L. Q., and Harris W. (1998). Background Concentrations of Trace Elements in Florida Surface Soils: *Annual Progress Report*, Soil and Water Science Dept, University of Florida, Gainesville.

Chino, M. (1981). *Heavy Metal Pollution of Soils in Japan* (K. Kitagishi and I. Yamane, Eds.), Japan Scientific Society Proceedings, Tokyo.

Culbard, E. B., Thornton, I., Watt, J., Wheatley, M., Moorcroft, S., and Thompson, M. (1988). Metal Contamination in British Suburban Dusts and Soils, *J. Environ. Qual.*, 17, 226–234.

Dudka, S. (1993). Factor Analysis of Total Element Concentration in Surface Soils of Poland. *Sci. Total Environ.*, 163, 161–172.

Forbes, E. A., Posner, A. M., and Quirk, J. P. (1976). The Specific Adsorption of Divalent Cd, Co, Cu, Pb and Zn on Goethite, *J. Soil Sci.*, 27, 154–166.

Graham, R. D. (1981). Absorption of Copper by Plant Roots. In *Copper in Soils and Plants* (J. F. Loneragan, A. D. Robson, and R. D. Graham, Eds.), Academic Press, Sydney, pp. 141–164, chap. 7.

Grant, C. A., Bailey, L. D., McLaughlin, M. J., and Singh, B. R. (1999). Management Factors which Influence Cadmium Concentrations in Crops. In *Cadmium in Soils and Plants* (M. J. McLaughlin and B. R. Singh, Eds.), Kluwer Academic Publishers, Dordrecht, The Netherlands, chap. 7.

Hayes, M. H. B., and Swift, R. S. (1978). The Chemistry of Organic Colloids. In *The Chemistry of Soil Constituents* (D.

J. Greenland and M. H. B. Hayes, Eds.), John Wiley & Sons, Chichester, England, pp. 179–320, chap 3.

Holmgren, C. G. S., Meyer, N. W., Chaney, R. L., and Daniels, R. B. (1993). Cadmium, Lead, Zinc, Copper and Nickel in Agricultural Soils in the United States, *J. Environ. Qual.*, 22, 335–348.

ISSS Working Group RB (1998). *World Reference Base for Soil Resources: Atlas* (E. M. Bridges, N. H. Batjes, and F. O. Nachtergaele, Eds.), ISRIC-FAO-ISSS-Acco, Leuven, Belgium.

Jenny, H. (1941). *The Factors of Soil Formation*, McGraw-Hill, New York.

Kabata-Pendias, A., and Pendias, H. (1992). *Trace Elements in Soils and Plants* (2nd edition), CRC Press, Boca Raton, FL.

Kinniburgh, D. G., Jackson, M. L., and Syers, J. K. (1976). Adsorption of Alkaline Earth Transition and Heavy Metal Cations by Hydrous Gels of Iron and Aluminum, *Soil Sci. Soc. Am. Proc.*, 40, 769–799.

Lindsay, W. L. (1979). *Chemical Equilibria in Soils*, John Wiley & Sons, New York.

McGrath, S. P., and Loveland, P. J. (1992). *Soil Geochemical Atlas of England and Wales*, Blackie Academic & Professional, Glasgow.

McKenzie, R. M. (1980). The Adsorption of Lead and Other Heavy Metals on Oxides of Manganese and Iron, *Aust. J. Soil Res.*, 18, 61–73.

Marschner, H. (1995). *Mineral Nutrition of Higher Plants* (2nd edition), Academic Press, San Diego, CA.

Mench, M., and Martin, E. (1991). Mobilisation of Cadmium and Other Metals from Two Soils by Root Exudates of *Zea Mays* L., and *Nicotiana rustica* L., *Plant Soil*, 132, 187–193.

Millstone, E. (1997). *Lead and Public Health*, Earthscan Publications, London.

Morgan, H., and Sims, D. L. (1988). Discussion and Conclusions (in "The Shipham Report"), *Sci. Total Environ.*, 75, 135–143.

Murray, K., and Lindler, P. W. (1983). Fulvic acids: Structure and Metal Binding. 1. A Random Molecular Model, *J. Soil Sci.*, 34, 511–523.

Okazaki, M., Takamidoh, K., and Yamane, I. (1986). Adsorption of Heavy Metal Cations on Hydrated Oxides of Iron and Aluminum with Different Crystallinities, *Soil Sci. Plant Nutr. (Tokyo)*, 32, 522–533.

Papadopoulos, P., and Rowell. D. L. (1988). The Reactions of Cadmium with Calcium Carbonate Surfaces, *J. Soil Sci.*, 39, 23–35.

Reimann, C., Siewers, U., Tarvainen, T., Bityukova, L., Eriksson, J., Glucis, A., Gregorauskiene, V., Lukashev, V., Matinian, N. N., and Pasieczna, A. (2000). Baltic Soil Survey: Total Concentrations of Major and Selected Trace Elements in Arable Soils from 10 Countries Around the Baltic Sea, *Science Total Environ.*, 257, 155–170.

Ross, S. A. (1989). *Soil Processes: A Systematic Approach*, Routledge, London.

Shimbo, S., Zhang, Z. W., Watanabe, T., Nakatsuka, H., Matsuda-Inoguchi, N., Higashikawa, K., and Ikeda, M. (2001). Cadmium and Lead Contents in Rice and Other Cereal Products in Japan in 1998–2000, *Sci. Total Environ.*, 281, 165–175.

Singh, M. V. (2001). Evaluation of Current Micronutrient Stocks in Different Agro-Ecological Zones of India for Sustainable Crop Production, *Fert. News (Delhi)*, 46(2), 25–42.

Sposito, G. (1983). The Chemical Form of Trace Elements in Soils. In *Applied Environmental Geochemistry* (I. Thornton, (Ed.), Academic Press, London, pp. 123–170.

Stevenson, F. J. (1979). In *Encyclopedia of Soil Science* (R. W. Fairbridge and C. W. Finkl, Eds.), Dowden, Hutchinson & Ross, Stroudsburg, PA.

Stuanes, A. (1976). Adsorption of Mn^{2+}, Zn^{2+}, Cd^{2+} and Hg^{+} from Binary Solutions by Mineral Material, *Acta Agrica Scand.*, 26, 243–250.

Tipping, E., and Hurley, M. A. (1992). A Unifying Model of Cation Binding by Humic Substances, *Geochem. Cosmochim.*, 56, 3627–3641.

Wei, F. S., Chen, J. S., Zheng, C. J., and Jiang, D. Z. (ed) (1990). *Elemental Background Concentrations of Soils in China*. Chain Environmental Scientific Publishing Ltd., Beijing (reported in Chen et al., 1998).

Welch, R. M., and Norvell, W. A. (1999). Mechanisms of Cadmium Uptake, Translocation and Deposition in Plants. In *Cadmium in Soils and Plants* (M. J. McLaughlin and B. R. Sing, Eds.), Kluwer Academic Publishers, Dordrecht, The Netherlands, pp. 125–150, chap. 6.

White, R. E. (1997). *Principles and Practice of Soil Science* (3rd edition), Blackwell Science, Oxford, England.

Wild, A., and Jones, L. H. P. (1988). Mineral Nutrition of Crop Plants, In *Russell's Soil Conditions and Plant Growth* (11th edition) (A. Wild, Ed.), Longman Scientific & Technical, Harlow, England, chap. 4.

Wixson, B. G., and Davies, B. E. (1993). *Lead in Soil: Recommended Guidelines*, Science Reviews, Northwood, England.

Zhang, Z. W., Moon, C. S., Watanabe, T., Shimbo, S., and Ikeda, M., (1996). Lead Content of Rice Collected from Various Areas of the World, *Sci. Total Environ.*, 191, 169–175.

Zhang, Z. W., Watanabe, T., Shimbo, S., Higahikawa, K., and Ikeda, M. (1998). Lead and Cadmium Contents in Cereals and Pulses in North-Eastern China, *Sci. Total Environ.*, 220, 137–145.

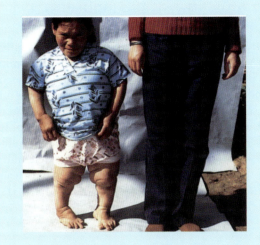

SELENIUM DEFICIENCY AND TOXICITY IN THE ENVIRONMENT

FIONA FORDYCE
British Geological Survey

CONTENTS

I. BACKGROUND

Selenium (Se) is a naturally occurring metalloid element, which is essential to human and other animal health in trace amounts but is harmful in excess. Of all the elements, selenium has one of the narrowest ranges between dietary deficiency ($<40\,\mu g\,day^{-1}$) and toxic levels ($>400\,\mu g\,day^{-1}$), which makes it necessary to carefully control intakes by humans and other animals, hence, the importance of understanding the relationships between environmental exposure and health. Geology exerts a fundamental control on the concentrations of selenium in the soils on which we grow the crops and animals that form the human food chain. The

selenium status of populations, animals, and crops varies markedly around the world as a result of different geological conditions. Because diet is the most important source of selenium in humans, understanding the biogeochemical controls on the distribution and mobility of environmental selenium is key to the assessment of selenium-related health risks. High selenium concentrations are associated with some phosphatic rocks, organic-rich black shales, coals, and sulfide mineralization, whereas most other rock types contain very low concentrations and selenium-deficient environments are far more widespread than seleniferous ones. However, health outcomes are not only dependent on the total selenium content of rocks and soils but also on the amount of selenium taken up into plants and animals—the bioavailable selenium. This chapter demonstrates that even soils containing adequate total amounts of selenium can still produce selenium-deficient crops if the selenium is not in a form ready for plant uptake.

The links between the environmental biogeochemistry of selenium and health outcomes have been documented for many years. Selenium was first identified in 1817 by the Swedish chemist Jons Jakob Berzelius; however, selenium toxicity problems in livestock had been recorded for hundreds of years previously although the cause was unknown. Marco Polo reported a hoof disease in horses during travels in China in the 13th century and similar problems were noted in livestock in Colombia in 1560 and in South Dakota (U. S.) in the mid-19th century where the symptoms were called "alkali disease." In 1931 this disease, which is characterized by hair and hoof loss and poor productivity, was identified as selenium toxicosis (selenosis). Since then, seleniferous areas have been reported in Ireland, Israel, Australia, Russia, Venezuela, China, the United States, and South Africa.

Conversely, selenium was identified as an essential trace element during pioneering work into selenium-responsive diseases in animals in the late 1950s and early 1960s. Selenium forms a vital constituent of the biologically important enzyme glutathione peroxidase (GSH-Px), which acts as an antioxidant preventing oxidative cell degeneration. In animals, selenium deficiency has been linked to muscular weakness and muscular dystrophy, but it also causes reduced appetite, poor growth and reproductive capacity, and embryonic deformities. These disorders are generally described as white muscle disease (WMD). Following these discoveries, selenium deficiencies in crops and livestock have been reported in all regions of the world including the United States, the UK, Finland, Denmark, Sri Lanka, New Zealand, Australia, India, Canada, Thailand, Africa, and China and selenium supplementation has become common practice in agriculture.

Selenium deficiency has also been implicated in the incidence of a heart disorder (Keshan disease) and bone and joint condition (Kashin-Beck disease) in humans in various parts of China. Recent research has shown that selenium deficiency also adversely affects thyroid hormone metabolism, which is detrimental to growth and development. Indeed, approximately 20 essential selenoproteins have now been identified in microbes, animals, and humans, many of which are involved in catalytic functions in the body. Selenium deficiency has also been implicated in a host of conditions including cancer, heart disease, immune system function, and reproduction. This chapter outlines some of the health problems in humans and animals that can arise as a result of selenium deficiency and toxicity in the natural environment. These links are more obvious in regions of the world where the population is dependent on local foodstuffs in the diet, but studies show that even in countries such as the United States where food is derived from a range of exotic sources, the local environment still determines the selenium status of the population. This fact should not be ignored because medical science continues to discover new essential functions for this biologically important element.

II. SELENIUM IN THE ENVIRONMENT

The naturally occurring element selenium belongs to group VIA of the periodic table and has chemical and physical properties that are intermediate between metals and non-metals (Table I). Selenium occurs in nature as six stable isotopes; however, it should be noted that although ^{82}Se is generally regarded as a stable isotope, it is actually a β^- emitter with a very long half-life of 1.4×10^{20} yr. The chemical behavior of selenium resembles that of sulfur and like sulfur, selenium can exist in the 2^-, 0, 4^+, and 6^+ oxidation states (Table II).

TABLE I.　Physical Properties of Selenium

Element name	Selenium
Chemical symbol:	Se
Atomic number:	34
Periodic table group:	VIA
Atomic mass:	78.96
Density:	$4808 \, kg \, m^{-3}$
Melting point:	220°C
Boiling point:	685°C
Vapor pressure:	1 mmHg @ 356°C
Natural isotopes:	Abundance:
^{74}Se	0.87%
^{76}Se	9.02%
^{77}Se	7.58%
^{78}Se	23.52%
^{80}Se	49.82%
^{82}Se	9.19%

From Jacobs (1989) and U.S.-EPA (2002a).

TABLE II.　Chemical Forms of Selenium in the Environment

Oxidative state	Chemical forms
Se^{2-}	Selenide (Se^{2-}, HSe^-, H_2Se_{aq})
Se^0	Elemental selenium (Se^0)
Se^{4+}	Selenite (SeO_3^{2-}, $HSeO_3^-$, H_2SeO_{3aq})
Se^{6+}	Selenate (SeO_4^{2-}, $HSeO_4^{2-}$, H_2SeO_{4aq})
Organic Se	Selenomethionine, Selenocysteine

From Jacobs (1989) and Neal (1995).

As a result of this complex chemistry, selenium is found in all natural materials on Earth including rocks, soils, waters, air, and plant and animal tissues (Table III) (see also Chapter 2, this volume).

At the global scale, selenium is constantly recycled in the environment via the atmospheric, marine, and terrestrial systems. Estimates of selenium flux indicate that anthropogenic activity is a major source of selenium release in the cycle, whereas the marine system constitutes the main natural pathway (Table IV) (Haygarth, 1994). Selenium cycling through the atmosphere is significant because of the rapidity of transport, but the terrestrial system is most important in terms of animal and human health because of the direct links with agricultural activities and the food chain.

Although the element is derived from both natural and man-made sources, an understanding of the links between environmental geochemistry and health is particularly important for selenium as rocks are the primary source of the element in the terrestrial system (Table V) (Fleming, 1980; Neal, 1995). Selenium is dispersed from the rocks through the food chain via complex biogeochemical cycling processes including weathering to form soils, rock-water interactions, and biological activity (Figure 1). As a result, selenium is not distributed evenly across the planet, rather concentrations differ markedly depending on local conditions and an understanding of these variations is essential to aid the amelioration of health problems associated with selenium deficiency and toxicity. The following sections of this chapter provide a brief summary of anthropogenic sources of the element before going on to discuss the important aspects of selenium in the natural biogeochemical cycle and impacts on health.

A. Man-Made Sources of Selenium

Following its discovery in 1817, little industrial application was made of selenium until the early 20th century when it began to be used as a red pigment and improver in glass and ceramic manufacture; however, it was not until the invention of the photocopier in the 1930s that demand for the element significantly increased due to its photoelectric and semi-conductor properties. Today selenium is widely used in a number of industries (Table VI); most commonly selenium dioxide is employed as a catalyst in organic synthesis and as an antioxidant in inks, mineral, vegetable, and lubricating oils. Selenium mono- and disulfide are also used in anti-dandruff and antifungal pharmaceuticals (WHO, 1987; Haygarth, 1994; U.S.-EPA, 2002b).

The world industrial output of selenium was over 2310 tonnes in 1995, and the largest producers are Japan, Canada, and the United States. In 1985, production of selenium in the United States alone reached 195 tonnes, which was used mainly in the electronic/photocopying and glass industries (Table VI). It is not economical to mine mineral deposits specifically for selenium, rather the element is recovered from the electrolytic refining of copper and lead and from the sludge accumulated in sulfuric acid plants (U.S.-EPA, 2002a,b).

Selenium compounds are released to the environment during the combustion of coal and petroleum fuels; during the extraction and processing of copper, lead, zinc, uranium, and phosphate; and during the manufacture of selenium-based products. According to monitoring data in the United States, between 1987 and 1993 over 460 tonnes of selenium were released to the environment, primarily from copper-smelting industries (Table VII) (U.S.-EPA, 2002a,b).

It is estimated that 76,000–88,000 tonnes yr^{-1} of selenium are released globally from anthropogenic activity, compared to natural releases of 4500 tonnes yr^{-1}, which gives a biospheric enrichment factor value of 17. This value is significantly higher than 1 indicating the important influence of man in the cycling of selenium (Nriagu, 1991). For example, long-term monitoring data from the Rothamstead Agricultural Experimental Station in the UK demonstrate the impact of anthropogenic activity on selenium concentrations in herbage. Samples collected between 1861 and 1990 bulked at 5-year intervals reveal that the highest concentrations occurred between 1940 and 1970 which coincided with a period of intensive coal use. Due to the move to fuel sources such as nuclear, oil, and gas in more recent decades, selenium concentrations in herbage are declining (Haygarth, 1994).

Selenium is also released inadvertently into the environment from the agricultural use of phosphate fertilizers, from the application of sewage sludge and manure to land, and from the use of selenium-containing pesticides and fungicides (Table V). For example, in the European Union (EU) it is no longer permissible to dump sewage sludge at sea, consequently the application to land has increased in recent years. To help avoid potential environmental problems, a maximum permissible concentration (MAC) of selenium in sewage sludge in the UK is set at 25 mg kg^{-1}, whereas the MAC in soil after application is 3 mg kg^{-1} in the UK and 10 mg kg^{-1} in France. Clearly, the application of sewage sludge to

TABLE III. Selenium Concentrations in Selected Natural Materials

Material	Total Se (mg kg^{-1})	Water-soluble Se (ng g^{-1})	Material	Total Se (mg kg^{-1})
Earth's crust:	0.05		Water (µg L^{-1}):	
			World freshwater	0.02
Igneous rocks (general):	0.35		Brazil River Amazon	0.21
Ultramafic (general)	0.05		U. S. (general)	<1
Mafic (general)	0.05		U. S. seleniferous	50–300
Granite (general)	0.01–0.05		U. S. Kesterson	<4200
			U. S. River Mississippi	0.14
Volcanic rocks (general):	0.35		U. S. River Colorado	10–400
United States	<0.1		U. S. River Gunnison	10
Hawaii	<2.0		U. S. Lake Michigan	0.8–10
Tuffs (general)	9.15		U. S. seleniferous gw	2–1400
			U. S. drinking water	0.0–0.01
Sedimentary rocks:			Spain freshwater	0.001–0.202
Limestone (general)	0.03–0.08		China Se-deficient sw	0.005–0.44
Sandstone (general)	<0.05		China Se-adequate sw	1.72
Shale (general)	0.05–0.06		China seleniferous sw	0.46–275
W. USA shales	1–675		Finland stream water	0.035–0.153
Wyoming shales	2.3–52		Canada stream water	1–5
South Korea shales	0.1–41		Norway groundwater	0.01–4.82
China Carbon-shale	206–280		Slovakia groundwater	0.5–45
Mudstones (general)	0.1–1500		Bulgaria drinking water	<2
Carbonates (general)	0.08		Sweden drinking water	0.06
Marine carbonates	0.17		Germany drinking water	1.6–5.3
Phosphates (general)	1–300		Ukraine surface water	0.09–3
U. S. Coal	0.46–10.65		Ukraine groundwater	0.07–4
Australia Coal	0.21–2.5		Argentina surface water	2–19
China stone-coal	<6471		Reggio, Italy dw	7–9
Oil (general)	0.01–1.4		Sri Lanka drinking water	0.056–0.235
			Greece drinking water	0.05–0.700
Soil:			Polar ice (general)	0.02
World (general)	0.4		Seawater (general)	0.09
World seleniferous	1–5000			
U. S. (general)	<0.1–4.3		Plants:	
U. S. seleniferous	1–10		U. S. grasses	0.01–0.04
England/Wales (general)	<0.01–4.7	50–390	U. S. clover and alfalfa	0.03–0.88
Ireland seleniferous	1–1200		Norway moss	0.8–1.23
China (general)	0.02–3.81		Canada tree bark	2–16
China Se-deficient	0.004–0.48	0.03–5	Norway grain	0.006–0.042
China Se-adequate	0.73–5.66		Norway forage	0.05–0.042
China seleniferous	1.49–59.4	1–254	Finland hay	0–0.04
Finland (general)	0.005–1.241		Finland grain	0.007
India Se-deficient	0.025–0.71	19–66		
India seleniferous	1–19.5	50–620	Algae:	
Sri Lanka Se-deficient	0.112–5.24	4.9–43.3	Marine (general)	0.04–0.24
Norway (general)	3–6		Freshwater (general)	<2
Greece Se-deficient	0.05–0.10			
Greece Se-adequate	>0.2		Fish:	
New Zealand (general)	0.1–4		Marine (general)	0.3–2
			Freshwater (general)	0.42–0.64
Stream sediments: Wales	0.4–83			
			Animal tissue (general):	0.4–4
Atmospheric dust (general):	0.05–10			
Air (ng m^{-3}) (general)	0.00006–30			

Note: gw = groundwater, sw = surface water, dw = drinking water.

From Fleming (1980); Thornton et al. (1983); Levander (1986); WHO (1987); Jacobs (1989); Nriagu (1989); Tan (1989); Fergusson (1990); Hem (1992); Haygarth (1994); Neal (1995); Rapant et al. (1996); Fordyce et al. (1998); Reimann and Caritat (1998); Vinceti et al. (1998); Oldfield (1999); British Geological Survey (2000); Fordyce et al. (2000a).

land increases the selenium content of the soil; however, relationships between elevated contents and increased uptake into plants have yet to be established (Haygarth, 1994). The application of selenium-bearing fertilizers to land has been used to remediate selenium deficiency in a number of countries and is discussed in Section V of this chapter. Environmental problems related to selenium emissions may also arise in areas surrounding selenium processing or fossil-fuel burning industries. Selenium concentrations in the air within 0.5–10 km of copper-sulfide ore processing plants have been reported to reach 0.15–6.5 $\mu g\,m^{-3}$ (WHO, 1987).

It is clear that man-made sources of selenium have a major impact upon the selenium cycle; despite this, the natural environment is still a very important source and pathway of selenium in animal and human exposure and requires careful consideration in selenium-related health studie (see also Chapter 3, this volume).

B. Selenium in Rocks

The most important natural source of selenium in the environment is the rock that makes up the surface of the planet. Selenium is classed as a trace element as average crustal abundances are generally very low (0.05–0.09 mg kg^{-1}; Taylor & McLennan, 1985). Average concentrations in magmatic rocks such as granites rarely exceed these values (Table III). Relationships with volcanic rocks are more complicated. Volcanoes are a major source of selenium in the environment and it is estimated that over the history of the Earth, volcanic eruptions account for 0.1 g of selenium for every cm^2 of the Earth's surface. Ash and gas associated with volcanic activity can contain significant quantities of selenium and, for example, values of 6–15 mg kg^{-1} have been reported in volcanic soils on Hawaii. Conversely,

TABLE IV. Global Selenium Fluxes

Cycle	Selenium flux (tonnes per year)
Anthropogenic	76,000–88,000
Marine	38,250
Terrestrial	15,380
Atmospheric	15,300

Modified from Haygarth (1994).

TABLE V. Main Sources of Selenium in the Environment

Natural sources	Comments
Volcanic activity	Important source
Weathering of rocks	Important source
Sea spray	Concentrations in ocean water are only an order of magnitude lower than those in rocks
Atmospheric flux	From the ocean surface to the atmosphere
Volatilization and recycling from biota	
Aerial deposition	For example, in the UK annual selenium deposition = 2.2–6.5 g ha^{-1}

Man-made sources	Comments
Selenium-based industries	
Metal processing industries	Important source
Burning of fossil fuels	Important source
Disposal of sewage sludge to land	Typical selenium contents 1–17 mg kg^{-1}
Agricultural use of pesticides	Potassium ammonium sulfide ([K(NH$_4$)S]$_5$Se)
Agricultural use of lime	Typical selenium contents 0.08 mg kg^{-1}
Agricultural use of manure	Typical selenium contents 2.4 mg kg^{-1}
Agricultural use of phosphate fertilizers	Typical selenium contents 0.08–25 mg kg^{-1}

From Fleming (1980); Haygarth (1994); and Neal (1995).

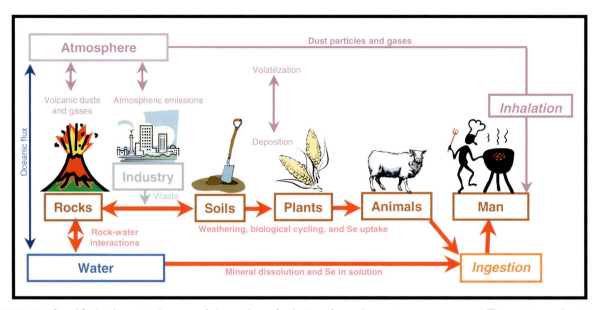

FIGURE 1 Simplified schematic diagram of the cycling of selenium from the environment to man. The main geochemistry and health pathways are shown in red.

TABLE VI. Industrial Uses of Selenium

Industry	Percentage of total production used by U. S. industry in 1985
Electrical components—semi-conductors, cables, and contacts	35%
Photocopier components	
Glass manufacture	30%
Photographic emulsions, printing, and graphics	
Pigments in plastics, paints, enamels, inks, rubber, textiles	25%
Pharmaceutical catalyst	
Additive to petroleum fuel and lubricant products	
Accelerator and vulcanizing agent in rubber manufacture	
Metal alloys	
Fungicides	
Pesticides	10%
Nutritional additive to livestock feed	
Anti-dandruff shampoo	
Medical use, e.g., dietary supplements	

From U.S.-EPA (2000a,b).

because selenium escapes as high-temperature gases during volcanic activity, selenium concentrations left behind in volcanic rocks such as basalts and rhyolites are usually very low (Fleming, 1980; Jacobs, 1989; Nriagu, 1989; Neal, 1995). In general terms, sedimentary rocks contain greater concentrations of selenium than igneous rocks, but even so, levels in most limestones and sandstones rarely exceed 0.1 mg kg^{-1} (Neal, 1995). Because these major rock types account for most of the Earth's surface, a picture should begin to emerge that selenium-deficient environments are far more widespread than selenium-adequate or selenium-toxic ones.

TABLE VII. Industrial Dispersion of Selenium in the U. S. Between 1987 and 1993

Source (tonnes)	Selenium release to land and water 1987–1993	
	Water	Land
Top 5 States:		
Utah	0.7	316
Arizona	0	118
Wisconsin	0	20
Indiana	2.4	0
Texas	0.16	2.2
Main industries:		
Copper smelting, refining	0.7	436
Metal coating	0	2
Petroleum refining	4	0.4
Totals	6.15	460

U.S.-EPA (2002b).

TABLE VIII. Most Common Mineral Forms of Selenium in Natural Rocks

Selenium–mineral	Chemical formula
Crookesite	$(Cu,Tl,Ag)_2Se$
Clausthalite	$PbSe$
Berzelianite	Cu_2Se
Tiemannite	$HgSe$
Elemental selenium	Se

Selenium is also commonly found in the sulfide host minerals

Pyrite	FeS_2
Chalcopyrite	$CuFeS_2$
Pyrrhotite	FeS
Sphalerite	ZnS

Typical mineral associations with selenium:

Polymetallic sulfide ores	Se-Hg-As-Sb-Ag-Cu-Zn-Cd-Pb
Copper-pyrite ores	Cu-Ni-Se-Ag-Co
Sandstone-uranium deposits	U-V-Se-Cu-Mo
Gold-silver selenide deposits	Au-Ag-Se

From Fleming (1980); Neal (1995); and Reimann and Caritat (1998).

Exceptions to the generally low concentrations occur in particular types of sedimentary rocks and deposits. Selenium is often associated with the clay fraction in sediments and is found in greater concentrations in rocks such as shales ($0.06\,mg\,kg^{-1}$) than limestones or sandstones. Very high concentrations ($\leq 300\,mg\,kg^{-1}$) of selenium have also been reported in some phosphatic rocks, probably reflecting similarities between organically derived PO_4^{3-} and SeO_4^{2-} anions (Fleming, 1980; Jacobs, 1989; Nriagu, 1989; Neal, 1995). Selenium concentrations in coal and other organic-rich deposits can be high relative to other rock types and typically range from 1 to $20\,mg\,kg^{-1}$ (although values of over $600\,mg\,kg^{-1}$ have been reported in some black shales) with selenium present as organoselenium compounds, chelated species, or adsorbed element (Jacobs, 1989). Selenium is often found in sulfide mineral deposits and has been used as a pathfinder for gold and other precious metals in mineral exploration (Boyle, 1979). In most situations, selenium substitutes for sulfur in sulfide minerals due to similarities in crystallography, however, elemental Se^0 is occasionally reported (Fleming, 1980; Neal, 1995; Tokunaga et al., 1996). The main mineral forms and common mineral associations of selenium are outlined in Table VIII.

Therefore, the distribution of selenium in the geological environment is highly variable depending on dif-ferent rock types. An illustration of the relationships between geology and selenium distribution is shown in the map of Wales (Figure 2). The highest selenium concentrations in stream sediment are associated with the mineralized areas of Parys Mountain, the Harlech Dome, and Snowdon in north Wales. In South Wales they are seen in the Forest of Dean and Pembrokeshire Coalfields and in the Permian Mercia Mudstone Group of the Welsh borderlands. In contrast concentrations over Devonian age sandstones in mid-Wales are extremely low ($<0.4\,mg\,kg^{-1}$) (see also Chapter 2, this volume).

C. Selenium in Soil

In the majority of circumstances there is a very strong correlation between the concentration of selenium in geological parent materials and the soils derived from them. The selenium content of most soils is very low at $0.01–2\,mg\,kg^{-1}$ (world mean is $0.4\,mg\,kg^{-1}$), but high concentrations of up to $1200\,mg\,kg^{-1}$ have been reported in some seleniferous areas (Table IX) (Fleming, 1980; Jacobs, 1989; Mayland, 1994; Neal, 1995). The relationships between geology, soil selenium

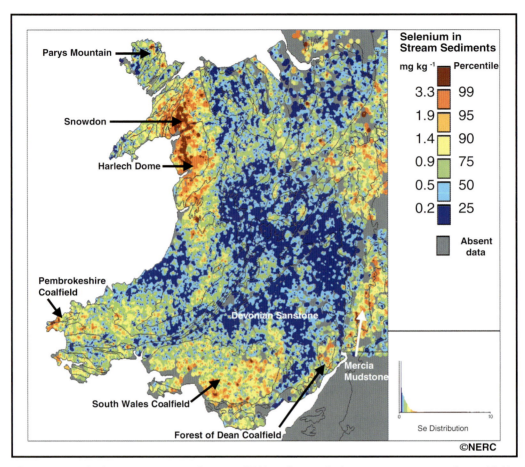

FIGURE 2 Distribution of selenium in stream sediments of Wales, showing high concentrations over the coalfields of south Wales, the mineralized areas (Parys Mountain, Snowdon, Harlech Dome) of north Wales, and the Mercia Mudstone in the Welsh Borderlands. Contrasting low values occur over the sandstones and siltstones of mid-Wales demonstrating a very strong relationship between selenium in the environment and geology. (Adapted with permission from British Geological Survey, 2000.)

concentrations, uptake into plants, and health outcomes in animals were first examined in detail in pioneering work carried out during the 1930s by Moxon (1937). Soils capable of producing selenium-rich vegetation toxic to livestock were reported over black shale and sandstone deposits of the Great Plains in the United States. Subsequent studies into selenium-deficiency-related diseases in animals lead to one of the first maps of the selenium status of soils, vegetation, and animals and the establishment of the classic Great Plain seleniferous soil types (Figure 3) (Muth & Allaway, 1963).

Although the underlying geology is the primary control on selenium concentrations in soils, the mobility and uptake of selenium into plants and animals, known as the bioavailability, is determined by a number of bio-physiochemical parameters. These include the prevailing pH and redox conditions, the chemical form or speciation of selenium, soil texture and mineralogy, and organic matter content and the presence of competitive ions. An understanding of these controls is essential to the prediction and remediation of health risks from selenium as even soils that contain adequate total selenium concentrations can result in selenium deficiency if the element is not in readily bioavailable form.

The principal controls on the chemical form of selenium in soils are the pH and redox conditions (Figure 4). Under most natural redox conditions, selenite (Se^{4+}) and selenate (Se^{6+}) are the predominant inorganic phases with selenite the more stable form. Selenite is adsorbed by ligand exchange onto soil particle surfaces with greater affinity than selenate. This process is pH

TABLE IX. Some Examples of Seleniferous Soils and Geological Parent Materials

Country	Parent material
U. S.	Cretaceous shale, Jurassic shales and sandstones, Triassic sandstones
Canada	Cretaceous shales
Colombia	Black slates
Puerto Rico	Volcanic soils
Ireland	Carboniferous shales and limestones
UK	Carboniferous and Ordovician shales and slates
Israel	Cretaceous limestone
South Africa	Cretaceous shales and sandstones
Australia	Cretaceous shales and limestones
Russia	Jurassic sandstones
China	Permian coal and shales

Modified from Fleming (1980).

dependent and adsorption increases with decreasing pH. In acid and neutral soils, selenite forms very insoluble iron oxide and oxyhydroxide complexes such as $Fe_2(OH)_4$ SeO_3. The low solubility coupled with stronger adsorption makes selenite less bioavailable than selenate. In contrast, selenate, the most common oxidation state in neutral and alkaline soils, is generally soluble, mobile, and readily available for plant uptake. For example, experiments have shown that addition of selenate to soils results in ten times more plant uptake than addition of the same amount of selenium as selenite (Jacobs, 1989; Neal, 1995).

Elemental selenium (Se^0), selenides (Se^{2-}), and selenium sulfide salts tend to exist in reducing, acid and organic-rich environments only. The low solubility and oxidation potential of these element species make them largely unavailable to plants and animals. However, the oxidation and reduction of selenium is closely linked to microbial activity, for example, the bacterium *Bacillus megaterium* is known to oxidize elemental selenium to

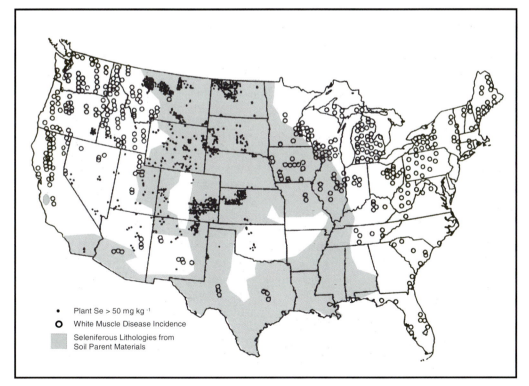

FIGURE 3 The geographic distribution of selenium-rich soils (shading), localities where plant selenium concentrations are known to exceed $50\,mg\,kg^{-1}$, and reported incidences of the selenium-deficiency-related disorder WMD in animals in the United States. (Adapted with permission from Muth and Allaway, 1963.)

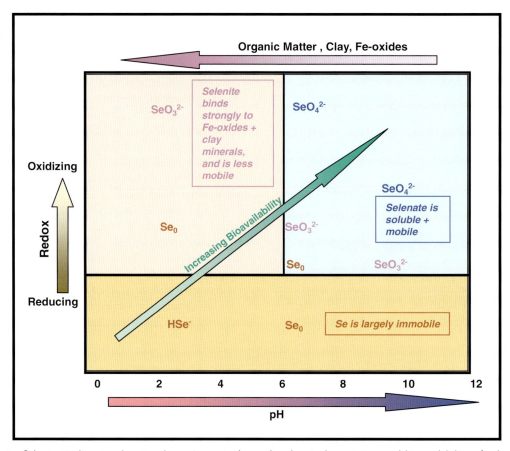

FIGURE 4   Schematic diagram showing the main controls on the chemical speciation and bioavailability of selenium in soils.

selenite. It is estimated that perhaps 50% of the selenium in some soils may be held in organic compounds, however, few have been isolated and identified. To date, selenomethionine has been extracted from soils and is two to four times more bioavailable to plants than inorganic selenite whereas selenocysteine is less bioavailable than selenomethionine (Jacobs, 1989; Mayland, 1994; Neal, 1995).

The bioavailability of the different selenium species in soils is summarized in Figure 5. In summary, selenate is more mobile, soluble, and less well adsorbed than selenite, thus, selenium is much more bioavailable under oxidizing alkaline conditions and much less bioavailable in reducing acid conditions (Figure 4) (Fleming, 1980; Jacobs, 1989; Neal, 1995) (see also Chapter 14, this volume).

In addition to the speciation of selenium in soils, other soil properties affect mobility. The bioavailability of selenium in soil generally correlates negatively with clay content due to increased adsorption on fine particles; indeed, the selenium uptake in plants grown on clay-loamy soils can be half that of plants grown on sandy soils. Iron also exerts a major control on selenium mobility as both elements are affiliated under oxidizing and reducing conditions and adsorption of selenium by iron oxides exceeds that of clay minerals. As mentioned above, the capacity of clays and iron oxides to adsorb selenium is strongly influenced by pH, reaching a maximum between pH 3–5 and decreasing with increasing pH (Jacobs, 1989; Neal, 1995). Soil organic matter also has a large capacity to remove selenium from soil solution possibly as a result of fixation by organometallic complexes. For example, plant uptake of selenate added to organic-rich soils can be ten times less than from mineral soils (Jacobs, 1989; Neal, 1995).

The presence of ions such as SO_4^{2-} and PO_4^{3-} can influence selenium uptake in plants by competing for fixation sites in the soil and plants. SO_4^{2-} inhibits the uptake of selenium by plants and has a greater effect on selenate than selenite. The addition of PO_4^{3-} to soils has been shown to increase selenium uptake by plants as the PO_4^{3-} ion is readily adsorbed in soils and displaces

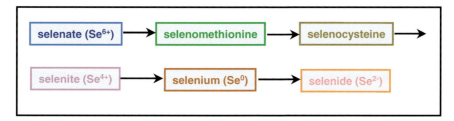

Sources: Jacobs (1989); Neal (1995)

FIGURE 5 Decreasing order of bioavailability of the different forms of selenium found in soils.

selenite from fixation sites making it more bioavailable. Conversely, increasing the levels of PO_4^{3-} in soils can dilute the selenium content of vegetation by inducing increased plant growth (Jacobs, 1989; Mayland, 1994; Neal, 1995).

Therefore, in any study of the selenium status of soil, consideration of the likely bioavailability is important. Several different chemical techniques are available to assess bioavailability but one of the most widely accepted indicators is the water-soluble selenium content (Jacobs, 1989; Tan, 1989; Fordyce et al., 2000b). In most soils, only a small proportion of the total selenium is dissolved in solution (0.3–7%) and water-soluble selenium contents are generally $<0.1\,mg$ kg^{-1} (Table III) (Jacobs, 1989).

The importance of soil selenium bioavailability and health outcomes is exemplified by seleniferous soils in the United States. Toxicity problems in plants and livestock have been reported in soils developed over the Cretaceous shales of the northern mid-West which contain $1–10\,mg\,kg^{-1}$ total selenium because up to 60% of the element is in water-soluble readily bioavailable form in the semi-arid alkaline environment. In contrast, soils in Hawaii with up to $20\,mg\,kg^{-1}$ total selenium do not cause problems in vegetation and livestock, because the element is held in iron and aluminium complexes in the humid lateritic soils of that region (Oldfield, 1999).

D. Selenium in Plants

Although there is little evidence that selenium is essential for vegetation growth, it is incorporated into the plant structure. Selenium concentrations in plants generally reflect the levels of selenium in the environment such that the same plant species grown over high and low selenium-available soils will contain concentrations reflecting the soil composition. However, an important factor that may determine whether or not selenium-related health problems manifest in animals and humans is the very wide-ranging ability of different plant species to accumulate selenium (WHO, 1987; Jacobs, 1989; Neal, 1995).

Rosenfield and Beath (1964) were the first to classify plants into three groups on the basis of selenium uptake when grown on seleniferous soils. Some examples of this scheme are outlined in Table X. Selenium accumulator plants grow well on high-selenium soils and can absorb $>1000\,mg\,kg^{-1}$ of the element, whereas secondary selenium absorbers rarely concentrate more than $50–100\,mg\,kg^{-1}$. The third group, which includes grains and grasses, usually accumulates less than $50\,mg\,kg^{-1}$ of selenium. Selenium concentration in plants can range from $0.005\,mg\,kg^{-1}$ in deficient crops to $5500\,mg\,kg^{-1}$ in selenium accumulators, but most plants contain $<10\,mg$ kg^{-1} selenium. Some species of the plant genera *Astragalus*, *Haplopappus*, and *Stanleya* are characteristic of seleniferous semi-arid environments in the western United States and other parts of the world and are often used as indicators of high-selenium environments. It should be noted, however, that other species in these genera are non-accumulators (WHO, 1987; Jacobs, 1989; Neal, 1995).

The reason why some plants are better at accumulating selenium than others depends upon selenium metabolism. Plants contain many different selenium compounds and the main form in non-accumulator species is protein-bound selenomethionine; however, selenocysteine and selenonium have also been reported (Jacobs, 1989; Neal, 1995). In contrast, the selenium metabolism in accumulator plants is primarily based on water-soluble, non-protein forms such as Se-methylselenomethionine. The exclusion of selenium from the proteins of accumulator plants is thought to be the basis of selenium tolerance (Jacobs, 1989; Neal, 1995). Plants also reduce selenate to elemental Se^0 and selenide Se^{2-} forming the volatile organic compounds dimethylselenide and dimethyldiselenide, which are released to the air during respiration giving rise to a

TABLE X. Examples of the Three Types of Selenium Accumulating Vegetation

Type	Examples (genus, family, or species)
Primary accumulator	G. *Astragalus* (e.g., milk vetch)
	G. *Machaeranthera* (woody aster, U. S.)
	G. *Haplopappus* (North and South American goldenweed)
	G. *Stanleya* (Prince's Plume)
	G. *Morinda* (rubiaceous trees and shrubs, Asia/Australia)
	F. *Lecythidaceae* (South American trees)
	Sp. *Neptunia* (Legume Asia/Australia)
Secondary accumulator	G. *Aster*
	G. *Astragalus*
	G. *Atriplex* (Saltbush)
	G. *Castilleja* (North and South American perennials)
	G. *Grindelia* (gummy herbs of western North and Central America)
	G. *Gutierrezia* (perennial herbs of western North and South America)
	G. *Machaeranthera*
	G. *Mentzelia* (bristly herbs of western America)
	Sp. *Brassica* (mustard, cabbage, broccoli, cauliflower)
Non-accumulator	Sp. *Pascopyrum* (wheat grass)
	Sp. *Poasecunda* (blue grass)
	Sp. *Xylorhiza* (Woody Aster)
	Sp. *Trifolium* (clover)
	Sp. *Buchloe* (buffalo grass)
	Sp. *Bouteloua* (North and South American tuft grass)
	Sp. *Beta* (sugar beet)
	Sp. *Horedeum* (barley)
	Sp. *Triticum* (wheat)
	Sp. *Avena* (oats)

From Rosenfield and Beath (1964); Jacobs (1989); and Neal (1995).

"garlic" odor characteristic of selenium-accumulating plants (Mayland, 1994).

Despite these coping mechanisms, plants can suffer selenium toxicity via the following processes (Jacobs, 1989; Fergusson, 1990; Mayland, 1994; Wu, 1994):

- Selenium competes with essential metabolites for sites in the plant biochemical structure.
- Selenium may replace essential ions, mainly the major cations (for example, iron, manganese, copper, and zinc).

- Selenate can occupy the sites for essential groups such as phosphate and nitrate.
- Selenium can be incorporated into analogues of essential sulfur compounds in plant tissues.

No phytotoxicity symptoms have been reported in nature in the United States, but experimental evidence has shown a negative correlation between increased selenium contents in soil and growth (plant dry weight, root length, and shoot height all decrease). For example, alfalfa yields have been shown to decline when extractable selenium exceeds $500 \, mg \, kg^{-1}$ in soil. Other symptoms include yellowing, black spots, and chlorosis of plant leaves and pink root tissue (Jacobs, 1989; Wu, 1994). However, phytotoxicity has been reported in nature in China, where high concentrations in soil caused pink discoloration of maize corn-head embryos; the pink color was attributed to the presence of elemental selenium. Levels of $>2 \, mg \, kg^{-1}$ and $>1.25 \, mg \, kg^{-1}$ selenium were detrimental to the growth and yield of wheat and pea crops, respectively (Yang et al., 1983). In addition to disturbances to the plant metabolism, a recent study has shown that at low concentrations, selenium acts as an antioxidant in plants inhibiting lipid peroxidation but at high concentrations (additions of $>10 \, mg \, kg^{-1}$), it acts as a pro-oxidant encouraging the accumulation of lipid peroxidation products, which results in marked yield losses (Hartikainen et al., 2000).

Food crops tend to have relatively low tolerance to selenium toxicity, and most crops have the potential to accumulate the element in quantities that are toxic to animals and humans. In general, root crops contain higher selenium concentrations than other plants (Table XI) and plant leaves often contain higher concentrations than the tuber. For example, Yang et al. (1983) noted that selenium concentrations in vegetables (0.3–$81.4 \, mg \, kg^{-1}$) were higher than in cereal crops (0.3–$28.5 \, mg \, kg^{-1}$ in rice and maize) in seleniferous regions of China. Turnip greens were particularly high in selenium, with an average of 457 and ranged up to $24,891 \, mg \, kg^{-1}$ compared to an average of $12 \, mg \, kg^{-1}$ in the tuber. In moderate to low selenium environments, alfalfa (Sp. *Medicago*) has been shown to take up more selenium than other forage crops, which may be due to deeper rooting accessing more alkaline conditions, hence more bioavailable selenium at depth. However, in general, crop species grown in very low-selenium soils show little difference in take up and changing the type of plants makes little impact on the selenium content of crops (Jacobs, 1989). An exception is reported in New Zealand (Section V of this chapter).

TABLE XI. Relative Uptake of Selenium in Agricultural Crops

Selenium accumulation	Plant species
Better accumulators	Cruciferae (broccoli, radish, cress, cabbage, turnip, rape, and mustard)
	Liliaceae (onion)
	Leguminosae (red and white clover, peas)
	Helianthus (sunflower)
	Beta (Swiss chard)
Poorer Accumulators	Compositae (lettuce, daisy, artichoke)
	Gramineae (cocksfoot, ryegrass, wheat, oats, barely)
	Umbelliferae (parsnip, carrot)

Average selenium mg kg^{-1} dry weight	U. S. crop type
0.407	Roots and bulbs
0.297	Grains
0.110	Leafy vegetables
0.066	Seed vegetables
0.054	Vegetable fruits
0.015	Tree fruits

Jacobs (1989).

E. Selenium in Water

It is estimated that the annual global flux of selenium from land to the oceans is 14,000 tonnes yr^{-1} via surface and groundwaters, which represent a major pathway of selenium loss from land in the selenium cycle (Nriagu, 1989). Approximately 85% of the selenium in most rivers is thought to be in particulate rather than aqueous form, however, the cycling of selenium from the land to the aqueous environment is poorly understood and requires further investigation (Haygarth, 1994).

The average concentration of selenium in seawater is estimated at 0.09 µg L^{-1} (Cutter & Bruland, 1984), but the mean residence time for selenium is thought to be 70 years in the mixed layer and 1100 years in the deep ocean, hence, the oceans constitute an important environmental sink for selenium (Haygarth, 1994). Biogenic volatilization from seawater to the atmosphere is estimated at 5000–8000 tonnes annually (Nriagu, 1989) and Amouroux et al. (2001) have demonstrated that the biotransformation of dissolved selenium in seawater during spring blooms of phytoplankton is a major pathway for the production of gaseous selenium emission into the atmosphere. This makes the oceans an important component of the selenium cycle.

Although the oceans via seafood do play a role in human selenium exposure, water used for drinking is more important. Selenium forms a very minor component of most natural waters and rarely exceeds 10 µg L^{-1}. Typically ranges are <0.1–100 µg L^{-1} with most concentrations below 3 µg L^{-1}. A garlic odor has been noted in waters containing 10–25 µg L^{-1}, whereas waters containing 100–200 µg L^{-1} selenium have an acerbic taste (WHO, 1987; Jacobs, 1989). In general, groundwaters contain higher selenium concentrations than surface waters due to greater contact times for rock–water interactions (Hem, 1992). Groundwaters containing 1000 µg L^{-1} selenium have been noted in seleniferous aquifers of Montana in the United States and up to 275 µg L^{-1} in China (Jacobs, 1989; Fordyce et al., 2000b) (Table III). Although rare in nature, concentrations of up to 2000 µg L^{-1} selenium have also been reported in saline lake waters in the United States, Venezuela, and Pakistan (Afzal et al., 2000). Anthropogenic sources of selenium can impact surface water quality as a result of atmospheric deposition from fossil fuel combustion, industrial processes, and sewage disposal. For example, concentrations of 400 µg L^{-1} in surface waters have been reported around the nickel–copper smelter at the Sudbury ore deposit in Ontario, Canada (Nriagu, 1989) and sewage effluents are known to contain 45–50 µg L^{-1} selenium (Jacobs, 1989). Irrigation practices can also affect the amount of selenium in water such as at Kesterson Reservoir in the San Joaquin Valley, California (Jacobs, 1989). (See Section VI of this chapter).

F. Atmospheric Selenium

The volatilization of selenium from volcanoes, soil, sediments, the oceans, microorganisms, plants, animals, and industrial activity all contribute to the selenium content of the atmosphere. It is estimated that natural background levels of selenium in non-volcanic areas are very low, around 0.01–1 ng m^{-3}; however, the residency time of selenium can be a matter of weeks, which makes the atmosphere a rapid transport route for selenium in the environment. Volatilization of selenium from the surface of the planet to the atmosphere results from microbial methylation of selenium from soil, plant, and water surfaces and is affected by the availability of selenium, carbon source, oxygen availability, and temperature (Haygarth, 1994).

The majority of gaseous selenium is thought to be in dimethylselenide form and it is estimated that terrestrial biogenic sources contribute 1200 tonnes of selenium per year to the atmosphere. Atmospheric dusts derived from volcanoes and wind erosion of the Earth's surface (180 tonnes per year) and suspended sea salts (550 tonnes per year) from the oceans also constitute significant sources of atmospheric selenium (Nriagu, 1989). It is suggested that particle-bound selenium can be transported several thousand kilometers before deposition back to the Earth's surface in both wet and dry forms. Wet deposition is thought to contribute 5610 tonnes per year to land (Haygarth, 1994). For example, in the UK it has been demonstrated that wet deposition (rain, snow, etc.) accounts for 76–93% of total deposition with >70% of selenium in soluble form. In the proximity of selenium sources (such as industrial emissions), atmospheric deposition can account for 33–82% of uptake in the leaves of plants (Haygarth, 1994).

G. Selenium Is All Around Us

From the descriptions above, it is clear that selenium is present in varying quantities in the environment all around us as a result of natural and man-made processes. Animals and humans are exposed to environmental selenium via dermal contact, the inhalation of air, and via ingestion of water and of plants and animals in the diet produced on soils containing selenium.

III. SELENIUM IN ANIMALS AND HUMANS

A. Selenium Exposure

In most non-occupational circumstances atmospheric exposure is insignificant as concentrations of selenium are so low ($<10\,ng\,m^{-3}$). However, occupational inhalation exposure may occur in the metal, selenium-recovery, and paint industries. In these circumstances, acute (short-term) exposure of humans to hydrogen selenide, the most toxic selenium compound, which exists as a gas at room temperature, results in irritation of the mucous membranes, pulmonary edema, severe bronchitis, and bronchial ammonia whereas inhalation of selenium dust can cause irritation of the membranes

TABLE XII. U.S.-EPA Inhalation Exposure Criterion for Selenium-Bearing Compounds

Regulation	Compound	Value
Lethal concentration	Hydrogen selenide	$12,700\,\mu g\,m^{-3}$
Permissible exposure limit	Se-hexafluoride	$400\,\mu g\,m^{-3}$
Permissible exposure limit	Se-compounds	$200\,\mu g\,m^{-3}$

From U.S.-EPA (2002a).

in the nose and throat, bronchial spasms, and chemical pneumonia. Selenium dioxide gas is the main source of problems in industrial situations as selenious acid is formed on contact with water or sweat causing irritation. Indigestion and nausea, cardiovascular effects, headaches, dizziness, malaise, and irritation of the eyes have also been reported in occupational selenium exposure (WHO, 1987, 1996) (see also Chapter 23 this volume).

As a result of these effects, hydrogen selenide gas is classed as a highly toxic substance and the common selenium-bearing compounds sodium selenite and sodium selenate are considered "high concern" pollutants (U.S.-EPA, 2002a). Some regulatory values for selenium compounds in air are presented in Table XII. Little information on the long-term (chronic) effects of selenium inhalation is available. In seleniferous areas of China, there is some evidence to suggest that the selenium-loading of the population is enhanced by inhalation of coal smoke from open fires used for cooking as concentrations have been known to rise to $160,000\,ng\,m^{-3}$ in air during combustion (Yang et al., 1983). However, it is difficult to assess the amount of exposure via this route compared to other sources (Fordyce et al., 1998). Smoking is an inadvertent inhalation exposure route to selenium as tobacco commonly contains $0.03–0.13\,mg\,kg^{-1}$ (WHO, 1987). Assuming a cigarette contains 1 g of tobacco and that all the selenium is inhaled, a person smoking 20 cigarettes could intake $1.6\,\mu g$ Se day^{-1}. The inhalation of locally grown selenium-rich tobacco in seleniferous regions of China may contribute to the loading of the local population (selenium concentration is $9.05\,mg\,kg^{-1}$, Fordyce et al., 1998). In general, however, inhalation is a less important exposure route than ingestion. For example, studies carried out on dogs found that only 52 and 73% of selenium in the form of metal and selenious acid aerosols

were adsorbed in the lungs compared to 73 and 96% absorption in the gut (Levander, 1986; WHO, 1987).

Very few studies have examined the effects of dermal exposure to selenium although sodium selenite and selenium oxychloride solutions have been proved to absorb into the skin of experimental animals. The insoluble compound selenium sulfide is used in anti-dandruff shampoos and is not normally absorbed through the skin, but elevated selenium concentrations in urine have been noted in people with open skin lesions who use these products. In an occupational setting, selenium dioxide gas can result in burns and dermatitis and an allergic body rash. In most normal circumstances, however, dermal contact is not an important exposure route (Levander, 1986; WHO, 1987, 1996) (see also Chapter 23, this volume).

In the majority of cases, water selenium concentrations are extremely low ($<10\,\mu g\,L^{-1}$) and do not constitute a major exposure pathway; however, aquatic life-forms are sensitive to selenium intoxication as soluble forms of selenate and selenite are highly bioavailable and cause reduced reproduction and growth in fish (Jacobs, 1989). For this reason, the U.S.-EPA has set a chronic ecotoxicity threshold of $5\,\mu g\,L^{-1}$ in surface water (Canton, 1999). Selenium is a bioaccumulator, which means that plants and animals retain the element in greater concentrations than are present in the environment (Table XIII) and the element can be bioconcentrated by 200–6000 times. For example, concentrations in most waters are approximately $1\,\mu g\,L^{-1}$ whereas freshwater invertebrates generally contain up to $4\,mg\,kg^{-1}$ of selenium (Jacobs, 1989). Phytoplanktons are efficient accumulators of dissolved selenomethionine and incorporate inorganic selenium into amino acids and proteins (estimated bioconcentration factors range from 100 to 2600) (Jacobs, 1989). However, the reported lethal doses of selenium in water for invertebrates ($0.34–42\,mg\,L^{-1}$) and fish ($0.62–28.5\,mg\,L^{-1}$) indicate that in most circumstances water alone is not a major environmental problem. It should be noted, however, that inorganic and organic selenium enter the food chain almost entirely via plants and algae and bioconcentration from high-selenium waters could cause problems, because selenium passes up the food chain from algae and larval fish to large fish, birds, and humans (WHO, 1987; Jacobs, 1989). A MAC of 10–$11\,mg\,kg^{-1}$ selenium in the diets of fish has been proposed in the United States to prevent toxicity and the uptake of too much selenium into the food chain (Jacobs, 1989). At concentrations $>50\,\mu g\,L^{-1}$ in water, selenium intake can contribute significantly to overall dietary intake in animals and humans and the U.S.-EPA

TABLE XIII. Average Concentrations of Selenium in Selected Animals

Animal	Selenium $mg\,kg^{-1}$
Fish, U. S.	0.5 wet weight
Terrestrial arthropods	1 fresh weight
Earthworms	2.2 (normal soil)–22 (sewage sludge amended soil) fresh weight
Bird livers	4–10 dry weight
Bird eggs	0.4–0.8 wet weight
Bird kidneys	1–3 dry weight
Mammal livers	<2 dry weight

From Jacobs (1989).

(2002b) currently recommends this as the MAC for selenium in drinking water. The World Health Organization currently sets a more precautionary MAC of $10\,\mu g\,L^{-1}$ selenium for drinking water (WHO, 1996).

However, the most important exposure route to selenium for animals and humans is the food we eat, as concentrations are orders of magnitude greater than in water and air in most circumstances (WHO, 1996). In terms of the human diet, organ meats such as liver and kidney are good sources of selenium and some seafoods contain almost as much. Muscle meats are also a significant source and garlic and mushrooms contain more than most other vegetables. Cereals are another important source, however, white bread and flour contain less selenium than whole meal by about 10–30% (Table XIV). Brazil nuts sold in the UK are high in selenium, indeed, cases of selenium poisoning in Amazon peoples following consumption of nuts of the Lecythidaceae family have been reported in Brazil. These incidents resulted in nausea, vomiting, chills, diarrhea, hair and nail loss, painful joints, and death in some cases (see Section III.F of this chapter). Cooking reduces the selenium contents of most foods, and studies have shown that vegetables that are normally high in selenium such as asparagus and mushrooms lose 40% during boiling. Other studies estimate 50% of the selenium content is lost from vegetables and dairy products during cooking especially if salt and low pH foods such as vinegar are added, whereas frying foods results in much smaller losses (Levander, 1986; WHO, 1987, 1996; Rayman, 2002) (see also Chapter 7, this volume).

Levels of dietary selenium intake show huge geographic variation and are dependent upon the geochemical conditions of the food source environments as well as differences in dietary composition. For example in 1995, cereals accounted for 75% of the total 149 μg day^{-1} selenium intake in Canada but only 10% of the 30 μg day^{-1} intake in Finland (WHO, 1996). In general terms, cereals grown in North America contain more selenium than European crops and concern is growing in Europe over declining selenium intakes. The UK traditionally imported large quantities of wheat from North America but since the advent of the EU, most cereals are now more locally derived and as a consequence, daily intakes of selenium in the UK have been falling. Marked declines are evident even over a 4-year period from intakes of 43 μg day^{-1} in 1991 to 29–39 μg day^{-1} in 1995 (Figure 6) (Rayman, 2002). This downward trend is also attributed to a reduction in cereal consumption in the UK, which fell from 1080 g person^{-1} week^{-1} in 1970 to 756 g person^{-1} week^{-1} in 1995 (MAFF, 1997). The selenium content of Irish bread is also significantly lower than in the United States and only marginally higher than the UK (Table XIV) (Murphy & Cashman, 2001). Other cereal crops such as rice generally contain low selenium contents (Table XIV) and can have a significant influence on overall dietary intake when consumed as the staple food as in most of Asia. Conversely Japanese diets can be very high in selenium (up to 500 μg day^{-1}) in areas where a large amount of seafood is consumed (WHO, 1987).

Some examples of daily dietary selenium intakes from around the world are listed in Table XV. On a global scale it is estimated that dietary intakes in adults range from 3 to 7000 μg day^{-1} and for infants in the first month of life from 5 to 55 μg day^{-1}. The wide ranges are attributed to selenium contents in the environment (WHO, 1996). The greatest variations in dietary intake are reported from selenium-deficient and seleniferous regions of China (Tan, 1989), but contrasts also occur in South America between high daily intakes (100–1200 μg) associated with foodstuffs grown on selenium-rich shales in the Andes and Orinoco River of Venezuela and widespread selenium deficiency in Argentina (WHO, 1987; Oldfield, 1999). Dietary intakes in countries such as New Zealand, Finland, and Turkey are also poor as a consequence of low-selenium soils, whereas intakes in Greece, Canada, and the United States are generally adequate (WHO, 1987). On the basis of selenium requirement studies, a range of 50–200 μg day^{-1} has been recommended by the U.S. National Research Council (NRC) for adults depending on various factors such as physiological status. Balance studies to more precisely determine the ratio of selenium inputs and outputs in human beings were attempted, however,

TABLE XIV. Concentrations of Selenium in Selected Foodstuffs From Around the World

Food type	Source	Selenium mg kg^{-1}
Whole meal flour	Ireland	0.077–0.099
White flour	Ireland	0.060–0.069
Wheat flour	Russia	0.044–0.557
Whole meal bread	Ireland	0.086–0.129
White bread	Ireland	0.066
Wheat	World	0.1–1.9
Wheat	Greece	0.019–0.528
Wheat	Colombia	180
Wheat	China	Deficient 0.001–0.105
Barley	U. S.	0.2–1.8
Oats	U. S.	0.15–1
Corn	China	Deficient 0.005–0.089
Corn	China	Seleniferous 0.5–28.5
Corn	Venezuela	14
Maize	China	Deficient 0.001–0.105
Maize	China	Seleniferous 0.017–9.175
Maize	China	Adequate 0.021–2.324
Maize	U. S.	0.136
Rice	Venezuela	18
Rice	China	Deficient 0.007–0.022
Rice	China	Seleniferous 0.3–20.2
Rice	Sri Lanka	0.0001–0.777
Cereals	World	0.1–0.8 wet weight
Cereals	Finland/New Zealand	0.01–0.07
Liver, kidney, seafood	World	0.4–1.5 wet weight
Liver, kidney, seafood	Finland/New Zealand	0.09–0.92 wet weight
Muscle meat	World	0.1–0.4 wet weight
Muscle meat	Finland/New Zealand	0.01–0.06 wet weight
Dairy products	World	0.1–0.3 wet weight
Dairy products	Finland/New Zealand	0.01 wet weight
Cow's milk	Turkey	11.28–36.05 mg L^{-1}
Human milk	World	0.013–0.018 mg L^{-1}
Human milk	New Zealand	0.005 mg L^{-1}
Dried milk	Russia	0.038–0.115
Egg whites	Chile	0.55–1.10
Fruit and vegetables	World	0.1 wet weight
Fruit and vegetables	Finland/New Zealand	0.01–0.07 wet weight
Vegetables	China	Seleniferous 2.0–475
Brazil nuts	UK	22.3–53
Soyabeans	China	Deficient 0.010
Soyabeans	China	Seleniferous 0.34–22.2

From Yang et al. (1983); Levander (1986); WHO (1987, 1996); Jacobs (1989); Tan (1989); Fordyce et al. (1998); Oldfield (1999); Fordyce et al. (2000a); Murphy and Cashman (2001).

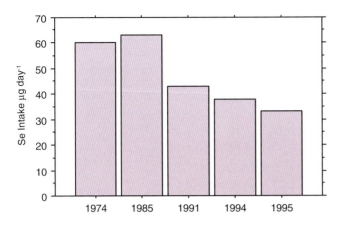

Data Source: MAFF (1997)

FIGURE 6 Bar chart showing the decline in the average daily dietary intake of selenium in UK adults 1974–1995.

these were not successful as humans have the ability to modify fecal and urinary excretion of selenium depending on levels of intake (WHO, 1996). Current recommended daily allowances (RDAs) of dietary selenium range from 55 μg in women to 75 μg day^{-1} in men and 8.7–10 μg day^{-1} in infants (Table XV) (WHO, 1996; MAFF, 1997).

Just as bioavailability is an important factor in terms of plant uptake of selenium, it is also an important factor in the diets of animals and humans. Dietary studies have shown that selenomethionine is more readily absorbed in the guts of animals and humans than selenate, selenite, or selenocysteine. More than 90% of ingested selenomethionine and selenate is absorbed, whereas the rate for sodium selenite is slightly lower (>80%). Selenides and elemental selenium are poorly absorbed. Very few studies have examined the chemical form of selenium in foodstuffs, but evidence suggests that 7.6–44% of selenium in tuna is in the form of selenate with the remainder present as selenite and selenide. In contrast, 50% of the selenium in wheat and 15% in cabbage is in the form of selenomethionine. These differences in the chemical forms of selenium are reflected in the rate of absorption and bioavailability of the element in foodstuffs. For example, it is estimated that over 90% of the selenium in Brazil nuts and beef kidney is bioavailable, compared to only 20–60% in tuna. However, other seafood, such as shrimp, crab, and Baltic herring, have higher bioavailability. As an indication of the diet in general, studies carried out in New Zealand have shown that 79% of selenium present in natural foods is bioavailable. In addition to foodstuffs, mineral supplements are a source of dietary selenium to

humans and animals. Chemical supplement tests show 97% adsorption of selenomethionine, 94% of selenate, and 60% of sodium selenite in this dietary form (Levander, 1986; WHO, 1987, 1996; Rayman, 2002).

In animals, 85–100% of dietary plant selenium is absorbed whereas only 20–50% of the selenium present in meat and fish is taken up by birds and mammals. In general terms, selenium in plant forms is more readily bioavailable than selenium in animal forms (Levander, 1986; WHO, 1987, 1996) (see also Chapter 20, this volume).

The bioavailability of selenium to humans and animals is not only dependent on the amount of absorption but also on the conversion of the ingested selenium to metabolically active forms. In humans, studies based on the activity of the selenium-dependent enzyme GSH-Px have shown that the bioavailability of selenium in wheat is >80% whereas the bioavailability of selenium in mushrooms is very low. In a comparison between wheat and mineral selenate supplements, while the latter were shown to enhance GSH-Px activity, patients fed wheat demonstrated greater increases and better long-term retention of selenium (WHO, 1987).

Much has still to be learned about the uptake of selenium in humans and animals, however, it is clear that in most normal circumstances food forms the major exposure route as selenium accumulates from the environment via plants and algae through the food chain to animals and man. Selenium in the form of selenomethionine and selenate is highly bioavailable to animals and humans and foodstuffs that contain high proportions of these forms, such as organ and muscle meats, Brazil nuts, and wheat, are good sources of the element in the diet (WHO, 1987; Rayman, 2002).

B. Selenium in the Body of Animals and Humans

Once ingested into the body, most selenium is absorbed in the small intestine of animals and humans, but the rates and mechanisms of selenium metabolism vary between different animal species. In general, single-stomached animals absorb more selenium than ruminants due to the reduction of selenite to insoluble forms by rumen microorganisms. Experiments on rats indicate very little difference in the process of absorption of different selenium forms; 92% of selenite, 91% of selenomethionine, and 81% of selenocysteine were absorbed primarily in the small intestine and none in the stomach. Approximately 95% of the total selenium intake was absorbed regardless of whether the rats were

TABLE XV. Some Daily Dietary Intakes of Selenium From Around the World (μg day^{-1})

Country	Year	Vegetables and fruit	Cereals	Dairy	Meat and fish	Men	Women	Total
New Zealand (general)	1981–1982	1–2	3–4	11	12–16			28–32
New Zealand infants	1987							0.5–2.1
Finland	1975–1979	1	3–25	7–13	19			30–60
UK	1978	3	30	5	22			60
UK	1991							43
UK	1995							29–39
Japan (general)	1975	6	24	2	56			88
Japan high seafood	1987							500
Canada	1975	1–9	62–133	5–28	25–90			98–224
Canada	1987							149
U.S. (general)	1974–1976	5	45	13	69			132
U.S. (general)	2002							71–152
U.S. South Dakota	1976	10	57	48	101			216
U.S. Maryland	1987							81
China (general)	1987			2–212				
China seleniferous	1983							240–6690
China Se-adequate	1983					19.1	13.3	42–232
China Se-deficient	1983					7.7	6.6	3–22
India, Mumbai	2001							61.9
Turkey	2001							20–53
Venezuela	1999			58				100–1200
Swedish Pensioners	1987							8.7–96.3
Greece	1999							110

Country	RDA range	RDA W	RDA M	RDA I				
U.S.	50–200	55	70	8.7				
UK	60–200	60	75	10				
China	40–600							

Note: RDA = Recommended Daily Allowance; W = Women; M = Men; I = Infants.

From Yang et al. (1983); WHO (1987, 1996); Tan (1989); MAFF (1997); Oldfield (1999); Aras et al. (2001); Mahapatra et al. (2001); U.S.-EPA (2002a).

fed a low- or high-selenium diet indicating that selenium intake is not under homeostatic control. This is true in general for intake in animals and humans. However, other studies have shown that oral doses of selenomethionine are retained more readily and turned over more slowly than selenite in humans, therefore unlike rats there is a difference in the metabolism of different forms of selenium. In fact, selenomethionine, the main form of uptake from plants to animals, becomes associated with protein tissues in the body whereas inorganic selenium is absorbed into other tissues (Levander, 1986; WHO, 1987, 1996).

Most of the ingested selenium is quickly excreted in the urine, breath, perspiration, and bile and the remainder becomes bound or incorporated into blood and proteins. Urine is the primary route of excretion (70–80%) in single-stomached animals, however, in ruminants selenium is mostly excreted in the feces and studies have shown that the majority of this selenium is in unavailable elemental form. Chemical selenium tracer experiments in humans suggest that the main extraction pathway is via urine, however, in studies using natural foods, excretion in feces was equal to that of urine; whereas minimal amounts of selenium were exuded in sweat and respiration and expulsion of volatile forms of selenium only occurred at very high exposures. Unlike selenium absorption, which is not homeostatically regulated, selenium excretion in animals and humans is

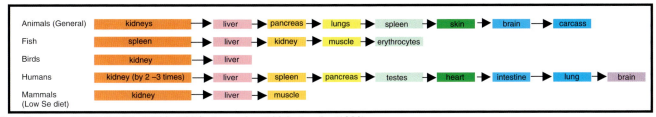

Data Sources: WHO (1987); WHO (1996); Levander (1986); Jacobs (1989)

FIGURE 7 Relative concentrations of selenium in the tissues of different animals, decreasing in concentration from left to right.

directly influenced by nutritional status: excretion rises as intake increases and decreases when selenium intakes are low (Levander, 1986; WHO, 1987, 1996) (see also Chapters 7, 8, and 20, this volume).

The remaining selenium is transported rapidly around the body and concentrates in the internal organs, which are rich in protein. This pattern is present in a number of animal species (Figure 7). Hence, in normal conditions in humans, selenium levels are highest in the liver and kidneys and lower in muscle tissues; the largest total amount of selenium in the human body is in the muscles as these form the main body mass. Total human body selenium contents are estimated at 3–14.6 mg (WHO, 1987).

In rats fed selenium-deficient diets, however, the pattern of selenium distribution is different with selenium reserved in the testes, brain, thymus, and spleen. Also in humans the supply to the testes has priority over the other tissues during selenium deficiency, because the element is found in the mitochondrial capsule protein (MCP) and is involved in biosynthesis of testosterone. Consequently, the selenium content of the testes increases considerably during puberty (Levander, 1986; WHO, 1987).

Both inorganic and organic selenium are converted by animals and humans to mono-, di-, or trimethylated forms by the main metabolic pathway, rarely reduction. Trimethylselenonium, is the main urinary excretion form. However, in cases of selenium toxicity, this pathway becomes overloaded and the volatile selenium metabolite dimethylselenide is produced and exhaled via the lungs, which results in the characteristic "garlic breath" symptom in animals and humans suffering selenosis. There is much debate over the form of selenium held in protein tissues. Non-ruminant animals and humans cannot synthesize selenite into selenomethionine, but there is evidence to suggest that selenomethionine can be incorporated into protein tissues directly. However, in the case of rats it is then converted to selenite or selenate. Rabbits and rats can also convert selenite into selenocysteine tissue proteins. Selenium may also be present in proteins in the selenotrisulfide and acid-labile form. Early work suggested that selenium intake in naturally occurring organic forms was retained in tissues to a greater extent than inorganic forms, however, experiments with mice using selenite, Se-methylselenocysteine, and selenomethionine showed that mice fed selenomethionine had greater quantities and better long-term retention of selenium than those fed selenite or Se-methylselenocysteine. Therefore, the distinction between inorganic and organic forms of selenium does not hold true as the metabolism of Se-methylselenocysteine and selenocysteine resemble that of selenite rather than selenomethionine. There is some evidence for metabolic pools of selenium in animals and humans. Studies with ewes fed selenium-adequate and then selenium-deficient diets showed that they were able to pass on adequate levels of selenium to their lambs even though the lambs were born 10 months into the selenium-deficient diet. Possible mechanisms for these pools may be the sequestration of selenomethionine or selenamino acids incorporated into protein structures and then released during protein turnover (Levander, 1986; WHO, 1987, 1996).

In most circumstances there is a close correlation between the levels of selenium in the diet of humans and animals and blood selenium content. On average, plasma levels vary from 0.079 to 0.252 mg L^{-1} depending on selenium intake, whereas the mean concentration of selenium in human whole blood is 0.2 mg L^{-1} (WHO, 1987, 1996). Human whole blood selenium levels show marked geographic variation depending on dietary intake. Ranges of 0.021–3.2 mg L^{-1} have been reported worldwide with highest concentrations in seleniferous areas of China and Venezuela and lowest concentrations in the selenium-deficient regions of Scandinavia, New Zealand, and China (Table XVI)

TABLE XVI. Examples of Selenium Concentrations in Human Tissues From Around the World

Country	Selenium (mg L^{-1}) Whole blood	Year	Selenium (mg L^{-1}) Serum	Year
Average (humans)	0.2			
Normal (humans)			0.06–0.105	
Canada, Ontario	0.182	1967		
China, high Se	1.3–7.5	1983		
China, high Se, no disease	0.44	1983		
China, mod Se	0.095	1983		
China, low Se, no disease	0.027	1983		
China, low Se, disease	0.021	1983		
Tibet, low Se			<0.005	1998
Egypt	0.068	1972		
Finland	0.056–0.081	1977		
Guatemala	0.23	1967		
New Zealand	0.083–0.059	1979		
Sweden		1987	0.86	
UK	0.32	1963		
U.S.	0.256–0.157	1968		
Russia	0.11–0.442	1976		
Venezuela seleniferous	0.355–0.813	1972		
Bulgaria			0.0548	1998
Hungary			0.0558	1998
Slovenia			0.0570	1998
Croatia			0.0642	1998
Russia			0.0718	1999
Italy, Lombardy	0.04–0.19	1986	0.033–0.121	1986
Spain, Barcelona			0.060–0.106	1995
Canary Islands			0.008–0.182	2001

	Selenium (mg kg^{-1}) Hair	Year		Selenium (µg L^{-1}) Urine
China, Se deficient	0.074	1983		0.007
China, Se deficient	0.170–0.853	1998		
China, Se deficient	0.094–0.359	1996		
China, low Se	0.16	1983		
China, Se adequate	0.343	1983		0.026
China, high Se	1.9–100	1983		0.04–6.63
China, high Se	0.566–141	1998		
Italy, Lombardy		1986		0.0002–0.068
Sri Lanka	0.104–2.551	1998		

Yang et al. (1983); Levander (1986); Akesson and Steen (1987); WHO (1987, 1996); Oldfield (1999); Fordyce et al. (2000a); Vinceti et al. (2000); Romero et al. (2001); Fordyce et al. (2000b)

(WHO, 1996; Oldfield, 1999). Similarly, concentrations in hair, nails, and urine vary according to differences in dietary intake, and some examples of the selenium composition of these tissues are given in Table XVI. Selenium levels in human milk are affected by maternal intake and infants and young children have a high requirement for the element during the rapid growth periods of early life. However, the age of mother and the concentration of selenium during pregnancy do not affect the weight of baby or the length of pregnancy.

Wide ranges of 2.6–283 mg L^{-1} in human milk have been reported from selenium-deficient and seleniferous regions in China, compared to ranges of 7–33 mg L^{-1} in the United States (Levander, 1986.)

In terms of biological function, approximately 20 essential selenoproteins containing selenocysteine have now been identified in microbes, animals, and humans, many of which are involved in redox reactions acting as components of the catalytic cycle (WHO, 1996). The selenoproteins found in mammals are listed in Table XVII (Rayman, 2002). Enzyme activity is attributed to the glutathione peroxidase, thioredoxin reductase, iodothyronine deiodinase, and selenophosphate synthetase groups. In complex interactions with vitamin E and polyunsaturated fatty acids, selenium plays an essential biological role as part of the enzyme GSH-Px, which protects tissues against peroxidative damage by catalyzing the reduction of lipid hydrogen peroxide or organic hydroperoxides. Together, GSH-Px, vitamin E and superoxide dismuthases form one of the main antioxidant defense systems in humans and animals. As such, selenium has been linked to enzyme activation, immune system function, pancreatic function, DNA repair, and the detoxification of xenobiotic agents such as paraquat, however, the exact mechanisms of immune function and detoxification are unknown (Combs & Combs, 1986; Levander, 1986; WHO, 1987, 1996). Selenium is found in the prosthetic groups of several metalloenzymes and appears to protect animals against the toxic effects of arsenic, cadmium, copper, mercury, tellurium, and thallium in most circumstances, but this is not always the case and the biological response depends on the ratio of selenium/metal involved (WHO, 1987; Fergusson, 1990). Selenium behaves antagonistically with copper and sulfur in humans and animals inhibiting the uptake and function of these elements. Selenium has been identified as a component of the cytochrome P-450 system in humans and animals, however, the exact biological role of this selenoprotein has yet to be established (WHO, 1987, 1996). Important developments in recent years have shown that selenium is beneficial to the thyroid hormone metabolism. There are three iodothyronine deiodinase (IDI) selenoenzymes. Types 1 and 2 are involved in the synthesis of active 3, 3′ and 5-triiodothyronine (T3) hormones, whereas type 3 IDI catalyzes the conversion of thyroxine (T4) to inactive T3(rT3). These hormones exert a major influence on cellular differentiation, growth, and development, especially in the fetus and child (Arthur & Beckett, 1994). Selenium also appears to be important in reproduction. In addition to aiding the biosynthesis of testosterone

TABLE XVII. List of the Approximately Fourteen Known Mammalian Selenoproteins

Name	Function
Glutathionine peroxidase GSH-PX	Antioxidant enzymes
Mitochondrial capsule selenoprotein	Protects sperm cells from oxidative damage
Sperm nuclei selenoprotein	Essential for male fertility and sperm maturation
Spermatid selenoprotein 34 kDa	May protect developing sperm
Iodothyronine deiodinases	Regulation and production of active thyroid hormones
Thioredoxin reductases	Reduction of nucleotides and binding of transcription factors in DNA
Selenophosphate synthetase SPS2	Required for selenoprotein synthesis
Selenoprotein P	Protects endothelial cells against perioxynitrite
Selenoprotein W	Skeletal and heart muscle metabolism
15-kDa selenoprotein	May protect prostate cells against carcinoma
18-kDa selenoprotein	Found in the kidney
Methionine sulfoxide reductase	May regulate lifespan
Selenoprotein N	If deficient may be linked to muscular dystrophy

From Rayman (2002).

(see above), the selenium contents of avian eggs are high whereas morphological deformities, immotility, and reduced fertility have been reported in sperm in selenium-deficient experimental animals (WHO, 1987, 1996; Rayman, 2002). Although many of the *in vivo* functions of selenium are still poorly understood, deficient and excessive dietary intakes of selenium have a marked effect on animal and human health, some of which are discussed below.

C. Selenium Deficiency—Effects in Animals

Due to the complementary role of selenium and Vitamin E, all selenium deficiency diseases in animals are concordant with vitamin E deficiency with the exception of neutrophil microbicidal activity reduction and the 5-deiodinase enzymes responsible for the production of triiodothyronine from thyroxine. Selenium is necessary for growth and fertility in animals and clinical signs of deficiency include dietary hepatic apoptosis in rats and pigs; exudative diathesis, embryonic mortality, and pancreatic fibrosis in birds; nutritional muscular dystrophy, known as white muscle disease, and retained placenta in ruminants and other species; and mulberry heart disease in pigs. Clinical signs of selenium deficiency in animals include reduced appetite, growth, production and reproductive fertility, unthriftyness, and muscle weakness (Levander, 1986; WHO, 1987, 1996; Oldfield, 1999).

White muscle disease is a complex condition that is multifactorial in origin and causes degeneration and apoptosis of the muscles in a host of animal species. This disease rarely affects adult animals but can affect young animals from birth. In lambs born with the disease, death can result after a few days. If the disease manifests slightly later in life, animals have a stiff and stilted gait, arched back, are not inclined to move about, lose condition, become prostrate and die. The disease responds to a combination of vitamin E and selenium supplementation (Levander, 1986; WHO, 1987, 1996; Oldfield, 1999).

Exudative diathesis in birds leads to massive hemorrhages beneath the skin as a result of abnormal permeability of the capillary walls and accumulation of fluid throughout the body. Chicks are most commonly affected between 3–6 weeks of age and become dejected, lose condition, show leg weakness, and may become prostrate and die. The disease responds to either vitamin E or selenium supplementation, but it will not respond to vitamin E alone if selenium is deficient (WHO, 1987).

Hepatic apoptosis in pigs generally occurs at 3–15 weeks of age and is characterized by necrotic liver lesions. Supplements of alpha-tocopherol and selenium can protect against death (Levander, 1986).

Low-selenium pastures containing 0.008–0.030 mg kg^{-1} are associated with a condition called "ill thrift" in lambs and cattle from New Zealand. The disease is characterized by subclinical growth deficits, clinical unthriftyness, rapid weight loss, and sometimes death but can be prevented by selenium supplementation with marked increases in growth and wool yields (Levander, 1986; WHO, 1987).

The level of dietary selenium needed to prevent deficiency depends on the vitamin E status and species of the host. For example, chicks receiving 100 mg of vitamin E require 0.01 mg kg^{-1} of selenium to protect against deficiency, whereas chicks deficient in vitamin

E require 0.05 mg kg^{-1} of selenium. Under normal vitamin E status, concentrations of 0.04–0.1 mg kg^{-1} (dry weight) in feedstuffs are generally adequate for most animals with a range of 0.15–0.20 mg kg^{-1} for poultry and 0.03–0.05 mg kg^{-1} for ruminants and pigs (Levander, 1986; WHO, 1987).

Selenium deficiency and WMD are known to occur in sheep when blood selenium levels fall below 50 μg L^{-1} and kidney concentrations below 0.21 mg kg^{-1} (dry weight). Blood levels of 100 μg L^{-1} selenium are needed in sheep and cattle and 180–230 μg L^{-1} in pigs to maintain the immunoresponse systems. Studies have shown that most farmland grazing in the UK is not able to provide enough selenium to support 0.075 mg L^{-1} in blood in cattle. Indeed, selenium deficiency in animals is very common and widespread around the globe affecting much of South America, North America, Africa, Europe, Asia, Australia, and New Zealand. Many western countries now adopt selenium supplementation programs in agriculture, but these are often not available in South America, Africa, and Asia and livestock productivity is significantly impaired by selenium deficiency in these regions (Levander, 1986; WHO, 1987, 1996; Oldfield, 1999) (see also Chapter 20, this volume).

D. Selenium Deficiency—Effects in Humans

No clear-cut pathological condition resulting from selenium deficiency alone has been identified in humans, however, the element has been implicated in a number of diseases (WHO, 1996) (see also Chapters 8 and 23, this volume).

1. Keshan Disease

Keshan disease (KD) is an endemic cardiomyopathy (heart disease) that mainly affects children and women of childbearing age in China. The disease has been documented for over 100 years, but the name is derived from a serious outbreak in Keshan County, northeast China in 1935. Outbreaks have been reported in a broad belt stretching from Heilongjiang Province in the northeast of China to Yunnan province in the southwest that transcends topography, soil types, climatic zones, and population types (Figure 8). This disease manifests as an acute insufficiency of the heart function or as a chronic moderate-to-severe heart enlargement and can result in death. Seasonal variations in outbreak were noted with peaks in the winter in the south and in the

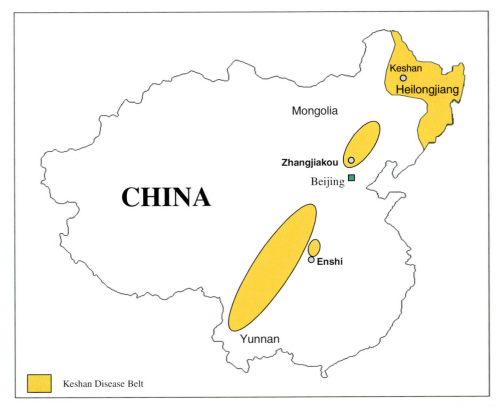

FIGURE 8 Distribution of the incidence of KD in China. (Adapted with permission from Tan, 1989.)

summer in the north. The worst affected years on record were 1959, 1964, and 1970 when the annual prevalence exceeded 40 per 100,000 with more than 8000 cases and 1400–3000 deaths each year (Tan, 1989).

Although the disease occurred in a broad belt across China, all of the affected areas were characterized by remoteness and a high proportion of subsistence farmers who were very dependent on their local environment for their food supply. Investigators noticed that WMD in animals occurred in the same areas and further studies demonstrated that the soils and crops were very low in selenium. KD occurred in areas where grain crops contained <0.04 mg kg^{-1} of selenium and dietary selenium intakes were extremely low, between 10 and 15 µg day^{-1}. Affected populations were characterized by very low selenium status indicated by hair contents of <0.12 mg kg^{-1} (Xu & Jiang, 1986; Tan, 1989; Yang & Xia, 1995). On the basis of these findings, large-scale mineral supplementation was carried out on 1- to 9-year-old children who were at high risk of the disease. In a trial carried out in Mianning County, Sichuan Province, from 1974 to 1977, 36,603 children were given 0.5- to 1.0-mg sodium selenite tablets per week whereas 9642 children were given placebo tablets.

During the 4 years of investigation, 21 cases of the disease and 3 deaths occurred in the selenium-supplemented group whereas 107 cases and 53 deaths occurred in the control group. By 1977 all the children were supplemented with selenium and the disease was no longer prevalent in either group. The results showed that supplements of 50-µg day^{-1} selenite could prevent the disease but if the disease was already manifest, selenium was of no therapeutic value (Anonymous, 2001).

Although the disease proved to be selenium-responsive, the exact biological function of the element in the pathogenesis was less clear and the seasonal variation in disease prevalence suggested a viral connection. Subsequent studies have demonstrated a high prevalence of the Coxsackie B virus in KD patients (see, for example, Li et al., 2000) and studies have proved increased cardiotoxicity of this virus in mice suffering from selenium and vitamin E deficiency. For a number of years it was thought that selenium deficiency impaired the immune function lowering viral resistance, however, more recent work by Beck (1999) has shown that a normally benign strain of Coxsackie B3 (CVB3/0) alters and becomes virulent in either selenium-deficient or vitamin E-deficient mice. Once the mutations are completed, even mice with normal nutritional status become susceptible to KD. These changes in the virus are thought to occur as a result of oxidative stress due to low vitamin E and low selenium status. This work demonstrates not only the importance of selenium deficiency in immunosuppression of the host but in the toxicity of the viral pathogen as well. Other studies have implicated moniliformin mycotoxins produced by the fungi *Fusarium proliferatum* and *F. subglutinans* in corn as a possible cause of KD (Pineda-Valdes & Bullerman, 2000). As with many environmental conditions, KD is likely to be multifactorial but even if selenium deficiency is not the main cause of the disease, it is clearly an important factor.

During the 1980s the prevalence of KD dropped to less than 5 per 100,000 with less than 1000 cases reported annually. The reason for this is twofold: first, widespread selenium supplementation programs have been carried out on the affected populations and secondly, economic and communication improvements in China as with the rest of the world mean that the population is increasingly less dependent on locally grown foodstuffs in the diet. In recent years the incidence of the disease has dropped still further so that it is no longer considered a public health problem in China (Burk, 1994).

2. Kashin-Beck Disease

Kashin-Beck disease (KBD), an endemic osteoarthropathy (stunting of feet and hands) causing deformity of the affected joints, occurs in Siberia, China, North Korea, and possibly parts of Africa. The disease is named after the Russian scientists who first described it between 1861 and 1899. It is characterized by chronic disabling degenerative osteoarthrosis affecting the peripheral joints and the spine with apoptosis of the hyaline cartilage tissues. Impairment of movement in the extremities is commonly followed by bone development disturbances such as shortened fingers and toes and in more extreme cases, dwarfism (Figure 9) (Levander, 1986; Tan, 1989; WHO, 1996). Indeed the main feature of KBD is short stature caused by multiple focal apoptosis in the growth plate of the tubular bones. In China, the pattern of disease incidence is concordant with KD in the north of the country, but the links with selenium-deficient environments are less clear (Tan, 1989).

Initial studies revealed that rats fed grain and drinking water from the affected areas in China suffered acute massive liver apoptosis and foodstuffs from the affected areas were found to be low in selenium. Children and nursing mothers were supplemented with

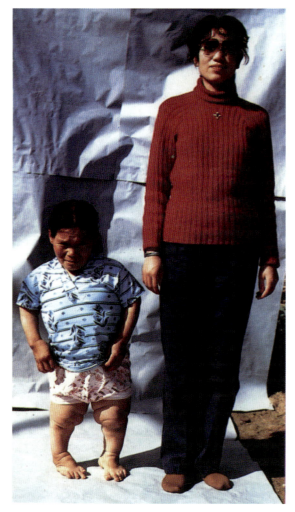

FIGURE 9 Kashin-Beck disease patient (left) and woman of the same age. (Reproduced with permission from Tan, 1989.)

0.5–2.0 mg sodium selenite per week for a period of 6 years and the disease prevalence dropped from 42% to 4% in children aged 3–10 years as a result (Tan, 1989). More recent studies carried out since the early 1990s demonstrated KBD-like cartilage changes and bone mineral density reduction in selenium-deficient rats (Sasaki et al., 1994; Moreno-Reyes et al., 2001). However, other factors have been implicated in the pathogenesis of KBD. The main theory proposed by Russian investigators was that the disease was a result of mycotoxins in the diet, and other work carried out in China has suggested ingestion of contaminated drinking water as a possible cause. In China, higher fungal contamination of grain in KBD areas has been known for a number of years. Other work suggests that the presence of humic substances in drinking water is a factor, and the mechanism of action is free radical generation from the oxy and hydroxyl groups of fulvic acid. Nonetheless, selenium was confirmed as a preventative factor in KBD in these studies (Peng et al., 1999).

There are similarities between KBD and the iodine-deficiency disorder cretinism. Several studies have considered the relationships between KBD and selenium and iodine deficiency. Work in the Yulin District of China (Zhang et al., 2001) carried out on 353 rural school children aged 5–14 years compared data between three endemic KBD villages (prevalence rates 30–45%) and a non-endemic village. Higher fungal contamination was recorded in cereal grain stores in KBD areas than in the non-endemic village, and hair selenium and urinary iodine concentrations were lower in families suffering from the disease than in control groups. However, iodine deficiency did not correlate significantly with increased KBD risk. Recent work into the disease has focused on Tibet and it does implicate iodine in the pathogenesis. Among 575 5- to 15-year-old children examined in 12 villages, 49% had KBD, 46% had the iodine deficiency disorder goiter, and 1% had cretinism. Of the examined population, 66% had urinary iodine contents of $<0.02\,\mu g\,L^{-1}$ and the content was lower in KBD patients ($0.12\,\mu g\,L^{-1}$) than in control subjects ($0.18\,\mu g\,L^{-1}$). Hypothyroidism was more frequent in the KBD group (23%) compared to 4% in the controls. Severe selenium deficiency was present in all groups with 38% of subjects with serum concentrations of $<5\,mg\,L^{-1}$ (normal range 60–105 $mg\,L^{-1}$). Statistical analyses revealed an increased risk of KBD in groups with low urinary iodine in the severe selenium-deficient areas. Here also, mesophilic fungal contamination in barley (*Alternaria* sp.) was higher in KBD areas than non-endemic areas and disease prevalence correlated positively with the humic content of drinking water. The results suggest that KBD is multifactorial and occurs as a consequence of oxidative damage to cartilage and bone cells when associated with decreased antioxidant defense. Another mechanism that may coexist is bone remodeling stimulated by thyroid hormones whose actions are blocked by certain mycotoxins (Suetens et al., 2001).

3. Iodine Deficiency Disorders

In addition to the links between selenium and iodine deficiency in KBD, the recent establishment of the role of the selenoenzyme, iodothyronine deiodinase (IDI), in thyroid function means that selenium deficiency is now being examined in relation to the iodine deficiency disorder (IDD) goiter and cretinism. Many areas around the world where IDD is prevalent are deficient in selenium including China, Sri Lanka, India, Africa, and South America (WHO, 1987, 1996). Concordant selenium and iodine deficiency are thought to account for the high incidence of cretinism in Central Africa, in Zaire and Burundi in particular (Kohrle, 1999), and selenium deficiency has been demonstrated in populations suffering IDD in Sri Lanka (Fordyce et al., 2000a). However, these links require further investigation to determine the role of selenium in these diseases.

4. Cancer

Following studies that revealed an inverse relationship between selenium in crops and human blood versus cancer incidence in the United States and Canada (Shamberger & Frost, 1969), the potential anti-carcinogenic effect of selenium has generated a great deal of interest in medical science. Many studies to examine the links between selenium and cancer in animal experiments and humans have been carried out; however, to date, the results are equivocal. There is some evidence to suggest that selenium is protective against cancer due to its antioxidant properties, the ability to counteract heavy metal toxicity, the ability to induce cell death, the ability to inhibit cell growth, and the ability to inhibit nucleic acids and protein synthesis (WHO, 1987, 1996; Clark et al., 1996; Varo et al., 1998). However, other studies have shown that selenium may promote cancer based on the pro-oxidant mutagenic and immunosuppressive actions of some selenium compounds. For example, the supplementation of sodium selenate, sodium selenite, and organic selenium have been shown to reduce the incidence of several tumor types in laboratory animals, but selenium sulfide has been shown to be carcinogenic in animals and has been classified as a Group B2 compound—a possible human carcinogen (WHO, 1987, 1996; U.S.-EPA, 2002a). Human studies have demonstrated low levels of selenium in the blood of patients suffering gastrointestinal cancer, prostrate cancer, or non-Hodgkin's lymphoma, but there is some evidence to suggest that selenium increases the risks of pancreatic and skin cancer (WHO, 1987, 1996; Birt, 1989).

An excellent review of the work into selenium and cancer is presented by Vinceti et al. (2000) and is summarized as follows. In a recent study of patients with a history of basal or squamous cell skin cancer, selenium intakes of 200 μg day^{-1} appeared to reduce mortality from all cancers and the incidence of lung, colorectal, and prostate cancers. However, it did not prevent the appearance of skin cancer. Indeed, some studies have shown an inverse relationship with melanoma risk but other studies have shown no relationship with non-melanoma skin cancer. However, Vinceti et al. (1998) carried out assessments of populations inadvertently exposed to high selenium in drinking water and reported higher mortality from lung cancer, melanoma, and urinary cancer among men and lymphoid neoplasm in women in the exposed group compared to controls. Other studies have shown an increased risk of colon and prostate cancer in populations taking selenium supplements in Iowa in the United States and Finland than in control populations. However, a study in Montreal carried out between 1989 and 1993 found no association between selenium status (measured by toenail selenium) and breast or prostate cancer but showed an inverse relationship with colon cancer (Vinceti et al., 2000).

In a study of stomach cancer in Finland and The Netherlands, an inverse relationship between selenium status and disease prevalence was found in Finnish men but not in Finnish women or in men or women from The Netherlands. No relationship between selenium status and stomach cancer was evident in studies carried out in Japan. A link between low-selenium status and pancreatic cancer was observed in Maryland in the United States and in Finland, but a similar relationship with bladder and oropharyngeal cancer was evident in Maryland only (Vinceti et al., 2000). Indeed Finland provides an interesting case because the government was so concerned about the low level of selenium intake in the Finnish diet that in 1984 a national program was initiated to increase the selenium content of Finnish foodstuffs by adding sodium selenate fertilizers to crops. Mean daily intakes rose from $45 \mu g \, day^{-1}$ in 1980 to $110-120 \mu g \, day^{-1}$ between 1987 and 1990 and $90 \mu g \, day^{-1}$ in 1992 (Varo et al., 1998). Studies of cancer incidence over this time carried out in Finland, Sweden, and Norway showed no reduction in colon cancer, non-Hodgkin's lymphoma, or melanoma in Finland whereas breast and prostrate cancer rates increased compared to the other two countries. Populations in Finland and New Zealand are known to have much lower selenium status than many other countries and yet no excess incidence of breast and colon cancer is evident. Similarly no relationship between cancer prevalence and selenium intakes as low as $14 \mu g \, day^{-1}$ was identified among rural farmers in China. Conversely, recent work by Finley et al. (2001) has demonstrated a link between consumption of high-selenium broccoli and reduced colon and mammary cancer prevalence. Based on the evidence presented above, it is fair to say that "the jury is still out" in terms of the beneficial effects or otherwise of selenium and cancer.

5. Cardiovascular Disease

Selenium deficiency has also been implicated in cardiovascular health and it is suggested that serum concentrations of less than $45 \mu g \, L^{-1}$ increase the risk of ischemic heart disease. Animal studies have demonstrated that selenium could play a protective role by influencing platelet aggregation and increasing production of thromboxane A2 while reducing prostacyclin activity. However, the epidemiological evidence and studies into selenium status and disease risk provide contradictory results (Levander, 1986; WHO, 1987, 1996).

6. Reproduction

The full role of selenium in reproduction has yet to be established; however, selenium deficiency has been shown to cause immotile and deformed sperm in rats (Wu et al., 1979; Hawkes & Turek, 2001) and studies into this field are ongoing in the medical profession at the current time.

7. Other Diseases

Selenium deficiency has been linked to a number of other conditions in man as the concentration of the element is decreased in the serum/plasma or erythrocytes of patients with AIDS, trisomy-21, Crohn's and Down syndrome, and phenylketonuria. The evidence of viral mutogeny under selenium deficiency established by Beck (1999) in the case of the Coxsackie B virus has major implications in terms of the toxicity and immunoresponse to many viral infections, particularly AIDS, in light of the widespread selenium-deficient environments of central and southern Africa where the disease has reached epidemic proportions in recent years (Longombe et al., 1994). The links between selenium status and AIDS require further investigation in that region.

Selenium deficiency has also been linked to muscular dystrophy (a similar human disease to WMD in animals) and muscular sclerosis, but again the medical evidence for the role of the element is equivocal. Selenium supplementation has proved beneficial to patients suffering renal disease and finally an inverse relationship between selenium status and asthma incidence has also been postulated (WHO, 1987).

Dietary intakes of $0.1–0.2\,mg\,kg^{-1}$ selenium are considered nutritionally generous converting to $50–100\,\mu g$ day^{-1} for a typical person and $0.7–2.8\,\mu g\,kg$ body $weight^{-1}$. Even in New Zealand and Finland, where selenium intake is $30–50\,\mu g\,day^{-1}$, compared with $100–250\,\mu g\,day^{-1}$ in the United States and Canada, overt clinical signs of selenium deficiency are rare among humans (WHO, 1987, 1996). Nonetheless, research increasingly shows the essential nature of selenium to human health and the potential for subclinical effects should not be underestimated. Concern is growing in many regions of the world over low levels of dietary selenium intake in human populations.

E. Selenium Toxicity—Effects in Animals

Experiments on laboratory animals have demonstrated that hydrogen selenide is the most toxic selenium compound by inhalation, sodium selenite the most toxic via ingestion, and elemental selenium in the diet has low toxicity as it is largely insoluble (U.S.-EPA, 2002a; WHO, 1987). Sodium selenite or seleniferous wheat containing $6.4\,mg\,kg^{-1}$ selenium causes growth inhibition and hair loss in animals and at concentrations of $8\,mg\,kg^{-1}$, pancreatic enlargement, anemia, elevated serum bilirubin levels, and death follow (Levander, 1986; WHO, 1987). In addition to food intake, the application of sodium selenate in drinking water has been shown to cause fetal deaths and reduced fertility in mice. Selenium sulfide is the only compound proven to be carcinogenic in animal studies which results in increased liver tumors in rats. Although it is used in anti-dandruff shampoos it is not normally found in food and water. The oral lethal dose for sodium selenite in laboratory animals has been shown to range from 2.3 to $13\,mg\,kg^{-1}$ body weight. Methylation of selenium is used as a detoxification mechanism by animals, and inorganic and organic forms of selenium are metabolized to form mono-, di-, or trimethylated selenium, of which monomethylated forms are most toxic. For example, dimethylselenide is 500–1000 times less toxic than selenide (Se^{2-}) (WHO, 1987) (see also Chapter 20, this volume).

In natural conditions, acute selenium intoxication is uncommon as animals are not normally exposed to high selenium forage and tend to avoid eating selenium accumulator plants. Abnormal posture and movement, diarrhea, labored respiration, abdominal pains, prostration, and death, often as a result of respiratory failure, characterize toxicity. The characteristic symptoms of selenium poisoning are the garlic odor due to exhalation of dimethylselenide, vomiting, shortness of breath, and tetanic spasms. Pathological changes include congestion of the liver and kidneys, and swelling and hemorrhages of the heart (Levander, 1986; WHO, 1987, 1996).

Chronic selenium intoxication is more common and leads to two conditions known as alkali disease and blind staggers in grazing animals. Alkali disease occurs after ingestion of plants containing $5–40\,mg\,kg^{-1}$ over weeks or months and is characterized by dullness, lack of vitality, emaciation, rough coat, sloughing of the hooves, erosion of the joints and bones, anemia, lameness, liver cirrhosis, and reduced reproductive performance. Although much of the work on alkali disease has focused on cattle, consumption of feeds containing $2\,mg\,kg^{-1}$ of selenium has also been shown to cause hoof deformation, hair loss, hypochromic anemia, and increased alkali and acid phosphatase activities in sheep (Levander, 1986; WHO, 1987).

Blind staggers occurs in cattle and sheep but not in horses and dogs and occurs in three stages:

- Stage 1: The animal wanders in circles, has impaired vision, and is anorexic.
- Stage 2: The stage 1 effects get worse and front legs weaken.
- Stage 3: The tongue becomes partially paralyzed and the animal cannot swallow and suffers blindness, labored respiration, abdominal pain, emaciation, and death.

Pathological changes include liver apoptosis, cirrhosis, kidney inflammation, and impaction of the digestive tract. Treatment of the condition involves drenching with large amounts of water and ingestion of strychnine sulfate. However, selenium may not be the main cause of blind staggers, which has similarities to thiamine deficiency. High sulfate intake has been implicated in the disease and may enhance the destruction of thiamine (WHO, 1987).

In addition to alkali disease and blind staggers, high selenium intakes in pigs, sheep, and cattle have been shown to interfere with normal fetal development. Selenosis has been known to cause congenital malformation in sheep and horses and reproductive problems in rats, mice, dogs, pigs, and cattle whereby females with high selenium intakes had fewer smaller young that were often infertile. Blood selenium levels of $>2\,mg\,L^{-1}$ in cattle and $>0.6–0.7\,mg\,L^{-1}$ in sheep are associated with selenosis with borderline toxicity at $1–2\,mg\,L^{-1}$ in cattle (Levander, 1986; WHO, 1987).

Although much of the work into selenium toxicity has focused on agricultural species, selenosis has also been reported in wild aquatic species and birds. Selenium concentrations of $47–53\,\mu g\,L^{-1}$ in surface waters results in anemia and reduced hatchability of trout whereas concentrations of $70–760\,\mu g\,L^{-1}$ in water are toxic to most aquatic invertebrates. Cranial and vertebral deformities occur in frogs exposed to $2000\,\mu g\,L^{-1}$ in surface waters. Selenium toxicity is also associated with embryonic deformities in birds; indeed, the hatchability of fertile eggs is a sensitive indicator of selenium intoxication. At concentrations of $6–9\,mg\,kg^{-1}$ in the diet, embryos suffer brain tissue, spinal cord, and limb bud deformities whereas $>7\,mg\,kg^{-1}$ causes reduced egg production and growth (WHO, 1987; Jacobs, 1989).

F. Selenium Toxicity—Effects in Humans

The toxicity of selenium compounds to humans depends on the chemical form, concentration, and on a number of compounding factors. The ingestion of selenious acid is fatal to humans, preceded by stupor, hypertension, and respiratory depression whereas the toxicity of methylated selenium compounds depends not only on the dose administered but also on the previous level of selenium intake. Higher selenium intake prior to dosing with methylated compounds has been shown to be protective against toxicity in animal experiments. Poor vitamin E status increases the toxicity of selenium and the nutritional need for the element, whereas sulfate counteracts the toxicity of selenate but not of selenite or organic selenium and increases selenium urinary excretion. Methylmercury enhances selenium deficiency but inorganic mercury increases methylated selenium toxicity. At intakes of $4–8\,mg\,kg^{-1}$, selenium increases the copper contents of the heart, liver, and kidneys but has a detoxifying or protective effect against cadmium and mercury (WHO, 1987; Bedwal et al., 1993). High selenium intake has also been shown to decrease sperm motility in healthy men (Hawkes & Turek, 2001) and has been related to increased incidence of some forms of cancer including pancreatic and skin cancer (see Section III.D in this chapter).

Overt selenium toxicity in humans is far less widespread than selenium deficiency. Following the discovery of seleniferous environments and the incidence of alkali disease in animals in the Great Plains in the United States during the 1950s, concern about potential adverse affects on the human population were raised. The health status of rural populations in seleniferous areas was examined. Results showed elevated urinary selenium levels in the population but no definite links to clinical symptoms of selenosis. However, a higher incidence of gastrointestinal problems, poor dental health, diseased nails, and skin discoloration were reported (Smith & Westfall, 1937). In similar studies in a seleniferous region of Venezuela, the prevalence of dermatitis, hair loss, and deformed nails among children was higher than in non-seleniferous areas. The hemoglobin and hematocrit values in children from the seleniferous areas were lower than in controls but did not correlate with blood or urine selenium levels and evidence of selenium toxicity effects was rather inconclusive. Nine cases of acute selenium intoxication due to the intake of nuts of the *Lecythis ollaria* tree in a seleniferous area of Venezuela have been reported to result in vomiting and diarrhea followed by hair and nail loss and the death of a two-year-old boy (WHO, 1987) (see also Chapter 23, this volume).

In China, an outbreak of endemic human selenosis was reported in Enshi District, Hubei Province, and in Ziyang County, Shanxi Province, during the 1960s. This condition was associated with consumption of high-selenium crops grown on soils derived from coal containing $>300\,mg\,kg^{-1}$ selenium. In the peak prevalence years (1961 to 1964) morbidity rates reached 50% in the worst affected villages, which were all located in remote areas among populations of subsistence farmers. Hair and nail loss were the prime symptoms of the disease but disorders of the nervous system, skin, poor dental heath, garlic breath, and paralysis were also reported. Although no health investigations were carried out at the time, subsequent studies in these areas carried out in the 1970s revealed very high dietary intakes of 3.2–6.8 mg with a range of selenium in the blood of $1.3–7.5\,mg\,L^{-1}$ and hair selenium levels of $4.1–100\,mg\,kg^{-1}$ (Yang et al., 1983; Tan, 1989).

Selenium toxicity related to mineral supplement intake has also been reported in the United States. In 1984, 12 cases of selenosis due to intakes over 77 days of tablets labeled to contain 0.15–0.17 mg selenium, but which actually were found to contain 27–31 mg selenium were reported. Patients suffered nausea, vomiting, nail damage, hair loss, fatigue, irritability, abdominal cramps, watery diarrhea, skin irritation, and garlic breath and had blood serum levels of $0.528\,mg\,L^{-1}$ (WHO, 1987). The U.S.-EPA recommends an upper limit of mineral supplementation of selenium of $0.1\,mg\,kg^{-1}$.

Indeed, a whole list of symptoms has been implicated in elevated selenium exposure including severe irrita-

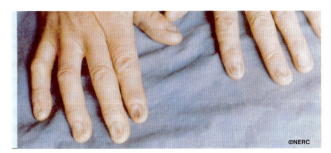

FIGURE 10 Nail deformities as a result of selenium poisoning in Enshi District, China. (Photo: Professor Mao Dajun. Reproduced with permission from the British Geological Survey, Keyworth, Nottingham.)

tions of the respiratory system, metallic taste in the mouth, tingling and inflammation of the nose, fluid in the lungs, pneumonia, the typical garlic odor of breath and sweat due to dimethylselenide excretion, discoloration of the skin, dermatitis, pathological deformation, and loss of nails (Figure 10), loss of hair (Figure 11), excessive tooth decay and discoloration, lack of mental alertness and listlessness, peripheral neuropathy, and gastric disorders. The links with dental health are somewhat equivocal and many of the studies indicating a possible link with selenium failed to take into account other factors such as the fluoride status of the areas of study (WHO, 1987).

Part of the problem in assessing high selenium exposure is that there is some evidence to suggest populations can adapt to or tolerate high selenium intakes without showing major clinical symptoms. Investigations are also hampered by the lack of a sensitive biochemical marker of selenium overexposure (WHO, 1996). Hair loss and nail damage are the most common and consistent clinical indications of the condition. Chinese studies carried out in the seleniferous areas of Hubei and Ziyang have demonstrated that these effects are evident above dietary intakes of $900\,\mu g\ day^{-1}$, blood plasma levels of $1\,mg\,L^{-1}$, and whole blood concentrations of $0.813\,mg\,L^{-1}$ (Yang et al., 1983). Interestingly, further work in China has shown a marked reduction in the ratio of selenium in plasma compared to that in erythrocytes at dietary intakes of $750\,mg\,kg^{-1}$. This is the first indication of a biochemical response to high selenium intakes prior to the development of clinical symptoms (Yang et al., 1989). There is still a great deal of uncertainty about harmful doses of selenium, but a maximum recommended dietary intake of $400\,\mu g\ day^{-1}$ has been proposed based on half the level of intake found in the Chinese studies (WHO, 1996).

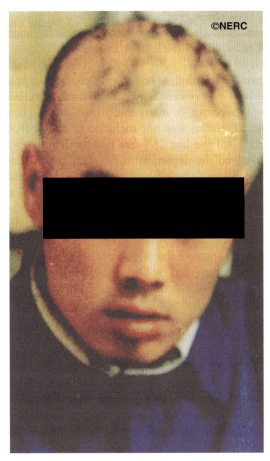

FIGURE 11 Hair loss as a result of selenium toxicity in Enshi District, China. (Photo: Professor Mao Dajun. Reproduced with permission from the British Geological Survey, Keyworth, Nottingham.)

IV. MEASURING SELENIUM STATUS

Thus far in this chapter selenium deficiency and toxicity in the environment, plants, animals, and human beings has been discussed, but in order to assess selenium status, it is important to consider how it is measured. Information about the chemical composition of the terrestrial environment is generally collated by national survey organizations concerned with geology, soil, water, agriculture, and vegetation. In terms of geology, for example, over 100 countries around the world carry out national geochemical mapping programs. These programs are based on systematic collection of materials such as soil, sediment, water, rock, and vegetation, which are then analyzed for a range of element compositions and used to produce maps of element distributions in the environment. This type of approach was pioneered in the disease and areas affected with selenosis (Tan, 1989). Determination of the selenium status of water is usually made using comparisons to the WHO maximum admissible concentration of $10 \mu g \, L^{-1}$.

In veterinary science, concentrations of $<0.04 \, mg \, L^{-1}$ selenium in animal whole-blood are considered deficient and are related to WMD in ruminant species whereas $0.07–0.10 \, mg \, L^{-1}$ is considered adequate, which highlights the extremely narrow range in selenium status between clinical and non-clinical outcomes (Levander 1986; WHO, 1987). Human selenium status is rather more difficult to categorize because of the lack of overt clinical symptoms in many populations exposed to supposedly deficient or toxic intakes, but based on work in China, deficiency and toxicity thresholds in human hair of 0.2 and $>3 \, mg \, kg^{-1}$ (Yang et al., 1983; Tan, 1989; Yang & Xia, 1995), respectively, have been suggested whereas dietary limits of 40 and $400 \mu g \, day^{-1}$ are proposed by the WHO as an indication of human selenium status (WHO, 1996).

V. REMEDIATION

A variety of methods have been used to try and counteract the impacts of selenium deficiency and toxicity in environments and within animals and humans as follows.

A. Remediating Selenium Deficiency

Methods to enhance selenium in the environment and uptake into agricultural crops and animals have been developed over a number of years. One approach is to alter the species of crops grown on deficient soil to plant types that take up more selenium. Switching from white clover production to certain grasses to increase the selenium content of fodder crops has been used successfully in New Zealand (Davis & Watkinson, 1966). Another approach is to apply selenium-rich fertilizers to the soil to increase the amount of selenium taken up by plants, animals, and humans. Some rock phosphate fertilizers are rich in selenium and can be used to enhance uptake; however, there is some risk associated with application of selenates to alkali and neutral soils because of high bioavailability. Use of selenate fertilizer results in much higher selenium contents of first cuts of crops or forage, which decrease sharply with subsequent cropping.

Addition of selenite to acid-neutral soils can result in some loss of selenium to soil adsorption, which decreases the effectiveness of the application, but in some cases this mechanism can ensure that levels of uptake are not toxic (Fleming, 1980; Jacobs, 1989). The selenium concentration of foods can also be increased by supplementing ordinary fertilizers with soluble selenium compounds. At the current time, only Finland, New Zealand, and parts of Canada and China allow selenium-enhanced fertilizers to be used for the cultivation of food crops. These countries use fertilizers based on sodium selenate (Oldfield, 1999). For example, in New Zealand 1% granular selenium is mixed with granulated fertilizer and is applied at a rate of $10 \, g \, Se \, ha^{-1}$ over about a quarter of the agricultural land in the country (in 1998 1.2 million of 4.5 million ha underwent selenium fertilization) (Jacobs, 1989; Oldfield, 1999).

Problems of uptake associated with retention of selenium in the soil can be circumvented by direct application of the fertilizer to the plants themselves. Foliar application of selenite to plants has been successfully used to increase the selenium content of crops and animals. Spraying selenium at $3–5 \, g \, ha^{-1}$ has been shown to increase the content in grain whereas sodium selenite applied at $50–200 \, g \, Se \, ha^{-1}$ maintained $>0.1 \, mg \, kg^{-1}$ contents in crops through three harvests. Studies have shown that the selenium content of crops is enhanced by mid-tillering spraying with selenium fertilizer but it cannot be applied successfully to seeds (Jacobs, 1989).

For example, Chinese workers have reported much better uptake of selenium in maize crops grown on aerated oxygenated soils than in rice grown in the same soils under waterlogged conditions due to reduction of selenium to insoluble forms. To avoid poor uptake of selenium from soils as a consequence of the waterlogged conditions, foliar spraying of sodium selenite at an early shooting stage of the rice plant growth was found to improve the selenium content of the grain and hull (Cao et al., 2000). In another study, the average wheat selenium contents in Kashin-Beck endemic areas were $0.009 \, mg \, kg^{-1}$ dry weight resulting in daily intakes of $12 \mu g$ in the local population. Following foliar selenium fertilizer application, wheat contents increased to $0.081 \, mg \, kg^{-1}$ dry weight and human daily dietary intakes rose to $47 \mu g$ (Tan et al., 1999).

In Finland, the bioavailability of soil selenium for plants is generally poor due to the relatively low selenium concentration, low pH, and high iron content of the soil as much of the country comprises very ancient hard crystalline granite and gneiss rocks. This is very similar to eastern Canada where selenium

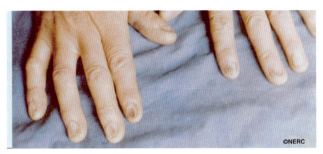

FIGURE 10 Nail deformities as a result of selenium poisoning in Enshi District, China. (Photo: Professor Mao Dajun. Reproduced with permission from the British Geological Survey, Keyworth, Nottingham.)

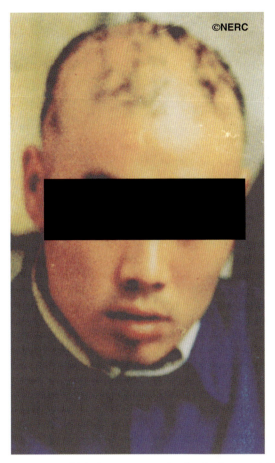

FIGURE 11 Hair loss as a result of selenium toxicity in Enshi District, China. (Photo: Professor Mao Dajun. Reproduced with permission from the British Geological Survey, Keyworth, Nottingham.)

tions of the respiratory system, metallic taste in the mouth, tingling and inflammation of the nose, fluid in the lungs, pneumonia, the typical garlic odor of breath and sweat due to dimethylselenide excretion, discoloration of the skin, dermatitis, pathological deformation, and loss of nails (Figure 10), loss of hair (Figure 11), excessive tooth decay and discoloration, lack of mental alertness and listlessness, peripheral neuropathy, and gastric disorders. The links with dental health are somewhat equivocal and many of the studies indicating a possible link with selenium failed to take into account other factors such as the fluoride status of the areas of study (WHO, 1987).

Part of the problem in assessing high selenium exposure is that there is some evidence to suggest populations can adapt to or tolerate high selenium intakes without showing major clinical symptoms. Investigations are also hampered by the lack of a sensitive biochemical marker of selenium overexposure (WHO, 1996). Hair loss and nail damage are the most common and consistent clinical indications of the condition. Chinese studies carried out in the seleniferous areas of Hubei and Ziyang have demonstrated that these effects are evident above dietary intakes of $900\,\mu g\,day^{-1}$, blood plasma levels of $1\,mg\,L^{-1}$, and whole blood concentrations of $0.813\,mg\,L^{-1}$ (Yang et al., 1983). Interestingly, further work in China has shown a marked reduction in the ratio of selenium in plasma compared to that in erythrocytes at dietary intakes of $750\,mg\,kg^{-1}$. This is the first indication of a biochemical response to high selenium intakes prior to the development of clinical symptoms (Yang et al., 1989). There is still a great deal of uncertainty about harmful doses of selenium, but a maximum recommended dietary intake of $400\,\mu g\,day^{-1}$ has been proposed based on half the level of intake found in the Chinese studies (WHO, 1996).

IV. MEASURING SELENIUM STATUS

Thus far in this chapter selenium deficiency and toxicity in the environment, plants, animals, and human beings has been discussed, but in order to assess selenium status, it is important to consider how it is measured. Information about the chemical composition of the terrestrial environment is generally collated by national survey organizations concerned with geology, soil, water, agriculture, and vegetation. In terms of geology, for example, over 100 countries around the world carry out national geochemical mapping programs. These programs are based on systematic collection of materials such as soil, sediment, water, rock, and vegetation, which are then analyzed for a range of element compositions and used to produce maps of element distributions in the environment. This type of approach was pioneered in the

1950s by Russian geochemists and the wide application of these methods has been made possible by improvements in rapid multi-element analysis techniques over the last 60 years.

However, selenium is not an easy element to analyze, partly because concentrations in natural materials are so low. Therefore, in many multi-element geochemical surveys, selenium was not included in the analysis. It was not until the last 25 years that analytical advances have allowed the detection of selenium at low enough concentrations to be of real interest to environmental studies but because these techniques are more expensive than routine analytical programs, selenium is still often missing from the group of determinants despite its environmental importance (Darnley et al., 1995). A summary of some of the selenium data available around the world is provided by Oldfield (1999).

Analytical methods that give good limits of detection ($<1\,mg\,kg^{-1}$) include colorimetry, neuron activation analysis (NAA), x-ray fluorescence spectrometry (XRF), atomic fluorescence spectrometry (AFS), gas chromatography (GC), inductively coupled mass spectrometry (ICP-MS), and inductively coupled atomic emission spectrometry (ICP-AES). Of these, AFS is the most widely used for natural materials such as foods, plants, and soils. NAA is often used to determine different selenium isotopes, especially in tracer studies using ^{75}Se, etc., and gives good detection limits, but it is a more specialized form of analysis. For studying stable isotopes of selenium (for example, ^{78}Se and ^{82}Se) ICP-MS may be used, but it requires enriched and expensive isotope materials. In more recent years, hydride-generation techniques have improved the detection limits of spectrometric methods such as ICP-AES. Ion exchange chromatography has been extensively used to determine selenium compounds in plants whereas gas chromatography is employed to determine volatile selenium compounds. Ion exchange or solvent extraction methods are used to distinguish selenate and selenite species in solution. Recently developed anion exchange high-performance liquid chromatography (HPLC) and ICP-dynamic-reaction–cell MS methods can be used to measure selenium isotopes and selenamino acids including selenocysteine and selenomethionine. Using these techniques it is now possible to measure relatively low selenium concentrations and selenium element species in a wide variety of environmental and biological materials (WHO, 1987).

In animals and humans a variety of bio-indicators of selenium status have been employed. Due to the close association between the level of dietary selenium intake and GSH-Px activity, the fact that the enzyme activity represents functional selenium and that assessments of this enzyme are easier to perform than selenium tests, GSH-Px activity has been used extensively to measure selenium status, especially in animals. However, this method requires caution because GSH-Px activity is influenced by other physiological factors and a non-selenium-dependent GSH-Px enzyme is also present in animals and humans. Furthermore, the enzyme activity may provide an indication of selenium status at lower levels of intake, but at higher concentrations of selenium the GSH-Px activity becomes saturated and the enzyme cannot be used to indicate toxic selenium status (WHO, 1996) (see also Chapters 23 and 30, this volume).

Other indicators of selenium status include whole blood, plasma, or serum; hair; toenail; and urine content. Of these, hair has been used extensively as it is easy to collect. However, caution is required to ensure that samples are not contaminated with residues from selenium-containing shampoos. It should also be noted that urinary selenium cannot be used to measure inhalation exposure to hydrogen selenide gas, selenium oxychloride, or organic-selenium compounds as severe damage to the lungs occurs before elevated selenium contents are evident in the urine (WHO, 1987). Dietary surveys are also commonly used as an indication of selenium intake. Single-day dietary surveys can give errors of up to 90% when used to estimate the real long-term exposure to selenium, because wide ranges in daily intake are commonplace ($0.6–221\,\mu g\,day^{-1}$). Comparisons of different methods have shown that three-week dietary observations give estimates of overall intake to within 20% and are a much more reliable indication of likely selenium status (WHO, 1987).

Regardless of the material sampled, whether it is soil, food, blood, water, hair, etc., selenium status is determined by comparison to a set of thresholds and normative values that have been determined by examining the levels at which physiological effects occur in plants, animals, and humans. Some of these thresholds are listed in Table XVIII. In general, total soil selenium contents of $0.1–0.6\,mg\,kg^{-1}$ are considered deficient as these are the concentrations of selenium found in regions where selenium-deficient livestock are commonplace such as New Zealand, Denmark, and the Atlantic Region of Canada. Work regarding Keshan disease in China suggests levels of $0.125\,mg\,kg^{-1}$ total selenium in soil cause selenium deficiency in the food chain (Yang et al., 1983; Yang & Xia, 1995). However, it should be kept in mind that the amount of total selenium in the soil is not necessarily the critical factor determining selenium status. Several studies have

TABLE XVIII. Deficiency and Toxicity Thresholds and Recommended Upper Limits for Selenium in Various Media

	Deficient $mg\,kg^{-1}$	Marginal $mg\,kg^{-1}$	Moderate $mg\,kg^{-1}$	Adequate $mg\,kg^{-1}$	Toxic $mg\,kg^{-1}$	Reference
Soils						
World Total Se	0.1–0.6					Various
China Total Se	0.125	0.175	0.400		>3	Tan (1989)
China Water-Soluble Se	0.003	0.006	0.008		0.020	Tan (1989)
Plants						
World Plants	0.1			0.1–1.0	3–5	Jacobs (1989)
China Cereal Crops	0.025	0.040	0.070		>1	Tan (1989)
Animals						
Fodder, Animals chronic	<0.04			0.1–3	3–15	Jacobs (1989)
Cattle and sheep liver	0.21					WHO (1987)
Cattle/sheep blood $mg\,L^{-1}$	<0.04	0.05–0.06		0.07–0.1		Mayland (1994)
Humans						
China Hair	0.200	0.250	0.500		>3	Tan (1989)
Urine Excretion $\mu g\,day^{-1}$				10–200		Oldfield (1999)
Ref. Dose $mg\,kg^{-1}day^{-1}$					0.005	US-EPA (2002a)
Food	<0.05				2–5	WHO (1996)
Diet $\mu g\,day^{-1}$	<40			55–75	>400	WHO (1996)

	Maximum Admissible Concentration $mg\,kg^{-1}$	Reference
UK Soil	35–8000	EA (2002)
France Soil	10	Haygarth (1994)
USA Sewage Sludge	100	US-EPA (2003)
UK Sewage Sludge	25	Haygarth (1994)
USA Air $mg\,m^{-3}$	0.2	US-EPA (2002a)
World Water $\mu g\,L^{-1}$	10	WHO (1996)
USA Water $\mu g\,L^{-1}$	50	US-EPA (2002b)
World Human Urine $mg\,L^{-1}$	0.1	WHO (1987)

demonstrated that the total selenium content of the soil can be considered "adequate" but if the selenium is not in bioavailable form, it is not taken up into plants and animals and selenium deficiency can result (see, for example, Fordyce et al., 2000a). Total selenium concentrations in soil can give an indication of likely selenium status but do not necessarily tell the whole story, and a selenium deficient environment is not necessarily one in which total concentrations of selenium in soil are the lowest. In recent years soil with water-soluble selenium has been used as an indicator of the bioavailable fraction and Chinese scientists have recommended soil deficiency and toxicity thresholds on this basis, 0.003 and 0.020 mg kg^{-1}, respectively (Tan, 1989).

Due to the many different factors that can influence the uptake of selenium from soil into plants, vegetation often provides a better estimate than soil of likely environmental status with regard to health problems in animals and humans. Feed crops containing more than 0.1 mg kg^{-1} of selenium will protect livestock from selenium deficiency disorders, whereas levels of >3–5 mg kg^{-1} in plants have been shown to induce selenium toxicity in animals (Levander 1986; Jacobs, 1989). The current MAC of selenium in animal feedstuffs in the United States is 5 mg kg^{-1}. In terms of cereal crops for human consumption, Chinese workers suggest deficiency and toxicity thresholds of 0.02 and 0.10 mg kg^{-1}, respectively, based on epidemiological studies in Keshan disease and areas affected with selenosis (Tan, 1989). Determination of the selenium status of water is usually made using comparisons to the WHO maximum admissible concentration of $10\,\mu g\,L^{-1}$.

In veterinary science, concentrations of $<0.04\,mg\,L^{-1}$ selenium in animal whole-blood are considered deficient and are related to WMD in ruminant species whereas 0.07–$0.10\,mg\,L^{-1}$ is considered adequate, which highlights the extremely narrow range in selenium status between clinical and non-clinical outcomes (Levander 1986; WHO, 1987). Human selenium status is rather more difficult to categorize because of the lack of overt clinical symptoms in many populations exposed to supposedly deficient or toxic intakes, but based on work in China, deficiency and toxicity thresholds in human hair of 0.2 and $>3\,mg\,kg^{-1}$ (Yang et al., 1983; Tan, 1989; Yang & Xia, 1995), respectively, have been suggested whereas dietary limits of 40 and $400\,\mu g\,day^{-1}$ are proposed by the WHO as an indication of human selenium status (WHO, 1996).

V. REMEDIATION

A variety of methods have been used to try and counteract the impacts of selenium deficiency and toxicity in environments and within animals and humans as follows.

A. Remediating Selenium Deficiency

Methods to enhance selenium in the environment and uptake into agricultural crops and animals have been developed over a number of years. One approach is to alter the species of crops grown on deficient soil to plant types that take up more selenium. Switching from white clover production to certain grasses to increase the selenium content of fodder crops has been used successfully in New Zealand (Davis & Watkinson, 1966). Another approach is to apply selenium-rich fertilizers to the soil to increase the amount of selenium taken up by plants, animals, and humans. Some rock phosphate fertilizers are rich in selenium and can be used to enhance uptake; however, there is some risk associated with application of selenates to alkali and neutral soils because of high bioavailability. Use of selenate fertilizer results in much higher selenium contents of first cuts of crops or forage, which decrease sharply with subsequent cropping.

Addition of selenite to acid-neutral soils can result in some loss of selenium to soil adsorption, which decreases the effectiveness of the application, but in some cases this mechanism can ensure that levels of uptake are not toxic (Fleming, 1980; Jacobs, 1989). The selenium concentration of foods can also be increased by supplementing ordinary fertilizers with soluble selenium compounds. At the current time, only Finland, New Zealand, and parts of Canada and China allow selenium-enhanced fertilizers to be used for the cultivation of food crops. These countries use fertilizers based on sodium selenate (Oldfield, 1999). For example, in New Zealand 1% granular selenium is mixed with granulated fertilizer and is applied at a rate of $10\,g\,Se\,ha^{-1}$ over about a quarter of the agricultural land in the country (in 1998 1.2 million of 4.5 million ha underwent selenium fertilization) (Jacobs, 1989; Oldfield, 1999).

Problems of uptake associated with retention of selenium in the soil can be circumvented by direct application of the fertilizer to the plants themselves. Foliar application of selenite to plants has been successfully used to increase the selenium content of crops and animals. Spraying selenium at 3–$5\,g\,ha^{-1}$ has been shown to increase the content in grain whereas sodium selenite applied at 50–$200\,g\,Se\,ha^{-1}$ maintained $>0.1\,mg\,kg^{-1}$ contents in crops through three harvests. Studies have shown that the selenium content of crops is enhanced by mid-tillering spraying with selenium fertilizer but it cannot be applied successfully to seeds (Jacobs, 1989).

For example, Chinese workers have reported much better uptake of selenium in maize crops grown on aerated oxygenated soils than in rice grown in the same soils under waterlogged conditions due to reduction of selenium to insoluble forms. To avoid poor uptake of selenium from soils as a consequence of the waterlogged conditions, foliar spraying of sodium selenite at an early shooting stage of the rice plant growth was found to improve the selenium content of the grain and hull (Cao et al., 2000). In another study, the average wheat selenium contents in Kashin-Beck endemic areas were $0.009\,mg\,kg^{-1}$ dry weight resulting in daily intakes of $12\,\mu g$ in the local population. Following foliar selenium fertilizer application, wheat contents increased to $0.081\,mg\,kg^{-1}$ dry weight and human daily dietary intakes rose to $47\,\mu g$ (Tan et al., 1999).

In Finland, the bioavailability of soil selenium for plants is generally poor due to the relatively low selenium concentration, low pH, and high iron content of the soil as much of the country comprises very ancient hard crystalline granite and gneiss rocks. This is very similar to eastern Canada where selenium

supplementation is also practiced (Jacobs, 1989). In 1984, the Finnish government approved a program of selenium supplementation in fodder and food crops. The program initially involved spraying a 1% selenium solution onto fertilizer granules giving an application rate of 6 g Se ha^{-1} for silage and 16 g ha^{-1} for cereal crops. Within two years a threefold increase of mean selenium intake in the human population was observed and human serum contents increased by 70%. The supplementation affected the selenium content of all major food groups with the exception of fish. In 1990 the amount of selenium that was supplemented was reduced to 6 g Se ha^{-1} for all crops and the mean human selenium intake fell by 30% and the serum selenium concentration decreased by 25% from the highest levels observed in 1989. According to data obtained, supplementation of fertilizers with selenium is a safe and effective means of increasing the selenium intake of both animals and humans and is feasible in countries like Finland with relatively uniform geochemical conditions (Aro et al., 1995). In other countries where the low level of selenium intake is currently of concern, such as in the UK, this kind of intervention would require careful planning and monitoring of the effects on both animal and human nutrition and the environment, because geochemical conditions vary markedly across the country as a result of a diverse geological environment.

In addition to attempts to enhance selenium in fodder crops and animal feeds supplemented with sodium selenite or selenate, selenium deficiency in animals is also prevented by veterinary interventions such as selenium injections to females during late gestation and/or to the young stock shortly after birth, dietary supplements, salt licks, and drenches (Levander, 1986).

In humans also, direct dietary supplementation methods have been used successfully to counteract selenium deficiency. Pills containing selenium alone or in combination with vitamins and/or minerals are available in several countries. Selenium supplements contain selenium in different chemical forms. In the majority of supplements, the selenium is present as 35 selenomethionine; however, in multivitamin preparations, infant formulas, protein mixes, and weight-loss products sodium selenite and sodium selenate are predominantly used. In other products, selenium is present in protein-chelated or amino acid chelated forms. Current animal studies and epidemiological evidence favors selenomethionine as the most bioavailable and readily taken up form of selenium in mineral supplements. A dosage of 200 μg day^{-1} is generally considered safe and adequate for adults of average weight consuming a North American diet (WHO, 1987). Studies carried out in KD

areas of China have shown that both selenite and selenium-yeast supplements were effective in raising GSH-Px activity of selenium-deficient populations, but selenium-yeast provided a longer lasting body pool of selenium (Alfthan et al., 2000). Altering the diets of humans to include selenium-rich foods has also proved successful in preventing selenium deficiency. In China, selenium-rich tea, mineral water, and cereal crops are now marketed in selenium-deficient areas.

B. Remediating Selenium Toxicity

One of the most common methods to reduce the effects of selenium in soil is phytoremediation. This practice is carried out by growing plant species, which accumulate selenium from the soil and volatilize it to the air to reduce levels in soil. For example, the hybrid poplar trees *Populus tremula x alba* can transfer significant quantities of selenium by volatilization from soil to air; the rate for selenomethionine is 230 times that of selenite and 1.5 times higher for selenite than selenate. These trees have been used successfully to reduce selenium contents in soil in the western areas of the United States (Oldfield, 1999).

There is some evidence to suggest that the presence of phosphate and sulfate in soils can inhibit the uptake of selenium in plants and application of these minerals as soil treatments could be beneficial against selenium toxicity in agricultural crops. Studies have shown a tenfold increase in sulfate content reduced uptake from selenate by >90% in ryegrass and clover, whereas a similar increase in phosphate content caused 30–50% decreases in selenium accumulation from selenite in ryegrass, but in clover such decreases only occurred in the roots. Therefore, sulfate-selenate antagonisms were much stronger than phosphate-selenite antagonisms. The addition of sulfur or calcium sulfate (gypsum) to seleniferous soils in North America was not successful in reducing uptake into plants probably because these soils already contain high quantities of gypsum. However, additions of calcium sulfate and barium chloride have been shown to markedly reduce the uptake of selenium in alfalfa in the United States (90–100%) probably due to the formation of $BaSeO_4$, which is barely soluble. The practicalities of this type of selenium remediation method are rather limited (Jacobs, 1989).

It is more common to counteract selenium toxicity with veterinary and medical interventions. Sodium sulfate and high protein intakes have been shown to reduce the toxicity of selenate to rats but not of selenite or selenomethionine in wheat. Arsenic, silver, mercury, copper, and cadmium have all been shown to decrease the toxicity of selenium to laboratory animals and they have been used to alleviate selenium poisoning in dogs, pigs, chicks, and cattle (Moxon, 1938; Levander, 1986; WHO, 1987). The protective effect of arsenic is thought to be a consequence of increased biliary selenium excretion. Laboratory evidence suggests that mercury, copper, and cadmium exert a beneficial effect due to reactions with selenium in the intestinal tract to form insoluble selenium compounds. However, consideration must be given to the toxic effects of these elements before they are applied as selenium prophylaxis. Linseed meal has also been found to counter selenium toxicity in animals by the formation of selenocyanates, which are excreted (WHO, 1987).

In terms of human diets, dietary diversification can also help reduce selenium toxicity. In China, high-selenium cereal crops are banned from local consumption and exported out of the seleniferous regions where they are mixed with grains from elsewhere before they are sold in selenium-deficient parts of the country.

VI. CASE HISTORIES

A. Selenium Toxicity in Animals— Kesterson Reservoir, United States

One of the best known and most studied incidences of selenium toxicity in animals has been recorded at Kesterson Reservoir, California, in the United States. The information summarized here is taken from Jacobs (1989), Wu et al. (2000), Wu (1994), and Tokunaga et al. (1996). These publications should be referred to for further details.

Due to a scarcity of wetlands in California, wildlife resource managers tried using irrigation runoff from subsurface agricultural drains to create and maintain wetland habitats at the Kesterson Reservoir. The reservoir comprises 12 shallow ponds acting as evaporation and storage basins for agricultural drain waters from the San Joaquin Valley. During part of the year, the water from the reservoir was to be discharged via the San Luis drain back into the Sacramento-San Joaquin River delta when river flows were high enough to dilute the contaminants present in the agricultural water. However, construction of the San Luis drain was halted in 1975 due to increased environmental concerns about the impact of the drain water on the river delta. During the 1970s surface water flow into the reservoir predominated, but into the 1980s almost all the flow was shallow subsurface agricultural drainage water. Selenium concentrations in agricultural drainage water entering the Kesterson Reservoir area between 1983 and 1985 were $300\,mg\,L^{-1}$ as a result of contact with seleniferous soils in the catchment area. In this arid alkaline environment, 98% of the selenium was in the most readily bioavailable selenate form with only 2% present as selenite. The effects of this water on plants and animals were relatively unknown prior to studies carried out between 1983 and 1985 by the U. S. Wildlife Service comparing Kesterson to the adjacent Volta Wildlife area, which was supplied with clean irrigation water with normal concentrations of selenium. The mortality of embryos, young and adult birds, survival of chicks, and embryonic deformities were compared between the two sites. The selenium content of the livers of snakes and frogs from the two areas were also examined in addition to tissues from 332 mammals of 10 species, primarily moles. Results of some of the comparisons between biota from the two sites are shown in Table XIX. In all cases, the levels of selenium in biological materials at Kesterson Reservoir exceeded those of the Volta Wildlife area several-fold. Concentrations of selenium

TABLE XIX. Comparison of Selenium Toxicity Effects in Biota from the Seleniferous Kesterson Reservoir and the Selenium-Normal Volta Wildlife Area, California, United States

Sample type	Kesterson $mg\,kg^{-1}$	Volta $mg\,kg^{-1}$
Algae and rooted aquatic plants	18–390	0.17–0.87
Emergent aquatic plants leaves	17–160	<2.0
Terrestrial plants leaves	0.5–27	<4.7
Plankton (geomean)	85.4	2.03
Aquatic insects	58.9–102	1.1–2.1
Mosquito fish	149–380	1.1–1.4
Reptiles (frogs, snakes) liver	11.1–45	2.05–6.22
Birds (coot, duck, stilt, grebe) liver	19.9–43.1	4.41–8.82
Voles liver (geomean)	119	0.228
No. of dead or deformed chick/embryo	22%	1%

From Jacobs (1989).

in water were compared to those in biota collected from the same site and bioaccumulation factors of more than 1000 for animals were found at Kesterson. Although no overt adverse health effects were noted in reptile or mammal species such as voles and raccoons in the area, the levels of selenium present were of concern in terms of bioaccumulation in the food chain. In contrast, the overt health effects on birds were very marked with 22% of eggs containing dead or deformed embryos as a result of selenium toxicity. The developmental deformities included missing or abnormal eyes, beaks, wings, legs and feet, and hydrocephaly, and were fatal. It is estimated that at least 1000 adult and juvenile birds died at Kesterson from 1983 to 1985 as a result of consuming plants and fish with 12–120 times the normal amount of selenium (Jacobs, 1989).

Following these revelations, Kesterson Reservoir was closed and a series of remedial measures were tested by a team of scientists who were able to provide a more thorough understanding of processes leading to selenium transport and biologic exposure in this environment. Some of the schemes proposed included the development of an *in situ* chemical treatment to immobilize soluble selenium in drained evaporation pond sediments by amendment with ferrous iron, which occludes selenate and selenite in ferric oxyhydroxide (FeOOH). Phytoremediation techniques were also tested. These included the growing of barley (*Hordeum vulgare* L.) and addition of straw to the soil, which contained $0.68\,mg\,kg^{-1}$ soluble selenium and $6.15\,mg\,kg^{-1}$ total selenium. Four treatments were evaluated: soil only, soil + straw, soil + barley, and soil + straw + barley. At the end of the experiments, selenium in barley represented 0.1–0.7% of the total selenium in the system, and volatilized selenium accounted for 0.2–0.5% of total selenium. In contrast, straw amendments were found to greatly reduce the amount of selenium in soil solution by 92–97% of the initial soluble selenium and represented a possible remediation strategy for the reservoir. The planting of canola (*Brassica napus*) was also evaluated but accumulated $50\,mg\,kg^{-1}$ (dry weight), which accounted for less than 10% of total selenium lost in the soil solution during the post-harvest period.

Bioremediation through the microbial reduction of toxic oxyanions selenite and selenate into insoluble Se^0 or methylation of these species to dimethylselenide was proposed as a potential bioremediation cleanup strategy. Field trials demonstrated that microorganisms, particularly *Enterobacter cloacae*, were very active in the reduction of selenium oxyanions in irrigation drainage water, into insoluble Se^0 and, by monitoring various environmental conditions and the addition of organic amendments, the process could be stimulated many times. Based upon the promising results of these studies, a biotechnology prototype was developed for the cleanup of polluted sediments and water at Kesterson.

A soil excavation plan had been proposed to remove selenium-contaminated material from the site; however, extensive monitoring of porewater in the vadose zone demonstrated that this plan would be ineffective in reducing the elevated selenium concentration in ephemeral pools present during the winter at Kesterson. Furthermore, extensive biological monitoring demonstrated that selenium concentrations in the dominant species of upland vegetation at Kesterson were near or equal to "safe" levels.

On the basis of these studies, a cost-effective remediation strategy was devised. First, the groundwater under Kesterson was protected from selenium contamination by naturally occurring biogeochemical immobilization. Secondly the contaminated soil and sediment was left in place but low-lying areas were infilled to prevent the formation of the ephemeral pools that attracted wildlife. The area was then planted over with upland grassland species. Monitoring studies carried out on soil and vegetation between 1989 and 1999 showed that selenium losses from soil via volatilization were approximately 1.1% per year. Soil selenium concentrations in the fresh soil fill sites increased in the top 15 cm, which indicated that the plants were able to effectively take up soluble soil selenium from the lower soil profile and deposit it at the land surface thus reducing the rate of leaching of soil selenium. In general plant tissue concentrations reflected the amount of soil water-soluble selenium present, which was low. In 1999 plant tissue concentrations averaged $10\,mg\,kg^{-1}$ (dry weight) and soil water-soluble selenium contents $110\,mg\,kg^{-1}$ giving an estimated bioaccumulation value for the upland grassland of less than 10% of the previous wetland habitat. It was concluded that the new Kesterson grassland did not pose a risk to the environment (Wu et al., 2000).

B. Selenium Toxicity and Drinking Water—Reggio, Italy

Examples of high selenium exposure related to intakes in water are very scarce. An exception is reported by Vinceti et al. (1998) and occurred in the town of Reggio, Italy, between 1972 and 1988 where the population in the Rivalta neighborhood was inadvertently exposed to wells containing $3–13\,\mu g\,L^{-1}$ selenium as selenate and resultant tap water containing $7–9 \mu g \, L^{-1}$ compared to selenium contents in the drinking water of adjacent neighborhoods of $<1 \mu g \, L^{-1}$. The wells were closed off in 1989 and the population was no longer exposed to water from this source. Apart from the selenium content, water quality between Rivalta and the other neighborhoods was the same. Using residency and water supply records, 2065 people (1021 men and 1044 women) were identified as having been exposed for at least 11 years to the elevated selenium content in the water between 1975 and 1988. This cohort was compared to a control population of non-exposed individuals from the same town. To examine the effects of this exposure on cancer incidence in the local population, all cases of pathologically confirmed primary invasive melanoma occurring during 1996 were collated for the entire town of Reggio as well as records on age, sex, educational level, and occupation. The exposed and non-exposed populations had similar educational and occupational profiles and once the data were corrected for age and sex, a higher prevalence of skin cancer was noted in the exposed population. On the basis of melanoma rates in the unexposed population, 2.06 cases would be expected in the exposed population whereas 8 cases were reported. Although other confounding factors could not be taken into account in this study, there is some evidence to suggest that the skin is a target organism in chronic selenium toxicity and that inorganic selenium can act as a pro-oxidant and mutagen and cell apoptosis suppressant, which may account for the higher prevalence of cancer in the exposed group. It should be noted that selenium is ineffective against melanoma although beneficial for other forms of cancer (Clark et al., 1996).

C. Selenium Deficiency in Humans—Zhangjiakou District, Hebei Province, China

Zhangjiakou District, Hebei Province, in China lies between Inner Mongolia to the north and Beijing to the south and is one of the remotest regions of China lying within the northeast-southwest KD belt (Figure 8). The area is underlain by Archaen metamorphic and Jurassic volcanic rocks, which are overlain by Quaternary loess and alluvial deposits, all of which contain low amounts of selenium. Within Zhangjiakou District, the KD belt follows the mountainous watershed between the two rock types, which reflects the fact that villages in the remotest locations where populations are most dependent on locally grown foodstuffs are most at risk from the

disease. However, within the KD belt, prevalence rates show marked variability between villages ranging from 0 to 10.8% between 1992 and 1996. In a study to examine why this variation may occur and to pinpoint the relationships between environmental selenium and disease, Johnson et al. (2000) examined soil, staple crop (wheat and oats), water, and human hair selenium levels in 15 villages in the region classified according to disease prevalence into three groups: (1) no KD 0% prevalence; (2) moderate KD, 0–3% prevalence; and high KD, >3% prevalence. Results showed that hair, grain, and water selenium concentrations showed an inverse relationship with disease prevalence as expected; the highest selenium contents were reported in villages with lowest prevalence of the disease. However, contrary to expectations, soil total selenium contents showed the opposite relationship and were highest in the villages with greatest disease prevalence (Figure 12). Indeed comparisons between the data collected from high prevalence villages for the study and selenium deficiency thresholds proposed by Tan (1989) indicated that the selenium contents of all sample types were very low, whereas hair (geometric mean $177 \, ng \, g^{-1}$, threshold 200 $ng \, g^{-1}$) and grain (geometric mean $7.8 \, ng \, g^{-1}$, threshold $25 \, ng \, g^{-1}$) contents would be classed as deficient, soil total selenium contents would not (geometric mean 171 $ng \, g^{-1}$, threshold $125 \, ng \, g^{-1}$). There was a strong correlation between the selenium content of grain and the selenium status of the local population determined by hair sampling, but relationships with local soils were less clear. Further examinations into the soil geochemistry demonstrated that soils in the high KD prevalence villages were black or dark brown with a high organic matter content and lower pH than other soils in the region. Although these soils contained high total selenium contents, it was not in a readily bioavailable form as it was held in the organic matter in the soil. Despite the higher total selenium contents, water-soluble selenium in the high prevalence villages was in fact lower than deficiency threshold values (geometric mean 0.06 $ng \, g^{-1}$, threshold $3 \, ng \, g^{-1}$). This study concluded that when environmental concentrations of selenium are low, any factor that is responsible for reducing the mobility of selenium may have a critical effect and emphasizes the importance of determining the bioavailability of selenium rather than the total selenium content when assessing impacts on human health. On the basis of this study, conditioning treatments to raise the soil pH thus increasing the bioavailability of selenium in the organic-rich soils or foliar application of selenium fertilizer to crops to avoid selenium adsorption in the soils were recommended as remediation

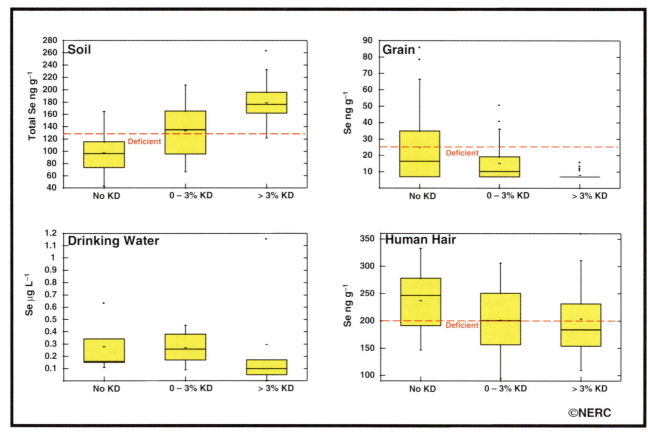

FIGURE 12 Box and whisker plots of the 10th, 25th, 50th, 75th, and 90th percentiles of selenium concentrations in soil, grain, drinking water, and human hair samples from Zhangjiakou District, Hebei Province, China where the selenium-deficiency related condition KD is prevalent in the local population. Samples were collected from three village groups classified into no KD (0% prevalence), moderate KD (0–3% prevalence), and high KD (>3% prevalence). Deficiency thresholds from Tan (1989) are shown as dashed lines on the plots for soil total selenium, grain selenium, and human hair selenium contents. (Reproduced with permission from the British Geological Survey, Keyworth, Nottingham.)

strategies to increase the levels of selenium in local diets. The study also demonstrates the importance of understanding the biogeochemical environment in the determination of selenium-deficient regions and appropriate remediation techniques. However, here as with elsewhere in China, no incidences of KD have been reported in the area since 1996 as economic and communication improvements diversify the diet and enhance the health of the population.

D. The Geological Impact of Selenium on Human Health—Deficiency and Toxicity, Enshi District, Hubei Province, China

If there is one place on Earth that demonstrates the importance of geological controls on selenium and

human health it is Enshi District, Hubei Province, in China, which lies approximately 100 km south of the Yangtze River Gorges and 450 km west-southwest of the provincial capital Wuhan (Figure 8). In Enshi District, selenium-deficiency related diseases (Keshan disease) and selenium toxicity (selenosis) occur within 20 km of each other; their incidence is controlled by geology. The area is very mountainous with little connectivity between villages; some of which can only be reached on foot, hence, populations are very dependent on the local environment for their food supply.

Jurassic sandstones, which contain low concentrations of trace elements including selenium, underlie the northwest part of Enshi District and KD is present in this area. Selenium toxicity, on the other hand, is associated with high environmental selenium derived from Permian age coal-bearing strata in the center and east of the Enshi District. Soils developed over Jurassic sandstones comprise red-purple sands whereas light-brown silts and clays containing many carbonaceous fragments are typical in areas underlain by the Permian strata.

Studies into the selenium balance of local populations were carried out during the 1960s and 1970s by the Enshi Public Health Department in response to outbreaks of selenium-related diseases in the area. Between 1923 and 1988, 477 cases of human selenosis were reported. Of these cases, 338 resulted in hair and nail loss and disorders of the nervous system. They occurred between 1959 and 1963 in Shadi, Xin Tang and Shuang He communities. In Yu Tang Ba village, Shuang He community, the population was evacuated after 82% (19 out of 23 people) suffered nail and hair loss and all livestock died from selenium poisoning. During the same period, 281 selenosis cases were reported in five villages in the Shadi area. Cases of selenosis in pigs reached peak prevalence between 1979 and 1987 when 280 out of 2238 animals were affected in Shatou, which resulted in 122 deaths. No human cases of selenium toxicity have been reported in recent years but animals commonly suffer hoof and hair loss as a result of the high environmental selenium.

During the late 1960s and 1970s, an area of selenium deficiency in Lichuan County to the northwest of Enshi District was also identified and lies within the KD belt across China. In total, 312 people have suffered KD in the county, an average incidence rate of 103 per 100,000. Among the 312 cases, 136 recovered, 163 died, and 13 persons still suffer from the disease. The village of Chang Ping was the worst affected with a total of 259 cases out of a population of 20,368 and 117 of those affected died. Children between the ages of 3 and 8 accounted for 83.4% of the total cases and 80% of the children affected by the disease died. Following peak prevalence in 1969 (106 cases), the number of cases has fallen dramatically and current prevalence rates are unknown as no medical investigations have been carried out in recent years.

Yang et al. (1983) were the first to compare levels of selenium in soil, crops, drinking water, human urine, blood, nail, and hair samples from the Enshi area with other regions of China and demonstrate that the endemic selenium intoxication of humans in Enshi was related to the occurrence of Permian selenium-enriched shaley coal, which contains up to 6471 mg kg^{-1} selenium. There is some evidence to suggest that selenium in these rocks is in the form of micro particles of elemental selenium in association with organic carbon and that the carbon content of the rock controls the selenium content. However, some selenium is also found in the lattice of pyrite minerals. Selenium concentrations in soil, food, and human samples from areas underlain by carbonaceous strata were up to 1000 times higher than in samples from selenium-deficient areas where KD was prevalent and dietary intakes of selenium greatly exceed the U. S. NRC and Chinese recommended standards (Table XX). It was estimated that locally grown crops constituted 90% of the diet in the Enshi area and cereal crops (rice and maize) accounted for 65–85% of the selenium intake, which indicated the importance of the local environment to selenium in the food chain. In addition to exposure via soils and foodstuffs, villagers in the selenosis region also mine the carbonaceous shale for fuel and use burnt coal residues as a soil conditioner. Although the epidemiological investigations revealed that selenosis occurred in areas of high environmental selenium associated with the carbonaceous strata, not all villages underlain by this strata were affected.

Further studies carried out by Fordyce et al. (2000b) into three groups of villages: one suffering KD, one with high environmental selenium but no selenosis, and one with high environmental selenium and selenosis. In villages with high selenium, concentrations in soils and foodstuffs could markedly vary from low to toxic within the same village, with these variations dependent on the outcrop of the coal-bearing strata. The wide range in geochemical conditions could in part explain why some villages suffered selenosis and others did not. Villagers were therefore advised to avoid cultivating

TABLE XX. Estimated Daily Dietary Intake of Selenium in Three Areas of Enshi District, China. Compared to Recommended Intakes Elsewhere

Source	Daily dietary intake of selenium (µg)
Enshi low Se and KD	62–70
Enshi high Se and no selenosis	194–198
Enshi high Se and selenosis	1238–1438
U.S. National Research Council	50–200 RDA
UK Ministry of Agriculture	60–200 RDA
China	40–600 RDA

RDA = Recommended Daily Allowance.
From Yang et al. (1983); Tan (1989); WHO (1996); and MAFF (1997).

TABLE XXI. Selenium in Soil, Grain, and Hair from Enshi District, China. Compared to Selenium Deficiency and Toxicity Threshold Values

Threshold	Soil total Se (mg kg⁻¹)	Soil water soluble Se (ng g⁻¹)	Grain Se (mg kg⁻¹)	Hair Se (mg kg⁻¹)
Deficient	0.125	3	0.025	0.2
Toxic	3	20	1	3
Enshi area village	Geometric mean Se		Concentration ranges (n = 15)	
Low Se	0.069–0.199	0.21–0.44	0.001–0.003	0.252–0.345
High Se	2.07–19.54	2–61	0.041–2.902	0.692–29.21

From Tan (1989) and Fordyce et al. (2002b)

fields underlain by the coal and were counseled against using coal-derived products, such as ash, to condition the soil. In the KD affected villages of Lichuan County, selenium concentrations in staple food crops (rice and maize), drinking water, and the human populations (measured in hair samples) were very low and soils in this area had lower pH contents than soils in the high-selenium villages, which would further inhibit the uptake of selenium into plants. Conditioning the soil with lime to increase the pH making selenium more mobile was suggested as a remediation strategy.

Although all the villages in the low-selenium area had a marginally deficient selenium status and the majority of villages in the high-selenium area had excessive amounts of selenium in the environment and human population using the thresholds defined by Tan (1989) (Table XXI), no new incidents of either KD or overt selenium poisoning have occurred in recent years. This suggests that the local population may have adapted to the high and low selenium intakes present in the different environments and that the historical occurrences of clinical effects related to selenium imbalances were caused by other factors. The outbreaks of human selenosis in Enshi during the late 1950s and early 1960s coincided with a drought and the failure of the rice crop. The crop failure had serious implications for the dietary intake and health of the local population with less food available, reduced protein intake, and higher dependence on vegetables and maize and natural plants. These factors may have lead to the severe outbreaks of selenosis in the Enshi area and demonstrate that in geologically controlled high- or low-selenium environments

additional stresses can lead to serious health outcomes in the local population.

E. The Geological Impact of Selenium on Animal Health—Deficiency and Toxicity, Queensland, Australia

Another example of the effects of geology on selenium and health has been reported in Queensland, Australia. Here seleniferous limestones and shales of the Tambo Formation cause selenium toxicity symptoms in livestock grazing plants in this area, whereas less than 100 km to the south selenium deficiency and WMD in grazing animals is a problem over Tertiary volcanic soils. Grain grown over the seleniferous limestone rocks contains >0.2 mg kg⁻¹ selenium whereas over the southern selenium-deficient region concentrations rarely exceed 0–0.05 mg kg⁻¹ in grain (Oldfield, 1999). This is another example of how geologically controlled geochemical variation can influence selenium status and health over relatively short distances.

F. Selenium Status in Western Countries—Is Environment Still Important?

With the exception of Italy, the human case studies presented in this chapter refer to developing countries where populations are very dependent on the local environment to provide the correct mineral balance. Under these circumstances it is easy to see why considerations of selenium status may be important. But what of the western world where people generally move around more during their lifetimes, products are derived from all around the world, and people buy food in large supermarkets rather than growing it in their own back yard? Under these conditions, the links between environment and health are less direct. Nonetheless, the impact of the selenium status of the environment on animals and humans is still evident. It has already been pointed out in this chapter that New Zealand is a generally selenium-deficient country compared to other areas of the world; indeed cereal grains grown in New Zealand contain 10 times less selenium than grain from Canada and the United States. In years when the crops in New Zealand are poor, wheat is imported from Australia and a corresponding notable increase in blood selenium levels is seen in the population. Studies have also shown that the average total body contents of selenium are only $3–6.1\,mg\,kg^{-1}$ in New Zealand compared to $14.6\,mg\,kg^{-1}$ in the United States, and studies into individual tissues of the body show that in New Zealand, concentrations are half that of the United States. A marked lowering of blood selenium levels has been noted in populations moving from selenium-adequate areas of the United States to New Zealand; however, the actual resultant values also depend on factors such as physiological status (WHO, 1987). There is clear evidence therefore of the influence of the geochemical environment on food and human selenium status when either food or people move from one area to another. But what about variation within a country? Perhaps the most compelling evidence that environmental differences are important even within western countries comes from the United States where, despite one of the most diverse and mobile food supply chains in the world, selenium concentrations in animals and humans reflect the surrounding environment. Studies have shown that despite the widespread use of agricultural management practices including selenium supplementation, the selenium content of skeletal muscle in cattle shows marked geographic variation concordant with selenium contents in soils and grasses and perhaps even more surprisingly, human blood selenium levels are higher in the seleniferous western United States than in selenium-poor areas. For example, serum selenium contents average $0.161\,\mu g\,L^{-1}$ in Ohio compared to $0.265\,\mu g\,L^{-1}$ in South Dakota (WHO, 1987). Hence, even in populations who now live one step removed from their natural environment, the cycling of selenium from nature into humans is still of fundamental importance to health.

VII. FUTURE CONSIDERATIONS

This chapter has demonstrated that human exposure to the biologically important element selenium is largely dependent on dietary intakes in food and water, which are significantly controlled by variations in the geology of the Earth's surface. Although much work has been done over the past 30 years to enhance our understanding of environmental selenium, over large areas of the globe information is still missing because until recently selenium was a difficult element to analyze. More work is required to understand not just the total amounts of selenium present but also the bioavailability of the element and cycling through the environment. For example, it is only recently that the importance of the oceans in the cycling of selenium has been recognized. The selenium status of the human and animal populations around the globe closely reflect environmental levels and although overt clinical symptoms of selenium toxicity and deficiency are rarely reported, the possible subclinical effects and implications of selenium status are poorly understood and should not be underestimated as medical science continues to uncover new essential functions for the element. In the future, closer collaboration between medical and environmental scientists will be required to evaluate the real environmental health impact of this remarkable element in diseases such as cancer, AIDS, and heart disease.

SEE ALSO THE FOLLOWING CHAPTERS

Chapter 2 (Natural Distribution and Abundance of Elements) · Chapter 6 (Biological Functions of the Elements) · Chapter 14 (Bioavailability of Elements in Soil) · Chapter 16 (Soils and Iodine Deficiency) · Chapter 20 (Animals and Medical Geology) · Chapter 23 (Environmental Pathology) · Chapter 31 (Modeling Groundwater Flow and Quality)

FURTHER READING

Afzal, S., Younas, M., and Ali, K. (2000). Selenium Speciation Studies from Soan-Sakesar Valley, Salt Range, Pakistan, *Water Int.*, 25, 425–436.

Akesson, B., and Steen, B. (1987). Plasma Selenium and Glutathione Peroxidase in Relation to Cancer, Angina Pectoris and Short-Term Mortality in 68-year-old Men, *Comprehens. Gerontol.*, A(1), 61–64.

Alfthan, G., Xu, G. L., Tan, W. H., Aro, A., Wu, J., Yang, Y. X., Liang, W. S., Xue, W. L., and Kong, L. H. (2000). Selenium Supplementation of Children in a Selenium-Deficient Area in China—Blood Selenium Levels and Glutathione Peroxidase Activities, *Biolog. Trace Element Res.*, 73(2), 113–125.

Amouroux, D., Liss, P. S., Tessier, E., Hamren-Larsson, M., and Donard, O. F. X. (2001). Role of the Oceans as Biogenic Sources of Selenium, *Earth Planet. Sci. Lett.*, 189(3–4), 277–283.

Anonymous (2001). Observations on the Effect of Sodium Selenite in the Prevention of Keshan Disease (reprinted from the *Chinese Medical Journal*, Vol. 92, pp. 471–476, 1979), *J. Trace Elements Exp. Med.*, 14(2), 213–219.

Aras, N. K., Nazli, A., Zhang, W., and Chatt, A. (2001). Dietary Intake of Zinc and Selenium in Turkey, *J. Radioanal. Nucl. Chem.*, 249(1), 33–37.

Aro, A., Alfthan, G., and Varo, P. (1995). Effects of Supplementation of Fertilizers on Human Selenium Status in Finland, *Analyst*, 120(3), 841–843.

Arthur, J. R., and Beckett, G. T. (1994). New Metabolic Roles for Selenium, *Proc. Nutr. Soc.*, 53, 615–624.

Beck, M. A. (1999). Selenium and Host Defense Towards Viruses, *Proc. Nutr. Soc.*, 58(3), 707–711.

Bedwal, R. S., Nair, N., Sharma, M. P., and Mathur, R. S. (1993). Selenium—Its Biological Perspectives, *Med. Hypotheses* 41(2), 150–159.

Birt, D. F., Pour, P. M., and Pelling, J. C. (1989). The Influence of Dietary Selenium on Colon, Pancreas and Skin Tumorigenesis. In *Selenium in Biology and Medicine* (A. Wendel, Ed.), Springer-Verlag, Berlin, pp. 297–304.

Boyle, R. W. (1979). The Geochemistry of Gold and its Deposits: Together with a Chapter on Geochemical Prospecting for the Element, Bulletin 280, Geological Survey of Canada, Ottawa.

British Geological Survey (2000). Regional Geochemistry of Wales and Part of West-Central England: Stream Sediment and Soil, British Geological Survey, Keyworth, Nottingham.

Burk, R. F. (Ed.) (1994). *Selenium in Biology and Human Health*, Springer-Verlag, New York.

Canton, S. P. (1999). Acute Aquatic Life Criteria for Selenium, *Environ. Toxicol. Chem.*, 18, 1425–1432.

Cao, Z. H., Wang, X. C., Yao, D. H., Zhang, X. L., Wong, M. H. (2000). Selenium Geochemistry of Paddy Soils in the Yangtze River Delta, *Environ. Int.*, 26(5–6), 335–339.

Clark, L. C., Combs, G. F., Turnbill, B. W., Slate, E. H., Chalker, D. K., Chow, J., Davies, L. S., Glover, R. A., Graham, G. F., Gross, E. G., Krongrad, A., Lesher, J. L., Park, H. K., Sander, B. G., Smith, C. L., and Taylor, J. R.

(1996). Effects of Selenium Supplementation for Cancer Prevention in Patients with Carcinoma of the Skin, *J. Am. Med. Assoc.*, 276, 1957–1963.

Combs, G. F., and Combs, S. B. (1986). *The Role of Selenium in Nutrition*, Academic Press, Orlando, FL.

Cutter, G. A., and Bruland, K. W. (1984). The Marine Biogeochemistry of Selenium: A Re-Evaluation, *Liminol. Oceanogr.*, 29, 1179–1192.

Darnley, A. G., Björklund, A., Bølviken, B., Gustavsson, N., Koval, P. V., Plant, J A., Steenfelt, A., Tauchid, M., and Xuejing Xie, X. (1995). A Global Geochemical Database for Environmental and Resource Management. Recommendations for International Geochemical Mapping, Earth Science Report 19, UNESCO Publishing, Paris.

Davies, E. B., and Watkinson, J. H. (1966). Uptake of Native and Applied Selenium by Pasture Species, *N. Z. J. Agric. Res.*, 9, 317–324.

EA (2002). *Soil Guideline Values for Selenium Contamination*, R+D Publication SGV 9, Environment Agency, Bristol.

Fergusson, J. E. (1990). *The Heavy Elements: Chemistry, Environmental Impact and Health Effects*, Pergamon, London.

Finley, J. W., Ip, C., Lisk, D. J., Davis, C. D., Hintze, K. J., and Whanger, P. D. (2001). Cancer-Protective Properties of High-Selenium Broccoli, *J. Agric. Food Chem.*, 49(5), 2679–2683.

Fleming, G. A. (1980). Essential Micronutrients II: Iodine and Selenium. In *Applied Soil Trace Elements* (B. E. Davis, Ed.), John Wiley & Sons, New York, pp. 199–234.

Fordyce, F. M., Johnson, C. C., Navaratne, U. R. B., Appleton, J. D., and Dissanayake, C. B. (2000a). Selenium and Iodine in Soil, Rice and Drinking Water in Relation to Endemic Goiter in Sri Lanka, *Science Total Environ.*, 263(1–3), 127–142.

Fordyce, F. M., Zhang, G., Green, K., and Liu, X. (1998). Soil, Grain and Water Chemistry and Human Selenium Imbalances in Enshi District, Hubei Province, China, Overseas Geology Series Technical Report WC/96/54, British Geological Survey, Keyworth, Nottingham.

Fordyce, F. M., Zhang, G., Green, K., and Liu, X. (2000b). Soil, Grain and Water Chemistry and Human Selenium Imbalances in Enshi District, Hubei Province, China, *Appl. Geochem.*, 15(1), 117–132.

Hartikainen, H., Xue, T. L., and Piironen, V. (2000). Selenium as an Anti-Oxidant and Pro-Oxidant in Ryegrass, *Plant Soil*, 225(1–2), 193–200.

Hawkes, W. C., and Turek, P. J. (2001). Effects of Dietary Selenium on Sperm Motility in Healthy Men, *J. Androl.*, 22(5), 764–772.

Haygarth, P. M. (1994). Global Importance and Cycling of Selenium. In *Selenium in the Environment* (W. T. Frankenberger and S. Benson, Eds.), Marcel-Dekker, New York, pp. 1–28.

Hem, J. (1992). Study and Interpretation of the Chemical Characteristics of Natural Water, Water Supply Paper 2254, U. S. Geological Survey, Reston, VA.

Jacobs, L. W. (Ed.) (1989). *Selenium in Agriculture and the Environment*, Soil Science Society of America Special Publication 23, SSSA, Madison, WI.

Johnson, C. C., Ge, X., Green, K. A., and Liu, X. (2000). Selenium Distribution in the Local Environment of Selected Villages of the Keshan Disease Belt, Zhangjiakou District, Hebei Province, People's Republic of China, *Appl. Geochem.*, 15(3), 385–401.

Kohrle, J. (1999). The Trace Element Selenium and the Thyroid Gland, *Biochimie*, 81(5), 527–533.

Levander, O. A. (1986). Selenium. In *Trace Elements in Human and Animal Nutrition* (W. Mertz, Ed.), Academic Press, London, pp. 139–197.

Li, Y., Peng, T., Yang, Y., Niu, C., Archard, L. C., and Zhang, H. (2000). High Prevalence of Enteroviral Genomic Sequences in Myocardium From Cases of Endemic Cardiomyopathy (Keshan Disease) in China, *Heart*, 83(6), 696–701.

Longombe, A. O., Arnaud, J., Mpio, T., and Favier, A. E. (1994). Serum Selenium in HIV-Infected Zairian Patients, *Trace Elements Electrolytes*, 11(2), 99–100.

MAFF (1997). Food Surveillance Information Sheet of the Joint Food Safety and Standards Group, 126, Ministry of Agriculture Fisheries and Food, London.

Mahapatra, S., Tripathi, R. M., Raghunath, R., and Sadasivan, S. (2001). Daily Intake of Selenium by the Adult Population of Mumbai, India, *Sci. Total Environ.*, 277(1–3), 217–223.

Mayland, H. F. (1994). Selenium in Plant and Animal Nutrition. In *Selenium in the Environment* (W. T. Frankenberger and S. Benson, Eds.), Marcel-Dekker, New York, pp. 29–47.

Moreno-Reyes, R., Egrise, D., Neve, J., Pasteels, J. L., and Schoutens, A. (2001). Selenium Deficiency-Induced Growth Retardation is Associated with an Impaired Bone Metabolism and Osteopenia, *J. Bone Miner. Res.*, 16(8), 1556–1563.

Moxon, A. L. (1937). Alkali Disease or Selenium Poisoning, Bulletin 311, South Dakota Agricultural Experimental Station, Vermillion.

Moxon, A. L. (1938). The Effect of Arsenic on the Toxicity of Seleniferous Grains, *Science*, 88, 81.

Murphy, J., and Cashman, K. D. (2001). Selenium Content of a Range of Irish Foods, *Food Chem.*, 74(4), 493–498.

Muth, O. H., and Allaway, W. H. (1963). The Relationship of White Muscle Disease to the Distribution of Naturally Occurring Selenium, *J. Am. Vet. Med. Assoc.*, 142, 1380.

Neal, R. H. (1995). Selenium. In *Heavy Metals in Soils* (B. J. Alloway, Ed.), Blackie Academic & Professional, London, pp. 260–283.

Nriagu, J. O. (1989). *Occurrence and Distribution of Selenium*, CRC Press, Boca Raton, FL.

Nriagu, J. O. (1991). Heavy Metals in the Environment, 1 CEP Consultants, Edinburgh.

Oldfield, J. E. (1999). Selenium World Atlas, Selenium-Tellurium Development Association, Grimbergen, Belgium.

Peng, A., Wang, W. H., Wang, C. X., Wang, Z. J., Rui, H. F., Wang, W. Z., and Yang, Z. W. (1999). The Role of Humic Substances in Drinking Water in Kashin-Beck Disease in China, *Environ. Health Perspect.*, 107(4), 293–296.

Pineda-Valdes, G., and Bullerman, L. B. (2000). Thermal Stability of Moniliformin at Varying Temperature, pH and Time in an Aqueous Environment, *J. Food Prot.*, 63(11), 1589–1601.

Rapant, S., Vrana, K., and Bodis, D. (1996). Geochemical Atlas of Slovakia: Part 1 Groundwater, Geological Survey of the Slovak Republic, Bratislava.

Rayman, M. (2002). Selenium Brought to Earth, *Chem. Br.*, October, 28–31.

Reimann, C., and Caritat, P. (1998). *Chemical Elements in the Environment*, Springer-Verlag, Heidelberg.

Romero, C. D., Blanco, F. L., Sanchez, P. H., Rodriguez, E., and Majem, L. S. (2001). Serum Selenium Concentration in a Representative Sample of the Canarian Population, *Sci. Total Environ.*, 269(1–3), 65–73.

Rosenfield, I., and Beath, O. A. (1964). *Selenium, Geobotany, Biochemistry, Toxicity and Nutrition*, Academic Press, New York.

Sasaki, S., Iwata, H., Ishiguron, N., Habuchi, O., and Miura, T. (1994). Low-Selenium Diet, Bone and Articular-Cartilage in Rats, *Nutrition*, 10(6), 538–543.

Shamberger, and Frost, (1969). Possible Protective Effect of Selenium Against Human Cancer, *Can. Med. Assoc. J.*, 100, 682.

Smith, M. I., and Westfall, B. B. (1937). Public Health Report 52, U. S. Public Health Department, Washington DC.

Suetens, C., Moreno-Reyes, R., Chasseur, C., Mathieu, F., Begaux, F., Haubruge, E., Durand, M. C., Neve, J., and Vanderpas, J. (2001). Epidemiological Support for a Multifactorial Etiology of Kashin-Beck Disease in Tibet, *Int. Orthop.*, 25(3), 180–187.

Tan, J. (Ed.) (1989). *The Atlas of Endemic Diseases and Their Environments in the People's Republic of China*, Science Press, Beijing.

Tan, J., Wang, W., Yang, L., Zhu, W., Li, R., and Hou, S. (1999). The Approaches to Amelioration of Selenium Eco-Cycle in Selenium-Deficient Biogeochemical Area and Its Effects on Selenium Dietary Intake and Health in China, In *Proceedings of the 5th International Conference on the Biogeochemistry of Trace Elements 2* (W. W. Wenzel, D. C. Adriano, B. E. Alloway, H. E. Doner, C. Keller, M. Lepp,

R. Mench, R. Naidu, and G. M. Pierzynski, Eds.), Vienna, Austria, 620–621.

Taylor, S. R., and McLennan, S. M. (1985). *The Continental Crust: Its Composition and Evolution*, Blackwell, Oxford, England.

Thornton, I., Kinniburgh, D. G., Pullen, G., and Smith, C. A. (1983). Geochemical Aspects of Selenium in British Soils and Implications to Animal Health. In *Trace Substances in Environmental Health XVII* (D. D. Hemphill, Ed.), University of Missouri, pp. 391–398.

Tokunaga, T., Pickering, I., and Brown, G. (1996). Selenium Transformations in Ponded Sediments, *Soil Sci. Soc. Am. J.*, 60, 781–790.

U.S.-EPA (2002a). Air Toxics Web-site Selenium and Compounds, http://www.epa.gov/ttn/atw/hlthef/selenium.html.

U.S.-EPA (2002b). Drinking Water and Health Consumer Fact Sheet on Selenium, http://www.epa.gov/safewater/dwh/t-ioc/selenium.html.

U.S.-EPA (2003). Biosolids Management Handbook, http://www.epa.gov/excotox.

Varo, P., Alfthan, G., Ekholm, P., and Aro, A. (1998). Effects of Selenium Supplementation Fertilizer on Human Nutrition and Selenium Status. In *Environmental Chemistry of Selenium* (W. T. Frankenberger and R. A. Engberg, Eds.), Marcel-Dekker, New York.

Vinceti, M., Rothman, K. J., Bergomi, M., Borciani, N., Serra, L., and Vivoli, G. (1998). Excess Melanoma Incidence in a Cohort Exposed to High Levels of Environmental Selenium, *Cancer Epidemiol. Biomarkers Prev.*, 7, 853–856.

Vinceti, M., Rovesti, S., Bergomi, M., and Vivoli, G. (2000). The Epidemiology of Selenium and Human Cancer, *Tumori*, 86, 105–118.

World Health Organization (1987). Environmental Health Criterion 58—Selenium, World Health Organization, Geneva.

World Health Organization (1996). Trace Elements in Human Nutrition and Health, World Health Organization, Geneva.

Wu, A. S., Oldfield, J. E., Shull, L. R., and Cheeke, P. R. (1979). Specific Effect of Selenium Deficiency on Rat Sperm, *Biol. Reprod.*, 20, 793–798.

Wu, L. (1994). Selenium Accumulation and Colonization of Plants in Soils with Elevated Selenium and Salinity. In *Selenium in the Environment* (W. T. Frankenberger and S. Benson, Eds.), Marcel-Dekker, New York, pp. 279–227.

Wu, L., Banuelos, G., and Guo, X. (2000). Changes of Soil and Plant Tissue Selenium Status in an Upland Grassland Contaminated by Selenium-Rich Agricultural Drainage Sediment After Ten Years Transformed From a Wetland Habitat, *Exotoxicol. Environ. Safety*, 47(2), 201–209.

Xu, G., and Jiang, Y. (1986). Selenium and the Prevalence of Keshan and Kashin-Beck Diseases in China. In *Proceedings of the First International Symposium on Geochemistry and Health* (I. Thornton, Ed.), Science Reviews Ltd., Northwood, England, pp. 192–205.

Yang, G., and Xia, M. (1995). Studies on Human Dietary Requirements and Safe Range of Dietary Intakes of Selenium in China and Their Application to the Prevention of Related Endemic Diseases, *Biomed. Environ. Sci.*, 8, 187–201.

Yang, G., Wang, S., Zhou, R., and Sun, S. (1983). Endemic Selenium Intoxication of Humans in China, *Am. J. Clin. Nutr.*, 37, 872–881.

Yang, G., Yin, S., Zhou, R., Gu, L., Yan, B., Lui, Y., and Liu, Y. (1989). Studies of Safe and Maximal Daily Dietary Selenium Intake in a Seleniferous Area in China. Part 2, *J. Trace Element Electrolytes Health Dis.*, 3, 123–129.

Zhang, W. H., Neve, J., Xu, J. P., Vanderpas, J., and Wang, Z. L. (2001). Selenium, Iodine and Fungal Contamination in Yulin District (People's Republic of China) Endemic for Kashin-Beck Disease, *Int. Orthop.*, 25(3), 188–190.

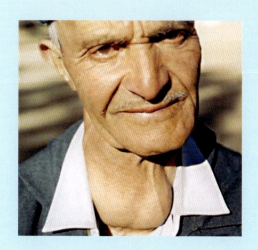

CHAPTER 16

SOILS AND IODINE DEFICIENCY

RON FUGE
University of Wales

CONTENTS

I. INTRODUCTION

Iodine has long been known as an essential element for humans, and mammals in general, where it is concentrated in the thyroid gland. It is a component of the thyroid hormone thyroxine. Deprivation of iodine results in a series of iodine deficiency disorders (IDD), the most common of which is endemic goiter, a condition where the thyroid gland becomes enlarged in an attempt to be more efficient. Iodine deficiency during fetal development and in the first year of life can result in endemic cretinism, a disease which causes stunted growth and general development along with brain damage. However, while these two diseases are easily recognizable, perhaps the more insidious problem is that iodine deficiency impairs brain development in children even when there is no obvious physical effect. Indeed it has been suggested that iodine deficiency is the most common preventable cause of mental retardation (see Geological Impacts on Nutrition, and Biological Responses of Elements, this volume).

Endemic goiter and cretinism along with related IDD have long been recognized as serious health problems and consequently much work has been carried out on the etiology and geographical distribution of these diseases. Many authors have suggested the involvement of several other elements in the etiology of these diseases, whereas a group of sulfur-containing compounds, thiocyanates and thiouracils, identified collectively as goitrogens, have been strongly implicated in some endemics. These compounds have been found to either inhibit iodine uptake by the thyroid gland or inhibit the formation of the thyroid hormones. However, it is generally agreed that the primary cause of IDD is a lack of iodine in the diet.

Iodine was the first element recognized as being essential to humans and the disease of endemic goiter was the first to be related to environmental geochemistry. Indeed, this disease appears to have been identified by the ancient Chinese. One Chinese medical writer from the 4th century AD noted the use of the seaweeds *Sargassum* and *Laminaria* (which are known to be very iodine rich) for treatment of goiter (Langer, 1960). However, there are many earlier records in ancient literature of seaweeds and burnt sea sponges used in the treatment of endemic goiter possibly from as early as 2700 BC (Langer, 1960). Iodine was discovered by the French chemist Bernard Courtois in 1811 when he accidentally added concentrated sulfuric acid to the seaweed *Fucus vesiculosus*, one of the seaweeds used in goiter treatment. It was soon realized that iodine was the active ingredient in the treatment. Iodine was later identified as an essential element in human nutrition. Despite this early recognition of the role of iodine in endemic goiter and related disorders, it is apparent that IDD is still affecting large numbers of people worldwide, with some estimates suggesting that around 30% of the world's population are at risk. Prior to the middle of the 20th century iodine-deficiency problems affected virtually every country (Kelly & Snedden, 1960). Although the problem has been eradicated in many countries, it is still a major concern in the developing world. In addition it has been shown recently that some more affluent countries in western Europe are also affected with IDD. Delange (1994) suggested that 50–100 million people in Europe are at risk.

The areas where IDD are concentrated tend to be geographically defined. Thus many of the most severe occurrences of endemic goiter and cretinism have been found to occur in high mountain ranges, rain shadow areas, and central continental regions (Kelly & Snedden, 1960). This distribution of IDD results from the unique geochemistry of iodine. Geochemists generally agree that little iodine in the secondary environment is derived from weathering of the lithosphere. Most iodine is derived from volatilization from the oceans with subsequent transport onto land (Goldschmidt, 1954). Therefore, to understand the role of soil iodine geochemistry in IDD it is essential to consider some aspects of the general geochemistry of iodine and its cyclicity in the environment.

II. IODINE GEOCHEMISTRY

Iodine has been shown to be concentrated in seawater, in the biosphere, and in the atmosphere. Because of this, it is classified as a hydrophile, biophile, and atmophile element (Goldschmidt, 1954). In addition iodine has been found to be concentrated in sulfur-containing minerals causing it to be classified as a chalcophile element.

A. The Lithosphere

In the lithosphere iodine is an ultra-trace element; its crustal abundance is estimated to be $0.25\,mg\,kg^{-1}$ by Fuge (1988) and more recently $0.3\,mg\,kg^{-1}$ by Muramatsu and Wedepohl (1998). Due to the large ionic radius of the iodide ion (220 pm), it is thought unlikely that iodine enters the crystal lattices of most rock-forming minerals. In fact the iodine content of the various rock-forming minerals has been shown to be fairly uniform and low, with some enrichment found in only the chlorine-containing minerals sodalite and eudialyte. Its distribution in the different igneous rocks is fairly uniform (Table I) with only volcanic glasses showing comparative enrichment. Some data suggest that carbonatites are also somewhat richer in iodine than average igneous rocks.

Sedimentary rocks show a greater range of iodine content with clay-rich or argillaceous rocks more enriched than the sand-rich, arenaceous rocks (Table I). The highest concentrations of iodine have been found in organic-rich shales, with concentrations as high as 44 $mg\,kg^{-1}$ recorded in some bituminous shales. The iodine content of carbonate rocks (limestones) and shales is highly variable but generally correlates with the amount of organic matter.

Recent sediments of marine origin can be extremely enriched in iodine with concentrations of up to 20,00 $mg\,kg^{-1}$ recorded in some surficial sediments (Wong, 1991).

It seems likely that the iodine content of metamorphic rocks is similar to that of igneous rocks. Recent

TABLE I. Iodine in Igneous and Sedimentary Rocks

Rock type	Mean iodine content ($mg\,kg^{-1}$)
Igneous rocks	
Granite	0.25
All other intrusives	0.22
Basalts	0.22
All other volcanics	0.24
Volcanic glasses	0.52
Sedimentary rocks	
Shales	2.3
Sandstones	0.80
Limestones	2.3
Organic-rich shales	16.7

analytical data produced by Muramatsu and Wedepohl (1998) suggest that metamorphic rocks have uniformly low iodine contents of <0.025 mg kg^{-1}. The same authors suggest that felsic igneous rocks are also very low in iodine with <0.009 mg kg^{-1}. It is suggested that the low concentrations are due to volatilization of iodine during the formation of these rocks.

B. The Marine Environment

Seawater is by far the biggest reservoir of iodine; its average concentration is about 60 µg L^{-1} (Wong, 1991). Iodine is thought to have fairly uniform concentrations with depth, but it is slightly depleted in surface ocean waters because it is concentrated by organisms. Iodine can exist in several forms in seawater and the transformations between the various forms and their mechanisms are vital for understanding the cycling of iodine in the general environment.

Inorganic iodine is essentially present in two forms, the iodide anion, I$^-$, and the iodate anion, IO$_3^-$. Iodate is the thermodynamically stable form of inorganic iodine in oxygenated, alkaline seawater, whereas iodide, the reduced species, is in a metastable state. There is considerable variation of the I$^-$/IO$_3^-$ ratio with depth with I$^-$ enriched in surface waters and depleted in deeper waters. There is also considerable geographic variation of the I$^-$/IO$_3^-$ ratio with waters from the shallower inner shelf areas containing more iodide than the deeper mid-shelf waters. The conversion of IO$_3^-$ to I$^-$ is thought to be due to biological activity, with the enzyme nitrate reductase implicated in the reaction. The increased iodide in surface and shallow shelf waters is thus thought to be due to the high biological activity in these zones. However, it is possible that some abiological mechanisms might also be involved in the conversion of iodate to iodide. Once formed, iodide is only slowly re-oxidized to iodate (Wong, 1991).

It has long been suggested that some iodine in seawater is present as an organic phase, and recently it has been found that in some coastal waters dissolved organic iodine can constitute up to 40% of the total iodine content (Wong & Cheng, 1998). Although Wong (1991) has indicated that few specific organo-iodine compounds have been identified in seawater, one of the substances identified, methyl iodide (CH$_3$I), is of interest due to its volatility and possible role in the transfer of iodine from the oceans to the atmosphere. Lovelock et al. (1973) were the first researchers to detect methyl iodide in seawater. This compound can be formed both biologically, by seaweeds and phytoplankton, and by photochemical reactions, with the likelihood that chemical formation of methyl iodide is more important in the open ocean. Recently it has been found that methyl iodide is oversaturated in surface ocean waters (Moore & Grosko, 1999).

The strong enrichment of iodine in marine organisms has long been recognized with Courtois discovering the element in the brown alga *F. vesiculosus*. Since that time brown algae have been shown to strongly concentrate iodine with an enrichment factor of alga/seawater having been estimated as being over 32,000 (Fuge & Johnson, 1986). Although concentration factors for red and green algae are lower, they are still enriched in iodine. Phytoplankton also concentrates iodine.

It is generally held that the iodide ion is preferentially incorporated into the organisms; however, it has been demonstrated that phytoplankton can take in iodate (Moisan et al., 1994). Whatever the mechanism of uptake, these marine organisms play an important role in transformation of iodine species and ultimately in its transfer from the oceanic environment to the atmosphere. It has been demonstrated that organo-iodine compounds are released from brown seaweeds and this could represent a source of atmospheric iodine.

C. Transfer of Iodine From the Marine to Terrestrial Environment Via the Atmosphere

Iodine has been shown to be concentrated in the atmosphere. This atmophile behavior is the most important part of its geochemical cycle, with the transfer of iodine from the oceans to the atmosphere ultimately governing its distribution in the terrestrial environment.

The mechanism of transfer of iodine from the oceans to the atmosphere has been a subject of some debate. Undoubtedly some iodine is transferred to the atmosphere as seawater spray. However, it has been shown that the amount of iodine relative to chlorine in the atmosphere is several hundred times that in seawater. Although some of the iodine is likely to be in aerosols, it has also been shown that a large percentage of atmospheric iodine is in a gaseous form (Duce et al., 1973). Therefore, the major mechanism of iodine transfer must reflect preferential volatilization of seawater iodine into the atmosphere.

The iodide ion can be converted to elemental iodine (I$_2$) by photochemical oxidation, and this has been proposed as a mechanism for volatilization of iodine from the oceans. More recently it has been proposed that I$_2$

(and possibly HOI) could be produced at the sea surface from oxidation of iodide by such species as ozone and nitrogen dioxide (Garland & Curtis, 1981; Heumann et al., 1987). However, some workers have suggested that volatilization of elemental iodine is not likely to be a major source of transfer of iodine from the oceans to the atmosphere.

As methyl iodide has been found to occur in the surface waters of the sea and as this compound is volatile, it could represent a significant source of atmospheric iodine. Yoshida and Muramatsu (1995) found that on the Japanese coast 90% of atmospheric iodine is gaseous with organically bound iodine as the dominant species. In addition, Gabler and Heumann (1993) found that organically bound iodine is the most abundant species of iodine in European air. In the atmosphere methyl iodide is broken down and has an estimated lifetime of about five days (Zafirou, 1974).

Whereas several authors have suggested that methyl iodide is the dominant form of iodine released from the oceans, it seems likely that there are several mechanisms of transfer of iodine from the oceans to the atmosphere and ultimately to the landmasses. From studies in Antarctica, Heumann et al. (1990) suggested that during short distance transport into coastal areas iodine is carried as I_2, HI, and sea spray; for long distance transport CH_3I is responsible. The residence time of inorganic iodine in the atmosphere has been estimated to be 10 days (Rahn et al., 1976; Chameides & Davis, 1980) and the total residence time of iodine has been estimated to be 15 days (Kocher, 1981).

Iodine is transferred from the atmosphere to the terrestrial environment by wet and dry deposition. On the basis of the marine origin of atmospheric iodine it might be expected that deposition should be highest at the coast and decrease inland. However, literature data are limited on the iodine content of rain from coastal and inland locations. Several workers have quoted data for the iodine content of rainfall in the UK, the whole of which has a strong maritime influence. Generally the samples were found to contain around $2 \mu g L^{-1}$ iodine with some higher values (up to $5 \mu g L^{-1}$) recorded in coastal rainfall. It has been shown that rain collected from an upland area 12 km inland from the mid-Wales coast contained over three times as much iodine as rainfall from 84 km inland. Although some studies of iodine contents of rainfall have failed to demonstrate significantly higher values in coastal rain, analyses of rainfall from the continental United States (Missouri) showed iodine contents of $<1.0 \mu g L^{-1}$, and it does seem likely from the limited data that the iodine content of rain is influenced by its proximity to the oceans.

The form of iodine in rainfall has been the subject of much discussion. In general it has been shown that iodide is the most common form of iodine in rain making up over 50% of rainfall iodine, and the iodate ion is the second major component. It has also been demonstrated that the iodate content of rainfall decreases inland with iodide content showing a parallel increase. In view of the likely importance of organically bound forms in atmospheric transport of iodine, it would seem likely that some iodine in rain should be present as organo-iodine, with an early study suggesting as much as 40% is organically bound (Dean, 1963). However, more recently, only small amounts of "non-ionic" dissolved iodine have been found to occur in Japanese rain; iodide and iodate are the dominant forms (Takagi et al., 1994). It seems likely that any atmospheric methyl iodide is converted to inorganic forms before deposition.

Dry deposition of iodine could be an important transfer mechanism for iodine from the atmosphere to the land surface. Few studies have attempted to quantify the amount of iodine deposited on land surfaces by dry deposition. Those studies suggest that dry deposition in marine influenced areas is a significant source of terrestrial iodine. However, there is no agreement on the relative importance of wet and dry deposition, and little is known with regard to the quantities of iodine in dry deposition in areas remote from the sea.

III. IODINE GEOCHEMISTRY OF SOILS

There is a considerable body of data for iodine in soils and this shows a very broad range of concentrations from <0.1–$150 \, mg \, kg^{-1}$. The iodine content of soils is generally considerably higher than the rocks from which they derive. Most geochemists agree that the majority of the iodine in soils is derived from the atmosphere and ultimately the marine environment. The proximity of an area to the sea therefore is likely to exert a strong influence on the iodine content of soils in that area. This also results in considerable geographic variation of soil iodine content. The other important feature of soil iodine geochemistry is the fact that the element can become strongly adsorbed by various soil components, and thus its concentration and behavior in soils is going to depend on soil composition. In this respect the nature of the composition of the soil parent material indirectly exerts a strong influence on the iodine chemistry of the soil. The iodine geochemistry

of soils then can be summarized as dependent on the quantity of iodine supplied coupled with the soil's ability to retain this iodine.

A. Factors Influencing the Supply of Iodine to Soils

It has generally been suggested that soils in close proximity to the coast are likely to be enriched in iodine with those far removed from the coast depleted of iodine (Goldschmidt, 1954) (see also Table II). However, in some cases such a relationship is not obvious and some workers have found no correlation between iodine and distance from the sea. This has led some workers to suggest that marine influence extends a considerable distance inland. Whitehead (1984), for instance, suggested that all soils from the UK are affected by the strong maritime influence over the whole country and that low iodine soils would only be found in the middle of continental areas. To some extent this is true, but it is perhaps pertinent to point out that in a soil traverse from the Welsh coast 120 km inland (Figure 1), it was found that samples from beyond 100 km (8 samples) contained between 2 and 3 mg kg^{-1} of iodine. Similarly, a traverse from the Irish coast showed that soil samples collected beyond 80 km inland contained between 1.7

and 2.8 mg kg^{-1}. These values are not much higher than concentrations in soils from continental United States (Table II). On this basis it is quite likely that the strong influence of the marine environment on soil iodine contents does not extend very far inland.

TABLE II. Iodine in Coastal and Inland Soils

Sample origin	Iodine Range Units	Content Mean Units	Ref.
Coastal areas			
Northwest Norway	5.4–16.6	9.0	Låg & Steinnes (1976)
Wales	1.5–149	14.7	Fuge (1996)
Ireland	4.2–54	14.7	Fuge (unpublished)
Inland areas			
East Norway	2.8–7.6	4.4	Låg & Steinnes (1976)
Wales/England	1.8–10.5	4.2	Fuge (1996)
Missouri, U. S.	0.12–5.6	1.3	Fuge (1987)
Whole of U. S.[a]	<0.5–9.6	1.2	Shacklette & Boerngen (1984)

[a]Includes some from coastal localities.

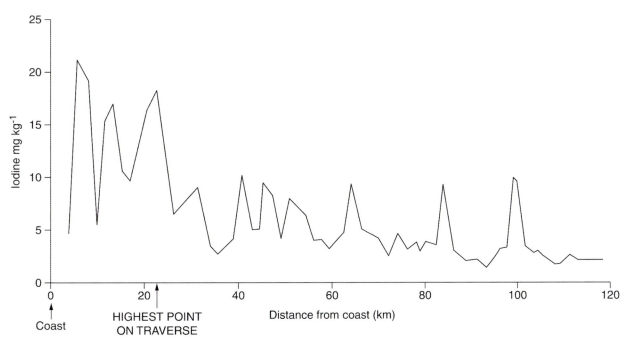

FIGURE 1 Iodine in topsoils on a traverse from the west Wales coast to the English Midlands. (After Fuge, 1996.)

TABLE III.   Iodine Content of Various Soil Types

Soil type	Iodine Range Units	mg kg⁻¹ Mean	Ref.
Peats (70% organic matter), UK	28–98	56	Johnson (1980); Fuge & Johnson (1986)
Peats, UK	18.7–98.2	46.8	Whitehead (1984)
Peaty tundra soils, Russian Plain	0.2–42	12.0	Vinogradov (1959)
Non-peaty soils, Russian Plain	0.3–9.8	2.8	Vinogradov (1959)
Iron-rich soils, UK	7.5–32.5	16.0	Fuge & Ander (1998) and unpublished information
Chalk and limestone parent material, UK	7.9–21.8	13.0	Whitehead (1984)
Over limestone, Derbyshire, UK	2.58–26.0	8.2	Fuge (1996)
Over shale, sandstone, and dolomite, Derbyshire, UK	1.88–8.53	3.44	Fuge (1996)
Carbonate-rich soils, Austria	1.64–5.63	3.75	Gerzabek et al. (1999)
Carbonate-free soils, Austria	1.08–4.80	2.58	Gerzabek et al. (1999)
Clay parent material, UK	2.1–8.9	5.2	Whitehead (1984)
Sand and sandstone parent material, UK	1.7–5.4	3.7	Whitehead (1984)
River and terrace alluvium parent material, UK	0.5–7.1	3.8	Whitehead (1984)
Marine estuarine alluvium parent material, UK	8.8–36.9	19.6	Whitehead (1984)

The influence of topography on the iodine concentration of soils is well illustrated in Figure 1. The highest values in the traverse occur in soils over the Welsh Mountains, which range up to 350 m (Ordinance Datum). In addition, several of the higher values in the latter half of the traverse occur on high ground. The greater precipitation that occurs in the upland areas causes a greater degree of washout of atmospheric iodine and hence a higher input of iodine. An additional factor is that in the area of the traverse, upland soils tend to be organic rich and thus are more able to retain the increased iodine input (see below).

The corollary of washout of iodine in upland areas is the low supply of iodine to rain shadow areas beyond. This is going to be particularly pronounced in the rain shadow areas of high mountain ranges such as the Himalayas and the Alps.

B. Factors Influencing the Retention of Iodine in Soils

As stated earlier soil iodine geochemistry reflects both the input of iodine and the ability of the soil to retain it. Many factors have been implicated in the retention of iodine in soil. Iodine is strongly enriched in organic-rich sediments and it seems likely that one of the most influential soil components with regard to retention of iodine is organic matter. Soils rich in organic matter are frequently enriched in iodine with the concentration of iodine correlated with the content of organic matter. Peaty soils are particularly enriched in iodine (see also Table III).

Whereas organic matter has been shown to be the major contributor to the retention of soil iodine, it has also been suggested that iron and aluminum oxides play an important role in soil iodine retention (Whitehead, 1974). With regard to iron oxide it is noteworthy that the weathered surfaces of iron meteorites have been found to be strongly enriched in iodine (Heumann et al., 1990). In addition, it has been found that iodine is concentrated in iron-rich soils (Table III). In the case of aluminum oxide, it has been demonstrated in experimental work with both aluminum and iron oxides that sorption of iodide by aluminum oxide is similar to that by iron oxide (Whitehead, 1974). The sorption of iodide by aluminum and iron oxides is strongly dependent on soil pH with sorption greatest in acid conditions, which is typical of anion adsorption. It has also been shown that iodate will be strongly sorbed by aluminum and iron oxides; however, this ion is not sorbed by organic matter.

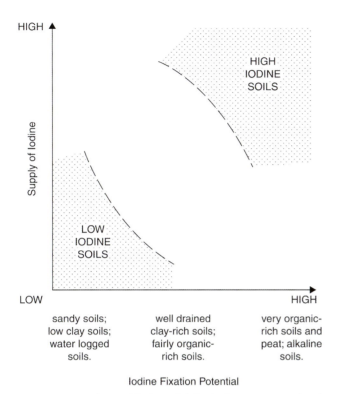

HIGH

Supply of Iodine

HIGH IODINE SOILS

LOW IODINE SOILS

LOW   HIGH

sandy soils;	well drained	very organic-
low clay soils;	clay-rich soils;	rich soils and
water logged	fairly organic-	peat; alkaline
soils.	rich soils.	soils.

Iodine Fixation Potential

FIGURE 2 A simplified model for iodine content of soils based on supply and fixation potential. (Modified from Fuge & Johnson, 1986.)

Clay minerals have been thought to be involved in retention of soil iodine with the suggestion that the sorption of iodine to clays is also pH dependent (Prister et al., 1977). However, it is generally held that clay minerals are relatively unimportant in soil iodine retention, with organic matter and aluminum and iron oxides providing the bulk of the retentive capacity of soils.

The ability of soil to retain iodine has been called the iodine fixation potential (IFP) by Fuge and Johnson (1986). Thus soils that are rich in one or more of organic matter, iron oxides, and aluminum oxides are likely to have a high IFP, whereas those with low amounts of the major fixation components have a low IFP (Figure 2).

Whereas little iodine is derived from the weathering of bedrock, the parent material governs the type of soil formed. In this context, the bedrock can exert a strong influence on the iodine retention capacity of soils (Table III). Thus in the case of soils derived from sand-rich parent material, iodine contents tend to be low as the sandy soils derived will have little ability to trap iodine. A particularly interesting lithological control on the iodine content of soils occurs in areas underlain by carbonate rocks. In such soils, which generally have alkaline or circum-neutral pH values, iodine contents are generally elevated. This association of iodine with carbonate bedrocks is particularly well illustrated when comparing the iodine content of soils overlying limestone with neighboring soils overlying non-limestone lithologies as illustrated for the Derbyshire area of the U.K. (Table III). A similar relationship has been demonstrated for carbonate-rich versus non-carbonate soils from Austria (Gerzabek et al., 1999) (see Table III). It has been shown that the soils overlying limestone have distinctly higher iodine contents and hence have a high fixation potential.

Whereas soils do not normally derive much iodine from their parent materials, where the parent material is sediment of fairly recent marine origin and likely to be iodine-rich, the soils derived could inherit some of their iodine from the parental source (Table III). In addition, it is likely that soils recently inundated by marine incursions or those that occur over reclaimed marine areas will be high in iodine.

As iodine is generally strongly sorbed within soils, very little is in a water-soluble form. Various researchers have found that up to 25% is in a soluble form with the water-soluble content of the majority of soils being less than 10% (Johnson, 1980). In most high-iodine, organic-rich soils water-soluble iodine accounts for much less than 10% of the total. More recently, a study of German soils revealed that 2.5–9.7% of the iodine was water soluble; however, in the same soils ^{129}I, recently added from a nuclear reprocessing plant, was considerably more soluble (21.7–48.7%). This suggests that natural iodine had become strongly bound through time (Schmitz & Aumann, 1994). For most soils it can be confidently predicted that water-soluble iodine will account for only a small percentage of the total, but in arid areas alkaline soils are likely to contain more elevated amounts of water-soluble iodine.

Whereas iodine sorbed in soils is generally strongly held and is not easily desorbed, it has been suggested that iodine is strongly desorbed in waterlogged soils. Thus Yuita et al. (1991) demonstrated that under flooding, and the resultant reducing conditions, 2–3 times as much iodine is solubilized from soils as in dry, oxidizing conditions. Such waterlogged, reducing conditions are typical of rice paddies and strong desorption of iodine has been found to occur in these soils (Muramatsu et al., 1996). Similarly lowland Japanese soils are low in iodine. It is suggested that this is due, in part, to flooding and desorption from the reducing conditions occasioned by microbial activity (Muramatsu & Yoshida, 1999).

C. Chemical Form of Iodine in Soils

As outlined in the previous section of this chapter, in most soils iodine is strongly bound to organic matter and iron and aluminum oxides. However, it is of interest to establish the form of the soluble or easily leached iodine in soils as this is the fraction which should be plant available. Several authors have found that the iodide ion is the dominant form of soluble soil iodine in acidic soils, particularly in waterlogged soils. However, in dry, oxidizing conditions iodate was found to be the dominant form of soluble iodine. The dominance of iodide in acidic soils, with iodate dominating in alkaline soils, demonstrates the importance of pH in governing the form of soluble iodine. However, Eh will also exert an important control on the form of soil iodine. It has also been demonstrated that Fe^{3+} and SO_4^{2-} reducing bacteria found in soils are capable of reducing iodate to iodide (Councell et al., 1997).

Despite some conflicting evidence in the literature, from the Eh-pH diagram for iodine (Figure 3), it seems likely that in acid soils soluble iodine will predominate as iodide, whereas in alkaline soils iodate will be dominant. As will be discussed later, this is important when considering the bioavailability and possible volatility of soil iodine.

D. Volatilization of Soil Iodine

Volatilization of iodine from soils as iodine gas has long been suggested to be an important process in the iodine cycle. Not all authors agree on the degree of volatilization of iodine from soils, but it is generally suggested that its volatilization is significant. It seems likely from consideration of the Eh-pH diagram for iodine (Figure 3), that in oxidizing conditions iodine gas is quite likely formed from iodide in acid soils. Perel'man (1977) has also suggested that the Fe^{3+} and Mn^{4+} ions could oxidize iodide under both acid and alkaline conditions.

In addition to the possible volatilization of gaseous elemental iodine from soils, several workers have demonstrated that soil iodine can be volatilized as methyl iodide, particularly under waterlogged and reducing conditions. Further, it has been suggested that volatilization of methyl iodide from the waterlogged soils may be a contributory cause of the low iodine contents

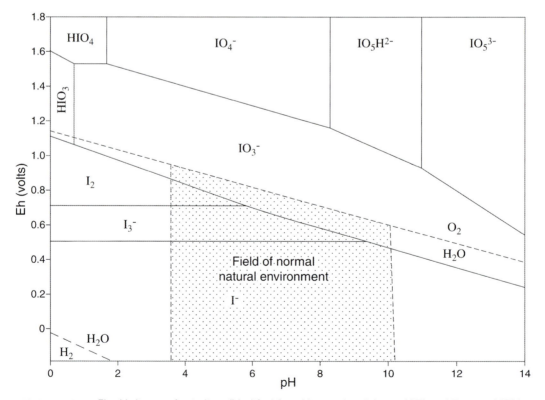

FIGURE 3   Eh-pH diagram for iodine. (Modified from Vinogradov & Lapp, 1971 and Bowen, 1979.)

of lowland soils in Japan (Muramatsu & Yoshida, 1999).

Therefore, it seems that volatilization of iodine from soils is quite likely and as such could play a very important role in the iodine cycle and the transfer of iodine into the biosphere. Fuge (1996) has suggested that only a relatively small proportion of iodine derived from the marine environment is transported into central continental regions and regions generally remote from the sea. Some of the iodine that occurs in environments far removed from marine influence could have been volatilized from soils, with iodine deposited on land by wet and dry precipitation being subsequently re-volatilized enriching the atmosphere in iodine. Such precipitation and re-volatilization could occur several times resulting in iodine migrating "stepwise" inland. In soils where iodine is strongly bound, such as organic-rich soils, it is likely that iodine is not available for re-volatilization. This could have a detrimental effect on the local iodine cycle by depriving plants of a potential source of iodine (see Section IV). It is also important to note that volatilization is going to be dependent on both Eh and pH, with volatilization under alkaline oxidizing conditions very unlikely, which is a possible explanation for the elevated iodine contents of the circum-neutral to alkaline soils occurring over limestone (Table III).

IV. Transfer of Iodine from Soil to Plants

When iodine is strongly sorbed in most soils it will not be readily bioavailable. Therefore, the presence of high iodine concentrations in soil does not necessarily mean that plants growing in the soil will incorporate large concentrations of iodine; indeed, it has been shown that there is no correlation between the iodine content of soils and the plants growing on them (Al-Ajely, 1985). This is particularly important when considering the distribution of IDD. An additional consideration is that iodine in high concentrations has been shown to be toxic to most plants (Sheppard & Evenden, 1995), with high iodine uptake in rice plants thought to be responsible for Akagare disease (Yuita, 1994b).

In most circumstances the major pathway of elements into plants is through the root system followed by translocation to the upper parts of the plant. For iodine it has been shown experimentally that it can be taken in through the root system of plants with the iodide ion more readily incorporated than iodate. However, it has also been demonstrated that there is little translocation from the roots to the aerial parts of the plant. In some circumstances high concentrations of iodine have been shown to occur in rice grown on flooded soil, which leads to Akagare disease. This has been suggested to be due to high soluble iodide in soils, which results in greater uptake by the roots; however, it has been demonstrated that there is only a relatively small increase in the iodine content of rice grown in flooded soil when compared with rice grown on drained soil, but submerged leaves when compared to other leaves showed dramatically increased iodine content.

From these considerations it seems likely that root uptake of iodine is relatively unimportant for the overall iodine content of plants. It is probable that the most important pathway into plants is from the atmosphere by direct absorption. Experiments utilizing radioactive isotopes of iodine have demonstrated that plant leaves can absorb this iodine, and it has been found that the absorption of gaseous iodine by leaves increases with increasing humidity. This is probably due to increased opening of the leaf stomata. Iodine absorbed through the leaf can be translocated through the rest of the plant, albeit slowly.

While the uptake of gaseous iodine has been demonstrated to be significant, doubt has been expressed that significant amounts of methyl iodide could be taken in through the stomata. However, even if little is absorbed, atmospheric iodine, whatever the form, could be deposited on plant surfaces and as such could represent a significant source of iodine to grazing animals, etc. The source of the atmospheric iodine in near-coastal areas will be mainly marine. However, the source in inland areas could be derived in part from iodine volatilized from soil, in some circumstances this being very significant.

The uptake of iodine into plants is, therefore, accomplished in two ways: through the roots and through leaf stomata. The latter is probably the most important. However, it must be stressed that the bioavailability of iodine in soils is low, whatever the preferred uptake route. Thus, strongly bound iodine will not be bioavailable for either root uptake or for volatilization from the soils. Several workers have quoted soil-to-plant concentration factors for iodine (iodine in plant/iodine in soil); these are generally very low and are in the range of 0.01–1.5 with most falling between 0.01 and 0.1 (Ng, 1982), while IAEA (1994) quoted a value of 0.0034 for grass. Variation of the plant concentration factor for different soil types has been demonstrated, and Muramatsu et al. (1993) quoted factors for brown rice grown on an andosol and gray lowland soil in Japan of 0.007 and 0.002, respectively.

Not surprisingly then, the iodine content of plants is generally low. Grass and herbage analyses from many different countries have shown iodine contents to be about 0.2 mg kg^{-1}; a typical example being for the UK with an estimated mean of 0.22 ± 0.16 mg kg^{-1} (Whitehead, 1984). In a study of Japanese plants Yuita (1994a) found the mean iodine content of different plant parts to be green leaves 0.46 mg kg^{-1}, fruit 0.14, edible roots 0.055, and seeds 0.0039.

V. IODINE SOURCES FOR HUMANS

The daily recommended dietary intake of iodine is variously quoted but is estimated to be 110–130 μg/day for children under the age of 1, 90–120 μg/day for children aged 1–10, and 150 μg/day for adults and adolescents, with higher concentrations required during pregnancy and lactation (see Geological Impacts on Nutrition, this volume). The traditional view is that humans derive their iodine from consumption of crops and vegetables, etc. In near-coastal areas such a source may provide sufficient iodine. However, in inland areas this will provide only relatively low quantities of iodine. In this context it has been demonstrated that vegetarian diets result in low iodine intake, which could lead to iodine deficiency (Davidsson, 1999; Remer et al., 1999) (see also Chapter 7, this volume).

Seafood is a potentially rich source of dietary iodine and, where such food is a major part of the diet as in Japan and Iceland, some problems of excess iodine in the diet have been described. A high iodine intake causes a decrease of thyroid hormone production resulting in formation of "high iodine goiter." However, at the present time the major source of dietary iodine in many developed countries is dairy produce such as milk, butter, and cheese, which are rich sources due to the addition of iodine to cattle feed and the use of iodine-containing disinfectants in the dairy industry. On this point, it is likely that the recent re-emergence of IDD in some affluent countries is due to the consumption of lower quantities of dairy produce in the move to "healthy eating."

It is also likely that even without iodine added to animal diets, animal products are probably enriched in iodine as grazing animals will take in iodine that has been deposited on the surfaces of grass and leaves; human preparation of food is likely to remove much of this surface-deposited iodine. In addition animals are known to inadvertently consume soil and this could add iodine that is not bioavailable to plants. This source of iodine has been shown to prevent iodine deficiency in grazing animals in some areas where pasture is iodine deficient as in New Zealand and Tasmania (see Geophagy and the Involuntary Ingestion of Soil, this volume), even though the soil contains only 1–2 mg kg^{-1}. In this context it is also of note that Lidiard (1995) found that farm animals in Exmoor, Somerset, UK, developed iodine deficiency symptoms when grazing areas of reclaimed land. These areas, which are about 40 km from the sea, were originally covered in peat and previous to reclamation no iodine-deficiency problems had occurred. Thus it seems likely that the animals had previously obtained iodine from inadvertent intake of peat.

It is possible that humans may obtain some iodine by inhalation, particularly in near-coastal environments. While it is indeed possible that some atmospheric iodine might be inhaled even in areas remote from the sea, this is unlikely to be a major component of iodine intake. It is of note that the Nordic Project Group (1995) suggested that an individual acquires only 0.5 μg per day from inhalation.

Although it has been suggested that some dietary iodine derives from drinking water, generally this source is unlikely to provide more than 10% of the daily adult iodine requirement. However, very high iodine drinking waters derived from groundwater have been recorded in some areas of China. In these areas it has been found that goiter incidence is negatively correlated with the iodine content of the drinking water. In some areas of China, such as around Bohai Bay in Hebai and Shandong Provinces and in some areas of Xinjiang and Shanxi Provinces, drinking waters contain between 100 and 200 μg L^{-1} and the incidence of goiter was found to be positively correlated with the iodine content of the water. The high iodine causes populations in the areas to suffer from high iodine goiter (Tan, 1989).

VI. THE GLOBAL DISTRIBUTION OF IODINE DEFICIENCY DISORDERS AND THE IODINE CYCLE

The most important parts of the iodine cycle, in terms of the environmental distribution of iodine and its impact on human health, are summarized in Figure 4.

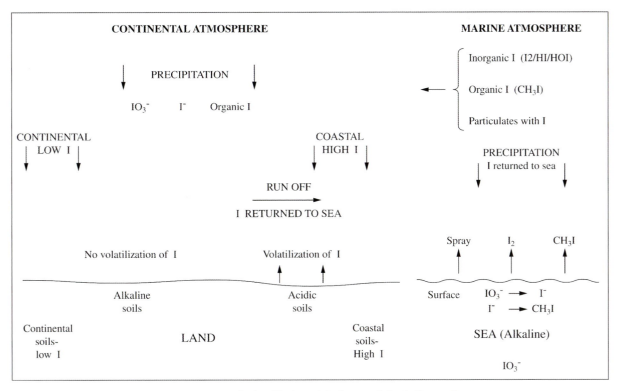

FIGURE 4 A simplified model of part of the iodine cycle.

Prior to about 1950 IDD had affected virtually every country in the world. Because the problem is known to be essentially due to the lack of iodine in the diet, it has been possible to introduce schemes for the mass treatment of affected populations such as the addition of iodine to the diet through the use of iodized salt and bread, etc., injections of iodized oil, or the addition of iodine to irrigation waters (see Cherian and Nordberg, this volume), which results in the alleviation of the symptoms in many countries. It was noted in the 1970s that IDD had been effectively eradicated in the developed world and endemic goiter was described as a disease of the poor that was largely confined to third world countries. However, it is important to point out that in the 1990s iodine deficiency has been reported to occur in several affluent countries in western Europe, probably as a result of dietary changes. This suggests that such changes could cause the reintroduction of IDD in many countries.

The global distribution of iodine-deficiency problems is shown in Figure 5. This figure is based on the data in Dunn and van der Haar (1990) and is limited as data are lacking for some areas such as parts of Africa and the Middle East.

As outlined by Kelly and Sneddon (1960) many of the areas affected by iodine deficiency are remote from marine influence. Many of the areas highlighted are the mountainous regions and their rain shadow areas such as the Himalayan region, the European Alps region, and the Andean Chain. Tan (1986) has indicated that many of the extremely seriously affected IDD areas of China are the mountainous and hilly regions over the whole country. Central continental regions, such as those of Africa and China, are also well-documented areas of iodine deficiency. In all of these situations IDD can be explained according to the classic explanations of low iodine supply and hence low iodine availability. Thus in continental areas such as the central United States, where iodine-deficiency problems were described prior to the 1950s (Kelly & Sneddon, 1960), soil iodine is relatively low with concentrations typically $1.3 \, mg \, kg^{-1}$ (Table II).

However, many endemias are not explicable in these simplistic terms and several countries and regions that are close to the coast have been known to suffer from IDD problems. For example, large regions of the UK have histories of IDD, despite the strong maritime influence on the country (see Figure 6). IDD in Sri Lanka

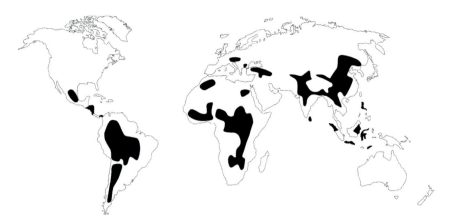

FIGURE 5   Global distribution of IDD. (Modified from Dunn & van der Haar, 1990.)

FIGURE 6   Distribution of IDD in the UK and Ireland. (Modified from Kelly & Snedden, 1960.)

occurs in coastal areas (Dissanyake & Chandrajith, 1996) while Tan (1986) had documented coastal areas of China where endemic goiter had been recorded.

In some endemias the involvement of the sulfur-containing goitrogens has been invoked as a reason. Some of these may derive from geological sources and be incorporated into drinking water or food (Gaitan et al., 1993), but others occur naturally in vegetables such as those of the *Brassica* genus and in such staple items as cassava (see also Chapter 8, this volume).

Several other elements such as fluorine, arsenic, zinc, magnesium, manganese and cobalt have been suggested to be involved in the etiology of IDD, but there is no real evidence to support their involvement. However, more recently in some areas of IDD selenium has been shown to be strongly implicated (Vanderpas et al., 1990) (see also Chapter 15, this volume).

However, even allowing for the involvement of other elements and compounds in some endemias, it seems likely that the causes of IDD problems in several areas are governed by the geochemistry of iodine and its bioavailability. In some cases sandy soils have been found to occur in goitrous areas such as the coastal regions of China. Here any iodine added to the soil from atmospheric sources is going to be leached from the soil rapidly, which results in low-iodine soils. This means little iodine is available for any plants or crops growing in the region, thus depriving humans and livestock of a source of iodine.

In sandy soils iodine is not retained, as outlined in Section III of this chapter. The iodine retention capacity of soils is related to composition with organic matter, and iron and aluminum oxides are the most important retentive components. So that in general, high-iodine soils are rich in one or more of these components. The iodine sorbed by these soil components is strongly held and little has been found to be easily leachable, so that in general terms this strongly held soil iodine is non-bioavailable. Thus Tan (1986) noted that some peaty areas of China were goitrous. Similarly, several peaty areas of the UK, even though some are in coastal regions, are listed as goitrous by Kelly and Sneddon (1960).

Another example of environmental control on IDD would seem to be the strong association of some goitrous regions with limestone bedrock as first mentioned by Boussingault (1831) for a region of Columbia, and subsequently by several other researchers such as Perel'man (1977). This association of limestone with goitrous areas led to the suggestion that calcium was a goitrogen, a claim that was subsequently disproved in clinical tests.

There are many examples of such limestone-associated endemias occurring in the UK and Ireland with the limestone regions of north Yorkshire historically renowned for its severe goiter and cretinism problems (see Figure 6). Similarly County Tipperary in Ireland, underlain by Carboniferous Limestone, was historically one of the major areas of IDD in that country. One of the most well-documented areas of IDD in the UK is the Derbyshire region of northern England, an area that is between 150 and 180 km from the west coast in the direction of the prevailing wind. Here endemic goiter was rife, known as "Derbyshire neck," with the endemia confined to areas underlain by limestone bedrock (see Figure 6). Analyses of soil in the former goitrous region have shown that iodine concentrations range up to 26 mg kg^{-1} with a mean value of 8.2 mg kg^{-1}. In neighboring areas underlain by non-limestone lithologies and with no history of IDD, soil iodine concentrations are lower with a mean of 3.44 mg kg^{-1}. Similarly, there is an area of north Oxfordshire, England, where endemic goiter was prevalent (Kelly & Sneddon, 1960) and where iodine-deficiency problems in school children were recorded as recently as the 1950s. This area is underlain by limestone and soil iodine ranges between 5 and 10 mg kg^{-1}.

IDD occurs in these limestone areas despite relatively high iodine in soils. This would imply that iodine is not bioavailable. Soils over the limestones would generally be well drained and circum-neutral to alkaline in nature. In these conditions any soluble soil iodine is likely to be present as the iodate anion (Figure 3). It has been shown that iodate uptake through plant roots is more limited than the uptake of iodide, thus in neutral to alkaline soils root uptake of iodine may be low.

In addition, as stated earlier, it is likely that plants derive much of their iodine from the atmosphere, through the upper parts of the plant. In the neutral to alkaline soils generally found overlying limestones where iodate is going to be the dominant soluble species of iodine, there is no possibility of conversion of this species to gaseous elemental iodine (see Figure 3). An additional factor is that in the well-drained soils there is going to be no conversion of iodine to methyl iodide. Thus in the limestone areas, plants will be deprived of a local source of atmospheric iodine.

The distribution of IDD reflects the geochemistry of iodine and, as stated above, with large areas of iodine deficiency occurring in central continental regions and mountainous and rain shadow areas, this distribution fits in with the classical explanation of IDD governed by the external supply of iodine from the marine environment via the atmosphere. However, the geochemistry of iodine is more complex than this simplistic approach, and from a closer scrutiny of the distribution of IDD it is apparent that, in many cases, iodine-deficiency problems are related to the bioavailability of iodine in soils and are not related directly to the external supply of iodine.

VII. RADIOACTIVE IODINE IN THE ENVIRONMENT

A relatively recent problem regarding iodine and human health is that of anthropogenically produced radioactive iodine, a topic that has received considerable attention during the last two decades. Although the problem of radioactive iodine that is of anthropogenic origin is perhaps out of place in this chapter, it is included because the distribution of natural iodine is likely to have a marked influence on the health effects of the radioactive iodine.

Whereas natural iodine is essentially mono isotopic and the one stable isotope is ^{127}I, over 20 radioactive isotopes have been identified ranging from ^{117}I to ^{139}I. Of these radioactive isotopes only ^{129}I has a significantly long half-life of 1.6×10^7 years. Extremely small amounts of ^{129}I are produced naturally by spontaneous fission of uranium and also by spallation of xenon in the upper atmosphere. However, comparatively large quantities of ^{129}I are produced from nuclear fission fallout and reactors. The pre-nuclear age ^{129}I/^{127}I ratio has been estimated as 10^{-12} while present day "background values" have been estimated to be about 10^{-10}.

Relatively high concentrations of ^{129}I have been found in the biosphere around nuclear plants, and elevated concentrations of this isotope have been shown to occur in the thyroid glands of animals in the vicinity of nuclear installations. The release of fairly large quantities of ^{129}I from nuclear sources is of concern, but it is the allied release of the shorter lived isotopes ^{125}I (half-life 60 days), ^{131}I (half-life 8.04 days), and ^{133}I (half-life 20.9 hours) that represents the greater threat to human health. Of these it is ^{131}I that has been suggested to pose the greatest risk because it is produced in fairly large quantities from fission of enriched uranium (the yield is about 3%). It has a particularly high specific activity, and hence is highly radioactive. It has been estimated that in the Chernobyl accident, in April 1986, 35 million curies of ^{131}I was released.

The problem of the short-lived radioactive isotopes of iodine was first encountered in the early years of the nuclear industry after an accident at Windscale, UK, in 1957. Radioactive iodine was found to enter the biosphere rapidly and was found in plants and in cow's milk shortly after the accident. Presumably as the radioactive iodine would have been released as a gas, it could be absorbed through plant stomata and deposited on foliage. Thus the contaminant iodine passes fairly rapidly through the food chain and into humans where it can be taken into the thyroid gland. Once there, radioactive breakdown would result in an increased risk of thyroid cancer.

In areas where natural iodine is deficient, the problem of large releases of ^{131}I would be potentially very serious, as a sudden increase of iodine would result in a high percentage of bioavailable iodine that is radioactive. Thus much of the iodine entering the human body would be radioactive. The area around Chernobyl is situated in a central continental region and as such has a history of iodine deficiency (Kelly & Sneddon, 1960) with goiter found to occur in schoolchildren in the area. Since the Chernobyl accident there has been an increased incidence of childhood thyroid cancers, and this has been shown to be causally linked to the release of radioactive iodine.

VIII. SUMMARY

The lithosphere is generally depleted in iodine and although it contributes to iodine in soils through weathering of bedrock, this is not the most important part of the iodine geochemical cycle. The oceans represent the largest reservoir of iodine on the Earth, and virtually all iodine in the terrestrial environment derives from the oceans by way of the atmosphere. Iodine is volatilized from the sea as methyl iodide (CH_3I), elemental iodine (I_2), and possibly as some other inorganic iodine compounds such as HI or HOI, with CH_3I probably the most important of these. This volatilized iodine is deposited on land by wet and dry precipitation and consequently soils from near-coastal environments are enriched in iodine and soils remote from the sea are depleted.

The iodine content of soils reflects not only the amount of iodine input from the atmosphere but is markedly dependent on soil composition. Thus organic matter, and iron and aluminum oxides in soils are able to strongly sorb iodine, and soils enriched in these components are frequently enriched in iodine.

It has generally been assumed that iodine in soils is transferred to plants and these, in turn, represent a major pathway of iodine into animals and humans. However, the soil-to-plant concentration factor for iodine has been shown to be low due to the strong sorption of iodine by soil components such as organic matter and iron and aluminum oxides. Little iodine in soils has been found to be easily leachable and no correlation of soil and plant iodine has been demonstrated. In addition, it has been shown that although iodine is taken into plant roots with the iodide ion more readily incorporated than iodate, little of this iodine is translocated from the roots of plants to the aerial parts. It is likely that most iodine in plants is taken in from the atmosphere through the stomata. Whereas in coastal areas such an atmospheric source of iodine is likely to be abundant, in inland areas such an atmospheric source would, to a large extent, be dependent on iodine volatilized from soils. Volatilization of soil iodine is likely to be of major importance in the iodine cycle, but in many areas such volatilization may be limited due to iodine being strongly bound in soil. This is likely to be particularly important in areas underlain by limestone where soils would be expected to be circum-neutral to alkaline, which results in any labile iodine being present as the iodate ion and hence unable to be converted to gaseous elemental iodine.

Traditionally, crops and vegetables have been suggested to be important sources of dietary iodine for humans. However, this is unlikely to be true in all but coastal regions as little iodine in soils is generally bioavailable. Seafood is generally a rich source of dietary iodine, whereas in some areas drinking waters are important sources. In developed countries, dairy products are a major source of dietary iodine due to the addition of iodine to cattle feed and use of iodine-containing sterilants in the dairy industry. However, even without the addition of iodine to dairy products animal products are likely to be richer sources of dietary iodine as grazing animals will take in iodine that has been deposited on the surfaces of grass and leaves. In addition, many grazing animals are known to inadvertently take in soil, which has been shown to provide more iodine.

The global distribution of IDD reflects the geochemistry of iodine with large areas of iodine deficiency occurring in central continental regions and mountainous and rain shadow areas, which reflect the supply of iodine from the marine environment via the atmosphere. However, the geochemistry of iodine is more complex than this simplistic approach, and from a closer scrutiny of the distribution of IDD it is apparent that, in many cases, iodine-deficiency problems are related to the bioavailability of iodine in soils and are not related directly to the external supply of iodine.

A relatively modern problem concerning iodine is the release of radioactive iodine from anthropogenic sources. Of the radioactive isotopes of iodine it is ^{131}I, with a half-life of 8.04 days, which has been suggested to pose the greatest risk as it is produced in fairly large quantities from fission of enriched uranium and is highly radioactive. In areas where natural iodine is deficient the problem of large releases of ^{131}I would be potentially very serious, as a sudden increase of iodine would result in a high percentage of bioavailable iodine being radioactive. This problem was highlighted in the Chernobyl accident which occurred in a central continental area where iodine-deficiency problems have been described. A high incidence of childhood thyroid cancer since the accident has been shown to be causally linked to the release of the radioactive iodine.

SEE ALSO THE FOLLOWING CHAPTERS

Chapter 6 (Biological Functions of the Elements) · Chapter 7 (Geological Impacts on Nutrition) · Chapter 8 (Biological Responses of Elements) · Chapter 14 (Bioavailability of Elements in Soil) · Chapter 15 (Selenium Deficiency and Toxicity in the Environment) · Chapter 17 (Geophagy and the Involuntary Ingestion of Soil)

FURTHER READING

Al-Ajely, K. O. (1985). Biogeochemical Prospecting as an Effective Tool in the Search for Mineral Deposits, Ph.D. thesis, University of Wales, Aberystwyth.

Boussingault, J. B. (1831). Recherches sur la cause qui produit le Goitre dans les Cordileres de la Novelle-Grenade, *Ann. Chim. (Phys.)*, 48, 41–69.

Bowen, H. J. M. (1979). *Environmental Chemistry of the Elements*, Academic Press, London.

Chameides, W. C., and Davis, D. D. (1980). Iodine: Its Possible Role in Tropospheric Photochemistry, *J. Geophys. Res.*, 85, 7383–7398.

Councell, T. B., Landa, E. R., and Lovley, D. R. (1997). Microbial Reduction of Iodate, *Water Air Soil Pollut.*, 100, 99–106.

Davidsson, L. (1999). Are Vegetarians an "At Risk Group" to Iodine Deficiency?, *Br. J. Nutr.*, 81, 3–4.

Dean, G. A. (1963). The Iodine Content of Some New Zealand Drinking Waters with a Note on the Contribution of Sea Spray to the Iodine in Rain, *N. Z. J. Sci.*, 6, 208–214.

Delange, F. (1994). The Disorders Induced by Iodine Deficiency, *Thyroid*, 4, 107–128.

Dissanayake, C. B., and Chandrajith, R. L. R. (1996). Iodine in the Environment and Endemic Goitre in Sri Lanka. In *Environmental Geochemistry and Health* (J. D. Appleton, R. Fuge, and G. J. H. McCall Eds), Geological Society Special Publication 113, 201–211.

Duce, R. A., Zoller, W. H., and Moyers, J. L. (1973). Particulate and Gaseous Halogens in the Antarctic Atmosphere, *J. Geophys. Res.*, 78, 7802–7811.

Dunn, J. T., and van der Haar, F. (1990). A Practical Guide to the Correction of Iodine Deficiency, International Council for the Control of Iodine Deficiency Disorders Technical Manual No. 3.

Fuge, R. (1987). Iodine in the Environment: Its Distribution and Relationship to Human health, *Trace Substances Environ. Health*, 21, 74–87.

Fuge, R. (1988). Sources of Halogens in the Environment, Influences on Human and Animal Health, *Environ. Geochem. Health*, 10, 51–61.

Fuge, R. (1996). Geochemistry of Iodine in Relation to Iodine Deficiency Diseases. In *Environmental Geochemistry and Health* (J. D. Appleton, R. Fuge, and G. J. H. McCall, Eds.), Geological Society Special Publication 113, 201–211.

Fuge, R., and Ander, L. (1998). Geochemical Barriers and the Distribution of Iodine in the Secondary Environment: Implications for Radio-Iodine. In *Energy and the Environment: Geochemistry of Fossil, Nuclear and Renewable Resources* (K. Nicholson, Ed.), MacGregor Science, Insch, Scotland, 163–170.

Fuge, R., and Johnson, C. C. (1986). The Geochemistry of Iodine-a Review, *Environ. Geochem. Health*, 8, 31–54.

Gabler, H. E., and Heumann, K. G. (1993). Determination of Iodine Species Using a System of Specifically Prepared Filters and IDMS, *Fresnius J. Anal. Chem.*, 345, 53–59.

Gaitan, E., Cooksey, R. C., Legan, J., Cruse, J. M., Lindsay, R. H., and Hill, J. (1993). Antithyroid and Goitrogenic Effects of Coal-Water Extracts From Iodine-Sufficient Areas, *Thyroid*, 3, 49–53.

Garland, J. A., and Curtis, H. (1981). Emission of Iodine From the Sea Surface in the Presence of Ozone, *J. Geophys. Res.*, 86, 3183–3186.

Gerzabek, M. H., Muramatsu, Y., Strebl, F., and Yoshida, S. (1999). Iodine and Bromine Contents of Some Austrian Soils and Relations to Soil Characteristics, *J. Plant Nutr. Soil Sci.*, 162, 415–419.

Goldschmidt, V. M. (1954). *Geochemistry*, Oxford University Press, London.

Heumann, K. G., Gall, M., and Weiss, H. (1987). Geochemical Investigations to Explain Iodine-Overabundances in Antarctic Meteorites, *Geochim. Cosmochim. Acta*, 51, 2541–2547.

Heumann, K. G., Neubauer, J., and Reifenhauser, H. (1990). Iodine Overabundances Measured in the Surface Layers of an Antarctic Stony and Iron Meteorite, *Geochim. Cosmochim. Acta*, 54, 2503–2506.

IAEA (1994). Handbook of Parameter Values for the Prediction of Radionuclide Transfer in Temperate Environments, International Atomic Energy Authority, Technical Report Series, No. 364.

Johnson, C. C. (1980). The Geochemistry of Iodine and a Preliminary Investigation into Its Potential Use as a Pathfinder Element in Geochemical Exploration, Ph.D. thesis, University of Wales, Aberystwyth.

Kelly, F. C., and Sneddon, F. W. (1960). Prevalence and Geographical Distribution of Endemic Goitre. In *Endemic Goitre*, World Health Organization, Geneva, 27–233.

Kocher, D. C. (1981). A Dynamic Model of the Global Iodine Cycle and Estimation of Dose to the World Population From Release of I-129 to the Environment, *Environ. Int.*, 5, 15–31.

Låg, J., and Steinnes, E. (1976). Regional Distribution of Halogens in Norwegian Forest Soils, *Geoderma*, 16, 317–325.

Langer, P. (1960). History of Goitre. In *Endemic Goitre*, World Health Organization, Geneva, 9–25.

Lidiard, H. M. (1995). Iodine in Reclaimed Upland Soil of a Farm in the Exmoor National Park, Devon, U.K., and Its Impact on Livestock Health, *Appl. Geochem.*, 10, 85–95.

Lovelock, J. E., Maggs, R. J., and Wade, R. J. (1973). Halogenated Hydrocarbons in and Over the Atlantic, *Nature*, 241, 194–196.

Moison, T. A., Dunstan, W. M., Udomkit, A., and Wong, G. T. F. (1994). The Uptake of Iodate by Marine Phytoplankton, *J. Phycol.*, 30, 580–587.

Moore, R. M., and Groszko, W. (1999). Methyl Iodide Distribution in the Ocean and Fluxes to the Atmosphere, *J. Geophys. Res.*, 104, 11163–11171.

Muramatsu, Y., and Wedepohl., K. H. (1998). The Distribution of Iodine in the Earth's Crust, *Chem. Geol.*, 147, 201–216.

Muramatsu, Y., and Yoshida, S. (1999). Effects of Microorganisms on the Fate of Iodine in the Soil Environment, *Geomicrobiol. J.*, 16, 85–93.

Muramatsu, Y., Uchida S., and Ohmomo, Y. (1993). Root-Uptake of Radioiodine by Rice Plants, *J. Radiat. Res.*, 34, 214–220.

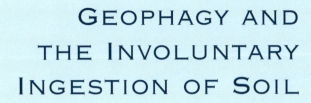

CHAPTER 17

GEOPHAGY AND THE INVOLUNTARY INGESTION OF SOIL

PETER W. ABRAHAMS
University of Wales

CONTENTS

Geophagy or geophagia can be defined as the deliberate ingestion of soil. This is a practice that is common among members of the animal kingdom, including people. Any person who studies geophagy undertaken by humans will invariably confront a problem during their research. Few people will believe them. This is perhaps understandable for members of a developed urban society that is educated, has ready access to modern pharmaceuticals, and which has increasingly, in both a physical and mental sense, become more remote from soils. Yet many of these people will readily accept that wild animals deliberately eat soil. For example, television programs which feature wildlife may show

animals consuming soil, with the presenter commonly stating that the soils are being eaten for their mineral nutrient content (although, as will be seen in the following sections, there are a variety of reasons why animals consume soil). But many people find it more difficult to accept that humans can deliberately eat soil. This ignorance of geophagy is not restricted to the layperson, because academic writers have used adjectives such as curious, odd, perverted, and strange when commenting on human geophagy. The use of such words demonstrates a misunderstanding of geophagy. The practice is common in certain human societies and can be readily found in many countries provided one has appropriate knowledge of the subject. An understanding of geophagy also allows an appreciation of the practice. There are perfectly sensible reasons as to why certain people deliberately eat soil, and the consumer can benefit from indulging in geophagy in a number of ways. It has been suggested that the practice should be considered within the normal range of human behavior (Vermeer, 1986), an enlightened viewpoint that I personally support. However, before the reader hurries away to indulge in geophagy, a word of warning is necessary. Aside from the benefits that eaten soils can impart to the consumer, very serious health problems may also result. These benefits and banes of soil consumption are considered in more detail later in Section

VI of this chapter, along with other aspects of geophagy undertaken by humans. This information follows a discussion about geophagy that is practiced by members of the animal kingdom other than humans. In addition to geophagy, many animals (including humans) also accidentally ingest soil. In order to appreciate geophagy in its proper context, this involuntary ingestion of soil is considered first (see also Chapter 14, this volume).

I. INVOLUNTARY SOIL INGESTION: DOMESTICATED ANIMALS

Grazing and browsing animals are especially prone to what is variously called accidental, involuntary, or incidental soil ingestion. Although both wild and domesticated animals can ingest soil involuntarily, the majority of research on this topic has concentrated on the latter largely because of the economic and health implications to humans. On farmland, pasture plants growing close to the surface are subject to soil contamination resulting from the effects of trampling by grazing animals, rain splash, or wind. Grazing animals will ingest soil that has adhered to vegetation because of these processes. Soil can also be licked from snouts and, on closely cropped pastures, can be ingested directly from the ground surface. The soil adhering to roots can be another source. For example, in a study undertaken on a semi-arid range located in Idaho in the United States, soil was ingested by cattle primarily with the roots of cheatgrass (*Bromus tectorum*) that were pulled up and consumed along with the aboveground plant parts (Mayland et al., 1975).

The amount of soil ingested can be quantified in a variety of ways (Healy, 1973). The ash content of feces from animals gives a measure of soil content, with a correction necessary for the ash contribution from undigested herbage. Treatment of ash with dilute acid gives an acid-insoluble residue (AIR) that allows a more accurate measure of soil content to be calculated. If the fecal output is known together with soil content of the feces, then the quantity of soil ingested can be determined. Alternatively, the titanium content of soil and feces can be used to estimate rates of soil ingestion (Miller et al., 1976). This method is based on the premise that titanium, which is abundant in soils (containing typically several thousand $mg\,kg^{-1}$), is present only in small quantities (usually $<10\,mg\,kg^{-1}$) in plants not contaminated with soil. Any titanium recorded in fecal samples can thus be assumed to originate from a soil source. With

animals absorbing a negligible amount of ingested soil titanium, soil ingestion can be calculated using the equation:

$$\% \text{ soil ingestion} = \frac{(1 - D_h)Ti_f \times 100}{Ti_s - D_h Ti_f} \qquad (1)$$

where D_h = digestibility of herbage, Ti_s = titanium in soil, and Ti_f = titanium in feces.

Research in New Zealand has indicated that sheep can ingest >75 kg of soil annually, and dairy cows can consume between 150 kg to 650 kg of soil per year (Healy, 1968). In New Zealand and elsewhere, seasonal variations of soil ingestion are marked, depending on soil type, weather, and management factors such as stocking rate and the use of supplementary feed. Within individual flocks and herds, significant differences in soil ingestion can be found between animals at any point in time, though there is some evidence to suggest that identical twins of cows have an inherited tendency to consume similar amounts of soil. In the UK, dairy cattle can ingest at certain times of the year >10% of their dry matter (DM) intake in the form of soil. For sheep, grazing typically closer to the ground, the figure may exceed 30% (Thornton, 1974). In countries such as New Zealand and the UK, the rates of soil ingestion are low during the summer months when there is an adequate supply of herbage. Soil ingestion is greater in the autumn, winter, and early spring months attributable to factors such as the low rates of herbage production.

There are a number of economic and health implications that are associated with the involuntary ingestion of soil. For example, animal production will be adversely affected if the consumption of soil reduces the digestible DM intake (Pownall et al., 1980). Also, as soil is highly abrasive to dentine, excessive tooth wear attributable to a high ingestion of soil can lead to culling at a comparatively early age because the animal can no longer graze efficiently (Healy & Ludwig, 1965). The abrasive effects on the alimentary tract could also prove irritating to animals, and may additionally increase their vulnerability to infections. The majority of research, however, has investigated the implications of ingested soils as a source of potentially beneficial mineral nutrients or of undesirable constituents such as pesticide residues, heavy metals, and radionuclides (Harrison et al., 1970; Beresford & Howard, 1991; Green et al., 1996; Lee et al., 1996; Abrahams & Steigmajer, 2003). As soil passes through the gastrointestinal tract of

Muramatsu, Y., Yoshida, S., Uchida S., and Hasebe, A. (1996). Iodine Desorption From Rice Paddy Soil *Water Air Soil Pollut.*, 86, 359–371.

Ng, Y. C. (1982). A Review of Transfer Factors for Assessing the Dose From Radionuclides in Agricultural Products, *Nucl. Safety*, 23, 57–71.

Nordic Project Group (1995). Risk Evaluation of Essential Trace Elements–Essential Versus Toxic Levels of Intake, Nordic Council of Ministers, Nord 1995, 18.

Perel'man, A. J. (1977). *Geochemistry of Elements in the Supergene Zone*, Keterpress Enterprises, Jerusalem.

Prister, B. S., Grigor'eva, T. A., Perevezentsev, V. M., Tikhomirov, F. A., Sal'nikov, V. G., Ternovskaya, I. M., and Karabin, T. (1977). Behaviour of Iodine in Soils, *Pochvovedenie*, 6, 32–40 (in Russian).

Rahn, K. A., Borys, R. D., and Duce, R. A. (1976). Tropospheric Jalogen Gases: Inorganic and Organic Components, *Science*, 192, 549–550.

Remer, T., Neubert, A., and Manz, F. (1999). Increased Risk of Iodine Deficiency with Vegetarian Nutrition, *Br. J. Nutr.*, 81, 45–49.

Schmitz, K., and Aumann, D. C. (1994). Why are the Soil-to-Pasture Transfer Factors, as Determined by Field-Measurements for I-127 Lower than for I-129, *J. Environ. Radioact.*, 24, 91–100.

Shacklette, H. J., and Boerngen, J. G. (1984). Element Concentrations in Soils and Other Surficial Materials of the Conterminous United States, United States Geological Survey Professional Paper 1270.

Sheppard, S. C., and Evenden, W. G. (1995). Toxicity of Soil Iodine to Terrestrial Biota with Implications for I-129, *J. Environ. Radioact.*, 27, 99–116.

Takagi, H., Iijima, I., and Iwashima, K. (1994). Determination of Iodine with Chemical Forms in Rain Water with Fractional Sampling NAA, *Bunseki Kagaku*, 43, 905–909.

Tan, J. (Ed.) (1989). *The Atlas of Endemic Diseases and Their Environments in the People's Republic of China*, Science Press, Beijing.

Vanderpas, J. B., Contempre, B., Duale, N. L., Gossans, W., and Bebe, N. G. O. (1990). Iodine and Selenium Deficiency Associated with Cretinism in Northern Zaire, *Am. J. Clin. Nutr.*, 52, 1087–1093.

Vinogradov, A. P. (1959). *The Geochemistry of Rare and Dispersed Chemical Elements in Soils*, 2nd edition, Consultants Bureau, New York.

Vinogradov, A. P., and Lapp, M. A. (1971). Use of Iodine Haloes to Search for Concealed Mineralisation, *Vestn. Leningr. Univ. Ser. Geolog. Geogr.*, No. 24, 70–76 (in Russian).

Whitehead, D. C. (1974). The Sorption of Iodide by Soil Components, *J. Sci. Food Agric.*, 25, 461–470.

Whitehead, D. C. (1984). The Distribution and Transformation of Iodine in the Environment, *Environ. Int.*, 10, 321–339.

Wong, G. T. F. (1991). The Marine Geochemistry of Iodine, *Rev. Aquat. Sci.*, 4, 45–73.

Wong, G. T. F., and Cheng, X. H. (1998). Dissolved Organic Iodine in Marine Waters: Determination, Occurrence and Analytical Implications, *Mar. Chem.*, 59, 271–281.

Yuita, K. (1994a). Overview and Dynamics of Iodine and Bromine in the Environment. 1. Dynamics of Iodine and Bromine in Soil-Plant System, *Jpn. Agric. Res. Q.*, 28, 90–99.

Yuita, K. (1994b). Overview and Dynamics of Iodine and Bromine in the Environment. 2. Iodine and Bromine Toxicity and Environmental Hazards, *Jpn. Agric. Res. Q.*, 28, 100–111.

Yuita, K., Tanaka, T., Abe, C., and Aso, S. (1991). Dynamics of Iodine, Bromine and Chlorine in Soil. 1. Effects of Moisture, Temperature and pH on the Dissolution of the Triad from Soil, *Soil Sci. Plant Nutr.*, 37, 61–73.

Yoshida, S., and Muramatsu, Y. (1995). Determination of Organic, Inorganic, and Particulate Iodine in the Coastal Atmosphere of Japan. *J. Radioanal. Nucl. Chem. -Articles*, 196, 295–302.

Zafiriou, O. C. (1974). Photochemistry of Halogens in the Marine Atmosphere, *J. Geophys. Res.*, 79, 2730–2732.

animals, en route it is exposed to digestive fluids that have the potential to extract soil elements thus contributing to the concentrations in these solutions. The release of elements such as cobalt, iodine, and selenium into the digestive fluids may be of benefit to the animal, because it is from the pool of elements in solution that essential mineral nutrients are absorbed into the bloodstreams of animals for distribution throughout their tissues. For example, in both New Zealand and Australia research has suggested that ingested soils can supplement iodine in potentially goitrous grazing sheep (Healy et al., 1972; Statham & Bray, 1975). Soils are a significant source of cobalt to grazing animals because they contain 100–1000 times more of this mineral nutrient than the pasture herbage they support. Cobalt is an essential constituent of vitamin B_{12}, the anti-pernicious anemia factor of liver that is produced by the synthesizing abilities of the symbiotic gastrointestinal bacteria of an animal. Research undertaken in the UK has shown that not only is cobalt extracted from soil in the rumen of sheep, but it is also synthesized into vitamin B_{12} as required by the animal (Brebner et al., 1987). Cobalt deficiency in sheep can thus be prevented by farmers dosing animals with soil (see also Chapter 20, this volume).

These examples demonstrate the importance of the direct soil–animal pathway of mineral nutrients that complements the soil-plant-animal route in agricultural systems. However, ingested soil can also be an important source of potentially harmful elements (PHEs) in geochemically anomalous areas such as mineralized and mined districts (Figure 1). This was found in a study

undertaken in southwest England, a province that is extensively contaminated because of mineralization and a long history of mining and smelting. Here concentrations of soil arsenic are high, typically several hundred $mg\,kg^{-1}$, yet the aboveground pasture herbage when free of soil contains only about $1\,mg\,kg^{-1}$ of this element (Abrahams & Thornton, 1994). Consequently, the soil–animal pathway is the dominant source of arsenic to grazing cattle in the province, with ingested soil contributing up to 97% of the total intake of this element (Table I). In contrast, the ingestion of soil is not

FIGURE 1 Sheep grazing mineralized ground disturbed by historical mining activity, Derbyshire, UK. At this locality the ingestion of soils containing high concentrations of fluorine and lead can contribute to health problems suffered by young livestock. (Photo: Peter W. Abrahams.)

TABLE I. Maximum and Minimum Soil and Washed Herbage Arsenic (As) and Copper (Cu) Concentrations, Rates of Soil Ingestion, and Calculated Intake by Cattle of the Two Elements

	Soil ingested (%)	Soil concentration ($mg\,kg^{-1}$)		Washed herbage concentration ($mg\,kg^{-1}$ DM)		Daily intake as soil ($mg\,day^{-1}$)		Daily intake as herbage ($mg\,day^{-1}$)		Total daily intake ($mg\,day^{-1}$)		% element ingested as soil	
		As	Cu	As	Cu	As	Cu	As	Cu	As	Cu	As	Cu
April	1.5–17.9			0.06–1.1	10–23	9–189	6–154	0.8–15	108–294	10–196	250–396	80–97	3.7–59
June	0.2–3.9	19–320	12–319	0.03–0.8	8–15	2–47	2–62	0.4–10	104–210	2.5–57	113–236	41–93	2.1–34
August	1.4–4.7			0.10–1.0	9–15	8–101	6–79	1.1–13	123–194	10–113	113–273	79–96	3.2–36

Data are taken from a study undertaken in the soil-contaminated province of southwest England.

Reprinted from Agriculture, Ecosystems and Environment, 48, Peter W. Abrahams and Iain Thornton, The contamination of agricultural land in the metalliferous province of southwest England: implications to livestock, 125–137, Copyright 1994, with permission from Elsevier Science.

so important in supplying copper to grazing cattle in southwest England. The mineralization and mining activity within this province has contaminated large areas of soil with this metal, but relative to arsenic, more copper is absorbed and transferred to the aboveground parts of the pasture herbage species. This reduces the importance of soil ingestion in supplying copper to the animals, although sites of particularly high rates of soil ingestion (e.g., 17.9%) and some contaminated locations can be areas where soil ingestion of this metal is significant (e.g., 59% of the total intake; Table I). To date, however, there is little information on the gut uptake of these elements and their transfer into animal products. In southwest England, much of the ingested arsenic is not available for uptake, and it is consequently found in animal feces. Still, some of the arsenic is known to be absorbed, and it has been reported that owners of arsenic-contaminated land in southwest England rent their fields to farmers wishing to present their livestock at agricultural shows. The resulting elevated intake and absorption of arsenic leads to a nice "bloom" on the coats of the animals and thus improving their appearance.

Ingested soils also have the potential to reduce the uptake of elements by animals. Adsorption on the organo-mineral cation exchange complex; sorption by hydrous oxides of aluminum, iron, and manganese; and the formation of stable complexes with soil organic matter are all mechanisms that can reduce the availability of elements to animals. In addition, antagonistic elements can be released from ingested soil. For example, the release of iron or molybdenum from soil in the stomach of sheep may interfere with copper metabolism, which leads to a disease known in the UK as swayback (Suttle et al., 1975, 1984). This disease affects the nervous system of newborn lambs. Its name is derived from the characteristic uncoordinated gait with the back legs swaying from side to side. In the UK this disease is more severe after mild winters, and forecasts as to the severity of the disease in lambs can be made on the basis of the number of days with snow cover. During mild, snow-free winters, soil ingestion rates are elevated and a high risk of swayback in spring is likely since intake of soil-contaminated herbage reduces the availability of copper to the ewe. Conversely, severe winters encourage supplementary feeding, limiting the intake of soil, and reducing the risk of swayback (although an additional factor here is that supplementary feeding of in-lamb ewes is likely to provide more copper than from pasture herbage alone) (see also Chapter 20, this volume).

II. GEOPHAGY IN THE ANIMAL KINGDOM

The deliberate ingestion of soil has been observed in both domesticated and wild animals, although research investigations have concentrated almost exclusively on the latter. Among terrestrial vertebrates, geophagy has been reported in many species of birds, reptiles, and mammals. Whereas humans may be geophagists and are a member of the latter class, this section of the chapter considers only other animal species. Table II illustrates that only carnivores have not been observed deliberately eating soil. The listing shown in this table, however, is certainly not complete with ongoing research adding knowledge to the number of species that indulge in geophagy. For example, avian geophagy was reported from the tropical island of New Guinea for the first time in 1999 when 11 bird species (all predominantly herbivores, and especially frugivores) were observed consuming soil (Diamond et al., 1999). It is also worth reporting that geophagy is not restricted to vertebrates; for example, isopods and butterflies are also known to deliberately consume soil.

Geophagy is a widespread practice that is reported from many parts of the globe. However, most observations on geophagical behavior come from North America and the savanna of Africa (Kreulen & Jager, 1984), perhaps reflecting a bias in the study of the ecosystems and animal species of these areas. Usually the soil intake is selective, with specific sites and sometimes even particular soil horizons being exploited (Figure 2). In the literature, these locations are variously referred to as mining sites, salt licks, natural salt licks, salines, mineral licks, natural mineral licks, or natural licks (Klaus & Schmid, 1998). The use of such terminology at times can be misleading, because the words "mineral" and "salt" suggest a chemical enrichment of soil, with animals indulging in geophagy to satisfy a mineral nutrient imbalance. This may be the case, but there are a variety of reasons as to why soils are deliberately consumed by animals, and not all are enriched in mineral nutrients. Consequently, the simple term "lick" is perhaps best, because its use does not imply a specific benefit that is gained from the soil.

The size of lick sites varies from small, unspectacular scrapes to large, treeless sites like those found within tropical rainforests. For example, in a study undertaken in Dzanga National Park (Central African Republic), the licks varied in size from 2000 m^2 to 55,000 m^2, with holes and caves excavated by the trunks, tusks, and front

TABLE II. Taxonomic Categories of Reptiles, Birds, and Mammals That Engage in Geophagy

Class	Families	Representatives
Reptilia	Iguanidae	Iguana
	Emydidae	Box turtle
	Testudinidae	Tortoise
Aves	Struthionidae	Ostrich
	Anatidae	Goose
	Aegypiidae	Palm-nut vulture
	Phasianidae	Pheasant
	Numididae	Guinea fowl
	Columbidae	Dove, pigeon
	Psittacidae	Parrot
	Musophagidae	Turaco
	Coliidae	Mousebird
	Sturnidae	Starling
	Ploceidae	Sparrow, weaver
	Fringillidae	Canary, bunting
Mammalia	Leporidae	Rabbit
	Sciuridae	Squirrel, woodchuck
	Erethizontidae	American porcupine
	Elephantidae	Elephant
	Equidae	Horse, ass, zebra
	Tapiridae	Tapir
	Rhinocerotidae	Black rhino
	Suidae	Bushpig, warthog, wild boar
	Tayassuidae	Peccary
	Camelidae	Camel
	Cervidae	Caribou, moose, mule deer, roe deer, sambar, white-tailed deer
	Giraffidae	Giraffe
	Bovidae	Antelope (e.g., duiker, gazelle), bighorn (e.g., Dall sheep), mountain goat, African buffalo, banteng, gaur, domestic ox, goat, sheep
	Indridae	Lemur
	Cercopithecidae	Baboon
	Colobidae	Colobus, langur
	Hylobatidae	Gibbon
	Pongidae	Chimpanzee, gorilla
	Hominidae	Man

After Kreulen & Jager, 1984.

FIGURE 2 A soil lick located in the Mkomazi Game Reserve, Tanzania. The reserve ranger is standing in an excavation mined by animals, and other holes made by elephant tusks can be seen. This extremely alkaline, highly calcareous and saline-sodic soil, referred to as site 2 in Table III, may provide a range of benefits (e.g., sodium supplementation, an antacid function) to animals if consumed in appropriate amounts. (From Abrahams, 1999; Plenum Publishing Corporation.)

legs of the forest elephant, *Loxodonta africana cyclotis* (Klaus et al., 1998). There is considerable variability in the use of licks. Sites are not necessarily used by all species in an area, and while a particular species may utilize a lick at one location, the same species may ignore licks in other areas. Some observations on ungulates have recorded no differences in lick use among different sex and age groups, whereas other reports observe geophagy mainly or exclusively in pregnant or lactating females and/or juveniles. In the Yankari Game Reserve (Nigeria), all ages of warthog (*Phaco-*

choerus aethiopicus) exploit licks, in contrast to (mainly) adolescent hartebeest (*Alcelaphus buselaphus*; Henshaw & Ayeni, 1971). A seasonal use of licks is also evident (Kreulen & Jager, 1984). In both North America and Europe, a peak in lick use by ungulates is linked to forage changes during leaf flush in the spring. Similarly, in the arid areas of southern Africa, seasonality of lick use is associated with leaf flush at the beginning of the wet season. For the humid tropics, information on the seasonality of lick use is limited, although red leaf monkeys (*Presbytis rubicunda*) have been observed consuming mineral nutrient enriched soil from termite mounds in the rainforests of northern Borneo from April to August (Davies & Baillie, 1988). In contrast, the forest elephants in Dzanga National Park visit licks throughout the year though lick use decreases during the main fruiting season (Klaus et al., 1998).

The selective use of lick sites indicates that soils of these locations have certain qualities that animals find desirable. In northeastern Peru, mustached tamarins (*Saguinus mystax*) have been observed eating soil from a broken mound of leaf-cutting ants (Heymann & Hartmann, 1991). Geochemical analysis of the fine-textured soil revealed elevated concentrations of a number of elements (e.g., iron and potassium), which are attributable to the ants constructing the mound with deeper soil materials that are less leached than the surrounding surface soil. Red leaf monkeys in northern Borneo (see above) and chimpanzees (*Pan troglodytes schweinfurthii*) in Tanzania (Mahaney et al., 1996) are examples of other primates that have been reported to feed on soils (in both cases from termite mounds) that are similarly enriched in potentially beneficial mineral nutrients (such as calcium, magnesium, potassium, and phosphorus) and clay minerals. In the Kalahari sandveld of Botswana, geomorphological processes are important in the formation of licks, with fine-textured and nutrient-rich material accumulating in depressions (called pans) by sheet flow from adjacent areas following periods of heavy rainfall (Kreulen & Jager, 1984). The properties of lick soils can vary a great deal. For example, in the 1500 km^2 Mkomazi Game Reserve located in Tanzania, the three known lick soils show considerable chemical and mineralogical variability (Abrahams, 1999) (Table III). Despite such differences, some common properties of lick soils are

- A high content of clay-sized particles
- A high salinity
- Among saline (halomorphic) licks, sodium chloride and/or sodium sulfate may predominate in neutral or slightly acidic soils, whereas sodium carbonate and sodium bicarbonate are associated with alkaline lick sites
- High quantities of calcium and/or magnesium carbonate
- Licks may be chemically enriched (e.g., in nitrogen, sulfur, potassium, and phosphorus) because of fecal and urinary contamination; the excreta from diseased animals will heavily laden sites with pests, cysts, and nematodes

A common practice by research investigators is to compare the properties of lick samples with non-lick soils found in the same region. Differences, if any, between the two soil types can then lead to suggestions as to why animals are indulging in geophagy. Sodium often appears to be the major attracting substance of many licks (e.g., site 2, Table III), and it is known that herbivores can seek extra sources of this macronutrient because they have a sodium-specific perception and hunger mechanism that is activated, among other things, by depletion of this element (Denton, 1969). It is also appreciated that terrestrial plants may not accumulate sufficient sodium to satisfy the nutrient demands of an animal. There are a number of factors that can account for a seasonal demand of sodium that matches the periods of use of licks as noted above. For example, a temporary large increase in urinary and fecal output of sodium (attributable to a dietary change caused by the sudden transition from winter or dry season roughage to lush grass/browsing plants at the onset of spring or the wet season) will create a seasonal demand for this macronutrient that may be satisfied by the ingestion of appropriate soils (Kreulen & Jager, 1984).

Even though there is strong evidence that sodium-rich soils are a cause of geophagy, there are licks that are not enriched in this element (e.g., site 1, Table III) and clearly other benefits are sought by the geophagists. Licks may be exploited for mineral nutrients other than sodium, and calcium, iron, phosphorus, and sulfur have all been suggested as target elements. However, most licks contain relatively low concentrations of phosphorus, and wild animals indulge in osteophagy (the consumption of bone) as a source of this element, rather than geophagy. At high altitudes in the tropics, it has been suggested that ingested soils supplement African buffaloes (*Syncerus caffer caffer* (Sparrman)) and mountain gorillas (*Gorilla gorilla beringei*) with iron (Mahaney & Hancock, 1990; Mahaney et al., 1990). These animals may require relatively large amounts of this mineral nutrient for erythrocyte formation, in the same way that it is known that humans living at high altitudes need iron-rich food to increase erythrocytes in the blood.

TABLE III. Selected Geochemistry[a] and Mineralogy of Three Lick Soils From the Mkomazi Game Reserve, Tanzania

Geochemistry:

		Calcium		Iron	Potassium		Magnesium		Sodium		Phosphorus	
Site	pH	Total	Extr.	Total	Total	Extr.	Total	Extr.	Total	Extr.	Total	Extr.
1	4.3	280	230	42,000	2,000	225	940	88	120	9	1,250	8
2	11.0	91,000	360	8,400	160	7.5	8,000	13	12,600	11,875	<100	4
3	8.2	7,450	2,400	29,400	3,300	188	5,400	1,500	5,040	3,188	1,600	<8

Mineralogy:

	Smectite	Illite	Kaolinite	Amphibole	Quartz	K Feldspar	Plagioclase	Calcite	Dolomite	Halite	Pyrite	Total
Site												
1	0.0	0.0	41.0	0.0	52.6	2.4	1.8	0.0	0.0	0.0	2.2	100
2	11.6	0.0	7.3	0.0	17.8	5.7	26.3	26.1	3.8	1.4	0.0	100
3	2.3	8.8	12.0	8.7	32.2	10.0	16.9	0.0	1.5	3.7	3.9	100

[a]pH measured in 1 : 2.5 w/v water suspension. Total and extractable (Extr.) concentrations in mg kg^{-1}. The extractable concentrations are the water-soluble and exchangeable (i.e., adsorbed on soil particle surfaces) fraction of the element in the soil.

From Abrahams, 1999, Plenum Publishing Corporation.

The consumption of soil to obtain calcium has been substantiated for reptiles (e.g., the desert tortoise, *Gopherus agassizii*; Marlow & Tollestrup, 1982) and birds, though for the latter the most common explanation for geophagy is to provide grit. Because birds lack teeth, many ingest pebbles or coarse soil particles for the grinding of food in their gizzards (reptiles and ruminants are also known to ingest soil for the breakdown of food). But licks used by parrots in Peru (Figures 3 and 4) are fine-textured, which strongly suggests that these soils are not consumed to aid digestion (Gilardi et al., 1999). Instead the lick soils have a higher cation exchange capacity (CEC) than those from non-preferred sites, and bioassays have shown the ability of the soils to adsorb toxins (such as quinine) associated with the birds' plant diet. Geophagy by seed-eating birds has also been observed elsewhere, and it would appear that this soil consumption represents one weapon in the escalating "biological warfare" between plants and animals (Diamond, 1999). From a plant's evolutionary perspective, a seed needs to be enriched in nutrients both to support germination and subsequent growth, while nutrient-rich fruits attract animals like birds that disperse seeds following plucking and consumption. However, chemical toxins in seeds and fruit will be repulsive to animals thus inducing regurgitation

or defecation of the former and deterring the harvest of the latter until the seed is viable. From an animal's evolutionary perspective, by overcoming the plant's toxin defenses a creature will obtain nutrients from seeds and fruit, and will outcompete other animals that find the diet repulsive and unpalatable. The ability of parrots (and other birds) to overcome plant toxins by indulging in geophagy would appear to suggest that they excel at the evolutionary "arms race" that exists between plants and animals. Many other animals may also benefit from geophagy and the ability of ingested soils to effectively detoxify plants. For example, this hypothesis has been proposed to explain the deliberate ingestion of soil undertaken by at least 14 species of non-human primates (Krishnamani & Mahaney, 2000), including apes (e.g., Sumatran orangutan, *Pongo pygmaeus abelii*), prosimians (mongoose lemur, *Eulemur mongoz*), New World monkeys (masked titi monkey, *Callicebus personatus melanochir*), and Old World monkeys (guereza monkey, *Colobus guereza*).

The research undertaken on the geophagous parrots of Peru has indicated that ingested soils may also serve a function other than the adsorption of toxins. Gastrointestinal cytoprotection results from the interaction of high surface area clays such as smectite and attapulgite with the gut lining (Gilardi et al., 1999). By

FIGURE 4 Red and Green Macaw group feeding at a lick site, Manu National Park, Peru (Frans Lanting/Minden/FLPA).

FIGURE 3 Red and Green Macaw eating a chunk of soil, Manu National Park, Peru (Frans Lanting/Minden/FLPA).

increasing mucus secretion and preventing mucolysis, clays in the gastrointestinal tract enhance the ability of the mucus barrier in protecting the gut lining from either chemical or biological insults, thereby alleviating the symptoms of diarrhea. The ingestion of soils containing clay minerals with a moderate to high surface area, together with the long time of passage of soil through the gastrointestinal tract, suggests the possibility of cytoprotection as an important function of geophagy undertaken by the Peruvian parrots.

Diarrhea (and other gastrointestinal upsets) can also be cured by geophagical practices because clay minerals are able to adsorb bacteria and their toxins. For example, chimpanzees in the Mahale Mountains National Park (Tanzania) have been observed consuming soil from termite mounds containing the clay minerals metahalloysite and smectite (Mahaney et al., 1996). This mineralogy makes the soil similar to the pharmaceutical Kaopectate that is used to treat minor

gastric ailments in humans. The soils consumed by the chimpanzees could also function as antacids, with the commonly alkaline termite mound soils acting as a buffering agent to counteract the effects of acidic foods. Acidosis can also afflict wild ruminants such as the giraffe (*Giraffa camelopardalis*) and wild ungulates such as the mountain goat (*Oreamnos americanus*). The problem of acidosis associated with these animals arises due to the sudden dietary changes and the lush growth of vegetation that is coincident with the onset of spring or the early wet season (Kreulen, 1985). A sudden lack of fiber and the increase in readily fermentable carbohydrates (e.g., sugars) and soluble proteins lead to a drop in stomach pH that causes several ailments such as anorexia, diarrhea, and gastrointestinal irritation. As a source of calcium carbonate, potassium carbonate, sodium bicarbonate, sodium chloride, and montmorillonite clays, ingested soils can avert acidosis by preventing a decline in stomach pH and by improving digestion efficiency through altering the sites of digestion and absorption of carbohydrates and proteins.

Other motives that may lead to the deliberate ingestion of soil have been suggested (Kreulen, 1985; Klaus & Schmid, 1998; Krishnamani & Mahaney, 2000). These include:

- The use of soil as a famine food to ease the pangs of hunger during periods of starvation
- Microbial inoculation, where the ingestion of feces-contaminated soil facilitates the transfer of bacteria between animals, thus accelerating digestive adaptation within a population during periods of dietary change
- A behavioral tradition, where animals ingest soil because others are doing likewise

With licks providing a number of potential benefits to consumers, it is not surprising that geophagy has such a wide distribution in the animal kingdom. However, there are a number of costs that are also associated with geophagy, which include:

- The adverse physical effects of excessive tooth wear, erosion of the mucosal surfaces of the stomach and intestines, and obstruction of the digestive tract. Soils enriched in silica or sodium bicarbonate may be responsible for the development of kidney stones.
- The adsorption of nutrients by (for example) clay minerals may cause deficiency symptoms, while an excessive intake of an element can lead to mineral nutrient imbalances or problems of toxicity.
- Feces and urine accumulation at lick sites may cause problems of parasitism and disease. Soil fungi produce antibiotics that may have a bacteriostatic effect in the stomachs of animals such as ungulates.
- The attraction of animals to licks is associated with energetic costs and time lost for foraging. Lick sites may also be focal points of disease transmission and predation (including poaching).

Clearly, licks must provide benefits that enhance both animal performance and resource utilization, which compensate for the costs and risks associated with their use. "Aversion learning" may also lessen some of the adverse effects that are linked with geophagy (Kreulen, 1985). This practice is important not only to individual animals but, by influencing population densities and structures, it also has broader ecological consequences. Yet the extent of research dedicated to this practice is, to date, relatively limited. Consequently, much of what we know about geophagy practiced by wild animals is speculative, and many questions relating to (for example) how animals find appropriate soils for consumption, why they ingest them, the quantities that are consumed, and the implications of the soil ingestion still remain to be fully answered.

III. INVOLUNTARY SOIL INGESTION: HUMANS

All members of an exposed human population will ingest at least small quantities of soil. Foods, for example, may be contaminated with soil particles that are then inadvertently ingested. This contamination is especially likely in the tropics because of the tradition of drying foodstuffs like cassava and millet outdoors. Soil can also be ingested via inhalation. Particles entrained in the aboveground air can be inhaled, but while some will reach and be retained in the lungs, the bulk is trapped and ultimately taken over the epiglottis into the esophagus before passing through the gastrointestinal tract. Soil particles adhering to the skin of fingers can also be involuntarily ingested by so-called hand-to-mouth activity. Young children in particular can ingest significant amounts of soil through this behavior; their hands are typically contaminated with soil through normal play activities (Figure 5).

Most research undertaken on the involuntary ingestion of soil by humans has concentrated on young children. This group of the population can be expected to ingest the greatest quantities of soil involuntarily, and it will be the most vulnerable to any health effects. Research is difficult to undertake on such people, because observations that do not disturb children are difficult to conduct. Attempts have been made to estimate soil ingestion rates through recording the amount of soil on a child's hand and estimating the frequency of finger or thumb sucking (Ferguson & Marsh, 1993). However, ingestion cannot be estimated reliably without some knowledge of how much soil is removed during each mouthing action. This information has not been well recorded, and there is further inadequate

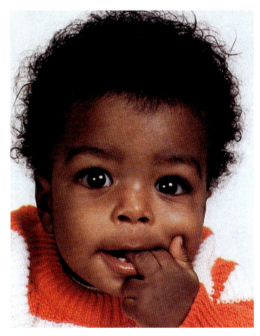

FIGURE 5 Hand-to-mouth activity displayed by a young child (Erika Stone; Photo Researchers, Inc.).

knowledge about the frequency of mouthing, and how much soil is retained on the hands following skin contact. Simple "soil-on-finger" estimates have therefore proved inadequate, and they have been superceded by more elaborate experiments using tracer elements. The ideal tracer element for estimates of soil ingestion by humans is one that is not present in food (or water or air or medications), is uniformly present in high concentrations in soil, and is poorly absorbed via the gastrointestinal tract (Binder et al., 1986). Another criterion for a gold standard tracer element is that the soil concentration should not vary significantly by particle size. No element exactly meets these ideal criteria with, for example, all tracer elements found to some extent in food. The mass-balance equation for a tracer element can be written:

$$I_a + I_{fo} + I_s + I_w = O_f + O_u \qquad (2)$$

where the subscripts refer to intakes (I) of the element in air, food, soil, and water, respectively, and outputs (O) in feces and urine. Because some of these inputs and outputs are negligible for an ideal tracer element, Eq. 2 can be simplified to:

$$I_{fo} + I_s = O_f \qquad (3)$$

that results in the soil ingestion estimate:

$$S_a = (O_f - I_{fo})/S_c \qquad (4)$$

where S_a is the mass of soil ingested and S_c is the concentration of the tracer element in the soil (Stanek & Calabrese, 1991).

Table IV records data from three tracer studies estimating soil ingestion by children, and the varying values illustrate the difficulties in interpretation. For example,

Study 1 made a number of assumptions including that the daily stool output averaged 15 g (dry weight) per child. A later adjusted recalculation using measured stool weights instead of the 15 g assumed in the original work gives lower estimates of soil ingestion. Table IV also illustrates the problem of poor intertracer consistency caused by errors in the mass-balance studies. For example, because titanium is widely used in inks, soil ingestion can be overestimated if a child eats printed paper or ingests ink residues sticking to the fingers. Such errors can be quantified, leading to adjustments in the soil ingestion estimates. This was done for the third study illustrated in Table IV. Using the original uncorrected data for six tracer elements, the mean soil ingestion estimates ranged from 21 to 459 mg day^{-1}. A marked improvement following adjustment led to a narrower range of 97–208 mg day^{-1}. Aluminum, silicon, and yttrium were considered to be the most reliable tracer elements in this particular study.

Despite efforts to improve the design of these soil ingestion studies, there still remains the problem of intertracer variability and the determination of which tracer element provides the best estimate of soil ingestion. There is a lack of information regarding the true variability of ingestion and the uncertainty of any average intake values. The limited number of investigations also provides little knowledge regarding factors such as the seasonal, regional, or ethnic variations in the rates of soil ingestion. Research undertaken in The Netherlands suggests that soil ingestion by children occurs mainly when the weather is dry and more time is spent outdoors (van Wijnen et al., 1990). The studies to date, however, do not address those children of tribal societies who live in a subsistence economy and who are

TABLE IV. Estimates of Soil Ingestion (mg day^{-1}) by Children

Tracer element	Study 1		Study 1 (adjusted)		Study 2		Study 3		Study 3 (adjusted)
	Mean	Median	Mean	Median	Mean	Median	Mean	Median	Mean
Aluminum	181	121	97	48	39	25	153	29	136
Silicon	184	136	85	60	82	59	154	40	133
Titanium	1834	618	1004	293	246	81	218	55	208
Vanadium							459	96	148
Yttrium							85	9	97
Zirconium							21	16	113

Note: Study 1: 59 children 1–3 years of age; Study 2: 101 children 2–7 years of age; Study 3: 64 children 1–4 years of age.
From Calabrese and Stanek, 1995.

most likely to inadvertently ingest the highest amounts of soil (Simon, 1998).

Young children with high hand-to-mouth activity will ingest more soil than older children, who in turn will involuntarily consume more soil than adults. Relative to studies investigating children, data quantifying the rates of soil ingestion by adults are more limited. The U. S. Environmental Protection Agency (EPA) concluded that adults could ingest 100 mg of soil day^{-1}, but this guidance figure would appear to be an overestimate. A pilot study undertaken on just six individuals, and using the tracers aluminum and yttrium, suggested a soil ingestion rate of approximately 50 mg day^{-1} (Calabrese et al., 1990). More recently a study on 10 adults used a tracer-based mass-balance study over 28 days of observation. The findings, representing the largest amount of data available on soil ingestion by adults, indicated an average estimate of 10 mg of soil day^{-1} (Stanek et al., 1997).

Regarding health, the implications of this involuntary soil ingestion to humans may prove to be beneficial. With foods in developing countries often contaminated with iron from soil residues on vegetables and cereals, it has been suggested that the contamination could be a good dietary source of this important mineral nutrient (Hallberg & Björn-Rasmussen, 1981). Considerably more work has emphasized the potentially deleterious effects of soil ingestion on health. In particular, ingested soils are likely to be a significant source of contaminants such as dioxins, and PHEs such as lead and the radionuclide isotopes, because there is only a limited uptake into the aerial parts of plants of these constituents from soils. In parts of Derbyshire (UK), soils are enriched in lead due to the weathering of mineral deposits and contamination associated with metalliferous mining. In a study undertaken in Derbyshire during the early 1970s, soil ingestion was found to be prevalent to an unexpected degree with 43% of children aged 2–3 years showing a pica for soil (Barltrop et al., 1974). An increased absorption of lead was found among children residing in villages near the extensive old mine workings, but the values found in blood and hair were still within the accepted normal range (Table V). A subsequent investigation in the same area showed that hand-wipe samples from children have relatively high lead concentrations, which suggested the importance of hand-to-mouth activity in transferring significant quantities of the metal to the child (Cotter-Howells & Thornton, 1991). The same investigation showed that many of the lead-rich soil grains were composed of pyromorphite, a stable soil–lead mineral that is only very slightly soluble and (presumably) has a low bioavailability to humans following ingestion.

TABLE V. Lead in Blood and Hair Samples Collected From Children Residing in High and Low Soil-Lead Areas of Derbyshire

	Geometric mean	
	Blood (μg 100 ml^{-1})	Hair (mg kg^{-1})
High soil-lead area:[a]		
No current pica (n = 27)	23.6	10.8
Present pica for soil (n = 16)	26.4	21.1
Low soil-lead area[b]		
No current pica (n = 17)	19.9	5.7
Present pica for soil (n = 16)	22.1	9.0

[a]Mean soil-lead concentration about 10,000 mg kg^{-1}.
[b]Mean soil-lead concentration about 500 mg kg^{-1}.
From Barltrop et al., 1974.

Lead poisoning is a very important issue in the United States with medical, learning, and social costs having broad and long-term implications. There has been a substantial decline in blood lead levels during the last decade, yet there are still 900,000 American children under 6 years old that have blood lead concentrations high enough to suggest impairment of intelligence, behavior, and development. Urban soils in large American cities form a reservoir of lead and other PHEs such as cadmium and zinc because of pollutants that include leaded gas and paint (Mielke et al., 1999). The ingestion of soil from gardens, school playgrounds, and other open spaces may therefore constitute a significant risk, especially because these urban soils will contain more soluble forms of lead (e.g., chloride and bromide) than the previously mentioned pyromorphite. Soil ingestion estimates are now routinely incorporated into all risk assessment procedures for contaminated sites in the United States (see also Chapters 8 and 23, this volume).

Other health issues may also be associated with involuntary soil ingestion. For example, doctors in the United States have recorded eosinophilia (a high count of a type of white blood cell that is usually found when a toxin or parasitic infection is present) in children admitted for treatment of lead poisoning following hand-to-mouth activity (Berger & Hornstein, 1980). The cause was attributable to infection with the larval form of the dog or cat parasite *Toxocara canis* or *T. cati*, respectively, which led to toxocariasis. Physicians managing children with lead intoxication following soil ingestion should therefore be aware of the possibility of concurrent parasitic infection. Toxoplasmosis is another disease associated with soil ingestion, attributable to the protozoan parasite *Toxoplasma gondii*, that sexually reproduces in cats who then release eggs in their feces to the soil. The ingestion of feces-contaminated soils by pigs, cattle, and sheep leads to their infection, and most people get toxoplasmosis from undercooked meat. However, the direct ingestion of soil by humans is a secondary source of infection. Medical opinion has insisted that *Toxoplasma* is nearly always harmless to people (Randerson, 2002), but recent research has suggested that by adversely affecting human behavior latent toxoplasmosis, the mildest form of *T. gondii* infection, might represent a serious and highly underestimated economic and public health problem (Flegr et al., 2002).

IV. HUMAN GEOPHAGY: HISTORICAL AND CONTEMPORARY PERSPECTIVES

The recognition that geophagy is widespread among non-human primates suggests that the deliberate ingestion of soil predates our evolution as a species. The oldest evidence of geophagy practiced by humans comes from the prehistoric site at Kalambo Falls on the border between Zambia and Tanzania (Root-Bernstein & Root-Bernstein, 2000). Here a calcium-rich white clay, believed to have been used for geophagical purposes, has been found alongside the bones of *Homo habilis* (the immediate predecessor of *Homo sapiens*). Migration transferred the practice from Africa to every other continent that has been permanently settled by humans, though there are some areas—Japan, Korea, much of Polynesia, Madagascar, and the south of South America—where geophagy is limited or unknown (Laufer, 1930). However, a lack of reporting on the practice is common, and because many geophagists are also reluctant to admit to soil consumption, undoubtedly the prevalence of geophagy is greater than suggested from literature sources.

Throughout history a large number of writers, including anthropologists, explorers, scientists and physicians have commented on geophagy. In the first century AD, both Dioscorides and Pliny mention a famous medicament known as *terra sigillata* (earth that has been stamped with a seal) otherwise known as Lemnian Earth (Thompson, 1913). This soil, derived

from the Greek island of Lemnos, was mixed with the blood of a sacrificed goat, shaped into tablets somewhat larger than a thumbnail, stamped with an impression of a goat, and dried. Lemnian Earth was used for many maladies (but most notably for poisoning), and so great was the demand from the 13th to the 14th centuries, that almost every country in Europe strove to find within its boundaries a source of supply. Thus varieties of *terra sigillata* emanated from numerous localities including Bohemia, England, Italy, Malta, Portugal, Sicily, and the Mediterranean island of Samos. Some of these rival medicaments, notably *terra sigillata strigoniensis* or Strigian Earth derived from Silesia, acquired a considerable reputation. So valuable and respected were these that false Earths were also sold, leading Thevet to comment in 1554: ". . . . . . adulterate it considerably when they sell it to people who have no knowledge of it" (Thompson, 1913, p. 438). The fame of *terra sigillata* reached a peak at the end of the 16th century, and throughout the following 200 years the medicament was mentioned in most of the official medical books published in Europe. Its last appearance in any important pharmacopoeia was in 1848.

In the New World, geophagy was widespread previous to its discovery by Columbus. The oldest written history of Native Americans is provided by the explorer Alvar Nuñez Cabeza de Vaca who for 8 years (1528–1536) traveled through what is now known as the southeast United States (Loveland et al., 1989). Cabeza de Vaca writes of a tribe that was often exposed to starvation and they ate as much as they could, including soil. In another passage, the same explorer states that the fruit of the mesquite tree (*Prosopis juliflora*) was eaten with soil, making the food sweet and palatable. The Portuguese colonist and chronicler Gabriel Soares de Sousa provides the earliest (i.e., 1587) account of geophagy in South America. Commenting on the Tupinamba of Brazil, Soares describes how members of the tribe would commit suicide by eating soil "when they are seized by disgust or when they are grieved to such a degree that they are determined to die." The association of geophagy and suicide became a tragic part of American history. Slaves shipped across the Atlantic were responsible for the large-scale transfer of the practice from Africa. These slaves ingested soils perhaps to fill their stomachs as well as for medical (including nutritional) and cultural reasons (Hunter, 1973). Additionally large numbers indulged in excessive soil consumption not only to become ill and to avoid work, but also to commit suicide in the belief that their spirit would return to the African homeland (Haller, 1972). There are records of mass suicides caused by geophagy among plantation workers, and some estates were abandoned because of this practice (this problem was not confined to the Americas, since in 1687 approximately 50% of slaves in Jamaica died because of geophagy). Methods to deter the practice were harsh and included the use of facemasks, iron gags, chaining to plank floors, whipping, and confinement. The dismemberment of bodies of those who perished as a result of geophagy also proved effective, possibly because slaves believed that the spirit of a mutilated body could not return to the African homeland.

It was generally believed that African-Americans would cease the practice of geophagy following the termination of the slave trade and through the influence of Christianity. However, although the amelioration of life in the New World during the 19th century led to a decline of their medical problems that were associated with geophagy, the practice persisted for reasons other than as a means of committing suicide. For three-quarters of a century following the Civil War, geophagy in the United States was mentioned for the most part only incidentally in a few articles on the poor Whites of the American South as one of their numerous eccentricities. An anonymous writer in 1897 commented on the "clay-eaters" of Winston County, Alabama, who consumed a "dirty white" or "pale yellow" colored soil found along the banks of a small mountain stream (Anonymous, 1897). The quantity eaten at any time varied from a pea-sized lump consumed by a child or beginner, to a piece the size of a man's fist for those who had eaten it for many years. Although the life expectancy of the clay-eaters was apparently not affected, they developed anemic, pale complexions. Several years later, Dr. Charles Wardell Stiles demonstrated that the anemia was attributable to hookworm disease that also caused the geophagy (ingested soils are reported to alleviate the gastric pain associated with hookworm disease).

A pioneering study reporting geophagy among African-Americans in 1941 proved to be the beginning of a number of important investigations on the subject (e.g., Dickins & Ford, 1942). Geophagy was found to be extensive among black children, (especially pregnant) women of the American South, and the U. S. postal system used to deliver soil to friends and relatives who had migrated to the North. By the early 1970s, the practice could still readily be found, and was recorded as a structured custom embedded in a well-defined system of beliefs and rituals. Within a decade it was reported that the forces of urbanization and modernization had caused a decline of geophagy among the African-Americans (additionally the same report noted the increasing consumption of baking soda and laundry starch substi-

tutes used instead of soil; Frate, 1984). Nevertheless the practice of geophagy can still be found relatively easily. A medical report in the early 1990s indicated that the prevalence of pica (of which geophagy was a predominant form) had stabilized among pregnant women, affecting about one-fifth of high-risk patients (defined as rural blacks, with a positive family history of pica; Horner et al., 1991).

Relative to the number of studies undertaken on African-Americans, investigations on North American Indians are very limited. Nevertheless, geophagy was described in the early 1980s as widespread among certain desert-dwelling Indian tribes of the American southwest (Fisher et al., 1981). Here the consumption of soil fluorine has been reported to lead to skeletal fluorosis. The problem was exacerbated because people of such tribes have a high prevalence of renal impairment that results in a decreased excretion of fluoride.

Elsewhere in the world, geophagy can be commonly found in particular areas and among certain societies. For example, the practice remains widespread throughout Africa (Abrahams & Parsons, 1996), and some recent research undertaken in Kenya indicates the contemporary prevalence of geophagy. In a cross-sectional study of 285 school children aged 5–18 years, 73% were reported to consume soil (Geissler et al., 1997). The prevalence decreased with age for both sexes up to the age of 15, then remained stable for girls between 15 and 18 years but continued to decrease for boys in that age range. The median amount consumed daily was 28 g, but it varied between individuals from 8 to 108 g. A cross-sectional survey of 275 pregnant women undertaken in the same country revealed that 56% consumed soil regularly (Geissler et al., 1998). The median estimated daily intake was 41.5 g (range 2.5–219.0 g). Namibia is an example of an African country where the practice had not been reported until recently. Then, in 1997, a study undertaken on 171 pregnant women in eastern Caprivi found that some 44% admitted to eating earth and utilized soil taken primarily from termite mounds (Thomson, 1997). This investigation suggests that a lack of reporting on the practice may be contributing to a significant underestimation of geophagy. This is likely not only in Africa but also elsewhere such as the Middle East and Southeast Asia where few contemporary reports on the practice have been published. Hawass et al. (1987) recorded the first cases of geophagy undertaken by adults in Saudi Arabia, while the widespread and reasonably frequent occurrence of geophagy in Indonesia, where the practice has been ongoing for generations, has recently been reported by Mahaney et al. (2000).

FIGURE 6 A 1-kg bag of sikor purchased from a shop in Birmingham, UK. Local pregnant Bengali women mainly consume the tablets that may be a significant source of iron. (Photo: Peter W. Abrahams.)

Although geophagy can be found relatively easily in many developing countries and among the more tribally oriented people (e.g., the Aborigines of Australia; Abrahams & Parsons, 1996), in many of the developed nations the forces of modernization and urbanization could be expected to lead to only infrequent reported cases. In the UK, pica undertaken by humans in the 1920s was described as very uncommon and confined almost entirely to the "abnormal cravings" of pregnant women and the "dirt-eating propensities" of children (Foster, 1927, p. 72). In the early 1970s, a study of 100 pregnant women revealed cravings for normal food and drink substances, but pica was not recorded (Dickens & Trethowan, 1971). Nevertheless, geophagy can be easily found in the UK by those with a desire to investigate the practice. Soil, variously known as sikor, mithi, patri, khuri, kattha, poorcha, or slatti, is traditionally taken by Asian ethnic groups as a remedy for indigestion and as a tonic during pregnancy (Figure 6). The soil is imported from the Bengal region of south Asia, and is sold by weight in shops throughout the country (e.g., Birmingham, Bristol, London, and Swansea). Geophagy appears to be so well established as a custom of Bengali society, that immigration into the UK has resulted in the cultural transfer of the practice. The likelihood is that immigration into other developed nations has also transferred the practice in recent years or decades.

It is clear that geophagy is not limited to any particular age group, race, sex, geographic area, or time period. Nevertheless, the practice is especially associated with certain regions and people (e.g., contemporary developing nations, people of low socioeconomic status, children, pregnant women). Typically only specific soils are consumed, and are selected for desired qualities of (for example) color, odor, flavor, texture, and plasticity. Often the material is a ferruginous clay, but other soils are certainly sought. In Africa, sand may be exploited both within the Sahara Desert of Mauritania and Ghana's Volta River delta (Vermeer, 1987). However, alluvial clay-enriched soil, from depths of 30–90 cm, are a common source of geophagical material. Scrapings from mudwalls of structures can also provide soil for occasional needs. In West Africa shales are mined and processed for geophagical use (Vermeer & Ferrell, 1985), and a field study in eastern Sierra Leone indicated that 50% of pregnant women regularly consume clay found in the interior of termite mounds (Hunter, 1984). In the same country another (less commonly used) geophagical source of clays are the nests of the mud-daubing wasp (genus *Synagris*).

Following collection, some rudimentary preparation of the soil may then occur. For example, tablets of sikor are made by compressing the soil prior to baking. The latter operation has the desired effect of destroying the eggs of potential intestinal parasites and produces the distinct smell of smoke to which the Bengali women are attracted. In eastern Guatemala, holy clay tablets known as pan del Señor (bread of the Lord) or *tierra santa* (sacred Earth) are produced following excavation (Hunter et al., 1989). Lumps of clay are pounded into small pieces before being crushed and passed through a 1-mm sieve. Water is added to produce a smooth dough-like material, and ceramic or wooden molds are then firmly pressed into the clay to produce tablets with a holy image. The tablets are sliced and trimmed with a carving knife and sun-dried for 24–48 hours before redistribution to retailers throughout Guatemala and the neighboring countries of Belize, Honduras, and El Salvador (Figure 7).

The qualitative and quantitative estimates of soil ingestion attributable to geophagy indicate the large quantities that can be consumed. For example, 3–4 tablets of sikor (about 64 g) may be ingested daily by pregnant Bengali women. The average daily consumption of soil by pregnant women in Africa has been reported as 30–50 g (Vermeer, 1987). Sometimes geomania is encountered, whereby people develop a craving and uncontrollable urge for eating soil (Halsted, 1968). This is evident from the quotations given to the interviewers of black women of the American South during the middle decades of the 20th century: "I feel awful, just about crazy when I can't get clay"; "I craves

FIGURE 7 A vendor selling tablets of *tierra santa* in Central America. The soil is used as a pharmaceutical, and with its religious associations it provides psychological comfort. Here, the woman is daubing candy makers' red dye onto the tablets to simulate the blood of Jesus. (Photo: John M. Hunter.)

it"; "I crave something sour like the taste of clay" (Edwards et al., 1959, p. 811; Ferguson & Keaton, 1950, p. 463). The cravings may be difficult to control. Avicenna, the Arabian philosopher and physician who lived about 1000 years ago, talks of the necessity of the whip in controlling boys from geophagy, while restraints or prison was used for older people. He also records how "incorrigible ones are abandoned to the grave," demonstrating the hazards of persistent and excessive soil consumption. Avicenna was the first to mention the benefit of iron preparations in treating geophagy, and the association between iron and the ingestion of soil is still of considerable debate today as will be discussed further in the next section.

V. THE CAUSES OF HUMAN GEOPHAGY

It may be thought that the causes of soil consumption can be easily established by interviewing geophagists. However, many people who undertake the practice are reluctant to admit to soil eating. Perhaps they have a sense of shame or guilt, and fear that the interviewers may view the trait negatively. In some societies, as commonly found in Africa, the practice is more overt and open discussion is possible (Vermeer, 1987), but even in such situations many geophagists are at a loss to explain their desire for soil. One evident feature is that they generally like the practice, since many speak positively about the qualities (e.g., taste, feel, or odor) of the soil that is consumed. Despite the difficulties of obtaining information from geophagists, the practice is known to have multiple causes including those listed below.

Soil as a food and food detoxifier—The use of soil as a food supplement during periods of famine has been frequently recorded, as the ingested soil gives a sensation of fullness to the stomach. For example, Alexander von Humboldt on his travels in South America at the beginning of the 19th century recorded the eating of clay by the Ottomac tribe in the Orinoco Valley (Ross, 1895). Local supplies of fish and turtles were curtailed during the time of annual flooding, and Humboldt records how 12.5- to 15-cm diameter balls of mainly alluvial material were prepared and eaten by the Ottomacs in "prodigious quantities." Such consumption of soil is not restricted to distant times. In China, a country where traditional knowledge of famine foods has been transmitted between generations, soils have been utilized during famine as recently as about 40 years ago (Aufreiter et al., 1997), while in 2002 food shortages were reported to be causing geophagy in Malawi.

Closely associated to the practice of geophagy undertaken during periods of famine is the use of soils by humans in plant detoxification. Many plants containing toxins are consumed during periods of food shortage, but the mixing of soil with such plants adsorbs the potentially harmful chemicals and renders the food palatable. In this way the most important African famine food, the wild yam *Dioscorea dumentorum* Pax., is detoxified through the use of clay. Elsewhere, Native American populations have been reported to mix soil with acrid acorns, tubers, and berries as a corrective of taste. Indeed, the Aymara and Quechua people of the Andes Mountains of Bolivia and Peru still continue to eat wild potatoes by dipping them in a thick slurry of clay that effectively adsorbs potentially toxic glycoalkaloids (Johns, 1986). The importance of these "potato clays" can be judged from the suggestion that because all wild potatoes are poisonous to humans, the domestication of the modern potato (the world's premier vegetatively propagated cultigen) may have required geophagy at first.

Perhaps as an extension of its use during times of famine, soil can also be consumed as a regular food item. A study of African-American women in the late 1950s reported that many ate soil as part of the menu, and the meal seemed incomplete without it (Edwards et al., 1959). In Turkey, clay has been reported as used for snacks in place of candy or chewing gum. The literature also reports the use of soil as a sort of relish, condiment, or delicacy.

Psychiatric and psychological causes of geophagy —Pica is common in institutionalized mentally retarded people, and soil is one of the nonfood items commonly sought. It may be considered that the mentally retarded cannot discriminate between food and nonfood items, but a study in the United States showed that individuals are often aggressive in seeking the nonfood item of their choice, and they are quite deliberate about what they ingest (Danford et al., 1982).

Other research undertaken in the United States has demonstrated that pica in early childhood is related to elevated, extreme, and diagnosable problems of bulimia nervosa in adolescents. This suggests that pica may be a symptom of a more general tendency to indiscriminant or uncontrolled eating behaviors.

A psychological cause for geophagy is also evident. For example, pregnant African-Americans have commented on feelings of anxiety and agitation that are overcome by experiencing a sense of satisfaction following the consumption of soil. Similarly, the holy clay tablets that are consumed in Central America (Figure 7) provide psychological comfort by helping to allay anxieties associated with ill health or pregnancy (Hunter et al., 1989).

The consumption of soil as a pharmaceutical— Soil may be the world's oldest medicine recorded as a pharmaceutical throughout history. The pre-modern Chinese extensively utilized soil in medicines with, for example, the pharmacologist Li Shi-Chen in 1590 listing 61 uses for soil materials in treating a variety of conditions (Root-Bernstein & Root-Bernstein, 2000). Such multiple applications of soil in treating ailments is still evident (e.g., Figure 8), although their effectiveness as a medicine for treating so many maladies must be questioned. The varieties of *terra sigillata* utilized throughout Europe for some 2000 years are recorded as

FIGURE 8   A vendor selling traditional soil/herbal remedies (the cylindrical-shaped objects) in the central market of Kampala, Uganda. The distinctive medicines, typically broken up and mixed with water before drinking, are used for treating a large variety of ailments such as asthma, nausea and vomiting, syphilis, poisoning, and anemia. (Photo: Peter W. Abrahams.)

used for a number of ailments including plague, the bites and stings of venomous animals, malignant ulcers, nose bleeds, gout, dysentery, and poisoning (Thompson, 1913). Certainly, *terra sigillata* would have proved effective in treating the latter due to the ion exchange capability of the soil constituents. Indeed, so effective are soils in treating cases of poisoning that fuller's earth and bentonite, both enriched in the clay mineral montmorillonite, are used in contemporary developed nations as an antidote for poisons such as the herbicides paraquat and diquat.

Contemporary modern societies also utilize kaolin and smectite clays for treating gastrointestinal disorders. It should be of no surprise, therefore, that the literature describing human geophagy mentions the effectiveness of ingested soils in treating gastrodynia (stomachache), dyspepsia (acid indigestion), nausea, and diarrhea. For example, the chief geophagical clay entering the well-developed West African market system comes from the village of Uzalla, Nigeria (Vermeer & Ferrell, 1985). The clay, called *eko* by the villagers who prepare the material, is obtained from the working of Paleocene shales, and it is used as a traditional medicine including its use for, among other treatments, stomach and dysenteric ailments. Some 400–500 tonnes of *eko* is reportedly produced each year and widely redistributed to markets greater than 1600 km from the source. Mineralogical analysis of *eko* indicates a kaolinitic composition similar to the clay in the modern pharmaceutical Kaopectate.

A cultural explanation for geophagy—For some people there are symbolic links between themselves, fertility, blood, ancestors, and graves that are strengthened through the ingestion of soil. For example, among the Luo people of western Kenya, soil is eaten openly by women of reproductive age (Geissler, 2000). There is a particular preference for soils from termite mounds that have a symbolic significance because of their red color (i.e., the color of blood), intense taste, fertility, the use of their material for building dwellings, and their location that may be coincident with sites of burials or former habitation. Luo children learn earth-eating by observation and imitation within the family, but girls and boys have different views on the practice. Boys have to stop indulging in geophagy in order to "become men."

The consumption of soil for physiological reasons—It is commonly assumed that humans ingest soils to satisfy a nutritional deficiency. Such a physiological explanation for geophagy is an attractive hypothesis, and certainly soils have the potential to supply mineral nutrients to the geophagist. Hydrochloric acid is secreted in the stomach and is a major component of gastric juice that is consequently strongly acidic (pH is dependent upon physiological parameters, varying from 1 to 2 when fasting to 2 to 5 when fed; Oomen et al., 2002). But the pancreatic secretion of bicarbonate ions that neutralizes acid from the stomach modifies conditions in the small intestine where a higher pH (i.e., less acidic) environment exists (the small intestine consists of three sections: duodenum, pH 4–5.5; jejunum, 5.5–7; and ileum, 7–7.5). The human digestive system can thus be considered to be a two-part, acid-alkaline extraction system that operates on any soil constituent passing through the gastrointestinal tract. For example, ingested clays encountering the acidity of the stomach will release elements by cation exchange reactions, while iron oxide and other minerals can be expected to be partially solubilized. Mineral nutrients released in the stomach may then be adsorbed by soil constituents as they enter the small intestine, because the adsorption of nutrients tends to increase with pH. Soil extractants such as 0.1 M hydrochloric acid can be used in simple laboratory experiments to simulate human digestion and its effects on the availability of mineral nutrients to the geophagists. Table VI provides a summary of such experimentation that has been undertaken on samples collected within Africa. This table illustrates the varying ability of the different soils in supplying mineral nutrients to the geophagist following their consumption.

As previously explained there is strong evidence that sodium-rich (saline) soils are a cause of soil ingestion by members of the animal kingdom. But salt has never been shown to be a stimulus in primate geophagy, and most soils utilized by humans are reported to be essentially salt free. This means that its craving is an improbable cause of the practice. Indeed geophagists can deliberately add salt to the soil that they then consume: a practice that is, for example, commonly undertaken by pregnant women in Nigeria. Calcium and iron are the two mineral nutrients that are frequently implicated in the physiological explanation for geophagy. Daily calcium needs increase for pregnant women from 800–1200 mg day^{-1}, primarily to provide the nutrient for fetal skeletal growth and development. In Africa, research has demonstrated that geophagy is especially common among women belonging to non-dairying tribes that have consequently a depleted intake of calcium (Wiley & Katz, 1998). The ingestion of calcium-rich soil therefore provides a plausible explanation for geophagy as was made, for example, in a study reported in 1966 on the non-milk-drinking Tiv tribe of Nigeria (see Table VI). However, such explanations remain speculative because there are no detailed, consistent, and well-controlled data to support any observation that human geophagy represents a craving generated by a nutritional deficiency (Feldman, 1986; Reid, 1992). Nevertheless, some soils do have the potential to supply various mineral nutrients to the geophagist in significant quantities, even if the soils are not consumed to satisfy a physiological requirement. Table VI shows the varying quantities of iron that can be extracted from geophagical soils following laboratory experimentation. Some soils appear to be poor providers of iron, but others are capable of contributing toward a significant proportion of the Reference Nutrient Intake (RNI) for this element. For example, research undertaken on Kenyan girls and boys in 1998 (see Table VI) indicated that on average ingested soils were providing 32 and 42%, respectively, of the RNI for iron. However, such findings have been criticized recently, because the laboratory extractions ignore the effect of changes in the Eh/pH regime and kinetics during passage of soil through the gastrointestinal tract. Consequently, new experimental methods for the estimation of the bioaccessible fraction of elements (defined as the fraction of a substance that is soluble in the gastrointestinal environment and is available for absorption) are being developed. One such method, the physiologically based extraction system (PBET), is an *in vitro* procedure that incorporates gastrointestinal tract parameters representative of a human (such as stomach and small intestine pH and chemistry, soil-to-solution ratio, stomach mixing, and stomach emptying rates). Recent experimental work simulating this extraction

TABLE VI. Extractable Concentrations (mg kg^{-1}) of Selected Macro- and Micronutrients Determined From Geophagical Materials Collected Within Africa

Date of study and origin of sample	Calcium	Copper	Iron	Potassium	Magnesium	Manganese	Sodium	Zinc
1966, Nigeria[a]	3910	—	—	53	2005	—	44	—
1971, Ghana	120	—	—	165	31	—	—	—
1973, Ghana[b]	1133	10	95	130	331	<1	—	15
1984, Nigeria[c]	265	0.6	134	41	179	29	—	30
1991, Cameroon	77	—	9	45	—	—	—	—
Gabon	68	—	4	87	—	—	—	—
Kenya 1	791	2	7	432	135	63	—	3
Kenya 2	220	1	12	793	112	349	—	5
Nigeria	19	2	10	102	9	nd	—	3
Togo	120	—	5	177	—	—	—	—
Zambia[d]	142	11	74	93	60	19	—	2
Zaire	16	—	497	84	—	—	—	—
1997, Uganda[e]	1341	2.1	528	763	458	59	186	6.7
1997, Uganda	1800	—	326	460	1180	50	143	4
Zaire	440	—	380	1730	4100	12	3140	2
1998, Kenya[f]	—	—	169	—	—	—	—	2.7
1998, Kenya[g]	—	—	103	—	—	—	—	1.7

[a]Mean of two samples collected from soil pits that are utilized by the Tiv tribe.
[b]Median concentrations determined from 12 samples.
[c]*Eko* clay.
[d]Sample from the Kalambo Falls archaeological site.
[e]Median concentrations determined from 12 samples used as traditional medicines.
[f]Mean concentrations determined from 48 samples of soil that are typically consumed by boys and girls.
[g]Mean concentrations determined from 27 samples of soil that are typically consumed by pregnant women.
nd = not detected.
Adapted from Abrahams & Parsons, 1996.

system has confirmed the bioaccessibility of iron from Ugandan soils including those from termite mounds. This indicates that their consumption will satisfy a major proportion of the geophagist's RNI for this nutrient (Smith et al., 2000). However, bioaccessibility estimates vary according to the type of *in vitro* procedure employed, and more research is required to establish which method most accurately reflects the human *in vivo* situation (Oomen et al., 2002).

VI. HUMAN GEOPHAGY: BENEFITS AND BANES

The preceding section reveals how ingested soils, as a medicament, food detoxifier, psychological comforter, and a supplier of mineral nutrients, can have a positive role in human society. Although people may seek a particular outcome by indulging in geophagy, at times the ingestion of soil can confer multiple benefits to the consumer. For example, there is a long association of the practice with pregnancy, and it has been suggested that during the first trimester ingested soils will adsorb dietary toxins that are potentially teratogenic to the embryo, while simultaneously quelling the common symptoms of pregnancy sickness. In the second trimester, when pregnancy sickness usually ends, soils may serve as a source of mineral nutrients, and calcium supplementation may aid in the formation of the fetal skeleton, and reduce the risk of pregnancy-induced hypertension (Wiley & Katz, 1998). Despite such benefits, problems may arise if inappropriate quantities or types of soil are ingested. Paradoxically, even though soils can be a source of mineral nutrients to the geophagist, the cation exchange and adsorption properties of soil constituents have been reported to result in deficiency symptoms of certain elements. In Turkey, iron-deficiency anemia has been linked to the consumption of clays (mainly sepiolite and montmorillonite) of high CEC (Minnich et al., 1968). Clinical trials confirmed the effectiveness of the clays in adsorbing iron, and the conclusion of the study indicated the prominent role of geophagy in contributing to the problem, although other nutritional and parasitic factors were also probably involved in the anemia.

The adsorption of potassium by soil constituents can induce hypokalemia in an individual that is reflected by abnormally low concentrations of this element in the blood. Literature from the mid-1800s indicates that the condition was, along with iron-deficiency anemia, common among black slaves. The concurrent iron and potassium deficiency was associated with a disease known as cachexia Africana, the symptoms of which could be relieved through the use of iron- and potassium-containing tonics (Cragin, 1836). Today, hypokalemia attributable to geophagy is only occasionally reported. For example, an isolated case was recorded in the late 1980s of an African-American woman who had a 25-year history of geophagy. An increase in her soil consumption produced a condition similar to that associated with cachexia Africana. However, symptoms abated following potassium replacement and cessation of soil consumption (Severance et al., 1988).

The association between geophagy and zinc deficiency has been noted in a number of countries. In Turkey, soil consumption can be commonly found among village women and children, and the practice is linked to a combined deficiency of iron and zinc, and the latter causes symptoms of growth retardation and delayed puberty (Çavdar et al., 1980). Physical growth and improved sexual maturation were observed in patients following zinc supplementation. There may be a number of causes of the pathogenesis of the zinc deficiency in these patients. The high cereal diet of the Turkish villagers provides little zinc because the phytate-rich food depresses the bioavailability of the metal (the phytate in cereals binds zinc to produce a highly insoluble complex that prevents its absorption). Thus in people who already have a low intake of zinc, geophagy can be considered as an accelerating factor leading to a deficiency of this metal, and ingested soil adsorbs significant quantities of zinc through cation exchange reactions. Additional factors may also be important; for example, of 300 Aboriginal people examined in a study located in the northwest of Australia, half of the individuals had low plasma zinc concentrations (hypozincemia). Geophagy and the high cereal diet of the Aborigines causes a decreased absorption of zinc, and an excessive loss of the metal from these individuals also occurs attributable to intestinal parasites and excessive perspiration. All the requirements are therefore present in the north of Australia for zinc deficiency to be widespread among Aboriginal people (Cheek et al., 1981).

Potentially life-threatening hyperkalemia, an abnormally high potassium concentration in the blood, has been associated with geophagy (Gelfand et al., 1975). This condition is attributable to the absorption of potassium released from ingested soils that are enriched in this element. However, hyperkalemia and its links with geophagy have only been occasionally reported, but the widespread contamination of soils with lead provides another example of toxicity that can be associated with the ingestion of soil, deliberate or otherwise. With lead being especially harmful to the developing brains and nervous systems of young people, there must be a concern if children are consuming soils enriched in this metal, especially with recent research suggesting that there may be no safe level of lead for children (Canfield et al., 2003). Furthermore, lead toxicity will not be restricted to children. As an example, Wedeen et al. (1978) reported on a 46-year-old American black woman who was found to have lead poisoning. The consumption of garden soil with a lead content of $700\,mg\,kg^{-1}$ resulted in damage to the patient's red blood cells, brain, and kidneys. Yet studies investigating the lead intoxication of geophagists remain limited, and bearing in mind that potentially deleterious quantities of this metal may be bioavailable even from soils that contain normal amounts of lead, further investigations are urgently required.

The biotic component of soils can pose hazards to geophagists because the eggs or larvae of parasitic worms (geohelminths) can be consumed, although infection is likely to be significantly reduced if subsoil or baked soil is utilized. Ascariasis and trichiuriasis are caused by the ingestion of *Ascaris lumbricoides* and *Trichuris trichiura* eggs, respectively, while toxocariasis occurs through infection with the larvae of *Toxocara canis* or *T. cati*. Hookworm infection can occur via the oral ingestion of *Ancylostoma duodenale* and *A. ceylanicum* (though skin contact with soil is the main cause of infection). It has been suggested that chronic liver disorders and cirrhotic changes may be associated with ingested soil bacteria and fungi.

Excessive tooth wear is another consequence of human geophagy, though the problem has been seldom reported in the literature (Abbey & Lombard, 1973). Rather, more attention has focused on the internal accumulation of soil that can lead to constipation, the reduction of the power of absorption of food materials by the body, severe abdominal pain, and obstruction and perforation of the colon. In pregnant women, this can lead to dysfunctional labor and maternal death (Key et al., 1982). Deleterious outcomes in fetuses and infants of mothers who practice geophagy are also likely, although the lack of research means that the quality of any evidence is poor. Clearly, the strong association of geophagy with pregnancy warrants further investigations on this important topic.

VII. Conclusions

This chapter considers both the involuntary and deliberate ingestion of soil by humans and other members of the animal kingdom. To many people, the word soil is commonly understood to be the material directly underfoot, as with organic-enriched topsoil. Although this material may be ingested involuntarily by humans and grazing animals, it is often strenuously avoided by geophagists. For example, human geophagists in Africa commonly exploit material from excavations that extend into clay-enriched subsoils or even further to underlying soft shales *in situ*. This mining zone is free of most organic matter and parasitic infestation. In such cases, clay eating may be a more accurate term for the geophagical practice, rather than soil consumption. With the close association between geophagy and pregnancy, sometimes the expression pregnancy clay may be appropriate (Hunter, 1993).

To date soil ingestion, whether involuntary or deliberate, has received relatively sporadic attention in the medical, sociologic, veterinary, or soil science literature. Yet such ingestion is demonstrably widespread and has important consequences for members of the animal kingdom. For example, it has been controversially suggested that modern urban human societies may experience health problems since their contact with (and ingestion of) soil is diminishing (Hamilton, 1998). Consequent decreasing human exposure to soil mycobacteria may contribute to the increasing prevalence of allergic and autoimmune diseases (e.g., asthma, diabetes, rheumatoid arthritis) that have been observed in affluent societies over the past 20 years (Figure 9). Similarly, a recent decline of intestinal worm infection (e.g., *Ascaris lumbricoides*) in people of developed societies may

FIGURE 9 Dietary soil supplements for sale. Behind the humor of the cartoon is the serious message that contemporary urban societies may be at risk for ill health because they are not ingesting soil that can afford them appropriate protection. (From Kate Charlesworth.)

be the cause of the increasingly common inflammatory bowel diseases that are now being recorded.

Relative to involuntary ingestion, geophagy is associated with a more substantial intake of soil. The functionality of non-human geophagy is not argued, whereas human geophagy has been typically viewed as a low-status, deviant, or highly suspect behavior, which is limited to marginal or deprived societies. A more enlightened appraisal is to realize that humans have frequently turned to geophagy as a useful way of overcoming problems that they experience. So important was the practice that it became embedded in the culture and customs of societies, which were perpetuated by learning rather than instinct. Although advances in education and medicine have caused a significant decline in the practice, geophagists still continue their indulgence in spite of the ill effects that can occur (perhaps some people may not connect any health problems with geophagy, or their beliefs are strong enough to overcome the fear of any ill effects).

There is evidence to suggest that geophagy is now attaining renewed and serious interest within the academic fraternity. Hopefully this will lead to future multidisciplinary research that will investigate the issues that have been hinted at in this chapter. For example, the role of ingested soil in either supplying mineral nutrients such as iron, or PHEs such as lead or radionuclides to humans needs to be quantified by undertaking properly controlled *in vitro* and/or *in vivo* studies. Such research will create a better understanding of the implications of soil ingestion that would benefit epidemiological and risk assessment studies. This should be considered as urgent bearing in mind the widespread nature of soil ingestion and human nutritional imbalances.

SEE ALSO THE FOLLOWING CHAPTERS

Chapter 5 (Uptake of Elements from a Biological Point of View) · Chapter 8 (Biological Responses of Elements) · Chapter 14 (Bioavailability of Elements in Soil) · Chapter 20 (Animals and Medical Geology) · Chapter 23 (Environmental Pathology)

FURTHER READING

Abbey, L. M., and Lombard, J. A. (1973). The Etiological Factors and Clinical Implications of Pica: Report of Case, *J. Am. Dent. Assoc.*, 87, 885–887.

Abrahams, P. W. (1999). The Chemistry and Mineralogy of Three Savanna Lick Soils, *J. Chem. Ecol.*, 25, 2215–2228.

Abrahams, P. W., and Parsons, J. A. (1996). Geophagy in the Tropics: A Literature Review, *Geogr. J.*, 162, 63–72.

Abrahams, P. W., and Steigmajer, J. (2003). Soil Ingestion by Sheep Grazing the Metal Enriched Floodplain Soils of Mid-Wales, *Environ. Geochem. Health*, 25, 17–24.

Abrahams, P. W., and Thornton, I. (1994). The Contamination of Agricultural Land in the Metalliferous Province of Southwest England: Implications to Livestock, *Agric. Ecosys. Environ.*, 48, 125–137.

Anonymous (1897). The Clay Eaters, *Sci. Am.*, 76, 150.

Aufreiter, S., Hancock, R. G. V., Mahaney, W. C., Stambolic-Robb, A., and Sanmugadas, K. (1997). Geochemistry and Mineralogy of Soils Eaten by Humans, *Int. J. Food Sci. Nutr.*, 48, 293–305.

Barltrop, D., Strehlow, C. D., Thornton, I., and Webb, J. S. (1974). Significance of High Soil Lead Concentrations for Childhood Lead Burdens, *Environ. Health Perspect.*, 7, 75–82.

Beresford, N. A., and Howard, B. J. (1991). The Importance of Soil Adhered to Vegetation as a Source of Radionuclides Ingested by Grazing Animals, *Sci. Total Environ.*, 107, 237–254.

Berger, O. G., and Hornstein, M. D. (1980). Eosinophilia and Pica: Lead or Parasites?, *Lancet*, 1, 553.

Binder, S., Sokal, D., and Maughan, D. (1986). Estimating Soil Ingestion: The Use of Tracer Elements in Estimating the Amount of Soil Ingested by Young Children, *Arch. Environ. Health*, 41, 341–345.

Brebner, J., Suttle, N. F., and Thornton, I. (1987). Assessing the Availability of Ingested Soil Cobalt for the Synthesis of Vitamin B_{12} in the Ovine Rumen, *Proc. Nutr. Soc.*, 46, 66A.

Calabrese, E. J., and Stanek, E. J. (1995). Resolving Intertracer Inconsistencies in Soil Ingestion Estimation, *Environ. Health Perspect.*, 103, 454–457.

Calabrese, E. J., Stanek, E. J., Gilbert, C. E., and Barnes, R. (1990). Preliminary Adult Soil Ingestion Estimates: Results of a Pilot Study, *Reg. Toxicol. Pharmacol.*, 12, 88–95.

Canfield, R. L., Henderson, C. R., Cory-Slechta, D. A., Cox, C., Jusco, T. A., and Lanphear, B. P. (2003). Intellectual Impairment in Children with Blood Lead Concentrations Below 10μg per Deciliter, *N. Engl. J. Med.*, 348, 1517–1526.

Çavdar, A. O., Arcasoy, A., Cin, S., and Gümüs, H. (1980). Zinc Deficiency in Geophagia in Turkish Children and Response to Treatment with Zinc Sulphate, *Hematologica*, 65, 403–408.

Cheek, D. B., Smith, R. M., Spargo, R. M., and Francis, N. (1981). Zinc, Copper and Environmental Factors in the Aboriginal Peoples of the North West, *Aust. N. Z. J. Med.*, 11, 508–512.

Cotter-Howells, J., and Thornton, I. (1991). Sources and Pathways of Environmental Lead to Children in a Derbyshire Mining Village, *Environ. Geochem. Health*, 13, 127–135.

Cragin, F. W. (1836). Observations on Cachexia Africana or Dirt-Eating, *Am. J. Med. Sci.*, 17, 356–364.

Danford, D. E., Smith, J. C., and Huber, A. M. (1982). Pica and Mineral Status in the Mentally Retarded, *Am. J. Clin. Nutr.*, 35, 958–967.

Davies, A. G., and Baillie, I. C. (1988). Soil Eating by Red Leaf Monkeys (*Presbytis rubicunda*) in Sabah, Northern Borneo, *Biotropica*, 20, 252–258.

Denton, D. A. (1969). Salt Appetite, *Nutr. Abstr. Rev.*, 39, 1043–1049.

Diamond, J. M. (1999). Dirty Eating for Healthy Living, *Nature*, 400, 120–121.

Diamond, J., Bishop, K. D., and Gilardi, J. D. (1999). Geophagy in New Guinea Birds, *Ibis*, 141, 181–193.

Dickens, G., and Trethowan, W. H. (1971). Cravings and Aversions During Pregnancy, *J. Psychosom. Res.*, 15, 259–268.

Dickins, D., and Ford, R. N. (1942). Geophagy (Dirt Eating) Among Mississippi Negro School Children, *Am. Sociol. Rev.*, 7, 59–65.

Edwards, C. H., McDonald, S., Mitchell, J. R., Jones, L., Mason, L., Kemp, A. M., Laing, D., and Trigg, L. (1959). Clay- and Cornstarch-Eating Women, *J. Am. Diet. Assoc.*, 35, 810–815.

Feldman, M. D. (1986). Pica: Current Perspectives, *Psychosomatics*, 27, 519–523.

Ferguson, C., and Marsh, J. (1993). Assessing Human Health Risks From Ingestion of Contaminated Soil, *Land Contamin. Reclam.*, 1, 177–185.

Ferguson, J. H., and Keaton, A. G. (1950). Studies of the Diets of Pregnant Women in Mississippi: I. The Ingestion of Clay and Laundry Starch, *New Orleans Med. Surg. J.*, 102, 460–463.

Fisher, J. R., Sievers, M. L., Takeshita, R. T., and Caldwell, H. (1981). Skeletal Fluorosis From Eating Soil, *Ariz. Med.*, 38, 833–835.

Flegr, J., Havlícek, J., Kodym, P., Maly, M., and Smahel, Z. (2002). Increased Risk of Traffic Accidents in Subjects with Latent Toxoplasmosis. *BMC Infect. Dis.*, www. biomedcentral. com/1471–2334/2/11.

Foster, J. W. (1927). Pica. *Kenya East Afr. Med. J.*, 4, 68–76.

Frate, D. A. (1984). Last of the Earth Eaters, *Sciences*, 24, 34–38.

Geissler, P. W. (2000). The Significance of Earth-Eating: Social and Cultural Aspects of Geophagy Among Luo Children, *Africa*, 70, 653–682.

Geissler, P. W., Mwaniki, D. L., Thiong'o, F., and Friis, H. (1997). Geophagy Among School Children in Western Kenya, *Trop. Med. Int. Health*, 2, 624–630.

Geissler, P. W., Shulman, C. E., Prince, R. J., Mutemi, W., Mnazi, C., Friis, H., and Lowe, B. (1998). Geophagy, Iron Status and Anaemia Among Pregnant Women on the Coast of Kenya, *Trans. R. Soc. Trop. Med. Hyg.*, 92, 549–553.

Gelfand, M. C., Zarate, A., and Knepshield, J. H. (1975). Geophagia: A Cause of Life-Threatening Hyperkalemia in Patients with Chronic Renal Failure, *JAMA*, 234, 738–740.

Gilardi, J. D., Duffey, S. S., Munn, C. A., and Tell, L. A. (1999). Biochemical Functions of Geophagy in Parrots: Detoxification of Dietary Toxins and Cytoprotective Effects, *J. Chem. Ecol.*, 25, 897–922.

Green, N., Johnson, D., and Wilkins, B. T. (1996). Factors Affecting the Transfer of Radionuclides to Sheep Grazing on Pastures Reclaimed From the Sea, *J. Environ. Radioact.*, 30, 173–183.

Hallberg, L., and Björn-Rasmussen, E. (1981). Measurement of Iron Absorption From Meals Contaminated with Iron, *Am. J. Clin. Nutr.*, 34, 2808–2815.

Haller, J. S. (1972). The Negro and the Southern Physician: A Study of Medical and Racial Attitudes 1800–1860, *Med. Hist.*, 16, 238–253.

Halsted, J. A. (1968). Geophagia in Man: Its Nature and Nutritional effects, *Am. J. Clin. Nutr.*, 21, 1384–1393.

Hamilton, G. (1998). Let Them Eat Dirt, *New Sci.*, 159 (No. 2143), 26–31.

Harrison, D. L., Mol, J. C. M., and Healy, W. B. (1970). DDT Residues in Sheep From the Ingestion of Soil, *N. Z. J. Agric. Res.*, 13, 664–672.

Hawass, N. D., Alnozha, M. M., and Kolawole, T. (1987). Adult Geophagia–Report of Three Cases with Review of the Literature, *Trop. Geogr. Med.*, 39, 191–195.

Healy, W. B. (1968). Ingestion of Soil by Dairy Cows, *N. Z. J. Agric. Res.*, 11, 487–499.

Healy, W. B. (1973). Nutritional Aspects of Soil Ingestion by Grazing Animals. In *Chemistry and Biochemistry of Herbage* (G. W. Butler and R. W. Bailey, Eds.), Academic Press, London, pp. 567–588.

Healy, W. B., and Ludwig, T. G. (1965). Wear of Sheep's Teeth. I. The Role of Ingested Soil, *N. Z. J. Agric. Res.*, 8, 737–752.

Healy, W. B., Crouchley, G., Gillett, R. L., Rankin, P. C., and Watts, H. M. (1972). Ingested Soil and Iodine Deficiency in Lambs, *N. Z. J. Agric. Res.*, 15, 778–782.

Henshaw, J., and Ayeni, J. S. O. (1971). Some Aspects of Big-Game Utilization of Mineral Licks in Yankari Game Reserve, Nigeria, *East Afr. Wildl. J.*, 9, 73–82.

Heymann, E. W., and Hartmann, G. (1991). Geophagy in Moustached Tamarins, *Saguinus mystax* (Platyrrhini: Callitrichidae), at the Río Blanco, Peruvian Amazonia, *Primates*, 32, 533–537.

Horner, R. D., Lackey, C. J., Kolasa, K., and Warren, K. (1991). Pica Practices of Pregnant Women, *J. Am. Diet. Assoc.*, 91, 34–38.

Hunter, J. M. (1973). Geophagy in Africa and the United States: A Culture-Nutrition Hypothesis, *Geogr. Rev.*, 63, 170–195.

Hunter, J. M. (1984). Insect Clay Geophagy in Sierra Leone, *J. Cultural Geogr.*, 4, 2–13.

Hunter, J. M. (1993). Macroterme Geophagy and Pregnancy Clays in Southern Africa, *J. Cultural Geogr.*, 14, 69–92.

Hunter, J. M., Horst, O. H., and Thomas, R. N. (1989). Religious Geophagy as a Cottage Industry: The Holy Clay Tablet of Esquipulas, Guatemala, *Natl. Geogr. Res.*, 5, 281–295.

Johns, T. (1986). Detoxification Function of Geophagy and Domestication of the Potato, *J. Chem. Ecol.*, 12, 635–646.

Key, T. C., Horger, E. O., and Miller, J. M. (1982). Geophagia as a Cause of Maternal Death, *Obstet. Gynecol.*, 60, 525–526.

Klaus, G., Klaus-Hügi, C., and Schmid, B. (1998). Geophagy by Large Mammals at Natural Licks in the Rain Forest of the Dzanga National Park, Central African Republic, *J. Trop. Ecol.*, 14, 829–839.

Klaus, G., and Schmid, B. (1998). Geophagy at Natural Licks and Mammal Ecology: A Review, *Mammalia*, 62, 481–497.

Kreulen, D. A. (1985). Lick Use by Large Herbivores: A Review of Benefits and Banes of Soil Consumption. *Mammal Rev.*, 15, 107–123.

Kreulen, D. A., and Jager, T. (1984). The Significance of Soil Ingestion in the Utilization of Arid Rangelands by Large Herbivores with Special Reference to Natural Licks on the Kalahari Pans, In *Herbivore Nutrition in the Subtropics* (F. M. C. Gilchrist and R. I. Mackie, Eds.), The Science Press, Johannesburg, pp. 204–221.

Krishnamani, R., and Mahaney, W. C. (2000). Geophagy Among Primates: Adaptive Significance and Ecological Consequences, *Anim. Behav.*, 59, 899–915.

Laufer, B. (1930). Geophagy. *Anthropology Series*, Field Museum of Natural History, Chicago, publication 280, 18, 97–198.

Lee, J., Rounce, J. R., Mackay, A. D., and Grace, N. D. (1996). Accumulation of Cadmium with Time in Romney Sheep Grazing Ryegrass-White Clover Pasture: Effect of Cadmium From Pasture and Soil Intake, *Aust. J. Agric. Res.*, 47, 877–894.

Loveland, C. J., Furst, T. H., and Lauritzen, G. C. (1989). Geophagia in Human Populations, *Food Foodways*, 3, 333–356.

Mahaney, W. C., and Hancock, R. G. V. (1990). Geochemical Analysis of African Buffalo Geophagic Sites and Dung on Mount Kenya, East Africa, *Mammalia*, 54, 25–32.

Mahaney, W. C., Hancock, R. G. V., Aufreiter, S., and Huffman, M. A. (1996). Geochemistry and Clay Mineralogy of Termite Mound Soil and the Role of Geophagy in Chimpanzees of the Mahale Mountains, Tanzania, *Primates*, 37, 121–134.

Mahaney, W. C., Milner, M. W., Mulyono, H., Hancock, R. G. V., Aufreiter, S., Reich, M., and Wink, M. (2000). Mineral and Chemical Analyses of Soils Eaten by Humans in Indonesia, *Int. J. Environ. Health Res.*, 10, 93–109.

Mahaney, W. C., Watts, D. P., and Hancock, R. G. V. (1990). Geophagia by Mountain Gorillas (*Gorilla gorilla beringei*) in the Virunga Mountains, Rwanda, *Primates*, 31, 113–120.

Marlow, R. W., and Tollestrup, K. (1982). Mining and Exploitation of Natural Mineral Deposits by the Desert Tortoise, *Gopherus agassizii*, *Anim. Behav.*, 30, 475–478.

Mayland, H. F., Florence, A. R., Rosenau, R. C., Lazar, V. A., and Turner, H. A. (1975). Soil Ingestion by Cattle on Semiarid Range as Reflected by Titanium Analysis of Feces, *J. Range Manag.*, 28, 448–452.

Mielke, H. W., Gonzales, C. R., Smith M. K., and Mielke, P. W. (1999). The Urban Environment and Children's Health: Soils as an Integrator of Lead, Zinc, and Cadmium in New Orleans, Louisiana, U. S. A., *Environ. Res.*, 81, 117–129.

Miller, J. K., Madsen, F. C., and Hansard, S. L. (1976). Absorption, Excretion, and Tissue Deposition of Titanium in Sheep, *J. Dairy Sci.*, 59, 2008–2010.

Minnich, V., Okçuoğlu, A., Tarcon, Y., Arcasoy, A., Cin, S., Yörükoğlu, O., Renda, F., and Demirağ, B. (1968). Pica in Turkey. II. Effect of Clay Upon Iron Absorption, *Am. J. Clin. Nutr.*, 21, 78–86.

Oomen, A. G., Hack, A., Minekus, M., Zeijdner, E., Cornelis, C., Verstraete, W., van de Wiele, T., Wragg, J., Rompelberg, C. J. M., Sips, A. J. A. M., and van Wijnen, J. H. (2002). Comparison of Five *in vitro* Digestion Models to Study the Bioaccessibility of Soil Contaminants, *Environ. Sci. Technol.*, 36, 3326–3334.

Pownall, D. B., Lucas, R. J., and Ross, A. D. (1980). Effects of Soil-Contaminated Feed on Dry Matter and Water Intake in Sheep, *Proc. N. Z. Soc. Anim. Produc.*, 40, 106–110.

Randerson, J. (2002). All in the Mind?, *New Sci.*, 176 (No. 2366), 40–43.

Reid, R. M. (1992). Cultural and Medical Perspectives on Geophagia, *Med. Anthropol.*, 13, 337–351.

Root-Bernstein, R. S., and Root-Bernstein, M. R. (2000). *Honey, Mud, Maggots, and Other Medical Marvels*, Pan, London.

Ross, T. (Trans.). (1895). *Personal Narrative of Travels to the Equinoctial Regions of America During the Years 1799–1804* by Alexander von Humboldt and Aime Bonpland, Vol. 2, Routledge, London.

Severance, H. W., Holt, T., Patrone, N. A., and Chapman, L. (1988). Profound Muscle Weakness and Hypokalemia Due to Clay Ingestion, *South. Med. J.*, 81, 272–274.

Simon, S. L. (1998). Soil Ingestion by Humans: A Review of History, Data, and Etiology with Application to Risk Assessment of Radioactively Contaminated Soil, *Health Phys.*, 74, 647–672.

Smith, B., Rawlins, B. G., Cordeiro, M. J. A. R., Hutchins, M. G., Tiberindwa, J. V., Sserunjogi, L., and Tomkins, A. M. (2000). The Bioaccessibility of Essential and Potentially Toxic Trace Elements in Tropical Soils From Mukono District, Uganda, *J. Geol. Soc.*, 157, 885–891.

Stanek, E. J., and Calabrese, E. J. (1991). A Guide to Interpreting Soil Ingestion Studies. 1. Development of a Model to Estimate the Soil Ingestion Detection Level of Soil Ingestion Studies, *Reg. Toxicol. Pharmacol.*, 13, 263–277.

Stanek, E. J., Calabrese, E. J., Barnes, R., and Pekow, P. (1997). Soil Ingestion in Adults—Results of a Second Pilot Study, *Ecotoxicol. Environ. Safety*, 36, 249–257.

Statham, M., and Bray, A. C. (1975). Congenital Goitre in Sheep in Southern Tasmania, *Aust. J. Agric. Res.*, 26, 751–768.

Suttle, N. F., Abrahams, P., and Thornton, I. (1984). The Role of a Soil x Dietary Sulphur Interaction in the Impairment of Copper Absorption by Ingested Soil in Sheep, *J. Agric. Sci., Cambridge*, 103, 81–86.

Suttle, N. F., Alloway, B. J., and Thornton, I. (1975). An Effect of Soil Ingestion on the Utilization of Dietary Copper by Sheep, *J. Agric. Sci., Cambridge*, 84, 249–254.

Thompson, C. J. S. (1913). *Terra sigillata*, a Famous Medicament of Ancient Times. In Proceedings of the 17th International Congress of Medical Sciences (Section 23, History of Medicine), London, pp. 433–444.

Thomson, J. (1997). Anaemia in Pregnant Women in Eastern Caprivi, Namibia, *S. Afr. Med. J.*, 87, 1544–1547.

Thornton, I. (1974). Biochemical and Soil Ingestion Studies in Relation to the Trace Element Nutrition of Livestock. In Trace Element Metabolism in Animals 2 (W. G. Hoekstra, J. W. Suttie, H. E. Ganther, and W. Mertz, Eds.), University Press, Baltimore, MD, pp. 451–454.

van Wijnen, J. H., Clausing, P., and Brunekreef, B. (1990). Estimated Soil Ingestion by Children, *Environ. Res.*, 51, 147–162.

Vermeer, D. E. (1986). Geophagy in the American South, *Bull. Shreveport Med. Soc.*, 37, 38.

Vermeer, D. E. (1987). Geophagy in Africa, *Bull. Shreveport Med. Soc.*, 38, 13–14.

Vermeer, D. E., and Ferrell, R. E. (1985). Nigerian Geophagical Clay: A Traditional Antidiarrheal Pharmaceutical, *Science*, 227, 634–636.

Wedeen, R. P., Mallik, D. K., Batuman, V., and Bogden, J. D. (1978). Geophagic Lead Nephropathy: Case Report, *Environ. Res.*, 17, 409–415.

Wiley, A. S., and Katz, S. H. (1998). Geophagy in Pregnancy: A Test of a Hypothesis, *Curr. Anthropol.*, 39, 532–545.

NATURAL AEROSOLIC MINERAL DUSTS AND HUMAN HEALTH

EDWARD DERBYSHIRE
University of London

CONTENTS

I. INTRODUCTION

Fine atmospheric dust (including fine mineral aggregates, fibrous minerals, and fibrous organic materials) reaches concentrations in many parts of the world sufficient to constitute a major influence upon both human and animal health.

The visible effect of dust in the atmosphere has been noted in written records since at least 1150 BC in China, since ancient times in the Mediterranean, and over the North Atlantic to leeward of the Sahara since at least the eighteenth century. Written records of Saharan dust falls in western Europe became increasingly common from the mid-nineteenth century. Following the "dust bowl years" of the 1930s, awareness of soil-derived atmospheric dusts increased considerably in the United States, particularly after 1945. Understanding of the complex role of atmospheric dust as a factor influencing climate and climatic change has made notable progress in the past 20 years, although there is still much to learn (Houghton et al., 2001). In contrast, the impact of high concentrations of natural dust on human and animal health has received relatively little attention when compared to work on artificially generated particulates, smoke and gases.

Aerosols include gases, liquids, and solid particles suspended in the atmosphere for varying lengths of time. Solid aerosols include particles injected into the atmosphere, such as mineral dust and sea salt, and those that form within the atmosphere, notably sulfates. Natural and man-made fires, including extensive burning of vegetation, generate smoke plumes that are often carried several thousands of kilometers from their sources, which contributes to regional air pollution and adds to atmospheric health hazards. Biomass burning yields black carbon which, together with mineral dust, is monitored by ultraviolet and other sensors on Earth-orbiting satellites. This provides increasingly detailed information on the incidence and seasonality of aerosol plumes over both land and water surfaces. Emphasis here is given to the release, transportation, and deposition of mineral particulate aerosols derived from soils, sediments, and weathered rock surfaces and their impact on human health when in suspension in the atmosphere. The finer components ($<10\,\mu m$) of respirable natural atmospheric dusts include single particles, aggregates of very fine mineral grains (notably

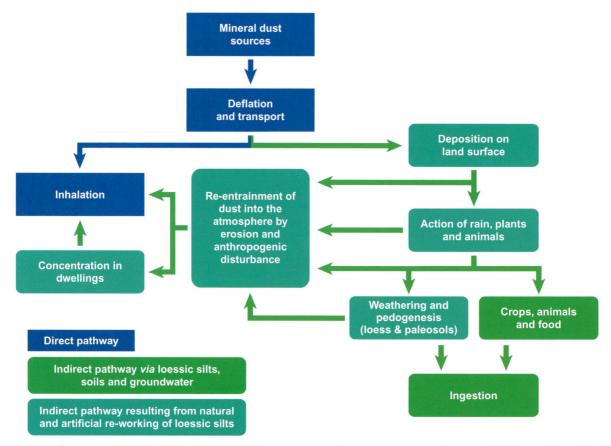

FIGURE 1   Some essential links in the direct and indirect pathways from dust sources to human inhalation and ingestion. Note: The wind-lain and variously weathered sediment known as loess commonly contains buried soils (paleosols) marking phases of relatively stable former land surfaces during the accumulation of the loess.

silica), fibrous minerals (e.g., the asbestos group), and fibrous organic materials.

Wind-borne dusts may affect human health by way of direct and indirect pathways. The elemental composition of dust, both when airborne and when accumulated on a land surface to form loess, can enhance the toxicity of the air breathed as well as that of the soil and the waters that drain through it. Inhaled dusts derived from fine-grained sediment sources such as seasonally dry rivers and dry lakebeds make up the direct pathway (Figure 1).

A significant indirect pathway arises from generation of respirable mineral dusts by both natural and human-induced re-working (erosion) of loess and loessic soils. A second indirect pathway (Figure 1), involving transfer and concentration of some toxic minerals by groundwater movement through thick loess accumulations, is not considered here. Some accumulations of mineral aerosols, including varying amounts of fine volcanic ash (tephra), contribute to the minerogenic dust

in the atmosphere, which often endows surface sediments and soils with a distinctive mineralogy and chemistry (Section II.B). Large volumes of ash and dust have been emitted since 1995 during eruptions of the Soufrière Hills volcano on the island of Montserrat in the eastern Caribbean. The finer dusts contain up to 24% of cristobalite (a form of silica), which poses a potential health threat to local populations in conditions of prolonged eruption.

The detachment of mineral dust from the ground surface and its entrainment and subsequent transport by the wind are functions of several variables, which include the wind speed (both mean regional wind speed and the critical wind speed or threshold velocity required to dislodge particles), the degree of instability of the atmosphere, the size of the particles, the roughness and moisture content of the land surface, and the degree of particle exposure.

Source environments of mineral aerosol dusts are diverse, and some dust takeup by the atmosphere (the

FIGURE 2 Dust storm in the upper Hunza valley, Karakoram Mountains, northern Pakistan, summer 1980. The thick pall is mixed fine sand and silts carried by a cold, dense, gravity-enhanced airflow (katabatic wind) from the 59-km long Batura Glacier (not visible from this viewpoint). Such glacially induced winds are frequent in summer in the dry mountains of High Asia, deflating the finer components of extensive, dried-out meltwater deposits that accumulate around glacier margins.

FIGURE 3 Bare, eroded slopes in thick (>200m) loess of the subhumid Xining Basin, northwestern China.

process of deflation) is a natural phenomenon that occurs at some time in most terrestrial environments. However, certain types of landscape, notably the sparsely vegetated terrains characteristic of the world's drylands, are particularly susceptible to the massive deflation that accounts for most of the atmospheric dust plumes thought to have a bearing on human and animal health. Seasonally deposited fine water-lain sediments (notably the finer grades in the silt and clay range carried in typically turbid glacial melt waters: Figure 2), actively aggrading alluvial fan deposits, and fine lake sediments exposed in extensive basins by climatic desiccation are important examples of terrain types serving as atmospheric dust sources. Mineral particles are released by a variety of surface processes grouped together under the general heading of "weathering." These processes include breaking up of rock surfaces by the action of frost, salt, and chemical reactions and the biochemical complex of processes involved in soil formation; the latter accounts for the presence in some aerosol dust of plant fibers, phytoliths (biogenic opal), pollens, and spores. The silt-rich wind-lain deposits known as loess, which accumulated to great thickness after about 2.5 million years ago in Eurasia and the Americas, and particularly in central and eastern Asia (Derbyshire, 2001; Derbyshire et al., 2000), are readily eroded in certain circumstances, thus constituting a secondary source of minerogenic atmospheric dust (Figure 3).

Silt-sized particles, especially those in the ~10–50 μm range, are readily entrained by the wind from dry, unvegetated surfaces, but the clay-size (<2 μm) component of soils and sediments is not readily detached by the wind as individual particles because of the high interparticle cohesive forces typical of such colloidal materials. Entrainment of material finer than 2 μm usually occurs in association with the coarser (silt-sized) grains, and also in the form of coarse or medium silt-sized aggregates made up of variable mixtures of fine silt and clay-grade particles (Figure 4). Critical wind speeds for dust entrainment (threshold velocities) vary notably; those for the semi-arid/subhumid, silt-covered terrains of northern China being approximately twice those required to initiate dust storms in the Sahara (Wang et al., 2000).

Silicon, making up more than one-quarter of the elements in the Earth's crust, is highly reactive and readily combines with oxygen to form free silica (SiO_2), the most common form of which is quartz. SiO_2 dominates the composition of dust from North Africa (60.95%) and China (60.26%). These values closely match the world mean (59.9%) and its average content in the world's rocks (58.98%). Silicon also combines with other elements in addition to oxygen to form the dominant mineral group known as the silicate family, which includes the group of fibrous amphibole minerals grouped together under the general term asbestos. In the finest (<2 μm) fractions of many dryland surface sediments and soils, quartz is an important, and sometimes a dominant mineral, ranging in type from lithic fragments to biogenic opal. Varying amounts of clay minerals (hydrous aluminous phyllosilicates) are also common (notably kaolinite, illite, chlorite, vermiculite,

A   B

FIGURE 4 Scanning electron micrographs of windblown dust aggregates. (A): Silt size aggregate made up of clay-grade mineral particles, as commonly found in the young (Last Glacial) loess deposits of central and eastern Asia. (B): Silt size aggregate taken from the dust on a house beam in Ladakh. Elemental composition of such beam dust is dominated by silica, with lesser amounts of oxygen, aluminum, sulfur, potassium, calcium, and iron. Both scale bars = 10 μm.

smectite, and several mixed layer clays), with varying amounts of calcite, gypsum, and iron compounds.

The natural process by which substantial volumes of mineral dust are injected into the atmosphere is usually periodic, sometimes strongly seasonal, and mainly located in the subtropical arid and semi-arid regions. These potential dust sources cover about 30% of the total land area of the Earth. The dominant dust source regions lie in the northern hemisphere continents and include the subtropical and temperate deserts stretching from the Sahara of North Africa, through the Middle East and the northwest of the Indian subcontinent, and into central and eastern Asia. More modest sources of atmospheric dust have been identified in the Great Basin (United States) and in the Southern Hemisphere (the Lake Eyre Basin, Australia; central and northern Argentina; and a small part of southern Africa).

The relative contribution to global dust palls from these dominant source regions may have changed over recent geological time. For example, some sedimentary records from both the continents and the oceans indicate that rates of minerogenic dust accumulation during the last glacial maximum (about 20,000 years ago) were up to 10 times greater than at present (Kohfeld & Harrison, 2001). Such variable rates reflect changes in the location and size of dust source regions as glaciers waxed and waned, changing wind regimes (especially those associated with the monsoons) and climatically driven fluctuations in the hydrological cycle that affected surface conditions including vegetation cover.

Considerable contrasts in dust accumulation, expressed as calculated dust fluxes (mass accumulation rates in $g/m^2/yr$; e.g., Derbyshire 2003), are beginning to emerge from studies of the Earth's loess deposits.

The process of entrainment and transport of mineral dusts varies from a local to a global scale. It is important to discriminate between source-proximal and source-distal dust plumes (Figure 5). In general, the size of particles entrained by the wind declines with transport distance. As a result, the proportion consisting of the respirable fractions (commonly regarded as <10 μm) makes up an increasing proportion of the dust plume with distance from the source, although the absolute mass of the respirable fraction is greatest close to the source, as suggested by the colloquial term "desert lung" to describe pneumoconiosis in North Africa and the Middle East. High atmospheric dust concentrations show considerable variety in terms of their periodicity (from days to decades) and extent and proximity to sources, as well as in the percentage of the dust in the respirable range. In some regions of the world, extensive dust plumes have become an integral part of regional culture, and the washing out of the brown-to-red mineral particles by precipitation is known as "loess rain" in China, and "blood rain" in Mediterranean Europe and further north. These terms also draw attention to the contrast between yellow Chinese dust and the red dust of North Africa.

The concentration of mineral dust in the atmosphere (the atmospheric aerosol loading) is both a function of,

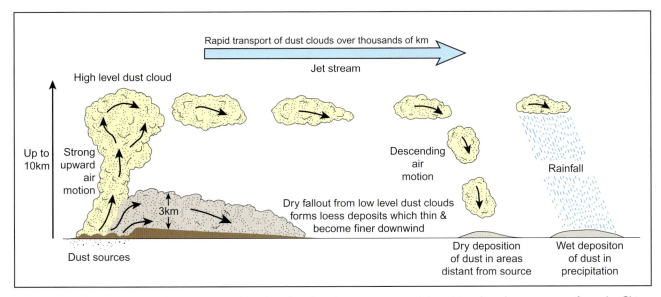

FIGURE 5 Sketch showing the two principal modes of aeolian dust transport and deposition, based on a transect from the Chinese drylands to the Loess Plateau and the North Pacific Ocean. Re-drafted from Pye and Zhou (1989).

and a factor influencing climatic change, as it affects physical and biogeochemical exchanges between atmosphere, land, and ocean. The presence of aerosols influences the chemistry of the troposphere including the proportion of ozone. Aspects of climate affected by atmospheric dust loading include the ability of dust to raise or lower air temperatures depending upon the differential effect of its particle size and chemistry, and upon the extent to which solar radiation is absorbed and scattered, which is an effect significantly modified by the amount, altitude, and thickness of any cloud cover. Deposition of dust may add notable volumes of certain nutrients to the world's oceans, including nitrates, ammonia, phosphates, and oxides of potassium and iron. It is considered that such inputs of iron to oceanic waters stimulate nitrogen fixation by plankton, thus enhancing productivity. The Sahara, commonly regarded as the world's greatest source of wind-transported mineral dust (Goudie & Middleton, 2001), has some influence on the nutrient dynamics and biogeochemical cycles of a region stretching from northern Europe to South America (Prospero, 1999) (Figure 6). In addition, it is claimed that Saharan dust storms sometimes transport bacteria and fungal spores that cause deterioration in Caribbean coral reefs (Shinn et al., 2000), events that have also been linked to reduced air quality and cases of asthma and other respiratory problems in residents of parts of the southeastern United States. Such an intimate relationship between aerosols and the global environment, taken together

with the probability that human actions in the past century or so have progressively enhanced the atmospheric dust loading, has implications for future climatic change (Harrison et al., 2001). The effect of such changes upon human societies is likely to include some notable health impacts.

II. HEALTH-IMPACTING MINEROGENIC AEROSOLS

A. Dust Storms

The type, size, and extent of dust plumes raised during dust storm events are fundamental factors influencing the degree to which naturally occurring atmospheric dust impacts upon the health of human and animal populations. Dust storms may be generated by local vortexes, generally known as "dust devils" (or willy-willies in Australia). Dust devils are only a few meters in diameter and raise dust to heights of 100–200 m (exceptionally 1000 m) for periods of a few minutes to a few hours. Similarly low altitude, though more regionally extensive dust-carrying wind systems in northwest Africa, for example, arise from the relatively shallow northeasterly trade winds and by squall lines associated with northward incursions of equatorial air (the West African summer monsoon). These raise dust

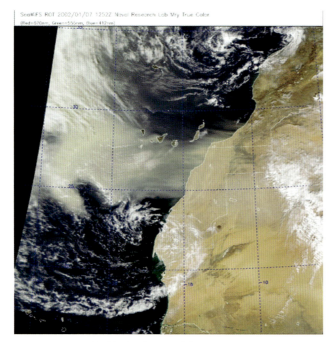

FIGURE 6   (Left): Outbreak of Saharan dust across the North Atlantic, January 2002. High pressure over and north of Morocco, with a depression centered off the West Sahara-Mauritania coast, indicates a strong easterly flow across the African coast and the Canary islands (center) and into the mid-Atlantic. A dense pall of dust, about 500 km wide, reduced visibility and enhanced sunset colors for several days in the southern Canaries and deposited red dust. (Right): African dust over western Europe, October 2001. High pressure over the Mediterranean basin and an extensive and vigorous depression west of North Africa and Spain (centered close to the island of Madeira) induced strong southerlies from Mauritania to Scandinavia. A high-level dust pall can be clearly seen running from off the Moroccan coast across western Iberia, the Bay of Biscay, western and central France, southern and central Great Britain, the Low Countries, North Germany, and Denmark. Both are NASA SeaWiFS images.

above the shallow trades and into the troposphere so that it crosses the Atlantic in winter as the dry northeasterly "harmattan." Such source-proximal transport involves a relatively high percentage of the coarser dusts (medium and coarse silts with some very fine sand), which are usually deposited at distances of only hundreds of kilometers downwind. Extensive regional atmospheric turbulence, however, arising from air mass frontal systems associated with the hemispherical wind regimes, notably the upper westerlies, carry the finer dust fractions at high levels within the troposphere. These source-distal events frequently transport terrestrial dust across oceans, including the Atlantic and Pacific, with deposition occurring some two or three weeks after initial entrainment (Pye, 1987).

An example of the relationship between landforms, surface sediments, and soils that are particularly susceptible to dust deflation and specific meteorological situations, on the one hand, and atmospheric dust loadings and their source-proximal and source-distal effects, on the other, are illustrated by a dust storm that

occurred in northwest China in May 1993 (Derbyshire et al., 1998). The highest wind velocities and most severe damage and loss of human life were felt in the Hexi Corridor (Figure 7), a WNW-ESE topographical constriction between the mountains bordering the northeastern edge of the Tibetan Plateau (the Qilian Shan; in Chinese *shan* means mountains) and the Mongolian Plateau. The wider impact was felt in the provinces of Gansu, Ningxia, Inner Mongolia, Shaanxi, and Hebei, a region equal to the combined area of France and Spain.

The meteorological situation on May 4, 1993, was controlled by a large high-pressure system (the Siberian High) over western Eurasia, with a depression centered on the northern Urals but with a trough extending far to the south. The cold front associated with this trough extended to the northern edge of the Tian Shan range. The constriction between the Tian Shan and the Altai Shan ranges resulted in increasingly convergent, and hence accelerating, air flow toward the east. By the next day, the cold front had reached

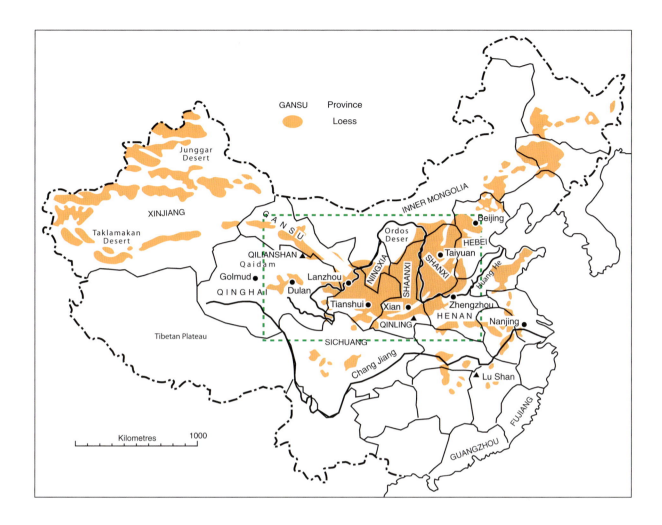

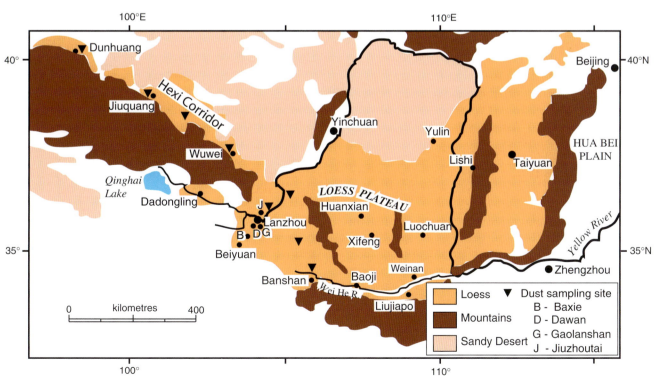

FIGURE 7 Upper: The Loess Plateau of North China in relation to the Hwang He (Yellow River) and the principal deserts. The box indicates area covered in lower half of the figure. Lower: Part of northern China, showing the Loess Plateau and the Hexi Corridor. See text. Re-drawn from Derbyshire et al. (1998).

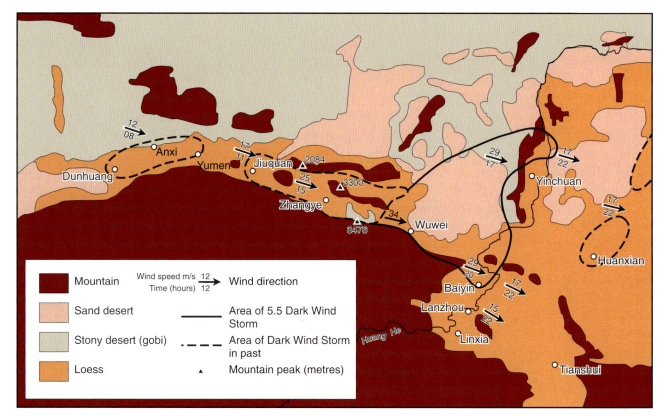

FIGURE 8 Terrain sediment cover types, dust storm zones, wind velocities, and timing (in hours) of the "dark storm" of May 5, 1993, in the Hexi Corridor and the western Loess Plateau, China. Data provided by the State Meteorological Service of China, and the Meteorological Bureau of Gansu Province. Re-drawn from Derbyshire et al. (1998).

Dunhuang (Figure 8), where the channeling effect of the western end of the Hexi Corridor sustained a wind velocity of 12 m s. In 3 hours the cold front had reached Jiuquan, with velocities of 17 m s, and within 5 hours it had reached the narrowest part of the Corridor (between Zhangye and Wuwei) where velocities peaked at 34 m s (Figure 8). These high velocities were sustained across the Tengger Desert, mobilizing sand as well as dust as far as Beiyin. With the opening out eastward of the Hexi Corridor, however, airflow became increasingly divergent and progressively slower; only the finer dust fractions were transported beyond the North China plain.

The effect of this "dark storm" in the proximal area of the Hexi Corridor was extremely serious. Visibility declined below 10 m in full daylight, and the depressed temperatures created severe frosts (minima −6.6°C) with some local snowfalls. The direct effects included 380 people and 120,000 farm animals killed and damage to about 3300 km² of crops. The particulate aerosol concentrations reached the "extensive dust pall" category (see below, Section II.D) in the center and east of the Hexi Corridor, and there was widespread loess rain in the more distal provinces to the east (Shaanxi and Hebei). The coarser suspension load in the lower atmosphere, including coarse silts, reached as far as the northern slopes of the 3700-m high Qinling Shan, south of the city of Xi'an. Five storms of similar magnitude occurred in the Hexi Corridor between 1952 and 1993. A satellite image of a dust storm that affected the Hexi Corridor and a broad region to the east of it on March 29, 2002, is shown as Figure 9. Comparison of this image with the regional details of the 1993 event (Figures 7 and 8) shows it to be very similar in source and extent, if not in destructive power.

This Hexi Corridor case study is an example of dust impact from a proximal source with the bulk of the visible dust pall consisting of relatively coarse silt particles in the lower few kilometers of the atmosphere. However, the finer components of such palls, traveling at higher levels, are known to be carried great distances across the Pacific Ocean. Such "distal source–high

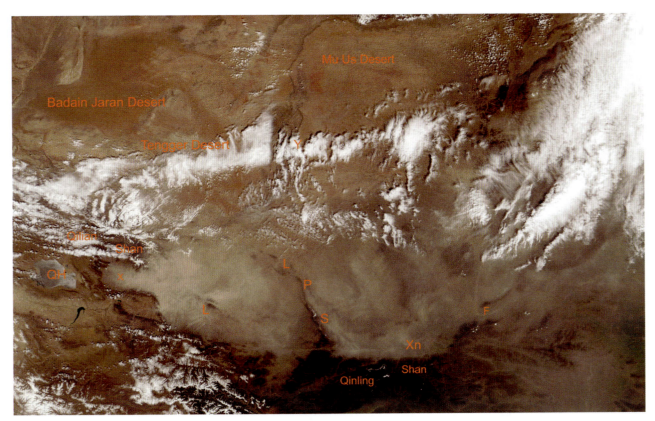

FIGURE 9 Part of a Terra Satellite image, using the MODIS sensor, taken on March 29, 2002, and showing a dust storm generated by winds from the west-northwest over northwest China. (For comparative location, compare with Figures 7 and 8). The air over the Badain Jaran and Tengger deserts is clear of dust, as it is over the Mu Us Desert (to the north of which can be seen the big bend of the Yellow River). However, the dust thickens rapidly with proximity to the cold-weather-front and the alluvial-fan-covered Hexi Corridor on the northern side of the Qilian Shan. The Xining Basin (X = city of Xining, just to the east of Qinghai Lake; QH) generates its own pulse of dust, but not the adjacent Qaidam or Gonghe basins in this case. Lanzhou city (L), near the outlet of the Hexi Corridor, has a thick dust pall over it as well as its own locally generated pollution cloud. The dense dust plume is split by the NNW-SSE aligned Liupan Shan (L-P-S). East of this mountain range, the plume completely covers the twin basins of the Jing and Luo rivers (draining the central and southern part of the Loess Plateau). This part of the plume just covers the city of Xi'an (Xn); its pollution pall is more modest than that of Lanzhou. The southeastern margin of the plume is very sharp as it comes up against the ~3700 m high Qinling Shan (on which several snow-covered areas can be seen). The plume extends eastward, crossing the sharp bend of the Yellow River at Fenglingdu (75 km west of the Sanmenxia Reservoir on the lower Yellow River). The dust plume over the green farmlands of Henan and Hebei provinces in the southeastern part of the image is much more diffuse, which suggests that it may be the product of a pre-frontal trough.

altitude–finer dust" systems have aroused considerable recent interest because they can be tracked using orbital imagery backed up by study of synoptic meteorological charts. One such example of a distal dust source of global significance is the Tarim Basin, a region that probably has the highest mean annual number of "dust days" in China. The Taklamakan Desert, occupying most of the Tarim Basin, is predominantly a sandy desert, but loess accumulations are found along extensive parts of its windward (southern and western)

mountain rim, which shows that the Tarim also has a functioning "proximal, low altitude–coarser dust" system (Figure 10).

Steep atmospheric pressure gradients associated with extensive Siberian–Mongolian ridges of high pressure, most notably between late winter and early summer, strengthen the easterlies around the southern flank of the seasonal high pressure cell over the Taklamakan Desert. This air flow is then subject to vigorous uplift as it comes up against the western Kunlun, the Pamir,

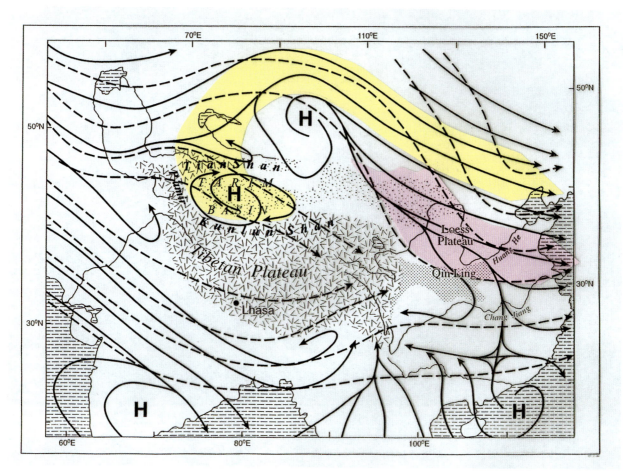

FIGURE 10 Relationship between generalized winter-spring pressure systems over central and eastern Asia, the major orographic features, the two dominant dust source regions, and the associated principal atmospheric dust pathways. The deserts, alluvial fans, and dry lake basins to the north and west of the Loess Plateau provide both source-proximal, coarser silts to the Loess Plateau, with much finer material being carried at higher atmospheric levels across the east China plain, and the Pacific Ocean and beyond (pink arrow). The Tarim Basin surrounded by high mountains (the Tian Shan, the Pamir, and the Kunlun Shan) concentrates fine sediments largely derived from glacial meltwater rivers and alluvial fans. These are re-worked by winds blowing from north of east to be deposited as loessic silts on the northern flanks of the Kunlun Shan. At over 5000 m above sea level, this may be the highest loess in the world (from Sun, 2002b). Finer dusts are lifted above this level and enter the westerly jet stream to be carried great distances, sometimes as far as Europe (yellow arrow).

and the western Tian Shan ranges, all of which have peaks around 7000 m above sea level. Current opinion (e.g., Sun, 2002a) is that the finest fractions of the Tarim dust cloud are uplifted into the upper troposphere to be carried northeastward across Outer Mongolia, eastern Siberia, and across the Pacific Ocean (Figure 10). Fine Chinese airborne dust is commonly recorded in western North America, and it is known to reach the eastern United States from time to time. This airborne dust was recorded over the Atlantic Ocean on April 20, 2001, and has recently been discovered in the French Alps, a distance from source of about 20,000 km (Grousset et al.,

2003). Preservation of mineral dust from both of the principal Chinese source regions in the Greenland ice cores confirms that the "distal–high level–fine dust" pathways are long established, persistent, and also of global scale.

B. Dust Sources

Interest in the detection, tracking, and measurement of distal (regional and global scale) mineral dust in the atmosphere has been greatly stimulated by the increasing availability of images and other data provided by Earth-orbiting satellites. Aerosol optical thickness (AOT), as estimated using the advanced very high resolution radiometer (AVHRR) of the United States' National Oceanic and Atmospheric Administration (NOAA), is based on backscatter radiation measurements made at an effective wavelength of 0.63 μm. In general, high AOT values indicate high atmospheric dust concentrations. However, because the AOT algorithm requires the surface below the dust plume to have a low and constant albedo, AOT can be estimated in this way only over the oceans. This restriction is generally true of satellite sensors operating in the visible spectrum. The situation was greatly improved by the advent, in 1980, of the total ozone mapping spectrometer (TOMS). This is used to detect absorbing aerosols based on the spectral contrast at 340 and 380 nm in the upwelling ultraviolet (UV) spectrum. TOMS is sensitive to a range of UV-absorbing aerosols such as mineral dust, volcanic ash, and black carbon from fossil-fuel combustion sources and biomass burning. The UV surface reflectivity is typically low and nearly constant over both land and water, which allows TOMS to detect aerosols over both continents and oceans. The UV spectral contrast is used in a non-quantitative way as an absorbing aerosol index (AAI). The temporal and spatial variability of this TOMS AAI has been matched to types of absorbing aerosols, as well as to known sources such as individual volcanic eruptions, forest fires, and large-scale dust events. The global distribution of the occurrence frequency of relatively high TOMS AAI values (January and July 1980–1992) is shown in Figure 11 (Prospero et al., 2002) (see also Chapter 27, this volume).

This image contains one huge, dominant area with high AOT values extending westward of the North African coast and eastward from the Middle East, indicative of dust plumes from the world's premier atmospheric dust source region. The plume off the west coast of South Africa, in contrast, is attributed to biomass burning.

The irregular timing and the variety of sources contributing to dust storms, as well as the technical limitations of the different sensors in use, complicate the determination of the location and extent of individual dust source regions or areas around the globe. The common association of dust-raising conditions with the cloudy conditions generated by pressure troughs and air-mass fronts is a case in point. The TOMS system is most sensitive to aerosols in the middle and upper troposphere and above (distal dust) and least sensitive in the boundary layer where aerosol residence times are

shorter (proximal dust). Aerosols below the altitude range of 1000–1500 m remain largely undetected. Thus, the source-proximal components of major destructive dust storms may not be detected on some visual imagery.

Despite such difficulties, the use of global distributions based on the month in the year that best represents the long-term (13 year) frequency of dust storm occurrence as indicated by TOMS AAI has yielded a map of major global dust sources that closely matches the information available from other types of observation (Prospero et al., 2002). The result (Figure 12) indicates sources in all continents except Europe and Antarctica. Most of the major sources are in surface depressions or adjacent to mountain fronts.

For example, the Ahaggar and Tibesti mountains in the Sahara are surrounded by what may be the greatest single regional dust source on Earth. There is a strong link between dust sources and extensive alluvial deposits, as well as ephemeral, saline, and dried out lakes throughout North Africa, the Middle East, the northwestern Indian subcontinent, Middle Asia (from the Caspian Sea to Kazakhstan), and across northwest China, as shown above (Section II.A). Sand dune deserts, as such, are not important consistent sources, although their sporadic drainage systems frequently provide abundant fine particles for deflation. There are many smaller, but important, sources outside these major regions. These include the Basin and Range province of the southwestern United States and northern Mexico; the Lake Eyre and Great Artesian Basin in Australia; Patagonia, the Andean footslopes in central Argentina, and the Altiplano of Bolivia and northern Argentina; and southern Africa. Secondary sources also include the major loess deposits, notably in parts of northern China. The "mountain deserts" of Iran and Pakistan, at the western end of the Himalayan tract, constitute an important regional dust source, notably in summer, which involves channeling of dust by down-valley (katabatic) winds (Figure 2).

The human impact, varying in both type and intensity of activity as well as in length of its history, further complicates assessment of "natural" dust sources. The major concentrations of fine-grained, poorly vegetated deposits that constitute important dust sources are those associated with floodplains, alluvial fans and lake depressions, and sites that are fed by seasonal perennial freshwater flows that also attract human communities and their animals. The dust sources in the Middle East include the Tigris-Euphrates basin where agriculture has been widespread on this rich alluvium for thousands

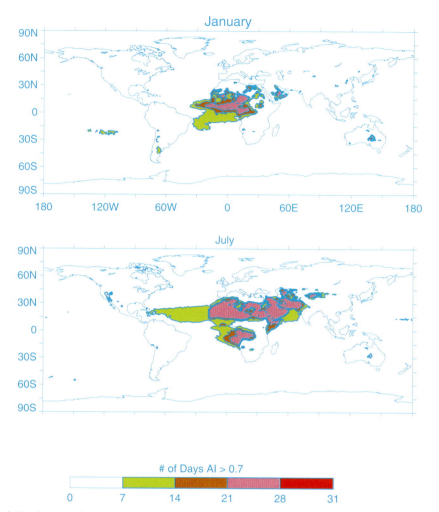

FIGURE 11   Global distribution of dust and smoke. Monthly frequency of TOMS absorbing aerosol product for January (top) and July from 1980 to 1992. Scale: number of days per month when the AAI equaled or exceeded 0.7. The large, dark area in southern Africa in July is a product of biomass burning, and there is also evidence of biomass burning in January just north of the Equator in Africa. Part of the plume over the Equatorial Atlantic is smoke. All other distributions shown are due to the presence of dust. After Prospero et al. (2002), with kind permission of the first author.

of years. There is also a long history of human use and interference with such water sources in the drylands of central Asia. Many small states and cities in western China have collapsed as a result of a failure of water supply due either to overuse or destruction of dams in warfare as well as severe periodic drought (Derbyshire et al., 2000). The present-day use of dung, wood, and, to a lesser extent, coal, as fuel sources in the drylands of western China is an important factor affecting the extent and composition of airborne dust.

The colonization, by sophisticated agricultural people, of the Loess Plateau of northern China, a mass of wind-deposited silt with an area >400,000 km² and an average thickness of 100 m, has had a notable impact.

Locally dense populations practicing hand agriculture and their grazing animals have played a major role in accelerating river erosion and slope failure. Some commentators take the view that the Loess Plateau and the Mongolian steppe lands to the north of it can be regarded as a secondary source of atmospheric dust in present climatic conditions (Figure 13), although this view is being challenged (see Section II.C).

Northern India, another region with a dense human population, injects the products of the burning of dung, wood, and fossil fuels into the atmosphere to an extent that makes it difficult to estimate the natural component in atmospheric dust palls. The definition of the main dust sources is also complicated by widespread

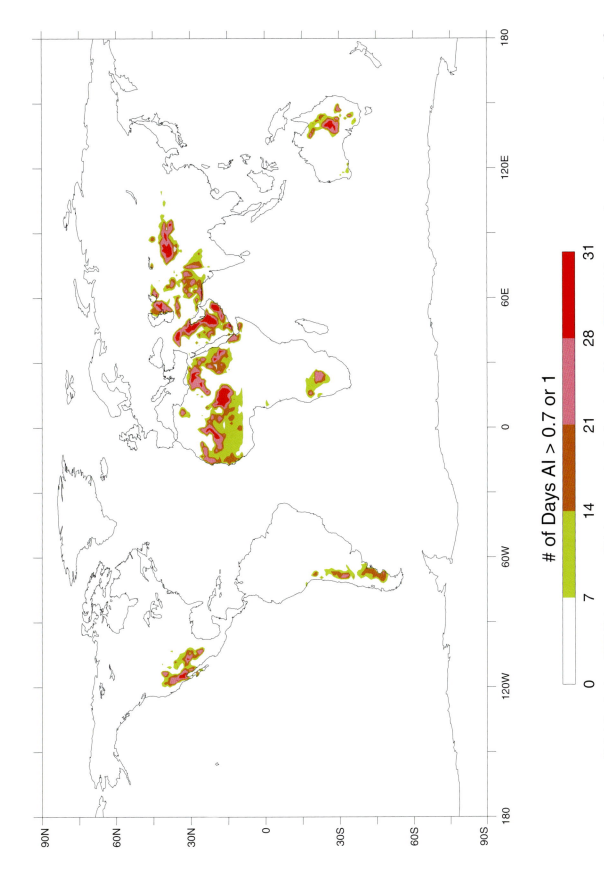

FIGURE 12 Global distribution of TOMS dust sources. This is a composite of selected monthly mean TOMS AAI frequency of occurrence distributions for specific regions using those months that best illustrate the configuration of specific dust sources. The distributions were computed using a threshold of 1.0 in the "global dust belt" (west African Saharan coast, through the Middle East, and central Asia to the Yellow Sea), and 0.7 elsewhere. After Prospero et al. (2002), with kind permission of the first author.

# of Days AI > 0.7 or 1

0   7   14   21   28   31

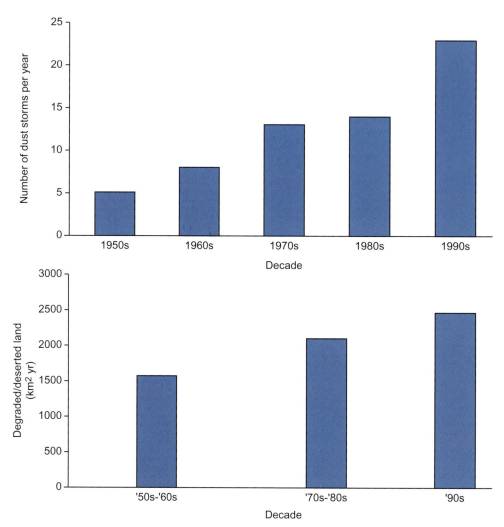

FIGURE 13 Graphs showing (top) the number of dust storms per year by decade, from the 1950s to the 1990s, in Beijing and (bottom) the annual rate of land degradation (in km² yr) in Inner Mongolia (lying northwest (windward) of Beijing). Data kindly supplied by Dr. Xingmin Meng from a Chinese-language Web site.

biomass burning in central and southern Africa. Land degradation arising from resettlement of quite large populations since the middle of the 20th century, and subsequent land clearing, agriculture, and/or animal grazing, has created new sources of dust in the Etosha Pan in Namibia (southern Africa; Bryant, 2003). This is also true of the Mongolian Plateau in eastern Asia. The recent history of human interference by damming rivers draining into the Caspian and Aral seas provides another such example. The diminution of the Aral Sea has become a classic case of the human generation of an atmospheric dust source, but smaller examples are known, such as Owens Lake in California in the United States (see Section III.B).

A number of studies have attempted to quantify global dust emissions, but estimates vary substantially, which reflects the evidently wide gap between modeling-based estimates and those based on the numerous available (but short-term) measurements. Equally diverse are estimates of the effect of land use and land use changes on the global atmospheric dust loading, which may be less than 20% (Prospero et al., 2002).

C. Dust Storm Frequencies

Dust storm frequency is usually measured by the annual number of "dust storm days," defined as a reduction of

visibility by dust to less than 1 km for all or part of a day. Allowing for the fact that the number of ground observations is low, the location of measuring sites is extremely uneven, and the measurement record very short for many sites, the range of dust storm days per year values is wide with figures reaching about 80 in southwest Asia, >60 in Turkmenistan and the Karakum, >45 in Kazakhstan, >30 in the Tarim Basin, the Hexi Corridor, and the Loess Plateau (China), ~30 in parts of North Africa, ~20 in the Northwestern Indian subcontinent, and >15 in central Australia (Middleton et al., 1986).

Measured dust storm frequencies are open to different interpretations, however, even in some densely populated regions of the world. For example, the rise in frequency in North China in the second half of the twentieth century referred to above (Figure 13) is currently under review in the context of global warming. It is argued that, while increased desertification might be expected to result in more frequent dust storms, the recent decline in the number of dust storms in the Beijing region, indicated by some records, is consistent with the current warming trend, although the effects on total seasonal dust volumes of any such decline may be offset by an increase in the vigor of individual dust events. Certainly, higher mean temperatures over North China would be consistent with a generally weaker winter-spring monsoon and fewer outbreaks of cold, dense air from the northwest. Thus, although there is some observational evidence suggesting a negative correlation between the number of spring dust storms and mean temperatures over this large region, the case for decline is inconclusive. This may reflect the complex cause-and-effect-relationships involved in dust storm analysis. Another more skeptical point of view is that governmental and public concern about dust storms is now much greater than it was in the social climate of the 1950s, 1960s, and 1970s.

D. Ambient Dust: Continental Concentrations

Dispersion of mineral aerosols during and following dust storms yields ambient atmospheric dust in the form of sets of discrete plumes and extensive regional sheets or palls. Discrete plumes, thought to derive from point sources, give rise to very high dust concentrations that include the coarsest particle fractions; such plumes persist for relatively short periods (hours to days). With increasing distance from source, atmospheric dust disperses as regional palls of finer particles in which the respirable fractions are dominant; these may persist for several days or longer.

Numerous measurements of ambient dust concentrations have been made, although most cover short periods only. In addition, the terrestrial sites concerned are very unevenly distributed. Both of these facts make it difficult to assess with any precision the terrestrial "dust climate" at a regional scale, as exemplified by the current controversial situation in North China mentioned above. Direct and regular measurement of dust concentration is rare or absent in many parts of the world. Most national meteorological services measure or estimate visibility. However, because visibility is a composite product of both atmospheric humidity and concentration of aerosols (both natural and anthropogenic), visibility estimates provide only a rough guide to dust concentrations (see below). Moreover, although satellite observations are very informative with respect to sources and transport paths in the troposphere, many dust events are not detected in this way (Section II.B). Further, the common assumption of global uniformity in the optical properties of dust is manifestly unrealistic. In regions in which frequent high concentrations of minerogenic dust constitute a health hazard for large human populations, such as northern China, existing data gathering is clearly inadequate, as indicated in the recent call within China for a nationwide state-owned network for the monitoring and analysis of atmospheric dust. Many of the existing measurements of atmospheric dust in continental locations are orders of magnitude greater than those obtained from measurements over the oceans. Maximum values of mineral dust cited in the literature for continental areas are around $10^5 \, \mu g \, m^{-3}$ (Pye, 1987), although concentrations may occasionally exceed this value in some parts of the world. The "normal background" lower atmosphere dust concentration in northwest China is $0.083 \, mg \, m^{-3}$, which reaches values of about $4 \, mg \, m^{-3}$ in "ordinary"

TABLE I. Categories of Dust Concentration Used to Describe Chinese Dust Storm Events

Dust concentration (mg m^3)	Description
0.083	Normal background
0.356	"Detachment mode"
1.206	Extensive dust pall
3.955	Ordinary dust storm

From Chinese Meteorological Bureau.

dust storms (Table I). However, concentrations of $69\,mg\,m^{-3}$ (April 1998) and $21.61\,mg\,m^{-3}$ (April 2003) have recently been recorded, and an extreme value of $1016\,mg\,m^{-3}$ occurred in May 1993 (Derbyshire et al., 1998).

Human settlement of continental drylands has undoubtedly served to enhance the frequency, magnitude, and impact on health of dust-entraining events and ambient dust levels. Activities such as arable farming, intensive grazing, industry, urbanization, and road and rail construction are frequently concentrated within natural dust source environments such as alluvial fans, river floodplains and terraces, and lake basin margins. Loess and silts deposited by rivers and lakes provide some of the most fertile and readily cultivated soils on Earth, as well as a widely used building material (known in parts of the Americas as adobe). Recent extension of agricultural activities along desert margins in several continents, and notably in Asia, has caused varying degrees of land degradation (often generalized as desertification). Hand cultivation and shallow plowing of such deposits certainly stimulates local dust palls, and dust concentrations in many adobe dwellings are often some orders of magnitude higher than those found in normal background conditions in regions such as Ladakh and northern China. In such situations, it is often difficult to discriminate between natural and anthropogenic dust, and so to attribute with any assurance human health effects exclusively to natural versus "occupational" dust events.

III. Pathological Effects of Aerosol Dust

Inhalation of mineral aerosol particles, followed by deposition in human pulmonary alveoli, varies with a number of factors, but particle size and composition and certain lung functions are particularly important. Most coarse particles in minerogenic dust (diameter $<100\,\mu m$) are abundant close to the dust sources, and they are deposited relatively quickly by both dry and wet depositional processes. Thick dust palls characteristic of locations relatively close to dust sources pose a number of hazards to human health and welfare, which include transport accidents, destruction of crops, and eye irritation. When inhaled, many of the larger dust particles are eventually rejected by expectoration. However, inhalation of large dust particles ($>10\,\mu m$) may constitute a health risk if the mineralogy is toxic, regardless of where the grains lodge in the respiratory system. Of the finer dust particles (diameter $<10\,\mu m$) that remain in suspension in the atmosphere for much longer periods (the respirable fraction), most between 10 and $5\,\mu m$ become trapped in the upper respiratory tract and are ultimately removed by coughing. Particles finer than $5\,\mu m$ frequently penetrate more deeply into the lungs to cause silicosis (Pendergrass, 1958), asbestosis, and other lung conditions. Recent studies suggest that about 75% of dust found in some Chinese post-mortem lung tissue is finer than $3\,\mu m$. Atmospheric dust finer than $2.5\,\mu m$ is considered to be of particular importance with respect to community health, as in the PM standard of the United States' Environmental Protection Agency.* Ambient dust may also absorb harmful gases, disease-generating bacteria, and even carcinogenic hydrocarbon compounds. Recent work in China, for example, has shown that the denser the ambient dust, the higher the rates of chronic respiratory disease and associated death rates. Respiratory disease may also exacerbate cardiac problems (see also Chapter 23, this volume).

A. Pneumoconioses

The pneumoconioses, lung diseases that include silicosis and asbestosis, are a result of prolonged inhalation of fine minerogenic aerosol dust. The condition is best documented from studies of workers in certain industries in which high mineral concentrations are generated (occupational pneumoconiosis). Much less attention has been accorded to cases of pneumoconiosis arising from non-occupational exposure to ambient mineral dust; one recent exception is the "cleanup" campaign at Libby, Montana in the United States (see below). Studies based on occupational cases show both pneumoconioses to be insidious in the early stages but then progressively noticeable when exercising, thus the symptoms are sometimes attributed to a patient's aging (Wagner, 1997). Many symptoms are nonspecific in the absence of radiography, and this may constitute an important factor influencing diagnosis in some developing countries. Radiographic diagnosis of silicosis is made with confidence only after the appearance of silicotic nodules 2–5 mm in size. Continued dust exposure leads to an increase in nodular size and number so that they eventually cover much of the lung, and the nodules sometimes coalesce to form conglomerate shadows often called progressive massive fibrosis (Saiyed, 1999).

---

*The PM (particulate matter) standard is based on the total mass of particles measuring 2.5 microns or less observed in a 24-hour period.

Some types of inhaled particulates are degraded by macrophages, but many are highly resistant to this process and persist in the lung cavity and lymph nodes. Some resistant particulates appear to cause no problems, but others stimulate fibroblastic cells to deposit collagen. In the case of asbestos, for example, detectable fibrosis appears only after a threshold number of particles have been retained (Bar-Ziv & Goldberg, 1974). Silica is a highly fibrogenic agent in lung tissue, and the reaction is very different from the granulomatous reaction to many other nondegradable grains; the fibrotic reaction has been associated with release of polysilicic acids. Fine-grained, sharply angular quartz grains are widely considered to enhance this process, although the precise nature of the pulmonary response to crystalline silica remains rather poorly known (Saiyed, 1999). Continued exposure to silica is thought to lead to increased rates of infection with pulmonary tuberculosis, a notable public health problem in many developing countries, with non-tuberculosis mycobacterial infection (involving intercellular bacterial parasites) also occurring. Many people with silicosis have been shown to be susceptible to tuberculosis (Snider, 1978), although present constraints on diagnosis, especially in poor and remote regions of the world, carry with them a continuing risk of confusing silicosis-related massive fibrosis with tuberculosis. Nevertheless, evidence exists that patients with silicosis carry a greater susceptibility to tuberculosis, and the World Health Organization (1997) has now listed crystalline silica as a human carcinogen. Silicosis has a number of deleterious effects upon the immune system. One important effect is a reduced ability of the macrophages to inhibit growth of tubercle bacilli. Some rheumatic, as well as chronic renal diseases also show higher than average incidence in individuals exposed to silica, and it is likely that such increased susceptibility of subjects to a suite of mycobacterial diseases is to some degree due to impaired function of macrophages in silicotic lungs (Snider, 1978).

A number of pathological conditions are associated with inhalation of asbestiform minerals. Asbestos is found in a wide variety of geological environments. For example, chrysotile is known to occur in hydrothermally altered ultramafic or carbonate rocks. Crocidolite is abundant in some metamorphosed iron formations, and may also occur as an authigenic mineral and as a hydrothermal alteration product in some carbonatite complexes. Although not covered by the term asbestos, the mineral erionite, a fibrous zeolite (group of hydrous aluminosilicates), is also known to cause asbestosis and related conditions as noted below (Section III.C). Natural release of asbestiform minerals from the host rock occurs by the processes of weathering and erosion, and the fibers are frequently concentrated by surface wash. In seasonally dry climates, such concentrations of fibers dry out and become susceptible to deflation. The health effects of asbestos inhalation include asbestosis, mesothelioma, and lung cancer. Some asbestos fibers penetrate body tissue and remain in the lungs, lung lining, and abdominal cavity. Radiographically visible fibrosis may take 15–20 years to appear following initial exposure (Wagner, 1997) (see also Chapter 22, this volume).

B. Case Studies of Non-Occupational Silicosis

Non-industrial deposition of silica in human lung tissue was first reported in three inhabitants of the Sahara Desert half a century ago. The autopsy results showed a high content of fine ($<3\,\mu m$) silica dust, but there was no sign of typical silicotic lesions (Policard & Collet, 1952). Other findings from different parts of North Africa include radiological evidence of multiple micronodules in reticular disposition scattered throughout the lungs, and this is considered to be consistent with silicosis. A radiographic survey of 18 asymptomatic Bedouin females in the Negev Desert by Hirsch et al. (1974) found positive indicators in all patients who were aged between 26 and 70 years, with 9 cases in people older than 50. Histological examination showed varying amounts of dust-laden macrophages in a perivascular and peribronchial distribution, mainly in the middle and lower lungs. No typical silicotic nodules with collagenization were found, and a 3- to 5-year radiological followup study showed no evidence of progression. The fact that older patients made up the largest proportion of the sample group, and the lack of any progression toward formation of fibrotic conglomeration, suggested a benign condition called simple siliceous pneumoconiosis, which was possibly attributable to long periods of work within the confined environment of the home tent. In a larger study involving 54 cases, siliceous particles were found both free and in macrophages, and the incidence of fibrosis was shown to be age related with a progression more noted in women (13 out of 22) than in men (only 4 out of 32) (Bar-Ziv & Goldberg, 1974).

Such "desert lung syndrome" has a long history and was even found in ancient Egyptian mummies (Tapp et al., 1975). In the past half century it has also been recorded in Pakistani farmers, Californian farm workers, Ladakh villagers, people in the Thar Desert of Rajasthan, northwest India, and residents of northern

China as well as the cases mentioned above from the greater Saharan region.

A survey of two villages situated at altitudes between 3200 and 3500 m in the western Himalayas, situated only 15 km from Leh (the capital city of Ladakh), was undertaken by Norboo et al. (1991). There are no mines or industries in Ladakh, but dust storms are frequent between late winter and summer and there is characteristic local variability in their incidence in this high mountain environment. Radiographic evidence was taken from an equal number (23) of men and women between the ages of 50 and 62 years. Of these, 8 men and 16 women showed varying grades of silicosis, with important differences resulting from the higher dust concentrations at the lower village (at 3200 m) compared to the higher one (at 3500 m). Three cases of progressive massive fibrosis were found in the lower village, which suggested the likelihood that silicosis here causes appreciable morbidity. Later augmentation with necropsy lung tissue samples revealed heavy dust deposition with abundant hard, 1- to 3-mm nodules and a lymph node largely replaced by hyaline collagenous nodules, a classic feature of silicosis. More than 20% of the mineral dust extracted from the lung tissue consisted of quartz, bulk chemical analyses yielding 54% elemental silica, and 19.2% aluminum. A larger study (Saiyed et al., 1991), involving a total of 449 patients aged over 50 years (245 women, 204 men) from three different villages in the Leh vicinity, showed typical cases of pneumoconiosis associated with progressive massive fibrosis and egg-shell calcification of the bronchial glands (indicative of high concentrations of free silica) in 101 cases (22.5% of the population sampled). A close correlation was found between frequency of dust storms and number of cases of pneumoconiosis: the village with low dust storm frequency recorded only a 2.0% incidence, the one with moderate frequency showed that 20.1% of the sample population was affected, and the village with severe dust storm incidence revealed a 45.3% incidence of pneumoconiosis. The existence of such a high proportion of pneumoconiosis cases in the populations of such remote villages, with no possibility of exposure to the products of mining or industrial activity, is striking evidence of the role of minerogenic dust. Although there is little doubt that the burning of brushwood and dung in the adobe dwellings places the women at higher risk of developing pneumoconiosis, as clearly indicated by Norboo et al. (1991), Saiyed et al. (1991) found no clear differences in incidence between the sexes. Mineral dust found on the upper surfaces of wooden roof beams in houses built of loess in Ladakh is all finer than 15 μm,

more than 25% by weight being <1 μm; the silica content is >60% (Figure 4B).

Several studies, both published and unpublished, have implied that the large number of people subject to the frequent dust storms characteristic of north China are potential silicosis cases. In one investigation 395 subjects (294 men and 101 women) from two communes in the middle of the Hexi Corridor, Gansu Province, were studied (mean dust concentration: 8.25–22.0 mg m^3), and 88 people (46 men and 42 women) were randomly chosen from a third commune with low dust storm exposure as a control group (mean dust concentration: 1.06–2.25 mg m^3). The incidence of silicosis was 7.09%, with no cases in the control group, but this rose to 21% in subjects over 40 years of age. There was no significant difference in incidence between the sexes, but comparative necropsy of the lungs of camels showed them to contain evidence of silicosis (Xu et al., 1993). In another, larger but unpublished radiological survey, involving 9591 residents in Gansu Province, a prevalence of 1.03% was found, which rose to 10% in subjects over 70 years old (Changqi Zou, personal communication).

In the past four decades, about half (36,000 km^2) of the former bed of the Aral Sea (western Uzbek Republic, middle Asia) has been exposed, providing a new and frequent source of fine dust (Figure 14). This situation has been exacerbated by the diversion, for irrigation purposes, of most of the waters of the two main rivers (Syr Darya and Amu Darya) that drain into the Aral Sea, adding to the regional desiccation. The fine-grained sediments (silts and clays) on the seabed are rich in agricultural chemicals, and they are readily deflated. Despite some reports of increasing respiratory illness in children, including some mention of interstitial lung disease in this region, there is little authoritative information about the link between the desiccation of the Aral Sea region and human health (Wiggs et al., 2003).

Owens Lake (California in the United States) was a shallow but perennial water body for most of the last million years, but diversion of water for use by the city of Los Angeles began in 1913 and, by 1926, the lake had dried up. The dry lakebed is now probably the greatest single source of respirable mineral dust particles in the United States. Palls of dust <10 μm in aerodynamic diameter occur on about 10 days per year across a wide area. With arsenic levels sometimes as high as 400 ng m^3 within air samples (Reid et al., 1994), and tests indicating high solubility of the arsenic in simulated lung fluids (G. Plumlee, personal communication), these events are viewed as a health hazard for residents of Owens valley (Reheis, 1997).

FIGURE 14 Orbital image of the Aral Sea area taken on June 30, 2001, from Space Station Alpha (Earth Sciences and Image Analysis Laboratory, Johnson Space Center). A major dust storm can be seen, driven by strong westerly winds. The sharp northern margin of the dust pall coincides with the Syr Darya River. This is beyond the area of exposed sea floor sediments, where soil moisture and vegetation cover impede deflation.

C. Case Studies of Non-Industrial Asbestosis

Asbestosis arises from inhalation of asbestos fibers, although conclusive risk assessment has been hampered because the microscopically detectable fibers make up only an insignificant proportion of the total dust burden in lung tissue (Eitner, 1988). Most studies of this interstitial lung disease have been concerned with the impact of asbestos on the health of workers in a wide range of occupations, including mining, manufacturing, and construction, as well as in users of the thousands of commercial products that contain asbestos primarily because of its insulating qualities. However, cases of non-occupational asbestosis have been reported in several countries in Europe and around the Mediterranean, including Czechoslovakia, Austria, Bulgaria, Greece, and Turkey.

In central Turkey, inhalation of agricultural soils rich in tremolite (a common fibrous amphibole found in contact-metamorphosed impure calcareous rocks) and erionite (most commonly arising from alteration of volcanic rocks) is responsible for an endemic malignant pleural mesothelioma. Incidence of this disease is specific to certain villages around which the soils contain one or both of these minerals.

Incidence of pleural plaques, associated with mesothelioma, was also found in residents of northern Corsica who had no history of occupational contact with asbestos. The percentage of 1721 subjects shown by radiographs to have bilateral pleural plaques was 3.7% for those born in northeastern Corsica compared to only 1.2% for those born in the northwest. The rocks of the northeast are rich in serpentine, asbestos, and chrysotile, but this region is separated from the northwest part of the island by a mountain barrier. A clear excess of subjects with bilateral plaques born in villages close to asbestos outcrops was shown (94.6% for affected subjects born in the northeast compared to only 5.4% of subjects born in unexposed villages; Boutin et al., 1986). Preliminary data indicated high levels of chrysotile fibers in the atmosphere, suggesting that incidence of the disease arises directly from inhalation, in an environment in which asbestos exposures and patients with plaques are juxtaposed.

In 1999, public concern led to investigation of a vermiculite mine in Libby, Montana, USA, following its closure in 1990 after more than a century of operation. It had been found that the vermiculite, a micaceous mineral widely used in the insulation of buildings, was contaminated with the tremolite-actinolite form of asbestos. In the alkalic intrusive complex at Libby, the amphiboles are a product of hydrothermal alteration of pyroxenites which also occur as hydrothermal veins cutting across the igneous rocks (G. Plumlee, personal communication). Investigations designed to determine the extent of the impact upon human health arising from occupational links as well as any non-occupational effects arising from activities such as gardening and use of unpaved roads included testing of more than 7000 people over 18 years old in the years 2000 and 2001. This involved interviews, medical history, chest x-ray, and lung function (spirometry) tests. The results showed that radiographic pleural and interstitial abnormalities were present in 51% of former mine workers. The risk of such abnormalities increased with age and with increasing length of residence in the Libby area. The odds of finding pleural abnormalities were stated to be 1.7–4.4 times greater (depending on age) in the case of former mine workers compared to residents with no mine connection, although the incidence of abnormalities in the latter group (3.8%) was higher than for groups within the United States with no known asbestos exposure (range 0.2%–2.3%; United States Environmental Protection Agency, 2003).

D. Tuberculosis

It has been suggested that the incidence of pulmonary tuberculosis in dryland environments may be linked to non-occupational silicosis. Sunlight and aridity are antipathetic to the tubercle bacilli and droplet transmission of pulmonary tuberculosis is favored by lack of sunlight, higher humidity, and overcrowding. However, data from the Thar Desert, India, presented by Mathur and Choudhary (1997), show a prevalence of tuberculosis in desert areas of Rajasthan about 25% higher than the non-desert parts. The presence of radiographically determined evidence of non-occupational silicosis in the desert people offers some support for the hypothesis that silicosis may be an important factor in the higher prevalence of tuberculosis in this desert.

IV. Conclusions

Aerosol mineral dusts affect human health as a result of inhalation and retention of the finest fractions derived directly from source sediments and indirectly from disturbance of surface layers of loess, a geological formation consisting primarily of wind-lain minerogenic dust. The geological and meteorological study of dust sources, sinks, transport, and geochemistry provides a foundation for improved understanding of the extent and magnitude of the impacts of natural minerogenic aerosols on human health.

The pathological effects of prolonged exposure to natural aerosol dust have been recognized in a general way since ancient times, but the number of modern studies of the pneumoconioses outside occupation-specific contexts remains small. The specific health effects of direct inhalation of high concentrations of fine minerogenic dusts, generated by natural deflation from loose, poorly bound soil surfaces, including those exposed by accelerated erosion of weak geological formations such as loess, thus remain rather poorly known and relatively little researched. Knowledge of many of the suspected linkages involved is incomplete, and so is inferential to varying degrees. The magnitude of the world's population affected by inhalation of fine mineral aerosols can only be estimated, although it is likely to number millions of people in the middle latitude desert zone especially across Eurasia between the eastern Mediterranean and the Yellow Sea. Given the progressive improvement and sensitivity of remotely sensed information derived from the several types of orbiting satellite platforms, there is a need for greater investment in improved "ground truth" systems. Obtaining the necessary data on the nature of the "dust climate" and degrees of dust exposure in susceptible environments requires the application of appropriate geological and meteorological methods of monitoring and analyzing dust. This should include regular, standardized measurement and collection and analysis of dust concentrations in the lower atmosphere as a routine component of the meteorological observation systems already operated by most countries. Systematic research programs designed to quantify the respiratory health status of people in the same environments but with contrasting dust exposure potential, and taking full account of other risk factors including those of anthropogenic origin (occupational conditions, cigarette smoking, lifestyle, etc.), will be needed to complement the environmental monitoring.

The impact of trace elements on human health by way of the indirect pathway through soils and groundwater, as found in the loess and loess-like sedimentary accumulations and associated soil types within and adjacent to the great dryland zones of the world, has received much more attention than the direct and indirect pathways considered here, as shown elsewhere in this volume.

Finally, some account of way of life must be considered as a factor in any assessment of the health impact of respirable mineral dust because it directly affects dust generation, re-suspension, and inhalation in many of the world's drylands. Loess and loessic alluvium are abundant and easily applied to building materials used widely in Asia; the predominantly flat-roofed dwellings require only small amounts of wood to complete them. Traditionally, many such houses use small interior kitchens for cooking in winter, with open fires and some primitive chimneys. Although the situation is now slowly changing, domestic burning of dried cattle-dung, wood, and (rarely) low-grade coal is still common, for example, in Ladakh and the Hexi Corridor in northwestern China. To the smoky atmosphere in such confined environments is added fine, re-suspended loessic dust raised by sweeping the dried loess floors. This is mixed with varying concentrations of cigarette smoke. Such high dust concentrations in the home place females at relatively higher risk than males. This may be further enhanced by additional exposure to field dust in areas, such as the tributary valleys of the Indus (northernmost Pakistan), in which females also play a primary role in cultivating the fine silty soils. Such complexity renders the design of a set of strategies for amelioration, if not prevention, a formidable task. It is yet another, perhaps unsung, addition to the challenges posed particularly by many countries in the developing world, with a bearing on the lives of many tens of millions of people.

SEE ALSO THE FOLLOWING CHAPTERS

Chapter 2 (Natural Distribution and Abundance of Elements) · Chapter 11 (Arsenic in Groundwater and the Environment) · Chapter 19 (The Ecology of Soil-Borne Human Pathogens) · Chapter 22 (Environmental Medicine) · Chapter 23 (Environmental Pathology)

FURTHER READING

Bar-Ziv, J., and Goldberg, G. M. (1974). Simple Siliceous Pneumoconiosis in Negev Bedouins, *Arch. Environ. Health*, 29, 121.

Boutin, C., Viallat, J. R., Steinbauer, D. G., Massey, D. G., and Mouries, J. C. (1986). Bilateral Pleural Plaques in Corsica: A Non-Occupational Asbestos Exposure Marker, *Eur. J. Resp. Dis.*, 69, 4–9.

Bryant, R. G. (2003). Monitoring Hydrological Controls on Dust Emissions: Preliminary Observations From Etosha Pan, Namibia, *Geogr. J.*, 169, 131–141.

Derbyshire, E. (2001). Geological Hazards in Loess Terrain, with Particular Reference to the Loess Regions of China, *Earth Sci. Rev.*, 54, 231–260.

Derbyshire, E. (Ed.) (2003). Loess and the Dust Indicators and Records of Terrestrial and Marine Palaeoenvironments (DIRTMAP) database, *Quat. Sci. Rev.*, 22 (18, 19).

Derbyshire, E., Meng, X. M., and Dijkstra, T. A. (Eds.) (2000). *Landslides in the Thick Loess Terrain of Northwest China*, Wiley, New York.

Derbyshire, E., Meng, X. M., and Kemp, R. A. (1998). Provenance, Transport and Characteristics of Modern Aeolian Dust in Western Gansu Province, China, and Interpretation of the Quaternary Loess Record, *J. Arid Environ.*, 39, 497–516.

Eitner, F. (1988). Substantiating Hazardous Exposure to Asbestos by Examination of Pulmonary Dust, *Int. Arch. Occup. Environ. Health*, 61, 163–166.

Goudie, A. S., and Middleton, N. J. (2001). Saharan Dust Storms: Nature and Consequences, *Earth Sci. Rev.*, 56, 179–204.

Grousset, F. E., Ginoux, P., Bory, A., and Biscaye, P. E. (2003). Case Study of a Chinese Dust Plume Reaching the French Alps, *Geophys. Res. Lett.*, 30, 6, 10.1029/2002GL016833.

Harrison, S. P., Kohfeld, K. E., Roelandt, C., and Claquin, T. (2001). The Role of Dust in Climate Changes Today, at the Last Glacial Maximum and in the Future, *Earth Sci. Rev.*, 54, 43–80.

Hirsch, M., Bar-Ziv, J., Lehmann, E., and Goldberg, G. M. (1974). Simple Siliceous Pneumoconiosis of Bedouin Females in the Negev Desert, *Clin. Radiol.*, 25, 507–510.

Houghton, J. T., Ding, Y., Griggs, D. J., Noguer, M., van der Linden, P. J., Dai, X., Maskell, K., and Johnson, C. A. (2001). *Climate Change 2001: The Scientific Basis*, Cambridge University Press, England.

Kohfeld, K. E., and Harrison, S. P. (2001). DIRTMAP: The Geological Record of Dust, *Earth Sci. Rev.*, 54, 81–114.

Mathur, M. L., and Choudhary, R. C. (1997). Desert Lung in Rural Dwellers of the Thar Desert, India, *J. Arid Environ.*, 35, 559–562.

Middleton, N. J., Goudie, A. S., and Wells, G. L. (1986). The Frequency and Source Areas of Dust Storms. In *Aeolian Geomorphology* (W. G. Nickling, Ed.), Allen and Unwin, New York.

Norboo, T., Angchuk, P. T., Yahya, M., Kamat, S. R., Pooley, F. D., Corrin, B., Kerr, I. H., Bruce, N., and Ball, K. P. (1991). Silicosis in a Himalayan Village Population: Role of Environmental Dust, *Thorax*, 46, 341–343.

Pendergrass, E. P. (1958). Silicosis and a Few Other Pneumoconioses: Observations on Certain Aspects of the Problems with Emphasis on the Role of the Radiologist, *Am. J. Roentgenol.*, 80, 1–41.

Policard, A., and Collet, A. (1952). Deposition of Silicosis Dust in the Lungs of the Inhabitants of the Saharan Regions, *Arch. Indust. Hyg. Occupat. Med.*, 5, 527–534.

Prospero, J. M. (1999). Long-Term Measurements of the Transport of African Mineral Dust to the Southeastern United States: Implications for Regional Air Quality, *J. Geophys. Res.*, 104, (15)917–(15)927.

Prospero, J. M., Ginoux, P., Torres, O., Nicholson, S. E., and Gill, T. E. (2002). Environmental Characterization of Global Sources of Atmospheric Soil Dust Identified with the NIMBUS-7 TOMS Absorbing Aerosol Product, *Rev. Geophys.*, 40, 1, 1002, doi: 10.1029/2000RG000095, 2002.

Pye, K. (1987). *Aeolian Dust and Dust Deposits*, Academic Press, London.

Pye, K., and Zhou, L. P. (1989). Late Pleistocene and Holocene Aeolian Dust Deposition in North China and the Northwest Pacific Ocean, *Palaeogeogr. Palaeoclimatol. Palaeoecol.*, 73, pp 11–23.

Reheis, M. C. (1997). Dust Deposition Downwind of Owens (Dry) Lake, 1991–1994: Preliminary Findings, *J. Geophys. Res.*, 102, (D22), 25,999–26,008.

Reid, J. S., Flocchini, R. G., Cahill, T. A., Ruth, R. S., and Salgado, D. P. (1994). Local Meteorological, Transport, and Source Aerosol Characteristics of Late Autumn Owens Lake (Dry) Dust Storms, *Atmos. Environ.*, 28, 1699–1706.

Saiyed, H. N. (1999). Silicosis–an Uncommonly Diagnosed Common Occupational Disease, *Indian Counc. Med. Res.*, 29, 1–17.

Saiyed, H. N., Sharma, Y. K., Sadhu, H. G., Norboo, T., Patel, P. D., Patel, T. S., Venkaiah, K., and Kashyap, S. K. (1991). Non-Occupational Pneumoconiosis at High Altitude Villages in Central Ladakh, *Br. J. Indust. Med.*, 48, 825–829.

Shinn, E. A., Smith, G. W., Prospero, J. M., Betzer, P., Hayes, M. L., Garrison, V., and Barber, R. T. (2000). African Dust and the Demise of Caribbean Coral Reefs, *Geophys. Res. Lett.*, 27, 3029–3032.

Snider, D. E. (1978). The Relationship Between Tuberculosis and Silicosis, *Am. Rev. Resp. Dis.*, 118, 455–460.

Sun, J. (2002a). Provenance of Loess Material and Formation of Loess Deposits on the Chinese Loess Plateau, *Earth Planet. Sci. Lett.*, 203, 845–859.

Sun, J. (2002b). Source Regions and Formation of the Loess Sediments on the High Mountain Regions of Northwestern China, *Quat. Res.*, 58, 341–351.

Tapp, E., Curry, A., and Anfield, C. (1975). Sand Pneumoconiosis in an Egyptian Mummy, *Br. Med. J.*, 3; 2(5965), 276.

United States Environmental Protection Agency (2003). http://www.epa.gov/region8/superfund/libby/lbybkgd.html.

Wagner, G. R. (1997). Asbestosis and Silicosis, *Lancet*, 349, 1311–1315.

Wang, Z., Ueda, H., and Huang, M. (2000). A Deflation Module for Use in Modelling Long-Range Transport of Yellow Sand Over East Asia, *J. Geophys. Res.*, 105(D), 26,947–26,959.

Wiggs, G. F. S., O'Hara, S. I., Wegerdt, J., van der Meers, J., Small, I., and Hubbard, R. (2003). The Dynamics and Characteristics of Aeolian Dust in Dryland Central Asia: Possible Impacts on Human Exposure and Respiratory Health in the Aral Sea Basin, *Geogr. J.*, 169, 142–157.

World Health Organization (1997). Monograph on the Evaluation of Carcinogenic Risks to Humans of Silica, some Silicates, Coal Dust and Para-Aramid Fibrils, IARC Monographs. 68, World Health Organization, Geneva, Switzerland.

Xu, X. Z., Cai, X. G., and Men, X. S. (1993). A Study of Siliceous Pneumoconiosis in a Desert Area of Sunan County, Gansu Province, China, *Biomed. Environ. Sci.*, 6, 217–222.

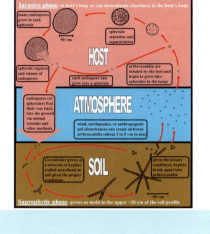

CHAPTER 19

THE ECOLOGY OF SOIL-BORNE HUMAN PATHOGENS

MARK W. BULTMAN
United States Geological Survey

FREDERICK S. FISHER
University of Arizona

DEMOSTHENES PAPPAGIANIS
University of California–Davis

CONTENTS

I. INTRODUCTION

The surface of the Earth, with the exception of the oceans and polar ice caps, is in large part covered with a marvelously complex layer of material called soil, from which we derive a host of useful products including fiber, fuels, building materials, animal forage, many mineral commodities, natural medicines (including antibiotics), and most of our food supply. Soil is teeming with life and is home for a huge array of living organisms. The vast majority of these living organisms are microbes that are ubiquitous on Earth (Table I). They occur in all soils, salt and fresh water, the harsh climates of the Artic and Antarctic, adjacent to deep-sea hydrothermal vents associated with spreading zones between tectonic plates, throughout the atmosphere, and deep below the surface of the Earth in oil wells where they have been isolated from the surface environment for millions of years (Staley, 2002, p. 13).

Some soil-dwelling microbes are pathogenic for humans, including protozoa, fungi, bacteria, and also viruses and the less well understood prions, both of which require a plant or animal host for their survival.

TABLE I.   Typical Numbers of Soil Organisms in Healthy Ecosystems

		Agricultural Soils	Prairie Soils	Forest Soils
Bacteria	*Per teaspoon (one gram dry)*	100 million to 1 billion.	100 million to 1 billion.	100 million to 1 billion.
Fungi		Several yards. (Dominated by vesicular-arbuscular mycorrhizal (VAM) fungi).	Tens to hundreds of yards. (Dominated by vesicular-arbuscular mycorrhizal (VAM) fungi).	Several hundred yards in deciduous forests. One to forty miles in coniferous forests (dominated by ectomycorrhizal fungi).
Protozoa		Several thousand flagellates and amoebae, one hundred to several hundred ciliates.	Several thousand flagellates and amoebae, one hundred to several hundred ciliates.	Several hundred thousand amoebae, fewer flagellates.
Nematodes		Ten to twenty bacterial-feeders. A few fungal-feeders. Few predatory nematodes.	Tens to several hundred.	Several hundred bacterial- and fungal-feeders. Many predatory nematodes.
Arthropods	*Per cubic foot*	Up to one hundred.	Five hundred to two thousand.	Ten to twenty-five thousand. Many more species than in agricultural soils.
Earthworms		Five to thirty. More in soils with high organic matter.	Ten to fifty. Arid or semi-arid areas may have none.	Ten to fifty in deciduous woodlands. Very few in coniferous forests.

From NRCS, 1999.

Helminths (which are in the mesofauna size class) are included in this chapter because of their importance as human pathogens and because of the numbers of pathogenic viruses and bacteria associated with them that can be introduced into the soil environment. Over 400 genera of bacteria have been identified with possibly as many as 10,000 species and, with the exception of viruses, they are in most cases more abundant than any other organism in soils. The number of bacteria that can be cultured in the laboratory is probably less than 1%; thus their actual diversity is probably much greater (Paul & Clark, 1996; Coyne, 1999). Fortunately, relatively few of this vast population of microbes are pathogenic for humans. For example, of the approximately 100,000 species of fungi currently recognized (University of Leicester, 1996), only about 300 are known to cause human disease (McGinnis, 1998).

Nonetheless, soil-borne human pathogens have extracted an unbelievable toll in disfigurement, suffering, blindness, death, and medical costs from the human race throughout history, and they will continue to do so for the foreseeable future. Examples are Ascariasis (roundworm) 60,000 deaths in 1993; Schistosomiasis, 200,000 deaths in 1993; and *Clostridium tetani*, killing 450,000 newborns and about 50,000 mothers each year (World Health Organization, 1996). Almost 3 million deaths a year, mostly in developing countries, are attrib-

uted to diarrheal diseases, and many are contracted from microbes introduced into the soil via fecal waste and then ingested (NIAID, 2000).

II. SOIL FUNDAMENTALS

Soil may be defined as that part of the regolith that is capable of supporting plant life. The regolith is the portion of unconsolidated rock material that overlies bedrock and forms the surface of most land. The upper boundary of soils is either air or shallow water, and horizontally, soil boundaries are bodies of deep water, rock outcrops, or permanent ice fields. The lower limit of soil layers is the underlying bedrock where biological activity is severely restricted. Microbes are abundant in the upper parts of soil layers, where organic material is more likely to be present, and decrease in numbers with depth. However, some microbes may occur in the deepest soil layers and in places within the fractured underlying bedrock. The physical properties of soils are described in depth by Brian Alloway in Bioavailability of Elements in Soil in this volume. In this chapter only additional material important to the understanding of soil dwelling microbes is presented.

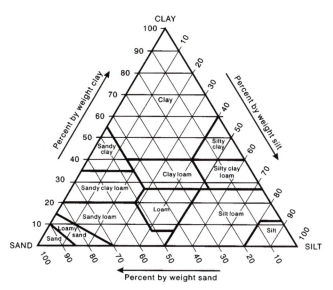

FIGURE 1 U.S. Department of Agriculture soil texture triangle. Sand, silt, and clay are defined by the size of their constituent particles. Sand is composed of particles with sizes ranging from 2.0 to 0.05 mm. Silt is composed of particles from 0.05 to 0.002 mm. Clay is composed of particles less than 0.002 mm in size.

A. Important Physical and Chemical Properties of Soil

Physical properties of soils—texture, porosity, permeability, water-holding capacity, and temperature—are of considerable importance in determining the characteristics and microhabitat utilization of a given soil by microbes. Texture is the term used to describe the relative proportions of sand, silt, and clay-sized particles in a given soil (Figure 1). Texture is important to microbes for several reasons: (1) it is the major factor controlling water holding capacity of a soil, (2) it determines the amount of pore space and the character (size and distribution) of the pore space, (3) it affects the rate of chemical reactions, (4) it is a major control on root penetration by plants, and (5) it controls soil aeration. All of these factors influence the types, distribution, and abundance of different microbes in any given soil profile.

Porosity is the percentage of the volume of a soil not occupied by solids; i.e., the interstices, isolated or interconnected, between individual solid soil particles (either mineral or organic). The availability of water and air are controlled largely by the amount, size, and interconnections between interstices, which are also the habitat for microbes.

Permeability is the ability of a soil to transmit water. Soils that contain a high proportion of sand will have larger continuous pores and will rapidly transmit water and air. In comparison, clay soils, which often have a high porosity because of the small size of individual particles, may have low permeability and transmit water slowly because of poor connectivity between soil interstices and swelling of individual clay particles. Soil water-holding capacity and soil temperature are discussed in the next section.

B. Soil-Forming Factors

Five major factors interact to form soils: parent material, climate, soil organisms, topography, and time.

1. Parent Material

Parent material may be residual or transported igneous, metamorphic, or sedimentary bedrock debris. The mineralogy of parent material is very important in determining the type and amount of clay minerals developed in a given soil profile, which in turn has a profound effect on the types and distribution of soil microbes. Texture and the degree of consolidation of the parent material directly control the movement of water and air within the soil, the rooting ability of plants, and consequently the mobility of many microbes.

2. Climate

Climate characteristics, temperature and moisture in particular, influence the kinds and amount of soil microorganisms, the rate of decomposition of organic material, weathering rates of mineral matter, rates of formation of secondary minerals, biological activity, and the removal, movement, and deposition of materials between different soil layers.

Soil temperatures vary with latitude, altitude, slope aspect, vegetation cover (shading), soil color, and moisture content. Typically the surface temperature of soils ranges from below freezing to as much as 60°C and displays daily and annual cycles. Soil temperature has an important influence on biological processes such as the germination of seeds, nutrient absorption, root growth, and microbial activity, which generally is enhanced at higher temperatures if adequate moisture is available.

Precipitation may result in water on the land surface, water within the soil horizons, or deeper below the surface in ground water. Water that infiltrates the soil will surround soil particles, fill pore spaces, and may eventually move downward into the water table. A quantitative measure of water availability is the soil water potential, which is defined as the amount of work that must be done per unit quantity of pure water in order to transport reversibly and isothermally an infinitesimal quantity of water from a pool of pure water, at a specified elevation and at atmospheric pressure, to the soil water (at the point under consideration) (Soil Science Society of America, 1998).

Informally, soil water potential is the amount of energy that must be expended to extract water from soil. It may be expressed mathematically as:

$$\Psi_{soil} = \psi_g + \psi_m + \psi_s$$

where Ψ_{soil} is the soil water potential, (ψ_g) is the gravitational potential, (ψ_m) is the matric potential, and (ψ_s) is the solute (or osmotic) potential. Water potentials are expressed in megapascals (MPa). One megapascal equals 1×10^6 pascals. One atmosphere equals 0.1013 MPa. The matric potential is related to the attraction of water molecules to solid surfaces. In unsaturated soils it is always negative and becomes more negative as the surface area of the soil increases. Solute potential is associated with the solutes in the soil and is always a negative value. It becomes more negative as soil solutes increase and the soils become more saline. The gravitational potential may be positive or negative depending on where in the soil it is measured relative to some arbitrary reference level. The reference level is generally set in the soil profile at some point (e.g., at the water table) below the soil profile considered, thus the gravitational potential is usually positive in reference to microbes. If the reference level is the soil surface then the gravitational potential is negative for any point below the surface. The osmotic and matric potential together determine the amount of energy that must be expended by microbes to extract water from soil.

In general, microbial growth rates are greatest for Ψ_{soil} near −0.01 MPa and decrease as soils become drier and have correspondingly larger, negative water potentials. Microbial activity also decreases as soils become waterlogged or saturated, which results in Ψ_{soil} values at or near zero. At field capacity the matric potential of the soil water will generally be in the range of −0.01 to −0.03 MPa and the wilting point will occur near −1.5 MPa. Soils vary widely in the size and shape of their pore space and this variation precludes a simple relation between the water content of any given soil and the associated water potential. However, when a soil is saturated and all pores are completely filled with water, the matric potential is zero.

Because few soils have solid particles that are uniform in size, shape, or composition, texture and structure have a large influence on the matric potential of unsaturated soils. For example, the amount and type of clay in a given soil is extremely important, as some types of clay will swell as water is added and shrink as they dry. Matric potential is complicated by these changes. As water drains from a saturated soil containing some clay, its removal is mostly influenced by the larger particles (most often sand-sized particles) and the matric potential determination given above is relatively straightforward. As drying progresses the behavior of the clay particles becomes more important. The clay particles carry a variable negative electrical charge and thus tend to repel one another, and work is necessary to bring them closer together. Drying also causes the clay to shrink due to the loss of water between layers in the clay structure. The force of repulsion between the particles thus determines the matric potential of the remaining water. Detailed discussions of water potential and its effects on soil microbes may be found in Griffin (1972), Harris (1981), and Brown (1990).

Many soil microbes obtain nutrients from water by diffusion and rely on water for mobility, either by swimming, movement by Brownian forces, or flow of water. Void spaces smaller than the diameter of a given organism and pores from which water has been removed act as barriers. As soils dry, the water films around solid particles become thinner, which impedes bacterial and protozoal mobility, limits nutrient availability, and slows nutrient diffusion through cell membranes. If films become discontinuous, then microbes may be trapped or have to move in much more tortuous paths in the remaining water films. In general, in wet unsaturated soils, all major groups of microbes are active and in competition. In drying soils with progressively lower matric potentials, unicellular organisms become less active and ultimately microbial activity in relatively dry soils is confined largely to filamentous organisms, such as fungi that can, through growth of hyphae, utilize water unavailable to bacteria.

3. Soil Organisms

Many organisms have long been recognized as having an important role in the genesis of soils. Microbes play an important role in many soil processes. They convert plant and animal residue into humus and produce compounds that help bind soil into aggregates. They create new organic compounds that are nutrients for other microbes. Many microbes have mutually beneficial associations with higher plant forms and may fix the nitrogen that is needed by these organisms. They may also inhibit the growth of disease causing soil-borne organisms and mediate many soil related chemical reactions.

Soil microbial populations are generally more abundant in surface horizons than in deeper horizons but are not uniformly distributed laterally or with depth (Coyne, 1999, p. 152). Typically, populations occur in localized concentrations associated with various favorable and unfavorable microenvironments throughout the soil profile (a vertical section of the soil which includes all the soil horizons). For example, the presence of plant roots is influential in determining the activity of microbes. The soil area influenced by a plant root is called the rhizosphere. This area (a few millimeters from the root) usually has chemical properties quite different from the bulk of the soil because of the uptake by the plant roots of moisture and nutrients and the secretions and exudates of amino and organic acids, sugars, proteins, and other chemical elements from the roots. Microbes may be up to 100 times more concentrated in the rhizosphere than in soil some distance from roots. Plants and microbes may form symbiotic relationships that are mutually beneficial. Examples are bacteria of the genera *Rhizobium* and *Bradyrhizobium* that convert (fix) atmospheric nitrogen into nitrogen compounds that can be utilized by the plant that in turn reciprocates and supplies carbohydrates to the bacteria. Another example are mycorrhizal fungi that attach to and grow on roots thereby acting as an extension of the root system which greatly facilitates the uptake of water and nutrients by the plant. In exchange, the fungus obtains sugars and other nutrients directly from the plant root.

4. Topography and Time

On the largest scale, climate and vegetation have great relevance to the distribution and formation of various soil types. On a smaller scale, topography modifies the climate and vegetation factors and has a major role in determining the character and development different soil types. The length of time that a given parent material has been exposed to the forces of climate, development of vegetation, and influence of animal life is believed by many soil scientists to be a strong influence on the nature of soil development. And, in general, the degree of development of differentiation between horizons in a soil is related to the age of a soil.

III. SOIL AS A COMPLEX SYSTEM

The study of complex systems has appeared as a separate scientific discipline only over the last 25 years and debate is ongoing as to its precise definition (Flake, 1998; Cowan et al., 1999). However, there seems to be general agreement about some features that all complex systems exhibit:

1. They are collections of many simple nonlinear units that operate in parallel and interact locally with each other to produce behavior that cannot be directly deduced from the behavior of its component units
2. There are numerous units that operate in layers or at different scales
3. Local interactions within layers sometimes produce a global behavior at another scale or layer, which is called emergent behavior
4. The result of emergent behavior is that simple components combine to form more complexity than the sum of its parts (Flake, 1998).

Soils are the interface between the lithosphere, atmosphere, hydrosphere, and biosphere. They are composed of an enormous number of individual constituents, both organic and inorganic, that are highly diverse in form, composition, and purpose. All the constituents adapt and interact with each other and respond chemically, biologically, and physically to processes specific to any of the abovementioned Earth domains. Rates of reaction to changes in environmental processes are not constant and rarely reach equilibrium. Many processes (perhaps most?) are nonlinear and thermodynamically irreversible. Soils are also open systems that exchange matter, energy, and organisms across numerous boundaries. This incredibly diverse, heterogeneous, dynamic, reactive, and adaptable nature makes the study and description of any aspect of soils a daunting task. Perhaps the overriding rule about soils is that they are constantly changing and adapting to their environment, and that the aggregate behavior of a soil cannot be predicted by summing up the behavior of its individual parts. These changes may be relatively rapid and readily visible or measurable (e.g., microbial growth, movement of soluble components, changes in water potential, seasonal changes in salinity, etc.) or they may be longer term taking place over years, decades, centuries, or millennia (e.g., weathering of silicate minerals and formation of clays, destruction of agricultural lands due to salt accumulation in irrigated areas, vegetation changes due to global warming or cooling, continental erosion, and deposition processes).

Most soil processes and microbial responses are gradational as are many boundaries between soil types, horizons, and chemical and physical properties. This lack of sharply defined volumes bedevils soil mapping and descriptions because spatial gradients in both process and products are at best poorly represented and most often ignored. One way of dealing with diverse data bounded by gradients is through the use of fuzzy logic; a method that allows the grouping of data into continuous sets in which membership can range between 0 and 1. Classes of data may then be expressed on an intermediate scale and spatial variations accounted for. The use of fuzzy logic (by the authors) to describe the habitat of a soil-borne human pathogen (*Coccidioides*) is described in the case study (Section XIII).

IV. PATHOGEN CLASSIFICATION

Classification is necessary to develop and discuss the connections and relationships between geology, soils, and soil-borne human pathogens. Soil fauna have been classified by several criteria including body size, degree of presence in soil, habitat preference, and activity (Wallwork, 1970). Soil-borne human pathogens include many biological entities, most of which contain widely diverse members. Thus, it is useful to organize different pathogenic microbes into groups on the basis of the character of their soil residency. Any type of pathogen that is present in the soil, for any reason or time period, is included in the following classification in which two factors are essential: presence in soil (for any reason), and human pathogenicity. The classification terms of permanent, periodic, and transient from Wallwork (1970) were adopted, but redefined. An additional term, incidental, was added and defined.

Included in our classification are several human pathogens that have been classified as water- or food-borne. Many pathogens classified as water-borne are the result of animal or human fecal, urine, or other wastes introduced first into the soil environment and subsequently washed into surface water or incorporated into groundwater. Infection then follows by contact with, or consumption of, contaminated water. Similar circumstances take place with some food-borne diseases. The pathogen is again introduced into the soil via defecation

or through some type of contaminated waste material and then may be consumed on unwashed raw fruit or vegetables.

Many pathogens have complex life cycles that may involve hosts in which they live and reproduce, biological vectors (insects, animals) and physical vectors (wind, water) for transport, and reservoirs to exist in during adverse environmental conditions. Soils provide these features to a wide variety of microbes and thus, the following classification to illustrate the importance of soil attributes and processes in the understanding the pathology of numerous diseases is offered.

In our classification some microbes may be classified as both transient and incidental. An example is *Giardia lamblia*, which can be introduced naturally into the soil through animal feces but can also be introduced anthropogenically via sewage systems, allowing for its dual classification. Four soil-borne pathogen residencies are defined below.

1. Permanent: Pathogenic organisms that are permanent soil inhabitants and can complete their entire life cycle within a soil environment. Examples are the bacteria *Clostridium botulinum, C. tetani, Burkholderia pseudomallei*, and *Listeria monocytogenes*. Also included are dimorphic organisms if one of their morphologic forms is capable of living and reproducing completely within the soil. Examples of permanent dimorphic soil pathogens are the fungi *Coccidioides* and *Histoplasma capsulatum*.

2. Periodic: Pathogenic organisms that require part of their life cycle to be completed within a soil environment on a regular, recurring basis. Examples are spores of *Bacillus anthracis* and the eggs laid in the soil by tick vectors that contain the bacterium *Rickettsia rickettsii*. Additional examples are eggs of the helminths *Ancylostoma duodenale* and *Necator americanus* (hookworms).

3. Transient: Pathogenic organisms that may naturally occur in soil, but the soil environment is not necessary for the completion of the organism's life cycle. Examples are cysts of a protozoan parasite *G. lamblia* and viruses in the genus *Hantavirus* that are introduced into soil environments worldwide via urine and feces of rodent vectors. Also included are *Leptospira*, a bacterium shed in urine of animals on soil, skin, and water and spores of the bacterium *Coxiella burnetii*.

4. Incidental: Pathogenic organisms introduced into the soil via anthropogenic means such as in sewage

sludge, waste water, septic systems, unsanitary living conditions, biologically toxic spills, dumping of biohazardous waste materials, and release of biological warfare agents. Examples of viruses are enterovirus poliovirus (etiological agent of polio), enterovirus Coxsackie A and -B, and enterovirus hepatitis A. Length of survival time and virulence depend on numerous physical and chemical factors of the soil and the effluent, and it can range from hours to years.

V. Gateways for Infection

A soil-borne pathogen must come in physical contact with and establish itself in a human to cause disease. There are many ways in which this can be accomplished. First, soil-borne pathogens need to start this process from the soil. From the pathogen residency classification given above, the pathogen may be in the soil on a permanent, periodic, transient, or incidental basis. Permanent and periodic soil-borne organisms occur in the soil through natural pathways. Their presence in the soil is a normal part of the life cycle of a particular organism. Transient soil-borne organisms generally are incorporated into the soil via the excrement of wild or domestic animals. Residency in the soil is not necessary for their survival, but they are deposited there by natural means. Man places incidental organisms in the soil, generally in the form of solid human waste.

Probably the most common method of introducing soil-borne pathogens into the human body is through ingestion. Ingestion of soil-borne microbes is generally accidental, but in some cultures, geophagia is practiced. This subject is covered extensively in Geophagy and the Involuntary Ingestion of Soil, this volume.

In many countries, human waste is a valuable commodity that cannot be wasted. The practice of spreading human waste, or night soil, is found in many cultures and is an efficient means of introducing many human pathogens into the soil. The concentration of viruses in feces can be very high: e.g., enterovirus, 10^6 virus particles per gram; hepatitis A, 10^9 virus particles per gram; and rotavirus, 10^9 virus particles per gram (Sobsey & Shields, 1987). In the United States and other industrialized nations, the application of human waste products in the form of treated sewage sludge is also practiced, but regulated. In the United States 24% of sludge is applied on or just below the land surface as fertilizer (Bertucci et al., 1987). Federal pollution control acts regulate the microbial content of the sludge and access to treated land by people, grazing animals, and use for crops grown for human consumption. In the United States, 90–99% of viruses are removed by primary and secondary sewage treatment (Bertucci et al., 1987). Raw sewage in the United States can contain concentrations as high as 100,000 infectious units per liter (Sobsey & Shields, 1987). A 99% reduction still leaves 1000 infectious units per liter. In countries where no such standard exists, pathogens from human waste have a much easier time entering soil and the environment. Many bacteria are also known to survive the sewage treatment process. If pathogens are ingested that became soil-borne as human waste, the gateway of infection will be referred to as the fecal-soil-oral gateway.

Other oral (ingested) gateways for soil-borne pathogens require an intermediate host. For example, the trematode *Schistosoma mansoni* has a transient soil residency. In soil its eggs hatch into larvae, which can infect an intermediate host. This intermediate host releases infective larvae into its environment, water. Consumption of infected water results in the pathogen moving to the human host. The cestode *Taenia saginata* also has a transient soil residency. Cattle ingest soil infected with this pathogen and develop cysts in their muscles. When humans consume incompletely cooked beef, they can become infected with the pathogen.

A respiratory gateway for soil-borne pathogens occurs when soil-borne pathogens are inhaled as airborne dust. Airborne dust is soil in motion. Human activity or natural forces can cause dust emission for every type of soil given the proper environmental conditions. Each year, several million tons of airborne soil makes its way from Africa to the Americas, Europe, and the Middle East as dust. Asian dust crosses the Pacific Ocean and dust from the southwestern United States can make its way to Canada (Raloff, 2001). Every place on Earth receives dust from both local and distant sources and along with the minerals that make up the inorganic part of the dust ride fungi, bacteria, and viruses. African dust in the Caribbean has been shown to contain the fungus *Aspergillus sedowii*. Between 1973 and 1996 the Queen Elizabeth Hospital in Barbados documented a 17-fold increase in asthma attacks (NASA, 2001). This time frame corresponds to an increased period of dust production in Africa due to drought conditions.

Microbes rarely penetrate the intact skin. It has been speculated that the bacterium *Francisella tularensis*, the etiological agent of tularemia, may be capable of penetrating unbroken skin, but there is little support for this. It may, however, enter through the thinner epithelium of the conjunctiva. The nematode *Strongyloides stercoralis* is capable of burrowing its way through healthy skin. The bacterium *C. tetani* enters the body if contaminated soil makes contact with a break in the skin.

VI. SOIL-BORNE HUMAN PATHOGENIC HELMINTHS AND MICROBES

There are enormous numbers of helminths and microbes in the soil but only a small number of these are pathogenic to humans. Helminths are multicellular parasitic worms with complex reproductive systems and life cycles. Microbes are microscopic or submicroscopic organisms. For our purposes these include protozoa, fungi, bacteria, viruses, and, possibly, the agents (prions) of transmissible spongiform encephalopathies (TSE).

A. The Importance of Soil-Borne Human Pathogenic Helminths and Microbes

Diseases where the responsible pathogen spends some or part of its life in the soil are a major concern. In 1995, the World Health Organization (WHO) estimated that there were 3.7 million deaths worldwide from food-, water-, and soil-borne pathogens. More than 2.4 million of these deaths were children under the age of 5. In that same year, there were over 4 billion (4×10^9) new cases of these diseases (World Health Organization, 1996). A soil-borne bacterium, *C. tetani*, was responsible for almost half a million deaths to newborns and 50,000 mothers each year due to tetanus (World Health Organization, 1996). Billions of infections and over a million deaths occur each year from soil-borne helminths and protozoa (MacLean, 2002).

Some soil-borne human pathogens are called frank pathogens because they are capable of infecting anyone. An example is *C. tetani*. Most soil-inhabiting pathogens are opportunistic pathogens; their main targets of opportunity are individuals with a suppressed immune system. These may include young children, the malnourished, HIV-positive individuals, individuals who have had transplant surgery, and the elderly.

B. The Distribution of Helminths and Microbes in Soils

The distribution of helminths in soils fluctuates greatly with season, climate, and amount of organic matter in the soil. Helminths typically prefer warm, moist soils with plentiful organic material. During favorable conditions, most helminths are found in the upper 10–15 cm of the soil profile. They may move vertically in the soil profile in response to seasonal weather changes.

As discussed in the introduction, microbes are found virtually everywhere on the planet. They cannot move very far on their own, but their small size allows them to be distributed globally by wind, water, animals, and humans.

On a small scale, the type of microbes in the soil are determined by the types of soils and the local climate. *Penicillium* is a fungus that is found in both warm and cold soils, whereas *Aspergillus* (a fungus) grows better in warm soils. *Fusarium*, a fungus that causes banana wilt, does not thrive in soils with the clay mineral smectite (Paul & Clark, 1996). *Coccidioides*, a soil-borne fungus and the etiological agent of coccidioidomycosis, is found in dry, alkaline soils with a soil texture that includes large percentages of silt and very fine sand.

Within a given soil, the distribution of microbes on soil surfaces is uneven or irregular. Microbes are found clustered where conditions are favorable for growth and there may be a relatively large distance between the clusters. The determining factors for the locations of these clumps include the size and distribution of pore spaces in the soil, the soil water potential, the types of gases present in the soil pore spaces, the distribution of organic debris, and the local mineralogy of the soil. The influence of water potential on microbes has been discussed previously. Soil gases tend to be enriched in carbon dioxide and depleted in oxygen due to biological activity. Also, even in well-aerated soils, water may be blocking many pore spaces limiting the diffusion of oxygen in and carbon dioxide out. Some carbon dioxide is dissolved in soil water and produces carbonic acid, which helps to dissolve soil minerals (Coyne, 1999). Many microbes have an affinity for clay minerals within soils. Clays have large surface areas, are chemically reactive, and have a net negative charge. They are a source of inorganic nutrients, such as potassium and ammonia, and modify the chemical and physical habitat immediately around them. Also, clays adsorb water making it less available for microbes.

Most soil-dwelling microbes are found in the upper 8 cm of the soil profile, and their numbers decrease significantly below 25 cm depth (Coyne, 1999). The main reason is that the soil organic content, including root density, tends to decline with soil depth. Also, in alluvial soils the microbial populations fluctuate with textural changes in the soil profile. Microbes are more numerous in clay layers than in sand or coarser size materials. There is an increase in the numbers of microbes in the unsaturated zone directly above water table (Coyne, 1999).

The population of soil-dwelling microbes is also affected by human activities. Global warming (anthropogenic and natural) is changing the characteristics of soils worldwide. Acid rain has changed the pH and mineralogy of soils and has affected the microbial populations of these soils. Clear-cutting can increase microbial populations in soils due to an increased supply of dead organic material (Paul & Clark, 1996). Microbial populations are lower in tilled soils and compacted soils. Tilled soils are less moist than non-tilled and compacted soils have reduced pore space and aeration (Coyne, 1999).

VII. SOIL-BORNE HUMAN PATHOGENIC HELMINTHS

Human pathogenic helminths are all parasites and have man as their definitive host. Most inhabit the human intestines at some point in their life cycle; however, some are systemic in the lymph system or in other tissue.

The exact taxonomy of the helminths is under continuing discussion and will not be addressed here. They can be grouped into the nematodes (including hookworms, roundworms, whipworms, and pinworms), the trematodes (flukes), and the cestodes (tapeworms). Human diseases caused by cestodes and trematodes are saprozoonoses. These are zoonotic diseases where the transmission of the disease requires a non-animal development site or reservoir. In many cases this site is the soil. Table II presents a summary of the soil-borne human pathogenic helminths discussed here.

A. Selected Soil-Borne Human Pathogenic Nematodes

Most human pathogenic helminths are nematodes. There are approximately 10,000 species of nematodes and approximately 1000 of these are found in soils. Most of the soil-dwelling nematodes are found in the

TABLE II. Selected Soil-Borne Human Pathogenic Helminths

Nematodes
Ancylostoma duodenale and *Necator americanus* (hookworm)
Ascaris lumbricoides (roundworm)
Enterobius vermicularis (pinworm)
Strongyloides stercoralis (roundworm)
Toxocara canis (roundworm)
Trichuris trichiura (whipworm)

Trematodes
Schistosoma spp. (fluke)

Cestodes
Taenia saginata (tapeworm)
Taenia solium (tapeworm)

upper 10 cm of the soil profile. Desert soils have the lowest population density of nematodes (about 400,000/m^2) with the highest densities occurring in permanent pastures (up to 10,000,000/m^2). Although there are large numbers of soil-dwelling nematodes, they do not contribute to a large percentage of the biomass (Coyne, 1999).

Nematodes are generally microscopic and transparent or translucent, ranging in length from about 0.05–2 mm. They have cylindrical unsegmented bodies with a bilateral symmetry. The body is covered with a tough cuticle. Nematodes have internal organs including a digestive, excretory, nervous, and muscular systems. They develop and grow by molting (shedding the cuticle). In almost all cases individual nematode species have sexual organs and separate sexes. Their life cycle begins with the development of an egg (most lay eggs in soil), followed by egg fertilization, embryonic growth in the egg, hatching and development of larvae, and molting and growth into an adult. Nematodes can produce five to six generations a year.

Nematodes are generally associated with water films in soils. These films partially fill the interstitial spaces between soil particles with water and are held in place by adhesion and cohesion. If soil becomes dry, many nematodes can form cysts or enter a dormant period allowing them to survive. They do well in warm organic-rich soils with a neutral pH, but can tolerate many soils. Most nematodes are predators or saprophytes. In this role they regulate microbial populations in soil by consuming up to 5000 bacteria per minute (Coyne, 1999).

The toll from nematode-caused human disease is staggering. Billions of people, rich and poor, are infected throughout the world each year, which causes much discomfort and suffering. There are over 130,000 deaths worldwide from nematode infections annually (MacLean, 2002). Many infections are the result of a lack of appropriate personal hygiene or from poor sanitation. Most nematodes infect a human host by being ingested and some infect by entering through the skin. Almost all soil-borne human pathogenic nematodes inhabit the intestines. One exception is the nematode that causes trichinosis. In this case the mature nematodes live in the small intestines but, after a short period of time, they release larvae that migrate to striated muscle tissue and form cysts. Nematodes are the etiological agent of numerous human diseases and have complex life cycles involving soil, water, and animals as illustrated in the following discussion.

1. Ancylostoma duodenale and Necator americanus

Two species of hookworms are capable of causing human intestinal infection, generally called ancylostomiasis. These are *Ancylostoma duodenale* and *Necator americanus*. *A. duodenale* is found in parts of southern Europe, North Africa, northern Asia, and parts of western South America. *N. americanus* is found in Central and South America, southern Asia, Australia, and the Pacific Islands. Worldwide, there are approximately 1.2 billion cases annually of human hookworm infections (Cambridge University Schistosomiasis Research Group, 2002). About 100 million of these involve a serious infection that creates a continuous loss of blood leading to chronic anemia. Less severe cases usually include mild diarrhea and cramps. Serious hookworm infection can create major health problems for newborns, children, pregnant women, and the malnourished.

Ancylostomiasis is a disease usually associated with unsanitary conditions. Hookworm eggs pass from the feces of infected humans to the soil. The eggs must be in the terrestrial environment to hatch and the soil residency of *A. duodenale* and *N. americanus* is periodic. Individuals with a major infection can excrete 2000 eggs per gram of feces (National Institutes of Health, 2001). Once exposed to air the eggs will develop rapidly in the upper few centimeters of moist warm soil. They hatch into larvae after a few days and feed on bacteria and organic matter. After about five days they molt and form the infectious form of the larvae. During cool damp periods, the larvae may come to the surface and extend their bodies into the air searching for a host. If they come into contact with human skin they attach and burrow in. The larvae are then transported in the blood to the lungs where they burrow into the airspace then migrate or are coughed up in the bronchi and trachea and are swallowed into the gut. Once in the intestine, they attach themselves to the wall of the small intestine and mature to adulthood which causes damage by blood ingestion. Thus, infection in conjunction with poor nutritional status can induce chronic anemia. Female *N. americanus* hookworms can produce 10,000 eggs daily and female *A. duodenale* can produce 20,000 eggs daily (Cambridge University Schistosomiasis Research Group, 2002).

2. Ascaris lumbricoides

Ascaris lumbricoides is a large roundworm that causes ascariasis, an infection of the small intestines. There are over 1.5 billion new cases of ascariasis annually; about 210 million of them are symptomatic (Cambridge University Schistosomiasis Research Group, 2002). Ascariasis is the most common helminthic infection and is distributed worldwide. The highest prevalence is in tropical and subtropical regions and in areas with inadequate sanitation.

A. lumbricoides is the largest nematode (roundworm) parasitizing the human intestine. Adult females can be 20–35 cm long and adult males can be 15–30 cm long (CDC, 2002). Although infections may cause stunted growth, acute symptoms are usually not caused by adult worms. High worm burdens may cause abdominal pain and intestinal obstruction. Migrating adult worms may cause symptomatic occlusion of the biliary tract or oral expulsion. During the lung phase of larval migration, pulmonary symptoms can occur (cough, dyspnea, hemoptysis, eosinophilic pneumonitis).

The female parasite may produce 240,000 eggs per day (CDC, 2002) which are passed in the feces. Fertile eggs may remain viable in the soil for many years if conditions are optimal. In warm, moist, shaded soil they begin to develop and can become infective after about 18 days. The soil residency of *A. lumbricoides* is classified as periodic. Infection occurs when the infective eggs are swallowed through the ingestion of contaminated raw food, such as fruit or vegetables, or through the incidental ingestion of soil. The eggs then hatch in the small intestine and the resulting larvae migrate to the lungs. In the lungs they molt twice and then migrate up through the air passages of the lungs to the trachea. They then enter the throat and are swallowed, finally ending up in the small intestine where they mature and mate, to complete their life cycle. Usually the roundworms only feed on the semi-digested contents of the gut. There is some evidence that they may also feed on blood and tissue taken from the intestinal mucous membrane. The worms can live for up to two years in the intestine (CDC, 2002).

3. Enterobius vermicularis

Enterobius vermicularis is a pinworm that causes enterobiasis, often referred to as human pinworm infection. *E. vermicularis* females grow up to 13 mm in length and males grow up to 5 mm. Enterobiasis is a common infection in children worldwide. It is the most common nematode parasite in temperate climates and in areas with modern sanitation. Symptoms are generally mild and vague. Most often anal itching is the only problem. It is estimated that over 200 million people are infected annually (Cambridge University Schistosomiasis Research Group, 2002). Enterobiasis is most common in soils associated with poor sanitation where human feces are distributed in yards or fields, generally, as fertilizer.

Adult pinworms live in the human colon. Eggs are deposited by the female in the perianal region and can enter the environment with the feces. Given the proper conditions, the eggs become infective about 4 hours after being laid. The eggs are resistant to drying and can remain infective in dust for several days (National Institutes of Health, 2001). The soil residency of *E. vermicularis* is classified as incidental. Person-to-person transmission can also occur through handling items contaminated by an infected person. Following ingestion of infective eggs, the larvae hatch in the small intestine and the adults establish themselves in the colon.

4. Strongyloides stercoralis

Strongyloides stercoralis is a small roundworm, 2 mm in length, which causes strongyloidiasis. It is parasitic in the mucosa of small intestines. Strongyloidiasis is generally found in tropical and subtropical areas with poor sanitation. However, it does occur in temperate areas, including the southern United States (CDC, 2002). In this area, it is often found in rural areas and in lower socioeconomic groups. Strongyloidiasis is often asymptomatic but sometimes causes chronic disease. In individuals who are immunosuppressed, it can be life-threatening.

The life cycle of *S. stercoralis* is complex. Unlike other helminths, its eggs can hatch in and re-infect the host. It is also capable of completing its life cycle and reproducing in the soil and thus is classified as a permanent soil resident. *S. stercoralis* eggs can hatch in the small

intestines and produce rhabditiform larvae that can be passed in the feces of an infected host and may enter the soil. If they molt twice, they become a form (filariform) that is infectious to humans. If the rhabditiform larvae molt four times, they become free-living males and females that can mate and produce rhabditiform larvae in the soil (CDC, 2002).

The parasitic cycle begins when filariform larvae penetrate human skin where they are then transported in the blood to the lungs and penetrate the airway. They are then swallowed and reach the small intestines where they molt twice and become adult, egg-laying females. The eggs hatch into rhabditiform larvae in the host. These larvae can either be passed in the feces or penetrate the intestinal mucosa which creates an internal autoinfection (CDC, 2002).

5. Toxocara canis

Toxocara canis is the most common cause of visceral larva migrans. The disease is found worldwide and most often afflicts children aged 1–4 (Pitetti, 2001). The definitive host for *T. canis* is the dog. Humans are called paratenic hosts, because the life cycle of *T. canis* cannot be completed in humans.

A heavily infected dog can pass millions of eggs each day in their feces (Pitetti, 2001). Humans contract *Toxocara* infections as accidental hosts by ingesting embryonated eggs in contaminated soil (CDC, 2002). The soil residency of *T. canis* is transient. The eggs hatch in the small intestine and the resulting larvae invade the mucosa and enter the bloodstream. The larvae can then disseminate to any organ in the body, provoke a granulomatous reaction, and die. Symptoms depend upon the degree of tissue damage and the associated immune system response. In the United States, 2–10% of children test positive for *Toxocara*, and international incidence is probably similar or slightly higher (Pitetti, 2001).

6. Trichuris trichiura

Trichuris trichiura is the whipworm responsible for trichuriasis (often called whipworm disease). Trichuriasis is an infection of large intestine and is caused by the accidental ingestion of *T. trichiura* eggs. Most infections are associated with tropical areas with poor sanitation and occur in children. The disease occurs worldwide and there are an estimated 800 million people infected (CDC, 2002). The soil residency of *T. trichiura* is classified as incidental.

Adult *T. trichiura* worms are approximately 4 cm in length. Female worms can shed 3000–20,000 eggs per day in the feces. In the preferred environment of warm, moist, shaded soil the eggs embryonate and become infective in 15–30 days. They are often ingested through soil-contaminated food or hands causing the eggs to hatch in the small intestines. This releases larvae that mature to adults and establish themselves in the large intestine. There, the larvae fix themselves into the mucosa to feed. They have a life span of about one year (CDC, 2002).

B. Selected Soil-Borne Human Pathogenic Trematodes

Trematodes are soft-bodied invertebrate animals with bilateral symmetry that are also called flukes. They can cause parasitic infections in humans. Trematodes have complex life cycles that always involve an intermediate host that is a mollusk. Trematodes and cestodes are members of a group of animals (the phylum Platyhelminths) that are commonly called flatworms, because most species are flattened dorsoventrally. This shape is due to the fact that they must respire by diffusion and no cell can be too far from the surface. Trematodes and cestodes have only one opening to the gut, which must both take in food and expel waste. The gut is often extensive and branched in order to provide nourishment to the entire animal.

As adults, trematodes are usually found in vertebrate animals including fish, amphibians, reptiles, birds, and mammals. These animals serve as the definitive hosts (a host where the adult parasite is able to reproduce sexually), but other intermediate hosts are usually involved in the trematode life cycle. Trematodes may cause highly severe infections of the lungs, bladder, blood, liver, and most often, the gastrointestinal tract. Several *Schistosoma* species are important human pathogens.

1. *Schistosoma mansoni, Schistosoma japonicum,* and *Schistosoma haematobium*

Schistosomiasis is a trematode infection that affects over 200 million people in 74 tropical countries. Up to 600 million people worldwide are at risk for this disease, which is spread by bathing or wading in infected rivers, lakes, and irrigation systems (World Health Organization, 1996). People who are repeatedly infected can face liver, intestinal, lung, and bladder damage.

Schistosoma mansoni, S. *japonicum,* and S. *haematobium* are the three species that cause the most prevalent form of the disease. Each species occupies a different geographic region and there are slight differences in the clinical presentation of the disease itself. S. *mansoni* is found in parts of South America, the Caribbean, Africa, and the Middle East; S. *haematobium* in Africa and the Middle East; and S. *japonicum* in the Far East. These trematodes are unusual in that they reproduce sexually (one mated pair of S. *japonicum* can produce about 3000 eggs per day) and that they live in the mesenteric veins that lie outside of the liver. The life expectancy of an adult is from 10 to 25 years. Some of these eggs can pass through the membranes of the bowels and are excreted in feces. The soil-residency of *Schistosoma* spp. is periodic. When fully wetted, the eggs hatch into free-swimming larvae call miracidia. These larvae infect amphibious snails that live in mud on the edges of bodies of water. These larvae are not infectious to humans at this stage of their development.

Once a snail is infected, the larvae reproduce asexually into another free-swimming larval stage and are called cercaria. The fork-tailed cercaria completes development in the snail, migrates to the surface of the snail's soft tissue, and enters the environment (water). If the cercaria comes into contact with the skin of a human host it releases enzymes that soften the skin and allow it to enter the host. Once in the host the cercaria sheds its tail and becomes a schistosomula and finds its way to the mesenteric veins where it matures into an adult worm capable of mating. Diseases caused by *Schistosoma* spp. are saprozoonoses. The disease is a zoonosis, but the eggs require an environmental residence to develop (CDC, 2002; MacLean, 2002; University of California, 2002b).

C. Selected Soil-Borne Human Pathogenic Cestodes

Cestodes are often referred to as tapeworms because of the shape of their long ribbon-like body. They resemble a colony of animals in that their bodies are divided into a series of segments each with its own set of internal organs. Adults of the species can reach 100m in length.

There are several parasitic tapeworm infections where man is the definitive host. Two examples of soil-borne tapeworms discussed below are *Taenia saginata* and *T. solium;* and both have a periodic soil residency. Adult parasitic human pathogenic cestodes live in the intestines. The bodies of cestodes can be divided into three regions: the scolex or head, the neck, and the strobila. The strobila is composed of a series of segments called proglottids and each proglottid has its own complete set of internal reproductive organs. As the organism grows proglottids are added, and the result may be thousands in a mature animal. This pattern of growth forms a long, ribbon-like body referred to as a tapeworm. Mature proglottids containing eggs are shed from the rear of the animal. The eggs, or proglottids, exit the body in feces and enter the environment.

1. Taenia saginata (the beef tapeworm)

Humans are the only definitive host for *T. saginata*, which causes about 50 million cases of tapeworm infection annually at locations spanning the globe. Infections are generally asymptomatic, but in some cases vitamin deficiency may be the result of excessive absorption of nutrients by the parasite. Occasionally mild symptoms like abdominal pain, digestive disturbances, excessive appetite, or loss of appetite, weakness, and loss of weight may accompany the infection.

Adult tapeworms live in the small intestines. They are generally 5 m or less long, but may reach lengths up to 25 m. Mature worms have over 1000 proglottids and mature proglottids contain 80,000–100,000 eggs each. Once mature, the proglottids separate from the tapeworm and can be passed in the feces and may enter the soil. The soil residency is classified as transient. The eggs can survive for months to years in the environment. Cattle become infected by ingesting vegetation contaminated with eggs (or proglottids). The eggs develop in the intestines and release onchospheres that evaginate and invade the intestinal wall. They then migrate to the striated muscles and develop into a cysticercus, which is capable of surviving for several years. Ingesting undercooked meat containing cysticercus infects humans. In the human intestine the cysticercus develops into an adult tapeworm that is capable of surviving for over 30 years (CDC, 2002; MacLean, 2002).

2. Taenia solium (the pork tapeworm)

Taenia solium, the pork tapeworm, can cause both taeniasis and cysticercosis in humans. As with beef tapeworm infections, the disease occurs worldwide with about 50 million annual cases. *T. solium* and *T. saginata* have very similar life cycles, but *T. solium* larvae can infect humans as well as swine. Humans are the definitive host for *T. solium* adults, which can live up to 25 years in the intestine. *T. solium* have less than 1000 proglottids and are 2–7 m in length. Their proglottids are less active than in *T. saginata* and each contain about 50,000 eggs. When the proglottids mature, eggs or proglottids are shed in the feces and into the soil. The

soil residency is classified as transient. If swine ingest them, they mature to onchospheres that move to the muscles and grow into the larval form of *T. solium*, *Cysticercus cellulosae*. Undercooked pork from infected swine is then infective for taeniasis in humans.

If humans swallow the eggs or proglottids, they can also develop *C. cellulosae* infection which results in cysticercosis, a disease that can be quite severe. It can also develop in humans infected with adult *T. solium* due to autoinfection from proglottids carried to the stomach by reverse peristalsis. In humans, *C. cellulosae* can develop in the striated muscles, the brain, the liver, and in other tissues. When infecting the central nervous system, the pressure produced by growing larvae can cause severe pain, paralysis, optical and/or psychic disturbances, and epileptic convulsions. Mortality due to cysticercosis is estimated at about 50,000 worldwide (CDC, 2002; Duckworth et al., 2002; MacLean, 2002).

VIII. Selected Soil-Borne Human Pathogenic Protozoa

Protozoa are single cell eukaryotic organisms that are phagotrophic, which means that they feed by engulfing and ingesting their prey inside a cell membrane. There are over 30,000 species of which only a small number are parasites of man. Most protozoa range in size from 0.01 to 0.1 mm. In soils, they feed on bacteria and algae. The protozoan life cycle ranges from binary fission in a single host to many morphological transformations in a series of hosts. There are no eggs, larva, or adults. There are about 10,000–100,000 protozoa per gram of upper soil surface (Coyne, 1999). Archeozoa are similar to protozoa except that they lack mitochondria.

A. Cryptosporidium parvum

Cryptosporidium parvum is a protozoan that causes a self-limiting diarrheal illness called cryptosporidiosis. It can be more serious in infants and in the immunosuppressed. The first documented human case occurred in 1976 and cryptosporidiosis is now considered a worldwide disease. It is especially common in developing countries. No effective specific treatment is known. The incidence of cryptosporidiosis is not known, but an outbreak in Milwaukee in 1993 infected over 400,000 people (Coyne, 1999).

Many animal species, including man, act as a reservoir for *C. parvum*. Infected hosts excrete sporulated oocysts in the feces. In this fashion, oocysts may enter the soil as animal or human waste products. Infection begins when oocysts are ingested, most often through contaminated water and food or by direct fecal-oral transmission. After ingestion, the oocysts mature to sporozoites and parasitize the epithelial cells of the gastrointestinal tract. These parasites then undergo asexual and then sexual reproduction. The soil residency of *C. parvum* is transient-incidental. Generally cryptosporidiosis is a water-borne disease, but *C. parvum* only reaches the environment through human and animal waste. The organism must survive in and travel through soils to become water-borne in many cases. Studies have indicated that *C. parvum* can survive in surface water for six months and in liquid manure tanks for many months (Cambridge University Schistosomiasis Research Group, 2002; CDC, 2002; Duckworth 2002; and Health Canada, 2002).

B. *Cyclospora cayetanensis*

Cyclospora cayetanensis causes cyclosporiasis, a diarrheal disease found worldwide. It is most commonly found in tropical and subtropical regions. This disease has a life cycle similar to *C. parvum* with the exception that when passed in the feces, the oocyst is not infective. This means that no direct fecal-oral transmission can occur. Freshly passed oocysts sporulate after spending days or weeks in the environment at temperatures of between 22 and 32°C and become infective. Soil residency is incidental. Little is known about possible animal reservoirs or environmental survival time for *C. cayetanensis*. The oocysts are thought to be able to survive for long periods of time in the environment if kept moist (CDC, 2002; Garcia, 2002).

C. *Entamoeba histolytica*

Entamoeba histolytica causes amebiasis, a disease characterized by diarrhea, with severe cases including dysentery or a serious invasive liver abscess. Amebiasis is a worldwide disease with an estimated 40,000,000 people infected annually. Around 40,000 die from the disease each year (University of Leicester, 2001). Man is the definitive host. As with *G. lamblia, E. histolytica* exists in two forms: the active parasite (trophozoite) and the dormant parasite (cyst). The trophozoites live in the intestine and feed on bacteria or on the wall of the intes-

tine. Trophozoites are expelled in feces and die rapidly. However, cysts expelled in the feces are very hardy and can survive days to weeks in the external environment. In areas where sanitation is poor, indirect transmission of the cysts is more common. The soil residency of *E. histolytica* is incidental (University of Leicester, 2001).

D. *Balantidium coli*

Balantidium coli is the largest protozoan found in humans. It causes balantidiasis, a disease with a worldwide distribution. It is capable of causing acute hemorrhagic diarrhea and ulceration of the colon. Pigs, large primates, humans, and dogs are the definitive hosts. The life cycle and methods of transmission of the parasite are similar to *E. histolytica*. The soil residency is transient-incidental. Under favorable temperature and humidity conditions, the cysts can survive in soil or water for weeks to months (CDC, 2002).

E. *Giardia lamblia*

Giardiasis is a disease that is especially common among children and in places where sanitation is poor worldwide. About 200 million people in Asia, Africa, and Latin America display symptoms and there are about 500,000 new cases annually (World Health Organization, 1996). It is also one of the most common parasitic diseases in developed nations. It is caused by *G. lamblia*, an archeozoan. Clinically, giardiasis presents as non-inflammatory diarrhea and associated abdominal cramps, bloating, fatigue, and weight loss. Infections can be asymptomatic or chronic.

G. lamblia trophozoites (the active stage of organism) live in the large intestine of infected humans or animals. At times they form cysts and millions of these cysts (and trophozoites) are released in the feces and may enter the soil. The soil residency of *G. lamblia* is transient-incidental. The cysts can persist for some time (up to many months) in the environment, which includes soil, food, water, or surfaces that have been contaminated. Infection results from ingestion of the cyst, usually in contaminated water or food (University of Leicester, 2001; CDC, 2002).

F. *Isospora belli*

Isospora belli is a protozoan that causes isosporiasis, an infection of the small intestine. Isosporiasis is found worldwide, but is most common in tropical and sub-tropical areas. It can cause chronic diarrhea, abdominal pain, and weight loss and is especially important in immunosuppressed individuals.

Large and football-shaped *I. belli* oocysts are passed in the feces. The soil residency of *I. belli* is transient-incidental. The oocysts contain a sporoblast (rarely two) that splits and develops cyst walls, thereby becoming sporocysts. Infection occurs by ingestion of sporocysts. *I. belli* has a complex asexual and sexually reproductive cycle within its host, both human and animal, which results in the production of oocysts that are then excreted in feces (University of Leicester, 2001; CDC, 2002).

G. *Toxoplasma gondii*

Toxoplasma gondii causes toxoplasmosis. *T. gondii* infection can produce flu-like symptoms in healthy people and severe disseminated disease in immunosuppressed individuals. It can also cause birth defects in infants when women are exposed during pregnancy. Toxoplasmosis occurs worldwide and is more common in warm climates and at low altitudes. Under some conditions, toxoplasmosis can cause serious pathology, including hepatitis, pneumonia, blindness, and severe neurological disorders.

Cats are the definitive host for *T. gondii*. They generally acquire the infection though consumption of infected rodents. After a cat consumes the tissue containing cysts or oocysts, viable tachyzoites invade the small intestine. These eventually form oocysts that are excreted. The soil residency of *T. gondii* is transient. The oocysts can remain infective in water or soil for about one year. Tachyzoites can also form cysts in tissue. Humans can become infected in several ways including the ingestion of cysts through contaminated food or soil or ingestion of undercooked meat (e.g., lamb, pork, or beef) infected with cysts.

Its life cycle includes two phases called the intestinal (or enteroepithelial) and extraintestinal phases. The intestinal phase occurs in cats only (wild as well as domesticated cats) and produces oocysts. The extraintestinal phase occurs in all infected animals (including cats) and produces tachyzoites and, eventually, bradyzoites or zoitocysts. The disease toxoplasmosis can be transmitted by ingestion of oocysts in cat feces or bradyzoites in raw or undercooked meat (University of Leicester, 2001; CDC, 2002; Health Canada, 2002).

H. *Dientamoeba fragilis*

Dientamoeba fragilis is a protozoan responsible for *D. fragilis* infection. Symptoms occur in only 15–25% of infected individuals and may result in mild, chronic gastrointestinal problems (abdominal pain, gas, diarrhea, etc.). The disease is found worldwide but has a higher prevalence in developing countries with poor sanitation.

D. fragilis trophozoites are one of the smallest human parasites and survive in the human gastrointestinal tract. No cyst stage has been reported, so a fecal-soil-oral infectious gateway is unlikely. The soil residency of *D. fragilis* is unknown. There is evidence that this organism is transmitted among humans in the eggs of human pinworms (*E. vermicularis*). Infection by *D. fragilis* may require infection by *E. vermicularis* (Mack, 2001; CDC, 2002).

IX. SELECTED SOIL-BORNE HUMAN PATHOGENIC FUNGI

There are over 100,000 species of fungi of which about 300 are known to be pathogenic (University of Leicester, 1996; McGinnis, 1998). Fungi are non-motile eukaryotic organisms with chitin-based cell walls that can be grouped into molds and yeasts. Molds are composed of branching filaments called hyphae that grow by elongation at their tips. Hyphae can be composed of one cell with continuous cytoplasm, called coenocytic hyphae, or are composed of cells separated by walls (septa) in which case they are called septate hyphae. The mass of hyphae of an individual organism is referred to as mycelium. Reproduction is through sexual or asexual spores and fragmentation of hyphae. Single cell non-filamentous fungi are called yeasts. They are generally spherical or ovoid in shape and reproduce by budding. Some fungi are dimorphic in that they can switch between filamentous or yeast growth.

Most fungi are saprophytes and must absorb nutrients from the environment. In this way they help decompose dead plants and animals. Fungi do not contain chlorophyll and are therefore not capable of photosynthesis. Most molds are aerobic and cannot survive in saturated soils. They need to be able to extend hyphae into air spaces that contain oxygen. Many yeasts are facultative anaerobes and some yeasts are capable of surviving in anaerobic environments. In moist soils, the largest fraction of the microbial biomass is made up of fungi. Soil-borne fungi are more tolerant of acidic soils,

TABLE III.   Selected Soil-Borne Human Pathogenic Fungi and Their Properties

Pathogen(s) and disease	Distribution and residency	Gateway(s) and incidence[a]	Comments and soil survival time
Coccidioides Disease: coccidioidomycosis	Southwestern U.S., Mexico, microfoci in Central and South America Residency: permanent	Respiratory, rarely trauma Incidence: 15/100,000 in Arizona in 1995	Please see the case study in Section XIII for complete information
Histoplasma capsulatum Disease: histoplasmosis	Locally in eastern and central U.S., microfoci in Central and South America, Africa, India, and southeast Asia Residency: permanent	Respiratory Incidence: About 80% of people living in endemic area have a positive skin test; mortality rate is about 10% in HIV-infected persons with disseminated disease	Found in soils contaminated with bird or bat feces
Blastomyces dermatitidis Disease: blastomycosis	South-central, southeastern and mid-western U.S., microfoci in Central and South America and Africa Residency: permanent	Respiratory Incidence: 1 to 2/100,000 in endemic areas	Found in soils enriched with decomposing organic debris
Aspergillus fumigatus; A. flavus; less commonly A. terreus, A. nidulans, A. niger Disease: aspergillosis	Worldwide; ubiquitous; found in soil, dust, plants, food, and water Residency: permanent	Respiratory, occasionally via contaminated biomedical devices Incidence: 1 to 2/100,000 is suggested	Found in soils, decomposing plant material, household dust, food, water, and plants
Sporothrix schenckii Disease: sporotrichosis	New World, Africa, and Europe Residency: permanent	Trauma to skin Incidence: disease is uncommon and sporadic	Most common in sphagnum moss, plants, baled hay

[a]Incidence is the annual rate of confirmed infection. An incidence of 15:100,000 means that there were 15 confirmed cases per 100,000 population. From CDC, 2002; DoctorFungus, 2002; Duckworth et al., 2002; Health Canada, 2002.

grow best between 6 and 50°C, and are usually found in the top 15 cm of the soil (Coyne, 1999).

The life cycles of the various fungi are generally not as complex or varied as helminths and protozoa. Table III illustrates selected soil-borne human pathogenic fungi and the diseases they cause. It also contains information on the geographic distribution, soil residency, gateway of infection, and disease incidence for each of the fungi.

X. SELECTED SOIL-BORNE HUMAN PATHOGENIC BACTERIA

Coyne (1999), presents an excellent introduction to bacteria in the soil that is summarized below. Bacteria are single-cell prokaryotic organisms that have existed on Earth for over 3 billion years. They are small, generally less than 50 µm in length and 4 µm in width. One bacterium weighs about 10^{-12} grams. Their small size affords them a high surface area to volume ratio, which allows them to maximize nutrient uptake through diffusion. They also possess a very high metabolism and the ability to reproduce through binary fission. Bacteria occur in a wide variety of shapes. Aerobic bacteria require oxygen for existence, whereas anaerobic bacteria do not tolerate gaseous oxygen. Some bacteria, called facultative anaerobes, prefer oxygen, but can grow without it. Heterotrophic bacteria use organic compounds in the environment for energy and for synthesis of cellular constituents. Autotrophs make use of energy from light or of reactions of inorganic chemicals to fix carbon dioxide and synthesize organic cellular components (University of California, 2002a).

There are up to one billion bacteria in one gram of soil (Table I). In general, they prefer warm, moist soils. Soil bacteria can be classified as autochthonous or allocthonous. Autochthonous organisms inhabit the bulk of the soil and are specialists at getting the most out of the available nutrients. Allocthonous microbes are more opportunistic. They are generally saprophytic or pathogenic and tend to be found in areas that are rich with nutrients, even if only for a limited time period. Allocthonous microbes maximize growth when conditions are right. They often are found in the rhizosphere and many of them are plant pathogens. Allocthonous microbes can cover 5–10% of root surfaces and there is a steep decrease in microbial populations just 5 mm from the plant root.

There are a large number of soil-borne human pathogenic bacteria that generally have similar life cycles. They grow and reproduce through binary fission in the proper conditions. When the conditions are unfavorable, they may die off or form a spore that can grow and reproduce again when conditions improve. Examples of several soil-borne human pathogenic bacteria are presented in Table IV.

Actinomycetes are prokaryotic bacterial organisms that display filamentous growth. They make up 10–50% of the total microbial population in soils (Coyne, 1999). Most actinomycetes are aerobic and prefer warm, dry soils. They tend to be spore formers and are adept at surviving droughts. Most are saprophytic. Due to their filamentous growth, they resemble fungi. The filaments of actinomycetes are much smaller than fungal hyphae, 0.5–1.0 µm as opposed to 3–8 µm for fungi. Although actinomycetes are an important component of the soil microbial population, there are few known to be important soil-borne human pathogens. Four pathogenic genera are included in Table IV. Many actinomycetes make antibacterial molecules. About 75% of the 5000 known antibacterial drugs are derived from actinomycetes (Coyne, 1999).

XI. Selected Soil-Borne Human Pathogenic Viruses

There are over 140 types of pathogenic enteric viruses transmitted from humans to the environment in human feces. For some (like the Norwalk virus and rotavirus), immunity is short-term; there is no life-long protection after recovering from an infection (Schwartzbrod,

1995). Viruses are the smallest pathogens, most having maximum dimension of less than 30 nm (Coyne, 1999). They are acellular organisms, have no cell membrane, and occur in many shapes including cubic, helical, and icosahedral. Most viruses have two basic structural components: a protein coating that can help the virus survive in the environment and a nucleic acid core. Viruses are so small that their genetic material (the nucleic acid core) contains only 10–200 genes (Coyne, 1999). There are viruses that infect animals, plants, fungi, protozoans, algae, and bacteria. They are always host specific.

Viruses are parasites that must use the chemical machinery and metabolism of a host cell to reproduce. In a host, viruses attach to a cell, use enzymes to break through the cell wall, and inject their nucleic acid core into the cell. Once in the cell, the genetic material from the virus begins making three types of proteins. It replicates its own genetic material, builds protein coating, and assembles proteins that will help it get out of the cell. These parts come together by chance to form and release a single or many new copies of the original virus. Outside of the host, viruses are inert. They do not grow or reproduce. Human pathogenic viruses with protective coatings can remain infectious in the environment for up to 6 months.

A. Viruses in Soils

Soil is not a natural reservoir for viruses. Viruses can only persist in soil in a dormant state but may retain their infectivity in this state. Plant viruses rarely survive in soils for long periods; however, some insect viruses remain infective for years (Coyne, 1999). Viruses are also known to infect many soil helminths and microbes.

There are a number of factors that influence the ability of a virus to survive and to move in unsaturated soils. Because viruses are added to soil as anthropogenic waste, the objective is to keep the viruses from the water table where generally cool water temperatures can keep the virus alive for long periods of time. Factors that affect survival include: temperature, soil moisture, soil microbial activity, soil type, virus type, soil organic matter, and adsorption of the virus, generally to clay minerals (Sobsey & Shields, 1987). Viruses generally survive longer in cooler, wetter, pH neutral soils with low microbial activity. Humic and fulvic organic material may cause reversible loss of infectivity, but, some other organic materials may complex with the virus and protect it from inactivation by preventing adsorption to

Pathogen(s) and disease	Distribution and residency	Gateway(s) and information on incidence, morbidity and/or mortality (IMM) if available	Comments and soil survival time if available
Nocardia spp., Rhodococcus spp. Disease: Nocardiosis	Worldwide, some species more likely in tropics Residency: permanent?	Cutaneous disease from skin trauma contaminated with soil, pulmonary and disseminated infections from inhalation IMM: in the U.S., there are an estimated 500–1000 new cases of Nocardia infection annually	Aerobic actinomycete found in soil and water; Nocardia asteroides is tolerant of 40–50°C
Rickettsia rickettsii and other Rickettsia spp. Disease: Rocky Mountain spotted fever, other fevers (African tick bite fever, Queensland tick typhus) and spotted fevers (Mediterranean, Japanese)	Worldwide, individual species are geographically contained by their mammalian reservoir Residency: periodic	Zoonosis-spread by bite of tick or by contamination of the skin with tick blood or feces; rodents are the main mammalian reservoir IMM: 3–5% of individuals who become ill with Rocky Mountain spotted fever still die from the infection; U.S. has 250–1200 cases of Rocky Mountain spotted fever annually	Vector in North America is the wood tick, American dog tick, or Lone Star tick; female ticks transfer the bacterium to their eggs that are infective as they mature and hatch in the soil (unlike many other tick vector diseases, i.e., Lyme disease); feces of infected ticks quickly lose their infectivity on drying
Salmonella spp. Disease: primarily salmonellosis (diarrhea), typhoid fever and paratyphoid fever	Worldwide, primarily a food-borne disease Residency: incidental	Fecal-soil-oral, ingestion of contaminated (often uncooked) food contaminated by infected soil and water; shed in human and animal feces IMM: 2–4 million cases in U.S. annually	Can survive in sludges and soils for many months given proper conditions; in sludge applied to arid soils survival may be 6–7 weeks.
Shigella spp. Disease: diarrhea, dysentery	Worldwide Residency: incidental	Fecal-soil-oral, ingestion of contaminated (often uncooked) food contaminated by infected soil and water; shed in human feces IMM: 300,000 cases annually in U.S.	May survive a few weeks in water below 10°C; soil survival unknown
Streptomyces spp. Disease: skin infection	Africa, India, Latin America Residency: permanent?	Skin trauma contaminated with soil IMM: invasive infection is extremely rare	An aerobic actinomycete and soil saprophyte
Thermoactinomyces spp.	Probably worldwide Residency: permanent?	Inhalation IMM: Farmer's lung can occur in 2–10% of farm workers but is regionally variable	Actinomycete found in soil, contaminated compost piles, silos; tolerant of 45–60°C heat.
Yersinia spp. Disease: diarrhea	Worldwide Residency: incidental	Fecal-soil-oral, ingestion of food (often uncooked) contaminated by infected soil and water; shed in human and animal feces	Disease most often occurs in infants and small children; Y. enterocolitica is known to survive in soil for 540 days.

From World Health Organization, 1996; Canadian Centre for Occupational Health and Safety, 1999; Rusin et al., 2000; Carey et al., 2001; Pennsylvania Environmental Network, 2001; Ania and Asenjo, 2002; CDC, 2002; Duckworth et al., 2002; Health Canada, 2002.

Pathogen(s) and disease	Distribution and residency	Gateway(s) and information on incidence, morbidity and/or mortality (IMM) if available	Comments and soil survival time if available
Clostridium spp. (other than two listed above) Disease: gas gangrene	Worldwide Residency: permanent? (in proper setting can complete life cycle in soil)	Fecal-soil-oral, humans and many animals; skin trauma (major or minor), burns, deep puncture wounds, ear infections, animal bites; spores introduced into the body through a wound contaminated with soil, street dust, feces, or injected street drugs; also through lacerations, burns, and trivial wounds. IMM: fairly common before general use of antibiotics to treat injuries; can still pose threat to those immunosuppressed	Before antibiotic treatments, about 5% of battlefield injuries were complicated by this bacterium
Coxiella burnetti Disease: Q fever	Worldwide with the exception of New Zealand and Antarctica; Residency: transient	Inhalation of infected aerosol, often produced from animal products and especially during parturition; also shed in urine and feces. IMM: morbidity from 5% in urban to 30% in rural areas worldwide	Highly infective, but unable to grow outside of host (commonly goats, sheep, and cattle); has a spore-like form that is very resistant to heat and desiccation and can last for months outside of host in soils
Escherichia coli several pathogenic strains Disease: diarrhea	Worldwide Residency: incidental	Fecal-soil-oral, ingestion of contaminated food. IMM: major cause of traveler's diarrhea of which there are some 5 million cases per year worldwide	Can survive for months in cool, dark, nutrient-rich soils
Francisella tularensis Disease: tularemia	Many areas of U.S. with most cases in Arkansas, Oklahoma, and Missouri; increasing numbers of cases in the Scandinavian countries, eastern Europe, and Siberia; also in the Middle East and Japan; rare in the UK, Africa, and Central and South America Residency: transient	Enters soil through tick feces and possibly other sources; humans acquire through contact of infected soil with broken skin (might be able to penetrate unbroken skin) and with mucous membranes; also tick and insect bites, inhalation, and ingestion. IMM: worldwide incidence not known; in the U.S. there are now less than 200 cases per year	A zoonosis; one of the most infectious agents known; highly infectious in both skin and aerosol routes; often found in rural areas; possible biological warfare agent; known to survive in water and moist soil for weeks
Leptospira spp. Disease: leptospirosis	Worldwide, but more common in temperate or tropical climates Residency: transient	Ingestion and skin contact, especially mucosal surfaces; contact with water, food, or soil contaminated with urine from infected animals. M: about 200 cases annually in U.S.; considered to be the most widespread zoonotic disease in the world	Outbreaks associated with heavy rainfall and flooding; known to survive many weeks in contaminated soil
Listeria monocytogenes Disease: listeriosis	Worldwide? Residency: permanent	Ingestion of food (often uncooked) contaminated by infected soil and water. IMM: 2500 serious cases per year in U.S. of which 500 are fatal	Found in soil, water, and fecal material of domestic animals; can grow at temperatures found in refrigerators

continued

TABLE IV. Selected Soil-Borne Human Pathogenic Bacteria and Their Properties

Pathogen(s) and disease	Distribution and residency	Gateway(s) and information on incidence, morbidity and/or mortality (IMM) if available	Comments and soil survival time if available
Actinomadura Spp. Disease: maduramycosis, actinomycetoma	Tropical regions, especially Africa, India, South and Central America Residency: permanent?	Skin trauma. IMM: Africa may have highest incidence; in the Sudan, 300–400 patients per year are seen; causes disfigurement, rarely fatal	An aerobic actinomycete that is a soil saprophyte
Bacillus anthracis Disease: anthrax	South and Central America, southern and eastern Europe, Asia, Africa, the Caribbean, and the Middle East Residency: periodic(?)	Respiration, skin trauma, ingested (gastrointestinal); often infected by above methods while handling contaminated animal products IMM: unknown to rare.	A spore-forming aerobic bacterium; spores can survive in soil environment for many years, possibly for decades; biological warfare agent. Possible biological warfare agent; soil saprophyte
Burkholderia (*Pseudomonas*) *pseudomallei* Disease: melioidosis	Worldwide, primarily in tropical and subtropical regions, especially in Southeast Asia and northern Australia, also in South Pacific Africa, India, and Middle East; isolated cases in Central and South America, Hawaii, and Georgia Residency: permanent	Direct contact with contaminated soil and water, inhalation of dust, ingestion of contaminated water, skin trauma, and contact with mucous membranes IMM: very important cause of morbidity and mortality in Thailand	No information on survival in soil; known to survive in 4°C stream water for over four months
Campylobacter jejuni Disease: diarrhea, gastroenteritis, Guillain-Barre syndrome	Worldwide Residency: incidental	Fecal-soil-oral; contaminated water, raw milk, and raw or undercooked meat, poultry, or shellfish IMM: *C. jejuni* along with rotaviruses and enterotoxigenic *Escherichia coli*, is a major cause of diarrhea worldwide	
Clostridium tetani Disease: tetanus	Worldwide, most frequently in densely populated regions in hot damp climates in soils rich in organic material especially manure Residency: permanent? (in proper setting it can complete life cycle in soil)	Fecal-soil-oral, humans and many animals; tetanus spores introduced into the body through a wound contaminated with soil, street dust, feces, or injected street drugs; also through lacerations, burns, and trivial wounds IMM: annual deaths: newborn = 450,000; maternal = 50,000.	Anaerobic (all *Clostridium* spp.) but can form spores that are very resistant to heat, many antiseptics, and chemical agents
Clostridium botulinum Disease: botulism	Worldwide Residency: permanent? (in proper setting can complete life cycle in soil)	Fecal-soil-oral, humans and many animals; ingestion of contaminated food containing toxin IMM: generally rare, associated with confined outbreaks	Obligate anaerobic bacterium that can live in oxygen-free pockets in soil as vegetative cells or spore; present in the soil and water; spores can be found on food that comes into contact with infected soil or water

continued

soil particles (Sobsey & Shields, 1987). Specific species of viruses have different survivability. Also, viruses tend to survive longer when clustered together.

Soil texture is also very important to virus survival in unsaturated soils. Clay minerals, which are found in fine-grained soils, prolong the virus's ability to survive through adsorption of the virus to these minerals. Adsorption by clay minerals can prolong virus survival because adsorption affords protection against inactivation (Gerba, 1987). They are essentially removed from the water film and physically protected by the clay minerals, but, they can desorb back into water with heavy rain and then be moved passively in the soil. Matson et al. (1987) showed that viruses have been recovered at distances from septic tanks of over 90 m horizontally and up to 67 m vertically in the soil. Schwartzbrod (1995), summarized the survival of viruses in soils. These results indicated that viruses have been shown to survive from 11 to 180 days in soils, and the length of survival depended on the type of soil, humidity, the soil moisture, and the soil temperature.

Sim and Chrysikopoulos (2000) presented an excellent review of virus sorption in an unsaturated soil. The ability of a soil to adsorb viruses is strongly correlated with the degree of soil moisture. Decreasing the moisture content enhances virus sorption onto the solid matrix by forcing viruses to move into a thin film of water surrounding soil particles. Virus adsorption at the liquid-solid interface is mainly from electrostatic double-layer interaction and van der Waals forces. Also, there is enhanced removal of viruses at low soil moisture because viruses are sorbed on the air-water interface as well as the water-mineral (solid) interface. The sorbtion at the air-liquid interface may be greater than the air-solid interface. Sorption at the air-liquid interface is primarily controlled by virus surface hydrophobicity, solution ionic strength, and particle charge. Even though viruses sorbed at the liquid-solid interface can remain infective, viruses sorbed at the air-liquid are deformed by interfacial tension to the degree that the protein coat of the virus is disrupted and the virus is inactivated.

Although the adsorption of viruses to clay minerals and water films in soils can prevent their movement and even kill them, the potential desorption of the viruses from clay minerals means that viruses can reach the ground and surface waters that may be used by man. This, in addition to viruses that may be directly ingested from contaminated soils, indicates that soil-borne viruses are important human pathogens. Table V lists the names of some viruses as well as the diseases they cause.

XII. TRANSMISSIBLE SPONGIFORM ENCEPHALOPATHIES

Transmissible spongiform encephalopathies (TSE) are fatal, degenerative diseases of animals and humans characterized by abnormal limb movements, progressive dementia, and the development of sponge-like holes in brain tissues. The accumulation of an abnormal protease-resistant protein in the brain is associated with all TSE. The nature of the causative agent of TSE is still being debated. One theory is that the agent is a biologically active, self-replicating, infectious protein called a prion, which accumulates in and destroys brain tissue. A second theory is that the agent resembles a virus in that it exists as different strains and causes infective, transmissible diseases and possesses nucleic acids, which carry genetic information (Council for Agricultural Science and Technology, 2000; Rabenau et al., 2001). TSE induced diseases of humans include: Creutzfeldt-Jakob disease, fatal familial insomnia, Gerstmann-Straussler-Scheinker disease, new variant Creutzfeldt-Jakob disease, and Kuru. Major examples in animals include scrapie in sheep, chronic wasting disease in deer and elk, bovine spongiform encephalopathy in cattle (mad cow disease), and TSE in cats, monkeys, and mink. Transmission of TSE is believed to be mostly by ingestion of infected animal parts or transplanted by use of contaminated medical instruments.

The relationship of TSE to the soil environment is unknown. However, some questions require consideration. First, can TSE agents, specifically bovine spongiform encephalopathy (BSE), be introduced into the soil by the natural death of an infected animal or possibly by anthropogenic activities? How long could the TSE agents exist in the soil? Concern has been raised in the UK by scientific advisors that of the nearly 500,000 cattle that were culled, killed, and buried as a result of foot-and-mouth disease, some were likely to be also infected with bovine spongiform encephalopathy and could spread the agent via soil and groundwater (Reuters Health Information, 2001). If this were the case then BSE would be classified as an incidental soil pathogen. TSE agents are found in brain and nervous tissue connected to the brain, and also in bone marrow (National Cattleman's Beef Association and Cattleman's Beef Board, 2001). This fact may be relevant to the possible natural transmission of the disease among cattle. Bone chewing is a relatively common trait of cattle worldwide and is believed to be related to dietary

TABLE V. Selected Soil-Borne Human Pathogenic Viruses and Their Properties

Pathogen(s) and disease	Distribution and residency	Gateway(s) and information on incidence, morbidity, and/or mortality (IMM) if available	Comments and soil survival time if available
Adenovirus spp. Disease: respiratory illness, conjunctivitis, diarrhea	Worldwide, specific viral species are found in specific locations Residency: incidental	Fecal-soil-oral IMM: Almost everyone is infected at some point; diarrheal disease blamed for at least 6 million deaths per year worldwide	Soil association with solid human waste; most infections are mild; serious infection possible in immunosuppressed
Arenavirus spp.: Lassa Fever virus, *A. junin*, *A. machupo*, *A. sabia*, *A. guanarito*, and others Disease: hemorrhagic fever (general), Lassa fever, Argentine, Bolivian, Brazilian, and Venezuelan hemorrhagic fevers	Worldwide, each species of virus has a different associated rodent(s) governing its geographic location Residency: transient	Respiratory and others IMM: The number of Lassa virus infections per year in West Africa is estimated at 100,000–300,000, with approximately 5000 deaths; may also have a nematode vector	Human zoonosis with the definitive reservoir being rodents; virus can enter soil from rodent urine, feces, and saliva; disturbance of feces, infected soil, or nesting materials can aerosolize; also spread by contact with contaminated surface
Astrovirus spp. Disease: diarrhea and gastroenteritis, mostly in children and immunosuppressed individuals	Worldwide Residency: incidental	Fecal-soil-oral IMM: diarrheal disease blamed for at least 6 million deaths per year worldwide	Soil association with solid human waste; most infections are mild; serious infection possible in immunosuppressed individuals
Caliciviruses spp. Including Hepatitis E virus Disease: diarrhea	Worldwide Residency: incidental	Fecal-soil-oral IMM: diarrheal disease blamed for at least 6 million deaths per year worldwide	Childhood diarrhea; most adults may be immune; Hepatitis E is very dangerous in developing countries and has a 20% mortality rate for pregnant women
Hantavirus spp. Including Sin Nombre, Puumala, Thailand, Prospect Hill, Khabarovsk, Thottapalayam, Tula, New York, Black Creek Canal, El Moro Canyon, Bayou, and others Disease: HPS-hantavirus pulmonary syndrome	Worldwide, specific virus species inhabit specific hosts in specific locations Residency: transient	Respiratory and others IMM: probably thousands of cases annually worldwide but quite variable	Human zoonosis with the definitive reservoir being rodents; each species has a different associated rodent(s); virus can enter soil from rodent urine, feces, and saliva; disturbance of feces, infected soil, or nesting materials can aerosolize; HPS has a 40–60% fatality rate

continued

Continued

Pathogen(s) and disease	Distribution and residency	Gateway(s) and information on incidence, morbidity and/or mortality (IMM) if available	Comments and soil survival time if available
Enterovirus poliovirus Disease: polio	Worldwide Residency: incidental	Fecal-soil-oral, respiratory IMM: disease eliminated in many parts of the world	Soil association with solid human waste; known to survive 91 days in unsaturated sand and humid conditions; known to survive 180 days in saturated sand and compost
Enterovirus Hepatitis A Disease: hepatitis	Worldwide Residency: incidental	Fecal-soil-oral IMM: in countries with poor sanitation, most children infected by age 9	Soil association with solid human waste; known to survive 91 days in sand and humid conditions
Enterovirus Coxsackievirus A Disease: diarrhea, hand-foot-and-mouth disease, respiratory infection	Worldwide Residency: incidental	Fecal-soil-oral IMM: diarrheal disease blamed for at least 6 million deaths per year worldwide	Soil association with solid human waste; known to survive 180 days in saturated sand and compost
Enterovirus Coxsackievirus B Disease: pleurodynia, aseptic meningitis, pericarditis myocarditis	Worldwide Residency: incidental	Fecal-soil-oral	Soil association with solid human waste; known to survive 180 days in saturated sand and compost
Enterovirus echovirus Disease: diarrhea, aseptic meningitis	Worldwide Residency: incidental	Fecal-soil-oral IMM: diarrheal disease blamed for at least 6 million deaths per year worldwide	Soil association with solid human waste; can survive 3–33 weeks depending on soil environment
Norwalk virus Disease: acute viral gastroenteritis, diarrhea	Worldwide Residency: incidental	Fecal-soil-oral IMM: diarrheal disease blamed for at least 6 million deaths per year worldwide	Soil association with solid human waste; very little is known about this virus
Orthopoxvirus variola Disease: smallpox	Worldwide Residency: incidental?	Mostly direct human-to-human transmission, but some respiratory environmental transmission known; humans are the only known reservoir IMM: last case acquired outside of a laboratory was in Somalia in 1977	Variola virus is unlikely to survive for more than 48 hours in environment; virus recovered in scabs on infected corpses after 13 years.
Rotavirus spp. Disease: diarrhea, gastroenteritis	Worldwide Residency: incidental	Fecal-soil-oral, respiratory(?) IMM: Kills 600,000 children worldwide annually; causes 2.7 million cases of gastroenteritis in children under 5 each year in U.S.	Soil association with solid human waste; major cause of death in the third world

From World Health Organization, 1996; Rusin et al., 2000; Toranzos & Marcos, 2000; CDC, 2002; Duckworth et al., 2002; Health Canada, 2002; Public Health Laboratory Service, 2002.

FIGURE 2   Areas endemic for *Coccidioides*. (After Valley Fever Center for Excellence, 2002.)

phosphorus deficiency, which is often associated with soils deficient in phosphorus. Decaying carcasses and bones are part of the surficial soil horizons and host a large array of microbes. If BSE can be transmitted among cattle by ingestion of contaminated bone material then it would also be classified as a transient soil pathogen.

XIII. COCCIDIOIDES CASE STUDY

A. Habitat of *Coccidioides*

Coccidioides is a dimorphic soil-inhabiting fungus, an important human pathogen, and the etiological agent of coccidioidomycosis (Valley Fever).

Coccidioides grows in the upper (5–20 cm) horizons of soils in endemic areas (Figure 2). This saprophytic phase

of the fungus is characterized by branching, segmented hyphae that form a network of mycelium. As the fungus matures, arthroconidia, 2–5 μm in size, are formed as barrel-shaped, rectangular segments of the hyphae that can be easily separated by soil disturbance (natural or anthropogenic) and consequently be dispersed by the wind. Arthroconidia are also very buoyant and may be readily moved by sheet-wash water during rainstorms only to be concentrated in fine sedimentary material some distance from the initial growth site. Under suitable environmental conditions the arthroconidia can germinate to form new hyphae and mycelium, which can repeat the cycle. If the airborne arthroconidia are inhaled by an appropriate host (humans, animals, even reptiles), then the parasitic phase of *Coccidioides* is initiated (Figure 3). In tissue the arthroconidia transform into spherules 10–80 μm in diameter that, when mature, are internally divided into endospores that are about 3–5 μm in diameter. The mature spherules then rupture and the

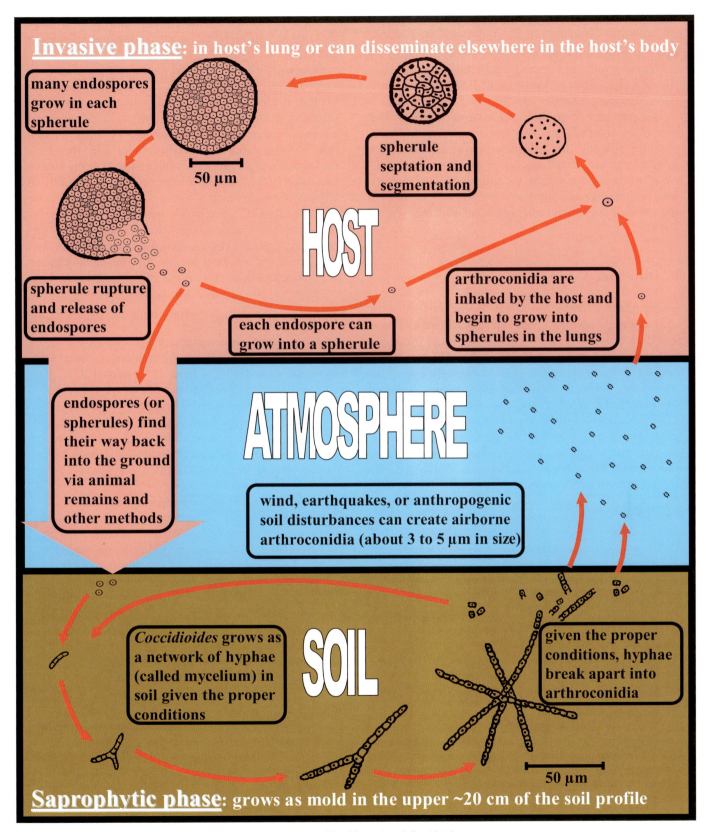

FIGURE 3 The life cycle of *Coccidioides*.

endospores are released into the surrounding tissue and spread the infection locally, or at times, into other organs by disseminating outside of the respiratory system. The epidemiology and human impacts of the disease are discussed in Pappagianis (1980, 1988, 1999) and Galgiani (1993, 1999).

The parasitic phase may end with the death of the host or by the expulsion of spherules outside of living tissue by sputum, pus, exudates, or degradation of an infected carcass. If this occurs in an acceptable environment then the spherules (or endospores) will germinate and hyphae and mycelium will be developed and the saprophytic phase will again be initiated (Fiese, 1958).

B. Distribution and Endemic Areas

In the United States *Coccidioides* is endemic in parts of Arizona, California, New Mexico, Nevada, Texas, and Utah (Figure 2). An outbreak at Dinosaur National Monument in northeastern Utah in 2001 (Figure 2) lies outside of the generally recognized endemic area. The site of this outbreak may represent a unique location where several things have come together, which include a favorable microclimate, to allow *Coccidioides* to survive in the soil. Outside of the United States it is endemic in parts of Argentina, Brazil, Colombia, Guatemala, Honduras, Mexico, Nicaragua, Paraguay, and Venezuela. With some exceptions endemic areas are generally arid to semi-arid with low to moderate rainfall (5–20 inches), mild winters, and long hot seasons. In 1993 the CDC declared that coccidioidomycosis was epidemic in parts of California (Kern County) and also issued a warning to physicians nationwide to watch for the disease in patients who may have become infected while traveling in endemic areas. The CDC also listed coccidioidomycosis as an example of one of the important disease threats to the United States and has called for expanded studies of the disease (Bryan et al., 1994).

C. Habitat Criteria Essential for the Growth and Survival of *Coccidioides*

Laboratory and site-specific field studies have shown that many physical, chemical, climatic, and biological factors influence the growth of *Coccidioides* in the soil and the consequent development and deployment of arthroconidia. Many of the following factors are closely interwoven, and the influence on the presence or growth of *Coccidioides* by any combination of, or single

factor, is an intricate balance that varies both in time (season) and in response to environmental changes at any given location.

Oxygen, carbon, nitrogen, phosphorus, sulfur, iron, and other trace elements along with water are necessary for the survival of *Coccidioides*. Furthermore, these raw materials must be available in a physical and chemical environment suitable for *Coccidioides* to satisfy its specific biological functions required for life. Based on measurements and observations gathered from known sites where *Coccidioides* is present in the soil and also on laboratory experiments where *Coccidioides* is grown under controlled conditions, several general conclusions can be made about the habitat parameters required for its life processes and also those parameters that, while not essential for the survival of *Coccidioides*, are favorable for its existence.

1. Important Criteria

1. Most known occurrences are in hyperthermic or thermic aridisols or entisols with mean annual soil temperatures ranging from 15°C to over 22°C.
2. The presence of soils with textures that provide adequate pore space in the upper (20 cm) parts of the profile, for moisture, oxygen, and growing room is very important. Soils in known occurrence sites are mostly fine sand to silt (0.002- to 0.2-mm particle size) with less than 10% clay-sized (<0.002 mm) material (Figure 1). Small amounts of clay foster water holding capacity, but large amounts of clay may be detrimental for *Coccidioides* growth. Smectite (a type of clay mineral) soils may be detrimental because their shrink and swell properties may provide room and water for bacterial growth that would compete with *Coccidioides*. Also, and perhaps, more important, they contain exchangeable cations that lower pH thereby enhancing bacterial growth at the expense of the growth of fungi.
3. The presence of some organic material is needed for carbon and nitrogen but in most known occurrences it is generally sparse, less than 2%. Large amounts of organic compounds may be detrimental because they would foster the growth of bacteria and other fungal species that would compete with *Coccidioides*.
4. Moisture is essential. Rainfall in endemic areas is generally seasonal with some areas receiving most of their precipitation in the winter months while precipitation in other areas may be split between winter rains and summer monsoons. In all cases,

annual precipitation ranges from less than 250 mm to 410 mm.

2. Favorable Criteria

1. Many *Coccidioides* growth sites have soils with elevated salinity. High soluble salts may act as an inhibitor of microbial competitors. Measured values of soluble salts in soils from known occurrence sites are sodium, 8–75% greater in positive soils than in negative soils; calcium, 2–5 times greater in positive than in negative soils; potassium, 2–5 times greater in positive than in negative soils; sulfates, 2–5 times greater in positive than in negative soils; borates, 3–25 times greater in positive than in negative soils; and chlorides, 10–240 times greater in positive than in negative soils (Elconin et al., 1964)

2. Several *Coccidioides* growth sites are in soils derived from marine sedimentary rocks. These rocks often contain elevated amounts of salts, and when weathered, provide material with textures favorable for *Coccidioides* growth. Also the elevated salinity of these derived soils inhibits microbial competition.

3. The presence of borates in the soil profile may act as antiseptics for bacteria that are competitive with *Coccidioides*.

4. Any environmental factor that reduces competition with other fungal, bacterial, and/or plant species is favorable for *Coccidioides* growth.

5. Parent material derived from aeolian deposits is a good source for the development of soil with favorable textures.

Habitat modeling of the saprophytic phase of the *Coccidioides* life cycle is difficult due to the limited number of known growth sites. This confounds the establishment of statistical relationships of the physical, chemical, and biological habitat parameters. Therefore, habitat modeling is accomplished using analysis of the physical properties of known *Coccidioides* sites and a spatial fuzzy system. A spatial fuzzy system is a system of spatial variables where some or all of the spatial variables are described with fuzzy sets. The fuzzy system is capable of translating structured knowledge into a flexible numerical framework and processing it with a series of if-then rules called fuzzy associative memory (FAM) rules.

Fuzzy systems can describe nonlinear numerical processes with linguistic common sense terms and can handle differing precision and accuracy in the data. They produce models that can be repeated and updated easily.

Fuzzy system analysis was applied to each 30 × 30 m spatial cell over the study area, Organ Pipe Cactus

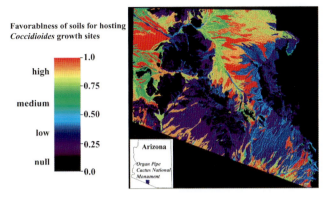

FIGURE 4 The fuzzy habitat suitability index of *Coccidioides* measured as the favorableness of soils for hosting *Coccidioides*, Organ Pipe Cactus National Monument, Arizona.

National Monument, Arizona. The resulting product is a map (Figure 4) depicting each cell's favorableness for hosting *Coccidioides* based on a scale of 0 to 1, which we define as its fuzzy habitat suitability index (FHSI). The fuzzy system allows modelers to change and update relationships among the variables as more is learned about *Coccidioides* habitat. An important property of this kind of analysis is that "what if" scenarios can be used to allow dynamic representation of climate related variables and may predict changes in habitat with changing climate.

Long-term climate fluctuations will undoubtedly have an effect on the distribution of *Coccidioides*. Changes in temperature and precipitation over time will directly influence soil characteristics critical to its growth and propagation. *Coccidioides* growth sites are believed to be relatively small and widely distributed throughout its endemic area and the fungus does not readily colonize outside of established growth sites. Laboratory studies show that *Coccidioides* is quite robust with respect to the physical and chemical factors of its habitat, but is very sensitive to competition from other microbes and vegetation. Therefore, climate change models that result in an increase in microbes in the soil profile and in the vegetation of a given area would result in decreased habitat for *Coccidioides* and scenarios that would reduce microbes in the soil profile and the vegetation in a given area may result in an increase in suitable habitat.

XIV. SOILS AND EMERGING DISEASES

Emerging infectious diseases may be defined as those that have newly appeared in a population and those whose incidence in humans has increased within the past two decades. A re-emergent disease is the reappearance of a known disease after a decline in incidence (Lederberg et al., 1992, pp. 34, 42). Many soils provide an ideal environment for the emergence of new infectious diseases due to their overall chemical and physical diversity, supply of essential nutrients for microbial growth, and their constantly evolving character in response to various soil forming factors, especially climate.

Disease emergence can be attributed to numerous factors such as the natural evolution and mutation of existing organisms; the spread of known diseases into new populations and/or new geographic areas; increasing human population; ecological changes that increase exposure of people to pathogens carried by insect or animal vectors; environmental changes that increase the exposure of people to contaminated dust, water, and soil; and exposure to as yet unknown pathogens (World Health Organization, 2002). The re-emergence of known soil-borne pathogens may occur in response to the breakdown or overtaxing of existing public health infrastructures as a result of refugee circumstances or other types of major demographic changes.

Examples of recently emerging and re-emerging soil-borne pathogens are *Clostridium* spp. bacteria, which cause a variety of diseases and are probably a permanent soil resident, transmitted by the fecal-oral route and through skin trauma; *Listeria monocytogenes*, a bacterium, which causes listeriosis and is a permanent soil resident transmitted by contact with soil contaminated with infective animal feces and also by inhalation of the organism; Sin Nombre virus, a *Hantavirus*, which causes hemorrhagic fever, is a transient soil resident, and is transmitted by inhalation of dust containing aerosolized rodent urine and feces; *Rotavirus* spp., which causes diarrhea and enteritis, an incidental or less commonly, transient soil resident, transmitted by the fecal-soil-oral route also by the fecal-respiratory route; *Coccidioides*, a fungus, which causes coccidioidomycosis, a permanent soil resident, transmitted by inhalation of *Coccidioides* arthroconidia (Bryan et al., 1994); and variant Creutzfeldt-Jakob disease, a TSE caused by prion infection, outbreaks in the UK in the late 1990s, possibly(?) a transient or incidental soil resident (Lederberg et al., 1992, Table 2.1, p. 36; World Health Organization, 1998).

Antimicrobial resistance, which is a natural consequence of the adaptation of microbes to exposure of drugs designed to kill them, may also cause re-emergence of infectious diseases. Even though resistance to antimicrobial agents is an irreversible, natural, and evolutionary process, it is exacerbated by several human activities including overuse (in developed countries) and under use (in developing countries) of antibiotic drugs; discharges of wastes from pharmaceutical production plants; disposal into landfills and sewage of wastes from common antibacterial household products (soaps, over-the-counter drugs, cosmetics, cleaning supplies, etc.); introduction into the human food chain by agricultural use of antibiotics for disease and pest control on plants and also by use of antibiotics on many types of livestock for therapeutic reasons and as growth enhancers; disposal of waste products from agricultural operations; disposal of sewage sludge in landfills and by application directly onto the land surface; and disposal of all types of household and industrial garbage into landfills (Standing Medical Advisory Committee, 1998, American Academy of Microbiology, 1999). Antibacterial drugs have received the most attention in regards to antimicrobial resistance; however, resistance is also developing to antiviral and antifungal drugs.

The ability of pathogenic microbes to respond and adapt quickly to new environmental conditions is fundamental to the development of antimicrobial resistance. Both the disposal of waste material in landfills and application of sewage sludge to the land surface create chemical and biological modifications of the natural soil environment in any given place and provide new environments that can foster microbial genetic change. In developed countries modern sanitary landfills and municipal sewage plants are closely regulated and designed to limit the escape of chemical and biological toxins. Homes in these countries, not connected to municipal sewage systems, in most cases utilize septic systems with leach fields that rely on the soils for sewage treatment. Nonetheless some sewage sludge and landfill leachates contain a variety of bacterial, parasitic, and viral pathogens derived from food waste, domestic animal feces, disposable diapers, and garden waste.

In developing countries raw sewage and untreated waste of all types are commonly disposed of directly into soils and at times are added directly to soils as fertilizer. Under these circumstances some bacterial pathogens in sludge and sewage may fail to adapt and die out, whereas others will adapt to the new environment and experience new growth.

Also of concern is the presence of residual amounts of antimicrobial agents (pharmaceuticals, heavy metals, toxic chemicals) that may select for the growth of new bacterial forms that are resistant to various antibiotics. Non-biological dispersal of pathogens from landfills are mainly by water (both surface runoff and groundwater in the vadose zone) and wind. Biological dispersal may be due to birds, rodents, insects, and humans.

XV. INTERCONNECTIONS: GEOLOGY/ SOIL/PATHOGENIC MICROBES

The importance of the soil environment for hosting human pathogens was recognized over 2000 years ago by Hippocrates who suggested that a physician, when arriving at an unfamiliar town, should examine the winds, sun aspect, sources of water, and ". . . the soil too, whether bare and dry or wooded and watered, hollow and hot or high and cold." (Jones, 1923). That wisdom has enormous room for development in the 21st century.

The study of the ecological habitats of soil microbes (both friendly and pathogenic) has been hampered for centuries by the inability to see, measure, count, and weigh organisms too small for the human eye to distinguish, especially *in situ*. It has also been hampered by the inadequate exchange of ideas and approaches among the wide diversity of scientific disciplines studying soils and microbes. Complete scientific descriptions of soil attributes, profiles, and classifications are rarely, if ever, given in the medical and microbiological literature focused on site-specific occurrences of soil-borne human pathogens. This makes it nearly impossible to conduct followup studies or further experiments in the same area or soil type, or to extrapolate results to other locations and studies. Commonly the only mention of the terrestrial environment of pathogenic organisms is described by phrases such as "soil" or "soil contaminated with bird feces," or "moist anaerobic soil," when describing the organism's habitat. This is not meant to fault prior research but instead to underline the need for multidisciplinary efforts. Many scientists are not knowledgeable about soils or soil attributes that affect microbes. Therefore, important geologic-soil-pathogen-process relationships are overlooked in many studies.

Geological features and processes are inherent in many soil attributes, which are, in turn, important controls over microbial activity and existence. For example, the abundance of ferromagnesian minerals and feldspars in a parent rock will determine the abundance and types of clay minerals formed (given the right climatic conditions) in soils by weathering processes. The presence of clay minerals strongly influences soil water potential, soil aggregation and pore size, microbe movement, virus adsorption, and the types of microbes present in any given soil.

Infection by soil-borne pathogens can be prevented or reduced by disrupting their life cycle. However, to do this a complete understanding of the infectious cycle is necessary to determine where interdiction will be most effective. For example, interdiction of the life cycle of soil-borne enteric pathogens is accomplished by the use of proper disposal and sanitation measures of human wastes. Another example is disruption of the hantavirus cycle of infection by controlling rodents in enclosed areas, thereby reducing or preventing exposure to contaminated aerosols from rodent feces and urine. These and similar examples require basic research into all aspects of the life processes and ecology of soil-borne pathogens and their interaction with the physical, chemical, and biological attributes of their habitat. These studies are best accomplished in the field and on-site using, whenever possible, noninvasive methods, some of which are reviewed by Madsen (1996). In the best circumstances, studies of soil-borne human pathogens would include a soil scientist familiar with field measurements and determinations of soil properties and classification. At the minimum, soil pathogen collection sites should be precisely located, soil textures should be determined, sand-silt-clay proportions should be estimated, organic content determined, hydrologic setting described, geomorphologic setting determined, pH and salinity (electrical conductivity) measured, and vegetation type and density described. These observations would go a long way to address Hippocrates' counsel to look at the soils.

Infectious diseases are a major cause of human suffering and mortality and account for an estimated 13 million deaths worldwide each year (World Health Organization, 1999), and that number is expected to grow. As indicated in previous sections, soil-borne human pathogens are important contributors to those numbers. Drug-resistant microbes are increasing at a dramatic rate and large urbanized areas in developing countries with dismal health care and sanitary facilities are magnets for displaced people. Deteriorating natural environments through urbanization, deforestation, and pollution of soils and waters coupled with the ease of human travel ensures breeding places and rapid transportation for many infectious agents. Increased understanding of the life cycles of pathogenic soil-borne microbes, the ecology of their habitats, and the environmental gateways they utilize for infectious transmissions will help break these cycles of infection. These problems are complex in character, global in distribution, and applicable to every human being. Their solutions are contingent on scientists from many disciplines working together to study the attributes and processes of complex soil ecosystems and communicating their results to public health officials.

SEE ALSO THE FOLLOWING CHAPTERS

Chapter 14 (Bioavailability of Elements in Soil) · Chapter 17 (Geophagy and the Involuntary Ingestion of Soil) · Chapter 18 (Natural Aerosolic Mineral Dusts and Human Health) · Chapter 27 (Investigating Vector-Borne and Zoonotic Diseases with Remote Sensing and GIS)

FURTHER READING

References cited and recommended readings are marked with an *. For references cited located at Web sites, see Appendix B.

Bertucci, J. J., Sedita, S. J., and Lue-Hing, C. (1987). Viral Aspects of Applying Sludges to Land. In *Human Viruses in Sediments, Sludges, and Soils* (V. C. Rao and J. L. Melnick, Eds.), CRC Press, Boca Raton, FL.

*Brady, N. C., and Weil, R. R. (2002). *The Nature and Properties of Soils*, Prentice Hall, Upper Saddle River, N. J.

*Brown, A. D. (1990). *Microbial Water Stress Physiology, Principles and Perspectives*, John Wiley & Sons, New York.

Bryan, R. T., Pinner, R. W., Gaynes, R. P., Peters, C. J., Aguilar, J. R., and Berkelman, R. L. (1994). Addressing Emerging Infectious Disease Threats: A Prevention Strategy for the United States Executive Summary, *CDC MMWR*, RR-5, 43, 1–18.

Council for Agricultural Science and Technology (2000). Transmissible Spongiform Encephalopathies in the United States: Task Force Report No. 136, p. 36.

*Cowan, G., Pines, D., Meltzer, D. (1999). Complexity, Metaphors, Models, and Reality: Santa Fe Institute, *Studies in the Sciences of Complexity, Proc., Vol. XIX*, Perseus Books, Cambridge, MA.

*Coyne, M. S. (1999). *Soil Microbiology: An Exploratory Approach*, Delmar Publishers, Albany, New York.

Elconin, A. F., Egeberg, R. O., and Egeberg, M. C. (1964). Significance of Soil Salinity on the Ecology of *Coccidioides immitis*, *J. Bacteriol.*, 87(3) 500–503.

Fiese, M. J. (1958). *Coccidioidomycosis*, Charles C Thomas, Springfield, IL.

Fisher, F. S., Bultman, M. W., and Pappagianis, D. (2000). Operational Guidelines for Geological Fieldwork in Areas Endemic for Coccidioidomycosis, U. S. Geological Survey Open File Report 00–348.

Flake, G. W. (1998). *The Computation Beauty of Nature: Computer Explorations of Fractals, Chaos, Complex Systems, and Adaptation*, MIT Press, Cambridge, MA.

*Galgiani, J. N. (1993). Coccidioidomycosis, *West. J. Med.*, 159, 153–171.

Galgiani, J. N. (1999). Coccidioidomycosis: A Regional Disease of National Importance, Rethinking Approaches for Control, *Ann. Int. Med.*, 130(4) (part 1), 293–300.

Gerba, C. P. (1987). Transport and Fate of Viruses in Soils: Field Studies. In *Human Viruses in Sediments, Sludges, and Soils* (V. C. Rao and J. L. Melnick, Ed.), CRC Press, Boca Raton, FL, pp. 141–154.

Goodman, R. A. (Ed.) (1994). Emerging Infectious Diseases, Update: Coccidioidomycosis—California, 1991–1993, *CDC MMWR*, 43(23), 421–423.

Griffin, D. M. (1972). *Ecology of Soil Fungi*, Syracuse University Press, New York.

Harris, R. F. (1981). Effect of Water Potential on Microbial Growth and Activity. In *Water Potential Relations in Soil Microbiology* (J. F. Parr, W. R. Gardner, and L. F. Elliott, Eds.), Soil Science Society of America, Special Publication, no. 9, p. 151.

Jones, W. H. S. (1923). *Hippocrates, Airs Waters Places, Vol. 1*, Harvard University Press, Cambridge, MA.

*Kosko, B. (1992). *Neural Networks and Fuzzy Systems*, Prentice Hall, Englewood Cliffs, NJ.

*Lederberg, J., Shope, R. E., and Oaks, S. C., Jr., (Eds.) (1992). Emerging Infections, Microbial Threats to Health in the United States, National Academy Press, Washington DC.

*Madsen, E. L. (1996). A Critical Analysis of Methods for Determining the Composition and Biogeochemical Activities of Soil Microbial Communities *in situ*. In *Soil Biochemistry* (G. Stotzky and Jean-Marc Bollag), Marcel Dekker, New York, pp. 287–370.

*Maier, R. M., Pepper, I. L., and Gerba, C. P. (Eds.) (2000). *Environmental Microbiology*, Academic Press, New York.

Matson, J. V., Lowry, C. L., Yee Ming C., and Whitworth, M. E. (1987). Physical and Chemical Characteristics of Sediments, Sludges, and Soils. In *Human Viruses in Sediments, Sludges, and Soils* (V. C. Rao and J. L. Menick, Eds.), CRC Press, Boca Raton, FL.

*Miller, R. W., and Gardiner, D. T. (2001). *Soils in Our Environment*, Prentice Hall, Upper Saddle River, N. J.

NASA (2001). Microbes and the Dust They Ride in Pose Potential Health Risks, Earth Observatory Release No. 01–120, NASA, Washington, D.C.

Pappagianis, D. (1980). Epidemiology of Coccidioidomycosis. In *Coccidioidomycosis* (D. A. Stevens, Ed.), Plenum Medical Book Company, New York, pp. 63–85.

*Pappagianis, D. (1988). Epidemiology of Coccidioidomycosis. *Curr. Top. Med. Mycol.*, 2, 199–238.

Pappagianis, D. (1999). *Coccidioides immitis* Antigen, Letter to the Editor, *J. Infect. Dis.*, 180, 243–44.

Pappagianis, D., Sun, R. K., Werner, S. B., Rutherford, G. W., Elsea, R. W., Miller, G. B., Jr., Egleston, M. D., and Hopkins, R. S. (1993). Coccidioidomycosis—United States, 1991–1992, *JAMA*, 269, 9, 1098.

*Paul, E. A., and Clark, F. E. (1996). *Soil Microbiology and Biochemistry*, Academic Press, New York.

Rabenau, H. F., Cinatl, J., and Doerr, H. W. (Eds.) (2001). *Prions: Contributions to Microbiology*, Karger, New York, vol. 7.

Raloff, J. (2001). Dust Storms Ferry Toxic Agents Between Countries and Even Continents, *Sci. News*, 160, 218–220.

Rusin, P., Enriquez, C. E., Johnson, D., and Gerba, C. P. (2000). Environmentally Transmitted Pathogens. In *Environmental Microbiology* (R. M. Maier, I. L. Pepper, and C. P. Gerba, Eds.), Academic Press, London, pp. 447–489.

Scogins, J. T. (1957). Comparative Study of Time Loss in Coccidioidomycosis and Other Respiratory Diseases. In Proceeding of a symposia on Coccidioidomycosis, Public Health Service Publication, n. 575, p. 132–135.

Sim, Y., and Chrysikopoulos, C. V. (2000). Virus Transport in Unsaturated Porous Media, *Water Resour. Res.*, 36(1), 173–179.

Sobsey, M. D., and Shields, P. A. (1987). Survival and Transport of Viruses in Soils: Model Studies. In *Human Viruses in Sediments, Sludges, and Soils* (V. C. Rao and J. L. Melnick, Eds.), CRC Press, Boca Raton, FL, pp. 155–177.

Staley, J. T. (2002). A Microbiological Perspective of Biodiversity. In *Biodiversity of Microbial Life* (J. T. Staley and Anna-Louise Reysenbach, Eds.), John Wiley & Sons, New York.

*Toranzos, G. A., and Marcos, R. P. (2000). Human Enteric Pathogens and Soil-Borne Disease. In *Soil Biochemistry*, Volume 10 (Jean-Marc Bollag and G. Stotzky, Eds.), Marcel Decker, New York, pp. 461–481.

Wallwork, J. A. (1970). *Ecology of Soil Animals*, McGraw-Hill, London.

World Health Organization (1996). Foodborne, Waterborne, and Soilborne Diseases, World Health Report, Fighting Disease Fostering Development, World Health Organization, Geneva.

CHAPTER 20

ANIMALS AND
MEDICAL GEOLOGY

BERNT JONES
Swedish University of Agricultural Sciences

CONTENTS

I. INTRODUCTION

It has been recognized for a very long time that animals can become sick after grazing in certain areas of the world. Eventually, it was determined that local geochemical anomalies were responsible, but in some cases the exact mechanisms behind the deficiencies and intoxications have only been known for the last 50 to 70 years. Local geochemical anomalies affect both domestic and wild herbivorous animals living in an area; however, wild animals will be especially susceptible to deficiency or intoxication due to the fact that they normally are totally dependent on feeds growing in a specific area.

Problems associated with restricted access to suitable feeds are seen in domestic animals in developing countries or among animals that have been reared under less intensive conditions, and where farmers rely on locally produced feeds. Extreme weather conditions, especially

drought, affect the availability of pasture and feed in areas with low groundwater tables and sensitive soils. The availability of nutrients is also affected by the seasonal development of forage plants eaten by the animals. The protein content of mature plants is often low, as is the amount of easily digestible carbohydrates and other important nutrients, but the fiber content is higher than in earlier stages of plant development. This seasonal shortage of available nutrients can be circumvented by migration between different areas and is believed to be one of the driving forces behind the seasonal migrations normally seen in many wild African herbivores. In some areas, this behavior is restricted, as some wildlife reservations are fenced to protect the animals from poachers and grazing competition from domestic animals (Maskall & Thornton, 1996). Migration between different grazing areas can also be applied to domestic animals and is a common practice for nomadic animal husbandry in Africa and continental parts of Asia. The nomadic behavior of caribou and semidomestic reindeer in the subarctic is primarily caused by climatic factors such as snow and ice, which can limit the availability of feed.

The increasing interest in organic farming in some European countries and North America exposes animals on such farms to higher risks of deficiency or dietary imbalance because of the use of feeds produced locally without conventional fertilizers. Farmers who are aware of these risks can overcome them by growing specific feeding plants selected to accumulate or exclude particular elements known to cause nutritional imbalances.

In the intensive type of animal husbandry seen in many parts of the world today, especially in Europe and North America, the effects of local geochemical anomalies are less apparent, as the content of nutrients in the rations fed to animals is controlled, and appropriate supplements or feedstuffs are utilized to compensate for potential deficiencies or excesses.

Generally, the metabolic effects of deficient or excessive intakes of nutrients due to the local geochemistry are the same in animals as in humans; that is, a change in activity of certain important enzymes is the major effect that compromises the health of animals. The symptoms exhibited by various animal species or humans suffering from the same pathological processes will therefore be identical or similar.

A. Recognition of Problems

Pathological changes caused by mineral imbalances, deficiency, or intoxication due to local geochemical anomalies are often difficult to detect in the living animal, especially in the free-living wild animal. The diffuse and ambiguous signs of these changes include retarded growth, decreased fertility, and decreased immunological capacity. In many cases, the only signs are suboptimal growth or low reproduction. Diagnosis is further complicated by interactions among the various elements already present in the soil. These interactions affect nutrient uptake by plants, making it difficult to evaluate results obtained from soil and plant analyses with regard to the nutritional value for the grazing animal or to estimate the animal's intake of required nutrients. Interactions also occur among various elements in the animal itself, both in the gastrointestinal tract and in different tissues of the body (see Table I). The gastrointestinal interactions are probably more important in herbivores, as the longer passage time through the tract and its larger volume provide greater opportunities for chemical reactions to occur among the many different compounds present. Often, extended analyses of blood and other tissues such as liver and kidney are needed to arrive at a definite diagnosis (see Section III.J).

TABLE I.   Important Interactions Between Various Elements in Animals

	F	Na	Mg	Al	P	S	K	Ca	Mn	Fe	Co	Cu	Zn	As	Se	Mo	Cd	I
F			+					+										+
Na							+											
Mg				+			+	+	+									
Al	+				+													
P			+	+				+	+	+		+	+			+		
S								+				+	+		+	+		
K		+	+															
Ca	+		+		+	+			+				+					
Mn			+		+			+			+							+
Fe					+			+			+	+	+					
Co								+		+								+
Cu					+	+				+			+		+	+		
Zn					+	+				+		+				+		
As															+			+
Se						+								+				
Mo					+	+						+						
Cd												+	+					
I	+							+			+			+				

Note: Interactions can occur in the gastrointestinal tract of the animal or on a cellular level in the tissues.
Source: Modified after Jacobson *et al.*, 1972.

B. Mineral and Trace Element Availability

Apart from general factors such as soil composition and pH, the plant species and stage of development will affect root uptake and hence the mineral content found in the plant. Alkaline soils will have more available molybdenum and selenium because these elements are present as anions, whereas most other metals are present as cations and their availability is favored by lower pH. The availability of minerals in soil eaten by animals is often difficult to estimate from common chemical analyses of mineral concentrations, as the actual gut absorption often is very low, depending on the chemical form of the mineral. The different types of interactions presented in Table I will also be of relatively large importance in these situations. Minerals taken up by different plants and present in plant tissues are generally more available as the root uptake requires solubility in soil water or intracellular water. Dust or other soil contaminations on plant surfaces, especially on pilose or sticky structures, will be as unavailable to absorption in the gut as pure soil eaten by an animal. Involuntary soil intake will be much higher in animals grazing hard and close to the ground, such as cattle and sheep, than goats and many wild ruminants that to a great extent feed on shrubs, bushes, and tree leaves (browse). Especially in situations of feed shortage, soil intake can be high (several kilograms per day in cattle).

C. Mineral and Trace Element Requirement and Excess

The needs for mineral and trace elements by common domestic animal species have been established in experiments using normal feeding procedures or specialized synthetic or semisynthetic feeds (see Tables II to VII). An animal's growth rate and normal development are followed in these types of experiments, as well as the health and reproductive functions of the animal. In many cases, the nutrient requirements are influenced by the selective breeding of animals to provide increased growth rates or improved milk or egg production. The complex interrelationships among minerals and trace elements and other feed components make it difficult to determine if nutrient requirements are being met or if adverse effects due to excess nutrients could be expected when unconventional or extreme feeding regimes are used. Some of the known requirements for various minerals and trace elements are discussed in Section III. The mineral and trace element requirements of wild or

TABLE II. Nutrient Requirements for Calcium and Phosphorus for Some Domestic Species

Animal	Calcium	Phosphorus	Ref.
Lactating dairy cows	4.8–8.5	3.1–5.3	NRC, 1989a
Growing calves	3.2–5.8	2.6–3.4	NRC, 1989a
Beef cattle	1.8–17	1.9–6.6	NRC, 1984a
Sheep	2.2–9.1	1.8–4.2	NRC, 1985b
Swine	5.6–10	4.4–7.8	NRC, 1988
Horses	2.7–7.6	1.9–4.2	NRC, 1989b
Dogs	6.6	4.9	NRC, 1985a

Note: The amounts are expressed as g/kg d.m. daily feed intake.

TABLE III. Amounts of Manganese Required for Normal Development and Health in Some Domestic Animal Species and Amounts Causing Adverse Effects

Animal	Requirements	Adverse Effects	Ref.
Dairy cattle	40	1000	NRC, 1989a
Beef cattle	40	1000	NRC, 1984a
Sheep	20–40	—	NRC, 1985b
Horses	40	400	NRC, 1989b
Swine	2–10	400	NRC, 1988
Chicken	30–60	2000	NRC, 1984b

Note: The amounts are expressed as mg/kg feed d.m.

TABLE IV. Nutrient Requirements for Copper and Amounts Causing Toxicity or Other Adverse Effects in Domestic Animals

Animal	Requirements	Toxic or Adverse Effects	Ref.
Dairy cattle	10	100	NRC, 1989a
Beef cattle	8	100	NRC, 1984a
Sheep	7–11	25	NRC, 1985b
Swine	3–6	250	NRC, 1988
Horses	10	—	NRC, 1989b
Dogs	2	—	NRC, 1985b
Chicken	6–8	500	NRC, 1984b

Note: The amounts are expressed as mg/kg feed d.m.

TABLE V. Nutrient Requirements for Zinc and Amounts Causing Adverse Effects in Domestic Animals

Animal	Requirements	Adverse Effects	Ref.
Dairy cattle	40	1000	NRC, 1989a
Beef cattle	30	1000	NRC, 1984a
Sheep	20–33	750	NRC, 1985a
Goats	10	—	NRC, 1979
Growing swine	50–100	1000	NRC, 1988
Breeding swine	50	—	NRC, 1988
Horses	40	—	NRC, 1989b
Chicken	35–65	1200	NRC, 1984b

Note: The amounts are expressed as mg/kg feed d.m.

TABLE VI. Nutrient Requirements for Selenium and Amounts Causing Adverse Effects in Domestic Animals

Animal	Requirements	Adverse Effects	Ref.
Dairy cattle	0.3	5	NRC, 1989a
Beef cattle	0.2	5	NRC, 1984a
Sheep	0.1–0.2	—	NRC, 1985b
Swine	0.1–0.3	20	NRC, 1988
Horses	0.1	5	NRC, 1989b
Dogs	0.11	—	NRC, 1985a
Chicken	0.1–0.15	—	NRC, 1984b

Note: The amounts are expressed as mg/kg feed d.m.

TABLE VII. Nutrient Requirements for Iodine for Some Domestic Species

Animal	Requirements	Ref.
Dairy cattle	0.5	NRC, 1984a
Beef cattle	0.5	NRC, 1989a
Sheep	0.1–0.8	NRC, 1985b
Swine	0.14	NRC, 1988
Horses	0.1	NRC, 1989b
Chicken	0.3	NRC, 1984b

Note: The amounts are expressed as mg/kg feed d.m.

exotic domestic species are not well known, and neither are the amounts needed to cause deleterious effects in these animals. Extrapolations from known needs of related domestic species should be done with great caution as many wild species have evolved in specific areas and have adapted to the geochemical situations in those particular habitats.

D. Salt and Mineral Licks

Both domestic and wild animals are regularly seen eating soil or earth. Under conditions of overgrazing or a shortage of feed, soil eating is believed to correct nutritional deficiencies. In certain situations, animals are exploiting natural sources of sodium- or phosphorus-rich minerals for this purpose. Other minerals are actively sought by animals, but the mechanisms behind this behavior are not known. It is a common belief that animals, and humans, will correct deficiencies in their mineral and trace element supply by eating soil (*i.e.*, pica). This behavior is seen in many cases of nutrient deficiencies, but no known mechanism induces or regulates it. More probable in animals is that this knowledge is passed down from mother to offspring as an important factor in the utilization of special sites with favorable mineral compositions. One illustrative example of this is the utilization of minerals found in caves in Kenya that are actively mined by elephants living in the area (Bowell *et al.*, 1996). The elephants use their tusks to mine the veins of calcite–zeolite in the roofs and walls of the caves; some of the rocks are eaten by the animals, but some are left on the cave floor. The environment on the cave floor (*e.g.*, water and debris from animals) induces the formation of salts on the surfaces of the rocks. This cave salt is utilized by local tribesmen for their cattle, and it is also used by other wild animals, such as baboons (*Papio cyancephalus*) and leopards (*Pantera pardus*). Bushbucks (*Tragelphus scriptus*) also seek the formed salt deposits in the caves.

II. SPECIES AND BREED DIFFERENCES

For a number of reasons, important differences can be seen between monogastric and ruminant species in the metabolic handling of several elements that cause deficiency or intoxication. One reason is the simple fact that the pH of the forestomachs of ruminant species ranges from 6 to 7, and the pH of the abomasum, the true stomach of ruminants, is 2 to 3, which renders most metallic elements less available for absorption as compared to most monogastric species, for which the stomach pH varies from 1 to 2. Another important reason for species differences is the simple chemical reactions that can occur in the forestomachs of ruminants. The anaerobic fermentation that takes place in the forestomachs can produce valence stages and complex compounds that are not available for intestinal absorption, as most minerals and trace elements are absorbed in cationic form. For a more comprehensive description of the digestive physiology of animals see, for example, Chapter 14 in Sjaastad *et al.* (2003).

In some cases, breed differences can exist within the same domestic species due to the development of breeding animals having special properties desirable in specific situations. The very fast growth rate seen in many modern domestic animals is a breeding effect that can affect an animal's sensitivity to insufficient or excessive intakes of minerals and trace elements. Because many modern breeds have a high demand for nutrients and therefore a high feed intake, they may not necessarily meet their mineral and trace element requirements. This is especially obvious in very young animals during the suckling period if they feed on milk that is poor in iron. The available iron is insufficient to meet the requirements for normal hemoglobin formation, thus causing anemia, which, in turn, often also leads to increased susceptibility to infection.

In most cases, attributing pathological conditions to local geochemical anomalies is complicated. The clinical picture and postmortem examination are often equivocal. Support from clinical chemical analyses of blood and other tissues from the living animal as well as chemical analyses from tissues obtained during a postmortem examination are necessary to obtain a definitive diagnosis.

III. Specific Elements

This section focuses on effects seen in normal environmental situations in both wild and domestic animals and only briefly discusses effects observed in areas polluted by mining or other industrial activities. The minerals calcium and phosphorus are discussed first, followed by the trace elements in order of increasing atomic weight.

A. Phosphorus and Calcium

The metabolism of phosphorus and the metabolism of calcium are so interrelated that they are discussed together. Utilization of these two elements is profoundly affected by the amount of vitamin D present in feeds consumed by the animal. Disturbances in the metabolism of phosphorus and calcium that are seen in young growing animals can affect bone formation when cartilage is being gradually replaced by bone tissue containing hydroxyapatite, a calcium phosphate compound that makes bone hard and mineralized. If this process is disturbed, both bone growth and function are affected. A well-known pathology associated with this mineralization is rickets, a disease known in both animals and humans since antiquity. Often, rickets is caused by a vitamin D deficiency. A typical case of rickets shows enlargement of the junction between the ribs and costal cartilages (costochondral junction) as well as the growth plates (metaphyses) of the long bones. The stability of the bone is reduced, resulting in fractures or bending of the legs, thus giving the animal the tell-tale appearance of rickets. In adult animals, where bone growth is completed, a similar defect is called *osteomalacia*.

Direct toxic effects of an excessive intake of calcium are not seen as the balance between calcium and phosphorus is crucial for the biological effects. An excessive intake of calcium, however, severely affects the availability of several other minerals and trace elements, as shown in Table I. High phosphorus intake also interferes with the metabolism of many other elements, as seen in the table. Normal calcium and phosphorus requirements for common domestic species are presented in Table II; these amounts are relevant in situations where the calcium–phosphorus ratio is balanced (ideally, 1–1.2).

Endemic areas deficient in phosphorus are found in South Africa and Latin America; cattle grazing on grassland in these areas often have a very low intake of both phosphorus and calcium. Clinical signs of phosphorus and calcium deficiency are more apparent in young animals than adults. Subnormal growth, low reproduction, and pica are observed before the bone pathology (osteomalacia) becomes apparent, in both young and adult animals. Deficiencies can also be seen in pigs in the form of poor growth, gait disturbances, lameness, bone deformation, and even fractures. The fractures occur in the vertebrae, which are more prone to demineralization than the long bones of the legs.

A special situation related to excessive calcium intake is calcinosis, which is seen in areas where plants producing substances related to vitamin D are growing. In

New Guinea, the European Alps, Argentina, Brazil, and Florida, enzootic calcinosis with calcium deposits in soft tissues has been described. In the Americas solanaceous plants produce compounds that can mimic vitamin D activity in the gastrointestinal tract, thus promoting an excessively high absorption of calcium and phosphates and disturbing the normal homeostatic regulation of these elements. In Argentina and Brazil, this condition (known as *enteque seco*) is caused by ingestion of leaves from the shrub *Solanum malacoxylon*, and in Florida a large ornamental plant (*Cestrum diurnum*) is the cause. In Central Europe, calcinosis is caused by consumption of the pasture grass *Trisetum flavescens*. The increased uptake of calcium results in calcification of soft tissues in the affected animals, and the often massive deposits of calcium seen in the walls of major arteries, especially the aorta, will eventually kill the animal due to disturbed blood circulation and heart failure.

B. Aluminum

Only trace amounts of aluminum are found in most biological organisms, although this element is abundant in most soils; however, some exceptional tropical plants do accumulate aluminum, which can reach levels of >1 g/kg of plant (dry matter). Aluminum can be toxic to plants as well as fish and other aquatic biota, but direct toxic effects of aluminum on livestock are unlikely at natural environmental levels. High levels of ingested aluminum will, however, decrease the availability of nutrients such as iron, phosphorus, and, to some extent, calcium (see Table I). Most of this aluminum would be present as external contamination on the plant material eaten by the animals, especially in a dry and dusty environment. The high levels of aluminum and iron seen in some acidic soils can form insoluble phosphate complexes that produce secondary phosphorus deficiency in animals. This could result in grass tetany and osteomalacia in exposed cattle. The amounts needed to cause such problems are estimated to be 0.5 to 8 g/kg plant material (d.m.).

C. Fluorine

Some species might need fluorine as an essential trace element, but it is generally known to be toxic to both humans and animals. Small amounts of fluorine absorbed in the gastrointestinal tract will quickly form CaF, which is incorporated into bone and teeth, result-

ing in more stable and less reactive bone tissue. Signs of fluorine toxicity were described very early (around the year 1000) in Iceland, where cattle grazed on grass contaminated with fallen volcanic ashes. Areas with volcanic activity are particularly prone to excessive fluorine exposure due to external contamination of pasture plants. Parts of the North African coastal plain have been known for centuries to be areas of endemic fluorine toxicity. Often, drinking water is the source of excessive fluorine intake, and endemic chronic fluorosis has been described in domestic ruminants and horses in Australia, India, Turkey, and several parts of Africa due to high levels of fluorine in their drinking water (see also Chapter 12, this volume).

Ruminants are more sensitive to fluorine than monogastric species, but even ruminants can tolerate only 80 to 100 μg/g d.m. in their feed (Mertz, 1987). Poultry is more resistant to the toxic effects of fluorine, and reports suggest that poultry is able to tolerate twice as much as mammals. A high intake of aluminum or calcium (see Table I) will protect species from the toxic effects of excessive fluorine in their feed or drinking water.

Clinical signs of fluorine toxicity are evident primarily in the teeth and bone. Teeth still under development are most sensitive to high fluorine intake and become mottled and discolored. More severe indications are modifications of size and shape of developing teeth and decreased strength resulting in fractures even after limited mechanical stress. Also, bone will show changes in size and shape as a result of fluorine toxicity, especially in growing animals. Lameness can also be seen as a result of bone pathology caused by excessive fluorine intake.

Most plants have low concentrations of fluorine, but external contamination is the primary route of toxic exposures; however, in South Africa, some plants do produce fluoroacetic acid, which can poison the grazing sheep. This fluoroacetic acid is converted *in vivo* to fluorocitrate, which specifically inhibits activity of the enzyme aconitase, which is necessary for the metabolism of citrate in the energy-producing citric acid cycle (or Krebs cycle) vital to all living cells. Fluoroacetate was used earlier as a rodenticide due to its potent toxic effect.

D. Manganese

Concentrations of manganese in plant material are highly variable, and the effects of a deficiency due to low manganese intake can be seen in ruminants, swine, and poultry fed normal feeding plants. The risk for low manganese intake is greater if the land used for feed production or grazing is excessively limed. Symptoms in affected animals are changes in lipid and carbohydrate metabolism and impaired growth and infertility, depending on the duration of the deficiency and the age of the animal. These symptoms are caused by the decreased activity of manganese-containing enzymes such as arginase, manganese–superoxide dismutase, and pyruvate carboxylase. The requirements of manganese in feeds are generally greater for poultry (see Table III), and feeds based on corn and barley could be deficient. Toxic effects of manganese are unlikely, as this element is considered one of the least toxic; however, continuous grazing on the volcanic soils of Costa Rica can result in a manganese intake of >200 mg per day and has been reported to produce reproductive changes in cattle. Adverse effects of excessive manganese intake are mainly caused by its interactions with iron and other elements (see Table I), resulting in secondary iron deficiency anemia that can be seen in both lambs and pigs.

E. Cobalt

A deficiency of cobalt *per se* is not described in carnivorous animals, but their diet should have this element incorporated into vitamin B_{12} (cyanocobalamin) for normal methionine synthesis and energy metabolism. Two different forms of vitamin B_{12} are responsible for these two metabolic pathways: methylcobalamin for methionine synthesis and adenosylcobalamin for energy metabolism. Methylcobalamin is of importance for both protein and lipid synthesis, and in humans a vitamin B_{12} deficiency typically causes a specific megaloblastic anemia (pernicious anemia) and neurological disorders due to progressive demyelination. Pernicious anemia is not seen in domestic animals in cases of vitamin B_{12} deficiency. Ruminants and other herbivorous species do not need preformed vitamin B_{12} in their feed, as the microbes present in their gastrointestinal tract, especially the rumen, can produce this complex compound if sufficient cobalt is present in the feed. The requirement in all domestic ruminant species is about 0.1 mg/kg feed (d.m.), as the uptake of the vitamin is comparatively low in ruminants and even less in monogastric animals.

Young, growing ruminants grazing cobalt-deficient pastures are most sensitive to deficient levels of the adenosylcobalamin moiety. Lambs especially will show symptoms of inanition in spite of access to grazing, but older animals can also be affected under certain circumstances. Underlying this problem is the fact that adenosylcobalamin converts methylmalonyl-CoA to succinyl-CoA, beginning with propionic acid formed in the forestomachs of ruminants. In monogastric species, this metabolic pathway is of minor significance, but the energy metabolism of ruminants is based on the volatile fatty acids (VFAs) acetate, butyrate, and propionate and not glucose, which is typical for most other species. The clinical effects of cobalt deficiency in ruminants grazing deficient pastures have been recognized for quite some time in many parts of the world. The disease, or rather the apparent symptoms, goes by various local names (*e.g.*, in New Zealand, bush sickness; in Australia, wasting disease; in the United States, salt sick or neck ail; in Brazil, *mal de colete* or *pest de secar*; in Great Britain, pining disease; in Norway, white liver disease), apart from more general terms such as muscular wasting or enzootic marasmus. Sheep grazing in deficient areas are more sensitive than cattle, and young animals are more sensitive than adults. The deficiency is often due to low levels of cobalt in peaty soils or otherwise highly organic soils. Signs of cobalt deficiency in ruminants are retarded development in lambs, slow growth rate, inappetence, emaciation, and weakness.

Eventually, animals suffering a deficiency of cobalt die of emaciation. At an early stage, the symptoms are nonspecific and other reasons for the observed pathology could be suspected. Cobalt deficiency has to be confirmed by determination of methylmalonic acid (MMA) in blood samples or determination of cobalt/vitamin B_{12} in tissue samples, preferably the liver. At autopsy, the carcass is generally pale, and the liver is also characteristically pale with a pathologic texture that gives the deficiency one of its more common names—ovine white liver disease (OWLD) (see Figures 1–3). In some situations, OWLD or cobalamin deficiency can be provoked by some additional nutritional factor suspected to be fructan or other water-soluble carbohydrate compounds present in some grasses early in the season. These carbohydrates are fermented to VFAs in the forestomachs which increases the demand for adenosylcobalamin in the energy metabolism of the animal.

Neurological symptoms due to central demyelination that have been reported in vitamin B_{12}-deficient lambs include blindness and locomotor disturbances (Ulvund & Pestalozzi, 1990). Recently, similar neurological symptoms observed for decades in Nova Scotia moose have been associated with cobalt deficiency (Frank *et al.*, 2003). Vitamin B_{12} deficiency and neurological prob-

FIGURE 1 The liver from an animal euthanized due to ovine white liver disease (OWLD); note pale irregular areas protruding above the surface. (Photograph courtesy of Martha J. Ulvund, Sandnes, Norway.)

FIGURE 2 Lamb showing nonspecific signs such as serous eye discharge and small crusts on the ears, typical of OWLD. This lamb was only half the size of its healthy flockmates. (Photograph courtesy of Martha J. Ulvund, Sandnes, Norway.)

lems among elderly humans have attracted considerable interest lately.

F. Copper

The effects of copper deficiency have been extensively studied in several parts of the world, because it is a widespread problem for sheep and cattle production. Copper deficiencies due to local geochemical anomalies produce significant losses for farmers worldwide. Studies of copper metabolism in various species have shown that there are important differences between ruminants and monogastric animals due to copper–molybdenum sulfate interactions in the forestomachs and biochemical handling of copper in the body after absorption. These biochemical differences are mainly due to the specific binding properties of metallothioneins in the liver and kidneys of ruminant species which differ from those seen in nonruminants. The copper–molybdenum sulfate interactions can make it difficult to evaluate the copper status of an animal because it is necessary to know the availability of all three compounds to make a reliable estimation of biologically active copper. To avoid an imbalance between copper and molybdenum, the ratio in the feeds should be between 1 to 2 or 1 to 4 (this important interaction is further discussed in Section III.J). These differences in the binding properties of metallothionein render ruminants very sensitive to chronic copper toxicity even after slight but prolonged excessive intake of copper.

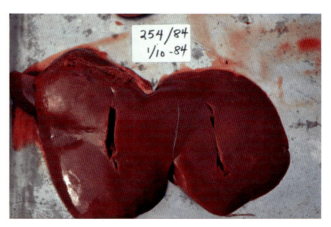

FIGURE 3 Swollen and soft liver from the animal shown in Figure 2; the color is paler than normal. (Photograph courtesy of Martha J. Ulvund, Sandnes, Norway.)

Copper deficiency has been described in cattle and sheep from many areas of the world, such as New Zealand, Australia, Northern Europe, and the United States. In Northern Europe, the problem is associated with grazing on peaty soils which often are also low in several other trace elements. The symptoms of copper deficiency are dominated by anemia and altered iron metabolism, bone disorders, diarrhea, and wasting, but also seen are infertility and depigmentation of wool and hair. In newborn lambs, dysfunctioning nerve development (swayback) can be seen when the ewe has been fed insufficient amounts of copper during the gestational period, when the central nervous system of the fetus develops. Swayback can also be seen in goats but rarely, as goats appear to be less sensitive to this type of deficiency. Cattle can sometimes show cardiovascular disorders (falling disease) in cases of copper deficiency, with sudden death after excitement or physical activity.

1. Chronic Intoxication

Chronic copper toxicity seen in grazing ruminants is very often due to very low molybdenum and sulfate intake. A suitable ratio between the content of copper and molybdenum in the feed is 1–2 to 1–4. Sheep and calves are more sensitive to this intoxication than adult cattle and goats. The excessive copper is bound to metallothioneins and initially accumulated in the liver. Symptoms of chronic copper intoxication are seen only when the animal is stressed by pregnancy and delivery, transportation, infection, or other disease. The stress induces a massive release of copper from metallothioneins in the liver that initiates a hemolytic crisis, with erythrocyte disruption and secondary liver and kidney damage. Only at this stage is the concentration of copper in blood elevated. Often, affected animals do not show any symptoms but are suddenly found dead. If an animal survives the acute hemolytic crisis, jaundice and symptoms of liver and kidney failure will be seen. A typical postmortem picture includes a discolored carcass and kidneys. If chronic copper intoxication is suspected in a flock of sheep, supplementation with molybdenum and sulfate could be used to bind excessive copper and facilitate excretion. In areas with known high levels of copper in feeds, special boluses containing molybdenum are placed in the rumen of grazing sheep and cattle to prevent chronic copper toxicosis. Goats are less sensitive and will seldom show chronic copper toxicosis unless the copper intake is very high or the molybdenum and sulfate intake is very low.

G. Zinc

Zinc is vital to a very large number of important enzymes, and inadequate amounts of zinc in feeds will produce many different effects on animals. Apart from the interactions shown in Table I, the intestinal uptake of zinc will decrease if feed levels of phytate are high or if clay substances are ingested. Clinical apparent zinc deficiency has been reported in various parts of the world (e.g., New Zealand, Australia, northwestern Europe). Early and very obvious signs are skin lesions (parakeratosis and hair loss), seen both in pigs and cattle

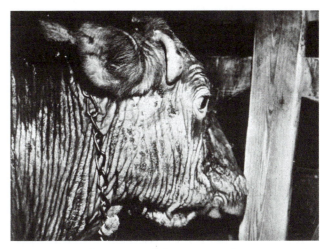

FIGURE 4 Severe zinc deficiency in a dairy cow causing parakeratosis and pronounced wrinkling of the neck skin.

(see Figure 4). Reduced appetite and retarded growth rate are other signs of deficiency, as are vomiting and scouring in pigs. In calves, a stiff gait and swelling of joints are seen as a result of disturbances in bone mineralization. This effect on bone metabolism is potentiated by a high calcium intake, which is often seen in young growing animals (see Table I).

In the Holstein–Friesian breed of cattle, an autosomal recessive trait has been described that interferes with intestinal absorption of zinc. In this condition, which parallels *acrodermatitis enterohepatica* in humans, affected calves show signs of zinc deficiency even though the feed intake of zinc is adequate.

H. Arsenic

Chronic arsenic poisoning is found in humans and animals in several parts of the world (e.g., Bangladesh, Argentina, Mexico, and the United States). In many cases, water from deep wells with high concentrations of arsenic is responsible for the toxic effects (see also Chapter 11, this volume); however, the use of arsenic compounds in herbicides, anthelmintics, insecticides, and rodenticides presents a greater risk for accidental or inadvertent exposure of animals to toxic amounts of arsenic. The symptoms described in cattle with chronic arsenic toxicity include diarrhea with weight loss, inflammation of the eyes and upper respiratory tract, gait incoordination, and changes in hair coat. More acute cases of arsenic intoxication are rapidly lethal.

I. Selenium

Diseases caused by a high selenium intake have been recognized for several hundred years in some selenium-rich areas of the world. Diseases caused by selenium deficiency are also well known, although selenium was one of the last elements to be recognized as an essential trace element, in 1957 (see also Chapter 15, this volume).

1. Selenium Deficiency

Problems associated with a low intake of selenium were recognized much later than the toxic effects of an excess of this element; however, several areas in New Zealand, China, Finland, Sweden, and North America do have insufficient selenium levels in their native feeds. Such a deficiency seems to be related to the growth rate of animals, as young, fast-growing animals such as piglets, foals, lambs, and calves are more sensitive to a low intake of selenium. The selenium-containing enzyme glutathione peroxidase (GSH-Px) has a cytoplasmic localization and acts together with fat-soluble vitamin E, which has a membrane localization in the cell. In fresh feeds, the supply of vitamin E is high but during storage the content decreases; therefore, in situations where stored feeds are used, it is especially important to ensure that the selenium requirements are met. Normally, GSH-Px and vitamin E will detoxify the free radicals produced by the continuous oxidative metabolism present in all cells of an animal. If feeds high in polyunsaturated fatty acids or other oxidative compounds are utilized, the nutritional requirements for selenium will be higher, especially if the feed has been stored for a long time. In pigs, liver degeneration and necrosis are prominent signs, but muscular degeneration is also seen in both skeletal and cardiac muscles. Muscular pathology, often precipitated by even minor physical activity, is the dominating symptom in other species and causes paralysis and acute heart failure. In many cases, the young animals are found dead in the pasture or in their boxes. The typical postmortem picture in these cases features degeneration of muscular tissue, which looks like fish meat. Similar changes can be seen at autopsy in skeletal muscles from slaughtered paralytic animals.

2. Acute Selenium Toxicity and Blind Staggers

If large quantities of selenium are ingested, an acute form of intoxication, called *blind staggers*, can be seen for which blindness and other symptoms from the nervous system are prominent. Other symptoms include diarrhea and prostration. Affected animals often die within hours. Animals with acute intoxication also develop a garlic-like breath odor due to the formation of volatile methylated selenium compounds. This type of acute intoxication can be seen in areas throughout the world where seleniferous plants accumulate selenium from the selenium-rich soil (*e.g.*, Australia, China, North America, Russia, and Western Europe). Normally, grazing animals avoid these plants, but when a shortage of feed occurs the animals will graze on these plants and become intoxicated. If smaller amounts of selenium are ingested, the neurological symptoms are more prominent and, in addition to blindness, paralysis, respiratory problems, abdominal pain, and other signs of pain such as grinding of the teeth and salivating will be seen. Eventually the animals die due to starvation or respiratory failure. Both the blindness and the locomotor problems contribute to the inability to graze and meet the metabolic needs of the sick animal.

3. Chronic Selenium Toxicity and Alkali Disease

Around 1295, in his travels in western China, Marco Polo described what is most probably chronic selenium intoxication in horses. Animals eating certain plants that accumulated selenium showed dramatic changes, such as loss of hoofs (exungulation) and dermatological changes. Similar toxic effects were later described in North America when horses in the U.S. Cavalry were affected. It is even claimed by some that the outcome of the battle at Little Bighorn was determined by the ingestion of seleniferous plants by the horses in General Custer's squadrons which made them sick and unfit for the ensuing battle. Cattle brought by settlers who came to the Great Plains also showed signs of selenosis. These farmers associated the disease with the high salt content of the water and alkali seeps in the area and named the problem alkali disease. It appears, however, that certain seleniferous plants are necessary for the appearance of alkali disease. Such plants can accumulate selenium several hundredfold in their tissues compared to the soil concentration. Seleniferous plants are also found in limited areas of India, Israel, Central Europe, and South America, where chronic selenosis is observed. In other areas of the world, such as Hawaii and Costa Rica, the volcanic soil is rich in selenium but no seleniferous plants grow, and alkali disease is not seen.

J. Molybdenum

1. Deficiency

The biochemical roles of molybdenum-containing enzymes vary according to the different valence states

(+4, +5, or +6) attainable for this element. Deficiencies caused by low molybdenum intake are seen in experimental situations but are rare under natural conditions. The nutritional requirement in ruminants is estimated to be <0.1 mg/kg feed (Mertz, 1987). The intricate interrelationships among molybdenum, copper, and sulfate make it difficult, however, to give reliable figures for molybdenum requirements and adverse effects in various animal species (see Section III.J.2). It has been suggested, though, that a molybdenum deficiency is the cause of the high incidence of urinary xanthine calculi in New Zealand sheep grazing on pasture low in molybdenum due to the resulting low activity of the molybdenum-containing enzyme xanthine oxidase. Other situations where molybdenum deficiencies have been suspected to occur spontaneously involve poultry, but these problems have not responded to simple molybdenum supplementation because the etiology is more complex. Most probably the earlier mentioned interactions among molybdenum, copper, and sulfate are involved, but other transition elements may also be involved as contributing factors in these cases.

2. Intoxication

The effects of excessive molybdenum intake (known collectively as molybdenosis) are closely linked to copper deficiency, and most of the symptoms seen in molybdenosis are linked to the decreased activity of copper-containing enzymes. This is especially true in ruminants, where the balance between these two elements and sulfur in the form of sulfate or sulfur-containing amino acids is very important but often difficult to estimate from analytical data on feed composition. The chemical reason for this interaction is the formation of oxythiomolybdate and tetrathiomolybdate (TTM) in the rumen. The TTM will react with copper and other elements in the gastrointestinal tract, making these elements unavailable for absorption. TTM will also be absorbed from the gut to the blood and distributed to various tissues in the body where it binds copper in enzymes and other proteins, thus rendering them dysfunctional. Cattle are most susceptible to molybdenosis, especially dairy cows and young animals. Sheep and goats are less sensitive, and horses as well as swine appear insensitive to high intakes of molybdenum. In the young, growing foal, however, the occurrence of rickets has been reported when the mare has been kept on high-molybdenum pasture. Defective bone metabolism with osteoporosis and joint abnormalities have also been reported as a prominent symptom in molybdenotic cattle and sheep.

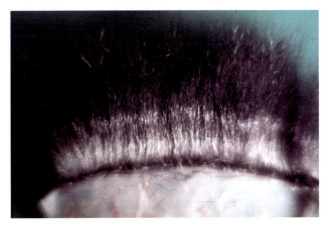

FIGURE 5 Discoloration of the black wool in a sheep suffering from molybdenosis.

Severe molybdenosis in cattle is widespread in the world. Plant uptake of molybdenum is favored by soil with a neutral or high pH. This problem has been known for a long time in England, where it is called *teart*, and in New Zealand, where the problem is known as *peat scours*. When cattle are put on pasture high in molybdenum, after only a few days they may exhibit typical signs of molybdenosis, such as profuse scouring and a harsh and discolored coat (see Figure 5). In milder cases, the diarrhea is less severe and the animals will recover rapidly if put on pasture with low molybdenum content. Less obvious effects are seen with regard to reproduction and thyroid function in cattle, for which a direct effect of molybdenum on the endocrine regulation in the hypothalamus is suspected. Anemia is also seen in cattle and sheep due to decreased hemoglobin synthesis, because the copper-containing enzyme ceruloplasmin is needed for the incorporation of iron into the heme molecule.

3. Molybdenosis in Moose

An example of the complexity of diagnosing trace element imbalances in animals, especially wild animals, is the occurrence of molybdenosis in moose (*Alces alces* L.) in an area in southwest Sweden (Älvsborg county) beginning around 1985 (Frank, 1998; Frank *et al.*, 2000). Affected animals that were found alive displayed a number of symptoms, including behavioral disturbances with apathy and pathological locomotion, loss of appetite, diarrhea, emaciation, discoloration and loss of hair, and opacities in the cornea and lens of one or both eyes. Postmortem examinations revealed edema, hyperemia, hemorrhage, and erosive lesions in the mucosa of the gastrointestinal tract. The heart was dilated and flabby, and myocardial lesions were seen macroscopically. Microscopic studies of different tissues showed pronounced hemosiderosis in the liver and spleen. Often, an animal showed some but not all of these symptoms or postmortem signs, making it difficult to classify a specific individual as a victim of this particular disease.

The gastrointestinal changes indicated a viral etiology of the disease, but other findings did not support this, and still no virus had been detected that could be responsible for the disease. It was suggested that starvation due to overpopulation or high age in the local moose population could account for the disease. Some of the chemical findings in samples from sick animals, however, were not compatible with starvation and lack of nutrients, energy, minerals, or trace elements. An indication of the probable etiology was found in measurements of trace elements in livers and kidneys from yearling moose sampled all over Sweden during the normal hunting periods in 1982, 1988, 1992, and 1994. These results showed that from 1982 (before the outbreak of the disease) to 1994 the hepatic concentration of selenium and especially molybdenum increased in the affected area, whereas the levels of copper, cadmium, and some other elements decreased in the livers of these young animals. These results were different from those for neighboring counties during the same period.

A plausible explanation for this discrepancy is the fact that the Älvsborg county was heavily limed to counteract the effects of acid rain, which was damaging the local forest and aquatic ecosystems. An increase in soil pH after liming increases the availability of molybdenum and selenium but decreases the availability of other metallic trace elements. The majority of symptoms and pathological signs found in the moose can be explained, then, by decreased activity of copper-containing enzymes such as tyrosinase, ceruloplasmin, superoxide dismutase, lysyloxidase, diaminoxidase, and methioninsyntase. Specific analysis of the copper enzyme cytochrome c oxidase in myocardial tissue showed lower activity in sick animals than in healthy controls. Determinations of the pancreatic hormone insulin in blood plasma from diseased moose showed concentrations about double those of apparently healthy animals, and still there was evidence of persistent hyperglycemia in these animals, as higher concentrations of furosine and pentosidine were found in kidney tissue compared to the levels determined in healthy animals. Furosine, pentosidine, and some other compounds are formed by nonenzymatic glucation of tissue proteins in the presence of high levels of blood glucose. The hyperinsulinemia and persistent hyperglycemia observed in the moose are similar to non-insulin-dependent dia-

betes (NIDD, or type 2 diabetes) often seen in humans, especially the elderly.

Some other endocrine disruptions could be suspected, because the direct effects of TTM on the hypothalamus have been demonstrated in experimental animals. Decreased reproductive capacity was suspected but difficult to detect in moose in the field; however, lowered thyroxin levels were measured in blood samples from the wild, free-living moose, which supported the idea of a central endocrine disruption. Similar pathological effects affecting central endocrine functions have recently been reported from clinical and experimental studies in sheep with chronic molybdenosis (Haywood *et al.*, 2004). Similar molybdenum-induced copper deficiency problems were suspected earlier and reported in moose (*Alces alces gigas*) in Alaska (Kubota, 1974; Flynn *et al.*, 1977) and in Grant's gazelle (*Gazelle granti*) in the Rift Valley in Kenya (Hedger *et al.*, 1964). The diagnostic biochemical investigations were, however, not complete in these earlier cases.

K. Iodine

Goiter, enlargement of the thyroid gland due to iodine deficiency, was described in ancient Egypt in humans and animals. It is still probably the most widespread trace element deficiency in world animal husbandry. Typical deficient areas are inland and mountain areas, such as the inner parts of Siberia, Africa, and South America; the slopes of the Himalayas, Alpine valleys, the Pyrenees; and the Andes, where wind and rain carry only limited amounts of iodine from the oceans. Some lowland and coastal areas also have soil in which leaching has caused iodine depletion (see also Chapter 16, this volume). The most sensitive and typical signs of iodine deficiency are enlargement of the thyroid gland and changes in the microscopic morphology (see Figure 6). Lambs and calves are often born with severe goiter when their mothers have been fed insufficient iodine during gestation. In severe cases, these newborn animals can suffocate shortly after birth due to the pressure of the enlarged thyroid gland on the trachea. If they survive the first critical period, they are often so weak that they have severe problems getting up and standing to suckle their mothers to get the antibodies and nutrients vital to the start of life.

In adult animals, iodine deficiency causes a generally depressed metabolism due to insufficient production of the thyroid hormone thyroxin. Lethargy, increased fat deposits, and impaired reproduction are other observed

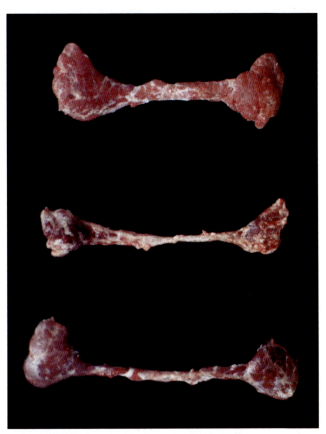

FIGURE 6 Thyroid glands from three steers fed different amounts of iodine. The gland in the middle is normal sized and comes from an animal with adequate iodine supply. The other two show different degrees of goiter. The upper gland shows a minor increase in size and the lower gland shows a moderate increase due to insufficient iodine intake.

effects. In breeding animals, irregular and suppressed estrous cycles are early signs of iodine deficiency, and the reduced fertility will have obvious effects on the economy of the farmer. If animals get pregnant, the fetus can suffer developmental problems that result in fetal death, abortion, or stillbirth.

Endemic iodine deficiency can be further aggravated by at least two factors operating in modern agriculture. First, plants containing goitrogenic substances are common in both cultivated land and growing on natural pasture. These goitrogenic substances are chemically glucosinolates (*e.g.*, isothiocyanates, thiocyanates, nitriles, and oxazolidinethiones), which block the uptake of iodine in the thyroid. Glucosinolates are present in the green parts and seeds of commonly cultivated crops of Cruciferae plants, such as various types of kale and rape. In temperate zones of the world,

cultivation of these glucosinolate-containing plants is common because they contribute valuable proteins to locally produced animal feeds. A second important factor contributing to iodine deficiency is the increasing interest in organic farming, especially in North America and Western Europe, where the limited use of supplements decreases the iodine supply in deficient areas.

Both of these problems can be easily alleviated by using appropriate supplements rich in iodine. The most common remedy is the use of iodized salt (NaCl) as a salt lick or a component of mineral supplements given to the animals. An important detail regarding iodine supplementation via salt licks or otherwise is that KI, often used as an inexpensive remedy, evaporates even at ambient temperature during storage; therefore, old salt licks may not contain the necessary amount of iodine supplementation. It is particularly important to use iodine supplementation for dairy production, as milk and other diary products are a major iodine source for humans in large parts of the world, especially in areas where marine products such as kelp, fish, or shellfish are not consumed regularly.

FURTHER READING

Bowell, R. J., Warren, A., and Redmond, I. (1996). Formation of Cave Salts and Utilization by Elephants in the Mount Elgon Region, Kenya. In *Environmental Geochemistry and Health* (J. D. Appleton, R. Fuge, and G. J. H. McCal, Eds.), Special Publ. No. 113, Geological Society, London, pp. 63–79.

Chaney, S. G. (2002). Principles of Nutrition II: Micronutrients. In *Textbook of Biochemistry with Clinical Applications* (T. M. Devlin, Ed.), 5th ed., Wiley-Liss, Philadelphia.

Flynn, A., Franzman, A. W., Arenson, P. D., and Oldemeyer, J. L. (1977). Indications of Copper Deficiency in a Sub-Population of Alaskan Moose, *J. Nutr.*, 107, 1182–1189.

Frank, A. (1998). "Mysterious" Moose Disease in Sweden: Similarities to Copper Deficiency and/or Molybdenosis in Cattle and Sheep. Biochemical Background of Clinical Signs and Organ Lesions, *Sci. Total Environ.*, 209, 17–26.

Frank, A., Danielsson, R., and Jones, B. (2000). The Mysterious Disease in Swedish Moose. Concentrations of Trace and Minor Elements in Liver, Kidneys and Ribs, Haematology and Clinical Chemistry. Comparison with Experimental Molybdenosis and Copper Deficiency in the Goat, *Sci. Total Environ.*, 249, 107–122.

Frank, A., MacPartlin, J., and Danielsson, R. (2003). Nova Scotia Moose Mystery—A Moose Sickness Related to Cobalt and Vitamin B_{12} Deficiency, *Sci. Total Environ.*, 318, 1–3, 89–100.

Haywood, S., Dincer, Z., Jasani, B., and Loughran, M. J. (2004). Molybdenum-Associated Pituitary Endocrinopathy in Sheep Treated with Ammonium Tetrathio-Molybdate, *J. Comp. Pathol.*, 130, 21–31.

Hedger, R. S., Howard, D. A., and Burdin, M. L. (1964). The Occurrence in Sheep and Goats of a Disease Closely Similar to Swayback, *Vet. Rec.*, 76, 493–497.

Jacobson, D. R., Hemken, R. W., Button, F. S., and Hatton, R. H. (1972). Mineral Nutrition, Calcium, Phosphorus, Magnesium, and Potassium Interrelationships, *J. Dairy Sci.*, 55, 935–944.

Kubota, J. (1974). Mineral Composition of Browse Plants for Moose, *Nature Can.*, 101, 291–305.

Markert, B., Kayser, G., Korhammer, S., and Oehlmann, J. (2000). Distribution and Effects of Trace Substances in Soil, Plants and Animals. In *Trace Elements—Their Distribution and Effects in the Environment* (B. Markert, and K. Friese, Eds.), Trace Elements in the Environment, Vol. 4, Elsevier, Amsterdam, pp. 3–33.

Maskall, J., and Thornton, I. (1996). The Distribution of Trace and Major Elements in Kenyan Soil Profiles and Implications for Wildlife Nutrition. In *Environmental Geochemistry and Health* (J. D. Appleton, R. Fuge, and G. J. H. McCal, Eds.), Special Publ. No. 113, Geological Society, London, pp. 47–62.

McDowell, L. R. (1992). *Minerals in Animal and Human Nutrition*, Academic Press, San Diego, CA.

Mertz, W. (1987). *Trace Elements in Human and Animal Nutrition*, 5th ed., Vols. 1 and 2, Academic Press, San Diego, CA.

NRC (1979). *Metabolic and Biologic Effects of Environmental Pollutants. Zinc*, University Park Press, Baltimore, MD.

NRC (1984a). *Nutrient Requirements of Domestic Animals: Nutrient Requirements of Beef Cattle*, 6th ed., National Academy of Sciences/National Research Council, Washington, D.C.

NRC (1984b). *Nutrient Requirements of Domestic Animals: Nutrient Requirements of Poultry*, 8th ed., National Academy of Sciences/National Research Council, Washington, D.C.

NRC (1985a). *Nutrient Requirements of Domestic Animals: Nutrient Requirements of Dogs*, 2nd ed., National Academy of Sciences/National Research Council, Washington, D.C.

NRC (1985b). *Nutrient Requirements of Domestic Animals: Nutrient Requirements of Sheep*, 5th ed., National Academy of Sciences/National Research Council, Washington, D.C.

NRC (1988). *Nutrient Requirements of Domestic Animals: Nutrient Requirements of Swine*, 9th ed., National Academy of Sciences/National Research Council, Washington, D.C.

NRC (1989a). *Nutrient Requirements of Domestic Animals: Nutrient Requirements of Dairy Cattle*, 6th ed., National Academy of Sciences/National Research Council, Washington, D.C.

NRC (1989b). *Nutrient Requirements of Domestic Animals: Nutrient Requirements of Horses*, 5th ed., National Academy of Sciences/National Research Council, Washington, D.C.

Sjaastad, Ö.V., Hove, K., and Sand, O. (2003). *Physiology of Domestic Animals*, Scandinavian Academic Press, Oslo, Norway.

Ulvund, M. and Pestalozzi, M. (1990). Ovine White-Liver Disease (OLWD) in Norway: Clinical Symptoms and Preventive Measures, *Acta Vet. Scand.*, 3, 53–62.

SECTION III

ENVIRONMENTAL TOXICOLOGY, PATHOLOGY, AND MEDICAL GEOLOGY

INTRODUCTION: JOSÉ A. CENTENO

In recent decades, there has been an increasing awareness of the importance of the interaction of mammalian systems with their natural environment. The primary focus has been on understanding exposure to hazardous agents in the natural environment through air, water, and soil. Such appreciation has led to a myriad of investigations focused on identifying those natural (and sometimes anthropogenic) environmental risk factors that may be involved in the development of diseases in humans and other animals.

Environmental medicine may be defined as the study of how environmental risk factors affect human health, which includes the practice of how to minimize and/or prevent any adverse effects. Humans are continually exposed to hazardous agents in the natural environment through air, water, soil, rocks, and even the workplace. The aim of environmental medicine is to better characterize exposure to a particular risk factor, to identify the type of adverse effects in tissues, and to determine the relationship between environmental exposure and genetic susceptibility for disease. It has been long thought that the genetic makeup of an individual plays a key role in the cellular response to environmental toxic agents such as chromium, arsenic, and nickel. Understanding the effects on humans from exposure to many of these elements present in the natural environment is a complex undertaking and one that requires an understanding of pathologic, toxicologic, speciation analysis, and epidemiological techniques. In this section, these scientific fields and techniques are summarized, with a focus on their application to the complexities of disease processes.

As indicated by Chapter 21, toxicology and epidemiology have significant roles in the study and implementation of environmental medicine. Environmental epidemiology is the study of associations between environmental exposures and the occurrence of disease within a population. The chapter presents common problems in environmental epidemiology, with exposure scenarios and case studies illustrating the application of epidemiology in environmental medicine. The objective of this chapter is to provide nonepidemiologists with basic skills to critically read and understand most epidemiological studies and recognize strengths, weaknesses, and biases related to design and exposure assessment. The chapter uses three recently published studies and illustrates different study designs, different outcome measures, and different types of exposure. At the end of the chapter, the authors provide a point-to-point checklist to guide the reader through reading and understanding an epidemiological paper.

In Chapter 22, the authors provide the reader with a comprehensive description of the principles and practices of environmental medicine focusing specifically on those aspects of the environment directly related to geological materials and geological processes. This chapter describes in detail several fundamental concepts in environmental medicine, including the characterization of the mechanism(s) of exposure and the study of the internal (cellular) response to hazardous substances. The importance of understanding individual variability in response to exposure to a particular environmental agent is emphasized and illustrated using clinical and epidemiological studies. This chapter describes a number of approaches to the study and implementation of environmental medicine practices, including toxicology, surveillance, and intervention. Finally, the case of chronic arsenic exposure from contaminated drinking water is illustrated as an example of environmental medicine and medical geology.

Several diseases are related to the accumulation of trace elements, metals, and metalloids within different organ tissues, either as the primary or as a secondary manifestation of diseases. The accumulation of metals is associated with varying degrees of organ injury. In the liver, for example, these disorders include Wilson's disease (hepatolenticular degeneration), primary biliary cirrhosis associated with copper accumulation, and hemochromatosis associated with excess iron accumulation. Chapter 23 provides a comprehensive overview of environmental pathology and exposure to toxic metals. The chapter is concentrated on major organ systems of the body and discusses pathologic states induced by various elements in the skin, brain, lungs, heart, and liver. This is by no means all inclusive but rather serves as an introduction to our current understanding of pathologic states in these organ systems. Most studies have concentrated on chronic high-level exposures, and much less work has been done on low-dose exposures to toxic elements. Much remains to be learned, and in the chapters that follow, technical analyses are discussed that assist in broadening our understanding.

In Chapter 24, toxicologic testing and the geosciences are reviewed. Toxicology is critical in medical geology, and toxicologic principles are applied through clinical toxicology, risk assessment and hazard control, monitoring, and surveillance. A general framework of these areas is provided by an examination of toxico kinetics and toxicodynamics.

Understanding the speciation of trace elements in environmental medicine, nutrition, and diseases may play a significant role in promoting legislative and regulatory actions in human health protection pursuant to the decrease of chronic diseases and/or amelioration of debilitating effects. Chapter 25 describes in detail the emerging scientific field of trace element speciation, including a discussion on a wide range of techniques for speciation analysis that can be used to establish the chemical form (oxidation state), morphology, and compounds formed by trace elements, metals, and metalloids.

CHAPTER 21

ENVIRONMENTAL EPIDEMIOLOGY

JESPER B. NIELSEN AND TINA KOLD JENSEN
University of Southern Denmark

Biological markers of exposure

Environmental exposure ┈┈▶ Black box ┈┈▶ Biological effect

Individual exposure

External dose ──▶ Internal dose

Target organ deposition

CONTENTS

I. INTRODUCTION

Epidemiology is the study of the occurrence of disease in populations. It originates from an observational discipline that describes changes in the prevalence or incidence of a specific disease—changes that may be observed over time, between geographical regions, or between populations. Thus, basic epidemiology delivers numbers with no explanation. Prevalences or incidences are, however, only really useful if associated with explanatory variables. These variables may relate to genetics, lifestyle, age, gender, occupation, environment, etc. Environmental epidemiology is therefore the study of associations between environmental exposures and the occurrence of disease within a population.

Few environmental diseases are pathognomonic in the sense that only one specified exposure may cause a certain disease. In most cases, several chemical exposures may cause the same disease, aggravate an existing disease, or in some situations even offer a certain degree of protection. Likewise, several sociodemographic factors and occupational exposures may affect exposure as well as disease. Proof of causation in epidemiological studies is therefore seldom, and associations between exposure and disease may often be biased.

This chapter does not replace epidemiological textbooks, but it is intended to introduce and discuss some more basic features related to study design and measures of exposure and outcome, as well as bias. The purpose is that non-epidemiologists should be able to critically read and understand most epidemiological studies, know strengths and weaknesses of different common study designs, and be able to recognize the more general types of bias occurring in health and exposure assessment.

The chapter will present some common problems related to environmental epidemiology and primarily use three exposure scenarios (case 1–3) based on recently published scientific articles to illustrate some of these problems. Most problems are general in nature and an inherent consequence of the chosen study design. The three examples have been chosen to illustrate different study designs, different outcomes or health effects, and different types of exposures. The studies selected are a case-control study of residential radon exposure and lung cancer (Barros-Dios et al., 2002), a cohort study on malignant mesothelioma and environmental exposure to asbestos (Metintas et al., 2002), and an ecological study on adverse pregnancy outcomes and exposure to arsenic in drinking water (Yang et al., 2003). All three articles are available as full text articles free of charge on the Internet.

II. STUDY DESIGN

Study designs can be broadly categorized according to whether they are describing distributions of a health outcome (descriptive studies) or elucidating its determinants (analytical studies). Descriptive studies describe general characteristics of the distribution of an outcome in relation to person, place, and time. Analytical studies are used to test specific hypotheses and infer that exposure precedes outcome. They can be categorized into case-control or cohort studies according to whether the study subjects are selected on the basis of outcome or exposure (Figure 1). This section will briefly introduce the different study designs and discuss their strengths and weaknesses. For further reading and more specified details, the reader is referred to epidemiological textbooks (e.g., Rothman & Greenland, 1998).

A. Descriptive Studies

Descriptive studies describe general characteristics of the distribution of an outcome in relation to person, place, and time. The identification of descriptive char-

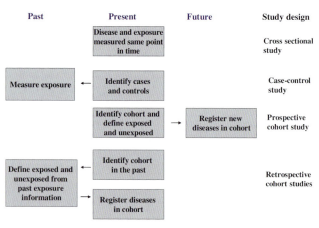

FIGURE 1   A schematic view of different study designs.

acteristics is an important first step in the search for determinants or risk factors for specific outcomes, and thereby for the formulation of hypotheses to be tested in analytical studies. Descriptive studies use information from diverse sources such as census data, disease registers, and vital and clinical records, as well as national figures on consumption of food, drinking water, etc. Because this information is already available, descriptive studies are generally far less expensive and time-consuming than analytical studies. Usually they preclude the ability to test epidemiological hypotheses. Descriptive studies can be categorized into ecological or cross-sectional studies.

1. Ecological Studies

In an ecological study data from entire populations are used to compare outcome frequencies between different groups during the same time period or in the same population at different points in time. It is not possible to link exposure information to the occurrence of outcome in a particular individual. Furthermore, the studies are unable to control for confounding. Therefore, they cannot be used to test hypotheses or infer causality. They are, however, quick and inexpensive and use already available information (Table I).

Case 1 is an ecological study where birthweight distributions in two different regions of Taiwan with different levels of arsenic in drinking water are compared. No individual exposure information was collected and the place of birth determined the exposure status. The women may, however, have moved to that area just before the delivery and therefore not have been exposed to the drinking water of that region in pregnancy at all. Furthermore, no information about the actual intake of water in pregnancy was available and it was not known whether the women with high water intake were those who delivered prematurely. In addition, no information on how much water the women drank at home was available, and the study was unable to control for confounding as, for example, the women's smoking habits. The study is, however, good for generating hypotheses to be tested in analytical studies (see also Chapters 11 and 22, this volume).

Case 1

Prevalence of adverse pregnancy outcome among 18,259 first-parity singleton live births in Taiwan was linked to place of birth. Two geographic regions with different median levels of arsenic in the drinking water

TABLE I. Strengths and Limitations in Different Types of Epidemiological Studies

Study type	Ecological study	Cross-sectional study	Case-control study	Cohort study
Strengths	Quick and inexpensive often using already available information	Quick and inexpensive Provide information about health status of great public health relevance	Optimal for evaluation of rare diseases Can examine multiple etiologic factors for a single disease Relatively quick and inexpensive compared to cohort studies Well suited for evaluation of diseases with long latency periods	Valuable for rare exposures Can examine multiple effects of a single exposure Can elucidate temporal relationship between exposure and disease Minimizes bias in exposure assessment Allows direct incidence rates to be calculated
Limitations	Unable to link exposure with disease in particular individuals Unable to control for confounding No individual exposure information	Cannot determine whether exposure preceded or resulted from the disease Considered prevalent and will reflect determinants of etiology as well as survival	Inefficient for evaluation of rare exposures Cannot compute incidence rates in exposed and unexposed individuals The temporal relationship between exposure and disease may be difficult to establish Prone to bias, particularly recall and selection bias	Inefficient for evaluation of rare diseases Prospective: extremely expensive and time-consuming Retrospective: requires the availability of adequate records Losses of followup can affect results

were included. Children from the arsenic-endemic area had on average a 30-g lower birthweight (statistical significant) and the rate of preterm deliveries was increased by 9% (insignificant). No data on individual exposures were available and the exposure to arsenic in drinking water in the arsenic-endemic area varied between <0.15 ppb and 3.6 ppm (20,000-fold), whereas the exposure in the area of comparison was below 0.9 ppb. In the arsenic-endemic area, 83% of drinking water resources had arsenic concentrations above 0.9 ppb.

Design: Ecological study
Outcome: Preterm delivery and birthweight
Exposure: Arsenic in drinking water

Reference: Yang et al., Environmental Research, 2003.

2. Cross-Sectional Studies

In cross-sectional studies the status of an individual with respect to the presence or absence of both exposure and outcome is assessed simultaneously (Figure 1). Thus, a cross-sectional study provides information about the frequency and characteristics of an outcome by a "snapshot" of the population at a specific time. Such data are of great value to public health administrators when assessing the health status or health-care needs of a population. However, as exposure and outcome are assessed at the same point in time, cross-sectional surveys cannot always distinguish whether the exposure preceded the outcome development or whether the presence of disease affected the individual's level of exposure. It is, in other words, not possible to determine whether the exposure preceded or was caused by the disease (Table I). Thus, cross-sectional studies have found that infertile couples report more psychological distress symptoms, which implies that stress therefore causes infertility. It is, however, not known whether the couples became infertile because of the stress or whether the infertility and its consequences and treatment caused the stress. Cross-sectional studies are like ecological studies, which are valuable for raising a question of the presence of an association rather than for testing a hypothesis.

B. Analytical Studies

1. Case-Control Studies

In a case-control study, subjects are selected on the basis of whether they have (cases) or do not have (controls) a specific outcome. In its most basic form, cases with the outcome of interest are selected from hospitals or the general population and compared with a group (controls) without the outcome. More refined study designs exist, and interested readers are referred to Rothman (1998, 2002). The proportions with the exposure of interest in each group are compared (Figure 1). The case-control design is a good way to study diseases with long latency periods, because investigators can identify affected and unaffected individuals and assess antecedent exposures rather than waiting a number of years for the disease to develop. Therefore, case-control studies are time and cost efficient. In addition, by selecting the cases on the basis of outcome, the study can identify an adequate number of affected and unaffected individuals. Consequently, this strategy is particularly well suited for rare diseases, which in cohort studies would need inclusion of very large numbers of individuals in order to accumulate a sufficient number of cases with the outcome of interest. Finally, case-control studies allow evaluation of a range of potential etiologic exposures and their effect on the outcome. The case-control design can therefore be used to test specific a priori hypotheses or explore the effect of a range of different exposures.

The major drawback of case-control studies is that both the exposure and the outcome have already occurred at the time when the participants enter the study. This may affect the motivation to participate and the way that participants remember and report their exposures. This study design is therefore particularly vulnerable to selection and information bias, especially recall bias (see below). Furthermore, case-control studies are not efficient for rare exposures, as too few cases would then be exposed. In addition, only one outcome can be studied because the cases and controls are selected on the basis of that outcome. In case-control studies, no absolute measures of risk or incidence can be calculated. Instead the odds ratio estimates the relative risk or incidence rate ratio. This is, however, not a reason for not conducting case-control studies, as they offer advantages mentioned before and provide answers to hypotheses relatively fast (Table I).

One of the first issues to be considered in the evaluation of a case-control study is the definition of disease or outcome of interest. It is important that the defini-

tion of disease (outcome) is as homogeneous as possible, because very similar manifestations of disease may have very different etiologies. For example, congenital malformations which encompass many different diseases such as congenital heart malformations, cleft-palate, or neural tube defects are often compiled into one outcome because of the rare nature of each of these disease categories. They do, however, have very different etiologies and combining them does not give clues to the risk factors of each particular outcome. It is therefore important to establish strict diagnostic criteria for the disease under study.

The selection of appropriate controls is perhaps the most difficult and critical issue in a case-control study. Controls are necessary to evaluate whether the exposure observed in the case group differs from what would have been expected in a comparable group of individuals without the disease. Controls must be selected, not to represent the entire non-diseased population, but the population of individuals who would have been identified and included as cases had they also developed the disease. They can be chosen from hospitals or the general population. Hospital controls are selected from people admitted to the same hospital as the cases but with a different disease. The advantage of this approach is that people admitted to hospitals are easy to identify, motivated, and more likely to be aware of antecedent exposures. The disadvantages are that there might be different selection factors leading to admission to that hospital for different diseases. Furthermore, they differ from healthy individuals and may therefore not represent the exposure distribution in the population from which the cases derived. Controls can also be chosen from the general population. This can be done in a number of ways including canvassing households in the targeted neighborhood, random digit telephone dialing, or identification from population registers or voting lists. This is, however, usually more costly and time-consuming. Furthermore, the quality of the information obtained and the participation rate from cases and controls may differ as healthy individuals from the general population do not recall exposures with the same level of accuracy and they are less motivated to participate.

It is often argued that cases should be representative of all persons with the disease. This is, however, not true and case-control studies can be restricted to a particular type of case from whom complete and reliable information on exposure and disease can be obtained (for example, in a limited age range). Then control subjects should be selected to be comparable to the cases. Such case-control studies will provide a valid estimate of the association between exposure and disease and a judgment of the generalizability of the findings can then be safely made.

Case 2

A total of 163 cases of primary lung cancer (response rate 70%) and 241 cancer-free controls (response rate 62%) were included. Cases were on average 8 years older than controls and had a 40% higher rate of cancer within the family. Close to 92% of cases were smokers as compared to 55% of the controls. Residential radon was measured in 98% of the homes for an average of 150 days. Residential radon exposure was close to 20% higher among cases than controls. This study concludes that residential radon exposure at levels below official guidelines of 148–200 Bq/m³ may lead to a 2.5-fold increase in lung cancer risk. Further, synergism (an effect greater than that expected by their separate actions) between residential radon exposure and smoking was demonstrated.

Design: Population-based case-control study
Outcome: Confirmed primary lung cancer
Exposure: Indoor radon concentration
Reference: Barros-Dios et al., 2002

Case 2 is an example of a population-based, case-control study where lung cancer patients were compared with healthy controls from the same area in Spain. Controls were proportionally stratified randomly but excluded if they had respiratory tract disease, lived in the area less than 5 years, or were younger than 35 years of age. Exposure information was obtained from next of kin if the case or control had died. Information about radon exposure was measured, so no recall bias was present. The length that the participants lived at their current address was, however, not taken into account. In addition the participation rates were 10% higher among cases than controls and cases were approximately 8 years older, which may have introduced selection bias (see below). Moreover, more than 90% of the cases were smokers as compared to 55% of controls.

2. Cohort Studies

In a cohort study, a group of individuals are defined on the basis of presence or absence of exposure. At the time of exposure classification, subjects must be free from the outcome under investigation. Participants are then followed over a period of time to assess the occurrence of the specified outcome among those who are exposed and unexposed (Figure 1). Most often, the followup period must be at least several years to allow an adequate number to develop the outcome so that meaningful comparisons of disease frequency between exposed and unexposed individuals can be made. As the participants by inclusion criteria are free of disease at the time when their exposure status is defined and the study initiated, the temporal sequence between exposure and disease can be more clearly defined.

For many exposures the proportion of exposed individuals with the outcome is too small to make meaningful comparisons between exposed and unexposed. Therefore, cohort studies are particularly well suited for assessing the effect of rare exposures. Thus, cohort studies can enroll participants on the basis of their exposures and thereby include a large number of exposed, for example, among a cohort of heavily exposed workers. Furthermore, cohort studies offer less potential for selection bias and direct measurement of association (incidence rates) can be calculated among the exposed and unexposed. Finally, cohort studies allow the examination of multiple outcomes of a single exposure.

As cohorts studies often involve a large number of individuals followed for many years, they are time-consuming and expensive. Furthermore, only a proportion of those eligible actually participate in the study, and they often differ from the non-participants in motivation and attitudes toward health. As outcome is compared among exposed and unexposed, this does not usually affect the relationship except when non-response is related to both exposure and outcome. A way to address this problem is by comparing participants with non-participants with respect to basic available information such as age and socioeconomic status. In addition, losses to followup may seriously affect the results, especially if it differs between exposed and unexposed individuals or is related to exposure or outcome or both. Losses to followup should therefore be minimized and for those lost to followup, attempts to gain information about outcome from independent sources should be made (for example, through death or disease registers).

Cohort studies are often categorized into prospective or retrospective studies according to whether the outcome of interest has occurred at the time the study is initiated (Figure 1). In a prospective cohort study, the cohort is identified and categorized according to exposure. After a followup period the frequency of the outcome among exposed and unexposed is compared. In a retrospective cohort study both exposure and outcome have occurred at the time of the start of the investigation. A historical cohort is identified at the start of the study and past exposures in the cohort are identified from already existing information. Then the frequency of outcomes (which has occurred) is determined. Prospective or retrospective solely refers to whether the outcome has occurred at the start of the study. Case-control studies can also be both prospective and retrospective, but they are most often retrospective, i.e., the outcome defining the case has occurred when the study is initiated. Retrospective cohort studies can usually be conducted more quickly and cheaply than their prospective counterparts, because all relevant events have already occurred at the time the study is initiated. They do, however, depend on routine availability of relevant exposure data in adequate details from pre-existing records. Because these data were collected for other purposes, the quality is often not optimal. Moreover, information on potential confounding factors is often unavailable (see also Chapter 18, this volume).

Case 3

In a cohort of 1886 villagers in a rural area in Turkey, the incidence of malignant pleural mesothelioma (MPM) was studied. The villagers were environmentally exposed to asbestos dust due to the use of asbestos-contaminated white soil. The soil was used as a whitewash or plaster material for walls, as insulation, and also in pottery. Exposure was assessed on a subgroup level through measurement of airborne fiber concentrations both indoors and outdoors. During a 10-year observation period, 24 cases of MPM were diagnosed within the cohort corresponding to an annual incidence rate close to 130/100,000. This incidence rate exceeds the expected in the general Turkish population by more than 100-fold, and is comparable to risks of MPM observed in occupational settings with much higher exposures.

Design: Cohort-study
Outcome: Malignant pleural mesothelioma
Exposure: Inhalation of dust from asbestos in soil
(Metintas et al. 2002)

Case 3 is an example of a retrospective cohort study examining the incidence of malignant mesothelioma among people in villages exposed to white soil containing asbestos compared to the incidence among the background population in Turkey. It is an example on how rare exposures can be studied if highly exposed cohorts are chosen. The authors sampled 11 out of 403

villages, and it is difficult to rule out if a selection bias is present. Exposure levels were measured in white soil and information on potential confounding factors was obtained by interview with relatives. As it is often the problem with retrospective studies, the information about confounders is limited and no information on smoking habits was obtained.

III. EXPOSURE ASSESSMENT

Outcome and exposure assessment are equal partners in a well-balanced epidemiological study. Accordingly, the very same questions on validity, bias, or confounding should be considered. Further, the representability of the exposure assessment with respect to individual, time, and place should be scrutinized. Thus, exposure assessments include qualitative as well as quantitative questions. This section will illustrate some common problems related to exposure assessment, but for a more in-depth discussion of this theme, readers are referred to more specific literature (a starting point is included in the Further Reading list).

A. What?

The qualitative questions relate to the validity of the analytical methods: (1) what is measured; (2) what is the specificity of the method, and (if the exposure is a mixture), (3) does the mixture change qualitatively over time or between areas included in the study. In relation to exposure to metals, the analysis of the total concentration of metal is often insufficient as different metal species will have specific toxicological profiles. Thus, for the assessment of intake of mercury or lead from soil by children, it would be relevant to know the species and salts occurring, as the intestinal absorption as well as toxicity of these metals depends on these features. Likewise, the authors in Case 3 use a well-validated method for fiber collection and only analyze and report the fraction of fibers (>5 um) relevant for the outcome (mesothelioma). This is a relevant approach, but it is not clear from the description in the article what proportion of the larger fibers analyzed were asbestos fibers. Neither is the exact type of asbestos described, which could be expected to influence the risk for the exposed individuals. These queries are especially important if the exposure is expected to vary qualitatively over time or between geographic regions included in a study.

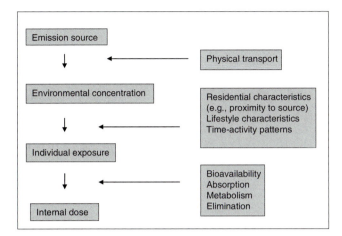

FIGURE 2 Sources of variability in exposure assessment.

B. Who?

Exposure may be assessed through measurements at different levels, beginning with very unspecific measurements at the emission source and ending with specific measurements of internal dose (Figure 2). Ecological studies (like Case 1) will use exposure assessments at a very crude level, which causes severe uncertainty with respect to the individual exposure. Thus, in Case 1, 83% of the drinking water resources in the arsenic-endemic region had arsenic concentrations above 0.9 ppb. It is, however, not known whether the women with adverse pregnancy outcomes actually consumed water from these wells or from the 17% with uncontaminated drinking water nor how much water they drank. Advancing the exposure assessment will clarify different factors of variability and increase the knowledge of individual exposure (Figure 2). An important point is, however, that the interindividual differences in exposure will not necessarily decrease with more specific exposure data. Instead, the uncertainty will be replaced by variability that may potentially be used in modeling or otherwise taken into account. In cohort or case-control studies, the possibility of getting more specific information on exposure is better as illustrated in Cases 2 and 3, where the exposure was assessed in the homes of the participants. However, the inherent problem that the exposure assessment will also have to be done retrospectively needs to be addressed. Prospective studies will have the advantage that exposure assessment can be planned in advance. Individual exposure and information on internal dose requires personal monitoring equipment or the use of biomarkers of exposure. These approaches are, however, often time-consuming and expensive, and they may not be an option in larger cohort studies.

C. How Much?

Using exposure assessments without individual exposure data raises some general questions relating to the representativity of the concentrations measured in the environment. At their best, these concentrations will represent an average exposure, but may over- as well as underestimate individual exposures. In Case 3, the concentration of airborne fibers was measured in a few homes in each village, but it is not known whether these exposure levels represent the exposure in the remaining houses. In Case 2, radon was measured in 98% of the homes, but not related to the time spent indoors. Besides averaging exposures between persons and locations, a measurement of a contaminant in the environment will be a snapshot in time. A single measurement or even a few will not be very informative regarding variation in exposure levels over time. The ideal exposure assessment in an epidemiological study is seldom achieved, but the key information that allows for a useful exposure assessment is quantitative information on changes in exposure over time and between locations. This will help exchange uncertainty with variability.

D. When?

Knowledge of variations in exposure over time is critical for several outcome measures. This information is needed if the health outcome depends on specific windows of susceptibility within the exposed population. Thus, for adverse pregnancy outcomes, neither information on exposures dating 20 years back nor information on exposure after birth is relevant. Likewise, when the outcome appears after a delay in time, e.g., cancer, current exposure is not relevant. Thus, in Case 3, the most relevant information on exposure to asbestos fibers probably dates 10–20 years back.

Information on present exposure levels is mainly useful if the outcome appears without delay, in prospective studies, or when the exposure can be assumed to be unchanged over time. However, most exposures change with time. If risk is a function of time of exposure, exposure profiles including information on variation will be valuable (Figure 3). In these profiles, the exposure concentration or dose is plotted as a function of time. Concentration versus time is used to describe the exposure, while amount versus time characterizes dose. If the elimination rate of the chemical is known, dose characterization may be used to estimate accumulation of contaminants. Further, exposure profiles may be used to identify more limited time periods with higher than

Exposure/dose profiles are important
- When risk is a function of time of exposure
  - Reprotoxicity
  - Neurotoxicity
- When risk may occur following short peak exposures
- When a chemical is accumulated
  - Lead, mercury, etc.
- When outcomes appear after delay
  - Cancer

FIGURE 3 Importance of exposure/dose profiles.

average exposures, which may be relevant for some outcomes. Thus, a single short-term exposure to very high concentrations may induce adverse effects, even if the average exposure is much lower than an apparent no-effect-level. Such short-term peak exposures clearly remain unidentified if only average values are available. In other exposure scenarios, however, average values are sufficient to perform a valid exposure assessment. Therefore, an epidemiological study protocol should include careful considerations concerning the most relevant collection of data on exposure.

E. Modeling Exposure

Several models, e.g., Monte Carlo simulations, have been developed through recent decades to estimate risk and exposure, to assess changes in exposure over time, or to identify worst case scenarios. These models are often very useful as they are able to accommodate and use vast amounts of information on parameters of importance for modeling individual exposures. If physiological and behavioral parameters are included, these models may even estimate target organ deposition. One of the major achievements of these models is that they have enabled risk and exposure assessors to replace the often very conservative estimates of worst case scenarios with more realistic scenarios based on probability functions. It is, however, important to remember, that a model never gets more valid than the validity of the exposure information obtained and entered into the model.

IV. BIAS

Two types of errors may occur in epidemiological studies: random and systematic errors. Random errors

are, as the word implies, random and are minimized when the study size or the precision of information is increased whereas systematic errors are unaffected by the size of the study. If a participant is weighed on an imprecise weighing scale, his or her weight may be over- or underestimated, but the error is random. Therefore, an increase in number of participants will reduce bias. If, however, the weighing scale is systematically overestimating the true weight of the participants, there is no effect of increasing the number of participants. It will still overestimate their weight, and a systematic error is introduced.

Systematic errors are often referred to as bias. Bias may be defined as any systematic error in an epidemiological study that causes an incorrect estimate of the association between exposure and outcome. Because epidemiological studies involve humans, even the most perfectly designed study will have the potential for one or more types of errors. Consequently, evaluating the role of bias as an alternative explanation for an observed association is a necessary step in the interpretation of any study. Therefore it is essential to discuss types of biases that might be present as well as the most likely direction and magnitude of their impact. A study can be biased because of the way in which the study subjects are selected (selection bias), the way the study variables are measured (information bias), or by the lack of measurement of other exposures related to the outcome (confounding).

A. Selection Bias

Selection bias is a systematic error in the study that occurs when the association between exposure and outcome differs for those who participate and those who do not participate in the study. The participation rate in a study is never 100%, therefore it is important to gain information about age and sociodemographic status from non-participants and compare these with the participants.

Selection bias may change the estimates both toward and away from the null hypothesis. Selection bias is of particular importance if the participation rate is low, varies between cases and controls or between exposed and unexposed. In Case 2, 70% of the lung cancer patients and 61% of the controls participated. Patients may be more interested in participating as they believe that the exposures studied may have caused their disease.

A special form of selection bias occurs when the prevalence of an outcome for a group of workers is compared with the prevalence for the general population. This comparison is biased because the general population includes many people who cannot work because they are too ill. Consequently, the outcome is more frequent in the general population. This bias is often referred to as healthy worker effect.

B. Information Bias

Information bias is caused by systematic differences in the way data on exposure or outcome are obtained from the various study groups. The participants are thereby misclassified with respect to either exposure or disease. This misclassification can be either differential or non-differential. Consider the smoking information in Case 2. If both cases and controls underestimate the number of cigarettes that they smoke on average, this would lead to a categorization of heavy smokers as light smokers. The classification of exposure (smoking) is unrelated to the outcome (lung cancer) as both cases and controls underreport to the same extent. The information bias is therefore a non-differential misclassification. A non-differential misclassification will produce estimates of the effect that are diluted and will tend to support the null hypothesis. Now imagine that lung cancer patients underreport their smoking to a greater extent than controls. The classification of exposure (smoking) is then related to the outcome (lung cancer), and the bias is differential and may under- as well as overestimate the effect.

A common type of information bias is recall bias, which may occur in case-control studies where a subject is interviewed to obtain exposure information after the outcome has occurred. Cases then tend to have a different recall than controls as a result of their disease. In Case 2 lung cancer patients and controls were interviewed. If lung cancer cases remember and report their smoking differently than controls, this causes a recall bias. Case 2 was further complicated by the fact that some cases had died by the time of the investigation and next of kin were interviewed instead. They may not remember exposures as precisely as the cases themselves.

V. CONFOUNDING

Confounding is a mixing of effects. A confounder is an exposure other than the one investigated, which is associated with outcome, but unequally distributed between the groups compared. Furthermore, it must not be an intermediate step in the causal pathway from exposure to disease. Confounding can cause bias in either direction.

In Case 2, smoking is a confounder. It is associated with the outcome (lung cancer) and unequally distributed among cases and controls (92% of cases and 55% of controls smoked). Furthermore, it is not an intermediate step in the causal pathway between radon exposure and lung cancer. Therefore, if the study does not take smoking into account, the effect of radon exposure would be overestimated as the estimate would really measure the aggregated effect of smoking and radon on lung cancer.

Confounding may be controlled by restricting the study population, thus all participants are equal with respect to a potential confounder (e.g., restricting the study to a specific age category). In Case 2, the study could have included only non-smokers. Another way to deal with confounding is by matching the study subjects with respect to the confounder. In Case 2, a smoking control could be included every time a smoking case was included. Matching poses special challenges and is not discussed further in this chapter, but readers are referred to epidemiological textbooks (see Further Reading section). Confounding control may also be addressed during analysis of the data by multiple regression analysis or by stratifying data, i.e., study lung cancer and radon exposure among smokers and non-smokers separately.

VI. STATISTICS

The majority of statistical analyses involves comparisons between groups of subjects. Initially, a hypothesis, called the null hypothesis, states that there is no difference in the outcome of interest between the groups of subjects. Statistical analysis is then an evaluation whether to accept or reject this hypothesis. The selected study subjects are only subsamples of the entire population, and probabilities are used to describe the certainty by which the null hypothesis is rejected. This probability is given as a p value in most statistics. Thus, a p value of 0.05 means that there is 95% certainty that the null hypothesis is not true and should be rejected. There is, however, a remaining probability of 5% that the rejected null hypothesis was actually true. This is known as a false positive result and termed a type I error. Thus, the risk of a type I error is determined by the size of the p value that is used as the level of rejection of the null hypothesis. False negative results may also occur, i.e., the acceptance of a null hypothesis that is not true, and this is called a type II error. In Case 1, the authors use a probability of $p < 0.001$ to conclude that the mean birthweight is different in the two regions. There is only a 0.1% risk of a type I error, e.g., that the difference observed is actually only a chance finding and not true. Type I and type II errors are interdependent. Thus, whenever the risk of a type I error is reduced, i.e., by decreasing the p value used as level of rejection of the null hypothesis, the risk of a type II error is increased and vice versa. For a thorough statistical explanation, readers are referred to statistical textbooks.

In many epidemiological studies large numbers of comparisons are made between different subgroups within the observed group of subjects. If a probability of $p = 0.05$ is used as the level for rejection of the null hypothesis, this means that for every 20 comparisons one will, just by chance, be a false positive finding. Therefore, most statistical packages include methods to reduce this risk of type I errors when doing large numbers of comparisons. Failing to apply these methods in multicomparison scenarios may invalidate conclusions.

VII. CHECK LIST FOR EVALUATING AN EPIDEMIOLOGICAL PAPER

Reading and understanding an epidemiological study report may be time-consuming and it is often difficult to evaluate its validity. We have tried to develop a checklist to guide the reader through the most pertinent questions relating to the validity, strengths, and weaknesses of an epidemiological paper. Answering the questions on the checklist will make the reader recognize possible problems in a paper.

1. What type of study design was used?

See strength and limitations for the different design in Table 1.

- Cross-sectional study—The problem with cross-sectional studies is that information about exposure and outcome is collected at the same point in time. It is therefore impossible to draw conclusions about causation.

- Ecological study—No individual exposure information is collected in an ecological study. It is therefore impossible to infer causation in these studies.
- Case-control study—Exposure among a group of cases selected on the basis of an outcome, i.e., disease, is compared to exposure among a control group without the outcome.

Study Design:

- How were the cases defined?
- How were the controls selected? Community or hospital controls?
- Were cases and controls from the same source population?
- Were cases and controls matched?
- What was the number of controls per case?

Study population:

- What was the target population?
- What recruitment procedures were used?
- Did the participation rate among cases and controls differ?

Validity:

- Over what time period was the study population recruited?
- Were the cases and controls comparable with respect to characteristics, response rates and time of recruitment?
- Was any information about non-responders obtained?

- Cohort study—A cohort is defined and categorized into exposed and unexposed, who are then followed in order to determine new occurrences of the outcome.

Study Design:

- Was the cohort defined retrospectively or prospectively?
- How large was the cohort?
- How many were exposed?
- How many observed events were there?

Study population:

- What was the target population?
- What recruitment procedures were used?
- How many were lost to followup (percentages)?

Validity:

- Over what time period was the study population recruited?

- Was there any description of the losses to followup?
- Do the losses to followup introduce bias?

2. What were the hypotheses of the study?

- Was the study originally designed to test these specific hypotheses?

3. How was the data quality?

- Were the data collected for the purpose of the study or were they obtained from other sources, for example, registers or hospital files?

4. How was exposure assessed?

- Who provided the information about exposure (subject, family or others)?
- Was the quality of exposure information assessed?
- Were the subjects and/or interviewer blinded to the hypothesis?
- How was the exposure information linked to the cases?
- Is risk a function of time, and were exposure profiles included?

5. How was information about outcome measured?

- Self-reported, by health personnel or from registers

6. Was adequate statistical analysis used?

- Did the analyses control for potential confounders?
- How wide were the confidence intervals?
- Is a type 1 or type 2 error possible?

7. Bias

Selection and information bias:

- It should be detected from the questions asked under case-control and cohort studies.

Confounders:

- Was adequate information about the confounders obtained?
- How accurate and adequate was the information about confounders?
- Did the study control for confounding by restriction, matching, stratification, or multiple regressions?

8. Have other studies reported similar findings (consistency)?

- Have a number of studies conducted by different investigators, in different geographical areas, and among different cultures at various points in time using different methodology found similar results?
- Lack of consistency should lead to a high degree of caution at any causal interpretation of the findings.

9. What is the strength of the association?

- The magnitude of the observed association is useful to judge the likelihood that the exposure itself affects the risk of developing the disease, and therefore, the likelihood of a cause-effect relationship. Specifically, the stronger the association—that is the greater the magnitude of the increased (or decreased) risk observed—the less likely that it is merely due to the effect of unexpected and uncontrolled confounding. This does not imply that a weak association cannot be causal, merely that it is more likely to exclude alternative explanations.

10. Is there a plausible biological mechanism of action and do experimental studies show similar results?

- Because what is considered biologically plausible or tested in animal studies at a given time depends on the current knowledge, the lack of these criteria do not necessary mean that a relationship is not causal.

11. Did the exposure precede outcome?

- Many lifestyle factors are likely to be altered as the first symptoms of a disease appear.

SEE ALSO THE FOLLOWING CHAPTERS

Chapter 2 (Natural Distribution and Abundance of Elements) · Chapter 11 (Arsenic in Groundwater and the Environment) · Chapter 18 (Natural Aerosolic Mineral Dusts and Human Health) · Chapter 22 (Environmental Medicine) · Chapter 23 (Environmental Pathology)

FURTHER READING

Basic epidemiology:
Rothman, K. J. (2002). *Epidemiology. An Introduction*, Oxford University Press, New York.
Rothman, K. J., and Greenland, S. (1998). *Modern Epidemiology*, second edition, Lippincott-Raven, Philadelphia, PA.
Exposure assessment:

Paustenbach, D. J. (2000). The Practice of Exposure Assessment: A State-Of-The-Art Review, *J. Toxicol. Environ. Health*, Part B, 3, 179–291.

REFERENCES FOR CASE-STUDIES

Barros-Dios, J. M., Barreiro, M. A., Ruano-Ravina, A., and Figueiras, A. (2002). Exposure to Residential Radon and Lung Cancer in Spain: A Population-Based Case-Control Study, *Am. J. Epidemiol.*, 156, 548–555.

Metintas, S., Metintas, M., Ucgun, I., and Oner, U. (2002). Malignant Mesothelioma Due to Environmental Exposure to Asbestos—Follow-Up of a Turkish Cohort Living in a Rural Area, *Chest*, 122, 2224–2229.

Yang, C.-Y., Chang, C.-C., Tsai, S.-S., Chuang, H.-Y., Ho, C.-K., and Wu, T.-N. (2003). Arsenic in Drinking Water and Adverse Pregnancy Outcome in an Arseniasis-Endemic Area in Northeastern Taiwan, *Environ. Res.*, 91, 29–34.

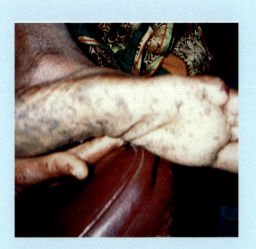

JEFFERSON FOWLES
Institute of Environmental Science and Research

PHILIP WEINSTEIN
University of Western Australia

CHIN-HSIAO TSENG
National Taiwan University Hospital

CONTENTS

ronment," so we will stick to the former term. The word environment also requires interpretation, because there are many subsets of environments that have special branches of environmental medicine associated with them. For example, occupational medicine (the study of the effects of the work environment on health) and social medicine (the study of the effects of social structures and dynamics on health). Of particular interest in this book are those aspects of the environment that relate directly to substances and processes of geological origin. This chapter will attempt to provide a brief outline of the principles and practice of environmental medicine by drawing largely on examples with geological relevance.

I. INTRODUCTION: WHAT IS ENVIRONMENTAL MEDICINE?

Abroad definition of environmental medicine is the study of how the environment affects health, including the practice of how to minimize any adverse effects. The term environmental medicine is effectively synonymous with the term environmental health, except that the latter term is often confused with "health of the envi-

II. EXTERNAL PROCESSES: HOW ARE WE EXPOSED TO HAZARDOUS SUBSTANCES?

A. Environmental Media

A medium (pl. media) is a vehicle by means of which exposure occurs. Thus, when humans are exposed to arsenic (for example), it is often by means of their drinking water, that is, water is the environmental medium of exposure. Other media by which substances of geological origin may come into contact with humans include air (e.g., inhalation of radon gas or airborne particulates), soil, and food. Although there is some direct ingestion of soil in food preparation or in conditions such as pica, contaminants in soil often enter humans by way of the food chain. Soil and food can therefore be conveniently thought of as one medium. For example, in regions where iodine is deficient in the soil, it is also deficient in the food chain, and goiter (or cretinism in children) may result as a deficiency disease in people dependent on local produce. The soil/food medium is complicated by the factor of bioavailability, which determines the ability of a substance to migrate, in its current chemical state, from the soil fraction to living organisms. (see also Chapter 25, this volume). These three environmental media: air, water, and soil/food form the basis of a useful analytic framework in environmental medicine, as illustrated in Table I.

B. Hazardous Substances

To carry the analytic framework further, the nature of the injurious agent conveyed by the above media can be examined. It is in this examination that environmental medicine most clearly shows itself as a broad science; the injurious effects of landslides are just as relevant in this context as are the toxins traditionally studied by toxicologists. The latter affect health because of their chemical nature, and are arguably most readily associated with the field of medical geology. A broader perspective was illustrated in Table I with physical as well as microbiological health effects that result, however indirectly, from geological processes.

C. Mechanisms of Exposure

Given a hazardous substance that has made its way to a human through an environmental medium, it remains for that substance to exert its influence by entering into the biochemical and metabolic functioning of the body at the cellular level. First and most obviously cellular function may be interrupted at the physical level by direct cellular injury in a rockfall, or by asphyxiation as the respiratory systems fails following a tsunami (Table I). Chemical and biological substances, on the other hand, enter the body through existing organ systems, by

TABLE I. A Simple Framework for Classifying Health Effects of the Environment

Environmental medium	Type of health effect	Example from medical geology
Air	Physical	Death by asphyxiation: absence of oxygen in a mine shaft; chronic obstructive pulmonary by disease caused inhalation of particles.
	Chemical	Neurotoxicity by inhalation: geothermal hydrogen sulfide accumulated in basement
	Biological	Pneumonia: infection by airborne anthrax from dried out soils
Water	Physical	Death by drowning: inundation by tsunami or other flooding
	Chemical	Skin cancer and vascular pathology: chronic ingestion of arsenic in drinking water
	Biological	Meningitis: infection by amoebae in hot springs
Soil/food	Physical	Direct injury: from displacement of natural or artificial structures in earthquake
	Chemical	Goiter: disruption of metabolism in areas with iodine deficient soils
	Biological	Tetanus: infection of cuts by contact with soil containing Clostridium spores

means of which they reach the cells and organ systems that they affect. Inhalation, ingestion, and absorption are such modes of penetration to the cellular level, with the systems responsible to include the respiratory, gastrointestinal, and integumentary (skin) systems respectively. Thus carbon monoxide from incomplete coal combustion is inhaled, passes through the lungs to red blood cells in the alveolar capillaries, and irreversibly binds hemoglobin so that cells are starved of oxygen. Lead from old paint is ingested with house dust, passes through the intestinal tract where it is taken up by the bloodstream, and transported to the brain where it exerts its neurotoxic effect on brain cells. Methylmercury accidentally spilled on the skin is absorbed directly through the skin and into the bloodstream, and is likewise transported to the brain where it exerts its neurotoxic effect. An understanding of these particular mechanisms of introducing hazardous substances to the body at the cellular level are of critical importance to the practice of environmental medicine: only with such an understanding can appropriate barriers be devised to protect individual as well as public health (see also Chapter 5, this volume).

D. Case Study: Soil, Water, and Amoebae

New Zealand is located in the Pacific Ring of Fire, on the boundary of two major tectonic plates. As these rub together, a number of geothermal phenomena are produced, including volcanism (see Volcanic Emissions and Health, this volume) and the geothermal heating of freshwater springs. Such hot springs are an integral part of New Zealand's cultural history; since the first humans colonized the islands some 1000 years ago, hot springs have been used for recreation, cooking, healing, and heating.

About 100 natural thermal springs are dotted around the country, and they are the delight of New Zealand children and tourists alike. The diving, jumping, and water sliding unfortunately took a serious blow in the late 1960s when isolated deaths began to occur from primary amoebic meningoencephalitis (PAM).

The amoeba responsible for PAM, *Naegleria fowleri*, invades the nasal mucosa and olfactory nerve endings, and then tracks up to the meninges and brain through a sieve-like weakness in the skull known as the cribriform plate. The infection is usually accompanied by a fever, sore throat, and headache and progresses in a few days to nausea, vomiting, neck stiffness, and sometimes olfactory hallucinations. Death is usual by day 5 or 6, and almost inevitable by day 10. Active and healthy young people are often the victims, which fuels public outrage disproportional to the public health impact of this disease (there have been less than a dozen deaths from PAM in New Zealand, and less than 200 reported cases globally).

An epidemiological association (see Section IV) was soon established between PAM and exposure to hot springs, and *N. fowleri* was isolated from water samples from suspect springs. Once the microbial hazard, environmental medium, and mechanism of introduction had all been established, appropriate barriers could be devised to protect the public from further infections. There was a publicity campaign requiring signs at hot pools to warn the public to keep their heads above water. This safety message persists in the New Zealand psyche, and many home spa users adhere to it even if they can no longer remember exactly why.

Interestingly, there is a geological link in this example because *N. fowleri* was only found in soil-contaminated waters. Free living in soil, the organism thrives in the warm waters of hot springs (and therefore not surprisingly also does well at body temperature). It is introduced into the water if the spring runs over bare soil before entering the pool, if the pool has unsealed edges or bottom allowing direct soil contact or runoff with rain, or if the pool surrounds are of exposed soil that enters the pool on bathers' feet. A second public health intervention was therefore to isolate the hydrological environment from the geological environment: standards were introduced for piping water into pools; improving pool construction; and installing wide, paved pool surrounds (Giddens, 1993).

There have now been no cases in New Zealand for over 20 years. This provides an excellent example of the effectiveness of the sound practice of environmental medicine.

III. INTERNAL PROCESSES: HOW DOES THE BODY RESPOND TO HAZARDOUS SUBSTANCES?

A. Concepts of Absorption, Distribution, Metabolism, and Repair

The basic processes of absorption, distribution, and metabolism are critical in gauging the body's response to any toxic chemical. In particular, these processes are crucial in understanding the significance of laboratory animal results in relation to human health risks, when modeling or estimating exposures in epidemiology studies, and when discussing the potential for increased susceptibility of subgroups within the population. Figure 1 depicts this relationship (see also Chapters 6 and 8, this volume).

1. Absorption

Chemicals are absorbed through oral ingestion, inhalation, or through the skin at a rate relating to their water solubility, size, ionization state, acid dissociation constant (pKa), and exposure concentration. In general, ionically neutral compounds are absorbed more readily than ionized compounds. Acids tend to be better absorbed in the acidic stomach, while basic compounds may be better absorbed under the more alkaline condi-

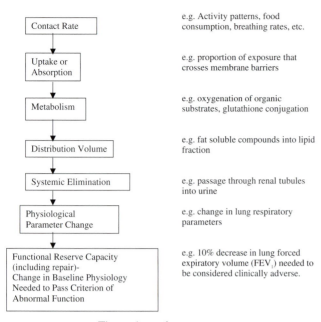

e.g. Activity patterns, food consumption, breathing rates, etc.

e.g. proportion of exposure that crosses membrane barriers

e.g. oxygenation of organic substrates, glutathione conjugation

e.g. fat soluble compounds into lipid fraction

e.g. passage through renal tubules into urine

e.g. change in lung respiratory parameters

e.g. 10% decrease in lung forced expiratory volume (FEV$_1$) needed to be considered clinically adverse.

FIGURE 1 The pathway from exposure to response.

tions in the small intestine. When considering metals, active uptake mechanisms play an important role, such as the use of calcium transport mechanisms serving to increase the uptake of lead. Some characteristics of compounds and their influence on absorption are shown in Table II.

It is thought that water-insoluble metal forms are not bioavailable for absorption by the stomach and small intestine. However, some insoluble forms become bioavailable in the acidic stomach. For example, the gastrointestinal effects for copper sulfate and copper oxide in humans are similar (Pizarro et al., 2001).

2. Distribution

The distribution of a chemical in the body is a function of its water solubility, ionization state, molecular size, and affinity for specific receptor sites in tissues. Generally, the more water soluble the compound, the more easily it can be excreted, and most metabolic processes work toward this end. Extreme cases, such as the case with DDT, can be illustrative. DDT is extremely lipophilic and therefore is retained in adipose tissue throughout the body and the fat found in the bloodstream. In cases such as this, the distribution affects the

TABLE II. Substance Characteristics That Influence Absorption or Bioavailability

Substance characteristic	Uptake mechanism	Example
pH of weak organic acids and bases	Ionically neutral states are more readily absorbed	Aniline absorption in the small intestine. Aspirin absorbed in the stomach
Valence states of metals	Specific active transport mechanisms	Chromium VI uptake into cells, while Chromium III is excluded. Inorganic arsenic transported by phosphate transport mechanisms
Cationic metal form may have competitive antagonists	Specific active transport mechanisms	Lead competes with iron for uptake and reduces iron bioavailability; molybdenum competes with copper for absorption. A child who gets enough iron and calcium will absorb less lead
Sorption status of metals in soils (bioavailability)	Ionic water soluble forms are more available	Cadmium is bound to organic counter ions in soil at high pH and therefore less bioavailable
Molecular weight	Smaller compounds tend to cross membranes more easily	Limited bioavailability of some large marine biotoxins (e.g., gymnodamine) through oral ingestion
Water solubility	Water-soluble liquids that are fat soluble cross membranes; water-soluble unionized compounds cross membranes more easily than ionized counterparts	

compound's toxicity and biological half-life (in this case, approximately 8 years) through storage reservoirs that prevent metabolism.

The chlorinated dibenzo-*p*-dioxins are another example of lipophilic compounds distributing into adipose tissue, and because those isomers with chlorines at the 2,3,7,8-positions are highly resistant to metabolism, the result is a biological elimination half-life of 7–11 years (U.S. EPA, 2000).

The tissues of highest concentration may or may not be the critical target organs. For example, copper is stored primarily in the liver, brain, heart, kidney, and muscles. About one-third of all the copper in the body is contained in the liver and brain, and the critical toxicological effect from chronic exposure is in the liver. Another one-third is contained in the muscles, where no toxic effect is known to occur. The remaining one-third is dispersed in other tissues.

Some compounds have deep storage depots, such as fluoride or lead in bone. When the chemical is in these depots, it is not bioavailable for activity at distal sites, and the elimination half-lives of such compounds are very long. The half-life for lead in bone is estimated to be about 25 years. Lead has a half-life of about 25 days in the blood. Methylmercury has a half-life for elimination of about 50 days in adult males and females. However, infants have poor elimination of organic and inorganic heavy metals in the first 6 months prior to development of metal transport systems, and it is possible that mercury is eliminated much more slowly in the child (Brown, 2001).

3. Metabolism

Metabolism of a substance is subject to enormous interspecies and interindividual differences, which in turn affects susceptibility to the chemical.

Metabolizing enzymes serve the functions of detoxification, intoxication, and facilitation of excretion of compounds. Key enzymes are the cytochrome P-450 family of heme-containing oxygenases, acetyltransferases, glutathione-S-transferases, and glucuronidases. These enzymes are needed to increase the polarity and water solubility and thus the rate of excretion of organic environmental contaminants.

Through metabolic enzyme differences, chemicals that are benign to humans can be highly toxic to some species and vice versa. The toxicity of paracetamol (acetaminophen) to cats is a classic example of how species-specific metabolism can influence toxicity. Cats are particularly susceptible to paracetamol intoxication because of their impaired glucuronic acid conjugation

mechanism and rapid saturation of their sulfate conjugation pathway, whereas humans rely heavily on a much more robust glucuronic acid conjugation system, which effectively detoxifies the critical metabolites.

For some metals, metabolism is also influential in determining bioavailability and toxicity. Arsenic methylation, the primary process by which the metal is metabolized in the body, has generally been considered a method of detoxification. Methylated arsenicals produced metabolically from inorganic arsenicals are excreted faster and have a lower affinity for tissue sulfhydryl groups than inorganic arsenicals, especially arsenites (Chan & Huff, 1997). It has also been shown that the incidence of skin cancer due to oral arsenic exposure is associated with individual methylating capacity (Hsueh et al., 1997). The epidemiological evidence indicates that the methylation detoxification mechanism does not become saturated even at high doses but that some inorganic arsenic always remains in human urine, regardless of the amount of arsenic exposure (Hopenhayn-Rich et al., 1993). Arsenic poisoning has been thought to occur only when the rate of exposure exceeds the rate of methylation (Le et al., 2000).

4. Repair

The body is constantly repairing itself from damage sustained from a myriad of environmental insults. Among the most significant of these mechanisms is with the repair of genetic lesions. Genetic lesions are central to the mechanisms of carcinogenesis and some developmental defects.

There are over 100 genes responsible for maintaining the integrity of our DNA. These include endonucleases, polymerases, and ligases. Each enzyme is important in one or more areas of DNA repair, which in turn affects individual susceptibility to particular agents. The clinical manifestations of defects in one or more of these enzymes can be seen in a number of genetic diseases. Xeroderma pigmentosa (XP) patients carry 1000-fold increase in skin cancer incidence, but no significant increase in internal cancers. UV-light-induced single DNA base point mutations are the most critical for these patients. XP patients have a defect in a key endonuclease required for carrying out excision-repair of point mutations (Hoffman, 1994). Ataxia telangiectasia patients, on the other hand, can repair point mutations induced by UV light, but are very susceptible to x-rays because of a defect in repair of strand breaks. Similarly, Bloom's syndrome patients carry a defect in DNA ligase I. This defect effectively increases chromosome fragility and results in huge increases in cancer rates (e.g., 28 out of 103 patients died of cancers at a mean age of 20.7 years).

A number of metals are carcinogens by mechanisms that are not absolutely clear, but some metals do appear to inhibit one or more DNA repair enzymes. This could be a mechanism of indirect action for several metal carcinogens.

B. Dose-Response Relationships

A critical aspect of toxicology is the description of a dose-response relationship. Conceptually, a dose-response relationship requires either the severity of a particular end point, or the incidence of the adverse effect in the population, to increase with increasing dose. The increase does not need to be monotonic, but there should be a region of the dose range that is linear. Once much higher doses are reached, additional toxic effects of a different nature may manifest and cloud the estimation of the response at lower doses (see also Chapter 24, this volume).

For most acute toxicity end points, such as acute lethality, the dose-response relationship is described in a relatively straightforward manner, and linear or log-linear models describe these relationships quite adequately (Fowles et al., 1999).

For chronic toxicity end points, such as cancer, the behavior of the dose-response curve at the low dose end is critical to interpretation and assessment of health risks. Some compounds have associated biological mechanisms that support a "hockey-stick" shaped dose-response curve for cancer. This is particularly relevant to the case of overloading of functional reserve or repair capacity of the body at high doses.

The dose-response relationship is important for the following reasons:

- Validation of hypothetical causal relationships to chemical exposure
- Provision of a measure of toxicological potency which facilitates prioritization of hazards and risks by risk managers
- Description of the range of variability in responses in the test population

C. Varieties of Effects: "Non-Cancer" Toxicity and Carcinogenicity

Regulatory toxicology and toxicological risk assessment generally divides responses into those of a carcinogenic and non-carcinogenic nature.

1. Carcinogenicity

It has been estimated that over 80% of all cancers, population wide, are environmentally induced (Doll & Peto, 1981). This estimate includes cancers from smoking, dietary carcinogens, and exposure to air and water pollution in addition to those from cosmic and solar radiation (see also Chapter 23, this volume).

One example of a type of cancer thought to have strong environmental links is prostate carcinoma (PC). In the United States, 10% of all 75-year-old men (black, white, and Japanese ethnicity) have latent PC. However, the active form of this carcinoma is 60:30:1 in these populations. This suggests that there are environmental, endocrine, or dietary factors that influence the progression of latent PC to active cancer.

Adult tissues, even those that are composed of rapidly replicating cells, maintain a constant size and cell number by regulating the rate of replication, by differentiation to assume specialized functions, and by programmed cell death (apoptosis). Cancers are diseases in which there are somatic mutations of genes critical to maintenance of control over cell division that lead to loss of control over cell replication, differentiation, and death. The International Agency for Research on Cancer (IARC) has defined chemical carcinogenesis as: "the induction by chemicals of neoplasms that are not usually observed, the earlier induction by chemicals of neoplasms that are commonly observed, and/or the induction by chemicals of more neoplasms than are usually found."

Carcinogenicity is thought to occur in a multistep or "multistage" process, with several key events occurring in sequence for a given normal cell to convert into a malignant cell with unregulated growth. The number of genes altered in a cancer cell compared to a normal cell is not known; recent evidence suggests that 3–10 genetic events are involved in common adult malignancies in humans. Two distinct classes of genes, proto-oncogenes and tumor-suppressor genes, are involved in the cancer process (Barrett, 1993).

In the current multistage model of carcinogenesis, development of a malignant tumor occurs in three stages: initiation, promotion, and progression. Initiation involves an irreversible change in a normal cell (usually an alteration of the genome) allowing for unrestricted growth. The initiated cell may remain latent for months or years. During this period of latency, the initiated cell is phenotypically indistinguishable from surrounding cells of the same type. Further development of the initiated cell into a neoplastic cell requires a period of promotion. Under the influence of a pro-

moter, tumor formation is accelerated through clonal expansion of initiated cells. Promoters, which do not interact directly with DNA, are a diverse group of agents believed to act by a variety of mechanisms most often resulting in increased cell proliferation. The process of promotion is considered reversible and requires prolonged and repeated exposure to promoter agents. Progression is the final step in which pre-neoplastic foci develop into malignant cells. In this stage, tumor development is characterized by kary-otypic changes, increased growth rate, and invasiveness. Progression may be spontaneous, influenced by environmental factors, or mediated by progressors. Resulting tumors may be either benign or malignant.

The mechanisms of metal-induced carcinogenesis are less clear than for genotoxic agents, and they probably involve a number of biochemical events that indirectly affect the integrity of the genome of particular cells. It has been found, for example, that nickel and arsenic exposure can induce DNA hypermethylation (Tang, 2000), and that arsenic inhibits DNA ligase I and II, which plays a role in DNA repair. Such non-mutational epigenetic changes could also result in suppression of tumor-suppressor genes, such as the p53 gene, triggering tumorigenesis. These "indirect" mechanisms of carcinogenesis often translate into what are believed to be sublinear, or hockey-stick type dose-response curves. If this occurs, extrapolations from high-dose effects are likely to overpredict risks at low doses (Rudel et al., 1996).

Background tumor formation is a normal observation in control animals from rodent carcinogenicity studies. The incidence of spontaneous tumors varies between tissues, and the susceptibility of a given tissue or organ varies between species and strain, and can be influenced by other factors including diet. For example, it has been found, in lifetime studies, that the incidence of testicular interstitial cell adenoma is 49% in F-344 rats compared with 9% among FBNF1 rats (Haseman et al., 1998). A 40% food restriction lowered incidences in these strains to 19 and 4%, respectively.

2. Non-Cancer Effects

The non-cancer end points are subdivided into acute or chronic exposures and effects with different levels of further organization relating to the target tissue, organ, or system, depending on the risk management need.

The critical biological target is identified from an exhaustive search of published and unpublished literature until a reliable study shows an effect that occurs at doses below those causing any other measured effect. As

TABLE III. Target Organs, tissues, or Systems Used by California Environmental Protection Agency for Air Toxicology Non-Cancer Risk Assessments

Eyes
Lung (upper and lower respiratory tracts)
Liver
Kidney
Immune system (e.g., reduced host-resistance)
Nervous system (peripheral and central)
Blood (i.e., anemia)
Endocrine
Reproductive and/or developmental
Gastrointestinal
Bones (i.e., fluorosis)

From OEHHA, 2000.

these studies are typically done on laboratory animals, the most sensitive species and sex is used as the basis for identifying the critical dose.

One organizational scheme for target organ systems is shown in Table III.

Generally speaking, the fewer the categories of classification, the more conservative or public health protective an assessment of the impacts of a mixture of chemicals.

D. Identifying Thresholds

The highest experimental "no observed adverse effect" level (NOAEL) is the basis for most practical thresholds of toxicological effects. The critical NOAEL is combined with uncertainty factors (UFs) to provide a margin of safety for the exposed population.

Acceptable Daily Intake/Air Quality Standard/
Water Quality Standard/etc.

= Experimental NOAEL ÷ Margin of Safety

or:

$$\text{NOAEL} \div (\text{UF}_A * \text{UF}_H * \text{UF}_T * \text{UF}_D * \text{UF}_L)$$

A = animal to human; H = human variability; T = temporal factors; D = data gaps/quality; and L = LOAEL* to NOAEL

UFs = 1–10, depending on source of data, with a maximum cumulative UF of 3000 (U. S. EPA, 2002).

TABLE IV. U.S. EPA—Interim Guidelines for Development of Inhalation Reference Concentrations (EPA 600/8–90/066A, 1990)[a]

Effect or no effect level	Rank	General effect	Cal/EPA severity ranking
NOEL	0	No observed effects	< Mild
NOAEL	1	Enzyme induction or other biochemical changes, consistent with possible mechanism of action, with no pathologic changes and no change in organ weights	< Mild
NOAEL	2	Enzyme induction and subcellular proliferation or other changes in organelles, consistent with possible mechanism of action, but no other apparent effects	< Mild
NOAEL	3	Hyperplasia, hypertrophy, or atrophy, but no change in organ weights	≤Mild
NOAEL/LOAEL	4	Hyperplasia, hypertrophy, or atrophy, with changes in organ weights	Mild
LOAEL	5	Reversible cellular changes including cloudy swelling, hydropic change, or fatty changes	Mild/severe
(LO)AEL[b]	6	Degenerative or necrotic tissue changes with no apparent decrement in organ function	Severe
(LO)AEL/FEL	7	Reversible slight changes in organ function	Severe
FEL	8	Pathological changes with definite organ dysfunction that are unlikely to be fully reversible	Severe
FEL	9	Pronounced pathologic changes with severe organ dysfunction with long-term sequelae	Severe
FEL	10	Death or pronounced life-shortening	Life-threatening

[a]Adapted from Hartung, 1987.

[b]The parentheses around the LO in the acronym LOAEL refer to the fact that any study may have a series of doses that evoke toxic effects of rank 5–7. All such doses are referred to as adverse effect levels (AELs). The lowest AEL is the (LO)AEL.

The definition of NOAELs and LOAELs (lowest observed adverse effect levels) are shown in Table IV.

The NOAEL concept and its implications for human health risk assessment is shown in Figure 2.

An alternative to the NOAEL is the benchmark dose (BMD) approach, which is favored by regulatory agencies when the data are of sufficient quality.

Some important characteristics of the BMD are

- Benchmark doses are used by the U.S. Environmental Protection Agency (U.S. EPA) and the World Health Organization (WHO) in many non-cancer risk assessments.
- BMDs use dose-response information.
- BMDs take into account statistical uncertainty and sample sizes.
- BMDs assume a distribution of responses rather than a point estimate.
- BMDs can assume a threshold and account for background responses as options.
- Dichotomous and continuous data can be used to calculate BMDs.

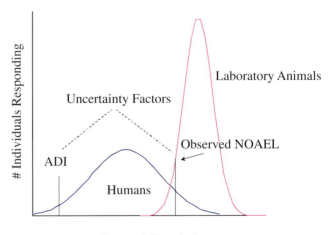

FIGURE 2 Concept of extrapolation from animal experimental NOAEL to humans in deriving an acceptable daily intake (ADI). The laboratory animals are assumed to be, on average, less sensitive to the toxicant than humans, and are also assumed to have less individual variability (indicated by the relatively narrow distribution on the right).

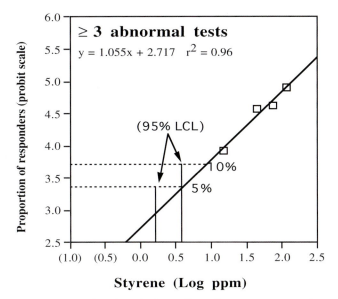

FIGURE 3 Benchmark dose effect of styrene on neurobehavioral tests in plastics workers. (Data from Mutti et al., 1984 and Rabovsky et al., 2001.) LCL is the statistical 95% lower confidence limit on the probit line.

A number of mathematical models exist for calculating BMDs. Public domain software developed by the U.S. EPA for benchmark dose calculations can be found at the U.S. EPA Web site (www.epa.gov).

An example of a BMD relationship is shown in Figure 3. A percentage response in the test population is taken to be the practical threshold for the adverse effect. Typically this is a 5–10% response incidence.

1. What Is An Adverse Effect?

Not all biological responses to a toxicant are considered adverse. Some effects are considered to be adaptive responses that have no short- or long-term consequences. This gray area in the definition of a threshold for adverse effects is one of considerable discussion and debate in regulatory toxicology. Table IV illustrates the types of definitions that have been assigned as severity qualifiers to adverse effects by the U.S. EPA.

Adverse effects from human clinical and epidemiological studies have also been defined in the context of regulatory purposes (Table V).

TABLE V. California Environmental Protection Agency Symptom and Sign Severity Rating for Human Studies

Severity rating	Symptoms	Signs/laboratory findings
Mild adverse	Mild subjective complaints with few to no objective findings:	Statistically significant findings of pre-clinical significance:
	Mild mucous membrane (eye, nose, throat) irritation	Mild conjunctivitis
		Mild lung function changes
	Mild skin irritation	Abnormal immunotoxicity test results
	Mild headache, dizziness, nausea	Mild decreases in hemoglobin concentration
Severe adverse	Potentially disabling effects that affect one's judgement and ability to take protective actions; prolonged exposure may result in irreversible effects:	Clinically significant findings: Findings consistent with central or peripheral nervous system toxicity
		Loss of consciousness
	Severe mucous membrane irritation	Hemolysis
	Blurry vision	Asthma exacerbation
	Shortness of breath, wheezing	"Mild" pulmonary edema
	Severe nausea	Clinically significant lung function changes
	Severe headache	Cardiac ischemia
	Incoordination	Some cardiac arrhythmias (e.g., atrial fibrillation)
	Drowsiness	Renal insufficiency
	Panic, confusion	Hepatitis
		Reproductive/developmental end points (e.g., infertility, spontaneous abortion, congenital anomalies)
Life-threatening		Potentially lethal effects:
		Severe pulmonary edema
		Respiratory arrest
		Ventricular arrhythmias
		Cardiac arrest

E. Variation in Effects: Genetic and Phenotypic Variability in Susceptibility

The variability in response to a chemical agent can create difficulties in establishing statistically significant associations in epidemiology. However, the variability in a given response often has biological roots that are increasingly important as regulatory and public health agencies try to determine ways to identify and protect the most sensitive individuals in society from adverse toxicological effects.

Toxicological risk assessment has traditionally relied on estimates of no-effect thresholds (i.e., NOAELs) combined with uncertainty factors, which are intended to account for the fact that individual variability in response exists. Default values between one and ten are typically used in these calculations because the precise amount of individual variability is not known.

More recently, investigators have been determining the degree of variability in physiological parameters. For example, the U.S. National Research Council (NRC) reports a range of elimination half-lives for 13 different drugs that are 0.7- to 17-fold greater in newborn infants than in adults (NRC, 1993). For a given rate of exposure, these drugs would remain for a longer time in an infant's body, thus likely increasing the infant's susceptibility.

The individual variability in some key toxicokinetic mechanisms have been described using clinical and epidemiological studies as shown in Table VI.

Table VI shows that although toxicological risk assessment relies on the assumption that a 10-fold uncertainty factor for individual variability is health protective, in some cases clearly a factor of 10 comes nowhere near the amount of variability that actually exists. However, this type of research with application to risk assessment is relatively recent, and more data are needed in order to more precisely define what are appropriate default values for this parameter.

Although the long neglected field of human susceptibility to environmental toxicants is currently receiving renewed attention, there is only scant literature on factors influencing susceptibility to heavy metals (Gochfeld, 1997).

IV. TOXICOLOGY AND EPIDEMIOLOGY: HOW IS ENVIRONMENTAL MEDICINE STUDIED?

Toxicology, risk assessment, and epidemiology all have important roles in the study of environmental medicine.

TABLE VI. Reported Ranges of Variability in Parameters Related to Susceptibility

Parameter	Width of 90% range for an average chemical or test
Systemic uptake pharmacokinetics	
Breathing rates	1.8- to 2.8-fold
Half-life or elimination	2.3- to 5.8-fold
Skin absorption	2.5-fold
Maximum blood concentration	2.3- to 11-fold
Area under concentration curve	3.0- to 8.1-fold
Blood concentration measurements	
Serum PCB concentrations	12-fold
Blood mercury levels	12-fold
Blood lead levels	13-fold
Pharmacodynamic or combined kinetic and dynamic parameters	
Cisplatin hearing loss	4.1-fold
Effects of methyl mercury in adults	12- to 78-fold
Fetal/developmental effects of methylmercury	460- to 10,000-fold
Hemodynamic responses to nitrendipine	8.3- to 17-fold
FEV1 response to cigarette smoke	8.3-fold
Salbutamol FEV1 increase (asthma)	128-fold
Acute toxicity	
Death from compounds metabolized by plasma cholinesterase	5.5-fold
Death from parathion	12-fold

Adapted from Hattis, 1996.

The scientific confidence in the public health actions that are taken in response to environmental contaminants is a function of how thoroughly each of these areas are addressed. History shows that heavy emphasis on one discipline alone can lead to actions that later are determined to be unfounded. The case of saccharin is one example. Saccharin was, for many years, considered to be a probable human carcinogen by the IARC and the U.S. National Toxicology Program due to its ability to induce bladder cancer in male rats. Many studies were done on rats to confirm this effect, and a dose-response relationship was developed and widely accepted. At one point, saccharin was one of the most studied compounds in terms of long-term cancer bioas-

TABLE VII. Basic Framework for the Identification, Risk Assessment, and Epidemiological Study of Environmental Contaminants

Discipline/Area of Research	Goals and Outcomes
Basic biochemical toxicological research	• Identification of biochemical mechanisms • Hypothesis of downstream physiological end points • Development of biomarkers
Toxicity testing	• Descriptive toxicology test battery (acute, chronic, mutagenicity, carcinogenicity, reproductive/developmental toxicity, sensitization) • Identification of critical effects starting from high doses and reducing dose until no effect is seen • Establishment of dose-response relationship • Multiple species and both genders
Risk assessment and management	• Identification of a critical dose (no observed effect level or benchmark dose) • Application of margin of safety (uncertainty factors) • Establishment of an acceptable chronic dose level
Epidemiological research	• Probe established toxicological limits and biochemical mechanisms determined from animal studies to determine if they are applicable to humans • Describe the probabilities of adverse effects in humans occurring following exposure • Signal for further mechanistic research or the need to develop human biomarkers for exposure or effect

says and dollars spent. This carcinogen designation was then driven by toxicity testing because the mechanism for cancer formation from saccharin was not known, nor was it known why only male rats were susceptible, and not mice, any other species, or female rats (see also Chapters 21 and 24, this volume).

The epidemiology research showed that there was no evidence for elevated incidence of cancers in humans using saccharin, but there were concerns that the latency period required for cancer to develop in people had not been allowed to mature. The mechanism of cancer formation was later found to be due to induction of a male-rat-specific protein (alpha-2-microglobulin) that caused chronic irritation of the bladder, which led to bladder cancer at high doses. This led regulatory agencies to de-list saccharin as a human carcinogen in 1999 (IARC, 1999). The relationship of these disciplines in environmental medicine is shown in Table VII.

A. Concepts of Dose and Duration

1. Cancer Potency

Chemicals that are carcinogenic to humans are identified through several authoritative bodies using established weight of evidence approaches. The IARC, the U.S. National Toxicology Program (NTP), and the U.S. EPA are the three most authoritative sources for identification of new and existing carcinogens.

There are 48 individual chemicals that are known human carcinogens under the IARC classification scheme.

The data from cancer bioassays is usually fit to a linearized multistage model, which is of the form originally described by Crump (1984).

Linearized multistage model:
$$P(d) = 1 - e^{-(q0 + q1d + q2d2 + \ldots + qkdk)}$$
$$P(0) = 1 - e^{-q0}$$

where P(d) is the probability of developing a tumor at a given dose rate and P(0) is the estimated background incidence. The q parameters are derived from the model.

Cancer potency is usually described by a dose-response slope factor (q1) and its respective 95% confidence limit (q1). The units of potency are usually in $(mg/kg/day)^{-1}$ which, when combined with a given dose level, gives a unitless risk factor (e.g., 10^{-6}, or 1 in a million). This concept applies to daily doses experienced over a chronic period, up to an entire lifetime. Much less is known about estimating risks from single acute exposures to carcinogens, and there are few animal studies through which to judge the difference in the dose-response relationship.

Often, a rodent cancer bioassay yields information only at a few doses that far exceed those found in the environment, and the dose-response is extended to very low doses.

Potency factors assume the absence of a threshold for cancer at low doses. Therefore, it may be inappropriate to apply potency estimates to carcinogens that are thought to have a threshold (e.g., non-genotoxic carcinogens, such as dioxin).

The utility of cancer potency factors lies in their use in cancer risk projections. When calculating cancer risks, if exposures to specific carcinogens can be quantified, then it is assumed that the risk of getting cancer from a long-term exposure is a function of exposure (i.e., mg/kg/day) multiplied by the respective cancer potency factor $(mg/kg/day)^{-1}$. The two most critical assumptions with this calculation are

1. The assumption that the basis for the cancer potency factor is a mechanism that applies to human physiology
2. The tumors seen at high doses in experimental studies are part of a linear or curvilinear function that extends to low doses that are more relevant to environmental exposures

The California Environmental Protection Agency (Cal/EPA) has an active program that identifies carcinogens and lists potency values for each on its Web site (www.oehha.ca.gov).

2. Dose-Response Slope

The slope of the response curve from a toxicology study imparts significant meaning to the causal relationship of the chemical and effect being studied. The dose-response relationship also helps in understanding the risks from exposure to low doses.

Some additional uses of the dose-response relationship include the characterization of individual variability in the measured response to the chemical. Table VIII shows the relationship of probit slope term to the individual variability in a battery of neurological tests in response to styrene exposure in the workplace (Rabovsky et al., 2001). In general, the more shallow the dose-response slope, the greater the variability in the test population. When using a log-normal model to

TABLE VIII. Expected Variability in the Abnormal Responses to Styrene

Number of abnormal neurological tests	Probit slope (1/log GSD)[a]	Fold differences between percentiles (5–95% range)[b]	(1–99% range)[b]
≥1	1.346	270	2860
≥2	1.225	485	7190
≥3	1.055	1300	25,540

[a]The probit slope was taken from the log-probit analysis using Tox-Risk V.3.5[30] described for Table III.
[b]The range of variability is obtained from the equation:

$$[\log(X) - \log(GM)]/\log(GSD) = t(\alpha)$$

where X is the distance on the dose axis from the mean to the 95th (or the 99th) percentile. To obtain the range of variability, the same distance from the mean to the 5th (or 1st) percentile is added to "X" and the bounds of the 5th–95th (or 1st–99th) percentile are the ranges of variability. Other abbreviations are GM, geometric mean; GSD, geometric standard deviation; and t (α), t distribution value at the desired test level (i.e., 0.05 or 0.01).
From Rabovsky et al., 2001.

describe a dose-response relationship, slope terms that approach 1.0 show very large individual variability (Table VIII).

B. Estimating Exposure (Analytical Chemistry, Biomarkers, and Modeling)

Exposures to metals can be measured or estimated in various ways. For chronic dietary exposures, blood samples for the metal may be the most direct and simple measure of exposure. For historical exposures, blood samples may not be appropriate if the body has had time to depurate the metal from the bloodstream. Lead has a half-life of about 25 days in the blood. Methylmercury has a half-life for elimination from the body of about 50 days in adult males and females. However, infants have poor elimination of organic and inorganic heavy metals in the first 6 months prior to development of metal transport systems. It is possible that mercury is eliminated much more slowly in the child. Cadmium is removed from the human body much more slowly, so that the elimination half-life is on the order of 20 years (Gochfeld & Syers, 2001).

Metallothioneins (MTs) are metal-binding proteins that are considered central in the intracellular regulation of metals such as copper, zinc, and cadmium. Variability in tissue MT levels influence susceptibility of tissues and species to the toxic effects of some metals, such as cadmium and mercury.

Metallothionein is the major protein thiol induced in cells exposed to cytokines and bacterial products (Schwarz et al., 1995). This protein is inducible by exposure to some metals, and it appears to impart protection from some adverse effects. For example, zinc exposure during fetal life results in MT induction (Mengheri et al., 1993). Zinc pretreatment lowers cadmium carcinogenicity in laboratory animals presumably through induction of MT (Coogan et al., 1992). Similarly, MT is thought to contribute to the placental barrier to the transfer of mercury to the fetus (Yamamoto et al., 2001).

C. Is There a Health Effect? Animal Models

The widespread use of animal models in toxicity testing continually raises the possibility that a given adverse effect may not be relevant to humans. The case of saccharin already described, is an example of this. The converse of the saccharin case, however, can also be true. Arsenic, for example, is a human carcinogen and a human neurotoxin, but it does not appear to cause cancer in laboratory animals at doses that are considered carcinogenic to humans. Thalidomide is a prominent case in which a relatively minor heart valve developmental defect in rodents translated into major deformities in human babies. Benzene is another example where humans appear to be the most sensitive species tested for its critical effect (leukemia). Thus, animal models show a number of examples of failure in predicting effects on humans. However, by and large, animal models, particularly when multiple species are tested, are thought to provide adequate evidence for an initial risk assessment of a substance to proceed, provided that adequate margins of safety are used. Epidemiological studies will always be needed to ascertain if the safe limits proposed in risk assessment are adequately protective.

In the case of arsenic, there is significant interspecies variability in metabolism. However, in most mammals, arsenic is metabolized and detoxified by the addition of methyl groups, and is eventually excreted, primarily in urine, in a mono- or dimethylated form. The methylation pathway is dependent on folate metabolism, which provides methyl groups through conversion of S-adenosylmethionine to S-adenosylhomocysteine. The availability of methyl groups from this pathway is thus essential to the metabolic detoxification of arsenic and may at least partially account for both inter- and intraspecies differences in sensitivity to this toxicant. Human serum, for example, is folate-deficient compared to rodent serum (unpublished observation), and this observation may explain experimental findings that arsenic, a known carcinogen in humans, does not induce cancer in rodents.

Given the critical role of methylation in the disposition of arsenic, further characterization of the enzymatic basis of arsenic methylation is required. To date, human arsenic methyltransferase has not been isolated, but transferases are generally polymorphic. Understanding the factors affecting human sensitivity would improve the arsenic risk assessment. The objective of this section is to evaluate variations in arsenic metabolism as reflected in variations in urinary metabolites or other biomarkers of exposure as associated with the exposure level, nutritional status, genetic factors, and other variables. Included in this area are studies to improve mass balance data on typical human metabolism of arsenic at various doses and chemical forms. There is a need for the development and refinement of assay procedures to characterize arsenic methyltransferases in human tissues. In addition, these studies would compare biomarkers of arsenic metabolism in individuals exposed to varying levels of arsenic with differences that include, but are not limited to, nutritional status, age, sex, and genetic variations.

D. Evidence at the Population Level

Further evidence (follows from animal studies section) of the relationship between an exposure and a health effect can be sought through the epidemiological study of human populations. Such epidemiological studies take three main forms: cross-sectional, case-control, and cohort. Each study design has its own strengths and weaknesses, and *all answer different questions*.

The cross-sectional study is in essence a survey to determine how common something is, or at what level it occurs. In Figure 4, for example, the prevalence of caries in the teeth of children was determined by cross-sectional survey. By carrying out such surveys in populations using different drinking water supplies, it was possible to answer the question: Are caries more common in populations with low fluoride exposure? An

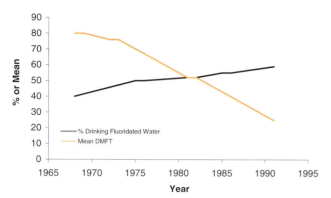

FIGURE 4 Percentage of population residing in areas with fluoridated community water systems and mean number of decayed, missing (because of caries), or filled permanent teeth (DMFT) among children aged 12 years, United States 1967–1992. (From CDC, 1992; National Center for Health Statistics, 1974, 1981; National Institute of Dental Research, 1989; and CDC, unpublished data, 1988–1944.)

advantage of the cross-sectional study is that it is relatively quick, easy, and cheap to administer. Disadvantages include that it cannot take into account individual exposures and is subject to confounding. For example, if the people drinking low-fluoride water are also poorer, they may have more caries for dietary reasons.

The case-control study starts with identified cases of a disease and compares their exposure to controls who do not have the disease. In the case study on amoebic meningitis in Section II, for example, children with the disease would be identified, and the frequency of their hot spring usage would be compared to that of children without the disease. With such a study it is possible to answer the question: Did children with amoebic meningitis swim in hot springs more frequently than children without amoebic meningitis? If yes, then hot springs are implicated by association, but note that this does not demonstrate a causal relationship.

An advantage of the case-control study is that risk factors for very rare events can be identified. A disadvantage is that they contain no denominator information: note that we have no idea how many children swam in the hot springs to give rise to the few rare cases that we are studying, and we therefore cannot say how dangerous that activity is.

The cohort study, seen by many as the epidemiological gold standard, starts with a group of people with a common exposure.

The risk factors of cases with the disease are then compared to the risk factors of controls without the disease, all still within this group or cohort of people.

For example, a cohort might be defined as all people living in a village with high levels of arsenic in the water supply. Some of these people would develop skin cancer and become "cases," whereas others would not develop skin cancer and would instead become "controls." The epidemiologist would record the water consumption habits of all members of the cohort. With such a study it is possible to ask the question: Do people who drink more water develop skin cancer more frequently? If yes, the arsenic-rich water would be implicated by association, but again, a causal relationship has not been established. The advantage of this study design is that it does provide denominator information, and it is therefore possible to directly calculate the risk of developing cancer from drinking the water. The disadvantage is that this approach is both very time-consuming and potentially very expensive as very large cohorts are often required to give the study enough power to achieve statistical significance.

As mentioned, all of these studies are epidemiological, and can at best establish association between exposure and effect. To establish a causal relationship between an exposure and a health effect, several other criteria are generally considered, including the following (Hill, 1965): a temporal relationship between exposure and effect; a biologically plausible relationship, including a dose-response effect; associations by epidemiological studies that are both strong and consistent; and reversibility of the association if the exposure is removed.

A number of approaches to the study of environmental medicine have been outlined in this chapter. Let's not forget that the aim of such a study is to devise interventions to reduce morbidity and mortality, which is the subject of the following section.

V. HEALTH PROTECTION: HOW CAN ADVERSE EFFECTS BE MINIMIZED?

In Section II (and Table I) we explored a useful framework for considering health effects from environmental exposures. Other than simply to satisfy the human compulsion to classify things, such a framework is a prerequisite to health protection: in order to minimize the adverse health effects of environmental exposure, we must first be able to consider all possible exposures. We can then prioritize surveillance and intervention so as to maximize health gain from the use of our (invariably limited) resources.

A. Risk Assessment

One common approach to risk assessment follows the four simple steps of hazard description, dose-response estimation, exposure assessment, and risk characterization. These can be simply illustrated here using an example such as exposure to volcanic gases during an eruption (see Volcanic Emissions and Health, this volume). The hazard description would list the gases concerned, e.g., CO_2, H_2S, and HF, and detail the volume, concentration, and duration of emission. The dose-response estimation (effect) would consider, usually graphically, the relationship between the amount of gas and the relevant health effect. Thus respiratory distress may increase in a linear fashion with the dose of gas inhaled, and a threshold may exist at which consciousness is lost.

The exposure assessment would take into account how many people are likely to suffer from such health effects, for how long, and if any particular groups are at greater risk. For example, the number of people living in a village downwind from the eruption will be relevant, as will be the number of children in cots (arguably more susceptible to suffocation from gas that is heavier than air).

The risk characterization summarizes all this information into a prediction about the likely outcome if the hazard goes unchecked. For example, 2 deaths in children and 100 adults with respiratory distress may be predicted for such a village if it is not evacuated. Medical and civil defense authorities will then make a decision about the appropriate deployment of resources based on this risk assessment. Should the two buses available be deployed to evacuate this village, or would they be better deployed to evacuate the village on the other side of the mountain that is potentially in the path of a lava flow?

The problem is that even with the best risk assessment, there is always a degree of uncertainty, and decisions about intervention can therefore be difficult to make. It seemed sensible at one time to recommend that well water be used in Bangladesh to avoid the risk of gastrointestinal disease from pathogens in surface waters. The well water, however, turned out to contain arsenic at levels that were not anticipated (see Arsenic in Groundwater and the Environment, this volume). It is sometimes unclear if a known exposure constitutes a health risk or not. For example, low-level chronic exposure to geothermal hydrogen sulfide has not been demonstrated to cause clear-cut pathology, but one might expect it to on the basis of respiratory and neurological toxicity at higher levels of exposure. In such cases, practitioners of environmental medicine often apply the "precautionary principle," which states that any substance suspected of adversely affecting health should be avoided (as far as possible) until proven otherwise.

B. Surveillance

Surveillance is the term used, in environmental medicine, to refer to the ongoing monitoring processes that inform public health intervention. Surveillance is the use of monitoring data to attempt to reduce morbidity and mortality: without the completion of this criterion, surveillance would be no more than data collection.

A good example of a surveillance system is the "notifiable disease" list, which compels medical practitioners in most countries to notify some authority responsible for disease control on each occasion that he makes a diagnosis of one of the diseases on the list. Lead poisoning causes learning difficulties and neurological complications, and is on the notifiable list in most countries. The relevant authority collects and analyzes the notified case data to determine the source of the exposure; for example, toddlers are commonly poisoned by ingesting flakes of old lead-based paint in poorly maintained houses, and adults may have occupational exposure in industries such as battery recycling.

The "surveillance loop" is completed when this information is used to impose recommendations for home improvement or factory practices. Note that the frequency of notified cases provides feedback as to the effectiveness of the interventions (at least on a regional or national scale), thereby forming a genuine "intelligence loop."

In the example above, health surveillance was carried out for the health effects of exposure to the element lead. It is also possible to carry out *hazard* surveillance, where the environmental levels of, e.g., lead are monitored directly, rather than (or as well as) monitoring the health effects caused by lead. The same requirements for surveillance hold, and the data on lead concentration and distribution are used to inform public health intervention: the lead level in the factory can now be kept below safety limits without the need for workers to develop symptomatic disease (which is obviously preferable). The implementation of one or both types of surveillance, health and hazard, are integral to the practice of environmental medicine. In conjunction with the environmental medicine framework discussed in Section II, it should now be possible for the reader to devise, at least at the theoretical level, surveillance systems to deal with most situations that might be encountered in medical geology.

C. Intervention

Intervention Success Story: The Fluoridation of Public Water Supplies

In the 1930s American dental epidemiologists noted a considerable regional variation in the rate of dental caries. By carrying out a cross-sectional study (as described in Section IV.D), Dean (1936) established an association between differing levels of fluoride (F⁻) in regional water supplies and caries rates in the populations drinking those waters. Although fluoride is a toxin that in high concentration can kill, it appeared in this case that a small amount in the drinking water was beneficial to dental health. In populations supplied with drinking water containing about 1 mg/L F⁻, dental caries rates were reduced by about 50%. The variation in fluoride levels in drinking waters in those days was entirely natural, and resulted from the dissolution of fluorine by surface waters as they coursed over fluorine-rich substrates such as geological deposits of marine origin. Population health was therefore directly affected by living in an area with particular geological characteristics, a situation that is in essence the theme of this book.

Public health authorities now began to ask the obvious question: For populations not living in fluoride-rich areas, could dental health be improved by artificially supplementing F⁻ levels in drinking water? The suggestion was examined from several perspectives including cost-benefit considerations and human health risk assessment (Section V.A). The risk was (and remains) one of balancing the fluoride delivery carefully so as to achieve the reduction in caries rate without also causing fluorosis—the condition of excessive chronic fluoride ingestion. Many middle-aged people in the western world have dental fluorosis (stained patches of brittle enamel) from receiving both fluoridated water and fluoride tablets as children. Industrial fluorosis is also known, largely historically, as an occupational hazard in the aluminum and fertilizer industries. In these cases, fluoride deposition into bone increases bone and cartilage density which can result in restricted flexibility and movement, especially around the lumbar spine (Derryberry et al., 1963). In the developing world, many people in specific geographic areas have a far more serious form of fluorosis, which is known as endemic fluorosis. This is a potentially crippling disease, the major manifestation of which is the overgrowth and distortion of bone, with tendinous, articular, and neurological involvement. Such severe disease only results after many decades of ingestion of drinking water with 10 mg/L F⁻ or more, a concentration one order of magnitude higher than the 1 mg/L observed to be beneficial by Dean (1936). Since the 1950s therefore, public drinking water supplies in most developed countries have been topped up to about 1 mg/L of F⁻, with minor adjustments to account for differing levels of climate-dependent water consumption. The improvement in the dental health of children has been remarkable since that time, with large reductions in caries rates.

The debate continues to this day about the relative contribution of water fluoridation to this reduction in dental caries rates, because there have been concurrent improvements in nutrition, dental hygiene, and dental services that also contribute to a reduction in the incidence of caries. The issue has become further blurred by the advent of fluoridated toothpaste, sports drinks, and other sources of fluoride that dilute the improvement attributed to fluoridated water alone. It is known, however, that populations receiving fluoridated drinking water show better dental health, on average, than do control populations without fluoridated water (WHO, 1994). Health surveillance data (caries rates) collected by school dental services show that this continues to be the case, and hazard surveillance data (water F⁻ levels) collected by water treatment plants continue to ensure that populations are not at risk of fluorosis. The fluoridation of public water supplies is therefore a good example of a successful public health intervention informed by ongoing surveillance.

VI. CASE STUDY: ARSENIC

A. Exposure to Arsenic

Arsenic is a metalloid element found ubiquitously in nature. It is present in the Earth's crust with an average concentration of 2 mg/kg. Arsenic can be found in soil, air, water, food, and some manufactured chemicals. Humans can be exposed to arsenic from either natural sources or anthropogenic sources. Natural sources of arsenic include rocks (soil), volcanic emissions, undersea smokers, and extraterrestrial material. Volcanic emission is the most important natural source of arsenic. Arsenic can be found in more than 200 mineral species, of which the most common is arsenopyrite. Anthropogenically, arsenic can be found in products of herbicides, fertilizers, pesticides, leather treatment, cotton desiccants, wood preservation, animal feeds as food additives, and pharmaceuticals (see also Chapters 11 and 23, this volume).

Humans can be exposed to arsenic through ingestion of arsenic-containing water, food, and drugs (such as Fowler's solution containing 1% of potassium arsenite used to treat psoriasis and arsenic trioxide used to treat leukemia). Airborne arsenic can be absorbed into the bloodstream in workers involved in the processing of copper, gold, and lead ores; in the production and use of arsenic-containing pesticides; in the manufacturing of glass, semi-conductors, and pharmaceutical substances; in using arsenic as pigments and dyes; in burning coal containing high levels of arsenic (Guizhou Province, China); in smoking high-arsenic-contaminated tobacco; and in chimney sweeping.

Water contamination is the most common source of arsenic exposure. Currently, Bangladesh and West Bengal, India, have the most serious problem of groundwater contamination with arsenic in the world. Tracing back the history of these areas, surface water was replaced by tubewell water in 30 years ago to fight against infectious diarrheal diseases. These programs to provide "safe" drinking water from underground unexpectedly brought up another health problem of arsenic hazards. It is estimated that more than 95% of the 120 million people in Bangladesh drink tubewell water and more than one-third of the tubewell water contains arsenic above 0.05 mg/L (the maximum allowable level recommended by the WHO). In 2001 the U. S. EPA lowered the maximum allowable level of arsenic in drinking water from 0.05 mg/L to 0.01 mg/L. High arsenic level in drinking water is also reported in countries such as Argentina, Australia, Chile, China, Hungary, Mexico, Peru, Taiwan, Thailand, and the United States. See Arsenic in Groundwater and the Environment, this volume.

B. Effects of Arsenic

Arsenic exists in four valence states: −3, 0, +3, and +5. Elemental arsenic and arsine (−3) exist in strongly reducing environments; arsenite (+3) is the dominant form in moderately reducing conditions; and arsenate (+5) is stable in oxygenated environments. Inorganic forms of arsenic are much more toxic than organic forms found abundant in seafoods, and, in general, inorganic arsenic of trivalent forms are more toxic than pentavalent forms. Immediate symptoms of an acute poisoning typically include vomiting, esophageal and abdominal pain, and bloody "rice water" diarrhea. However, a variety of symptoms and signs involving the gastrointestinal, dermal, nervous, renal, hepatic, hematopoietic, cardiovascular, respiratory, and ophthalmic systems can be observed (Table IX) (Chen et al., 1999; Tseng, 1999). Treatment with chelating agents such as dimercaprol or dimercaptosuccinic acid during acute intoxication is classical but may have varying effects. Chelating agents may not be effective in chronic poisoning. See Biological Responses of Elements, this volume.

Long-term exposure to arsenic can cause a variety of cancers involving the skin (squamous cell and basal cell carcinoma), lung, bladder, kidney, and liver. Although arsenic does not induce point mutations, it can cause chromosomal aberrations, affect methylation and repair of DNA, induce cell proliferation, transform cells, and promote tumors.

A wide spectrum of non-cancerous diseases and clinical problems are also reported in long-term arsenic exposure (Table IX). Arsenic skin lesions are characterized by the coexistence of hyper- and hypopigmentation giving rise to a raindrop pattern (Figure 5) and hyperkeratosis of the palms and soles (Figure 6). In recent years, long-term exposure to arsenic from drinking water has also been found to be highly associated with hypertension and diabetes mellitus (Tseng et al., 2000, 2002). Preclinical microcirculatory defects (Tseng et al., 1995) and arterial insufficiency (Tseng et al., 1994) can also been demonstrated in subjects exposed to arsenic. Arsenic could also cause lower IQs in children exposed to arsenic in Thailand. The symptoms and signs that arsenic causes appear to differ between individuals, population groups, and geographic areas.

C. The Study of Arsenic Intoxication: The Example of Blackfoot Disease

Exposure to arsenic from drinking water in Taiwan has been shown to cause a severe peripheral vascular disease, which might progress from intermittent claudication, ulceration, gangrene, and spontaneous or surgical amputation (a case with spontaneous amputation is shown in Figure 7). The disease has been named

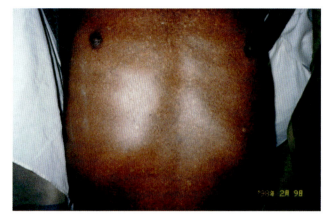

FIGURE 5 Raindrop pattern of hyper- and hypopigmented skin lesions in a patient with long-term arsenic intoxication.

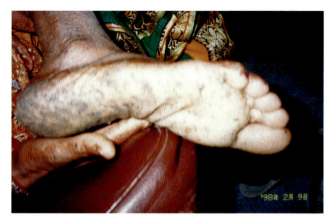

FIGURE 6 Hyperkeratosis of the sole in a patient with long-term arsenic intoxication.

TABLE IX. Non-Cancerous Effects of Arsenic on Humans

Organ system	Diseases or symptoms/signs
Dermal	Hypo- and hyperpigmentation (raindrop pattern), hyperkeratosis of palms and soles, exfoliative dermatitis, Bowen's disease (pre-cancerous lesions), facial edema, non-pitting pedal edema
Cardiovascular	Arrhythmia, pericarditis, ischemic heart disease, peripheral vascular disease, cerebral infarction, hypertension, microcirculatory defects
Gastrointestinal	Abdominal discomfort, anorexia, malabsorption, body weight loss
Nervous	Peripheral neuropathy involving sensory and motor systems, cranial nerve involvement, hearing loss, mental retardation, encephalopathy
Renal	Nephritis and proteinuria
Hepatic	Fatty degeneration, non-cirrhotic portal fibrosis, cirrhosis, hepatomegaly
Hematopoietic	Bone marrow hypoplasia, aplastic anemia, anemia, leukopenia, thrombocytopenia, impaired folate metabolism, karyorrhexis
Respiratory	Rhino-pharyngo-laryngitis, tracheobronchitis, pulmonary insufficiency (emphysematous lesions)
Ophthalmic	Conjunctivitis
Reproductive	High perinatal mortality, low birth weight, spontaneous abortions, stillbirths, pre-eclampsia, congenital malformation
Metabolic	Diabetes mellitus, goiter

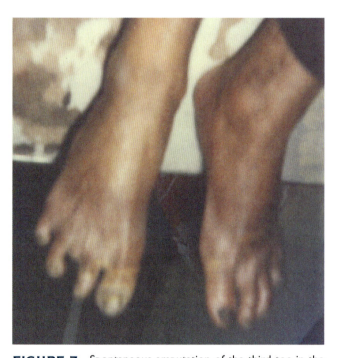

FIGURE 7 Spontaneous amputation of the third toe in the right foot of a subject living in the blackfoot disease endemic area in Taiwan.

blackfoot disease after its clinical appearance (Tseng 1999, 2002). This disease was first reported in the early twentieth century and was confined to the southwestern coast of Taiwan where people used artesian well water from as deep as 100–300 m underground (Tseng et al., 1996). The prevalence ranged from 6.51 to 18.85 per 1000 people in different villages. A series of epidemiologic studies and surveillance of the arsenic concentrations of the artesian wells carried out during the mid-twentieth century revealed the association between blackfoot disease and the consumption of high-arsenic-containing artesian well water. Besides arsenic intake from well water, residents in the endemic area could also be exposed to arsenic from a variety of other sources, because the artesian well water was extensively used for agricultural and piscicultural purposes. The amount of arsenic ingested by the residents of the endemic area was estimated to be as high as 1 mg per day. The lethal dose in humans is estimated to be 1 mg/kg/day.

Although studies in several other countries have demonstrated that arsenic exposure can be associated with some forms of peripheral vascular disease, similar endemic occurrence of severe blackfoot disease has not been observed. It is possible that nutritional status, coexistence of other factors, and interaction with other trace elements determine the development of the various clinical manifestations.

There are two main pathways of arsenic metabolism: the reduction reactions and the oxidative methylation reactions. Pentavalent arsenic is reduced to trivalent arsenic, followed by the methylation reactions to form mono-, di-, and trimethylated products. S-adenosyl methionine is the methyl donor and glutathione is an essential cofactor. Low amounts of methionine or protein in the diet decrease the methylation of inorganic arsenic in animals and similar nutritional deficiency was observed in the residents of blackfoot disease areas. Vitamin B_{12} is needed in the methylation process and insufficient intake of this vitamin in poor people and/or increased requirement during the reproduction ages in the women may put these people at higher risk for the development of arsenic-related health problems. Zinc and selenium may provide protective effects against the toxic effects of arsenic, and residents in the blackfoot disease areas were found to have deficiency of these elements in their diet. Lower levels of beta-carotene have also been shown to carry a higher risk of developing vascular disease and skin cancers in residents of the blackfoot disease endemic areas (Hsueh et al., 1997, 1998).

The absorption, distribution, and metabolism of arsenic differ significantly across species. Animals are less sensitive to the toxic effect of arsenic and most of the effects of long-term arsenic exposure on humans are not observed in animals. Genetic factors may play important roles on these metabolic cascades of arsenic; and thus, may also be involved in the development of the clinical effects of arsenic.

D. Public Health Intervention

The arsenic-related health problems in emerging endemic areas can be critical issues in public health. Drinking water poses the greatest threat to public health from arsenic. However, exposure from coal-burning, working environment, mining, and industrial emissions may also be significant in some areas. There is no universal definition of the disease caused by arsenic and there is no way to differentiate pathologically those vascular or cancerous lesions caused by arsenic from other etiologies. All of these complicate the assessment of the burden of arsenic on health. However, the use of interventional measures to terminate the hazards associated with arsenic should not wait until all these ambiguities are clarified. Up to now, there is no magic bullet for the treatment of the diseases associated with arsenic intoxication. The best strategy is prevention and avoidance of exposure. New sources of water and coal with low arsenic contents, techniques for arsenic removal from drinking water, decreasing industrial arsenic emissions, improving working environments, and promoting health education among the affected people are necessary.

As for the conditions in Bangladesh, a collaborative approach is required for scientists, health workers, policy makers, and members of the community to work together to plan and implement a sustainable and environmentally friendly water supply system. The government and professional people should have reliable, timely, and easily available information on the status of knowledge about the problems and what can be done to tackle the problems. Large-scale and concerted action is required from all sectors to take effective and practical remedial measures at affordable cost.

People need to be educated with correct knowledge about arsenic. Arsenic-related health problems are not infectious and they are manageable with a change of water-consumption pattern and adequate intake of food nutrients. Absorption of arsenic through the skin is minimal and thus hand-washing, bathing, laundry, etc., with water containing arsenic do not pose significant risk. However, arsenic-containing water is readily absorbed into the human body by the gastrointestinal tracts.

People at risk of arsenic exposure should be warned about the health hazards associated with arsenic. Tube-well water can be replaced by surface water. However, people should be educated to boil surface water before drinking to avoid the affliction of infectious diseases. Utilization of domestic arsenic removal devices are encouraged to obtain clean water. Adding alum or ferrous salts to arsenic-contaminated water to convert arsenic into insoluble substances is one of the methods. Rainwater harvesting can be helpful at a low cost during the monsoon season. Handy and low-cost technology to detect arsenic components in water can be applied to identify safe water sources.

The successful eradication of blackfoot disease in Taiwan set an exemplar in the public health approach for the prevention of arsenic-related health hazards. Because of the link between the etiology of these endemic diseases with the artesian well water, the Provincial Government of Taiwan began to implement tap water supply systems to replace the artesian well water in the endemic areas. Programs that moved villagers to other residential areas had even been carried out in some seriously affected villages. Since the 1970s, the incidence rates of blackfoot disease decreased dramatically after the implementation of these public health measures. The eradication of blackfoot disease by changing the water supply system also demonstrated an excellent example that many environmental diseases can be successfully eradicated by removal of their vectors, even when the real etiology remains controversial.

SEE ALSO THE FOLLOWING CHAPTERS

Chapter 5 (Uptake of Elements from a Biological Point of View) · Chapter 8 (Biological Responses of Elements) · Chapter 11 (Arsenic in Groundwater and the Environment) · Chapter 21 (Environmental Epidemiology) · Chapter 23 (Environmental Pathology) · Chapter 24 (Toxicology) · Chapter 25 (Speciation of Trace Elements)

FURTHER READING

Barrett, J. C. (1993). Mechanisms of Multistep Carcinogenesis and Carcinogen Risk Assessment, *Environ. Health Perspect.*, 100, 9–120.

Brown, D. (2001). Theoretical Estimates of Blood Mercury Levels From Thiomersal Injections Using a One-Compartment Biokinetic Model, www.iom.edu/iom/iomhome.nsf/WFiles/Brown/%24file.

CDC (1988). Unpublished data, Third National Health and Nutrition Examination Survey, 1988–1994.

CDC (1993). Fluoridation Census 1992, U. S. Department of Health and Human Services, Public Health Service, CDC,

National Center for Prevention Services, Division of Oral Health, Atlanta, GA.

Chan, P. C., and Huff, J. (1997). Arsenic Carcinogenesis in Animals and in Humans: Mechanistic, Experimental and Epidemiological Evidence, *J. Environ. Sci. Health Environ Carcinog. Ecotoxicol. Rev.*, C15(2), 83–122.

Chen, C. J., Hsu, L. I., Tseng, C. H., Hsueh, Y. M, and Chiou, H. Y. (1999). Emerging Epidemics of Arseniasis in Asia. In *Arsenic Exposure and Health Effects* (W. R. Chappell, C. O. Abernathy, and R. L. Calderon, Eds.), Elsevier Science Ltd, Oxford, UK, pp. 113–121.

Coogan, T. P., Bare, R. M., and Waalkes, M. P. (1992). Cadmium-Induced DNA Strand Damage in Cultured Liver Cells: Reduction in Cadmium Genotoxicity Following Zinc Pretreatment, *Toxicol. Appl. Pharmacol.*, 113(2), 227–233.

Crump, K. S. (1984). A New Method for Establishing Allowable Daily Intakes. Fundamental & Applied Toxicology 4, 860–866.

Dean, H. T. (1936). Chronic Endemic Dental Fluorosis (Mottled Enamel), *JAMA*, 107, 1269–1272.

Derryberry, O. M., Bartholomew, M. D., and Fleming, R. B. L. (1963). Fluoride Exposure and Worker Health—The Health Status of Workers in a Fertilizer Manufacturing Plant in Relation to Fluoride Exposure, *Arch. Environ. Health*, 6, 503–514.

Doll, R., and Peto, R. (1981). The Causes of Cancer: Quantitative Estimates of Avoidable Risks of Cancer in the United States Today, *J. Natl. Cancer Inst.*, 66(6), 1192–1308.

Fowles, J. R., Dodge, D., and Alexeeff, G. V. (1999). The Use of Benchmark Dose Methodology with Acute Inhalation Lethality Data, *Reg. Toxicol. Pharmacol.*, 29, 262–278.

Gochfeld, M. (1997). Factors Influencing Susceptibility to Metals, *Environ. Health Perspect.*, 105(Suppl. 4), 817–822.

Gochfeld, M., and Syers, K. (2001). Scientific Committee on Problems of the Environment, http://www.icsu-scope.org/index.html.

Hartung, R. (1987). Dose-response Relationships. In *Toxic Substances and Human Risk*, Tardiff, R. G. and Rodricks, J. U. (eds) New York, Plenum Press, pp. 29–46.

Haseman, J. K., Hailey, J. R., and Morris, R. W. (1998). Spontaneous Neoplasm Incidences in Fischer 344 Rats and B6C3F1 Mice in Two-Year Carcinogenicity Studies: A National Toxicology Program Update, *Toxicol. Pathol.*, 26(3), 428–441.

Hattis, D. (1996). Variability in Susceptibility–How Big, How Often, for What Responses to What Agents?, *Environ. Toxicol. Pharmacol.*, 2, 135–145.

Hill, A. B. (1965). The Environment and Disease: Association or Causation?, *Proc. Roy. Soc. Med.*, 58, 295–300.

Hoffman, G. (1994). Genetic Toxicology. In *Cassarett and Doull's Toxicology: The Basic Science of Poisons, 5th edition* (C. Klaassen, Ed.), McGraw-Hill, New York.

Hopenhayn-Rich, C., Smith, A. H., and Goeden, H. M. (1993). Human Studies Do Not Support the Methylation Threshold Hypothesis for the Toxicity of Inorganic Arsenic, *Environ. Res.*, 60(2), 161–177.

Hsueh, Y. M., Chiou, H. Y., Huang, Y. L., Wu, W. L., Huang, C. C., Yang, M. H., Lue, L. C., Chen, G. S., and Chen, C. J. (1997). Serum Beta-Carotene Level, Arsenic Methylation Capability, and Incidence of Skin Cancer, *Cancer Epidemiol. Biomarkers Prev.*, 6(8), 589–596.

Hsueh, Y. M., Wu, W. L., Huang, Y. L., Chiou, H. Y., Tseng, C. H., and Chen, C. J. (1998). Low Serum Carotene Level and Increased Risk of Ischemic Heart Disease Related to Long-Term Arsenic Exposure, *Atherosclerosis*, 141, 249–257.

International Agency for Research on Cancer (IARC). (1999). Sacchurin and its salts. IARC Monograph 73. WHO, Geneva.

Le, C., Lu, X., Ma, M., Cullen, W., and Aposhian, V. (2000). Speciation of Key Arsenic Metabolic Intermediates in Human Urine, *Anal. Chem.*, 72, 5172–5177.

Mengheri, E., Murgia, C., Vignolini, F., Nobili, F., and Gaetani, S. (1993). Metallothionein Gene is Expressed in Developing Rat Intestine and is Induced by Zinc but not by Corticosteroids, *J Nutr.*, 123(5), 817–822.

Moolgavkar, S. H., and Luebeck, E. G. (1995). Incorporating Cell Proliferation Kinetics into Models for Cancer Risk Assessment, *Toxicology*, 102(1–2), 141–147.

Mutti, A., Mazzucchi, A., Rusticelli, P., Frigeri, G., Arfini, G., and Franchini, I. (1984). Exposure-effect and Exposure-Response Relationships Between Occupational Exposure to Styrene and Neuropsychological Functions. American Journal of Industrial Medicine 5, 275–286.

National Center for Health Statistics (1974). Decayed, Missing, and Filled Teeth Among Youth 12–17 Years, U. S. Department of Health, Education, and Welfare, Public Health Service, Health Resources Administration, Vital and Health Statistics, Vol. 11, no. 144, DHEW publication no. (HRA) 75–1626, Rockville, MD.

National Center for Health Statistics (1981). Decayed, Missing, and Filled Teeth Among Persons 1–74 Years, U. S. Department of Health and Human Services, Public Health Service, Office of Health Research, Statistics, and Technology, Vital and Health Statistics, Vol. 11, no. 223, DHHS publication no. (PHS)81–1673, Hyattsville, MD.

National Institute of Dental Research (1989). Oral Health of United States Children: The National Survey of Dental Caries in U. S. School Children, 1986–1987, U. S. Department of Health and Human Services, Public Health Service, National Institutes of Health, NIH publication no. 89–2247, Bethesda, MD.

National Research Council (NRC). (1993). Guidelines for Developing Community Emergency Exposure Levels for Hazardous Substances. Natl. Academy Press. Washington, D.C.

Office of Environmental Health Hazard Assessment (OEHHA) (2000). Air toxics Hot Spots Program Risk Assessment Guidelines; Part III: Non-cancer Chronic Reference Exposure Levels. Feb. 2000. California Environmental Protection Agency (www.OEHHA.ca.gov).

Pizarro, F., Olivares, M., Araya, M., Gidi, V., and Uauy, R. (2001). Gastrointestinal Effects Associated with Soluble and Insoluble Copper in Drinking Water, *Environ. Health Perspect.*, 109(9), 949–952.

Rabovsky, J., Fowles, J., Hill, M., and Lewis, D. (2001). A Health Risk Benchmark for the Neurological Effects Of Styrene: Comparison with the NOAEL/LOAEL Approach, *Risk Anal.*, 21(1), 117–126.

Rudel, R., Slayton, T. M., and Beck, B. D. (1996). Implications of Arsenic Genotoxicity for Dose Response of Carcinogenic Effects, *Reg. Toxicol. Pharmacol.*, 23(2), 87–105.

Schwarz, M. A., Lazo, J. S., Yalowich, J. C., Allen, W. P., Whitmore, M., Bergonia, H. A., Tzeng, E., Billier, T. R., Robbins, P. D., and Lancaster, J.R. Jr. (1995). Metallothiouein Protects Against the Cytotoxic and DNA-damaging Effects of Nitric Oxide, *Proceedings of the National Academy of Sciences* 92(10), 4452–4456.

Tang, E. M. (2000). Epigenetic Effects on Susceptibility to Heavy Metal and PAH Induced DNA Damage, Crisp Data Base National Institutes of Health, Bethesda, MD.

Tseng, C. H. (1999). Chronic Arsenic Intoxication in Asia: Current Perspectives, *J. Intern. Med. Taiwan*, 10, 224–229.

Tseng, C. H. (2002). An Overview on Peripheral Vascular Disease in Blackfoot Disease-Hyperendemic Villages in Taiwan, *Angiology*, 53(5), 529–537.

Tseng, C. H., Chen, C. J., Lin, B. J., and Tai, T. Y. (1994). Abnormal Response of Ankle Pressure After Exercise in Seemingly Normal Subjects Living in Blackfoot Disease-Hyperendemic Villages in Taiwan, *Vasc. Surg.*, 28, 607–617.

Tseng, C. H., Chong, C. K., Chen, C. J., and Tai, T. Y. (1996). Dose-Response Relationship Between Peripheral Vascular Disease and Ingested Inorganic Arsenic Among Residents in Blackfoot Disease Endemic Villages in Taiwan, *Atherosclerosis*, 120, 125–133.

Tseng, C. H., Tai, T. Y., Chong, C. K., Tseng, C. P., Lai, M. S., Lin, B. J., Chiou, H. Y., Hsueh, Y. M., Hsu, K. H., and Chen, C. J., (2000). Long-Term Arsenic Exposure and Incidence of Non-Insulin-Dependent Diabetes Mellitus: A Cohort Study in Arseniasis-Hyperendemic Villages in Taiwan, *Environ. Health Perspect.*, 108, 847–851.

Tseng, C. H., Tai, T. Y., Lin, B. J., and Chen, C. J. (1995). Abnormal Peripheral Microcirculation in Seemingly Normal Subjects Living in Blackfoot Disease-Hyperendemic Villages in Taiwan, *Int. J. Microcirc.*, 15, 21–27.

Tseng, C. H., Tseng, C. P., Chiou, H. Y., Hsueh, Y. M., Chong, C. K., and Chen, C. J. (2002). Epidemiologic Evidence of Diabetogenic Effect of Arsenic, *Toxicol. Lett.*, 133, 69–76.

United States Environmental Protection Agency (U. S. EPA) (2000). Exposure and Human Health Reassessment of 2,3,7,8-tetrachlorodibenzo-p-dioxin (TCDD) and Related Compounds, EPA/600/P-00/001Ag.

United States Environmental Protection Agency (U.S. EPA) (2002). A review of the reference dose and reference concentration processes. EPA/630/P-02/002F. www.epa.gov/iris/Rfd.final.pdf

WHO (1994). Fluorides and Oral Health, WHO Technical Report Series 846, WHO, Geneva.

Yamamoto, E., Shimada, A., Morita, T., Yasutake, A., Yoshide, M., Nishimura, N., Suzuki, J. S., Satok, M., and Tokyama, C. (2001). Localization and Role of Placental Metallothioueins as a Barrier Against Maternal to Fetal Transfer of Mercury. Toxicologist 60(1), 357.

CHAPTER 23

ENVIRONMENTAL PATHOLOGY

José A. Centeno, Florabel G. Mullick,
Kamal G. Ishak, Teri J. Franks,
and Allen P. Burke
The Armed Forces Institute of Pathology

Michael N. Koss
University of Southern California Medical Center

Daniel P. Perl
Mount Sinai School of Medicine

Paul B. Tchounwou
Jackson State University

Joseph P. Pestaner
East Carolina University

CONTENTS

This chapter is dedictated to the memory of Kamal G. Ishak, M.D. Ph.D. (1928–2004).

I. INTRODUCTION

Humans are constantly exposed to hazardous pollutants in the environment—for example, in the air, water, soil, rocks, diet, or workplace. Trace metals are important in environmental pathology because of the wide range of toxic reactions and their potential adverse effects on the physiological function of organ systems. Exposures to toxic trace metals have been the subject of numerous environmental and geochemical investigations, and many studies have been published on the acute and/or chronic effects of high-level exposures to these types of agents; however, much fewer data are available concerning the health effects of low-dose chronic exposure to many trace metals. Chronic low-dose exposures to toxic elements such as cadmium and arsenic have been shown to cause these metals to accumulate in tissues over time, leading to multiple adverse effects in exposed individuals.

Exposure to toxic trace metals occurs via three principal routes: percutaneous absorption, ingestion, or inhalation. The toxic effects may affect specific target organ components, resulting in immunological-induced injury or specific functional changes. The diseases caused by metals can be genetic or acquired, and the effects can be acute or chronic. This chapter provides a review of some of these pathologies and discusses the critical organ systems that are affected. Examining such toxicities is a medical challenge in that a number of metallic elements, such as iron, copper, and manganese, are essential to life. Distinguishing normal and pathologic states is critical to our understanding of the pathogenesis of metal-induced diseases. The toxic properties of certain metals, such as lead and mercury, have been acknowledged since ancient times, but enhanced pathologic analyses have allowed us to learn much about how metals can affect specific organ systems. Also reviewed in this chapter are the pathologic states caused by metals in the skin, brain, lung, heart, and liver.

II. THE SKIN

The list of metals exhibiting dermal toxicity has been well catalogued. Such metals include compounds used in medicinal products, industrial processes, pesticides, cosmetics, dyes, and jewelry (Lansdown, 1995). Of major concern is exposure to metals and metalloids through contaminated water and other environmental and geological media. Dermal toxicity is a result of local tissue responses to direct contact of a metal with skin or, alternatively, it may represent a manifestation of systemic toxicity following ingestion or inhalation. Allergic contact dermatitis induced by nickel is one such example of a local tissue response. The adverse cutaneous reactions resulting from chronic ingestion or inhalation of arsenical compounds exemplify systemic toxicity.

A variety of pathologic responses in the skin are associated with both acute and chronic exposures to metals. Categorization of these responses presents a challenge to the environmental pathologist, as the histologic features associated with metal-induced skin lesions may mimic virtually any known morphologic skin disease. The more frequently encountered morphologic changes include spongiotic dermatitis (allergic contact dermatitis and primary irritant dermatitis), granulomatous inflammation, pigmentation disorders, and cancer. A pertinent exposure history correlated with pathologic findings should be done to establish a precise diagnosis.

The occurrence of hyper- and hypopigmentation of the skin has been reported worldwide in populations chronically exposed to arsenic from contaminated drinking water (see also Chapter 11, this volume). Because of its widespread presence in the environment, arsenic has become one of the most studied elements in environmental toxicology and public health. The ensuing discussion centers on arsenic and arsenic-related skin diseases, which are considered by many to be the prototype of the development of disease following exposure to a metal.

A. Arsenic and Metal-Induced Cancer of the Skin

The most widely recognized toxic element affecting the skin is arsenic. Arsenic is the twentieth most abundant element in the Earth's crust. It is odorless and tasteless and exhibits both acute and chronic health effects in humans. In nature, arsenic can occur as metalloid alloys or in a variety of chemical compounds. In geological media such as rocks, arsenic is commonly found as a sulfide such as orpiment (As_2S_3) or realgar (As_2S_2) in the form of arseno–pyrite or mixed sulfides (AsFeS). Significant amounts of arsenic may also be found bound to gold, silver, copper, lead, zinc, and cobalt ores. Mining of these minerals may result in the mobilization and/or transport of arsenic into drinking water. Arsenic has also been used in a variety of agricultural applications (*e.g.*, pesticides, insecticides), industrial applications (*e.g.*, manufacturing of solid-state detectors), and medical applications (*e.g.*, drugs and medical treatments).

As with all toxic metals, the toxic effects of arsenic are related to the chemical and physical forms in which it appears: metallic, As(0); inorganic, As(III) and As(V); and organic, As(III) and As(V). Although arsenic exhibits both organic and inorganic forms, the inorganic trivalent arsenic compounds are considered to demonstrate the greatest toxicity. The molecular basis by which arsenic compounds may induce their toxicity in humans has been described. Impairment of cellular respiration through inhibition of various mitochondrial enzymes and uncoupling of oxidative phosphorylation is one of the major mechanisms by which arsenic exerts its toxic effects. At the molecular level, the toxicity of arsenic results from its ability to interact with sulfhydryl groups of proteins and enzymes and to substitute phosphorus in a variety of biochemical reactions (Li *et al.*, 1989). *In vitro* experiments have demonstrated that arsenic reacts with protein sulfhydryl groups to inactivate enzymes such as dihydrolipoyl dehydrogenase and thiolase, thereby producing inhibited oxidation of pyruvate and beta-oxidation of fatty acids (Belton *et al.*, 1985).

Humans are exposed to inorganic arsenic mainly through the oral and inhalation routes. Direct dermal exposure also occurs, but to a lesser extent. The oral route includes contaminated drinking water, food, drugs (including Chinese herbal medications), and tobacco. Inhalation occurs primarily in occupational settings; workers may be exposed to arsenic in the air as a by-product of copper and lead smelting, pesticide production, manufacturing of glass, and production of semiconductors (Chan & Huff, 1997). Arsenic tends to concentrate in ectodermal tissues, including the skin, hair, and nails. Biomethylation is considered the major metabolic pathway for inorganic arsenic in humans. Historically, the enzymatic conversion of inorganic arsenic to mono- and dimethylated species has been considered a primary detoxification mechanism of inorganic arsenic; however, compelling experimental evidence obtained from several laboratories suggests that biomethylation, particularly the production of methylated metabolites that contain trivalent arsenic, is a process that can activate arsenic as a toxin and a carcinogen (Styblo *et al.*, 2002; Wei *et al.*, 2002).

Epidemiological studies have confirmed the role of arsenic in the induction of cancers of the skin. Of the metals known to exhibit dermal toxicity, only arsenic has been shown conclusively to be carcinogenic (Chen *et al.*, 1992). Squamous cell carcinomas *in situ* (Bowen's disease) and basal cell carcinomas of the skin have been associated with chronic inorganic arsenic ingestion with a latency period of 2 to 20 years after exposure (Maloney, 1996; Tsai *et al.*, 1999). In addition, epidemiological studies have provided suggestive evidence linking arsenic exposure to various internal cancers, including angiosarcoma of the liver, lung cancer, and bladder cancer. In the majority of these cases, in which the internal cancer is ascribed to arsenic exposure, some dermatologic hallmark of arsenic poisoning (such as hyper- or hypopigmentation) is identified.

Arsenical keratosis is a well-established clinical entity resulting from chronic exposure. The lesions are usually most pronounced on the palms and soles, although they can occur on the trunk and other areas of the extremities (see Figure 1). Arsenical keratoses are characterized by several specific pathologic features, including hyperkeratosis, parakeratosis, and acanthosis (see Figure 2). Nuclear atypia is sometimes present as well. In severe atypia, squamous cells exhibit hyperchromatic nuclei and a disorderly arrangement within the epithelium. Within the spectrum of keratotic lesions, arsenical keratosis may be differentiated from the more commonly diagnosed actinic keratosis by the absence of epidermal atrophy and basophilic degeneration of the upper

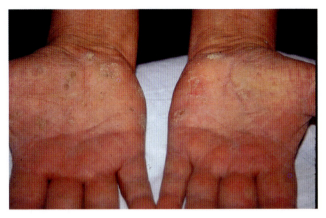

FIGURE 1 Arsenic-induced hyperkeratosis of the hands.

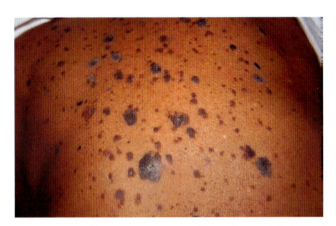

FIGURE 2 Hyperpigmentation and hyperkeratosis lesions of the back induced by chronic exposure to arsenic from consumption of contaminated drinking water.

dermis. All arsenical skin changes, including keratoses, tend to occur in non-exposed sites with an absence of dermal solar elastosis noted histologically.

As a human carcinogen, inorganic arsenic remains an enigma because arsenic-related tumors occur in humans but not in laboratory animals. The biological mechanisms by which arsenic induces chronic effects including cancer are not well understood and are the subject of considerable research efforts. Arsenic does not directly damage DNA but rather causes chromosome aberrations, aneuploidy, cell transformation, and gene amplification in many cell types. Recently, arsenic-induced carcinogenesis has been studied using molecular biological methods. Among the oncogenes evaluated, the tumor-suppressor gene *p53* appears to play a role in arsenic-induced carcinogenesis. In one study from an endemic area, 48 cases representing a variety of arsenic-induced skin cancers (including Bowen's disease, squamous cell carcinoma, and basal cell carcinoma) were examined (Chung *et al.*, 1998). All of the specimens demonstrated positive *p53* immunostaining. Positive *p53* staining was identified in all perilesional normal skin as well, which suggested that the *p53* mutation may be an early event in arsenic-related carcinogenesis. Another study demonstrated that *p53* mutation rates, sites, and types in arsenic-related skin cancer are significantly different from those in ultraviolet-induced cancer, which implicates arsenic as the etiologic agent and suggests a possible mechanism of action (Hsu *et al.*, 1999).

In the absence of animal models, *in vitro* studies become particularly important in providing information on the carcinogenic mechanisms of arsenic toxicity. Arsenic and arsenical compounds have been reported to induce morphological changes in cultured cells. Experimental studies have indicated that arsenic is cytotoxic and able to transcriptionally activate a significant number of stress genes in transformed human liver cells (Tchounwou *et al.*, 2000, 2001). Arsenic and arsenic-containing compounds have also been shown to be potent clastogens both *in vivo* and *in vitro*. Arsenical compounds have been reported to induce sister chromatid exchanges and chromosome aberrations in both human and rodent cells in culture (Nakamuro & Sayato, 1981; Barrett *et al.*, 1989). Arsenical compounds have also been shown to induce gene amplification, arrest cells in mitosis, inhibit DNA repair, and induce expression of the *c-fos* gene and the oxidative stress protein heme oxygenase in mammalian cells (Ramirez *et al.*, 1997; Jingbo *et al.*, 2002), and they have been implicated as promoters and comutagens for a variety of agents (Cavigelli *et al.*, 1996).

Research has also indicated that inorganic arsenic does not act through classic genotoxic and mutagenic mechanisms but may be a tumor promoter that modifies signal transduction pathways involved in cell growth and proliferation (Simeonova & Luster, 2000; Kitchin, 2001). Modulation of gene expression and/or DNA-binding activities of several key transcription factors, including nuclear factor kappa B (NFκB), tumor-suppressor protein (p53), and activating protein-1 (AP-1) has been associated with arsenic exposure (Barchowsky *et al.*, 1996). Mechanisms of AP-1 activation by trivalent arsenic include stimulation of the mitogen-activated protein kinase (MAPK) cascade with a consequent increase in the expression and/or phosphorylation of the two major AP-1 constituents, c-jun and c-fos (Simeonova & Luster, 2000).

Much remains to be learned about arsenic and its toxicity to skin, and similar challenges remain when examining arsenic and its toxicity to other important organ systems in the body. As is discussed in the next section, arsenic damage to the nervous system is relatively acute, and the extent of damage is proportional to the exposure dose (see also Chapters 11 and 21, this volume).

III. THE BRAIN

A. Introduction

It should be recognized that a number of metallic elements, such as iron, copper, and manganese, are essential to life and play an important role in the functioning of the central nervous system (CNS); nevertheless, that certain metals, such as lead and mercury, have neurotoxic properties has been acknowledged since ancient times. The brain must be viewed in a somewhat different light than other organ systems because of its position behind the protective shield of the blood–brain barrier. Because of this critical protective barrier, one must recognize that, for a metal to induce neurologic damage, it must be able to cross the blood–brain barrier, enter the CNS, and gain access to target cells in sufficient quantity to produce pathologic damage. Compounds in which the metal is linked to a lipophilic organic compound tend to be particularly neurotoxic as they can readily cross the lipid membranes that comprise the blood–brain barrier. One example is mercury, which, in its inorganic form, is relatively nontoxic to the CNS; however, when mercury is methylated to form methylmercury, the compound rapidly crosses the blood–brain barrier, is readily taken up by neurons, and produces massive cellular destruction. This results in the severe parenchymal damage that occurred in the outbreak of severe CNS damage in Minamata, Japan (see discussion below). Finally, it should be recognized that the effects of toxins that damage neurons are particularly serious because of the inability of these cells to regenerate.

Certain metals function as classic toxins to the nervous system, in which damage is relatively acute following exposure, and the extent of damage is proportional to the exposure dose. One such example is arsenic. As noted earlier, arsenic is generally found as an impurity of ores containing copper, lead, gold, and zinc. Exposure to arsenic is a relatively rare event but does occur due to accidental ingestion, suicide, or murder and most commonly is related to occupational activities and consumption of contaminated drinking water (see Chapter 11, this volume). Arsenic is an active ingredient of herbicides, insecticides, and other pesticides. In cases of acute arsenic poisoning, sudden fatal circulatory collapse may follow ingestion of a single large dose. When smaller doses are involved, gastrointestinal symptoms will predominate initially to be followed after a period of 2 to 3 weeks by the development of a rapidly progressive peripheral neuropathy. The extent and severity of the neurologic symptomatology will depend on the arsenic dosage. In cases of chronic exposure to lower doses of arsenic, a sensory neuropathy is generally observed. Relatively little is known of the mechanism by which arsenic produces its toxic damage, but recent studies suggest that the process involves primary axonal damage.

B. Lead Poisoning: Acute and Chronic Effects

Lead is perhaps the most important metallic neurotoxin. Adults are relatively resistant to its effects, and only in high doses is a peripheral neuropathy encountered. In children, however, the effects of relatively low doses can be much more devastating. High doses in children can cause acute lead encephalopathy, a life-threatening condition characterized by generalized cerebral edema with increased intracranial pressure, which leads to transtentorial and cerebellar tonsillar herniation. Clinical features of acute lead encephalopathy may include ataxia, seizures, stupor, coma, and often death. Almost always other associated systemic signs of lead exposure, such as anemia and the presence of lead lines on x-rays of the long bones, are noted in affected children. At autopsy, the brain is markedly swollen with compressed gyri, obliterated sulci, and collapsed lateral ventricles. Uncal and cerebellar tonsillar herniation are commonly encountered. Microscopically, there is a breakdown of the blood–brain barrier with transudation of fluid into the pericapillary space and ischemic damage to cerebral cortical neurons. Once again, the adult is apparently resistant to such toxic changes, and it is extremely rare to see acute encephalopathy in an adult even following extremely high-exposure doses. More typically, adults exposed to lead develop a peripheral neuropathy with wrist drop and/or ankle drop. Exposure in children leading to encephalopathy is primarily through the eating of peeling paint chips in houses dating from an era when lead-based paints were employed. In general, exposure to lead can come from

the air, in the form of lead fumes (*e.g.*, firing range operators, stained glass workers, solderers); from water contaminated by lead plumbing components (particularly in acid conditions); the use of lead-containing vessels (lead crystal, lead glazes of pottery in which acid liquids have been stored); and airborne particulates, such as is found in restorers of old homes and bridge workers. Other workers who are particularly prone to high lead exposures are battery production workers and bronze workers.

Largely through the work of Needleman and colleagues (1979, 1988, 1990), the long-term effects of lower doses of lead exposure on children have been increasingly recognized. This relates to the adverse effects of lead on intellectual functioning as well as its association with behavioral problems in children exposed to what had been previously considered to be relatively low lead burdens. Documenting low-level, chronic lead exposure has not been easy to accomplish. It is clear that determining a single blood lead level in a child can be completely misleading. In the initial studies of Needleman *et al.*, lead exposure was estimated through the calculation of lead levels in the dentine of deciduous teeth. These studies suggested that exposure to relatively low levels of lead in childhood is associated with a slight, but significant, drop in IQ. Furthermore, such low-level exposures have been associated with a tendency toward disturbed classroom behavior. Follow-up studies on children with elevated lead levels showed a markedly high dropout level from high school, increased absenteeism, and impaired eye–hand coordination. The neuropathologic substrate of all such effects of low-level, chronic lead exposure remains unclear, but these data have resulted in a progressive lowering of what is considered an "acceptable" blood lead level.

C. Mercury Poisoning: Inorganic Versus Organic Forms

The hazards of inorganic mercury poisoning have been known since Roman days, and the dangers of working in the mercury mines of Almedon, Spain, have been widely acknowledged for centuries. Central nervous system symptoms related to such exposure are generally related to the presence of high concentrations of inorganic mercury fumes. These produce tremor and irritability which typically are reversible upon removal from the exposure source. The behavioral disturbances that result gave rise to the notion of being "mad as a hatter," a saying that refers to the practice of employing mercury rather indiscriminately in the felt tanning industry of Victorian England and the subsequent frequency of cases of inorganic mercury poisoning among hatters. Nevertheless, oral mercury-containing calomel medicinals have been used relatively safely over the centuries, and inorganic mercury has a relatively low level of toxicity. On the other hand, exposure to organic mercury, such as methylmercury, causes dramatic nervous system destruction (Hunter & Russell, 1954).

The importance of this distinction was dramatically demonstrated in the tragic outbreak of methylmercury poisoning among Japanese villagers living along Minamata Bay (Marsh, 1979). In the small town of Minamata, Japan, on the southern portion of the island of Kyushu, a large chemical factory employed a significant amount of inorganic mercury as a catalyst for the production of the raw ingredients of plastic. Some of this mercury was discharged into Minamata Bay. The mercury deposited in the sediment of the bay was subsequently taken up by certain bacteria, which are capable of methylating it to form methylmercury (Jensen & Jernelov, 1969). These bacteria entered the food chain to eventually deposit methylmercury in the fish, the major protein source of the local villagers.

The methylated form of mercury readily crosses the blood–brain barrier, inducing severe neuronal destruction in extremely small doses. Indeed, the solubility and neurotoxicity of methylmercury are so great that fatal poisoning has been reported in a research chemist who was exposed to minute amounts of dimethylmercury that gained access to his skin directly through his latex gloves. Clinically, exposed patients initially show a constricted visual field, paresthesias, gait ataxia, and impairment of vibration sense (stereognosis) and two-point discrimination. Subsequently, more profound visual loss (frequently progressing to total blindness), ataxia, and sensory–motor signs are noted. The neuropathologic lesions of methylmercury poisoning are quite distinct, with acute necrosis of the calcarine and pre-central cortex and the cerebellum (Shiraki, 1979). Nerve cell loss is severe, with a profound glial response. In the cerebellum, the neuronal loss predominates in the internal granular layer, with sparing of the Purkinje cells. Although blindness is a common complication of the disease, the retina appears to be intact, and the loss of vision is thought to be directly related to destruction of the primary visual (calcarine) cortex. In patients with prolonged survival after exposure to methylmercury, dramatic atrophy and glial scarring of the abovementioned regions are observed. Among the most severely affected victims of the Minamata outbreak were newborns who were exposed *in utero*. The methylmercury readily crossed the placenta and produced dramatic widespread nervous system destruction.

D. Tin Poisoning: An Example of Selective Toxicity of Organic Metal Complexes

A phenomenon similar to that of enhanced neurotoxicity of methylmercury also occurs following exposure to tin-containing compounds. Metallic tin and its inorganic salts have been included in medicinal preparations since the 16th century. These were widely administered in relatively high doses without any apparent adverse health effects; however, organic tin compounds, similar to organic mercury compounds, are highly lipid soluble and are also highly toxic to the nervous system (Cavanagh & Nolan, 1994). Unfortunately, these characteristics were learned following a serious outbreak of organic tin poisoning in 1954 in France due to the introduction of the drug Stalinon for the treatment of staphylococcal infections. Each Stalinon capsule contained 15 mg of diiodide–diethyltin. Of those exposed to this drug, approximately half died, one-third recovered, and the remainder suffered from a wide variety of chronic neurologic deficits. Autopsies on the acute fatal cases showed evidence of an unusual form of diffuse edema that was confined to the white matter. Subsequent experimental work has shown that triethyltin produces a characteristic and selective white matter edema, presumably through its toxic effects on the functioning of oligodendroglial membranes. Interestingly, animals exposed to trimethyl tin fail to show evidence of white matter edema but instead develop a selective necrosis of neurons of the hippocampus. The mechanism behind this dramatic difference in the selective toxicity of these two extremely similar compounds remains unclear; nevertheless, the organic moiety on each form allows for its rapid uptake into the CNS, thus providing access to the different cellular targets for its toxicity.

E. Manganese-Induced Parkinsonism

Manganese is a neurotoxic metal that is capable of readily entering the CNS when it reaches the general circulation. Upon reaching the brain, manganese accumulates selectively in the globus pallidus, where it can destroy local nerve cells. Manganese is the twelfth most abundant element in the Earth's crust and the fourth most widely used metal in the world. Eight million tons of manganese metal are extracted annually, of which 94% is employed in the manufacture of steel. Manganese is also used in the manufacture of batteries. Potassium permanganate is a widely employed bactericidal and fungicidal agent in water purification processes. Additionally, methylcyclopentadienyl manganese tricarbonyl (MMT) is an organic manganese compound that has been used as an anti-knock additive to gasoline. Medically significant manganese toxicity is almost exclusively encountered in an industrial setting, either in association with manganese mining or in smelting operations. The classic papers of Mena and coworkers (Mena *et al.*, 1967; Mena, 1979) documented the development of psychiatric and parkinsonian features in a group of manganese miners in Chile, and more recent papers from Taiwan have documented cases seen in association with smelting (Olanow *et al.*, 1994).

The clinical features of manganese neurotoxicity closely resemble Parkinson's disease but have more prominent dystonic features. Initial stages typically include psychiatric disturbances, such as behavioral abnormalities, hallucinations, and, at times, frank psychosis. This syndrome is referred to as "manganese madness" or *locura manganica*. Not all cases pass through the psychiatric phase of the disease and present with the extrapyramidal manifestations. The extrapyramidal features consist of bradykinesia, gait disturbance, postural instability, rigidity, micrographia, masked facies, and speech disturbances. Although tremors may be seen, they usually are not the "pill-rolling" tremors seen at rest that are typical of Parkinson's disease but are described as more of an intension tremor. Dystonic features can be rather prominent with facial grimacing and plantar flexion of the foot, which results in a very characteristic gait disturbance. The gait of manganese poisoning consists of the patient keeping the foot dorsoflexed and walking with elbows flexed and spine erect. This rather unique and characteristic gait is commonly referred to as *coq au pied* or "cock walk," as it resembles the strutting of a rooster.

The manganese that produces this form of neurotoxicity is generally thought to enter the body through the lungs. Experimental studies have shown that airborne manganese is readily taken up into the systemic circulation and is selectively deposited in the globus pallidus of the brain, where the major site of damage occurs (Newland *et al.*, 1989; Olanow *et al.*, 1996). Pathological studies of both experimental models and human cases demonstrate evidence of selective neuronal loss and gliosis in globus pallidus (primarily the medial segment) and striatum. Accompanying the neuronal loss is a dramatic increase in the amount of stainable iron within the damaged regions (Olanow *et al.*, 1996). Whether the increased iron participates in the neuronal damage through oxidative damage or is secondary to the parenchymal destruction remains unclear. The substantia nigra pars compacta is the location of the dopaminergic cells that project to the striatum and are the main targets of damage in Parkinson's disease, but they remain intact in manganese poisoning. Lewy bodies, the characteristic microscopic feature of Parkinson's disease, are not encountered in cases of manganese-induced parkinsonism. Presumably dysfunction of the extrapyramidal system occurs secondary to damage to the globus pallidus and striatum, the major targets of the striatonigral projections. Because the postsynaptic targets of dopaminergic transmission are largely lost, treatment by levodopa is generally considered to be ineffective in manganese-induced parkinsonism. This represents a major defining characteristic that distinguishes manganese-induced parkinsonism from Parkinson's disease.

F. Metals and Age-Related Neurodegenerative Diseases

The accumulation of metals in the brain has also been associated with several of the age-related neurodegenerative disorders. The two metals most often cited are aluminum and iron. Excess amounts of both metals have been identified in microprobe studies within the characteristic neurofibrillary tangles of cases of Alzheimer's disease (Perl & Brody, 1980; Good *et al.*, 1992a). Similarly, increased concentrations of iron and aluminum have been noted in the neurons of the substantia nigra in cases of Parkinson's disease (Good *et al.*, 1992b). It is not considered likely that such elemental accumulations play an etiologic role in these diseases, although they may certainly have the capacity to contribute to their pathogenesis (Perl & Good, 1992). Iron, through its ability to donate electrons in energy transfer reactions, may actively support oxidative damage. Evidence for oxidative damage in the brains of Alzheimer's disease and Parkinson's disease victims is well established (Markesbery, 1997). Aluminum is a highly charged element that binds strongly to proteins and acts as a cross-linking stabilizer. Many of the proteins comprising the intraneuronal inclusions that characterize the neurodegenerative diseases (the neurofibrillary tangle being the most prominent example) are highly cross-linked and thus resistant to degradation. Whether aluminum serves this role in the natural history of the disorder remains unclear, although recent evidence appears to support this concept. These findings have raised concerns regarding the source of such elemental accumulations and whether aspects of environmental and geological exposures might play a role in the further onset of these age-related disorders. These questions remain largely unresolved.

One question that is resolved is the critical role oxygen plays in maintaining the tissues in the brain. Effects of metals on the lungs can secondarily affect respiratory effectiveness, which will have secondary effects on the brain. Toxicity of metals and the lung are reviewed in the next section.

IV. INHALATION INJURY

A. Introduction

Zenker (1867) first proposed the term "pneumonokoniosis" as a name for lung disease resulting from the inhalation of dust. While many forms of pneumoconioses are known today, each with its own etiologic agent, the definition of pneumoconiosis varies. Some authors restrict the term to non-neoplastic reactions of the lung to inhaled minerals or organic dusts, excluding asthma, bronchitis, and emphysema. Others use the term more broadly to define the accumulation of abnormal amounts of dust and the resulting pathologic reactions (Gibbs, 1996). These reactions range from minimal responses to inert dust particles, such as interstitial dust macules, to lethal scarring associated with fibrogenic dusts (see also Chapters 9 and 18, this volume). Damage caused by inhaled particles depends on a variety of factors, including the number, size, and physiochemical properties of the particles; the deposition, clearance, and retention of particles in the respiratory tract; the host's inflammatory response to the inhaled particles; and the duration of exposure and interval since initial exposure, as well as interactions with other inhaled particles, particularly cigarette smoke (Roggli & Shelburne, 1994).

B. Deposition, Clearance, and Retention

The respiratory system acts like a filter by removing particles and bacteria from inhaled air and leaving exhaled air essentially free of contaminants (Brain &

Valberg, 1979). With each breath, air is drawn through the nares into the nasopharynx and trachea and on into the conducting system of the lungs. Air then passes by way of the bronchi, terminal bronchioles, respiratory bronchioles, and alveolar ducts to the alveoli. In healthy individuals, the tracheobronchial tree is covered by a thin, watery layer of mucus, which is continuously moved up and out of the lungs by ciliated bronchial epithelium. The velocity of inspired air decreases over the course of the airways. Flow rates decrease slightly in the trachea and are markedly reduced in the third through fifth generations of bronchi. At the level of the terminal bronchiole, flow rates are not more than 2 to 3 cm per second (Morgan & Seaton, 1975).

Particles that are nearly spherical in shape are called *compact particles*, whereas those with a length-to-width ratio of 3 : 1 or greater are called *fibers*. Compact particles with an aerodynamic equivalent diameter of 1 to 5 μm are the most likely to deposit in the lung parenchyma (Roggli & Shelburne, 1994; Gibbs, 1996), whereas particles with extreme shapes, such as fibers and plates, tend to deposit in the airway walls (Gibbs, 1996). Fiber deposition is primarily a function of diameter and less a function of length (Gibbs, 1996). Fibers with diameters below 3 μm, even when several hundred micrometers in length, can align axially with the airstream (Timbrell et al., 1970) and can deposit in the alveoli; however, the longer the fiber, the less its statistical chance of reaching the alveolus (Churg, 1998a).

Deposition refers to the fraction of particles in inspired air that are trapped in the lung and fail to exit with expired air. Deposition of particles occurs in the respiratory tract as a consequence of the physical processes of inertial impaction, sedimentation, and diffusion. Inertial impaction occurs when the air current carrying a particle changes direction and the momentum of the particle carries it along its original path. This frequently results in particle deposition at bifurcation points in the respiratory tract (Brain & Valberg, 1979). Sedimentation of particles occurs secondary to the force of gravity, whereas very small particles are deposited by diffusion and Brownian motion. In general, a high percentage of particles greater than 1 μm in diameter is filtered in the nasopharynx by inertial impaction and sedimentation. A smaller percentage of particles is deposited by the same mechanisms in the trachea and bronchi. When particles are under 0.5 μm in diameter, they deposit in the alveoli by sedimentation and diffusion (Morgan & Seaton, 1975). Breathing rate, tidal volume, and nose versus mouth breathing influence the pattern of deposition (Roggli & Shelburne, 1994; Gibbs, 1996).

Clearance is the output of particles previously deposited in the respiratory tract. Mechanisms such as dissolution and absorption, coughing, sneezing, and tracheobronchial and alveolar transport systems keep the lungs relatively free of foreign material (Brain & Valberg, 1979). The tracheobronchial system, also known as the mucociliary escalator, extends from the terminal bronchiole to the glottis. Particles deposited in the conducting system adhere to a thin layer of mucus secreted by the mucous glands and goblet cells of the bronchial tree; subsequently, particles are swept toward the glottis by ciliary action and are swallowed or expectorated. The rate of transport is approximately 3 mm per minute, with 80 to 90% of particles removed within 2 hours. The bronchial mucus is arranged in two layers: an outer viscous gel, which does not absorb water, and an inner liquid solid phase in which the cilia beat. The function and integrity of the inner layer are affected by dehydration. Ciliary function is influenced by ionic charges, the presence of oxidants, high concentrations of oxygen, low concentrations of adenosine triphosphate, and the presence of oxidants and cigarette smoke. In the alveolar transport system, particles are dissolved, removed to lymphatics, engulfed by macrophages, or transported by surface fluid movement to the mucociliary escalator (Morgan & Seaton, 1975). Retention is defined as the amount of particulate matter found in the lungs at any time and is dependent on the rates of deposition and clearance.

The diseased and blackened lungs of miners who die of coal worker's pneumoconiosis contain less than 2% of the dust originally deposited in the lungs (Brain & Valberg, 1979). This illustrates the efficiency of the respiratory tract as a filter; however, the normal defense mechanisms of the lung can be overwhelmed by high exposure levels to inhaled foreign material, which increases the probability of a pathologic response.

C. Pathologic Responses to Inhaled Particles

Inhaled particles can produce a variety of injury patterns in the lung, which include diffuse alveolar damage (fume exposure), dust macules with or without small airway fibrosis, diffuse interstitial fibrosis, alveolar proteinosis, granulomatous interstitial pneumonitis, giant-cell and desquamative interstitial pneumonitis, and lung cancer. Generally, these injury patterns lack specific features that implicate a causative agent.

Diffuse alveolar damage (DAD), the histologic correlate to adult respiratory distress syndrome (ARDS), is the most common serious reaction to inhaled gases and fumes (Wright & Churg, 1998). Microscopically, DAD is characterized by hyaline membranes lining edematous alveolar septa. Virtually any noxious gas or fume (e.g., beryllium and cobalt metal) inhaled in sufficient concentration can potentially cause DAD (Wright & Churg, 1998); however, numerous other agents, including infectious organisms, drugs, ingestants, shock, and sepsis, cause DAD that is microscopically indistinguishable from that produced by gases and fumes.

Dust macules are nonpalpable, peribronchiolar interstitial aggregates of pigmented dust and dust-laden macrophages. Initially, macules may have little associated fibrosis; however, with sufficient exposure, peribronchiolar fibrosis may occur. This pattern of injury generally has little functional deficit, but due to the radiodensity of the dust it is usually associated with an abnormal chest radiograph (Churg & Colby, 1998). Agents that produce dust macules include antimony, barium, chromium ore, iron, rare earths, tin, titanium, and tungsten.

A diffuse interstitial fibrosis pattern of injury, an uncommon complication of metal exposure, can be seen with iron (mild fibrosis), aluminum, hard metal (containing cobalt), copper, rare earths, and silicon carbide exposure. Asbestos and mixed dusts containing silicates can also produce interstitial fibrosis.

Alveolar proteinosis pattern is characterized by a relatively uniform filling of alveoli with granular, eosinophilic exudate containing dense bodies and acicular clefts accompanied by a variable amount of chronic interstitial inflammation and fibrosis. This pattern of injury is usually associated with acute exposure to high levels of silica dust and rarely to aluminum dusts.

Giant-cell (GIP) and desquamative interstitial pneumonitis (DIP) patterns of injury can result from exposure to "hard metals." Patchy interstitial fibrosis with mild chronic inflammatory cell infiltrates, accompanied by striking intraalveolar accumulations of macrophages, characterize these two injury patterns. The presence of enlarged, multinucleated alveolar macrophages, which may contain engulfed inflammatory cells, distinguishes GIP from DIP. GIP is characterized microscopically by interstitial fibrosis accompanied by noncaseating granulomas. This pattern can be seen in chronic beryllium disease.

Asbestos (particularly when associated with asbestosis), arsenic, beryllium, cadmium, chloromethyl ether, hexavalent chromium compounds, nickel, and radon have been linked to lung cancer. Lung cancer occurring in occupationally exposed individuals is histologically indistinguishable from cancer in unexposed individuals (see also Chapter 18, this volume).

Although many forms of pneumoconioses are known, only a handful are seen by the surgical pathologist with any regularity (Table I). The ensuing discussion centers on asbestos and asbestos-related disease, which is considered by many to be the prototype inhalation injury.

D. Asbestos

Asbestos is a group of naturally occurring mineral fibers with the economically useful properties of flexibility, high tensile strength, acoustic insulation, and corrosion, thermal, and electrical resistance. These properties led to the widespread incorporation of asbestos into many products including fireproofing and insulating materials, cloth, cement, plastics, floor tiles, paper, paints, and brake, clutch, furnace, and kiln linings. Virtually everyone in the general population is exposed to a low level of asbestos fibers, and normal lungs can contain small numbers of them (Gibbs, 1996).

Asbestos fibers occur as hydrated fibrous silicates that are mined directly from the Earth. The world's principal, commercially exploited mines are found in Canada, South Africa, Western Australia, and Russia (Roggli, 1994). Based on physical and chemical features, asbestos is classified into two major mineralogic groups: serpentine fibers (containing only chrysotile) and amphiboles, comprised of crocidolite, amosite, tremolite, anthophyllite, and actinolite. Serpentine fibers are curly and pliable, whereas the amphiboles are straight, rigid, and brittle (see also Chapter 18, this volume).

Until the late 1970s, chrysotile (white asbestos) accounted for 95% of the asbestos used commercially (Becklake, 1983), while crocidolite (blue asbestos) and amosite (brown asbestos) made up the remaining 5%. Current usage is almost 100% chrysotile asbestos.

Tremolite, anthophyllite, and actinolite are encountered most commonly as contaminants of other minerals, including chrysotile ore and probably most forms of processed chrysotile products (Dupres *et al.*, 1984). These forms of amphibole asbestos have seen little commercial use, in part because of their physical and chemical properties but primarily due to the lack of commercially useful deposits (Churg, 1998a).

It is generally accepted that exposure to the amphiboles is far more pathogenic than exposure to chrysotile, due to the differential rate of fiber clearance between these groups (Gibbs, 1996). Evidence indicates that amphibole fibers longer than 20 μm cannot be cleared from the peripheral lung (Morgan, 1980). The straight, broad amphibole forms do not readily fragment, whereas the long fibers of chrysotile fragment into short, straight fibers that are cleared (Churg, 1998a). Moreover, chrysotile is thought to be chemically unstable and likely to dissolve, whereas amphibole fibers are stable in the environment of the lung (Hume & Roe, 1992). Zielhuis (1977) broadly classified exposure to asbestos into *direct*, *indirect* (bystander), *paraoccupational* (women washing contaminated work clothes), *neighborhood* (living in the vicinity of asbestos mines or factories), and *ambient* (atmospheric contamination).

Asbestos bodies are the most characteristic feature of asbestos exposure (see Figure 3). These structures are composed of a clear asbestos core surrounded by a golden yellow coating of iron and mucopolysaccharides. The clear core distinguishes asbestos bodies from other ferruginous bodies. The coat may be continuous or bead-like with terminal bulbs. Fiber dimensions are

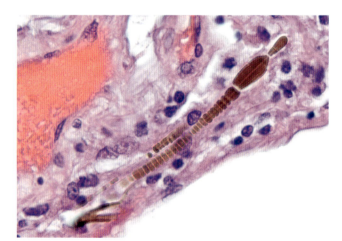

FIGURE 3 Asbestos body (oil-immersion photomicrograph, ×400) with the characteristic clear core that is surrounded by an iron and mucopolysaccharide coating with terminal bulbs.

important factors in determining whether a fiber becomes coated (Roggli, 1992). Longer thicker fibers are more likely to become coated than shorter thinner ones (Morgan & Holmes, 1985), and fibers less than 20 μm in length rarely become coated (Morgan & Holmes, 1980). Because alveolar macrophages are unable to phagocytize long fibers completely within their cell cytoplasm, they coat the asbestos fibers with an iron–protein matrix (Roggli, 1994). For many years, asbestos bodies were thought to be markers of asbestos exposure only in primary asbestos workers. Subsequent studies indicate that asbestos bodies are present in virtually everyone in the general population. Of the asbestos bodies identified in these studies, 98% contained amphibole cores, while the cores in the remaining 2% were chrysotile asbestos (Churg & Warnock, 1981); however, asbestos bodies are present in such small numbers from background exposure that an observer should not see an asbestos body in more than 1 in 100 iron-stained, routine-sized paraffin sections (Roggli & Pratt, 1983).

E. Asbestos-Related Diseases

A variety of benign and malignant diseases of the pleura and lung are associated with asbestos exposure. The benign diseases consist of pleural plaques, diffuse pleural fibrosis, pleural effusion, rounded atelectasis, and asbestosis. Carcinoma of the lung in the presence of asbestosis and malignant mesothelioma of the pleura comprise the malignant diseases. Although asbestos fibers are present in virtually everyone in the general population, disease always occurs with fiber burdens greater than those seen in the general population (Wagner et al., 1988). The various asbestos-related diseases occur at different fiber burdens; in general, the fiber burden required to induce disease is greater for chrysotile, with its tremolite contaminant, than for amosite and crocidolite (Churg, 1998b; Churg et al., 1993).

1. Benign Pleuropulmonary Disease

Pleural plaques, the most common form of benign asbestos-related pleuropulmonary disease, are discrete, raised, irregular foci of dense fibrous tissue. Thin plaques are grayish-white and smooth, whereas thicker ones are pearly-white with either a smooth or bosselated surface. They vary in size from a few millimeters to 10 cm across, and they vary in consistency from leathery to heavily calcified and brittle (Roberts, 1971).

Pleural plaques occur most frequently on the posterolateral parietal pleura and on the domes of the hemidiaphragms. Generally, the apices, anterior chest wall, and costophrenic angles are not involved. Pleural plaques are seen predominantly in persons exposed to asbestos, and bilateral plaques are almost pathognomonic of asbestos exposure (Roggli, 1994). There is evidence of a dose–response relationship between the presence of pleural plaques and the number of asbestos bodies in the lung (Roberts, 1971). Microscopically, pleural plaques consist of dense, paucicellular, hyalinized collagen arranged in a basket-weave pattern in which asbestos bodies are not seen.

In contrast to pleural plaques, diffuse pleural fibrosis typically involves the visceral pleura. The lung may be encased in a thick rind of fibrotic pleura that bridges the major fissures and distorts the edges of the lung (Churg, 1998b). In its most severe form, fusion of the visceral and parietal pleura obliterates the pleural cavity, which results in a condition known as fibrothorax. Microscopically, the thickened fibrous pleura consists of dense collagenous tissue admixed with varying numbers of chronic inflammatory cells.

Diagnosis of benign asbestos effusion is based on: (1) a history of exposure to asbestos, (2) confirmation of the effusion by radiographs or thoracentesis, (3) absence of another disease that could produce an effusion, and (4) no malignant tumor developing within three years after the effusion (Epler et al., 1982). Asbestos-induced pleural effusion is characteristically a serous or serosanguineous exudate with increased numbers of eosinophils. Most effusions are small (usually less than 500 mL) and asymptomatic and may persist from 2 weeks to 6 months (Fraser et al., 1999a). In 1982, Epler et al. reported that the prevalence of asbestos-induced pleural effusion was dose related and, with a shorter latency period than other asbestos-related disease, it is the most common disorder in the first 20 years after asbestos exposure.

Rounded atelectasis is most often an incidental radiographic finding and is resected for suspicion of neoplasm (Churg, 1998b). Radiographically, rounded atelectasis is a unilateral, rounded, pleural-based mass in the lower lobe of the lung, with one or more curvilinear densities radiating from the mass toward the hilum of the lung (Comet's tail sign). Grossly, the visceral pleura is irregularly thickened and invaginated into the underlying lung. Pleural fibrosis and folding, accompanied by a variable degree of parenchymal atelectasis and fibrosis, are seen microscopically. The majority of cases have a history of asbestos exposure; however, other causes of chronic pleuritis, such as congestive heart failure, pulmonary infarct, tuberculosis, and histoplasmosis, have also been associated with rounded atelectasis.

Roggli and Pratt (1992) describe asbestosis as the prototype of disease caused by the inhalation of mineral fibers. Asbestosis is defined as interstitial fibrosis of the lung parenchyma in which asbestos bodies or fibers may be demonstrated, and it is the only asbestos-related disease to which this term should be applied (American Thoracic Society, 1986). Asbestosis shows a dose–response relationship, and there appears to be a threshold below which asbestosis is not seen (Browne, 1994). The time from initial exposure to the appearance of asbestosis—the latency period—is inversely proportional to the exposure level and is generally several decades (Roggli & Pratt, 1992). Studies suggest that cigarette smoke acts synergistically with asbestos by increasing the incidence of asbestosis (Barnhart et al., 1990; Roggli & Pratt, 1992).

The clinical, physiologic, radiologic, and pathologic findings in asbestosis vary with the severity of the disease. The clinical and physiologic findings are not pathognomonic of asbestosis, and they can be seen in diffuse interstitial fibrosis of any cause. Patients with well-established disease usually present with shortness of breath, basilar end-inspiratory crepitations, and a restrictive defect with decreased diffusing capacity on pulmonary function testing. Small, irregular linear opacities, most prominent in the lower lobes, are seen on plain films (Fraser et al., 1999a).

The earliest microscopic manifestation of asbestosis is fibrosis of the walls of respiratory bronchioles. As the disease progresses, fibrosis involves the walls of terminal bronchioles and alveolar ducts, and ultimately it extends into the adjacent alveolar septae. The minimal histologic criteria for the diagnosis of asbestosis can be defined as the presence of peribronchiolar fibrosis and asbestos bodies, with or without accompanying alveolar fibrosis (Craighead et al., 1982). As peribronchiolar fibrosis occurs with inhalants other than asbestos, it is suggested that, in the absence of alveolar septal fibrosis, the histologic diagnosis of asbestosis be restricted to cases in which the majority of bronchioles are involved (Roggli, 1989). The report from the College of American Pathologists suggests that there must be a minimum of two asbestos bodies in areas of fibrosis for the histologic diagnosis of asbestosis (Craighead et al., 1982).

2. Malignant Pleuropulmonary Disease

Numerous epidemiologic studies demonstrate an association between asbestos and an excess risk of lung cancer (McDonald, 1980). The clinical features—anatomic distribution within the lung and histologic subtypes of asbestos-related lung cancer—are identical to carcinoma in non-exposed individuals. The latency period for asbestos-related lung cancer is lengthy; it peaks at 30 to 35 years (Selikoff et al., 1980). A linear dose–response relationship exists between exposure and the risk of lung cancer at high-exposure levels (McDonald, 1980), but at low-exposure levels Browne (1986) suggests that a threshold exists below which the risk of cancer is not increased. This threshold is thought to be in the range of exposure required to produce asbestosis (Churg, 1998b). Available data on asbestos workers indicate that the interaction between cigarette smoke and asbestos in increasing the risk of lung cancer is synergistic, or multiplicative, rather than additive (Greenberg, 1992). Although persons with asbestosis are at excess risk of developing lung cancer (Doll, 1993), investigators are uncertain whether this risk is directly related to the asbestos or to the pulmonary fibrosis. Several studies (Sluis-Cremer & Bezuidenhout, 1989; Hughes & Weill, 1991) demonstrate an increased rate of lung cancer only in the presence of asbestosis, supporting the view that asbestos is carcinogenic because of its fibrogenicity. Churg (1998b) offers the following approach when evaluating cancer in a given case:

> If asbestosis is present, then the carcinoma is ascribed to asbestos exposure. If the patient smokes or has a history of smoking, then smoking is considered a contributing factor to causation. In the absence of asbestosis, the cancer should not be ascribed to asbestos exposure; the most common causative agent in this instance is cigarette smoke.

Malignant mesothelioma may be idiopathic, or it may occur secondary to ionizing radiation, chemical carcinogens, chronic inflammation and scarring of the pleura, and erionite exposure (Browne, 1994; Churg, 1998b); however, it is most strongly associated with asbestos exposure. The latency period for asbestos-related mesothelioma is long, peaking at 30 to 40 years, and virtually never occurring before 15 years (McDonald & McDonald, 1987). Asbestos fiber types differ in ability to produce mesothelioma, with the amphiboles posing greater risk than chrysotile (Browne, 1994). Characteristically, mesothelioma grows over the surface of the lung in thick sheets or as nodular masses, which may progress to encase the lung in a hard white rind. Mesotheliomas spread by direct extension into adjacent structures or by lymphatic or hematogenous metastases. Peripheral lung cancers may spread in a similar fashion, which makes the distinction from mesothelioma difficult. Generally, the presence of metastatic disease at clinical presentation favors carcinoma. Mesotheliomas are subclassified by microscopic appearance into epithelioid, sarcomatoid, and biphasic patterns. The diagnosis of diffuse malignant mesothelioma can be very difficult, and it must be distinguished from benign reactive pleural lesions as well as primary and metastatic neoplasms of the pleura.

Frequently, disorders of the lung can lead to cardiovascular dysfunction due to added stress from increased vascular pressures. In the next section, toxicity from metals to the heart is discussed.

V. CARDIOVASCULAR SYSTEM

Deficiencies of trace elements as well as toxic exposures of metals may be involved in physiologic changes in the cardiovascular system. The clinical, pathologic, and epidemiologic data that support or refute an association between metals and three common cardiovascular disorders—dilated cardiomyopathy, atherosclerosis, and systemic hypertension—are discussed in detail in this section.

A. Dilated Cardiomyopathy

1. Selenium Deficiency

Selenium is an essential nutrient in trace quantities. It combines with cysteine as a component of selenoproteins, many of which have antioxidant properties. For example, thioredoxin reductase, iodothyronine deiodinases, and glutathione peroxidases are selenium-dependent enzymes (Rayman, 2000). Selenoproteins are believed to be especially important in relation to the immune response and cancer prevention; however, the role of selenium in the maintenance of the cardiovascular system is less clear (Rayman, 2000).

It has been suggested that two levels of selenium deficiency are involved in the causation of human disease (Rayman, 2000). Rare endemic diseases occur where the soil is extremely low in selenium, specifically parts of China. These diseases include Keshan cardiomyopathy and Kashin-Beck disease, a deforming arthritis. (See Chapter 15, this volume.) High-prevalence diseases, such as cancer and heart disease, may have as a risk factor a relatively mild deficiency of selenium, such as has been reported in Europe. The role of mildly low selenium intake in the development of reproductive

disease, mood disorders, thyroid function, inflammatory disease, and cancer has been recently reviewed (Rayman, 2000).

Markedly inadequate dietary intake of selenium in areas of China with endemic selenium deficiency is associated with a form of dilated cardiomyopathy referred to as Keshan disease (Ge & Yang, 1993). It has been demonstrated that selenium levels in soil and rocks vary greatly in China and that areas with low levels have higher rates of the disease (Ge & Yang, 1993). Intervention studies that have shown a prophylactic effect of sodium selenite provide further evidence that selenium plays a role in Keshan disease. Morphologically, the condition is characterized by multifocal necrosis and replacement fibrosis of the myocardium, which results in acute or chronic heart failure (Ge et al., 1983) (see Figures 4–6). Some patients with Keshan disease show the clinical features of dilated cardiomyopathy, but autopsy studies from China demonstrate distinct pathologic features that separate Keshan disease from sporadic dilated cardiomyopathy (Li et al., 1985). Typical dilated cardiomyopathy does not demonstrate frequent necrosis or fibrosis, which are considered hallmarks of Keshan cardiomyopathy. Ultrastructural observations have shown mitochondrial abnormalities (Ge et al., 1983), which have recently been expanded to include biochemical defects and proteinaceous granular deposits (Yang et al., 1988).

Activities of succinate dehydrogenase, succinic oxidase, and cytochrome c oxidase, H(+)-ATPase and its sensitivity to oligomycin, as well as the response of membrane potential to energization by ATP, are decreased, and affected mitochondria had markedly

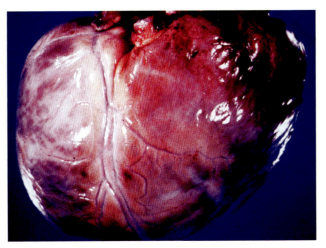

FIGURE 4 Gross appearance of Keshan cardiomyopathy demonstrating globular configuration of the external aspect of the heart.

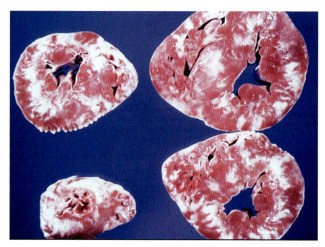

FIGURE 5 Gross appearance of cross sections of heart muscle demonstrating multiple areas of white tan scar tissue.

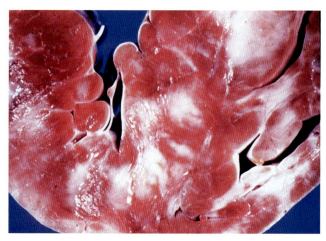

FIGURE 6 Gross appearance of scar tissue involving the right and left ventricles of the heart.

decreased selenium content (Yang *et al.*, 1988). The sole role of selenium in the causation of Keshan disease is debated, and it is generally believed that other cofactors are responsible for the pathogenesis of Keshan cardiomyopathy, including possible genetic and other environmental factors (Li *et al.*, 1985). Recently, enteroviral infection as a cofactor has been proposed based on the finding of viral genomes in the myocardial tissue of affected patients. (See Chapter 15, this volume.)

An association between sporadic dilated cardiomyopathy and selenium deficiency is far less clear than the association between selenium and Keshan disease. Serum levels of selenium may be decreased in patients with dilated and postpartum cardiomyopathy in Chinese and African populations as well as Indians (Vijaya *et al.*, 2000), although the data in idiopathic dilated cardiomyopathy are conflicting in Western and Middle Eastern countries (Raines *et al.*, 1999). There have been anecdotal reports of reversible cardiomyopathy due to selenium deficiency in patients with total parenteral nutrition. Selenium deficiency may play a role in the cardiopathy associated with chronic renal failure, but some studies do not support any association (Li & Nan, 1989; Chou *et al.*, 1998). The association between dilated cardiomyopathy and dystrophic epidermolysis bullosa does not appear to be mediated by selenium deficiency. The mechanism of cardiomyopathy in selenium deficiency is unclear but is likely related to protection of low-density lipoproteins against oxidative modification, regulation of glutathione peroxidase (Li & Nan, 1989; Chou *et al.*, 1998), modulation of prostaglandin synthesis and platelet aggregation,

and protection against toxic heavy metals (Oster & Prellwitz, 1990; Neve, 1996).

a. Cobalt Toxicity

An epidemic of cardiomyopathy related to cobalt toxicity was reported in 1967 in Canada (Kesteloot *et al.*, 1968). Cobalt ingestion in this epidemic was related to the addition of cobalt sulfate or cobalt chloride to beer as a stabilizer of beer foam. The mechanism of myocyte damage is unclear but may involve increased cardiac vulnerability to oxygen free radicals and may be enhanced in patients with dietary deficiencies. The histologic features of cobalt cardiomyopathy differ somewhat from idiopathic cardiomyopathy, in that greater myofibrillar loss and atrophy and less fibrosis and myocyte hypertrophy are observed in cobalt cardiomyopathy as compared to idiopathic dilated cardiomyopathy (Centeno *et al.*, 1996) (see Figure 7). Based on heavy metal analysis of cardiac tissues, there is little evidence that elevated cobalt levels are a significant factor in sporadic dilated cardiomyopathy (Centeno *et al.*, 1996).

b. Mercury-Induced Cardiomyopathy

Dilated cardiomyopathy has been described in Greenlanders with high mercury levels in their blood, presumably from eating seal meat contaminated with high levels of mercury. In addition, evidence suggests that, in cases of sporadic dilated cardiomyopathy, there are thousand-fold elevations of mercury in heart tissue (Frustaci *et al.*, 1999). This latter observation remains to be duplicated, and the mechanism of potential

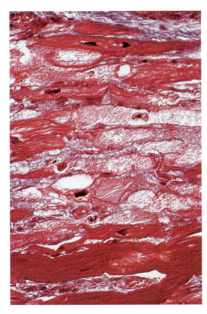

FIGURE 7 Photomicrograph of myocytes with decreased numbers of myocytes (H&E ×400).

mercury-induced cell damage must yet be elucidated. It is believed that mercury ions may act as calcium antagonists at the actin–myosin junction, inhibiting sarcomere contraction, or that they may disrupt microtubule structure. It has been postulated that a source of mercury in patients with dilated cardiomyopathy may, in addition to occupational exposure, be amalgam fillings, although this remains conjectural.

c. Iron Overload

Iron overload may adversely affect the heart in patients with hemosiderosis or primary hemochromatosis. The most common cause of death in patients with hemochromatosis is from cardiac failure and arrhythmias followed by cirrhosis and hepatocellular carcinoma. Histologically, these changes are similar to dilated cardiomyopathy with abundant stainable iron present within myocytes. Most patients with hemochromatosis-related cardiomyopathy have a genetic predisposition to iron overload (primary hemochromatosis); significant cardiac failure in patients with hemosiderosis secondary to blood transfusions is rare. There is little evidence that elevated serum iron plays a role in idiopathic dilated cardiomyopathy, although the data are conflicting (Oster, 1993; Chou et al., 1998). A recently described mutation in the hemochromatosis gene may contribute to the development of dilated cardiomyopathy that possibly is not mediated by a rise in serum iron (Mahon et al., 2000).

d. Aluminum Toxicity

Aluminum levels may be elevated in patients on hemodialysis. Although multifactorial, the mechanism of cardiac hypertrophy in patients with end-stage renal disease may involve direct aluminum cardiotoxicity (Reusche et al., 1996). Aluminum may be demonstrated by special stains in the myocardium of such patients. There is little reported evidence that aluminum toxicity plays a role in idiopathic dilated cardiomyopathy.

e. Arsenic Toxicity

Acute arsenic toxicity may result in a toxic myocarditis, and chronic exposure to arsenic, such as in occupational exposure, may cause chronic cardiomyopathy characterized by small vessel disease and interstitial inflammation with fibrosis (Hall & Harruff, 1989).

f. Miscellaneous

Elevations of other potentially toxic metals have been demonstrated in heart tissue from patients with dilated cardiomyopathy. In addition to cobalt and mercury, these metals include antimony, gold, and chromium (Frustaci et al., 1999). The role of these elements in the pathogenesis of dilated cardiomyopathy remains to be elucidated.

B. Atherosclerosis

1. Iron Excess

The body has no method of controlling iron excretion; therefore, dietary excesses may lead to iron overload (Schumann, 2001). The most likely excesses result from dietary supplements or alcoholic beverages brewed in iron containers (Schumann, 2001). Genetic factors are also important in chronic iron overload (hemochromatosis). The medical literature covering the experimental and epidemiological relationship between excess iron stores and coronary artery disease has been recently reviewed (de Valk & Marx, 1999). It has long been appreciated that males have a higher incidence of atherosclerosis than females and have greater body iron stores; however, a causative link between iron and atherosclerosis has not been fully established, and a recent prospective study has shown no relationship between iron stores (serum ferritin) and the development of ischemic heart disease (Sempos et al., 2000).

Experimental data support the role of iron in the process of lipid peroxidation, which is hypothesized to be involved in early stages of atherosclerosis. The exact mechanism by which endothelial cells and macrophages interact with iron and low-density lipoprotein are still unknown. There is speculation that catalytically active iron, in the form of nontransferrin-bound plasma iron and hemoglobin, oxidizes low-density lipoprotein to interact with the macrophage oxidized low-density lipoprotein receptor (Evans *et al.*, 1995).

At a clinical level, the role of iron in lipid oxidation has been questioned. Some studies have failed to show an association between serum ferritin or dietary iron intake with markers of low-density lipoprotein oxidation (Iribarren *et al.*, 1998). Several studies correlating body iron stores with clinically measured atherosclerosis have been positive (Kiechl *et al.*, 1997), which demonstrates a synergistic effect with low-density lipoprotein levels. Other studies have been negative (Rauramaa *et al.*, 1994). A recent review has demonstrated that the majority of studies correlating body iron stores with clinical assessments of atherosclerosis have not shown a significant correlation between clinical indices of coronary, carotid, or aortic atherosclerosis and body iron stores (Meyers, 1996).

2. Selenium Deficiency

Like dilated cardiomyopathy, ischemic heart disease has been linked to low levels of serum selenium. The mechanism of selenium deficiency resulting in an increased risk of coronary artery disease is unknown but may be related to increased oxidative stress, as is the case with iron, or increased platelet aggregation (Salonen *et al.*, 1988).

Epidemiological studies have provided some evidence for the role of selenium deficiency in the etiology of atherosclerotic disease. Angiographic studies suggest that low levels of selenium are associated with coronary stenosis (Yegin *et al.*, 1997), and patients with acute myocardial infarction tend to have lower serum selenium compared to controls (Navarro-Alarcon *et al.*, 1999). Plasma, red blood cell, and urine selenium concentrations have been shown to be decreased in patients with acute myocardial infarction. The results of longitudinal studies within populations are conflicting (Huttunen, 1997), however, and the risk of decreased levels of selenium appears to correlate with elevated low-density lipoprotein cholesterol. A nested case-control study among participants in the Physicians' Health Study showed no effect of selenium on the risk for acute myocardial infarction (Salvini *et al.*, 1995). Other studies using toenail selenium concentration, which is a better measure of chronic selenium levels, have shown an inverse relationship between selenium and myocardial infarction only in Germany, which has relatively low levels of selenium (Kardinaal *et al.*, 1997).

Salonen *et al.* (1982) showed an increase in overall cardiovascular morbidity and mortality for individuals with low serum selenium (less than 45 μg/L), whereas Virtamo's group (1985) found no significant associations except for stroke mortality. A large prospective study in a Danish population demonstrated that men with the lowest tertile of blood selenium had a mildly increased risk for developing acute ischemic events of stroke or myocardial infarct (risk ratio of 1.7); this risk remained significant to a low level of probability ($p = .05$) when other risk factors were considered. This study also demonstrated a correlation between selenium and traditional risk factors, including tobacco smoking, social class, alcohol consumption, total cholesterol, hypertension, age, and physical inactivity (Suadicani *et al.*, 1992). Other studies have found that selenium levels are affected by cigarette smoking and alcohol use, and that selenium is a negative acute phase reactant and may therefore be secondarily decreased by the inflammatory component of atherosclerosis. The current consensus is that the effect of selenium on ischemic heart disease is small and likely mediated by an association with other risk factors (Rayman, 2000).

3. Magnesium

The majority of research related to magnesium and cardiovascular disease addresses magnesium physiology after acute myocardial infarction. Specifically, the role of intravenous magnesium in the treatment of acute myocardial infarction and the effects of magnesium levels or intake on the risk of acute myocardial infarction are controversial areas. Less attention has been paid to the effects of chronic intake of low magnesium because of environmental and natural geological factors and the effect of chronic magnesium supplementation on the incidence of acute coronary events. Low levels of magnesium and fluoride in drinking water were found to be associated with increased risk for acute myocardial infarction in Finland and Sweden. No such effect was found in England, where there was a small protective effect of magnesium in drinking water against death due to ischemic heart disease, but no protection against deaths due to acute myocardial infarction. Oral magnesium supplementation does not decrease the rate of subsequent cardiac events after myocardial infarction, and the level of serum magnesium is not related to morbidity in acute myocardial infarction patients.

4. Calcium

Calcium has a complex relationship with atherosclerosis, and it is frequently deposited in the extracellular

matrix of advanced atherosclerotic plaques. There is no clear association between dietary calcium or serum levels of calcium and the risk for development of coronary atherosclerosis. High levels of serum calcium may have a mildly protective effect, because there is a borderline inverse relationship with serum calcium and the extent of coronary atherosclerosis (Narang *et al.*, 1997). However, there is no clear association between calcium in drinking water and the incidence of deaths due to acute myocardial infarction deaths, as demonstrated in studies in England and Scandinavia (Maheswaran *et al.*, 1999). (See Chapter 13, this volume.) An exception to these results is a study from Finland that demonstrated a protective effect of high levels of calcium in drinking water to fight against acute myocardial infarction in postmenopausal women (Rubenowitz *et al.*, 1999).

5. Other Metals

The effects of trace elements on atherosclerosis are difficult to determine because of the complex interplay between metal levels and traditional risk factors for cardiovascular disease, such as disorders of blood lipids, blood pressure, coagulation, glucose tolerance, and circulating insulin. In addition, the relevance of serum metal levels, metal concentrations within blood cells, and analyses of metal content of portions of atherosclerotic plaque is often unclear.

Like iron, copper ions may exert oxidative stress on the arterial wall. Elevated copper within circulating white cells has been associated with atherosclerosis. Elevated serum copper has been related to peripheral vascular disease, especially when related to serum zinc levels. No consistent data show a link between zinc and atherogenesis. Pathologic studies have demonstrated elevated levels of zinc within fibrous plaques of the aorta (Mendis, 1989). There is evidence that zinc requirements of the vascular endothelium are increased during inflammatory conditions such as atherosclerosis, where apoptotic cell death is also prevalent. Theoretically, zinc deficiency may exacerbate the detrimental effects of specific fatty acids and inflammatory cytokines on vascular endothelial functions, but an association between decreased dietary zinc and an increased risk for coronary atherosclerosis has yet to be demonstrated.

A mortality study of mercury miners in Europe demonstrated no increased deaths due to ischemic heart disease (Boffetta *et al.*, 2001); however, in a prospective study among 1014 men, high hair mercury content was one of the strongest predictors of the 4-year increase in the mean intima-media thickness, which is a measurement of carotid atherosclerosis (Salonen *et al.*, 2000).

A study of chromium welders has suggested that welders exposed to chromium have increased plasma lipid oxidation (Elis *et al.*, 2001). Although arsenic has been only indirectly associated with coronary heart disease, increased serum arsenic may lead to persistent oxidative stress, thereby theoretically leading to atherosclerosis.

C. Hypertension

1. Lead

The toxic effects of lead poisoning are well known; however, the effects of lower levels of lead exposure are uncertain. Although the finding is controversial, chronic low-level lead exposure has been linked to hypertension in both clinical and experimental studies. The exact pathogenic mechanisms that underlie the actions of lead in the cardiovascular system have yet to be elucidated definitively (Kopp *et al.*, 1988).

There are a number of difficulties in proving a link between low-level exposure to lead (blood levels of $<1.45\,\mu m/L$ or $<30\,\mu g/dL$) and the development of high blood pressure. A number of possible covariates must be considered in epidemiologic studies, such as body mass index, alcohol consumption, other metals, race, and gender. Measurements of blood levels may not accurately reflect body stores, and there may be discrepant results depending on whether blood lead, bone lead, or urinary lead levels are obtained after chelation testing is performed. Several cross-sectional epidemiologic studies have demonstrated an association between blood lead levels and hypertension in individuals without known occupational exposure and in men with occupational risk. In most of these studies, the association persists after correction for other variables. A similar number of studies have shown no association or only a very weak association between lead and hypertension, after adjustment for age and body mass index, in populations with or without known lead exposure. A study from the United States demonstrated an association only in African-Americans, who had higher lead levels (Sokas *et al.*, 1997). In a study of postpartum women in Los Angeles, an association between lead and hypertension was demonstrated only in immigrants who had, in general, higher levels than nonimmigrants. A study in Germany demonstrated a link only in heavy drinkers of alcohol; results from a cross-sectional analysis of alcoholics in the United States also demonstrate a link between hypertension and blood lead levels. A metaanalysis of pooled data from several studies suggested only a weak positive association between blood pressure and lead exposure (Staessen *et al.*, 1994).

Cross-sectional epidemiologic studies using bone indices have often shown a positive correlation between lead exposure and hypertension with an estimated relative risk of about 1.5 times for individuals with increased skeletal lead stores in men and women. These results, however, are not always confirmed by blood testing in the same patients, and results vary by bone site measured (*e.g.*, patella vs. tibia). Studies estimating body stores of lead in measurements of blood and urine after a chelating challenge have also suggested a link between hypertension and increased lead stores, especially in the presence of renal failure (Sanchez-Fructuoso *et al.*,1996).

In the United States and Western countries, mild blood pressure elevations due to moderate increases in lead blood levels translate into potentially large numbers of patients dying with coronary artery disease. This link between lead and coronary artery disease is corroborated by data showing a direct association between blood lead and electrocardiographic changes of left ventricular hypertrophy. There is evidence that occupational exposure to lead is associated with increased risk of dying from hypertension-related illness. A prospective study in Belgium, however, in which patients were followed for 6 years did not show any association between blood levels and hypertension, a finding that refutes a significant link between coronary heart disease and lead exposure (Staessen *et al.*, 1996). The mechanisms by which lead may induce hypertension include increased vascular responsiveness to catecholamines mediated by effects on calcium channels and increased expression of endothelin and generation of reactive oxygen species.

Sources of lead exposure are numerous and include occupational exposure (workers in battery factories, smelters), environmental exposure (traffic exhaust, dust, paint chips, drinking water), cosmetics, food supplements, food preparation utensils, and illegal alcoholic beverages ("moonshine").

2. Mercury

Acute mercury poisoning may result in hypertension, but few data suggest that chronic low-level exposure to mercury is an etiologic factor in essential hypertension. Mortality statistics from mercury miners in several regions across Europe demonstrated no consistent increase in deaths due to hypertension (Boffetta *et al.*, 2001), with the exception of Spain and Ukraine, which had elevated mortality ratios of 2.8 and 9.4, respectively, for hypertension-related diseases.

3. Other Metals

Chronic arsenic exposure from drinking water in contaminated wells has been implicated in a modest increased risk for hypertension in China and Bangladesh. Patients with hypertension, who are not on dialysis, may have elevated levels of serum aluminum, but an etiologic link has yet to be shown between hypertension and aluminum toxicity.

The effects of hypertension can ultimately lead to cardiac failure, resulting in increased venous backpressures. Depending on the severity of the failure, severe chronic congestive heart failure may lead to fibrosis of the liver. Direct toxicity of metal ions on the liver also may lead to injury to the liver and fibrosis of the liver, as is discussed in the final section.

VI. HEPATOTOXICITY OF METAL IONS

Several metals can injure the liver, but by far the most important are iron and copper. The diseases caused by metals may be genetic or acquired, and the effects can be acute or chronic (see Table II). Acute metal toxicity produces hepatocellular and cholestatic injury:

1. *Hepatocellular injury*—Ferrous sulfate poisoning in children can lead to coagulative degeneration in zone 1 of the hepatic acinus, and phosphorus poisoning induces lytic necrosis in that zone, as well as steatosis. Copper toxicity causes zone 3 necrosis.
2. *Cholestatic injury*—Intrahepatic cholestasis has been reported with acute arsenical toxicity and as an idiosyncratic reaction to the use of gold salts for treatment of rheumatoid arthritis.

In chronic metal toxicity the injuries listed below are seen:

1. *Vascular injury*—Hepatoportal sclerosis is a recognized complication of chronic arsenical toxicity, for example, from ingestion of high levels of arsenic in drinking water (India, Bangladesh).
2. *Chronic hepatitis*—Chronic hepatitis is a stage in the evolution of Wilson's disease; hemosiderin is an accumulation in the liver in chronic hepatitis C which may lead to a poor response to interferon therapy.

TABLE II. Hepatotoxicity of Metals

Metal	Circumstances	Histopathology	Comments
		Acute Toxicity	
Iron	Accidental ingestion by children (usually FeSO₄)	Zone 1 necrosis	Also severe gastrointestinal injury
Copper	Suicidal or accidental	Zone 3 necrosis	Also cholestasis
		Chronic Toxicity	
Iron	Genetic hemochromatosis	Hepatic hemosiderosis, fibrosis, cirrhosis, HCC	C282Y and H63D gene mutations
		Other (see Table II)	
Copper	Wilson's disease	Copper overload, chronic hepatitis, cirrhosis, fulminant liver failure with necrosis, HCC (rare)	ATP7B gene mutations
	Indian childhood cirrhosis	Copper overload, Mallory bodies, fibrosis, cirrhosis	Excess ingestion of copper in milk from copper or brass containers
	Endemic infantile Tyrolean cirrhosis	Copper overload, cirrhosis	Excess ingestion of copper in milk plus genetic factor
Arsenic	Excess ingestion of arsenic via drinking water, drugs, or exposure to insecticides	Hepatoportal sclerosis, angiosarcoma	

Note: HCC = hepatocellular carcinoma.

3. *Fibrosis and cirrhosis*—Fibrosis and cirrhosis occur in genetic hemochromatosis, Wilson's disease, Indian childhood cirrhosis, acquired copper toxicosis, neonatal hemochromatosis, and sub-Saharan hemosiderosis; the presence of iron in nonalcoholic steatohepatitis may increase the hepatic injury and contribute to the fibrosis.

4. *Granulomas*—Chronic beryllium toxicity (berylliosis) in the past was associated with a sarcoidosis-like disease; exposure of vineyard sprayers to copper sulfate in Portugal has been reported to lead to granulomas in the lungs and liver.

5. *Malignant tumors*—Hepatocellular carcinoma is a dreaded complication of genetic hemochromatosis and, rarely, of Wilson's disease; angiosarcoma has been reported after chronic exposure to arsenic and rarely to iron or copper.

The hepatotoxicity of several metals, in alphabetical order, is discussed in more detail in the following paragraphs.

A. Arsenic

1. Acute Toxicity

Toxicity from inorganic arsenicals may result from attempts at homicide or suicide; exposure to insecticides, herbicides or rodenticides; industrial exposure; exposure to naturally contaminated materials; or from treatment with arsenical medications. The clinical features, mechanism of action, and other aspects of arsenical toxicity are comprehensively discussed in a review by Schoolmeester and White (1980). Hepatic injury is usually overshadowed by the effects of damage to other organ systems, such as the gastrointestinal, cardiovascular, neurologic, and hematologic ones. Observed histologic changes in the liver have included steatosis, cholestasis, and zone 3 necrosis. A striking increase in hepatocytic mitotic activity was noted in two cases reported by Brenard *et al.* (1996). Hepatic veno-occlusive disease following severe arsenic poisoning was described by Labadie *et al.* (1990). Predominantly cholestatic injury has been reported with the antiamebicidal agent carbarsome and the anti-

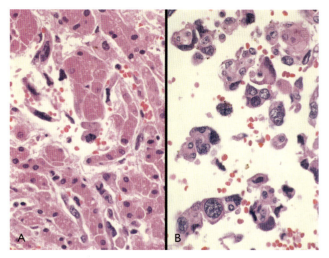

FIGURE 8 Arsenic-related angiosarcoma. (A) The malignant endothelial cells infiltrating the sinusoids have led to disruption of the liver cell plates (B); higher magnification of the angiosarcoma cells (B) shows marked nuclear pleomorphism and hyperchromasia.

syphilitic drug arsphenamine. Acute arsenic and mercury intoxication has also resulted from ingestion of Indian ethnic remedies (Kew *et al.*, 1993) and Chinese herbal balls (Espinoza *et al.*, 1995). Experimental work with arsenates in rabbits was reported to cause periportal focal necrosis and cirrhosis of an unspecified type (Van Glahn *et al.*, 1938).

2. Chronic Toxicity

Chronic arsenical injury to the liver can lead to hepatoportal sclerosis or noncirrhotic portal hypertension (Datta *et al.*, 1979; Mazumder *et al.*, 1988), cirrhosis (Denk *et al.*, 1959; Lunchtrath, 1972), systemic arterial disease (Rosenberg, 1974), hepatocellular carcinoma (Jhaveri, 1959), and angiosarcoma (Falk *et al.*, 1981) (see Figure 8). The chronic exposure was the result of therapy (*e.g.*, the use of Fowler's solution [potassium arsenite] for the treatment of psoriasis), high levels of arsenic in drinking water, use of arsenical preparation as insecticides in vineyards, or industrial or environmental exposure (*e.g.*, among workers in copper smelters or residents living nearby). The histologic changes that have been described in hepatoportal sclerosis include portal area expansion and fibrosis, an increase in the number of vessels in portal areas, thickening of portal vein branches (intimal thickening and muscular hypertrophy, but no thrombosis), and perisinusoidal fibrosis. The cirrhosis reported with chronic arsenicism has been described as "post-necrotic" or macronodular in type.

B. Copper

1. Acute Copper Toxicity

Acute copper toxicity may result from: (1) ingestion (either suicidal or accidental) of copper sulfate, (2) its use as an emetic or for the treatment of burns, (3) the release of copper ions from equipment made of copper, (4) acid pH conditions (*e.g.*, malfunctioning hemodialysis equipment or vending machines), or (5) high levels of copper in drinking water (Blomfield *et al.*, 1971; Spitainy *et al.*, 1984; Holtzman *et al.*, 1986). Copper poisoning leads to a hemolytic anemia related to several effects on erythrocytes that include inhibition of glucose-6-phosphate dehydrogenase activity, inhibition of glycolysis, denaturation of hemoglobin (with Heinz body formation), and oxidation of glutathione. The largest series of cases of acute poisoning with copper sulfate, ingested for suicidal purposes, was reported from New Delhi, India (Chuttani *et al.*, 1965). Jaundice appeared on the second or third day in 11 of 48 patients (23.0%). In five patients, the jaundice was deep, the liver was enlarged and tender, the average total serum bilirubin was 11.2 ± 809 mg/dL, the AST averaged 252.4 ± 142 units, and the prothrombin time was markedly prolonged. In the remaining six patients, the jaundice was mild, the serum bilirubin was about 3.0 mg/dL, and the AST averaged 144 ± 80 IU. Histologic changes in biopsy or autopsy material showed zone 3 necrosis and cholestasis in the deeply jaundiced patients and focal necrosis or no changes in the mildly jaundiced patients. When measured, the serum copper levels have been markedly elevated in acute copper intoxication (Chuttani *et al.*, 1965; Blomfield *et al.*, 1971; Holtzman *et al.*, 1986), as well as in the liver (Blomfield *et al.*, 1971).

2. Chronic Copper Toxicity

Chronic copper toxicity can be hereditary or acquired. In humans, hereditary copper overload is exemplified by Wilson's disease and in animals by the copper toxicosis of the Long-Evans cinnamon rat; both conditions have an autosomal recessive inheritance and are not discussed in this review. Acquired forms of copper overload also occur in humans and animals. Sheep are particularly susceptible to copper intoxication, as they can be chronically exposed to excessive copper by grazing in sprayed orchards or by eating contaminated feed. The changes in sheep livers include steatosis, focal necrosis, swelling of liver and Kupffer cells, and cytochemically demonstrable copper in Kupffer cells (incorporated in lysosomes with lipofuscin), but Mallory bodies have not been observed. An interesting copper toxicosis was reported in North Ronaldsay sheep by Haywood *et al.* (2001). These sheep are a primitive breed that has adapted to a copper-impoverished environment in the Orkney Islands. Sheep transferred to copper-replete pastures on the mainland developed hepatic copper toxicity resembling that of Indian childhood cirrhosis and idiopathic copper toxicosis.

a. Acquired Chronic Copper Toxicity

In humans, acquired chronic copper toxicity may result from occupational environmental, or domestic exposure. An example of occupational exposure is the chronic exposure of vineyard workers to fungicide sprays containing copper sulfate. The resultant hepatic injury includes noncaseating granulomas, intraacinar and periportal fibrosis, cirrhosis, and, rarely, angiosarcoma (Pimental & Menezes, 1977). The geological environment is the setting for Indian childhood cirrhosis, and a similar disease reported from outside the Indian subcontinent occurred in children who had ingested large quantities of copper in their drinking water.

b. Indian Childhood Cirrhosis

Indian childhood cirrhosis (ICC) is associated with marked hepatic copper overload. The copper storage has been demonstrated histochemically (by staining tissue sections with orcein for copper-binding protein and the rhodanine method for copper) and by quantitative techniques such as atomic absorption spectrophotometry. It is generally accepted that copper accumulation is directly responsible for the histopathologic lesions in ICC; these include the presence of numerous Mallory bodies, intraacinar and portal inflammation, copper accumulation (beginning in zone 1), periportal ductular proliferation, intraacinar and periportal fibrosis, occlusive lesions of terminal hepatic venules, and, eventually, the development of a micronodular cirrhosis (Smetana *et al.*, 1961; Joshi, 1987; Bhagwat *et al.*, 1983) (see Figure 9).

According to Bhagwat *et al.* (1983), cases of ICC have been reported from 17 countries as well as from the United States (Lefkowitch *et al.*, 1982). The non-Indian cases also are characterized by marked copper overload and histopathologic alterations indistinguishable from those of ICC. In 1983, Tanner *et al.* (1983) proposed that increased dietary copper (from copper-contaminated milk stored in brass and copper containers) could be of etiologic significance in ICC. Subsequently, O'Neill and Tanner (1989) demonstrated experimentally that copper (but not zinc) is avidly taken up from brass and bound to casein from which it is com-

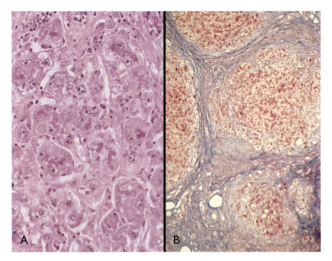

FIGURE 9 Indian childhood cirrhosis. (A) High magnification shows Mallory bodies in liver cells. (B) This cirrhotic liver is composed of nodules separated by fibrous septa (Masson).

pletely removable by picolinate chelation. They concluded that milk is an effective carrier of copper from a brass utensil to the infant enterocyte. Lending support to a direct cytopathic effect of copper are recent reports of clinical recovery, improved survival, and reversal of the hepatic histological lesions by penicillamine therapy of infants with ICC (Tanner *et al.*, 1987; Bhusnurmath *et al.*, 1991).

Of great interest are reports of several children who developed an illness, clinically and histopathologically resembling ICC, from the chronic ingestion of well water contaminated with high levels of copper (Müller-Höcker *et al.*, 1985, 1987; Schramel *et al.*, 1988). In all cases, the copper was leached from copper pipes into drinking water delivered to the children's homes. Early exposure to copper appears to be crucial, because siblings exposed after nine months of age and the parents who drank the same water did not develop the disease (Müller-Höcker *et al.*, 1985; Schramel *et al.*, 1988). Other cases of "non-Indian childhood cirrhosis," also referred to as "idiopathic hepatic copper toxicosis," not specifically related to increased dietary copper intake have been reported (Maggiore *et al.*, 1987; Horslen *et al.*, 1994; Ludwig *et al.*, 1995). To date, most of the affected children have succumbed to liver failure. It is conceivable that some cases of ICC occurring outside the Indian subcontinent are examples of environmental chronic copper toxicity, but other cases are not. The occurrence of the disease in siblings and parental consanguinity in two cases suggest a genetic abnormality, possibly an autosomal recessive disorder (Scheinberg & Sternlieb, 1994). Lack of support for excess dietary copper as a cause of non-Indian childhood cirrhosis comes from a study of Scheinberg and Sternlieb (1994) in which no liver-related deaths were found in children less that six years of age in three Massachusetts towns with high copper concentrations in drinking water. The study covered 64,124 child-years of exposure. Of relevance to these observations are cases of endemic Tyrolean infantile cirrhosis, which is transmitted by autosomal recessive inheritance but requires an additional risk factor such as excess dietary copper (from cows' milk contaminated with copper leached from untinned copper or brass containers) for its expression (Müller *et al.*, 1996). The disease was eradicated when the untinned copper containers were replaced by modern industrial vessels.

Before leaving this section, it is worth noting that chronic acquired copper toxicity in the adult is very rare. There is one report of chronic toxicity from massive ingestion of copper coins; 275 copper coins were found at autopsy in the stomach of a mentally deranged person (Yelin *et al.*, 1987). Corrosion of the coins led to absorption of copper from the stomach, resulting in liver and kidney injury. The changes in the liver resembled those of ICC and childhood hepatic copper toxicosis. It is important to remember that accumulation of copper occurs in a number of chronic cholestatic conditions, most frequently in primary biliary cirrhosis and primary sclerosing cholangitis. Although demonstration of the stored copper, mainly periportal, is of diagnostic value, it does not appear to be a risk factor for progression of the disease, and attempts at removing the copper by penicillamine therapy have been abandoned.

C. Iron

Injury to the liver by iron can be either acute or chronic. Genetic hemochromatosis, the most common inherited metabolic disease, is not discussed here; instead, the focus of this section is on acquired iron toxicity, which may be acute or chronic.

1. Acute Iron Toxicity

Accidental overdose of iron-containing drugs and dietary supplements (in tablet or capsule form) is a leading cause of fatal poisoning of children under 6 years of age (Nightingale, 1997), but occasional cases have been reported in adults (Monoguerra, 1976). Since 1986, U.S. poison control centers have received reports of more than 110,000 incidents of children younger

than 6 years accidentally swallowing iron tablets. Almost 17% of pediatric deaths reported to poison control centers between 1988 and 1992 were due to iron poisoning, compared with 12% between 1984 and 1987. Death has occurred from ingesting as little as 200 mg to as much as 5.85 g of iron (Nightingale, 1997).

A large number of products available in pharmacies, foodstores, and discount stores contain iron, and the problem is compounded by the attractiveness of dosage forms, high availability, and ambiguous labeling (Krenzelok & Hoff, 1979). The Food and Drug Administration (FDA) issued a regulation, effective July 15, 1997, with labeling and packaging requirements to protect children from accidental poisoning from iron-containing drugs and dietary supplements in tablet or capsule form. Under the regulation, these products must display—in a prominent and conspicuous place set off by surrounding lines—the following warning statement: "WARNING: Accidental overdose of iron-containing products is a leading cause of fatal poisoning in children under 6. Keep this product out of reach of children. In case of accidental overdose, call a doctor or poison control center immediately" (Nightingale, 1997). In addition, the agency requires most products that contain 30 mg or more of iron per dosage, such as iron tablets or capsules for pregnant women, to be packaged as individual doses (for example, in blister packages). The FDA has concluded that this packaging will limit the number of unit doses that a child may consume once access is gained to the product, thus significantly reducing the likelihood of serious injury.

Although the morbidity of acute iron poisoning is high, the mortality in two large series was low—none in the series of 66 patients reported by James (1970), and 5.16% of the 172 cases reported by Westlin (1966). The typical victim of iron poisoning vomits within 10 minutes to 1.5 hours after swallowing a toxic dose of iron; consequently, it is often difficult to state what the toxic dose is. In a severe case that does not prove fatal, the three stages of poisoning are:

- First stage—This stage lasts approximately six hours and begins with abdominal pain, nausea, and vomiting and goes on to hematemesis, melena, and subsequently shock and possibly coma.
- Second stage—This stage is one of apparent recovery during which toxic iron compounds are formed.
- Third stage—This stage begins about 24 hours after ingestion of the toxic dose. Acidosis and hyperglycemia develop, and there may be convulsions, coma, bleeding, and evidence of

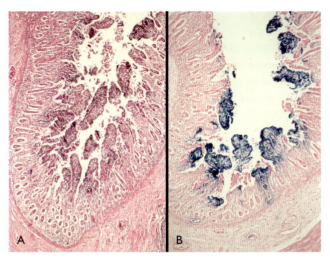

FIGURE 10 Acute ferrous sulfate toxicity: (A) small intestine showing marked necrosis of the villi; (B) iron encrustation (blue) of the necrotic intestinal villi (Prussian blue stain).

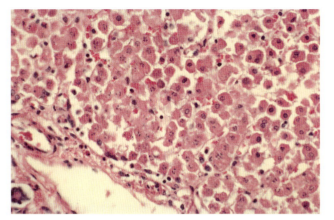

FIGURE 11 Acute ferrous sulfate toxicity; section of the liver of the same patient illustrated in Figure 10 shows coagulative degeneration and dissociation of liver cells in zone 1. A portal area is present in the lower left corner.

hepatic damage. Invariably, edema and necrosis of the gastric mucosa are found in fatal cases. If the child recovers, there may be fibrosis of the gastric mucosa leading to pyloric stenosis.

Gandhi and Robarts (1962) reviewed 11 such cases. The interval between the time of ingestion of the ferrous sulfate and the symptoms of obstruction ranged from 13 to 40 days. Pyloric stenosis was present in five patients. "Hourglass" stricture of the stomach was found in the other six children, two of whom had pyloric stenosis as well; one child also had a penetrating gastric ulcer that involved the liver.

Nearly 90% of deaths from iron poisoning occur in the first 48 hours after ingestion. Pathologic findings include widespread necrosis of the upper gastrointestinal tract (see Figure 10), thrombosis of intestinal vessels, periportal (zone 1) necrosis of hepatic cells, hemorrhagic pneumonia, and pericardial effusion. Liver cells in zone 1 show coagulative degeneration with nuclear pyknosis or karyorrhexis, and there is variable drop out of cells (Pestaner *et al.*, 1999) (see Figure 11). The inflammatory response is typically minimal, but there may be a sprinkling of neutrophils and lymphocytes. Histochemical stains for iron often show positive staining of the coagulated cells and damaged sinusoids.

2. Chronic Iron Toxicity

Numerous conditions, other than genetic hemochromatosis, can lead to chronic accumulation of iron in the liver. Among these are the hereditary anemias listed in

Table III. In transfusion-dependent β-thalassemia major, the iron overload is severe and, over a period of months or years, can lead to congestive heart failure, hypothyroidism, diabetes mellitus, hepatomegaly, fibrosis, and cirrhosis (Modell, 1975; Schafer *et al.*, 1981). In transfusional siderosis, iron initially accumulates in reticuloendothelial cells but eventually storage also occurs in parenchymal cells, beginning in zone 1. A large amount of iron also is found in the spleen, in contrast to genetic hemochromatosis (Oliver, 1959).

Iron overload has been reported in a number of inherited metabolic diseases, including hereditary tyrosinemia, Zellweger's syndrome, congenital atransferrinemia, aceruloplasminemia, Wilson's disease after penicillamine therapy (Shiono *et al.*, 2001), porphyria cutanea tarda, and alpha-1 antitrypsin (AAT) deficiency. In the last condition, no association between genetic hemochromatosis and AAT deficiency was found in one study (Fargion *et al.*, 1996). In another study, the association of AAT deficiency and genetic hemochromatosis led to an earlier onset of cirrhosis in genetic hemochromatosis, but it did not increase the risk of hepatocellular carcinoma (Elzouki *et al.*, 1995). In porphyria cutanea tarda, 60 to 70% of patients have mild to moderate iron overload, and approximately 10% have increases in the range of genetic hemochromatosis; the major cause is the presence of mutations of the HFE gene (Bonkovsky & Lambrecht, 2000). Hepatic iron accumulation can occur in a number of acquired diseases of the liver (Table III). The ones discussed briefly here are chronic hepatitis C, nonalcoholic and alcoholic steatohepatitis, cirrhosis, and post-portacaval shunt surgery.

TABLE III. Causes of Hepatic Iron Overload Not Related to Genetic Hemochromatosis

Etiology	Ref.
Hereditary Anemias	
Sickle cell anemia	Searle et al., 2001
Thalassemia major	
Sideroblastic anemia	
Hereditary spherocytosis	Barry et al., 1968
Chronic Liver Disease	
Chronic hepatitis C	Hezode et al., 1999; Ganne-Carrie et al., 2000
Alcoholism and alcoholic liver disease	Kageyama et al., 2000
Nonalcoholic steatohepatitis	George et al., 1998; Younossi et al., 1999; Bonkovsky & Lambrecht, 2000
Cirrhosis or diverse etiology	Deugnier et al., 1997; Ludwig et al., 1997
Post-portacaval shunt	Bonkovsky et al., 1996
Increased Oral or Parenteral Iron Loading	
African iron overload (dietary overload with genetic factor)	Gordeuk et al., 1986, 1992; Gangaidzo & Gordeuk, 1995; Mandishona et al., 1998
African-American iron overload (similar to African iron overload)	Barton et al., 1995
Medicinal iron ingestion	Hennigar et al., 1979
Transfusional siderosis	Oliver et al., 1959; Schafer et al., 1981
Hemodialysis	Gokal et al., 1979
Inherited Metabolic Diseases	
α-1 Antitrypsin deficiency with genetic hemochromatosis	Elzouki et al., 1995; Farigon et al., 1996
Porphyria cutanea tarda	Bonkovsky & Lambrecht, 2000
Hereditary tyrosinemia	
Zellweger's syndrome	
Congenital atransferrinemia	Searle et al., 2001
Aceruloplasminemia	Andrews, 1999
Wilson's disease, especially after penicillamine therapy	Shiono et al., 2001
Neonatal hemochromatosis?	Bonkovsky et al., 1996

Hepatic iron accumulation occurs in chronic hepatitis C. In one study from Italy, 10% of patients with chronic hepatitis C had an elevated hepatic iron content (Riggio et al., 1997). Stainable iron in sinusoidal cells and portal tracts reflects an increased hepatic iron concentration, and elevated serum iron values have been linked to poor responses of chronic hepatitis C to interferon therapy. In one study, hepatic iron accumulation was found to be significantly associated with histologic activity and cirrhosis (Hezode et al., 1999). Barton et al. (1995) have suggested that the percentage of portal areas staining for iron should be included in the pathology report of liver biopsies; these investigators found that patients who respond to interferon therapy have less than 40% portal areas positive for iron. Enhanced iron accumulation and lipid peroxidation were improved by interferon therapy in one study (Kageyama et al., 2000). Pirisi et al. (2000) believed that interface hepatitis (piecemeal necrosis) and its sequelae (sinusoidal capillarization and microshunting) are major factors in iron deposition in chronic hepatitis C. HFE mutations are not responsible for the iron deposition but could favor the progression of virus-induced damage independently from interference with iron metabolism. The topic of iron and its relationship to chronic viral hepatitis is reviewed in detail by Bonkovsky et al. (1996, 1999).

Patients with heavy alcohol consumption and genetic hemochromatosis (GH) have a higher prevalence of cirrhosis than those who do not. Ludwig et al. (1997) found that subjects with GH who drank more than 60 g of alcohol per day were approximately nine times more likely to develop cirrhosis than those who drank less than that amount. The range of the hepatic iron concentration associated with cirrhosis (in the absence of other cofactors) was 237 to 675 μmol/g dry weight.

Iron overload in cirrhosis was studied in 447 native livers by Ludwig et al. (1997) and Deugnier et al. (1997). It is very common in many types of nonbiliary cirrhosis and appears to be acquired and to occur rapidly once cirrhosis develops (Ludwig et al., 1997). Quantitative iron determinations cannot be relied on to differentiate such cases from GH. Patients with nonalcoholic cirrhosis can accumulate iron (and zinc) after end-to-side portacaval shunting, but quantitative iron analyses have shown that such patients have far less iron than patients with GH, and they are unlikely to have tissue injury resulting from the iron overload (Adams et al., 1994).

Nonalcoholic steatohepatitis (NASH), now sometimes referred to as nonalcoholic fatty liver disease, may be accompanied by iron overload. In one study of 65 patients from the United States, iron accumulation, when present, was not associated with increased overall mortality, liver-related mortality, or cirrhosis (Younossi *et al.*, 1999). Bonkovsky *et al.* (1999) found increased prevalence of HFE mutations in 57 subjects with NASH. Those with C282Y mutations had significantly higher levels of serum ALT and greater hepatic fibrosis than those without. In another study of 51 patients with NASH from Australia, the C282Y mutation was found to be responsible for most of the mild iron overload, and it was significantly associated with hepatic damage in these patients (George *et al.*, 1998). It was suggested that phlebotomy therapy to remove the increased iron stores could delay or even reverse the liver damage.

African iron overload (formerly referred to as Bantu siderosis) is now rare in South Africa but remains a public health issue in sub-Saharan Africa (Gordeuk *et al.*, 1986). It appears to be caused by an interaction between the amount of dietary iron (consumed in traditional beer brewed in steel drums) and a gene distinct from any HLA-linked gene (Gordeuk *et al.*, 1992). African iron overload is considered a risk factor for hepatocellular carcinoma in black Africans (Gangaidzo & Gordeuk, 1995; Mandishona *et al.*, 1998). Iron overload in African-Americans (that is unexplained by dietary or medicinal iron excess, transfusions, or sideroblastic anemia) is considered to be similar to that in sub-Saharan Africans (Barton *et al.*, 1995).

Miscellaneous effects of iron overload that should be briefly mentioned include impaired cellular immune function and increased susceptibility to bacterial infections (Bonkovsky *et al.*, 1996), as well as hepatocellular carcinoma (HCC). The occurrence of HCC in patients with genetic hemochromatosis and cirrhosis is too well recognized to require comment, but iron may play a putative role in HCC that occurs rarely in a non-cirrhotic liver (Turlin *et al.*, 1995).

ACKNOWLEDGEMENTS

The authors would like to express their appreciation to Drs. Elena Ladich, Todor Todorov, Leonor Martinez, Elizabeth Meze, and Linda Murakata for their valuable contributions during the preparation of this chapter.

FURTHER READING

Adams, P. C., Bradley, C., and Frei, J. V. (1994). Hepatic Iron and Zinc Concentrations After Portacaval Shunting for Nonalcoholic Cirrhosis, *Hepatology*, 19, 101–105.

American Thoracic Society (1986). The Diagnosis of Non-malignant Diseases Related to Asbestos, *Am. Rev. Respir. Dis.*, 134, 363–368.

Andrews, N. (1999). Disorders of iron Metabolism, *N. Engl. J. Med.*, 341, 1886–1995.

Barchowsky, A., Dudek, E. J., Treadwell, M. D., and Wetterhahn, K. E. (1996). Arsenic Induces Oxidant Stress and NFKB Activation in Cultured Aortic Endothelial Cells, *Free Radic. Biol. Med.*, 21, 783–790.

Barnhart, S., Thornquist, M., Omenn, G. *et al.* (1990). The Degree of Roentgenographic Parenchymal Opacities Attributable to Smoking Among Asbestos-Exposed Subjects, *Am. Rev. Respir. Dis.*, 141, 1102–1106.

Barrett, J. C., Lamb, P. W., Wang, T. C., and Lee, T. C. (1989). Mechanisms of Arsenic-Induced Cell Transformation, *Biol. Trace Element Res.*, 21, 421–429.

Barry, M., Scheuer, P. J., Sherlock, S. *et al.* (1968). Hereditary Spherocytosis with Secondary Haemochromatosis, *Lancet*, 2, 481–485.

Barton, A. L., Banner, B. F., Cable, E. E. *et al.* (1995). Distribution of Iron in the Liver Predicts the Response of Chronic Hepatitis C Infection to Interferon Therapy, *Am. J. Clin. Pathol.*, 103, 419–424.

Barton, J. C., Edwards, C. Q., Bertoli, L. F. *et al.* (1995). Iron Overload in African Americans, *Am. J. Med.*, 99, 616–623.

Becklake, M. (1983). Occupational Lung Disease: Past Record and Future Trend Using the Asbestos Case as an Example, *Clin. Invest. Med.*, 6, 305–337.

Belton, J. C., Benson, N. C., Hanna, M. L., and Taylor, R. T. (1985). Growth Inhibition and Cytotoxic Effects of Three Arsenic Compounds on Cultured Chinese Hamster Ovary Cells, *J. Environ. Sci. Health*, 20A, 37–72.

Bhagwat, A. G., Walia, B. N., Koshy, A., and Banerji, C. K. (1983). Will the Real Indian Childhood Cirrhosis Please Stand Up?, *Clev. Clin. Q.*, 50, 323–237.

Bhusnurmath, S. R., Walia, B. N. S., Singh, S. *et al.* (1991). Sequential Histopathologic Alterations in Indian Childhood Cirrhosis Treated with D-Penicillamine, *Hum. Pathol.*, 22, 653–658.

Blomfield, J., Dixon, S. R., and McCredie, D. A. (1971). Potential Hepatotoxicity of Copper in Recurrent Hemodialysis, *Arch. Intern. Med.*, 128, 555–560.

Boffetta, P., Sallsten, G., Garcia-Gomez, M., Pompe-Kirn, V., Zaridze, D., Bulbulyan, M., Caballero, J. D., Ceccarelli, F., Kobal, A. B., and Merler, E. (2001). Mortality From Cardiovascular Diseases and Exposure to Inorganic Mercury, *Occup. Environ. Med.*, 58, 461–466.

Bonkovsky, H. L., and Lambrecht, R. W. (2000). Iron-Induced Liver Injury, *Clin. Liver Dis.*, 4, 409–429.

Bonkovsky, H. L., Banner, B. F., Lambrecht, R. W., and Rubin, R. B. (1996). Iron in Diseases Other Than Hemochromatosis, *Semin. Liver Dis.*, 16, 65–82.

Bonkovsky, H. L., Jawaid, Q., Tortorelli, K. *et al.* (1999). Non-Alcoholic Steatohepatitis and Iron; Increased Prevalence of Mutations of the HFE Gene in Non-Alcoholic Steatohepatitis, *J. Hepatol.*, 31, 421–429.

Brain, J., and Valberg, P. (1979). Deposition of Aerosol in the Respiratory Tract, *Am. Rev. Respir. Dis.*, 120, 1325–1373.

Brenard, R., Laterre, P. F., Reynaert, M. *et al.* (1996). Increased Hepatocytic Mitotic Activity as a Diagnostic Marker of Acute Arsenic Intoxication, *J. Hepatol.*, 25, 218–220.

Browne K. (1986). A Threshold for Asbestos Related Lung Cancer, *Br. J. Ind. Med.*, 43, 556–558.

Browne, K. (1994). Asbestos-Related Disorders. In *Occupational Lung Disorders* (W. Parkes, Ed.), Butterworth-Heinemann, Oxford.

Cavanagh, J. B., and Nolan, C. C. (1994). The Neurotoxicity of Organolead and Organotin Compounds. In *Handbook of Clinical Neurology*, Vol. 20 (F. A. de Wolff, Ed.), Elsevier, Amsterdam, pp. 129–150.

Cavigelli, M., Li, W. W., Lin, A., Su, B., Yushioka, K., and Karin, M. (1996). The Tumor Promoter Arsenite Stimulates AP-1 Activity by Inhibiting a JNK Phosphatase, *EMBO J.*, 15, 6269–6279.

Centeno, J. A., Pestaner, J. P., Mullick, F. G., and Virmani, R. (1996). An Analytical Comparison of Cobalt Cardiomyopathy and Idiopathic Dilated Cardiomyopathy, *Biol. Trace Element Res.*, 55, 21–30.

Chan, P., and Huff, J. (1997). Arsenic Carcinogenesis in Animals and in Humans: Mechanistic, Experimental, and Epidemiological Evidence, *Environ. Carcinog. Ecotoxicol. Rev.*, 15, 83–122.

Chen, C. J., Chen, C. W., Wu, M. M., and Kuo, T. L. (1992). Cancer Potential in Liver, Lung, Bladder, and Kidney Due to Ingested Inorganic Arsenic in Drinking Water, *Br. J. Cancer*, 66, 888–892.

Choie, D. D., and Richter, G. W. (1972). Lead Poisoning: Rapid Formation of Intranuclear Inclusions, *Science*, 177(55), 1194–1195.

Chou, H. T., Yang, H. L., Tsou, S. S., Ho, R. K., Pai, P. Y., and Hsu, H. B. (1998). Status of Trace Elements in Patients with Idiopathic Dilated Cardiomyopathy in Central Taiwan, Chung Hua I Hsueh Tsa Chih (Taipei), 61, 193–198.

Chung, C., Tsai, R., Chen, G., Yu, H., and Chai, C. (1998). Expression of bcl-2, p53 and Ki-67 in Arsenical Skin Cancers, *J. Cutan. Pathol.*, 25, 457–462.

Churg, A. (1998a). Non-Neoplastic Disease Caused by Asbestos. In *Pathology of Occupational Lung Disease* (A. Churg and F. Green, Eds.), Williams & Wilkins, Baltimore, MD.

Churg, A. (1998b). Neoplastic Asbestos-Induced Disease. In *Pathology of Occupational Lung Disease* (A. Churg and F. Green, Eds.), Williams & Wilkins, Baltimore, MD.

Churg, A., and Colby, T. (1998). Diseases Caused by Metals and Related Compounds. In *Pathology of Occupational Lung Disease* (A. Churg and F. Green, Eds.), Williams & Wilkins, Baltimore, MD.

Churg, A., and Warnock, M. (1981). Asbestos and Other Ferruginous Bodies, *Am. J. Pathol.*, 102, 447–456.

Churg, A., Wright, J., and Vedal, S. (1993). Fiber Burden and Patterns of Asbestos-Related Disease in Chrysotile Miners and Millers, *Am. Rev. Respir. Dis.*, 148, 25–31.

Chuttani, H. K., Gupta, P. S., Gulati, S., and Gupta, D. N. (1965). Acute Copper Sulfate Poisoning, *Am. J. Med.*, 39, 849–854.

Craighead, J., Abraham, J., Churg, A. *et al.* (1982). The Pathology of Asbestos-Associated Diseases of the Lungs and Pleural Cavities: Diagnostic Criteria and Proposed Grading Schema (Report of the Pneumoconiosis Committee of the College of American Pathologists and the National Institute for Occupational Safety and Health), *Arch. Pathol. Lab. Med.*, 106, 544–596.

Datta, D. V., Mitra, S. K., Chhuttani, P. N., and Chakravarti, R. N. (1979). Chronic Oral Arsenic Intoxication as a Possible Aetiological Factor in Idiopathic Portal Hypertension (Non-Cirrhotic Portal Fibrosis) in India, *Gut*, 20, 378–384.

de Valk, B., and Marx, J. J. (1999). Iron, Atherosclerosis, and Ischemic Heart Disease, *Arch. Intern. Med.*, 159, 1542–1548.

Denk, R., Holzmann, H., Lange, H. J., and Greve, D. (1959). Uber Arsenspatschaden bei obduzierten Moselwinzern, *Die. Med. Welt.*, 11, 557–567.

Deugnier, Y., Turlin, B., Le Quilleue, D. *et al.* (1997). A Reappraisal of Hepatic Siderosis in Patients with End-Stage Cirrhosis: Practical Implications for the Diagnosis of Hemochromatosis, *Am. J. Surg. Pathol.*, 21, 669–675.

(1999). Diseases of Immunity. In *Pathologic Basis of Disease* (R. Cotran, V. Kumar, and T. Collins, Eds.), W. B. Saunders, Philadelphia, pp. 188–259.

Doll, R. (1993). Mortality from Lung Cancer in Asbestos Workers 1955 (classical article), *Br. J. Ind. Med.*, 50, 485–490.

Dupres, J., Mustard, J., and Uffen, R. (1984). *Report of the Royal Commission on Matters of Health and Safety Arising from the Use of Asbestos in Ontario*, Queen's Printer for Ontario, Toronto.

Elis, A., Froom, P., Ninio, A., Cahana, L., and Lishner, M. (2001). Employee Exposure to Chromium and Plasma Lipid Oxidation, *Int. J. Occup. Environ. Health*, 7, 206–208.

Elzouki, A.-N. Y., Hulterantz, R., Stäl, P. *et al.* (1995). Increased PiZ Frequency for α-1 Antitrypsin in Patients with Genetic Hemochromatosis, *Gut*, 36, 922–926.

Epler, G., McLoud, T., and Gaensler, E. (1982). Prevalence and Incidence of Benign Asbestos Pleural Effusion in a Working Population, *JAMA*, 247, 617–622.

Evans, P. J., Smith, C., Mitchinson, M. J., and Halliwell, B. (1995). Metal Ion Release From Mechanically Disrupted Human Arterial Wall. Implications for the Development of Atherosclerosis, *Free Radic. Res.*, 23, 465–469.

Espinoza, E. O., Mann, M. J., and Bleasell, B. (1995). Arsenic and Mercury in Traditional Chinese Herbal Balls (letter), *N. Engl. J. Med.*, 333, 803–804.

Falk, H., Caldwell, G. G., Ishak, K. G. *et al.* (1981). Arsenic-Related Hepatic Angiosarcoma, *Am. J. Ind. Med.*, 2, 43–50.

Fargion, S., Bissoli, F., Fracanzani, A. L. *et al.* (1996). No Association Between Genetic Hemochromatosis and α-1 Antitrypsin Deficiency, *Hepatology*, 24, 1161–1164.

Fraser, R., Muller, N., Colman, N. *et al.* (1999a). Pleural Effusion. *Fraser and Pare's Diagnosis of Diseases of the Chest*, W. B. Saunders, Philadelphia.

Fraser, R., Muller, N., Colman, N. *et al.* (1999b). Inhalation of Inorganic Dust (Pneumoconiosis), *Fraser and Pare's Diagnosis of Diseases of the Chest*, W. B. Saunders, Philadelphia.

Frustaci, A., Magnavita, N., Chimenti, C., Caldarulo, M., Sabbioni, E., Pietra, R., Cellini, C., Possati, G. F., and Maseri, A. (1999). Marked Elevation of Myocardial Trace Elements in Idiopathic Dilated Cardiomyopathy Compared with Secondary Cardiac Dysfunction, *J. Am. Coll. Cardiol.*, 33, 1578–1583.

Gandhi, R. K., and Robarts, F. H. (1962). Hourglass Stricture of the Stomach and Pyloric Stenosis Due to Ferrous Sulphate Poisoning, *Br. J. Surg.*, 49, 613–617.

Gangaidzo, I. T., and Gordeuk, V. K. (1995). Hepatocellular Carcinoma and African Iron Overload, *Gut*, 37, 727–730.

Ganne-Carrié, N., Christidis, C., Chastang, C. *et al.* (2000). Liver Iron Is Predictive of Death in Alcoholic Cirrhosis: A Multivariate Study of 229 Consecutive Patients with Alcoholic and/or Hepatitis C Virus Cirrhosis: A Prospective Followup Study, *Gut*, 46, 277–282.

Ge, K., and Yang, G. (1993). The Epidemiology of Selenium Deficiency in the Etiological Study of Endemic Diseases in China, *Am. J. Clin. Nutr.*, 57, 259S–263S.

Ge, K., Xue, A., Bai, J., and Wang, S. (1983). Keshan Disease—An Endemic Cardiomyopathy in China, *Virchows Arch. A Pathol. Anat. Histopathol.*, 401, 1–15.

George, D. K., Goldwurm, S., MacDonald, G. A. *et al.* (1998). Increased Hepatic Iron Concentration in Non-Alcoholic Steatohepatitis Is Associated with Increased Fibrosis, *Gastroenterology*, 114, 311–318.

Ghadially, F. N. (1979). Ultrastructural Localization and *in situ* Analysis of Iron, Bismuth, and Gold Inclusions, *CRC Crit. Rev. Toxicol.*, 6(4), 303–50.

Gibbs, A. (1996). Occupational Lung Disease. In *Spencer's Pathology of the Lung* (P. Hasleton, Ed.), McGraw-Hill, New York.

Gokal, R., Millard, P. R., Weatherall, D. J. *et al.* (1979). Iron Metabolism in Haemodialysis Patients, *Q. J. Med.*, 48, 369–391.

Good, P. F. *et al.* (1992a). Selective Accumulation of Aluminum and Iron in the Neurofibrillary Tangles of Alzheimer's Disease: A Laser Microprobe (LAMMA) Study, *Ann. Neurol.*, 31, 286–292.

Good, P. F., Olanow, C. W., and Perl, D. P. (1992b). Neuromelanin-Containing Neurons of the Substantia Nigra Accumulate Iron and Aluminum in Parkinson's Disease: A LAMMA Study, *Brain Res.*, 593, 343–346.

Gordeuk, V., Mukübi, S. J., Hasstedt, S. J. *et al.* (1992). Iron Overload in Africa: Interaction Between a Gene and Dietary Iron Content, *N. Engl. J. Med.*, 326, 95–100.

Gordeuk, V. R., Boyd, R. D., and Brittenham, G. M. (1986). Dietary Iron Overload in Rural Sub-Saharan Africa, *Lancet*, 1, 1310–1313.

Greenberg, S., and Roggli, V. (1992). Carcinoma of the Lung. In *Pathology of Asbestos-Associated Disease* (V. Roggli, S. Greenberg, and P. Pratt, Eds.), Little Brown & Company, Boston.

Hall, J. C., and Harruff, R. (1989). Fatal Cardiac Arrhythmia in a Patient with Interstitial Myocarditis Related to Chronic Arsenic Poisoning, *South. Med. J.*, 82, 1557–1560.

Hanada, K., Hashimoto, I., Kon, A., Kida, K., and Mita, R. (1998). Silver in Sugar Particles and Systemic Argyria, *Lancet*, 351, 960.

Haywood, S., Müller, T., Müller, W. *et al.* (2001). Copper-Associated Liver Disease in North Ronaldsay Sheep: A Possible Animal Model for Non-Wilsonian Hepatic Copper Toxicosis of Infancy and Childhood, *J. Pathol.*, 195, 264–269.

Helwig, E. (1951). Chemical (Beryllium) Granulomas of the Skin, *Mil. Surg.*, 109, 540–558.

Hennigar, G. R., Greene, W. B., Walker, E. M. *et al.* (1979). Hemochromatosis Caused by Excessive Vitamin Iron Intake, *Am. J. Pathol.*, 96, 611–624.

Hezode, C., Cazeneuve, C., Coué, O. *et al.* (1999). Liver Iron Accumulation in Patients with Chronic Active Hepatitis C: Prevalence and Role of Hemochromatosis Gene Mutations and Relationship with Hepatitis Histological Lesions, *J. Hepatol.*, 31, 979–984.

Holtzman, N. A., Elliot, D. A., and Heller, R. H. (1986). Copper Intoxication: Report of a Case with Observations on Ceruloplasmin, *N. Engl. J. Med.*, 275, 347–352.

Horslen, S. P., Tanner, M. S., Lyon, T. D. B. *et al.* (1994). Copper Associated Childhood Cirrhosis, *Gut*, 35, 1497–1500.

Hsu, C. H., Yang, S. A., Wang, J. Y., Yu, H. S., and Lin, S. R. (1999). Mutational Spectrum of p53 Gene in Arsenic-Related Skin Cancers from the Blackfoot Disease Endemic Area of Taiwan, *Br. J. Cancer*, 80, 1080–1086.

Hughes, J. M., and Weill, H. (1991). Asbestosis as a Precursor of Asbestos Related Lung Cancer: Results of a Prospective Mortality Study, *Br. J. Ind. Med.*, 48, 229–233.

Hume, L., and Roe, M. (1992). The Biodurability of Chrysotile Asbestos, *Am. Mineralogist*, 77, 1125–1128.

Hunter, D., and Russell, D. S. (1954). Focal Cerebral and Cerebellar Atrophy in a Human Subject Due to Organic Mercury Compounds, *J. Neurol. Neurosurg. Psychol.*, 17, 235–241.

Hutchinson, J. (1888). On Some Examples of Arsenic-Keratosis of the Skin and Arsenic-Cancer, *Trans. Pathol. Soc. Lond.*, 39, 352–363.

Huttunen, J. K. (1997). Selenium and Cardiovascular Diseases—An Update, *Biomed. Environ. Sci.*, 10, 220–6.

Iribarren, C., Sempos, C. T., Eckfeldt, J. H., and Folsom, A. R. (1998). Lack of Association Between Ferritin Level and Measures of LDL Oxidation: The ARIC Study. Atherosclerosis Risk in Communities, *Atherosclerosis*, 139, 189–195.

James, J. A. (1970). Acute Iron Poisoning: Assessment of Severity and Prognosis, *J. Pediatr.*, 77, 117–119.

Jensen, S., and Jernelov, A. (1969). Biological Methylation of Mercury in Aquatic Organisms, *Nature*, 223, 753–756.

Jhaveri, S. S. (1959). A Case of Cirrhosis and Primary Carcinoma of the Liver in Chronic Arsenical Intoxication, *Br. J. Ind. Med.*, 16, 248–250.

Jingbo, P. I., Hiroshi, Y., Yoshito, K., Guifan, S., Takahiko, Y., Hiroyuki, A., Claudia, H. R., and Nobuhiro, S. (2002). Evidence for Induction of Oxidative Stress Caused by Chronic Exposure of Chinese Residents to Arsenic Contained in Drinking Water, *Environ. Health Perspect.*, 110(4), 331–336.

Johnson, R. A., Baker, S. S., Fallon, J. T., Maynard, E. P., 3rd, Ruskin, J. N., Wen, Z., Ge, K., and Cohen, H. J. (1981). An Occidental Case of Cardiomyopathy and Selenium Deficiency, *N. Engl. J. Med.*, 304, 1210–1212.

Joshi, V. V. (1987). Indian Childhood Cirrhosis, *Perspect. Paediatr. Pathol.*, 11, 175–192.

Kageyama, F., Kobayashi, Y., Kawasaki, T. *et al.* (2000). Successful Interferon Therapy Reverses Enhanced Hepatic Iron Accumulation and Lipid Peroxidation in Chronic Hepatitis C, *Am. J. Gastroenterol.*, 95, 1041–1050.

Kardinaal, A. F., Kok, F. J., Kohlmeier, L., Martin-Moreno, J. M., Ringstad, J., Gomez-Aracena, J., Mazaev, V. P., Thamm, M., Martin, B. C., Aro, A., Kark, J. D., Delgado-Rodriguez, M., Riemersma, R. A., van't Veer, P., and Huttunen, J. K. (1997). Association Between Toenail Selenium and Risk of Acute Myocardial Infarction in European Men. The EURAMIC Study. European Antioxidant Myocardial Infarction and Breast Cancer, *Am. J. Epidemiol.*, 145, 373–379.

Kesteloot, H., Roelandt, J., Willems, J., Class, J. M., and Joosens, J. V. (1968). An Inquiry into the Role of Cobalt in the Heart Disease of Chronic Beer Drinkers, *Circulation*, 37, 854–864.

Kew, J., Morris, C., Aihie, A. *et al.* (1993). Arsenic and Mercury Intoxication Due to Indian Ethnic Remedies, *Br. Med. J.*, 306, 506–507.

Kiechl, S., Willeit, J., Egger, G., Poewe, W., and Oberhollenzer, F. (1997). Body Iron Stores and the Risk of Carotid Atherosclerosis: Prospective Results from the Bruneck Study, *Circulation*, 96, 3300–3307.

Kitchin, K. T. (2001). Recent Advances in Arsenic Carcinogenesis: Modes of Action, Animal Model Systems, and Methylated Arsenic Metabolites, *Toxicol. Appl. Pharmacol.*, 172(3), 249–261.

Koch, P., and Bahmer, F. A. (1999). Oral Lesions and Symptoms Related to Metals Used in Dental Restorations: A Clinical, Allergological, and Histologic Study, *J. Am. Acad. Dermatol.*, 42, 422–430.

Kopp, S. J., Barron, J. T., and Tow, J. P. (1988). Cardiovascular Actions of Lead and Relationship to Hypertension: A Review, *Environ. Health Perspect.*, 78, 91–99.

Krenzelok, E. P., and Hoff, J. V. (1979). Accidental Iron Poisoning. A Problem of Marketing and Labeling, *Pediatrics*, 63, 591–596.

Labadie, H., Stoessel, P., Callard, P., and Beaugrand, M. (1990). Hepatic Veno-Occlusive Disease and Perisinusoidal Fibrosis Secondary to Arsenic Poisoning, *Gastroenterology*, 99, 1140–1143.

Lansdown, A. (1995). Physiological and Toxicological Changes in the Skin Resulting from the Action and Interaction of Metal Ions, *Crit. Rev. Toxicol.*, 25, 397–462.

Lefkowitch, J. H., Honig, C. L., King, M. E., and Hagstrom, J. W. C. (1982). Hepatic Copper Overload and Features of Indian Childhood Cirrhosis in an American Sibship, *N. Engl. J. Med.*, 307, 271–277.

Li, G. S., Wang, F., Kang, D., and Li, C. (1985). Keshan Disease: An Endemic Cardiomyopathy in China, *Hum. Pathol.*, 16, 602–609.

Li, J. H., and Rossman, T. C. (1989). Inhibition of DNA Ligase Activity by Arsenite: A Possible Mechanism of Its Comutagenesis, *Mol. Toxicol.*, 2, 1–9.

Li, Y., and Nan, B. S. (1989). Correlation of Selenium, Glutathione Peroxidase Activity and Lipoperoxidation Rates in Dilated Cardiomyopathy, *Chin. Med. J.*, 102, 670–671.

Ludwig, J., Farr, G. H., Freese, D. K., and Sternlieb, I. (1995). Chronic Hepatitis and Hepatic Failure in a 14-Year-Old Girl, *Hepatology*, 22, 1874–1879.

Ludwig, J., Hashimoto E., Poroyko, M. D. *et al.* (1997). Hemosiderosis in Cirrhosis: A Study of 447 Native Livers, *Gastroenterology*, 112, 888–890.

Lunchtrath, H. (1972). Cirrhosis of the Liver in Chronic Arsenical Poisoning of Vintners, *Ger. Med. Mon.*, 2, 127–128.

Maggiore, G., De Giacomo, C., Sessa, F., and Burgio, G. R. (1987). Idiopathic Hepatic Copper Toxicosis in a Child, *J. Pediatr. Gastroenterol. Nutr.*, 6, 980–983.

Maheswaran, R., Morris, S., Falconer, S., Grossinho, A., Perry, I., Wakefield, J., and Elliott, P. (1999). Magnesium in Drinking Water Supplies and Mortality from Acute Myocardial Infarction in North West England, *Heart*, 82, 455–460.

Mahon, N. G., Coonar, A. S., Jeffery, S., Coccolo, F., Akiyu, J., Zal, B., Houlston, R., Levin, G. E., Baboonian, C., and McKenna, W. J. (2000). Haemochromatosis Gene Mutations in Idiopathic Dilated Cardiomyopathy, *Heart*, 84, 541–547.

Maloney, M. (1996). Arsenic in Dermatology. *Dermatol. Surg.*, 22, 301–304.

Mandishona, E., MacPhail, A. P., Gordeuk, V. R. *et al.* (1998). Dietary Iron Overload as a Risk Factor for Hepatocellular Carcinoma in Black Africans, *Hepatology*, 27, 1563–1566.

Markesbery, W. R. (1997). Oxidative Stress Hypothesis in Alzheimer's Disease, *Free Radic. Biol. Med.*, 23, 134–147.

Marsh, D. (1979). Organic Mercury: Methylmercury Compounds. In *Handbook of Clinical Neurology*, Vol. 36 (P. J. Vinken and G. W. Bryun, Eds.), North-Holland, Amsterdam, pp. 73–81.

Mazumder, D. N. G., Chakraborty, A. K., Ghosh, A. *et al.* (1988). Chronic Arsenic Toxicity from Drinking Tube Well Water in Rural West Bengal, *Bull. WHO*, 66, 499–506.

McDonald, J. C. (1980). Asbestos and Lung Cancer: Has the Case Been Proven?, *Chest*, 78, 374–376.

McDonald, A., and McDonald, J. (1987). Epidemiology of Malignant Mesothelioma, In *Asbestos-Related Malignancy* (K. Antman and J. Aisner, Eds.), Harcourt Brace Jovanovich, Orlando, FL.

Mena, I. (1979). Manganese Poisoning. In *Handbook of Clinical Neurology*, Vol. 36 (P. J. Vinken and G. W. Bryun, Eds.), North-Holland, Amsterdam, pp. 217–237.

Mena, I. *et al.* (1967). Chronic Manganese Poisoning: Clinical Picture and Manganese Turnover, *Neurology*, 17, 128–136.

Mendis, S. (1989). Magnesium, Zinc, and Manganese in Atherosclerosis of the Aorta, *Biol. Trace Element Res.*, 22, 251–256.

Meyers, D. G. (1996). The Iron Hypothesis—Does Iron Cause Atherosclerosis?, *Clin. Cardiol.*, 19, 925–929.

Modell, C. B. (1975). Transfusional haemochromatosis. In *Iron Metabolism and Its Disorders* (H. Kief, Ed.), Excerpta Medica, Amsterdam, pp. 230–239.

Monoguerra, A. S. (1976). Iron Poisoning: Report of a Fatal Case in an Adult, *Am. J. Hosp. Pharmacol.*, 33, 1088–1090.

Morgan, A. (1980). Effect of Length on the Clearance of Fibers from the Lung and on Body Formation. In *Biologic Effects of Mineral Fibers* (J. Wagner, Ed.), International Agency for Research on Cancer, Lyon.

Morgan, A., and Holmes, A. (1980). Concentrations and Dimensions of Coated and Uncoated Asbestos Fibres in the Human Lung, *Br. J. Ind. Med.*, 37, 25–32.

Morgan, A., and Holmes, A. (1985). The Enigmatic Asbestos Body: Its Formation and Significance in Asbestos-Related Disease, *Environ. Res.*, 38, 283–292.

Morgan, W., and Seaton, A. (1975). *Occupational Lung Diseases*, W. B. Saunders, Philadelphia.

Muehrcke, R. C., and Pirani, C. L. (1968). Arsine-Induced Anuria. A Correlative Clinicopathological Study with Electron Microscopic Observations, *Ann. Intern. Med.*, 68(4), 853–866.

Müller, T., Feichinger, H., Berger, H., and Muller, W. (1996). Endemic Tyrolean Infantile Cirrhosis: An Ecogenetic Disease Disorder, *Lancet*, 347, 877–880.

Müller-Höcker, J., Meyer, U., Wiebecke, B., and Hübner, G. (1985). Copper Storage Disease of the Liver and Chronic Dietary Copper Intoxication in Two Further German Infants Mimicking Indian Childhood Cirrhosis, *Pathol. Res. Pract.*, 183, 39–45.

Müller-Höcker, J., Weib, M., Meyer, U. *et al.* (1987). Fatal Copper Storage Disease of the Liver in a German Infant Resembling Indian Childhood Cirrhosis, *Virchows Arch. (A)*, 411, 379–385.

Murphy, G. (1995). Skin. In *Pathology of Environmental and Occupational Disease* (J. Craighead, Ed.), Mosby, St. Louis, MO, pp. 437–453.

Murphy, M., Hunt, S., McDonald, G. S. A. *et al.* (1991). Intrahepatic Cholestasis Secondary to Gold Therapy, *Eur. J. Gastroenterol. Hepatol.*, 3, 855–859.

Nakamuro, K., and Sayato, Y. (1981). Comparative Studies of Chromosomal Aberration Induced by Trivalent and Pentavalent Arsenic, *Mutat. Res.*, 8873–8880.

Narang, R., Ridout, D., Nonis, C., and Kooner, J. S. (1997). Serum Calcium, Phosphorus and Albumin Levels in Relation to the Angiographic Severity of Coronary Artery Disease, *Int. J. Cardiol.*, 60, 73–79.

Navarro-Alarcon, M., Lopez-Garcia de la Serrana, H., Perez-Valero, V., and Lopez-Martinez, C. (1999). Serum and Urine Selenium Concentrations in Patients with Cardiovascular Diseases and Relationship to Other Nutritional Indexes, *Ann. Nutr. Metab.*, 43, 30–36.

Needleman, H. L. (1988). The Persistent Threat of Lead: Medical and Sociological Issues, *Curr. Probl. Pediatr.*, 18, 697–744.

Needleman, H. L., Gunnoe, C., and Leviton, A. (1979). Deficits in Psychologic and Classroom Performance in

Children with Elevated Dentine Lead Levels, *N. Engl. J. Med.*, 300, 689–695.

Needleman, H. L., Schell, A., and Bellinger, D. (1990). The Long-Term Effects of Exposure to Low Doses of Lead in Childhood. An 11th Year Follow-Up Report, *N. Engl. J. Med.*, 322, 83–88.

Neve, J. (1996). Selenium as a Risk Factor for Cardiovascular Diseases, *J. Cardiovasc. Risk*, 3, 42–47.

Newland, M. C. *et al.* (1989). Visualizing Manganese in the Primate Basal Ganglia with Magnetic Resonance Imaging, *Eur. Neurol.*, 11106, 251–258.

Nightingale, S. L. (1997). Action To Prevent Accidental Iron Poisonings in Children, *JAMA*, 277, 1343.

Olanow, C. W. *et al.* (1994). Manganese-Induced Neurotoxicity. In *Advances in Research on Neurodegeneration*. Vol. II. *Etiopathogenesis* (Y. Mizuno, D. B. Calne, and R. Horowski, Eds.), Birkhauser, Boston, pp. 53–62.

Olanow, C. W. *et al.* (1996). Manganese Intoxication in the Rhesus Monkey: A Clinical, Imaging, Pathologic, and Biochemical Study, *Neurology*, 46, 492–498.

Oliver, R. A. M. (1959). Siderosis Following Transfusions of Blood, *J. Pathol. Bacteriol.*, 77, 171–194.

O'Neill, N. C., and Tanner, M. S. (1989). Uptake of Copper from Brass Vessels in Bovine Milk and Its Relevance to Indian Childhood Cirrhosis, *J. Pediatr. Gastroenterol. Nutr.*, 9, 167–172.

Oster, O. (1993). Trace Element Concentrations (Cu, Zn, Fe) in Sera from Patients with Dilated Cardiomyopathy, *Clin. Chim. Acta*, 214, 209–218.

Oster, O., and Prellwitz, W. (1990). Selenium and Cardiovascular Disease, *Biol. Trace Element Res.*, 24, 91–103.

Patrizi, A., Rizzoli, L., Vincenzi, C., Trevisi, P., and Tosti, A. (1999). Sensitization to Thimerosal in Atopic Children, *Contact Derm.*, 40, 94–97.

Perl, D. P., and Brody, A. R. (1980). Alzheimer's Disease: X-Ray Spectrographic Evidence of Aluminum Accumulation in Neurofibrillary Tangle-Bearing Neurons, *Science*, 208, 297–299.

Perl, D. P., and Good, P. F. (1992). Aluminum and the Neurofibrillary Tangle: Results of Tissue Microprobe Studies. In *Ciba Symposium: Aluminum in Biology and Medicine* (R. M. P. Williams, Ed.), John Wiley & Sons, New York, pp. 217–236.

Pestaner, J. P., Ishak, K. G., Mullick, F. G., and Centeno, J. (1999). Ferrous Sulfate Toxicity. A Review of Autopsy Findings, *Biol. Trace Element Res.*, 70, 1–8.

Pimental, J. C., and Menezes, A. P. (1977). Liver Disease in Vineyard Sprayers, *Gastroenterology*, 72, 275–289.

Pirisi, M., Scott, C. A., Avellini, C. *et al.* (2000). Iron Deposition and Progression of Disease in Chronic Hepatitis C. Role of Interface Hepatitis, Portal Inflammation, and HFE Missense Mutations, *Am. J. Clin. Pathol.*, 113, 546–554.

Raines, D. A., Kinsara, A. J., Eid Fawzy, M., Vasudevan, S., Mohamed, G. E., Legayada, E. S., Al-Rawithi, S., and El-Yazigi, A. (1999). Plasma and Urinary Selenium in Saudi Arabian Patients with Dilated Cardiomyopathy, *Biol. Trace Element Res.*, 69, 59–68.

Ramirez, P., Eastmond, D. A., Laclette, J. P., and Ostrosky-Wegman, P. (1997). Disruption of Microtubule Assembly and Spindle Formation for the Induction of Aneuploid Cells by Sodium Arsenite and Vanadium Pentoxide, *Mutat. Res.*, 386, 291–298.

Rauramaa, R., Vaisanen, S., Mercuri, M., Rankinen, T., Penttila, I., and Bond, M. G. (1994). Association of Risk Factors and Body Iron Status to Carotid Atherosclerosis in Middle-Aged Eastern Finnish Men, *Eur. Heart J.*, 15, 1020–1027.

Rayman, M. P. (2000). The Importance of Selenium to Human Health, *Lancet*, 356, 233–241.

Reusche, E., Koch, V., Friedrich, H. J., Nunninghoff, D., Stein, P., and Rob, P. M. (1996). Correlation of Drug-Related Aluminum Intake and Dialysis Treatment with Deposition of Argyrophilic Aluminum-Containing Inclusions in CNS and in Organ Systems of Patients with Dialysis-Associated Encephalopathy, *Clin. Neuropathol.*, 15, 342–347.

Riggio, O., Montagnese, F., Fiore, P. *et al.* (1997). Iron Overload in Patients with Chronic Viral Hepatitis: How Common Is It?, *Am. J. Gastroenterol.*, 92, 1298–1301.

Roberts, G. (1971). The Pathology of Parietal Pleural Plaques, *J. Clin. Pathol.*, 24, 348–353.

Roggli, V. (1989). Pathology of Human Asbestosis: A Critical Review. In *Advances in Pathology* (C. Fenoglio-Preiser, Ed.), Year Book Medical Publishers, Chicago.

Roggli, V. (1992). Asbestos Bodies and Nonasbestos Ferruginous Bodies. In *Pathology of Asbestos-Related Disease* (V. Roggli, S. Greenberg, and P. Pratt, Eds.), Little Brown & Company, Boston.

Roggli, V. (1994). The Pneumoconioses: Asbestosis. In *Pathology of Pulmonary Disease* (M. Saldana, Ed.), J. B. Lippincott, Philadelphia.

Roggli, V., and Pratt, P. (1983). Numbers of Asbestos Bodies on Iron-Stained Sections in Relation to Asbestos Body Counts in Lung Tissue Digests, *Hum. Pathol.*, 14, 355–361.

Roggli, V., and Pratt, P. (1992). Asbestosis. In *Pathology of Asbestos-Associated Diseases* (V. Roggli, S. Greenberg, and P. Pratt, Eds.), Little Brown & Company, Boston.

Roggli, V., and Shelburne, J. (1994). Pneumoconioses, Mineral and Vegetable. In *Pulmonary Pathology* (D. Dail and S. Hammar, Eds.), Springer-Verlag, New York.

Rosenberg, H. G. (1974). Systemic Arterial Disease and Chronic Arsenicism in Infants, *Arch. Pathol.*, 97, 360–365.

Rubenowitz, E., Axelsson, G., and Rylander, R. (1999). Magnesium and Calcium in Drinking Water and Death from

Acute Myocardial Infarction in Women, *Epidemiology*, 10, 31–36.

Russell, M., Langley, M., Truett, A., King, L., and Boyd, A. (1997). Lichenoid Dermatitis After Consumption of Gold-Containing Liquor, *J. Am. Acad. Dermatol.*, 36, 841–844.

Salonen, J. T., Alfthan, G., Huttunen, J. K., Pikkarainen, J., and Puska, P. (1982). Association Between Cardiovascular Death and Myocardial Infarction and Serum Selenium in a Matched-Pair Longitudinal Study, *Lancet*, 2, 175–179.

Salonen, J. T., Salonen, R., Seppanen, K., Kantola, M., Parviainen, M., Alfthan, G., Maenpaa, P. H., Taskinen, E., and Rauramaa, R. (1988). Relationship of Serum Selenium and Antioxidants to Plasma Lipoproteins, Platelet Aggregability and Prevalent Ischaemic Heart Disease in Eastern Finnish Men, *Atherosclerosis*, 70, 155–160.

Salonen, J. T., Seppanen, K., Lakka, T. A., Salonen, R., and Kaplan, G. A. (2000). Mercury Accumulation and Accelerated Progression of Carotid Atherosclerosis: A Population-Based Prospective 4-Year Follow-Up Study in Men in Eastern Finland, *Atherosclerosis*, 148, 265–273.

Salvini, S., Hennekens, C. H., Morris, J. S., Willett, W. C., and Stampfer, M. J. (1995). Plasma Levels of the Antioxidant Selenium and Risk of Myocardial Infarction Among U.S. Physicians, *Am. J. Cardiol.*, 76, 1218–1221.

Sanchez-Fructuoso, A. I., Torralbo, A., Arroyo, M., Luque, M., Ruilope, L. M., Santos, J. L., Cruceyra, A., and Barrientos, A. (1996). Occult Lead Intoxication as a Cause of Hypertension and Renal Failure, *Nephrol. Dial. Transplant.*, 11, 1775–1780.

Schafer, A. I., Cheron, R. G., Dluchy, R. *et al.* (1981). Clinical Consequences of Acquired Transfusional Iron Overload in Adults, *N. Engl. J. Med.*, 304, 319–324.

Scheinberg, I. H., and Sternlieb, I. (1994). Is Non-Indian Childhood Cirrhosis Caused by Excess Dietary Copper?, *Lancet*, 344, 1002–1004.

Schoolmeester, W. L., and White, D. R. (1980). Arsenic Poisoning. *South. Med. J.*, 73, 198.

Schramel, P., Müller-Höcker, J., Meyer, U. *et al.* (1988). Nutritional Copper Intoxication in Three Carcinoma Infants with Severe Liver Cell Damage (Features of Indian Childhood Cirrhosis), *J. Trace Element Electrolyte Health. Dis.*, 2, 85–89.

Schumann, K. (2001). Safety Aspects of Iron in Food, *Ann. Nutr. Metab.*, 45, 91–101.

Searle, J., Leggett, B. A., Crawford, D. H. G., and Powell, L. W. (2001). Iron Storage Disease. In *Pathology of the Liver* (R. N. M. MacSween, A. D. Burt, B. C. Portmann, K. G. Ishak, P. J. Scheuer, and P. P. Anthony, Eds.), Churchill Livingstone, London, pp. 257–272.

Selikoff, I. J., Hammond, E. C., and Seidman, H. (1980). Latency of Asbestos Disease Among Insulation Workers in the United States and Canada, *Cancer*, 46, 2736–2740.

Sempos, C. T., Looker, A. C., Gillum, R. E., McGee, D. L., Vuong, C. V., and Johnson, C. L. (2000). Serum Ferritin and Death from All Causes and Cardiovascular Disease: The NHANES II Mortality Study. National Health and Nutrition Examination Study, *Ann. Epidemiol.*, 10, 441–448.

Shiono, Y., Wakusawa, S., Hayashi, H. *et al.* (2001). Iron Accumulation in the Liver of Male Patients with Wilson's Disease, *Am. J. Gastroenterol.*, 96, 3147–3151.

Shiraki, H. (1979). Neuropathological Aspects of Organic Mercury Intoxication, Including Minamata Disease. In *Handbook of Clinical Neurology*, Vol. 36 (P. J. Vinken and G. W. Bryun, Eds.), North-Holland, Amsterdam, 83–145.

Simeonova, P. P., and Luster, M. I. (2000). Mechanisms of Arsenic Carcinogenicity: Genetic or Epigenetic Mechanisms, *J. Environ. Pathol. Toxicol. Oncol.*, 19, 281–286.

Sluis-Cremer, G. K., and Bezuidenhout, B. N. (1989). Relation Between Asbestosis and Bronchial Cancer in Amphibole Asbestos Miners (see comments), *Br. J. Ind. Med.*, 46, 537–540.

Smetana, H. F., Hadley, G. G., and Sirsat, S. M. (1961). Infantile Cirrhosis. An Analytic Review of the Literature and a Report of 50 Cases, *Pediatrics*, 28, 107–127.

Sokas, R. K., Simmens, S., Sophar, K., Welch, L. S., and Liziewski, T. (1997). Lead Levels in Maryland Construction Workers, *Am. J. Ind. Med.*, 31, 188–194.

Spitainy, K. C., Brondum, J., Vogt, R. L. *et al.* (1984). Drinking-Water-Induced Copper Intoxication in a Vermont Family, *Pediatrics*, 74, 1103–1106.

Staessen, J. A., Bulpitt, C. J., Fagard, R., Lauwerys, R. R., Roels, H., Thijs, L., and Amery, A. (1994). Hypertension Caused by Low-Level Lead Exposure: Myth or Fact?, *J. Cardiovasc. Risk*, 1, 87–97.

Staessen, J. A., Roels, H., and Fagard, R. (1996). Lead Exposure and Conventional and Ambulatory Blood Pressure: A Prospective Population Study. PheeCad Investigators, *J. Am. Med. Assoc.*, 275, 1563–1570.

Styblo, M., Drobna, Z., Jaspers, I., Lin, S., and Thomas, D. J. (2002). The Role of Biomethylation in Toxicity and Carcinogenicity of Arsenic: A Research Update, *Environ. Health Perspect.*, 110(5), 767–771.

Suadicani, P., Hein, H. O., and Gyntelberg, F. (1992). Serum Selenium Concentration and Risk of Ischaemic Heart Disease in a Prospective Cohort Study of 3000 Males, *Atherosclerosis*, 96, 33–42.

Tanner, M. S., Bhave, S. A., Prodham, A. M., and Pandit, A. N. (1987). Clinical Trials of Penicillamine in Indian Childhood Cirrhosis, *Arch. Dis. Child.*, 62, 118–1124.

Tanner, M. S., Kantarjian, A. H., Bhave, S. A., and Pandit, A. N. (1983). Early Introduction of Copper-Contaminated Animal Milk Feed as a Possible Cause of Indian Childhood Cirrhosis, *Lancet*, 2, 992–995.

Tchounwou, P. B., Wilson, B. A., Ishaque, A., and Schneider, J. (2001). Atrazine Potentiation of Arsenic Trioxide-Induced Cytotoxicity and Gene Expression in Human Liver Carcinoma Cells (HepG2), *Mol. Cell. Biochem.*, 222, 49–259.

Tchounwou, P. B., Wilson, B. A., Schneider, J., and Ishaque, A. (2000). Cytogenetic Assessment of Arsenic Trioxide Toxicity in the Mutatox, Ames II, and CAT-Tox Assays, *Metal Ions Biol. Med.*, 6, 89–91.

Timbrell, V., Pooley, F., and Wagner, J. (1970). Characteristics of Respirable Asbestos Fibers. In *Pneumoconioses: Proceedings of the International Conference, Johannesburg* (H. Shaprio, Ed.), Oxford University Press, London.

Tsai, S., Wang, T., and Ko, Y. (1999). Mortality for Certain Diseases in Areas with High Levels of Arsenic in Drinking Water, *Arch. Environ. Health*, 54, 186–193.

Turlin, B., Juguet, F., Moirand R. *et al.* (1995). Increased Liver Iron Stores in Patients with Hepatocellular Carcinoma Developed on a Noncirrhotic Liver, *Hepatology*, 22, 446–450.

Van Glahn, W. C., Flinn, F. B., and Keim, W. F. (1938). Effect of Certain Arsenates on the Liver, *Arch. Pathol.*, 25, 488–505.

Van Heck, E., Kint, A., and Temmerman, L. (1981). A Lichenoid Eruption Induced by Penicillamine, *Arch. Dermatol.*, 117, 676.

Vijaya, J., Subramanyam, G., Sukhaveni, V., Abdul Latheef, S. A., Gupta, S. R., Sadhasivaiah, G., and Salam, N. M. (2000). Selenium Levels in Dilated Cardiomyopathy, *J. Indian Med. Assoc.*, 98, 166–169.

Virtamo, J., Valkeila, E., Alfthan, G., Punsar, S., Huttunen, J. K., and Karvonen, M. J. (1985). Serum Selenium and the Risk of Coronary Heart Disease and Stroke, *Am. J. Epidemiol.*, 122, 276–282.

Wagner, J., Newhouse, M., Corrin, B. *et al.* (1988). Correlation Between Fibre Content of the Lung and Disease in East London Factory Workers, *Br. J. Indust. Med.*, 45, 305–308.

Wei, M., Wanibuchi, H., Morimura, K., Iwai, S., Yoshida, K., Endo, G., Nakae, D., and Fukushima, S. (2002). Carcinogenicity of Dimethylarsinic Acid in Male F344 Rats and Genetic Alterations in Induced Urinary Bladder Tumors, *Carcinogenesis*, 23(8), 1387–1397.

Westlin, W. F. (1966). Deferoxamine in the Treatment of Acute Iron Poisoning. Clinical Experiences with 172 Children, *Clin. Pediatr.*, 5, 531.

Wright, J., and Churg, A. (1998). Diseases Caused by Gases and Fumes. In *Pathology of Occupational Lung Disease* (A. Churg and F. Green, Eds.), Williams & Wilkins, Baltimore, MD.

Yang, F. Y., Lin, Z. H., Li, S. G., Guo, B. Q., and Yin, Y. S. (1988). Keshan Disease—An Endemic Mitochondrial Cardiomyopathy in China, *J. Trace Element Electrolytes Health Dis.*, 2, 157–163.

Yegin, A., Yegin, H., Aliciguzel, Y., Deger, N., and Semiz, E. (1997). Erythrocyte Selenium–Glutathione Peroxidase Activity Is Lower in Patients with Coronary Atherosclerosis, *Jpn. Heart J.*, 38, 793–798.

Yelin, G., Taff, M. L., and Sadowski, G. E. (1987). Copper Toxicity Following Massive Ingestion of Coins, *Am. J. Forensic Med. Pathol.*, 8, 78–85.

Younossi, Z. M., Gramlich, T., Bacon, B. R. *et al.* (1999). Hepatic Iron and Nonalcoholic Fatty Liver Disease, *Hepatology*, 30, 847–850.

Zenker, F. (1867). Iron Lung: Sclerosis Pulmonum, *Dtsch. Arch. Klin. Med.*, p. 116.

Zielhuis, R. (1977). *Public Health Risks of Asbestosis*, Pergamon Press, Elmsford, NY.

CHAPTER 24

TOXICOLOGY

TEE L. GUIDOTTI
The George Washington University

CONTENTS

Geosciences and chemistry and the scientific discipline of toxicology itself have been on parallel, often intertwined paths for many years. Issues related to toxic substances from natural sources, such as arsenic, lead, and other metals, and from the contamination of soil and groundwater, have been recognized from the historical beginnings of the discipline. The Society of Toxicology groups areas of activity in toxicologic research and the biomedical fields with which they interact as seen in Table I. Other categories of specialties within toxicology exist and are equally valid.

Toxicology has a long and colorful history. Initially, it was developed as a forensic science. Later it became a subdiscipline of pharmacology as the mechanisms of drug effects (many of the drugs were derived from classical toxins) were elucidated. From its early preoccupation with particularly toxic chemicals, from which it gained its essential definition as the science of poisons, toxicology has expanded its scope to include biological mechanisms of toxicity and host defenses (or resistance) against toxicity. In the 20th century, momentum for its development as an independent discipline has come (in roughly historical order) from food safety, chemical warfare, defense, product safety (especially cosmetics and food additives but also industrial chemicals), radiation biology, pesticide research, concern for environmental quality, environmental medicine, recent refinements in methodology of epidemiology and risk assessment, materials science and biocompatibility, molecular genetics and carcinogenesis research, and immunology. Toxicology has become highly specialized in the area of risk assessment, which identifies the level of hazard peculiar to a particular chemical exposure and the limits of acceptably safe exposure. These issues go far beyond characterizing the effects of poisons, because most of the chemicals of modern concern are not classically poisons in the sense of being potentially lethal at low doses.

For convenience in terminology, all substances not normally present in the body and introduced from outside are referred to as "xenobiotics" (from the Greek xeno-, meaning foreign). Xenobiotics may be drugs, food constituents, natural chemical exposures, or anthropogenic environmental chemical exposures. Because the delineation of safe levels of exposures assumes a socially determined level of acceptable risk (implicit in the definition of safety), toxicology has been adapted in the form of risk assessment to provide guidance to regulatory bodies.

TABLE I.   Scope of Toxicology

Field of specialization in toxicology	Interdisciplinary with
Clinical toxicology	
Drug adverse effects	Medicine, pharmacy, pediatrics, psychiatry
Drug abuse	Emergency medicine, pharmacy, forensic medicine, sports medicine
Natural products (venoms, toxins)	Pharmacology, pharmacy, pharmacognosy, emergency medicine
Suicide or accident prevention	Forensic medicine, pathology
Environmental and environmental toxicology	
Environmental toxicology	Medical geology, environmental medicine, epidemiology, agriculture, forestry
Environmental (media) toxicology	Medical geology, environmental health, epidemiology, agriculture, forestry
Risk assessment	Medical geology, political science, economics, law, public policy, epidemiology
Exposure assessment and biological monitoring	Medical geology, industrial hygiene, epidemiology
New product testing and product safety	Chemical engineering, cosmetology, food science, business, law, genetics, consumer protection, pharmaceuticals, agriculture, forestry
Basic toxicology	
Toxicokinetics	Pharmacology
Metabolism of xenobiotics	Pharmacology
Toxicodynamics	Biochemistry and molecular biology, carcinogenesis
Dermatotoxicology, ocular toxicology	Dermatology, ophthalmology, cosmetic
Target organ toxicology	(Following terminology of the Society of Toxicology)
Inhalation toxicology	Pulmonary medicine
Hematotoxicology	Hematology
Hepatotoxicity	Liver disease (hepatology)
Neurotoxicology	Neurology
Renal toxicology	Nephrology
Immunotoxicology	Immunology
Reproductive toxicology	Reproductive health
Carcinogenicity and genotoxicity	Oncology

Toxicology obviously plays a central role in medical geology. The scientific principles of toxicology are applied to medical geology in three broad areas: clinical toxicology, risk assessment, and hazard control and monitoring. Clinical toxicology is the recognition, diagnosis, and management of human toxicity, and in environmental medicine it reflects the outcome of environmental chemical exposures. Risk assessment, as the term is used here, is the identification and characterization of the level of risks resulting from exposure to hazards, including the uncertainties. Toxicology plays an essential role in risk assessment both in characterizing the potential toxicity of a chemical hazard, the first step in the process, and in providing the conceptual framework upon which quantitative risk assessment is based. The application of toxicology to risk assessment is particularly evident in the background to public policy and regulatory decisions.

Risk management is the general term for how the effects of these toxic substances are reduced. Hazard control is a term for measures that isolate, mitigate, or remove the toxic agent, and requires an understanding of the physiochemical characteristics of the chemical hazard. Here toxicology provides the essential information needed to design a control system and to set priorities for control.

In this chapter, the basic principles of toxicology will be briefly presented followed by a general framework for clinical toxicology and a general framework for toxicology as applied to risk assessment and to hazard control. The science of toxicology can be divided into toxicokinetics, the study and description of how xenobiotics enter and are handled by the body, and toxicodynamics, the study and description of what the xenobiotic does to the body (see also Chapters 8, 21, and 22, this volume).

I. TOXICOKINETICS

Regardless of their effect or origin, the behavior of xenobiotics in the body can be described by general terms and models reflecting the mechanisms by which exposure occurs and the body handles the chemical. From the standpoint of evolutionary biology, it is supposed that these mechanisms developed in response to selection pressures reflecting either of two biological needs: to detoxify and excrete harmful substances ingested in foods (especially in spoiled or putrefied foodstuffs) and to metabolize endogenous chemical compounds (such as steroid hormones).

Toxicokinetics is the toxicological analogy to pharmacokinetics and is based on identical concepts. It is therefore often useful to think of the disposition and metabolism of common drugs in thinking through the behavior of a toxic chemical or other xenobiotics. Four terms describe the disposition of xenobiotics: absorption, distribution, metabolism, and excretion. Modeled together, the terms describe the entry, local and overall accumulation, transformation, and removal from the body of the xenobiotic. Because tissue levels depend on transport of the xenobiotic to the target organ and the degree to which the xenobiotic partitions or is sequestered into the tissue, the kinetics of the xenobiotics determines the presentation of the xenobiotic to the target organ at the receptor level, where the toxic effect occurs. Figure 1 is an illustration of the principles of toxicokinetics.

A. Absorption

Xenobiotics may enter the body through any of several "portals" or routes of entry. In workplace situations, by far the most common opportunities for exposure are skin contact and breathing in the agent. In environmental medicine, the most significant portals of entry are therefore absorption through skin and inhalation. Ingestion, resulting from eating or placing objects such as cigarettes in the mouth in a situation where the object or the hands may have been contaminated, or in suicide attempts, is not a common problem in environmental medicine but appears from time to time. Splashes into the eyes are more often associated with local eye irritation and only rarely with absorption and systemic toxicity. Other routes of exposure, such as intravenous infusion or implantation of soluble agents, are artificial and seldom seen outside of medical care and experimental studies.

The toxicity of the xenobiotic may or may not involve the organ of first contact or site of entry. For example, carbon monoxide enters the body by the inhalation route but causes no toxicity to the lung. Other chemicals may cause local toxicity without significant absorption into the body, such as strong irritants applied to the skin. These routes of entry are not mutually exclusive. Inhalation of poorly soluble dusts such as silica, for example, may result in ingestion of the same material because of clearance from the lung bringing the material up the mucociliary escalator where it is swallowed or expectorated.

The rate at which a xenobiotic enters the bloodstream is determined by absorption across the barrier presented by the given route of exposure. Absorption of xenobiotics across membranes is determined for the most part by the chemical and physical properties of the agent. In general, lipid-soluble (lipophilic) substances are absorbed more readily than water-soluble substances across barriers such as skin. The rate of absorption is the most important determinant of the peak levels that will be reached in plasma. For many toxic substances, this is the prime determinant of acute toxicity.

The skin is sufficiently permeable to be a major route of entry of many chemicals into the body, particularly those that are readily lipid-soluble. Absorption across the skin is highly variable, depending on skin characteristics and the solubility of the xenobiotic in fat. Most transdermal absorption occurs directly across the superficial layers of the skin such as the stratum corneum, which consists of nonliving, keratinized cells, and the other living cell layers of the epidermis, where it is absorbed in the capillary bed of the dermis. Some chemicals applied to the skin may gain entry through a shortcut by passing more rapidly through hair follicles and sebaceous gland ducts. When the skin is injured with open wounds or abrasions, or in the presence of a skin rash, absorption across the skin is much faster. Transcutaneous absorption is generally a problem in the toxicology of pesticides, solvents, and halogenated hydrocarbons. Some agents may be significantly metabolized by enzyme systems in the skin, but most gain entry into the bloodstream unchanged.

Exposure by inhalation is relatively efficient absorption and the lung itself is vulnerable to damage from inhaled xenobiotics. The lungs are the organ of gas exchange and are in the circulation just before the heart. The organ receives venous blood from the body, oxygenates it, and returns it to the heart which pumps it out via arteries. Thus, blood reaching the lungs is initially low in oxygen and consists of mixed blood from

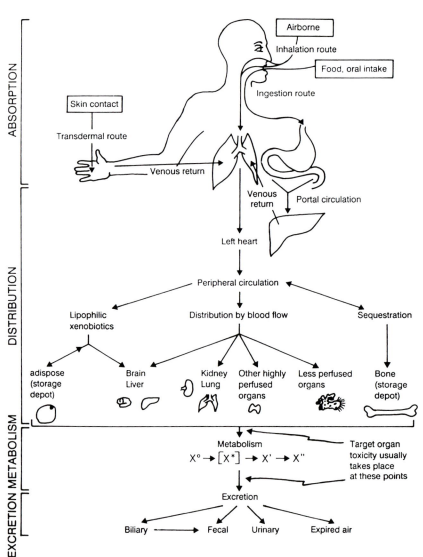

FIGURE 1   Principles of toxicokinetics: absorption, distribution, metabolism, and excretion. (Reproduced with permission from Guidotti, 1994.)

many tissues, but the oxygen tension in lung tissue itself is very high, which makes the organ susceptible to inflammatory chemicals derived from oxygen. Exposure by inhalation results in relatively efficient absorption of gases if the gas can penetrate the alveolar-capillary bed. Whether the gas will penetrate efficiently depends on its solubility in water, which reflects clearance rates in the bronchial tree. Once having penetrated the alveolar level, however, gases are readily absorbed across the alveolar-capillary bed by simple passive diffusion. Absorption across the alveolar membrane in the lung is usually very efficient and complete entry into the bloodstream is limited only by the xenobiotic's solubility in

plasma, which is an aqueous medium. Particles, on the other hand, are subject to a number of host defense mechanisms in the respiratory tract, which limit the efficiency of penetration to the alveolar level. Once at the alveolar level, their size prevents them from passing directly into the bloodstream and they must dissolve or be digested by macrophages before their constituent chemical contents can be absorbed and enter the bloodstream. Particles may contribute to systemic toxicity if they are composed of a soluble material, such as lead or polycyclic aromatic hydrocarbons. For this reason, inhalation of toxic gases is usually associated with acute systemic toxicity or vascular injury to the lung (resulting in pulmonary edema), but particle deposition in the lung is usually associated with localized pulmonary effects and chronic systemic toxicity.

Ingestion is an important route of exposure for water and food and sometimes soil. Absorption through the gastrointestinal (GI) tract for many organic compounds depends on pH (because passage is increased when they are in a non-ionized state) and therefore on location in the GI tract: the stomach is acid and the small intestine is basic. There are specialized transport mechanisms in the GI tract. Among them is facilitated diffusion to absorb glucose and a divalent-metal ion transporter that increases absorption of metals such as calcium and iron, as well as electrochemically similar ions. The GI route of exposure is unique in another important respect. Absorbed xenobiotics do not pass directly into the systemic circulation, as they do by transcutaneous and inhalation exposure, to be returned to the heart (via the lungs). Rather, veins draining the stomach and intestine conduct the blood to the liver by a specialized circuit (the portal circulation). The liver then metabolizes many xenobiotics before they pass into the systemic circulation and stores many xenobiotics. The veins draining the liver conduct blood to the main vein of the lower body and into the systemic circulation. Thus, when a xenobiotic is ingested it may produce a toxic effect on the GI tract; produce a toxic effect on the liver; be metabolized, sometimes to a more toxic product; and pass in an altered form into the general circulation (see also Chapters 5, 8, and 23, this volume).

B. Distribution

Once the xenobiotic is absorbed and enters the bloodstream, it is transported to the capillary level in tissues of the body where it becomes available for uptake by the target organ. After one pass through the circulation the xenobiotic is uniformly mixed in arterial blood regardless of its route of entry. When a bolus is absorbed, the peripheral tissues are therefore presented with an increasing concentration in the blood which peaks and then declines as the xenobiotic is distributed to tissues throughout the body and removed by metabolism, excretion, or storage.

When a xenobiotic is dissolved in plasma, some fraction of the total usually binds to circulating proteins, particularly albumin (which binds many organic compounds as well as calcium, copper, and zinc). Metals may also be bound to specialized proteins in the plasma, such as ceruloplasmin (copper) and transferrin (iron).

Binding occurs quickly and an equilibrium is established between the fraction of the xenobiotic bound to plasma protein, which cannot leave the vascular space, and that dissolved in the plasma, which is free to diffuse or be taken up by tissues. As the concentration of free xenobiotic falls in plasma, some molecules will separate from their binding sites until a new equilibrium is reached. Binding therefore acts as both a storage and distribution mechanism. It maintains a more even blood concentration than would otherwise be the case and reduces the peak concentration that would otherwise be presented to tissues. Bound xenobiotics may be displaced by other xenobiotics. Some xenobiotics, such as barbiturates or sulfonamides, compete with others for binding sites and may increase the concentration of free xenobiotic in the plasma and therefore increase toxicity. As a practical matter, this is of greatest significance in drug-related toxicology as a mechanism of drug interaction and overdose and is seldom a consideration in environmental toxicology.

The persistence of a xenobiotic in the bloodstream is an important determinant of the duration of its action and the penetration that may occur into tissues less avid in their uptake of the particular agent. However, the most important determinant of uptake by the target organ is the uptake of the xenobiotic from plasma into the tissue.

Uptake of a xenobiotic by an organ from the plasma depends on the blood flow to the organ and the avidity of the tissue for the material. Special transport mechanisms exist at the cellular level for some xenobiotics. As mentioned above in the context of absorption into the body, absorption of a xenobiotic from the bloodstream into the tissue depends on the solubility of the xenobiotic in fat; lipophilic agents will be accumulated in adipose tissue or lipid-rich organs such as the nervous system or liver. Where the physicochemical properties of the organ attract and bind metals, as in bone, a metal xenobiotic will be sequestered and will accumulate over time.

Entry into some tissues is restricted by special barriers to passage, such as the blood-brain barrier and the placenta. In most cases, however, delivery of a xenobiotic depends on the blood supply to a tissue relative to its weight. When the xenobiotic is neither particularly lipophilic nor sequestered nor preferentially taken up by some organ-specific mechanism, it is largely distributed on the basis of blood flow to the target organ. Organs with greater perfusion will tend to accumulate the xenobiotic because of the increased total amount presented to it. The lung, a very lightweight organ, is the only organ of the body to receive 100% of the cardiac output at a tissue level. (The heart, functioning as a pump, moves blood in bulk but is itself nourished by a much smaller coronary artery system.) Not surprisingly, the lung is a principal target organ for blood-borne as well as airborne xenobiotics. The liver and kidneys each receive massive fractions of the cardiac output and are therefore presented with circulating xenobiotics in quantity. The brain also receives a disproportionate fraction of the cardiac output but is partly protected by the blood-brain barrier; this barrier works well for most polar xenobiotics but is permeable to lipophilic compounds.

In the liver, the portal circulation also delivers ingested xenobiotics at high concentrations directly from the stomach and small intestine. This provides an opportunity for metabolism to take place before the xenobiotic enters the general circulation. The liver is the principal metabolic site for xenobiotics, as it is for nutrients. Xenobiotics metabolized in the liver may even be taken up and reprocessed through biliary excretion and reabsorption through the enterohepatic circulation, such as kepone. Some tissues have an affinity for xenobiotics with certain characteristics. Organs with a high adipose or lipid content accumulate much larger concentrations of highly lipophilic xenobiotics, such as the PCBs, than occurs in plasma or in other organs. This is useful scientifically as a means of measuring body burden, because subcutaneous fat biopsies are not difficult to perform and other fatty substances that can be easily recovered, such as cerumen, do reflect tissue levels. When an obese individual in whose adipose tissue is stored a high level of a fat-soluble toxic chemical rapidly loses weight as a result of dieting, food deprivation, unaccustomed exercise, or cachexia, the xenobiotic may be mobilized and a rapidly climbing circulating level of the agent may rise to toxic levels. In general, however, the principal significance of adipose and intracellular lipid is as a storage depot, in that the blood concentration comes into an equilibrium with release from the tissue where it is stored, remaining fairly constant for the remaining life of the individual. The xenobiotic can rarely be effectively purged from the body in this situation because of the extent of the storage, although strategies exist to steadily reduce the body burden over time by vigorous removal from plasma to force mobilization. Another important implication of storage in fatty tissues is accumulation in breast tissue and subsequent excretion into breast milk. This is the major route of exposure to a variety of xenobiotics for newborns who breastfeed. Metals such as lead are also sequestered in bone; mobilization from depots in bone by chelating agents may substantially increase blood levels and create a risk of renal toxicity (see also Chapter 23, this volume).

C. Metabolism

Many xenobiotics are substrates for intracellular enzyme systems, most of which appear to have evolved as mechanisms for clearing endogenous, mainly steroid, hormones or foreign substances taken in with food. These enzyme systems transform the xenobiotic in a series of steps from the original compound to a series of stable metabolites, often through intermediate unstable compounds. For many xenobiotics there are many pathways of metabolism, which result in numerous metabolites. These transformations may have the effect of either "detoxifying," by rendering the agent toxicologically inactive, or of "activation," by converting the native agent into a metabolite that is more active in producing the same or another toxic effect. An active xenobiotic may be transformed into an inactive metabolite, which effectively removes the agent from the body in its toxicologically active form. However, an inactive precursor may also be transformed into an active metabolite.

In general, the enzyme systems available for the metabolism of xenobiotics tend to convert non-polar, lipid-soluble compounds into polar water-soluble products that are more easily excreted in urine or bile. The general pattern consists of two phases. These are illustrated in Figure 2.

Phase I of the metabolic process involves the attachment of functional chemical groups to the original molecule. This usually results in activation, especially in the very important "mixed function oxidase" (MFO) system, and results in a metabolite capable of interacting with macromolecules, such as DNA in the early steps of carcinogenesis. The mixed function oxidase system requires a great deal of metabolic energy and is closely linked with the cytochrome oxidase system, which provides it. Because the particular cytochrome most closely linked with the system has a spectral absorption peak at 450 nm, there is frequent reference in the literature to P-450 as an indicator of MFO activity.

Most important of the metabolizing systems, the MFO system also is known by other names: aryl hydrocarbon hydroxylase, arene oxidase, epoxide hydroxylase, and cytochrome oxidase. It is a complex of membrane-associated enzymes closely linked to the cytochrome P-450 system (and other cytochromes) that acts on

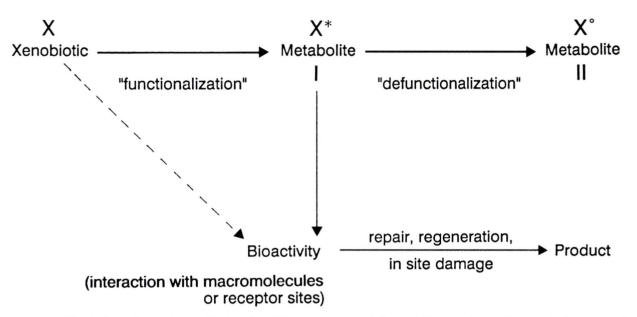

FIGURE 2 Metabolism of organic xenobiotics often follows a pattern of chemical "functionalization" in which the intermediate product may be more reactive and toxic than the original xenobiotic. This is followed by de-activation and defunctionalization or conjugation to make the product more water soluble.

organic compounds with aromatic or double bonds. The system attacks these bonds, which creates first an epoxide and then an alcohol, and in the process it first activates the compound and then deactivates it and renders it more easily excreted. The MFO system is virtually ubiquitous in the body but activity is particularly concentrated in liver and lung, and it can be found and conveniently studied in circulating lymphocytes. The MFO system has a huge capacity and acts on a wide variety of substrates. It is also inducible; when presented with suitable substrate, the cell synthesizes more MFO enzymes, which increases the capacity of the system and prepares itself for a greater load. The degree of inducibility and the level of baseline activity in a given tissue is genetically determined, so that at any one time MFO activity in a particular tissue reflects heredity combined with exposure in recent past.

Phase II involves the removal or conversion of chemical groups in such a way as to render the molecule more polar and therefore more easily excreted by the kidney (and less easily diffused back across the renal tubular epithelium after filtration). In the process, the activated xenobiotic metabolite from phase I usually becomes inactivated. This process frequently involves "conjugation," the attachment of a functional group such as sulfonate or glucoronic acid that makes the molecule much more hydrophilic.

The most complicated metabolic pathways are those for organic compounds. Metals may also be metabolized, however. The methylation of mercury and arsenic, especially, plays a major role in their toxicity. The methylation pathway of arsenic is species specific and this is thought to be the reason why arsenic is a carcinogen in humans but not in animals.

D. Excretion

The xenobiotic or its metabolites would accumulate and remain within the body if there were no mechanisms for excretion. Elimination is the term used for removal of the xenobiotic from the bloodstream, whether by excretion, metabolism, or sequestration (storage).

The kidney is the major route of excretion for most xenobiotics. Those that are water soluble may be filtered or excreted unchanged. The reserve capacity of the kidney is very great and this mechanism is rarely saturated in healthy people, but individuals with renal insufficiency may show accumulation and persistence of the xenobiotic and, consequently, prolonged and more severe toxicity. Other xenobiotics may be metabolically transformed into more water-soluble metabolites before renal clearance occurs. Xenobiotics that are themselves nephrotoxic may injure the kidney and reduce their own clearance thereby enhancing their own toxicity by further accumulation.

The liver, besides being an important metabolizing organ, secretes some xenobiotics into bile. These include heavy metals such as lead and mercury. These may recirculate by the enterohepatic circulation, persisting in the body much longer than otherwise, or they may pass out of the body in feces. Forced biliary excretion is not presently possible but interruption of the enterohepatic circulation by binding agents such as cholestyramine is a practical clinical intervention to hasten excretion and reduce the body burden of xenobiotics excreted in the bile and reabsorbed in the gut. This was first demonstrated for kepone. Although hepatotoxic agents may interfere with their own excretion by the liver, they are more likely to interfere with metabolism and as a practical matter this effect is rarely significant.

Volatile gases are readily excreted by the lungs through passive diffusion from the blood while crossing the alveolar-capillary barrier in "reverse" direction. Gases that are poorly soluble in blood, such as ethylene, are rapidly and efficiently eliminated by this route. Those that are readily soluble in blood, such as chloroform, are less efficiently eliminated and may be detectable in expired air for days or even weeks.

Xenobiotics and their metabolites are also eliminated by various minor routes that matter little with respect to reduction of the total body burden but may have toxicological implications. Lipid-soluble agents may be secreted in breast milk; this is a major route of exposure of neonates and young children to substances such as the organohalides, which include PCBs. Water-soluble agents are excreted in saliva and tears and are filtered through sweat glands; the latter function much like miniature nephrons. Lipid-soluble agents may also be found in cerumen and sebum. These minor elimination pathways permit non-invasive monitoring techniques for the detection of the agent but are rarely reliable enough to quantify exposure.

E. Kinetics

Metabolism and excretion define the rates of elimination and the change in the concentration of the xenobiotic in the plasma with time. Elimination may occur either because the xenobiotic is excreted, because it is converted to something else by metabolism, or because

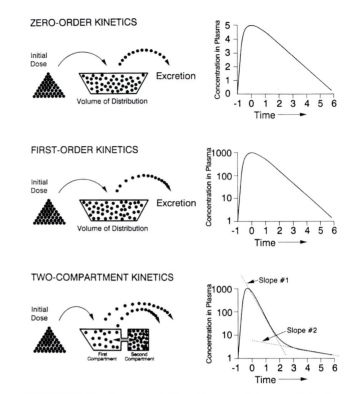

FIGURE 3　Kinetics of elimination are determined by the behavior of the xenobiotic in different toxicokinetic phases and can be modeled.

it is stored somewhere inaccessible to the bloodstream. The description of the rates of elimination of the agent is an important tool in understanding its behavior in the body. Each phase of the kinetics of a xenobiotic is governed by rates determined by properties of the agent and characteristics of the biological system, as illustrated in Figure 3. Each rate is described by a rate constant (k) that determines how rapidly the process proceeds.

Rate constants are described by their "order." A zero-order rate constant describes an elimination curve in which the rate is limited intrinsically by the fixed ability of the body to eliminate the agent, regardless of its concentration. In practice, the only important example of this is, ironically, the most common metabolizing system of toxicological concern: alcohol dehydrogenase, which metabolizes ethanol and other alcohols. Regardless of how much alcohol a person ingests, elimination will occur at the same rate and the rate of elimination will occur at a fixed rate regardless of the dose or plasma concentration.

A first-order rate constant describes a process in which the rate of elimination is independent of the dose and the concentration of the xenobiotic in plasma is proportional to the concentration of the agent in tissue. This is the most common situation. The concentration of the xenobiotic in plasma decreases over time by a constant proportion per unit time. This is called a "one-compartment" model, because the agent behaves as if it is restricted to one compartment of the body, the vascular space. In reality, the agent may not remain in the vascular space but it may equilibrate freely with it from its tissue depots. Water-soluble xenobiotics usually show first-order kinetics, except for alcohols.

A multicompartment, or "multiexponential," function of elimination suggests that the agent equilibrates in more than one compartment and is eliminated at different rates from each. The rate of elimination varies with the concentration in plasma and the initial dose and is biphasic. The elimination will not fit a simple exponential decay (or straight line on a logarithmic scale), but it will be described by a more complex equation with two rate constants: a "fast" rate constant and a "slow" one. Organohalides typically show two-compartment kinetics because of their storage and slow release from fatty tissue. (The term second order is not used because it would imply that elimination rate is a function of the square of the concentration, which is not the case.) Increasingly, the behavior of such xenobiotics is modeled using "physiologically based pharmacokinetic" (PBPK) models, so-named because they were first worked out for drugs. Metals often have multiple compartments and complicated toxicokinetics.

First-order kinetics are most common for water-soluble xenobiotics. In such systems elimination of the agent that is proportional to concentration results in an exponential decay or reduction in plasma concentration over time. The period required for the plasma concentration to drop by half is called the half-life ($t_{1/2}$). The $t_{1/2}$ can be calculated easily and accurately and is related to the elimination rate for first-order systems by the following equation:

$$t_{1/2} = 0.693/k_{el}.$$

II. TOXICODYNAMICS

The mechanisms of toxic injury are too numerous to accommodate simple classification or generalizations. There are, however, a few general principles that are useful in understanding toxic effects.

A. Mechanisms of Toxicity

Xenobiotics exert toxic effects by interfering with the normal functions of the body. These effects occur at the molecular and cellular level. Thus, an understanding of normal function and biochemistry is essential for understanding toxicodynamics. The toxic effect is an interaction between the xenobiotic and the cellular and biochemical mechanism. This usually involves the interaction between the xenobiotic and a macromolecule, as illustrated in Figure 2. This has not always been understood. For centuries, poisons were considered to be a special class of chemicals and the toxicity of poisons were understood to be intrinsic properties of the chemical, or magic (see also Chapter 5, this volume).

The liver, kidney, lungs, skin, and bladder are particularly susceptible to toxic effects. They are in harm's way because they may be the first to encounter a toxic exposure. These organs are highly metabolically active, actively metabolize xenobiotics themselves, concentrate toxic substances or their metabolites, or have biochemical characteristics that render them vulnerable.

Although there are as many potential mechanisms of toxic effects as there are reactions in biochemistry and functions in physiology, there are a few processes that play special roles. These process are listed below.

1. Inflammation: The body has natural mechanisms to repair and limit damage. Many xenobiotics are irritating to human tissues and induce local inflammation. For others, inflammatory reactions may contribute to systemic toxicity. A particularly important phenomenon associated with inflammation is the production of reactive oxygen and nitrogen free radicals, which cause intracellular damage as a by-product of inflammation.
2. Immune responses: The body also has natural mechanisms to respond to specific foreign substances (antigens) or cells by producing antibodies or by mobilizing special cells that destroy the foreign material. In the process the body sets off inflammatory responses. These immune responses require that the body sees the antigen first or that it is persistent, and they are triggered by subsequent exposure to low levels of the antigen. When the immune response is dysfunctional, it may cause allergies, diseases collectively called "immunopathies" and self-directed autoimmune diseases.
3. Carcinogenesis: Cancer is the prototype for "stochastic" or probabilistic toxic effects, in which

the response depends on the probability of an interaction rather than the magnitude of exposure and degree of response. This is discussed below. There is some evidence that certain other classes of toxic response, such as neurotoxicity, follow similar patterns.

4. Endocrine mimics: Many xenobiotics interact with hormonal receptors, sometimes by simulating the effect of hormones and sometimes by inhibiting them.

B. Exposure-Response Relationships

The exposure-response relationship is a concept fundamental to an understanding of toxicology. Paracelsus, the great medieval toxicologist, first said "it is the dose that makes the poison" and thereby established that poisons were not a mystically benighted form of matter but that all chemicals had toxic properties that become apparent as increasing quantities are consumed or absorbed. It follows from this simple observation that there may be "safe" levels of exposure to even the most toxic substances, which is a much more controversial assertion. Obviously, there are several dimensions to this seemingly straightforward concept. There are three distinct varieties of the exposure-response relationship that need to be distinguished conceptually. These are the toxicological dose-response relationship, the clinical dose- or exposure-response relationship, and the epidemiological exposure-response relationship. These are illustrated in Figure 1.

Dose is generally understood to mean the total quantity of a toxic substance administered; exposure is generally considered to be the level of concentration available for absorption by any or all routes at or over a given period of time. Thus, dose is best understood as total or cumulative exposure over a relevant time period. If the dose is given all at once, the dose-response relationship is most meaningful, as it is when the toxic substance is accumulated in the body. If the exposure takes place over a prolonged period of time, the internal dose at any given time tends to vary and it is more useful to think of an exposure-response relationship. When a xenobiotic accumulates and persists in the body, such as lead over a period of weeks or dioxin and pesticides over a period of months and years, cumulative exposure approximates dose in toxicological terms. When a xenobiotic does not readily accumulate and is quickly eliminated, cumulative exposure over a long period of time does not equate to effective dose in tox-

icological terms, although there may be cumulative effects if each exposure produces permanent injury.

The most fundamental building block of toxicology is the dose-response relationship demonstrable in the laboratory, often called the "toxicological" dose- or exposure-response relationship. The fundamental principle is that the physiological response depends on the amount of the agent presented to the tissue. In a given individual, exposure to an increasing amount of a toxic substance leads to the progressive appearance of new and usually more severe health problems leading to death, a sort of stepladder to lethality.

This gives rise to another type of dose- or exposure-response relationship, which might be termed the "clinical" exposure-response relationship. At a given level of exposure, often referred to clinically (if colloquially) as a "threshold," one can usually expect a given constellation of symptoms and signs. This clinical exposure-response relationship depends on the strength of the host defenses of the individual (which can be very variable) and whether the individual has an acquired condition or genetically determined phenotype that renders him or her more vulnerable than others. In a given exposure situation, one person may show one symptom and another a different symptom, based on personal susceptibility. At relatively low levels of lead toxicity, some patients show elevated uric acid levels because of reduced renal clearance, but most do not. As well, the detection of the expected clinical response depends on the sensitivity of clinical examination and laboratory tests. Clinical tests are often inadequate for early detection of equivocal cases because they were designed for making specific diagnoses in people known to be sick in a way that strongly suggested a particular type of disease.

The third type of exposure-response relationship relates exposure levels to the frequency of the response in a population. If one is interested in what personal characteristics of those exposed render them vulnerable to a toxic effect or in how frequently a response is associated with a given level of exposure in a population, one may do a "nose-count" of the observed response among individuals exposed. This is the essential method of epidemiology and yields what is usually called the "epidemiological" exposure-response relationship. To be meaningful, however, the outcome must be experimentally or clinically detectable. This removes from study many types of response that cannot be directly measured and which are usually considered "subclinical" or "adaptive responses." In this system, a threshold means the level of exposure associated with the first appearance of an excess of the health outcome representing the toxic response. It is this threshold for response that generates the most controversy in regulatory policy. However, interpretation of this type of threshold depends on understanding the basis for selecting and detecting the health outcome.

At higher levels of exposure, the exact shape of these exposure-response relationships are not critical and the general relationship is usually obvious. At lower levels of exposure, however, interpretation of the population response is very dependent on an interpretation of the general mechanism of the toxic effect and extrapolation to low exposures is very sensitive to the biological model applied.

A particularly important, if confusing, term in toxicology is threshold. This means the level of exposure at which an effect is first observed. The existence of thresholds for certain types of response (particularly carcinogenicity) are controversial and arguments surrounding identification of a threshold for response frequently neglect to specify the type of threshold under consideration.

C. Interaction

Some xenobiotics interact with others to produce disproportionate effects. For example, exposure to sulfur oxides in the presence of particulate air pollution in combination causes worse lung irritation than would be predicted by the individual effects of each added together. This is because the sulfur oxide adsorbs onto the surface of the particle and is carried deeper into the lung than it would be as a gas.

Interaction may be positive (often called synergy) or negative (often called inhibition). Different models of interaction are applied to the interpretation of data. When the effects of two or more are additive, no interaction occurs. This may suggest that both are acting by the same pathway or mechanism and are simply adding their proportionate share to the total magnitude of the effect. When the effects multiply, this is strong evidence for extensive interaction, and suggests that the xenobiotics are acting by different pathways that potentiate each other's effects. When the effects are less than additive, it is evidence that in some way the effects of one exposure are reducing or blocking the effects of the other or are acting by a similar pathway that has only limited capacity.

Toxicologists are very concerned that exposure to mixtures, such as cigarette smoke or heavily contaminated groundwater, could present the potential for numerous interactions and unpredictable effects. In practice, relatively few examples of interaction producing significant health effects have been documented. Some of the most important involve cigarette smoking and persistent carcinogens, which include asbestos, silica, and radon daughters. These greatly increase the risk of lung cancer compared to the sum of the separate risks of either smoking alone or exposure to the other carcinogen alone.

D. Carcinogenesis

Much of environmental toxicology is oriented toward the etiology and prevention of cancer. Carcinogenesis is not a straightforward, deterministic process. Rather, at each step in the sequence there is a finite probability of events leading to the next step. Chemical carcinogenesis is thus a stochastic, or probabilistic process, like a roulette wheel or radioactive decay, not a certain prediction based on chemical structure and properties. In any one individual, an exposure may increase the odds of getting cancer but does not make it certain in absolute terms that this will happen.

Chemical carcinogens are demonstrable by their effect in increasing the frequency of cancers observed in exposed subjects as compared to unexposed. They may produce malignant tumors that are often different in tissue type and wider in diversity than those usually observed among unexposed subjects, produce malignant tumors at characteristic or unusual sites in the body, and produce these malignant tumors earlier in the life span of subjects than they would otherwise be seen. Often, however, chemical carcinogens produce malignancies identical in tissue type, location, and onset to those seen in unexposed populations. The only clue is an increased frequency of cancers in exposed groups.

Recent advances in research on carcinogenesis, especially the identification of the oncogene, may have identified new and rather complicated mechanisms, but the effect has been to simplify our understanding by providing common pathways and unitary, comprehensible mechanisms by which many causes may act. The principal categories of causes of cancer are fairly conclusively identified as heredity, chemical exposure, viral infection, and radiation exposure. Other categories of causes may be identified, but these appear to be primary. Specific causes within each category may act by similar mechanisms such as by activation of oncogenes.

As understanding of the basic mechanisms of cancer improves, concepts of chemical carcinogenesis have also grown more refined. A deep understanding of the biology of cancer helps to explain many of the phenomena critical to regulation and control, such as latency periods and cancer promotion. For example, low-dose radiation and radiomimetic health effects have been difficult to unravel because three competing theoretical models exist for low-dose extrapolation (linear, quadratic, and linear-quadratic). The divergence in goodness-of-fit to available response data results from differences in the underlying assumptions involving adaptive mechanisms, threshold effects, receptor behavior, and transport to the target organ. Similarly, the population response to exposure to chemical carcinogens at low exposure levels depends on whether a "one-hit" model or an interactive model is operative. (One-hit refers to a single interaction with DNA in a single cell as theoretically sufficient to cause cancer, no matter how improbable; an interactive model assumes that more than one hit is required to sustain the carcinogenic process.) The discovery of the various oncogenes and emerging evidence as to their distribution in the genome among individuals in the general population and, perhaps, high-risk subgroups has led to a rethinking of our concepts of cancer-risk and susceptibility.

The contemporary model for induction of cancer by chemicals that is most consistent with available evidence for most chemicals and for radiation is the "two-stage model of carcinogenesis." (The model is insufficient to account for some other types of cancer induction and these are discussed below.) The two-stage model assumes the introduction of a carcinogen into the body (or the metabolic activation of a pro-carcinogen) and its distribution in the body in such a way as to be presented to a tissue at levels in which it is likely to be taken up intracellularly and to react with cellular constituents, most specifically nuclear DNA. This process may result in activation of proto-oncogenes into oncogenes or may, by other means, redirect the cell's set of instructions. Transformation of the cell may override normal regulation of cell growth and mutual adherence instead activates more primitive or fetal-like genes that result in less contact inhibition, greater migratory potential, and less surface adhesion, which are the basis for tumor growth, invasion, and metastasis. This does not occur with every interaction between a carcinogen and DNA, however. Only in a relatively small fraction of such interactions will the critical sites on DNA be affected, which results in a probabilistic phenomenon. When it occurs, the process is called "initiation" because those cancers that may ultimately result are initiated at this step. In many cases, things presumably go no further than initiation. The mechanism for much, if not most, initiation activity is oncogene activation.

Among these interactions with DNA are a handful that may cause the cell to behave in a manner more appropriate to a primitive, embryo-like state, and these are thought to be the mechanism for transforming normal cells into neoplastic cells. Oncogenes are capable of being activated by chemical exposure. They are latent within the genetic structure of all humans (and probably all advanced life) and at least some probably play a physiological role in normal embryonic and fetal growth and development. Activated in the absence of regulation; however, the oncogenes trigger malignant transformation of the cell, which causes a previously differentiated cell to regress to a more primitive state abnormal for that stage of the life of the organism.

The derepressed oncogene comes to life, and expresses itself by the production of proteins (many of them enzymes, others messenger molecules or receptors) for which it codes and that have the effects required to transform the cell. These "oncogene products" are not only important in the transforming process, but may serve as very early markers that initiation has occurred. Theoretically, this would allow early identification of workers at risk for subsequent development of cancer. This raises the possibility of prophylactic treatment such as chemoprevention or other interventions designed to avoid promotion or reduce cancer risk generally.

Next in sequence is the growth of a clone of transformed cells from a single cell altered in its growth characteristics to a small focus *in situ*. The transformed cell, having been altered in its DNA blueprint, does not necessarily begin to multiply at once. Rather, it may be held in check by host factors or cell-specific factors, such as those needed for further DNA reorganization or oncogene activation to take place. The abnormal cell may rest for a very long time contributing to the greater part of the latency period before appearance of the clinically evident tumor. Additional exposures may trigger the conversion of the initially abnormal cell into a transformed or pre-neoplastic cell capable of giving rise to a tumor. This process may be facilitated by exposure to chemicals that also have genotoxic potential, either simultaneously or after the action of the primary carcinogen. This ancillary process is called "co-carcinogenesis," which implies that the second or combination exposure may not be the initiator but it may participate in the genotoxic cell events that either lead to expression of the critical event, resulting in oncogene activation, or override mechanisms that would otherwise inhibit oncogene activation and cell transformation. In general, the same chemicals that are primary carcinogens are likely also to be co-carcinogens. The distinguishing feature is not which chemical reaches the DNA first or which exposure preceded which, but which chemical actually participated in the critical event that specifically altered the DNA in such a way as to activate the oncogene.

At this stage, exposure may occur to chemicals that are capable of triggering proliferation by removing the inhibitory factors that are suppressing the transformed cell. This is called "promotion" and it is the second stage in the two-stage model of carcinogenesis. Promoters are sometimes primary carcinogens themselves and probably act through genetic mechanisms, such as the polycyclic aromatic hydrocarbons, but others are either weakly or not carcinogenic and presumably act by nongenetic mechanisms. The most well known are the phorbol esters (specifically tetradecanoyl phorbol acetate, TPA), which are constituents of croton oil that are chemically extremely complex and seem to act at least in part pharmacologically by activating certain specific receptors on the cell surface. Chlorinated hydrocarbon species are often potent promoters, including the PCBs, DDT, PBBs, and certain dioxins. They seem to act by nongenetic means and have variable primary carcinogenic activity, depending on the species.

By whatever mechanism, promotion results in deregulation and progression of the neoplasm by proliferation into a clone of cells. The transformed cell has now become a cancer cell with the essential features of a malignancy: unresponsiveness to regulation, loss of contact inhibition, potential for sloughing and migration of cells, and the potential for inducing growth of new nutrient blood vessels.

To metastasize, malignant cells must digest or displace the matrix binding them (especially basement membranes), migrate through the degraded tissue, gain access to blood or lymphatic vessels for transport, and be deposited in a tissue favorable to growth. This does not occur until most tumors have reached a size of at least 1 cm^3, representing a population of 10^9 cells. This takes time, since the doubling time of a cancer is rarely less than six months and cells are being continually killed by host factors (especially natural killer, or NK, lymphocytes) or local nutrient inadequacy.

Because this all takes time at each step, there is a delay between the initiation (commonly assumed to be at first exposure) and earliest clinical presentation of a tumor. This is called the "latency period." For most chemically induced cancers it is on the order of 20–30 years but may be as long as 50 or more (in the case of mesothelioma and asbestos) or as short as 5 years (for radiation or radiomimetic exposures and some bladder carcinogens).

Meanwhile, back at the primary, local invasion and mechanical effects that lead to clinical detection are largely a function of tumor size and, therefore, are not obvious until the mass of the primary cancer has passed through some number of doubling times. Thus, there is a further delay between malignant proliferation and detection of the tumor clinically, which contributes to the latency period at the end of the process. The latency period is also influenced by the intensity of exposure and can be shortened by intense exposure at initiation or during promotion.

It is only at this late phase that screening programs for clinically apparent cancers have a role. Coming so late, cancers that are usually aggressive and metastasize early (such as lung cancer and melanoma) or that are difficult to detect because of their location (pancreas and ovary) do not lend themselves to effective management by early detection and treatment, because it is already too late in the great majority of cases by the time the tumor is detected. Less aggressive and more accessible malignancies, such as breast and cervical cancer, are more readily dealt with by these means.

In time, earlier screening methods for the detection of antigens reflecting oncogene expression may allow identification of persons at risk for cancer following exposure to carcinogens. The past interest in using cancer-related embryonic antigen (CEA) as a screening test may be revived with the introduction of more specific oncogene products that can be determined in urine or blood. How oncogenes and their products will become incorporated into occupational and environmental medicine in practice is not yet apparent, but it is clear that they will someday become a part of population health monitoring and part of the mainstream.

Not all chemically induced cancers act by this genetic mechanism. "Epigenetic" refers to the actions of cancer-inducing agents and exposures that do not directly interact with DNA. At least some probably act by inducing intracellular free radicals that damage DNA in a nonspecific manner, however. Others are more obscure in their mechanisms. None are adequately explained by the conventional two-stage model of carcinogenesis, but subsequent refinements in theory will almost certainly result in a unitary model demonstrating a final common mechanism for most cancers.

Because epigenetic mechanisms are associated with important occupational exposures (benzene), laboratory reagents (dioxane), consumer products (nitriloacetic acid, NTA), medical devices (foreign body), and pharmaceuticals (hormones), epigenetic carcinogens are of particular concern to occupational physicians. They represent a class of carcinogens of particular concern in risk assessment because they do not behave as typically genotoxic agents in the usual *in vitro* assays and are therefore more difficult to anticipate. With respect to new products, the risk of foreign body and hormonal induction of cancer demands particular attention because of, respectively, the development of new biomedical technology and the weakly estrogenic effects of many substituted hydrocarbon compounds, including some pesticides. Metal-induced carcinogenesis occurs by a variety of mechanisms and often strongly depends on the chemical composition, redox state, and solubility: arsenic (lung, bladder, and skin), beryllium (lung), cadmium (lung), chromium (hexavalent ion: lung), and nickel (subsulfide; lung).

Armed with a better comprehension of the process, it should become possible for society to set standards of exposure that provide greater assurance of protection and to anticipate problems with newly synthesized or introduced chemicals. It may even be possible, using interventions on the horizon or presently available, to reduce the risk of cancer once exposure to a carcinogen has occurred. When this is done by the administration of a drug it is called chemoprevention. There is a great deal of interest in applying chemoprevention in populations of workers occupationally exposed to carcinogens. Clinical trials were conducted in the United States to determine the efficacy of cis-retinoic acid (a vitamin A derivative) in blocking steps in the sequence of events leading to certain cancers. Unfortunately, they were not only unsuccessful but for reasons that are not clear there was actually an increased incidence of cancer in the treated group, which forced the study to be terminated. A successful strategy of chemoprevention could help thousands in high-risk groups that have already sustained exposure, but to date there is no good model.

III. TOXICITY TESTING

The usual approach to assessing the toxicity of a new chemical or an agent that has recently come under suspicion is to conduct a sequence of studies, each level of which is called a "tier." A tier-one study, for example, may involve the use of *in vitro* studies, such as the Ames assay (described below), in an effort to identify potential carcinogens early and to exclude them from further consideration as a possible product. A tier-two study might involve determination of LD_{50} or LC_{50} in animals. Higher tiers may involve "subchronic" studies (90-day

exposures, which result in sacrifice of the animals to examine sublethal effects), "chronic" studies of 6 months or a year, lifetime studies (to evaluate carcinogenicity over two to three years), and special studies to examine teratology, reproductive effects, toxicokinetics and metabolism, allergenicity, phototoxicity, and behavioral effects. As the tests become more sophisticated and the outcomes become more difficult to detect, they become much more expensive. Alternatives to animal studies are becoming available for specific purposes, but they cannot replace *in vivo* testing for all needs.

A major issue in selecting any kind of animal model is the biological relevance of the model to the application intended. The experiment must be at least comparable to human routes of exposure, metabolic pathways (if applicable), and the potential for expression of the effect. Strain differences within species are as important as species differences. Inbreeding has resulted in considerable differences among rat strains in response to longer term effects. The longevity of animals species places constraints on what can be studied. Animals that survive less than two years in confinement, such as mice, are difficult to use for long-term exposure studies. Rats do survive this long but full expression of the effects of exposure may require the animal to live out its life span rather than be sacrificed after an arbitrary time period. The age and sex of the animals are also important considerations. Although it is difficult to generalize, females are sometimes more susceptible to the effects of toxic exposures involving metabolism of the agent, especially if there is a possible parallel with hormonal effects as in the case of certain aromatic hydrocarbons. Young animals may differ from older animals in their degree of resistance to toxic effects; for example, neonate mice are relatively resistant to oxidant gases compared to older animals. (The dose or concentration, respectively, is sufficient to kill 50% of test animals as calculated from the dose-responsive curve.)

SEE ALSO THE FOLLOWING CHAPTERS

Chapter 5 (Uptake of Elements from a Biological Point of View) · Chapter 8 (Biological Responses of Elements) · Chapter 21 (Environmental Epidemiology) · Chapter 22 (Environmental Medicine) · Chapter 23 (Environmental Pathology)

SPECIATION OF TRACE ELEMENTS

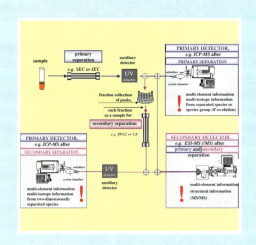

BERNHARD MICHALKE
GSF National Research Center for Environment and Health

SERGIO CAROLI
Istituto Superiore di Sanità

CONTENTS

I. Introduction
II. Speciation Analysis: Sampling, Storage, Preparation, Separation, Detection
III. Quality Control in Speciation
IV. Selected Examples of Speciation Studies in the Life Sciences
V. Summary

I. INTRODUCTION

The determination of trace elements has assumed a place of prominence in the life sciences. Elements present at even minimal concentrations in biological and environmental matrices can have a significant influence on vital functions, depending on the amount present. The study of, for example, pathophysiological processes in the human body requires the determination of elements at concentrations measured in $\mu g\,L^{-1}$, $ng\,g^{-1}$, and even $pg\,g^{-1}$. The higher concomitant amounts of organic and inorganic components make it difficult to determine the presence of trace elements. Moreover, it is a complex process that progresses from an initial trace element analysis to the final statement of

biological implications, one that requires close collaboration between the analytical chemist and life scientist. Furthermore, it should be kept in mind that the concept of *zero tolerance* for potentially toxic elements has been replaced by the more scientific notions of *safe ranges of exposure* and *range of safe intake*.

Over the last two decades, analytical chemists have come to realize that, in general, the total concentrations of chemical elements cannot provide information about their mobility, bioavailability, and eventual impact on ecological systems and biological organisms. Only knowledge of the chemical species of an element can provide information regarding possible chemical and biochemical reactions and thus lead to a greater understanding of toxicity or essentiality. It is also worth stressing, in this context, that new trace elements are continually being added to the list of those that are known or suspected to be essential.

Until now, already established separation and detection methods had to be combined in novel ways and modified according to particular speciation problems. Combination and hyphenation of separation techniques and element- or molecule-selective detection systems are generally the approaches of choice for speciation analysis; however, further methodological developments are still necessary, primarily for hyphenation and quality-control strategies. Investigations on quality control have shown that changes in the original species information can easily occur during sampling, sample

---

TABLE II.   Cell-Entering Mechanisms for Elemental Species

Diffusion	Diffusion is dependent on the size and lipophilic nature of the element species. It is fast and efficient for lipophilic molecules and is associated with high toxicity. Ionophores, which have such an increased lipophilic nature, may form complexes with (lipophobic) metals. These "excluded metals" are then transported across the cell membrane by the ionophores. In the cell, the metals are set free again and recomplexed by proteins or other ligands.
Active transport by ATPase-driven ion pumps	The ATPase-driven uptake mechanism has been proven for some essential elements such as Cu^{2+}, Zn^{2+}, and Ni^{2+}.
Carrier/shuttle transport	This transport mechanism is typically shown by proteins and hormones.
Uptake via ion-selective channels/active transport by electrical potential	Ion-selective channels have been tested for cations such as Ca^{2+} and K^+. Transportation across the membrane is dependent on the D_μ^{-H+}-membrane potential. A potential higher than $-70\,mV$ opens voltage-gated channels.

Source: Morrison, G. M. P., In *Trace Element Speciation: Analytical Methods and Problems*, G. E. Batley, Ed., CRC Press, Boca Raton, FL, 1989, pp. 25–42. With permission.

---

organisms experiencing such a metal load, a genetic MT transcription is induced that increases the concentration of the "offered" ligand to the toxic metals. The generated MT-metal complex is excreted via kidneys, thus protecting the reactive centers of enzymes.

- *Availability of many elements*—Elements must be available for the organism as well-defined species and at suitable quantities to guarantee a normal health status. A good example of this is chromium, which is essential in the trivalent oxidation state and highly toxic and carcinogenic in the hexavalent oxidation state.

F. Reference Values and Ranges

The achievement of reference concentration values and intervals for elements in biological and environmental matrices is of paramount importance in the detection of imbalances that can adversely affect human health and ecosystems. An exhaustive overview of problems and applications related to this issue was published several years ago (Caroli *et al.*, 1994). In this context, "normal values" are provided as tolerance limits for those elements that may be undetectable in human organs. Elements essential for life, on the other hand, are homeostatically regulated, and their concentrations are expected to fluctuate within narrow limits for each species under normal conditions. Although doing so is still far from feasible, there is no doubt that for chemical speciation determining reference figures will become even more crucial than obtaining knowledge regarding the total amount of a particular element.

II. SPECIATION ANALYSIS: SAMPLING, STORAGE, PREPARATION, SEPARATION, DETECTION

The direct determination of trace elements in samples is an important problem in technology, industry, and research, because decomposition and preconcentration procedures, as well as the storage of trace analytes in solutions, are often sources of concern. The accuracy of analytical results can, in fact, be threatened by these pretreatments. Only a few methods exist for direct analytical determinations in solids. In many cases, the detection power and reproducibility of spectral analyses are inadequate to meet the needs of analysts. This less-than-ideal performance is of particular concern in speciation analysis, which requires a series of carefully planned steps among which chemical and/or physical pretreatments of the material being tested are almost always mandatory. In this context, sampling and sample preparation are of prime importance. Without proper sampling and sample treatment procedures, there is little chance that any speciation analysis will be able to provide reliable data upon which human health or environmental decisions can be safely based. Thiers (1957) stated that, unless the complete history of samples is known with certainty, the analyst is well advised not to

TABLE I. Impact of Various Species of Elements

Element	Impact of Species
Arsenic	As(III) and As(V) are toxic; arsenobetaine is nontoxic.
Chromium	Cr(III) is essential; Cr(VI) is highly toxic and cancer-promoting.
Copper	Ionic Cu(II) is toxic in aquatic systems; humic complexes of Cu are almost nontoxic.
Iodine	Thyroid hormones influence an extended range of biochemical reactions in organisms and play an active role in immune defense. Triiodothyronine (T_3) shows about a fivefold effectivity compared to thyroxin (T_4), but it comprises only about 20% of the total iodine hormones.
Iron	Absorption capacity for Fe(II) is lower compared to Fe(III), but only Fe(II) is effective against Fe deficiency. This is important for supplementation; however, Fe(III) is utilized efficiently following reduction by ceruloplasmin.
Mercury	Inorganic Hg salts are less dangerous than methylated forms; these are more toxic and can be enriched (e.g., up to 10,000-fold in fish).
Platinum	Pt(0) is nearly completely insoluble in water. After emission from car catalysts as Pt(0), Pt species transformation occurs, and solubility in water as well as its availability are significantly increased.

Source: Data from Florence (1989) and Lustig *et al.* (1998).

D. Useful Fields for Speciation

Speciation is particularly relevant to the environmental field, as well as to biology and medicine. Food chemistry and nutrition, in turn, can also greatly benefit from the speciation approach, which can act as an interface between these two fields. The nature and amount of manmade species are altering natural species formation and equilibria; consequently, trace element mobility and bioavailability may be influenced and modified. Bioavailability is directly linked to biochemical mechanisms within the organism. Thus, the fate of a species— such as adsorption to membranes, transport and incorporation into larger molecules (*e.g.*, enzymes), and enrichment or excretion—may be modified and result in an unbroken path from environmentally changed species to toxicity, deficiency, or growth in biological systems.

E. Species Impact and Mechanism in Biological Systems

A necessary prerequisite for an elemental species to interact with an organism is that the species must be able to cross the cell membrane and participate in biochemical paths and reactions (Morrison, 1989). Several intake mechanisms are known, as detailed in Table II. These uptake mechanisms result in an enrichment of element species in the organism by a factor of 10^2 to 10^5. In some cases, toxic concentrations are reached even when the original species concentration in the environment is low (Morrison, 1989). The uptake and subsequent metabolization of element species is obviously dependent on the nature of the species itself, as are the consequences of that uptake and metabolization:

- *Immediate excretion without any interaction*—This action is considered beneficial for species having toxic potential and adverse for essential element species.
- *Interaction with the organism and participation in metabolic paths*—This result is considered to be beneficial for essential element species and adverse when toxification and a reduction in enzymatic selectivity and turnover rate occur. The replacement of an essential element by another one in the reactive center of the enzyme can sometimes cause enzyme damage.
- *Intracellular toxicity*—Such toxicity often appears when intracellular species transformations occur. The displacement of essential elements in the reactive enzyme centers results in inactive enzyme–metal complexes (*i.e.*, new species). Conversely, metal exchange at a protein can also be a detoxification reaction (*e.g.*, via metallothioneins [MTs]).
- *Metallothionein transcription*—Metallothioneins, 7- to 10-kDa proteins with about 30% S amino acids (~30% cystein) have a high affinity to metals such as cadmium, copper, mercury, and zinc. In

TABLE II.   Cell-Entering Mechanisms for Elemental Species

Diffusion	Diffusion is dependent on the size and lipophilic nature of the element species. It is fast and efficient for lipophilic molecules and is associated with high toxicity. Ionophores, which have such an increased lipophilic nature, may form complexes with (lipophobic) metals. These "excluded metals" are then transported across the cell membrane by the ionophores. In the cell, the metals are set free again and recomplexed by proteins or other ligands.
Active transport by ATPase-driven ion pumps	The ATPase-driven uptake mechanism has been proven for some essential elements such as Cu^{2+}, Zn^{2+}, and Ni^{2+}.
Carrier/shuttle transport	This transport mechanism is typically shown by proteins and hormones.
Uptake via ion-selective channels/active transport by electrical potential	Ion-selective channels have been tested for cations such as Ca^{2+} and K^+. Transportation across the membrane is dependent on the D_μ^{-H+}-membrane potential. A potential higher than $-70\,mV$ opens voltage-gated channels.

Source: Morrison, G. M. P., In *Trace Element Speciation: Analytical Methods and Problems*, G. E. Batley, Ed., CRC Press, Boca Raton, FL, 1989, pp. 25–42. With permission.

organisms experiencing such a metal load, a genetic MT transcription is induced that increases the concentration of the "offered" ligand to the toxic metals. The generated MT-metal complex is excreted via kidneys, thus protecting the reactive centers of enzymes.

- *Availability of many elements*—Elements must be available for the organism as well-defined species and at suitable quantities to guarantee a normal health status. A good example of this is chromium, which is essential in the trivalent oxidation state and highly toxic and carcinogenic in the hexavalent oxidation state.

F. Reference Values and Ranges

The achievement of reference concentration values and intervals for elements in biological and environmental matrices is of paramount importance in the detection of imbalances that can adversely affect human health and ecosystems. An exhaustive overview of problems and applications related to this issue was published several years ago (Caroli *et al.*, 1994). In this context, "normal values" are provided as tolerance limits for those elements that may be undetectable in human organs. Elements essential for life, on the other hand, are homeostatically regulated, and their concentrations are expected to fluctuate within narrow limits for each species under normal conditions. Although doing so is still far from feasible, there is no doubt that for chemical speciation determining reference figures will become even more crucial than obtaining knowledge regarding the total amount of a particular element.

II. SPECIATION ANALYSIS: SAMPLING, STORAGE, PREPARATION, SEPARATION, DETECTION

The direct determination of trace elements in samples is an important problem in technology, industry, and research, because decomposition and preconcentration procedures, as well as the storage of trace analytes in solutions, are often sources of concern. The accuracy of analytical results can, in fact, be threatened by these pretreatments. Only a few methods exist for direct analytical determinations in solids. In many cases, the detection power and reproducibility of spectral analyses are inadequate to meet the needs of analysts. This less-than-ideal performance is of particular concern in speciation analysis, which requires a series of carefully planned steps among which chemical and/or physical pretreatments of the material being tested are almost always mandatory. In this context, sampling and sample preparation are of prime importance. Without proper sampling and sample treatment procedures, there is little chance that any speciation analysis will be able to provide reliable data upon which human health or environmental decisions can be safely based. Thiers (1957) stated that, unless the complete history of samples is known with certainty, the analyst is well advised not to

Speciation of Trace Elements

Bernhard Michalke

GSF National Research Center for Environment and Health

Sergio Caroli

Istituto Superiore di Sanità

Contents

I. Introduction

The determination of trace elements has assumed a place of prominence in the life sciences. Elements present at even minimal concentrations in biological and environmental matrices can have a significant influence on vital functions, depending on the amount present. The study of, for example, pathophysiological processes in the human body requires the determination of elements at concentrations measured in $\mu g \, L^{-1}$, $ng \, g^{-1}$, and even $pg \, g^{-1}$. The higher concomitant amounts of organic and inorganic components make it difficult to determine the presence of trace elements. Moreover, it is a complex process that progresses from an initial trace element analysis to the final statement of biological implications, one that requires close collaboration between the analytical chemist and life scientist. Furthermore, it should be kept in mind that the concept of *zero tolerance* for potentially toxic elements has been replaced by the more scientific notions of *safe ranges of exposure* and *range of safe intake*.

Over the last two decades, analytical chemists have come to realize that, in general, the total concentrations of chemical elements cannot provide information about their mobility, bioavailability, and eventual impact on ecological systems and biological organisms. Only knowledge of the chemical species of an element can provide information regarding possible chemical and biochemical reactions and thus lead to a greater understanding of toxicity or essentiality. It is also worth stressing, in this context, that new trace elements are continually being added to the list of those that are known or suspected to be essential.

Until now, already established separation and detection methods had to be combined in novel ways and modified according to particular speciation problems. Combination and hyphenation of separation techniques and element- or molecule-selective detection systems are generally the approaches of choice for speciation analysis; however, further methodological developments are still necessary, primarily for hyphenation and quality-control strategies. Investigations on quality control have shown that changes in the original species information can easily occur during sampling, sample preparation and storage, separation, and detection. For specific details on various analytical techniques, see Chapter 29, this volume.

A. Definitions of Terms Related to Speciation

The use of concepts and terms related to chemical speciation in recent years still reflects a certain degree of inconsistency within the scientific community. In recognition of the importance of standard terminology from the viewpoint of both interdisciplinary communication and constructive interaction with decision makers, the International Union of Pure and Applied Chemistry (IUPAC) has undertaken a collaborative effort of three of its divisions and reached a consensus on some basic definitions that can be used by specialists in the discipline of speciation (Templeton *et al.*, 2000). One of the major conclusions of this working group was that the term "speciation" should be restricted to the distribution of an element among well-defined chemical forms. A clear distinction was also made between *speciation* and *fractionation*. As it is quite crucial for those working in this field to fully abide by such definitions, they are reported below verbatim:

- *Chemical species* (of a chemical element)—Specific form of an element defined as to isotopic composition, electronic or oxidation state, and/or complex or molecular structure.
- *Speciation analysis* (in analytical chemistry)—Analytical activities of identifying and/or measuring the quantities of one or more individual chemical species in a sample.
- *Speciation of an element*—Distribution of an element among defined chemical species in a system.
- *Fractionation*—Process of classification of an analyte or a group of analytes from a certain sample according to physical (*e.g.*, size, solubility) or chemical (*e.g.*, bonding, reactivity) properties.

The analytical activity of identifying and measuring species includes a clear identification of the species (elements and possibly the binding partners) and exact quantification in representative samples, as well as quality control (Caroli, 1996; Michalke, 1999a). If the identification and quantification of a chemical species cannot be performed, the analytical procedure can only lead to operationally defined species characterization.

B. Operationally and Functionally Defined Characterization of Species (Groups)

Operationally and functionally defined characterization of species has to be distinguished from chemical speciation analysis (Ure *et al.*, 1993). The former gives a characterization of molecule groups (not single species) that show a similar behavior during an analytical procedure (operation) such as extraction. Characterization of the molecule groups is strongly dependent on the selected analytical procedure, and usually the original species information (identity) is lost. In this sense, fractionation is regarded as an operationally defined characterization. The functionally defined characterization of species (groups) provides information about the function of species groups (not singly identified species) in organisms and their impact on living systems (Caroli, 1996; Mota & Simaes Gonçalves, 1996). Neither is considered to be a real chemical speciation analysis, as identification of a single species does not actually take place.

C. Necessity and Reasons for Element Speciation

The quality and quantity of the relevant element species in a matrix, rather than the total element concentration, are greatly responsible for the mobility, bioavailability, and ecotoxicological or toxicological impact of an element (Florence, 1989; Templeton *et al.*, 2000). Elements usually interact as parts of macromolecules (*e.g.*, proteins, enzymes, hormones) or according to their oxidation state; therefore, only current knowledge about a species provides a reason for further assessing whether or not it is toxic, without (known) impact at a specific concentration, or essential. Problem-related speciation analysis appears to play a key role in effectively assessing the risk posed by elements in the environment. Also, from a health viewpoint, the consequences of a trace element's essentiality, depletion, or toxicity can be better determined, and the development of diagnoses and potential therapies is improved. Some examples are provided in the following text to illustrate how different species of a given element can have different impacts (see Table I). It must be realized that, in speciation analysis, the analytical procedure generally interacts with the separation and detection of a species. These interactions usually shift equilibria among the species and possibly change some of the species themselves. The nature and extent of these alterations, as well as a critical discussion of the results achieved, should therefore be an essential part of well-conducted speciation analysis.

spend time analyzing them. The container in which the sample is stored is itself a potential source of contamination, as is the sample pretreatment procedure or manipulation or the analyst. Volatilization is another source of error. If unexpected changes to the form of the element occur, such as oxidation state, extent of chelation, or organometallic state, then clearly the species has been changed, and the original species identity and amount cannot be ascertained, thus defeating the purpose of the experiment. On the other hand, highly reliable data can be obtained with careful evaluation of the potential chemical changes in the sampling and sample preparation process.

A. Sampling

The sampling step is all the more critical in speciation analysis and usually shows uncontrolled and irreversible interferences in the species equilibria (Dunemann & Begerow, 1995). Sampling should be designed to preserve the original information about native species; however, existing techniques are often inadequate for the problem at hand. They must be adapted to the actual situation with regard to the element species of interest in the given matrix (Kersten et al., 1989). Several problems have been identified; for example, wall adsorption effects have been described, as well as contamination from the sampler and alteration of biological or chemical equilibria or oxidation by atmospheric oxygen (Caroli, 1996). When sampling natural water, unintended contamination of the probe frequently occurs in the surface microlayer, which is usually enriched by trace metal species. This contamination can be avoided by taking water samples 0.5 to 1.0 m below the surface. Also, materials such as dust and airborne particulate matter may be available in only very limited amounts (0.02–0.05 g); therefore, their analysis requires the use of methods with high detection power.

In biological samples, contamination from syringes, metal scalpels, and other metal tools can alter the species pattern of other elements. Species alterations due to bacterial activity have also been observed (Dunemann & Begerow, 1995). The extent to which the sample is representative of a chemical species is often not confirmed. Each of the thousands of possible major biological and environmental materials suffers from different matrix effects; even the urine of a given individual can differ substantially in its concentration of salts from one day to another. An overview of the sampling problems encountered under critical conditions in an extreme environment has recently been published (Caroli et al., 2000).

B. Sample Storage and Processing

Samples typically cannot be analyzed on site; therefore, storage becomes necessary and should be as short as possible. For replicate measurements on a given sample and for many applications, longer storage times become necessary. Proven storage techniques typically used for trace element analysis may be inappropriate for element speciation studies; for example, acidification cannot be used for element speciation studies because pH changes affect species composition and thus alter native speciation information.

Unreliable data can be found in the analytical literature due to the ubiquitous nature of certain elements and unaware analysts. Such elements are present in gloves, rubber stoppers, and anticoagulants. Among the many potential sources of contamination are dust, dirt, cosmetics, disinfectants, talc and dust on gloves, and metallic corrosion products. Laboratory dust can contain up to 0.3% aluminum, 0.3% calcium, 0.3% iron, 0.8% potassium, 0.2% magnesium, 0.3% sodium, 0.2% lead, 2% sulfur, and 0.2% zinc. Preservatives are also rich in elements; heparin contains calcium and zinc, and formalin contains iron, manganese, and zinc. These contaminants are likely to disturb not only trace element analysis but also element speciation. Such contaminations shift species pattern and may rearrange complexes in a sample. Even when the element species of interest is not the contaminating element, a shifting of species equilibria or changes in complexes are possible. Drying of samples can result in the loss of element species (predominantly volatile species). Problems of subsampling must also be considered. In most regulations and directives, analytical methods and sampling are only briefly addressed. Sampling and analysis should be described as accurately as possible to profit from the excellent spectrometric methods available.

Samples should not be contaminated or destabilized during storage and sample handling. Clean-room conditions and precleaned vials, among other precautions, must be used throughout. Sample preparation should be as simple as possible to reduce possible alterations. For this reason, storage time should be kept to a minimum, preferably at 4°C. For long-term storage, freeze-drying or shock freezing at −80°C is recommended; however, when the freeze-drying procedure is used, the process should allow for control of the sample temperature.

Even then, volatile species may be lost. No acid additions or other pH-changing agents are allowed, nor are repetitive, slow thaw/freeze cycles. Glassware can have ion-exchange properties; thus it is less suitable than polymer materials (Urasa, 1996). Before storing water samples, particulate matter must be removed to avoid species condensation and a shift in elemental speciation. On the other hand, colloids must not be removed, as they are considered to be a species group or fraction themselves. Mercury is a well-known example of the problems that can be encountered. The possibility of random errors in the determination of mercury due to migration phenomena is primarily associated with the sampling, pretreatment, and storage of samples. Mercury concentrations in water samples stored for a long time are strongly affected by several physical phenomena (*e.g.*, sorption and desorption), dissolvation, and passage through the walls of the container. The errors due to these processes are, however, in most cases independent of the analytical method used.

For biological samples, storage and sample preparation should be as short as possible, preferably at 4°C. Recently, solid biological samples were investigated for enzymatic extraction employing protease (different types) or lysozyme. Also, hotwater extractions were compared to diluted HCl leaching. When yeast was used for sample extraction, the efficiencies were quite high (*e.g.*, reaching, 80–100% of the total Se content of the sample after a protease treatment) (Potin-Gautier *et al.*, 1997). HCl leaching was used successfully for mimicking gastric juice digestion (Crews *et al.*, 1996), but it was found to be less suitable for speciation in a bacterial sample (Michalke *et al.*, 2002). Species stability in soils and sediments during storage can be a problem, as volatile species may be lost. Generally, extraction procedures are required for further analysis. This can obviously cause species alteration, depending on the chosen extractants, and bring subsequent analysis into the field of "operationally defined" approaches. As Pickering (1981) wrote: "The gained results often are bound to errors and limitations, finally leading to wrong or misunderstanding data."

Usually, extractions are single extractions aimed at availability studies or sequential extractions. Single extractions often use leaching agents such as H_2O, $NaNO_3$, NH_4NO_3, KNO_3, $CaCl_2$, CH_3COONH_4, EDTA, or CH_3COOH. A promising single extraction method recently was described for tin (Sn) species using low-power, short-time, microwave-assisted leaching of soil or sediment. Here, CH_3COOH or diluted HNO_3 were used as extractants, and microwaves were applied at only 30 W at approximately 60°C for a few minutes.

TABLE III.   Example of Steps in Sequential Extraction

Steps	Remarks
1. Exchangeable, adsorbed	—
2. Carbonate fraction	Subsequent fractions are usually not interfered.
3. Reducible fraction	This step is usually accomplished by using hydroxylamine–HCl/acetic acid; however, selectivity is doubtful.
4. Oxidizable fraction	Possible binding mechanisms are adsorption, chelating, and complexation; differentiation from step 3 is often impossible.
5. Residual fraction	For example, aqua-regia.

The stability of organotin was preserved, and extraction efficiency approached 100% (Rodriguez-Pereiro *et al.*, 1997).

Sequential extraction procedures attack the sample by utilizing consecutive leaching agents of increasing strength and ability to interact with the sample matrix. Typical steps are summarized in Table III. Representative extraction procedures have been described in a number of papers (Zeien & Bruemmer, 1989; Ure *et al.*, 1993). These procedures try to leach the elements from the different compounds and complexes in a stepwise fashion. Usually, however, the selectivity is not high enough. Problems may arise because of the pH dependence of the extraction step. Sample matrices can also alter pH and therefore selectivity.

C. Speciation Approaches: Direct Speciation Methods or Combined (Hyphenated) Techniques

After sampling, storage, and sample preparation, species are identified and analyzed. Direct speciation approaches can provide full information about the species in a sample without any additional (separation) method; that is, they can directly quantify the species. Such methods include nuclear magnetic resonance (NMR) or (in special cases) anodic stripping voltammetry (ASV) and cathodic stripping voltammetry (CSV). The concentration ranges detectable by NMR, however, are far too high for real-world samples of biological or ecological relevance. With ASV and CSV,

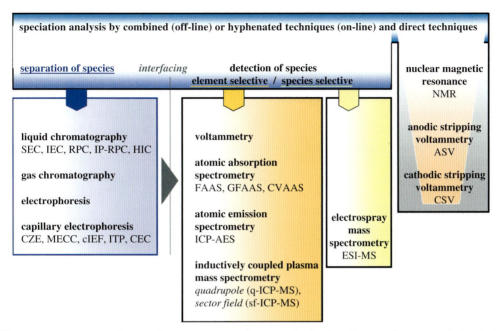

speciation analysis by combined (off-line) or hyphenated techniques (on-line) and direct techniques

separation of species *interfacing* **detection of species** **nuclear magnetic**
element selective / species selective **resonance**
NMR

liquid chromatography **voltammetry** **anodic stripping**
SEC, IEC, RPC, IP-RPC, HIC **voltammetry**
atomic absorption ASV
gas chromatography **spectrometry**
FAAS, GFAAS, CVAAS **cathodic stripping**
electrophoresis **voltammetry**
atomic emission CSV
capillary electrophoresis **spectrometry**
CZE, MECC, cIEF, ITP, CEC ICP-AES

electrospray
mass
inductively coupled plasma **spectrometry**
mass spectrometry ESI-MS
quadrupole (q-ICP-MS),
sector field (sf-ICP-MS)

FIGURE 1 Two speciation approaches are shown: one using direct methods, the other employing coupled techniques that include separation, interfacing/coupling, and selective detection. The latter are commonly used for element speciation due to flexibility and sensitivity.

quantification is rarely possible in samples with high organic loads.

The most usual approach for element speciation is based on combined (or hyphenated) systems. Here, species are selectively separated and then the elements in the various chemical forms are selectively detected. For increased quality control, molecule-selective detection is coupled to separation devices. These combinations provide extended flexibility and a broad applicability. Disadvantages include the increased complexity and thus increased risk for failure of the systems. Also, species equilibria can be drastically altered during separation due to dilution, and some components can be removed from the chemical equilibrium. Species transformation and destruction are likely consequences; therefore, the total separation time should be short compared to the transformation rate of species. Figure 1 gives an overview of the various speciation methods. In the following section, separation mechanisms are described and their features are taken into account from the viewpoint of element speciation. The hyphenation of separation techniques to element detectors is also discussed.

1. Liquid Chromatography

One of the most important advantages of liquid chromatography (LC) is the ample selection of separation mechanisms and the use of various mobile and stationary phases which provide nearly all the necessary techniques for separation of element species; therefore, specific problem-related speciation analysis is often possible that meets the requirements for species stability and efficient separation. Preservation of original species information is at least as necessary as good separation and influences the choice of separation mechanisms and reagents. Many stationary phases and buffers or organic modifiers can denature native species. Chelating eluents or ion exchangers may cause recomplexation of free or labile bound metal species (Dunemann & Begerow, 1995). A more general disadvantage is seen in the existence of a stationary phase, as compared, for example, to capillary electrophoresis (CE); the large surface allows adsorption effects, contamination, or miscellaneous alterations of the species to occur (Harms & Schwedt, 1994).

The mobile phases, too, can cause severe alterations of species, even if they may be very effective in separation. Although doubly distilled water is highly pure, some denaturing effects on biomolecules have been observed. Buffers can stabilize biomolecules, but they may also alter species equilibria. Complexing tendencies or input from metal contamination may also occur (Arnaud *et al.*, 1992).

The commonly used LC separation techniques in speciation are size exclusion chromatography (SEC), ion-exchange chromatography (IEC), reversed-phase liquid chromatography (RPLC), and ion-pairing chromatography (IPC). Advantages of SEC are seen in the fact that samples of unknown molecular masses are characterized in a mass-calibrated chromatographic system. SEC is a gentle method of performing chromatographic separation and normally does not result in a loss of element species or on-column alterations, although the column exhibits limited peak capacity. For complex multicomponent samples, complete resolution of the peaks is normally not achieved. Szpunar (2000b) reported shifting retention times for some compounds beyond the final elution volume. The stationary phase is not totally uncharged; for example, electrostatic effects have been observed when analyzing cadmium species. Adsorption, hydrophobic interactions, and species-specific affinities or H-bridging have also been observed. IEC shows high separation efficiency and wide applicability, thus solving many speciation problems. The relative retention of the ionic species is determined by three variables: namely, pH, ionic strength of the mobile phase, and nature of the ion exchanger. On the other hand, hydrophobic interactions between the sample and the non-ionic carbon backbone of the stationary phase cause organic ions to be retained in a way typical for RPLC (Mikes, 1988). The pore size of the resin particles is an important parameter for the success of the separation. Often, loosely bound metal ions are lost or replaced by other metals originating from the buffer (Quevauviller *et al.*, 1996); thus, IEC is a good fit for the separation of covalent bound element species in different valence states, such as Cr(III)/Cr(VI) or Sb(III)/Sb(V)/methyl-Sb (Lintschinger *et al.*, 1998). Another frequent application is for covalent bound Se amino acids (Michalke *et al.*, 2002) or a high number of arsenicals (Gössler *et al.*, 1997).

The predominant advantage of RPLC is the wide analytical spectrum available; this very effective separation technique provides high resolution of species, and the flexibility offered by the multiple mobile phases allows the addition of ion-pairing reagents for analysis of ionized molecules. Obtaining results is usually easy and fast. The normally high reproducibility is a significant advantage in speciation analysis (Mikes, 1988). In practice, the stationary phase can exhibit ion-exchange properties or undesired adsorption effects, especially for basic analytes. At pH values higher than 4, these basic analytes can be adsorbed tightly. Usually, two different eluents are necessary, at least one of them containing a considerable amount of organic modifier. The polar eluent often shows high complexing tendencies. Organic solvents and acids may easily change element

species such as protein–metal complexes. The structure of proteins may be unfolded, and complex-bound elements are subsequently released. Released metals are likely to be recomplexed by other ligands, which results in species transfer reactions; hence, only those analytes with no loosely bound metals may be separated by RPLC technique. Species with covalent element-ligand bondings are best suited to this method; therefore, RPLC is often used in parallel or in combination with other smoother techniques as a part of the so-called orthogonal speciation concepts (Szpunar, 2000a).

When performing RPLC, care should be taken to keep the pH below 7.5 to avoid hydrolysis of the stationary phase. Hyphenation to an element-selective technique, such as inductively coupled plasma–mass spectrometry (ICP–MS), causes further problems, particularly when RPLC is coupled online to ICP–MS. These problems are detailed in Section II.C.3. The use of ion-pairing reagents allows RPLC to analyze ionic species. Additional advantages of this method are self-evident; for example, flexibility and variability are drastically increased. Also, very efficient separation is possible for a wide range of analytes, and the separation conditions may be tailored for the specific separation task.

Problems and limitations of ion-pairing RPLC are also increased in element speciation and hyphenation to, for example, ICP–MS. Again, organic solvents and acids might be used that could easily change element species. The structure of proteins may be destroyed, and complex-bound elements can be released to be recomplexed by other ligands; thus, species transfer reactions are likely. Covalent element-ligand bondings are best suited for analysis by this method. The use of ion-pairing reagents intensifies these problems even more, and analyte stability during separation becomes more difficult to maintain with IPC than with RPLC alone.

2. Capillary Electrophoresis

One of the most powerful separation techniques, capillary electrophoresis provides a very efficient separation of species and is often superior to LC separation techniques. CE is able to separate positive, neutral, and negative ions in a single run with high separation efficiencies. A single CE instrument can even offer several different modes of separation: capillary zone electrophoresis (CZE), micellar electrokinetic chromatography (MEKC), isoelectric focusing (IEF), isotachophoresis (ITP), and capillary electrochromatography (CEC). All of these modes rely on the application of high voltage (Kuhn & Hofstetter, 1993); however, the separation principles are quite different, and they each provide a variety of characterization and identification mechanisms for elemental species. The latter point is of great importance, as species identification is rarely achieved by a single method; rather, it requires multidimensional strategies. Furthermore, species integrity is thought to be affected less than when HPLC is used. Limitations are seen in the very small sample volume used, thus giving rise to concerns with regard to the representativeness and homogeneity of the sample and instrumental detection limits (Michalke, 1999b).

3. Interfacing LC to ICP–MS

Combining separation methods with element-selective detection methods leads to hyphenated systems. These online systems give easier and faster results; thus the risk for contamination or loss is reduced. On the other hand, collected fractions allow for several quality control checks and application of the orthogonal identification concepts (Michalke & Schramel, 1997). Interfacing LC with ICP–MS is achieved by a nebulizer. Several pneumatic nebulizers are available, such as the Meinhard (MN), cross-flow (CFN), microconcentric (MCN), ultrasonic (USN), direct injection (DIN), and hydraulic high pressure (HHPN) nebulizers. Nebulization efficiency ranges from around 1 to 5% for the MN and CFN to quite high values for the DIN and USN (Dunemann & Begerow, 1995). The MCN and DIN are used for low flow rates between about 30 and 150 $\mu L\,min^{-1}$ and are believed to achieve nebulization efficiency up to 100%. These are suitable systems for interfacing with microbore LC.

The major problems with online hyphenation include:

1. The salt load of eluents can cause problems, such as crusting and changes in ionization.
2. The use of high amounts of organic solvents in RPLC cools the plasma and increases the reflected radiofrequency (RF) power. This results in plasma extinction already at relatively low organic solvent concentrations. Further, the ionization characteristics of the argon plasma are altered, which affects the sensitivity of the detector for element species. The extremely high carbon intake induces polyatomic interferences and carbon precipitation on the torch and cones. The conductivity of carbon can cause flashovers from the coil to the carbon-coated torch.

3. Sample transfer to ICP–MS; This may be influenced by dead volume of the interface, or affected by peak broadening and flow rate.

Developments to address the first issue listed above are typically based on novel column technology that provides high separation efficiency even at low (buffer) salt concentrations. The easiest way to overcome the problems caused by the second issue is post-column dilution of the eluent; however, this results in dilution of the separated analytes. The most common method to stand high organic modifier concentrations is to reduce the evaporation pressure of the modifier by cooling the transfer line and/or the spray chamber below 10°C. Methanol concentrations up to 60 to 80% in the eluent are tolerated at flow rates of about 1.5 ml min⁻¹. For improved burning of carbon from MeOH, the nebulizer gas is added with small amounts of oxygen. Desolvating systems are also used that can tolerate methanol concentrations up to 100% (Lustig et al., 1998); however, one should be aware that some species may be removed in the desolvator. Further aspects of nebulizer behavior are discussed by Montaser and Golightly (1992).

4. Interfacing CE to ICP–MS

Much effort has been devoted to interfacing CE with ICP–MS. It has been demonstrated that such hyphenated CE techniques can provide sub-$\mu g\,L^{-1}$ detection limits for the analysis of many types of environmental samples and are also capable of multiple-element monitoring of various metal functionalities. At present, an efficient interface is still unavailable. Furthermore, the tiny amounts of sample emerging from the capillary often give rise to detection limits that are usually higher than those of conventional chromatographic methods. An exhaustive discussion on CE–ICP–MS is given by Olesik (2000), who covers theory, application, and instrumentation of the procedure. One requirement for the interface is being able to close the electrical circuit from CE at the end of the capillary. The flow rate of CE generally does not match the flow rate needed for an efficient nebulization, but one possible way to circumvent this problem is to close the electrical circuit of CE during nebulization by applying a coaxial electrolyte flow around the CE capillary. The grounded outlet electrode is, in all cases, in contact with this electrolyte flow. The sheath flow has also been used to adapt the flow rate to a suitable nebulization efficiency. Optimization of nebulization efficiency was achieved by adjusting the flow rate when using MCN- or DIN-based systems or by exactly positioning the CE capillary to the point of nebulization (*e.g.*, by employing a micrometer screw).

5. Interfacing CE to ESI–MS

Electrospray ionization (ESI) interfaces for CE are commercially available. The closing of the electrical circuit from CE during ion evaporation is provided by an electrolyte sheath flow. Effective ion production is made possible by the use of a suitable spray voltage easily controlled by the instrument software. For older instruments, a laboratory-made device for reproducibly optimized positioning of the CE capillary to the ESI tip is still necessary.

6. Gas Chromatography

The choice of an adequate separation technique is determined by the physicochemical properties of the analyte (volatility, charge, polarity), whereas that of the detection technique is determined by the analyte level in the sample. The combination of gas chromatography (GC) with ICP–MS has become an effective method for the speciation of organometallic compounds in complex environmental samples. GC separates volatile and gaseous element species employing primarily capillary columns with bonded phases. The big advantage is seen in the fact that species no longer have to be transferred into an aerosol or gaseous phase. Sample input into, for example, ICP–MS is about 100%, thus improving detection limits. Very often, however, species are not sufficiently volatile, and in such cases time-consuming and tedious sample preparation, extraction, and derivatization procedures are necessary, especially when carrying out a Grignard derivatization (Quevauviller *et al.*, 1996). This approach, on the other hand, removes the matrix components and makes separation easier.

Major problems are often encountered with derivatization, which is sometimes less selective than expected (Quevauviller *et al.*, 1996). The detector response is species-derivative specific, an important aspect to consider in quantification; however, some derivatives have nearly no detector response. The separation typically is carried out at elevated temperature; thus, only thermally stable species can be handled by GC. Another problem arises when coupling to element-selective detectors such as ICP–AES or ICP–MS. This coupling is not as straightforward as with LC and is affected by several limitations; for example, analytes have to be maintained in the gas phase during their transport from the GC to the ICP–MS to avoid condensation effects. The transfer line to the plasma should also be heated, either by a pre-heated sheath gas or by electrical

heating. The proximity of metal parts (heating wire) to the generator coil is problematic. Also, the effluent from the gas chromatograph (a few milliliters per minute) requires an additional carrier gas to achieve sufficient flow in the central channel of the plasma. For these reasons, most of the published papers devoted to the use of GC–ICP–MS address the construction and development of adequate interfaces. Typical applications of GC in speciation analysis are quantification of alkylated species of arsenic, mercury, lead, selenium, and tin.

7. Element-Selective Detection

Spectrometry-based techniques for the quantification of elements have long since taken root. Spectrochemical methods are frequently used because of their speed, detection power, sensitivity, and specificity. The impact that atomic spectrometric techniques in general have had on governmental institutions and international organizations has been carefully discussed by Minderhoud (1983), who made particular reference to the analysis of chemical wastes, sewage sludges, surface waters, and airborne particulates. The role of spectrochemistry in bioinorganic chemistry is still growing and will be even greater when chemical speciation can be fully accomplished.

Atomic absorption spectrometry (AAS) systems are comparatively inexpensive element-selective detectors that recently also matured from mono-elemental to multiple-element systems. There are flame (FAAS), cold vapor (CVAAS), hydride generation (HGAAS), and electrothermal atomization (ETA-AAS) AAS systems. The detection power of FAAS is often insufficient for normal environmental or physiological concentrations (Dunemann & Begerow, 1995). The sample intake is high (4–5 mL min^{-1}), which complicates online hyphenations with HPLC (optimized flow rates at about 0.5–2 mL min^{-1}); therefore, it becomes necessary to have an auxiliary flow, which results in analyte dilution. CVAAS and HGAAS use selective derivatization for matrix separation and detectability of relevant species. The detector response is strongly species dependent and often easily interfered. As an example, As species should be mentioned: if As(III) is assumed to have the highest response (100%), then As(V) is only about 85%, whereas arsenobetaine shows absolutely no response. ETA-AAS needs samples of only a few microliters and provides really low detection limits of between 0.1 and 5 µg L^{-1}. Matrix interferences are widely eliminated via Zeeman correction and matrix modifiers. However, quantification errors are still possible, as the final atomization temperature is only up to

2900°C. An optimized temperature program is part of the determination, which is the reason for a discontinuous measurement; therefore, ETA–AAS is unsuitable for online hyphenation to HPLC, as the chromatographic data points are gained only at intervals of a few minutes. This is too slow for peak description in a chromatogram. Moreover, during the temperature increase for sample drying, volatile species may be lost before being atomized.

Commonly used alternatives to AAS-centered approaches include ICP–AES and ICP–MS detectors. The special diagnostic advantage of plasma-based techniques is their rapid screening ability, which has often confirmed suspected heavy metal poisoning (Ure *et al.*, 1993), mineral deficiencies, or storage diseases. Equally as often, however, these techniques have identified completely unsuspected etiologic factors. Barnes (1991) surveyed the potential of ICP–AES combined with flow injection analysis, direct sample introduction and vaporization systems, electrothermal vaporization, hydride, metal vapor and gas generation, and chromatographic techniques. The big advantage of ICP–AES is offered by its multiple-element capability and sensitivity. Online hyphenation is easily set up. The ionization source is an inductively coupled plasma, and the temperature is around 5000 to 9000 K. The plasma (mostly argon) is formed within a quartz torch made up of three concentric quartz tubes, with the gas flowing at different rates through each. The outer flow is the highest and is known as the "plasma," "coolant," or "support" gas flow. It is tangentially introduced into the plasma torch to provide a helical flow that sustains the plasma itself. The central gas flow, known as the "auxiliary" gas flow, keeps the plasma away from the edge of the quartz torch. The inner gas flow, commonly called the "nebulizer" gas flow, transports the nebulized samples to the plasma. An RF field provides the energy to sustain the argon plasma, and the plasma transfers energy to the analyte(s) for excitation and ionization. Excellent discussions of plasma theory, mechanism, and applications are given by Montaser and Golightly (1992). Chemical interferences such as molecular emissions cause no major problems, but background correction should be applied in any case. Sample introduction is performed via a nebulizer and spray chamber. Nebulizer types, related problems, and some solutions have already been discussed earlier. Spectrometers might be of the sequential or simultaneous type. For on-time, multiple-element monitoring, simultaneous ICP–AES systems are used, as the sequential type results in a loss of chromatographic datapoint resolution (similar to ETA–AAS). Recently, ICP–AES systems equipped with

a charge-coupled device (CCD) helped to overcome this problem.

Inductively coupled plasma–mass spectrometry is now the technique of choice for a wide range of samples with element concentrations in the $\mu g\,g^{-1}$ to sub-$ng\,g^{-1}$ range. It has become a highly versatile technique with low detection limits and high sensitivity. Also, thanks to its element specificity, it is a technique of choice for chromatographic detection, including GC, LC, SFC, and CE. The striking advantages of ICP–MS techniques are their detection selectivity, multiple-element capability, and high sensitivity. Isotopic and elemental information of species is obtained, and species not totally resolved but pertaining to different elements are distinguished by the selective detector.

Sample introduction is performed through an interface connecting the LC system to the ICP–MS. The availability and pros and cons of such interfaces (various types of nebulizers) have been discussed previously. In hyphenated systems, detection limits for element concentrations in elemental species were reported to be in the range of 10 to $100\,ng\,L^{-1}$ (Olesik, 2000; Michalke *et al.*, 2001). ICP–MS can be of the quadrupole type (Q-ICP–MS) or the highly resolving sector field type (SF-ICP–MS). The latter is reported to improve detection ability by a factor of 10 to 100, or even up to 1000 when equipped with a guard electrode (Prange & Schaumlöffel, 1999). The quadrupole mass filter provides a resolution of only 1 amu; therefore, polyatomic interferences are likely to occur, especially in the mass range of 40 to 80 amu. Well-known interferences are those of the $^{40}Ar^{35}Cl^+$ double ion on monoisotopic ^{75}As or of $^{40}Ar^{12}C^+$ on ^{52}Cr (Tittes *et al.*, 1994). They are produced in the argon plasma when chlorine and carbon are introduced into the plasma (sample and buffer components, respectively). The highly resolving SF-ICP–MS can distinguish between the interference and the element isotope, as the mass resolution is 7500 to 10,000 amu (as compared with 300 amu for Q-ICP–MS). An example for such separation is given in Figure 2.

When using SF-ICP–MS at a high mass resolution, the detection power is reduced, and it falls into the same range of Q-ICP–MS. With element-selective detectors such as ICP–MS, one has to realize that only the element in the species is detected, not the entire molecule, which is advantageous because the separation of molecules bound to different elements does not need to be complete. They may be screened by the detector responding to different isotopes; however, it must be kept in mind that the molecule itself is not seen, and the species identification is only possible by comparing

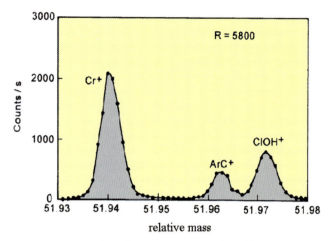

FIGURE 2 ICP–MS detection may be affected by, for example, polyatomic interferences. Here, the resolution of $^{40}Ar^{12}C^+$ interference and ^{52}Cr by sf-ICP–MS is shown. (From Tittes, W. et al., *Reduction of Some Spectral Interferences in ICP–MS*, Finnigan MAT Elemental Mass Spectrometry Technical and Application Note 3, Finnigan MAT GmbH, Bremen, Germany, 1994. With permission.)

retention times. In natural samples, this is not always achieved beyond doubt; therefore, very often speciation analysis with only, for example, LC–ICP–MS is not enough for unequivocal speciation results. Multidimensional analytical concepts are strongly indicated in such cases (Michalke, 1999a).

For the identification of polyatomic interferences, the monitoring of several isotopes of a given element can be helpful. Only when the natural isotope ratio is measured can interferences be ruled out. Unsatisfactory sensitivity is still a problem for very low concentration samples; hence, monitoring the most abundant isotopes of an element is recommended, except when these isotopes are interfered (e.g. poly-atomic interferences). On the other hand, ICP–MS is a sequential detector that monitors the programmed isotopes for several milliseconds. If too many isotopes are programmed for subsequent determinations, the detector operates too slowly to allow for highly resolved and fast-appearing peaks on one specific mass. This causes a loss in chromatographic resolution of the hyphenated system. The recently available time-of-flight (TOF) ICP–MS can overcome this drawback, as the mass filter does not jump in a time-consuming manner from one mass to another. Instead, the different isotopes are distinguished as a function of their individual (m/z-dependent) time to reach the detector.

8. Species-Selective Detectors: Electrospray Ionization Mass Spectrometry Detection

Electrospray ionization is a process that may preserve the whole species intact under optimal circumstances. ESI is suitable for extremely low flow rates. It is based on the so-called ion evaporation principle, where charged droplets of the analytes are transferred into the gas phase. A volatile buffer consisting of considerable amounts of, for example, methanol supports the ion evaporation process. In fact, the high volatilization capability of CE electrolytes is mandatory. The success of this detection method is based on the ability to produce multicharged ions from high molecular element species, such as metalloproteins, thus making the analysis of these compounds feasible up to molecular weights of 150,000 to 200,000 amu. The possibility of coupling this detector to LC or CE systems makes it extremely valuable. The soft ionization of element species finally allows preservation of the entire molecule (element species) when it is transferred into the gas phase and subsequent analysis by mass spectrometry (Cole, 1997). Structural changes normally do not appear as long as covalent bonds are present. In special cases, stable element–organic molecules can be analyzed.

When applying collision-induced dissociation (CID) together with an MS/MS system, further structural information can be gained as the parent ions are fragmented into molecule-specific daughter ions, which are then selected by the second quadrupole. No other detection technique is able to provide such detailed information about molecular weight and even structure of the analyzed compounds. On the other hand, significant problems have been described for ESI. One problem arises due to the ion–solvent clusters. During the transfer of gas-phase ions into the high-vacuum (10^{-9} bar) zone, condensation of solvent molecules (*e.g.*, methanol, water) from the gas-phase ions is likely to happen. This is what is referred to as an ion–solvent cluster and is caused by the cooling-down phenomena that occur when a gas expands into a vacuum (*i.e.*, free jet expansion, adiabatic expansion). Ion–solvent cluster production results in the splitting up of one species into multiple signals, worsening detection limits, and increasing spectral complexity. Electrolytic processes at the metallic ESI tip needle that can generate new species or transformations of species (*e.g.*, by metal exchange) are also observed. When analyzing free metal ions such as Cu(II), multiple ion–solvent cluster signals are again detected. Most important, however, is the fact that native counter-ions of the metal ion are replaced by H_2O and/or methanol, independent of the counter-ion initially present (*e.g.*, $[Cu(MeOH)]^+$) instead of $Cu^{2+}2Cl^-$). This implies a total loss of the original species information.

III. QUALITY CONTROL IN SPECIATION

Causes of disagreement may be traced back to poor methodology, improper instrument calibration, faulty experimental techniques, impure reagents, or a combination thereof. Because of the lack of reliable data on trace elements in biological fluids, the reported diagnostic significance of some elements is controversial. The same holds for inaccurate measurements of pollutants in environmental matrices such as sediments, water, and particulate matter. Extremely sensitive instrumentation is readily available in laboratories not equipped to control contamination, and many users of a particular technique do not fully understand the limitations of a methodology. All of these factors introduce a great deal of questionable data. Much work on trace elements in human body fluids and tissues has suffered from methodology deficiencies, but accuracy is needed to reach rational conclusions upon which basic health-care decisions are built. A rigorous program of quality control/quality assurance is needed to confirm the reliability of results for trace elements in biological materials.

A. General Aspects of Quality Control

Quality control plays a crucial role in element speciation. General hypotheses and analytical models depend on the reliability of data (Quevauviller *et al.*, 1996). The key to successful speciation is the preservation of species information during the whole analytical procedure from sampling to final analysis. Actually, this is rarely guaranteed, and the range of errors is extremely high; thus, quality control is needed in the planning stages of an experiment. Pertinent literature must be sought and general criteria must be adapted to the actual problem relevant to the elements and matrices of interest.

All trace and ultratrace analyses require clean workplaces and the continuous use of appropriate control materials. These prerequisites are absolutely necessary for reliable results, and determination of the blank level will be that much more accurate. If these rules are not

strictly followed, the money and time spent are completely wasted. For work with trace elements (and their chemical species) at very low concentration in biological materials, clean-room techniques may become mandatory. Other prerequisites are listed below:

1. *Calibration*—A necessary prerequisite of reliable analytical work is correct calibration using calibrants of each investigated species with known stoichiometry. Preparation of two parallel calibrant solutions should be done according to weight and not volume. High purity of chemicals is compulsory. New lots of calibrants must be verified. Calibration graphs of the single species must be generated.

2. *Quantitative speciation*—Species quantification is the basis for obtaining mass balances and gaining information about losses and contaminations. Quantification must be done using the relevant calibration curve (or standard for standard addition). If unknown compounds are monitored, quantification is not possible. Species can only be estimated by relating peak area calibration graphs to those of closely eluting known species.

3. *Certified reference materials*—One of the favored approaches for quality control is the use of reference materials (RMs) or certified reference materials (CRMs), although in environmental analysis there is a dismal lack of RMs to match real-world situations. The somewhat restricted market is certainly partially responsible for this. The introduction of more stringent regulations will enhance the demand for RMs. Operational complexity associated with the production of environmental RMs and the amounts required (5000–1000 kg) are also paramount factors.

Blending is an alternative procedure to prepare RMs of intermediate concentrations. Strategies to ensure accuracy include: (1) building accuracy into RMs under the guidance of a few centrally operating agencies, (2) transfer of accuracy from RMs to the measurement system, and (3) safeguarding the accuracy levels by continuous measurement quality control. Political and economical decisions concerning the environment are based on the correctness of analytical data. The last 30 years are testament to a growing awareness of the mistakes and pitfalls that accurate RMs should help eliminate. Suitable RMs and CRMs should be included at the earliest possible stage in any speciation analysis process. Confirmation of the certified value can prove the reliability of the results for unknown samples (Quevauviller *et al.*, 1996); however, doing so never gives full evidence

TABLE IV. General Criteria for Speciation Analysis

Sampling	Extent to which the samples are representative should be verified (Quevauviller *et al.*, 1996).
	Sampling time should be kept short.
	Contamination phenomena should be minimized.
	Volume-to-surface ratio should be high to reduce wall effects.
	Use of stainless steel tools is not advisable when sampling biological materials (Dunemann & Begerow, 1995).
Sample preparation	Short storage at °C or freeze-drying (possible loss of volatile species) is recommended.
	Extractions typically result in operationally defined speciation.
	Mass balances and recovery rates (spiking of species) should be determined; species spiked can exhibit different extraction behavior.
	Species spikes endanger native equilibria in the sample and could lead to a changed species pattern; comparison of different extractions is the best way to get reliable information (Quevauviller *et al.*, 1996).
Derivatization	Derivatization should be avoided; it increases detection power but can affect the species.
	Selectivity is much lower than expected unless other options are available.
	Different efficiencies of derivatization and different detector responses of the derivatives are observed to some extent (Quevauviller *et al.*, 1996).
Separation	Stationary phases may cause contamination and retain undesirable reactive groups.
	Contamination or stability problems of species occur; these undesirable effects should be monitored by mass balances or checked by reinjection experiments (Michalke & Schramel, 1997); potential species transfers should be investigated and possibly excluded.
	Identification problems arise when the identity of a species is only attributed by retention times in a low resolving separation system.
	Multidimensional analysis using various independent methods is the best alternative (Michalke, 1999a; Szpunar, 2000b).
Detection	Nonspecific detection should be avoided.
	Calibration should be done properly.
	Suitability of the detector for the problem at hand should be checked.
	The discontinuous measuring nature of ETA–AAS results in too low chromatographic resolution; quadrupole ICP–MS exhibits maximum mass interference in the *m/z* range of 40–80.
	Several isotopes of the same element should be monitored to check polyatomic interferences.
Multidimensional analytical concepts	Combining various separation techniques and detection systems is recommended.
	Two schemes are generally employed:
	1. Separation methods based on different separation principles are combined. After the first separation, fraction collection provides aliquots to be used for element determination and as samples for the second separation, which is again followed by online ICP–MS monitoring. This procedure provides an orthogonal characterization of molecules.
	2. Separation is monitored in parallel by ICP–MS and ESI–MS. In this case, ICP–MS provides the element information and ESI–MS (or MS/MS) gives molecular or structural information; hence, obtaining maximum species information and sometimes identification becomes possible. These schemes can also be combined in different ways (Michalke, 1999a; Szpunar, 2000b).

of the correctness of results, as sample matrices are rarely identical to the CRM matrix. The concentration range of the species or the species pattern may be different, which results in variation in their behavior during the analysis (Quevauviller *et al.*, 1996). Many CRMs are available—for example, from the National Institute of Standards and Technology (NIST) in the United States, Institute for Reference Materials and Measurements (IRMM) in the EU, and National Research Council (NRC) in Canada. In practice, however, many other precautions are necessary, as summarized in Table IV.

Preservation of samples for future controls and investigations into their content with regard to unsuspected chemical species are well-recognized needs. Sample banking is a new concept that involves the preservation

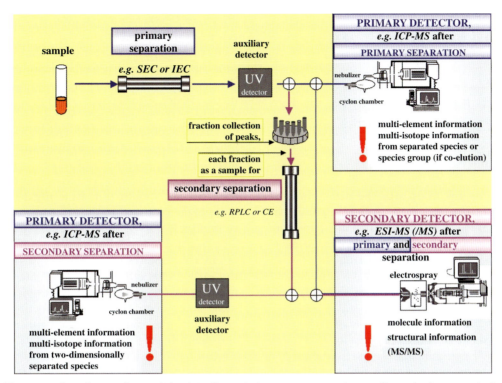

FIGURE 3 The principles of an orthogonal (analytical) speciation concept are shown. General advantages include the ability to obtain multi-element/multi-isotope information using element-selective detection after the first species separation combined with molecular (and probably structural) information provided by a second molecule-specific detector. In case of coelution after the primary separation step, further purification is achieved by a secondary separation step, again providing multi-element/multi-isotope and molecule-selective information. Clear species identification is usually possible.

of important samples under unequivocal conditions that ensure their integrity over extended periods of time, but sample banking is expensive.

B. Orthogonal Analytical Concepts

Analytical strategies that employ combinations of various separation and/or detection methods are referred to as *orthogonal analytical concepts*. They are an indispensable means for quality control in speciation and offer the best opportunity for obtaining accurate speciation results and even identification of unknown species. In analytical systems with (only) one separation and one detection system, the risk of coelution, impossibility of species identification, or misidentification is high (McSheehy *et al.*, 2002). This problem can be solved by employing different systems in various ways.

In short, these multidimensional analytical concepts rely upon combinations of various separation technologies and detection systems. Two schemes are primarily employed:

1. Two separation methods based on different separation principles are combined in a series. After the primary separation, fraction collection provides aliquots to be used for element determination and as samples for the secondary separation, which is again followed by an (online) ICP–MS monitoring. This results in an orthogonal characterization of molecules.
2. The separation is monitored in parallel by ICP–MS and ESI–MS. Here, ICP–MS provides the element information while ESI–MS (or ESI–MS/MS) delivers the molecular or structural information; hence, maximized species information and sometimes identification are possible.

Often, it is necessary to combine the two schemes in different ways (Michalke *et al.*, 1997; Chassaigne *et al.*, 1998; Michalke, 1999a; Szpunar, 2000a). Figure 3 provides an orthogonal flow chart, and a current example has been published by McSheehy *et al.* (2002). Similarly, Nischwitz *et al.* (2003) checked species preserving extraction techniques using an orthogonal scheme, first using SEC with ICP–MS, fraction collection with subsequent RPLC–ICP–MS, and, finally, after collecting cleaned fractions from the latter, ESI–MS detection.

IV. SELECTED EXAMPLES OF SPECIATION STUDIES ON THE LIFE SCIENCES

A. Environmental Applications

Environmental analysis deals with the detection of a variety of substances naturally or artificially present in our total environment and known or thought to exert adverse effects on human health. Environmental measurements are a special class of determinations with common problems. Ordinarily, data have large uncertainties, and the resulting decisions are controversial. Accuracy in environmental analysis is still a problem, notwithstanding the growing awareness of its importance. Trace element measurement quality is still far from satisfactory. Environmental data are obtained for the purpose of information and/or action. Since the industrial revolution, mankind has continued to pollute the Earth's biosphere with toxic substances such as heavy metals and halogenated organic compounds. Many of these substances are not amenable to processes that would lead to their removal; as a result, many industrial chemicals build up in the environment. During the past 100 years, the environment has been severely polluted by arsenic, cadmium, mercury, lead, and thallium. The increasing interest of governmental agencies, particularly the Environmental Protection Agency (EPA), in ICP–AES as a novel and useful alternative to established procedures for monitoring elements in environmental media was stressed 20 years ago in a report on the potential future of this technique (Barnes, 1983).

Large amounts of impurities (both natural and industrial) can be found in the atmosphere. Some of them form flying ashes and other gaseous emissions with numerous elements, the toxicity of which depends on their chemical species. Limit values in regulations must be checked regularly, and levels of pollutants must be known in order to identify systematic changes and pollution sources.

The well-documented risks posed by concentrations of mercury in the environment are caused by human activity. Although the annual world consumption of mercury by industry is estimated to be around 10,000 tonnes, the total worldwide release of mercury as a result of human activities has been estimated to be

20,000 to 35,000 tonnes per year. It is unlikely that such releases can significantly increase the average mercury concentrations in the oceans, although increased levels of mercury can occur regionally and locally because of the release of mercury compounds into the environment. Water courses passing through areas where mercury-rich minerals are found (the main mercury minerals are cinnabar and metacinnabar, the two polymorphs of HgS) are known to result in elevated mercury levels. In the aquatic food chain, mercury plays an important role. Because of bioaccumulation processes, the mercury content in marine organisms is normally quite high compared to the content in water, and concentration factors of 10^4 and higher have been reported. The health hazards associated with mercury pollution of environmental waters were first brought to light in the early 1960s due to the Minamata mass poisoning catastrophe of several hundred people (mainly fishermen and their families) by methylmercury. The poisoning was caused by the consumption of fish and shellfish caught in Minamata Bay, which had been contaminated by industrial mercury discharge (see also Chapter 23, this volume).

Three categories of methodologies for environmental analysis can be defined as: (1) definitive techniques of high precision and accuracy—and high cost; (2) more routine techniques to be used for continuous monitoring of exposure to pollutants; and (3) field methods for preliminary semiquantitative assessment under emergency circumstances requiring urgent countermeasures.

Marine- and freshwater have been significantly contaminated by organotin compounds, such as mono-, di-, and trisubstituted butyltin and phenyltin, because of their use as agrochemicals, biocides, and domestic products. The performance of four specific devices—namely, flame photometric detector, pulsed flame photometric detector, microwave-induced plasma atomic emission spectrometer, and inductively coupled plasma mass spectrometer—was compared in terms of ability to quantify the species of interest after separation by means of solid-phase microextraction and GC (Aguerre et al., 2002). The different approaches showed that the determination of ultratrace tin species is possible in natural waters even for routine purposes. Detection limits as good as $0.5 \, \text{ng} \, \text{L}^{-1}$ tin could be reached.

A detailed description of the complexation chemistry of copper in natural waters, with particular reference to water quality criteria and to possible effects on biota, was reported by Allen and Hansen (1986). It was concluded that the bioavailability and toxicity of Cu depend primarily on pH and alkalinity. When these two parameters are constant in a given system, then bioavailability and toxicity are proportional to the concentration of free copper ions and inorganic Cu complexes.

B. Nutritional Applications

It is a generally valid principle that each nutrient at excessive concentrations can be toxic; conversely, any of the trace elements now known for their toxicity might be shown in the future to have an essential function at low concentrations. Among the trace elements of nutritional interest are arsenic, chromium, manganese, molybdenum, nickel, selenium, silica, tin, vanadium, and perhaps cadmium. These elements present serious problems of analysis in the concentration ranges that are of interest to the nutritionist, either as toxic species even at low concentrations (*e.g.*, As (III)) or as essential element species.

The initial recognition that a micronutrient is essential for animal species raises questions about its nutritional and public health importance for humans. Nutritionists approach these questions by working through some specific tasks, such as identification of the metabolic parameters in humans that might be affected by the micronutrient, detection of disease states that can be prevented or cured by supplementation, determination of the human requirement for the micronutrient, assessment of the risk for dietary imbalances in population groups, and, finally, development of methods for assessing the nutritional status in individuals. All of these tasks, except for the last one, have been more or less accomplished.

Zinc, for example, plays a key role in erythrocyte carbonic anhydrase, an enzyme catalytically involved in the transport of CO_2 in blood. From a nutritional viewpoint, it is interesting that zinc deficiency is typically seen in population groups living in poverty. Consumption of vegetarian phytate-rich food is common and provides zinc–phytate complexes (species) of reduced bioavailability (Brätter *et al.*, 1992). An analogous situation is known from formulas for newborns as compared to human milk. Formulas based on cows' milk contain higher amounts of casein, calcium, and phosphorus. These compounds together associate to casein–phosphorus–zinc micelles of low availability for zinc. Contrary to this, human milk shows zinc–citrate complexes that are easily accessed by the newborn's gut.

The ability of surfactants to differentiate between methylmercury and inorganic Hg(I) in fish-egg oil was ascertained by using an online, time-based injection system in conjunction with CVAAS (Burguera *et al.*,

1999). An advantageous flow could be obtained from the highly viscous sample by injecting it into a three-phase surfactant (Tween 20®)–oil–water emulsion. Quantities of mercury as low as $0.1\,\mu g\,L^{-1}$ could still be measured. The range of organic mercury concentrations in the catfish-egg oil sample was found to be 2.0 to $3.3\,\mu g\,L^{-1}$.

C. Biomedical Applications

The overall importance of analytical atomic spectrometry in biology and medicine was reviewed by Dawson (1986). While this author recognized the fundamental role played by ICP–AES and other spectroscopic techniques in generating accurate and reliable data, he also emphasized that the next step in the analysis of elements of well-established biological importance would be a growing emphasis on the identification and quantification of their biologically active species through coupling with adequate online separation procedures.

The importance of trace elements and the identification and quantification of their chemical forms cannot be overestimated. Some of the elements essential to humans (cobalt, chromium, copper, iron, iodine, manganese, molybdenum, nickel, selenium, silica, and zinc; see Brätter *et al.*, 1992) have a vitamin character. Others are present in pharmaceuticals as active agents (aluminum, arsenic, gold, bismuth, copper, iron, mercury, lithium, platinum, and zinc). Uncontrolled use of pharmaceuticals can lead to undesired effects and intoxications, thus requiring immediate control therapy.

The human requirement for chromium is lower than that for any of the essential trace elements, except for cobalt as a part of vitamin B_{12}. Chromium is not only an essential ultratrace element but also a potent carcinogen. Many other essential trace elements are also toxic at excessive exposures, but chromium is unique in that its essentiality is limited to one valence state and toxicity to another, and transformation from the essential to the toxic valence state does not occur in the living organism. Those and other properties of chromium that influence its interaction with biological systems reside in its chemical behavior. Chromium in the trivalent state is an essential element for animal species and humans; in the form of certain hexavalent compounds it is a potent carcinogen. Hexavalent compounds are manmade and do not occur naturally in living organisms. They penetrate biological membranes and are reduced by organic matter, which leads to oxidative damage of cell structures. While there is little evidence for a role of Cr(III) in enzyme systems, its interactions with nucleic acids and with the functions of the thyroid gland and of insulin have been demonstrated. The lack of a simple diagnostic procedure, which chromium shares with many other essential trace elements, is the main impediment to the wide application of chromium in medicine and nutrition. Without diagnosis of chromium status, the response of an individual to supplementation is unpredictable.

Among the essential trace elements, selenium is receiving increasing attention as a natural cancer-preventing agent. The anticarcinogenic effects of selenium have been demonstrated in numerous animal tumor model systems as well as under conditions simulating human dietary conditions (Patching & Gardiner, 1999). Because the positive effects of selenium on Kashin-Beck disease are known, its protective mechanisms on several heart diseases, predominantly cardiovascular damage or congestive cardiomyopathy, have been widely investigated (see also Chapter 15, this volume). Detoxification effects of selenium have been proven and described for various metals such as arsenic, thallium, silver, cadmium, and mercury. Selenium deficiency is most critical for the brain and growth of infants. Furthermore, the thyroid metabolism may be impaired, because many deiodinases are Se proteins. These positive effects of selenium species led to several studies on selenium speciation in supplements, food (Quijano *et al.*, 2000), and body fluids such as human milk (Michalke & Schramel, 1997) and serum. The selenium speciation in serum is expected to mirror the available concentrations of relevant protective selenium species. As an example, Figure 4 shows various chromatograms monitoring ^{82}Se after SAX separation from children's sera. Although at present not fully explained, these chromatograms from the sera of children with absorption abnormalities in the gastrointestinal tract show different patterns of selenium species as well as different total selenium levels. Gaining a full understanding of possible interrelations between alterations in the gut and the selenium species pattern will be a demanding task in the future.

Iodine, long known to act beneficially in human health as an essential micronutrient, is utilized by the thyroid gland for the biosynthesis of the thyroid hormones thyroxin (T_4) and triiodothyronine (T_3) (Keller, 1991) (see also Chapter 5, this volume). These hormones strongly influence an extended range of biochemical reactions. Immune defense and antibody production are dependent on reliable thyroid function and the availability of T_4 and T_3 hormones. The speciation of various iodine species in serum or urine provides information about malfunction of the thyroid gland and can also explain other T_4/T_3-influenced metabolic abnormalities. The superiority of hyphenated techniques with ICP–MS (iodine-) detection over monoclonal antibody systems has recently been demonstrated, as the monoclonal antibody systems were generally unable to distinguish between iodinated (active) and non-iodinated hormones (Michalke *et al.*, 2000). Recently, iodine speciation was reported during investigations into the disruption of normal thyroid function by xenobiotic chemicals.

The increased rate at which zinc-containing metalloenzymes have been identified in the past is largely due to the development of highly precise, rapid, and convenient methods for determining this element. Definitive knowledge that zinc is indispensable to living matter has emerged only in the last few decades. In succession, the biological effects of zinc have been viewed as mostly harmful, then questionable, then essential; however, it is now well established that zinc is essential for the growth and development of all living forms. There is a lack of reliability in the techniques for aluminum determination that can affect a large number of samples encountered in clinical practice. Better measures for the quality control of such determinations have progressively led to improved detection limits for aluminum in biological material, thus paving the way to more reliable figures for the concentrations of their species of clinical relevance. Human serum was incubated *in vitro* with the radiotracers ^{51}Cr(III), ^{191}Pt(I), or carrier-free ^{48}V(V) (Lustig *et al.*, 1999), and the protein-bound metals were measured by flatbed electrophoresis followed by autoradiography with laser densitometry, followed by subsequent detection of the proteins through silver staining. At this stage of its development, however, this proposed technique, although highly promising, does not possess adequate detection power.

V. Summary

The future of speciation analysis depends to a large extent on three major factors: (1) development of instrumentation actually designed for this purpose (and not simply assembled from apparatuses originally conceived for another purpose), (2) production of a substantial number of CRMs especially prepared for chemical species (and not only the total content of a given element), and (3) ability to transfer the relevant know-how from the expert laboratory to the routine

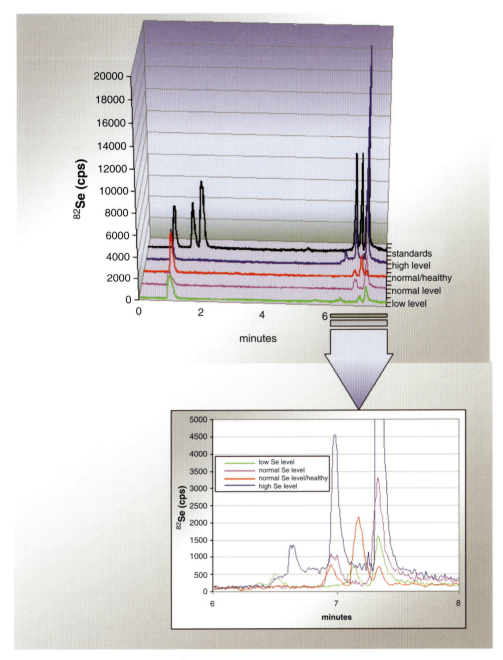

FIGURE 4 Chromatograms of children's sera [82]Se after SAX separation are shown (with the region between 6 and 8 minutes retention time additionally magnified). These sera from children having different absorption abnormalities in the gastrointestinal tract show different patterns of Se species as well as different total Se levels. For comparison, a 20-μg Se/L standard chromatogram is also plotted.

laboratory. Currently, the most suitable speciation analysis combines systems primarily based on the hyphenation of separation technologies online with element- or molecule-specific detectors. The variety of separation principles available today allows the separation of most species present in liquid samples. The selective detection ability of ICP–MS provides only signals of interest (not all compounds can be resolved) with very high detection sensitivity; however, there are still several pitfalls, so strict quality control and quality assurance schemes must be implemented. In speciation analysis, one of the most promising approaches is based on the orthogonal scheme. The achievement of the above three goals will allow decision makers to develop regulations incorporating more effective, current knowledge on chemical elements and their role in the life sciences to the full benefit of human health and the environment.

FURTHER READING

Aguerre, S., Lespes, G., Desauziers, V., and Potin-Gautier, M. (2002). Speciation of Organotin in Environmental Samples by SPME–GC: Comparison of Four Specific Detectors: FPD, PFPD, MIP-AES and ICP–MS, *J. Anal. Atom. Spectrom.*, 16, 263–269.

Allen, H. E., and Hansen, D. J. (1986). The Importance of Trace Metal Speciation to Water Quality Criteria, *Water Environ. Res.*, 68, 42–54.

Arnaud, J., Andre, D., Bouillet, M. C., Kia, D., and Favier, A. (1992). Problems Associated with the Use of Exclusion Diffusion Chromatography for Identification of Zinc Ligands in Human Milk, *J. Trace Elements Electrolytes Health Dis.*, 6, 81–90.

Barnes, R. M. (1983). Frontiers in Inductively Coupled Plasma Spectroscopy, *Chem. Anal. Warsaw*, 28, 179–198.

Barnes, R. M. (1991). Inductively Coupled and Other Plasma Sources: Determination and Speciation of Trace Elements in Biomedical Applications. In *Biological Trace Element Research—Multidisciplinary Perspectives* (K. S. Subramanian, G. V. Iyengar, and K. Okamoto, Eds.), American Chemical Society, Washington, D.C., pp. 158–180.

Brätter, P., Forth, W., Fresenius, W., Holtmeier, H. J., Hoyer, S., Kruse-Jarres, J., Liesen, H., Mohn, L., Negretti de Brätter, V., Reichlmayr-Lais, A. M., Sitzer, G., and Tölg, G. (1992). *Mineralstoffe und Spurenelemente, Leitfaden für die ärztliche Praxis*, Verlag Bertelsmann Stiftung, Gütersloh.

Burguera, J. L., Quintana, I. A., Salager, J. L., Burguera, M., Rondón, C., Carrero, P., Anton de Salager, R., and Petit de Peña, Y. (1999). The Use of Emulsions for the Determination of Methylmercury and Inorganic Mercury in Fish-Eggs Oil by Cold Vapor Generation in a Flow Injection System with Atomic Absorption Spectrometric Detection, *Analyst*, 124, 593–599.

Caroli, S. (1996). Chemical Speciation: A Decade of Progress. In *Element Speciation in Bioinorganic Chemistry* (S. Caroli, Ed.), John Wiley & Sons, New York, pp. 1–18.

Caroli, S., Alimonti, A., Coni, E., Petrucci, F., Senofonte, O., and Violante, N. (1994). The Assessment of Reference Values for Elements in Human Biological Tissues and Fluids: A Systematic Review, *CRC Crit. Rev. Anal. Chem.*, 24, 363–398.

Caroli, S., Cescon, P., and Walton, D. (Eds.) (2000). *Environmental Contamination in Antarctica: A Challenge to Analytical Chemistry*, Elsevier Science, Amsterdam.

Ceulemans, M., and Adams, F. C. (1996). Integrated Sample Preparation and Speciation Analysis for the Simultaneous Determination of Methylated Species of Tin, Lead and Mercury in Water by Purge-and-Trap Injection–Capillary Gas Chromatography–Atomic Emission Spectrometry, *J. Anal. Atom. Spectrom.*, 11, 201–206.

Chassaigne, H., and Lobinski, R., (1998). Speciation of Metal Complexes with Biomolecules by Reversed-Phase HPLC with Ion-Spray and Inductively Coupled Plasma Mass Spectrometric Detection, *Fresenius' J. Anal. Chem.*, 361, 267–273.

Cole, R. B. (Ed.) (1997). *Electrospray Ionization Mass Spectrometry—Fundamentals, Instrumentation and Applications*, John Wiley & Sons, New York.

Cornelis, R. (1996). Involvement of Analytical Chemistry in Chemical Speciation of Metals in Clinical Samples, *Ann. Clin. Lab. Sci.*, 26(3), 252–263.

Crecelius, E., and Yagen, J. (1997). Intercomparison of Analytical Methods for Arsenic Speciation in Human Urine. *Environ. Health Perspect.*, 105(6), 650–653.

Crews, H. M., Clarke, P. A., Lewis, D. J., Owen, L. M., Srutt, P. R., and Izquierdo, A. (1996). Gastrointestinal Extracts of Cooked Cod by High Performance Liquid Chromatography–Inductively Coupled Plasma Mass Spectrometry and Electrospray Mass Spectrometry, *J. Anal. Atom. Spectrom.*, 11, 1177–1182.

Dawson, J. B. (1986). Analytical Atomic Spectroscopy in Biology and Medicine, *Z. Anal. Chem.*, 324, 463–471.

Dunemann, L., and Begerow, J. (Eds.) (1995). *Kopplungstechniken zur Elementspeziesanalytik*, VCH, Weinheim, Germany.

Ferrarello, C. N., Montes Bayón, M., Fernández de la Campa, R., and Sanz-Medel, A. (2000). Multi-Elemental Speciation Studies of Trace Elements Associated with Metallothionein-Like Proteins in Mussels by Liquid Chromatography with Inductively Coupled Plasma Time-of-Flight Mass Spectrometric Detection, *J. Anal. Atom. Spectrom.*, 15, 1558–1563.

Florence, T. M. (1989). Electrochemical Techniques for Trace Element Speciation in Waters. In *Trace Element Speciation: Analytical Methods and Problems* (G. E. Batley, Ed.), CRC Press, Boca Raton, FL, pp. 77–116.

Francesconi, K. A., Edmonds, J. S., and Morita, M. (1994). Determination of Arsenic and Arsenic Species in Marine Environmental Samples. In *Arsenic in the Environment*, Part

I: *Cycling and Characterization* (J. O. Nriagu, Ed.), John Wiley & Sons, New York, pp. 189–219.

Harms, J., and Schwedt, G. (1994). Applications of Capillary Electrophoresis in Element Speciation Analysis of Plant and Food Extracts, *Fresenius' J. Anal. Chem.*, 350, 93–100.

Goessler, W., Schlagenhaufen, C., Kuehnelt, D., Greschonig H., and Irgolic K. (1997). Can Humans Metabolize Arsenic Compounds to Arsenobetaine?, *Appl. Organomet. Chem.*, 11(4), 327–335.

Heitkemper, D. T., Vela N. P., Stewart, K. R., and Westphal, C. S. (2001). Determination of Total and Speciated Arsenic in Rice by Ion Chromatography and Inductively Coupled Plasma Mass Spectrometry, *J. Anal. Atom. Spectrom.*, 16, 299–306.

Keller, H. (1991). *Klinische-chemische Labordiagnostik für die Praxis. Analyse, Befund, Interpretation*, Georg Thieme Verlag, Stuttgart.

Kersten, M., and Förstner, U. (1989). Speciation of Trace Elements in Sediments. In *Trace Element Speciation: Analytical Methods and Problems* (G. E. Batley, Ed.), CRC Press, Boca Raton, FL, pp. 245–318.

Kuhn, R., and Hofstetter-Kuhn, S. (1993). *Capillary Electrophoresis: Principles and Practice*, Springer-Verlag, Berlin.

Lewis, R. A., Klein, B., Paulus, M., and Horras, C. (1992). Environmental Specimen Banking. In *Hazardous Metals in the Environment* (M. Stoeppler, Ed.), Elsevier, Amsterdam, pp. 19–48.

Lintschinger, J., Schramel, O., and Kettrup, A. (1998). The Analysis of Antimony Species by Using ESI–MS and HPLC–ICP–MS, *Fresenius' J. Anal. Chem.*, 361, 96–102.

Lustig, S., Lampaert, D., De Cremer, K., De Kimpe, J., Cornelis, R., and Schramel, P. (1999). Capability of Flatbed Electrophoresis (IEF and Native PAGE) Combined with Sector Field ICP–MS and Autoradiography for the Speciation of Cr, Ca, Ga, In, Pt and V in Incubated Serum Samples, *J. Anal. Atom. Spectrom.*, 14, 1357–1362.

Lustig, S., Michalke, B., Beck, W., and Schramel, P. (1998). Platinum Speciation with Hyphenated Techniques: Application of RP–HPLC–ICP–MS and CE–ICP–MS to Aqueous Extracts from a Platinum Treated Soil, *Fresenius' J. Anal. Chem.*, 360, 18–25.

McSheehy, S., Szpunar, J., Lobinski, R. *et al.* (2002). Characterization of Arsenic Species in Kidney of the Clam *Tridacna derasa* by Multidimensional Liquid Chromatography–ICPMS and Electrospray Time-of-Flight Tandem Mass Spectrometry, *Anal. Chem.*, 74(10), 2370–2378.

Michalke, B. (1999a). Quality Control and Reference Materials in Speciation, *Fresenius' J. Anal. Chem.*, 363, 439–445.

Michalke, B. (1999b). Potential and Limitations of Capillary Electrophoresis–ICP–Mass Spectrometry, *J. Anal. Atom. Spectrosc.*, 14, 1297–1302.

Michalke, B., and Schramel, P. (1997). Selenium Speciation in Human Milk with Special Respect to Quality Control, *Biol. Trace Element Res.*, 59, 45–56.

Michalke, B., Witte, H., and Schramel, P. (2000). Iodine Speciation in Human Serum by Reversed Phase Liquid Chromatography–ICP–Mass Spectrometry, *Biol. Trace Element Res.*, 78, 81–92.

Michalke, B., Witte, H., and Schramel, P. (2001). Developments of a Rugged Method for Selenium Speciation, *J. Anal. Atom. Spectrom.*, 16, 593–597.

Michalke, B., Witte, H., and Schramel, P. (2002). Effect of Different Extraction Procedures on the Yield and Pattern of Se Species in Bacterial Samples, *Anal. Bioanal. Chem.*, 372, 444–447.

Mikes, O. (1988). *High-Performance Liquid Chromatography of Biopolymers and Biooligomers*, Elsevier, Amsterdam.

Minderhoud, A. (1983). Atomic Spectrometry for Trace Determinations in the Environment with a View to Government Regulations, *Spectrochim. Acta*, 38B, 1525–1532.

Montaser, A., and Golightly, D. W. (1992). *Inductively Coupled Plasmas in Analytical Atomic Spectrometry*, 2nd ed., VCH Publishers, New York.

Morrison, G. M. P. (1989). Trace Element Speciation and Its Relationship to Bioavailability and Toxicity in Natural Waters. In *Trace Element Speciation: Analytical Methods and Problems* (G. E. Batley, Ed.), CRC Press, Boca Raton, FL pp. 25–42.

Mota, A. M., and Simaes Gonçalves, M. L. (1996). Direct Methods of Speciation of Heavy Metals in Natural Waters. In *Element Speciation in Bioinorganic Chemistry* (S. Caroli, Ed.), John Wiley & Sons, New York, pp. 21–96.

Nischwitz, V., Michalke, B., and Kettrup, A. (2003). Investigations on Species-Preserving Extraction from Liver Samples, *Anal. Bioanal. Chem.*, 375, 145–156.

Olesik, J. W. (2000). Capillary Electrophoresis for Elemental Speciation Studies. In *Element Speciation: New Approaches for Trace Element Analysis* (J. A. Caruso, K. L. Sutton, and K. L. Ackley, Eds.), Elsevier, Amsterdam, pp. 151–211.

Patching, S. G., and Gardiner, P. H. E. (1999). Recent Developments in Selenium Metabolism and Chemical Speciation: A Review, *J. Trace Element Med. Biol.*, 13(4), 193–214.

Pickering, W. F. (1981). Selective Chemical Extraction of Soil Components and Bound Metal Species, *CRC Crit. Rev. Anal. Chem.*, 12, 233–238.

Potin-Gautier, M., Gilon, N., Astruc, M., de Gregori, I., and Pinochet, H. (1997). Comparison of Selenium Extraction Procedures for Its Speciation in Biological Materials, *Int. J. Environ. Anal. Chem.*, 67, 15–25.

Prange, A., and Schaumlöffel, D. (1999). Determination of Element Species at Trace Levels Using Capillary Electrophoresis–Inductively Coupled Plasma Sector–Field Mass Spectrometry, *J. Anal. Atom. Spectrom.*, 14, 1329–1332.

Quevauviller, P., Maier, E. A., and Griepink, B. (1996). Quality Control of Results of Speciation Analysis. In *Element Speciation in Bioinorganic Chemistry* (S. Caroli, Ed.), John Wiley & Sons, New York, pp. 195–222.

Quijano, M. A., Moreno, P., Gutierrez, A. M., Perez-Conde, C., and Camara, C. (2000). Selenium Speciation in Animal Tissues After Enzymatic Digestion by HPLC Coupled to ICP–MS, *J. Mass Spectrom.*, 35, 878–884.

Rodriguez-Pereiro, I., Wasik, A., and Lobinski, R. (1997). Trace Environmental Speciation Analysis for Organometallic Compounds by Isothermal Multicapillary Gas Chromatography—Microwave Induced Plasma Atomic Emission Spectrometry (MC GC MIP AES), *Anal. Chem.*, 42, 799–808.

Silva da Rocha, M., Soldado, A. B., Blanco-González, E., and Sanz-Medel, A. (2000). Speciation of Mercury Compounds by Capillary Electrophoresis Coupled On-Line with Quadrupole and Double-Focusing Inductively Coupled Plasma Mass Spectrometry, *J. Anal. Atom. Spectrom.*, 15, 513–518.

Szpunar, J. (2000a). Trace Element Speciation Analysis of Biomaterials by High Performance Liquid Chromatography with Inductively Coupled Plasma Mass Spectrometric Detection (HPLC–ICP MS), *Trends Anal. Chem.*, 19/2–3, 127–137.

Szpunar, J. (2000b). Bio-Inorganic Speciation Analysis by Hyphenated Techniques, *Analyst*, 125, 963–988.

Templeton, D. M., Ariese, F., Cornelis, R., Danielsson, L.-G., Muntau, H., van Leeuwen, H. P., and Lobinski, R. (2000). Guidelines for Terms Related to Chemical Speciation and Fractionation of Elements. Definitions, Structural Aspects and Methodological Approaches, *Pure Appl. Chem.*, 72(8), 1453–1470.

Thiers, R. E., and Vallee, B. L. (1957). Distribution of Metals in Subcellular Fractions of Rat Liver, *J. Biol. Chem.*, 226, 911–920.

Tittes, W., Jakubowski, H., and Stüwer, D. (1994). *Reduction of Some Spectral Interferences in ICP–MS*, Finnigan MAT Elemental Mass Spectrometry Technical and Application Note 3, Finnigan MAT GmbH, Bremen, Germany.

Urasa, I. T. (1996). Developments of New Methods of Speciation Analysis. In *Element Speciation in Bioinorganic Chemistry* (S. Caroli, Ed.), John Wiley & Sons, New York, pp. 121–154.

Ure, A. M., Quevauviller, P., Muntau, H., and Griepink, B. (1993). Speciation of Heavy Metals in Soils and Sediments. An Account of the Improvement and Harmonization of Extraction Techniques Undertaken Under the Auspices of the BCR of the Commission of the European Communities, *Int. J. Environ. Anal. Chem.*, 51, 135–151.

Zeien, H., and Bruemmer, G. W. (1989). Chemische Extraktionen zur Bestimmung von Schwermetallbimdungsformen in Böden, *Mitt. Dtsch. Bodenk. Gesellsch.*, 59, 505–510.

SECTION IV

TECHNIQUES AND TOOLS

INTRODUCTION: ROBERT B. FINKELMAN*

Geoscientists and medical researchers bring to medical geology an arsenal of valuable techniques and tools that can be applied to health problems caused by geologic materials and processes. Although some of these tools may be common to both disciplines, practitioners of these disciplines commonly apply them in novel ways or with unique perspectives. In this section, some of these tools and techniques are examined.

The geographic information system (GIS) is one of the more powerful techniques that is used by both the geoscience and public health/biomedical fields. GIS can help researchers to visualize the relationships between health problems and the physical environment. The value of GIS is dependent on having appropriate data to input, store, manipulate, and analyze.

In Chapter 26, an introduction is provided to a wide range of GIS databases that can be used for human health studies. The authors describe the types and features of databases that currently exist (with the caveat that this is a dynamic field with new databases appearing almost daily). They then provide several case studies that illustrate how geospatial data and GIS have been used in medical geology research. The examples include using GIS to determine the types of soil that Lyme disease-bearing ticks prefer and demonstrating the relationship of dental fluorosis to domestic coal combustion in southern China.

One of the most important types of data used in GIS is the information transmitted from Earth-orbiting satellites. Chapter 27 reviews fundamental aspects of

*The views expressed by the author are his own and do not represent the views of the United States Geological Survey or the United States.

remote sensing and describes the types of information that are available from many of the current remote-sensing satellites. It also describes how these geospatial data are used to gain insights into the occurrence and distribution of vector-borne diseases.

Classical mineralogical techniques, such as optical microscopy, X-ray diffraction, and microbeam analytical tools, have been used to characterize the complex biominerals formed in the human body that affect human health and well-being in many ways. Chapter 28 focuses on the largest biomineral in the human body, the bioapatites, which constitute the bulk of bones and teeth. It shows how these tools and techniques can be used to better understand these materials that are vital to our health. It also shows how classical geoscience tools and techniques have been used to gain and improve understanding of osteoporosis, a bone disorder afflicting millions of elderly people throughout the world.

Accurate, comprehensive determination of the composition of geologic materials with which we come in contact and of natural materials that gain entry to the human body through ingestion, inhalation, or physical contact is essential for the protection of public health. In Chapters 29 and 30, we can see the similarities and differences in the approaches that geoscientists and medical researchers adapt in characterizing these materials. In Chapter 29, a detailed overview is provided of several of the more important analytical techniques used for inorganic geochemical characterization of solids and liquids as well as several techniques for the characterization of organic matter. The authors describe the principles of each method, how each is applied, and the strengths and weaknesses of each.

Chapter 30 describes how biomedical researchers treat human tissue and body fluid samples to determine the impact of exposure to inorganic materials. In addition to histochemical techniques, a range of analytical tools used in identifying minute inclusions in human tissues are described. Some of these tools are similar to those described in Chapter 29.

Water is one of the most important media for transmission of disease. Hydrogeologists have developed powerful computer-based tools for modeling groundwater flow and quality. Chapter 31 provides a review of computer models of groundwater flow, solute transport, and geochemical reaction processes and describes how they can be applied to medical geology issues.

From images of Earth taken from thousands of miles above our heads to images of submicroscopic inclusions within human tissues, from the most sophisticated analytical equipment and computer programs to methods that have been used for 100 years, scientists are using their full arsenal of tools and techniques to solve a wide range of medical geology problems.

CHAPTER 26

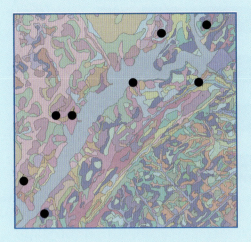

GIS IN HUMAN HEALTH STUDIES

JOSEPH E. BUNNELL, ALEXANDER W. KARLSEN,
AND ROBERT B. FINKELMAN*
United States Geological Survey

TIMOTHY M. SHIELDS
The Johns Hopkins University

CONTENTS

I. INTRODUCTION TO DATABASES AND GEOGRAPHIC INFORMATION SYSTEMS

Databases used in the field of medical geology are generally comprised of geospatial and/or temporal elements. Although these are not requirements for all medical geology research projects, much of the discussion in this chapter will be focused on databases incorporated into geographic information systems (GIS). GIS are computer-based (or manual) methods that allow a user to input, store, retrieve, manipulate,

*The views expressed by the author are his own and do not represent the views of the United States Geological Survey or the United States.

analyze, and output spatial data (Aronoff, 1989). There are four major systems of GIS: engineering mapping systems (computer-aided design/computer-assisted mapping; CAD/CAM), geographic base file systems, image processing systems, and generalized thematic mapping systems. Various software packages are available that perform one or more of these systems, and the relative ability to move data back and forth between them can be critical to the needs and success of a particular GIS. Relational databases are the most commonly used types of databases in GIS (Cromley & McLafferty, 2002). Relational database management models are convenient for linking formerly disparate databases together in a GIS. The databases to be joined must share one common attribute, usually an identifier such as coded patient number, sample site, or latitude/longitude. Other database management structures, such as hierarchical and network systems, are not as well suited to health GIS applications, although they may be useful for extremely large databases.

The capability to quickly and easily link large medical or public health databases with equally large geospatial databases represents an important technological breakthrough. Due to advances in computational power and speed, studies may be conducted today that could not have been done in reasonable time frames even just a few years ago. By linking disparate databases in a visually accessible manner (i.e., with maps), researchers are able to recognize relationships or discern patterns of

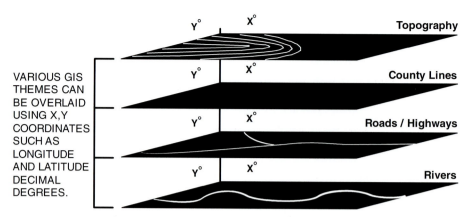

VARIOUS GIS THEMES CAN BE OVERLAID USING X,Y COORDINATES SUCH AS LONGITUDE AND LATITUDE DECIMAL DEGREES.

Topography

County Lines

Roads / Highways

Rivers

FIGURE 1 GIS conceptual diagram. Note that different data layers (covers or themes) are overlaid such that queries about individual point locations reveal numerous attributes from a variety of source databases. (Figure courtesy of Eric Morrissey, U.S. Geological Survey, Reston, VA.)

disease that can lead to an understanding of causality that was previously not apparent. The value in mapping disease occurrence is appreciated when doing so can illuminate the underlying cause of an outbreak, which may then enable mitigation measures to be taken to prevent further spread of a disease.

The earliest example of using such a spatioanalytical approach to solving an epidemiological riddle is generally credited to a physician named John Snow and the Broad Street Pump of London in 1854 (Cameron & Jones, 1983). Dr. Snow mapped a major outbreak of cholera, in a time before the germ theory was well accepted, and hypothesized that there was a causal association between the putative source of the contagion and locations of cases. He convinced city officials to remove the handle of the pump dispensing contaminated water—an intervention that promptly quelled the outbreak. Although modern tools are much more sophisticated than those at Dr. Snow's disposal, our goal remains the same in applying GIS to public health issues.

Currently, databases used in GIS applications are often developed by the user or are available on the Internet or from other sources. Databases are typically stored in electronic media formats such as a hard drive, floppy disk, and compact disc-read only memory (CD-ROM). GIS databases contain fields in columns and records in rows. A field, or item, is an element of a database record in which one piece of information is stored and represented as a column in a geodatabase table or spreadsheet (Kennedy, 2001). *Records* represent different entities with different values for the attributes represented by the fields. (Kennedy, 2001). *Attributes* are information about geographic features, and they are

contained within GIS data layers, or themes (Figure 1). For example, a climate data layer (the feature) may contain the attributes of temperature, rainfall, and relative humidity for a specific geographic point location or region. Attribute data and spatial data comprise the two critical types of data in a GIS. A more in-depth look at how to assemble databases into a project is presented in Other Sources, part C.

Along with the individual databases, *metadata* are also included. Metadata, also called data dictionaries, are simply data about data. Metadata contain information such as the time/place the database was created, field and record identifier information (attributes), data development process (lineage), and individual(s) to contact regarding the data. If the data are displayed in a geographic environment, the metadata must also include additional information such as map scale and projection. Guidelines for what should be included in metadata are provided by the National Spatial Data Infrastructure, which is maintained by the Federal Geographic Data Committee (FGDC) (see Other Sources, part A). An interagency U.S. government organization, the FGDC sets guidelines for all aspects of spatial data, and works with offshore/international partners to develop the global spatial data infrastructure.

II. TYPES OF DATABASES AND THEIR FEATURES

There are enormous numbers of databases available in digital format, but the types of data likely to be used by medical geologists fall into two broad categories: Earth science/geospatial databases and biomedical/health databases (see Other Sources, part A). What makes the field of medical geology innovative and unique is that it, by definition, brings together in a coherent manner databases from these two general areas in specific applications. This approach leads to fresh perspectives enabling recognition of connections between environmental factors and human health outcomes that may have previously gone unnoticed. Medical geological research can identify mechanistic connections that in turn can lead to new practices or policies. This may result in novel solutions to public health problems, which ultimately benefit large numbers of people. Two case studies are presented in Sections IV and V that illustrate the utility of such an approach.

Spatial data are represented in two models, vector and raster. In the vector models attribute data are attached or linked to one of three features: point, line, or polygon. Simply defined, a point is an x,y coordinate such as a mountain peak or soil sample location. The point feature does not have any length or area. A line is defined by the connection of two or more vertices (x,y coordinate pairs). A polyline is made up of numerous lines that represent the same feature, such as a road or river. The line or polyline feature has a length associated that is considered too narrow at the given scale to have an area. The polygon feature is defined by a series of lines that start and end at the same place, such as a state or country. Perimeter and area can be calculated for these features. In summary a line is a set of connected points whereas a polygon is a set of connected lines that have the same beginning and end points.

Features in a vector-based GIS can be linked or joined with attribute data from one or more databases provided a common identifier exists. The National Climatic Data Center (NCDC) generates climate data from ground-based observations. They supply a table of weather stations along with the stations' identifiers and x,y coordinates. This table can be imported into a GIS and point features can easily be made which correspond with weather station locations. Once these features exist in a GIS, they can be linked to other NCDC tables that contain station identifiers and hourly, daily, weekly, monthly, and/or annual climate data. Climatic conditions can be analyzed from the temporal frame of a single point in time or a complete historical compilation.

A key function of a vector-based GIS is topology. Topology is the spatial linkage between vector features. It stores the spatial relationship of the features with respect to each other. Topology enables the user to determine where a feature is in relation to other features, which parts of different features are shared, and how features are connected. Functions such as which sample points are located within a specific watershed or which counties does a river intersect can be performed (see Figure 1). Topology also reduces the amount of information that must be stored. If two polygons are adjacent, that is they share a common line as a border, the common line needs to be stored only once. It will be saved as the right side of one polygon and as the left side of the other.

Raster systems present data in a regular grid of squares or cells. These cells are often called "pixels," which is an abbreviation of the words picture elements. Each pixel is defined by a row and column number and in GIS these can be converted to x,y coordinates. Pixels contain a single attribute value relating to the feature they represent. In an image that represents soil, a unique value would be given for each type of soil.

The differences in the manner in which vector and raster systems present and store geospatial data often lead people to contrast the two in order to determine which is better. Each model has its advantages and disadvantages. Vector systems, with points, lines, and polygons produce maps that are more like those drawn by hand and therefore are more aesthetically pleasing. Raster systems tend to represent continuous surfaces such as elevation or vegetation well. At the same time they demand large storage capacity because they require a value for every pixel in the image. The answer to the question of which system is better is dependent upon the application. Both systems are effective, but the nature of the application or task generally determines which one should be utilized.

III. Software, Computational Technology, and Technical Issues

Developments in software are facilitating the integration of these two forms of data. Vector-based systems now incorporate some raster-based system functionality and vice versa. The crossover of functionality has helped the user immensely. No longer do entire layers of data need to be converted from one data structure to the other. The integration of the two forms has made data management faster and less expensive while providing better quality of data generated.

Getting data to be used in GIS is often a challenge. Manual efforts such as digitizing (using a flat digitizing

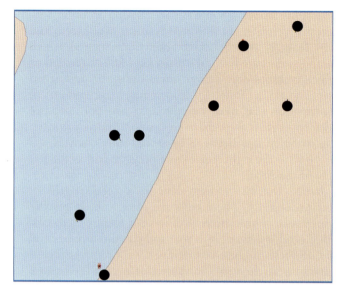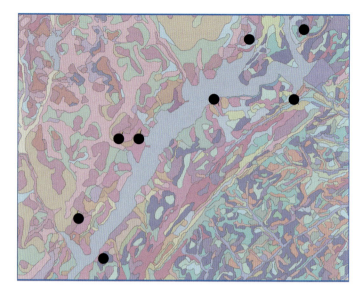

FIGURE 2   Example of the effect of using databases of different scale. Both panels represent soil types displayed from digital databases for the exact same field sites (locations of individual transects are indicated with black circles, and latitudes and longitudes have been determined with a hand-held global positioning system [GPS] device). Left panel shows level of resolution available with STATSGO data (from the U.S. Department of Agriculture, Natural Resources Conservation Service, National Soil Survey Center, Lincoln, NE), 1:250,000. Right panel reveals much greater detail available with SSURGO data (from the U.S. Department of Agriculture, Natural Resources Conservation Service, National Soil Survey Center, Lincoln, NE), 1:24,000. (Data from Bunnell et al., 2003.)

table to draw or copy a map into an electronic format), geocoding (using an electronic basemap to match an address to), or scanning (using a scanner to convert a paper map to electronic format) are time intensive. Using existing digital sources is often preferable, but these sources also have their share of issues pertaining to quality. These are outlined below.

Scale specifies the level at which real-world features have been reduced to be represented. It is usually stated as a ratio or fraction such as 1:1000 or 1/1000, where one unit on the map represents one thousand units in the area represented. An often confusing term for maps is small- or large-scale. Small-scale maps actually represent large areas but the ratio or fraction is a small number. Conversely, large-scale maps represent small areas. The map of a neighborhood would be considered large-scale whereas a map of Asia would be small-scale.

Resolution refers to the amount of detail in the features of the map. The scale of the map determines its resolution. It is the level at which features can be distinguished. On a small-scale map local features, such as ponds and small lakes, will not be represented. The terms fine (high) resolution and large-scale are synonymous; they contrast with coarse (low) resolution, and small-scale.

In GIS, issues of scale and resolution can determine which functions are appropriate as well as the level at which results can be stated (see Figure 2). Ecological fallacy occurs when statements or predictions are made at one level based upon observations made at another. A soil survey conducted in 1 of the 24 counties of Maryland would be insufficient to predict soil characteristics throughout the state. Likewise, a small-scale digital elevation model would be inappropriate to use as part of a local water drainage study.

Accuracy is an important concept when working with GIS. Spatial accuracy, or how well the mapped features are located, must be stated and understood. If the map you are using states a spatial accuracy of ±1000 feet, this will not be accurate enough to select a point to dig a well. Likewise, temporal accuracy must be stated. Population data from 1980 will not be accurate enough for current demographic studies. These are just a couple of ways error can be introduced in GIS.

Projections were developed to represent the curved surface of the Earth on a flat map. Projections are utilized to preserve local angles and shapes or relative size of areas. Every projection distorts the map in some manner and should be carefully chosen with respect to the study. Most GIS can convert data from one projection to another, which enables data layers from different sources to be compiled.

Metadata is information about data. All data sets utilized in GIS should have metadata accompanying them. This information should include the origin and characteristics of the data set, the purpose of the data set, and any problems the data set may have. This information is critical for the proper utilization of GIS data.

Remote sensing technology provides satellite imagery and high-resolution, high-altitude aerial photography. Such image data are also becoming more common in the context of GIS. Raster data can exist either simply or with multiple values, for example, representing spectral bands. Geodatabase features with the same type of geometry make up simple or topological feature classes (Zeiler, 1999). Object classes retain descriptive information related to geographic features, but they are not elements found on a map (Zeiler, 1999). Depths of wells could make up an object class, for example, in a medical geology GIS examining proximity of drinking water wells to sources of arsenic in Bangladesh (see also Chapters 11 and 27, this chapter). Due to advances in microcomputer technology, large raster data sets can now be manipulated and spatial data rigorously analyzed statistically with relative ease, thus making incorporation into epidemiological frameworks feasible (Robinson, 2000).

Medical geologists can now move beyond simply noting spatial coincidences of environmental features and disease patterns. By taking advantage of increasing computational speed and capabilities, sophisticated spatial statistics can be used in conjunction with GIS to reduce bias and correct for such potentially confounding effects as non-constant variance and autocorrelation (Haining, 1998). Moreover, spatial statistical models can rigorously test for clustering versus random distributions, and can incorporate a fourth, temporal dimension to better assess correlation and offer clues into disease etiology (Kulldorff, 1998).

IV. CASE STUDY 1: LYME DISEASE

The first is a study designed to identify environmental determinants of tick abundance in the Mid-Atlantic region of the United States (Bunnell et al., 2003). Lyme disease is the most commonly reported vector-borne disease in the United States and it is still rapidly growing with over ten thousand new cases annually (Centers for Disease Control and Prevention, 2001). In this part of the country, the black-legged, or deer, tick (*Ixodes scapularis*) transmits the microbial agents, that cause Lyme disease, ehrlichiosis and babesiosis, and other human and veterinary ailments. Tick abundance is likely a more reliable measure of the effect of landscape features on Lyme disease risk than human case data for several reasons, most notably because the location of Lyme disease cases will be, at best, the patients' home address, while many if not most of the cases are actually acquired elsewhere. Furthermore, by using tick distribution patterns, one avoids potentially misleading interpretations resulting from over- or underreporting due to the challenges in accurately diagnosing Lyme disease in humans. By better understanding the effects of environmental parameters on tick distribution, public health intervention strategies will likely be improved. Note that the word "vector" has a specialized meaning in the context of GIS (a mathematical definition, as above); in the biomedical community, a vector is an insect or other arthropod that actively transmits a pathogen from an infected reservoir host animal to another individual. Chapter 27 offers a thorough discussion of GIS technology applied to vector-borne diseases.

Environmental factors (covariates) that were previously known or suspected to correlate to tick abundance patterns included elevation, land cover, forest distribution, watersheds, and soil type (Glass et al., 1995; Ostfeld et al., 1996; Kitron & Kazmierczak, 1997; Jensen et al., 2000). Digital elevation model (DEM) databases were obtained in a raster format (U.S. Geological Survey, Reston, VA). Land cover attributes were obtained from the remotely sensed multiresolution landscape characterization (MRLC) Landsat thematic mapper (TM) source data, managed by the Earth Resources Observation Systems (EROS), an entity within the U.S. Geological Survey (USGS). Initially, soils data were obtained from the state soil geographic database (STATSGO) maintained by the Natural Resources Conservation Service (U.S. Department of Agriculture, National Soil Survey Center, Lincoln, NE).

From a five-state region, 320 field sites were randomly selected and ticks were collected along transects at each site. Latitude and longitude were recorded at the beginning and end of each transect. At first, locations of the transects were noted on a paper map, 7.5-minute topographical quadrangles were digitized by hand, and the sites were matched up. Digitizing a paper map requires a digitizing table and specialized hardware and computer software. The digitizing process was not required when latitudes and longitudes were recorded directly to a global positioning system (GPS) device, and then entered into a desktop personal computer. Newer GPS devices can further streamline the process by coupling directly with a laptop or desktop computer. This obviates the need for data entry by hand. Each field site, or transect, became a unique identifier, and the latitude and longitude were the link to the environmental data downloaded. DEM data were used as is so that elevation in meters was obtained at each transect. Land cover data were of limited usefulness, as they only indicated whether a given field site was located in forest, low-intensity residential, open water, etc. Because tick population densities are positively associated with forest edges (ecotones), and because ticks are not found in open water, the GIS was used to calculate the distance from the midpoint of each transect to the nearest specific type of forest and body of water. The databases were thus manipulated from their native state, which resulted in new databases of particular utility to this specific research project.

Newly developed spatial statistics that incorporated spatial autocorrelation were applied to the data, and multiple regression analysis was performed (Das et al., 2002). These techniques revealed significant associations between tick abundance and certain environmental covariates that included soil type. This latter finding was particularly intriguing, and since the inception of the project, some soil data at much higher resolution was made available digitally. The soil survey geographic database (SSURGO) was released county by county as it became available (U.S. Department of Agriculture, Natural Resources Conservation Service, National Soil Survey Center, Lincoln, NE). Only 75 of the 320 field sites happened to be located in counties with SSURGO data available at the time of analysis, and a second analysis was conducted on that subset of field sites. Because of the much finer resolution (1:24,000 vs. 1:250,000 for SSURGO and STATSGO, respectively), interpretations were made with greater precision (Figure 2).

The enhanced resolution available with SSURGO data, combined with the more extensive set of attributes of this database, revealed some surprising results. For example, with STATSGO data, well-drained soils were found to be positively associated with tick abundance, in keeping with previous reports in the literature. However, upon analysis using SSURGO data, it was found that poorly drained soils, too, could be positively associated with tick abundance. This seeming contradiction was apparently resolved by considering precip-

itation factors and water-holding capacity of the soil. This example demonstrates the power of a GIS approach to examining environmental influences on factors controlling human disease risk. Observation of previously obscured patterns has enabled a generation of hypotheses now being tested to explain factors responsible for spatioanalytical trends in biological terms. Important advances in our understanding of basic Lyme disease ecology are likely to follow from this application of newly acquired and improved computational technology.

V. Case Study 2: Fluorosis in China

Elucidating the causes of fluorosis in the People's Republic of China offers another example of how GIS can be used to address the relationship between human health problems and geologic materials. Fluorosis, an abnormal condition of bones and teeth caused by exposure to excessive amounts of fluorine, affects millions of people throughout China. There are three principal pathways of exposure: drinking high-fluorine water, drinking tea made from tea leaves rich in fluorine, and exposure to fumes from residential combustion of high-fluorine coal or briquettes made with fluorine-rich clays as a binder (Zhang & Cao 1996; Ando et al., 1998) (see also Chapter 12, this chapter).

Until recently, only general information existed on the epidemiology of fluorosis in China. For example, it was known that Kazakhs in the Xinjiang Autonomous Region in northwestern China were exposed to high levels of fluorine due to their preference for "brick tea" made from tea leaves rich in fluorine (Ben et al., 2000).

To determine where fluorosis was likely to be caused by exposure to fluorine-rich coal or coal briquettes, two GIS layers were required: the distribution of fluorosis and the distribution of coal deposits in China. No digital versions of either layer could be located. A map of the distribution and prevalence rates of dental fluorosis by county in China was located (Jianan, 1989) and a map of the coal deposits of China was obtained (Ruiling et al., 1996). Both paper maps were digitized into electronic format at a computer workstation. Once in electronic form, the individual features of the maps (i.e., areas with the same prevalence rate of dental fluorosis) were assigned unique identifiers (attributes, in this case different colors). Parameters that control the way the map is displayed (projection) were adjusted so that the digital maps, representing the same geographic

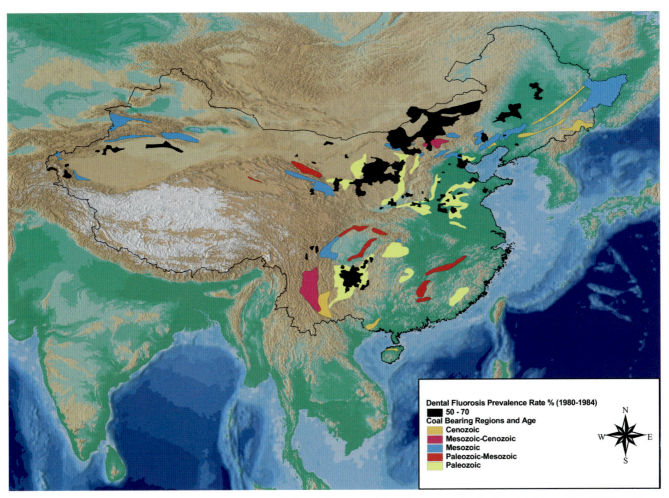

FIGURE 3 Relationship of high prevalence rates of dental fluorosis and coal deposits in the People's Republic of China. (From Karlsen et al., 2001.)

areas, would perfectly overlap each other (Karlsen et al., 2001).

In a GIS environment, the digital map of prevalence rates for dental fluorosis was overlain with the digital map showing coal distribution (Figure 3). The combined maps confirm the association of fluorosis and coal in Guizhou Province, where more than 10 million people are known to suffer from fluorosis (Zheng & Huang, 1989). Figure 3 indicates that the high incidence of fluorosis in north central China (Shaanxi Province, which is the largest coal-producing province in China) may not be related to coal use.

In Guizhou Province the fluorosis is caused primarily by combining moderately high fluorine coals (average of about 200 ppm) with clays having very high fluorine contents (average of about 800 ppm) to form briquettes

(Belkin et al., 1999). The high-fluorine clays are the residual products from intensive leaching of the limestone substrate that formed the beautiful karst landforms for which the region is noted. In Shaanxi Province the substrate is primarily loess, a silicate-rich, wind-deposited sediment that is unlikely to have high-fluorine contents. Therefore, unless the coals in Shaanxi Province have exceptionally high fluorine contents, it is unlikely that the high incidence of dental fluorosis in that region is due to residential coal use.

Surprisingly, in the Xinjiang Autonomous Region the incidence of fluorosis also parallels the coal deposits, perhaps this indicates that the distribution of fluorosis may be controlled by sedimentary rocks that favor the growth of the trees from which the fluorine-bearing tea leaves are obtained.

VI. Other Case Studies

A number of other examples of GIS applications to medical geology problems may be found. These include the examples below.

African trypanosomiasis—Environmental factors that influence temporal and spatial distributions of African trypanosomiasis (sleeping sickness) were analyzed in a GIS that incorporated temporal Fourier analysis and discriminating analytical techniques such as Mahalanobis distance metric to aid in data interpretation (Rogers, 2000). A relationship between the climate covariate included in the analysis was found to exist with vegetation patterns, which in turn influenced suitability of grazing for cattle, the main reservoir hosts for the tsetse fly (*Glossina* spp.) vector of the trypanosome parasites. Sequential statistical modeling of the tsetse fly populations and trypanosome disease transmission, when linked to biological modeling based on known differences due to the different species of tsetse, helped explain differences in the patterns of disease observed in different regions in Africa.

Hurricane Mitch—In 1998, one of the most powerful and deadly hurricanes in recorded history struck Central America. At least 6500 people died and over 11,000 went missing in Honduras alone as a result of Mitch's fury. Many cases of human disease and death were caused by flooding, even in areas not directly hit by the hurricane itself. Some of these floods, in turn, triggered lethal outbreaks of waterborne infectious diseases such as cholera, leptospirosis, Dengue fever, and malaria. GIS was used to predict high-risk areas for flood potential based on themes including river network configurations, elevation, and slope. This tool may have helped keep the casualty count low following this major disaster. Lessons learned from this experience may be helpful in using GIS to plan and execute preparedness and relief efforts before and after future catastrophic events (PAHO, 2000).

Cadmium in The Netherlands—From 1892 until 1973, a zinc works in the Kempen area of The Netherlands discharged zinc and cadmium (Cd) into the environment in an uncontrolled fashion and seriously contaminated the soil with up to 8 ppm Cd (Stein et al., 1995). Cleanup efforts undertaken in the 1980s made use of a GIS and geostatistics to contour the cadmium distribution and improve sampling efficiency. Data from more than 1700 soil samples were used as point data, and semi-variograms were created to compare stratified and ordinary kriging methods to interpolate the Cd concentrations. Neither mapping technique was uni-

formly superior; depending on the application (e.g., proximity to urban centers), one or the other map proved more useful. Had the GIS been used interactively, Stein et al. (1995) concluded that the number of soil samples necessary for testing could have been reduced approximately tenfold.

Malaria—A GIS analysis of malaria conducted in the Chiapas region, Mexico, and Peten, Guatemala, provides an example of how this analytical tool can be useful in active and iterative generation of testable research hypotheses. A priori hypotheses pertaining to environmental factors that influence malaria incidence by their impacts on the *Anopheles* spp. mosquito vectors invoked altitude, temperature, rainfall, land use, and vegetation type. In the course of developing the GIS, researchers observed that high-risk areas were often in close proximity to agricultural lands (SHA, 2000). Further analysis led to the generation of a novel hypothesis that relates malaria risk to deforestation, a potential linkage that is presently being investigated by several groups.

VII. Conclusions

In this chapter, the conceptual framework for GIS databases has been described, as have strategies and tools for conducting medical geology research projects. We have explained how recently developed GIS technology has truly revolutionized the study of human disease systems, which makes possible the simultaneous analysis of numerous interrelated factors that may exert unapparent and synergistic effects. Previously, such complex systems could only be addressed looking at one (or just a few) variable(s) at a time. With the case studies provided, the reader has seen examples of how to approach rigorous investigation causes of human disease patterns that have strongly suspected environmental influences. Finally, we have supplied an extensive, if admittedly selective, set of database resources that should at least provide a starting point for researchers wishing to conduct GIS studies of their own.

See Also the Following Chapters

Chapter 11 (Arsenic in Groundwater and the Environment) · Chapter 12 (Fluoride in Natural Waters) ·

Chapter 27 (Investigating Vector-Borne and Zoonotic Diseases with Remote Sensing and GIS)

FURTHER READING

Ando, M., Tadano, M., Asanuma, S., Matsushima, S., Wanatabe, T., Kondo, T., Sakuai, S., Ji, R., Liang, C., and Cao, S. (1998). Health Effects of Indoor Fluoride Pollution From Coal Burning in China, *Environ. Health Perspect.*, 106, 239–244.

Aronoff, S. (1989). *Geographic Information Systems: A Management Perspective*, WDL Publications, Ottawa, Canada.

Belkin, H. E., Finkelman, R. B., and Zheng, B. S. (1999). Geochemistry of Fluoride-Rich Coal Related to Endemic Fluorosis in Guizhou Province, China, Pan-Asia Pacific Conference on Fluoride and Arsenic Research, Abstract 45, p. 47.

Ben, K., Hua, L., and Hongchao, H. (2000). The Current State of Epidemic Tea-Induced Fluorosis and Its Control Countermeasures in Urumqi County, Xinjiang. In *Metal Ions in Biology and Medicine*, Vol. 6, (J. A. Centeno, P. Collery, G. Vernet, R. B. Finkelman, H. Gibb, and J.-C. Etienne, Eds.), John Libby Eurotext, Paris, pp. 303–305.

Bunnell, J. E., Price, S. D., Lele, S. R., Das, A., Shields, T. M., and Glass, G. E. (2003). Geographic Information Systems and Spatial Analysis of *Ixodes scapularis* (Acari: Ixodidae) in the Middle Atlantic Region of the U. S. A., *J. Med. Entomol.*, 40, 570–576.

Cameron, D., and Jones, I. G. (1983). John Snow, the Broad Street Pump and Modern Epidemiology, *Int. J. Epidemiol.*, 12, 393–396.

Centers for Disease Control and Prevention (2001). Lyme Disease–United States, 1999, *MMWR*, 50(10), 181–185.

Cromley, E. K., and McLafferty, S. L. (2002). *GIS and Public Health*, The Guilford Press, New York, p. 340.

Das, A., Lele, S. R., Glass, G. E., Shields, T. M., and Patz, J. A. (2002). Modeling a Discrete Spatial Response Using Generalized Linear Mixed Models: Application to Lyme Disease Vectors, *Int. J. Geog. Inform. Sci.*, 16, 151–166.

Glass, G. E., Schwartz, B. S., Morgan, III, J. M., Johnson, D. T., Noy, P. M., and Israel, E. (1995). Environmental Risk Factors for Lyme Disease Identified with Geographic Information Systems, *Am. J. Public Health*, 85, 944–948.

Haining, R. (1998). Spatial Statistics and the Analysis of Health Data. In *GIS and Health, GIS Data VI* (A. C. Gatrell and M. Löytönen, Eds.), Taylor & Francis, London, pp. 29–47.

Hock, R. (2001). *The Extreme Searcher's Guide to Web Search Engines*, 2nd edition, CyberAge Books, Information Today, Inc., Medford, NJ, p. 241.

Jensen, P. M., Hansen, H., and Frandsen, F. (2000). Spatial Risk Assessment for Lyme Borreliosis in Denmark, *Scand. J. Infect. Dis.*, 32, 545–550.

Jianan, T. (Ed.) (1989). *The Atlas of Endemic Diseases and Their Environments in the People's Republic of China*, Science Press, Beijing.

Karlsen, A. W., Schultz, A. C., Warwick, P. D., Podwysocki, S. M., and Lovern, V. S. (2001). Coal Geology, Land Use, and Human Health in the People's Republic of China, U. S. Geological Survey Open File Report 01–318 (CD-ROM).

Kennedy, H. (Ed.) (2001). *Dictionary of GIS Terminology*, ESRI Press, Redlands, CA.

Kitron, U., and Kazmierczak, J. J. (1997). Spatial Analysis of the Distribution of Lyme Disease in Wisconsin, *Am. J. Epidemiol.*, 145, 558–566.

Kulldorff, M. (1998). Statistical Methods for Spatial Epidemiology: Tests for Randomness. In *GIS and Health, GIS Data VI*, (A. C. Gatrell and M. Löytönen, Eds.), Taylor & Francis, London, pp. 49–62.

Ostfeld, R. S., Hazler, K. R., and Cepeda O. M. (1996). Temporal and spatial dynamics of *Ixodes scapularis* (Acari: Ixodidae) in a rural landscape, *J. Med. Entomol.*, 33, 90–95.

PAHO (Panamerican Health Organization) (2000). Geographic Information Systems in Health, Special Program for Health Analysis, PAHO, Washington DC.

Robinson, T. P. (2000). Spatial Statistics and Geographical Information Systems in Epidemiology and Public Health. In *Remote Sensing and Geographical Information Systems in Epidemiology, Advances in Parasitology 47* (S. I. Hay, S. E. Randolph, and D. J. Rogers, Eds.), Academic Press, San Diego, CA, pp. 81–128.

Rogers, D. J. (2000). Satellites, Space, Time and the African Trypanosomiases. In *Remote Sensing and Geographical Information Systems in Epidemiology, Advances in Parasitology 47* (S. I. Hay, S. E. Randolph, and D. J. Rogers, Eds.), Academic Press, San Diego, CA, pp. 129–171.

Ruiling, L., Tianyu, H., and Jianping, W. (Compilers) (1996). Coalfield Prediction Map of China. Surveying and Mapping Institute of Jilin Province, Publishing House of Surveying and Mapping, 9 map sheets, scale 1:2,500,000.

SHA (Special Program for Health Analysis) (2000). Incidence of Malaria and Land Use in Chiapas, Mexico and Peten, Guatemala, PAHO, Scientific paper No. 104.

Sherman, C., and Price, G. (2001). *The Invisible Web*, CyberAge Books, Information Today, Inc., Medford, NJ.

Stein, A., Staritsky, I., Bouma, J., and van Groenigen, J. W. (1995). Interactive GIS for Environmental Risk Assessment, *Int. J. Geogr. Inf. Syst.*, 9(5), 509–525.

Zeiler, M. (1999). *Modeling Our World*, ESRI Press, Redlands, CA.

Zhang, Y., and Cao, S. R. (1996). Coal Burning Induced Endemic Fluorosis in China, *Fluoride*, 29(4), 207–211.

Zheng, B., and Huang, R. (1989). *Human Fluorosis and Environmental Geochemistry in Southwest China, Developments in Geoscience, Contributions to 28th International Geologic Congress*, Washington DC Science Press, Beijing, China, pp. 171–176.

Suggested Reading

Bernhardsen, T. (1999). *Geographic Information Systems: An Introduction*, John Wiley & Sons, New York.

Briggs, D. J., and Elliott P. (1995). The use of geographical information systems in studies on environment and health, *World Health Stat. Q.*, 48, 85–94.

Burrough, P. A., and McDonnell, R. (1998). *Principles of Geographic Information Systems*, Oxford University Press, Oxford, England.

Clarke, K. C. (1998). *Getting Started With Geographic Information Systems*, third edition, Prentice Hall, Upper Saddle River, NJ.

DeMers, M. N. (2000). *Fundamentals of Geographic Information Systems*, second edition, John Wiley & Sons, New York.

Glass, G. E. (2000). Spatial Aspects of Epidemiology: The Interface with Medical Geography, *Epidemiol. Rev.*, 22(1), 136–139.

Green, K. (1992). Spatial Imagery and GIS: Integrated Data for Natural Resource Management, *J. Forestry*, Nov., 32–36.

Lang, L. (2000). *GIS for Health Organizations*, ESRI Press, Redlands, CA, p. 100 plus CD-ROM.

Longley, P. A., Goodchild, M. F., Maguire, D. J., and Rhind, D. W. (Eds.) (1999). *Geographical Information Systems*, second edition, John Wiley & Sons, New York, p. 1101 (2 volumes).

Meade, M. S, and Earickson, R. J. (2000). *Medical Geography*, second edition, The Guilford Press, New York.

Melnick, A. L. (2002). *Introduction to Geographic Information Systems in Public Health*, Aspen Publishers, Gaithersburg, MD.

Moore, G. S. (2002). *Living With the Earth: Concepts in Environmental Health Science*, second edition, Lewis Publishers, Boca Raton, FL.

de Savigny, D., and Wijeyaratne, P. (Eds.) (1995). *GIS for Health and Environment*, International Development Research Centre, Ottawa, Canada.

Vine, M. F., Degnan, D., and Hanchette, C. (1997). Geographic Information Systems: Their Use in Environmental Epidemiological Research, *Environ. Health Perspect.*, 105, 598–605.

Other Sources

A. Sources Of Earth Science/ Geospatial Information

FGDC database clearinghouse sources (all databases listed are accessible through the Internet at http://www.fgdc.gov). Information presented is current at the time of publication (accessed March 2002) and was prepared with sources readily accessible to Internet users in the United States. Additional sources are available internationally in various languages and formats.

- Alaska State Geospatial Data Clearinghouse (ASGDC)
- America Central Clearinghouse Nicaragua
- Australia–WALIS Interrogator–Spatial Data
- Australia–IndexGeo Pty Ltd–Eco Companion Catalogue
- Bureau of Land Management Spatial Data Clearinghouse
- Canada–Ecological Monitoring and Assessment Network Data Set Library (hosted by Environment Canada)
- Canada–National Forest Health Database–Archive of Insects and Diseases Found in Canadian Forests
- Canada–Newfoundland and Labrador Community Base Maps (1 : 2500 and 1 : 5000)
- Canada–Newfoundland and Labrador Geodetic Survey Data
- Canada–Purple Pages Business Directory
- Canada–RADARSAT Inventory held by CCRS
- Caribbean Environment Programme
- Columbia Environmental Research Centers Metadata Node
- Connecticut–Geospatial Data Clearinghouse
- Costa Rica Biological Resource Maps (KU)
- Earth Data Analysis Center, UNM–Prototype Hydrology WMS
- El Salvador, CNR Instituto Geografico Nacional
- Forest, aquatic, and rangeland Ecosystems in the Western United States
- Geography Network
- Geological Survey of Alabama Geospatial Data Clearinghouse Node
- Global Change Master Directory
- Illinois Natural Resources Geospatial Data Clearinghouse
- Kansas Ecological Reserves Clearinghouse
- Michigan GIS
- Minnesota Land Management Information Center
- Montana State Library
- NOAA Environmental Satellite, Data, and Information Services (SAT) Node
- NOAA NCDC Library Historical Data Sets (FDL) Node
- NOAA National Climatic Data Center (NCDC) Node

- National Biological Information Infrastructure Metadata Clearinghouse
- Natural Resources Conservation Service
- Nevada Dataworks Spatial Data Warehouse
- New Mexico Resource Geographic Information System
- Peru–Instituto Geografico Nacional
- Republica Dominicana–Nodo Nacional de Datos Geoespaciales
- Space Imaging 5-m Digital Ortho Quads
- Texas Natural Resources Information System
- Transboundary (U.S.–Mexico) Metadata Clearinghouse
- U.S. Geological Survey Advanced Very High Resolution Radiometer
- U.S. Geological Survey CORONA Satellite Photographs
- U.S. Geological Survey DOQ
- U.S. Geological Survey Digital Elevation Model 15-minute
- U.S. Geological Survey Digital Raster Graphics
- U.S. Geological Survey Geoscience data
- U.S. Geological Survey MAPS
- U.S. Geological Survey National Aerial Photography Program
- UK Node–British Geological Survey
- Uruguay–Clearinghouse Nacional de Datos Geograficos (Espanol)

For selected online earth science/geospatial journals see Appendix 2. For selected biomedical/health information see Appendix 2.

B. Libraries (for Further Research)

U.S. Geological Survey Library
950 National Center 12201 Sunrise Valley Drive
Reston, VA 20192
e-mail: library@usgs.gov
U.S. Library of Congress
101 Independece Avenue,
S. E. Washington DC 20540
e-mail: lcweb@loc.gov
U.S. National Library of Medicine
National Institutes of Health
8600 Rockville Pike
Bethesda, MD 20894
e-mail: NIHInfo@OD.NIH.GOV

C. Where Does One Start a Search for Relevant Databases?

The Federal Geographic Data Committee (FGDC) Web site http://www.fgdc.gov provides access to over 38 simultane-ously searchable data clearinghouses in the United States and internationally. These include databases related to Earth sciences, geography, landform information, ecosystem health, biological resources, and satellite imagery (see Other Sources, part A). The focus of this chapter has been on electronic sources, but don't forget to check your library's reference section, trade journals, or other specialized periodicals and books. To find databases on the Internet, you might use a search engine. Be aware that different search engines work in different ways, and that what may be overlooked by one search engine might be found by another one. There are also metasearch engines that use several search engines simultaneously (Hock, 2001). Of course, another very efficient way to find out what databases are used by experts in a given field is to simply ask them. Contact information for university professors is often listed on their institution's Web site, which can be found on any search engine. Even if the principal investigator is hard to reach in person, his postdoctoral fellows, graduate students, and technicians may be willing to help.

An efficient strategy when starting out on medical geology research projects is to seek out relevant database clearinghouses. Using clearinghouses also offers some protection from rapidly changing unique or uniform resource locators (URLs), which are often referred to as Web site addresses. One example, is the U.S. Geological Survey (USGS), which is a major clearinghouse for Earth science data. The URL for some particular databases contained therein may change, but the URL for a clearinghouse such as USGS generally remains stable over time. The primary clearinghouse organization will maintain proper internal links and keep access to all of their individual databases current.

The geographic and temporal range of the data needed must be ascertained at the outset of any medical geology GIS. It is better to err on the side of obtaining more information then deleting unnecessary elements, because it can be difficult to add data later if it is decided to examine additional parameters. But the initial cost of the data and the cost in resources to store data must also be taken into account. Because different databases will likely contain data archived in a variety of formats, it is advisable to store the initial downloaded data as is and make copies of it before any subsequent manipulation. The text-only ASCII file format is a "common denominator" useful for merging data from different sources into a single data set.

The use of Internet-derived databases can be made frustrating and difficult by two realities of this medium. One reality is that URLs change quickly, and so the Web site address that worked in the past may not take you to the same page today. This potential pitfall can be avoided by using the "gatekeeper" URL to a database clearinghouse as mentioned above, rather than by using direct URLs to individual databases. For example, one is advised to use a main clearinghouse Web site rather than a more specific URL for some individual database, such as for water table depths in India. The other major problem is the so-called Invisible Web (Sherman & Price, 2001). There are a great many databases accessible via the Internet with no easy way to find them or to find out about them. Many databases can only be accessed after registering and entering a password. Search engines will miss these and other relevant sites, and they will often come up with totally irrelevant sites. Investigators must be mindful, too, of the reliability of database sources accessible via the Internet. *If associated metadata are not available, that database should not be used.*

Once a database of interest has been identified and the legitimacy of the organization that maintains it is verified, you are ready to download. Appendix 2 of this volume lists a number of earth science/geospatial and biomedical/human health databases as examples. Make sure you have the minimum requirements and sufficient memory space on your computer before proceeding. As always when downloading any software or data to a personal computer, remember to have some tool in place for screening computer viruses. It is critical to ensure that once downloaded, the data were not corrupted in the process. You must examine the source data carefully and confirm that they match the data in the form that has been downloaded. Problems can arise, for example, if the source data are tab delimited and your default download is space delimited. As soon as you have downloaded the source data, you should make a backup copy before doing anything with the data. It is generally convenient to keep such files on a compact disc (CD). Now you are ready to open up your data with your spreadsheet software package and import it to your GIS application or to a statistical analysis package. In a GIS environment, you can easily query the data. That is, by clicking with a mouse on a location visibly displayed on a map, you can extract attributes of that point.

You will need to join data from different databases for use in a GIS project. Get to know the raw data well as you must always maintain quality assurance/quality control (QA/QC). It is easy to mix up or somehow corrupt data when manipulating it. For instance, if you sort the data for some reason, make sure you keep a copy of the original unsorted data. Also be careful not to sort only one field, but rather keep your unique identifiers tied to the data in the proper order. If your ultimate aim is to do some statistical analysis of the data, you should work closely with a statistician right from the start. The statistician will help you determine the appropriate data you need to answer the questions you are asking.

CHAPTER 27

INVESTIGATING VECTOR-BORNE AND ZOONOTIC DISEASES WITH REMOTE SENSING AND GIS

STEPHEN C. GUPTILL
United States Geological Survey

CHESTER G. MOORE
Colorado State University

CONTENTS

I. INTRODUCTION

For centuries, people have been intuitively aware of the relationships between human health and the environment. Today, geographic information systems (GIS), remote sensing satellites, and other technologies are providing scientists with the tools and the data to make clear the geographic relationships between the habitats of disease agents, their vectors and vertebrate hosts, and the occurrence of disease in the human population. Although the utility of the foregoing tools as an aid to epidemiology was pointed out 30 years ago (Cline, 1970), the medical community has been slow to put them to use.

In this discussion, we provide an introduction to remote sensing technology and then use vector-borne and zoonotic diseases to demonstrate some current and potential uses of remote sensing, GIS, and related technologies for the surveillance, prevention, and control of disease. Vector-borne is defined as those diseases that are transmitted from one vertebrate host to another by an invertebrate, usually an insect, a tick, or a snail. A zoonosis (pl. zoonoses) is a disease that normally exists in a non-human host, or reservoir. For example, passerine birds are the natural hosts of several viruses that can infect humans. Many vector-borne diseases are also zoonoses (for example, yellow fever, Lyme disease, and plague).

There are several reasons for choosing this group of diseases as our examples. First, these diseases account for a large portion of the annual global morbidity and mortality. It is estimated that 41% of the world's population (about 2.3 billion people) live in areas with malaria risk (Gratz, 1999). Of approximately 300–500 million people who become infected each year, some 1.5–2.7 million die from this disease. Dengue, a virus related to yellow fever, attacks as many as 50–100 million people each year, and there are more than 250,000 cases of the more severe forms of dengue hemorrhagic fever and dengue shock syndrome. Rabies, a zoonosis, is responsible for at least 35,000–45,000 deaths each year. Second, these diseases—especially the zoonoses—are naturally occurring systems that are impacted by a wide variety of physical and biotic factors that may be susceptible to remote measurement (e.g., Ostfeld et al., 1996). Finally, there is considerable concern about the potential impact of global change on the dynamics and spread of these diseases (e.g., Shope, 1992; Reeves et al., 1994; Martens et al., 1995; Jetten & Focks, 1997).

II. FUNDAMENTALS OF REMOTE SENSING

Remote sensing is a technology that involves the analysis and interpretation of images gathered through techniques that do not require direct contact with the object. Electromagnetic radiation sensors are used to record images of the environment. The sensors used in these devices can detect radiation from the ultraviolet through the visible and infrared spectra to microwave radar. In studying the Earth, the remote sensing devices are usually deployed in aircraft or Earth-orbiting satellites. For convenience, in the remainder of this paper, we will generally refer to remote sensing satellites, but the concepts and principles also apply to remote sensing instruments carried aboard aircraft.

Remote sensing satellites are designed to collect various types of information about the Earth's surface and atmosphere. The combination of spatial resolution of the sensor, the wavelengths detected by that sensor, and the frequency of data collection determine the types of applications for which the satellite will collect useful information. These design parameters must be weighed one against another, as they are somewhat mutually exclusive. Each of these parameters will be examined below along with descriptions of how they have been implemented in various operational systems.

A. Electromagnetic Spectrum

Electromagnetic radiation (EMR) extends over a wide range of energies and wavelengths (frequencies). A narrow range of EMR extending from 0.4 to 0.7 µm, the interval detected by the human eye, is known as the *visible region* (also referred to as *light*, but physicists often use that term to include radiation beyond the visible). White light contains a mix of all wavelengths in the visible region. The distribution of the continuum of all radiant energies can be plotted either as a function of wavelength or of frequency in a chart known as the electromagnetic spectrum.

Using spectroscopes and other radiation detection instruments, scientists have arbitrarily divided the electromagnetic spectrum into regions or intervals and applied descriptive names to them. At the very energetic (high frequency and short wavelength) end are gamma rays and x-rays, whose wavelengths are normally measured in angstroms (Å), which in the metric scale are in units of 10^{-8} cm. Radiation in the ultraviolet extends from about 300–4000 Å. It is convenient to measure the mid-regions of the spectrum in one of two units: micrometers (µm), which are multiples of 10^{-6} m or nanometers (nm), based on 10^{-9} m. The visible region occupies the range between 0.4 and 0.7 µm, or its equivalents of 4000–7000 Å or 400–700 nm. The infrared region (IR), spanning between 0.7 and 100 µm, has four subintervals of special interest: (1) reflected IR (0.7–3.0 µm); (2) its film responsive subset, the photographic IR (0.7–0.9 µm); and (3) thermal bands at 3–5 µm and (4) 8–14 µm. Longer wavelength intervals are measured in units ranging from millimeters to centimeters to meters. The microwave region spreads across 0.1–100 cm, which includes the entire interval used by radar systems. These systems generate their own active radiation and direct it toward targets of interest. The lowest frequency–longest wavelength region beyond 100 cm is the radio bands, from VHF (very high frequency) to ELF (extremely low frequency). Within any region, a collection of continuous wavelengths can be partitioned into discrete intervals called *bands*.

Most remote sensing is conducted above the Earth either within or above the atmosphere. The gases in the atmosphere interact with solar irradiation and with radiation from the Earth's surface. The atmosphere itself is excited by EMR and thus becomes another source of released photons. Figure 1 (Short, 2003) is a generalized diagram which shows relative atmospheric radiation transmission of different wavelengths.

Shaded zones mark minimal passage of incoming and/or outgoing radiation, whereas white areas denote atmospheric windows in which the radiation does not interact much with air molecules and hence, is not absorbed.

Most remote sensing instruments on air or space platforms operate in one or more of these windows by

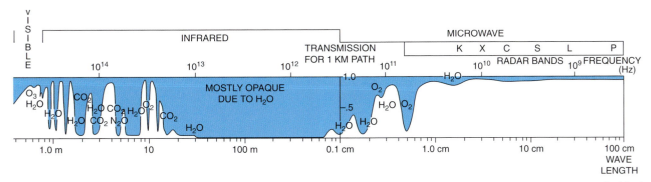

FIGURE 1   Electromagnetic spectrum showing areas of transmission (in white) and areas of atmospheric absorption (in blue).

making their measurements with detectors tuned to specific frequencies (wavelengths) that pass through the atmosphere. However, some sensors, especially those on meteorological satellites, directly measure absorption phenomena, such as those associated with carbon dioxide (CO_2) and other gaseous molecules. Note that the atmosphere is nearly opaque to electromagnetic radiation in part of the mid-IR and all of the far-IR regions. In the microwave region, by contrast, most of this radiation moves through unimpeded, so radar waves reach the surface.

Remote sensing of the Earth traditionally has used reflected energy in the visible and infrared and emitted energy in the thermal infrared and microwave regions to gather radiation that can be analyzed numerically or used to generate images whose variations represent different intensities of photons associated with a range of wavelengths that are received at the sensor. This gathering of a (continuous or discontinuous) range(s) of wavelengths is the essence of what is usually termed multispectral remote sensing (Short, 2003).

B. Spatial, Temporal, and Spectral Resolution

Spatial, temporal, and spectral resolution are three parameters that largely determine the characteristics of the data collected by a remote sensing instrument and thus to some degree the applications for which those data can be used.

The orbital characteristics of the satellite are the primary determinant of its temporal characteristics, that is, the time required for the sensor to re-image the same geographic location on the Earth. This can range from a daily revisit for the advanced very high resolution radiometer (AVHRR) sensor on the National Oceanic and Atmospheric Administration (NOAA) satellite to a

16-day revisit time for Landsat 7. Most satellites image the area directly below the sensor (nadir). However, some satellites are designed to point the sensor off-nadir, allowing the satellite to image the target area more frequently.

The imaging sensors are designed to collect energy from a certain portion or portions of the electromagnetic spectrum. For example a "panchromatic" sensor collects one set of data (i.e., a grayscale image) across a broad spectrum of visible and near infrared energy (450–900 nm). In contrast a hyperspectral sensor, like Hyperion, collects 250 bands of data, each 10 nm in width from 43–2400 nm. Figure 2a–d shows the ways in which different features are highlighted or hidden depending on the band combinations used to make the image.

Spatial resolution refers to the smallest unit of area within which the sensor integrates EMR. More commonly it is referred to as "pixel size." This size is basically determined by the height of the orbit above the Earth and the magnification power of the optics on the sensor. As of the fall of 2003 the QuickBird satellite obtains the highest spatial resolution image with a 61-cm (approximately 2 feet) pixel size panchromatic (black and white) image collected in the 450- to 900-nm band. The detail available at this resolution can be seen in Figure 3 which is an image of the Eiffel Tower taken on April 9, 2002. The coarsest image commonly used in our applications is from the AVHRR sensor, with a ground resolution of approximately 1 km.

Using digital image processing, satellite data of varying resolutions can be combined. In the example below, QuickBird 61-cm panchromatic imagery is combined with 2.4-m multispectral imagery (collected at the same time) to create a color image with 61-cm resolution. This image of Prague (Figure 4) was taken on August 17, 2002, and shows the extent of flood damage to the city.

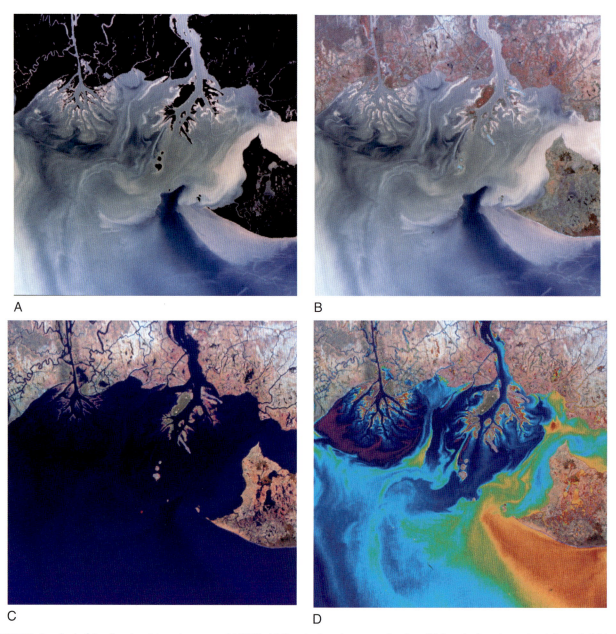

FIGURE 2   Atchafalya Bay, Louisiana (courtesy NASA). (a) Bands showing natural color, (b) bands showing color-infrared, (c) bands showing middle-infrared, and (d) bands showing thermal radiation.

III. REMOTE SENSING SATELLITES

Remote sensing instruments fall naturally into three groups based on their principal applications: land observation, meteorology, or oceanography. However, many of the satellites provide useful information for more than one set of applications. Here, compiled by Nicholas Short (2003), are the principal remote sensing spacecraft flown by the United States and other nations (identified in parentheses) along with the launch date (if more than one in a series, this date refers to the first one put successfully into orbit):

1. *Land observation:* Government satellites—Landsat 1–7 (1973); Seasat (1978); HCMM (1978); RESURS (Russia) (1985); *IRS* 1A–1D (India)

FIGURE 3   Panchromatic image of the Eiffel Tower, Paris, France, with resolution of 61 cm. (Courtesy DigitalGlobe.)

(1986); ERS 1–2 (1991); JERS 1–2 (Japan) (1992); Radarsat (Canada) (1995); ADEOS (Japan) (1996); Terra (1999)

Commercial satellites—SPOT (France) (1986); Resurs-01 series (Russia) (1989; became commercial in the 1990s); Orbview-2 (U.S.) (1997); SPIN-2 (Russia) (1998); IKONOS (U.S.) (1999); QuickBird (U.S.) (2001); EROS A (ImageSat International; Israel) (2002)

2. *Meteorological observation*: TIROS 1–9 (1960); Nimbus 1–7 (1964); ESSA 1–9 (1966); ATS (g = geostationary) 1–3 (1966); DMSP series I (1966); the Russian Kosmos (1968) and Meteor series (1969); ITOS series (1970); SMS(g) (1975); GOES(g) series (1975); NOAA 1–5 (1976);

DMSP series 2 (1976); GMS (Himawari) (g) series (Japan) (1977); Meteosat(g) series (Europe) (1978); TIROS-N series (1978); Bhaskara(g) (India) (1979); NOAA (6–14) (1982); Insat (1983); ERBS (1984); MOS (Japan) (1987); UARS (1991); TRMM (U.S./Japan) (1997); Envisat (European Space Agency) (2002); Aqua (2002)

3. *Oceanographic observations*: Seasat (1978); Nimbus 7 (1978) included the *CZCS*, the Coastal Zone Color Scanner that measures chlorophyll concentration in seawater; Topex-Poseidon (1992); SeaWiFS (1997)

The Center for Health Applications of Aerospace Related Technologies (CHAART), a part of NASA,

FIGURE 4   Panchromatic/multispectral merged image of Prague Czechoslovakia with effective resolution of 61 cm. (Courtesy DigitalGlobe.)

maintains a comprehensive list that shows the basic characteristics of current and future planned remote sensing satellite systems. A dynamic version of this table can be found at http://geo.arc.nasa.gov/sge/health/sensor/cfsensor.html.

So what applications are addressed by the data collected from these instruments? As part of NASA's Earth Observing System program, Michael King (2000) has created a table that shows a variety of physical measurements and characterizations that can be constructed from remote sensing data. The measurements/characteristics are shown on the left. The satellite/sensor that collects the raw information used in this process is on the right.

Atmosphere

Cloud properties (amount, optical properties, height)	MODIS, GLAS, AMSR-E, MISR, AIRS, ASTER, SAGE III
Radiative energy fluxes (top of atmosphere, surface)	CERES, ACRIM III, MODIS, AMSR-E, GLAS, MISR, AIRS, ASTER, SAGE III
Precipitation	AMSR-E
Tropospheric chemistry (ozone, precursor gases)	TES, MOPITT, SAGE III, MLS, HIRDLS, LIS
Stratospheric chemistry (ozone, ClO, BrO, OH, trace gases)	MLS, HIRDLS, SAGE III, OMI, TES
Aerosol properties (stratospheric, tropospheric)	SAGE III, HIRDLS MODIS, MISR, OMI, GLAS
Atmospheric temperature	AIRS/AMSU-A, MLS, HIRDLS, TES, MODIS
Atmospheric humidity	AIRS/AMSU-A/HSB, MLS, SAGE III, HIRDLS, Poseidon 2/JMR/DORIS, MODIS, TES
Lightning (events, area, flash structure)	LIS

Solar radiation

Total solar irradiance	ACRIM III, TIM
Solar spectral irradiance	SIM, SOLSTICE

Land

Land cover and land use change	ETM+, MODIS, ASTER, MISR
Vegetation dynamics	MODIS, MISR, ETM+, ASTER
Surface temperature	ASTER, MODIS, AIRS, AMSR-E, ETM+
Fire occurrence (extent, thermal anomalies)	MODIS, ASTER, ETM+
Volcanic effects (frequency of occurrence, thermal anomalies, impact)	MODIS, ASTER, ETM+, MISR
Surface wetness	AMSR-E

Ocean

Surface temperature	MODIS, AIRS, AMSR-E
Phytoplankton and dissolved organic matter	MODIS
Surface wind fields	SeaWinds, AMSR-E, Poseidon 2/JMR/DORIS
Ocean surface topography (height, waves, sea level)	Poseidon 2/JMR/DORIS

Cryosphere

Land ice (ice sheet topography, ice sheet volume change, glacier change)	GLAS, ASTER, ETM+
Sea ice (extent, concentration, motion, temperature)	AMSR-E, Poseidon 2/JMR/DORIS, MODIS, ETM+, ASTER
Snow cover (extent, water equivalent)	MODIS, AMSR-E, ASTER, ETM+

In the sections below, a few of the most widely available and useful systems for analyzing environmental influences on human health will be characterized.

A. Landsat

The Landsat program has been in operation since the early 1970s. Since then, many different satellites have been sent into orbit. Beginning with Landsat 1 in 1972 and most recently Landsat 7, these satellites have taken thousands of images of the Earth and documented the rapidly changing landscape of the planet.

Landsat 7 was launched on April 15, 1999, from the Western Test Range aboard a Delta II expendable launch vehicle. At launch, the satellite weighed approximately 4800 pounds (2200 kg). The spacecraft is about 14 feet long (4.3 m) and 9 feet (2.8 m) in diameter. It consists of a spacecraft bus that is provided under a NASA contract with Lockheed Martin Missiles and Space in Valley Forge, Pennsylvania, and the Enhanced Thematic Mapper Plus (ETM+) instrument, procured under a NASA contract with Raytheon (formerly Hughes) Santa Barbara Remote Sensing in California.

The ETM+ instrument is an eight-band multispectral scanning radiometer capable of providing high-resolution imaging information of the Earth's surface. It detects spectrally filtered radiation at visible, near-infrared, short-wave, and thermal infrared frequency bands from the sun-lit Earth in a 115-mile (183 km) wide swath when orbiting at an altitude of 438 miles (705 km). Nominal ground sample distances or pixel sizes are 49 feet (15 m) in the panchromatic band; 98 feet (30 m) in the six visible, near- and short-wave infrared bands; and 197 feet (60 m) in the thermal infrared band. A Landsat WorldWide-Reference System has catalogued the world's landmass into 57,784 scenes, each 115 miles (183 km) wide by 106 miles (170 km) long. The ETM+ produces approximately 3.8 gigabits of data for each scene. The seven bands have the following characteristics:

Band no.	Wavelength interval (μm)	Spectral response	Resolution (m)
1	0.45–0.52	Blue-green	30
2	0.52–0.60	Green	30
3	0.63–0.69	Red	30
4	0.76–0.90	Near-IR	30
5	1.55–1.75	Mid-IR	30
6	10.40–12.50	Thermal-IR	120
7	2.08–2.35	Mid-IR	30

The satellite orbits the Earth at an altitude of approximately 438 miles (705 km) with a sun-synchronous, 98-degree inclination and a descending equatorial crossing time of 10 a.m. The orbit will be adjusted upon reaching orbit so that its 16-day repeat cycle coincides with the Landsat Worldwide Reference System. This orbit will be maintained with periodic adjustments for the life of the mission. A three-axis attitude control subsystem will stabilize the satellite and keep the instrument pointed toward Earth to within 0.05 degrees. A state-of-the-art solid-state recorder capable of storing 380 gigabits of data (100 scenes) is used to store selected scenes from around the world for playback over a U. S. ground station. In addition to stored data, real-time data from ETM+ can be transmitted to cooperating international ground stations and to the U.S. ground stations.

These spectral bands allow ETM+ to detect subtle variations in surface characteristics. For example, on the border between Chile and the Catamarca province of Argentina lies a vast field of currently dormant volcanoes. Over time, these volcanoes have laid down a crust of magma roughly 2 miles (3.5 km) thick. It is tinged with a patina of various colors that can indicate both the age and mineral content of the original lava flows. This is shown in Figure 5.

The U. S. Geological Survey operates the Landsat 7 satellite. For more information on Landsat 7 including how to order data, go to http://landsat7.usgs.gov/index.php.

B. AVHRR

AVHRR is a broad-band, four- or five-channel (depending on the model) scanner, sensing in the visible, near-infrared, and thermal infrared portions of the electromagnetic spectrum. This sensor is carried on NOAA's Polar Orbiting Environmental satellites (POES), beginning with TIROS-N in 1978 and most recently on the NOAA-15 (launched in 1998) and NOAA-16 (launched in 2000) satellites.

The AVHRR sensor provides for global (pole to pole) on-board collection of data from all spectral channels. Each pass of the satellite provides a 1491-mile (2399-km) wide swath. The satellite orbits the Earth 14 times each day from 517 miles (833 km) above its surface.

The average instantaneous field-of-view (IFOV) of 1.4 milliradians (mrad) yields a LAC/HRPT ground resolution of approximately 1.1 km at the satellite nadir from the nominal orbit altitude of 517 miles (833 km).

The GAC data are derived from an on-board sample averaging of the full resolution AVHRR data. Four out of every five samples along the scan line are used to compute one average value and the data from only every third scan line are processed, yielding a 1.1 × 4 km resolution at nadir.

The current sensors cover five spectral bands as shown below:

Band	Wavelength (µm)	IFOV (mrad)
1	0.58–0.68	1.39
2	0.725–1.10	1.41
3	3.55–3.93	1.51
4	10.3–11.3	1.41
5	11.5–12.5	1.30

AVHRR data provide opportunities for studying and monitoring vegetation conditions in ecosystems including forests, tundra, and grasslands. Applications include agricultural assessment, land cover mapping, producing image maps of large areas such as countries or continents, and tracking regional and continental snow cover. AVHRR data are also used to retrieve various geophysical parameters such as sea surface temperatures and energy budget data.

Online requests for these data can be placed via the U.S. Geological Survey Global Land Information System (GLIS) interactive query system. The GLIS system contains metadata and online samples of Earth science data. With GLIS, you may review metadata, determine product availability, and place online requests for products. Additional data sets include the Alaska twice-monthly AVHRR and the U.S. Conterminous bi-weekly composites. These comprehensive time series data sets are calibrated, georegistered daily observations and twice-monthly maximum NDVI composites for each annual growing season. Global experimental bi-weekly normalized difference data, computed from Global Vegetation Index (GVI) data, are analyzed to monitor global vegetation and are a potential tool in global climatic studies.

Figure 6 shows an AVHRR image of the Mississippi River basin soon after the summer floods of 1993. This figure shows the extent of flooding and demonstrates the value of daily observations.

FIGURE 5 Landsat ETM+ image of Chile and Argentina. (Courtesy USGS.)

C. EOS Terra Spacecraft

On December 18, 1999, NASA launched the Earth Observing System (EOS) "flagship"—EOS Terra—to begin collecting a new 18-year global data set on which to base future scientific investigations about our complex home planet.

Physically, the EOS Terra spacecraft is roughly the size of a small school bus. It carries a payload of five state-of-the-art sensors that will study the interactions among the Earth's atmosphere, lands, oceans, and radiant energy (heat and light). Each sensor has unique design features that will enable EOS scientists to meet a wide range of science objectives.

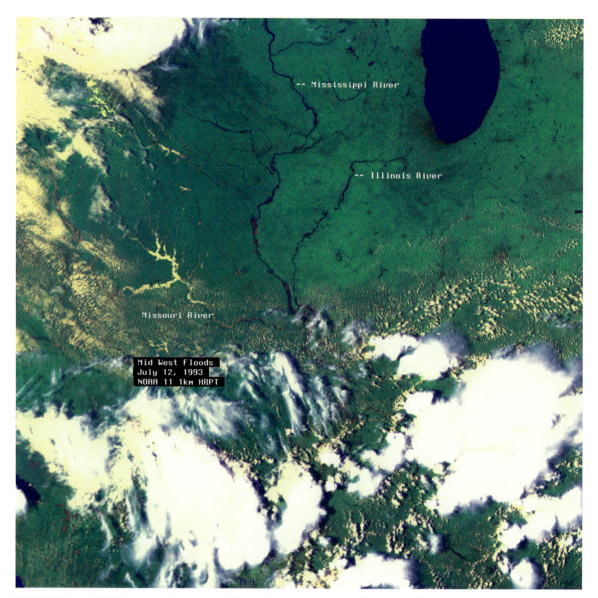

FIGURE 6   AVHRR image of the central United States showing the Mississippi floods of 1993. (Courtesy NOAA.)

EOS Terra orbits the Earth from pole to pole descending across the equator in the morning when cloud cover is minimal and its view of the surface is least obstructed. The satellite's orbit will be perpendicular to the direction of Earth's spin, so that the viewing swaths from each overpass can be compiled into whole global images. Over time, these global images will enable scientists to show and tell the stories of the causes and effects of global climate change.

The sensors on EOS Terra will not actively scan the surface (such as with laser beams or microwave pulses). Rather, the sensors work much like a camera. Sunlight that is reflected by Earth, and heat that is emitted from

Earth, will pass through the apertures of Terra sensors. This radiant energy will then be focused onto specially designed detectors that are sensitive to selected regions of the electromagnetic spectrum that range from visible light to heat. The information produced by these detectors will then be transmitted back to Earth and processed by computers into images that we can interpret.

The five Terra onboard sensors are

1. ASTER, or Advanced Spaceborne Thermal Emission and Reflection Radiometer
2. CERES, or clouds and earth's radiant energy system

3. MISR, or multi-angle imaging spectroradiometer
4. MODIS, or moderate-resolution imaging spectroradiometer
5. MOPITT, or measurements of pollution in the troposphere

MODIS provides continuous global coverage every one to two days and collects data from 36 spectral bands. Two bands (1–2) have a resolution of 250 m. Five bands (3–7) have a resolution of 500 m. The remaining bands (8–36) have a resolution of 1000 m. The swath width for MODIS is 2330 km. In its application, MODIS can be viewed in some ways as a higher resolution version of AVHRR. The MODIS image mosaic shown in Figure 7a–c is a "greenness" map of the United States in January, April, and June 2001. The NDVI index (or greenness) is calculated from several of the MODIS bands and the seasonal variation in vegetation vigor can be seen in this image sequence.

ASTER provides fourteen spectral bands with 15- to 90-m resolution, depending on bands. ASTER does not acquire data continuously, and its sensors are activated only to collect specific scenes upon request. The instrument consists of three separate telescopes, each of which provides different spectral range and resolution. The VNIR (visible and near-infrared) sensor provides 4 bands at 15-m resolution. The SWIR (short-wave infrared) sensor provides 6 bands at 30-m resolution. The TIR (thermal infrared) sensor provides 5 bands at 90-m resolution. The swath width for all sensors is 60 km.

ASTER data is generally available in the universal transverse mercator (UTM) projection, although some individual scenes may be cast to an alternative projection. The data is referenced to the World Geodetic Survey (WGS) System of 1984 (WGS84). Files are in the HDF-EOS format, and are distributed on CD-ROM, DVD, DLT, 8-mm tape, and file transfer protocol (FTP).

The ASTER data have about the same spatial resolution as Landsat, but they provide more spectral bands, which provide scientists with a greater ability to characterize various surface phenomena. Figure 8 shows an ASTER image of Washington, DC in color-infrared.

The life expectancy of the EOS Terra mission is 6 years. It will be followed in later years by other EOS spacecraft that take advantage of new developments in remote sensing technologies.

IV. IMAGE PROCESSING AND GEOGRAPHIC INFORMATION SYSTEMS

Since the early days of monitoring the Earth by orbiting spacecraft, the development of computer-aided

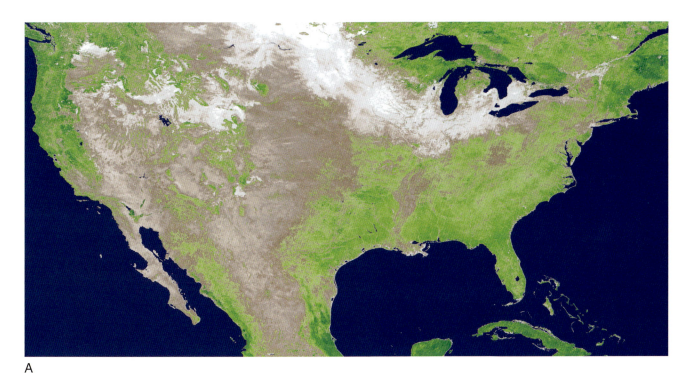

A

FIGURE 7 MODIS NDVI image of the United States (courtesy NASA). (a) January 2001,

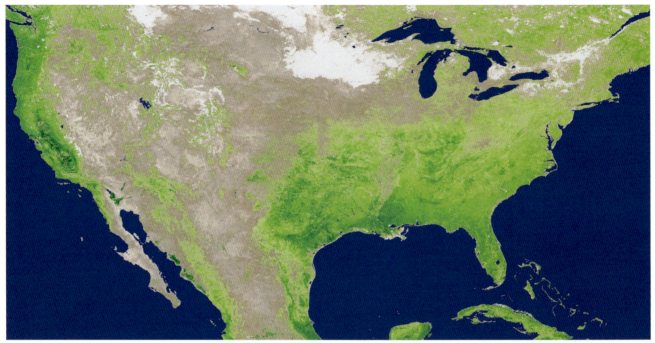

B

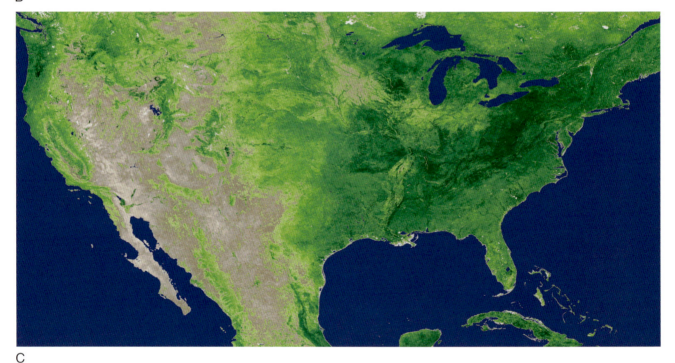

C

FIGURE 7 *Continued*   (b) April 2001, and (c) June 2001.

techniques for reliably identifying many categories of surface features within a remotely sensed scene, either by photo interpretation of enhanced images or by classification, ranks in itself as an outstanding achievement. Numerous practical uses of such self-contained information are made without strong dependence on other sources of complementary or supporting data. Thus, automated data processing assists in recognizing and mapping, for example, major crop types, estimating their yields, and spotting early

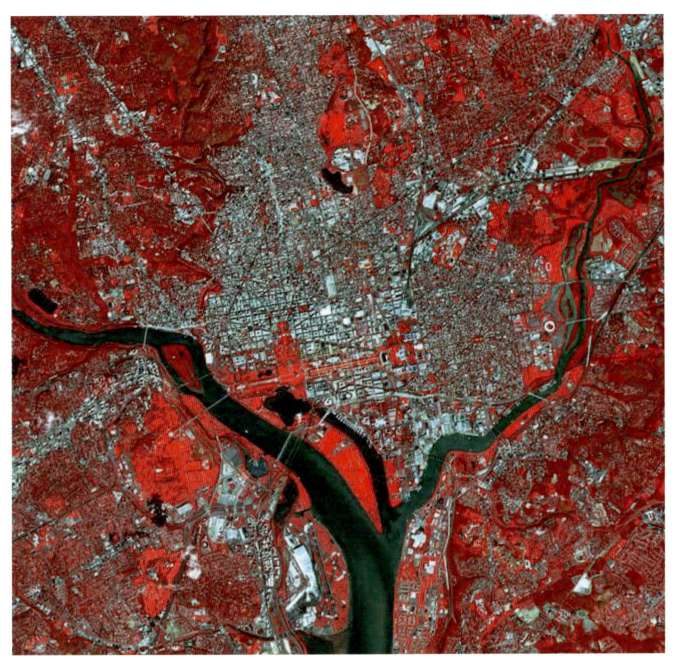

FIGURE 8   ASTER image of Washington, DC. (Courtesy NASA.)

warning indicators of potential disease or loss of vigor.

However, many applications, particularly those involving control of dynamic growth or change systems, decision making in management of natural resources, or exploration for nonrenewable energy or mineral deposits, require a wide variety of input data (from multiple sources) not intrinsic to acquisition by space-borne sensors such as those on Landsat, the commercial satellite SPOT, and others of similar purpose. Data from remote sensing satellites combined in a geographic information system with geospatial data on themes such as soils, terrain, geology, and hydrology provide the means for characterizing the land surface over extended areas.

Some data are essentially fixed or time-independent—slope, aspect, rock types, drainage patterns, archaeological sites, etc.—in the normal span of human events. These

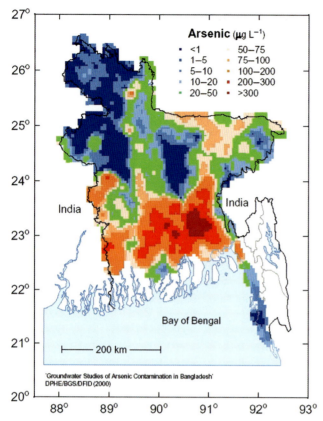

FIGURE 9 Arsenic concentrations in groundwater in Bangladesh. (Courtesy BGS)

the coastal flood plains of the country (see also Chapter 11, this volume).

V. LANDSCAPE ECOLOGY AND DISEASE SYSTEMS

Vector-borne diseases usually have complex life cycles (Figure 11). For any given system, the host(s), vector(s), and pathogen are each subjected to a variety of "pressures" within the ecosystem. Many environmental factors drive or constrain the system: weather and climate, food and space resources, predators, and parasites. For example, vertebrate hosts are affected by food quantity and quality, availability of nesting sites, and exposure to predators or parasites. The vector is affected by temperature, humidity, food resources (which may differ between adult and immature stages), and by predators and parasites. The pathogen is affected by host immune status and the frequency and timing of contact between vector and host. Temperature has a major impact on the development rate of the pathogen when it is developing in the vector. With the exception of vector-borne diseases that have humans as the primary or only vertebrate host (e.g., malaria, dengue, and bancroftian filariasis), humans often become involved in the transmission cycle by accident, and do not develop sufficient parasitemia or viremia to infect additional vectors; that is, they are "dead-end" hosts (see also Chapter 19, this volume).

Landscape ecology deals with the mosaic structure of landscapes and ecosystems, and considers the spatial heterogeneity of biotic and abiotic components as the underlying mechanism that determines the structure of ecosystems (Forman & Godron, 1986; Kitron, 1998). Vector-borne diseases are complex in their spatial and temporal distribution (Figure 12). All components of the system, pathogen, vector, and host must occur together in time and space for epizootics or epidemics to occur. Variations in landscape structure create a patchwork of suitable and unsuitable habitats, which lead to focal disease activity. Barriers, such as water bodies, deserts, or mountain ranges, may prevent the occurrence of a pathogen in an otherwise suitable location. Remotely sensed images of the land, coupled with other geographic data that characterize the landscape, help us to understand the ecosystem structure and to identify areas where risk of disease is greater.

data are usually collected and shown on maps. When digitized, they are included as layers in a GIS. Other data come from measurements or inventories conducted by people on the ground or in the air—such as weather information, population censuses, soil types and so forth. These too can be incorporated as GIS data layers. However, many vital data are transient or ephemeral—crop growth, flood water extent, insect infestation, limits of snow cover, etc.—and must be collected in a timely sense. Remote sensing data play a key role in this last instance, and in fact satellite monitoring is often the only practical and cost-effective way to acquire frequent data over large regions.

Using these tools we can relate GIS data layers and features shown on remote sensing imagery. For example, Figure 9 (BGS and DPHE, 2001) shows a map of arsenic concentrations in shallow water wells in Bangladesh. Comparing this map with the Landsat image of Bangladesh (Figure 10), one can see that the areas of high arsenic concentrations correspond with

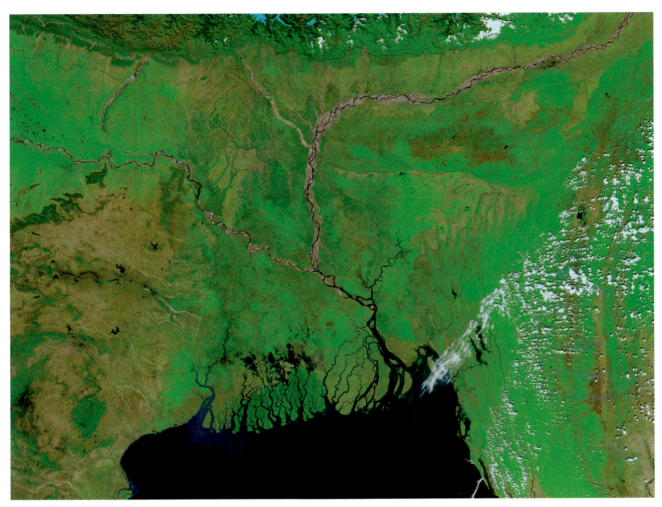

FIGURE 10   Landsat image of Bangladesh. (Courtesy USGS.)

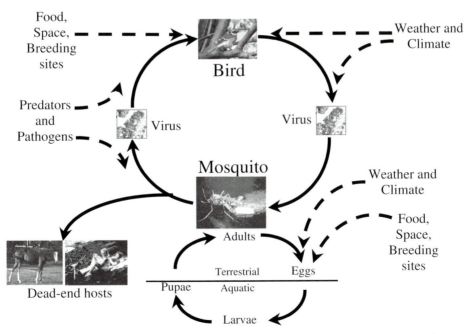

FIGURE 11   A typical arbovirus cycle showing sources of environmental pressure on the system.

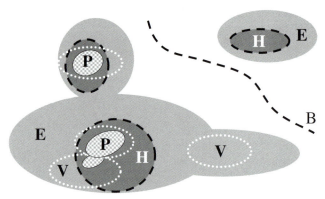

FIGURE 12 Spatial relationships in the arbovirus-vector-host ecosystem. All components are limited by a suitable environment (E). Populations of the vertebrate host (H) do not necessarily overlap with the distribution of the vector (V). It is only in those regions of E where both H and V overlap that the pathogen (P) can survive. Barriers (B) to dispersal may limit the presence of V or H.

VI. Relating the Disease Cycle to the Underlying Physical Environment

GIS, global positioning systems (GPS), remote sensing, and spatial statistics provide techniques and methods that we can use to analyze and integrate the spatial component into studies of the ecology and epidemiology of vector-borne disease (Kitron, 1998). These spatial tools, together with the concepts of landscape ecology, can help us to better understand emerging infectious diseases as well as the potential impact of global change on vector-borne diseases.

Several studies have utilized satellite-derived estimates of ground temperature and moisture to predict the distribution or risk of vector-borne disease. Malone et al. (1997) used an annual time series of diurnal temperature difference (dT, derived from day-night pairs of AVHRR data) to map high and low prevalence zones of the parasite, *Schistosoma mansoni* in Egypt. The dT values were associated with the depth of the water table, a major environmental determinant of the distribution of this water and snail-associated parasite.

Lindsay and Thomas (2000) characterized the climate at sites in Africa where surveys for lymphatic filariasis had taken place by using computerized climate surfaces. Logistic regression analysis of the climate variables predicted with 76% accuracy whether sites had microfilaremic patients or not. A map of the risk of lymphatic filariasis infection across Africa, built from the logistic equation in a GIS, compared favorably with expert opinion. A further validation, using a quasi-independent data set, showed that the model correctly predicted 88% of the infected sites. They then used a similar procedure to map the risk of microfilaremia in Egypt, where the dominant vector species differs from those in sub-Saharan Africa. By overlaying risk maps on a 1990 population grid, and adjusting for recent population increases, they estimated that around 420 million people would be exposed to this infection in Africa in the year 2000. The approach described by these authors could be used to produce a sampling frame for conducting filariasis surveys in countries that lack accurate distribution maps.

American and Mexican researchers collaborated in a study to map habitats of the malaria vector, *Anopheles albimanus*, using Landsat Thematic Mapper (TM) imagery and extensive ecological and epidemiological data (Beck et al., 1994). Pixel categories were associated with landscape type both by ground surveys and by comparison to color-infrared photography of the study area. Since the relationship between landscape type and suitability as larval habitat for *Anopheles albimanus* was very strong, it was possible to correctly distinguish between villages with high and low vector abundance with an overall accuracy of 90%. These analyses indicated that the most important landscape elements in terms of explaining mosquito abundance were the proportions of transitional wetlands and unmanaged pasture. Using these two landscape elements as predictors, they were able to correctly distinguish villages with high and low mosquito abundance with an overall accuracy of 90%.

Robinson et al. (1997) mapped the distribution of tsetse (*Glossina* spp.) habitat by using climate and remotely sensed vegetation data. Coarse-resolution (7.6 km) AVHRR images were combined with smoothed climate surfaces derived from continent-wide, long-term weather station records. Predictions were improved by subdividing habitats prior to classification. This system might be improved by using satellite-derived weather data and by using finer grain imagery (e.g., 1.1 km AVHRR).

Several groups have used remotely sensed data to improve our understanding of Lyme disease. Glass et al. (1992), for example, conducted a case-control study of Lyme disease in Baltimore County, Maryland. Land use/land cover maps (derived from Landsat TM imagery) were combined with soils, elevation, geology, and watershed maps to evaluate risk of exposure to Lyme disease and its vector, *I. scapularis*. The risk of disease was significantly lower in highly developed areas, and risk decreased with increasing distance from forests. Dister et al. (1997) used Landsat TM images to evaluate Lyme disease exposure risk on 337 residential premises in two communities in Westchester County, New York. Premises were categorized as no, low, or high risk based on seasonally adjusted densities of *I. scapularis* nymphs (previously determined by sampling). Spectral indices from the TM scene provided relative measures of vegetation structure and moisture (wetness) as well as vegetation abundance (greenness). They used GIS to spatially quantify and relate the landscape variables to risk category. A comparison of the two communities showed that the community with more high-risk premises was significantly greener and wetter than the community with fewer high-risk premises. Furthermore, high-risk premises were significantly greener and wetter than lower risk premises in the high-risk community. The high-risk sites appeared to contain a greater proportion of broadleaf trees, while lower risk sites were interpreted as having more non-vegetative cover or open lawn. The ability to distinguish these fine-scale differences among communities and individual properties illustrates the efficiency of a remote sensing/GIS-based approach for identifying peridomestic risk of Lyme disease over large geographic areas.

More recently, Moncayo et al. (2000) combined Landsat TM images with aerial videography to generate a map of landscape elements around 15 human and horse cases of eastern equine encephalomyelitis (EEE) in southeastern Massachusetts. EEE exists in enzootic foci transmitted between birds and the mosquito, *Culiseta melanura* (the enzootic vector). In Massachusetts and surrounding states, epidemic/epizootic transmission may involve as many as six additional mosquito species (*Aedes canadensis*, *Aedes vexans*, *Culex salinarius*, *Coquillettidia perturbans*, *Anopheles quadrimaculatus*, and *Anopheles punctipennis*), each with its characteristic larval and adult habitat requirements. Stepwise regression analysis showed that wetlands and more specifically, deciduous wetlands, were the most important major class element. Deciduous wetlands accounted for up to 72.5% of the observed variation in the host-seeking populations of *A. canadensis*, *A. vexans*, and *C. melanura*. The authors propose combining habitat mapping with street maps to identify and prioritize areas in need of vector mosquito control.

Even coarse-grain satellite imagery can be used to advantage. Daniel and Kolář (1990) analyzed a 25-year database on the distribution and abundance of *I. ricinus*, the vector of tick-borne encephalitis in Europe, in relation to land cover types derived from a 41 × 41 km section of a Landsat multispectral scanner (MSS) scene. They showed that *I. ricinus* is associated with specific land cover types, which allowed them to generate risk maps that could be used in public education and other prevention programs.

VII. TECHNICAL ISSUES AND LIMITATIONS

A. Analytical and Statistical Issues

Although a picture may be worth a thousand words, it is still necessary to know whether the patterns we think we see are, in fact, statistically significant rather than the result of random noise. Better integration of GIS and spatial statistics software would be advantageous, but it is probably not desirable or economically feasible for each GIS package to integrate a spatial statistics module. Rather, it would be helpful for developers from the two fields to agree upon common import/export formats to permit the rapid movement of data between the two types of software. Bailey (1994) provides a good review of progress in the integration of spatial statistics and GIS. As the focus of GIS and remote sensing in vector-borne disease moves from description to prediction, there will be a greater need for analytical techniques that simultaneously deal with space and time. Cressie (1996) identified several issues and approaches to dealing with spatiotemporal processes.

B. Modeling and Simulation

Perhaps the greatest potential for applying remotely sensed information to the surveillance, prediction, and control of vector-borne and zoonotic diseases is the ability to use these data in predictive models. Models come in a great variety, and it is crucial to use the right type of model to answer a specific question. Thus, models designed to elucidate the mechanism of a particular ecological process may not be good predictive models and vice versa. An area where considerable progress is being made is in landscape ecology, where issues of sustainable harvestable resources and preservation of endangered species habitats have become crucial. A good survey of theory and applications in this area can be found in Turner and Gardner (1991).

C. Temporal and Spatial Resolution

The spatial scale of satellite imagery should match the scale of the object of study at ground level. Early remote sensing studies (e.g., Hayes et al., 1985) were limited to Landsat MSS images, with a pixel resolution of about 80 m. This placed severe limitations on the size of habitats that could be detected. Today, with the advent of commercial imagery with 1-m resolution, the lack of sufficient high-resolution, ground-level data has become a limiting factor in the application of remote sensing to solving issues in vector-borne disease ecology.

D. The Issue of Scale

Ecological phenomena exhibit patterns at different scales (e.g., Turner et al., 1991; Quattrochi & Goodchild, 1997). Vector-borne diseases are no exception to this phenomenon (Korenberg, 1989). Thus, a virus-vector-vertebrate host system, such as LaCrosse encephalitis, can exhibit distributional pattern at the continental scale, the regional scale, and at the local scale (Figure 13). For a given question or hypothesis about any disease, there is probably an optimal scale at which to measure the system components. Thus, it is crucial to clearly understand the scale(s) at which different components of the disease system operate.

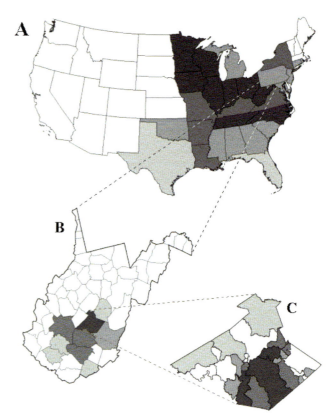

FIGURE 13 Effect of scale on pattern and distribution of LaCrosse encephalitis in the United States; patterns are evident at several spatial levels. Darker colors indicate more viral activity.

E. Imagery Issues

The resolution of remotely sensed imagery may not be satisfactory for studies at a particular scale. Resolution here can refer to space, time, or spectral composition. In particular, studies of some diseases, such as dengue and LaCrosse encephalitis, would benefit from the ability to identify water-holding containers on individual premises. This can be accomplished by using high-resolution aerial photography, but temporal coverage will be sparse due to the high cost of flying such missions. Also, the spectral coverage will be very narrow, which reduces the utility of the imagery.

Cloud cover and other disturbances are a perennial problem. One anticipated use of remote sensing is in evaluating the threat of disease outbreaks following floods, hurricanes, or other natural disasters. Because many of these disasters are weather related, it is not uncommon for the areas of interest to be completely covered by clouds for weeks following the event. Perhaps additional studies on the use of real-time radar in conjunction with historical imagery in other spectral regions or similar systems could improve prospects for using remote sensing in this setting (e.g., Imhoff & McCandless, 1988).

Landscape feature classification is an important tool for understanding and modeling vector-borne and zoonotic diseases. One or more specific land cover types (e.g., scrub, piñon-juniper, oak-hickory forest, etc.) often define these diseases. Thus, it is important to be able to distinguish between cover types. All of the problems related to slope, aspect, and shading as they impact the remotely sensed image also need to be dealt with. Soil type, elevation, slope, and aspect often have a major impact on the distribution of vertebrate hosts of these disease agents, especially small mammals. It should be possible, at least in theory, to infer some of these qualities from the remotely sensed spectral signals, particularly from hyperspectral imagery.

F. Ground Data Issues

One of the first things that strikes the entomologist or epidemiologist when looking at remotely sensed imagery is the enormous amount of information in comparison to the available data on the diseases we would like to study. For many diseases, such as cancer and birth defects, there are well-developed registries that give detailed information about cases. Such information allows the epidemiologist to search for associations between environmental indicators (from the remotely sensed data) and the location of cases. Unfortunately, reporting systems for vector-borne diseases are generally poorly developed or entirely absent. For example, dengue which, as reported above, is responsible for as many as 50–100 million cases each year, is not even reported in many countries. Only the more severe cases are entered into the reporting system. Even in the United States, the vector-borne diseases are generally grossly underreported. To successfully apply remote sensing to the surveillance, prevention, and control of vector-borne and zoonotic diseases, we need well-designed surveillance systems to provide the "ground truth" data to validate the models that are being developed.

G. Privacy Issues

A separate but extremely important issue relating to the collection of disease surveillance data is the issue of privacy. There is concern on the part of some citizens that their personal privacy would be violated by having the precise location of their residence, school, workplace, etc., entered into a database, particularly if that information in also associated with health-related information. Thus, many reporting systems show only the city or county of residence. This severely limits the utility of the data for remote sensing studies.

One way to bypass this problem, at least with the vector-borne and zoonotic diseases, is to monitor disease activity in the vector or the wild vertebrate host (in some situations, susceptible domestic animals can also be monitored). Detailed protocols have been developed for collecting the appropriate data from these parts of the disease system (e.g., Moore et al., 1993; Moore & Gage, 1996). Data from these field-oriented surveillance systems can be used to predict disease activity in the human (or domestic animal) population. Because these surveillance systems are generally removed in space (often rural) and time (activity pre-

cedes disease in humans) from the human population at risk, there is often a shortage of funding and staffing for these programs. Thus, the surveillance agency wants to collect as much useful information as possible in the shortest number of hours with the smallest possible work force. To do this, programs are increasingly moving to automated field data recording systems, GPS, and similar equipment to reduce paperwork, data entry error, and similar time-consuming problems. The Centers for Disease Control is currently funding the development of an integrated database, GIS, and decision support system that will speed the collection of vector-borne disease related data. Once completed, the system, called the National Electronic Arbovirus Reporting System (NEARS), will be accessible via the Worldwide Web. Data at differing spatial resolutions can be downloaded for use with remotely sensed data.

VIII. PROSPECTS FOR THE FUTURE

The utility of remotely sensed data for improving our understanding, prevention, and control of vector-borne and zoonotic diseases has been abundantly demonstrated. Even greater opportunities exist for applying the imagery from current and future remote-sensing platforms. Much of the eventual progress in applying that imagery to the solution of human problems, including the diseases of interest to us, will depend on the priorities set by universities, governments, and funding agencies. There is a need for cross-disciplinary programs in universities to provide qualified and broadly trained researchers for this growing field. Collaborative research, within and between agencies and departments, should be encouraged. It is no longer possible for a single researcher or a small group within a single discipline to "do it all." By sharing valuable resources and insights (expertise), progress can be more rapid and costs can be reduced.

SEE ALSO THE FOLLOWING CHAPTERS

Chapter 11 (Arsenic in Groundwater and the Environment) · Chapter 19 (The Ecology of Soil-Borne Human Pathogens) · Chapter 26 (GIS in Human Health Studies)

FURTHER READING

Bailey, T. C. (1994). A Review of Statistical Spatial Analysis in Geographical Information Systems. In *Spatial Analysis and GIS* (S. Fotheringham and P. Rogerson, Eds.), London, Taylor & Francis, pp. 13–44.

Beck, L. R., Lobitz, B. M., and Wood, B. L. (2000). Remote Sensing and Human Health: New Sensors and New Opportunities, *Emerg. Infect. Dis.*, 6(3), 217–227.

Beck, L. R., Rodriguez, M. H., Dister, S. W., Rodriguez, A. D., Rejmankova, E., Ulloa, A., Meza, R. A., Roberts, D. R., Paris, J. F., Spanner, M. A., Washino, R. K., Hacker, C., and Legters, L. J. (1994). Remote Sensing as a Landscape Epidemiologic Tool to Identify Villages at High Risk for Malaria Transmission, *Am. J. Trop. Med. Hyg.*, 51, 271–280.

BGS and DPHE. (2001). *Arsenic contamination of groundwater in Bangladesh*. Kinniburgh, D. G. and Smedley, P. L. (eds). British Geological Survey Technical Report WC/00/19, British Geological Survey, Keyworth.

Clarke, K. C., McLafferty, S. L., and Tempalski, B. J. (1996). On Epidemiology and Geographic Information Systems: A Review and Discussion of Future Directions, *Emerg. Inf. Dis.*, 2, 85–92.

Cline, B. L. (1970). New Eyes for Epidemiologists: Aerial Photography and Other Remote Sensing Techniques, *Am. J. Epidemiol.*, 92, 85–89.

Colwell, R. R. (1996). Global Climate and Infectious Disease: The Cholera Paradigm, *Science*, 274, 2025–2031.

Cressie, N. (1996). Statistical Modeling of Environmental Data in Space and Time, In *Spatial Accuracy Assessment in Natural Resources and Environmental Sciences: Second International Symposium* (H. T. Mowrer, R. L. Czaplewski, and R. H. Hamre, Eds.), U. S. Department of Agriculture, Forest Service, GTR RM-GTR-277, Fort Collins, CO, pp. 1–3.

Daniel, M., and Kolář, J. (1990). Using Satellite Data to Forecast the Occurrence of the Common Tick *Ixodes ricinus* (L.), *J. Hyg. Epidemiol. Microbiol. Immunol.*, 34, 243–252.

Dister, S. W., Fish, D., Bros, S. M., Frank, D. H., and Wood, B. L. (1997). Landscape Characterization of Peridomestic Risk for Lyme Disease Using Satellite Imagery. *Am. J. Trop. Med. Hyg.*, 57(6), 687–692.

Estes, J. E., and Loveland, T. R. (1999). Characteristics, Sources, and Management of Remotely-Sensed Data. In *Geographical Information Systems, Second Edition*, (P. A. Longely, M. F. Goodchild, D. J. Maguire, and D. W. Rhind, Eds.), John Wiley & Sons, New York, pp. 667–675.

Forman, R. T. T., and Godron, M. (1986). *Landscape Ecology*, John Wiley & Sons, New York.

Glass, G. E., Morgan, J. M., III, Johnson, D. T., Noy, P. M., Israel, E., and Schwartz, B. S. (1992). Infectious Disease Epidemiology and GIS: A Case Study of Lyme Disease, *GeoInfoSystems*, November/December.

Gratz, N. G. (1999). Emerging and Resurging Vector-Borne Diseases, *Annu. Rev. Entomol.*, 44, 51–75.

Hayes, R. O., Maxwell, E. L., Mitchell, C. J., and Woodzick, T. L. (1985). Detection, Identification and Classification of Mosquito Larval Habitats Using Remote Sensing Scanners in Earth-Orbiting Satellites, *Bull. WHO*, 63, 361–374.

Imhoff, M. L., and McCandless, S. W. (1988). Flood Boundary Delineation Through Clouds and Vegetation Using L-Band Space-Borne Radar: A Potential New Tool for Disease Vector Control Programs, *Acta Astronaut.*, 17, 1003–1007.

Jetten, T. H., and Focks, D. A. (1997). Potential Changes in the Distribution of Dengue Transmission Under Climate Warming, *Am. J. Trop. Med. Hyg.*, 57, 285–297.

King, M. (2000). EOS Measurements, National Aeronautics and Space Administration, Goddard Space Flight Center, http://eospso.gsfc.nasa.gov/ftp_docs/measurements.pdf.

Kitron, U. (1998). Landscape ecology and epidemiology of vector-borne diseases: tools for spatial analysis. *J. Med. Entomol.*, 35(4), 435–445.

Korenberg, E. I. (1989). Population Principles in Research into Natural Focality of Zoonoses, *Sov. Sci. Rev. F. Physiol. Gen. Biol.*, 3, 301–351.

Lindsay, S. W., and Thomas, C. J. (2000). Mapping and estimating the population at risk from lymphatic filariasis in Africa. *Trans. R. Soc. Trop. Med. Hyg.*, 94(1), 37–45.

Linthicum, K. J., Anyamba, A., Tucker, C. J., Kelley, P. W., Myers, M. F., and Peters, C. J. (1999). Climate and Satellite Indicators to Forecast Rift Valley Fever Epidemics in Kenya, *Science*, 285, 397–400.

Malone, J. B., Abdel-Rahman, M. S., El Bahy, M. M., Huh, O. K., Shafik, M., and Bavia, M. (1997). Geographic Information Systems and the Distribution of *Schistosoma mansoni* in the Nile Delta, *Parasitol. Today*, 13, 112–119.

Martens, W. J. M., Niessen, L. W., Rotmans, J., Jetten, T. H., and McMichael, A. J. (1995). Potential Risk of Global Climate Change on Malaria Risk, *Environ. Health Perspect.*, 103, 458–464.

Moncayo, A. C., Edman, J. D., and Finn, J. T. (2000). Application of Geographic Information Technology in Determining Risk of Eastern Equine Encephalomyelitis Virus Transmission, *J. Am. Mosq. Control Assoc.*, 16(1), 28–35.

Moore, C. G., and Gage, K. L. (1996). Collecting Methods for Vector Surveillance. In: *The Biology of Disease Vectors* (B. J. Beaty and W. C. Marquardt, Eds.), University Press of Colorado, Niwot, CO, pp. 471–491.

Moore, C. G., McLean, R. G., Mitchell, C. J., Nasci, R. S., Tsai, T. F., Calisher, C. H., Marfin, A. A., Moore, P. S., and

Gubler, D. J. (1993). Guidelines for Arbovirus Surveillance in the United States, U. S. Dept. of Health and Human Services, Centers for Disease Control, Fort Collins, CO.

Ostfeld, R. S., Jones, C. G., and Wolff, J. O. (1996). Of Mice And Mast: Ecological Connections in Eastern Deciduous Forests, *BioScience*, 46, 323–330.

Pope, K. O., Rejmankova, E., Savage, H. M., Arredondo-Jimenez, J. I., Rodriguez, M. J., and Roberts, D. R. (1994). Remote Sensing of Tropical Wetlands for Malaria Control in Chiapas, Mexico, *Ecol. Appl.*, 4, 81–90.

Quattrochi, D. A., and Goodchild, M. F. (Eds.) (1997). *Scale in Remote Sensing and GIS*, Lewis Publishers, Boca Raton, FL.

Reeves, W. C., Hardy, J. L., Reisen, W. K., and Milby, M. M. (1994). Potential Effect of Global Warming on Mosquito-Borne Arboviruses, *J. Med. Entomol.*, 31, 324–332.

Robinson, T., Rogers, D., and Williams, B. (1997). Mapping Tsetse Habitat Suitability in the Common Fly Belt of Southern Africa Using Multivariate Analysis of Climate and Remotely Sensed Vegetation Data, *Med. Vet. Entomol.*, 11, 235–245.

Shope, R. E. (1992). Impacts of Global Climate Change on Human Health: Spread of Infectious Diseases. In *Global Climate Change: Implications, Challenges and Mitigation Measures* (S. K. Majumdar, L. S. Kalkstein, B. Yarnal, E. W. Miller, and L. M. Rosenfeld, Eds.), Pennsylvania Academy of Science, Easton, PA, pp. 361–370.

Short, N. M. (Ed.) (2003). The Remote Sensing Tutorial, National Aeronautics and Space Administration, Goddard Space Flight Center, Greenbelt, MD.

Srivastava, A., Nagpal, B. N., Saxena, R., and Sharma, V. P. (1999). Geographic Information System as a Tool to Study Malaria Receptivity in Nadiad Taluka, Kheda District, Gujarat, India, *Southeast Asian J. Trop. Med. Public Health*, 30(4), 650–656.

Star, J. L., Estes, J. E., and Davis, F. (1991). Improved Integration of Remote Sensing and Geographic Information Systems: A Background to NCGIA Initiative 12, *Photogrammetr. Eng. Remote Sensing*, 57, 643–645.

Turner, M. G., and Gardner, R. H. (Eds.). (1991). *Quantitative Methods in Landscape Ecology*, Springer-Verlag, New York.

Turner, S. J., O'Neill, R. V., Conley, W., Conley, M. R., and Humphries, H. C. (1991). Pattern and Scale: Statistics for Landscape Ecology. In *Quantitative Methods in Landscape Ecology* (M. G. Turner and R. H. Gardner, Eds.), Springer-Verlag, New York, pp. 17–49.

MINERALOGY OF BONE

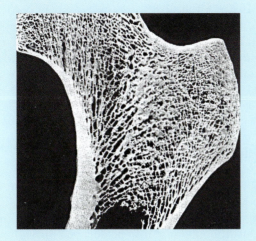

H. CATHERINE W. SKINNER
Yale University

CONTENTS

I. INTRODUCTION

Medical geology encompasses many scientific endeavors with global activities and impact, but it also includes aspects that are very personal and individual. Local environment is sampled through what is ingested and inhaled whether or not it is salubrious, marginal, or downright unhealthy. Human bodies react to the remarkable range of natural and man-made chemicals that they are exposed to every day. It is common knowledge that a range of nutrients is required, but some of the chemicals can be hazardous to our health. This chapter focuses on bones, and specifically the mineral portion in these tissues, the component essential to the functions of these organs. These discussions illustrate several attributes of the emerging field of medical geology. The scientific information outlined herein is drawn from a diversity of disciplines and expertise, from biophysical and biochemical sciences, from physicians and dentists, and from geologists, mineralogists, and engineers. This knowledge enables us to address the many individual and collective roles that minerals play in the body. A disorder that affects people on every continent, osteoporosis, is presented as an example of how information on minerals can be applied. It is only one of the possible targets of opportunity where mineralogical/geological expertise has, and continues to have, the potential to ameliorate suffering and promote better global and personal health. Selected classic and recent references are included to whet the appetite of those who will make contributions to our knowledge in the future.

II. THE SKELETON AND MINERALIZED TISSUES

Humans can be distinguished from all other mammalian species by their skeletons which are composed of over 200 bones and 32 teeth. Each of these skeletal components is a separate organ composed of mineralized tissues. Created by the actions of distinct cell systems, the tissues are true composites, an intimate association of extracellular macro- and microbioorganic molecules, and inorganic mineral materials. The tissues are in dynamic equilibrium with a highly controlled fluid whose chemical composition resembles that of the ocean: highly oxidized, with a pH around 7 with sodium, Na^+, and chlorine, Cl^-, as the dominant ionic species.

The mineral of bones and teeth is a calcium phosphate that closely resembles the naturally occurring mineral, hydroxylapatite (Gaines et al., 1997), and because of its intimate association with biological activity and molecules it is known as bioapatite. The mineral formed and maintained in the soft tissue or matrix of bones has peculiar attributes. Each mineralized tissue is a remarkable and unique chemical repository whose maintenance is only beginning to be comprehended as part of the dynamic skeletal system.

A. Normal Mineralized Tissues: Bones and Teeth

The relative amounts of mineral to bioorganic components and the distinct spatial aggregations or structures of the mineral-matrix combination have been objects of investigation for over 250 years. Four different types of normal human mineralized tissues have been described (Glimcher, 1976; Mann, 2000). With increased sensitivity and availability of analytical techniques, the four types have been shown to have discrete cell systems, chemical components, and spatial expressions that change during growth, with age, or with disease. Studies, especially those from different life stages, have allowed us to identify essential participants and gain some understanding of the expected or "normal" state and the reactions required to maintain function appropriate for the skeleton.

Three of the four tissues are found in teeth: enamel, dentine, and cementum (Figure 1) (Miles, 1967), and the fourth is in bone. Subsets of all four types have been described and amplified with each new and more sophisticated analytical technique. Optical examination and specialized methodology on thin sections of mineralized tissues at high resolution has defined the anatomy of their components, the cells, extracellular products, or the typical textures (Figure 1A). Tissues typical of bones (Figure 1B) have woven, lamellar, and haversian textures with mineral distribution and content that varies with

tissue age, with nutrition, or with disease (Albright & Skinner, 1987). All three textures may occur in cortical bone, the heavily mineralized portions of bone, or in trabecular bone, the porous and spongy segments of the organ, and are expressions of their dynamic nature. But all bioapatite deposits in humans are not normal and expected.

B. Pathological Apatitic Deposition

In addition to the normal bones and teeth, bioapatite may deposit pathologically, that is, in the tissues of other organs that would normally not mineralize. Nodular apatitic spherules may be detected throughout the body. One site where bioapatites may occur is in tumors, both benign and cancerous, where rapid cell production may cause accumulations of dead cells. When a cell dies it releases any phosphate bound to bioorganic molecules into the surrounding fluid. The elevated calcium concentration in the circulating serum is much higher than the calcium concentration within the cell and bioapatite may nucleate. Another likely site for bioapatite mineralization is at scar tissues, the sequelae to tissue trauma, where excessive amounts of the fibrous protein collagen accumulate as a result of the normal cellular repair systems that occur throughout the body. Collagen is the most common protein in the body; a normal biomolecular component of all connective tissues and the dominant protein (approximately 90 wt%) in the organic fraction of bone tissues. Cells that produce collagen are known as fibroblasts; in bone they are called osteoblasts. Many biochemical varieties of the collagen molecule with slightly different amino acid compositions and intramolecular cross links have been described (Miller, 1973; Skinner, 1987), but type 1 collagen is typical of bone, dentine, and cementum. Collagen is not found in enamel.

Pathological bioapatite may be found in the arteries, an expression of cardiovascular disease, and in kidney "stones" (Skinner, 2000a). It may occur in association with other calcium phosphate species (Table I). For example, calcium pyrophosphate, $Ca_2P_2O_7 \cdot H_2O$, occurs in joints (Skinner, 2000a), and octacalcium phosphate, $Ca_8H_2(PO_4)_6 \cdot 5\ H_2O$, and/or whitlockite, $CaMg(PO_3OH)(PO_4)_6$, in dental calculus, which is a mineral deposit that forms just below the gum line in the soft tissue around teeth (see Driessens & Verbeeck, 1990, for a complete discussion of dental plaque mineral constituents).

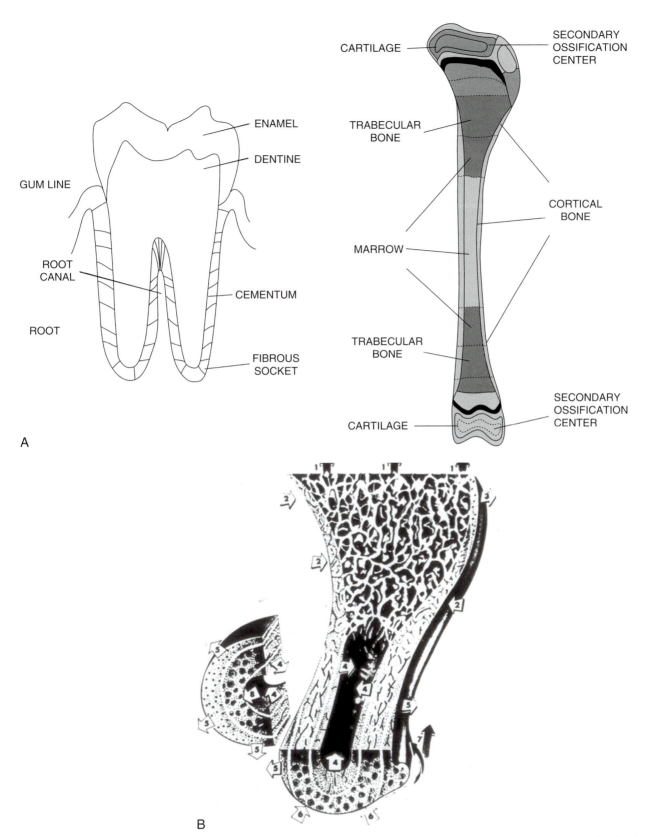

FIGURE 1 (A) Sketch of a tooth and a long bone indicating the typical tissues found in these organs. (From Skinner, 2000a, Figure 1, p. 356.) (B) Sketch of a longitudinal section and a cross section through a long bone illustrating the tissue types and textures that occur and the changes that take place during the development of a bone. Arrows indicate the directions of growth and remodeling. (From Albright and Skinner, 1987, Figure 5–16, part A, p. 175.) Numbers at specific sites illustrate the following changes: 1, length increase, typical growth direction of a long bone; 2, thickness or diameter increase that also takes place as the organ changes shape and size. Some initial trabecular tissues become cortical bone tissues. The textures depicted in 2 and 5 are typical of heavily mineralized lamellar bone that results from the remodeling of trabecular tissues into cortical tissues. 3, all bones require local remodeling to achieve final organ size and shape in order to function as part of the skeleton; 4, remodeling is also needed to maintain the internal (marrow) cavity size and shape; 5, extensive remodeling in the mid-shaft of long bones where previous trabeculae have been remodeled into heavily mineralized cortical tissue; 6, the circular patterns of haversian bone express the sites of resorption and re-deposition of mineralized tissues which continue throughout life; 7, an outer protuberance on the bone, probably the site of muscle attachment, which will change as the bone responds to growth and development; and 8, a cross section showing distinct layers of haversian and lamellar bone tissues surrounding the marrow cavity.

TABLE I. Formulae of Hydroxylapatite, Bioapatite, and Other Calcium Phosphate Phases and the Ca/P Ratios for Their Ideal Compositions

Name	Formula	Ca/P (wt .7%)
Hydroxylapatite	$Ca_5(PO_4)_3(OH)$	2.15
Bioapatite[a]	$(Ca, Na, [\ ])_{10}(PO_4, HPO_4, CO_3)_6(OH, F, Cl, H_2O, CO_3O, [\ ])_2$	1.33–2.25
Monetite	$CaHPO_4$	1.25
Brushite	$CaHPO_4 \cdot 2\,H_2O$	1.25
Octacalcium phosphate	$Ca_8H_2(PO_4)_6 \cdot 5\,H_2O$	1.33
Tricalcium phosphate	$Ca_3(PO_4)_2$	1.875
Whitlockite	$Ca_9\,Mg(PO_3OH)\,(PO_4)_6$	1.11–1.456
Calcium pyrophosphate dihydrate	$Ca_2P_2O_7 \cdot 2\,H_2O$	1.25
Tetracalcium phosphate	$Ca_4P_2O_9$	2.50
Dahllite	$(Ca, X)_{10}(PO_4, CO_3)_6(O, OH)_{26}{}^{b}$ $CO_2 > 1\%\ F < 1\%$	2.08
Francolite	$(Ca, X)_{10}(PO_4, CO_3)_6(O, F)_{26}{}^{b}$ $F > 1\%\ CO_2 > 1\%$	2.08

[a]An approximate formula containing vacancies [ ].
[b]Formula and composition from McConnell (1973), where X indicates a range of cations.

To cover all calcium phosphate mineral species that may be found in the human body is beyond the scope of this presentation. An introduction to the usual and predominant mineralizing system (Section III), the submicron characteristics of the mineral (Section IV), and the methodology and techniques used to determine the composition, concentration, and distribution of the mineral in mineralized tissues are presented (Section V). The final section (VI) on the disease osteoporosis illustrates how knowledge of the mineral, distinguishing normal from the abnormal, and mineral dynamics have become foci for medical research.

Although the techniques for studying bones and mineral are similar to those employed for all solid materials, mineralized tissues present special problems. Foremost are the difficulties of obtaining sufficient sample when the object of study is a living human. Initial diagnosis of osteoporosis usually employs noninvasive techniques such as transmission X-ray analyses (radiology) of a bone or bones. Once a clinical diagnosis is made it may be followed, especially today, by examination of mineralized tissue samples in spite of the difficulties of procurement and preparation (Sections V.A, B, C, D). The tiny amounts of tissues can provide the information that is essential in defining and adequately treating the disease in many patients. Whenever mineralized tissue samples are studied, either normal or pathological, the physical and chemical characteristics of the tissues and the mineral fraction benefit the health of future patients and populations.

III. Crystal Chemistry of the Mineral in Mineralized Tissues

The normal and much of the abnormal or pathological mineral deposits in humans is a calcium phosphate, a member of the apatite group of minerals (Gaines et al., 1997). The bio-deposits conform crystal chemically most closely to the mineral species hydroxylapatite in the ideal formula $Ca_5(PO_4)_3(OH)$, (hereafter abbreviated as HA), which is described in most introductory mineralogical texts (Klein & Hurlbut, 1985). However, the precise composition and crystal structure of the bioapatite mineral has proved difficult to pinpoint so the term "apatitic" is often used. These difficulties relate to the chemical variability reported from many analyses of the mineral from different mineralized tissues. Further, X-ray diffraction analyses, the criteria for accurate identification of any mineral species, applied to biomineral samples is only partially successful primarily because of the very small grain size of the mineral materials. What has been shown from extensive investigations is that the biomineral, although apatitic, is not the ideal or stoichiometric chemical compound whose formula is presented above. The following formula is a more appropriate presentation:

$$(Ca, Na, Mg[\ ])_{10}(PO_4, HPO_4, CO_3)_6(OH, F, Cl, CO_3, O, [\ ])_2$$

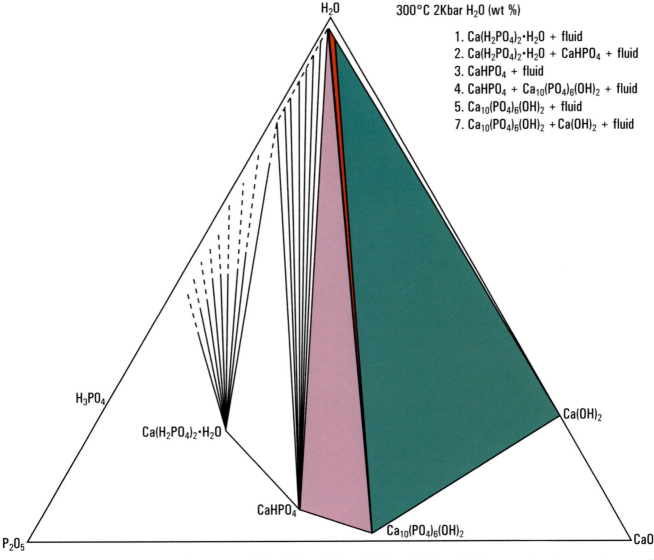

300°C 2Kbar H₂O (wt %)

1. $Ca(H_2PO_4)_2 \cdot H_2O$ + fluid
2. $Ca(H_2PO_4)_2 \cdot H_2O$ + $CaHPO_4$ + fluid
3. $CaHPO_4$ + fluid
4. $CaHPO_4$ + $Ca_{10}(PO_4)_6(OH)_2$ + fluid
5. $Ca_{10}(PO_4)_6(OH)_2$ + fluid
7. $Ca_{10}(PO_4)_6(OH)_2$ + $Ca(OH)_2$ + fluid

FIGURE 2 The phase diagram of the system $CaO—P_2O_5—H_2O$ at 300°C and 2 Kbars (H_2O pressure) to illustrate the several calcium phosphate phases that may occur with hydroxylapatite under equilibrium conditions. (From Skinner, 1973, Figure 1A.)

The brackets indicate vacancies in some lattice sites of the solid to achieve a charge-balanced solid phase.

This complicated chemical solid can be described as follows: bioapatite is predominantly a calcium phosphate mineral most closely resembling the species hydroxylapatite but usually contains many elements and molecular species other than calcium and phosphate that probably contribute to its physical attributes and reactivity, and should be part of any identification. The crystal structure of all apatite minerals is dominated by tetrahedral anions. In the case of hydroxylapatite and bioapatite it is phosphate, the phosphorous-oxygen tetrahedral anionic groups $(PO_4)^{3-}$, that forms the back-

bone of the structure and bonds predominantly to calcium, but other cation and anion species, vacancies, and complex molecular groups are usually detected on analysis.

A. The Mineralizing System: CaO—P₂O₅—H₂O

The chemical system that defines bioapatite can be simplified to $CaO—P_2O_5—H_2O$. Figure 2 is an experimentally determined phase diagram. It is a portion of the chemical system that describes the mineralization of bones and teeth, which is a summary of the solid and liquid phases that exist at equilibrium at a specific temperature and pressure in the simplified system (Skinner, 1973a). The following solids—monetite, $CaHPO_4$, brushite, $CaHPO_4 \cdot 2 H_2O$, and $Ca_2P_2O_7$, all with Ca/P weight ratios that equal 1.25 and mole ratios that equal 1.0—may occur in stable association with HA when the fluid phase is pure water. This phase diagram is the result of experiments at elevated temperatures and pressures to accelerate reaction times and provide crystalline products for accurate analyses. It shows the stability fields of the solid phases and fluid compositions for a range of bulk compositions. Brushite and octacalcium phosphate identified in some biomineral analyses are not found in this diagram as they occur under different temperature and pressure conditions. It should be noted that HA is the stable phosphate mineral phase throughout most of this diagram, but it coexists with other minerals from pH 4 to 12. Interestingly, only at pH 7 is HA the only solid phase associated with a variable composition fluid. At lower pH, the solid phases that occur with HA have higher phosphate content. At a pH greater than 7, the second mineral phase in this simplified experimental system is a mineral known as portlandite, $Ca(OH)_2$. Although the mineral analyzed after low temperature extraction from tissues (discussed in Section VI) may show a Ca/P higher than that of hydroxylapatite, portlandite has not been identified (Skinner, 1973b).

B. Composition of the Mineral in Mineralized Tissues

The bulk chemical composition of the tissues and of the included bioapatites found in them, the dominant human mineralized tissues, are listed in Table II. Table IIA lists the bulk composition of the major components in these tissues, and Table IIB presents the range of major elements and the different Ca/P ratios of the bioapatites from three different sources. The measured Ca/P wt% range, from 1.3 to 2.2, differs from the ideal or stoichiometric value for hydroxylapatite of 2.15. Because the values are both above and below the stoichiometric value, several hypotheses for such variations have been proposed.

1. Variations Due to Multiple Minerals

One group of investigators suggested that the biomineral matter was not a single mineral phase. Table I shows that there are a number of calcium phosphate mineral species with different, and in many cases lower,

TABLE II. The Composition of Normal Mineralized Tissues

A. Bulk Composition of Bone, Dentine, and Enamel

	Bone[a]		Dentine[a]		Enamel[a]	
	wt%	vol%	wt%	vol%	wt%	vol%
Inorganic	70	49	70	50	96	90
Water	6	13	10	20	3	8
Organic	24	38	20	30	1	2
Denisty (avg.) (g/cm^3)	2.35		2.52		2.92	

B. Composition of Major Elements and the Ca/P Ratio of the Bioapatites in Three Tissues (Wt% on a dry, fat-free basis)[b]

ASH	57.1	70.0	95.7
Ca	22.5	25.9	35.9
P	10.3	12.6	17.0
Ca/P	2.18	2.06	2.11
Mg	0.26	0.62	0.42
Na	0.52	0.25	0.55
K	0.089	0.09	0.17
CO_2	3.5	3.19	2.35
Cl	0.11	0.0	0.27
F	0.054	0.02	0.01

[a]From Driessens and Verbeeck, 1990, Table 8.2, p. 107; Table 9.4, p. 165; and Table 10.5, p. 183.
[b]From Zipkin, 1970, Table 17, p. 72.

Ca/P ratios than HA. Most of these minerals may nucleate and are stable in the body fluid-tissue environment. The presence, and a variable amount, of a second mineral with lower Ca/P that might occur with HA in bio-deposits has been cited as a possible reason for the compositional variability measured in tissues. The higher Ca/P ratio may reflect the substitution of another anion, e.g., CO_3^{2-}, for a portion of the PO_4^{3-} (discussed in Section IV.D).

2. Variations Due to Nucleation and Maturation

Another possible reason why the analyses of bioapatite may not show uniform and stoichiometric composition is that the mineral precipitate changes over time. It is well known that phosphate ions ($H_2PO_4^{2-}$, HPO_4^-) are usually detected in the fluid phase at initial mineral nucleation. This led to the suggestion that nucleation was a discrete process from growth and maturation of the mineral phase and the composition of the fluid at least locally varied over time as mineralization proceeded (Roberts et al., 1992). Glimcher (1976) suggested that nucleation could be induced by charges from phosphate groups associated with matrix collagen molecules. Alternatively, because the initial solid was so poorly crystalline, Posner (1977) suggested a separate phase called "amorphous" calcium phosphate mineral, octacalcium phosphate was suggested by Brown et al. (1987), and a third suggestion of brushite was made by Neuman and Neuman (1958). All three mineral materials were considered intermediates in the mineralization process before a mature biomineral phase that more closely resembled hydroxylapatite was produced. Even during the investigations of the basic mineralizing system, the synthetic calcium-phosphate-water investigations that delineated the formation of HA showed this stable and ubiquitous phase which was most often associated with other solid phases and had incongruent solubility (Van Wazer, 1958; Skinner, 1973a).

A variety of chemical techniques for determining the Ca/P ratio in the synthetic chemical system and early mineral deposition confirmed that initial precipitates produced by adding calcium to phosphate-rich solutions, or vice versa, was a calcium phosphate mineral with higher phosphorous (P) content than that of stoichiometric hydroxylapatite (and therefore a lower Ca/P). Although the presence of amorphous and brushite as initial phases has not been confirmed, this question remains under study (Brown, et al., 1987; Roberts et al., 1992; Kim et al., 1995; Aoba et al., 1998).

3. Variations Due to Substitutions Within the Mineral

Putting aside the nucleation and maturation processes and identification of the initial precipitates, biomineral has been mostly thought of as a single calcium phosphate phase whose variations in Ca/P may be accommodated via substitutions and/or vacancies within the solid. The opportunity for elements and molecular species other than calcium (Ca), phosphorus (P), oxygen (O), and hydroxyl (OH) to occur in normal mineralized tissues is a most important consideration from the perspective of medical geology.

A multiplicity of chemicals exist in the natural environment, and it is probable that at least some of them will become part of the dynamic mineral systems found in bones and teeth. Detailed investigations of the apatite crystal structure, presented in Section IV, enable us to discuss the ability of this species to incorporate a variety of chemical species. Each of these incorporations, known as solid solution, could alter the Ca/P ratio of the solid. For example, in geological environments, sodium (Na), lead (Pb), or strontium (Sr) may substitute for some of the calcium which means an accurate analysis for Ca/P that would be slightly lower than stoichiometric HA. Alternatively, if an anionic group, such as sulfate (SO_4), became incorporated in place of some of the phosphate (PO_4), the Ca/P would be above the stoichiometric HA value. Calculation of a Ca/P ratio, or a cation/anion ratio, from a chemical analysis will depend not only on which chemical elements and species were available when the mineral formed but also on the completeness of the data used in the calculation. If only calcium and phosphate are measured, for example, no matter how accurately, calculation of a Ca/P for bioapatite may be misleading. The following section on crystal structural details of mineral apatites will allow investigators to more fully appreciate the possible uptake and incorporation of specific elements, or molecular species, into bioapatites.

IV. THE CRYSTAL STRUCTURE OF CALCIUM APATITES

The structure of apatites, a common group of naturally occurring minerals in many rock types throughout the world, is distinguished by hexagonal symmetry (Gaines et al., 1997). Figure 3A depicts one projection of the three-dimensional arrangement of the atoms that make up the apatite structure. This view is down the unique c-axis and depicts a plane perpendicular to c. The rhombohedral outlines are the disposition of two of the three a axes (directions at 120 degrees apart) and the atoms are precisely placed conforming to the crystallographic characteristics of the repeating unit, or unit cell, of the apatite structure.

To further illustrate the distribution and importance of the tetrahedral orthophosphate groups (PO_4^{3-}) that form the backbone of the apatite structure, Figure 3B presents another view of the structure. It is a projection 90° to the c-axis along one of the a axes. The yellow phosphorus atom is in tetrahedral coordination surrounded by two white oxygen atoms in the same plane as the phosphorus and two light purple oxygen atoms, one above the plane, the other below it. Phosphate

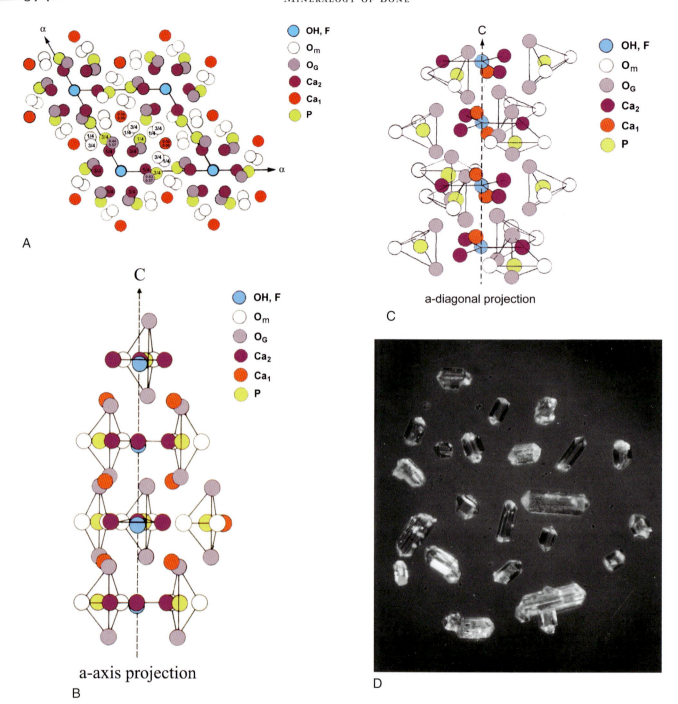

FIGURE 3 The crystal structure of hydroxylapatite, ideal formula $Ca_5(PO_4)_3(OH)$. (A) Projection down the c-axis, showing the distribution of all the atoms in the unit cell. (B) Projection down the a axis of the unit cell, note the tetrahedral orthophosphate groups. (C) Projection down the a diagonal, a different view of the three-dimensional arrangement of atoms in the unit cell. (D) Crystals of hydroxylapatite, note the hexagonal prismatic morphology.

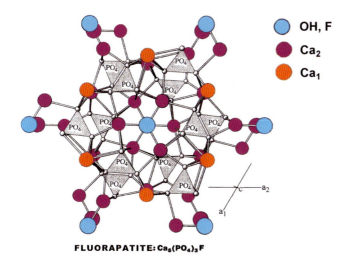

OH, F
Ca₂
Ca₁

FLUORAPATITE: Ca₅(PO₄)₃F

FIGURE 4 Distribution of the cations in two sites, Ca_1 and Ca_2, in fluorapatite, view down unique c-axis, similar to Figure 3A but with F in place of OH.

anionic groups coordinate with calcium cations to produce a charge neutral solid when a singly charged species, OH^{1-} or F^{1-}, the blue atom site, is added. Figure 3C is the view down the diagonal between two a axes. Note the distribution of the calcium atoms and the channelways where the OH^{1-} or F^{1-} occur parallel to the c-axis direction. The prismatic hexagon morphology typical of many apatite mineral samples shows up in Figure 3D. It is a selection of single crystals of mineral apatite produced during investigations of the simplified chemical system. These crystals are synthetic hydroxylapatites and are up to several millimeters in length.

There are two distinct sites for cations and they are discretely colored orange and raspberry red. One is connected to the tetrahedral phosphate (PO₄) backbone oxygens and the atoms in channelways (OH, F, or Cl) that parallel the c-axis, and the other is related to the trigonal a axes in the center of the cell. The two calcium sites are designated Ca1 and Ca2 in Figure 4. When the channels are occupied by OH^{1-} the mineral is known as hydroxylapatite, and when occupied by F^{1-} the name is fluorapatite, which is another member of the calcium apatite mineral group.

An amazing number of different composition minerals, each a separately named mineral species, may form with the apatite structure type. These minerals show mixed chemistries with very minor variations in structural detail that are related to the amount and exact placement of he particular elements. Because of the importance of the chemical variations to our medical purview and the fact that many of these naturally occurring minerals have been studied in great detail, the sec-

tions below present examples to show what elements find their way into bioapatites and where they may be located. Bioapatite composition mirrors elemental bioavailability.

A. Substitutions in the Phosphate Backbone

Though the most common minerals are probably the phosphate apatites, there are arsenate or vanadate minerals that form with virtually identical apatite-like crystalline structures. The AsO_4^{3-}, and VO_4^{3-} anions have a charge and size very similar to the PO_4^{3-} group. Naturally occurring arsenate or vanadate apatites usually contain lead (Pb) rather than calcium. Lead phosphate apatite, the mineral pyromorphite with ideal formula $Pb_5(PO_4)_3Cl$, has all cation sites occupied by lead, but chlorine, Cl^{1-}, takes the place of the OH in the channelways (see Section IV.C).

Other tetrahedral anionic groups, sulfate SO_4^{2-} and silicate SiO_4^{4-}, although of different charge, may also form minerals with apatite structure type, and small amounts of these groups as well as arsenate and vanadate may be found in the phosphate apatite minerals. Britholite-Y, a naturally occurring mineral, is one of a series of phosphate apatites that contain silicon substituting for phosphorus, but there is a second cation in addition to calcium, yttrium (Y). Both OH and fluorine (F) are present in the channelways.

B. Substitutions in the Cation Sites

There are two different cation sites, or distinct lattice locations, known as Ca1 and Ca2 in hydroxylapatite (Figure 4). The site designated Ca1 bonds to nine oxygen atoms of the tetrahedral phosphate groups while Ca2 bonds to six phosphate oxygens and one OH ion. Many other cations, especially those with double positive charge similar to Ca^{2+} such as Sr^{2+}, can be accommodated in these sites. Table III lists the elements by charge and by size that could substitute in apatites.

Lead apatite, the mineral pyromorphite, has an apatite structure but lead is a large ion relative to calcium. However, small amounts of lead occur in the predominantly calcium bioapatites. Lead isotope [207]Pb has been used to discriminate between the possible sources. Lead from leaded gasoline, or uranium, for example, might be taken up in human tissues. Lead may also occur in high concentrations indoors. It might be found as a component of dust distributed about the living areas with the possible source local soils tracked

TABLE III. Ions That can Substitute in Calcium Phosphate Apatites, Their Charge and Ionic Radii Depending on Coordination Number, i.e. VII, IX . . . .

Ion	Ionic Radius
Ca^{2+}	$1.06^{a,b}$(VII)
	1.18^{a}(IX)
Cd^{2+}	1.14^{a}(VII)
Mg^{2+}	0.79^{a}(VII)
	0.89^{b}(VIII)
Sr^{2+}	1.21^{a}(VII)
	1.31^{b}(IX)
Ba^{2+}	1.38^{b}(VII)
	1.47^{b}(IX)
Mn^{2+}	0.90^{b}(VII)
	0.96^{b}(VIII)
Na^{1+}	1.12^{b}(VII)
	1.24^{b}(IX)
K^{1+}	1.46^{b}(VII)
	1.55^{b}(IX)
Pb^{2+}	1.23^{b}(VII)
	1.35^{b}(IX)
P^{5+}	$0.17^{a,b}$(IV)
As^{5+}	0.335^{a}(IV)
Si^{4+}	$0.26^{a,b}$(IV)
V^{5+}	0.54^{a}(VI)
S^{6+}	0.12^{b}(IV)
	0.29^{b}(VI)
Sb^{5+}	0.61^{b}(VI)
Al^{3+}	0.39^{a}(IV)
U^{3+}	0.98^{a}(VII)
Ce^{3+}	1.07^{a}(VII)
	1.146^{b}(IX)

[a]From Shannon and Prewitt, 1969.
[b]From Shannon, 1976.

inside a residence or be from lead-based paint. Some from either source could be ingested and become sequestered in bones or teeth. These exposures make lead a "silent" hazard, a potential danger especially for young children who constantly put their fingers in their mouths. The term silent is used because Pb is not visible, and detection and amount determined in the environment or in the body is impossible without specialized analytical methods and tools. Therefore biological uptake and sequestration of lead in bones, which

may be cumulative, is not monitored. Lead exposure goes unrecognized unless some distinctive disease or signal, as described in a study on schoolchildren (Mielke, 2003), is recognized, evaluated, and related to lead levels.

Strontium apatite, ideal formula $(Sr,Ca)_5(PO_4)_3(OH)$, has been designated a separate mineral species and a member of one of the subsets within the apatite group, because the amount of Sr is greater than 50% of the total cations present. Studies on naturally occurring rare-earth-element-substituted apatites show coordinated substitution: when rare earth elements (REE) with a charge of 3^+ are incorporated in the structure, there may be a substitution of Si^{4+} for part of the phosphorus. With both substitutions, a charge balance comparable to the original association of Ca^{2+} and P^{5+} in the calcium apatite structure can be maintained. Another coordinated substitution is when an REE^{3+} is incorporated along with Na^{1+}. Charge balance may be achieved because the cations are distributed at both sites in calcium apatites (Hughes et al., 1991b).

Some elements prefer one cation site over the other. A study of REE-containing apatite minerals by Hughes et al. (1991b) demonstrated that some of the REE preferred the Ca^2 site while others preferred the Ca^1 site, provided charge balance was maintained by substitutions at the anionic sites or with additional and differently charged cations. Site designation for particular elements can be determined using high-resolution X-ray diffraction analyses, paramagnetic resonance, thermoluminescence, or infrared spectroscopy (Suitch et al., 1985).

Apatite samples can have different compositions in spite of coming from similar sources or sites. Although it is possible with modern techniques to show elements at specific sites in the crystal structure of geological samples, the tiny crystallites of bioapatite are too small for such detailed investigations. It is worthwhile to reiterate that bioapatite composition in one bone may not be identical to another bioapatite forming elsewhere at the same moment or at different times. The fluid-cell-matrix-mineral system composition is unlikely to be constant from one moment to another, much less from year to year as the human ages, resides in different geographic localities, and ingests water or food from different sources.

There is another set of concerns that relate directly to cation substitution in bioapatites: bone seeking α-emitting radionuclides and ionizing radiation exposures of humans. During atomic bomb tests a half century ago in the southwest United States there was a scare related to fallout of radioactive nuclides, especially ^{90}Sr. The anxiety was based on the similarities in behavior of calcium and strontium, the half-life of the nuclide (28 years), the prevailing wind direction toward the more heavily populated east, and the fact that American dietary calcium came mostly from milk products with the largest consumers being children who were actively putting down new bone. The worldwide average of ^{90}Sr was shown to be about 0.12 microcuries per gram of calcium in man or 1/10,000 of the acceptable permissible level at that time. This suggested that the atomic bomb circulating ^{90}Sr was not a global hazard. A remarkable study on baby teeth in mid-western communities of the United States compared the strontium concentrations with adult bones the late 1950s. These investigations showed that bioapatites discriminated against strontium during formation and concluded that only those individuals who obtained their total food supply from restricted areas (with low calcium in the rocks, soils, or waters) were at risk (Eckelman et al., 1957). The furor over the nuclide hazard eventually collapsed when it was realized that strontium bioavailability was overshadowed by calcium and no one, especially children, was likely to be at risk in the United States or abroad (Eckelman et al., 1954; Fowler, 1960).

Massive doses of radioactive elements from nuclear explosions such as the Chernobyl disaster are locally extremely hazardous as they spread the radionuclides in the soils and the plants that animals and humans ingest. However, the potential for incorporation in mineralized tissues at a hazardous exposure level of several potentially harmful radioactive materials has yet to be documented in spite of extensive surveys (Fabrikant, 1988).

C. Substitutions in the Hydroxyl Site

The hydroxyl (OH) site in the calcium apatite crystal structure can be fully occupied by fluorine or chlorine. As end members in that chemical system, they form the independent minerals fluorapatite and chlorapatite. These halogen species are the predominant forms of apatite found in sedimentary, metamorphic, and igneous rocks (Hughes et al., 1989). Bromine (Br) or iodine (I) can be incorporated in the apatite structure type, but mineral species in which the halogen site is filled entirely by bromine or iodine have not been found naturally.

Bioapatites originally precipitate and remain mostly as hydroxylapatite in human bones and teeth because of the predominance of aqueous fluid and OH concentration relative to the halogens in the human body. When higher amounts of fluorine become available the element can become incorporated substituting for part, at least, of the OH. For example, $1 \, mg \, L^{-1}$ fluorine added to drinking water, an amount that has become standard for many of the reservoirs that supply water to populations across the United States, appears to reduce caries and may lead to a reduction in the incidence of osteoporosis (Watts, 1999). On the other hand, the regular ingestion of greater than $100 \, mg \, L^{-1}$ of fluorine over a long period of time by humans leads to disease. Such high amounts of fluorine in local waters and agricultural products grown in soils irrigated with high-fluorine-containing water, or through industrial exposure, may result in fluorosis (Vischer, 1970; Finkelman et al., 1999). Whether the mineral matter in the bone and tooth tissues of such exposed human populations is partially fluorapatite, i.e., a mixture of the two separate apatite species, or whether each apatite crystallite has both fluorine and hydroxyl in its channel sites is unknown. Fluoride has been used to treat osteoporosis (see Section VI.B.1 and Chapter 12, this volume).

Chlorapatite, the calcium phosphate apatite mineral in which all the channelways are filled with chlorine, has not been identified in mineralized tissues in spite of the high concentration of chlorine (Cl) in body fluids, but small amounts of chlorine can be detected on analyses of bioapatites.

D. Carbonate (CO_3^{2-}) in Apatites

One additional chemical constituent that is often detected in apatite analyses becomes important when discussing bioapatites, and that is carbonate. Two carbonate-containing calcium apatite minerals, dahllite and francolite (Table I), have been described from phosphorites, fine-grained sedimentary deposits mined for fertilizer on many continents. These phosphate minerals are associated with the common calcium carbonate minerals calcite and aragonite (Gaines et al., 1997). Neither calcite nor aragonite has been identified in bone tissues, but because many bioapatites show higher than stoichiometric Ca/P ratio and CO_3^{2-} on analysis, the suggestion is that CO_3^{2-} substitutes either for PO_4^{3-} or for OH^{1-} in the apatite. The carbonate ion CO_3^{2-} has a different charge and size than the dominant phosphate groups (PO_4^{3-}). It is a planar trigonal ion with a diameter of 0.24 nm and does not easily fit in the crystal structure. The amount and disposition of CO_3^{2-} within the apatite crystal structure is, and has been, a topic of great interest for some time (McConnell, 1973, Skinner, 1989; Elliott, 1984, 1994).

If carbonate, CO_3^{2-}, is present in the crystal lattice other ions, such as triply positively charged cations, it might take the place of calcium so that local charge disruption could be balanced by coupled substitution or by vacancies in the lattice. The inclusion of CO_3 could compromise the ideal architecture of the apatite backbone and the channelways. Such destabilization may account for the very fine-grained nature of bioapatites.

The association of CO_2 with bioapatite is not particularly surprising. The molecular species CO_2 or bicarbonate, HCO_3^{1-}, are produced along with many others during cell metabolism and could adsorb on the high surface area of the tiny crystallites. Early deposition of mineral takes place under acidic conditions where orthophosphate species may aid the nucleation of bioapatite whereas other ions in the fluid, such as bicarbonate, may be inhibitory (Glimcher, 1998).

The carbonate ion and its distribution is not of major concern for this medical geology purview except that its presence makes us aware of the necessity to consider both the physical and chemical aspects of the mineralizing system. Carbonate probably aids incorporation of other elements into the lattice. Bioapatite precipitates are aggregates of crystallites, which means that the mineral mosaic of many crystallites can each present a slightly different composition and size. The lower crystallinity, and variable composition, of carbonate-containing-bioapatites may be an irritation preventing precise designation of the mineral phase, but it is a positive advantage for the biological system. The high surface area of the crystallites facilitates their dissolution as required for the dynamic bone mineral formation-resorption system. Nature has utilized a solid phase that fulfills several functional roles required for bone (Skinner, 2000b). Bioapatites record exposures of living creatures to the environment and particularly the bioavailability of elemental species in our diets.

This brief summary does not do justice to investigations with a variety of techniques which include electron and X-ray diffraction analysis, infrared, polarized infrared and Raman spectroscopy, solid state carbon-13 nuclear magnetic resonance spectroscopy, and most recently atomic force microscopy. These have been and are used to detect and quantitate the amount of carbonate within the crystal lattice of a single-phase mineral or bioapatite or as an adsorbed species.

From the above selected examples and Table III, the very wide range of elements and molecular species that can be accommodated within the apatite crystal structure is summarized. Table IV lists the levels of elements essential to proper body function. The match is quite

TABLE IV. The Essential Elements, and Their Recommended Daily Intake (RDI)

Element	RDI
Boron**	(1.7–7.0 mg)
Bromine	0.3–7.0 mg
Calcium**	0.8–1.3 g
Cesium**	0.1–17.5 mcg
Chromium**	130 mcg
Cobalt	15–32 mcg
Copper**	1–2 mg
Fluorine**	1.0 mgt
Iodine	70–150 mcg
Iron**	10–18 mg
Lithium	730 mcg
Magnesium**	200–450 mg
Manganese**	3.5 mg
Molybdenum	160 mcg
Nickel**	(35–700)
Phosphorus**	0.8–1.3 G
Potassium**	3500 mg
Selenium	70 mcg
Silicon**	(21–46 mg)
Tin**	0.13–12.69 mcg
Vanadium	(12.4–30.0 mcg)
Zinc**	8–15 mg

Note: Additional elements detected include the following non-essential elements: Ag, Af, Cd, Pb, and Sp.

**indicates those detected in bone tissues and ( ) indicates non-essential elements.

From the Food and Nutrition Board, National Research Council, Recommended Dietary Allowances 10th edition, National Academy Press, Washington DC, 1989, the Federal Register #2206, and for elements in bone Bronner, 1996 and Skinner et al., 1972.

remarkable. Many elements and chemicals entering the human body may become associated with, or become part of, the apatitic mineral matter.

V. ANALYSIS OF APATITIC BIOMINERALS

To ascertain the ranges of included elements and species in bioapatites, the mineral matter must be extracted and concentrated. The techniques devised for separating mineral from the associated organic and cellular materials can, in many cases, further complicate the assays. The other option is to analyze the tissues keeping both the mineral and organic fractions associated. Either way there are special techniques required to prepare the sample for analysis. Preparation of the mineral phase and the main methods of analysis, diffraction, will be presented followed by the sample preparations necessary for examining whole tissues. Histology is the general name for the host of techniques using optical and/or scanning electron microscopic (SEM) analyses to investigate thin sections of tissues.

A. Sample Preparation: Mineral

The opportunity to examine the mineral separately from surrounding organic materials, whether examining pathological aggregates from arteries or from normal bone tissues, is a non-trivial undertaking (Kim et al., 1995). Bioapatites have individual mineral grains of the order of $1 \times 2 \times 25$ nm and are loosely associated one with another as porous aggregates with random crystallite distribution or alternatively on and in a fibrous protein matrix. The latter is characteristic of bone tissues where the crystallites often align parallel to the length of the collagen molecules (Skinner, 1987). Aware that there is a range of composition of bioapatites and mindful of their very small grain size, care must be exercised. In addition it is necessary to record the specific tissue and site in the organ or exactly where a pathological deposit is located. The age of the individual and the date of sampling are also critical because different cell systems and tissue textures are encountered in every bone (Weiner & Wagner, 1998).

In all normal bone the intimate association of mineral with matrix proteins and other proteins shows variations dependent on source (Miller, 1973) Table II gives average amounts, but the mineral concentration as well as distribution also varies at submicron levels (Rey et al., 1996). The chemical variations detailed above reinforce the possibility that the mineral itself may also vary with growth and maturation, especially in bone where the entire structures, e.g., haversian bone, are constantly resorbed and re-deposited over time (Skinner, 1987). The chemical analysis of mineral that may attract and adsorb transient ions from the surrounding fluids is only an indication of the compositional range of the tissue at a specific moment in time.

Analyses of mineralized tissues are not only compromised by the poor crystallinity of the mineral but by the presence of non-mineral components. A sample of enamel with over 96 wt% mineral, less than one percent protein (enamelin), and the most highly mineralized normal tissue in the human body, is the preferred choice for mineral analyses. Enamel is also examined because it has a very restricted period of formation, and because no cells remain at maturity the final tissue is not reworked. Dentine, the mineralized tissue adjacent to enamel and the major tissue in the tooth, is maximally 75% mineral when fully mature, with collagen about 20% of the total, plus some small molecular protein species, and fluid. Bone tissues, especially in the first stages of formation, and spicular, or trabecular bone, found adjacent to the marrow cavity (Figure 1B), may contain less than 50% mineral per unit area. To accurately ascertain the mineral composition and structure, it is important to extract the mineral portion from these organic moieties.

The usual way to separate the mineral fraction has been to immerse the entire sample in sodium hypochlorite or bleach. Most of the organic matter will eventually dissolve, and the time it takes depends on the size and porosity of a particular sample relative to the amount of bleach. Smaller sample size allows for a more rapid dissociation of the mineral from the intimately associated organic molecules. To concentrate the mineral fraction from the heavily mineralized cortex of a long bone will require a long soak and much decanting and re-suspending of the tissue sample in the bleach. This procedure could alter the amount of mineral, especially as smaller crystallites are likely to be more soluble.

More exotic systems of extracting the organic moieties from the mineral phase using chemical methods have been suggested. Refluxing with ethylene diamine, (Skinner et al., 1972), but it is a lengthy procedure and possibly dangerous because the solution pH is 12 or greater. An interesting observation on careful chemical extractions is that the morphology of the organ or sample, whether bone or tooth, will likely remain after virtually all of the organics (>95%) have been removed. This is an illustration of the permeation of the biological macromolecules and how difficult it is to completely extract all of the intimately associated organic phases. Pathologic spherulitic apatitic samples also have ultra-small size, not only of the aggregates, but of the crystallites within them.

Several other methods of extraction have been employed. One technique, called "ashing" subjects tissues to elevated temperatures, above 500°C and often to 1000°C, for at least an hour. Researchers who wish to assure complete disappearance of the bioorganic portions advocate up to 10 or more hours at elevated temperatures. The extended high temperature treatment certainly eliminates any organic components, but it also re-crystallizes the mineral phase. An X-ray diffraction analysis of the ashed sample (see Section V.B) provides a sharp diffraction pattern of hydroxylapatite that is comparable to well-crystallized geologically obtained mineral. Occasionally, two mineral phases are detected with the identity of the second phase dependent on the bulk composition of the sample, the temperature, and duration of ashing. For example, above about 200°C adsorbed water and any carbonate will be lost and other calcium phosphates, $CaHPO_4$ or $Ca_2P_2O_7$, appear (Table I). This indicates that the bulk Ca/P composition was lower than stoichiometric HA. At greater than 1000°C, tricalcium phosphate, $Ca_3(PO_4)_2$, may take the place of the pyrophosphate. The relative amounts of the two species at a known temperature can be used to calculate a Ca/P for the sample. If the Ca/P is greater than 2.15, tetracalcium phosphate $Ca_4(PO_4)_2O$ may form. The appearance of different mineral phases reflects the experimental conditions that the mineralized tissue sample was subjected to, not the presence of the second phase in the low-temperature-produced tissues. An alternative, low-temperature ashing using activated oxygen, has also been employed to remove the organic constituents from mineralized tissues. X-ray or electron diffraction techniques on the extracted materials will be used to determine the crystal chemical characteristics of the mineral phase (Kim et al., 1995).

B. X-ray and Electron Diffraction

Unambiguous identification of most mineral materials utilizes diffraction techniques that can easily determine the species based on the unique crystal structural characteristics of the compound (Klug & Alexander, 1954). The powder diffraction method may be used for the identification of any crystalline materials. Either X-ray or electron diffraction may be employed, and the choice is usually dependent on instrument availability and the specifics of the sample. Vast databases have accumulated over the past hundred years since the techniques were elucidated. The International Union Committee of Diffraction has a compendium of X-ray diffraction data on crystalline compounds, both organic and inorganic, which contains a subset for natural and synthetic mineral materials including the apatites.

The tiny crystallite size and variable composition of bioapatites are fully expressed in the X-ray diffraction analysis of the extracted materials. Well-crystallized (without vacancies and >0.5 µm in average size) hydroxylapatite geological samples give many discrete diffraction maxima from which one easily calculates unit cell parameters. Comparing a mineral apatite X-ray powder diffraction pattern (Figure 5) with bioapatite, e.g., cortical bone mineral, the latter has few and broad maxima. A poorly crystalline compound makes it difficult to determine any compositional details other than to say

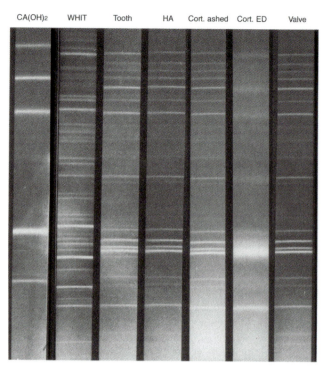

FIGURE 5 Comparison of X-ray powder diffraction patterns of mineral and tissue bioapatites: $Ca(OH)_2$, portlandite; WHIT, whitlockite; Tooth, bioapatite; HA, hydroxylapatite, synthetic hydrothermal sample (from Skinner, 1973); Cort. ashed and Cort. ED, cortical bone ashed and ethylene diamine extracted cortical bone; Valve, human aortic valve pathological bioapatite deposit.

the sample gives a pattern consistent with an apatite (Skinner, 1968) and focuses attention on the detection of any additional solids. Enamel is perhaps the only biomaterial that provides sufficient diffraction detail to calculate lattice parameters of a calcium apatite. Table V presents the results of calculations from powder diffraction data of several different apatites. Electron diffraction does not necessarily afford more precise results over X-ray diffraction as the level of crystallinity is the important criterion for producing diffraction. The advantage of electron diffraction is that the beam may be more finely focused, and a small mineral crystal aggregate within a thin section might be separately examined rather than extracting the mineral to obtain a pure mineral sample for powder diffraction.

C. Sample Preparation: Mineralized Tissues

To examine the mineral phase in its biological surroundings requires different sample preparation. The

TABLE V. Calculated Unit Cell (Lattice) Parameters for Synthetic Apatites, and Bioapatites

	a-Axis	c-Axis	Ref.
Hydroxylapatite (Synthetic)			
315°C, 2 Kbars H$_2$O pressure			
CaO/P$_2$O$_5$ 1.61	9.421	6.882	Skinner (1968)
CaO/P$_2$O$_5$ 1.12	9.416	6.883	Skinner (1968)
600°C, 2 Kbars H$_2$O pressure			
CaO/P$_2$O$_5$ 1.61	9.4145	6.880	Skinner (1968)
CaO/P$_2$O$_5$ 1.12	9.4224	6.8819	Skinner (1968)
100°C Precipitation	9.422	6.883	Bell & Mika (1979)
(Reitveld calc.)	9.4174	6.8855	Young & Holcomb (1982)
Mineral	9.418	6.875	Gaines et al. (1997), p. 856
Carbonate Apatites (Synthetic)[a]			
Low temperature (<100°C precipitates, air-dried)			
Direct (adding calcium acetate into a solution of ammonium carbonate+phosphate) 13.8% CO$_3$	9.373	6.897	Labarthe et al. (1973)
Inverse (adding ammonium solution into calcium acetate) 10.4% CO$_3$	9.354	6.897	Labarthe et al. (1973)
Direct using sodium carbonate	9.440	6.880	LeGeros et al. (1968)
Direct using sodium carbonate and fluoride 22.1% CO$_3$	9.268	6.924	LeGeros et al. (1986)
High temperatures 1.455 CO$_3$, the maximum determined	9.367	6.934	Rey et al. (1991)
Bone			
Ashed cortical	9.419	6.886	Skinner et al. (1972)
Enamel			
Human	9.421	6.881	Carlstrom (1955)

[a]The wide variations in the values may be partially attributed to differences in composition of the starting materials and to the methods of preparation, e.g., direct versus inverse, under low temperatures for carbonate apatite synthesis.

See Elliott, 1994, p. 234–248 for details on precipitation mechanisms and discussion of IR, X-ray diffraction analyses, and possible locations of the carbonate ion in the apatite lattice.

techniques must minimize any alteration of the tiny crystallites while maintaining the organic and cellular framework in which the mineral matter is distributed. Thin sections are prepared for examination with optical and electron microscopic techniques so that morphological relationships typical of the different tissue types and chemical observations on the mineral phase can be assayed (Section V.D).

The techniques to prepare biological tissues as thin sections parallel the methods employed in petrology (the study of rocks) (Blatt & Tracy, 1996). The mineralized tissue sample is embedded to obtain a planoparallel thin section, from 50 to less than 10 μm thick. The thickness will depend on the information desired and the method of analysis. Embedding preferentially employs plastics, rather than paraffin, the typical media

for soft tissue sections examined in pathology laboratories (Malluche et al., 1982). Plastic is required because the juxtaposition of tiny crystals of hard mineral with the soft organic matter and cells minimizes differential hardness of the medium put to sectioning.

After obtaining a fresh tissue sample the first act is to soak the pieces in alcohol solution about 10 times the sample volume for a few hours or overnight refreshing the solution several times. Occasionally formaldehyde is used but formalin has to be buffered or the solution will be acidic and at least some mineral crystallites may be lost or dissolve during immersion. The alcohol soak effectively lowers the water and fat content in the tissue, stops further biological degradation, and stabilizes the organic components. The second stage is to embed the whole sample with not too viscous epoxy materials such

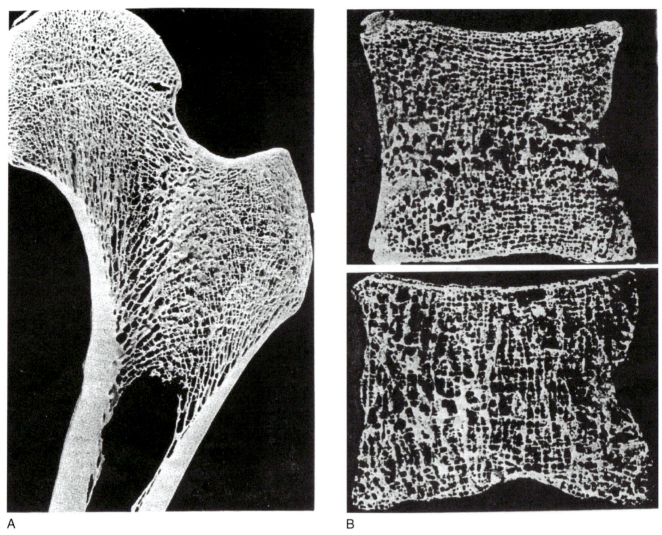

A   B

FIGURE 6 Transmission X-ray photos of excised samples of bone. (A) Longitudinal section through the upper femur of normal 29-year-old male (from Albright and Skinner, 1987, Figure 5–1). Note the arcuate pattern of the trabeculae. (B) Mid sagittal sections of first lumbar vertebra of a 20-year-old female (above) and 50 year old (below) (from Arnold et al., 1966, Figure 1.16). The trabeculae in the older bone show thickening and "reinforcement" accentuating their vertical orientation in the skeleton.

as methylmethacrylate, or other commercially available plastics such as Spurr™ or Epon™. The viscosity must be appropriate to facilitate penetration throughout the sample, and probably will require a vacuum to maximize efficiency. Most mineralized tissue laboratories will have an automated embedding system that takes the tissue through the extraction and embedding procedures and applies vacuum and heat and a "hardener" to ensure a fully homogeneous block of embedded sample. Any holes, effectively pockets of air that might incorporate dust or foreign materials, must be avoided especially if the analysis will use SEM coupled with energy dispersive analysis (SEM/EDXA) to measure the elemental composition of the mineral. Once embedded

the whole tissue piece is ready for sectioning and polishing.

Sectioning requires a sharp knife, usually diamond or carborundum blades, to cut the plastic embedded tissues without tearing the crystallites from their organic matrix. Once cut the thin section is mounted on a glass or plastic slide for microscopic study, usually without the addition of a cover slip. Grinding to assure plano-parallel surface for high-resolution electron microprobe elemental analysis may be essential, but it almost always results in smearing and disrupting the crystallites, so this level of the preparation procedure again requires care. It is wise to constantly view the section under a microscope with at least 40× magnification. Any surface roughness will cause interference in elemental analyses, although background corrections can be applied to the raw data (Reed, 1993).

Every laboratory that performs mineralized tissue section analyses, usually the pathology departments of hospitals, has their own sterile procedures. To prepare uniformly parallel, non-artifact-containing hard tissue thin sections is an art, not a science. Once made, sections may be stained to identify osteoid, cell walls, nuclei, etc., or the calcified portions for specialized investigations, but such treatments would compromise analysis of the mineral phase. Careful, sensitive attention to the preparation of any section is necessary for accurate documentation of the tissue components, their structures, and elemental components.

D. Histomorphometry

To study the variations in the textures found in all mineralized tissues, bone, teeth, or pathological materials, optical investigations at different magnifications are usually required. Optical light microscopy using transmitted light at approximately 40× magnification is sufficient to determine areas of mineralization in a tissue section. The mineral matter will be non-translucent and easily distinguished from organic components. By employing a polarizing light microscope, with the crossed polarizers in place, the mineral portions will show variations in birefringence on rotation of the stage, which is the characteristic used to identify any crystalline mineral (Nesse, 2004). The mineral calcite, for example, can be easily differentiated from hydroxylapatite because calcite has a much higher birefringence. However, to thoroughly document the tissue components and study the remodeling of bone tissue, combinations of SEM and optical and electron microscopy are usually employed.

Some of the distinct morphologies observed in normal and pathological mineralized tissues are next. Figure 6A is tissue section cut longitudinally through the upper end of a long bone. Dense mineral matter and cortical bone surrounds the shaft and the open marrow cavity. It also outlines the trabecular tissue in what appears to be the porous head of the femur. The arcuate patterning of trabeculae throughout the head is an expression of mineral deposition conforming to stress in this organ; the bone tissue distribution responds to mechanical strain (Albright & Skinner, 1987). In Figure 6B two tissue sections through vertebrae of different ages illustrate changes in trabecular thickness and patterning. The vertical struts thicken with age as the tissues respond to the effects of gravity from our vertical posture.

Figure 7A is an x-ray transmission microradiograph of cortical bone tissue from a dog tibia. It illustrates variations in mineral density in the several osteons (the circular patterns), typical of haversian bone. Variation in the levels of gray in this section reflect different amounts of mineral per unit area and detected at higher magnification because of the mineral impeding the transmission of x-ray energy. Figure 7B is another view of the same area but illuminated with UV light. Three of the osteons show two circular dark lines around the central vascular opening, the pattern resembling an archery target. The lines are due to the incorporation of the antibiotic tetracycline as the mineral deposits and the molecule glows under UV light. The two lines mark two separate doses of tetracycline allowing the rate of mineral deposition, and the growth of osteons in the cortex, to be determined (Skinner and Nalbandian, 1975).

Figure 8A is an image that shows spherulitic (pathologic) calcium phosphate deposits in breast tissue. The higher resolution (×1200) and use of back scattered electron imaging (one of the modes available with SEM) shows these tiny deposits. Figure 8B is an electron micrograph at still higher magnification (×25,000) of the dentine-enamel junction illustrating the different size, shape, and aggregation of hydroxylapatite crystallites in these two tissues. At magnifications greater than ×200,000, transmission electron micrographs have shown hexagonal outlines of the early-formed enamel crystals.

Weiner and Wagner (1998) recently reviewed the multiple levels of structural organization seen with histological examination and some of the physical contributions of the mineral crystallites. Bone organs require strong and flexible tissues, so they do not buckle and break when subjected to torsion, tension, compression, and instant and consistent responses to applied stresses for proper organ function. Although histology demonstrates the physical textures and mineral distribution during growth, development, and aging, the incorporation of chemicals may well alter the reactions of the mineral, and hence the mineralized tissues (Bronner, 1996; Skinner et al., 2004). A brief discussion of osteoporosis and outlining some treatments for this disease will illustrate how our present understanding of bone physical properties and bone mineralogy are benefiting medical care.

VI. OSTEOPOROSIS

Osteoporosis is a metabolic bone disorder characterized by reduction in the volume of bony tissue per unit

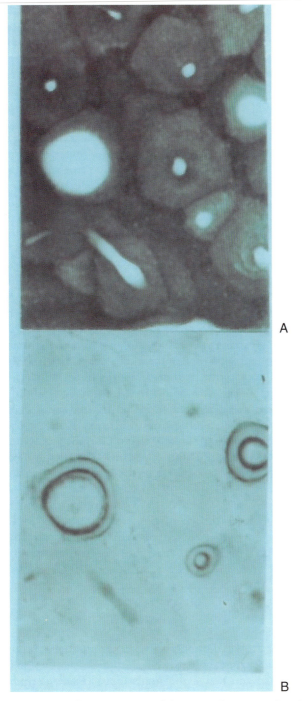

A

B

FIGURE 7 Ground section of the cortical portion of a rib from a dog given two doses of the antibiotic tetracycline. The tetracycline is incorporated at the time of deposition of bone tissue (Skinner and Nalbandian, 1975, Figure 3). Magnification 175x

A: In transmitted light, the darker the color the more mineral present, an expression of decreased transmission of light. Note variations in bone mineral density between the several haversian systems in the section. Some contain more mineral than others.

B: Same area in ultraviolet light (tetracycline glows in UV light) which shows two fluorescent rings (concentric dark circles) in the haversian systems marking the times when two doses of tetracycline were added to the diet as bone tissues were developing.

A

B

FIGURE 8 Microradiographs taken with the scanning electron microprobe, backscattered images of breast tissue samples. (A) Calcium phosphate spherules deposited in breast tissue. Note the size of the spherules at this high resolution (magnification ×1200) relative to the heavily mineralized calcium phosphate deposit (white areas) on the right. Section thickness is 6 μm. (From Poggi, et al., 1998, Figure 3.) (B) The dentin-enamel junction in a tooth showing the typical small crystallites in dentine (left) and the larger crystallites in enamel (right). (From Goose and Appleton, 1982, Figure 2.6.) Enamel crystallites are about 100 μm in length.

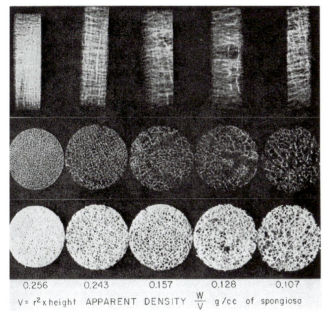

0.256   0.243   0.157   0.128   0.107

$V = r^2 \times height$ APPARENT DENSITY $\frac{W}{V}$ g/cc of spongiosa

A

FIGURE 9 Osteoporosis examined by transmission X-ray analysis. The first lumbar vertebra from five autopsy subjects illustrates the possibility of quantifying the level of mineral in a bone (from Barzel, 1970, Figure 2B). (A) Top row vertical sections through the lumbar vertebrae showing changes in the distribution of mineral (white) from homogeneous fully mineralized tissues on left part of the figure to more porous on the right where the trabeculae thicken and appear more vertically oriented in the samples from older individuals. Middle row transverse section through the vertebrae shows increased porosity and disorder of the vertical sections. Bottom row transverse sections have been extracted and are now fat-free and dry. The porosity increase and the density of the mineral per unit area can be measured. Formula used to calculate the apparent density of columnar sections is weight/volume, g/cc. (B) Transmission X-ray of the entire vertebrae illustrating the apparent density differences that might be visible on radiologic examination (from Barzel, 1970, Figure 4, left half). The several vertebrae depicted in this figure coincide with the mineral density calculated for the sections in Figure 9A.

volume of bone (Figure 9). The disease has been a topic of investigation for over a hundred years (Arnold, 1966; Barzel, 1970; Riggs & Melton, 1995). The tissue reduction leads to fragility of the bone organs and may cause an osteoporotic individual to sustain a fracture, perhaps without obvious trauma, pain, or other warnings. The main concern is that the usually elderly patient will be at high risk for additional fractures. In osteoporosis there is a normal mineral/collagen ratio, just less min-

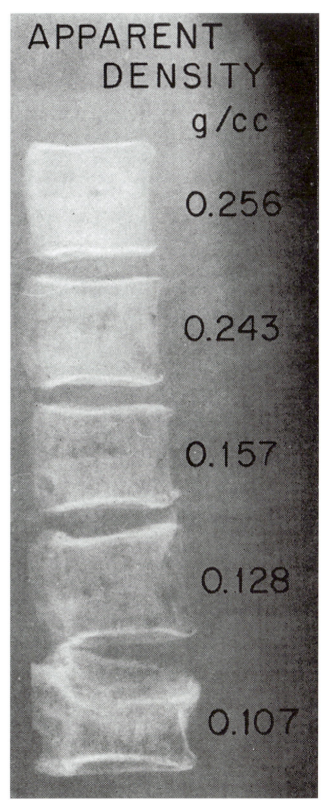

B

eralized tissue per unit area which distinguishes the disorder from osteomalacia where mineral is reduced in amount relative to the bioorganic matrix. There is, as might be expected, no known cure for osteoporosis. It is part of the normal aging processes and pervasive in postmenopausal women and older men. The disease is considered a health issue in the United States and in other developed countries. It has become an important area for basic and clinical research with the incidence of fracture considered a signal of disease (Figure 9). Onset of different types of osteoporosis, the relationship to nutrition, exercise, or other potential contributing factors, and the results of a variety of treatments are the factors considered in the many epidemiological studies in several countries.

Bone mass is genetically programmed for an individual but may be modulated qualitatively and quantitatively by environmental factors. Reduction of bone tissue density in the vertebrae, a hallmark of osteoporosis, and subsequent vertebral fracture is estimated to be as high as 1 out of 4 women by age 65–70 (Wasnich, 1996; Eastell, 1999). This expression of compromised bone strength results in debilitation and pain with normal body movements and often premature death. Medical attention now must go beyond bone quantity (mass) into bone quality: architecture and rates of turnover (Chestnut et al., 2001) that involve molecular biochemistry and the relationships of the inorganic with organic constituents.

Osteoporosis, a focus of public attention when bone loss was recorded for the astronauts subjected to weightlessness, is also a consideration from long periods of inactivity such as for disabled people in wheelchairs or bedridden. The dynamic bone tissues respond to normal wear and tear. A cadre of metabolic disorders that result in bone loss in the young as well as in the old have been identified (Avioli, 2000). The accumulated knowledge from abnormal situations has illuminated the complicated array of physical and chemical interactions necessary to maintain a viable skeleton. However, what constitutes effective treatment or, best of all, prevention of osteoporosis, remains elusive. A multiplicity of approaches that include not only pharmacologic intervention, but genetics (Econs, 2000), diet (Marcus et al., 1996), and exercise (Riggs & Melton, 1995) are addressing the disease.

A. Detection of the Disease

Throughout the skeleton and its interrelated body systems there are dynamic changes but probably none are so obvious a sign of aging as the bend of the spine, also known as dowagers hump, because it is often typical of older women. Figure 6B and Figures 9A and B depict the remarkable differences at the tissue level in the distribution of trabeculae in affected vertebrae. A diagnosis of osteoporosis is inferred from a clinical examination and confirmed by radiologic examination using X-ray radiographic transmission analyses. A radiograph of the forearm, or leg bones, which is available as a result of an accidental fracture, may present poorly mineralized, inhomogeneous, or "porous" bone and tissues that are reminiscent of these spine photos.

The density level, or mineral content, of bones can be measured using these radiographs, and such examinations are noninvasive (Johnston et al., 1996). Digital radiographic techniques can quantify the mass of any bone or portion thereof. The results are compared with results from persons of like race, stature, and age to estimate the degree of osteoporosis and the potential risk for future fractures for a specific patient. What is actually measured is the mineral concentration per mass of bone (Figure 10).

High-resolution radiographic analysis methodology, computerized axial tomography (CAT) scans, has also been used to show local differences in mineral density and distribution in the cortical bone, or the number, size, and organization of the trabeculae in a bone. These measurements are a reminder that each bone has its own biomineralization system, which is independent during formation and must be maintained at a certain level to be a contributing and effective part of the skeleton.

Radiological and densitometry analyses are useful in aiding diagnosis, but treatment for an individual depends on the patient's distinctive metabolic status. These data may inform on whether the osteoporosis is due to lowered amount or to overproduction of certain hormones that might influence the level of circulating calcium or phosphorus, for example. There are many other factors that can impact the several cells unique to the bone tissue system that must be considered in mineral formation and maintenance.

An initial designation of osteoporosis via radiographic survey and density of mineral amount per unit area may be followed with a bone biopsy. A small portion of tissue usually from the iliac crest (hip) is extracted by syringe and prepared for histological examination. Histomorphometry techniques examine texture and quantify tissue components, such as the number of the essential cells, specific hormones, and proteins, and estimate the level of bone formation and resorption. Does the tissue show normal amounts and distribution of mineral? Are there sufficient osteoblasts, the bone forming cells

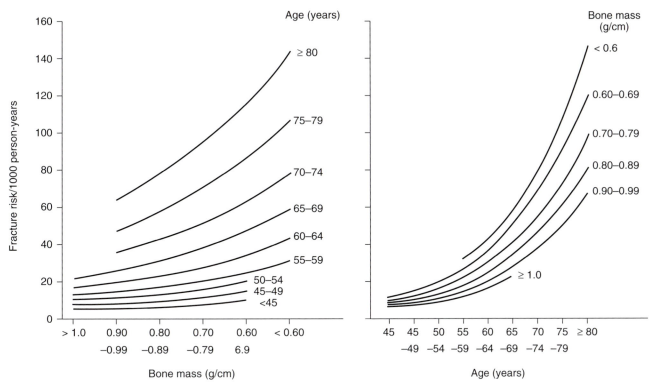

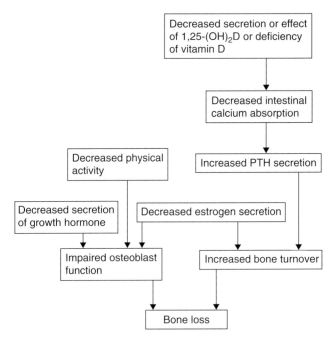

FIGURE 10 Graphs comparing the average (over 1000 person years) of fracture risk versus bone mass and fracture risk versus age. (From Johnston et al., 1996, Figure 1.)

present, or alternatively, bone tissue resorption accelerated with a high concentration of osteoclasts (Frost, 1973)? These two options are the yin and yang of bone remodeling, the dynamic system that underlies the viability of all skeletal organs (Urist, 1965; Mundy, 1999).

The onset of osteoporosis is asymptomatic and prevention is at least partially associated with an adequate diet and the intake, transport, and uptake of calcium, protein, and calories as well as vitamin D. The evaluation of personal choices, such as consumption of calcium-containing foods or specific trace elements via supplements may not be easily assessed for a particular patient. A variety of elements, hormones, enzymes, and proteins are essential in the complicated integration of other organs and cycles of the body system for maintaining a functional set of dynamic skeletal tissues (Figure 11). From the production of Vitamin D, or parathyroid hormone, to the absorption of calcium at the intestines, and recycling of phosphorus by the kidney, research has shown that all are important and must be properly integrated in order to arrive at the best treatment for a particular patient with osteoporosis (Avioli, 2000).

FIGURE 11 Sketch illustrating the proposed contributions to age-related bone loss in women. The interactions of vitamin D, hormones, and activity level affect the uptake of calcium, bone turnover, and cell function, which if not in balance, may cause bone loss. (From Eastell, 1999, Figure 1.)

B. Treatments for Osteoporosis

Over the past 50 years investigations from the cell level to animal model systems have been utilized to assess treatment regimens. Although some therapies may appear promising in animals, they may not translate to effective treatments for humans (Jee, 1995). Much of the research has involved tests using pharmacologic agents that act on several body systems and effect the balance of circulating calcium and phosphorus. This seems a sensible approach applicable to all the bones in the body rather than attempting to construct treatment for specific bones.

The goal of osteoporosis treatments is to assist the body toward normal bone growth and repair and specifically to provide for normal mineral formation and retention. Because the mineral is our focus, but only one component in a complicated and multifaceted system, the following three sections on pharmacologic treatments focus on the composition and relative amounts of mineral to bring out some of the contributions provided by this basic science presentation.

1. Fluoride

The first thrust for making fluorine a therapeutic agent was put forward by dental scientists. Their studies led them to suggest that if the mineral in dental tissues was fluorapatite it would be less susceptible to caries as fluorapatite was the more stable apatite (Vischer, 1970). To increase the formation of fluorapatite would require incrementing the amount of fluorine ingested that would then become part of the bone mineral as the inorganic portion of the tissues should respond to the nutrients supplied. However, it was known that some geographic areas with high fluorine in the waters led to a disease known as fluorosis. The simple hypothesis of more fluorine belies other aspects of the complicated bone tissue system. Fluoride may not only form fluorapatite, it may affect bone cells and the formation of the organic matrix. To obtain the optimum amount of fluorine to benefit human bones requires not only bioapatite production and composition, but also consideration of tissue recycling.

In the presence of small amounts of fluorine there was an increase in mineral matter and therefore the density of bone tissues. One study showed that fluorine was mitogenic for osteoblasts and a clinical investigation produced dramatic sustained (years) increases in bone mineral density when administered in doses of 20–30 mg fluoride daily (Riggs & Melton, 1995). However,

other clinical studies showed no effect and the cortical portion of the long bones decreased in density relative to an increase in trabecular bone density. The disturbance of cortical tissues could be modulated by increasing Ca and Vitamin D intake or by arranging F-free periods during treatment (Watts, 1999). These interesting insights into the use of fluoride supplements as a possible means to increase fluorapatite in bone tissues also showed that only some of the fluorine found its way into bioapatite. When new mineral was formed little if any fluorine was partitioned into already existing bone mineral. The predominant bioapatite phase on bulk analyses remained hydroxylapatite.

Fluoride may be rapidly absorbed from the stomach but 50% will be eliminated via the kidneys within a few hours. Serum fluoride levels between $0.1–0.25\,mg\,L^-$ and doses greater than $70\,mg$ daily ($35\,mg$ fluorine) produce grossly abnormal bone (Riggs & Melton, 1995). In areas of China with naturally high fluorine in the environment, telltale signs of brown spots (fluorosis) on the teeth and bent legs and bodies attest to its interference in the strength and architectural character of normal mineralized tissues (Finkelman et al., 1999). There is a narrow therapeutic window for fluorine. Domestic water levels at $1\,\mu g–L^{-1}$ achieve dental benefits and may also provide some osteoporotic relief, but what is the level and length of treatment, or which cooperative treatments might be paired with fluorine ingestion to ensure no abnormal demineralization, and strong bone?

In shark teeth whose enameloid tissue shows the highest fluorine mineral content for any vertebrate, the uptake unfortunately appears to be related to genetics and phylogeny and not to the fluoride concentration in the aqueous environment (Aoba et al., 1997). The appropriate dose and best schedule for human fluorine ingestion have not been fully determined. The possible impact of fluoride on bone and on other body systems will require much longer and more comprehensive investigations before fluoride can be part of routine treatment of postmenopausal or other types of osteoporosis.

2. Bisphosphonates

Bisphosphonates are a group of compounds, analogues of the pyrophosphates, where the phosphorus atom of the phosphate group connects directly to a carbon atom. The chemical arrangement of several pharmacological bisphosphonates that are used to treat osteoporosis approximates the following general structure:

$$OH \quad R_1 \quad OH$$
$$O = P — C — P = 0$$
$$OH \quad R_2 \quad OH$$

where R1 and R2 represent other chemical substituents (Fleisch, 2000). These phosphate-containing chemicals adsorb onto mineral surfaces and are not readily degraded chemically or enzymatically, an attribute that makes them useful markers of mineral materials. Specific formulations with easily detectable elements in the "R" groups (such as radioactive species in these adducts) have been employed in nuclear medicine experimentation. However, other, non-nuclear-containing bisphosphonates are useful in treatment of osteoporosis because they act locally by preventing the resorption of calcium phosphates by osteoclasts. Some bisphosphonates may inhibit osteoclast activity or cause apoptosis (cell death), but the major effect is to slow down the dissolution of bioapatite and therefore reduce bone remodeling, although the action is usually transient.

Only 5% of an oral dose of a bisphosphonate is absorbed from the gut, and the amount will be lower in the presence of calcium or other divalent ions. The drugs, usually taken in the morning before any food is consumed, will mostly be excreted by the kidney. Because bisphosphonates bind to the mineral they may become buried in the tissue becoming inactive. It is estimated that roughly 25% of the amount bound will be lost from the skeleton after 10 years, an indication of the slow rate of bone tissue turnover.

There are side effects of bisphosphonate ingestion: some tissues may not mineralize properly, and an increased amount of non-mineralized matrix may be the locus for fracture of the bone. Oral administration may also cause some soft tissue side effects but intravenous bisphosphonates have been effective in patients with Paget's disease (Watts, 1999). Current research using animal model systems and high-resolution transmission electron microscopy show that bisphosphonates may increase the width of individual apatitic crystallites in the tissues but not markedly increase the mineralization level. Such results beg additional questions. One might be phrased as follows: In the composite tissues typical of bone where the physical and chemical properties of all the components must be maintained to ensure proper and continued interactions and organ function, is the major effect thus far noted with bisphosphonates, increasing crystallite size, an appropriate and useful contribution for long-term treatment of osteoporosis?

3. Hormones

The most familiar pharmacologic treatment for postmenopausal osteoporosis is to augment the lowered production of estrogens in older women. Although the maximum level of skeletal bone density is affected by nutrition and usually achieved by the age of maturity (25–30 years), estrogen was shown to be important in producing and keeping the calcium levels in the circulation adequate for proper skeletal mineralization. For aging individuals the addition of estrogens, originally prescribed to relieve hot flashes and vaginal dryness, was shown to increase bone mineral density in the spine by 5–10% after age 65. Progesterone, or one of the progestins, was added to the estrogens in some formulations. Long-range clinical studies on the combined pharmaceuticals have shown no bone tissue benefits. Most recently these clinical studies have noted an increase in cardiovascular disease and breast cancer rates. This is most unfortunate and a sad commentary on hormonal replacement therapy that has been followed for many years by hundreds of thousands of women.

Although linked with skeletal health, the mechanisms for a direct action of the hormones on bone tissue remains under study. Early clinical investigations indicated reduced bone loss perhaps through suppression of osteoclastic bone resorption. Estrogens bind at receptors on the nuclei of target cells in both men and women and activate genes that affect the several different pathways required to maintain adequately mineralized bone tissues. For example, the normal production of growth factors and cytokines involved in calcium homeostasis may be altered without appropriate amounts of estrogens and other hormones. Estrogen acts at the kidneys and in the bowel, and via feedback mechanisms may sensitize the remodeling system bone cells to be more receptive to mechanically induced electrical signals (Watts, 1999).

The "normal dose" of 0.625 mg daily of conjugated estrogens, usually equine derived, was lowered to 0.3 mg to prevent bone loss (Watts, 1999). With a half-life of 10–18 hours, estrogen compounds are easily metabolized by several different tissues once absorbed. From a series of investigations it became known that the stable equilibrium for the hormone between estrone, the dominant form, estradiol, the active form, and other conjugated and esterified forms, is rapid. Estrogens circulate by binding to sex hormone-globins or albumin, the dominant circulating protein in blood and in tissues, but only the unbound form is biologically active. Estrogen compounds are excreted in bile, resorbed in the small intestine, and become less active in the process.

The side effects of tenderness, fluid retention, weight gain, occasional generation of deep vein thrombosis, and pulmonary emboli are some of the reasons that discouraged many women from starting or continuing the use of the estrogens post menopause. Raloxifene (Evista) acts to modulate the effect of estrogens on the surface of cells. It is an estrogen-acceptor modulator, and thus far it is the only drug approved that has been shown to prevent bone loss by causing differential expression of estrogen-regulating genes. The actions of estrogen and future use are cloudy at present and their potential contribution to minimizing bone tissue loss needs further study (Chlebowski et al., 2003).

4. Summary

Bisphosphonates, possibly useful for reducing bone mineral resorption; hormones, particularly estrogen, certainly important in calcium-mineral dynamics; and fluoride, thought to increase the stability of the mineral phase; all have shown modest success for retaining or restoring mineral in bones by a variety of clinical investigations. Together with exercise and a diet that includes vitamins and certain trace elements, the battle to understand and relieve if not prevent osteoporosis continues. The lack of consistent and sustained benefits from the proscribed regimens on osteoporosis-affected populations makes it clear that additional experimental protocols are needed and will require lengthy clinical evaluation. Any new techniques or pharmacologic agents in these ongoing research efforts will provide useful data on the mineral, its roles and reactions.

VII. MINERAL AND MINERALIZED TISSUE RESEARCH AND MEDICAL GEOLOGY

In this chapter the several roles that mineral plays in mineralized tissues were discussed. Bioapatite is not only the stiffening agent for our skeleton but it is also constantly recycled in bone tissues. It acts as a filter recording the many different chemicals ingested in our diet, some essential nutrients, and others perhaps hazardous to our health. Mineralized tissues play essential roles in growth, development, and maintenance of all bodily functions, organs, and tissues.

The fact that all naturally occurring elements and man-made chemicals are distributed throughout the environment where food and drink are obtained has caused us to ask whether they are impacting our health. The mineral composition of bone is suggested as a signal to identify potential environmental problems. By exploring in detail the special physical and chemical characteristics of bioapatites, we increase our understanding of how the human body and its tissues react. Although the addition of fluorine to domestic waters may benefit human dental health, it can also be one of the pharmacologic agents used in the treatment of osteoporotic patients. The importance of minerals to the complicated and interactive metabolism of human systems provides an illustration of the information and sophisticated analyses from a host of scientific disciplines that forge links between geological/mineralogical sciences and anthropological/health activities. Maximum integration between these disciplines is sought as a benefit for personal health and for populations at risk and to allow the contemplation and eventually practice of preventative medicine.

SEE ALSO THE FOLLOWING CHAPTERS

Chapter 5 (Uptake of Elements from a Biological Point of View) · Chapter 29 (Inorganic and Organic Geochemistry Techniques)

FURTHER READING

Albright, J. A., and Skinner, H. C. W. (1987). Bone: Structural Organization and Remodeling Dynamics. In *The Scientific Basis of Orthopaedics* (J. A. Albright and R. Brand, Eds.), Appleton & Lange, Norwalk, CT, pp. 161–198.

Aoba, T., Komatsu, H., Shimazu, Y., Yagishita, H., and Taya, Y. (1998). Enamel Mineralization and an Initial Crystalline Phase, *Connect. Tissue Res.*, 38, 129–134.

Arnold, J. S., Bartley, M. H., Tont, S. A., and Jenkins, D. P. (1966). Skeletal Changes in Aging and Disease, *Clin. Orthopaed.*, 49, 17–34.

Avioli, L. V. (Ed.) (2000). *The Osteoporotic Syndrome: Detection, Prevention and Treatment*, 4th edition, Academic Press, San Diego, CA.

Barzel, U. S. (Ed.) (1970). *Osteoporosis*, Grune & Stratton, New York.

Bell, L. C., and Mika, H. (1979). The pH Dependence of the Surface Concentration of Calcium and Phosphorus on

Hydroxylapatite in Aqueous Solutions, *J. Soil Sci.*, 30, 247–258.

Blatt, H., and Tracy, R. J. (Eds.) (1996). *Frontiers in Petrology*, W. H. Freeman, San Francisco, CA.

Bronner, F. (1996). Metals in Bone: Aluminum, Boron, Cadmium, Chromium, Lead, Silicon and Strontium. In *Principles of Bone Biology* (J. P. Bilezikian, L. G. Raisz, and G. A. Rodan, Eds.), Academic Press, San Diego, CA, pp. 295–303.

Brown, W. E., Eidelman, N., and Tomazic, B. (1987). Octacalcium Phosphate as a Precursor in Biomineral Formation, *Adv. Dental Res.*, 1, 306–313.

Carlstrom, D. (1955). X-Ray Crystallographic Studies on Apatites and Calcified Tissues, *Acta Radiolog. Suppl.*, 121, p.

Chlebowshi, R. T., Hendrix, S. L., Langer, R. D. et al. (2003). Influence of Estrogen Plus Progestin on Breast Cancer and Mammography in Healthy Postmenopausal Women. The Women's Health Randomized Trial, *JAMA*, 289, 3243–3253.

Chestnut, C. H., III, Rosen, C. J., and Bone Quality Discussion Group (2001). Reconsidering the Effects of Antiresorbtive Therapies in Recycline Osteoporotic Fracture, *J. Bone Miner. Res.*, 12, 2163–2172.

Driessens, F. C. M., and Verbeeck, R. M. H. (1990). *Biominerals*, CRC Press, Boca Raton, FL.

Eastell, R. C. (1999). Pathogenesis of Postmenopausal Osteoporosis. In *Primer on the Metabolic Diseases and Disorders of Mineral Metabolism* (M. J. Favus, Ed.), 4th edition, Lippincott, Williams & Wilkins, Philadelphia, PA, pp. 260–262.

Eckelmann, W. R., Kulp, J. L., and Schulert, A. R. (1957). Strontium–90 in Man, *Science*, 125, 219.

Econs, M. J. (2000). (Ed.) *The Genetics of Osteoporosis and Metabolic Bone Disease*, Humana Press, Torowa, New Jersey.

Elliott, J. C. (1984). Infrared and Raman Spectroscopy of Calcified Tissues. In *Methods of Calcified Tissue Preparation* (G. R. Dickson, Ed.), Elsevier, Amsterdam, Netherlands, pp. 413–434.

Elliott, J. C. (1994). *Structure and Chemistry of the Apatites and other Calcium Phosphates*, Elsevier, Amsterdam, Netherlands.

Fabrikant, J. I. (Chairman) (1988). *Health Risks of Radon and other Internally Deposited α Emitters. BER IV*, National Academy Press, Washington DC.

Finkelman, R. B., Belkin, H. E., and Zheng, B. (1999). Health Impacts of Domestic Coal Use in China, *Proc. Natl. Acad. Sci. U. S. A.*, 96, 3427–3431.

Fleisch, H. (Ed.) (2000). *Bisphosphonates in Bone Disease*, Academic Press, San Diego, CA.

Food and Nutrition Board, National Research Council, Recommended Daily Allowances, 10th edition 1989, National Academy Press, Washington, DC, Federal Register 2206.

Fowler, J. M. (1960). *Fallout: A study of Superbombs, Strontium 90 and Survival*, Basic Books, Inc., New York.

Frost, H. M. (1973). *Bone Remodeling and its Relationship to Metabolic Bone Disease, Orthopaedic Lectures Vol. III*, Charles C Thomas, Springfield, IL.

Gaines, R. V., Skinner, H. C. W., Foord, E. E., Mason, B., and Rosensweig, A. (1997). *Dana's New Mineralogy*, John Wiley & Sons, New York.

Glimcher, M. J. (1976). Composition, Structure, and Organization of Bone and Other Mineralized Tissues and the Mechanism of Calcification. In *Handbook of Physiology: Endocrinology. Vol. 7* (R. O. Greep and E. B. Astwood, Eds.), American Physiological Society, Washington DC, pp. 25–116.

Glimcher, M. J. (1998). The Nature of the Mineral Phase in Bone: Biological and Clinical Implications. In *Metabolic Bone Disease and Clinically Related Disorders* (L. V. Avioli and S. M. Crane, Eds.), 3rd edition, San Diego, CA, pp. 23–51.

Halstead, L. B. (1974). *Vertebrate Hard Tissues*, Wykeham Publications, London and Springer-Verlag, New York.

Hughes, J. M., Cameron, M., and Crowley, K. D. (1989). Structural Variations in Natural F, OH and Cl Apatites, *Am. Mineral.*, 74, 870–876.

Hughes, J. M., Cameron, M., and Crowley, K. D. (1991a). Ordering of Divalent Cations in the Apatite Structure: Crystal Structure Refinements of Natural Mn- and Sr-Bearing Apatite, *Am. Mineral.*, 76, 1857–1862.

Hughes, H. M., Cameron, M., and Mariano, A. N. (1991b). Rare Earth Element Ordering and Structural Variations in Natural Rare-Earth Bearing Apatites, *Am. Mineral.*, 96, 1165–1173.

Jee, W. S. S. (Ed.) (1995). Proceedings of the International Conference on Animal Models in the Prevention and Treatment of Osteopenia, *Bone*, 17(Suppl.), 1–466.

Johnston, Jr., A., Conrad, C., Slemenda, C. W., and Melton, III, L. J. (1996). Bone Density Measurement and the Management of Osteoporosis. In *Primer on the Metabolic Bone Diseases and Disorders of Mineral Metabolism* (M. J. Favus, Ed.), American Association of Bone and Mineral Research, Lippincott Raven, Philadelphia, PA, pp. 142–151.

Kim, H.-M., Rey, C., and Glimcher, M. J. (1995). Isolation of Calcium Phosphate Crystals of Bone by Non-Aqueous Methods at Low Temperature, *J. Bone Miner. Res.*, 10, 1589–1601.

Klein, C., and Hurlbut, C. S. (1985). *Manual of Mineralogy*, 20th edition, John Wiley & Sons, New York.

Klug, H. W., and Alexander, W. L. (1954). *X-Ray Diffraction Procedures*, John Wiley & Sons, New York.

Labarthe, J.-C., Bonel, G., and Montel, G. (1973). Sur la structure et les proprietes des apatites carbonates de type B phospho-calcique, *Ann. Chim. (Paris)*, 8, 289–301.

LeGeros, R. Z., Trautz, O. R., LeGeron, J. P., and Klein, E. (1968). Carbonate Substitution in the Apatite Structure, *Bull. Soc. Chim. France* (Special No.), 1712–1718.

Malluche, H. H., Sherman, D., Meyer, R., Massry, S. G. (1982). A New Semi-Automatic Method for Quantitative State and Dynamic Bone-Histology, *Calcif. Tissue Int.*, 34, 439–448.

Mann, S. (2000). *Biomineralization Principles and Concepts in Bioinorganic Materials Chemistry*, Oxford University Press, Oxford, England.

Marcus, R., Felman, U., and Kelsey, J. (1996). *Osteoporosis*, Academic Press, New York.

McConnell, D. (1973). *Apatite*, Springer-Verlag, New York.

Mielke, H. (2003). Anthropogenic Distribution of Lead. In *Geology and Health: Closing the Gap* (H. C. W. Skinner and A. N. Berger, Eds.), Oxford University Press, New York, pp. 119–124.

Miller, E. J. (1973). A Review of Biochemical Studies on the Genetically Distinct Collagens of the Skeletal System, *Clin. Orthopaed.*, 92, 260–280.

Miles, A. E. W. (Ed.) (1967). *Structural and Chemical Organization of Teeth*, Academic Press, New York.

Mundy, G. R. (1999). Bone Remodeling. In *Primer on Metabolic Bone Diseases and Disorders of Mineral Metabolism* (M. J. Favus), 4th edition, American Society for Bone and Mineral Research, Lippincott Williams & Wilkins, Philadelphia, PA, pp. 30–38.

Nesse, W. D. (2004). *Introduction to Optical Mineralogy*, 3rd edition, Oxford University Press, New York.

Neuman, W. F., and Neuman, M. W. (1958). *Chemical Dynamics of Bone Mineral*, University of Chicago Press, Chicago, IL.

Poggi, S. H., Skinner, H. C. W., Ague, J. J., and Carter, D. (1998). Using Scanning Electron Microscopy to Study Mineral Deposits in Breast Tissues, *Am. Mineral.*, 83, 1122–1126.

Posner, A. (1977). The Relation of Synthetic Amorphous Calcium Phosphate to Bone Mineral Structure, *Orthopaed. Trans.*, 1, 78–95.

Reed, S. J. B. (1993). *Electron Microprobe Analysis*, Cambridge University Press, Cambridge, England.

Rey, C., Kim, H. M., Gerstenfeld, L., and Glimcher, M. J. (1996). Characterization of the Apatite Crystals of Bone and Their Maturation in Osteoblast Cell Culture: Comparison with Native Bone Crystals, *Connect. Tissue Res.*, 35, 343–349.

Riggs, B. L., and Melton, L. J., III (Eds.) (1995). *Osteoporosis: Etiology, Diagnosis and Management*. Raven Press, New York.

Roberts, J. E., Bonar, L. C., Griffin, R. G., and Glimcher, M. J. (1992). Characterization of Very Young Mineral Phases of Bone by Solid State ^{31}phosphorus Magic Angle Sample Spinning Nuclear Magnetic Resonance and X-Ray Diffraction, *Calcif. Tissue Int.*, 50, 422–448.

Shannnon, R. D. (1976). Revised Effective Ionic Radii and Systematic Studies of Interatomic Distances in Halides and Chalcogenides, *Acta Crystallogr. A*, 32, 751–767.

Shannon, R. D., and Prewitt, C. T. (1969). Effective Ionic Radii in Oxides and Fluorides, *Acta Crystallogr. B*, 25, 925–946.

Skinner, H. C. W. (1968). X-Ray Diffraction Analysis Techniques to Monitor Composition Fluctuations Within the Mineral Group Apatite, *Appl. Spectrosc.*, 22, 412–414.

Skinner, H. C. W. (1973a). Phase Relations in the CaO-P_2O_5-H_2O System From 300–600°C at 2 kb H_2O Pressure, *Am. J. Sci.*, 273, 545–560.

Skinner, H. C. W. (1973b). Studies in the Basic Mineralizing System CaO-P_2O_5-H_2O, *Calcif. Tissue Res.*, 14, 3–14.

Skinner, H. C. W. (1987). Bone: Mineral and Mineralization. In *The Scientific Basis of Orthopaedics* (J. A. Albright and R. Brand, Eds.), 2nd edition, Appleton & Lange, Norwalk, CT, pp. 199–211.

Skinner, H. C. W. (1989). Low Temperature Carbonate Phosphate Materials or the Carbonate Apatite Problem. A review. In *Origin, Evolution and Modern Aspects of Biomineralization in Plants and Animals* (R. Crick, Ed.), Plenum Press, New York, pp. 251–264.

Skinner, H. C. W. (2000a). Minerals and Human Health. In *Environmental Mineralogy. European Mineralogical Union Notes in Mineralogy* (D. J. Vaughan and R. A. Wogelius, Eds.), Vol. 2, Eotvos University Press, Budapest, Hungary, pp. 383–412.

Skinner, H. C. W. (2000b). In Praise of Phosphates or Why Vertebrates Chose Apatite to Mineralize Their Skeletons, *Int. Geol. Rev.*, 42, 232–240.

Skinner, H. C. W., Kemper, E., and Pak, C. Y. C. (1972). Preparation of the Mineral Phase of Bone Using Ethylene Diamine Extraction, *Calcif. Tissue Res.*, 10, 257–268.

Skinner, H. C. W., and Nalbandian, J. (1975). Tetracyclines and Mineralized Tissues: Review and Perspectives, *Yale J. Biol. Med.*, 48, 377–397.

Skinner, H. C. W., Nicolescu, S., and Raub, T. D. (2004). A Tale of Two Apatites. In *Environment & Progress 2* (I. Petrescu, Ed.), Editura Fundatiel de Studii Europene, Cluj-Napoca, Romania, pp. 283–288.

Suich, P. R., LaCout, J. L. Hewat, A., and Young, R. A. (1985). The Structural Position and Type of Mn^{2+} Partially Substituted for Ca^{2+} in Fluorapatite, *Acta Crystallogr. B*, 41, 173–179.

Urist, M. R. (1965). Bone: Formation by Autoinduction, *Science*, 150, 893–899.

Van Waser, J. R. (1958). *Phosphorus and Its Compounds*, Interscience Publishers, John Wiley & Sons, New York.

Vischer, T. L. (1970). *Fluoride in Medicine*, Han Huber Publications, Bern, Switzerland.

Wasnich, R. D. (1996). Epidemiology of Osteoporosis. In *Primer on the Metabolic Bone Diseases and Disorders of Mineral Metabolism* (M. J. Favus, Ed.), 3rd edition, Lippincott Williams & Wilkins, Philadelphia, PA, pp. 249–252.

Watts, N. B. (1999). Pharmacology of Agents to Treat Osteoporosis. In *Primer on the Metabolic Bone Diseases and Disorders of Mineral Metabolism* (M. J. Favus, Ed.), 4th Edition, Lippincott Williams & Wilkins, Philadelphia, PA, pp. 278–283

Weiner, S., and Wagner, H. D. (1998). The Material Bone: Structure-Mechanical Function, *Ann. Rev. Mater. Res.*, 28, 271–298.

Zipkin, I. (1970). Inorganic Composition of bone. In *Biological Calcification: Cellular and Molecular Aspects* (H. Schraer, Ed.), Appleton-Century-Crofts, New York, pp. 69–104.

INORGANIC AND ORGANIC GEOCHEMISTRY TECHNIQUES

MITKO VUTCHKOV AND GERALD LALOR
University of the West Indies

STEPHEN MACKO
University of Virginia

CONTENTS

I. INTRODUCTION

Inorganic and organic geochemistry are applications of analytical chemistry to solve Earth science problems by analyzing the composition of geological and biological systems. Measurements and characterization of unknown substances are used for making policy decisions, exploring mineral resources, pollution prevention, and management of environmental hazards. Geochemical exploration methods use trace element geochemistry to identify the "fingerprints" of certain minerals and rocks and to discover new deposits. Environmental geochemistry has become an essential tool for determining where man-made or natural distributions of elements can become a potential environmental health hazard. Organic geochemistry adds a valuable contribution to determining the nature, origin, and distribution of organic constituents in the environment.

This chapter brings together descriptions of the most common inorganic and organic geochemistry techniques. The section below provides an introduction to the analytical geochemistry techniques, reference sources on the subject, and glossary of the technical terms and acronyms. The analytical techniques are described in a standardized format to facilitate the comparison between them and include a brief introduction, principle of operation, what each one does, and how it is used. Typical application examples are also provided.

The most common inorganic and organic geochemistry techniques comprise methods for analysis of major, minor, and trace elements; age dating; stable isotopes; etc. The data derived from these methods are used for interpretation of the global and local element cycles as well as the distribution of elements in the environment. Figure 1 compares the most common analytical techniques used for analysis of major, minor, and trace concentrations of the elements in the Earth's crust.

Inorganic geochemistry techniques for solids and liquids include a suite of analytical methods for measuring chemical parameters of materials that can be fundamentally divided into spectroscopic techniques and non-spectroscopic or classical, wet chemistry methods.

Spectroscopic techniques use the interaction of electro-magnetic (EM) radiation with a sample to perform an analysis, while the classical methods utilize physical means of detecting analytes, such as mass, volume, density, color, refraction, conductivity or electric charge. Figure 2 illustrates the different regions of the EM spectrum based on the origin of photons, i.e., radio waves, microwaves, infrared, visible, ultraviolet (UV), X-rays, gamma, and cosmic rays, and the associated analytical techniques.

The EM radiation can be described in terms of a stream of photons traveling in a wave-like pattern with the speed of light. The photons can be expressed in terms of energy (E), wavelength (λ), or frequency (ν) that are mathematically related by the fundamental equation of Max Planck

$$E = h\nu = hc/\lambda \qquad (1)$$

where, h = Plank's constant ($4.136 - 10^{-15}$ eV sec) and c is the velocity of light ($3 - 10^{8}$ m/sec).

The physical principles and mathematical description of radiation across the EM spectrum is the same, but due to the different energy levels, different mechanisms of interaction with matter occur. Low-energy photons behave more like waves, while higher energy photons such as X-rays and gamma rays behave more like particles, i.e., they exhibit wave-like particle duality. This is an important factor used in designing detectors and instruments for spectroscopy measurements.

The energy of the photons in various regions of the EM spectrum corresponds to different types of transitions in the atoms and molecules, which can be detected and measured using specific spectroscopy techniques. For example, microwaves and infrared spectroscopy methods are associated with molecular rotations and vibrations; UV, visible, and atomic absorption/fluorescence spectroscopy are based on outer electron transitions, whereas X-ray fluorescence (XRF) spectroscopy is related to inner electrons of the atom and gamma rays to nuclear transitions.

A variety of analytical instruments that are commonly used in inorganic and organic biogeochemistry are based on spectroscopy, e.g., UV-visible, fluorescence, atomic, infrared, X-ray, and nuclear spectroscopy. The interpretation of the spectra thus produced can be used for chemical analysis and examining atomic energy levels and molecular structures. Qualitative and quantitative analysis of the elemental composition using these

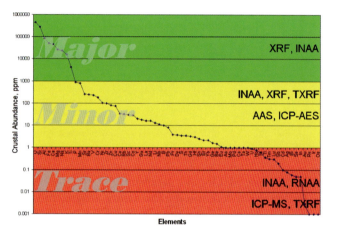

FIGURE 1 Analytical methods selection chart for major, minor, and trace elements concentration in the Earth's crust. XRF: X-ray fluorescence spectrometry, INAA: instrumental neutron activation analysis, TXRF: total reflection XRF, AAS: atomic absorption spectroscopy, ICP-AES: inductively coupled plasma atomic emission spectroscopy, RNAA: radiochemical neutron activation analysis, and ICP-MS: inductively coupled plasma mass spectrometry.

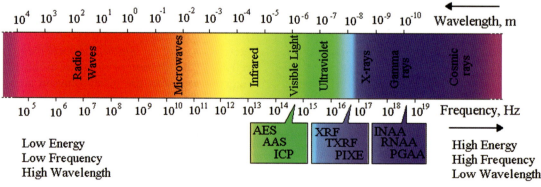

FIGURE 2 EM spectrum and related spectroscopy techniques.

techniques is based on absorption, emission, or scattering of electromagnetic radiation by matter, e.g., atoms, molecules, atomic or molecular ions, or solids. The analysis of concentrations of elements relies on the relationship between the measured EM radiation and concentration of the analyte. The unknown concentrations are determined from calibration curves using appropriate standard reference materials (SRM).

The various analytical techniques require analytical samples to be either in a solid or liquid form. Traditional geochemical analysis essentially involves getting a sample into solution and then using an appropriate method to measure the elemental concentration in the solution. These methods are generally referred to as destructive methods (sample dissolution) and include the classical wet chemistry methods: Atomic spectroscopy (AS), inductively coupled plasma (ICP) spectroscopy, and some other methods such as the radiochemical neutron activation analysis (RNAA). Nuclear and related analytical techniques depend on the physical properties of the atomic nucleus and do not need preparation of solutions, i.e., they are non-destructive. The most common non-destructive techniques include instrumental neutron activation analysis (INAA) and XRF spectrometry.

For detailed information on chemical analysis in pathology see Chapter 30, this volume.

II. INORGANIC GEOCHEMISTRY TECHNIQUES FOR SOLIDS

A. Neutron Activation Analysis

NAA is a nuclear analytical technique used for non-destructive multi-element analysis of solids, liquids, or gases. Theoretically, about 70% of naturally occurring elements can be analyzed by NAA at concentration levels down to $1\,\mu$/kg. Analysis can be performed instrumentally without chemical pretreatment of the sample, thus avoiding problems related to incomplete dissolution or loss of volatile elements. Since neutrons interact directly with the atomic nuclei, the NAA measures the "total" amount of an element, regardless of oxidation state or chemical form.

The neutron activation method was discovered in 1936 by Hevesy and Levi, and commercialized in the mid-1950s and 1960s. In the last three decades, NAA has been widely applied to mineral exploration, envi-

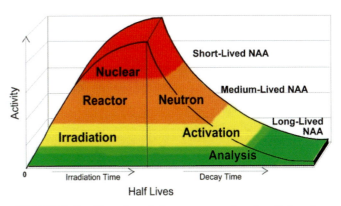

FIGURE 3 Growth and decay curves of radioactivity of a natural sample exposed to neutrons.

ronmental biogeochemistry, and health-related studies. NAA is well recognized as a referee method of choice in the certification of new reference materials or quality control trials.

1. How It Works

a. Principle of NAA

Neutron activation is a physical method for analysis of elemental composition that uses neutron irradiation of a sample to convert the elements into radioactive isotopes. The radioactive elements can then be detected and quantified by gamma ray spectroscopy. The basic principle of NAA, shown in Figure 3, illustrates how the radioactivity of atomic species grows and decays after exposure to neutrons.

The induced activity during the irradiation follows an exponential growth curve and depends on the half-life $(t_{1/2})$ of the particular nuclide. Therefore, the isotopes with longer half-lives $(t_{1/2})$ will need longer irradiation times compared to those of short-lived species. The induced activity of the sample will decrease with time according to the disintegration rates of isotopes present in the sample. The short-lived isotopes, used in NAA, have half-lives of less than an hour; medium-lived isotopes range from an hour to several days; and long-lived isotopes range from several weeks to 2–3 months.

The most common type of nuclear reaction used for NAA is the neutron capture or (n, gamma) reaction. When a naturally occurring stable isotope of an element (target nucleus) absorbs a neutron, it is transformed into higher mass unstable nucleus as shown in Figure 4.

The excited nucleus instantaneously decays through emission of prompt gamma rays or, in most cases, is converted to a radioactive nucleus, which decays primarily by emission of beta particles and/or gamma rays.

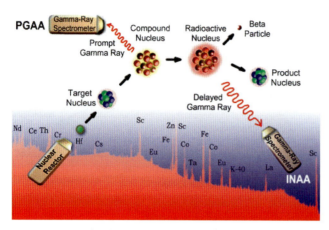

FIGURE 4 A schematic diagram illustrating the principle of neutron activation.

TABLE I. Classification of Most Common NAA Methods

	Neutron activation analysis			
Gamma ray spectroscopy				*Neutron counting*
Prompt gamma	*Delayed gamma*			*Delayed neutron*
PGAA	INAA & RNAA	ENAA	FNAA	DNC

The fundamental equation for activation, decay, and counting of a radionuclide with activity A and mass M (g) can be described as follows (Soete et al., 1972):

$$A = \frac{N\theta\sigma\gamma\rho W\phi\varepsilon}{M\lambda}\left(1 - e^{-\lambda t_i}\right)\left(e^{-\lambda t_d}\right)\left(1 - e^{-\lambda t_c}\right) \quad (2)$$

where: t_i, t_d and t_c = irradiation, cooling, and counting times, sec; φ = thermal neutron flux, neutrons/cm²/sec, N = Avogadro's number, $6.023 - 10^{23}$ atoms/g atom; θ = abundance of the activated nuclide; σ = absorption cross-section of the irradiated species, cm²; γ = gamma emission probability; ρ = concentration of analyte, µg/kg; W = sample weight, g; Φ = effective thermal neutron flux, neutrons/cm²/s; ε = efficiency of the counting system; M = molecular mass of the target element; λ = decay constant of the radioisotope produced, sec⁻¹.

The basic instrumentation necessary to carry out NAA includes a source of neutrons, gamma ray spectrometer, and data processing software. Nuclear reactors are most commonly used for NAA, providing high and well thermalized neutron fluxes. The neutron energy spectrum consists of three principal components sorted in increasing energy order: thermal, epithermal, and fast neutrons. The thermal neutrons comprise about 90–95% of the total neutron flux and induce the main (n, gamma) reactions of target nuclei used in conventional NAA. The epithermal and fast neutrons induce nuclear reactions, which release gamma rays or nuclear particles and the technique associated with them is called epithermal neutron activation analysis (ENAA) and fast neutron activation analysis (FNAA), respectively.

In principle, with respect to the underlying nuclear reactions and their measurement, neutron activation analysis methods are classified into three categories (Table I): (1) prompt gamma NAA (PGAA), (2) delayed gamma NAA (DGNAA), and (3) delayed neutron counting (DNC).

Neutron-induced PGAA uses prompt gamma rays released from excited nuclei during sample irradiation (Figure 4) and is manly applicable to elements that do not form radioactive products after irradiation (e.g., hydrogen and boron) or elements whose half-life is too short or long to be measured by NAA. DGNAA technique measures the delayed gamma rays obtained during the radioactive decay of nucleus and includes the conventional NAA, ENAA, and FNAA techniques. Depending on the treatment of the samples, the NAA technique is generally referred to as INAA when analysis is done without chemical processing and as RNAA when post-irradiated radiochemical separation of analyte or interfering nuclide is applied. DNC is used for quick determination of uranium and other fissionable radionuclides with improved detection limits.

b. Irradiation Facilities

The irradiation of samples is typically performed in nuclear reactors with neutron fluxes ranging from 10^{11} n.cm⁻².sec⁻¹ to 10^{16} n.cm⁻².sec⁻¹. The low power nuclear reactors such as TRIGA (training research isotopes general atomics) and Slowpoke (safe low power kritical experiment) with typical neutron fluxes 10^{13}–10^{12} n.cm⁻².sec⁻¹ are most commonly used in NAA. Figure 5a and b show the Slowpoke-2 reactor facility in Jamaica, which is used for NAA (Lalor, 1995).

For neutron activation analysis samples are irradiated in or near the core of the nuclear reactor for an appropriate period of time, from seconds (short-irradiations) to several weeks (long-irradiations). This period of time

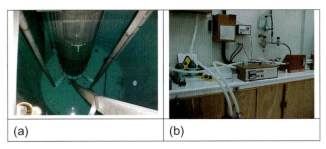

(a)   (b)

FIGURE 5 Neutron activation analysis using Slowpoke-2 nuclear reactor in Jamaica: (a) vertical cross-section view of the reactor core-reflector configuration and (b) irradiation-transfer system for NAA.

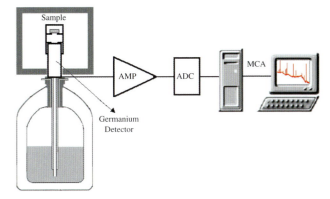

FIGURE 6 Schematic representation of a gamma ray counting system.

depends on the half-lives ($t_{1/2}$) of the elements of interest. The irradiation sites designed for short-lived radionuclides are equipped with computer-controlled pneumatic transfer systems for quick transport and measurement of the samples. Activation of long-lived nuclides usually require much higher flux levels and longer irradiation times. Samples are loaded/unloaded automatically or manually depending on the application and irradiation facility. Neutron irradiation for PGAA is limited to reactors with external neutron beam facilities designed to perform simultaneous irradiation and counting of the samples.

c. Gamma Ray and Neutron Counting

Simultaneous neutron irradiation of natural materials generates numerous isotopes of elements that decay with different half-lives. Elements of interest are usually measured in several steps with different cooling times to allow the interfering elements to decay. The typical gamma ray counting equipment comprises a high-resolution germanium detector (<2 keV for ^{60}Co 1.33 MeV line), pulse processing electronics including an amplifier, an analog-to-digital converter (ADC), and a multichannel analyzer (MCA) interfaced to a desktop computer (Figure 6).

Measurement of prompt gamma rays is performed using a high-purity germanium detector placed nearby to the neutron beam port. The geometry arrangement and shielding of the prompt gamma ray counting system is designed to minimize the background due to neutrons and gamma rays.

The delayed neutrons are detected with boron trifluoride (BF_3) proportional detectors surrounded by double cadmium sheets filled with paraffin wax. The paraffin wax is used to thermalize the fast neutrons while cadmium absorbs the thermal neutrons, thus ensuring low-neutron background around the BF_3 detectors.

TABLE II. Basic Steps for Neutron Activation Analysis

Step	INAA	RNAA	DNC	PGAA
1. Sample preparation	Yes	Yes	Yes	Yes
2. Irradiation	Yes	Yes	Yes	Yes
3. Separation	No	Yes	No	No
4. Cooling periods	Multiple	Multiple	Single	No
5. Measurement(s)	Multiple	Multiple	Single	Single
6. Quantitative analysis	Yes	Yes	Yes	Yes

d. Data Processing

Data processing software for neutron activation analysis includes programs for gamma ray spectrum analysis and qualitative and quantitative analysis. Spectrum analysis programs perform identification of nuclides (qualitative analysis), background subtraction, and calculation of the net peak areas of elements. The quantitative analysis software employs the net peak areas to calculate the elemental concentrations using standards.

2. What It Does

The principal objective of the neutron activation analysis, qualitative and quantitative multi-element analysis, can be completed in several basic steps, which are summarized in Table II.

a. Sample Preparation

Solid samples are analyzed by NAA as received without any chemical pretreatment, which avoids losses of volatile elements (e.g., arsenic, selenium, cadmium, and mercury) and contamination of the sample. Small volumes of liquid samples can be analyzed directly provided that their boiling point is greater than 60°C, or alternatively they are can be evaporated to dryness before analysis. The common sample preparation procedure for NAA involves weighing about 100- to −500-mg samples into heat-sealed polyethylene or quartz vials. As a rule, the vials should not be filled more than three quarters of the volume to prevent pressure buildup. The encapsulated samples are packed together with standards and a flux monitor into larger vials used for irradiation. All sample containers should not be handled with bare hands to avoid contamination of the vial surface with sodium and chlorine.

b. Irradiation of Samples

The "unknown" samples are activated in the nuclear reactor for different irradiation times and neutron flux levels taking into consideration nuclear properties of the element of interest and the expected concentration of elements. As a rule, the irradiation time should be several times greater than the half-life of the short-lived radionuclides and as long as practically possible for the long-lived isotopes. There are two main irradiation schemes employed in NAA, short- and long-irradiations. Short-irradiations are used for NAA of isotopes with half-lives less than one day, whereas long-irradiations are applied for medium- and long-lived nuclides. The overall activation time varies from seconds to several days and weeks. Table III lists the typical irradiation, decay, and counting times employed for activation of the very-short, short-, medium-, and long-lived nuclides using the Slowpoke-2 reactor (Lalor et al., 2000; Vutchkov et al., 2000).

As shown, the complete analysis of the elements listed in Table III can be performed after two irradiations, for short- and long-lived elements, respectively.

c. Radiochemical Separation

Radiochemical NAA involves isolation of one or a group of elements from the irradiated sample before measurement by gamma ray spectroscopy. Chemical separation is usually done after irradiation of the sample in order to remove interfering elements and/or to concentrate the analyte isotope(s). This procedure can improve the detection limits of elements by several orders of magnitude. RNAA requires special facilities and well-trained personnel in handling open radioactive materials. For this reason it is mainly used for certification of new reference materials, analysis of human tissue samples, and other special projects.

d. Cooling

Samples removed from the reactor are frequently highly radioactive due to the activation of elements with high-neutron cross-sections and/or high concentrations. The irradiated samples are stored in shielded lead containers for a specific period of time referred to as "cooling," before gamma ray spectrometry measurement. The cooling period allows the short-lived radioactive species that might interfere with the analyzed element to decay and to therefore reduce the overall radioactivity of the sample. If the half-life of the analyte is much longer than that of the interfering nuclide, after sufficient cooling time the interference can be completely eliminated. Because the production and decay rates of gamma ray radiation are dependent on the half-life of the nuclide, measurement of the elements of interest can be optimized by varying irradiation and cooling times. As shown in Table III, the sequential cooling and counting after several periods of time can optimize the determination of about 30–50 elements in various materials. Depending on the decay time of the elements of interest, the complete analysis report can be obtained from 10 minutes to 30 days after irradiation.

e. Measurement

In most cases, measurement of delayed gamma ray spectra of irradiated samples is performed repeatedly after different cooling times as indicated in Table III. The counting time of different groups of elements depends on their half-lives and the overall activity of the sample. Typical gamma ray spectra of Jamaican soil sample measured for very-short, short-, medium-, and long-lived elements are shown in Figure 7.

As seen from Figure 7a and b, elements such as aluminum, magnesium, vanadium, manganese, cooper, and dysprosium can be determined in two measurements after 5 and 30 minutes of cooling. Medium-lived nuclides of sodium, potassium, bromine, arsenic, cadmium, gold, and tungsten (Figure 7c) are counted after 2–5 days cooling time. Very-long lived elements such as chromium, iron, barium, cesium, rubidium, selenium, scandium, rare earth elements, etc. (Figure 7d) are typically counted after 3–4 weeks decay time.

Measurement of prompt gamma ray spectra is performed during irradiation of the sample with neutrons. In prompt gamma activation the multi-element analysis is performed in one measurement, but the counting times may range from minutes to several hours per sample. The detection limits of PGNAA can be improved by increasing the sample mass or by using longer measurements and irradiation times.

TABLE III. Typical Irradiation-Decay-Counting Scheme for Multi-Element NAA of Geological and Biological Materials Using Slowpoke-2 Reactor in Jamaica

Nuclides	Analyzed elements			
	Very short	Short	Medium	Long
Product half-life	<12 min	12–1440 min	1–3 days	>3 days
Neutron flux (n.cm^{-2}.sec^{-1})	2×10^{11}	5×10^{11}	10×10^{11}	10×10^{11}
Irradiation time	5 min	5 min	4 hours	4 hours
Cooling time	3–6 min	15 min	2–5 days	20–30 days
Counting time	5 min	30 min	1 hour	3 hour
Elements	Magnesium, calcium, aluminum, cooper, titanium, and vanadium	Chlorine, potassium, sodium, iodine, dysprosium, and manganese	Arsenic, bromine, sodium, cadmium, gold, lanthanum, samarium, etc.	Selenium, scandium, chromium, cesium, iron, europium, zinc, uranium, thorium, hafnium, tantalum, cerium, barium, etc.

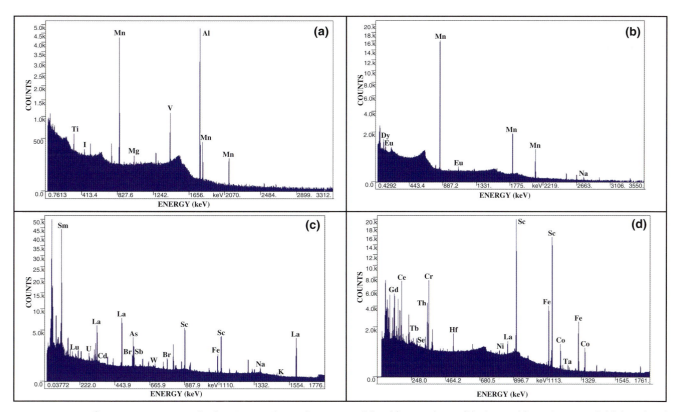

FIGURE 7 Gamma ray spectra of a Jamaican soil sample measured for (a) very-short, (b) short-, (c) medium-, and (d) long-lived elements.

Counting of delayed neutrons is performed after a decay time of 10–20 seconds for about 1 minute. The DNC method using a pneumatic transfer system can be fully automated and has the highest turnaround time among the NAA methods.

f. Qualitative Analysis

Qualitative neutron activation analysis involves processing of the gamma ray spectrum and identification of activated nuclides. Spectrum analysis normally performs smoothing of spectra, background subtraction, peak searching, determination of energies of the photo peaks, multiplet deconvolution, and net peak area determination. The qualitative NAA is accomplished by library matching of the energies of identified nuclides in the sample.

g. Quantitative Analysis

Quantitative NAA is generally accomplished by using the relative comparator method. The most common quantitative analysis approach is the comparator method in which the unknown sample (unk) is co-irradiated with comparator standards (std) and measured under identical counting conditions. The main advantage of this procedure is that it eliminates the nuclear and instrumental parameters and simplifies Eq. 2 as follows:

$$C_{unk} = \left(C_{st} \frac{W_{std}}{W_{unk}} \right) \left(\frac{A_{unk} e^{-\lambda t_{std}}}{A_{std} e^{-\lambda t_{unk}}} \right) \qquad (3)$$

Calculations of elemental concentrations include corrections for decays of unknown and standards, sample weights, and other parameters as indicated in Eq. 3. The comparator single and multi-element standards are prepared from certified solutions of a known concentration of the elements of interest. Each sample analysis batch also includes a certified standard reference material similar to the matrix of the analyzed samples such as soil, sediments, rock, coal, plant tissue, etc. In multi-element analysis mode the comparator method is a time-consuming procedure and requires the measurement of a large amount of standards.

An alternative quantitative analysis approach gaining large popularity is the so-called k0-method, which eliminates the need for co-irradiation of comparator standards. By measuring a single comparator (e.g., gold, zinc, zirconium, etc.) and the efficiency of the counting system, quantitative analysis can be performed using a library with k0-factors of elements. The k0-method has a great advantage in multi-element analysis of samples with various origin.

h. Accuracy and Detection Limit

The accuracy of NAA depends on several factors such as the counting statistics and geometry, gamma ray self-absorption, neutron self-shielding, spectral and nuclear interferences, etc. The principal error in the analysis is the counting statistics of the total and net peak area determination. In most cases, this error is better than 5–10%. The errors due to sample preparation in NAA are usually negligible because the samples do not require any special treatment. Errors due to neutron self-shielding and gamma ray self-absorption can be neglected when using small sample masses. Spectral interferences are negligible because most spectrum analysis programs perform accurate peak deconvolution and calculation of net peak area. Nuclear interferences are important for samples with elevated uranium concentrations and include corrections for the ^{235}U fission products while analyzing lanthanum, cerium, neodymium, molybdenum, and zirconium. The accuracy of quantitative NAA of trace metals is generally less than 5–10% for most elements, depending on the concentration of the element of interest and the sample matrix.

The overall detection limits for most elements range from less than one microgram per kilogram to several hundreds of milligrams per kilogram. The lowest detection limits are observed for gold, iridium, and samarium, which are important to the exploration geology. The second group of elements includes a number of potentially hazardous substances (e.g., arsenic, antimony), the radioactive elements uranium and thorium, and some of the rare earths. The third and fourth groups include the majority of the essential and toxic elements and some precious metals. The elements in the fifth group—calcium, potassium, and magnesium—exhibit poor detection limits, but due to their usually high concentrations in geological materials, these elements are readily quantified. Some elements such as silicon, tin, and sulfur have very poor detection limits and they are usually analyzed using other techniques.

The detection limits of elements in biological materials are much lower than those obtained for soils (Lalor et al., 2000) because the matrix of biological materials consists mainly of elements that are not activated with the conventional analytical schemes (e.g., carbon, nitrogen, oxygen, and sulfur).

3. Summary and Conclusions

The neuron activation analysis method is currently well recognized as a referee method because of the traceability of analytical results. The uncertainty in NAA can

TABLE IV. Advantages and Disadvantages of NAA

Advantages	Disadvantages
• Multi-element analysis of >50 elements	• Long turnaround times for multi-element NAA
• Nondestructive, no need for sample dissolution	• Some common elements, e.g., lead, sulfur, and silicon cannot be determined by NAA or have poor detection limits
• Analysis of solids, liquids, and gases	
• Analysis of "total" concentrations	
• Low detection limits for most elements	• Analyzed samples should be treated as a radioactive waste
• Dynamic concentration ranges, from µg/kg to 100%	

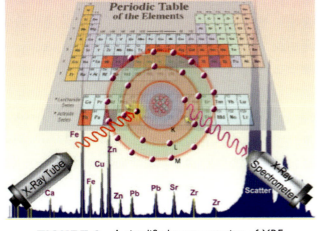

FIGURE 8 A simplified representation of XRF.

be fully assessed because the analytical procedures including activation, decay, and measurement are mathematically well described and defined. The main advantages and disadvantages of the neutron activation methods are summarized in Table IV.

B. X-Ray Fluorescence Spectrometry

X-ray fluorescence is a spectroscopic method for multi-element analysis of solids, powders, and liquids using the X-ray fluorescence radiation of elements present in a sample. X-ray spectroscopy dates back to 1913, when Moseley photographed the X-ray spectrum of several elements and demonstrated the relationship between the atomic number of elements and the wavelength of their emission lines. This relationship not only underlined the foundation of the quantitative XRF analysis, but was also useful in the discovery of new elements in the periodic table, such as hafnium (Bertin, 1975). Bragg constructed the first X-ray analytical device and received the Nobel Prize in physics in 1915. The XRF technique did not become widespread until the 1950s when Friedman and Birks built a prototype of the modern wavelength-dispersive XRF (WDXRF) spectrometer with multi-element capabilities and the sensitivity to analyze various materials. Twenty years later, the advances in solid-state detectors made possible the construction of the energy-dispersive XRF (EDXRF) spectrometers. In the early 1980s an ultratrace XRF analyzer was commercially introduced using the principle of the total reflection XRF (TXRF). Today, the lab-

oratory XRF spectrometers can analyze all elements with atomic number Z = 4 (beryllium) to Z = 92 (uranium) in various materials. Portable XRF instruments are widely used for *in situ* analysis of lead and other heavy metal contamination; the alpha-proton X-ray spectrometer on the Mars Pathfinder was used to analyze chemical composition of Martian soil and rocks.

1. How It Works

a. Principle of the XRF

X-rays cover an approximately 1- to 100-keV region of the EM spectrum and are associated with atomic electron transitions between different shells of the atom. The primary interaction of the X-rays with matter includes photoelectric absorption and incoherent and coherent scattering. The high-energy excitation radiation interacts with inner-shell orbitals of the atom as illustrated in Figure 8. If the energy of the incident radiation E_p is greater than the binding energy (E_b) of an inner-shell electron, an electron is ejected in the form of a photoelectron. The atom remains ionized for a very short time (about 10^{-14} sec) until the vacancy is filled by a higher energy electron from outer orbits. The difference between the two energy levels of the transferred electron results in emission of a photoelectron of a specific wavelength or energy. This photon will either escape from the atom in the form of characteristic X-rays or will be absorbed within the atom by ejection of an electron called an Auger electron.

In the stable atom, electrons occupy discrete energy levels that are labeled in order of decreasing binding energy as K, L, M, . . . , and the corresponding charac-

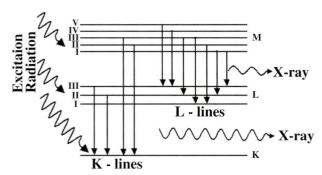

FIGURE 9   Principal K and L XRF lines and their electronic transitions.

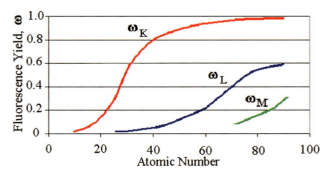

FIGURE 10   Fluorescence yield for K, L, and M lines as a function of atomic number.

teristic X-ray lines in these shells are known as the K, L, M, . . ., series of lines. Figure 9 shows the electronic transitions of the principal K and L emission lines used in XRF.

The probability of emission of the characteristic K, L, M, . . . lines, called fluorescence yield, increases with the atomic number of elements and decreases in order K > L > M series as shown in Figure 10. Because production of Auger electrons is the only other competing reaction, the yield of X-ray photons will be primarily reduced by the Auger effect.

The X-ray scattering process occurs in two modes, with and without changes in incident photon energy or wavelength, referred to as incoherent (Compton) and coherent (Rayleigh) scattering, respectively. Compton scattering is due to collision between an X-ray photon with a loosely bound outer shell electron in which the incident photon transfers a portion of its energy to the electron and scatters at angle θ following the relationship (Jenkins, 1981):

$$E_{Com} = \frac{E_{exc}}{1 + \dfrac{E_{exc}}{511}(1 - \cos\theta)} \qquad (4)$$

where E_{Com} and E_{exc} are the Compton and excitation photon energies in keV, respectively.

In Rayleigh scattering the incident photons interact coherently with all the electrons of the atom without loss of energy. The probability of either types of scattering depends on the X-ray energy and sample composition. Thus, the effect of Compton scattering increases with decreasing atomic number, whereas Rayleigh scattering is most prominent for low photon energies and high atomic number elements.

The absorption of X-rays through a sample with thickness d is described by an expression similar to Lambert-Beer law, i.e.,

$$I_d = I_o e^{-\mu\rho d} \qquad (5)$$

where, I_o and I_d are the intensities of incident and transmitted X-rays and ρ is the density (g.cm^{-2}) of the specimen. The mass absorption coefficients μ of absorber (cm^{2}g^{-1}) is a function of the atomic number and the energy (wavelength) of the X-rays, and exhibits sharp discontinuities at particular energy levels, known as K, L, M, etc., absorption edges. It accounts for the major interactions that can occur in the absorber, i.e.,

$$\mu(E) = \tau_{ph}(E) + \sigma_{inc}(E) + \sigma_{coh}(E) \qquad (6)$$

where $\tau_{ph}(E)$ is the photoelectric mass absorption coefficient and $\sigma_{inc}(E)$ and $\sigma_{coh}(E)$ are the incoherent and coherent mass scattering coefficients.

The mass attenuation coefficient μ_{mix} for a multi-component mixture of pure elements is calculated as the weighted average of the absorption coefficients μ_i of the different constituents i, i.e., $\mu_{mix} = \Sigma(W_i \cdot \mu_i)$, with W_i = weight fraction of component i.

b. X-ray Fluorescence Spectrometry

X-ray fluorescence works by irradiation of a sample with a primary EM radiation generated with radioisotopes, X-ray tubes, or charged particles (electrons, protons, and alpha particles), which in turn causes an element to emit characteristic X-ray lines. By measuring the intensities of the X-ray fluorescent lines against their wavelengths or energies, one can identify and quantify the concentrations of elements in a sample.

A typical X-ray fluorescent spectrometer includes an excitation source and apparatus for separation, detection, and measurement of the fluorescent X-rays. The intensities of the secondary X-rays are typically reduced by two to three orders of magnitude compared to the primary excitation radiation. Therefore, to perform an XRF analysis a high-intensity excitation source and a

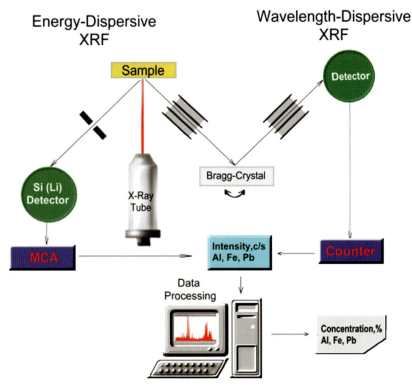

FIGURE 11 Schematic diagram of a simplified wavelength-dispersive and energy-dispersive XRF spectrometer.

sensitive X-ray detector are necessary. There are two common types of spectrometers in use, the wavelength-dispersive (WD) and energy-dispersive (ED) XRF. The term dispersion here refers to different methods of separation and measurement of the characteristic X-ray lines of elements by means of wavelength or energy dispersion.

X-ray fluorescence generally uses two types of excitation sources: radioisotopes with discrete gamma or X-ray emission lines, and sources of continuous radiation generated by X-ray tubes. WDXRF spectrometers use X-ray tubes, whereas the EDXRF spectrometers employ both radioisotope and X-ray tube excitations. The basic components of a typical wavelength-dispersive and energy-dispersive XRF spectrometer are shown in Figure 11.

The X-ray tube with selected anode materials (chromium, molybdenum, tungsten, etc.) excites the characteristic lines of elements present in the sample, which are then separated and dispersed into a pulse-height distribution spectrum called a characteristic X-ray spectrum. Collimators are placed on both optical paths of the sample, including the primary (excitation) and secondary (fluorescent) radiation. In WDXRF the separation of wavelengths of secondary X-rays is done

by Bragg's diffraction from natural or synthetic crystals according to the equation

$$n\lambda = 2d.\sin(\theta) \qquad (7)$$

where n is the diffraction order, λ is the wavelength of the characteristic X-ray, d is the lattice spacing of the analyzing crystal, and θ is the angle of incidence.

By rotating a crystal with specific d spacing at an angle θ along with the detector and associated collimators at diffraction angle (2θ), it is possible to separate and measure the characteristic wavelengths of different elements. To cover the entire range of X-ray wavelengths of elements, different diffraction crystals are usually used such as lithium fluoride (LiF), quartz (SiO$_2$), penta-erythritol (PET), NH$_6$PO$_3$ (ADP), and acid phthalate (KAP). The intensity of the wavelengths are detected by either NaI(Tl) scintillation or flow-proportional counters or both. Gas-flow proportional detectors are filled with a mixture of two gases such as helium, argon, krypton, or xenon and used for long-wavelength X-rays, but the scintillation detector is used for counting of short wavelengths.

In energy-dispersive XRF spectrometry, the wavelength-dispersive crystal-detector system is replaced by a solid-state energy-dispersive detector that can simultaneously separate and detect all X-rays from a sample. The common solid-state detectors used in XRF include lithium-drifted silicon semi-conductor detectors Si(Li), HgI_2, and silicon pin diode. Among those detectors, the Si(Li) has the best resolution (less than 150 eV) but needs to be operated at $-273°C$ to maintain the diffusion of lithium ions. The Si(Li) semi-conductor detector detects the X-ray photons, which are then converted by an ADC and an MCA directly into an X-ray spectrum.

Compared to wavelength-dispersive crystal-detector systems, the solid-state detectors in EDXRF have much higher dispersive efficiency because they perform direct separation and measurement of the secondary X-ray spectrum. This makes possible the use of low-powered X-ray sources, such as radioisotopes or X-ray tubes with secondary targets. The most common radioisotope sources employed in X-ray spectrometry are ^{55}Fe, used for analysis of elements with atomic number $12 < Z < 23$; ^{109}Cd, for analysis of elements with $20 < Z < 42$; and ^{241}Am, for elements with $47 < Z < 58$. The secondary-target excitation mode allows modification of broadband radiation of an X-ray tube to narrow band energy, suitable for efficient excitation of a particular group of elements.

The conventional WDXRF measurement electronics are generally referred to as a single-channel pulse-height analyzer in contrast to the MCA used for EDXRF spectrometry. Thus, the wavelength-dispersive XRF measures the intensity of the characteristic X-ray lines of elements in sequential mode, whereas the EDXRF records the entire X-ray spectrum at once. The EDXRF spectrum is further processed with a computer program that performs background subtraction, correction for escape peaks and other interferences, and calculates the net peak intensities of elements. In both cases, wavelength- and energy-dispersive XRF spectrometry provide information about the intensities of the characteristic X-ray emission lines, which is used to calibrate the equipment and calculate the concentrations of unknown samples.

2. What It Does

The X-ray excitation sources used in wavelength- and energy-dispersive XRF provide adequate analytical sensitivity for quantitative analysis across the elements sodium to uranium. The elements of interest are identified by their spectral wavelengths or energies for qualitative analysis and the intensities of the emitted spectral lines are used for quantitative analysis. The quantity of the characteristic X-rays is proportional to the amount of elements in the sample. The quantitative XRF analysis of unknown samples is usually performed using calibrations with matrix-matched standards.

a. Sample Preparation

Most solid or liquid samples can be analyzed by XRF as received, without any special sample preparation. However, for quantitative analysis the samples and standards must be homogeneous with an optically flat surface. Table V summarizes the most common specimen preparation procedures for solids, liquids and gases.

Solid geological and biological samples are previously oven- or freeze-dried and milled to a fine powder (<200 mesh) before preparation for XRF analysis. For the pressed powder pellet about 3–7 g of sample is homogenized with a binding agent and pressed into a solid pellet using a hydraulic press. For preparation of the glass discs the sample material is weighed and fused at a temperature of 900–1200°C using a suitable flux, e.g., lithium tetraborate.

Sample preparation is a critical step in XRF analysis. By choosing the most appropriate sample preparation procedure (e.g., fusion, pelletizing, etc.) some problems associated with sample homogeneity, particle size effects, nonlinear working range of concentrations, etc., could be greatly overcome (Buhrke et al., 1997).

b. Qualitative XRF Analysis

The qualitative XRF analysis is an important step in geochemical studies, which makes it possible to determine the overall composition of samples. The problems

TABLE V. Some Common Specimen Preparation Methods for XRF Analysis

Sample type	Sample preparation
Powders (geological)	Bulk sample in a sample cup
	Press pellet
	Fused glass beads
Powders (biological)	Bulk sample in a sample cup
	Press pellet
	Dry ashing and palletizing
Metals	Cutting, grinding, and polishing
Liquids	Measurement in a sample cup
	Preconcentration
	Filtering through a 0.45-μ filter
Air particulates	Aspiration through a filter paper
	Dust wipes with filters

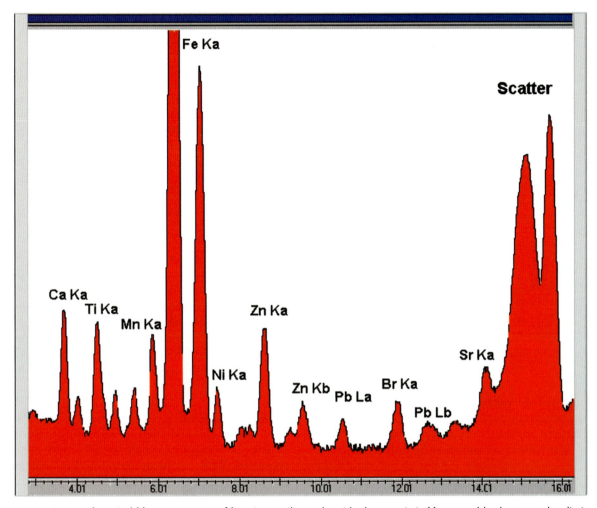

FIGURE 12 A typical X-ray spectrum of Jamaican soil sample with characteristic X-rays and backscattered radiation.

associated with qualitative analysis depend on the X-ray dispersion method. The qualitative WDXRF analysis is performed by stepwise scanning over the entire wavelength range by rotating the crystal detectors in small increments of angle θ. Modern WDXRF software includes advanced peak search routines to analyze the peak shape and to locate the true position of the peaks. The EDXRF spectrometer simultaneously collects the X-ray spectrum of elements for a short period of time; however, to cover the entire range of elements, usually two to three different excitation conditions and measurements are applied.

There are several interferences common to wavelength and energy-dispersive XRF that need to be considered in qualitative analysis. These include the incoherent and coherent scattering, K, L, and M line overlaps, tube emission lines, background distribution, higher order of diffraction lines in WDXRF, and the presence of escape and sum peaks in EDXRF. Figure 12 shows the X-ray spectrum of a typical Jamaican soil sample with characteristic peaks and scattered radiation (Johnson et al., 1996).

c. Quantitative XRF Analysis

Quantitative XRF analysis using wavelength and energy-dispersive spectrometry involves a series of steps designed to prepare, measure, and process the X-ray spectral data. Table VI outlines the basic procedures used to perform a quantitative XRF analysis.

TABLE VI.   Basic for Quantitative X-Ray Analysis

Step	Procedure
1. Preparation	• Perform qualitative analysis of unknown samples • Select calibration standards with composition similar to unknowns • Prepare samples and standards using similar procedures
2. Measurement	• Measure the samples and standards under identical operating conditions
3. Data Processing	• Perform spectrum analysis to obtain the net intensities of elements • Perform quantitative calibration for each element • Calculate concentrations of unknown samples • Carry out appropriate quality control checks

Although the measurement principle of the wavelength and energy-dispersive XRF spectrometers differ, the quantitative analysis process, converting the intensities into elemental concentration, is similar for both techniques.

In the case of monochromatic excitation of homogeneous samples, the relationship between the concentration C_i of an element i and intensity I_i can be generalized in the form

$$C_i = K_i \cdot I_i \cdot M_i \qquad (8)$$

where K_i is a sensitivity factor that depends upon instrumental parameters and physical constants of the analyte i, and M_i is a matrix correction factor accounting for the absorption and enhancement of the primary and secondary radiation.

Equation 8 shows that the concentration of an element can be determined by measuring the intensity I_i of the analyte and by calculation of K_i and M_i factors from standards. The accuracy of the analytical results to a large extent depends on the efficiency of the interelement correction factor M_i. The interelement or matrix interferences result from variations in the physical and chemical properties of the sample, such as particle size, mineralogy, chemical composition of major constituents, etc. The most common matrix effects encountered in XRF are the absorption and tertiary fluorescence or enhancement of X-rays, which can be corrected by using appropriate quantitative analysis methods (Jenkins, 1981; Tertian & Claisse, 1982).

d. Thin-Film Calibration

In thin-film samples the interelement effects associated with the absorption and enhancement effects are reduced significantly because each atom interacts with the primary and secondary (characteristic X-ray) radiation independently. A thin specimen is defined as satisfying the relationship $\rho d < 1/\mu$ (Eq. 8), where ρd is the mass thickness of the sample per unit area. The intensities of the analyte lines for thin samples are directly proportional to analyte concentrations.

e. Empirical Calibration Method

The empirical calibration is performed using matrix-similar standards or samples already analyzed by other techniques. The standards should include the elements of interest and cover the expected concentration range of elements. Calibration curves are typically generated using 10–15 calibration standards. The calibration coefficients are calculated through a least-squared regression analysis with a correlation coefficient greater than 0.95.

f. Dilution Methods

The sample dilution method is used to minimize the matrix differences between samples and comparator standards. This can be accomplished by adding either an element with a very high (e.g., lanthanum, barium, etc.) or a very low (e.g., lithium borate) absorption coefficient. This changes the overall mass absorption coefficient of the sample to a level that the effect of matrix elements on the intensity of the analyte line are negligible. Dilution methods also allow for obtaining linear calibrations over wider concentration ranges.

g. Matrix Correction Methods

The common approach for compensating the matrix effect includes the use of the incoherent (Compton) scattering of the primary radiation, which is measured in each XRF spectrum. The Compton method applies the inverse relationship between the Compton peak intensity and the mass absorption coefficient of the sample for correction of the matrix effects. The ratio of elemental intensities to the Compton peak is used to generate calibration curves. An alternative approach for matrix correction is the internal standard method, i.e., addition of an element to a sample that is affected in the same way as the analyte element by the matrix. The ratio of the line intensities of analyte element to the internal standard will then be independent of the matrix.

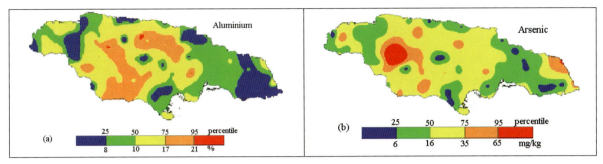

FIGURE 13 (a and b): Map of aluminum and arsenic distributions in Jamaican soils.

b. Mathematical Methods

Mathematical correction methods were introduced in an attempt to develop absolute methods for quantitative XRF analysis. In 1955 Sherman published the fundamental relationship between X-ray intensities and sample composition, but due to the lack of adequate computational power this method did not find practical application. Bettie and Brissey used the regression analysis to solve equations relating measured intensities of elements with corresponding concentrations. Other approaches include the intensity-correction model of Lucas-Touth and Price, the alpha-correction model of Lachance and Trail, the Rasberry and Heinrich enhancement correction model, and the fundamental parameters method.

3. Summary and Conclusions

Compared to NAA methods, XRF has the advantage of being a cost-effective technique providing good detection limits across the elements of the periodic table and applicable to a wide range of concentrations from milligram per kilogram to 100%. WD and EDXRF are not the methods of choice for sub-trace elements, nor are they for those in which calibration standards with similar matrix composition cannot be obtained.

The choice of a wavelength- or energy-dispersive spectrometer depends on several factors, such as cost, speed, and application type. Thus for high throughput quantitative analysis of a limited number of elements where the initial cost is not important, the WD XRF instruments are probably the best choice, whereas for multi-element analysis with some limitations in the detection limits or accuracy, the EDXRF is preferable.

C. Applications of NAA and XRF

Neutron activation and XRF analysis are widely used in environmental geochemistry and health. Some examples are shown below.

1. Geochemical Mapping of Soils

Surface soil samples were collected on an 8×8 km grid across Jamaica which resulted in some 204 samples, including field duplicates. Samples were prepared by wet sieving to $-150\,\mu$, drying at 80°C, and grinding to less than $50\,\mu$. The soil samples were analyzed by NAA and XRF for about 40 elements and the results were used to produce geochemical maps of soils (Lalor, 1995).

Figure 13a shows the geochemical map of aluminum in Jamaican soils. The highest concentrations of aluminum in soils (Al > 15%) overlap the areas of known bauxite deposits. As this is also an agricultural area the soil map provides useful information on possible aluminum intake by plants.

The arsenic distribution map in Figure 13b shows an anomalous area with arsenic concentrations as high as $373\,\mathrm{mg\,kg^{-1}}$. A followup study of the "hot spot" defines the boundaries of the anomalous area and indicates that under the present land use conditions arsenic does not present environmental hazard problems (Lalor et al., 1999).

2. Lead Pollution and Health

EDXRF spectrometry was used to provide data for mapping of lead contamination from an old mine. Figure 14a shows the distribution of lead in soils. The highest lead levels were around a school for preschool children 3–6 years old. Blood lead (PbB) screening of the children showed a mean level of $37\,\mu\mathrm{g\,dL^{-1}}$. The environmental intervention included isolation of lead contamination by means of paving, nutritional supplements of food rich in iron and calcium, and a lead-safe education campaign for teachers, parents, and children. The upper curve in Figure 14b shows the blood lead levels of children sorted in increasing order. Ten months after intervention the PbB levels of children were greatly reduced as shown by the lower curve in the same

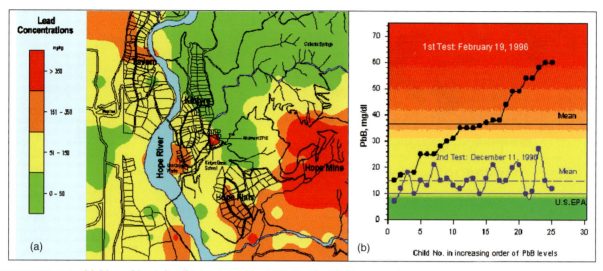

FIGURE 14 (a) Map of lead distribution in contaminated soils; (b) blood levels of children before and after remediation.

figure (Lalor et al., 2001). Five years later a new generation of children attending this school showed a mean PbB level of 8.3 µg dL^{-1}. This illustrates the effectiveness of the mitigation campaign undertaken in the lead-contaminated area.

III. Inorganic Geochemistry Techniques for Liquids

A. Introduction

There are a number of methods used to analyze inorganic elements in liquid samples. The most widely used are atomic spectroscopy (AS), mass spectroscopy (MS), and electrochemical techniques (ET). According to the principle of interaction of atoms with light, the AS methods are classified as absorption, emission, and fluorescence (Figure 15).

- Atomic absorption spectroscopy (AAS) measures the intensity of absorbed radiation of atomic species. Two AAS techniques are commonly used: flame (FAAS) and graphite-furnace (GFAAS).
- Atomic emission spectroscopy (AES) involves excitation of the analyte atoms and measurement of their emission as they revert to their ground state. The classical emission spectroscopy, known as flame emission photometry (FEP), has been further developed to accommodate multi-element capabilities using inductively coupled plasma (ICP) as the atomization source.

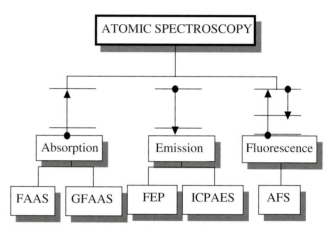

FIGURE 15 Block Diagram of atomic spectroscopy techniques.

- Atomic fluorescence spectroscopy (AFS) is based on measurement of radiation given off as excited atoms relax to a state higher than the ground state. Even though AFS is the most sensitive of the above techniques, it has not gained widespread use for routine analysis, because it is based on complex and expensive instrumentation technology.

Among the different spectroscopy techniques for liquids, the most commonly used are AAS, ICP-AES, and ICP-MS, each of which has particular advantages and disadvantages. ICP mass spectrometry (ICP-MS) utilizes hot plasma as an ionization source and has been extensively developed as a multi-element spectroscopic technique capable of analyzing both elemental concentrations and isotopic ratios. Along with modern spectroscopic techniques, classical electrochemical methods are increasingly used due to their exceptionally low detection limits for many elements and possibility of *in situ* analysis.

1. Sample Dissolution

Liquid analytical techniques require prior dissolution of the solid samples using appropriate digestion procedures. The choice of a particular dissolution technique depends on the sample type, the elements of interest, and their concentration levels. Metals in biological materials can be digested more easily than those in geological matrices. Water samples usually require relatively minimal treatment prior to analysis.

2. Geological Materials

The decomposition of geological samples is commonly classified as selective, partial, or total, according to the mineral composition of the sample and analytical requirements.

- *Selective extraction* is used to extract specific elements leaving the remaining sample matrix in the residue. This procedure usually involves single or sequential attack of the sample by weak acids. The most popular procedure involves the use of sodium acetate, which dissolves only alkali metals and carbonates. Sodium pyrophosphate is also used for liberating organically bound heavy metals.
- *Partial digestion* is usually accomplished using aqua-regia (3:1 mixture of concentrated hydrochloric and nitric acids). Attack with aqua-regia causes loosely bound metals to enter into solution while those embedded in mineral particles remain in the solid residue. Generally, about 30 elements can be digested using aqua-regia, with approximately half of them fully dissolved.
- *Total digestion* and analysis of minerals is important for geological exploration and industrial purposes. It usually involves oxidation of the organic matter with nitric acid, followed by hydrofluoric acid (HF) digestion, which attacks the silicate matrix. The excess silicon is removed during evaporation of volatile silicon fluoride (SiF_4). For digestion of easily volatilized elements closed digestion vessels are required.

Certain minerals such as barite and some refractory elements (e.g., zircon) will dissolve only partially under these conditions. In such cases, sodium peroxide (Na_2O_2) decomposition is required, which provides basic chemical attack. Generally, samples with high sulfide content require higher reaction temperatures for full dissolution, which can be achieved using microwave high-pressure bomb digestion (Kingston, 1988). This method, however, cannot be used for determination of rare earth elements (REE) and some other resistant minerals. The only attack in which REE, major oxides such as silicon dioxide (SiO_2), and refractory minerals in geological samples are fully dissolved is lithium metaborate/tetraborate fusion. This technique involves preliminary drying of the sample at 105°C, followed by an attack of the geological matrix with a high-temperature (about 1050°C) molten flux of lithium metaborate/tetraborate. The fusion process produces a homogenous amorphous solid solution (bead), which then may be easily dissolved in dilute acid and used for analysis.

3. Biological Materials

The inorganic content of most biological materials is a minor constituent and in most instances the organic matter is removed prior to analysis. Organic matter is eliminated by oxidation of carbon and hydrogen to carbon dioxide and water, which is done by heating in the presence of oxygen (dry ashing), or by the action of oxidizing acids (wet ashing), such as sulfuric, nitric, and perchloric acids. The dry ashing procedure is usually carried out at about 450–600°C for 4–12 hours. The inorganic residue (ash) is easily dissolved and used for analysis. This procedure is not applicable for some volatile elements such as arsenic, selenium, lead, and cadmium for which closed vessel microwave digestion in Teflon containers is recommended. Samples intended for analysis of mercury are oven dried at 60°C or freeze-dried.

4. Water Samples

Water samples are usually analyzed without sample digestion, unless they contain high concentrations of suspended solids. The standard methods of water analysis require prior filtration of the sample through a 0.45-μ filter. Freshwater samples are usually filtered and acidified in the field to prevent loss of metals. Wastewater samples generally require digestion, as they contain high concentrations of metals in a variety of matrices.

B. Atomic Absorption Spectroscopy

AAS is the most frequently used method for trace element analysis of liquid samples. This method involves an introduction of an atomized sample into a beam of appropriate EM wavelength, which causes the atoms to absorb light and become excited. The measured absorption of the light beam is compared with that of known calibration standards, and the concentration of the elemental species is obtained.

1. How It Works

Historically, two main types of AAS instruments have been developed, FAAS and GFAAS. These two AAS techniques have similar measurement principles, but they differ in the methods of sample introduction and atomization. The analysis is normally completed in a single element mode because of the limitation imposed by the excitation source (Welz, 1999).

a. Flame Atomic Absorption Spectroscopy

FAAS was developed by the Australian scientist Alan Walsh in the mid-1950s (Walsh, 1955). FAAS is theoretically applicable to most metals, provided that a light source is available for that element. A schematic representation of an FAAS system is shown in Figure 16.

The liquid sample is initially converted to a fine aerosol by a nebulizer, which increases the surface area of the solution sample and facilitates evaporation and volatilization. The sample aerosol is transported to the flame, where most of the metal ions are reduced to elemental atoms. Air-acetylene and nitrogen oxide-acetylene are the most commonly used flame mixtures. The universal source of EM radiation used in AAS is the hollow cathode lamp, which consists of a cathode filament made of the element to be analyzed and filled with argon or neon gas. When a high voltage is applied, the lamp emits characteristic radiation of the analyte element. To isolate the radiation of the lamp from the radiation emitted by the atoms in the sample cell, a mechanical device (chopper) located between the light source and the sample cell is used.

The radiation from the burner is directed to the monochromator or wavelength selector, which isolates the resonant spectral line from the background radiation of the sample. The spectral line is measured by a detector, usually a photomultiplier tube, which converts incident EM energy into an electrical signal. Modern instruments are operated by a computer, which acts as an output device and is used to process the analytical information.

b. Graphite Furnace Atomic Absorption Spectroscopy

GFAAS was introduced in the early 1970s as a method with greatly improved sensitivity compared to FAAS (L'vov, 1961). It is typically applicable to most elements analyzed by FAAS. A schematic representation of a GFAAS system is shown in Figure 17.

GFAAS instrumentation uses an electrically heated polycrystalline furnace, mounted in the graphite tube, which allows the sample to be heated by radiant heat rather than convective energy. This delays atomization and minimizes some non-spectral interferences. The electrothermal atomization allows almost full atomization of the sample, which makes this technique more sensitive than FAAS. The sample is placed directly on the platform of the graphite furnace, which is heated stepwise to dry the sample, ash the organic matter, and vaporize the atoms. At the end of the heating cycle, the graphite tube is externally flushed with argon gas to prevent the tube from burning away.

2. What It Does

Flame and graphite AAS techniques are primarily used for chemical analysis of metallic elements in a variety of materials. The quantitative analysis can be accomplished by using a calibration curve or standard addition method. The choice of quantification method

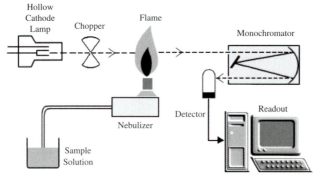

FIGURE 16 General setup of FAAS.

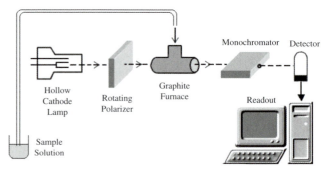

FIGURE 17 General setup of GFAAS.

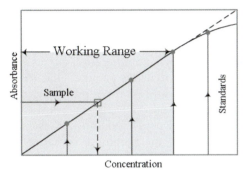

FIGURE 18 A typical calibration curve used for AAS analysis.

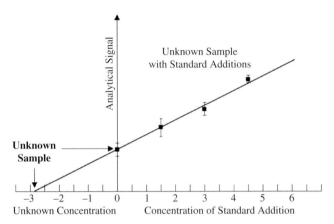

FIGURE 19 Standard addition calibration curve.

depends on the sample type, potential interferences, and matrix effects.

A calibration curve is a plot of measured analytical signal as a function of known concentrations of standards. Typically three standards, covering the expected concentration range of the sample, as well as a blank, are used to produce a working calibration curve (Figure 18). The working range is defined within the linear relationship of the calibration curve. If the sample concentration is too high, there are three alternatives that can help bring the absorbance into the linear working range; namely sample dilution, alternative wavelength with a lower absorptivity, or reduction of the path length by rotating the burner head.

The standard addition method is usually applied to quantification of samples that have substantial matrix effects. In this method, a known amount of analyte is added (spiked) to an aliquot of the sample. The concentrations and absorbance of the original and spiked samples are plotted on a graph, which is used to determine the concentration of the unspiked sample, as shown in Figure 19. Because the sample aliquots differ only in analyte concentration and have relatively no matrix differences, this method is often referred to as an "internal" calibration method, and it is more accurate than using external calibrations.

There are five major types of interferences encountered in atomic absorption spectroscopy: physical, chemical, spectral, ionization, and background. The physical, chemical, and ionization interferences influence the atomization process, whereas spectral and background interferences change the atomic absorption signal.

- *Physical* characteristics of dissolved samples, such as viscosity or density, may influence the aspiration rates or droplet size and therefore the measurement signal. This may change the slope of the calibration

curve and produce lower or higher analytical results.

- *Chemical* interferences occur when the atoms are not completely free or in their ground state. This may be due to the presence of certain chemical compounds in a sample matrix, which can change the sensitivity and the response between samples and standards.
- *Spectral* interferences exist when an emission line of atomic or molecular species other than the element being analyzed is present within a band pass of the wavelength that the analyte absorbs.
- *Ionization* interferences occur when the flame or furnace causes complete removal of atomic electrons. The ionized atoms of the analyte element will not absorb light, thereby lowering the concentrations of that element.
- *Background* interferences are due to undissociated molecules of the matrix that can scatter or absorb the light. In either case, the apparent concentration of the analyte will be much higher. Most GFAAS perform automatic background corrections using the Zeeman method.

The common procedures used to correct for interferences include: (1) serial dilution and re-analysis of the sample to determine whether the interference can be eliminated, (2) use of matrix modifiers to compensate for potential interferences, and (3) use of standard addition calibration method (Schlemmer & Radziuk, 1999).

C. Atomic Emission Spectroscopy

Atomic emission spectroscopy uses optical emission of excited atoms for quantitative elemental analysis. The

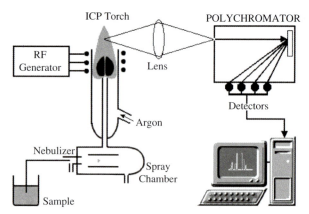

FIGURE 20   Schematic diagram of simultaneous ICP-AES.

earliest emission technique, flame emission photometry, was limited mainly to sodium and potassium, as most atoms of non-alkali metals remain in their ground state at normal flame temperatures. The emission source should ideally be able to vaporize the sample, break-down all compounds, and excite the atoms and ions. To achieve this, temperatures much higher than those of the flame are required. Thus, the use of alternative atomization sources, such as electrical arcs and plasma, are required. ICP has revolutionized AES by making it applicable to a wide range of elements. Because all atoms in the sample can be excited at the same time, they can be detected simultaneously, giving ICP-AES multi-element capabilities.

1. How It Works

ICP-AES was introduced in 1975 as a multi-element trace element technique utilizing argon plasma for atomization and excitation. The ICP source consists of a torch with three concentric quartz tubes through which ionized argon gas flows at high speed, reaching temperatures between 6000 to 10,000 K (Figure 20). The liquid sample is introduced into the plasma in the form of an aerosol. The high temperature of the plasma excites the atoms and they emit light of characteristic wavelengths. Because emission lines are quite narrow, there are less chances of interelement overlaps, and multiple elements can be determined simultaneously. A polychromator, which isolates all wavelengths of the analyte, is used in conjunction with multiple detectors that measure the wavelengths and intensities of analyte lines.

2. What It Does

ICP-AES is mainly used for analysis of dissolved/suspended samples, as solid samples require costly intro-

duction techniques such as laser ablation. The element range spans about 60 elements, including some non-metals such as sulfur and some halogens, which cannot be analyzed by AAS. The calibration curves are generally linear in wide dynamic ranges. Due to the high atomization temperatures, even most refractory elements—such as boron, phosphorous, tungsten, uranium, zirconium, and niobium—are atomized efficiently. As a result, the minimum detection limits of these and most other elements can be orders of magnitude lower with ICP-AES than with FAAS.

Spectral interference problems are fairly common in ICP-AES due to the line-rich spectra produced by the hot plasma source. High-resolution spectrometers can be used to minimize spectral interferences or an alternative analyte emission line can be used to detect the element. Chemical matrix effects are not as profound in ICP-AES as in other techniques, and they can be moderated by use of internal standards. Detection limits are typically less than 1 μg/L, which is about at least ten times higher than those of obtained by GFAAS.

D. Inductively Coupled Plasma Mass Spectrometry

ICP-MS was developed in the early 1980s and has become increasingly popular for the analysis of geological, environmental, and medical materials. The ICP-MS technique uses high-temperature plasma discharge to generate positively charged ions, which can be separated and quantified based on their mass-to-charge ratios by means of a mass spectrometer. Like AES, atomic mass spectrometry is inherently a multi-element analytical technique, which can also provide information about isotopes present in the sample. Conventional ICP-MS analyzes liquid samples, but techniques for direct analysis of solid samples such as laser ablation ICP-MS have also been developed.

1. How It Works

The main steps in ICP-MS operation involve (1) ionization of the liquid or solid material introduced into the ICP using a stream of argon carrier gas, (2) separation of the ions based on their mass-to-charge ratio using a mass spectrometer, and (3) measurement of the relative quantities of ions by an electron multiplier detector and counting system. A schematic diagram of an ICP-MS is shown in Figure 21. It consists of a sample introduction system, an excitation source, ion

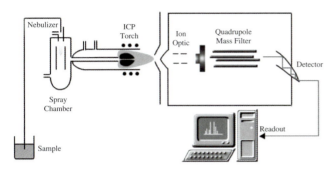

FIGURE 21 Schematic diagram of an ICP-MS.

transport system, mass separation device, detector, and computer.

The sample introduction system includes a nebulizer and a spray chamber. The nebulizer converts liquid samples into a fine aerosol by mixing it with argon in a spray chamber. The primary function of the spray chamber is to isolate fine aerosols to enter into the plasma. The sample aerosol is injected into the middle of the high-temperature plasma (5000–10,000 K), where it is vaporized, atomized, and ionized. Recombination of sample, matrix, and argon ions may occur in the cooler parts of the plasma, which leads to the formation of various molecular species.

The ions produced in the central part of the argon plasma are transferred into a mass spectrometer by an ion transport system consisting of an interface region and an ion lens system. The interface region extracts the ions from the plasma and directs the ion beam to the front section of the mass spectrometer chamber using rotary and turbomolecular vacuum pumps. The ion lens focuses and collimates the positive ions from the sample ion beam into the mass spectrometer, which separates them according to their mass-to-charge ratio. The most frequently used mass separation device is a quadrupole mass spectrometer, which consists of four parallel conducting rods to which radiofrequency (RF) and direct current (DC) potentials are applied. By changing the RF and DC potentials on the quadrupole rods, the mass spectrometer is capable of sequential scanning across the mass-to-charge ratio range of all naturally occurring elements. The ions sorted by the mass spectrometer are detected and measured by a photomultiplier tube detector.

The computer is the final constituent of an ICP-MS instrument. It controls all hardware components of the equipment, sample handling, measurement, data acquisition, and processing of analytical information.

2. What It Does

In ICP-MS, the intensities of the analytes are measured by continuous scanning of the mass spectrometer, which generates spectra of signal intensity versus mass. The peak positions of the spectra are used for identification of the elements present in a sample, whereas the signal intensity is used for quantitative analysis. Semi-quantitative multi-element analysis is often used for rapid screening of the unknown sample composition. A single multi-element standard containing most of the elements of interest is generally used for this purpose. The quantitative analysis is generally achieved by internal, external, or standard addition calibration. An internal standardization is used for correction of instrument drift by normalization of all analyte data to non-analyte isotopes present in known concentrations in both the samples and standards. External calibration uses intensities of analyte isotopes in a number of standards that cover the concentration range of interest. The blank-corrected intensity of standards versus known concentrations is used to produce a calibration curve applied for calculation of unknown concentrations. The standard addition calibration includes spiking of sample solutions with known concentrations of analyte elements. An advantage of this method is that it also provides an effective way to minimize any matrix effects of the sample.

Interferences in ICP-MS are generally low compared to the other atomic spectroscopy methods. Spectral interferences occur when equal mass isotopes of different elements are detected (e.g., ^{64}Ni on ^{64}Zn) or when ions with twice the mass of the analyte are produced. They can be avoided by using alternative interference-free analyte isotopes or corrected by measuring the intensity of isotopes of the interfering element and calculating a correction factor. A common matrix effect in ICP-MS takes place when abundant and easily ionized elements at high concentrations are present in the sample. These effects may be generally overcome by sample dilution or separation of the interfering matrix component.

ICP-MS instruments may also be used to determine the natural isotopic ratios of specific elements such as lead and strontium. Although no standards are required for ratio determinations, the application of the isotope dilution method can significantly improve the accuracy of analysis.

E. Summary and Conclusions

Atomic and mass spectroscopy techniques are widely used for analysis of trace metals in various materials.

TABLE VII.   Summary of the Key Advantages and Disadvantages of Atomic and Mass Spectroscopy Techniques

	FAAS	GFAAS	ICP-AES	ICP-MS
Elements analyzed	60+ metals	50+ metals	70+ metals, some non-metals	75+ metals, non-metals
Multi-element	No	No	Yes	Yes
Sample throughput	Fast (for <5 elements per sample)	Slow (3–5 min. per element)	Fast (multi-element analysis)	Fast (multi-element analysis)
Semi-quantitative analysis	No	No	Yes	Yes
Isotopic analysis	No	No	No	Yes
Detection limit	Good	Excellent	Very good	Excellent
Dynamic range	10^3	10^2	10^5	$10^5–10^8$
Precision	<1%	<5%	<2%	<3%
Sample volumes	Large	Small	Small	Small
Dissolved solids	<5%	<20%	<20%	<0.5%
Interferences				
Spectral	Very few	Very few	Many	Few
Chemical	Many	Many	Few	Some
Physical	Some	Very few	Very few	Some
Method development	Easy	Difficult	Moderately easy	Difficult
Ease of use	Very easy	Moderate	Easy	Moderate
Capital and running costs	Low	Medium	High	Very high

Selecting the most appropriate method for a particular analysis depends on many factors, such as sample type, elements of interest, required accuracy, sample volume, speed, cost, etc. This task is further complicated by the fact that the capabilities of the current analytical techniques such as FAAS, GFAAS, ICP-AES, and ICP-MS overlap to a great extent. Table VII summarizes the key advantages and disadvantages of these techniques, which might aid the selection process.

FAAS is the most commonly used analytical technique, mainly due to the lower capital and running cost of the instrument and the simplicity of operation. It is the technique of choice for single element analyses of small batches of samples. The GFAAS technique has a greater sensitivity and lower detection limits than FAAS and is preferred for analysis of few elements at very low levels using small sample sizes. The main advantages of ICP-AES and ICP-MS over AAS techniques are their multi-element capabilities, larger linear dynamic ranges, and reduced interferences. ICP-AES detection limits are comparable to those obtained by FAAS, and the technique is preferred when a high sample throughput with moderate sensitivity is required. ICP-MS is best for simultaneous analysis of a large number of elements with very low concentrations. However, sometimes the analytical requirements will not be fully satisfied using a single technique, and a combination of two techniques is used to meet all requirements.

IV. ORGANIC GEOCHEMISTRY TECHNIQUES

A. Chromatography

1. Principles

Chromatography is essentially a mechanism for the separation of compounds within a mixture. Originally, chromatography was used for the isolation of plant pigments (hence chroma or color). Different components of a mixture will migrate over a surface at different rates, depending on the chemical reactivity, or other interactions with the substrate. Depending on the interactions, the components will move at different speeds, with greater interactions resulting in slower speeds. Essentially, chromatography involves the use of two phases, one stationary and the other mobile (Figure 22). The mobile phase can be either liquid or gas, whereas the stationary phase is typically solid. The stationary phase can have a wide range of polarities that range from nonpolar to strongly ionic in nature, and may chemically

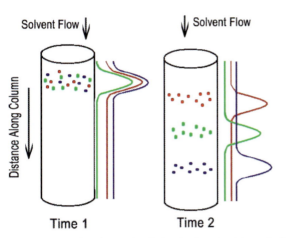

Time 1   Time 2

FIGURE 22 A schematic of the basic principles of chromatography, in which mixtures of compounds can be separated into individual molecular components (the compounds that travel more quickly through the column interact less strongly with the phase in the column).

bind to compounds in the mixture. Those compounds are then released by the introduction of more of the mobile phase, or may utilize chemical changes in the mobile phase. If the mobile phase is liquid, the technique is liquid chromatography (LC); if the mobile phase is gas, it is gas chromatography (GC). As the compounds are eluted from the stationary phase, they are detected by a wide array of systems that include, among others, spectrophotometric or refractive index detection, electron capture detectors (ECD), flame ionization detectors (FID), thermal conductivity detectors (TCD), or mass spectrometers.

Environmental samples are extracted from the native matrix using water or organic solvents, and they are concentrated through a process like evaporation of the solvent. The concentrated material is introduced onto the column, which is essentially a tube filled or coated with the stationary phase. The tube can be a few centimeters in length to 50 m or more, and from millimeters in diameter to tens of micrometers. Very small volumes are typically placed on the column, perhaps fractions of microliters in the case of GC or LC. In LC, the mobile phase may be water, or an organic solvent like hexane or benzene. In GC the mobile phase is typically helium. The chromatographic separation scheme is optimized for the kinds of components analyzed, with the stationary phase altered to enhance the resolution of the individual components. Typical separations of a mixture (in these cases polychlorinated biphenyls; PCBs, and fatty acids captured from an aerosol) are shown in Figures 23a and b, respectively.

B. Mass Spectrometry

1. Principles

Mass spectrometers are used both for the identification and characterization of organic compounds as well as the determination of the stable isotopic abundances of both bulk and molecular components. They are essentially instruments that have the capability to separate charged ions of either atoms or molecules based on their masses, through the use of electrical or magnetic fields (Waples, 1981). In the 1920s, F. W. Aston reported the first use of mass spectrometers in the precise determination of the masses of neon isotopes. Many of the developments in the early days of mass spectrometry were aimed at cataloguing and measuring the isotopic compositions of the elements. Following World War II, the use of mass spectrometers was extended to determine the molecular weights of compounds, and the instruments were coupled to chromatographic technologies in order to isolate single components.

Mass spectrometers used in geochemical research generally have four basic components (Figure 24): an inlet system, an ion source, a magnet, and ion collectors. All of these parts are housed in a system that allows for the maintenance of a high vacuum (typically 10^{-6}–10^{-9} mmHg pressure). High vacuums can be achieved by turbomolecular pumping, or diffusion pumps, which are backed by standard rotary pumps. The high vacuum is important for a number of reasons including the diminishment of interferences, enhancement of ion separation, etc. The inlet system can allow for either static or dynamic introduction of samples. Samples can be converted offline, for example, through acidification, or high temperature combustion, for the introduction of the products as purified gases. The mass spectrometer can also be coupled to a chromatographic system like a GC, which allows for the continuous introduction of species as they are separated.

The source is where the gas is ionized during static introduction of the sample, and it is also the location where the molecules are struck by electrons, which can cause them to fragment into ions. The electron source, which must produce electrons with sufficient energy to accomplish the ionization, is a heated filament.

Ions are deflected by the magnetic field and the radius of that deflection can be precisely calculated using the mass of the ions, the accelerating voltage in the source, and the charge on the ion.

All of these facets are carefully controlled for in the mass spectrometer, and the result is very precise measurements of the mass, as well as the ability for an

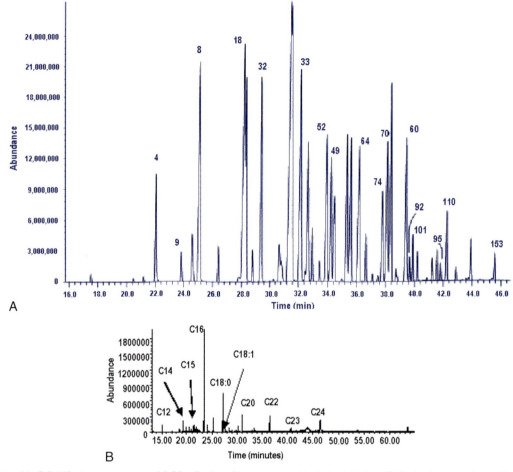

A

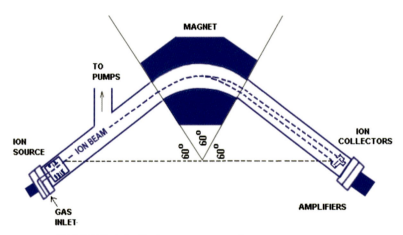

B

FIGURE 23 (a) GC-MS chromatogram of PCBs. Peaks of interest are labeled with their Ballschmiter numbers. (From Yanik et al., 2003a.) (b) GC-MS chromatogram of fatty acid methyl esters from an aerosol. Peaks are identified by their carbon chain length. (From Billmark et al., 2003.)

FIGURE 24 Basic components of a mass spectrometer.

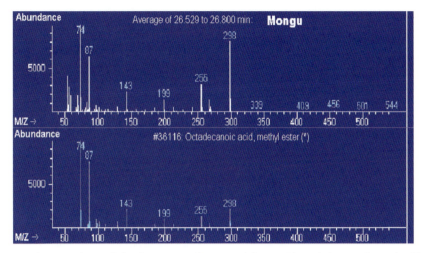

FIGURE 25 Mass spectral comparison of the fragments of a natural fatty acid methyl ester from the library of compounds that shows the match between the two. (From Billmark et al., 2003.)

abundance of the ions to be made. The magnetic field, produced by a permanent or an electromagnet, can either be carefully maintained or rapidly varied in order to accomplish the desired separation. By varying the magnetic field used in coupled GC/MS, the ions of different masses can be directed into a single collector and counted. In such a system, an entire mass spectrum can be observed (Figure 25) for the molecules and their fragments, which can be used to identify the parent molecule. With a static magnetic field, used in stable isotope mass spectrometry, the ions can be directed into multiple collectors, which have been positioned inside the flight tube of the mass spectrometer so that the ion masses are uniquely identified, and are able to be quantified.

The ions are focused and are counted in a collector, which is essentially a metal cup. The detector cup is connected to amplifiers that feed into a device for reporting the signal (typically a computer). In stable isotope mass spectrometry, the signals are observed relative to a standard signal for each cup, and are reported as ratios; hence the name isotope ratio mass spectrometry (IR-MS). More recent modifications of IR-MS technology include the coupling of a GC to the front end of the MS through a combustion system that converts the compounds coming from the GC separation into pulses of gas, which is isotopically assessed (Figure 26).

2. Stable IR-MS

Stable carbon and nitrogen isotope analysis of bulk organic materials is a well-established method for

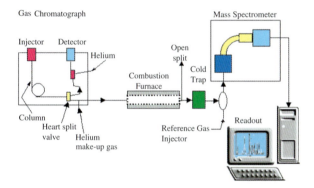

FIGURE 26 Schematic of the combustion interface between the gas chromatograph and a mass spectrometer designed for stable isotope characterization.

tracing biosynthesis as well as the sources and history of organic matter in the geosphere (Engel & Macko, 1993). For example, carbon isotopes have been routinely used to distinguish the biosynthetic pathway of the carbon to be either C3 or C4 synthesis. Stable nitrogen isotopes, sometimes in combination with carbon isotopes, have been used to assess trends in early diagenesis and to elucidate conditions on the early Earth, and to assess the origins of organic nitrogen in extraterrestrial materials as well as to establish trophic orders in modern and fossil food chains (Macko et al., 1991; Macko et al., 1999; Engel & Macko, 2001; MacAvoy et al., 2002).

3. Bulk Isotope Techniques

Samples for bulk organic isotope analysis are usually refluxed in distilled dichloromethane prior to isotope analysis in order to remove lipids. The lipid-extracted samples are converted to carbon dioxide and nitrogen gas for isotope analysis using an elemental analyzer, which was coupled to a stable isotope ratio mass spectrometer. Generally, this conversion is a high-temperature combustion involving strong oxidation at high temperature (1020°C) followed by reduction at lower temperature (650°C). The effluent gases are introduced into the mass spectrometer using a continuous flow interface. The stable isotopic ratio is reported as follows:

$$\delta^N E = [R_{sample}/R_{standard} - 1]10^3 (\text{‰}) \qquad (9)$$

where N is the heavy isotope of the element E and R is the abundance ratio of the heavy to light isotopes (for example, $^{13}C/^{12}C$ or $^{15}N/^{14}N$) of that element. The standard for carbon is the Pee Dee Belemnite limestone (PDB) and for nitrogen the standard is atmospheric nitrogen gas. These are assigned $\delta^N E$ values of 0.0‰. The reproducibility of the measurement is typically better than ±0.2‰ for the elements using the continuous flow interface on the mass spectrometer. In the laboratory, the samples are commonly measured against tanks of carbon dioxide and nitrogen gases that have been calibrated against the standards NBS 22 and atmospheric nitrogen, respectively.

4. Compound Specific Isotope Analysis

Stable isotopic determinations made on bulk materials are the weighted averages of the isotopic compositions of mixtures of hundreds to thousands of chemical compounds, each of which has its own isotopic abundances. The relative contribution of each of these materials to the isotopic content of the bulk material could theoretically be quantified through mass-balance or isotopic-mixing equations. The stable isotope analysis of individual molecular components holds great potential as a method of tracing the source, biochemistry, diagenesis, or indigeneity of a material.

The compounds studied to date include hydrocarbons, tetrapyrroles (chlorophyll derivatives), fatty acids, carbohydrates, and amino acids. In recent studies, nitrogen and carbon isotope analyses of components of petroleum and hydrocarbon extracts of sediments have indicated the preservation of original source materials. Isotope analyses of individual amino acids using both carbon and nitrogen isotopes have been useful in detail-ing indigeneity of organic matter in meteorites and fossils as well as helping to understand diagenesis. Inscribed in the isotopic signature is an indication of the biosynthetic pathway used in the formation of the compound. The transfer of nitrogen and carbon within the organism forming the component can thus be better understood. These pathways in turn imprint the signature of the organism in the rocks and sediments from which the compounds can later be isolated.

Over the years, numerous attempts have been made to isolate individual molecular components using liquid and gas techniques in order to better interpret or trace the source of history of an organic material. The possibility of comparative biochemistry in modern or fossil organisms has been suggested through the assessment of the isotopic differences between compounds of a family of components. Such differences are the result of enzymatic fractionation effects during synthesis or metabolism of the compound. Isotopic compositions of individual hydrocarbons have the potential for establishing sources for the materials, bacterial or otherwise, and have been useful in correlation techniques both in the petroleum industry and in pollution assessment. Individual carbohydrate isotope compositions also show great potential in metabolic and diagenetic studies. Depletions in the isotopic compositions of the products of reactions permit calculations to be done that quantify use and production of new organic materials and resolve them from native materials, even though the chemical compositions of the substances are identical. The studies to date have yielded important information regarding the source and history of the compounds characterized.

Through recent technological advancements, GC effluents can be combusted and the resulting carbon dioxide directly introduced into an IR-MS. This modification, GC-IRMS, allows for rapid analysis of the carbon and nitrogen isotopes on components in a mixture, and with increased sensitivity, on the order of 0.5 nmol/L of each compound. GC-based systems are presently constrained by the volatility of the components investigated. Compounds that do not have this constraint include hydrocarbons, the analyses of which have already clearly demonstrated the power of such technology and measurements in the assessments of source and the history of organic materials.

Nonvolatile, multifunctional molecules, including carbohydrates, fatty acids, and amino acids, require derivatization prior to GC analysis to increase volatility. Through the derivatization, additional carbon (but not nitrogen) is thus added to the parent compound. This addition, as well as fractionations associated with derivatization procedures involving bond rupture and formation in, for example, esterification and acylation, needs to be corrected in order to ascertain the original isotopic composition of the compound. The original carbon and nitrogen isotopic compositions of individual amino acids and their stereoisomers have been able to be computed, however, through analysis of standards prepared in a similar fashion. Through GC-IRMS analysis of amino acids, enrichments of ^{13}C and ^{15}N for both stereoisomers of the same amino acid from a meteorite have confirmed, for example, the extraterrestrial origin of those components and supported the lack of contamination by terrestrial compounds in the absolute concentrations and stereoisomer relationships.

Lipids that originate from tree waxes are more depleted (have less of the heavy isotope) than those derived from marine plants. Thus, in sediments, resolution of chemically indistinguishable sources is now possible. Additionally, pollutants, including PCBs and phenylalanine hydroxylase (PAHs), are ideally suited for the source and history analysis through the GC-IRMS analysis of extracts of natural materials (Zepp & Macko, 1997; Yanik et al., 2003a,b). Hydrocarbon components from fossil fuels have isotopic compositions that are readily resolvable from those of "natural" lipids, and thus have potential in the analysis and tracking of pollutants. Atmospheric contaminants from biomass burning have been able to be tracked to source materials because of the selective isotope signals of both the PAHs produced and the volatilization of fatty acids.

5. GC and GC-IRMS Procedures

The chromatographic separation techniques have a wide field of applicability in biochemistry, such as some of the examples given below (see also Knapp, 1979).

a. Amino Acids
Solutions of amino acids are extracted from mineral matrices using 6N HCl and dried under a stream of N_2 at 40°C. The dried samples are esterified with acidified 2-propanol for 1 hour at 110°C. The solvent is removed by evaporation under a gentle stream of nitrogen gas at 25°C. The amino acid isopropyl esters are acylated with trifluoroacetic anhydride (TFAA) at 110°C. Next, the excess TFAA and CH_2Cl_2 are removed by evaporation under N_2. The derivatives are redissolved in CH_2Cl_2. The TFAA isopropyl esters of the individual enantiomers of amino acids are analyzed directly for their stable carbon or nitrogen isotope compositions by using the GC-IRMS system (Macko et al., 1997).

b. Carbohydrates
Alditol acetates of the individual sugars are prepared with the sample reacted with sodium borohydride to reduce the aldehyde group. Following neutralization with acetic acid, which destructs the carbohydrate-borate complex, and vacuum rotary evaporation, the carbohydrate mixture is reacted with acetic anhydride. The acetylated products are then rotary evaporated and subsequently washed with methanol and chloroform and lastly, filtered prior to analysis by GC (Macko et al., 1998).

c. Fatty Acids and Hydrocarbons
The fatty acids are derivatized to fatty acid methyl esters (FAME) with BF_3 in methanol and extracted with hexane. The excess solvent is evaporated. Hydrocarbons can be analyzed directly in the GC-IRMS system without derivatization.

d. GC-MS Techniques
For the separation of the individual components by GC, the columns and conditions are optimized for the types of components. As an example, amino acid stereoisomers are separated using a chiral phase column and a temperature program. The GC conditions are as follows: ~1–2 nmol/L of each derivative is injected and subsequently introduced directly into the source of the MS; the carrier gas is ultrapure He (99.9999%) at a head pressure of 80 kPa; the injector temperature is 200°C; the GC temperature program is 45°C for 3 minutes, 45–90°C at 45 minutes, 90°C isothermal for 15 minutes, 90–190°C at 3°C per minute, and then 190°C isothermal for 30 minutes. The solvent (ethyl acetate) peaks are removed from the effluent of the GC through a heart split valve, which is open at the time of injection. The valve is programmed to close to allow the column effluent to be directed to the mass spectrometer.

e. GC-IRMS Techniques
For isotope analysis, an IR-MS is interfaced to a GC through a combustion furnace (copper/nichrome wire, at 850°C) and a cryogenic trap (–90°C). The GC is equipped with a fused silica capillary chromatographic column with the appropriate phase, temperature, and flow rates again optimized for the separation. The performance of the GC-IRMS system, including the combustion furnace, is evaluated by injection of a laboratory standard of known isotopic composition. For all runs, background subtraction is performed using the parameters supplied by the GC-IRMS software. Two to five replicate GC-IRMS runs are performed for each sample. The reproducibility ranges between 0.1 and

0.5% (1 standard deviation). The accuracy of the samples' analyses is also monitored by co-injection of a laboratory standard of known isotopic composition. The isotopic shift due to the carbon introduced, for example, in the fatty acids methylation is corrected by the following relationship (MacAvoy et al., 2002):

$$\delta^{13}C(FAME)$$
$$= f(FA)\delta^{13}C(FA) + f(MeOH)\delta^{13}C(MeOH) \quad (10)$$

where $\delta^{13}C(FAME)$, $\delta^{13}C(FA)$, and $\delta^{13}C(MeOH)$ are the carbon isotope compositions of the fatty acid methyl ester, the fatty acid, and the methanol used for methylation of the fatty acid, respectively; $f(FA)$ and $f(MeOH)$ are the carbon fractions in the fatty acid methyl ester in the underivatizad fatty acid and methanol, respectively. Amino acids and other less volatile components require derivatization for GC analysis, and will have similar correction strategies to obtain the naturally occurring isotopic composition.

Analysis of the stable nitrogen or carbon isotope composition of each component is accomplished by comparison to reference gas pulses introduced at the start of the run and following the opening of the heart split valve at the end of each run, i.e., after 4500 sec.

V. Conclusion

The potential for application of organic geochemical analysis to any multitude of environmental, ecological, or biochemical research areas is only beginning to be realized. Extension of compound-specific isotope analytical data derived from modern organisms and settings to yield interpretations of ancient depositional environments certainly appears possible. Further application of the technologies to understand the cycling of carbon and nitrogen, the identification and alteration of pollutants, or resolve metabolic relationships between compounds in living or extinct organisms are all within the scope of future research.

See Also the Following Chapters

Chapter 23 (Environmental Pathology) · Chapter 25 (Speciation of Trace Elements) · Chapter 28 (Mineralogy of Bone) · Chapter 30 (Histochemical and Microprobe Analysis in Medical Geology)

Further Reading

Bertin, E. P. (1975). *Principles and Practice of X-Ray Spectrometric Analysis*, Plenum Press, New York.

Billmark, K. A., Swap, R. J., and Macko, S. A. (2003). Characterization of Sources for Southern African Aerosols Through Fatty Acid and Trajectory Analyses, *J. Geophys. Res.*, 108D13, 8503 SAF 39, 1–9.

Buhrke, V. E., Jenkins, R., and Smith, D. K. (1997). *A Practical Guide for the Preparation of Specimens for X-Ray Fluorescence and X-Ray Diffraction Analysis*, John Wiley & Sons, New York.

Engel, M. H., and Macko, S. A. (1993). *Organic Geochemistry, Principles and Applications*, Plenum Press, New York.

Engel, M. H., and Macko, S. A. (2001). The Stereochemistry of Amino Acids in the Murchison Meteorite, *Precambrian Res.*, 106, 35–46.

Jenkins, R., Gould, R. W., and Gedcke, D. (1981). *Quantitative X-Ray Spectrometry*, Marcel Dekker, New York.

Johnson, A., Lalor, G. C., Robotham, H., and Vutchkov, M. K. (1996). Analysis of Jamaican Soils and Sediments by Energy-Dispersive X-Ray Fluorescence Spectrometry, *J. Radioanal. Nucl. Chem.*, 209(1), 101–111.

Kingston, H. M., and Jassie, L B. (Eds.) (1988). *Introduction to Microwave Sample and Preparation: Theory and Practice*, American Chemical Society, Washington DC.

Knapp, D. R. (1979). *Handbook of Analytical Derivatization Reactions*, Wiley & Sons, New York.

Lalor, G. C. (1995). *A Geochemical Atlas of Jamaica*, Canoe Press, University of the West Indies.

Lalor, G., Rattray, R., Simpson, P., and Vutchkov, M. K. (1999). Geochemistry of an Arsenic Anomaly in St. Elizabeth, Jamaica, *Environ. Geochem. Health*, 21(1), 3–11.

Lalor, G., Rattray, R., Vutchkov, M., Campbell, B., and Lewis-Bell, B. (2001). Blood Lead Levels in Jamaican School Children, *Sci. Total Environ.*, 269, 171–181.

Lalor, G. C., Vutchkov, M. K., Grant, C., Preston, J., Figueiredo, A. M. G., and Favaro, D. I. T. (2000). INAA of Trace Elements in Biological Materials Using the SLOWPOKE-2 Reactor in Jamaica, *J. Radioanal. Nucl. Chem.*, 244(2), 263–266.

L'vov, B. V. (1961). *Spectrochim. Acta*, 17, 761.

MacAvoy, S. E., Macko, S. A., and Joye, S. B. (2002). Fatty Acid Carbon Isotope Signatures in Chemosynthetic Mussels and Tube Worms From the Gulf of Mexico Hydrocarbon Seep Communities, *Chem. Geol.*, 185, 1–8.

Macko, S. A., and Engel, M. H. (1991). Assessment of Indigeneity in Fossil Organic Matter: Amino Acids and Stable Isotopes, *Phil. Trans. R. Soc. Lond. B*, 333, 367–374.

Macko, S. A., Engel, M. H., Andrusevich, V., Lubec, G., O'Connell, T. C., and Hedges, R. E. M. (1999). Documenting the Diet in Ancient Human Populations Through Stable Isotope Analysis of Hair, *Phil. Trans. R. Soc.*, 353, 1–12.

Macko, S. A., Ryan, M., and Engel, M. H. (1998). Stable Isotopic Analysis of Individual Carbohydrates by GC/C/IRMS, *Chem. Geol.*, 152, 205–210.

Macko, S. A., Uhle, M. E., Engel, M. H., and Andrusevich, V. (1997). Stable Nitrogen Isotope Analysis of Amino Acid Enantiomers by Gas Chromatography/Combustion/Isotope Ratio Mass Spectrometry, *Anal. Chem.*, 69, 926–929.

Schlemmer, G., and Radziuk, B. (1999). *Analytical Graphite Furnace Atomic Absorption Spectrometry: A Laboratory Guide*, Springer-Verlag, Berlin.

Soete, D., De., Gijbels, R., and Hoste, J. (1972). *Neutron Activation Analysis*, John Wiley & Sons, New York.

Tertian, R., and Claisse, F. (1982). *Principles of Quantitative X-Ray Fluorescence Analysis*, Heyden & Son, Philadelphia, PA.

Vutchkov, M., Grant, C., Lalor, G. C., and Preston, J. (2000). Standardization of the SLOWPOKE-2 Reactor in Jamaica for Routine NAA, *J. Radioanal. Nucl. Chem.*, 244(2), 355–359.

Walsh, A. (1955). The Application of Atomic Absorption Spectra to Chemical Analysis. *Spectrochim. Acta*, 1955; 7, 108–117.

Waples, D. (1981). *Organic Geochemistry for Exploration Geologists*, Burgess Publishers, New York.

Welz, B., and Sperling, M. (1999). *Atomic Absorption Spectrometry*, Weinheim, Germany.

Yanik, P. J., O'Donnell, T. H., Macko, S. A., Qian, Y., and Kennicutt, M. C. (2003a). Source Apportionment of Polychlorinated Biphenyls Using Compound Specific Isotope Analysis, *Organ. Geochem.*, 34, 239–251.

Yanik, P. J., O'Donnell, T. H., Macko, S. A., Qian, Y., and Kennicutt, M. C. (2003b). The Isotopic Composition of PAHs During Biodegradation, *Organ. Geochem.*, 34, 165–183.

Zepp, R. G., and Macko, S. A. (1997). Polycyclic Aromatic Compounds in Sedimentary Records of Biomass Burning. In *Sediment Record of Biomass Burning and Global Change* (J. Clark, H. Cachier, and J. G. Goldammer, Eds.), Springer-Verlag, Berlin, Germany, pp. 145–168.

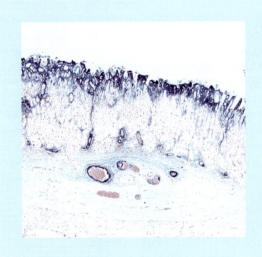

CHAPTER 30

HISTOCHEMICAL AND MICROPROBE ANALYSIS IN MEDICAL GEOLOGY

José A. Centeno, Todor Todorov, Joseph P. Pestaner, and Florabel G. Mullick
The Armed Forces Institute of Pathology

Wayne B. Jonas
Samueli Institute for Information Biology

CONTENTS

I. INTRODUCTION

Understanding the detrimental effects that geochemical processes and environmental pollutants may have on the health of humans and animals has been the subject of extensive study in medical geology and environmental pathology. For example, information obtained from the chemical analysis of mineral deposits in tissues may provide insight into a particular disease state and assist in the development of new treatments and therapy. In many cases, diseases related to the environment can be directly linked to the presence and distribution of toxic chemical elements in the soil, air, or water, such as chronic exposure to arsenic through contaminated drinking water (Centeno *et al.*, 2002). To demonstrate such links, it is necessary to analyze minerals and geo-environmental toxins to obtain information on the possible origins of such diseases. Because of the ever-growing complexity of geological sources and toxic environmental, biological, and chemical agents, accurate, rapid, and nondestructive techniques for qualitative and quantitative analysis of these materials are essential.

The identification and quantification of biogeochemical mineral deposits in human tissues are often valuable adjuncts to defining the diagnosis or understanding the pathogenesis and etiology of a disease. In the chemical pathology laboratory, analyses of tissues may proceed in a variety of ways but generally can be classified into three types: (1) histochemical/light microscopy, (2) ultrastructural methods (*e.g.*, scanning electron microscopy, energy dispersive x-ray microanalysis), and (3) spectroscopic techniques.

Utilizing light microscopy with histochemical stains on paraffin block tissue sections is often the simplest and least expensive approach for rendering a histopathological diagnosis. This process involves the use of selective stains that are sensitive to the presence of particular components within tissues; a well-known example is the Prussian blue stain used to highlight the presence and location of iron deposits in tissues. Because histochemical methods often utilize morphological abnormalities rather than biochemical changes, however, it is frequently difficult to use such methods to completely assess the identities of complex foreign materials and endogenous cellular components in tissues.

On the other hand, ultrastructural and microprobe methods, such as scanning electron microscopy (SEM) combined with energy-dispersive x-ray analysis (EDXA), are useful for identification of mineral deposits (Goldstein *et al.*, 2002). Despite the wide use of these electron-based optical systems, the SEM and EDXA techniques do have some disadvantages. Both techniques require extensive sample preparation (*e.g.*, paraffin sections must be mounted on pure carbon disks and coated with a layer of gold or carbon to avoid electron beam charging effects), and, although these techniques provide useful information on the elemental composition and morphological features of the sample, they may lack the required structural, molecular, and chemical sensitivity for *in situ* identification and localization of foreign substances in tissues.

Other physical and chemical techniques that have been traditionally used for the study of foreign materials and mineral deposits in tissues include x-ray diffraction, x-ray absorption, ultraviolet absorption, fluorescence, interference microscopy, polarization microscopy, and autoradiography. The powder method of x-ray diffraction employing the Gandolfi (1967) camera has been particularly useful for the analysis of crystals that have been separated from tissues or obtained from tissue biopsy specimens. In contrast to the combined SEM–EDXA technique, which provides information on the elemental composition of mineral specimens, x-ray diffraction has been widely used for the identification of individual chemicals or mixture of chemicals (Willard *et al.*, 1988).

Recent technological advances in optical instrumentation have led to the development of new microprobe techniques that are capable of providing *in situ* analysis, chemical distribution (mapping), and quantitative information of tissues. Chemical and molecular spectroscopic techniques, such as Fourier transform infrared (FTIR) microspectroscopy and Raman microprobe spectroscopy, provide accurate, rapid, and selective identification by virtue of a molecule's characteristic spectrum of vibrational frequencies (Gillie *et al.*, 2000; Mulvaney & Keating, 2000). Raman microprobe spectroscopy is a light-scattering technique in which a laser

of known wavelength is scattered off a molecule of interest (*e.g.*, mineral particle) to produce a spectrum of bands representing movements of atoms in the molecule (McCreery, 2000). A Raman microprobe instrument provides 1-µm spatial resolution on histological sections and does not require staining for identification of chemical species (Centeno *et al.*, 1999; Hanlon *et al.*, 2000). Analysis is nondestructive and is easily correlated with morphological features to aid in the evaluation of a disease process. The Raman microprobe can be used to identify the microscopic contents of various organelles and diseased versus healthy tissue through determination of differences in vibrational spectra of intrinsic molecules; as a result, it is now possible to provide full chemical mapping of specimens of interest without the use of histochemical stains.

Infrared microspectroscopy, another vibrational technique, in principle has characteristics similar to those of Raman scattering (Diem, 1999). In comparison with Raman microprobe spectroscopy, however, infrared microscopy suffers from two disadvantages: Spectral resolution is one order of magnitude poorer, and aqueous-based systems are intense absorbers. In contrast, water is a weak scatterer in Raman spectroscopy, which makes it an excellent medium for studying biological systems. Both infrared microscopy and Raman microprobes allow the analysis of individual particles as seen in tissue sections or of digested bulk samples in cases where digestion of tissues is required (Centeno *et al.*, 1994).

This chapter describes the use of histochemical, microspectroscopic, ultrastructural, and microprobe tools available to biomedical scientists that are essentially identical to those used by geoscientists. The preferred technique in a particular situation depends on the analyte; thus having some idea of what is being examined is important. The primary focus here is on the Raman microprobe because of its nondestructive operation and simple sample preparation. The Raman microprobe can be used to assist in the identification of such materials as calcium oxalate inclusions in tissues as well as for myriad other applications that are discussed in detail later in this chapter. For detailed information on chemical analyses in the geosciences, see Chapter 29.

II. HISTOCHEMICAL TECHNIQUES

Histochemical examination of tissue sections is the simplest and least expensive technique for analyzing minerals in tissue. The process involves the use of selective stains that are sensitive to the presence of particular components, such as minerals within tissues. This short review discusses the most prominent histochemical techniques; for greater detail on the various staining techniques, see *Armed Forces Institute of Pathology Special Staining Techniques* (1995). The three methods used today for analyzing minerals in tissue section are microincineration, digestion of tissues, and the use of special stains.

A. Microincineration

Microincineration is a procedure employed to study mineral content without the interference of organic elements; in particular, iron, calcium, magnesium, and silicon can be identified in incinerated material. Special stains allow dyes to bind to tissue elements to enhance microscopic details of specific elements in the tissue and to assist in ruling out other possibilities, such as infection. Incineration is accomplished by increasing the temperature of the tissue on a slide to as high as 650°C; the sample is then cooled. A Bunsen burner or appropriate furnace may be used for the incineration. Incinerated organic material (*i.e.*, ashes) appears gray to colorless; illumination and transmitted light microscopy are reduced. Calcium and magnesium oxides in tissues appear as a white ash, silicon is a white birefringent crystalline material, and iron appears as a red to yellow oxide.

B. Digestion of tissues

Digestion of tissues in 1-*N* sodium hydroxide is another histochemical technique used to separate foreign materials from fresh or fixed tissues. The procedure involves heating the tissue in NaOH for approximately 1 hour, after which the digested material is rinsed in absolute ethanol and washed with distilled deionized water. The resulting sediment is examined using light microscopy, which includes the use of polarized light and darkfield microscopy. In addition, residues can also be placed on a carbon disk and examined by SEM–EDXA.

C. Special Stains

The evaluation of toxic tissue reactions to foreign or apparently foreign materials in tissues is a frequent challenge to pathologists. Typically, such reactions range from minimal hemorrhage to extensive inflammatory reaction, and the question becomes one of determining the etiology of the reaction. Generally, infectious agents and exogenous minerals and elements generate similar inflammatory reactions; thus the pathologist's first approach to such histological analysis is to stain for infectious agents so that prompt and proper medical therapy may be administered. When inflammatory reactions are due to infectious agents, stains that are helpful in this diagnostic endeavor include hematoxylin and eosin (H&E), periodic acid–Schiff reaction (PAS stain), Gomori's methenamine silver, and Brown–Brenn Gram stain, as well as acid-fast staining techniques. Such staining techniques are beyond the scope of this chapter, but indepth treatments of these staining techniques can be found in staining manuals (see, for example, Prophet *et al.*, 1995). The exclusion of infectious agents is clinically important before more extensive evaluation for minerals is performed.

D. Selected Histological Techniques for Iron, Calcium, Copper, and Zinc

A deficiency or excess of iron in developing countries has been associated with the occurrence of a wide range of human diseases and health conditions, including anemia, growth impairment, and even liver diseases. In individuals affected with diseases caused by excess iron, iron is found in tissue in both ferric and ferrous forms. Perl's Prussian blue reaction (see Prophet *et al.*, 1995), which is useful for identifying ferric ions, is a very sensitive histochemical technique that forms a stable pigment that is preserved even after other stains are performed. In addition, single granules of iron can be identified that might otherwise remain undetected. Perl's stain for ferric iron involves treatment of tissue sections with acid solutions of potassium ferrocyanide, which reacts with ferric ions in the tissue to form ferric ferrocyanide which possesses a blue color (*i.e.*, the Prussian blue pigment). An example of the use of this stain is shown in Figure 1, taken during the study of a case of ferric poisoning. The ferric ions have penetrated through the stomach wall and are observed in the adjacent blood vessels (blue stains around numerous vessels). Figure 1 demonstrates that, in this particular case, the excess iron has been absorbed through the stomach wall surrounding the blood vessels, which probably resulted in internal hemorrhage. In contrast, Turnbull's blue

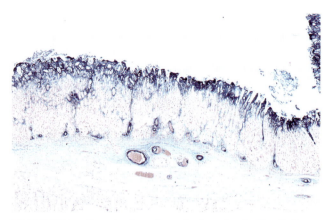

FIGURE 1 Examination of gastric tissue by the Pearl's Prussian blue histochemical stain for the presence and location of ferric iron (magnification 10×).

reaction utilizes potassium ferricyanide to demonstrate the presence of ferrous ions in tissue. The ferricyanide combines with any ferrous ions present in the tissue, resulting in Turnbull's blue pigment.

A link between calcium and cardiovascular disease has been the subject of many studies (see also Chapter 13, this volume). Because calcium levels in drinking water correlate with the extent of cardiovascular disease, identifying calcium ions in tissues is a helpful diagnostic tool. Alizarin red S and Von Kossa are special stains that may be used to demonstrate calcium in tissues. The Alizarin red reaction is not specific for calcium, as magnesium, manganese, barium, strontium, and iron may show similar reactions; positive staining depends on the chelation process with the dye, and calcium appears as an orange-red precipitate. The Von Kossa silver test is a metal substitution reaction where tissue sections are treated with a silver nitrate solution and silver is deposited in place of the calcium.

Copper is generally demonstrable only in diseases where it is present in relatively high concentrations, such as in Wilson's disease. Copper can be demonstrated by its reaction with rubeanic acid. Rubeanic acid reacts with copper to form a dark precipitate consisting of copper rubeanate. Nickel and cobalt have similar reactions, but these metals are soluble when acetate is in the staining solution; thus, using acetate in the staining solution makes the reaction with rubeanic acid more specific for copper. Nevertheless, variability in staining may be attributed to the protein binding of copper ions in tissue.

Zinc is an important element in cell metabolism, and its deficiency affects rapidly growing cells. Zinc toxicity is an unusual occurrence and usually has mild toxic effects. Zinc may be demonstrated by the dithizone reaction, where dithizone reacts with zinc to form an insoluble red-purple complex, although other metals can interfere with the reaction.

III. Microprobe (Vibrational) Spectroscopy

Definitive identification of foreign material for geologic and medical investigations can be problematic. Raman microprobe spectroscopy is a well-established vibrational technique for the study and identification of organic, polymeric, and inorganic materials (McCreery, 2000). Its utility in mineralogical analysis has increased significantly, as sample preparation is minimal, the test results are obtained rapidly, and the sample is not destroyed. In chemical pathology, the Raman microprobe has been used for more than 10 years for identification of chemical species of foreign inclusions and metabolic accumulations in histological sections of patients biopsied in order to understand their disease state. The microprobe has been fairly successful in identifying the chemical species of inclusions of unknown origin. Chemical mapping of such samples not only confirms the original identification but also provides additional information for the clinician regarding morphology, location, and distribution of species that can be useful in understanding the causes of diseases.

A. Raman Microspectroscopy

In Raman spectroscopy, the Raman effect or scattering signal arises due to the interaction of an incident laser beam (photons) (hv_0) that is focused onto a sample with the molecular vibrational modes (v_1) of the interrogating chemical species (Figure 2A). If no interaction takes place, the resulting Raman signal, known as Rayleigh scattering, occurs at the same frequency (hv_0) as the incident laser beam; hence, no information can be obtained from the chemical system. The result of the interaction, though, may produce Raman signals at frequencies higher (anti-Stokes) and lower (Stokes) than the frequency of the incident beam. At room temperature, the Stokes scattering is more intense and is used to characterize the chemical composition of the sample. The peaks or lines observed in a Raman spectrum are Raman active vibrations attributed to structurally active molecular bonds. The shift from the frequency of the

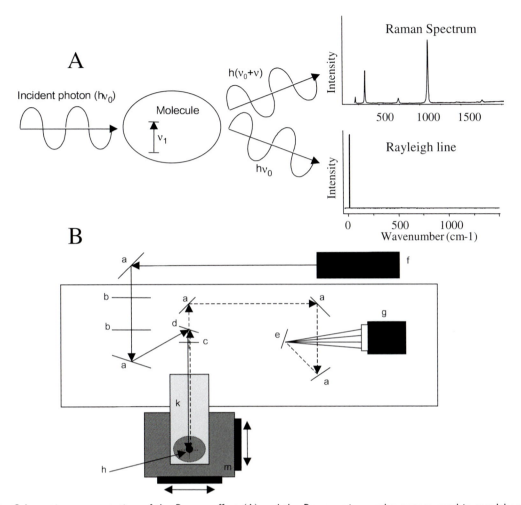

FIGURE 2 Schematic representation of the Raman effect (A) and the Raman microprobe system used in our laboratory (B). The excitation of the molecule by incident photons produces the Rayleigh line (bottom trace) and Raman scatter (top trace). Components of the Raman microscope setup (B) include: a, mirrors; b, beam-focusing lenses; c, holographic notch filter; d, beamsplitter; e, holographic grating; f, laser light source; g, low noise charge-coupled device (CCD) camera; h, specimen to be analyzed; k, microscope; and m, X–Y motorized stage.

incident wavelength is known as the Raman shift (*x*-axis). Although Raman spectroscopy has been widely used for the study of macrochemical systems, only recently has wide use of the technique coupled to high-resolution (1 μm) optical microscopes been demonstrated. In contrast to SEM and EDXA, which require complex processing and handling of the tissue, Raman microprobe spectroscopy can be used with fresh tissues as well as paraffin-embedded tissue sections without having to destroy, decompose, or extract the tissue to study the foreign material.

The Raman microprobe experiments reported in this chapter were conducted employing the system shown schematically in Figure 2B. Briefly, the system consists of a spectrograph (model LabRam, Jobin Yvon, Edison,

NJ) equipped with holographic gratings blazed at 1800 and 600 grooves/mm and interfaced to an Olympus microscope (Model BH-40). Specimens can be visualized using white light or laser source at magnification of 100× objective (N.A. = 0.90), 50× objective (N.A. = 0.75), or 10×. For light excitation, several laser lines may be used with wavelengths at 514 nm, 632 nm, and 785 nm. As demonstrated in Figure 2B, during the Raman microprobe measurement, a laser is focused onto the specimen using one of the objectives. The backscattered Raman signal is collected by the same objective, collimated, and directed into the spectrometer to determine the frequencies or wavelengths. The minimum size of the laser focus spot is 1 μm. The Raman scatter signal is then analyzed by a detector

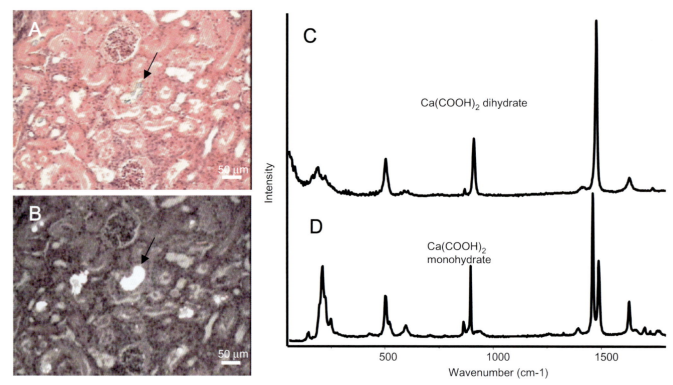

FIGURE 3 Raman spectroscopic investigation of Ca(COOH)$_2$ inclusions in kidney tissues. Panel A shows the light microscopy (H&E) and panel B the crossed polarized (H&E) images of calcium oxalate crystals in renal tissues. The arrow in the white light image (A) indicates the particle for which spectra were taken. Both forms of calcium oxalate—the monohydrate (Ca(COOH)$_2 \cdot$ H$_2$O) and dihydrate (Ca(COOH)$_2 \cdot$ 2H$_2$O)—were identified on the same crystal, which reveals the sensitive nature of Raman microprobe. The spectra were obtained using a 632-nm laser excitation and 100× magnification.

system consisting of a charged coupled device (CCD). The following examples illustrate the use of Raman microprobe spectroscopy.

B. Identification of Mineral Materials in Inorganic Particulate Tissues and Geo-Environmental Samples by Raman Microprobe Spectroscopy

The chemical, elemental, and/or molecular identification of mineral deposits in human tissues is an important step in defining a diagnosis or understanding the pathogenesis of a disease. Raman microprobe provides a useful analytical and diagnostic approach for the qualitative and quantitative identification of mineral deposits in human tissues. The formation of minerals such as calcium oxalate crystals in tissues serves as a simple example for the *in situ* nondestructive and noninvasive application of this technique in human pathology. The occurrence of calcium oxalate crystals in tissues is frequently associated with the formation of

chronic granulomatous inflammatory lesions from a wide range of causes. As such, these crystals may be regarded as products of inflammatory cells. Identification of these crystals in tissues has been demonstrated by the use of histochemical methods based on the microincineration of oxalate to produce carbonate (Johnson & Pani, 1962). SEM–EDXA has also been used to demonstrate the presence of calcium and oxygen. Each of these techniques requires extensive sample preparation procedures, and none of these methods provides the capability of studying the localization *in situ* of these crystals within tissues. The Raman microprobe, on the other hand, not only is capable of providing information about the localization of crystals but also provides multidimensional information on the chemical nature of the crystal. Figure 3 demonstrates the use of the Raman microprobe for the study of calcium oxalate crystals in kidney tissue (Pestaner *et al.*, 1996). Under ordinary light microscopy and crossed polarized image (Figures 3A and 3B, respectively; H&E stain), crystals of calcium oxalate are observed. In this case, Raman spectroscopy proved to be very useful in the identification of not one but two forms of calcium oxalate present in the mineral inclusion: the monohydrate and dihydrate forms. These species are difficult to distinguish by classical histochemical methods. The monohydrate is characterized by a doublet at 1463/1491 cm^{-1}. The two bands merge into one band at 1476 cm^{-1} for dihydrate calcium oxalate. This spectral difference provides rapid identification of calcium monohydrate and dihydrate mineral inclusions in tissue.

The characterization and identification of inorganic atmospheric particulate that may be involved in the development of respiratory diseases is an increasing concern for environmental pathologists and geoscientists. Formation of the so-called "intercontinental dust" or "African dust" has been described (Prospero, 2001); however, its composition and impact on human health have not been fully characterized. Applying the Raman microprobe to improve our understanding of the composition and behavior of this type of contaminant is expected to play an increasing role in medical geology and environmental pathology. Figure 4 demonstrates the histological and *in situ* analysis of dust particulates in lung tissues employing the confocal Raman microprobe. The histological appearance of dust particles in this particular case is characterized by the presence of opaque granular black material and is morphologically consistent with carbon and silica (Figure 4A). The opaque black granules are easily discernible under white light image (Figure 4B). Using the Raman microprobe, the foreign material was identified as a carbon-based dust particle with characteristic peaks at 1326, 1574, and 2656 cm^{-1} (Figure 4C).

In addition to point acquisition, as demonstrated in Figures 3 and 4, Raman microspectroscopy can be used to generate maps or images based on different Raman spectral features (*i.e.*, vibrations from the various components) (Carden & Morris, 2000; Ling *et al.*, 2002; Shafer-Peltier *et al.*, 2002; Uzunbajakava *et al.*, 2003). Figure 5 shows an example of Raman-based images obtained from a commercially available asbestos reference material. The particular type of asbestos reference material used in this study was amosite, with a representative spectrum as indicated in Figure 5G. The Raman chemical imaging analysis demonstrated the presence of two other components within the area investigated. The first component (upper right corner of the white light image, Figure 5D) is identified as a dust/coal particle. The Raman image generated based on the spectrum of the particle (Figure 5E) is shown in Figure 5A, and it can be observed that dust/coal is present only in that region. A second impurity was also

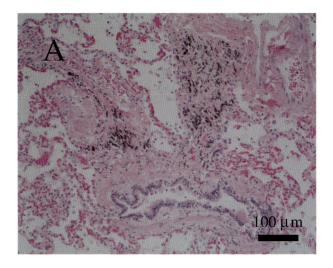

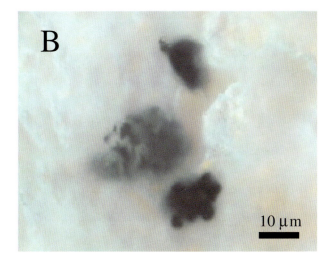

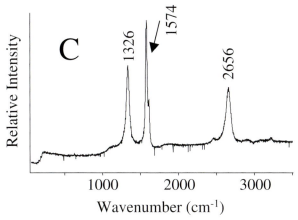

FIGURE 4 Analyses of inorganic particulate in lung tissues by Raman microprobe spectroscopy. (A) Histological section (H&E stain) at 10× magnification; (B) light microscopy image of unstained tissue demonstrating the black opaque dust inclusions (100× magnification); (C) Raman spectra of a carbon-based dust inclusion in the tissue section shown in (B).

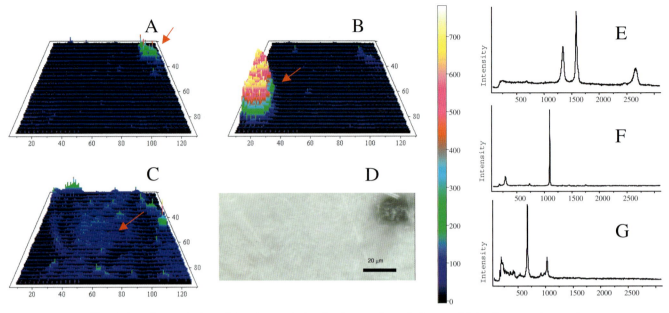

FIGURE 5 Chemical and morphological imaging analysis of asbestos (amosite) material by Raman microprobe spectroscopy. (A) Three-dimensional Raman image corresponding to spectrum D (dust/charcoal particle); (B) three-dimensional Raman image corresponding to spectrum E (unidentified material); (C) three-dimensional Raman image corresponding to spectrum F (the asbestos component); (D) white light image of area analyzed by Raman spectroscopy. The laser excitation line at 632 nm was used with a 50× magnification objective.

found (lower left corner of the white light image, Figure 5D). This unidentified particle is difficult to distinguish from the asbestos material based on white light microscopy, but it is readily distinguished by Raman spectroscopy (Figures 5B and F). The remaining part of the image area shows amosite crystals as observed both in the Raman and the white light images.

C. Infrared Spectroscopy

In addition to Raman microprobe spectroscopy, infrared (IR) spectroscopy is another useful vibrational technique for the *in situ* identification of mineral materials in tissues. IR spectroscopy is based on the absorption of energy by vibrating chemical bonds (primarily stretching and bending motions). As discussed earlier in this chapter, Raman scattering results from the same type of transitions, but the selection rules are different so the weak bands in the IR may be strong in the Raman and *vice versa*; thus, IR and Raman spectroscopies are complementary and when used together can provide a powerful tool for chemical compositional analysis.

IV. ELECTRON MICROSCOPY

Ultrastructural analysis allows minerals to be analyzed in an efficient and sensitive manner. The scanning electron microscope is a versatile approach for the study of micro-inclusions in tissues, as it has the capability of examining objects at low and high magnifications with substantial depth of field and a three-dimensional appearance. SEM combined with EDXA is a very useful procedure for identifying mineral substances even though EDXA provides information on elemental composition rather than specific compounds.

To prepare samples for SEM–EDXA of tissue sections, 4- to 6-μm-thick sections are placed on pure carbon disks. Prior to mounting the section on the disc, the carbon disk is washed with concentrated H_2SO_4 (to remove iron), thoroughly rinsed with distilled deionized water, placed in acetone, and ultrasonicated for 2 hours. The carbon disk is then washed with distilled deionized water again and placed in a vacuum oven to dry for 12 hours. Prior to SEM–EDXA, tissue sections mounted on the disks are deparaffinized with xylene and absolute ethyl alcohol. A Hitachi S-3500N scanning electron microscope with one energy-dispersive spectrometer

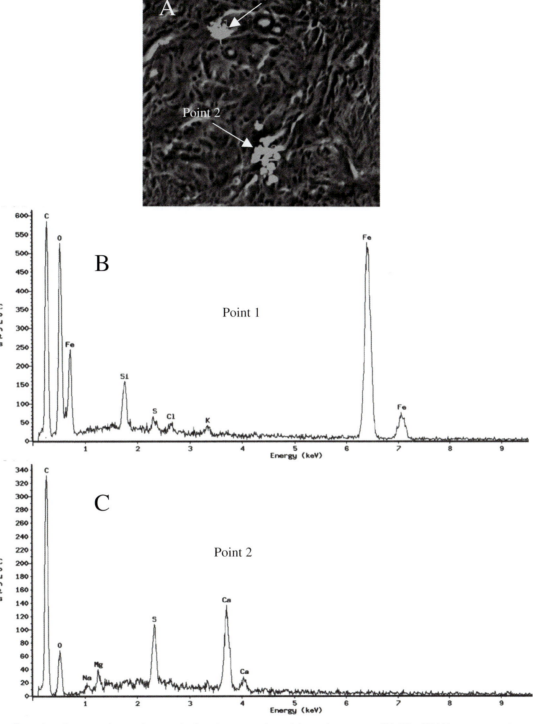

FIGURE 6 Scanning electron photomicrograph showing several particles in lung tissue (A). The EDXA spectrum demonstrates the presence of exogenous silicon (B), and endogeneous sulfur and calcium elements were found within the lung tissue (C). This case is from a 37-year-old male with a history of occupational exposure as a coal miner.

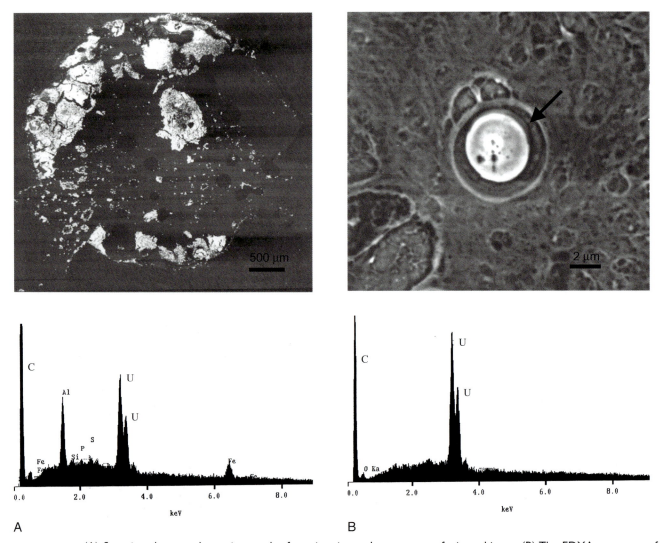

FIGURE 7 (A) Scanning electron photomicrograph of uranium in a subcutaneous soft tissue biopsy. (B) The EDXA spectrum of uranium at an energy of approximately 3.8 keV, as well as other elements, including aluminum and iron (magnification 20×). When using a higher magnification (200×), only the presence of uranium was detected within the subcutaneous soft tissue section.

(EDS) and KEVEX software were used to examine the tissue sections discussed here. An accelerating voltage of 20 keV and a 90-nA current were used for most of the measurements. Backscattered electron images at magnifications from 60 to 6000× were used to observe morphologic characteristics of tissue and to record the composition of the inorganic material.

Figure 6A is an SEM image of foreign material in lung tissues in which two clusters of particles can be seen. The EDXA spectrum is shown for clusters 1 and 2. Although the two particle clusters are similar in appearance in the SEM image, their elemental compositions as obtained by EDXA are different. Cluster 1 shows the presence of silicon, in addition to iron, sulfur,

chlorine, potassium, oxygen, and carbon (Figure 6B). The silicon, potassium, and iron suggest the possibility of the presence of silicates, which are exogenous materials (*i.e.*, they are not produced by living organisms). Cluster 2 is composed of calcium, sulfur, sodium, magnesium, oxygen, and carbon. All of these elements participate in a variety of biological processes or act as building blocks of tissue; thus cluster 2 is composed of endogenous material (*i.e.*, it might have been produced by the living tissue). Another example is shown in Figure 7A, which demonstrates the inclusion of uranium in a case of a subcutaneous soft tissue. The SEM–EDXA was obtained at two magnifications: 20× and 200×. At the lower magnification, the x-ray microanalysis demonstrated the presence of not only uranium but also aluminum and iron. At higher magnification, only uranium was detected. Energy-dispersive x-ray microanalysis allows us to examine the elemental composition of samples and to identify whether a material is of foreign or natural origin.

V. CONCLUSION

Each of the techniques described in this chapter—histochemical, Raman and infrared spectroscopies, and ultrastructural characterization with elemental analysis—has its advantages and limitations in chemical characterization of geochemical and medically relevant inclusions in tissue. Histochemical methods offer an inexpensive and simple approach, but they provide limited chemical information for complex biological systems. Light microspectroscopies have proven to be an excellent source for the chemical identification of tissue samples and other materials with or without minimal preparation, but instrumentation is expensive and requires the attention of well-trained technicians. SEM with x-ray microanalysis is extremely useful in morphological and elemental analyses, but sample preparation is extensive and instrumentation is expensive. No one technique, then, is highly superior when compared to the others; the choice of a method will depend on the particular sample, the instrumentation available, the type of information needed, and the expertise of the researcher.

SEE ALSO THE FOLLOWING CHAPTERS

Chapter 13 (Water Hardness and Health Effects) • Chapter 29 (Inorganic and Organic Geochemistry Techniques)

FURTHER READING

Carden, A., and Morris, M. D. (2000). Application of Vibrational Spectroscopy to the Study of Mineralized Tissues (review), *J. Biomed. Opt.*, 5(3), 259–268.

Centeno, J. A., Ishak, K. G., Mullick, F. B., Gahl, W. A., and O'Leary, T. J. (1994). Infrared Microspectroscopy and Laser Raman Microprobe in the Diagnosis of Cystinosis, *Appl. Spectrosc.*, 48(5), 569–572.

Centeno, J. A., Kalasinsky, V. F., Johnson, F. B., Vihn, T. N., and O'Leary, T. J. (1992). Fourier Transform Infrared Microspectroscopy Identification of Foreign Materials in Tissue Sections, *Lab. Invest.*, 292(2), 624–628.

Centeno, J. A., Mullick, F. G., Martinez, L., Page, N. P., Gibb, H., Longfellow, D., Thompson, D., and Ladich, E. R. (2002). Pathology Related to Chronic Arsenic Exposure, *Environ. Health Perspect.*, 110(5), 883–886.

Centeno, J. A., Mullick, F. G., Panos, R. G., Miller, F. W., and Valenzuela-Espinoza, A. (1999). Laser–Raman Microprobe Identification of Inclusions in Capsules Associated with Silicone Gel Breast Implants, *Mod. Pathol.*, 12(7), 714–721.

Diem, M. (1999). *Introduction to Modern Vibrational Spectroscopy*, 1st ed., John Wiley & Sons, New York.

Gandolfi, G. (1967). Discussion on Methods To Obtain X-Ray Powder Patterns from a Single Crystal, *Miner. Petrogr. Acta*, 13, 67–74.

Gillie, J. K., Hocklowski, J., and Arbuckle, G. A. (2000). Infrared Spectroscopy, *Anal. Chem.*, 72(2), 71R–79R.

Goldstein, J. I., Newbury, D., Joy, D., Lyman, C., Echlin, P., Lifshin, E., Sawyer, L., and Michael, J. (2002). *Scanning Electron Microscopy and X-Ray Microanalysis*, 1st ed., Plenum Press, New York.

Hanlon, E. B., Manoharan, R., Koo, T.-W., Shafer, K. E., Motz, J. T., Fitzmaurice, M., Kramer, J. R., Dasari, R. R., and Feld, M. S. (2000). Prospects for *In Vivo* Raman Spectroscopy, *Phys. Med. Biol.*, 45, R1–R59.

Johnson, F. B., and Pani, K. (1962). Histochemical Identification of Calcium Oxalate, *Arch. Pathol.*, 74, 347–350.

Ling, J., Weitman, S. D., Miller, M. A., Moore, R. V., and Bovik, A. C. (2002). Direct Raman Imaging Techniques for Study of the Subcellular Distribution of a Drug, *Appl. Opt.*, 41(28), 6006–6017.

McCreery, R. L. (2000). *Raman Spectroscopy for Chemical Analysis*, 1st ed., John Wiley & Sons, New York.

Mulvaney, S. P., and Keating, C. D. (2000). Raman Spectroscopy, *Anal. Chem.*, 72(2), 145R–157R.

Pestaner, J. P., Mullick, F. G., Johnson, F. B., and Centeno, J. A. (1996). Calcium Oxalate Crystals in Human Pathology: Molecular Analysis with the Laser Raman Microprobe, *Arch. Pathol. Lab. Med.*, 120, 537–540.

Prophet, E. B., Mills, B., Arrington, J. B., Sobin, J. B., and Sobin, L. H. (Eds.) (1995). *Armed Forces Institute of Pathology Special Staining Techniques*, American Registry of Pathology, Washington DC.

Prospero, J. M. (2001). African Dust in America, *Geotimes*, 46(11), 24–27.

Shafer-Peltier, K. E., Haka, A. S., Motz, J. T., Fitzmaurice, M., Dasari, R. R., and Feld, M. S. (2002). Model-Based Biological Raman Spectral Imaging, *J. Cell. Biochem. Suppl.*, 39, 125–137.

Uzunbajakava, N., Lenferink, A., Kraan, Y., Willekens, B., Vrensen, G., Greve, J., and Otto, C. (2003). Nonresonant Raman Imaging of Protein Distribution in Single Human Cells, *Biopolymers*, 72(1), 1–9.

Willard, H. H., Merritt, L. L., Dean, J. A., and Settle, F. A. (1988). *Instrumental Methods of Analysis*, 7th ed., Wadsworth Publishing, Belmont, CA.

LEONARD F. KONIKOW AND
PIERRE D. GLYNN
United States Geological Survey

CONTENTS

I. INTRODUCTION

In most areas, rocks in the subsurface are saturated with water at relatively shallow depths. The top of the saturated zone—the water table—typically occurs anywhere from just below land surface to hundreds of feet below the land surface. Groundwater generally fills all pore spaces below the water table and is part of a continuous dynamic flow system, in which the fluid is moving at velocities ranging from feet per millennia to feet per day (Figure 1). While the water is in close contact with the surfaces of various minerals in the rock material, geochemical interactions between the water and the rock can affect the chemical quality of the water, including pH, dissolved solids composition, and trace elements content. Thus, flowing groundwater is a major mecha-nism for the transport of chemicals from buried rocks to the accessible environment, as well as a major pathway from rocks to human exposure and consumption. Because the mineral composition of rocks is highly variable, as is the solubility of various minerals, the human health effects of groundwater consumption will be highly variable.

Groundwater provides about 40% of the public water supply in the United States. Also, most of the rural population in the United States, more than 40 million people, supply their own drinking water from domestic wells (Alley et al., 1999). Consequently, groundwater is considered an important source of drinking water in every state (Figure 2). Groundwater also is the source of much of the water used for irrigation, especially in areas with arid to semi-arid climates. Nearly all surface-water features (streams, lakes, reservoirs, wetlands, and estuaries) interact with groundwater. Groundwater is the source of base flow to streams and rivers (see Figure 1) and often is the primary source of water that sustains a wetland habitat.

It is long recognized that the chemical content of drinking water can have an adverse or beneficial affect on human health (Keller, 1978). Although the potential side effects associated with some trace elements of natural origin (e.g., arsenic, selenium) or anthropogenic origin (e.g., hexavalent chromium, organic compounds) present in concentrations exceeding public health standards for human consumption of drinking water has stituents dissolved in flowing water will tend to migrate with the water—the faster the water moves, the faster and further the solutes will migrate. This entrainment of dissolved chemicals is called advective transport.

The specific discharge calculated from Darcy's law represents a volumetric flux per unit cross-sectional area, but flow does not pass through the solid grains of the rock, only through the void spaces. Thus, to calculate the actual seepage velocity of groundwater, one must account for the actual cross-sectional area through which flow is occurring. The latter is done by dividing the specific discharge by the effective porosity of the porous medium. The effective porosity of fractured crystalline rocks can be less than 0.01, whereas for unconsolidated sands and gravels it can exceed 0.30.

C. Hydrodynamic Dispersion

Controlled laboratory and field experiments show that observed solute concentrations in a flow field cannot be predicted adequately just on the basis of seepage velocity, even for nonreactive constituents. Instead, it is observed that some solute will arrive at a given location sooner than predicted by the mean seepage velocity, whereas some solute arrives later than the mean velocity would indicate. That is, there is a spreading about the mean arrival time. Similarly, solute distribution will spread spatially with time and travel distance. This spreading and mixing phenomenon is called hydrodynamic dispersion. It results from molecular and ionic diffusion, and from mechanical dispersion arising from small-scale variations in the velocity of flow that cause the paths of solutes to diverge or spread from the average direction of groundwater flow (Bear, 1979). The outcome is a transient, irreversible, mixing (or dilution) process affecting the concentration distribution of a solute species in an aquifer.

The rate of solute flux caused by hydrodynamic dispersion is expressed in a form analogous to Fick's law of diffusion. This Fickian model assumes that the driving force is the concentration gradient and that the dispersive flux occurs in a direction from higher concentrations toward lower concentrations at a rate related to a constant of proportionality—the coefficient of hydrodynamic dispersion. However, this assumption is not always consistent with field observations and is the subject of much ongoing research and field study (see, for example, Gelhar et al., 1992). The coefficient of hydrodynamic dispersion is defined as the sum of mechanical dispersion and molecular diffusion (Bear,

1979). Mechanical dispersion is a function both of the intrinsic properties of the porous medium (expressed as a dispersivity coefficient, which is related to variability in hydraulic conductivity and porosity) and of the fluid flow (specifically, the fluid velocity). Molecular diffusion in a porous medium will differ from that in free water because of the effects of tortuous paths of fluid connectivity in porous media.

In most groundwater transport model applications, the dispersivity is defined in terms of just two unique constants: the longitudinal dispersivity and the transverse dispersivity of the medium. In practice, however, dispersivity values appear to be dependent on and proportional to the scale of the measurement. Field-scale dispersion (commonly called macrodispersion) results from large-scale spatial variations in hydraulic properties and seepage velocity. Consequently, the use of values of dispersivity determined for one scale of transport in a model designed to predict concentration changes over a different scale of travel probably is inappropriate. Overall, the more accurately and precisely a model can represent or simulate the true velocity distribution in space and time, then the uncertainty concerning representation of dispersion processes will be less of a problem.

D. Solute-Transport Equation

A generalized form of the solute-transport equation is presented by Bear (1979). The governing partial differential equation relates the change in concentration over time in a groundwater system to (1) hydrodynamic dispersion, (2) advective transport, (3) the effects of mixing with a source fluid that has a different concentration than the groundwater at the location of the recharge or injection, and (4) all of the physical, chemical, geochemical, and biological reactions that cause transfer of mass between the liquid and solid or air phases or conversion/decay of dissolved chemical species from one form to another. The chemical attenuation of inorganic chemicals can occur by sorption/desorption, precipitation/dissolution, or oxidation/reduction; organic chemicals can adsorb or degrade by microbiological processes and/or volatilization.

There has been considerable progress over the last 10–15 years in modeling reactive-transport processes; however, the complexity and computational requirements for solute-transport models and reaction models are intense and, therefore, applications of coupled multispecies reactive-transport models are rare. Although received the most attention in recent years, it is important to realize that many trace elements are greatly beneficial to human health (Hopps & Feder, 1986).

It is clear that the chemical content of natural waters varies greatly from place to place. Geology has a stronger and more direct effect on the quality of groundwater than on surface water. Surface water sources include a much greater component of direct precipitation and represent the rapid integration of water derived from large and diverse source areas within a drainage basin than do groundwater sources. Surface water also includes a smaller component which reflects the geochemical environment of the watershed. Furthermore, public water supply systems based primarily on surface water sources typically include a large distribution network and centralized treatment and monitoring facilities. Conversely, groundwater has much greater direct contact with the geochemical environment (e.g., mineral surfaces) during its slow migration and long residence time through the void spaces of the rocks that compose an aquifer system. Therefore, prior to its collection and distribution into a water supply system, the chemical content of groundwater will have been strongly affected by the geochemical environment in the rocks along the flow paths feeding wells or springs to which groundwater discharges. Many groundwater supply systems are small domestic systems designed to supply individual homes; these systems often are monitored on a minimal basis for chemical constituents in the water, especially for trace elements.

Understanding the pathway of dissolved minerals from the source rock to the environment or to human consumption is critical for evaluating and remediating possible toxic hazards. Evaluation and remediation, in turn, requires an understanding of the processes and parameters that control rock–water interactions and groundwater flow and solute transport. Conceptual knowledge of these processes and parameters can be quantified and incorporated into generic deterministic models, which can be applied to site-specific problems and be used to predict the fate and transport of dissolved chemicals.

The purpose of this chapter is to review the state-of-the-art in deterministic numerical modeling of groundwater flow, solute transport, and geochemical reaction processes. This chapter is intended to describe the types of models that are available and how they may be applied to complex field problems. However, as this chapter is only a review, it cannot offer comprehensive and in-depth coverage of this complex topic: instead, it guides the reader to references that provide more details. Other chapters in this book covering elements in groundwater are, for example, Chapters 10, 11, 12, and 13.

II. PHYSICAL PROCESSES

A. Groundwater Flow

It generally is assumed that the process of groundwater flow is governed by the relation expressed in Darcy's law, which was derived in 1856 on the basis of the results of laboratory experiments on the flow of water through a sand column. Darcy's law states that the groundwater flow rate (or specific discharge) is proportional to the hydraulic gradient (related to pressure and elevation differences) and to hydraulic conductivity, a property that depends on the characteristics of the porous media (such as grain size distribution or fractures) and the fluid (such as density and viscosity) (see Bear, 1979).

Darcy's law, however, has limits on its range of applicability. It was derived from experiments on the laminar flow of water through porous material. Flow probably is turbulent or in a transitional state from laminar to turbulent near the intakes of large-capacity wells. Turbulent flows also may occur in rocks as a result of the development of fractures, joints, or solution openings. What commonly is done in determining flow in such situations is to ignore local or small-scale turbulence and assume that flow behaves as if it were laminar flow through porous media on the regional scale, and, thus, that Darcy's law applies at that scale.

In some field situations, fluid properties such as density and viscosity may vary appreciably in space or time. This variation may occur where water temperature or dissolved-solids concentration changes greatly. When the water properties are heterogeneous and/or transient, the relations among water levels in monitoring wells, hydraulic heads, fluid pressures, and flow velocities are not straightforward. In such cases, the flow equation is written and solved in terms of fluid pressures, fluid densities, and the intrinsic permeability of the porous media.

B. Advective Transport

The migration and mixing of chemicals dissolved in groundwater obviously will be affected by the velocity of the flowing groundwater. That is, chemical constituents dissolved in flowing water will tend to migrate with the water—the faster the water moves, the faster and further the solutes will migrate. This entrainment of dissolved chemicals is called advective transport.

The specific discharge calculated from Darcy's law represents a volumetric flux per unit cross-sectional area, but flow does not pass through the solid grains of the rock, only through the void spaces. Thus, to calculate the actual seepage velocity of groundwater, one must account for the actual cross-sectional area through which flow is occurring. The latter is done by dividing the specific discharge by the effective porosity of the porous medium. The effective porosity of fractured crystalline rocks can be less than 0.01, whereas for unconsolidated sands and gravels it can exceed 0.30.

C. Hydrodynamic Dispersion

Controlled laboratory and field experiments show that observed solute concentrations in a flow field cannot be predicted adequately just on the basis of seepage velocity, even for nonreactive constituents. Instead, it is observed that some solute will arrive at a given location sooner than predicted by the mean seepage velocity, whereas some solute arrives later than the mean velocity would indicate. That is, there is a spreading about the mean arrival time. Similarly, solute distribution will spread spatially with time and travel distance. This spreading and mixing phenomenon is called hydrodynamic dispersion. It results from molecular and ionic diffusion, and from mechanical dispersion arising from small-scale variations in the velocity of flow that cause the paths of solutes to diverge or spread from the average direction of groundwater flow (Bear, 1979). The outcome is a transient, irreversible, mixing (or dilution) process affecting the concentration distribution of a solute species in an aquifer.

The rate of solute flux caused by hydrodynamic dispersion is expressed in a form analogous to Fick's law of diffusion. This Fickian model assumes that the driving force is the concentration gradient and that the dispersive flux occurs in a direction from higher concentrations toward lower concentrations at a rate related to a constant of proportionality—the coefficient of hydrodynamic dispersion. However, this assumption is not always consistent with field observations and is the subject of much ongoing research and field study (see, for example, Gelhar et al., 1992). The coefficient of hydrodynamic dispersion is defined as the sum of mechanical dispersion and molecular diffusion (Bear,

1979). Mechanical dispersion is a function both of the intrinsic properties of the porous medium (expressed as a dispersivity coefficient, which is related to variability in hydraulic conductivity and porosity) and of the fluid flow (specifically, the fluid velocity). Molecular diffusion in a porous medium will differ from that in free water because of the effects of tortuous paths of fluid connectivity in porous media.

In most groundwater transport model applications, the dispersivity is defined in terms of just two unique constants: the longitudinal dispersivity and the transverse dispersivity of the medium. In practice, however, dispersivity values appear to be dependent on and proportional to the scale of the measurement. Field-scale dispersion (commonly called macrodispersion) results from large-scale spatial variations in hydraulic properties and seepage velocity. Consequently, the use of values of dispersivity determined for one scale of transport in a model designed to predict concentration changes over a different scale of travel probably is inappropriate. Overall, the more accurately and precisely a model can represent or simulate the true velocity distribution in space and time, then the uncertainty concerning representation of dispersion processes will be less of a problem.

D. Solute-Transport Equation

A generalized form of the solute-transport equation is presented by Bear (1979). The governing partial differential equation relates the change in concentration over time in a groundwater system to (1) hydrodynamic dispersion, (2) advective transport, (3) the effects of mixing with a source fluid that has a different concentration than the groundwater at the location of the recharge or injection, and (4) all of the physical, chemical, geochemical, and biological reactions that cause transfer of mass between the liquid and solid or air phases or conversion/decay of dissolved chemical species from one form to another. The chemical attenuation of inorganic chemicals can occur by sorption/desorption, precipitation/dissolution, or oxidation/reduction; organic chemicals can adsorb or degrade by microbiological processes and/or volatilization.

There has been considerable progress over the last 10–15 years in modeling reactive-transport processes; however, the complexity and computational requirements for solute-transport models and reaction models are intense and, therefore, applications of coupled multispecies reactive-transport models are rare. Although

Modeling Groundwater Flow and Quality

Leonard F. Konikow and
Pierre D. Glynn
United States Geological Survey

Contents

I. Introduction

In most areas, rocks in the subsurface are saturated with water at relatively shallow depths. The top of the saturated zone—the water table—typically occurs anywhere from just below land surface to hundreds of feet below the land surface. Groundwater generally fills all pore spaces below the water table and is part of a continuous dynamic flow system, in which the fluid is moving at velocities ranging from feet per millennia to feet per day (Figure 1). While the water is in close contact with the surfaces of various minerals in the rock material, geochemical interactions between the water and the rock can affect the chemical quality of the water, including pH, dissolved solids composition, and trace elements content. Thus, flowing groundwater is a major mecha-

nism for the transport of chemicals from buried rocks to the accessible environment, as well as a major pathway from rocks to human exposure and consumption. Because the mineral composition of rocks is highly variable, as is the solubility of various minerals, the human health effects of groundwater consumption will be highly variable.

Groundwater provides about 40% of the public water supply in the United States. Also, most of the rural population in the United States, more than 40 million people, supply their own drinking water from domestic wells (Alley et al., 1999). Consequently, groundwater is considered an important source of drinking water in every state (Figure 2). Groundwater also is the source of much of the water used for irrigation, especially in areas with arid to semi-arid climates. Nearly all surface-water features (streams, lakes, reservoirs, wetlands, and estuaries) interact with groundwater. Groundwater is the source of base flow to streams and rivers (see Figure 1) and often is the primary source of water that sustains a wetland habitat.

It is long recognized that the chemical content of drinking water can have an adverse or beneficial affect on human health (Keller, 1978). Although the potential side effects associated with some trace elements of natural origin (e.g., arsenic, selenium) or anthropogenic origin (e.g., hexavalent chromium, organic compounds) present in concentrations exceeding public health standards for human consumption of drinking water has

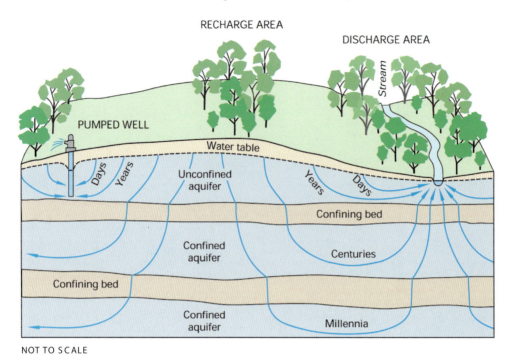

FIGURE 1   Groundwater flow paths vary greatly in length, depth, and travel time from points of recharge to points of discharge in the groundwater system. Flow lines typically are perpendicular to lines (or surfaces) of equal hydraulic head. (From Winter et al., 1998.)

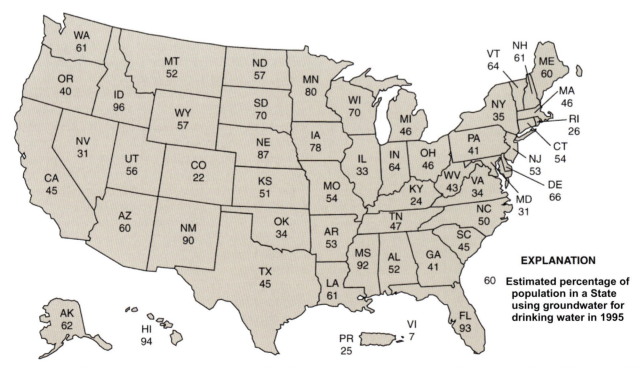

EXPLANATION

60   **Estimated percentage of population in a State using groundwater for drinking water in 1995**

FIGURE 2   Groundwater is an important source of drinking water for every state in the United States. (From Alley et al., 1999.)

some research tools are documented (e.g., see Lichtner et al., 1996; Ibaraki & Therrien, 2001), they still are at the leading edge of the state-of-the-art and usually require computational resources and input data that are beyond that available for most applications. Thus, for the analysis of field problems, it is much more common to apply only groundwater flow and solute-transport models or only multispecies geochemical reaction models, or to apply both sequentially.

In summary, the mathematical solute-transport model requires at least two partial differential equations. One is the equation of flow, from which groundwater flow velocities are obtained, and the second is the solute-transport equation, which describes the chemical concentration in groundwater. If the properties of the water are affected significantly by changes in solute concentration, as in a saltwater intrusion problem, then the flow and transport equations should be solved simultaneously (or at least iteratively). If the properties of the water remain constant and uniform, then the flow and transport equations can be decoupled and solved sequentially, which is simpler numerically.

III. GEOCHEMICAL PROCESSES

A. Basic Concepts

Thermodynamic models describing chemical reactions within an aqueous phase or between the aqueous phase and other phases (solid, gas, or surficial phases) use various common basic principles. Thermodynamic models assume that ion activities, which can be considered the "thermodynamically effective" concentrations determining the progress and direction of reactions, can be calculated from measured (or calculated) ion concentrations.

In addition to the calculation of ionic activities, thermodynamic models typically are based on the law of mass action. This law states that as the activity of the reactants is increased in a chemical reaction, the activity of the reaction products also will increase if equilibrium is maintained. For example, if the following chemical reaction is at equilibrium, $i\text{A} + j\text{B} \Leftrightarrow k\text{C} + l\text{D}$ where i, j, k, and l are stoichiometric coefficients, the law of mass action states that the following mathematical equation must hold:

$$K = \frac{a_C^k a_D^l}{a_A^i a_B^j} \qquad (1)$$

where K is a constant (at a fixed temperature and pressure) and a_A, a_B, a_C, and a_D represent the activities of A, B, C, and D at thermodynamic equilibrium. The law of mass action applies to any kind of chemical reaction at thermodynamic equilibrium, not just to aqueous reactions.

B. Aqueous Speciation, Hydrolysis, and Oxidation/Reduction Reactions

The calculation of solution ionic strength, ionic activity coefficients, and ionic activities must be conducted iteratively with the solution of the mass-action equations that determine the speciation of the aqueous solution. Aqueous speciation is the partitioning of chemical constituents present in a solution (whose total concentrations are typically measured) into various aqueous species that represent the different molecular forms assumed by the constituents in the aqueous solution. Aqueous speciation reactions are homogeneous reactions; all reactants and products are aqueous species. Two examples of speciation reactions are

$$\text{Al}^{3+} + 3\text{H}_2\text{O} \Leftrightarrow \text{Al(OH)}_2^+ + 2\text{H}^+ + \text{H}_2\text{O}$$
(hydrolysis of aluminum) and $\qquad (2a)$

$$6\text{CN}^- + \text{Fe}^{2+} \Leftrightarrow \text{Fe(CN)}_6^{4-}$$
(cyanide complexation by ferrous iron) $\qquad (2b)$

Simple mass-action equations describing the relative aqueous activities of products and reactants can be written for any of the above reactions. Aqueous speciation reactions control the concentrations of individual aqueous species, and, thereby, may appreciably affect the toxicity of a solution. Some aqueous species can be much less toxic than others. For example, cyanide present in the complexed Fe(CN)_6^{4-} form is much less toxic than cyanide in the CN^- or HCN^0 forms. Aqueous speciation reactions also affect the total concentration of constituents in solution through their control of general solution characteristics such as ionic strength, acidic nature (pH), and redox potential (pe) of the solution, and through their control of individual aqueous species involved in mass-transfer reactions (e.g., mineral precipitation/dissolution, surface sorption/desorption, ingassing/exsolution reactions).

The pH and pe of an aqueous solution are sometimes described as the *master variables*, which control the speciation of aqueous solutions. The pH of a solution simply relates to the acidic nature of a solution, and more specifically to the activity of protons (H^+) or equivalently of hydronium ions (H_3O^+ or $H_9O_4^+$) in the solution, as $pH = -\log a_{H^+}$. The pe of a solution is related to the ratio of the activity of an aqueous species present in oxidized form to that present in reduced form. Mathematically, the definition of the pe of a solution is analogous to that of the pH variable, except that it is defined in terms of the activity of free electrons in the solution, as $pe = -\log a_e$. For all practical purposes, however, any electrons produced by an oxidation reaction always must be consumed by a reduction reaction. Nevertheless, the redox potential of a solution does have practical relevance in describing the degree to which aqueous species are in oxidized or reduced form.

The Eh of a solution is the redox potential measured in the field. It is directly related to pe by the relation:

$$pe = Eh \frac{F}{2.303RT} \qquad (3)$$

where F is the Faraday constant, R is the gas constant, and T is the temperature in Kelvin. At 25°C, the relation is $pe = 16.904\ Eh$, where Eh is expressed in volts.

Field measurements of Eh are problematic. First, the redox-active species present in a water often are not in redox equilibrium and, therefore, the measurement may be meaningless; that is, more than one redox potential may be present for the solution depending on which redox couple (e.g., Fe^{3+}/Fe^{2+}) is chosen, and the measurement may at best represent some sort of "mixed" potential. Second, the only redox-active species to which platinum electrodes (typically used for Eh measurements) have been demonstrated to respond quickly, and, therefore, reflect their electrochemical equilibrium, are Fe^{2+}, Fe^{3+}, and S^{2-}, and only when these species are present at concentrations of 10^{-5} mol/L or greater (Nordstrom & Munoz, 1994). In general, redox equilibria and disequilibria in groundwater are best assessed through the measurement of individual redox couples, rather than through measurements of Eh.

Oxidation/reduction reactions can occur either in the aqueous phase only (homogeneous reactions) or between the aqueous phase and other phases (heterogeneous reactions). In most groundwaters, the presence of organic carbon commonly drives a sequence of redox reactions as water migrates from the unsaturated zone and water table to greater depths. Typically, organic carbon reduces dissolved oxygen in the water then reduces nitrate to nitrogen gas (and sometimes to ammonia). Dissolved organic carbon also may react with manganese oxide minerals producing Mn^{2+} in the water; at slightly lower oxidation potentials the organic carbon will react with ferric-iron minerals (typically oxides) and generate dissolved ferrous iron in the water. At greater depths, water becomes sulfidic as the carbon starts to reduce dissolved sulfate to sulfide. Finally, when no further electron acceptors (such as SO_4, O_2, and NO_3) are present, water typically becomes methanic, i.e., any remaining organic matter decomposes through a process of fermentation to methane and carbon dioxide. In this process, hydrogen generally is produced as an intermediate product. Despite strong thermodynamic potentials for their occurrence, rates of redox reactions are often slow unless microbially catalyzed. Most redox reactions in natural and contaminated environments are catalyzed microbially.

Redox reactions are important in medical geology, whether for natural or contaminated environments, because they affect the relative toxicity of various dissolved constituents in water. For example, Cr(IV) is a suspected carcinogen, whereas Cr(III) is an essential trace element for humans. Redox reactions also affect the solubility of various compounds (e.g., metal sulfides present in the rock materials).

C. Geochemical Mass-Transfer Processes

Mineral dissolution and precipitation processes are important in controlling the chemical evolution of groundwater. These processes strongly affect the overall chemical characteristics of the water through their effect on pH and pe conditions, ionic strength, and complexant concentrations (dissolved carbonate, sulfate, chloride, etc.). For example, the pH of natural waters often is buffered by the dissolution of calcite and described as $CaCO_{3,S} + H^+ \Leftrightarrow Ca^{2+} + HCO_3^-$.

Mineral precipitation processes also commonly limit the concentrations of many constituents in water. For example, barium and aluminum concentrations in water often are limited by the precipitation of barite and aluminum hydroxide:

$$Ba^{2+} + SO_4^{2-} \Leftrightarrow BaSO_{4,s} \qquad (4a)$$

$$Al(OH)^{2+} + 2H_2O \Leftrightarrow Al(OH)_{3,s} + 2H^+. \qquad (4b)$$

Although the above reactions are written for pure minerals, mineral phases invariably contain foreign ions and impurities, which were entrained as occlusion pockets during the formation of the minerals or are substituting as an integral part of the mineral lattice. In either case, minerals can take up and/or release these impurities through recrystallization processes. The thermodynamic theory describing the uptake and release of substitutional impurities in minerals (also known as solid solutions) is fairly complex and remains an area of active research (Glynn, 2000), but it is increasingly implemented in geochemical modeling codes. Examples of solid-solutions reactions that may control trace element concentrations include (1) the uptake of copper, nickel, cobalt, and zinc by precipitating manganese oxides; (2) the uptake of chromate by barite recrystallization in contaminated waters; and (3) the control of fluoride concentrations through dissolution and recrystallization uptake of apatites. Biogenic apatites, such as found in fossil bones and teeth, commonly are initially rich in hydroxylapatite and slowly recrystallize upon contact with groundwater to fluoroapatite. In certain groundwater systems, however, the reverse process also has been demonstrated to occur. For example, Zack (1980) has shown that the exchange of hydroxide ions for fluoride present in fossil shark teeth is responsible for anomalously high fluoride concentrations in the Atlantic Coastal Plain aquifers of South Carolina (this is discussed in more detail below).

Sorption reactions often are important in controlling the concentrations of constituents in groundwater and often may even affect the observed pH. A typical porous medium aquifer (with a porosity of 0.2 and a cation exchange capacity of 5 mEq/100 g) has about 500 mEq of cation exchange capacity per liter of water (Drever, 1997). This value is more than two orders of magnitude greater than the concentration of dissolved ions in dilute groundwater, and consequently can be expected to have a large effect on the chemistry of the water.

Sorption reactions generally are described either through an ion-exchange model (primarily affecting cations) or through a surface-complexation model. Ion-exchange models typically apply to mineral surfaces and interlayers with constant surface charge (e.g., clays and zeolites), and they usually only consider cation exchange reactions such as $[\text{Na} - \text{clay} + \text{K}^+ \Leftrightarrow \text{Na}^+ + \text{K} - \text{clay}]$.

Surface-complexation models commonly are used to describe the sorption of aqueous species on surfaces with variable charge (e.g., iron and manganese oxides, silica, organic matter, and clay edges). These surfaces become more negatively charged with increasing pH, and, therefore, their cation sorption capacity increases and their anionic sorption capacity decreases. At any pH, the surfaces are considered to contain a mix of positively charged, negatively charged, and neutral sites. This mix and the ensuing chemical reactions among the various sites and aqueous ions and complexes are fully described by a speciation of the surface in a manner analogous to that for an aqueous solution. This speciation describes the surface as a series of various surface-complexes and bare-surface sites of different charges.

Gas dissolution/exsolution/volatilization reactions can affect the concentrations of organic (aromatic and light aliphatic compounds) and inorganic (O_2, CO_2, N_2, noble gases) constituents in groundwater near the water table. These reactions can strongly affect the general chemistry of water through their effect on pH, redox potentials, and ionic strength; in some cases (e.g., volatilization of HCN and light organic compounds), the reactions also can directly affect the concentrations of contaminants in water.

D. Biodegradation/Biotransformation

Microorganisms are important in the chemical evolution of waters, and, for all practical purposes, can be considered present in almost every groundwater environment, even under extreme conditions. Microorganisms have been found underground at depths of more than a kilometer (Pedersen, 1993), at temperatures as high as 110°C (Stetter, 1998), and in waters with up to 30% salinity (Grant et al., 1998). Most microorganisms are heterotrophic and use organic carbon as a primary energy source. However, chemolithotrophic organisms can use reduced inorganic substrates, such as NH_4, H_2, H_2S, and CH_4 to derive energy in both aerobic and anaerobic environments.

Microbes also are essential in the degradation of organic molecules, generally of complex molecules to simpler ones, ultimately to inorganic compounds and forms of C, N, H, S, Cl and other elements. Complete transformation to inorganic compounds (mineralization) involves multiple, successive, biologically mediated reactions, which may proceed at different rates. Although the initial degradation rate of an organic molecule may be fast, degradation of some of its metabolites may be slow, which can be a problem if the metabolite is associated with a health risk. Although microbial "remediation" is an important issue considered in investigations of anthropogenic groundwater contaminants (such as pesticides, herbicides, and petroleum products), microbes also are likely to affect the "natural contamination" of groundwater through their catalysis of coal-water interactions and their consequent mobilization of soluble polar aromatic and polynuclear aromatic hydrocarbons. These compounds are thought to be an important factor in the observed incidences of Balkan endemic nephropathy (Feder et al., 1991).

Microbes also are known to catalyze many inorganic reduction reactions by generally using organic carbon as a reducing agent. Examples from naturally occurring subsurface constituents, which when modified by microorganisms may become more mobile, include: solid Fe(III)-oxyhydroxides to dissolved Fe(II), nitrate to N_2 (denitrification) or further to ammonia, sulfate to sulfide, As(V) to As(III) (Dowdle et al., 1996), Se(VI) to Se(IV) and Se(0) (Switzer Blum et al., 2001), and U(VI) to U(IV) (Lovley et al., 1991). Microbes also are known to catalyze many of the reverse oxidation reactions in the list above. Redox reactions usually are associated with large changes in free energy, and therefore provide microbes with an energy source.

Although the occurrence of these redox reactions can be predicted from thermodynamic considerations, in practice the kinetics of the reactions would be orders of magnitude slower if they were not mediated by microbes. Numerical modeling of biodegradation/biotransformation reactions reduces the complexity of the multiple chemical, enzymatic, biological, and ecological processes that are mediating the transformation of a constituent of interest to a simple mathematical description of the overall transformation kinetics. The mathematical model chosen often considers not just the degradation or transformation of a particular compound, but also keeps track of the effect of the transformation on the size and productivity of the microbial community responsible for the catalysis of the transformation. Monod and Michaelis-Menten kinetic models are used to describe microbial utilization of chemical substrates and microbial growth kinetics (Schwarzenbach et al., 1993). The computer codes BIOMOC (Essaid & Bekins, 1997) and RT3D (Clement, 1997) are examples of groundwater flow and transport codes that allow the simulation of biodegradation and transformation reactions using a variety of kinetic model formulations.

IV. Models

A. Overview

The word *model* has many definitions. A model perhaps is most simply defined as a representation of a real system or process. A *conceptual model* is a hypothesis for how a system or process operates. This hypothesis can be expressed quantitatively as a mathematical model. *Mathematical models* are abstractions that represent processes as equations, physical properties as constants or coefficients in the equations, and measures of state or potential in the system as variables.

Most groundwater models in use are deterministic mathematical models. *Deterministic models* are based on conservation of mass, momentum, and energy and describe cause-and-effect relations. The underlying assumption is that, given a high degree of understanding of the processes by which stresses on a system produce subsequent responses in that system, the system's response to any set of stresses can be predetermined, even if the magnitude of the new stresses falls outside of the range of historically observed stresses.

Deterministic groundwater models generally require the solution of partial differential equations. Exact solutions often can be obtained analytically, but *analytical models* require that the parameters and boundaries be highly idealized. Some deterministic models treat the properties of porous media as lumped parameters (essentially, as a black box), but this precludes the representation of heterogeneous hydraulic properties in the model. Heterogeneity, or variability in aquifer properties, is characteristic of all geologic systems and now is recognized as critical in affecting groundwater flow and solute transport. Thus, it often is preferable to apply distributed-parameter models, which allow the representation of more realistic distributions of system properties. Numerical methods yield approximate solutions to the governing equation (or equations) through the *discretization* of space and time. Within the discretized problem domain, the variable internal properties, boundaries, and stresses of the system are approximated. Deterministic, distributed-parameter, *numerical models* can relax the rigid idealized conditions of analytical models or lumped-parameter models, and, therefore, they can be more realistic and flexible for simulating field conditions (if applied properly).

The number and types of equations to be solved are determined by the concepts of the dominant governing processes. The coefficients of the equations are the parameters that are measures of the properties, boundaries, and stresses of the system; the dependent variables of the equations are the measures of the state of the system and are determined mathematically by the solution of the equations. When a numerical algorithm is implemented in a computer code to solve one or more partial differential equations, the resulting computer code can be considered a *generic model*. When the grid dimensions, boundary conditions, and other parameters (such as hydraulic conductivity and storativity), are specified in an application of a generic model to represent a particular geographical area, the resulting computer program is a *site-specific model*. The capability of generic models to solve the governing equations accurately typically is demonstrated by example applications to simplified problems. This does not guarantee a similar level of accuracy when the model is applied to a complex field problem.

B. Numerical Methods

The partial differential equations describing groundwater flow and transport can be solved mathematically by using either analytical solutions or numerical solutions. In general, obtaining the exact analytical solution to the partial differential equation requires that the properties and boundaries of the flow system be highly and perhaps unrealistically idealized. Many of the limitations of applying analytical methods to complex field problems can be overcome by using analytical element methods, which apply analytical methods to subareas of the problem domain (see Haitjema, 1995).

Alternatively, for problems where the simplified analytical models no longer describe the physics of the situation, the partial differential equations can be approximated numerically. In numerical approaches, the continuous variables are replaced with discrete variables that are defined at grid blocks or nodes. Thus, the continuous differential equation, which defines hydraulic head or solute concentration everywhere in the system, is replaced by a finite number of algebraic equations that defines the hydraulic head or concentration at specific points. This system of algebraic equations generally is solved using matrix techniques. This approach constitutes a numerical model.

Two major classes of numerical methods have come to be well accepted for solving the groundwater flow equation. These are finite-difference methods and finite-element methods. Each of these two major classes of numerical methods includes a variety of subclasses and implementation alternatives. An overview of the application of these numerical methods to groundwater problems is presented by Wang and Anderson (1982). Both of these numerical approaches require that the area of interest be subdivided by a grid into a number of smaller subareas (cells or elements) that are associated with nodal points (either at the centers or peripheries of the subareas).

Finite-difference methods approximate the first derivatives in the partial differential equations as difference quotients (the differences between values of the independent variable at adjacent nodes with respect to the distance between the nodes, and at two successive time levels with respect to the duration of the time-step increment). Finite-element methods use assumed func-

tions of the dependent variable and parameters to evaluate equivalent integral formulations of the partial differential equations. Huyakorn and Pinder (1983) present a comprehensive analysis of the application of finite-element methods to groundwater problems. In both numerical approaches, the discretization of the space and time dimensions allows the continuous boundary-value problem for the solution of the partial differential equation to be reduced to the simultaneous solution of a set of algebraic equations. These equations then can be solved using either iterative or direct matrix methods.

Each approach has advantages and disadvantages, but there are few groundwater problems for which one approach clearly is superior. In general, the finite-difference methods are simpler conceptually and mathematically, and are easier to program. They typically are keyed to a relatively simple, rectangular grid, which also eases data entry. Finite-element methods generally require the use of more sophisticated mathematics but, for some problems, may be more accurate numerically than standard finite-difference methods. A major advantage of the finite-element methods is the flexibility of the finite-element grid, which allows a close spatial approximation of irregular boundaries of the aquifer and/or of parameter zones within the aquifer when they are considered. However, the construction and specification of an input data set are much more difficult for an irregular finite-element grid than for a regular rectangular finite-difference grid. Thus, the use of a graphical model preprocessor that includes a mesh generator should be considered. A hypothetical aquifer system with impermeable boundaries and a well field (Figure 3A) has been discretized using finite-difference (Figure 3B) and finite-element (Figure 3C) grids. Grids can be adjusted to use a finer mesh spacing in selected areas of interest. The rectangular finite-difference grid approximates the aquifer boundaries in a step-wise manner, which results in some nodes or cells outside of the aquifer, whereas sides of the triangular elements of the finite-element grid can closely follow the outer boundary using a minimal number of nodes.

C. Groundwater Flow Models

A major revolution in the quantitative analysis of groundwater flow systems came in the early 1970s with the introduction and documentation of two-dimensional, deterministic, distributed-parameter, digital computer simulation models. These models rep-

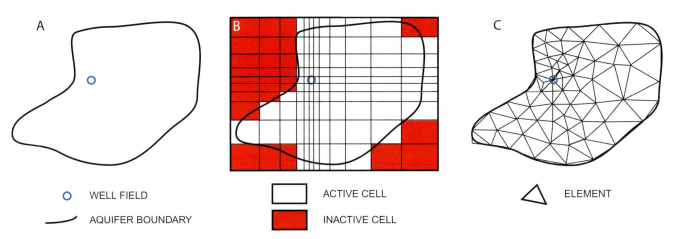

FIGURE 3   Application of a numerical model to simple hypothetical problem, showing (A) an irregularly bounded aquifer discretized using (B) a finite-difference grid and (C) a finite-element grid. (From Konikow & Reilly, 1998.)

resented a major improvement over analytical methods because they allowed the representation of heterogeneous properties, complex boundary conditions, and time-varying stresses. As improved numerical methods were developed and more powerful computers became widely available, three-dimensional modeling became standard practice by the early 1990s. Practical aspects of applying groundwater models are reviewed by Anderson and Woessner (1992).

Groundwater flow models solve a governing partial differential equation. The solution defines the hydraulic-head distribution at every point within the boundaries of the problem domain. When this is accomplished using numerical methods, the solution inherently also provides the fluid fluxes throughout the discretization grid. Solving the flow equation requires the specification of the properties of the groundwater system (and their spatial variability), the boundary conditions, and, for transient problems, the initial conditions.

The knowledge of the heads (or water levels or fluid pressures) and the direction and rate of flow provides much insight into the nature of the groundwater flow system, and allows inferences to be made about (1) potential source areas for toxic substances detected in groundwater and (2) potential discharge areas or receptors for flow and transport of dissolved toxic constituents away from known sources of soluble toxic substances. When problems are detected and analyzed using a simulation model, the model can be used as a management tool to help evaluate alternative decisions for reducing risks to public health or the environment.

The major difficulty in groundwater modeling is accurately defining the properties of the system and the boundary conditions for the problem domain. The subsurface environment is a complex, heterogeneous, three-dimensional framework. To determine the unique parameter distribution for a field problem, so much expensive field testing would be required that it is seldom feasible either economically or technically. Therefore, the model typically represents an attempt, in effect, to solve a large set of simultaneous equations with more unknowns than equations. It inherently is impossible to obtain a unique solution to such a problem. Therefore, limited sampling and understanding of the geological heterogeneity causes uncertainty in the model input data (aquifer properties, sources and sinks, and boundary and initial conditions). This uncertainty leads to non-uniqueness in the model solution.

Uncertainty in parameters logically leads to a lack of confidence in the interpretations and predictions that are based on a model analysis, unless the model can be demonstrated to be a reasonably accurate representation of the real system. To demonstrate that a deterministic groundwater simulation model is realistic, usually field observations of aquifer responses (such as changes in water levels for flow problems or changes in concentration for transport problems) are compared to corresponding model-calculated values. The objective of this calibration procedure is to minimize differences between the observed data and calculated values. The minimization is accomplished by adjusting parameter values within their ranges of uncertainty until a best fit is achieved between the calculated values of dependent variables and the corresponding observations. Thus, model calibration often is considered a parameter-estimation procedure. Usually, the model is considered calibrated when it reproduces historical data within some acceptable level of accuracy. The level of acceptability is, of course, determined subjectively. Although a poor match provides evidence of model errors, a good match does not necessarily prove the validity or adequacy of the model (Konikow & Bredehoeft, 1992).

The calibration of a deterministic groundwater model often is accomplished through a trial-and-error adjustment of the model input data to modify model output. Because a large number of interrelated factors affect the output, trial-and-error adjustment may be a highly subjective and inefficient procedure. Advances in automatic parameter-estimation procedures help to eliminate some of the subjectivity inherent in model calibration. The newer procedures generally treat model calibration as a statistical procedure using multiple regression approaches. Parameter-estimation procedures allow simultaneous model construction, application, and calibration using uncertain data, so that the uncertainties in model parameters and in predictions and assessments can be quantified.

Automated parameter-estimation techniques improve the efficiency of model calibration and have two general components: one that calculates the best fit (sometimes called automatic history matching) and a second that evaluates the statistical properties of the fit. These techniques also are called *inverse models*, as they treat the system parameters as unknowns. The minimization procedure uses sensitivity coefficients that are based on the change in calculated value divided by the change in the parameter (for example, the change in head with changing transmissivity). The sensitivity coefficients may be useful in the consideration of additional data collection. Hill (1998) provides an overview of methods and guidelines for effective model calibration using inverse modeling.

One of the most popular and comprehensive deterministic groundwater models available today is the U.S. Geological Survey's (USGS) MODFLOW code (McDonald & Harbaugh, 1988; Harbaugh et al., 2000). This model actually is an integrated family of compatible codes that centers on an implicit finite-difference solution to the three-dimensional flow equation. The basic model uses a block-centered, finite-difference grid that allows variable spacing of the grid in three dimensions. Flow can be steady or transient. Aquifer properties can vary spatially and hydraulic conductivity (or transmissivity) can be anisotropic. Flow associated with external stresses, such as wells, distributed recharge in areas, evapotranspiration, drains, lakes, and streams, can also be simulated through the use of specified head, specified flux, or head-dependent flux boundary conditions. The implicit finite-difference equations can be solved using any one of several solution algorithms. Although the input and output systems of the program were designed to permit maximum flexibility, usability and ease of interpretation of model results can be enhanced by using one of several commercially available preprocessing and post-processing packages; some of these operate independently of MODFLOW, whereas others are directly integrated into reprogrammed and/or recompiled versions of the MODFLOW code.

A variety of other MODFLOW accessory codes, packages, and features are available. Most of these were developed by the USGS; examples include coupled surface water and groundwater flow, aquifer compaction, transient leakage from confining units, rewetting of dry cells, horizontal flow barriers, alternative interblock transmissivity conceptualizations, cylindrical flow to a well, a statistical processor, a data-input program, and a program that calculates water budgets. Other packages have been developed by non-USGS sources to work with MODFLOW; one example is the advective-dispersive solute-transport model MT3D (Zheng & Wang, 1999). The latest version of MODFLOW (MODFLOW-2000) has an inverse modeling capability built in to the code, which allows the user to do parameter estimation and sensitivity analyses directly (Hill et al., 2000).

The utility of groundwater flow modeling is illustrated by its application to a selenium problem in California, where more than 2 million acres of agricultural land is irrigated in the western San Joaquin Valley (see Chapter 15). Since 1967, imported surface water has been the primary source for irrigation; hence, groundwater pumping simultaneously declined (Belitz & Phillips, 1992). This combination caused increased recharge to the underlying aquifers and subsequent water-table rises. By the early 1990s, the water table was high (within 10 feet of the land surface) over more than half of the area. Because such areas are prone to soil salinization and other problems, subsurface tile drains have been used to keep the water table deep enough to minimize these problems. However, the agricultural drainage water was high in selenium and eventually flowed into the Kesterson Wildlife Refuge, which lead to deaths and deformities of waterfowl and aquatic biota (Deverel et al., 1984; Presser & Barnes, 1985). These problems led to the closure of drains contributing selenium, which left considerable concern about how to manage the groundwater flow system in a manner that maintained agricultural productivity yet precluded selenium transport.

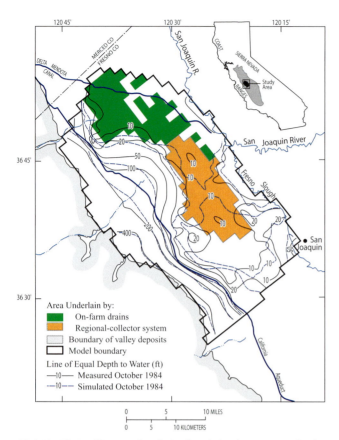

FIGURE 4 Measured and simulated depths to water in the central part of the western San Joaquin Valley, California, October 1984. (Modified from Belitz et al., 1993.)

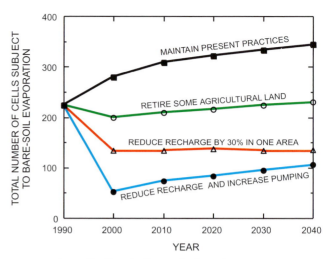

FIGURE 5 Predicted effects of selected water management alternatives on area of high water table. (Modified from Belitz & Phillips, 1992.)

According to Presser et al. (1990), the source of the selenium is from weathering of pyritic marine shales in the Coast Ranges just west of the San Joaquin Valley. They believe that selenium is mobilized by oxidative weathering in an acidic environment, which concentrates the soluble selenate form of selenium. Selenate is transported readily in flowing groundwater and surface runoff.

A transient, three-dimensional, finite-difference model of the regional groundwater flow system was developed to assess water-table responses to alternative management that would affect groundwater recharge and discharge (Belitz et al., 1993). The model was calibrated using hydrologic data collected from 1972 to 1988. The model results indicate good agreement between measured and simulated depths to water (Figure 4).

The calibrated model was used to evaluate the possible effects of various management practices on the depth of the water table (see Figure 5 for representative

results). The number of cells (each having an area of one square mile) subject to bare-soil evaporation is an indicator of the depth to water because only cells in which the water table is less than 7 feet deep will fall into this category. Higher water-table elevations also yield greater discharge to drains. If present practices are maintained, the area underlain by a high water table will continue to increase, as will discharge to drains. Reducing recharge (by increasing irrigation efficiency), increasing pumping, and removing land from agricultural production all will help to mitigate the problem. Thus, the groundwater flow model provides a powerful tool to help water managers mitigate the selenium problem while considering cost-benefit ratios.

D. Groundwater Pathline Models

Pathline models simulate the process of advective transport. They use calculated velocities to compute where and how fast water and nonreactive dissolved chemicals migrate. This requires the specification of an additional physical parameter—the porosity of the groundwater system (and its spatial variability). Also, the hydraulic-head gradients must be known, typically from the output of a groundwater flow model. It is useful for estimating where fluid and dissolved solutes are moving, how fast they are moving, and their source. They also can be useful for cross-checking age dates estimated from isotopic analyses. Pathline models, however, cannot calculate solute concentrations because dilution

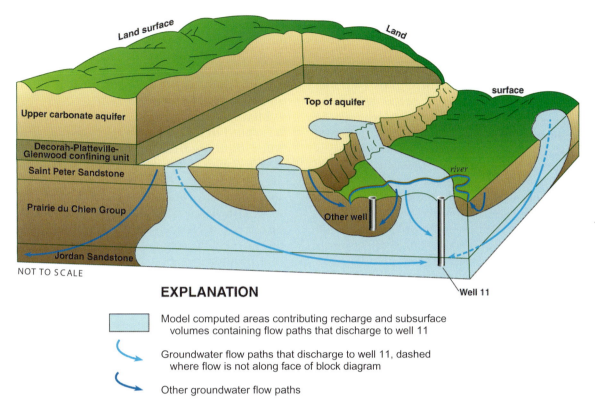

EXPLANATION

Model computed areas contributing recharge and subsurface volumes containing flow paths that discharge to well 11

Groundwater flow paths that discharge to well 11, dashed where flow is not along face of block diagram

Other groundwater flow paths

FIGURE 6 Long-term (steady-state) contributing recharge areas for well 11 near Rochester, Minnesota, calculated using pathline model. Complex three-dimensional patterns of groundwater flow yield irregularly shaped volumes in the subsurface containing the many flow paths that originate at the water table and discharge at well 11. (From Franke et al., 1998.)

or reaction mechanisms are not included in these models. Pathline models usually are much more efficient to run than transport models.

A widely used pathline program is MODPATH (Pollock, 1989), which uses MODFLOW model output and determines paths and travel times of water movement under steady-state and transient conditions. The semi-analytical, particle-tracking method assumes that each directional velocity component varies linearly within a grid cell in its own coordinate direction. For example, MODPATH was used to delineate source areas contributing recharge to a well in Minnesota (Figure 6). The results depict the complicated and discontinuous spatial patterns of contributing recharge areas to wells in highly developed aquifer systems and would be extremely difficult to derive without the aid of such a model. If toxic constituents are detected in a particular well (such as well 11 in Figure 6), the pathline model would help delineate the volume of rock material with which groundwater was in contact at earlier times. This is an invaluable aid in a search for the source of the toxic constituents.

E. Advection-Dispersion Models

The purpose of a model that simulates solute transport in groundwater is to compute the concentration of a dissolved chemical species in an aquifer at any specified time and place. The theoretical basis for the equation describing solute transport has been well documented in the literature (e.g., Bear, 1979). Zheng and Bennett (2002) provide a conceptual framework for analyzing and modeling solute-transport processes in groundwater as well as guidelines and examples for applications to field problems.

The mathematical solute-transport model requires at least two partial differential equations. One is the equation of flow, from which groundwater flow velocities are obtained, and the second is the solute-transport equation, whose solution gives the chemical concentration in groundwater. If the properties of the water are affected significantly by changes in solute concentration, as in a saltwater-intrusion problem, then the flow and transport equations should be solved simultaneously (or at least iteratively). If the properties of the water remain constant, then the flow and transport equations can be decoupled and solved sequentially, which is simpler numerically.

The solute-transport equation is more difficult to solve numerically than the groundwater flow equation, largely because the mathematical properties of the transport equation vary depending upon which terms in the equation are dominant in a particular situation. When solute transport is dominated by advection, as is common in many field problems, then the governing equation approximates a hyperbolic type of equation (similar to equations describing the propagation of a wave or of a shock front). But if transport is dominated by dispersive fluxes, such as might occur where fluid velocities are low and/or hydrodynamic dispersion is relatively high, then the governing equation becomes more parabolic in nature (similar to the transient groundwater flow equation).

The numerical methods that work best for parabolic partial differential equations are not best for solving hyperbolic equations and vice versa. Thus, no one numerical method or simulation model will be ideal for the entire spectrum of groundwater transport problems likely to be encountered in the field. Further compounding this difficulty is that the seepage velocity of groundwater in the field is highly variable, even if aquifer properties are homogeneous, because of the effects of complex boundary conditions. Thus, in low permeability zones or near stagnation points, the velocity may be close to zero and the transport processes will be dominated by dispersion processes; in high permeability zones or near stress points (such as pumping wells), the velocity may be up to several meters per day and the transport processes will be advection dominated. In other words, the governing equation may be more hyperbolic in one area (or at one time) and more parabolic in another area (or at another time). Therefore, regardless of which numerical method is chosen as the basis for a simulation model, it will not be accurate or optimal over the entire domain of the problem. The transport modeling effort must recognize this inherent difficulty and strive to minimize and control the numerical errors.

Additional complications arise when the solutes of interest are reactive. Simple reaction terms do not necessarily represent the true complexities of many reactions (see, for example, Glynn, 2003). Also, particularly difficult numerical problems arise when reaction terms are highly nonlinear, or if the concentration of the solute of interest is strongly dependent on the concentration of other chemical constituents. For field problems where reactions appreciably affect solute concentrations, simulation accuracy is less limited by mathematical constraints than by data constraints. That is, the types and rates of reactions for the specific solutes and minerals in the particular groundwater system of interest rarely are known and require an extensive amount of data to assess accurately.

Finite-difference and finite-element methods also can be applied to solve the transport equation, particularly when dispersive transport is large compared to advective transport. However, numerical errors, such as numerical dispersion and oscillations, may be large for some problems. The numerical errors generally can be reduced by using a finer discretization (either shorter time steps or finer spatial grid), but this discretization will increase the computational work load. An example of a documented three-dimensional, transient, finite-difference model that simultaneously solves the fluid pressure, energy-transport, and solute-transport equations for non-homogeneous miscible fluids is HST3D (Kipp, 1997). An example of a finite-element transport model is SUTRA (Voss, 1984).

Although finite-difference and finite-element models commonly are applied to transport problems, none of the standard numerical methods is ideal for a wide range of transport problems and conditions. Thus, there currently is much research on developing better mixed or adaptive methods that aim to minimize numerical errors and combine the best features of alternative standard numerical approaches. Examples of other types of numerical methods that also have been applied to transport problems include method of characteristics, random walk, Eulerian-Lagrangian methods, and adaptive grid methods. All these numerical methods have the capability to track sharp fronts accurately with a minimum of numerical dispersion. Documented models based on variants of these approaches include Konikow et al. (1996) and Zheng and Wang (1999).

As an example, the public domain MOC3D model (Konikow et al., 1996) is integrated fully with the MODFLOW-2000 code. The model computes changes in concentration over time caused by the processes of advective transport; hydrodynamic dispersion; mixing or dilution from fluid sources; matrix diffusion; a first-order irreversible-rate reaction, such as radioactive decay; and reversible equilibrium-controlled sorption with a linear isotherm. The model couples the groundwater flow equation with the solute-transport equation. MOC3D uses the method of characteristics to solve the solute-transport equation, by which a particle-tracking procedure represents advective transport and a finite-difference procedure calculates concentration changes caused by hydrodynamic dispersion.

There are many examples in the literature illustrating the application of solute-transport models to problems involving contaminant plumes emanating from point sources of contamination, but few dealing with natural sources. One of the few involves the application of MODFLOW and MT3D to the arsenic problem in Bangladesh, wherein the sustainability of groundwater development is evaluated on the basis of constraining increases in arsenic concentrations in supply wells calculated by the coupled models (Cuthbert et al., 2002).

F. Aqueous Speciation Modeling

Geochemical speciation models, such as WATEQF (Plummer et al., 1976) and WATEQ4F (Ball & Nordstrom, 1991), calculate the distribution of chemical elements among different aqueous species (bare ions, complexes, and ion pairs) at a given temperature and pressure, determine whether the aqueous solution is supersaturated or undersaturated with respect to various solid mineral phases, and calculate the partial pressure of gases that would be in equilibrium with the calculated solution composition. Speciation models also calculate the total dissolved inorganic concentration (TDIC) of a solution given its measured alkalinity, or conversely calculate its alkalinity given a measured TDIC concentration.

Speciation codes solve a set of algebraic equations that are basically of two types: mass-balance and mass-action. Mass-balance equations relate the total dissolved concentration (a user-provided measured quantity) of given elements or components to the sum of the concentrations of their aqueous species multiplied by the stoichiometric coefficient of the element/component in each species. Mass-action equations provide thermodynamic relations describing the dependence of the activity (i.e., the thermodynamically effective concentration) of a given aqueous species on the activities of other aqueous species, pH, and redox potential.

To solve the equations described above, speciation codes require that the user provide a complete chemical analysis of the water, including not only the total dissolved concentrations of major and minor elements, but also the pH of the solution and some indication of its redox potential. The redox potential either can be indicated by an Eh or pe value, or alternatively by one or more redox couples. If only one redox couple (e.g., the Fe(II)/Fe(III) couple) is entered, it typically will be used to define the redox potential for all redox-active elements in the solution. More advanced codes, such as PHREEQC (Parkhurst, 1995; Parkhurst & Appelo,

1999), allow the specification of more than one redox couple and allow the user to apply each couple to control the redox distribution of specific redox-active elements.

In addition to the solution-specific data that must be entered by the user, speciation codes also require a thermodynamic database that provides equilibrium constants, as a function of temperature and pressure, for the various aqueous-speciation and complexation reactions considered, and for potential mineral and gas dissolution and precipitation/exsolution reactions. The quality of a speciation code's output will, in large part, be determined both by the quality of the user-entered data and by the quality of the thermodynamic database associated with the code. Ideally, thermodynamic databases should be internally consistent, should consider all major aqueous species, and should be based on accurate measurements. Thermodynamic consistency has many meanings (Nordstrom & Munoz, 1994) such as (1) the data are consistent with basic thermodynamic relations; (2) common scales are used for temperature, energy, atomic mass, and fundamental physical constants; (3) the same mathematical and chemical models were used to fit different data sets; (4) conflicts among measurements were resolved; and (5) appropriate choices of standard states were made and used for all similar substances. In practice, there is considerable uncertainty in thermodynamic data and judging the extent of the uncertainty for different elements, conditions, and calculated results requires geochemical expertise and experience in using the database.

Thermodynamic databases typically consider few organic species, even though organic species are important constituents in both natural and contaminated waters. Most codes and associated thermodynamic databases also are limited to modeling the speciation of relatively dilute waters, with ionic strengths (or salinity) lower than seawater. The few codes that are available to model the speciation of saline waters and brines usually have little or no data available to model the speciation of minor elements, metals, or radionuclides or redox states. Finally, most speciation codes assume that the aqueous species present are at equilibrium with each other. Although most "homogeneous" aqueous-speciation reactions are fast, this is not always the case for reactions involving redox-active species and elements, and/or strong aqueous complexes and polymerized species. The kinetics of formation/dissociation of those species can be slow and the kinetics of redox reactions often depend on microbial catalysis.

Speciation models help in understanding the speciation of an aqueous solution and its thermodynamic state, particularly with respect to the potential dissolution/precipitation/exsolution of various minerals and gases, and to the potential for the exchange or sorption of ions and aqueous species on mineral surfaces. Also, the results of speciation codes can provide valuable insight into the potential toxicity of a natural or contaminated water. For example, dissolved Cr species are more toxic in the +6 oxidation state than in the +3 oxidation state, and strongly complexed cyanide species such as ferro- and ferricyanides also are less harmful than CN^- and HCN^0 species. The speciation of a water immediately reveals the predominant forms of potentially toxic elements in a water (assuming that proper thermodynamic data are available) in addition to total concentrations. Finally, aqueous speciation codes often form the core of other geochemical modeling codes, such as "inverse" geochemical modeling codes, sometimes also confusingly referred to as mass-balance models, and also "forward" modeling codes (mass-transfer codes and mass-transport codes).

G. Inverse Geochemical Modeling

Inverse geochemical modeling uses available chemical and isotopic analyses, which are assumed to be representative of the chemical and isotopic evolution of groundwater along a given flow path, and attempts both to identify and quantify the heterogeneous reactions that may have been responsible for that chemical and isotopic evolution. A speciation code typically is run, as part of the inverse modeling process, to help the user determine the set of reactions that is thermodynamically feasible, to convert alkalinity measurements into TDIC concentrations, and to calculate the redox state (not the redox potential) of the waters considered. Establishing the redox state of the waters is a convention-based process and simply allows the user to ensure that an electron mass balance is maintained, and that no free electrons are created or destroyed as a result of the reactions considered. Nevertheless, apart from the above considerations, the inverse modeling approach does not require that reactions proceed to thermodynamic equilibrium and indeed some of the early inverse modeling codes did not contain a speciation code or a thermodynamic database.

Inverse modeling codes essentially solve a set of algebraic mass-balance equations describing the changes in chemistry and isotopic composition between two waters (or more in the case of "mixtures") and relate those changes to lists of potential reaction sets and reaction

amounts. "Initial" waters represent source waters prior to mixing and reactions considered by the model. "Final" water represents measured composition after mixing and reaction processes. Typically, the user specifies a list of plausible reactions (sometimes called "phases") and also provides a list of components (chemical or isotopic) that will be used to set up and solve the set of mass-balance equations. The inverse modeling code calculates one or more possible "models" (i.e., reaction sets and amounts) that obey the specified mass balances. Glynn and Brown (1996) provide a detailed description of inverse geochemical modeling, its requirements and limitations, and the relative capabilities of the two most commonly used codes, NETPATH (Plummer et al., 1994) and PHREEQC (Parkhurst & Appelo, 1999).

The PHREEQC code has an advantage over NETPATH in that it accounts for the uncertainties in the analyses provided and, therefore, avoids consideration of reactions with small mass transfers that instead could be explained by uncertainties in the basic data. To do this analysis, PHREEQC assumes that the charge balance error on each given aqueous solution is caused by errors in the analytical data provided, and attempts to "adjust" the analytical data to correct for the charge balance error without exceeding uncertainty limits provided by the user for each analytical datum (total concentrations of each element, pH, isotopic data). PHREEQC also does a more complete accounting of redox balances than NETPATH, and allows redox balances to be maintained not only among the overall redox states of the different waters and reactions, but also among specified redox states for individual elements. Finally, PHREEQC also solves a water-balance equation, an alkalinity balance equation, and a mass balance on inorganic carbon.

NETPATH has some capabilities that are not matched by the inverse modeling capabilities of PHREEQC. Foremost, NETPATH incorporates ^{14}C dating capabilities using various literature-based models, or alternatively and preferably, using reaction-based inverse models. Additionally, NETPATH incorporates isotopic fractionation factors to calculate the ^{13}C, ^{34}S, and ^{15}N compositions of the final water. In its isotopic calculations, NETPATH also solves differential equations, which account for the progressive isotopic evolution of a water as various phases dissolve into it and various phases precipitate or exsolve from it with differing and evolving isotopic compositions. In contrast to NETPATH, PHREEQC inverse modeling only considers isotopic mass-balance constraints, as posed by the user. The PHREEQC user is required to provide the measured isotopic compositions and their uncertainties for the initial and final waters, for dissolving phases, and for precipitating/exsolving phases.

Inverse geochemical modeling is used to *explain* and help understand the observed chemical and isotopic evolution of natural (or contaminated) waters, rather than to *predict* future compositions (as is done by forward geochemical modeling). A minimum amount of data is required to use an inverse geochemical modeling code, namely the compositions of at least one "initial" water and a final water. Inverse geochemical modeling is best used early in the data-acquisition process because it forces the user to think and evaluate the nature and extent of knowledge gaps and uncertainties. Therefore, inverse geochemical modeling can be used to guide the field-data acquisition process. As should be the case for most hydrological and geochemical modeling, inverse geochemical modeling should be used as part of a continuous iterative cycle between data acquisition and data interpretation and modeling until some desired level of detail is obtained in understanding the system investigated.

Although inverse geochemical codes at a minimum require two chemical analyses, one for each water, the modeling process requires appreciable knowledge and expertise. The user has to postulate a list of possible reactions that may be responsible for the observed evolution, and, therefore, needs to have a mineralogical knowledge of the system to be able to make reasonable guesses as to what minerals and gases might be dissolving, precipitating, or exsolving. The user also needs to consult the speciation results to determine which reactions are thermodynamically feasible. For example, if both the initial and final waters are undersaturated with respect to a given mineral, it is unlikely that a reaction model that requires precipitation of the mineral would be valid.

The user needs to have some understanding of the relative kinetics of various reaction processes to be able to judge whether a given reaction process is likely to occur to the extent calculated for a given reaction model, given the estimated travel and evolution time of the water. Establishing a plausible hydrologic relation between the initial and the final waters and estimating a likely travel time between sampling points requires hydrological knowledge of the system and may involve application of a groundwater flow model. Conversely, the inverse geochemical modeling process may result in an improved, or sometimes radically modified, hydrologic understanding of the groundwater system. For example, if all available models predict that a chloride-containing phase needs to precipitate, a thermodynamically unrealistic conclusion in most cases, it is likely instead that either (1) the initial and the final waters are not hydrologically related or (2) the inverse geochemical modeling process perhaps should consider the diluting effect of an additional initial water to explain the lower chloride concentration of the final water.

Inverse geochemical models can account for the possibility of having more than one initial water responsible for the evolution to a final water composition. Inverse geochemical codes do not consider the various possible mechanisms responsible for the "mixing" of the various initial waters: hydrodynamic dispersion, solute diffusion, mixing of various waters as a result of the sampling process (long screens, temporal variations in water chemistry), and other possibilities. It is the responsibility of the model user to assess the hydrological situation and consider the likelihood of the various processes that might cause this "mixing."

A primary value of inverse geochemical modeling is to force the model user to put all available hydrological, chemical, isotopic, and mineralogical data within a conceptual framework. This action should (1) result in an improved understanding of the chemical and isotopic reactions that may be responsible for the observed evolution of the waters, (2) help refine and improve the user's hydrological understanding of the system, and, most importantly, (3) help assess the nature of some of the remaining uncertainties in the constructed conceptual framework.

H. Forward Geochemical Modeling: Overview

Forward geochemical modeling differs conceptually from inverse geochemical modeling. Inverse modeling uses available aqueous-solution data and calculates the mass-transfer amounts of various reactions suspected of accounting for the evolution of an initial water to a final water. Inverse modeling is most useful when abundant chemical, isotopic, mineralogical, and hydrologic data are available, and when the user's objective is to explain the past chemical evolution of a groundwater system.

In contrast, forward modeling attempts to predict the future chemical composition of an aqueous solution given an initial solution and given certain postulated reactions, some of which usually are considered to go to thermodynamic equilibrium. Forward modeling is most useful when the amount of chemical and isotopic data available for a given groundwater system is limited and when the modeler's objective is to predict the future evolution of the system.

I. Forward Modeling: Mass-Transfer Codes

Mass-transfer geochemical codes are used to predict the possible evolution of a water as it contacts, forms, and/or reacts with other phases such as minerals, gases, surface phases, organic matter, and non-aqueous-phase liquids (NAPLs). Most currently available geochemical codes consider only interactions with minerals, gases, and surfaces. A mass-transfer code essentially is an extension of a speciation code. The main difference is that a mass-transfer code uses thermodynamic (and sometimes kinetic) information to calculate not only the speciation of the aqueous solution (i.e., the aqueous-phase reactions), but also to calculate the effect of heterogeneous reactions (reactions between the aqueous phase and other phases) on the composition and speciation of the aqueous phase and on the composition of contacting phases. Many possible reactions and processes can be simulated, including mineral dissolution and precipitation, gas dissolution and exsolution, gas bubble formation, ion exchange on fixed-charge surfaces, ion sorption on variable charge surfaces, evaporation, dilution and mixing of aqueous solutions, precipitation and dissolution of solid-solution phases, boiling, temperature and pressure changes, radioactive decay, and biodegradation reactions. Most commonly, the user makes the assumption that the processes go to full (or partial) thermodynamic equilibrium, but the most recent codes, such as PHREEQC (Parkhurst & Appelo, 1999) and EQ3/6 (Wolery, 1992) also can consider reaction kinetics (given appropriate rate law and kinetic constant information from the user) and can calculate changes in composition as a function of time.

Similar to a speciation code, a mass-transfer numerical model solves a set of algebraic mass-balance and mass-action equations. The mass-balance equations impose conservation of mass for the various components of the system across all phases. The mass-action equations provide for specification of thermodynamic equilibrium for both homogeneous (aqueous-only) and heterogeneous (mass-transfer) reactions. If reaction kinetics are simulated, a set of one or more ordinary differential equations also is solved. Mass-transfer codes have all the limitations of speciation codes (uncertainties, errors, and gaps in thermodynamic and analytical data). In addition, numerical convergence problems tend to occur more frequently in mass-transfer codes than in speciation codes. These problems usually are caused by the extreme changes in the concentrations of individual species that can result from even minor heterogeneous-reaction-driven changes in the pH or pe conditions.

Mass-transfer geochemical codes are useful tools in understanding and predicting the effects of reaction processes in groundwater systems. They can be used to predict the minimum and maximum concentrations that may be expected, as a function of varying physicochemical conditions, for various chemical elements and constituents that may be either toxic or essential to human health. The accuracy of the predictions will be much greater for major constituents (Ca, Na, Mg, K, Cl, SO_4, C, SiO_2) than for minor and trace elements, which often are of concern in water-quality studies. Multiple competing processes often control the concentrations of minor and trace elements; these elements usually are associated and heterogeneously distributed across many different mineral phases and surfaces. Considerable uncertainties and gaps exist in the available thermodynamic data for minor and trace element processes. Finally, many of the processes are kinetically controlled and are not adequately described by the assumption of thermodynamic equilibrium.

Despite the above uncertainties, geochemical mass-transfer codes have the potential to improve the understanding of minor and trace element geochemistry. These codes are essential in determining and predicting the effects of the major reaction processes that are responsible for the evolution of pH, pe, and major element and complexant concentrations. Understanding and predicting the dominant chemical characteristics of the groundwater system is key to understanding and predicting the effects of reaction processes that control the concentrations of minor and trace elements.

Inverse and forward geochemical modeling codes can be complementary, as illustrated by an example describing a fluoride water-quality problem. Groundwater from the Black Creek aquifer in the coastal region of South Carolina has elevated concentrations of fluoride. The general geochemistry of the groundwater, the occurrence and causes of elevated fluoride concentrations, and the public-health aspects of the fluoride problem have been discussed by Zack (1980) and Zack and Roberts (1988). Fluoride concentrations in groundwater in this region generally range from 0.5 mg/L in shallow upgradient (younger) waters to 5.5 mg/L in downgradient (older) waters (compared to a recommended limit of 2.0 mg/L). Dentists in the area have noted a high occurrence of dental fluorosis (mottling of dental enamel) among people who have lived since childhood in the area. The problem seems to be mainly cosmetic; epidemiological studies have indicated no significant long-term health risk for fluoride concentrations of 10 mg/L or less.

TABLE I. Water Chemical Compositions Used in Inverse and Forward Geochemical Modeling of Groundwaters From the Black Creek Aquifer in South Carolina.

	Geo-113	Geo-117	Seawater
Temp	20	20	25
pH	8.5	8.5	8.22
HCO_3	390	626	142
Ca	2	3.4	410
Mg	1.8	1.4	1350
Na	170	320	10500
K	7	0.9	390
Fe	0.02	0.3	0.003
Cl	51	83	19000
SO_4	9.2	4.2	2700
F	0.5	4.6	1.3
SiO_2	13	19	6.4
PO_4	0.09	0.15	0.28

Units in mg/L.
Seawater composition from Hem, 1992.

Zack (1980) and Zack and Roberts (1988) gave a thorough and reasonable explanation of the sources and factors affecting fluoride concentrations and the general geochemistry of the groundwater in the region. Our numerical modeling analysis supports their conclusions and provides further insight, which may allow a better understanding of fluoride geochemistry in groundwaters of the Atlantic Coastal Plain and elsewhere.

Geochemical speciation of a typical high-fluoride water (Table I; Well Geo-117) indicates that the water is undersaturated with respect to fluorite (CaF_2), a mineral with fast reaction kinetics. If fluorite were the source of the dissolved fluoride, speciation of the high-fluoride groundwater likely would be close to thermodynamic equilibrium with respect to fluorite. Although some phosphate nodules found in the Black Creek aquifer also contain fluoride, tests have shown that water drawn from phosphatic deposits in eastern North Carolina contain relatively little fluoride (0.4 mg/L), and, consequently, are not thought to be a source of fluoride in the Black Creek aquifer. Instead, Zack (1980) and Zack and Roberts (1988) suggest that fossil shark teeth found in the most hydraulically conductive layers of the aquifer are the source of fluoride. Shark teeth consist of almost pure cryptocrystalline fluoroapatite.

The Black Creek Formation consists of fine to very fine sands interbedded with laminated clays. The shark teeth were deposited during the Upper Cretaceous in a marine environment and now are abundant in thin, relatively continuous layers of calcite-cemented quartz sands. These layers are present in the most transmissive upper third of the aquifer. Zack (1980; 2002, personal communication) notes that there is a strong linear correlation between dissolved fluoride concentration and alkalinity (mainly dissolved bicarbonate).

Zack suggests that the geochemical evolution of the groundwater is controlled primarily by the reaction of dilute low-pH, CO_2-rich recharge waters that, while flowing, dissolve calcite cement and exchange Ca for Na present in the initially Na-rich marine clays. The cation exchange reaction causes more dissolution of calcite than otherwise would occur. In turn, dissolution of calcite cement exposes surfaces of shark teeth. The high pH environment found at the calcite-dissolution interface causes substitutional exchange of hydroxide for fluoride in the apatite of the shark teeth, which results in high dissolved fluoride concentrations. Zack argues convincingly that substitutional exchange rather than fluoroapatite dissolution is the cause of the high fluoride concentrations.

Inverse geochemical simulations were used to identify and quantify reactions that could explain the geochemical evolution of a recharge water into a more saline high-fluoride water. The simulations were conducted first with the NETPATH code, and, subsequently, with the PHREEQC code. PHREEQC has the advantage because it accounts for data uncertainties and keeps a mass balance on H, O, charge, and alkalinity. NETPATH, which gave similar results to PHREEQC, is easier to use in the initial exploration of possible reaction models. The simulation results (Table II) largely confirm Zack's conceptual model, although it was found that proton exchange reactions occurring on disseminated organic materials could offer an additional control on the pH values of the groundwater without invoking a source of dissolved CO_2. Among reactions that were not considered in our preliminary modeling, silicate mineral weathering reactions could provide a sink for protons, whereas pyrite and marcasite oxidation could provide a source of protons. Additional information, such as isotopic data, would be required for further determination of the most likely reactions controlling the evolution of the groundwater.

Mass-transfer modeling (Figure 7) with PHREEQC indicates that thermodynamic consideration of reactions with a fluoroapatite-hydroxyapatite solid-solution series is essential in explaining the fluoride concentrations, pH values, and other geochemical characteristics observed in the groundwater. Reacting the

TABLE II. Three Inverse Models Determined with PHREEQC for the Evolution of a Low-Fluoride Water (Well Geo-113) to a High-Fluoride Water (Well Geo-117)

	CO_2 Diss.	H^+ release	H^+ Uptake
Geo-113 fraction	0.9984	0.9984	0.9984
Seawater fraction	0.0016	0.0016	0.0016
CO_2	1.904		4.134
Calcite	2.23	4.135	
HX		3.808	−4.461
CaX_2	−2.214	−4.118	0.0168
KX	−0.173	−0.173	−0.173
MgX_2	−0.110	−0.110	−0.110
NaX	4.819	4.819	4.819
CH_2O	0.223	0.223	0.223
Goethite	0.104	0.104	0.104
FeS(ppt)	−0.099	−0.099	−0.099
$SiO2$	0.100	0.100	0.100
Fluoroapatite	0.216	0.216	0.216
Hydroxyapatite	−0.216	−0.216	−0.216

Note: Minor mixing of seawater is included. Mass transfers in millimoles per kilogram of H_2O. Positive numbers indicate mass transfer into the aqueous phase.

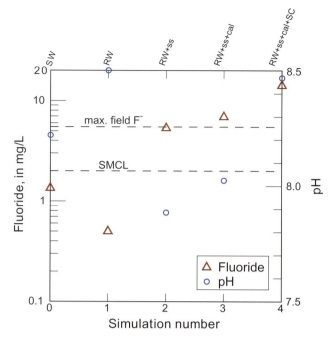

FIGURE 7 Simulated pH values and fluoride concentrations resulting from reactions between a low-fluoride recharge water (RW) sampled from well Geo-113 in the Black Creek aquifer, and combinations of the following: a fluoroapatite solid-solution (ss), calcite (cal), and a proton buffering surface (SC) modeled using a goethite surface-complexation model. Seawater (SW), pH, and fluoride values also are provided for reference, as are fluoride values corresponding to the U.S. Environmental Protection Agency's Secondary Maximum Contaminant Level (SMCL) and the maximum observed in the aquifer (dashed lines).

low-fluoride water (well Geo-113) with calcite and a 99.9% fluoroapatite solid solution resulted in a fluoride concentration slightly above the maximum observed in the Black Creek aquifer, although the pH was lower than observed by about half a pH unit. Adding a proton buffering surface (previously equilibrated with seawater) increased the pH to the observed field value of 8.5, but also resulted in an increase in the fluoride concentration to 17 mg/L, three times the maximum observed in the field. Modifications of the number of surface-complexation sites and of the zero-point-of-charge of the surface proton buffer could be attempted to obtain a better fit of the field data. Further modeling analyses (incorporating other reactions) should be conducted to provide a better understanding of the factors controlling fluoride concentrations in the Black Creek aquifer.

J. Forward Modeling: Mass-Transport Codes

Geochemical mass-transport codes are used to simulate (1) the movement of groundwater, (2) the transport of dissolved constituents, and (3) their reactions both within the water phase and with other phases. In addition to solving sets of algebraic mass-balance and mass-action equations, mass-transport codes also solve sets of partial differential equations describing, as a function of space and time, the distribution of groundwater potentials and velocities, and the advective and dispersive movement of solutes.

Geochemical mass-transport codes incorporate all the limitations and uncertainties associated with the use of geochemical mass-transfer codes and nonreactive solute-transport codes. Geochemical mass-transport codes commonly have convergence problems and other numerical problems (e.g., numerical oscillations, numerical dispersion) associated with the numerical solution of partial differential equations. Also, the description and simulation of physicochemical processes in geochemical transport codes suffers from a dichotomy of scale. Physical transport processes are described at a much larger scale than the molecular level based scale applicable to chemical reactions. This dichotomy generates conceptual and numerical errors and uncertainties in the application and use of geochemical mass-transport codes.

Additionally, running geochemical transport codes can require large computer time and memory, even on today's computers. Increases in computer processing speeds have been matched by the increasing sophistication and simulation capabilities of geochemical transport codes. Possible increases in the "realism" offered by more sophisticated and complex codes, however, are counterbalanced by increased data requirements and associated increases in the uncertainties relating both to the data entered and to the mathematical representation of the simulated processes. Sensitivity analyses, where simulations are run multiple times to test the effects of the data and process uncertainties, are crucial in any intelligent use of geochemical transport codes, but commonly are hampered by computer time requirements.

Geochemical mass-transport codes can be used to predict "best case" and "worst case" scenarios of contaminant transport, but in most cases they are not exact predictive tools. Both geochemical mass-transfer and mass-transport codes are useful tools that can be used to improve conceptual understanding and to gain an appreciation of the relative quantitative importance of processes controlling the chemical evolution (and transport) of natural or contaminated waters.

The use of geochemical transport modeling is illustrated by an application to an arsenic problem in Oklahoma. The Central Oklahoma aquifer underlies about 8000 km^2 of central Oklahoma and is a major source of drinking water for the region. The aquifer is composed mostly of fine-grained sandstones interbedded with siltstone and mudstone. Schlottmann et al. (1998) describe its mineralogy and geochemistry, and they also recognize the occurrence of arsenic as a problem. Recharge to the aquifer occurs mainly in its unconfined eastern area and most streams are gaining (see Figure 8). To the west, the aquifer is confined by low-permeability rocks.

Concentrations of dissolved arsenic in the Central Oklahoma aquifer exceed the 1986 federal drinking water limit of 50 µg/L in about 7% of 477 analyses and even more frequently exceed the more recent standard of 10 µg/L. The highest dissolved arsenic concentrations are found primarily in the western confined part of the aquifer (Figure 8).

Mineralogical and sequential extraction analyses have shown that iron oxides (goethite and hematite) in the sandstones are the primary mineral sources of arsenic.

Arsenic sorbs strongly to iron oxides, particularly at pH values below 8. Discrete arsenic mineral phases were not found, although some evidence was found of high arsenic concentrations in pyrite grains. Pyrite only is found in isolated, poorly conductive, low redox zones. Indeed, waters in the Central Oklahoma aquifer generally are oxic, with dissolved oxygen concentrations above 1 mg/L, and there is little organic matter or iron sulfides present. Iron oxide minerals predominate instead.

Extensive geochemical modeling of the Central Oklahoma aquifer has succeeded in elucidating the factors controlling dissolved arsenic concentrations and the general geochemical evolution of the waters (Parkhurst et al., 1993; Parkhurst, written communication, 2002). Both inverse and forward geochemical modeling was conducted, including a three-dimensional geochemical transport model using the USGS code PHAST (based on coupling HST3D with PHREEQC). Parkhurst's geochemical model assumes that the aquifer initially is filled with sodium chloride brine equilibrated with calcite and dolomite minerals, a cation-exchanger (clays), and a hydrous iron oxide surface with complexed arsenic.

In Parkhurst's model, fresh recharge water, equilibrated with calcite, dolomite, and with carbon dioxide at a partial pressure close to 100 times atmospheric (typical soil CO_2 partial pressure) enters the unconfined part of the aquifer in the east. The recharge water reacts with the initially Na-rich exchanger clays and with the As-rich and proton-depleted hydrous ferric oxide surfaces. As the groundwater flows through the porous media, ion exchange gradually changes the calcium-magnesium bicarbonate recharge water into a sodium bicarbonate water. The initial dissolution of soil carbon dioxide keeps the pH of the recharge water relatively low (between 7.0 and 7.5). After loss of contact with the soil CO_2 reservoir, however, the pH of the recharging water gradually increases from the further dissolution of calcite and dolomite (because of uptake of calcium and magnesium on exchange sites) and also from the protonation of the initially proton-depleted hydrous ferric oxide surface. The calculated pH of the resulting sodium bicarbonate water ranges from 8.5 to 9.2, which is close to the observed values. Under these higher pH conditions, sorption sites on the hydrous ferric oxide surface become predominantly negatively charged and, consequently, desorption of arsenic occurs, resulting in higher dissolved As concentrations.

In addition to simulating geochemical reactions, the PHAST code applied by Parkhurst to the Central Oklahoma aquifer also simulates groundwater flow and

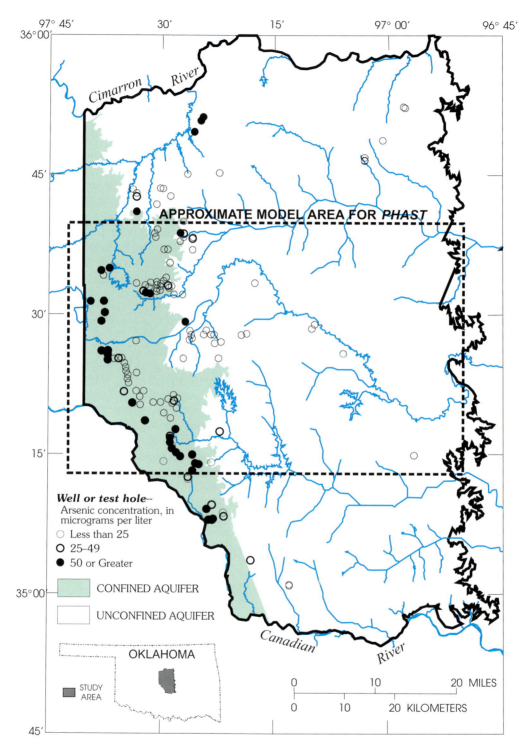

FIGURE 8   Areal distribution of arsenic in water from deep (>100 m) wells and test holes from the Central Oklahoma aquifer, showing the area where the *PHAST* model was applied. (Modified from Schlottmann et al., 1998, and D. Parkhurst, written communication, 2002.)

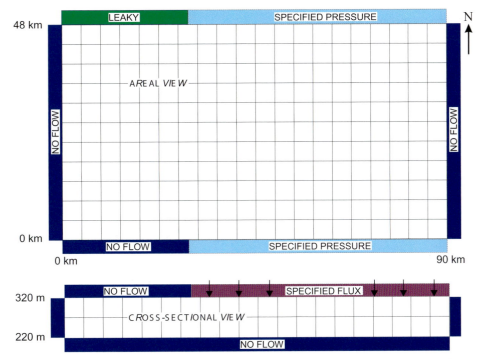

FIGURE 9 Schematic representation of grid and boundary conditions for application of *PHAST* model to Central Oklahoma aquifer. (From D. Parkhurst, written communication, 2002.)

solute-transport processes. Boundary conditions for the simulation domain (Figure 9) included specified pressures along the eastern part of the northern and southern boundaries to represent the hydrologic effects of adjacent rivers, which are the primary sinks for water discharge from the aquifer. A specified-flux boundary condition also was placed over the unconfined eastern part of the aquifer to simulate recharge.

The use of the PHAST model enabled Parkhurst and his coworkers to analyze the magnitude and sensitivity of various factors affecting groundwater flow, solute transport, and geochemical evolution observed in the Central Oklahoma aquifer. Their integrated model was successful in matching general hydrological and geochemical observations and in explaining the occurrence of high arsenic concentrations in the western part of the aquifer.

V. MODEL DESIGN AND APPLICATION

A. Overview

The first step in model design and application is to define the nature of the problem and the purpose of the model (Figure 10). This step is linked closely with the formulation of a conceptual model, which is required prior to development of a mathematical model. In formulating a conceptual model, one must evaluate which processes are important for the particular problem at hand. Some processes may be important to consider at one scale of study, but negligible or irrelevant at another scale. Good judgment is required to evaluate and balance the trade-offs between accuracy and cost, with respect to model development, model use, and data requirements. The key to efficiency and accuracy in modeling a system probably is more affected by the formulation of a proper and appropriate conceptual model than by the choice of a particular numerical method or code.

Once a decision to develop a model has been made, a code (or generic model) must be selected (or modified or constructed) that is appropriate for the given problem. Next, the generic code must be adapted to the specific site or simulated region. Development of a numerical deterministic, distributed-parameter, simulation model involves selecting or designing spatial grids and time increments that will yield an accurate solution for the given system and problem. The analyst must then specify the properties of the system (and their distributions), boundary conditions, initial conditions (for

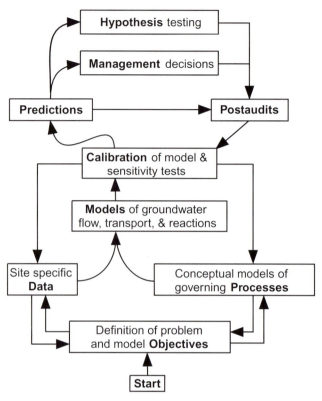

FIGURE 10 The use of models in the analysis of ground-water systems. (Modified from Konikow & Reilly, 1998.)

developed. In a field environment, perhaps the single most important key to understanding a transport or reaction problem is the development of an accurate definition (or model) of flow. In highly heterogeneous systems, the head distribution and flow directions often can be simulated fairly accurately, whereas the calculated velocity field still may be greatly in error, which results in considerable errors in simulations of transport.

B. Grid Design

The dimensionality of a flow or transport model (i.e., one, two, or three dimensions) should be selected during the formulation of the conceptual model. If a one- or two-dimensional model is selected, then it is important that the grid be aligned with the flow system so that there is no unaccounted flux into or out of the line or plane of the grid. For example, if a two-dimensional areal model is applied, then there should be no major vertical components of flow and any vertical leakage or flux must be accounted for by boundary conditions. If a two-dimensional profile model is applied, then the line of the cross section should be aligned with an areal streamline, and there should not be any major lateral flow into or out of the plane of the cross section.

To minimize a variety of sources of numerical errors, the model grid should be designed using the finest mesh spacing and time steps that are possible, given limitations on computer memory and computational time. The boundaries of the grid also should be aligned, to the extent possible, with natural hydrologic and geologic boundaries of the aquifer. Where it is impractical to extend the grid to a natural boundary, then an appropriate boundary condition should be imposed at the grid edge to represent the net effects of the continuation of the aquifer beyond the grid. These boundaries also should be placed as far away as possible from the area of interest and areas of stresses on the system to minimize any effect of conceptual errors associated with these artificial boundary conditions.

In specifying boundary conditions for a particular problem and grid design, care must be taken not to overconstrain the solution. That is, if dependent values are fixed at too many boundary nodes, at either internal or external nodes of a grid, the model may have too little freedom to calculate a meaningful solution (Franke & Reilly, 1987).

To optimize computational resources in a model, it sometimes is advisable to use an irregular (or variably-

transient problems), and geochemical processes/reactions. All of the parameter specifications and boundary conditions really are part of the overall conceptual model of the system.

Any model is a simplified approximation of a very complex reality, but the model should capture the essential features and processes relative to the problem at hand. The selection of the appropriate model and appropriate level of model complexity remains subjective and dependent on the judgment and experience of the analysts, the objectives of the study, the level of prior information available for the system of interest, and the complexity of the modeled system. The trade-off between model accuracy and model cost always will be difficult to resolve, but always will have to be made and may affect model reliability.

Because the groundwater seepage velocity is determined from the head distribution and because both advective transport and hydrodynamic dispersion are functions of the seepage velocity, a model of groundwater flow typically is calibrated before a pathline, solute-transport, or geochemical reaction model is spaced) mesh in which the grid is finest in areas of point stresses, where gradients are steepest, where data are most dense, where the problem is most critical, and/or where greatest numerical accuracy is desired. Similarly, time steps often can be increased geometrically during a transient simulation.

C. Model Calibration and Refinement

Model calibration may be viewed as an evolutionary process in which successive adjustments and modifications to the model are based on the results of previous simulations. Overviews on the philosophy of applying and testing groundwater flow and geochemical models are presented by Konikow and Bredehoeft (1992) and Nordstrom (1994). In general, it is best to start with a simple model and add complexity or refine the grid in small increments as needed and justified.

In applying and evaluating a model, one must decide when sufficient adjustments have been made to the representation of parameters and processes and at some time accept the model as adequately calibrated (or perhaps reject the model as inadequate and seek alternative approaches). This decision often is based on a mix of subjective and objective criteria. The achievement of a best fit between values of observed and computed variables is a regression procedure and can be evaluated as such. That is, the residual errors should have a mean that approaches zero and the deviations should be minimized. There are various statistical measures that can be used to assess the reliability and "goodness of fit" of groundwater models. The accuracy tests should be applied to as many dependent variables as possible.

The use of deterministic models in the analysis of groundwater problems is illustrated, in a general sense, in Figure 10. Perhaps the greatest value of the modeling approach is its capability to integrate site-specific data with equations describing the relevant processes as a quantitative basis for predicting changes or responses in a groundwater system. One objective of model calibration should be to improve the conceptual model of the system. The improvement in understanding of a system derived from a model application and calibration exercise for hypothesis testing often is of greater value than the predictive value for management purposes. Another objective should be to define inadequacies in the database and help set priorities for the collection of additional data.

D. Model Error

Discrepancies between observed and calculated responses of a groundwater system are the manifestation of errors in the conceptual or mathematical model. In applying groundwater models to field problems, there are three sources of error, and it may not be possible to distinguish among them (Konikow & Bredehoeft, 1992). One source is conceptual errors, that is, misconceptions about the basic processes that are incorporated in the model. Conceptual errors include both neglecting relevant processes as well as inappropriate representation of processes. Examples of conceptual errors include the use of a two-dimensional model where significant flow or transport occurs in the third dimension, or the application of a model based upon Darcy's law to media or environments where Darcy's law is inappropriate. A second source of error involves numerical errors arising in the equation-solving algorithm, such as truncation errors, round-off errors, and numerical dispersion. A third source of error arises from measurement errors and from uncertainties and inadequacies in the input data that reflect our inability to describe comprehensively and uniquely the properties, stresses, and boundaries of the groundwater system. In most model applications, conceptualization problems and data uncertainty are the most common sources of error.

In solving advection-dominated transport problems in which a sharp front (or steep concentration gradient) is moving through a groundwater system, it is difficult numerically to preserve the sharpness of the front. Obviously, if the width of the front is narrower than the node spacing, then it is inherently impossible to calculate the correct values of concentration in the vicinity of the sharp front. Even in situations where a front is less sharp, the numerical solution technique can calculate a greater dispersive flux than would occur by physical dispersion alone or would be indicated by an exact solution of the governing equation. That part of the calculated dispersion (or spreading of solute about the center of mass) introduced solely by the numerical solution algorithm is called numerical dispersion. Numerical dispersion can be controlled most easily by reducing the grid spacing, although that will increase computational costs proportionately.

One measure of numerical accuracy is how well the model conserves mass. This can be measured by comparing the net fluxes calculated or specified in the model (e.g., inflow and sources minus outflow and sinks) with changes in storage (accumulation or depletion). Mass-balance calculations always should be performed and checked during the calibration procedure to help assess the numerical accuracy of the solution.

As part of the mass-balance calculations, the hydraulic and chemical fluxes contributed by each distinct hydrologic component of the flow and transport model should be itemized separately to form hydrologic and chemical budgets for the modeled system. The budgets are valuable assessment tools because they provide a measure of the relative importance of each component to the total budget.

E. Geochemical Model Design

The guidelines for the design and application of geochemical models are similar to those outlined above, especially concerning geochemical transport modeling, which depends on the establishment of flow and solute-transport models. Other types of geochemical models do not depend on the establishment of a spatial grid and on the attendant issues of grid spacing and boundary conditions. The accuracy of a geochemical mass-transfer model may depend on time step size if reaction kinetics are simulated. Other types of geochemical calculations (equilibrium mass-transfer calculations, speciation calculations, inverse geochemical modeling) do not depend on the numerical value of any time-step increments, but can, in some cases, depend on the specification of an initial system state and also on the order in which different isolated geochemical processes or sets of processes are applied/simulated in the system. In all cases of geochemical modeling, the quality of results obtained strongly depend on the quality of the input chemical data and on the quality of any thermodynamic data used by the model.

Inverse geochemical modeling can be used as a first step in helping to construct a geochemical transport model (e.g., Glynn & Brown, 1996). The idea is to use inverse geochemical modeling to determine all the possible sets of reaction processes that potentially could explain the observed chemical and isotopic evolution of one groundwater into another. Consequently, different sets of reaction processes, and different geochemical characteristics, can be considered in a suite of transport simulations, thereby allowing the modeler to assess: (1) the dependence of the movement of a particular contaminant front or concentration on the reaction processes considered and (2) the need for additional specific field data that potentially could eliminate some of the uncertainties regarding the applicable geochemical processes. As stated by Glynn and Brown (1996):

"Identifying knowledge gaps and critical data needs, preventing us from more accurately determining the identity and importance of the reactions ... was one of the most important results of the inverse and reactive transport modeling simulations conducted."

VI. Obtaining Model Codes

A large number of generic deterministic groundwater models, based on a variety of numerical methods and a variety of conceptual models, are available. In selecting a model that is appropriate for a particular application, it is most important to choose one that incorporates the proper conceptual model; one must avoid force-fitting an inappropriate model to a field situation solely because of model convenience, availability, or familiarity to the user. Usability also is enhanced by the availability of graphical preprocessing and post-processing programs or features, and by the availability of comprehensive yet understandable model documentation.

A large number of public and private organizations distribute public domain and/or proprietary software for groundwater modeling. Some Internet sites allow computer codes to be downloaded at no cost whereas other sites provide catalog information, demonstrations, and pricing information. The International Groundwater Modeling Center in Golden, Colorado (www.mines.edu/research/igwmc/software/), maintains a clearinghouse and distribution center for groundwater simulation models. Many of the U. S. Geological Survey public domain codes are available from links on their Web sites at water.usgs.gov/nrp/models.html and water.usgs.gov/software/. The U. S. Environmental Protection Agency's Center for Subsurface Modeling Support (www.epa.gov/ada/csmos.html) also provides public domain groundwater modeling software.

See Also the Following Chapters

Chapter 11 (Arsenic in Groundwater and the Environment) · Chapter 12 (Fluoride in Natural Waters) · Chapter 15 (Selenium Deficiency and Toxicity in the Environment)

Further Reading

Alley, W. M., Reilly, T. E., and Franke, O. L. (1999). Sustainability of Ground Water Resources, U.S. Geological Survey Circular 1186.

Anderson, M. P., and Woessner, W. W. (1992). *Applied Groundwater Modeling: Simulation of Flow and Advective Transport*, Academic Press, San Diego.

Ball, J. W., and Nordstrom, D. K. (1991). User's Manual for WATEQ4F, with Revised Thermodynamic Database and Test Cases for Calculating Speciation of Major, Trace and Redox Elements in Natural Waters, U.S. Geological Survey Open-File Report 91–183.

Bear, J. (1979). *Hydraulics of Groundwater*, McGraw-Hill, New York.

Belitz, K., and Phillips, S. P. (1992). Simulation of Water-Table Response to Management Alternatives, Central Part of the Western San Joaquin Valley, California, U.S. Geological Survey Water Resources Investigations Report 91–4193.

Belitz, K., Phillips, S. P., and Gronberg, J. M. (1993). Numerical Simulation of Ground Water Flow in the Central Part of the Western San Joaquin Valley, California, U.S. Geological Survey Water Supply Paper 2396.

Clement, T. P. (1997). RT3D—A Modular Computer Code for Simulating Reactive Multi-Species Transport in 3-Dimensional Groundwater Aquifers, Report PNNL-11720, Pacific Northwest National Laboratory, Richland, Washington.

Cuthbert, M. O., Burgess, W. G., and Connell, L. (2002). Constraints on Sustainable Development of Arsenic-Bearing Aquifers in Southern Bangladesh. Part 2: Preliminary Models of Arsenic Variability in Pumped Groundwater. In *Sustainable Groundwater Development* (K. M. Hiscock, M. O. Rivett, and R. M. Davison, Eds.), Geological Society Special Publication 193, The Geological Society of London.

Deverel, S. J., Gilliom, R. J., Fujii, R., Izbicki, J. A., and Fields, J. C. (1984). Areal Distribution of Selenium and Other Inorganic Constituents in Shallow Ground Water of the San Luis Drain Service Area, San Joaquin Valley, California—A Preliminary Study, U.S. Geological Survey Water Resources Investigations Report 84–4319.

Dowdle, P. R., Laverman, A. M., and Oremland, R. S. (1996). Bacterial Dissimilatory Reduction of Arsenic (V) to Arsenic (III) in Anoxic Sediments, *Appl. Environ. Microbiol.*, 62, 1664–1669.

Drever, J. I. (1997). *The Geochemistry of Natural Waters: Surface and Groundwater Environments*, third edition, Prentice Hall, New Jersey.

Essaid, H. I., and Bekins, B. A. (1997). BIOMOC, a Multispecies Solute-Transport Model with Biodegradation, U.S. Geological Survey Water-Resources Investigations Report 97–4022.

Feder, G. L., Radovanovic, Z., and Finkelman, R. B. (1991). Relationship Between Weathered Coal Deposits and the Etiology of Balkan Endemic Nephropathy, *Kidney Int.*, 40, Suppl. 34, S9–S11.

Franke, O. L., and Reilly, T. E. (1987). The Effects of Boundary Conditions on the Steady State Response of Three Hypothetical Ground Water Systems—Results and Implications of Numerical Experiments, U.S. Geological Survey Water-Supply Paper 2315.

Franke, O. L., Reilly, T. E., Pollock, D. W., and LaBaugh, J. W. (1998). Estimating Areas Contributing Recharge to Wells: Lessons from Previous Studies, U.S. Geological Survey Circular 1174.

Gelhar, L. W., Welty, C., and Rehfeldt, K. R. (1992). A critical review of data on field-scale dispersion in aquifers, *Water Resour. Res.*, 28, 1955–1974.

Glynn, P. D. (2000). Solid-Solution Solubilities and Thermodynamics: Sulfates, Carbonates and Halides. In *Sulfate Minerals, Crystallography, Geochemistry and Environmental Significance* (C. N. Alpers, J. L. Jambor, and D. K. Nordstrom, Eds.), *Reviews in Mineralogy and Geochemistry*, Vol. 40, pp. 481–511, chap. 10.

Glynn, P. D. (2003). Modeling Np and Pu Transport with a Surface Complexation Model and Spatially Variant Sorption Capacities: Implications for Reactive Transport Modeling and Performance Assessments of Nuclear Waste Disposal Site, *Comput. Geosci.*, 29, 331–349.

Glynn, P. D., and Brown, J. G. (1996). Reactive Transport Modeling of Acidic Metal-Contaminated Groundwater at a Site with Sparse Spatial Information, In *Reactive Transport in Porous Media* (C. I. Steefel, P. Lichtner, and E. Oelkers, Eds.), Mineralogical Society of America, *Reviews in Mineralogy*, Vol. 34, pp. 377–438, chap. 9.

Grant, W. D., Gemmell, R. T., and McGenity, T. J. (1998). Halophiles. In *Extremophiles: Microbial Life in Extreme Environments* (K. Horikoshi and W. D. Grant, Eds.), Wiley Series in Ecological and Applied Microbiology, Wiley-Liss, New York.

Haitjema, H. M. (1995). *Analytic Element Modeling of Groundwater Flow*, Prentice Hall, Englewood Cliffs, NJ.

Harbaugh, A. W., Banta, E. R., Hill, M. C., and McDonald, M. G. (2000). MODFLOW-2000, The U.S. Geological Survey Modular Groundwater Model–User Guide to Modularization Concepts and the Ground-Water Flow Process, U.S. Geological Survey Open-File Report 00–92.

Hem, J. D. (1992). Study and interpretation of the chemical characteristics of natural water, third edition, U.S. Geological Survey Water-Supply Paper 2254.

Hill, M. C. (1998). Methods and guidelines for effective model calibration, U.S. Geological Survey Water-Resources Investigations Report 98–4005.

Hill, M. C., Banta, E. R., Harbaugh, A. W., and Anderman, E. R. (2000). MODFLOW-2000, The U.S. Geological Survey Modular Ground Water Model–User Guide to the Observation, Sensitivity, and Parameter-Estimation Processes and Three Post-Processing Programs, U.S. Geological Survey Open-File Report 00–184.

Hopps, H. C., and Feder, G. L. (1986). Chemical Qualities of Water that Contribute to Human Health in a Positive Way, *Sci. Total Environ.*, 54, pp. 207–216.

Huyakorn, P. S., and Pinder, G. F. (1983). *Computational Methods in Subsurface Flow*, Academic Press, New York.

Ibaraki, M., and Therrien, R., (Eds.) (2001). Practical Applications of Coupled Process Models in Subsurface Environments, *J. Contam. Hydrol.*, 52.

Keller, W. D. (1978). Drinking Water: A Geochemical Factor in Human Health, *Geol. Soc. Am. Bull.*, 89, 334–336.

Kipp, K. L., Jr. (1997). Guide to the Revised Heat and Solute Transport Simulator, HST3D–Version 2, U.S. Geological Survey Water-Resources Investigations Report 97–4157.

Konikow, L. F., and Bredehoeft, J. D. (1992). Groundwater Models Cannot be Validated, *Adv. Water Resour.*, 15, 75–83.

Konikow, L. F., and Reilly, T. E. (1998). Groundwater Modeling. In *The Handbook of Groundwater Engineering* (J. W. Delleur, Ed.), CRC Press, Boca Raton, FL.

Konikow, L. F., Goode, D. J., and Hornberger, G. Z. (1996). A Three-Dimensional Method of Characteristics Solute-Transport Model (MOC3D), U.S. Geological Survey Water-Resources Investigations Report 96–4267.

Lichtner, P. C., Steefel, C. I., and Oelkers, E. H. (Eds.) (1996). Reactive Transport in Porous Media. In *Reviews in Mineralogy*, Vol. 34, Mineralogical Society of America, Washington, DC.

Lovley, D. R., Phillips, E. J. P., Gorby, Y., and Landa, E. R. (1991). Microbial Reduction of Uranium, *Nature*, 35, 413–416.

McDonald, M. G., and Harbaugh, A. W. (1988). A Modular Three-Dimensional Finite-Difference Ground Water Flow Model, U.S. Geological Survey Techniques of Water-Resources Investigations, chap. A1.

Nordstrom, D. K. (1994). On the Evaluation and Application of Geochemical Models. In Proc. of 5th CEC Natural Analogue Working Group and Alligator Rivers Analogue Project, Report EUR 15176EN, 375–385.

Nordstrom, D. K., and Munoz, J. L. (1994). *Geochemical Thermodynamics*, second edition, Blackwell Scientific Publications, Boston, MA.

Parkhurst, D. L. (1995). User's Guide to PHREEQC—A Computer Program for Speciation, Reaction-Path, Advective-Transport, and Inverse Geochemical Calculations, U.S. Geological Survey Water-Resources Investigations Report 95–4227.

Parkhurst, D. L., and Appelo, C. A. J. (1999). User's Guide to PHREEQC (Version 2)—A Computer Program for Speciation, Batch-Reaction, One-Dimensional Transport, and Inverse Geochemical Calculations, U.S. Geological Survey Water-Resources Investigations Report 99–4259.

Parkhurst, D. L., Christenson, S., and Breit, G. N., (1993). Ground Water-Quality Assessment of the Central Oklahoma Aquifer, Oklahoma: Geochemical and Geohydrologic Investigations, U.S. Geological Survey Open-File Report 92–642.

Pedersen, K. (1993). The Deep Subterranean Biosphere, *Earth Sci. Rev.*, 34, 243–260.

Plummer, L. N., Jones, B. F., and Truesdell, A. H. (1976). WATEQF—A FORTRAN IV Version of WATEQ, A Computer Program for Calculating Chemical Equilibrium of Natural Waters, U.S. Geological Survey Water-Resources Investigations Report 76–13 (revised 1984).

Plummer, L. N., Prestemon, E. C., and Parkhurst, D. L. (1994). An Interactive Code (NETPATH) for Modeling NET Geochemical Reactions Along a Flow PATH—Version 2.0, U.S. Geological Survey Water-Resources Investigations Report 94–4169.

Pollock, D. W. (1989). Documentation of Computer Programs to Compute and Display Pathlines Using Results from the U.S. Geological Survey Modular Three-Dimensional Finite-Difference Ground Water Flow Model, U.S. Geological Survey Open-File Report 89–381.

Presser, T. S., and Barnes, I. (1985). Dissolved Constituents Including Selenium in Waters in the Vicinity of Kesterson National Wildlife Refuge and the West Grassland, Fresno and Merced Counties, California, U.S. Geological Survey Water-Resources Investigations Report 84–4220.

Presser, T. S., Swain, W. C., Tidball, R. R., and Severson R. C. (1990). Geologic Sources, Mobilization, and Transport of Selenium from the California Coast Ranges to the Western San Joaquin Valley: A Reconnaissance Study, U.S. Geological Survey Water-Resources Investigations Report 90–4070.

Schlottmann, J. L., Mosier, E. L., and Breit, G. N. (1998). Arsenic, Chromium, Selenium, and Uranium in the Central Oklahoma Aquifer. In *Ground Water-Quality Assessment of the Central Oklahoma Aquifer, Oklahoma: Results of Investigations* (S. Christenson and J. S. Havens, Eds.), U.S. Geological Survey Water-Supply Paper 2357-A, pp. 119–179.

Schwarzenbach, R. P., Gschwend, P. M., and Imboden, D. M. (1993). *Environmental Organic Chemistry*, Wiley, New York.

Stetter, K. O. (1998). Hyperthermophiles: Isolation, Classification and Properties. In *Extremophiles: Microbial Life in Extreme Environments* (K. Horikoshi and W. D.

Grant, Eds.), Wiley Series in Ecological and Applied Microbiology, Wiley-Liss, New York.

Switzer Blum, J., Stolz, J. F., Oren, A., and Oremland, R. S. (2001). Selenihalanaerobacter shriftii gen. nov., sp. nov., a Halophilic Anaerobe from Dead Sea Sediments that Respire Selenate, *Arch. Microbiol.*, 175, 208–219.

Voss, C. I. (1984). SUTRA—Saturated Unsaturated Transport—A Finite-Element Simulation Model for Saturated-Unsaturated Fluid-Density-Dependent Ground Water Flow With Energy Transport or Chemically-Reactive Single-Species Solute Transport, U.S. Geological Survey Water-Resources Investigations Report 84–4369.

Wang, J. F., and Anderson, M. P. (1982). *Introduction to Groundwater Modeling*, Freeman, San Francisco, CA.

Winter, T. C., Harvey, J. W., Franke, O. L., and Alley, W. M. (1998). Ground Water and Surface Water: A Single Resource, U.S. Geological Survey Circular 1139.

Wolery, T. J. (1992). EQ3/6, A Software Package for Geochemical Modeling of Aqueous Systems: Package Overview and Installation Guide (Version 7.0), UCRL-MA-110662 pt. 1, Lawrence Livermore National Laboratory.

Zack, A. L. (1980). Geochemistry of Fluoride in the Black Creek Aquifer System of Horry and Georgetown Counties, South Carolina—and its Physiological Implications, U.S. Geological Survey Water-Supply Paper 2067.

Zack, A. L., and Roberts, I. (1988). The Geochemical Evolution of Aqueous Sodium in the Black Creek Aquifer, Horry and Georgetown Counties, South Carolina, U.S. Geological Survey Water-Supply Paper 2324.

Zheng, C., and Bennett, G. D. (2002). *Applied Contaminant Transport Modeling* (second edition), Wiley-Interscience, New York.

Zheng, C., and Wang, P. P. (1999). MT3DMS: A Modular Three-Dimensional Multispecies Model for Simulation of Advection, Dispersion and Chemical Reactions of Contaminants in Groundwater Systems: Documentation and User's Guide, Report SERDP-99–1, U.S. Army Engineer Research and Development Center, Vicksburg, MS.

Appendix A (International Reference Values) *Soils*

Peter Bobrowsky
Geological Survey of Canada

Roger Paulen
Alberta Geological Survey

Brian J. Alloway
The University of Reading

Pauline Smedley
British Geological Survey

Maximum Permissible Concentrations of Heavy Metals and Metalloids in Soils (mg kg^{-1})

Element	UK (1)	UK (2)	Netherlands		USA	AUS	NZ	Europe
			Targ	Int				
Arsenic	20	50	29	55	—	20	10	—
Cadmium	1 (pH 6) 2 (pH 7) 8 (pH 8)	3 (pH > 5)	0.8	12	20	1 (3 SA)	3	1–3
Chromium	130	(400 prov)	100	380	1500	100	600	—
Copper	—	80 (pH 5–5.5) 100 (pH 5.5–6) 135 (pH 6–7) 200 (pH > 7)	36	190	750	100 (200 SA)	140	50–140
Mercury	8	1 (pH > 5)	0.3	10	8	1	1	1–1.5
Nickel	50	50 (pH 5–5.5)	35	210	210	60	35	30–75
Lead	450	300	85	530	150	150 (200 SA)	300	50–300
Zinc	—	200 (pH 5–7) 300 (pH > 7)	720	140	1400	200 (250 SA)	300	150–300

Notes and references: UK (1)—Contaminated Land Exposure Assessment (CLEA) guidance values (to be used as part of a risk assessment for contaminated sites), Department for Environment, Food and Rural Affairs, R & D Publications SGV 1, 3, 4, 5, 7, and 10, Environment Agency, Bristol, 2002. UK (2)—for normal agricultural soils and values for zinc and copper in all types of soil. The Soil Code: Code of Good Agricultural Practice for the Protection of Soil PB0617, MAFF, London, 1998. Netherlands—"Dutch Limits" Targ = target values (which it is intended that soil should reach) and Int = intervention values (when site needs to be cleaned up). These values are for assessing the need for remediation of land suspected of being contaminated and apply to a "standard soil" containing 10% organic matter and 25% clay. VROM (2000) circular on target values and intervention values for soil remediation. Ministry of Housing, Spatial Planning and Environment, Department of Soil Protection (VROM) The Hague, The Netherlands. DBO/1999226863. USA—Maximum concentrations for soils treated with biosolids (sewage sludge) McGrath et al., 1994. Land application of sewage sludge: scientific perspective if heavy metal loading limits in Europe and the United States. *Environmental Reviews*, 2, 108–118. Australia and New Zealand—Guidelines for controlling metal concentrations in soils for reuse of biosolids (SA = values used in the state of South Australia). McLaughlin et al. (2000). Review: A bioavailability-based rationale for controlling metal and metalloid contamination of agricultural land in Australia and New Zealand. *Australian Journal of Soil Research*, 38, 1037–1086. Europe—for countries of the European Union for soils receiving sewage sludge (assumes soil pH 6–7), lower value is guideline value, upper value is the mandatory limit. Commission of the European Communities (1986) Council Directive (86/278/EEC) on the protection of the environment, and in particular of the soil, when sewage sludge is used in agriculture. *Official Journal of the European Communities*, 15, 69–81.

APPENDIX A (INTERNATIONAL REFERENCE VALUES) *WATER*

PETER BOBROWSKY
Geological Survey of Canada

ROGER PAULEN
Alberta Geological Survey

BRIAN J. ALLOWAY
The University of Reading

PAULINE SMEDLEY
British Geological Survey

Regulations and Guidelines: Inorganic Trace Constituents in Drinking Water ($\mu g\, L^{-1}$)

Country or institution	Nature of standards	Comments	Date	Al	Ag	As	B	Ba	Be	Cd
Australia	Guidelines	Health-based guidelines	1996		100	7	300	4,000[b]	—	2
	Guidelines	Aesthetic guidelines	1996	200						
Canada	Guidelines	Maximum acceptable concentrations (MACs)	2003	100[a]				1,000	—	5
			2003			25	5,000			
	Guidelines	Interim maximum acceptable concentrations (IMACs)	2003							
	Guidelines	Aesthetic objectives (AOs)								
Japan	Standards	Health-based	1993	—		10	—	—	—	10
	Standards	Acceptability-based	1993							
EC (European Commission)	Directive	Maximum permissible values	1998			10	1,000	—	—	5
	Directive	Indicator parameters	1998	200						
U. S. EPA	Regulations	Primary standards (maximum contaminant levels, MCLs)	2002			10	—	2,000	4	5
	Guidelines	Secondary standards	2002	50–200	100					
WHO	Guidelines	Guideline values	2004					700	—	3
		Provisional guideline values	2004			10	500			
		Acceptability-based guidelines	2004	200						

[a]Water treatment outlets.
[b]Update 2001.
[c]WHO (2004). *Guidelines for Drinking-water Quality*, 3rd Ed., WHO, Geneva, In press.
[d]Short-term exposure (NO_2 long-term exposure value is $200\,\mu g\, L^{-1}$); provisional.

Cr	Cu	F	Fe	Hg	I	Mn	Mo	NH_3	Ni	NO_2	NO_3	Pb	Sb	Se	Tl	U	V	Zn
								as NH_3		as NO_2	as NO_3							
50	2,000	1,500		1	100	500	50		20	3,000	50,000	10	3	10	—	—	—	
	1,000		300			100		500										3,000
50		1,500		1	—		—	—			45,000	—		10	—		—	
	1,000		300			50							6			20		5,000
50		800		0.5	—		—	—	—	44,300	44,300	50	—	10	—	—	—	
	1,000		300			50												1,000
50	2,000	1,500		1	—		—	—	20	500	50,000	10	5	10	—	—	—	
			200			50		500										
100	1,300	4,000		2	—		—	—		4,430	44,300	15	6	50	2	30	—	
	1,000	2,000	300			50												5,000
		1,500		1	—		70			3,000[d]	50,000[d]	10	18	10	—		—	
50	2,000					400			20							9		
	1,000		300			100	1,500											3000

APPENDIX B (WEB LINKS)

ANNOTATED URLs FOR CHAPTER-RELATED WEB SITES OF INTEREST

Chapter 19

American Academy of Microbiology, accessed 1999. Antimicrobial resistance-an ecological perspective: Report of AAM colloquium, July 16, 1999, San Juan, Puerto Rico, American Academy of Microbiology, Washington DC, http://www.asmusa.org/acasrs/pdfs/Antimicrobial.pdf

Ania, B.J. and Asenjo, M., accessed 2002. Mycetoma, *eMedicine Journal*, http://www.emedicine.com/MED/topic30.htm

Cambridge University Schistosomiasis Research Group, accessed 2002. Helminth infections of man, University of Cambridge, Cambridge CB2 1TN, UK, http://www.path.cam.ac.uk/~schisto/General_Parasitology/Hm.helminths.html

Canadian Centre for Occupational Health and Safety, accessed 1999. What is Farmer's Lung, Canadian Centre for Occupational Health and Safety, Hamilton, Ontario, Canada, http://www.ccohs.ca/oshanswers/diseases/farmers_lung.html

Carey, J., Motyl, M., and Perlman, D.C., 2001. Catheter-related bacteremia due to *Streptomyces* in a patient receiving holistic infusions, *in* Emerging Infectious Diseases, Vol. 6, No. 6, November–December, 2001, Centers for Disease Control and Prevention, Atlanta, Georgia, http://www.cdc.gov/ncidod/eid/index.htm

CDC, accessed 2002. U.S. Centers for Disease Control and Prevention, Atlanta, Georgia, http://www.cdc.gov/

DoctorFungus, accessed 2002. DoctorFungus.org, http://www.doctorfungus.org/thefungi/index.htm

Duckworth, D.H., Crandall, R., and Rathe, R., accessed 2002. Medical microbiology and infectious disease "BUGS" program, University of Florida, Gainesville, Florida, http://www.medinfo.ufl.edu/year2/mmid/bms5300/bugs/index.html

Garcia, L.S., accessed 2002. *Cyclospora cayetanensis:* Epidemiology, Para-site Online, http://www.med-chem.com/Para/New/cc-epid.htm

Health Canada, accessed 2002. Material Safety Data Sheets, Health Canada Population and Public Health Branch, Ottawa, Canada, http://www.hc-sc.gc.ca/pphb-dgspsp/msds-ftss/index.html#menu

MacLean, J.D., accessed 2002. Clinical Parasitology, McGill University Centre for Tropical Diseases, McGill University, Montreal, Canada, http://www.medicine.mcgill.ca/tropmed/txt/lecture1.htm

Mack, D.R., 2001, *Dientamoeba fragilis* Infection, eMedicine World Medical Library, http://www.emedicine.com/

McGinnis, M.R., 1998. Introduction to mycology: *in* Baron, Samuel, editor, Medical Microbiology, http://www.md.huji.ac.il/microbiology/book

NASA, 2001a. Soil science education home page, Goddard Space Flight Center Laboratory for Terrestrial Physics, NASA, Greenbelt, Maryland, http://ltpwww.gsfc.nasa.gov/globe/tbf/tbfguide.htm

National Cattleman's Beef Association and Cattleman's Beef Board, 2001. Transmissible Spongiform Encephalopathies (TSE): BSE Info Resource, 2/14/2002, http://www.bseinfo.org/resource/tse_fact.htm

National Institutes of Health, 2001. Parasitic Roundworm Diseases Fact Sheet, National Institute of Health, Washington DC, http://www.niaid.nih.gov/factsheets/roundwor.htm

National Institute of Allergy and Infectious Diseases, 2000. Antimicrobial resistance: Fact Sheet, Office of Communications and Public Liaison, U.S. Department of Health and Human Services, 4 p., http://www.niaid.nih.gov/factsheets/antimicro.htm

NRCS, 1999. Soil biology primer, National Resources Conservation Service, U.S. Department of Agriculture, Washington DC, 53p., http://www.gsfc.nasa.gov/gsfc/earth/toms/microbes.htm

Pennsylvania Environmental Network, 2001. Survival of *E. coli* in sludge amended soil, the Pennsylvania Environmental Network, http://www.penweb.org/issues/sludge/ecoli-survival.htm

Pitetti, R.D., 2001. Visceral Larva Migrans, eMedicine World Medical Library, http://www.emedicine.com/

Public Health Laboratory Service, accessed 2002. Disease Facts, Secretary of State for Health, London, NW9 5DF, United Kingdom, http://www.phls.co.uk/facts/index.htm

Reuters Health Information, 2001. Cattle cull could present new risk of variant Creutzfeldt-Jakob disease: Reuters Medical News, May 24, 2001, http://www.reutershealth.com/archive/2001/05/24/professional/links/20010524publ007.htm

Schwartzbrod, Louis, 1995. Effect of human viruses on public health associated with the use of wastewater and sludge in agriculture and aquaculture, World Health Organization Publication 95.19, World Health Organization, Geneva, http://www.who.int/environmental_information/Information_resources/worddocs/Human_viruses.html

Soil Science Society of America, 1998. Internet Glossary of Soil Science Terms, Soil Science Society of America, Madison, Wisconsin, http://www.soils.org/sssagloss/

Soil Survey Staff, 1998. Keys to soil taxonomy, Natural Resource Conservation Service, U. S. Department of Agriculture, Washington, D.C., 40 p., http://soils.usda.gov/classification/keys/main.htm

Standing Medical Advisory Committee, 1998. The path of least resistance: Report of the Sub-Group on Antimicrobial Resistance, Department of Health, London, U.K., http://www.doh.gov.uk/smacful.htm

University of California-a, accessed 2002. Bacteria: Life History and Ecology, Copyright 1994–2002 by The University of California Museum of Paleontology, Berkeley, and the Regents of the University of California, http://www.ucmp.berkeley.edu/bacteria/bacterialh.html

University of California-b, 2002. Schistosomiasis in China, University of California, Berkeley, California, the Regents of the University of California, http://ehs.sph.berkeley.edu/china/

University of Leicester, 2001. Protozoa as Human Parasites, Department of Microbiology and Immunology, University of Leicester, Leicester UK, http://www-micro.msb.le.ac.uk/224/Parasitol.html

University of Leicester, 1996. Pathogenic Fungi, Department of Microbiology and Immunology, University of Leicester, Leicester UK, http://www-micro.msb.le.ac.uk/MBChB/6a.html

Valley Fever Center for Excellence, accessed 2002. What is valley fever?, University of Arizona and Southern Arizona Veteran's Administration Healthcare System, Tucson, Arizona, http://vfce.arl.arizona.edu/

World Health Organization, 1998. Emerging and re-emerging infectious diseases: Fact sheet #97, revised August 1998, World Health Organization, Geneva, http://www.who.int/inf-fs/en/fact097.html

World Health Organization, 1999. Removing Obstacle to healthy development, World Health Organization, Geneva, http://www.who.int/infectious-disease-report/index-rpt99.html

World Health Organization, 2002. Antimicrobial resistance: Fact sheet #194, revised January 2002, World Health Organization, Geneva, http://www.who.int/inf-fs/en/fact194.htm

Chapter 26

Selected Online Earth Science/Geospatial Journals

- *Canadian Journal of Remote Sensing* Provides index to journal issues dating back to 1992 (http://www.ccrs.nrcan.gc.ca/ccrs/cjrs/cjrsndxe.html).
- ESRI Digital Chart of the World & Data Quality Project Downloadable papers related to using ESRI's Digital Chart of the World data series (http://ilm425.nlh.no/gis/dcw/dcw.html#DOC).
- ESRI White Papers: Papers related to using ESRI products as well as GIS in general (http://www.esri.com/base/common/whitepapers/whitepapers.html).
- *GeoInformatica*—An International Journal on Advances of Computer Science for Geographic Information Systems (http://kapis.www.wkap.nl/kapis/CGI-BIN/WORLD/journalhome.htm?1384-6175).
- geoinformatik online (Uni Münster) (http://gio.uni-muenster.de/).
- Geo-Informations-Systeme (Wichmann/Huethig) (http://www.huethig.de/zeitschr/gis/gis.html).
- GIS World Magazine (http://www.gisworld.com/)
- Grassclippings: The *Journal of Open Geographic Information Systems* (http://deathstar.rutgers.edu/grassclip/grassclip.html).
- *International Journal of GIS* (Taylor & Francis, London) (http://www.tandf.co.uk/jnls/gis.htm).
- National Research Council, Board on Earth Sciences and Resources Online reports (http://www2.nas.edu/besr/22e2.html).
- *Photogrammetric Engineering & Remote Sensing* (American Society of Photogrammetry and Remote Sensing) (http://www.asprs.org/asprs/publications/journal/pers.html).
- The Harlow Report: Geographic Information

Systems Newsletter covering current GIS-related topics (http://www.geoint.com/).

Biomedical/Health Information
Selected Biomedical/Health Data Resources

- National Library of Medicine Developed by the U. S. National Library of Medicine, this program offers access to most of the MEDLARS databases including: MEDLINE, HealthSTAR, AIDSLINE, BIOETHICSLINE, HISTLINE (History of Medicine) (http://igm.nlm.nih.gov/).
- NCBI PubMed Sponsored by the National Center for Biotechnology Information, this search interface covers all citations covered in MEDLINE and PreMEDLINE (http://www.ncbi.nlm.nih.gov/PubMed/).
- NCI CANCERLIT Produced by the National Cancer Institute's International Cancer Information Center, CANCERLIT indexes over 1.3 million citations and abstracts from over 4,000 sources, including biomedical journals, books and doctoral theses (http://cnetdb.nci.nih.gov/cancerlit.html).
- Agency for Toxic Substances and Disease Registry (ATSDR): An agency of the U. S. Department of Health and Human Services. Monitors exposure to hazardous substances from waste sites, unplanned releases, and other sources of pollution present in the environment. Access to the *HazDat Database* and full-text to *Public Health Assessments* (http://atsdr1.atsdr.cdc.gov:8080/).
- Centers for Disease Control and Prevention (CDC): Includes the latest health information and news, publications (such as the *Morbidity and Mortality Weekly Report*), statistics, funding information and public domain computer software for working with public health data (http://www.cdc.gov/).
- National Center for Chronic Disease Prevention and Health Promotion: Clearinghouse for information on chronic disease prevention. Access to various full-text publications and reports (http://www.cdc.gov/nccdphp/index.htm).
- National Center for HIV, STD, and TB Prevention: Comprehensive guide to the prevention, treatment, and elimination of HIV, STDs, and TB. Full-text of the *HIV/AIDS Surveillance Report* and the *STD Treatment Guidelines* as well as access to several databases covering news releases, funding opportunities, and health services (http://www.cdc.gov/nchstp/od/nchstp.html).
- National Center for Health Statistics (NCHS): The nation's principal health statistics agency. Includes several statistical publications for download, including several fact sheets, news releases, and reports (http://www.cdc.gov/nchs).
- National Center for Infectious Diseases: Develops programs to evaluate and promote prevention and control strategies for infectious diseases. Includes information about many infectious diseases as well as online access to the publications *Emerging Infectious Diseases* and *Health Information for International Travel* (http://www.cdc.gov/ncidod/).
- National Institute of Allergy and Infectious Disease (NIAID): Includes news for consumers and professionals, full-text newsletters, and consumer fact sheets and booklets, as well as a list of research activities and clinical trials (http://www.niaid.nih.gov/).
- National Institutes of Health (NIH): Gateway to clinical and consumer oriented resources including health information, funding opportunities, and scientific resources (http://www.nih.gov/).
- U.S. Census Bureau: Social, demographic and economic information. Includes full-text to the *Statistical Abstract of the United States* (http://www.census.gov/).
- Center for International Health Information Provides timely, reliable, and accurate information on the Population, Health, and Nutrition (PHN) sector in developing countries assisted by USAID. Full-text of *Country Health Profile Reports*, and *Population, Health and Nutrition Indicators* (http://www.cihi.com/).
- World Health Organization (WHO): Promotes technical cooperation for health among nations, carries out programs to control and eradicate disease and strives to improve the quality of human life. Includes full text to the *World Health Reports* and the *Weekly Epidemiological Record* (http://www.who.ch/).
- Agency for Toxic Substances and Disease Registry (ATSDR): An agency of the U. S. Department of Health and Human Services. Monitors exposure to hazardous substances from waste sites, unplanned releases, and other sources of pollution present in the environment. Access to the *HazDat Database* and full-text to *Public Health Assessments* (http://atsdr1.atsdr.cdc.gov:8080/).

Appendix C (Glossary)

a-axis: a vector direction defined by the space group and crystal structure for a particular crystalline form; a term used in crystallography.

absorption: the process by which a substance or a xenobiotic is brought into a body (human or animal) or incorporated into the structure of a mineral.

acanthosis: increase in thickness of stratum spinosum (specific layer in epidermis/skin).

acid rain: contamination of rain by artificial pollutants or natural emissions (such as sulfur dioxide from volcanic activity) which produces an acid composition.

activity: the thermodynamically effective concentration of a chemical species or component.

acute myocardial infarction (AMI): gross necrosis of the heart muscle as a result of interruption of the blood supply to the area.

adsorption: the binding of a chemical compound to a solid surface.

advection: a transport process in which dissolved chemicals move with flowing groundwater.

albedo: the percentage of the incoming solar radiation reflected back by different parts of the Earth's surface.

aldosterone: a steroid hormone produced by the adrenal gland that participates in the regulation of water balance by causing sodium retention and potassium loss from cells.

aliquot: a known amount of a homogeneous material, assumed to be taken with negligible sampling error. When a sample is "aliquoted", or otherwise subdivided, the portions may be called split samples.

alkali disease: disease affecting animals that ingest feed with a high selenium concentration, characterized by dullness, lack of vitality, emaciation, rough coat, sloughing of the hooves, erosion of the joints and bones, anemia, lameness, liver cirrhosis, and reduced reproductive performance.

alkalinity: the capacity of solutes in a solution to react with and neutralize acid; determined by titration with a strong acid to an end point at which virtually all solutes contributing to the alkalinity have reacted. In general the alkalinity in water equates with the bicarbonate concentration.

allergy: immunologic state induced in a susceptible subject by an antigen (allergen).

alluvial: deposited by rivers.

alteration (Earth science): a process due to high-temperature fluids and gases that occurs within the Earth's crust and results in the formation of new mineral suites that are in equilibrium with their environment. Alteration can also occur at low temperatures.

aluminosilicate: a mineral composed dominantly of aluminum, silicon, and oxygen, and lesser amounts of cations such as sodium, potassium, calcium, magnesium, and iron.

amorphous: a lack of crystallinity or the regular extended three-dimensional order of the atoms in a solid.

anaerobic/aerobic: environmental conditions in which oxygen is absent/present.

analyte: any substance whose identity or concentration is being determined.

anemia: any of several conditions in which the oxygen-carrying capacity of the blood is below normal due to reductions in the number of red blood cells (hypocytic) and/or the amount of hemoglobin per red blood cell (hypochromic).

aneuploidy: cellular state where there is an abnormal number of chromosomes, not a multiple of the haploid number of chromosomes.

aneurysm: localized ballooning of the aorta or an artery, potentially causing pressure on adjacent structures and liability to rupture.

angiotensin: a vasoconstrictive hormone.

antisense: nucleic acid that has a sequence exactly opposite an mRNA molecule made by the body; binds to the mRNA molecule to prevent a protein from being made.

apo: without, especially metalloproteins without the metal/metals.

apoptosis: programmed cell death, in which a cell brings about its own death and lysis, signaled from outside or programmed in its genes, by systematically degrading its own macromolecules.

aqueous speciation: the partitioning of chemical components between various aqueous species in a solution: free species (e.g., Ca^{2+}), ion pairs (e.g., $CaCO_3^0$), and complexes (e.g., $Fe(CN)_6^{3-}$).

aquifer: a water-bearing rock formation.

aquitard: a rock formation with poor permeability and hence a poor water-bearing unit.

archaea: prokaryotes lacking a nucleus as bacteria, but they are as different from bacteria as are humans; they represent their own evolutionary pathway; they live in extreme places with high temperatures.

arenosols: sandy soils with >65% sand-sized (0.05–2 mm) particles; these soils have low moisture and low concentrations of most elements and are highly prone to causing deficiencies of micronutrients in crops.

aridisol: soils found in arid and semi-arid environments; characterized by a light color, poorly developed soil horizons, high soluble salt content, little organic material, and a coarse texture.

arrhythmia: irregularity of the heart beat.

arthroconidia: fungal spores released by fragmentation or separation of the cells of a hypha.

asbestos: a commonly used term for a group of fibrous silicate minerals that includes extremely fibrous serpentine (chrysotile) and the amphibole minerals crocidolite, amosite, tremolite, actinolite, and anthophyllite.

asbestosis: degenerative fibrosis of the lung resulting from chronic inhalation of asbestos fibers.

ascariasis: an infection caused by the parasitic worm *Ascaris lumbricoides* that is found throughout temperate and tropical regions. Intestinal infection may result in abdominal cramps and obstruction, while passage through the respiratory tract causes symptoms such as coughing and wheezing. In children, migration of the adult worms into the liver, gallbladder, or peritoneal cavity may cause death.

ascidian: any minute marine invertebrate animal of the class Ascidiacea, such as the sea squirt.

ash: fine particles of pulverized rock ejected from volcanoes.

asphyxiant: gas which produces suffocation by replacing oxygen in the respiratory system.

ataxia: lack of coordination of muscle for voluntary movement.

atelectasis: absence of gas in lung tissue from nonexpansion.

atherosclerosis: irregularly distributed intimal deposits of lipid.

atomization: the dispersion of fluids into fine particles.

atrium: the upper chamber of each half of the heart.

atrophy: diminished cellular proliferation.

attribute: information about geographic features contained within GIS data layers, or *themes*.

auger effect: phenomenon occurring when an electron is released from one of the inner orbiting shells, thereby creating two electron vacancies of the residual atom and repeated as the new vacancies are filled or X-rays are emitted.

autosome: a chromosome not involved in sex determination. The diploid human genome consists of 46 chromosomes, 22 pairs of autosomes and 1 pair of sex chromosomes (the X and Y chromosomes).

auxotroph: a microorganism possessing a mutation in a gene that affects its ability to synthesize a crucial organic compound.

atypia: reactive cellular state, which does not correspond to normal form.

background: the property, as applied to a location or measurements from such locations, of being due to natural processes alone and unaffected by anthropogenic processes. In some instances the term natural background is used to reinforce the non-anthropogenic aspect. With the global atmospheric transport of anthropogenic contaminants, e.g., persistent organic pollutants (POPs), it is a moot point whether background sites exist for some substances.

basal cell carcinoma: slow growing, locally invasive neoplasm derived from basal cells of epidermis or hair follicles.

baseline: a measure of the natural background or ambient level of an element/substance. Some people also suggest that baseline is the current background which could include natural and anthropogenic components.

basolateral membrane: part of the plasma membrane that includes the basal end and sides of the cell.

basophilic degeneration: pathologic change in tissue noted by blue staining of connective tissue with hematoxylin-eosin stain.

beneficiation: process of concentrating ores.

benign: usual or normal; the opposite of cancerous when applied to cells or tumors.

bioaccumulation: process by which an element is taken into an organism, possibly transformed into another chemical species, and retained so that the element's concentration in the biota is greater than its concentration in the media in which the biota is sustained.

bioapatite: the name given to the complex calcium phosphate mineral that forms in biological tissues and is characterized by extremely small crystallite size; maximum dimension is typically less that 20×10^{-9} m (200 Å). Generalized chemical formula: $(Ca,Na,Mg, \ldots [\ ])_{10}(PO_4, HPO_4, CO_3, SO_4 \ldots)_6(OH, F, Cl, CO_3, O, [\ ])_2$ where ... indicates the possible addition of other cations and [ ] indicates vacancies in the crystal structure at the cation or halogen sites.

bioavailability: the property of a substance that makes its chemical uptake by biota possible.

bioessential/bioessentiality: present in sufficient amounts to support essential biochemical processes imperative for sustaining life.

biogeochemical cycle: model encompassing the movement of elements (and some compounds) from the lithosphere through the hydrosphere, atmosphere, and biosphere.

biosphere: the sum of all organisms on Earth.

birefringence: the ability of anisotropic (non-isometric) crystalline materials to split plane polarized light into two non-equal rays of distinct velocities depending on the direction of the transmission relative to the orientation of the atomic structure of the compound. When the two rays emerge from the crystal, one is retarded relative to the other. Precise measurements of the interference colors of the rays define the optical characteristics and identify the compound.

bisphosphonates: a group of phosphorus- and carbon-containing compounds that have carbon connected to the phosphorus atom in place of one of the oxygen atoms of the tetrahedral phosphate (PO$_4$) groups.

blind staggers: blind staggers occurs in cattle and sheep ingesting high concentrations of selenium and is characterized by impaired vision leading to blindness, anorexia, weakened legs, paralyzed tongue, labored respiration, abdominal pain, emaciation, and death.

bombs (volcanic): clots of lava that are ejected in a molten or semi-molten state and congeal before striking the ground.

bone: a term applied to one of the many individual organs that make up vertebrate skeletons, or alternatively, to the fragments or the tissues that are found within these organs.

Bowen's disease: an intraepidermal carcinoma characterized as a small, circumscribed elevation on the skin.

buffer: a chemical compound that controls pH by binding to hydrogen ions.

bulk analysis: chemical analysis of an entire body/substance of rock or soil or a subpart with little or no segregation of specific areas or components.

c-axis: a vector direction defined by the space group and structure of a particular crystalline form. A crystallographic term.

calcisols: soils with a high content of free calcium carbonate either developed on limestones, or which have become calcified by the deposition of calcium carbonate in pores and voids as a result of the evaporation of soil solution in arid environments. These soils generally have neutral or alkaline pHs and can adsorb some trace elements very strongly.

calcitonin: hormone secreted by the thyroid gland; important in the homeostatic regulation of serum calcium levels.

capillary electrophoresis: electrophoretic separation technique performed in a small fused silica capillary.

carbon dioxide: a colorless odorless gas; in high concentrations CO$_2$ acts as an inert asphyxiant in humans.

carbonatite: an igneous rock composed of carbonate minerals.

carcinogen: a substance that can directly or indirectly cause a cell to become malignant.

carcinogenesis: the mechanism by which cancer is caused.

cardiomyopathy: disease of the heart muscle (myocardium).

cardiovascular disease (CVD): disease pertaining to the heart and blood vessels, including, for example, both AMI and cerebrovascular disease (stroke).

catecholamines: category of compounds including the neurotransmitters adrenaline and noradrenaline.

cation exchange: exchange of cations between a solution and a negatively-charged solid phase (e.g., a clay mineral) in response to a change in solution conditions; this is especially important in geochemistry for major cations such as calcium and sodium.

cation exchange capacity (CEC): the ability of a soil or soil constituent (e.g., clay mineral or humus) to adsorb cations on permanent, or pH-dependent, negatively charged sites on surfaces. Cations of different elements can replace each other as counter ions to the negative charges.

cDNA: complementary DNA: a DNA molecule copied from an mRNA template by the enzyme reverse transcriptase.

cementum: the thin tissue that forms the outer covering of a tooth below the gum line, similar in composition to dentine.

chaperones: proteins that help in folding proteins correctly and that discourage incorrect folding. Metallochaperones assist in the delivery of metal ions to target proteins or compartments.

chelate: the complex formed through the bonding of a metal ion with two or more polar groupings within a single molecule.

chitin: a tough white to semi-transparent substance that forms the major structural component of arthropod exoskeletons and the cell walls of certain fungi.

chloroplast: chlorophyll-containing photosynthetic organelle in some eurkaryotic cells.

choroid plexus: a network of intersecting blood vessels of the cerebral ventricles that regulate intraventricular pressure.

chromatin: the complex of DNA and proteins that make up eukaryotic chromosomes.

chromatography: the separation of a mixture of compounds using solid, liquid, or gas phases based on affinity of molecules for the phase.

chromosome aberrations: any deviation from the normal number or morphology of chromosomes.

clay minerals: phyllosilicate minerals with a small grain size, commonly <4 μm but ranging down to colloidal dimensions. When mixed with a limited amount of water they develop plasticity. Clay minerals are formed by high-temperature hydrothermal alteration processes, e.g., kaolinite in altered granitic rocks; or by low-temperature weathering processes, e.g., montmorillonite, smectite, chlorite, kaolinite, and illite.

clearance: output of particles previously deposited in the respiratory tract.

coccidioidomycosis: a respiratory disease of humans and animals caused by inhalation of arthroconidia of the soil-inhabiting fungus *Coccidioides immitis*. Fever, cough, weight loss, and joint pains characterize the disease, also called valley fever.

code (biological): the presentation of the content (of a molecule) in terms of symbols such as ATC and G for the DNA code where ATC and G are nucleotide bases.

codon: the fundamental unit of the genetic code consisting of a triplet sequence of nucleotide bases which specifies the ribosomal binding of a specific amino-acid-bearing tRNA during protein synthesis or the termination of that process.

coenzyme: a small molecule which binds to a protein to create a catalytic center.

collagen: protein making up the white fibers (collagenous fibers) of skin, cartilage, and all connective tissue.

collimator: a device for producing a beam of parallel rays.

compartment: a separated solution volume of a cell by an enclosing membrane not at equilibrium with any other separated volume.

complex system: natural or man-made system composed of many simple nonlinear agents that operate in parallel and interact locally with each other at many different scales. The behavior of the system cannot be directly deduced from the behavior of the component agents and the system sometimes produces behavior at another scale, which is called emergent behavior.

composite: a mixture of several components or parts blended together to form a functional whole.

condensation polymer: a polymer formed by loss of water molecules from monomers.

confined aquifer: aquifer over- and underlain by impermeable or near-impermeable rock strata.

cooling: the decrease of the activity of a radioactive material by nuclear decay.

coordination: the association of one atom with another in three-dimensional arrays. The coordination number reflects the atomic size of an atom. Octahedral or sixfold coordination is typical of metal atoms with oxygen.

coronary heart disease (CHD): disease caused by deficiency of blood supply to the heart muscle due to obstruction or constriction of the coronary arteries.

cortical: the tissue that forms the external portions of bones heavily mineralized with bioapatite-containing cells and exhibiting a variety of textures.

Cretaceous/Tertiary (K/T) boundary: the Cretaceous period was the last in the Mesozoic era and was succeeded 64 million years ago by the Tertiary period of the Cenozoic era. It is marked by the sudden extinction of genera of living organisms, most famously the dinosaurs.

crust: the outermost solid layer of a planet or moon.

crystallinity: the three-dimensional regular array typical of solids with definite chemical composition and crystal structure.

crystalline basement: solid igneous, sedimentary, or metamorphic rock; may crop out at the ground surface or be overlain by superficial deposits (unconsolidated sediments or soils).

crystallite: a general term applied to very small size materials, usually minerals, in which a crystal form or crystal faces may be observed, usually with magnification. The morphology of a crystallite suggests a material with a regular crystal structure and may be used to identify a specific compound or mineral species.

cytochrome P-450: iron-containing proteins important in cell respiration as catalysts of oxidation-reduction reactions.

cytoplasm: the central compartment of all cells that contains genes and DNA as well as synthetic systems.

database: a structured set of persistent data, that in a GIS context, contains information about the spatial locations and shapes of geographic features, and their *attributes*.

decay (radioactive): the disintegration of the nucleus of an unstable atom by spontaneous fission or emission of an alpha particle or beta particle.

deconvolution: a mathematical procedure used for separation of overlapping peaks.

definitive host: the host in which a parasite reaches sexual maturity and reproduces.

dental calculus: calcium phosphate mineral materials deposited around the teeth at and below the gumline, probably the result of bacterial action.

dental caries: cavities in teeth arising from tooth decay.

dentine: the tissue composed of greater than 70% bioapatite that forms the predominant segment of a tooth. This tissue is capped by enamel.

deposition: fraction of particles in inspired air that are trapped in the lung and fail to exit with expired air. In geology it is the laying down of sediments.

derivatization: the chemical modification of a naturally occurring compound so that it may be more volatile for gas chromatographic separation.

dermis: inner aspects of skin that interdigitate with epidermis and contain blood and lymphatic vessels, nerves, glands, and hair follicles.

desorption: release of a bound chemical compound from a solid surface (the opposite of adsorption).

detection limit: minimum amount of the characteristic property of an element that can be detected with reasonable certainty under specific measuring conditions.

diagenesis: changes to the original organic composition of a material caused by low-temperature processes, often involving bacterial action. It can occur in sediments where minerals are altered as well as organic matter. It changes the original chemistry of many minerals and bone when they are buried.

dioxygenase: a class of oxidoreductases that catalyze the binding of diatomic oxygen to a product of the reaction.

DOC (dissolved organic compounds, or dissolved organic carbon): the soluble fraction of organic matter in soils and ground and surface waters comprising low molecular weight organic compounds which have the ability to complex many elements and render them more available to plants and more prone to leaching down the soil profile.

dose: a general term for the quantity of radiation. The *absorbed dose* is the energy absorbed by a unit mass of tissue whereas the *dose equivalent* takes account of the relative potential for damage to living tissue of the different types of radiation. It is also the quantity of a substance taken in by the body in general.

dose response: the relationship between an exposure dose and a measurable biological effect.

dowagers hump: the abnormal concave bending of the upper or thoracic spine as a result of osteomalcia or osteoporosis often obvious in older women.

drift: (analysis) a slow change in the response of an analytical instrument; (geology) it is a superficial sediment.

dry matter (d.m.): remaining solid material after evaporation of all water. Often used to express concentration of minerals and trace elements to eliminate variation due to differences in water content of plant material.

ectodermal: relating to ectoderm, the outer layer of cells in the embryo.

eco-district/eco-classification: a relatively ecologically homogeneous area of the Earth's surface, an element of a classification based on climatic, biological, pedological, and geological criteria that becomes more specific from eco-zones, through eco-provinces and eco-regions to eco-districts.

effluent: the material that is coming from a chromatographic separation. Can also be the waste outfall from industries and is also the term for sewage (sewage effluent).

eggshell calcification: a thin calcified layer surrounding an intrathoracic lymph node.

elastosis: degenerative changes of collagen fibers with altered staining properties.

electromagnetic spectrum: the full range of frequencies, from radio waves to cosmic rays.

electrospray ionization (ESI): ionized molecules by application of a high voltage (approximately 5 kV) to the spray needle.

elimination: how xenobiotics are removed from the bloodstream, either by metabolism or excretion.

emissions (volcanic): any liquid, solid, or gaseous material produced by volcanic activity.

enamel: the tissue composed of greater than 96% bioapatite that forms the outer surface of teeth.

enantiomer: one of two indistinguishable forms of a compound that differ only in the orientation in space; a stereoisomer.

endemic: Where a disease is confined to specific geographical areas.

endocytosis: the process in which the plasma membrane engulfs extracellular material, forming membrane-bound sacs that enter the cytoplasm and thereby move material into the cell.

endosome: a small vesicle resulting from the invagination of the plasma membrane transporting components of the surrounding medium deep into the cytoplasm.

endospore: an asexual spore formed by some bacteria, algae, and fungi within a cell and released.

endothelium: a tissue consisting of a single layer of cells that lines the blood, lymph vessels, heart, and some other cavities.

enterovirus: group of viruses transient in the intestine which includes poliovirus, echovirus, and Coxsackievirus.

entisol: entisols are soils that formed recently and are often found on floodplains, deltas, or steep slopes where soil development is inhibited. They are weakly developed and lack distinct soil horizons. Entisols have a wide geographic and climatic distribution.

enzootic: a disease that affects animals in a specific area, locale, or region.

enzyme: proteins that act as catalysts driving plant and animal metabolism.

eosinophils: a specific type of white blood cell.

epidemiology: the study of the prevalence and spread of disease in a community.

epidermis: outer aspect of skin with multiple layers.

eruption (volcanic): the ejection of tephra, gas, lava, or other materials onto the Earth's surface as a result of volcanic or geothermal activity.

erythrocyte: a mature red blood cell. Erythrocytes are the major cellular element of the circulating blood, and transport oxygen as their principal function. An increase in the number of cells normally occurs at altitudes greater than 3000 m.

erythron: a collective term describing the erythrocytes and their predecessors in the bone marrow.

erythropoiesis: the formation of erythrocytes in the bone marrow.

estrogen: category of steroid hormones produced by ovarian and adipose tissues that can effect estrus and a number of secondary sexual characteristics and is involved in bone remodeling.

etiological: the cause of a disease determined by etiology, the branch of medical science which studies the causes and origins of disease. The etiological agent of coccidioidomycosis is *Coccidioides immitis*.

etiology: the process underlying development of a given disease.

eubacteria: true bacteria so named to differentiate them from archaea (earlier known as Archaebacteria).

eukaryote: cells of organisms of the domain Eukarya (kingdoms Protista, Fungi, Plantae, and Animalia). Eukaryotic cells have genetic material enclosed within a membrane-bound nucleus and contain other membrane-bound organelles.

eutrophication: nutrient enrichment of waters that stimulates phytoplankton and plant growth and can lead to deterioration in water quality and ecosystems.

evapotranspiration: transfer of water from the soil to the atmosphere by combined evaporation and plant transpiration. It results in a concentration of solutes in the remaining water.

excretion: excretion is the mechanism whereby organisms get rid of waste products.

exon: a DNA sequence that is ultimately translated into protein.

exposure response: the relationship between how much of a xenobiotic is presented to a person or animal and what happens in their body.

extracellular: space in tissue that is outside of cells.

FAO/*Unesco* Soil classification system: the soil classification system developed for the joint project by the UN Food and Agriculture Organization and UNESCO to produce the Soil Map of the World (1:5,000,000) published from 1974 onward.

felsic: igneous rock rich in feldspar and siliceous minerals (typically light-colored).

ferralsols: reddish iron oxide-rich soils characteristic of the tropical weathering and soil-forming environment (humid tropics). These soils generally have a low fertility with low CECs and nutrient contents. Also called oxisols (U. S. Soil Taxonomy), ferralitic, or lateritic soils.

ferritin: a soluble protein storage form of iron containing as much as 23% iron.

ferromagnesian: a silicate mineral dominated by iron, magnesium, sometimes with aluminum.

fibroblastic cells: secretionary cells of connective tissue.

fibroblasts: cells that produce collagen molecules.

fibrosis: formation of fibrous tissue.

fluorapatite: a mineral, ideal formula $Ca_5(PO_4)_3F$. One of the members of the calcium apatite mineral group.

fluoride: F^-, the dominant form of fluorine found in water.

fluorite: the dominant fluorine mineral, CaF_2; occurs as an accessory mineral in some sediments and igneous rocks and in some hydrothermal mineral veins.

fluorosis: disease affecting bones and teeth, caused at least in part by exposure to high doses of fluoride. Dental fluorosis causes weakening and possible loss of teeth, and skeletal fluorosis causes bone deformation and disability.

fluvial: pertaining to rivers and streams.

forestomachs: two or three sac-like dilations of the esophagus seen in ruminants and kangaroos. The physiological function of these structures is to serve as fermentation tanks to make cellulose and other carbohydrates in the feed available for absorbtion in the gastrointestinal tract of the animal.

fraction: in this context, a term used in sedimentology, pedology, and other physical sciences to describe the mechanical size range of a material.

fuzzy system: a system that uses fuzzy sets and if-then rules to store, compress, and relate many pieces of information and/or data in order to build a model free estimator.

gamma ray: a distinct quantity of electromagnetic energy, without mass or charge, emitted by a radionuclide.

genome: the DNA (or for some viruses, RNA) that contains one complete copy of all the genetic information of an organism or virus.

genotoxic: the ability of a substance to cause damage to DNA.

geothermal: pertaining to the internal heat of the Earth. Geothermal zones are areas of high heat flow, where hot water and/or steam issue at the Earth's surface. They are found close to tectonic plate boundaries or associated with volcanic systems within plates. Heat sources for geothermal systems may be from magmatism, metamorphism, or tectonic movements.

gleys: soils under reducing conditions caused by permanent or intermittent waterlogging; characterized by pale colors and low concentrations of iron oxides.

gliosis: a chronic reactive process in neural tissue.

glutathione peroxidase: a detoxifying enzyme in humans and animals that eliminates hydrogen peroxide and organic peroxides; it has a selenocysteine residue in its active site.

glycolysis: the energy-yielding metabolic conversion of glucose to lactic acid in muscle and other tissues.

gneiss: banded, usually coarse-grained metamorphic rock, having been modified from its original mineralogy and texture by high heat and pressure (high-grade regional metamorphism).

goitrogen: a substance which causes or enhances the symptoms of iodine deficiency, e.g., goiter formation.

granite: a coarse-grained igneous rock, composed mainly of quartz, alkali, feldspar, and mica. Accessory minerals may also include apatite, zircon, magnetite, and sphene. Granite characteristically has a high proportion of silica (>70% SiO_2) with high concentrations of sodium and potassium.

granitization: a metamorphic process by which sedimentary and metamorphic rocks with a chemistry similar to granites (granitoids) are transformed mineralogically into rocks that look like the granites formed by igneous intrusive processes.

granulomatous inflammation: inflammatory reaction where tissue cells of monocyte/macrophage cells predominate.

granulomatous reaction: reaction leading to the formation of granuloma, or chronic inflammatory lesions.

grazing: feeding behavior of cattle, sheep, and horses; consumption of grass and other plants from the ground, mostly rather indiscriminately.

groundwater: subsurface water in the zone of saturation in which all pore spaces are filled with liquid water (although sometimes the term groundwater is used inclusively for all water below the land surface, to distinguish it from surface water).

half-life: the time in which one-half of the atoms of a particular radioactive substance decay to another nuclear form.

hardness water: the content of metallic ions in water, predominantly calcium and magnesium, which react with sodium soaps to produce solid soaps or scummy residue and which react with negative ions to produce scale when heated in boilers.

haversian bone: the tissue type found throughout the skeleton in humans that signifies sites of resorption and remodeling. Characterized in cross section by a circular outline and a lamellar distribution of cells and mineralized tissue around a central blood vessel, which is called the haversian canal.

heavy metal: a metal with a density more than $4500\,kg\,m^{-3}$.

helminth: a multicellular worm, generally parasitic, often with a complex reproductive system and life cycle. Generally 50 to 2000 µm in length, but may be longer.

heme: the protoporhyrin component of hemoglobin (in erythrocytes) and myoglobin (in myocytes), the proteinaceous chelation complexes with iron that facilitate transport and binding of molecular oxygen to and in cells.

hemolysis: lysis of erythrocytes that potentially causes anemia.

hemorrhage: profuse bleeding from ruptured blood vessels.

hemosiderin: an insoluble iron-protein complex that comprises a storage form of iron mainly in the liver, spleen, and bone marrow.

hepatolenticular: hepato, means belonging to the liver; lenticular means lens shaped and refers to the basal ganglia of the brain.

herbivores: animals normally feeding on plant material such as cattle, horses, sheep, antelope, deer, and elephants, but also rodents like mice, rabbits, and hares. As vertebrates lack enzymes in the gastrointestinal tract that can digest cellulose and other complex carbohydrates present in plants, they utilize microorganisms living in their gastrointestinal tract for this process. See also Ruminants and Large Intestine Fermenters.

hexagonal: a description of a specific crystallographic form in which the c-axis is perpendicular to three axes, usually designated as a axes, which are 120 degrees relative to each other. Apatite crystals often show hexagonal prisms with a 60 degree angle measured between adjacent vertical or prism crystal faces.

histology: science concerned with the minute structure of cells, tissue, and organs, utilizing light microscopy.

histomorphometry: the study of the textures of tissues using sections of samples embedded in paraffin or epoxy. The sections cut from the embedded blocks may be stained to assist in the identification of specific tissue components, i.e., collagen or special components in the nucleus of a cell.

histones: the family of five basic proteins that associate tightly with DNA in the chromosomes of eukaryotic DNA.

homeostasis: the state of equilibrium in the body with respect to various functions and the chemical compositions of fluids and tissues, including such physiological processes as temperature, heart rate, blood pressure, water content, blood sugar, etc., and the maintenance of this equilibrium.

homeostatic control: the ability or tendency of an organism or cell to maintain internal equilibrium by adjusting its physiological processes.

homologue: a member of a chromosome pair in diploid organisms or a gene that has the same origin and functions in two or more species. To an organic chemist this is series of compounds that are similar in structure. For instance methanol, ethanol, and the other alcohols represent a homologous series of compounds.

hormone: a circulating molecule released by one type of cell or organ to control the activity of another over the long term, e.g., thyroxine.

host: a human or animal in which another organism, such as a parasite, bacteria, or virus, lives.

humus: the fraction of the soil organic matter produced by secondary synthesis through the action of soil microorganisms; it comprises a series of moderately high molecular weight compounds that have a high adsorptive capacity for many metal ions.

hydraulic conductivity: the volume of water that will move in unit time under a unit hydraulic gradient through a unit cross-sectional area normal to the direction of flow.

hydraulic gradient: the change in static head (elevation head + pressure head) per unit distance in a given direction. It represents the driving force for flow under Darcy's law.

hydrodynamic dispersion: the irreversible spreading of a solute caused by diffusion and mechanical dispersion (which, in turn, is caused by indeterminate advective transport related to variations in velocity about the mean).

hydroxylapatite: name of the mineral, ideal chemical formula $Ca_5(PO_4)_3(OH)$, one of the members of the calcium apatite mineral group. Hydroxylapatite occurs naturally throughout the different types of rocks on the surface of the Earth and closely resembles the mineral deposits in normal and pathological tissues. See also Bioapatite.

hyperchromatic: excessive dark staining.

hyperkeratosis: hyperplasia of the stratum corneum (specific layer in epidermis/skin), the outermost layer in the epidermis.

hyperplasia: an increase in the number of cells in tissue or an organ.

hypertension: high blood pressure.

hyphae: the branching threadlike filaments, generally 2–10 μm across, characteristic of the vegetative stage of most fungi.

hyphenated techniques: generally, two analytical methods connected in series, e.g., a chromatographic technique directly connected to a spectroscopic technique.

hypoxia: less than the physiologically normal amount of oxygen in organs/tissues.

idiopathic: describing a disease of unknown cause.

igneous rocks: formed from the cooling and solidification of molten rock originating from below the Earth's surface, includes volcanic rocks.

incidence: quantifies the number of new cases/events that develop in a population at risk during a specified time interval.

inductively coupled plasma (ICP): an argon plasma with a temperature of approximately 7000–10,000 K, produced by coupling inductively electrical power to an Ar stream with a high-frequency generator (transmitter). Then plasma is used as an emission source (atomic emission spectrometry) or as an ionization source (mass spectrometry).

inselberg: an isolated peak of hard rocks that has stubbornly resisted erosion; most commonly found in the tropics.

integrin: a membrane protein that conveys information in both directions across the plasma membrane.

internal dose: amount of an agent penetrating the absorption barriers via physical or biological processes.

iodothyronine deiodinase: selenoproteins responsible for the production and regulation of the active thyroid hormone from thyroxine.

ischemia: ischemia occurs due to the disruption of the supply of blood and oxygen to organs and cells.

isoform: the descriptor for a specific form of a protein that exists in multiple molecular forms; also, for enzymes, isozyme.

isotachophoresis: separation mode in capillary electrophoresis, separating according to analyte conductivity.

isotope: one of two or more atoms with the same atomic number but with different atomic weights.

Kashin-Beck disease: an endemic osteoarthropathy (stunting of feet and hands) causing deformity of the affected joints; occurs in Siberia, China, and North Korea.

keratinocytes: cells of the epidermis that produce the protein keratin.

Keshan disease: an endemic cardiomyopathy (heart disease) that mainly affects children and women of childbearing age in China.

kinase: an enzyme catalyzing the conversion of a proenzyme, or zymogen, to its metabolically active form, frequently via phosphorylation or proteolytic cleavage.

K_m: the Michaelis constant in enzyme kinetics.

lahar: a hot or cold flow of water-saturated volcanic debris flowing down a volcanic slope.

lamellar bone: the tissue that shows sequential layers of mineralized matrix, cells, and the blood system required to maintain its viability. This tissue probably represents a second stage after the initial deposition of woven bone.

large intestine fermenters: different animal species utilizing bacteria and protozoa in their large intestine

(cecum or colon) to digest cellulose and starch in plants eaten so the nutrients can be absorbed in the gut of the animal. Horses, donkeys, zebras, rabbits, and hares are examples of animal species utilizing large intestine fermentation to facilitate digestion.

lattice: an array with nodes repeating in a regular three-dimensional pattern. A crystal lattice is the array distinctive for the chemical and physical structure of the crystalline compound.

lava: magma which erupts onto the Earth's surface; lava may be emitted explosively, as lava fountains, or by oozing from the vent as lava flows.

leachate: a liquid that carries dissolved compounds from a material through which it has percolated (e.g., water which carries adsorbed elements from settled volcanic ash into soil or water).

Lewis acid: a chemical center which accepts electron pair donation from a donor base, e.g., M^{2+} is a Lewis acid in the complex $M^{2+} \leftarrow OH_2$.

Lewy bodies: intracytoplasmic inclusion seen in Parkinson's disease.

lichenoid: accentuation of normal skin markings.

ligand: a binding unit attached to a central metal ion.

limestone: a sedimentary rock composed of calcium carbonate.

lithosphere: the solid Earth.

lOAEL: the lowest dose at which adverse effects are observed to occur in an experimental setting.

loess: natural sedimentary formation made up of wind-lain mineral dust, mainly in the silt size range (1–60 μm), most of which accumulated, often in great thickness, during the Quaternary (the last about 2.6 million years).

lumen: a cavity of passage in a tubular organ; the lumen of the intestine.

lymph nodes: small nodes along the bronchi that drain the tissues of lymph fluid.

lymphatic: vascular channel that transports lymph, a clear fluid with predominantly lymphocytes.

lysis: destruction of a cell's plasma membrane or of a bacterial cell wall, releasing the cellular contents and killing the cell.

macronutrient: general term for dietary essential nutrients required in relatively large quantities (hundreds of milligrams to multiple grams) per day; includes energy (calories), protein, calcium, phosphorus, magnesium, sodium, potassium, and chloride.

macrophage: mononuclear phagocytes (large leukocytes) that travel in the blood and can leave the bloodstream and enter tissues protecting the body by digesting debris and foreign cells.

magma: any hot mobile material within the Earth that has the capacity to move into or through the crust.

marine black shales: sedimentary rocks formed from organic-rich muds which have developed under strongly reducing conditions and are generally enriched in a wide range of trace elements.

matrix: the basis or collection of materials within which other materials develop. The organic matrix is the base in which mineral materials are deposited to form bone.

matrix effect: the combined effect of all components of the sample other than analyte on the measurement of quantity.

melanin: dark pigment that provides color to hair, skin, and the choroid of the eye.

mesothelioma: a highly malignant type of cancer, usually arising from the pleura, which is the lining of the thoracic cavity, and characteristically associated with exposure to asbestos.

messenger (transmitter): a molecule or ion used to convey information rapidly in or between cells, e.g., Ca^{2+}.

metabolism: the enzymatic chemical alteration of a substance. In toxicology, how xenobiotics are converted chemically; in life sciences generally, the pathways of chemical reactions that occur in the body.

metabolome: the small organic molecule composition in concentration units of a cell or compartment.

metadata: data about data, typically containing information such as time and place of database creation, *field* and *record* identifier information (attributes), data development process, map projection, and person to contact regarding the database; also known as data dictionary.

metalliferous: rich in metals.

metalloid: an element which behaves partly as a metal and partly as a non-metal, sometimes referred to as a "semi-metal."

metallome: the element composition in concentration units of a whole or a part of a cell where the element may be in free or combined form.

metamorphic rocks: rock formed from the alteration of existing rock material due to heat and/or pressure.

micellar electrokinetic chromatography: separation mode in capillary electrophoresis, separating according to the ability of apolar analytes to enter the (apolar) core of surface charged micelles.

micronutrient: general term for dietary essential nutrients required in relatively small amounts (less than multiple milligrams) per day; includes the vitamins and trace elements.

microradiograph: a picture produced using X-rays or rays from a radioactive source showing the minute internal textures of a planar thin section of a mineralized tissue sample.

mineral: a naturally occurring compound with definite chemical composition and crystal structure, of which there exist over 4000 officially defined species.

mineral elements: equal to elements. This term is used by nutritionists.

mineral group: an aggregate of mineral species that shares structural and chemical affinities.

mineral nutrient: a metal, non-metal, or radical that is needed for proper body function and maintenance of health; also used in reference to plant nutrition.

mineralization: the presence of ore and non-ore (gangue) minerals in host rocks, concentrated as veins, or as replacements of existing minerals or disseminated occurrences; typically gives rise to rocks with high concentrations of some of the rarer elements.

mitochondrion: subcellular organelle containing the electron transport chain of cytochromes and the enzymes of the tricarboxylic acid cycle and fatty acid oxidation and oxidative phosphorylation, thus, constituting the cell's primary source of energy.

mitogenic: a factor that causes mitosis of cells.

mitosis: the division of a cell into two daughters with identical complements of the nucleic material (chromosomes) characteristic of the species.

model: a conceptual, physical, or mathematical representation of a real system or process.

monoclinic: the description of a special crystallographic form for the structure of a compound in which the three axes are not mutually perpendicular.

monooxygenase: a class of oxidoreductases that catalyze the dissociation of molecular (diatomic) oxygen such that single oxygen atoms are bound to different products of the reaction.

mT: metallothionein.

mucosal cell: cell of the mucous membranes of the gastrointestinal tract.

multichannel analyzer (MCA): an instrument that collects, stores, and analyzes time- or energy-correlated events.

multistage carcinogenesis model: a mathematical model that assumes a sequential series of DNA-damaging events is necessary for a single cell to become malignant. The model also assumes linearity at low doses.

mycelium: the vegetative part of a fungus (or in some cases bacteria), consisting of a mass of branching, threadlike hyphae.

mycorrhizae: symbiotic fungi which colonize the outer layers of the roots of many plant species and whose external mycelium effectively increases the effective absorptive surface area of the roots.

myocyte: a muscle cell.

myxedematous cretinism: form of mental retardation caused by perinatal iodine deficiency.

natural background: a term used to describe the geochemical variability and the range of data values due to natural processes, that characterize a particular geological or geochemical occurrence. See also Background and Baseline.

nebulizer: interface at plasma detectors for aerosol production.

necrosis: cell death.

nephrotoxin: cytotoxin specific for cells of kidney.

neurotransmitter: any of several compounds released by neurons to stimulate other neurons.

neutrophil: a specific type of white blood cell.

nOAEL: the highest dose at which no observed adverse effects occur in an experimental setting.

nuclide: a general term applied to any atom with data on the number of protons and neutrons in its nucleus.

odds: probability of disease divided by probability of no disease (p/1-p) within a study group (e.g., exposed individuals).

odds-ratio: ratio between odds for exposed and odds for non-exposed ($odds_{+exp}/odds_{-exp}$).

oligonucleotide: a DNA polymer composed of only a few nucleotides.

omnivores: animals normally feeding on both plant and animal material. Species considered omnivores are humans, dogs, and swine.

oncogene: a gene that controls growth and when aberrant or when activated inappropriately may permit cancer to develop.

operon: a cluster of genes with related functions that are under the control of a single operator and promoter, thereby allowing transcription of these genes to be turned on and off.

organ systems: part of body performing a specific function.

organelle: a compartment found in eukaryotes derived from captured bacteria and with residual independent genes, e.g., *mitochondria* which create useful energy from oxidation of sugars and *chloroplasts* which create useful energy from light-generating oxygen.

organization: a managed flow of material and energy in contrast with static order.

orthogonal (analytical) speciation concept: analytical strategies which employ combinations of various separation and/or detection methods are called orthogonal analytical concepts.

ortholog: a gene in two or more species that has evolved from a common ancestor.

osteoblasts: a bone-forming cell; function with bone-removing cells (osteoclasts) in the normal process of bone remodeling.

osteoclasts: multinucleate cells that destroy bone tissue.

osteomalacia: impaired mineralization of bone tissues resulting in areas where mineral is missing. One possi-

ble cause of osteomalacia is a deficiency of vitamin D, the hormone required for adequate calcium absorption and deposition as bioapatite in bone tissues.

osteon: the bulls-eye pattern of concentric rings of lamellar bone around a vascular canal. This structure is detected in tissue sections that form as a result of bone tissue remodeling. See Haversian Bone.

osteoporosis: a generalized term for the loss of bone tissues in bone organs. There are multiple possible causes of osteoporosis and the loss may occur at any age, but it is more prevalent in older individuals. The variations of osteoporosis remain active areas for investigation.

osteosclerosis: disease characterized by abnormal hardening of bone due to excessive calcification.

oxalic acid: a dicarboxylic acid (ethane dioic acid, $C_2H_2O_4$) found in some plants and produced by molds; forms stable chelation complexes with divalent cations (Ca^{2+}, Mg^{2+}, Fe^{2+}, Zn^{2+}, Cu^{2+}) rendering them unavailable from the diet.

oxidation: chemical process which can lead to the fixation of oxygen or the loss of hydrogen, or the loss of electrons; the opposite of reduction.

oxidoreductase: an enzyme that catalyzes an oxidation-reduction reaction.

p53 gene: a tumor-suppressor gene that codes for a transcription factor involved in preventing genetically damaged cells from proliferating.

Paget's disease: a disorder in which the normal resorption and sculpting of bone is compromised and superfluous or more dense mineralized tissue is deposited.

parakeratosis: retention of nuclei in the cells of the stratum corneum.

parasitimia: the condition of having parasites within the bloodstream. Usually the parasite is a protozoan.

parathyroid hormone: hormone secreted by the parathyroid gland; important in the homeostatic regulation of serum calcium levels.

parent material: the weathered rock material on which a soil is formed. Can be either fragments of the underlying solid geology or transported drift material overlying the solid geology.

parenteral: administration of substance into organism not through gastrointestinal tract but through intramuscular, subcutaneous, or intravenous injection.

parkinsonism: clinical syndrome characterized by diminished facial expression, slowness of voluntary movement, rigidity, tremor, and stooped posture.

pedogenesis: the process of soil formation involving various physical and chemical processes which give rise to the formation of a soil profile. The nature of soil formed is determined by the interactions of the climate, vegetation, parent material, topography, and time.

periodic table: a tabular classification of the chemical elements whereby they are organized into (vertical) groups based on progessive increases in numbers of electron shells surrounding the atomic nucleus and (horizontal) rows based on changes in the internal complexities of the electron shells. Elements within any group have similar chemical properties.

periplasm: a secondary enclosed compartment of a prokaryote outside the cytoplasm and surrounding it.

permafrost: permanently ice-bearing frozen ground, found in the Arctic, Antarctic, and some high-altitude regions.

pH: a measure of the acidic (or alkaline) nature of an aqueous solution, expressed as the negative base −10 logarithm of the activity of protons in the solution. Solutions with pH values below 7 are considered acidic; values greater than 7 indicate basic (or alkaline) conditions.

phagocytosis: a type of endocytosis in which extensions of a plasma membrane engulf extracellular particles and transport them into the interior of the cell.

pharmacognosy: the study of the useful drug effects of natural products.

phase: a volume of space, solid, liquid, or gas in equilibrium with other volumes and described by a boundary. A homogeneous, distinct portion of a chemical system.

phase diagram: a graphical representation of the stability relationships between phases in a chemical/physical system usually representing states at equilibrium. The presentation usually depicts relationships based on changes in composition, temperature, or pressure.

phenotype: the physical characteristics of an organism that can be defined as outward appearance (such as flower color), as behavior, or in molecular terms (such as glycoproteins on red blood cells).

phosphorite: a sedimentary rock with a high percentage of phosphate materials, shell, or bone fragments that may be mined for use as fertilizer. Prominent textural features are often nodules and pellets of extremely fine-grained calcium phosphate.

photoelectron: electron that is ejected from the surface when light falls on it.

phyllosilicate: a group of aluminosilicate minerals that have a sheeted crystal structure which permits cations to be trapped between the sheets and around the sheet edges. Because of these properties some are capable of sequestering geochemically significant amounts of cations, metals.

phytic acid: inositolhexaphosphoric acid ($C_6H_6O_6$ $[H_2PO_3]_6$) found in plants; forms stable chelation complexes with divalent cations (Ca^{2+}, Mg^{2+}, Fe^{2+}, Zn^{2+}, Cu^{2+}) rendering them unavailable from the diet.

phytoavailability: a specific instance of bioavailability with reference to plants. In some instances it is useful to differentiate between phyto- and bioavailability along the food chain. Phytoavailability controls the transfer of a trace element from soil to a plant, and bioavailabilty controls the transfer of the trace element from the plant material to the receptor organism; the transfer factors are unlikely to be the same.

phytosiderophores: organic compounds released by the roots of some plants suffering from a deficiency of iron or certain other micronutrients. They mobilize iron and elements co-precipitated onto iron oxides and render them available for uptake by the plant.

phytotoxic: toxic to plants.

pica: a craving for unnatural articles of food. The name pica comes from the Latin for magpie, a bird that picks up a variety of things either to satisfy hunger or out of curiosity. Geophagy, the deliberate ingestion of soil, is a form of pica.

placer deposits: alluvial deposits which contain ore minerals (commonly native gold, platinum, diamond, cassiterite) in economic quantities; these are heavy minerals which are concentrated by reworking of primary ore bodies. They typically concentrate in low-energy environments such as floodplains and deltas. Many important placer deposits occur also as beach placers where they have been concentrated by seawater movement.

platelet: a non-nucleated, hemoglobin-free cellular component of blood that functions in clotting; also called a thrombocyte.

platform: a term used in geology to describe a large stable section of the Earth's crust that is unaffected by current mountain building. Commonly formed over long periods of time by the erosion of the Earth's surface to relatively low relief.

plaque: the unwanted deposition of mineral materials in tissue areas such as in the vascular system or around teeth within the gum tissues.

pleiotropy: a situation in which a single gene influences more than one phenotypic characteristic.

pleural plaques: a fibrous thickening of the parietal pleura which is characteristically caused by inhalation of the fibers of asbestiform minerals.

pM standard: the PM (particulate matter) standard is based on the total mass of particles measuring $2.5\,\mu m$ or less observed in a 24-hour period.

pneumoconiosis: a chronic fibrosing lung disease from contact with respirable mineral dusts; examples include silicosis and asbestosis.

podsol: a type of soil which can be found in cool, humid environments on freely drained parent materials usually under coniferous trees or ericaceous vegetation. Typically has an iron pan as a result of leaching. Also called spodosols in the USDA Soil Taxonomy classification.

polymorph: a term applied in mineralogy to describe minerals with the same composition that can crystalize in multiple crystallographic forms. Possibly the most well-known polymorphic minerals are calcite and aragonite; both have the chemical composition $CaCO_3$.

primary: term used to describe position in the biogeochemical cycle; refers to bedrock.

primitive cell: a cell thought to have existed some 3 to 4 billion years ago, although a related form can be found in extreme anaerobic conditions today.

prions: an infectious microscopic protein that lacks nucleic acid thought to be responsible for degenerative diseases of the nervous system called transmissible spongiform encephalopathies (TSE); transmissible within and between species.

progesterone: the steroid hormone produced by the corpus luteum, adrenal cortex, and placenta that prepares the uterus for reception and development of the fertilized ovum.

progestins: a general term for the natural or synthetic progestinal agents.

prokaryote: cells of the domains Bacteria or Archaea. Prokaryotic cells have genetic material that is not enclosed in a membrane-bound nucleus; they lack other membrane-bound organelles.

proteome: the full complement of proteins produced (expressed) by a particular genome.

protista: eukaryotic one-celled living organisms distinct from multicellular plants and animals: protozoa, slime molds, and eukaryotic algae.

protozoa: comprise flagellates, ciliates, sporozoans, amoebas, and foraminifers.

pulmonary alveoli: out-pouchings on the fine lung passages in which oxygen exchange between the alveoli and the bloodstream occurs.

pump (in the context of organisms): a mechanical protein-based device in a cell membrane for transferring material from one compartment to another.

Purkinje cells: large nerve cells found in the cerebellum, a large portion of the posterior aspect of the brain.

pyrite: iron sulfide (FeS_2), otherwise known as fool's gold; occurs commonly in zones of ore mineralization and in sediments under strongly reducing conditions.

pyroclastic flow: a fast-moving heated cloud of gas and volcanic particles produced by explosive eruptions or volcanic dome collapse.

Quaternary: the most recent period of geological time, spanning 0–2 million years before Present; divided into the earliest period, the Pleistocene (ending with the last glacial maximum), and the subsequent Holocene (the last 13,000 years).

quaternary structure: the three-dimensional structure of a multisubunit protein; particularly the manner in which the subunits fit together.

radioactivity: atoms (known as radionuclides) which are unstable and will change naturally into atoms of another element accompanied by the emission of ionizing radiation. The change is called radioactive decay.

radionuclide: a radioactive nuclide.

radon: a colorless radioactive element; comprises the isotope radon-222, a decay product of radium. ^{222}Rn (radon) is a gas. It occurs in the uranium-238 decay series and provides about 50% of the total radiation dose to the average person.

radon potential map: a map showing the distribution of radon prone areas delineated by arbitrary grid squares, administrative or geological boundaries. The radon potential classification may be based on radon measurements in existing dwellings, measurements of radon in soil gas, or proxy indicators such as airborne radiometric measurements.

raman microprobe: vibrational spectroscopic technique where light scatter allows for characteristic spectra of materials to be obtained.

raster: a model of spatial data using an x,y coordinate system, rows and columns, and representing features as cells, or pixels, within.

reactive oxygen species: general descriptor for the superoxide (O_2), singlet oxygen (O), and hydrogen peroxide (H_2O_2), each of which has a much greater chemical reactivity with intracellular nucleophiles (proteins, DNA) than molecular oxygen from which it is derived metabolically.

recessive: a mode of inheritance in which a gene must be present from both parents for the trait to become manifest in an offspring.

recharge: process by which water is added from the atmosphere or ground surface to the saturated zone of an aquifer, either directly into the aquifer, or via another formation.

record: a unique entity, commonly in GIS a location, that possesses different values for its attributes in fields.

redox potential (pe or Eh): pe and Eh are related variables that express a measure of the ratio of the aqueous activity of an oxidized species (an electron acceptor, such as Fe^{3+}) to that of a reduced species (an electron donor, such as Fe^{2+}). The redox potential of a solution can provide a sense of the oxidizing or reducing nature of a solution or aqueous environment (oxic, suboxic, sulfidic, methanic).

redox reactions: coupled chemical oxidation and reduction reactions involving the exchange of electrons; many elements have changeable redox states in groundwater the most important redox reactions involve the oxidation or reduction of iron and manganese, introduction or consumption of nitrogen compounds (including nitrate), introduction or consumption of oxygen (including dissolved oxygen), and consumption of organic carbon.

reducing condition: anaerobic condition, formed where nearly all of the oxygen has been consumed by reactions such as oxidation of organic matter or of sulfide; reducing conditions commonly occur in confined aquifers.

reduction: chemical process leading to the loss of oxygen or increase of electrons by a compound; the opposite of oxidation.

reference nutrient intake (RNI): the daily dietary value of a nutrient above which the amount will almost certainly be adequate for everybody.

regolith: a deposit of physically and/or chemically weathered rock material which has not developed into a soil due to the absence of biological activity and the presence of organic matter.

reitfield refinement: a method of calculating the three-dimensional structure of compounds.

relational database: database where data are organized according to the relationships between entities.

relative risk (RR): a risk is the number of occurrences out of the total. Relative risk is the risk given one condition versus the risk given another condition; used in epidemiology.

repair (DNA): the action of biological machinery to fix damage, especially referring to maintenance of DNA integrity.

reservoir (biological): a host, carrier, or medium (such as soil), that harbors a pathogenic organism, without injury to itself in the case of carriers, and can directly or indirectly transmit that pathogen to individuals.

residence time: period during which water, solutes, or particles remain within an aquifer or organisms as a component part of the hydrological cycle.

respiratory distress: impairment of lung function, often resulting in uncomfortable respiratory symptoms, lowered oxygenation and/or elevated carbon dioxide levels in the blood.

retention time: elution time of a compound in a chromatographic system depending on its interaction at the stationary phase.

rheumatoid: indefinite term applied to conditions with symptoms related to the musculoskeletal system.

rhizosphere: the zone around plant roots (2 mm thick) in which there is intense microbial activity due to root exudates and which has chemical properties different from the bulk of the soil.

ribozyme: rNA molecule with catalytic activity.

rickets: disease of children characterized by under-mineralization of growing bone, leading to physical deformities of the weight-bearing bones most notably of the legs, wrists, and arms.

risk assessment: a systematic way of estimating the probability of an adverse outcome based on the known properties of a hazard such as a chemical.

ruminants: several groups of animal species utilizing bacteria, fungi, and protozoa in their forestomachs to digest cellulose and starch in plants eaten so the nutrients can be absorbed in the gut of the animal. Cattle, sheep, goats, antelope, deer, and camels are examples of ruminants.

saline intrusion: phenomenon occurring when a body of salt water invades a body of fresh water; it can occur either in surface water or groundwater bodies.

saprophyte: an organism, often a fungus or bacterium, that obtains its nourishment from dead or decaying organic matter.

saprozoonoses: zoonotic diseases where transmission requires a non-animal development site or reservoir. Soil can often serve as the reservoir.

sarcoidosis: a systemic granulomatous disease of unknown cause.

sarcomatoid: resembling a sarcoma, a neoplasm of soft tissue.

scanning electron microscope (SEM): a method employing an electron microscope and a finely-focused beam of electrons that is moved across a sample allowing the surficial textures to be examined at high resolution and the image displayed. By collecting the emitted electrons from a single spot (size 1–10 μm) chemical analysis of portions of the sample, i.e., a specific mineral species, can be made using energy dispersive X-ray analysis (SEM/EDXA).

screw axis: a specific translational and rotational characteristic of a lattice direction (axis) defined as part of one of the known 230 space groups. The calcium apatite group has a screw axis designated as 6_3. The c-axis has sixfold-symmetry with a screw. The screw rotates 120 degrees around the sixfold-axis with each one-third translation along the axis, part of the space group designation of the apatite unit cell.

secondary: terms used to describe position in the biogeochemical cycle; refers to weathering products and processes resulting from, or acting on, primary rock material.

sedimentary rock: rock formed by compression of material derived from the weathering or deposition of pre-existing rock fragments, marine or other organic debris, or by chemical precipitation.

selenocysteine: an unusual amino acid of proteins, the selenium analog of cysteine, in which a selenium atom replaces sulfur.

selenomethionine: 2-amino-4-(methylseleno) butanoic acid.

selenosis: selenium toxicity.

sesquioxide: oxide mineral containing three atoms of oxygen and two atoms of another chemical substance. Iron and aluminum oxides are the most important in the natural environment.

shale: a sedimentary rock composed of fine particles, mainly made up of clay.

silicate: a mineral composed dominantly of silicon and oxygen, with or without other elements such as magnesium, iron, calcium, sodium, and potassium.

silicosis: a form of pneumoconiosis produced by inhalation of fine silica particles.

smectite: a group of clay minerals (phyllosilicates) that includes montmorillonite and minerals of similar chemical composition. They possess high cation exchange capacities, and are therefore capable of sequestering labile cations.

soil profile (solum): the vertical section of a soil from the surface to its underlying parent material. It comprises distinct layers (horizons) differing in appearance or texture and chemical properties. The soil profile is the basis of soil classification (soils with characteristic combinations of horizons).

soil texture: the relative proportions of sand (0.05–2 mm), silt (0.002–0.05 mm), and clay (<0.002 mm) sized particles in a soil which affects both its physical and chemical properties.

solubility: equilibrium concentration of a solute in water at a given temperature and pressure when the dissolving solid is in contact with the solution.

sorption: the retention of ions on solid surfaces in the soil by a combination of mechanisms: ion exchange, specific adsorption, precipitation, and organic complexation.

space group: a mathematical expression that uniquely defines the three-dimensional array typical of a crystalline material.

spallation: splitting off, particularly applied to splitting off parts of the nucleus of an atom, resulting in the formation of a different element.

spherule: a small spherical structure of the invasive phase of *Coccidioides immitis* that fills with endospores as it matures. The spherule ruptures at maturity releasing the infective endospores into the host.

spongiosis: intercellular edema of epidermis.

spray chamber (chemical analysis): part of sample introduction system, connected to a nebulizer. Droplets from the aerosol that are too big are discarded.

squamous cell carcinoma: malignant neoplasm derived from stratified squamous epithelium.

stable isotope: isotope that does not undergo radioactive decay.

standardized mortality ratios: a statistical method for comparing the mortalities of different population groups by separating data according to sex and then age band.

steatosis: general term describing fatty degeneration; **t-RNA:** transfer ribonucleic acid; any of a number of such intracellular factors involved in protein synthesis by transferring in sequence individual amino acids to the ribosome.

stereoisomer: one of two forms of a compound that is indistinguishable from the other outside of the orientation in space. An enantiomer.

stoichiometric: a term applied when a phase or compound has the charge balance and chemical proportions expected in the ideal formula.

swayback: neonatal ataxia, a clinical manifestation of copper deficiency in lambs. The condition is characterized by incoordination of movement and high mortality. The disease is known as lamkrius in South Africa, kipsiepsiep in Kenya, and enzootic ataxia in several other countries, including the former Soviet Union.

symbiosis: the cohabiting of more than one organism which supply one another with vital material and energy.

synergy: a positive interaction.

tachycardia: rapid heart beat.

tachypnea: rapid breathing.

tephra: any solid material produced and made airborne by volcanic activity (including bombs, blocks, ash, and dust).

termite mounds: a common source of geophagical material in the tropics. The edible part of a termite mound is the extremely mineraliferous, soft, protected interior comprising the queen's chamber, nursing galleries, and fungus gardens.

tetrahedral orthophosphate group: the three-dimensional atomic array in which four oxygen atoms are distributed at the apices of the tetrahedron around the phosphorus atom.

theme (GIS): a GIS data layer, or coverage used in an overlay analysis with spatial referencing.

threshold: in biology it is a dose level, below which, no adverse effect is expected. In Earth science it represents the upper or lower limit of background—above or below which is anomalous.

thylakoid: a disk-shaped, membranous sac found in chloroplasts, the membranes of which contain the photosystems and ATP-synthesizing enzymes used in the light-dependent reactions of photosynthesis.

thyroxine: also referred to as $3:5,3':5'$ tetra-iodothyronine (T_4) is the major hormone secreted by the thyroid gland. T_4 is involved in controlling the rate of metabolic processes in the body and influencing physical development.

TNF: tumor necrosis factor.

tomography: a method employing transmission X-radiological analysis to visualize the bones or bony portions of the skeleton. The X-ray source moves relative to the patient.

tonsillar herniation: physical displacement of cerebellar tonsil into foramen magnum, a large opening at base of the brain.

toxicity: state of being poisonous and disturbing organ function.

toxicodynamics: the mechanisms by which xenobiotics induce their effects in the body; the mechanisms of the toxic response.

toxicokinetics: the mechanisms by which xenobiotics are handled in the body, comprising the steps *absorption*, *distribution*, *metabolism*, and *excretion*.

toxicology: originally, the study of poisons and now the general science of the handling by, and response of, the body to xenobiotics and the patterns of adverse effects that result.

toxocariasis: also called visceral larva migrans (VLM), toxocariasis is caused through infection with the larvae of *Toxocara canis* or *T. cati* (the common roundworm of dogs and cats, respectively). After infection, the eggs hatch into larvae and are carried into the circulation and to various tissues. Respiratory symptoms develop, and there is a swelling of body organs such as the liver. A complication of VLM is epilepsy and ocular larva migrans, the latter caused by microscopic worms entering the eye.

toxoplasmosis: a disease attributable to the ingestion of *Toxoplasma gondii*, one of the most common human parasites that infect 30–60% of the global population. Commonly caused by eating of undercooked meat with soil ingestion as secondary source. Recent research has suggested that human behavior can be adversely affected following *T. gondii* infection.

trace elements (in medicine): general term for the nutritionally essential mineral elements that are required at levels of intake less than about $50\,mg\,d^{-1}$; includes iron, copper, zinc, iodine, selenium, manganese, molybdenum, chromium, fluoride, and cobalt.

transcription: the act of producing RNA from DNA leading to *translation*, protein production.

transfection: the uptake and expression of a foreign DNA sequence by cultured eukaryotic cells or the introduction of foreign DNA into a host cell.

transposon: a segment of DNA that can become integrated at many different sites along a chromosome (especially a segment of bacterial DNA that can be translocated as a whole).

trichiuriasis: infestation with the roundworm *Trichuris trichiura* that may cause nausea, abdominal pain, diarrhea, and occasionally anemia and rectal prolapse.

triiodothyronine: also referred to as 35,3′ triiodothyronine (T_3) produced in the thyroid gland and involved in controlling the rate of metabolic processes in the body and physical development.

trabecular: the porous tissues forming the internal sectors of bones. The trabeculae are bone tissue spicules. This type of tissue is often adjacent to the hollow core or within the marrow cavity.

transmissible spongiform encephalopathies (TSE): rare forms of progressively degenerative diseases of the nervous system that affect both humans and animals. They are caused by agents called prions and generally produce spongiform changes in the brain. Examples include chronic wasting disease (CWD) in deer and elk, bovine spongiform encephalopathy (BSE) in cattle, and Creutzfeldt-Jakob disease (CJD) in humans.

type 1 collagen: the special variety of the collagen molecule typically found in the matrix of tissues that will become mineralized as bone.

ultramafic rock: igneous rock composed substantially of ferromagnesian silicate minerals and metallic oxides and sulfides, with <45% silica, and almost no quartz or feldspar.

ultrastructure: morphometry of particles and cell structure based on electron microscopy.

unconfined aquifer: aquifer containing unconfined groundwater, i.e., having a water table and an unsaturated zone.

unit cell: the smallest geometric volume that uniquely defines the composition and precise structure of a crystalline compound. The basis for the repetitive pattern that completely characterizes a compound, its chemistry, and three-dimensional arrangements of all the constituent atoms.

USDA Soil Taxonomy: the soil classification system devised by the United States Department of Agriculture (published in 1975).

vasodilation: expansion of the blood vessels.

vadose zone: also known as the "unsaturated zone" is the part of the Earth's surface extending down to the water table.

vector (GIS): model of spatial data using points, lines, and polygons to represent geospatial features and boundaries. Vector in the entomological sense, is typically an arthropod that transmits disease-causing pathogens to humans.

vector-borne disease: disease that is transmitted from one vertebrate host to another by an invertebrate, usually an insect, tick, or snail.

viremia: the existence of virus or viral particles in the bloodstream.

virulence: the capacity of a microorganism for causing disease.

v_{max}: the maximum velocity (never attained) in enzyme kinetics.

volatile fatty acids (VFA): common name for acetic acid, butyric acid, and propionic acid normally formed under anaerobic conditions in the forestomachs and large intestine of herbivores. After absorption from the gastrointestinal tract, VFA can be further metabolized and used mainly for energy production. In ruminants, VFAs are the dominating energy source equivalent to glucose in the metabolism of other species.

volcanic gas: gas produced by volcanic activity or geothermal processes. Steam is the most common gas; those of relevance to health include the inert asphyxiants, irritant gases, or noxious asphyxiants.

volcanic monitoring: geological and epidemiological testing and surveillance prior to, surrounding, and subsequent to an eruptive event or degassing episode; includes the period of post-disaster recovery and rehabilitation.

volcano: an opening in the crust from which gases, lava, and/or tephra are expelled.

voltammetry: an electrochemical determination method based on the characteristic redox potential of the measured compound.

weathering: a process at or near the Earth's surface caused by the interaction of water, oxygen, carbon dioxide, and organic acids with the minerals present; includes hydrolysis and oxidation reactions. Weathering can result in the formation of new mineral suites that are in equilibrium with their environment. In Arctic and high mountainous regions chemical weathering may be limited, and weathering is largely limited to mechanical breakdown due to frost action that liberates fragments of the pre-existing minerals.

white muscle disease: a complex medical condition, which is multifactorial in origin but linked to selenium deficiency, causes degeneration of the muscles in animal species. In lambs born with the disease, death can result after a few days. Later in life, animals have a stiff and stilted gait, arched back, are not inclined to move about, lose condition, and die.

world Reference Base for Soil Resources: a classification system, database, and atlas produced by the working group RB International Society of Soil Science in 1998.

Woven bone: the first deposited bone tissue that may display a haphazard distribution of matrix, cells, vascular channels, and mineral and which is usually later reworked into lamellar or haversian bone over time.

xenobiotic: a chemical substance foreign to the body or introduced to the body in higher quantities or by a different pathway than occurs in normal metabolism.

X-ray diffraction maxima: the periodic coherent scattering of X-rays that arises from crystalline materials. These data are used to determine the coordinates from which the space group and unit cell of the compound can be determined.

X-ray/electron diffraction: the method employed to examine the crystallinity and crystal structure of materials.

zoonotic/zoonosis: a disease which has a natural reservoir in an animal or non-human species that can be transmitted to humans.

Note: The italicized *f* or *t* following a page number denotes a figure or table on that page. The italicized *ff* or *tt* following a page number denotes multiple figures or tables on that page